D1428918

HANDBOOK OF
INTEGRAL EQUATIONS

Andrei D. Polyanin and
Alexander V. Manzhirov

HANDBOOK OF
INTEGRAL
EQUATIONS

CRC Press
Boca Raton London New York Washington, D.C.

2116034 l

Library of Congress Cataloging-in-Publication Data

Polianin, A. D. (Andreĭ Dmitrievich)
   Handbook of integral equations / Andrei D. Polyanin, Alexander
V. Manzhirov.
        p. cm.
   Includes bibliographical references (p.   –   ) and index.
   ISBN 0-8493-2876-4 (alk. paper)
      1. Integral equations--Handbooks, manuals, etc. I. Manzhirov, A.
V. (Aleksandr Vladimirovich) II. Title.
QA431.P65 1998
515′.45—dc21
                                                                98-10762
                                                                CIP

© 1998 by CRC Press LLC

No claim to original U.S. Government works
International Standard Book Number 0-8493-2876-4
Library of Congress Card Number 98-10762
Printed in the United States of America   2 3 4 5 6 7 8 9 0
Printed on acid-free paper

ANNOTATION

More than 2100 integral equations with solutions are given in the first part of the book. A lot of new exact solutions to linear and nonlinear equations are included. Special attention is paid to equations of general form, which depend on arbitrary functions. The other equations contain one or more free parameters (it is the reader's option to fix these parameters). Totally, the number of equations described is an order of magnitude greater than in any other book available.

A number of integral equations are considered which are encountered in various fields of mechanics and theoretical physics (elasticity, plasticity, hydrodynamics, heat and mass transfer, electrodynamics, etc.).

The second part of the book presents exact, approximate analytical and numerical methods for solving linear and nonlinear integral equations. Apart from the classical methods, some new methods are also described. Each section provides examples of applications to specific equations.

The handbook has no analogs in the world literature and is intended for a wide audience of researchers, college and university teachers, engineers, and students in the various fields of mathematics, mechanics, physics, chemistry, and queuing theory.

FOREWORD

Integral equations are encountered in various fields of science and numerous applications (in elasticity, plasticity, heat and mass transfer, oscillation theory, fluid dynamics, filtration theory, electrostatics, electrodynamics, biomechanics, game theory, control, queuing theory, electrical engineering, economics, medicine, etc.).

Exact (closed-form) solutions of integral equations play an important role in the proper understanding of qualitative features of many phenomena and processes in various areas of natural science. Lots of equations of physics, chemistry and biology contain functions or parameters which are obtained from experiments and hence are not strictly fixed. Therefore, it is expedient to choose the structure of these functions so that it would be easier to analyze and solve the equation. As a possible selection criterion, one may adopt the requirement that the model integral equation admit a solution in a closed form. Exact solutions can be used to verify the consistency and estimate errors of various numerical, asymptotic, and approximate methods.

More than 2100 integral equations and their solutions are given in the first part of the book (Chapters 1–6). A lot of new exact solutions to linear and nonlinear equations are included. Special attention is paid to equations of general form, which depend on arbitrary functions. The other equations contain one or more free parameters (the book actually deals with families of integral equations); it is the reader's option to fix these parameters. Totally, the number of equations described in this handbook is an order of magnitude greater than in any other book currently available.

The second part of the book (Chapters 7–14) presents exact, approximate analytical, and numerical methods for solving linear and nonlinear integral equations. Apart from the classical methods, some new methods are also described. When selecting the material, the authors have given a pronounced preference to practical aspects of the matter; that is, to methods that allow effectively "constructing" the solution. For the reader's better understanding of the methods, each section is supplied with examples of specific equations. Some sections may be used by lecturers of colleges and universities as a basis for courses on integral equations and mathematical physics equations for graduate and postgraduate students.

For the convenience of a wide audience with different mathematical backgrounds, the authors tried to do their best, wherever possible, to avoid special terminology. Therefore, some of the methods are outlined in a schematic and somewhat simplified manner, with necessary references made to books where these methods are considered in more detail. For some nonlinear equations, only solutions of the simplest form are given. The book does not cover two-, three- and multidimensional integral equations.

The handbook consists of chapters, sections and subsections. Equations and formulas are numbered separately in each section. The equations within a section are arranged in increasing order of complexity. The extensive table of contents provides rapid access to the desired equations.

For the reader's convenience, the main material is followed by a number of supplements, where some properties of elementary and special functions are described, tables of indefinite and definite integrals are given, as well as tables of Laplace, Mellin, and other transforms, which are used in the book.

The first and second parts of the book, just as many sections, were written so that they could be read independently from each other. This allows the reader to quickly get to the heart of the matter.

We would like to express our deep gratitude to Rolf Sulanke and Alexei Zhurov for fruitful discussions and valuable remarks. We also appreciate the help of Vladimir Nazaikinskii and Alexander Shtern in translating the second part of this book, and are thankful to Inna Shingareva for her assistance in preparing the camera-ready copy of the book.

The authors hope that the handbook will prove helpful for a wide audience of researchers, college and university teachers, engineers, and students in various fields of mathematics, mechanics, physics, chemistry, biology, economics, and engineering sciences.

A. D. Polyanin
A. V. Manzhirov

SOME REMARKS AND NOTATION

1. In Chapters 1–11 and 14, in the original integral equations, the independent variable is denoted by x, the integration variable by t, and the unknown function by $y = y(x)$.

2. For a function of one variable $f = f(x)$, we use the following notation for the derivatives:

$$f'_x = \frac{df}{dx}, \quad f''_{xx} = \frac{d^2 f}{dx^2}, \quad f'''_{xxx} = \frac{d^3 f}{dx^3}, \quad f''''_{xxxx} = \frac{d^4 f}{dx^4}, \quad \text{and} \quad f_x^{(n)} = \frac{d^n f}{dx^n} \quad \text{for } n \geq 5.$$

Occasionally, we use the similar notation for partial derivatives of a function of two variables, for example, $K'_x(x, t) = \dfrac{\partial}{\partial x} K(x, t)$.

3. In some cases, we use the operator notation $\left[f(x) \dfrac{d}{dx} \right]^n g(x)$, which is defined recursively by

$$\left[f(x) \frac{d}{dx} \right]^n g(x) = f(x) \frac{d}{dx} \left\{ \left[f(x) \frac{d}{dx} \right]^{n-1} g(x) \right\}.$$

4. It is indicated in the beginning of Chapters 1–6 that $f = f(x)$, $g = g(x)$, $K = K(x)$, etc. are arbitrary functions, and A, B, etc. are free parameters. This means that:

(a) $f = f(x)$, $g = g(x)$, $K = K(x)$, etc. are assumed to be continuous real-valued functions of real arguments;*

(b) if the solution contains derivatives of these functions, then the functions are assumed to be sufficiently differentiable;**

(c) if the solution contains integrals with these functions (in combination with other functions), then the integrals are supposed to converge;

(d) the free parameters A, B, etc. may assume any real values for which the expressions occurring in the equation and the solution make sense (for example, if a solution contains a factor $\dfrac{A}{1-A}$, then it is implied that $A \neq 1$; as a rule, this is not specified in the text).

5. The notations $\operatorname{Re} z$ and $\operatorname{Im} z$ stand, respectively, for the real and the imaginary part of a complex quantity z.

6. In the first part of the book (Chapters 1–6) when referencing a particular equation, we use a notation like 2.3.15, which implies equation 15 from Section 2.3.

7. To highlight portions of the text, the following symbols are used in the book:

▶ indicates important information pertaining to a group of equations (Chapters 1–6);

⊙ indicates the literature used in the preparation of the text in specific equations (Chapters 1–6) or sections (Chapters 7–14).

---

* Less severe restrictions on these functions are presented in the second part of the book.

** Restrictions (b) and (c) imposed on $f = f(x)$, $g = g(x)$, $K = K(x)$, etc. are not mentioned in the text.

AUTHORS

Andrei D. Polyanin, D.Sc., Ph.D., is a noted scientist of broad interests, who works in various fields of mathematics, mechanics, and chemical engineering science.

A. D. Polyanin graduated from the Department of Mechanics and Mathematics of the Moscow State University in 1974. He received his Ph.D. degree in 1981 and D.Sc. degree in 1986 at the Institute for Problems in Mechanics of the Russian (former USSR) Academy of Sciences. Since 1975, A. D. Polyanin has been a member of the staff of the Institute for Problems in Mechanics of the Russian Academy of Sciences.

Professor Polyanin has made important contributions to developing new exact and approximate analytical methods of the theory of differential equations, mathematical physics, integral equations, engineering mathematics, nonlinear mechanics, theory of heat and mass transfer, and chemical hydrodynamics. He obtained exact solutions for several thousands of ordinary differential, partial differential, and integral equations.

Professor Polyanin is an author of 17 books in English, Russian, German, and Bulgarian. His publications also include more than 110 research papers and three patents. One of his most significant books is *A. D. Polyanin and V. F. Zaitsev, Handbook of Exact Solutions for Ordinary Differential Equations, CRC Press, 1995.*

In 1991, A. D. Polyanin was awarded a Chaplygin Prize of the USSR Academy of Sciences for his research in mechanics.

Alexander V. Manzhirov, D.Sc., Ph.D., is a prominent scientist in the fields of mechanics and applied mathematics, integral equations, and their applications.

After graduating from the Department of Mechanics and Mathematics of the Rostov State University in 1979, A. V. Manzhirov attended a postgraduate course at the Moscow Institute of Civil Engineering. He received his Ph.D. degree in 1983 at the Moscow Institute of Electronic Engineering Industry and his D.Sc. degree in 1993 at the Institute for Problems in Mechanics of the Russian (former USSR) Academy of Sciences. Since 1983, A. V. Manzhirov has been a member of the staff of the Institute for Problems in Mechanics of the Russian Academy of Sciences. He is also a Professor of Mathematics at the Bauman Moscow State Technical University and a Professor of Mathematics at the Moscow State Academy of Engineering and Computer Science. Professor Manzhirov is a member of the editorial board of the journal "Mechanics of Solids" and a member of the European Mechanics Society (EUROMECH).

Professor Manzhirov has made important contributions to new mathematical methods for solving problems in the fields of integral equations, mechanics of solids with accretion, contact mechanics, and the theory of viscoelasticity and creep. He is an author of 3 books, 60 scientific publications, and two patents.

CONTENTS

Part I. Exact Solutions of Integral Equations

CONTENTS

Part I. Exact Solutions of Integral Equations

6. Nonlinear Equations With Constant Limits of Integration **371**

Part II. Methods for Solving Integral Equations

Supplements

Part I

Exact Solutions of Integral Equations

Chapter 1

Linear Equations of the First Kind With Variable Limit of Integration

▶ *Notation:* $f = f(x)$, $g = g(x)$, $h = h(x)$, $K = K(x)$, and $M = M(x)$ *are arbitrary functions (these may be composite functions of the argument depending on two variables x and t); A, B, C, D, E, a, b, c, α, β, γ, λ, and μ are free parameters; and m and n are nonnegative integers.*

▶ **Preliminary remarks.** For equations of the form

$$\int_a^x K(x,t)y(t)\,dt = f(x), \qquad a \le x \le b,$$

where the functions $K(x,t)$ and $f(x)$ are continuous, the right-hand side must satisfy the following conditions:

1°. If $K(a,a) \ne 0$, then we must have $f(a) = 0$ (for example, the right-hand sides of equations 1.1.1 and 1.2.1 must satisfy this condition).

2°. If $K(a,a) = K'_x(a,a) = \cdots = K_x^{(n-1)}(a,a) = 0$, $0 < \left|K_x^{(n)}(a,a)\right| < \infty$, then the right-hand side of the equation must satisfy the conditions

$$f(a) = f'_x(a) = \cdots = f_x^{(n)}(a) = 0.$$

For example, with $n = 1$, these are constraints for the right-hand side of equation 1.1.2.

3°. If $K(a,a) = K'_x(a,a) = \cdots = K_x^{(n-1)}(a,a) = 0$, $K_x^{(n)}(a,a) = \infty$, then the right-hand side of the equation must satisfy the conditions

$$f(a) = f'_x(a) = \cdots = f_x^{(n-1)}(a) = 0.$$

For example, with $n = 1$, this is a constraint for the right-hand side of equation 1.1.30.

For unbounded $K(x,t)$ with integrable power-law or logarithmic singularity at $x = t$ and continuous $f(x)$, no additional conditions are imposed on the right-hand side of the integral equation (e.g., see Abel's equation 1.1.36).

In Chapter 1, conditions 1°–3° are as a rule not specified.

1.1. Equations Whose Kernels Contain Power-Law Functions

1.1-1. Kernels Linear in the Arguments x and t

1. $\displaystyle\int_a^x y(t)\,dt = f(x).$

   Solution: $y(x) = f'_x(x)$.

2. $\displaystyle\int_a^x (x - t)y(t)\, dt = f(x).$

Solution: $y(x) = f''_{xx}(x).$

3. $\displaystyle\int_a^x (Ax + Bt + C)y(t)\, dt = f(x).$

This is a special case of equation 1.9.5 with $g(x) = x$.

1°. Solution with $B \neq -A$:

$$y(x) = \frac{d}{dx}\left\{ [(A + B)x + C]^{-\frac{A}{A+B}} \int_a^x [(A + B)t + C]^{-\frac{B}{A+B}} f'_t(t)\, dt \right\}.$$

2°. Solution with $B = -A$:

$$y(x) = \frac{1}{C}\frac{d}{dx}\left[\exp\left(-\frac{A}{C}x\right) \int_a^x \exp\left(\frac{A}{C}t\right) f'_t(t)\, dt \right].$$

1.1-2. Kernels Quadratic in the Arguments x and t

4. $\displaystyle\int_a^x (x - t)^2 y(t)\, dt = f(x),\qquad f(a) = f'_x(a) = f''_{xx}(a) = 0.$

Solution: $y(x) = \frac{1}{2} f'''_{xxx}(x).$

5. $\displaystyle\int_a^x (x^2 - t^2)y(t)\, dt = f(x),\qquad f(a) = f'_x(a) = 0.$

This is a special case of equation 1.9.2 with $g(x) = x^2$.

Solution: $y(x) = \dfrac{1}{2x^2}\left[x f''_{xx}(x) - f'_x(x) \right].$

6. $\displaystyle\int_a^x \left(Ax^2 + Bt^2\right)y(t)\, dt = f(x).$

For $B = -A$, see equation 1.1.5. This is a special case of equation 1.9.4 with $g(x) = x^2$.

Solution: $y(x) = \dfrac{1}{A + B}\dfrac{d}{dx}\left[x^{-\frac{2A}{A+B}} \int_a^x t^{-\frac{2B}{A+B}} f'_t(t)\, dt \right].$

7. $\displaystyle\int_a^x \left(Ax^2 + Bt^2 + C\right)y(t)\, dt = f(x).$

This is a special case of equation 1.9.5 with $g(x) = x^2$.

Solution:

$$y(x) = \operatorname{sign}\varphi(x)\,\frac{d}{dx}\left\{ |\varphi(x)|^{-\frac{A}{A+B}} \int_a^x |\varphi(t)|^{-\frac{B}{A+B}} f'_t(t)\, dt \right\},\qquad \varphi(x) = (A + B)x^2 + C.$$

8. $\displaystyle\int_a^x \left[Ax^2 + (B - A)xt - Bt^2\right]y(t)\, dt = f(x),\qquad f(a) = f'_x(a) = 0.$

Differentiating with respect to x yields an equation of the form 1.1.3:

$$\int_a^x [2Ax + (B - A)t]y(t)\, dt = f'_x(x).$$

Solution:

$$y(x) = \frac{1}{A + B}\frac{d}{dx}\left[x^{-\frac{2A}{A+B}} \int_a^x t^{\frac{A-B}{A+B}} f''_{tt}(t)\, dt \right].$$

9. $\displaystyle\int_a^x \left(Ax^2 + Bt^2 + Cx + Dt + E\right)y(t)\,dt = f(x).$

This is a special case of equation 1.9.6 with $g(x) = Ax^2 + Cx$ and $h(t) = Bt^2 + Dt + E$.

10. $\displaystyle\int_a^x \left(Axt + Bt^2 + Cx + Dt + E\right)y(t)\,dt = f(x).$

This is a special case of equation 1.9.15 with $g_1(x) = x$, $h_1(t) = At + C$, $g_2(x) = 1$, and $h_2(t) = Bt^2 + Dt + E$.

11. $\displaystyle\int_a^x \left(Ax^2 + Bxt + Cx + Dt + E\right)y(t)\,dt = f(x).$

This is a special case of equation 1.9.15 with $g_1(x) = Bx + D$, $h_1(t) = t$, $g_2(x) = Ax^2 + Cx + E$, and $h_2(t) = 1$.

1.1-3. Kernels Cubic in the Arguments x and t

12. $\displaystyle\int_a^x (x-t)^3 y(t)\,dt = f(x), \qquad f(a) = f'_x(a) = f''_{xx}(a) = f'''_{xxx}(a) = 0.$

Solution: $y(x) = \frac{1}{6} f''''_{xxxx}(x).$

13. $\displaystyle\int_a^x (x^3 - t^3) y(t)\,dt = f(x), \qquad f(a) = f'_x(a) = 0.$

This is a special case of equation 1.9.2 with $g(x) = x^3$.

Solution: $y(x) = \dfrac{1}{3x^3}\left[x f'''_{xxx}(x) - 2f'_x(x)\right].$

14. $\displaystyle\int_a^x \left(Ax^3 + Bt^3\right)y(t)\,dt = f(x).$

For $B = -A$, see equation 1.1.13. This is a special case of equation 1.9.4 with $g(x) = x^3$.

Solution with $0 \le a \le x$: $y(x) = \dfrac{1}{A+B}\dfrac{d}{dx}\left[x^{-\frac{3A}{A+B}}\displaystyle\int_a^x t^{-\frac{3B}{A+B}} f'_t(t)\,dt\right].$

15. $\displaystyle\int_a^x \left(Ax^3 + Bt^3 + C\right)y(t)\,dt = f(x).$

This is a special case of equation 1.9.5 with $g(x) = x^3$.

16. $\displaystyle\int_a^x (x^2 t - xt^2)y(t)\,dt = f(x), \qquad f(a) = f'_x(a) = 0.$

This is a special case of equation 1.9.11 with $g(x) = x^2$ and $h(x) = x$.

Solution: $y(x) = \dfrac{1}{x}\dfrac{d^2}{dx^2}\left[\dfrac{1}{x}f(x)\right].$

17. $\displaystyle\int_a^x (Ax^2 t + Bxt^2)y(t)\,dt = f(x).$

For $B = -A$, see equation 1.1.16. This is a special case of equation 1.9.12 with $g(x) = x^2$ and $h(x) = x$.

Solution:

$$y(x) = \frac{1}{(A+B)x}\frac{d}{dx}\left\{x^{-\frac{A}{A+B}}\int_a^x t^{-\frac{B}{A+B}}\frac{d}{dt}\left[\frac{1}{t}f(t)\right]dt\right\}.$$

18. $\int_a^x (Ax^3 + Bxt^2)y(t)\,dt = f(x).$

This is a special case of equation 1.9.15 with $g_1(x) = Ax^3$, $h_1(t) = 1$, $g_2(x) = Bx$, and $h_2(t) = t^2$.

19. $\int_a^x (Ax^3 + Bx^2 t)y(t)\,dt = f(x).$

This is a special case of equation 1.9.15 with $g_1(x) = Ax^3$, $h_1(t) = 1$, $g_2(x) = Bx^2$, and $h_2(t) = t$.

20. $\int_a^x (Ax^2 t + Bt^3)y(t)\,dt = f(x).$

This is a special case of equation 1.9.15 with $g_1(x) = Ax^2$, $h_1(t) = t$, $g_2(x) = B$, and $h_2(t) = t^3$.

21. $\int_a^x (Axt^2 + Bt^3)y(t)\,dt = f(x).$

This is a special case of equation 1.9.15 with $g_1(x) = Ax$, $h_1(t) = t^2$, $g_2(x) = B$, and $h_2(t) = t^3$.

22. $\int_a^x \left(A_3 x^3 + B_3 t^3 + A_2 x^2 + B_2 t^2 + A_1 x + B_1 t + C\right)y(t)\,dt = f(x).$

This is a special case of equation 1.9.6 with $g(x) = A_3 x^3 + A_2 x^2 + A_1 x + C$ and $h(t) = B_3 t^3 + B_2 t^2 + B_1 t$.

| **1.1-4. Kernels Containing Higher-Order Polynomials in x and t** |

23. $\int_a^x (x - t)^n y(t)\,dt = f(x), \qquad n = 1, 2, \ldots$

It is assumed that the right-hand of the equation satisfies the conditions $f(a) = f'_x(a) = \cdots = f_x^{(n)}(a) = 0$.

Solution: $y(x) = \dfrac{1}{n!} f_x^{(n+1)}(x).$

Example. For $f(x) = Ax^m$, where m is a positive integer, $m > n$, the solution has the form

$$y(x) = \frac{Am!}{n!\,(m - n - 1)!}\,x^{m-n-1}.$$

24. $\int_a^x (x^n - t^n)y(t)\,dt = f(x), \qquad f(a) = f'_x(a) = 0, \qquad n = 1, 2, \ldots$

Solution: $y(x) = \dfrac{1}{n}\dfrac{d}{dx}\left[\dfrac{f'_x(x)}{x^{n-1}}\right].$

25. $\int_a^x \left(t^n x^{n+1} - x^n t^{n+1}\right)y(t)\,dt = f(x), \qquad n = 2, 3, \ldots$

This is a special case of equation 1.9.11 with $g(x) = x^{n+1}$ and $h(x) = x^n$.

Solution: $y(x) = \dfrac{1}{x^n}\dfrac{d^2}{dx^2}\left[\dfrac{f(x)}{x^n}\right].$

1.1-5. Kernels Containing Rational Functions

26. $\displaystyle\int_0^x \frac{y(t)\,dt}{x+t} = f(x).$

$1°$. For a polynomial right-hand side, $f(x) = \displaystyle\sum_{n=0}^{N} A_n x^n$, the solution has the form

$$y(x) = \sum_{n=0}^{N} \frac{A_n}{B_n} x^n, \qquad B_n = (-1)^n \left[\ln 2 + \sum_{k=1}^{n} \frac{(-1)^k}{k} \right].$$

$2°$. For $f(x) = x^\lambda \displaystyle\sum_{n=0}^{N} A_n x^n$, where λ is an arbitrary number ($\lambda > -1$), the solution has the form

$$y(x) = x^\lambda \sum_{n=0}^{N} \frac{A_n}{B_n} x^n, \qquad B_n = \int_0^1 \frac{t^{\lambda+n}\,dt}{1+t}.$$

$3°$. For $f(x) = \ln x \left(\displaystyle\sum_{n=0}^{N} A_n x^n \right)$, the solution has the form

$$y(x) = \ln x \sum_{n=0}^{N} \frac{A_n}{B_n} x^n + \sum_{n=0}^{N} \frac{A_n I_n}{B_n^2} x^n,$$

$$B_n = (-1)^n \left[\ln 2 + \sum_{k=1}^{n} \frac{(-1)^k}{k} \right], \quad I_n = (-1)^n \left[\frac{\pi^2}{12} + \sum_{k=1}^{n} \frac{(-1)^k}{k^2} \right].$$

$4°$. For $f(x) = \displaystyle\sum_{n=0}^{N} A_n (\ln x)^n$, the solution of the equation has the form

$$y(x) = \sum_{n=0}^{N} A_n Y_n(x),$$

where the functions $Y_n = Y_n(x)$ are given by

$$Y_n(x) = \left\{ \frac{d^n}{d\lambda^n} \left[\frac{x^\lambda}{I(\lambda)} \right] \right\}_{\lambda=0}, \qquad I(\lambda) = \int_0^1 \frac{z^\lambda\,dz}{1+z}.$$

$5°$. For $f(x) = \displaystyle\sum_{n=1}^{N} A_n \cos(\lambda_n \ln x) + \sum_{n=1}^{N} B_n \sin(\lambda_n \ln x)$, the solution of the equation has the form

$$y(x) = \sum_{n=1}^{N} C_n \cos(\lambda_n \ln x) + \sum_{n=1}^{N} D_n \sin(\lambda_n \ln x),$$

where the constants C_n and D_n are found by the method of undetermined coefficients.

$6°$. For arbitrary $f(x)$, the transformation

$$x = \tfrac{1}{2}e^{2z}, \quad t = \tfrac{1}{2}e^{2\tau}, \quad y(t) = e^{-\tau} w(\tau), \quad f(x) = e^{-z} g(z)$$

leads to an integral equation with difference kernel of the form 1.9.26:

$$\int_{-\infty}^{z} \frac{w(\tau)\,d\tau}{\cosh(z-\tau)} = g(z).$$

27. $\displaystyle\int_0^x \frac{y(t)\,dt}{ax+bt} = f(x), \qquad a > 0, \quad a+b > 0.$

1°. For a polynomial right-hand side, $f(x) = \displaystyle\sum_{n=0}^N A_n x^n$, the solution has the form

$$y(x) = \sum_{n=0}^N \frac{A_n}{B_n} x^n, \qquad B_n = \int_0^1 \frac{t^n\,dt}{a+bt}.$$

2°. For $f(x) = x^\lambda \displaystyle\sum_{n=0}^N A_n x^n$, where λ is an arbitrary number ($\lambda > -1$), the solution has the form

$$y(x) = x^\lambda \sum_{n=0}^N \frac{A_n}{B_n} x^n, \qquad B_n = \int_0^1 \frac{t^{\lambda+n}\,dt}{a+bt}.$$

3°. For $f(x) = \ln x \left(\displaystyle\sum_{n=0}^N A_n x^n \right)$, the solution has the form

$$y(x) = \ln x \sum_{n=0}^N \frac{A_n}{B_n} x^n - \sum_{n=0}^N \frac{A_n C_n}{B_n^2} x^n, \qquad B_n = \int_0^1 \frac{t^n\,dt}{a+bt}, \qquad C_n = \int_0^1 \frac{t^n \ln t}{a+bt}\,dt.$$

4°. For some other special forms of the right-hand side (see items 4 and 5, equation 1.1.26), the solution may be found by the method of undetermined coefficients.

28. $\displaystyle\int_0^x \frac{y(t)\,dt}{ax^2+bt^2} = f(x), \qquad a > 0, \quad a+b > 0.$

1°. For a polynomial right-hand side, $f(x) = \displaystyle\sum_{n=0}^N A_n x^n$, the solution has the form

$$y(x) = \sum_{n=0}^N \frac{A_n}{B_n} x^{n+1}, \qquad B_n = \int_0^1 \frac{t^{n+1}\,dt}{a+bt^2}.$$

Example. For $a = b = 1$ and $f(x) = Ax^2 + Bx + C$, the solution of the integral equation is:

$$y(x) = \frac{2A}{1-\ln 2} x^3 + \frac{4B}{4-\pi} x^2 + \frac{2C}{\ln 2} x.$$

2°. For $f(x) = x^\lambda \displaystyle\sum_{n=0}^N A_n x^n$, where λ is an arbitrary number ($\lambda > -1$), the solution has the form

$$y(x) = x^\lambda \sum_{n=0}^N \frac{A_n}{B_n} x^{n+1}, \qquad B_n = \int_0^1 \frac{t^{\lambda+n+1}\,dt}{a+bt^2}.$$

3°. For $f(x) = \ln x \left(\displaystyle\sum_{n=0}^N A_n x^n \right)$, the solution has the form

$$y(x) = \ln x \sum_{n=0}^N \frac{A_n}{B_n} x^{n+1} - \sum_{n=0}^N \frac{A_n C_n}{B_n^2} x^{n+1}, \qquad B_n = \int_0^1 \frac{t^{n+1}\,dt}{a+bt^2}, \qquad C_n = \int_0^1 \frac{t^{n+1} \ln t}{a+bt^2}\,dt.$$

29. $\displaystyle\int_0^x \frac{y(t)\,dt}{ax^m + bt^m} = f(x)$, $\qquad a > 0$, $\quad a + b > 0$, $\quad m = 1, 2, \ldots$

1°. For a polynomial right-hand side, $f(x) = \displaystyle\sum_{n=0}^{N} A_n x^n$, the solution has the form

$$y(x) = \sum_{n=0}^{N} \frac{A_n}{B_n} x^{m+n-1}, \qquad B_n = \int_0^1 \frac{t^{m+n-1}\,dt}{a + bt^m}.$$

2°. For $f(x) = x^\lambda \displaystyle\sum_{n=0}^{N} A_n x^n$, where λ is an arbitrary number ($\lambda > -1$), the solution has the form

$$y(x) = x^\lambda \sum_{n=0}^{N} \frac{A_n}{B_n} x^{m+n-1}, \qquad B_n = \int_0^1 \frac{t^{\lambda+m+n-1}\,dt}{a + bt^m}.$$

3°. For $f(x) = \ln x \left(\displaystyle\sum_{n=0}^{N} A_n x^n \right)$, the solution has the form

$$y(x) = \ln x \sum_{n=0}^{N} \frac{A_n}{B_n} x^{m+n-1} - \sum_{n=0}^{N} \frac{A_n C_n}{B_n^2} x^{m+n-1},$$

$$B_n = \int_0^1 \frac{t^{m+n-1}\,dt}{a + bt^m}, \qquad C_n = \int_0^1 \frac{t^{m+n-1}\ln t}{a + bt^m}\,dt.$$

1.1-6. Kernels Containing Square Roots

30. $\displaystyle\int_a^x \sqrt{x - t}\, y(t)\,dt = f(x)$.

Differentiating with respect to x, we arrive at Abel's equation 1.1.36:

$$\int_a^x \frac{y(t)\,dt}{\sqrt{x-t}} = 2f'_x(x).$$

Solution:

$$y(x) = \frac{2}{\pi} \frac{d^2}{dx^2} \int_a^x \frac{f(t)\,dt}{\sqrt{x-t}}.$$

31. $\displaystyle\int_a^x \left(\sqrt{x} - \sqrt{t} \right) y(t)\,dt = f(x)$.

This is a special case of equation 1.1.44 with $\mu = \frac{1}{2}$.

Solution: $y(x) = 2\dfrac{d}{dx}\left[\sqrt{x}\, f'_x(x) \right]$.

32. $\displaystyle\int_a^x \left(A\sqrt{x} + B\sqrt{t} \right) y(t)\,dt = f(x)$.

This is a special case of equation 1.1.45 with $\mu = \frac{1}{2}$.

33. $\displaystyle\int_a^x \left(1 + b\sqrt{x-t}\right) y(t)\,dt = f(x).$

Differentiating with respect to x, we arrive at Abel's equation of the second kind 2.1.46:

$$y(x) + \frac{b}{2} \int_a^x \frac{y(t)\,dt}{\sqrt{x-t}} = f'_x(x).$$

34. $\displaystyle\int_a^x \left(t\sqrt{x} - x\sqrt{t}\right) y(t)\,dt = f(x).$

This is a special case of equation 1.9.11 with $g(x) = \sqrt{x}$ and $h(x) = x$.

35. $\displaystyle\int_a^x \left(At\sqrt{x} + Bx\sqrt{t}\right) y(t)\,dt = f(x).$

This is a special case of equation 1.9.12 with $g(x) = \sqrt{x}$ and $h(t) = t$.

36. $\displaystyle\int_a^x \frac{y(t)\,dt}{\sqrt{x-t}} = f(x).$

Abel's equation.
    Solution:

$$y(x) = \frac{1}{\pi}\frac{d}{dx}\int_a^x \frac{f(t)\,dt}{\sqrt{x-t}} = \frac{f(a)}{\pi\sqrt{x-a}} + \frac{1}{\pi}\int_a^x \frac{f'_t(t)\,dt}{\sqrt{x-t}}.$$

⊙   Reference: E. T. Whittacker and G. N. Watson (1958).

37. $\displaystyle\int_a^x \left(b + \frac{1}{\sqrt{x-t}}\right) y(t)\,dt = f(x).$

Let us rewrite the equation in the form

$$\int_a^x \frac{y(t)\,dt}{\sqrt{x-t}} = f(x) - b\int_a^x y(t)\,dt.$$

Assuming the right-hand side to be known, we solve this equation as Abel's equation 1.1.36. After some manipulations, we arrive at Abel's equation of the second kind 2.1.46:

$$y(x) + \frac{b}{\pi}\int_a^x \frac{y(t)\,dt}{\sqrt{x-t}} = F(x), \qquad \text{where} \quad F(x) = \frac{1}{\pi}\frac{d}{dx}\int_a^x \frac{f(t)\,dt}{\sqrt{x-t}}.$$

38. $\displaystyle\int_a^x \left(\frac{1}{\sqrt{x}} - \frac{1}{\sqrt{t}}\right) y(t)\,dt = f(x).$

This is a special case of equation 1.1.44 with $\mu = -\frac{1}{2}$.
    Solution: $y(x) = -2\left[x^{3/2}f'_x(x)\right]'_x, \qquad a > 0.$

39. $\displaystyle\int_a^x \left(\frac{A}{\sqrt{x}} + \frac{B}{\sqrt{t}}\right) y(t)\,dt = f(x).$

This is a special case of equation 1.1.45 with $\mu = -\frac{1}{2}$.

40. $\displaystyle\int_a^x \frac{y(t)\,dt}{\sqrt{x^2 - t^2}} = f(x).$

Solution: $\displaystyle y = \frac{2}{\pi}\frac{d}{dx}\int_a^x \frac{t f(t)\,dt}{\sqrt{x^2 - t^2}}.$

⊙ Reference: P. P. Zabreyko, A. I. Koshelev, et al. (1975).

41. $\displaystyle\int_0^x \frac{y(t)\,dt}{\sqrt{ax^2 + bt^2}} = f(x),\qquad a > 0,\quad a + b > 0.$

$1°$. For a polynomial right-hand side, $f(x) = \sum\limits_{n=0}^N A_n x^n$, the solution has the form

$$y(x) = \sum_{n=0}^N \frac{A_n}{B_n} x^n,\qquad B_n = \int_0^1 \frac{t^n\,dt}{\sqrt{a + bt^2}}.$$

$2°$. For $f(x) = x^\lambda \sum\limits_{n=0}^N A_n x^n$, where λ is an arbitrary number $(\lambda > -1)$, the solution has the form

$$y(x) = x^\lambda \sum_{n=0}^N \frac{A_n}{B_n} x^n,\qquad B_n = \int_0^1 \frac{t^{\lambda+n}\,dt}{\sqrt{a + bt^2}}.$$

$3°$. For $f(x) = \ln x \left(\sum\limits_{n=0}^N A_n x^n \right)$, the solution has the form

$$y(x) = \ln x \sum_{n=0}^N \frac{A_n}{B_n} x^n - \sum_{n=0}^N \frac{A_n C_n}{B_n^2} x^n,\qquad B_n = \int_0^1 \frac{t^n\,dt}{\sqrt{a + bt^2}},\qquad C_n = \int_0^1 \frac{t^n \ln t}{\sqrt{a + bt^2}}\,dt.$$

$4°$. For $f(x) = \sum\limits_{n=0}^N A_n (\ln x)^n$, the solution of the equation has the form

$$y(x) = \sum_{n=0}^N A_n Y_n(x),$$

where the functions $Y_n = Y_n(x)$ are given by

$$Y_n(x) = \left\{ \frac{d^n}{d\lambda^n}\left[\frac{x^\lambda}{I(\lambda)} \right] \right\}_{\lambda=0},\qquad I(\lambda) = \int_0^1 \frac{z^\lambda\,dz}{\sqrt{a + bz^2}}.$$

$5°$. For $f(x) = \sum\limits_{n=1}^N A_n \cos(\lambda_n \ln x) + \sum\limits_{n=1}^N B_n \sin(\lambda_n \ln x)$, the solution of the equation has the form

$$y(x) = \sum_{n=1}^N C_n \cos(\lambda_n \ln x) + \sum_{n=1}^N D_n \sin(\lambda_n \ln x),$$

where the constants C_n and D_n are found by the method of undetermined coefficients.

1.1-7. Kernels Containing Arbitrary Powers

42. $\displaystyle\int_a^x (x-t)^\lambda y(t)\,dt = f(x), \qquad f(a) = 0, \quad 0 < \lambda < 1.$

Differentiating with respect to x, we arrive at the generalized Abel equation 1.1.46:

$$\int_a^x \frac{y(t)\,dt}{(x-t)^{1-\lambda}} = \frac{1}{\lambda}f'_x(x).$$

Solution:

$$y(x) = k\frac{d^2}{dx^2}\int_a^x \frac{f(t)\,dt}{(x-t)^\lambda}, \qquad k = \frac{\sin(\pi\lambda)}{\pi\lambda}.$$

⊙ Reference: F. D. Gakhov (1977).

43. $\displaystyle\int_a^x (x-t)^\mu y(t)\,dt = f(x).$

For $\mu = 0, 1, 2, \ldots$, see equations 1.1.1, 1.1.2, 1.1.4, 1.1.12, and 1.1.23. For $-1 < \mu < 0$, see equation 1.1.42.

Set $\mu = n - \lambda$, where $n = 1, 2, \ldots$ and $0 \le \lambda < 1$, and $f(a) = f'_x(a) = \cdots = f_x^{(n-1)}(a) = 0$. On differentiating the equation n times, we arrive at an equation of the form 1.1.46:

$$\int_a^x \frac{y(t)\,d\tau}{(x-t)^\lambda} = \frac{\Gamma(\mu - n + 1)}{\Gamma(\mu + 1)}f_x^{(n)}(x),$$

where $\Gamma(\mu)$ is the gamma function.

> **Example.** Set $f(x) = Ax^\beta$, where $\beta \ge 0$, and let $\mu > -1$ and $\mu - \beta \ne 0, 1, 2, \ldots$ In this case, the solution has the form $y(x) = \dfrac{A\Gamma(\beta + 1)}{\Gamma(\mu + 1)\Gamma(\beta - \mu)}x^{\beta-\mu-1}.$

⊙ Reference: M. L. Krasnov, A. I. Kisilev, and G. I. Makarenko (1971).

44. $\displaystyle\int_a^x (x^\mu - t^\mu)y(t)\,dt = f(x).$

This is a special case of equation 1.9.2 with $g(x) = x^\mu$.

Solution: $y(x) = \dfrac{1}{\mu}\left[x^{1-\mu}f'_x(x)\right]'_x.$

45. $\displaystyle\int_a^x \left(Ax^\mu + Bt^\mu\right)y(t)\,dt = f(x).$

For $B = -A$, see equation 1.1.44. This is a special case of equation 1.9.4 with $g(x) = x^\mu$.

Solution: $y(x) = \dfrac{1}{A + B}\dfrac{d}{dx}\left[x^{-\frac{A\mu}{A+B}}\int_a^x t^{-\frac{B\mu}{A+B}}f'_t(t)\,dt\right].$

46. $\displaystyle\int_a^x \frac{y(t)\,dt}{(x-t)^\lambda} = f(x), \qquad 0 < \lambda < 1.$

The generalized Abel equation.

Solution:

$$y(x) = \frac{\sin(\pi\lambda)}{\pi}\frac{d}{dx}\int_a^x \frac{f(t)\,dt}{(x-t)^{1-\lambda}} = \frac{\sin(\pi\lambda)}{\pi}\left[\frac{f(a)}{(x-a)^{1-\lambda}} + \int_a^x \frac{f'_t(t)\,dt}{(x-t)^{1-\lambda}}\right].$$

⊙ Reference: E. T. Whittacker and G. N. Watson (1958).

47. $\displaystyle\int_a^x \left[b + \frac{1}{(x-t)^\lambda} \right] y(t)\,dt = f(x), \qquad 0 < \lambda < 1.$

Rewrite the equation in the form

$$\int_a^x \frac{y(t)\,dt}{(x-t)^\lambda} = f(x) - b \int_a^x y(t)\,dt.$$

Assuming the right-hand side to be known, we solve this equation as the generalized Abel equation 1.1.46. After some manipulations, we arrive at Abel's equation of the second kind 2.1.60:

$$y(x) + \frac{b\sin(\pi\lambda)}{\pi} \int_a^x \frac{y(t)\,dt}{(x-t)^{1-\lambda}} = F(x), \qquad \text{where} \quad F(x) = \frac{\sin(\pi\lambda)}{\pi} \frac{d}{dx} \int_a^x \frac{f(t)\,dt}{(x-t)^{1-\lambda}}.$$

48. $\displaystyle\int_a^x \left(\sqrt{x} - \sqrt{t} \right)^\lambda y(t)\,dt = f(x), \qquad 0 < \lambda < 1.$

Solution:

$$y(x) = \frac{k}{\sqrt{x}} \left(\sqrt{x}\, \frac{d}{dx} \right)^2 \int_a^x \frac{f(t)\,dt}{\sqrt{t}\left(\sqrt{x} - \sqrt{t} \right)^\lambda}, \qquad k = \frac{\sin(\pi\lambda)}{\pi\lambda}.$$

49. $\displaystyle\int_a^x \frac{y(t)\,dt}{\left(\sqrt{x} - \sqrt{t} \right)^\lambda} = f(x), \qquad 0 < \lambda < 1.$

Solution:

$$y(x) = \frac{\sin(\pi\lambda)}{2\pi} \frac{d}{dx} \int_a^x \frac{f(t)\,dt}{\sqrt{t}\left(\sqrt{x} - \sqrt{t} \right)^{1-\lambda}}.$$

50. $\displaystyle\int_a^x \left(Ax^\lambda + Bt^\mu \right) y(t)\,dt = f(x).$

This is a special case of equation 1.9.6 with $g(x) = Ax^\lambda$ and $h(t) = Bt^\mu$.

51. $\displaystyle\int_a^x \left[1 + A(x^\lambda t^\mu - x^{\lambda+\mu}) \right] y(t)\,dt = f(x).$

This is a special case of equation 1.9.13 with $g(x) = Ax^\mu$ and $h(x) = x^\lambda$.
   Solution:

$$y(x) = \frac{d}{dx} \left\{ \frac{x^\lambda}{\Phi(x)} \int_a^x \left[t^{-\lambda} f(t) \right]_t' \Phi(t)\,dt \right\}, \qquad \Phi(x) = \exp\left(-\frac{A\mu}{\mu+\lambda} x^{\mu+\lambda} \right).$$

52. $\displaystyle\int_a^x \left(Ax^\beta t^\gamma + Bx^\delta t^\lambda \right) y(t)\,dt = f(x).$

This is a special case of equation 1.9.15 with $g_1(x) = Ax^\beta$, $h_1(t) = t^\gamma$, $g_2(x) = Bx^\delta$, and $h_2(t) = t^\lambda$.

53. $\displaystyle\int_a^x \left[Ax^\lambda(t^\mu - x^\mu) + Bx^\beta(t^\gamma - x^\gamma) \right] y(t)\,dt = f(x).$

This is a special case of equation 1.9.45 with $g_1(x) = Ax^\lambda$, $h_1(x) = x^\mu$, $g_2(x) = Bx^\beta$, and $h_2(x) = x^\gamma$.

54. $\int_a^x \left[A x^\lambda t^\mu + B x^{\lambda+\beta} t^{\mu-\beta} - (A+B) x^{\lambda+\gamma} t^{\mu-\gamma} \right] y(t)\, dt = f(x).$

This is a special case of equation 1.9.47 with $g(x) = x$.

55. $\int_a^x t^\sigma (x^\mu - t^\mu)^\lambda y(t)\, dt = f(x), \qquad \sigma > -1, \quad \mu > 0, \quad \lambda > -1.$

The transformation $\tau = t^\mu$, $z = x^\mu$, $w(\tau) = t^{\sigma-\mu+1} y(t)$ leads to an equation of the form 1.1.42:

$$\int_A^z (z - \tau)^\lambda w(\tau)\, d\tau = F(z),$$

where $A = a^\mu$ and $F(z) = \mu f(z^{1/\mu})$.

Solution with $-1 < \lambda < 0$:

$$y(x) = -\frac{\mu \sin(\pi\lambda)}{\pi x^\sigma} \frac{d}{dx} \left[\int_a^x t^{\mu-1} (x^\mu - t^\mu)^{-1-\lambda} f(t)\, dt \right].$$

56. $\int_0^x \frac{y(t)\, dt}{(x+t)^\mu} = f(x).$

This is a special case of equation 1.1.57 with $\lambda = 1$ and $a = b = 1$.

The transformation

$$x = \tfrac{1}{2} e^{2z}, \quad t = \tfrac{1}{2} e^{2\tau}, \quad y(t) = e^{(\mu-2)\tau} w(\tau), \quad f(x) = e^{-\mu z} g(z)$$

leads to an equation with difference kernel of the form 1.9.26:

$$\int_{-\infty}^z \frac{w(\tau)\, d\tau}{\cosh^\mu(z-\tau)} = g(z).$$

57. $\int_0^x \frac{y(t)\, dt}{(a x^\lambda + b t^\lambda)^\mu} = f(x), \qquad a > 0, \quad a + b > 0.$

$1°$. The substitution $t = xz$ leads to a special case of equation 3.8.45:

$$\int_0^1 \frac{y(xz)\, dz}{(a + b z^\lambda)^\mu} = x^{\lambda\mu-1} f(x). \qquad (1)$$

$2°$. For a polynomial right-hand side, $f(x) = \sum_{m=0}^n A_m x^m$, the solution has the form

$$y(x) = x^{\lambda\mu-1} \sum_{m=0}^n \frac{A_m}{I_m} x^m, \qquad I_m = \int_0^1 \frac{z^{m+\lambda\mu-1}\, dz}{(a + b z^\lambda)^\mu}.$$

The integrals I_m are supposed to be convergent.

$3°$. The solution structure for some other right-hand sides of the integral equation may be obtained using (1) and the results presented for the more general equation 3.8.45 (see also equations 3.8.26–3.8.32).

$4°$. For $a = b$, the equation can be reduced, just as equation 1.1.56, to an integral equation with difference kernel of the form 1.9.26.

58. $\int_a^x \frac{\left(\sqrt{x} + \sqrt{x-t}\right)^{2\lambda} + \left(\sqrt{x} - \sqrt{x-t}\right)^{2\lambda}}{2t^\lambda \sqrt{x-t}} y(t)\, dt = f(x).$

The equation can be rewritten in terms of the Gaussian hypergeometric functions in the form

$$\int_a^x (x-t)^{\gamma-1} F\left(\lambda, -\lambda, \gamma; 1 - \frac{x}{t}\right) y(t)\, dt = f(x), \qquad \text{where} \quad \gamma = \tfrac{1}{2}.$$

See 1.8.86 for the solution of this equation.

1.2. Equations Whose Kernels Contain Exponential Functions

1. $\displaystyle\int_a^x e^{\lambda(x-t)} y(t)\, dt = f(x).$

Solution: $y(x) = f'_x(x) - \lambda f(x)$.

> **Example.** In the special case $a = 0$ and $f(x) = Ax$, the solution has the form $y(x) = A(1 - \lambda x)$.

2. $\displaystyle\int_a^x e^{\lambda x + \beta t} y(t)\, dt = f(x).$

Solution: $y(x) = e^{-(\lambda+\beta)x}\left[f'_x(x) - \lambda f(x)\right]$.

> **Example.** In the special case $a = 0$ and $f(x) = A\sin(\gamma x)$, the solution has the form $y(x) = A e^{-(\lambda+\beta)x} \times [\gamma\cos(\gamma x) - \lambda\sin(\gamma x)]$.

3. $\displaystyle\int_a^x \left[e^{\lambda(x-t)} - 1\right] y(t)\, dt = f(x),\qquad f(a) = f'_x(a) = 0.$

Solution: $y(x) = \frac{1}{\lambda} f''_{xx}(x) - f'_x(x)$.

4. $\displaystyle\int_a^x \left[e^{\lambda(x-t)} + b\right] y(t)\, dt = f(x).$

For $b = -1$, see equation 1.2.3. Differentiating with respect to x yields an equation of the form 2.2.1:

$$y(x) + \frac{\lambda}{b+1}\int_a^x e^{\lambda(x-t)} y(t)\, dt = \frac{f'_x(x)}{b+1}.$$

Solution:

$$y(x) = \frac{f'_x(x)}{b+1} - \frac{\lambda}{(b+1)^2}\int_a^x \exp\left[\frac{\lambda b}{b+1}(x-t)\right] f'_t(t)\, dt.$$

5. $\displaystyle\int_a^x \left(e^{\lambda x + \beta t} + b\right) y(t)\, dt = f(x).$

For $\beta = -\lambda$, see equation 1.2.4. This is a special case of equation 1.9.15 with $g_1(x) = e^{\lambda x}$, $h_1(t) = e^{\beta t}$, $g_2(x) = 1$, and $h_2(t) = b$.

6. $\displaystyle\int_a^x \left(e^{\lambda x} - e^{\lambda t}\right) y(t)\, dt = f(x),\qquad f(a) = f'_x(a) = 0.$

This is a special case of equation 1.9.2 with $g(x) = e^{\lambda x}$.

Solution: $y(x) = e^{-\lambda x}\left[\dfrac{1}{\lambda} f''_{xx}(x) - f'_x(x)\right].$

7. $\displaystyle\int_a^x \left(e^{\lambda x} - e^{\lambda t} + b\right) y(t)\, dt = f(x).$

For $b = 0$, see equation 1.2.6. This is a special case of equation 1.9.3 with $g(x) = e^{\lambda x}$.

Solution:

$$y(x) = \frac{1}{b} f'_x(x) - \frac{\lambda}{b^2} e^{\lambda x}\int_a^x \exp\left(\frac{e^{\lambda t} - e^{\lambda x}}{b}\right) f'_t(t)\, dt.$$

8. $\displaystyle\int_a^x \left(Ae^{\lambda x} + Be^{\lambda t}\right) y(t)\, dt = f(x).$

For $B = -A$, see equation 1.2.6. This is a special case of equation 1.9.4 with $g(x) = e^{\lambda x}$.

Solution: $\displaystyle y(x) = \frac{1}{A+B} \frac{d}{dx} \left[\exp\left(-\frac{A\lambda}{A+B}x\right) \int_a^x \exp\left(-\frac{B\lambda}{A+B}t\right) f_t'(t)\, dt \right].$

9. $\displaystyle\int_a^x \left(Ae^{\lambda x} + Be^{\lambda t} + C\right) y(t)\, dt = f(x).$

This is a special case of equation 1.9.5 with $g(x) = e^{\lambda x}$.

10. $\displaystyle\int_a^x \left(Ae^{\lambda x} + Be^{\mu t}\right) y(t)\, dt = f(x).$

For $\lambda = \mu$, see equation 1.2.8. This is a special case of equation 1.9.6 with $g(x) = Ae^{\lambda x}$ and $h(t) = Be^{\mu t}$.

11. $\displaystyle\int_a^x \left[e^{\lambda(x-t)} - e^{\mu(x-t)} \right] y(t)\, dt = f(x), \qquad f(a) = f_x'(a) = 0.$

Solution:
$$y(x) = \frac{1}{\lambda - \mu} \left[f_{xx}'' - (\lambda + \mu) f_x' + \lambda\mu f \right], \qquad f = f(x).$$

12. $\displaystyle\int_a^x \left[Ae^{\lambda(x-t)} + Be^{\mu(x-t)} \right] y(t)\, dt = f(x).$

For $B = -A$, see equation 1.2.11. This is a special case of equation 1.9.15 with $g_1(x) = Ae^{\lambda x}$, $h_1(t) = e^{-\lambda t}$, $g_2(x) = Be^{\mu x}$, and $h_2(t) = e^{-\mu t}$.

Solution:
$$y(x) = \frac{e^{\lambda x}}{A+B} \frac{d}{dx} \left\{ e^{(\mu-\lambda)x} \Phi(x) \int_a^x \left[\frac{f(t)}{e^{\mu t}} \right]_t' \frac{dt}{\Phi(t)} \right\}, \qquad \Phi(x) = \exp\left[\frac{B(\lambda - \mu)}{A+B} x \right].$$

13. $\displaystyle\int_a^x \left[Ae^{\lambda(x-t)} + Be^{\mu(x-t)} + C \right] y(t)\, dt = f(x).$

This is a special case of equation 1.2.14 with $\beta = 0$.

14. $\displaystyle\int_a^x \left[Ae^{\lambda(x-t)} + Be^{\mu(x-t)} + Ce^{\beta(x-t)} \right] y(t)\, dt = f(x).$

Differentiating the equation with respect to x yields

$$(A + B + C)y(x) + \int_a^x \left[A\lambda e^{\lambda(x-t)} + B\mu e^{\mu(x-t)} + C\beta e^{\beta(x-t)} \right] y(t)\, dt = f_x'(x).$$

Eliminating the term with $e^{\beta(x-t)}$ with the aid of the original equation, we arrive at an equation of the form 2.2.10:

$$(A + B + C)y(x) + \int_a^x \left[A(\lambda - \beta)e^{\lambda(x-t)} + B(\mu - \beta)e^{\mu(x-t)} \right] y(t)\, dt = f_x'(x) - \beta f(x).$$

In the special case $A + B + C = 0$, this is an equation of the form 1.2.12.

15. $\displaystyle\int_a^x \left[Ae^{\lambda(x-t)} + Be^{\mu(x-t)} + Ce^{\beta(x-t)} - A - B - C\right]y(t)\,dt = f(x),$ $f(a) = f_x'(a) = 0.$

Differentiating with respect to x, we arrive at an equation of the form 1.2.14:

$$\int_a^x \left[A\lambda e^{\lambda(x-t)} + B\mu e^{\mu(x-t)} + C\beta e^{\beta(x-t)}\right]y(t)\,dt = f_x'(x).$$

16. $\displaystyle\int_a^x \left(e^{\lambda x + \mu t} - e^{\mu x + \lambda t}\right)y(t)\,dt = f(x),$ $f(a) = f_x'(a) = 0.$

This is a special case of equation 1.9.11 with $g(x) = e^{\lambda x}$ and $h(t) = e^{\mu t}$.
    Solution:
$$y(x) = \frac{f_{xx}'' - (\lambda + \mu)f_x'(x) + \lambda\mu f(x)}{(\lambda - \mu)\exp[(\lambda + \mu)x]}.$$

17. $\displaystyle\int_a^x \left(Ae^{\lambda x + \mu t} + Be^{\mu x + \lambda t}\right)y(t)\,dt = f(x).$

For $B = -A$, see equation 1.2.16. This is a special case of equation 1.9.12 with $g(x) = e^{\lambda x}$ and $h(t) = e^{\mu t}$.
    Solution:
$$y(x) = \frac{1}{(A + B)e^{\mu x}}\frac{d}{dx}\left\{\Phi^A(x)\int_a^x \Phi^B(t)\frac{d}{dt}\left[\frac{f(t)}{e^{\mu t}}\right]dt\right\}, \qquad \Phi(x) = \exp\left(\frac{\mu - \lambda}{A + B}x\right).$$

18. $\displaystyle\int_a^x \left(Ae^{\lambda x + \mu t} + Be^{\beta x + \gamma t}\right)y(t)\,dt = f(x).$

This is a special case of equation 1.9.15 with $g_1(x) = Ae^{\lambda x}$, $h_1(t) = e^{\mu t}$, $g_2(x) = Be^{\beta x}$, and $h_2(t) = e^{\gamma t}$.

19. $\displaystyle\int_a^x \left(Ae^{2\lambda x} + Be^{2\beta t} + Ce^{\lambda x} + De^{\beta t} + E\right)y(t)\,dt = f(x).$

This is a special case of equation 1.9.6 with $g(x) = Ae^{2\lambda x} + Ce^{\lambda x}$ and $h(t) = Be^{2\beta t} + De^{\beta t} + E$.

20. $\displaystyle\int_a^x \left(Ae^{\lambda x + \beta t} + Be^{2\beta t} + Ce^{\lambda x} + De^{\beta t} + E\right)y(t)\,dt = f(x).$

This is a special case of equation 1.9.15 with $g_1(x) = e^{\lambda x}$, $h_1(t) = Ae^{\beta t} + D$, and $g_2(x) = 1$, $h_2(t) = Be^{2\beta t} + De^{\beta t} + E$.

21. $\displaystyle\int_a^x \left(Ae^{2\lambda x} + Be^{\lambda x + \beta t} + Ce^{\lambda x} + De^{\beta t} + E\right)y(t)\,dt = f(x).$

This is a special case of equation 1.9.15 with $g_1(x) = Be^{\lambda x} + D$, $h_1(t) = e^{\beta t}$, and $g_2(x) = Ae^{2\lambda x} + Ce^{\lambda x} + E$, $h_2(t) = 1$.

22. $\displaystyle\int_a^x \left[1 + Ae^{\lambda x}(e^{\mu t} - e^{\mu x})y(t)\,dt = f(x).\right.$

This is a special case of equation 1.9.13 with $g(x) = e^{\mu x}$ and $h(x) = Ae^{\lambda x}$.
    Solution:
$$y(x) = \frac{d}{dx}\left\{e^{\lambda x}\Phi(x)\int_a^x \left[\frac{f(t)}{e^{\lambda t}}\right]_t'\frac{dt}{\Phi(t)}\right\}, \qquad \Phi(x) = \exp\left[\frac{A\mu}{\lambda + \mu}e^{(\lambda + \mu)x}\right].$$

23. $\int_a^x \left[A e^{\lambda x}(e^{\mu x} - e^{\mu t}) + B e^{\beta x}(e^{\gamma x} - e^{\gamma t}) \right] y(t)\, dt = f(x).$

This is a special case of equation 1.9.45 with $g_1(x) = A e^{\lambda x}$, $h_1(t) = -e^{\mu t}$, $g_2(x) = B e^{\beta x}$, and $h_1(t) = -e^{\gamma t}$.

24. $\int_a^x \big\{ A \exp(\lambda x + \mu t) + B \exp[(\lambda + \beta)x + (\mu - \beta)t]$
$$- (A + B) \exp[(\lambda + \gamma)x + (\mu - \gamma)t] \big\} y(t)\, dt = f(x).$$

This is a special case of equation 1.9.47 with $g_1(x) = e^{\lambda x}$.

25. $\int_a^x \left(e^{\lambda x} - e^{\lambda t} \right)^n y(t)\, dt = f(x), \qquad n = 1, 2, \ldots$

Solution:
$$y(x) = \frac{1}{\lambda^n n!} e^{\lambda x} \left(\frac{1}{e^{\lambda x}} \frac{d}{dx} \right)^{n+1} f(x).$$

26. $\int_a^x \sqrt{e^{\lambda x} - e^{\lambda t}}\; y(t)\, dt = f(x), \qquad \lambda > 0.$

Solution:
$$y(x) = \frac{2}{\pi} e^{\lambda x} \left(e^{-\lambda x} \frac{d}{dx} \right)^2 \int_a^x \frac{e^{\lambda t} f(t)\, dt}{\sqrt{e^{\lambda x} - e^{\lambda t}}}.$$

27. $\int_a^x \frac{y(t)\, dt}{\sqrt{e^{\lambda x} - e^{\lambda t}}} = f(x), \qquad \lambda > 0.$

Solution:
$$y(x) = \frac{\lambda}{\pi} \frac{d}{dx} \int_a^x \frac{e^{\lambda t} f(t)\, dt}{\sqrt{e^{\lambda x} - e^{\lambda t}}}.$$

28. $\int_a^x (e^{\lambda x} - e^{\lambda t})^\mu y(t)\, dt = f(x), \qquad \lambda > 0, \quad 0 < \mu < 1.$

Solution:
$$y(x) = k e^{\lambda x} \left(e^{-\lambda x} \frac{d}{dx} \right)^2 \int_a^x \frac{e^{\lambda t} f(t)\, dt}{(e^{\lambda x} - e^{\lambda t})^\mu}, \qquad k = \frac{\sin(\pi \mu)}{\pi \mu}.$$

29. $\int_a^x \frac{y(t)\, dt}{(e^{\lambda x} - e^{\lambda t})^\mu} = f(x), \qquad \lambda > 0, \quad 0 < \mu < 1.$

Solution:
$$y(x) = \frac{\lambda \sin(\pi \mu)}{\pi} \frac{d}{dx} \int_a^x \frac{e^{\lambda t} f(t)\, dt}{(e^{\lambda x} - e^{\lambda t})^{1-\mu}}.$$

1.2-2. Kernels Containing Power-Law and Exponential Functions

30. $\int_a^x \left[A(x - t) + B e^{\lambda(x-t)} \right] y(t)\, dt = f(x).$

Differentiating with respect to x, we arrive at an equation of the form 2.2.4:
$$B y(x) + \int_a^x \left[A + B\lambda e^{\lambda(x-t)} \right] y(t)\, dt = f'_x(x).$$

31. $\int_a^x (x-t)e^{\lambda(x-t)}y(t)\,dt = f(x), \qquad f(a) = f'_x(a) = 0.$

Solution: $y(x) = f''_{xx}(x) - 2\lambda f'_x(x) + \lambda^2 f(x).$

32. $\int_a^x (Ax + Bt + C)e^{\lambda(x-t)}y(t)\,dt = f(x).$

The substitution $u(x) = e^{-\lambda x}y(x)$ leads to an equation of the form 1.1.3:

$$\int_a^x (Ax + Bt + C)u(t)\,dt = e^{-\lambda x}f(x).$$

33. $\int_a^x (Axe^{\lambda t} + Bte^{\mu x})y(t)\,dt = f(x).$

This is a special case of equation 1.9.15 with $g_1(x) = Ax$, $h_1(t) = e^{\lambda t}$, and $g_2(x) = Be^{\mu x}$, $h_2(t) = t.$

34. $\int_a^x \left[Axe^{\lambda(x-t)} + Bte^{\mu(x-t)}\right]y(t)\,dt = f(x).$

This is a special case of equation 1.9.15 with $g_1(x) = Axe^{\lambda x}$, $h_1(t) = e^{-\lambda t}$, $g_2(x) = Be^{\mu x}$, and $h_2(t) = te^{-\mu t}.$

35. $\int_a^x (x-t)^2 e^{\lambda(x-t)}y(t)\,dt = f(x), \qquad f(a) = f'_x(a) = f''_{xx}(a) = 0.$

Solution: $y(x) = \frac{1}{2}\left[f'''_{xxx}(x) - 3\lambda f''_{xx}(x) + 3\lambda^2 f'_x(x) - \lambda^3 f(x)\right].$

36. $\int_a^x (x-t)^n e^{\lambda(x-t)}y(t)\,dt = f(x), \qquad n = 1, 2, \ldots$

It is assumed that $f(a) = f'_x(a) = \cdots = f^{(n)}_x(a) = 0.$

Solution: $y(x) = \dfrac{1}{n!}e^{\lambda x}\dfrac{d^{n+1}}{dx^{n+1}}\left[e^{-\lambda x}f(x)\right].$

37. $\int_a^x (Ax^\beta + Be^{\lambda t})y(t)\,dt = f(x).$

This is a special case of equation 1.9.6 with $g(x) = Ax^\beta$ and $h(t) = Be^{\lambda t}.$

38. $\int_a^x (Ae^{\lambda x} + Bt^\beta)y(t)\,dt = f(x).$

This is a special case of equation 1.9.6 with $g(x) = Ae^{\lambda x}$ and $h(t) = Bt^\beta.$

39. $\int_a^x (Ax^\beta e^{\lambda t} + Bt^\gamma e^{\mu x})y(t)\,dt = f(x).$

This is a special case of equation 1.9.15 with $g_1(x) = Ax^\beta$, $h_1(t) = e^{\lambda t}$, $g_2(x) = Be^{\mu x}$, and $h_2(t) = t^\gamma.$

40. $\int_a^x e^{\lambda(x-t)}\sqrt{x-t}\,y(t)\,dt = f(x).$

Solution:

$$y(x) = \frac{2}{\pi}e^{\lambda x}\frac{d^2}{dx^2}\int_a^x \frac{e^{-\lambda t}f(t)\,dt}{\sqrt{x-t}}.$$

41. $\displaystyle\int_a^x \frac{e^{\lambda(x-t)}}{\sqrt{x-t}} y(t)\,dt = f(x).$

Solution:

$$y(x) = \frac{1}{\pi} e^{\lambda x} \frac{d}{dx} \int_a^x \frac{e^{-\lambda t} f(t)\,dt}{\sqrt{x-t}}.$$

42. $\displaystyle\int_a^x (x-t)^\lambda e^{\mu(x-t)} y(t)\,dt = f(x), \qquad 0 < \lambda < 1.$

Solution:

$$y(x) = k e^{\mu x} \frac{d^2}{dx^2} \int_a^x \frac{e^{-\mu t} f(t)\,dt}{(x-t)^\lambda}, \qquad k = \frac{\sin(\pi\lambda)}{\pi\lambda}.$$

43. $\displaystyle\int_a^x \frac{e^{\lambda(x-t)}}{(x-t)^\mu} y(t)\,dt = f(x), \qquad 0 < \mu < 1.$

Solution:

$$y(x) = \frac{\sin(\pi\mu)}{\pi} e^{\lambda x} \frac{d}{dx} \int_a^x \frac{e^{-\lambda t} f(t)}{(x-t)^{1-\mu}}\,dt.$$

44. $\displaystyle\int_a^x \left(\sqrt{x} - \sqrt{t}\right)^\lambda e^{\mu(x-t)} y(t)\,dt = f(x), \qquad 0 < \lambda < 1.$

The substitution $u(x) = e^{-\mu x} y(x)$ leads to an equation of the form 1.1.48:

$$\int_a^x \left(\sqrt{x} - \sqrt{t}\right)^\lambda u(t)\,dt = e^{-\mu x} f(x).$$

45. $\displaystyle\int_a^x \frac{e^{\mu(x-t)} y(t)\,dt}{\left(\sqrt{x} - \sqrt{t}\right)^\lambda} = f(x), \qquad 0 < \lambda < 1.$

The substitution $u(x) = e^{-\mu x} y(x)$ leads to an equation of the form 1.1.49:

$$\int_a^x \frac{u(t)\,dt}{\left(\sqrt{x} - \sqrt{t}\right)^\lambda} = e^{-\mu x} f(x).$$

46. $\displaystyle\int_a^x \frac{e^{\lambda(x-t)}}{\sqrt{x^2 - t^2}} y(t)\,dt = f(x).$

Solution: $\displaystyle y = \frac{2}{\pi} e^{\lambda x} \frac{d}{dx} \int_a^x \frac{t e^{-\lambda t}}{\sqrt{x^2 - t^2}} f(t)\,dt.$

47. $\displaystyle\int_a^x \exp[\lambda(x^2 - t^2)] y(t)\,dt = f(x).$

Solution: $y(x) = f'_x(x) - 2\lambda x f(x).$

48. $\displaystyle\int_a^x [\exp(\lambda x^2) - \exp(\lambda t^2)] y(t)\,dt = f(x).$

This is a special case of equation 1.9.2 with $g(x) = \exp(\lambda x^2)$.

Solution: $\displaystyle y(x) = \frac{1}{2\lambda} \frac{d}{dx}\left[\frac{f'_x(x)}{x \exp(\lambda x^2)}\right].$

49. $\displaystyle\int_a^x \left[A\exp(\lambda x^2) + B\exp(\lambda t^2) + C\right] y(t)\, dt = f(x).$

This is a special case of equation 1.9.5 with $g(x) = \exp(\lambda x^2)$.

50. $\displaystyle\int_a^x \left[A\exp(\lambda x^2) + B\exp(\mu t^2)\right] y(t)\, dt = f(x).$

This is a special case of equation 1.9.6 with $g(x) = A\exp(\lambda x^2)$ and $h(t) = B\exp(\mu t^2)$.

51. $\displaystyle\int_a^x \sqrt{x-t}\,\exp[\lambda(x^2 - t^2)] y(t)\, dt = f(x).$

Solution:
$$y(x) = \frac{2}{\pi}\exp(\lambda x^2)\frac{d^2}{dx^2}\int_a^x \frac{\exp(-\lambda t^2)}{\sqrt{x-t}} f(t)\, dt.$$

52. $\displaystyle\int_a^x \frac{\exp[\lambda(x^2 - t^2)]}{\sqrt{x-t}}\, y(t)\, dt = f(x).$

Solution:
$$y(x) = \frac{1}{\pi}\exp(\lambda x^2)\frac{d}{dx}\int_a^x \frac{\exp(-\lambda t^2)}{\sqrt{x-t}} f(t)\, dt.$$

53. $\displaystyle\int_a^x (x-t)^\lambda \exp[\mu(x^2 - t^2)] y(t)\, dt = f(x), \qquad 0 < \lambda < 1.$

Solution:
$$y(x) = k\exp(\mu x^2)\frac{d^2}{dx^2}\int_a^x \frac{\exp(-\mu t^2)}{(x-t)^\lambda} f(t)\, dt, \qquad k = \frac{\sin(\pi\lambda)}{\pi\lambda}.$$

54. $\displaystyle\int_a^x \exp[\lambda(x^\beta - t^\beta)] y(t)\, dt = f(x).$

Solution: $y(x) = f'_x(x) - \lambda\beta x^{\beta-1} f(x).$

1.3. Equations Whose Kernels Contain Hyperbolic Functions

1.3-1. Kernels Containing Hyperbolic Cosine

1. $\displaystyle\int_a^x \cosh[\lambda(x-t)] y(t)\, dt = f(x).$

Solution: $y(x) = f'_x(x) - \lambda^2 \displaystyle\int_a^x f(x)\, dx.$

2. $\displaystyle\int_a^x \left\{\cosh[\lambda(x-t)] - 1\right\} y(t)\, dt = f(x), \qquad f(a) = f'_x(a) = f''_{xx}(x) = 0.$

Solution: $y(x) = \dfrac{1}{\lambda^2} f'''_{xxx}(x) - f'_x(x).$

3. $\displaystyle\int_a^x \{\cosh[\lambda(x-t)] + b\}y(t)\,dt = f(x).$

For $b = 0$, see equation 1.3.1. For $b = -1$, see equation 1.3.2. For $\lambda = 0$, see equation 1.1.1.
Differentiating the equation with respect to x, we arrive at an equation of the form 2.3.16:

$$y(x) + \frac{\lambda}{b+1}\int_a^x \sinh[\lambda(x-t)]y(t)\,dt = \frac{f'_x(x)}{b+1}.$$

1°. Solution with $b(b+1) < 0$:

$$y(x) = \frac{f'_x(x)}{b+1} - \frac{\lambda^2}{k(b+1)^2}\int_a^x \sin[k(x-t)]f'_t(t)\,dt, \qquad \text{where} \quad k = \lambda\sqrt{\frac{-b}{b+1}}.$$

2°. Solution with $b(b+1) > 0$:

$$y(x) = \frac{f'_x(x)}{b+1} - \frac{\lambda^2}{k(b+1)^2}\int_a^x \sinh[k(x-t)]f'_t(t)\,dt, \qquad \text{where} \quad k = \lambda\sqrt{\frac{b}{b+1}}.$$

4. $\displaystyle\int_a^x \cosh(\lambda x + \beta t)y(t)\,dt = f(x).$

For $\beta = -\lambda$, see equation 1.3.1.
Differentiating the equation with respect to x twice, we obtain

$$\cosh[(\lambda+\beta)x]y(x) + \lambda\int_a^x \sinh(\lambda x + \beta t)y(t)\,dt = f'_x(x), \tag{1}$$

$$\{\cosh[(\lambda+\beta)x]y(x)\}'_x + \lambda\sinh[(\lambda+\beta)x]y(x) + \lambda^2\int_a^x \cosh(\lambda x + \beta t)y(t)\,dt = f''_{xx}(x). \tag{2}$$

Eliminating the integral term from (2) with the aid of the original equation, we arrive at the first-order linear ordinary differential equation

$$w'_x + \lambda\tanh[(\lambda+\beta)x]w = f''_{xx}(x) - \lambda^2 f(x), \qquad w = \cosh[(\lambda+\beta)x]y(x). \tag{3}$$

Setting $x = a$ in (1) yields the initial condition $w(a) = f'_x(a)$. On solving equation (3) with this condition, after some manipulations we obtain the solution of the original integral equation in the form

$$y(x) = \frac{1}{\cosh[(\lambda+\beta)x]}f'_x(x) - \frac{\lambda\sinh[(\lambda+\beta)x]}{\cosh^2[(\lambda+\beta)x]}f(x)$$
$$+ \frac{\lambda\beta}{\cosh^{k+1}[(\lambda+\beta)x]}\int_a^x f(t)\cosh^{k-2}[(\lambda+\beta)t]\,dt, \qquad k = \frac{\lambda}{\lambda+\beta}.$$

5. $\displaystyle\int_a^x [\cosh(\lambda x) - \cosh(\lambda t)]y(t)\,dt = f(x).$

This is a special case of equation 1.9.2 with $g(x) = \cosh(\lambda x)$.

Solution: $\displaystyle y(x) = \frac{1}{\lambda}\frac{d}{dx}\left[\frac{f'_x(x)}{\sinh(\lambda x)}\right].$

6. $\displaystyle\int_a^x [A\cosh(\lambda x) + B\cosh(\lambda t)]y(t)\,dt = f(x).$

For $B = -A$, see equation 1.3.5. This is a special case of equation 1.9.4 with $g(x) = \cosh(\lambda x)$.

Solution: $\displaystyle y(x) = \frac{1}{A+B}\frac{d}{dx}\left\{[\cosh(\lambda x)]^{-\frac{A}{A+B}}\int_a^x [\cosh(\lambda t)]^{-\frac{B}{A+B}}f'_t(t)\,dt\right\}.$

7. $\displaystyle\int_a^x \big[A\cosh(\lambda x) + B\cosh(\mu t) + C\big]y(t)\,dt = f(x).$

This is a special case of equation 1.9.6 with $g(x) = A\cosh(\lambda x)$ and $h(t) = B\cosh(\mu t) + C$.

8. $\displaystyle\int_a^x \big\{A_1\cosh[\lambda_1(x-t)] + A_2\cosh[\lambda_2(x-t)]\big\}y(t)\,dt = f(x).$

The equation is equivalent to the equation

$$\int_a^x \big\{B_1\sinh[\lambda_1(x-t)] + B_2\sinh[\lambda_2(x-t)]\big\}y(t)\,dt = F(x),$$

$$B_1 = \frac{A_1}{\lambda_1}, \quad B_2 = \frac{A_2}{\lambda_2}, \quad F(x) = \int_a^x f(t)\,dt,$$

of the form 1.3.41. (Differentiating this equation yields the original equation.)

9. $\displaystyle\int_a^x \cosh^2[\lambda(x-t)]y(t)\,dt = f(x).$

Differentiation yields an equation of the form 2.3.16:

$$y(x) + \lambda\int_a^x \sinh[2\lambda(x-t)]y(t)\,dt = f'_x(x).$$

Solution:

$$y(x) = f'_x(x) - \frac{2\lambda^2}{k}\int_a^x \sinh[k(x-t)]f'_t(t)\,dt, \qquad \text{where} \quad k = \lambda\sqrt{2}.$$

10. $\displaystyle\int_a^x \big[\cosh^2(\lambda x) - \cosh^2(\lambda t)\big]y(t)\,dt = f(x), \qquad f(a) = f'_x(a) = 0.$

Solution: $\displaystyle y(x) = \frac{1}{\lambda}\frac{d}{dx}\left[\frac{f'_x(x)}{\sinh(2\lambda x)}\right].$

11. $\displaystyle\int_a^x \big[A\cosh^2(\lambda x) + B\cosh^2(\lambda t)\big]y(t)\,dt = f(x).$

For $B = -A$, see equation 1.3.10. This is a special case of equation 1.9.4 with $g(x) = \cosh^2(\lambda x)$.
Solution:

$$y(x) = \frac{1}{A+B}\frac{d}{dx}\left\{[\cosh(\lambda x)]^{-\frac{2A}{A+B}}\int_a^x [\cosh(\lambda t)]^{-\frac{2B}{A+B}}f'_t(t)\,dt\right\}.$$

12. $\displaystyle\int_a^x \big[A\cosh^2(\lambda x) + B\cosh^2(\mu t) + C\big]y(t)\,dt = f(x).$

This is a special case of equation 1.9.6 with $g(x) = A\cosh^2(\lambda x)$, and $h(t) = B\cosh^2(\mu t) + C$.

13. $\displaystyle\int_a^x \cosh[\lambda(x-t)]\cosh[\lambda(x+t)]y(t)\,dt = f(x).$

Using the formula

$$\cosh(\alpha - \beta)\cosh(\alpha + \beta) = \tfrac{1}{2}[\cos(2\alpha) + \cos(2\beta)], \quad \alpha = \lambda x, \quad \beta = \lambda t,$$

we transform the original equation to an equation of the form 1.4.6 with $A = B = 1$:

$$\int_a^x [\cosh(2\lambda x) + \cosh(2\lambda t)]y(t)\,dt = 2f(x).$$

Solution:

$$y(x) = \frac{d}{dx}\left[\frac{1}{\sqrt{\cosh(2\lambda x)}}\int_a^x \frac{f'_t(t)\,dt}{\sqrt{\cosh(2\lambda t)}}\right].$$

14. $\displaystyle\int_a^x [\cosh(\lambda x)\cosh(\mu t) + \cosh(\beta x)\cosh(\gamma t)]y(t)\,dt = f(x).$

This is a special case of equation 1.9.15 with $g_1(x) = \cosh(\lambda x)$, $h_1(t) = \cosh(\mu t)$, $g_2(x) = \cosh(\beta x)$, and $h_2(t) = \cosh(\gamma t)$.

15. $\displaystyle\int_a^x \cosh^3[\lambda(x - t)]y(t)\,dt = f(x).$

Using the formula $\cosh^3\beta = \frac{1}{4}\cosh 3\beta + \frac{3}{4}\cosh\beta$, we arrive at an equation of the form 1.3.8:

$$\int_a^x \left\{\tfrac{1}{4}\cosh[3\lambda(x - t)] + \tfrac{3}{4}\cosh[\lambda(x - t)]\right\}y(t)\,dt = f(x).$$

16. $\displaystyle\int_a^x \left[\cosh^3(\lambda x) - \cosh^3(\lambda t)\right]y(t)\,dt = f(x), \qquad f(a) = f'_x(a) = 0.$

Solution: $\displaystyle y(x) = \frac{1}{3\lambda}\frac{d}{dx}\left[\frac{f'_x(x)}{\sinh(\lambda x)\cosh^2(\lambda x)}\right].$

17. $\displaystyle\int_a^x \left[A\cosh^3(\lambda x) + B\cosh^3(\lambda t)\right]y(t)\,dt = f(x).$

For $B = -A$, see equation 1.3.16. This is a special case of equation 1.9.4 with $g(x) = \cosh^3(\lambda x)$. Solution:

$$y(x) = \frac{1}{A + B}\frac{d}{dx}\left\{\left[\cosh(\lambda x)\right]^{-\frac{3A}{A+B}}\int_a^x \left[\cosh(\lambda t)\right]^{-\frac{3B}{A+B}} f'_t(t)\,dt\right\}.$$

18. $\displaystyle\int_a^x \left[A\cosh^2(\lambda x)\cosh(\mu t) + B\cosh(\beta x)\cosh^2(\gamma t)\right]y(t)\,dt = f(x).$

This is a special case of equation 1.9.15 with $g_1(x) = A\cosh^2(\lambda x)$, $h_1(t) = \cosh(\mu t)$, $g_2(x) = B\cosh(\beta x)$, and $h_2(t) = \cosh^2(\gamma t)$.

19. $\displaystyle\int_a^x \cosh^4[\lambda(x - t)]y(t)\,dt = f(x).$

Let us transform the kernel of the integral equation using the formula

$$\cosh^4\beta = \tfrac{1}{8}\cosh 4\beta + \tfrac{1}{2}\cosh 2\beta + \tfrac{3}{8}, \quad \text{where} \quad \beta = \lambda(x - t),$$

and differentiate the resulting equation with respect to x. Then we obtain an equation of the form 2.3.18:

$$y(x) + \lambda\int_a^x \left\{\tfrac{1}{2}\sinh[4\lambda(x - t)] + \sinh[2\lambda(x - t)]\right\}y(t)\,dt = f'_x(x).$$

20. $\displaystyle\int_a^x [\cosh(\lambda x) - \cosh(\lambda t)]^n y(t)\,dt = f(x), \qquad n = 1, 2, \ldots$

The right-hand side of the equation is assumed to satisfy the conditions $f(a) = f'_x(a) = \cdots = f_x^{(n)}(a) = 0.$

Solution: $\displaystyle y(x) = \frac{\sinh(\lambda x)}{\lambda^n n!}\left[\frac{1}{\sinh(\lambda x)}\frac{d}{dx}\right]^{n+1} f(x).$

21. $\displaystyle\int_a^x \sqrt{\cosh x - \cosh t}\, y(t)\, dt = f(x).$

Solution:

$$y(x) = \frac{2}{\pi} \sinh x \left(\frac{1}{\sinh x} \frac{d}{dx} \right)^2 \int_a^x \frac{\sinh t\, f(t)\, dt}{\sqrt{\cosh x - \cosh t}}.$$

22. $\displaystyle\int_a^x \frac{y(t)\, dt}{\sqrt{\cosh x - \cosh t}} = f(x).$

Solution:

$$y(x) = \frac{1}{\pi} \frac{d}{dx} \int_a^x \frac{\sinh t\, f(t)\, dt}{\sqrt{\cosh x - \cosh t}}.$$

23. $\displaystyle\int_a^x (\cosh x - \cosh t)^\lambda y(t)\, dt = f(x), \qquad 0 < \lambda < 1.$

Solution:

$$y(x) = k \sinh x \left(\frac{1}{\sinh x} \frac{d}{dx} \right)^2 \int_a^x \frac{\sinh t\, f(t)\, dt}{(\cosh x - \cosh t)^\lambda}, \qquad k = \frac{\sin(\pi\lambda)}{\pi\lambda}.$$

24. $\displaystyle\int_a^x (\cosh^\mu x - \cosh^\mu t) y(t)\, dt = f(x).$

This is a special case of equation 1.9.2 with $g(x) = \cosh^\mu x$.

Solution: $\displaystyle y(x) = \frac{1}{\mu} \frac{d}{dx} \left[\frac{f'_x(x)}{\sinh x \cosh^{\mu-1} x} \right].$

25. $\displaystyle\int_a^x \left(A \cosh^\mu x + B \cosh^\mu t \right) y(t)\, dt = f(x).$

For $B = -A$, see equation 1.3.24. This is a special case of equation 1.9.4 with $g(x) = \cosh^\mu x$.

Solution:

$$y(x) = \frac{1}{A+B} \frac{d}{dx} \left\{ \left[\cosh(\lambda x) \right]^{-\frac{A\mu}{A+B}} \int_a^x \left[\cosh(\lambda t) \right]^{-\frac{B\mu}{A+B}} f'_t(t)\, dt \right\}.$$

26. $\displaystyle\int_a^x \frac{y(t)\, dt}{(\cosh x - \cosh t)^\lambda} = f(x), \qquad 0 < \lambda < 1.$

Solution:

$$y(x) = \frac{\sin(\pi\lambda)}{\pi} \frac{d}{dx} \int_a^x \frac{\sinh t\, f(t)\, dt}{(\cosh x - \cosh t)^{1-\lambda}}.$$

27. $\displaystyle\int_a^x (x-t) \cosh[\lambda(x-t)] y(t)\, dt = f(x), \qquad f(a) = f'_x(a) = 0.$

Differentiating the equation twice yields

$$y(x) + 2\lambda \int_a^x \sinh[\lambda(x-t)] y(t)\, dt + \lambda^2 \int_a^x (x-t) \cosh[\lambda(x-t)] y(t)\, dt = f''_{xx}(x).$$

Eliminating the third term on the right-hand side with the aid of the original equation, we arrive at an equation of the form 2.3.16:

$$y(x) + 2\lambda \int_a^x \sinh[\lambda(x-t)] y(t)\, dt = f''_{xx}(x) - \lambda^2 f(x).$$

28. $\displaystyle\int_a^x \frac{\cosh\left(\lambda\sqrt{x-t}\right)}{\sqrt{x-t}}\, y(t)\, dt = f(x).$

Solution:

$$y(x) = \frac{1}{\pi}\frac{d}{dx}\int_a^x \frac{\cos\left(\lambda\sqrt{x-t}\right)}{\sqrt{x-t}}\, f(t)\, dt.$$

29. $\displaystyle\int_0^x \frac{\cosh\left(\lambda\sqrt{x^2-t^2}\right)}{\sqrt{x^2-t^2}}\, y(t)\, dt = f(x).$

Solution:

$$y(x) = \frac{2}{\pi}\frac{d}{dx}\int_0^x t\,\frac{\cos\left(\lambda\sqrt{x^2-t^2}\right)}{\sqrt{x^2-t^2}}\, f(t)\, dt.$$

30. $\displaystyle\int_x^\infty \frac{\cosh\left(\lambda\sqrt{t^2-x^2}\right)}{\sqrt{t^2-x^2}}\, y(t)\, dt = f(x).$

Solution:

$$y(x) = -\frac{2}{\pi}\frac{d}{dx}\int_x^\infty t\,\frac{\cos\left(\lambda\sqrt{t^2-x^2}\right)}{\sqrt{t^2-x^2}}\, f(t)\, dt.$$

31. $\displaystyle\int_a^x \left[Ax^\beta + B\cosh^\gamma(\lambda t) + C\right]y(t)\, dt = f(x).$

This is a special case of equation 1.9.6 with $g(x) = Ax^\beta$ and $h(t) = B\cosh^\gamma(\lambda t) + C$.

32. $\displaystyle\int_a^x \left[A\cosh^\gamma(\lambda x) + Bt^\beta + C\right]y(t)\, dt = f(x).$

This is a special case of equation 1.9.6 with $g(x) = A\cosh^\gamma(\lambda x)$ and $h(t) = Bt^\beta + C$.

33. $\displaystyle\int_a^x \left(Ax^\lambda \cosh^\mu t + Bt^\beta \cosh^\gamma x\right)y(t)\, dt = f(x).$

This is a special case of equation 1.9.15 with $g_1(x) = Ax^\lambda$, $h_1(t) = \cosh^\mu t$, $g_2(x) = B\cosh^\gamma x$, and $h_2(t) = t^\beta$.

1.3-2. Kernels Containing Hyperbolic Sine

34. $\displaystyle\int_a^x \sinh[\lambda(x-t)]y(t)\, dt = f(x), \qquad f(a) = f_x'(a) = 0.$

Solution: $y(x) = \dfrac{1}{\lambda}f_{xx}''(x) - \lambda f(x).$

35. $\displaystyle\int_a^x \left\{\sinh[\lambda(x-t)] + b\right\}y(t)\, dt = f(x).$

Differentiating the equation with respect to x, we arrive at an equation of the form 2.3.3:

$$y(x) + \frac{\lambda}{b}\int_a^x \cosh[\lambda(x-t)]y(t)\, dt = \frac{1}{b}f_x'(x).$$

Solution:

$$y(x) = \frac{1}{b}f_x'(x) + \int_a^x R(x-t)f_t'(t)\, dt,$$

$$R(x) = \frac{\lambda}{b^2}\exp\left(-\frac{\lambda x}{2b}\right)\left[\frac{\lambda}{2bk}\sinh(kx) - \cosh(kx)\right], \quad k = \frac{\lambda\sqrt{1+4b^2}}{2b}.$$

36. $\displaystyle\int_a^x \sinh(\lambda x + \beta t)y(t)\,dt = f(x).$

For $\beta = -\lambda$, see equation 1.3.34. Assume that $\beta \neq -\lambda$.

Differentiating the equation with respect to x twice yields

$$\sinh[(\lambda + \beta)x]y(x) + \lambda \int_a^x \cosh(\lambda x + \beta t)y(t)\,dt = f_x'(x), \tag{1}$$

$$\left\{\sinh[(\lambda + \beta)x]y(x)\right\}_x' + \lambda \cosh[(\lambda + \beta)x]y(x) + \lambda^2 \int_a^x \sinh(\lambda x + \beta t)y(t)\,dt = f_{xx}''(x). \tag{2}$$

Eliminating the integral term from (2) with the aid of the original equation, we arrive at the first-order linear ordinary differential equation

$$w_x' + \lambda \coth[(\lambda + \beta)x]w = f_{xx}''(x) - \lambda^2 f(x), \qquad w = \sinh[(\lambda + \beta)x]y(x). \tag{3}$$

Setting $x = a$ in (1) yields the initial condition $w(a) = f_x'(a)$. On solving equation (3) with this condition, after some manipulations we obtain the solution of the original integral equation in the form

$$y(x) = \frac{1}{\sinh[(\lambda + \beta)x]}f_x'(x) - \frac{\lambda \cosh[(\lambda + \beta)x]}{\sinh^2[(\lambda + \beta)x]}f(x)$$
$$- \frac{\lambda\beta}{\sinh^{k+1}[(\lambda + \beta)x]}\int_a^x f(t)\sinh^{k-2}[(\lambda + \beta)t]\,dt, \qquad k = \frac{\lambda}{\lambda + \beta}.$$

37. $\displaystyle\int_a^x [\sinh(\lambda x) - \sinh(\lambda t)]y(t)\,dt = f(x), \qquad f(a) = f_x'(a) = 0.$

This is a special case of equation 1.9.2 with $g(x) = \sinh(\lambda x)$.

Solution: $\displaystyle y(x) = \frac{1}{\lambda}\frac{d}{dx}\left[\frac{f_x'(x)}{\cosh(\lambda x)}\right].$

38. $\displaystyle\int_a^x [A\sinh(\lambda x) + B\sinh(\lambda t)]y(t)\,dt = f(x).$

For $B = -A$, see equation 1.3.37. This is a special case of equation 1.9.4 with $g(x) = \sinh(\lambda x)$.

Solution: $\displaystyle y(x) = \frac{1}{A + B}\frac{d}{dx}\left\{[\sinh(\lambda x)]^{-\frac{A}{A+B}}\int_a^x [\sinh(\lambda t)]^{-\frac{B}{A+B}}f_t'(t)\,dt\right\}.$

39. $\displaystyle\int_a^x [A\sinh(\lambda x) + B\sinh(\mu t)]y(t)\,dt = f(x).$

This is a special case of equation 1.9.6 with $g(x) = A\sinh(\lambda x)$, and $h(t) = B\sinh(\mu t)$.

40. $\displaystyle\int_a^x \left\{\mu\sinh[\lambda(x - t)] - \lambda\sinh[\mu(x - t)]\right\}y(t)\,dt = f(x).$

It is assumed that $f(a) = f_x'(a) = f_{xx}''(a) = f_{xxx}'''(a) = 0$.

Solution:

$$y(x) = \frac{f_{xxxx}'''' - (\lambda^2 + \mu^2)f_{xx}'' + \lambda^2\mu^2 f}{\mu\lambda^3 - \lambda\mu^3}, \qquad f = f(x).$$

41. $\displaystyle\int_a^x \big\{ A_1 \sinh[\lambda_1(x-t)] + A_2 \sinh[\lambda_2(x-t)] \big\} y(t)\, dt = f(x), \quad f(a) = f'_x(a) = 0.$

1°. Introduce the notation

$$I_1 = \int_a^x \sinh[\lambda_1(x-t)]y(t)\, dt, \quad I_2 = \int_a^x \sinh[\lambda_2(x-t)]y(t)\, dt,$$

$$J_1 = \int_a^x \cosh[\lambda_1(x-t)]y(t)\, dt, \quad J_2 = \int_a^x \cosh[\lambda_2(x-t)]y(t)\, dt.$$

Let us successively differentiate the integral equation four times. As a result, we have (the first line is the original equation):

$$A_1 I_1 + A_2 I_2 = f, \qquad f = f(x), \tag{1}$$

$$A_1 \lambda_1 J_1 + A_2 \lambda_2 J_2 = f'_x, \tag{2}$$

$$(A_1 \lambda_1 + A_2 \lambda_2)y + A_1 \lambda_1^2 I_1 + A_2 \lambda_2^2 I_2 = f''_{xx}, \tag{3}$$

$$(A_1 \lambda_1 + A_2 \lambda_2)y'_x + A_1 \lambda_1^3 J_1 + A_2 \lambda_2^3 J_2 = f'''_{xxx}, \tag{4}$$

$$(A_1 \lambda_1 + A_2 \lambda_2)y''_{xx} + (A_1 \lambda_1^3 + A_2 \lambda_2^3)y + A_1 \lambda_1^4 I_1 + A_2 \lambda_2^4 I_2 = f''''_{xxxx}. \tag{5}$$

Eliminating I_1 and I_2 from (1), (3), and (5), we arrive at the following second-order linear ordinary differential equation with constant coefficients:

$$(A_1 \lambda_1 + A_2 \lambda_2)y''_{xx} - \lambda_1 \lambda_2(A_1 \lambda_2 + A_2 \lambda_1)y = f''''_{xxxx} - (\lambda_1^2 + \lambda_2^2)f''_{xx} + \lambda_1^2 \lambda_2^2 f. \tag{6}$$

The initial conditions can be obtained by substituting $x = a$ into (3) and (4):

$$(A_1 \lambda_1 + A_2 \lambda_2)y(a) = f''_{xx}(a), \quad (A_1 \lambda_1 + A_2 \lambda_2)y'_x(a) = f'''_{xxx}(a). \tag{7}$$

Solving the differential equation (6) under conditions (7) allows us to find the solution of the integral equation.

2°. Denote

$$\Delta = \lambda_1 \lambda_2 \frac{A_1 \lambda_2 + A_2 \lambda_1}{A_1 \lambda_1 + A_2 \lambda_2}.$$

2.1. Solution for $\Delta > 0$:

$$(A_1 \lambda_1 + A_2 \lambda_2)y(x) = f''_{xx}(x) + Bf(x) + C \int_a^x \sinh[k(x-t)]f(t)\, dt,$$

$$k = \sqrt{\Delta}, \quad B = \Delta - \lambda_1^2 - \lambda_2^2, \quad C = \frac{1}{\sqrt{\Delta}}\big[\Delta^2 - (\lambda_1^2 + \lambda_2^2)\Delta + \lambda_1^2 \lambda_2^2\big].$$

2.2. Solution for $\Delta < 0$:

$$(A_1 \lambda_1 + A_2 \lambda_2)y(x) = f''_{xx}(x) + Bf(x) + C \int_a^x \sin[k(x-t)]f(t)\, dt,$$

$$k = \sqrt{-\Delta}, \quad B = \Delta - \lambda_1^2 - \lambda_2^2, \quad C = \frac{1}{\sqrt{-\Delta}}\big[\Delta^2 - (\lambda_1^2 + \lambda_2^2)\Delta + \lambda_1^2 \lambda_2^2\big].$$

2.3. Solution for $\Delta = 0$:

$$(A_1 \lambda_1 + A_2 \lambda_2)y(x) = f''_{xx}(x) - (\lambda_1^2 + \lambda_2^2)f(x) + \lambda_1^2 \lambda_2^2 \int_a^x (x-t)f(t)\, dt.$$

2.4. Solution for $\Delta = \infty$:

$$y(x) = \frac{f''''_{xxxx} - (\lambda_1^2 + \lambda_2^2)f''_{xx} + \lambda_1^2 \lambda_2^2 f}{A_1 \lambda_1^3 + A_2 \lambda_2^3}, \qquad f = f(x).$$

In the last case, the relation $A_1 \lambda_1 + A_2 \lambda_2 = 0$ is valid, and the right-hand side of the integral equation is assumed to satisfy the conditions $f(a) = f'_x(a) = f''_{xx}(a) = f'''_{xxx}(a) = 0$.

42. $\int_a^x \left\{ A \sinh[\lambda(x-t)] + B \sinh[\mu(x-t)] + C \sinh[\beta(x-t)] \right\} y(t)\, dt = f(x).$

It assumed that $f(a) = f_x'(a) = 0$. Differentiating the integral equation twice yields

$$(A\lambda + B\mu + C\beta)y(x) + \int_a^x \left\{ A\lambda^2 \sinh[\lambda(x-t)] + B\mu^2 \sinh[\mu(x-t)] \right\} y(t)\, dt$$
$$+ C\beta^2 \int_a^x \sinh[\beta(x-t)]y(t)\, dt = f_{xx}''(x).$$

Eliminating the last integral with the aid of the original equation, we arrive at an equation of the form 2.3.18:

$$(A\lambda + B\mu + C\beta)y(x)$$
$$+ \int_a^x \left\{ A(\lambda^2 - \beta^2) \sinh[\lambda(x-t)] + B(\mu^2 - \beta^2) \sinh[\mu(x-t)] \right\} y(t)\, dt = f_{xx}''(x) - \beta^2 f(x).$$

In the special case $A\lambda + B\mu + C\beta = 0$, this is an equation of the form 1.3.41.

43. $\int_a^x \sinh^2[\lambda(x-t)]y(t)\, dt = f(x), \qquad f(a) = f_x'(a) = f_{xx}''(a) = 0.$

Differentiating yields an equation of the form 1.3.34:

$$\int_a^x \sinh[2\lambda(x-t)]y(t)\, dt = \frac{1}{\lambda} f_x'(x).$$

Solution: $y(x) = \frac{1}{2}\lambda^{-2} f_{xxx}'''(x) - 2f_x'(x).$

44. $\int_a^x \left[\sinh^2(\lambda x) - \sinh^2(\lambda t) \right] y(t)\, dt = f(x), \qquad f(a) = f_x'(a) = 0.$

Solution: $y(x) = \dfrac{1}{\lambda} \dfrac{d}{dx} \left[\dfrac{f_x'(x)}{\sinh(2\lambda x)} \right].$

45. $\int_a^x \left[A \sinh^2(\lambda x) + B \sinh^2(\lambda t) \right] y(t)\, dt = f(x).$

For $B = -A$, see equation 1.3.44. This is a special case of equation 1.9.4 with $g(x) = \sinh^2(\lambda x)$.
    Solution:

$$y(x) = \frac{1}{A+B} \frac{d}{dx} \left\{ \left[\sinh(\lambda x) \right]^{-\frac{2A}{A+B}} \int_a^x \left[\sinh(\lambda t) \right]^{-\frac{2B}{A+B}} f_t'(t)\, dt \right\}.$$

46. $\int_a^x \left[A \sinh^2(\lambda x) + B \sinh^2(\mu t) \right] y(t)\, dt = f(x).$

This is a special case of equation 1.9.6 with $g(x) = A \sinh^2(\lambda x)$ and $h(t) = B \sinh^2(\mu t)$.

47. $\int_a^x \sinh[\lambda(x-t)] \sinh[\lambda(x+t)] y(t)\, dt = f(x).$

Using the formula

$$\sinh(\alpha - \beta) \sinh(\alpha + \beta) = \frac{1}{2}[\cosh(2\alpha) - \cosh(2\beta)], \quad \alpha = \lambda x, \quad \beta = \lambda t,$$

we reduce the original equation to an equation of the form 1.3.5:

$$\int_a^x [\cosh(2\lambda x) - \cosh(2\lambda t)]y(t)\, dt = 2f(x).$$

Solution: $y(x) = \dfrac{1}{\lambda} \dfrac{d}{dx} \left[\dfrac{f_x'(x)}{\sinh(2\lambda x)} \right].$

48. $\int_a^x \left[A \sinh(\lambda x) \sinh(\mu t) + B \sinh(\beta x) \sinh(\gamma t) \right] y(t)\, dt = f(x).$

This is a special case of equation 1.9.15 with $g_1(x) = A \sinh(\lambda x)$, $h_1(t) = \sinh(\mu t)$, $g_2(x) = B \sinh(\beta x)$, and $h_2(t) = \sinh(\gamma t)$.

49. $\int_a^x \sinh^3[\lambda(x - t)] y(t)\, dt = f(x),$ $f(a) = f'_x(a) = f''_{xx}(a) = f'''_{xxx}(a) = 0.$

Using the formula $\sinh^3 \beta = \frac{1}{4} \sinh 3\beta - \frac{3}{4} \sinh \beta$, we arrive at an equation of the form 1.3.41:

$$\int_a^x \left\{ \tfrac{1}{4} \sinh[3\lambda(x - t)] - \tfrac{3}{4} \sinh[\lambda(x - t)] \right\} y(t)\, dt = f(x).$$

50. $\int_a^x \left[\sinh^3(\lambda x) - \sinh^3(\lambda t) \right] y(t)\, dt = f(x),$ $f(a) = f'_x(a) = 0.$

This is a special case of equation 1.9.2 with $g(x) = \sinh^3(\lambda x)$.

51. $\int_a^x \left[A \sinh^3(\lambda x) + B \sinh^3(\lambda t) \right] y(t)\, dt = f(x).$

This is a special case of equation 1.9.4 with $g(x) = \sinh^3(\lambda x)$.
    Solution:

$$y(x) = \frac{1}{A + B} \frac{d}{dx} \left\{ [\sinh(\lambda x)]^{-\frac{3A}{A+B}} \int_a^x [\sinh(\lambda t)]^{-\frac{3B}{A+B}} f'_t(t)\, dt \right\}.$$

52. $\int_a^x \left[A \sinh^2(\lambda x) \sinh(\mu t) + B \sinh(\beta x) \sinh^2(\gamma t) \right] y(t)\, dt = f(x).$

This is a special case of equation 1.9.15 with $g_1(x) = A \sinh^2(\lambda x)$, $h_1(t) = \sinh(\mu t)$, $g_2(x) = B \sinh(\beta x)$, and $h_2(t) = \sinh^2(\gamma t)$.

53. $\int_a^x \sinh^4[\lambda(x - t)] y(t)\, dt = f(x).$

It is assumed that $f(a) = f'_x(a) = \cdots = f''''_{xxxx}(a) = 0.$
    Let us transform the kernel of the integral equation using the formula

$$\sinh^4 \beta = \tfrac{1}{8} \cosh 4\beta - \tfrac{1}{2} \cosh 2\beta + \tfrac{3}{8}, \quad \text{where} \quad \beta = \lambda(x - t),$$

and differentiate the resulting equation with respect to x. Then we arrive at an equation of the form 1.3.41:

$$\lambda \int_a^x \left\{ \tfrac{1}{2} \sinh[4\lambda(x - t)] - \sinh[2\lambda(x - t)] \right\} y(t)\, dt = f'_x(x).$$

54. $\int_a^x \sinh^n[\lambda(x - t)] y(t)\, dt = f(x),$ $n = 2, 3, \ldots$

It is assumed that $f(a) = f'_x(a) = \cdots = f_x^{(n)}(a) = 0.$
    Let us differentiate the equation with respect to x twice and transform the kernel of the resulting integral equation using the formula $\cosh^2 \beta = 1 + \sinh^2 \beta$, where $\beta = \lambda(x - t)$. Then we have

$$\lambda^2 n^2 \int_a^x \sinh^n[\lambda(x - t)] y(t)\, dt + \lambda^2 n(n - 1) \int_a^x \sinh^{n-2}[\lambda(x - t)] y(t)\, dt = f''_{xx}(x).$$

Eliminating the first term on the left-hand side with the aid of the original equation, we obtain

$$\int_a^x \sinh^{n-2}[\lambda(x-t)]y(t)\,dt = \frac{1}{\lambda^2 n(n-1)}\left[f_{xx}''(x) - \lambda^2 n^2 f(x)\right].$$

This equation has the same form as the original equation, but the exponent of the kernel has been reduced by two.

By applying this technique sufficiently many times, we finally arrive at simple integral equations of the form 1.1.1 (for even n) or 1.3.34 (for odd n).

55. $\displaystyle\int_a^x \sinh\left(\lambda\sqrt{x-t}\right)y(t)\,dt = f(x).$

Solution:

$$y(x) = \frac{2}{\pi\lambda}\frac{d^2}{dx^2}\int_a^x \frac{\cos\left(\lambda\sqrt{x-t}\right)}{\sqrt{x-t}}f(t)\,dt.$$

56. $\displaystyle\int_a^x \sqrt{\sinh x - \sinh t}\,y(t)\,dt = f(x).$

Solution:

$$y(x) = \frac{2}{\pi}\cosh x\left(\frac{1}{\cosh x}\frac{d}{dx}\right)^2\int_a^x \frac{\cosh t\,f(t)\,dt}{\sqrt{\sinh x - \sinh t}}.$$

57. $\displaystyle\int_a^x \frac{y(t)\,dt}{\sqrt{\sinh x - \sinh t}} = f(x).$

Solution:

$$y(x) = \frac{1}{\pi}\frac{d}{dx}\int_a^x \frac{\cosh t\,f(t)\,dt}{\sqrt{\sinh x - \sinh t}}.$$

58. $\displaystyle\int_a^x (\sinh x - \sinh t)^\lambda y(t)\,dt = f(x),\qquad 0 < \lambda < 1.$

Solution:

$$y(x) = k\cosh x\left(\frac{1}{\cosh x}\frac{d}{dx}\right)^2\int_a^x \frac{\cosh t\,f(t)\,dt}{(\sinh x - \sinh t)^\lambda},\qquad k = \frac{\sin(\pi\lambda)}{\pi\lambda}.$$

59. $\displaystyle\int_a^x (\sinh^\mu x - \sinh^\mu t)y(t)\,dt = f(x).$

This is a special case of equation 1.9.2 with $g(x) = \sinh^\mu x$.

Solution: $y(x) = \dfrac{1}{\mu}\dfrac{d}{dx}\left[\dfrac{f_x'(x)}{\cosh x \sinh^{\mu-1} x}\right].$

60. $\displaystyle\int_a^x \left[A\sinh^\mu(\lambda x) + B\sinh^\mu(\lambda t)\right]y(t)\,dt = f(x).$

This is a special case of equation 1.9.4 with $g(x) = \sinh^\mu(\lambda x)$.

Solution with $B \neq -A$:

$$y(x) = \frac{1}{A+B}\frac{d}{dx}\left\{[\sinh(\lambda x)]^{-\frac{A\mu}{A+B}}\int_a^x [\sinh(\lambda t)]^{-\frac{B\mu}{A+B}}f_t'(t)\,dt\right\}.$$

61. $\displaystyle\int_a^x \frac{y(t)\,dt}{(\sinh x - \sinh t)^\lambda} = f(x), \qquad 0 < \lambda < 1.$

Solution:

$$y(x) = \frac{\sin(\pi\lambda)}{\pi}\frac{d}{dx}\int_a^x \frac{\cosh t\, f(t)\,dt}{(\sinh x - \sinh t)^{1-\lambda}}.$$

62. $\displaystyle\int_a^x (x-t)\sinh[\lambda(x-t)]y(t)\,dt = f(x), \qquad f(a) = f_x'(a) = f_{xx}''(a) = 0.$

Double differentiation yields

$$2\lambda\int_a^x \cosh[\lambda(x-t)]y(t)\,dt + \lambda^2\int_a^x (x-t)\sinh[\lambda(x-t)]y(t)\,dt = f_{xx}''(x).$$

Eliminating the second term on the left-hand side with the aid of the original equation, we arrive at an equation of the form 1.3.1:

$$\int_a^x \cosh[\lambda(x-t)]y(t)\,dt = \frac{1}{2\lambda}\left[f_{xx}''(x) - \lambda^2 f(x)\right].$$

Solution:

$$y(x) = \frac{1}{2\lambda}f_{xxx}'''(x) - \lambda f_x'(x) + \tfrac{1}{2}\lambda^3\int_a^x f(t)\,dt.$$

63. $\displaystyle\int_a^x \left[Ax^\beta + B\sinh^\gamma(\lambda t) + C\right]y(t)\,dt = f(x).$

This is a special case of equation 1.9.6 with $g(x) = Ax^\beta$ and $h(t) = B\sinh^\gamma(\lambda t) + C$.

64. $\displaystyle\int_a^x \left[A\sinh^\gamma(\lambda x) + Bt^\beta + C\right]y(t)\,dt = f(x).$

This is a special case of equation 1.9.6 with $g(x) = A\sinh^\gamma(\lambda x)$ and $h(t) = Bt^\beta + C$.

65. $\displaystyle\int_a^x \left(Ax^\lambda\sinh^\mu t + Bt^\beta\sinh^\gamma x\right)y(t)\,dt = f(x).$

This is a special case of equation 1.9.15 with $g_1(x) = Ax^\lambda$, $h_1(t) = \sinh^\mu t$, $g_2(x) = B\sinh^\gamma x$, and $h_2(t) = t^\beta$.

1.3-3. Kernels Containing Hyperbolic Tangent

66. $\displaystyle\int_a^x \left[\tanh(\lambda x) - \tanh(\lambda t)\right]y(t)\,dt = f(x).$

This is a special case of equation 1.9.2 with $g(x) = \tanh(\lambda x)$.

Solution: $y(x) = \dfrac{1}{\lambda}\left[\cosh^2(\lambda x)f_x'(x)\right]_x'.$

67. $\displaystyle\int_a^x \left[A\tanh(\lambda x) + B\tanh(\lambda t)\right]y(t)\,dt = f(x).$

For $B = -A$, see equation 1.3.66. This is a special case of equation 1.9.4 with $g(x) = \tanh(\lambda x)$.

Solution: $y(x) = \dfrac{1}{A+B}\dfrac{d}{dx}\left\{[\tanh(\lambda x)]^{-\frac{A}{A+B}}\int_a^x [\tanh(\lambda t)]^{-\frac{B}{A+B}}f_t'(t)\,dt\right\}.$

68. $\displaystyle\int_a^x \big[A\,\tanh(\lambda x) + B\,\tanh(\mu t) + C\big]y(t)\,dt = f(x).$

This is a special case of equation 1.9.6 with $g(x) = A\,\tanh(\lambda x)$ and $h(t) = B\,\tanh(\mu t) + C$.

69. $\displaystyle\int_a^x \big[\tanh^2(\lambda x) - \tanh^2(\lambda t)\big]y(t)\,dt = f(x).$

This is a special case of equation 1.9.2 with $g(x) = \tanh^2(\lambda x)$.

Solution: $\displaystyle y(x) = \frac{d}{dx}\left[\frac{\cosh^3(\lambda x)f_x'(x)}{2\lambda \sinh(\lambda x)}\right].$

70. $\displaystyle\int_a^x \big[A\,\tanh^2(\lambda x) + B\,\tanh^2(\lambda t)\big]y(t)\,dt = f(x).$

For $B = -A$, see equation 1.3.69. This is a special case of equation 1.9.4 with $g(x) = \tanh^2(\lambda x)$.

Solution: $\displaystyle y(x) = \frac{1}{A+B}\frac{d}{dx}\left\{\big[\tanh(\lambda x)\big]^{-\frac{2A}{A+B}}\int_a^x \big[\tanh(\lambda t)\big]^{-\frac{2B}{A+B}}f_t'(t)\,dt\right\}.$

71. $\displaystyle\int_a^x \big[A\,\tanh^2(\lambda x) + B\,\tanh^2(\mu t) + C\big]y(t)\,dt = f(x).$

This is a special case of equation 1.9.6 with $g(x) = A\,\tanh^2(\lambda x)$ and $h(t) = B\,\tanh^2(\mu t) + C$.

72. $\displaystyle\int_a^x \big[\tanh(\lambda x) - \tanh(\lambda t)\big]^n y(t)\,dt = f(x), \qquad n = 1, 2, \ldots$

The right-hand side of the equation is assumed to satisfy the conditions $f(a) = f_x'(a) = \cdots = f_x^{(n)}(a) = 0$.

Solution: $\displaystyle y(x) = \frac{1}{\lambda^n n!\,\cosh^2(\lambda x)}\left[\cosh^2(\lambda x)\frac{d}{dx}\right]^{n+1} f(x).$

73. $\displaystyle\int_a^x \sqrt{\tanh x - \tanh t}\; y(t)\,dt = f(x).$

Solution:
$$y(x) = \frac{2}{\pi \cosh^2 x}\left(\cosh^2 x\,\frac{d}{dx}\right)^2 \int_a^x \frac{f(t)\,dt}{\cosh^2 t\,\sqrt{\tanh x - \tanh t}}.$$

74. $\displaystyle\int_a^x \frac{y(t)\,dt}{\sqrt{\tanh x - \tanh t}} = f(x).$

Solution:
$$y(x) = \frac{1}{\pi}\frac{d}{dx}\int_a^x \frac{f(t)\,dt}{\cosh^2 t\,\sqrt{\tanh x - \tanh t}}.$$

75. $\displaystyle\int_a^x (\tanh x - \tanh t)^\lambda y(t)\,dt = f(x), \qquad 0 < \lambda < 1.$

Solution:
$$y(x) = \frac{\sin(\pi\lambda)}{\pi\lambda \cosh^2 x}\left(\cosh^2 x\,\frac{d}{dx}\right)^2 \int_a^x \frac{f(t)\,dt}{\cosh^2 t\,(\tanh x - \tanh t)^\lambda}.$$

76. $\displaystyle\int_a^x (\tanh^\mu x - \tanh^\mu t) y(t)\, dt = f(x).$

This is a special case of equation 1.9.2 with $g(x) = \tanh^\mu x$.

Solution: $\displaystyle y(x) = \frac{1}{\mu}\frac{d}{dx}\left[\frac{\cosh^{\mu+1} x\, f'_x(x)}{\sinh^{\mu-1} x}\right].$

77. $\displaystyle\int_a^x \left(A\tanh^\mu x + B\tanh^\mu t\right) y(t)\, dt = f(x).$

For $B = -A$, see equation 1.3.76. This is a special case of equation 1.9.4 with $g(x) = \tanh^\mu x$.
Solution:

$$y(x) = \frac{1}{A+B}\frac{d}{dx}\left\{ [\tanh(\lambda x)]^{-\frac{A\mu}{A+B}} \int_a^x [\tanh(\lambda t)]^{-\frac{B\mu}{A+B}} f'_t(t)\, dt \right\}.$$

78. $\displaystyle\int_a^x \frac{y(t)\, dt}{[\tanh(\lambda x) - \tanh(\lambda t)]^\mu} = f(x), \qquad 0 < \mu < 1.$

This is a special case of equation 1.9.42 with $g(x) = \tanh(\lambda x)$ and $h(x) \equiv 1$.
Solution:

$$y(x) = \frac{\lambda \sin(\pi\mu)}{\pi}\frac{d}{dx}\int_a^x \frac{f(t)\, dt}{\cosh^2(\lambda t)[\tanh(\lambda x) - \tanh(\lambda t)]^{1-\mu}}.$$

79. $\displaystyle\int_a^x \left[Ax^\beta + B\tanh^\gamma(\lambda t) + C\right] y(t)\, dt = f(x).$

This is a special case of equation 1.9.6 with $g(x) = Ax^\beta$ and $h(t) = B\tanh^\gamma(\lambda t) + C$.

80. $\displaystyle\int_a^x \left[A\tanh^\gamma(\lambda x) + Bt^\beta + C\right] y(t)\, dt = f(x).$

This is a special case of equation 1.9.6 with $g(x) = A\tanh^\gamma(\lambda x)$ and $h(t) = Bt^\beta + C$.

81. $\displaystyle\int_a^x \left(Ax^\lambda \tanh^\mu t + Bt^\beta \tanh^\gamma x\right) y(t)\, dt = f(x).$

This is a special case of equation 1.9.15 with $g_1(x) = Ax^\lambda$, $h_1(t) = \tanh^\mu t$, $g_2(x) = B\tanh^\gamma x$, and $h_2(t) = t^\beta$.

1.3-4. Kernels Containing Hyperbolic Cotangent

82. $\displaystyle\int_a^x \left[\coth(\lambda x) - \coth(\lambda t)\right] y(t)\, dt = f(x).$

This is a special case of equation 1.9.2 with $g(x) = \coth(\lambda x)$.

Solution: $\displaystyle y(x) = -\frac{1}{\lambda}\frac{d}{dx}\left[\sinh^2(\lambda x) f'_x(x)\right].$

83. $\displaystyle\int_a^x \left[A\coth(\lambda x) + B\coth(\lambda t)\right] y(t)\, dt = f(x).$

For $B = -A$, see equation 1.3.82. This is a special case of equation 1.9.4 with $g(x) = \coth(\lambda x)$.

Solution: $\displaystyle y(x) = \frac{1}{A+B}\frac{d}{dx}\left\{ [\tanh(\lambda x)]^{\frac{A}{A+B}} \int_a^x [\tanh(\lambda t)]^{\frac{B}{A+B}} f'_t(t)\, dt \right\}.$

84. $\displaystyle\int_a^x \big[A\coth(\lambda x) + B\coth(\mu t) + C\big]y(t)\,dt = f(x).$

This is a special case of equation 1.9.6 with $g(x) = A\coth(\lambda x)$ and $h(t) = B\coth(\mu t) + C$.

85. $\displaystyle\int_a^x \big[\coth^2(\lambda x) - \coth^2(\lambda t)\big]y(t)\,dt = f(x).$

This is a special case of equation 1.9.2 with $g(x) = \coth^2(\lambda x)$.

Solution: $y(x) = -\dfrac{d}{dx}\left[\dfrac{\sinh^3(\lambda x)f_x'(x)}{2\lambda\cosh(\lambda x)}\right].$

86. $\displaystyle\int_a^x \big[A\coth^2(\lambda x) + B\coth^2(\lambda t)\big]y(t)\,dt = f(x).$

For $B = -A$, see equation 1.3.85. This is a special case of equation 1.9.4 with $g(x) = \coth^2(\lambda x)$.

Solution: $y(x) = \dfrac{1}{A+B}\dfrac{d}{dx}\left\{[\tanh(\lambda x)]^{\frac{2A}{A+B}}\displaystyle\int_a^x [\tanh(\lambda t)]^{\frac{2B}{A+B}} f_t'(t)\,dt\right\}.$

87. $\displaystyle\int_a^x \big[A\coth^2(\lambda x) + B\coth^2(\mu t) + C\big]y(t)\,dt = f(x).$

This is a special case of equation 1.9.6 with $g(x) = A\coth^2(\lambda x)$ and $h(t) = B\coth^2(\mu t) + C$.

88. $\displaystyle\int_a^x \big[\coth(\lambda x) - \coth(\lambda t)\big]^n y(t)\,dt = f(x),\qquad n = 1, 2, \ldots$

The right-hand side of the equation is assumed to satisfy the conditions $f(a) = f_x'(a) = \cdots = f_x^{(n)}(a) = 0$.

Solution: $y(x) = \dfrac{(-1)^n}{\lambda^n n!\,\sinh^2(\lambda x)}\left[\sinh^2(\lambda x)\dfrac{d}{dx}\right]^{n+1} f(x).$

89. $\displaystyle\int_a^x (\coth^\mu x - \coth^\mu t)y(t)\,dt = f(x).$

This is a special case of equation 1.9.2 with $g(x) = \coth^\mu x$.

Solution: $y(x) = -\dfrac{1}{\mu}\dfrac{d}{dx}\left[\dfrac{\sinh^{\mu+1} x f_x'(x)}{\cosh^{\mu-1} x}\right].$

90. $\displaystyle\int_a^x (A\coth^\mu x + B\coth^\mu t)y(t)\,dt = f(x).$

For $B = -A$, see equation 1.3.89. This is a special case of equation 1.9.4 with $g(x) = \coth^\mu x$.
Solution:

$$y(x) = \dfrac{1}{A+B}\dfrac{d}{dx}\left\{|\tanh x|^{\frac{A\mu}{A+B}}\int_a^x |\tanh t|^{\frac{B\mu}{A+B}} f_t'(t)\,dt\right\}.$$

91. $\displaystyle\int_a^x \big[Ax^\beta + B\coth^\gamma(\lambda t) + C\big]y(t)\,dt = f(x).$

This is a special case of equation 1.9.6 with $g(x) = Ax^\beta$ and $h(t) = B\coth^\gamma(\lambda t) + C$.

92. $\displaystyle\int_a^x \big[A\coth^\gamma(\lambda x) + Bt^\beta + C\big]y(t)\,dt = f(x).$

This is a special case of equation 1.9.6 with $g(x) = A\coth^\gamma(\lambda x)$ and $h(t) = Bt^\beta + C$.

93. $\int_a^x \left(Ax^\lambda \coth^\mu t + Bt^\beta \coth^\gamma x \right) y(t)\, dt = f(x).$

This is a special case of equation 1.9.15 with $g_1(x) = Ax^\lambda$, $h_1(t) = \coth^\mu t$, $g_2(x) = B \coth^\gamma x$, and $h_2(t) = t^\beta$.

1.3-5. Kernels Containing Combinations of Hyperbolic Functions

94. $\int_a^x \left\{ \cosh[\lambda(x - t)] + A \sinh[\mu(x - t)] \right\} y(t)\, dt = f(x).$

Let us differentiate the equation with respect to x and then eliminate the integral with the hyperbolic cosine. As a result, we arrive at an equation of the form 2.3.16:

$$y(x) + (\lambda - A^2 \mu) \int_a^x \sinh[\mu(x - t)] y(t)\, dt = f'_x(x) - A\mu f(x).$$

95. $\int_a^x \left[A \cosh(\lambda x) + B \sinh(\mu t) + C \right] y(t)\, dt = f(x).$

This is a special case of equation 1.9.6 with $g(x) = A \cosh(\lambda x)$ and $h(t) = B \sinh(\mu t) + C$.

96. $\int_a^x \left[A \cosh^2(\lambda x) + B \sinh^2(\mu t) + C \right] y(t)\, dt = f(x).$

This is a special case of equation 1.9.6 with $g(x) = \cosh^2(\lambda x)$ and $h(t) = B \sinh^2(\mu t) + C$.

97. $\int_a^x \sinh[\lambda(x - t)] \cosh[\lambda(x + t)] y(t)\, dt = f(x).$

Using the formula

$$\sinh(\alpha - \beta) \cosh(\alpha + \beta) = \tfrac{1}{2} \left[\sinh(2\alpha) - \sinh(2\beta) \right], \quad \alpha = \lambda x, \quad \beta = \lambda t,$$

we reduce the original equation to an equation of the form 1.3.37:

$$\int_a^x \left[\sinh(2\lambda x) - \sinh(2\lambda t) \right] y(t)\, dt = 2 f(x).$$

Solution: $y(x) = \dfrac{1}{\lambda} \dfrac{d}{dx} \left[\dfrac{f'_x(x)}{\cosh(2\lambda x)} \right].$

98. $\int_a^x \cosh[\lambda(x - t)] \sinh[\lambda(x + t)] y(t)\, dt = f(x).$

Using the formula

$$\cosh(\alpha - \beta) \sinh(\alpha + \beta) = \tfrac{1}{2} \left[\sinh(2\alpha) + \sinh(2\beta) \right], \quad \alpha = \lambda x, \quad \beta = \lambda t,$$

we reduce the original equation to an equation of the form 1.3.38 with $A = B = 1$:

$$\int_a^x \left[\sinh(2\lambda x) + \sinh(2\lambda t) \right] y(t)\, dt = 2 f(x).$$

99. $\int_a^x \big[A \cosh(\lambda x) \sinh(\mu t) + B \cosh(\beta x) \sinh(\gamma t)\big] y(t)\, dt = f(x).$

This is a special case of equation 1.9.15 with $g_1(x) = A \cosh(\lambda x)$, $h_1(t) = \sinh(\mu t)$, $g_2(x) = B \cosh(\beta x)$, and $h_2(t) = \sinh(\gamma t)$.

100. $\int_a^x \big[\sinh(\lambda x) \cosh(\mu t) + \sinh(\beta x) \cosh(\gamma t)\big] y(t)\, dt = f(x).$

This is a special case of equation 1.9.15 with $g_1(x) = \sinh(\lambda x)$, $h_1(t) = \cosh(\mu t)$, $g_2(x) = \sinh(\beta x)$, and $h_2(t) = \cosh(\gamma t)$.

101. $\int_a^x \big[\cosh(\lambda x) \cosh(\mu t) + \sinh(\beta x) \sinh(\gamma t)\big] y(t)\, dt = f(x).$

This is a special case of equation 1.9.15 with $g_1(x) = \cosh(\lambda x)$, $h_1(t) = \cosh(\mu t)$, $g_2(x) = \sinh(\beta x)$, and $h_2(t) = \sinh(\gamma t)$.

102. $\int_a^x \big[A \cosh^\beta(\lambda x) + B \sinh^\gamma(\mu t)\big] y(t)\, dt = f(x).$

This is a special case of equation 1.9.6 with $g(x) = A \cosh^\beta(\lambda x)$ and $h(t) = B \sinh^\gamma(\mu t)$.

103. $\int_a^x \big[A \sinh^\beta(\lambda x) + B \cosh^\gamma(\mu t)\big] y(t)\, dt = f(x).$

This is a special case of equation 1.9.6 with $g(x) = A \sinh^\beta(\lambda x)$ and $h(t) = B \cosh^\gamma(\mu t)$.

104. $\int_a^x \big(A x^\lambda \cosh^\mu t + B t^\beta \sinh^\gamma x\big) y(t)\, dt = f(x).$

This is a special case of equation 1.9.15 with $g_1(x) = A x^\lambda$, $h_1(t) = \cosh^\mu t$, $g_2(x) = B \sinh^\gamma x$, and $h_2(t) = t^\beta$.

105. $\int_a^x \big\{(x - t) \sinh[\lambda(x - t)] - \lambda(x - t)^2 \cosh[\lambda(x - t)]\big\} y(t)\, dt = f(x).$

Solution:

$$y(x) = \int_a^x g(t)\, dt,$$

where

$$g(t) = \sqrt{\frac{\pi}{2\lambda}} \frac{1}{64\lambda^5} \left(\frac{d^2}{dt^2} - \lambda^2\right)^6 \int_a^t (t - \tau)^{\frac{5}{2}} I_{\frac{5}{2}}[\lambda(t - \tau)] f(\tau)\, d\tau.$$

106. $\int_a^x \left\{\frac{\sinh[\lambda(x - t)]}{x - t} - \lambda \cosh[\lambda(x - t)]\right\} y(t)\, dt = f(x).$

Solution:

$$y(x) = \frac{1}{2\lambda^4} \left(\frac{d^2}{dx^2} - \lambda^2\right)^3 \int_a^x \sinh[\lambda(x - t)] f(t)\, dt.$$

107. $\int_a^x \big[\sinh\big(\lambda\sqrt{x - t}\big) - \lambda\sqrt{x - t} \cosh\big(\lambda\sqrt{x - t}\big)\big] y(t)\, dt = f(x), \quad f(a) = f'_x(a) = 0.$

Solution:

$$y(x) = -\frac{4}{\pi\lambda^3} \frac{d^3}{dx^3} \int_a^x \frac{\cos\big(\lambda\sqrt{x - t}\big)}{\sqrt{x - t}} f(t)\, dt.$$

108. $\displaystyle\int_a^x \left(Ax^\lambda \sinh^\mu t + Bt^\beta \cosh^\gamma x\right) y(t)\, dt = f(x).$

This is a special case of equation 1.9.15 with $g_1(x) = Ax^\lambda$, $h_1(t) = \sinh^\mu t$, $g_2(x) = B\cosh^\gamma x$, and $h_2(t) = t^\beta$.

109. $\displaystyle\int_a^x \left[A\tanh(\lambda x) + B\coth(\mu t) + C\right] y(t)\, dt = f(x).$

This is a special case of equation 1.9.6 with $g(x) = A\tanh(\lambda x)$ and $h(t) = B\coth(\mu t) + C$.

110. $\displaystyle\int_a^x \left[A\tanh^2(\lambda x) + B\coth^2(\mu t)\right] y(t)\, dt = f(x).$

This is a special case of equation 1.9.6 with $g(x) = \tanh^2(\lambda x)$ and $h(t) = B\coth^2(\mu t)$.

111. $\displaystyle\int_a^x \left[\tanh(\lambda x)\coth(\mu t) + \tanh(\beta x)\coth(\gamma t)\right] y(t)\, dt = f(x).$

This is a special case of equation 1.9.15 with $g_1(x) = \tanh(\lambda x)$, $h_1(t) = \coth(\mu t)$, $g_2(x) = \tanh(\beta x)$, and $h_2(t) = \coth(\gamma t)$.

112. $\displaystyle\int_a^x \left[\coth(\lambda x)\tanh(\mu t) + \coth(\beta x)\tanh(\gamma t)\right] y(t)\, dt = f(x).$

This is a special case of equation 1.9.15 with $g_1(x) = \coth(\lambda x)$, $h_1(t) = \tanh(\mu t)$, $g_2(x) = \coth(\beta x)$, and $h_2(t) = \tanh(\gamma t)$.

113. $\displaystyle\int_a^x \left[\tanh(\lambda x)\tanh(\mu t) + \coth(\beta x)\coth(\gamma t)\right] y(t)\, dt = f(x).$

This is a special case of equation 1.9.15 with $g_1(x) = \tanh(\lambda x)$, $h_1(t) = \tanh(\mu t)$, $g_2(x) = \coth(\beta x)$, and $h_2(t) = \coth(\gamma t)$.

114. $\displaystyle\int_a^x \left[A\tanh^\beta(\lambda x) + B\coth^\gamma(\mu t)\right] y(t)\, dt = f(x).$

This is a special case of equation 1.9.6 with $g(x) = A\tanh^\beta(\lambda x)$ and $h(t) = B\coth^\gamma(\mu t)$.

115. $\displaystyle\int_a^x \left[A\coth^\beta(\lambda x) + B\tanh^\gamma(\mu t)\right] y(t)\, dt = f(x).$

This is a special case of equation 1.9.6 with $g(x) = A\coth^\beta(\lambda x)$ and $h(t) = B\tanh^\gamma(\mu t)$.

116. $\displaystyle\int_a^x \left(Ax^\lambda \tanh^\mu t + Bt^\beta \coth^\gamma x\right) y(t)\, dt = f(x).$

This is a special case of equation 1.9.15 with $g_1(x) = Ax^\lambda$, $h_1(t) = \tanh^\mu t$, $g_2(x) = B\coth^\gamma x$, and $h_2(t) = t^\beta$.

117. $\displaystyle\int_a^x \left(Ax^\lambda \coth^\mu t + Bt^\beta \tanh^\gamma x\right) y(t)\, dt = f(x).$

This is a special case of equation 1.9.15 with $g_1(x) = Ax^\lambda$, $h_1(t) = \coth^\mu t$, $g_2(x) = B\tanh^\gamma x$, and $h_2(t) = t^\beta$.

1.4. Equations Whose Kernels Contain Logarithmic Functions

1.4-1. Kernels Containing Logarithmic Functions

1. $\displaystyle\int_a^x (\ln x - \ln t) y(t)\, dt = f(x).$

This is a special case of equation 1.9.2 with $g(x) = \ln x$.
    Solution: $y(x) = x f''_{xx}(x) + f'_x(x)$.

2. $\displaystyle\int_0^x \ln(x - t) y(t)\, dt = f(x).$

Solution:

$$y(x) = -\int_0^x f''_{tt}(t)\, dt \int_0^\infty \frac{(x-t)^z e^{-Cz}}{\Gamma(z+1)}\, dz - f'_x(0) \int_0^\infty \frac{x^z e^{-Cz}}{\Gamma(z+1)}\, dz,$$

where $C = \lim\limits_{k \to \infty}\left(1 + \dfrac{1}{2} + \cdots + \dfrac{1}{k+1} - \ln k\right) = 0.5772\ldots$ is the Euler constant and $\Gamma(z)$ is the gamma function.

⊙ References: M. L. Krasnov, A. I. Kisilev, and G. I. Makarenko (1971), A. G. Butkovskii (1979).

3. $\displaystyle\int_a^x [\ln(x - t) + A] y(t)\, dt = f(x).$

Solution:

$$y(x) = -\frac{d}{dx} \int_a^x \nu_A(x-t) f(t)\, dt, \qquad \nu_A(x) = \frac{d}{dx} \int_0^\infty \frac{x^z e^{(A-C)z}}{\Gamma(z+1)}\, dz,$$

where $C = 0.5772\ldots$ is the Euler constant and $\Gamma(z)$ is the gamma function.
    For $a = 0$, the solution can be written in the form

$$y(x) = -\int_0^x f''_{tt}(t)\, dt \int_0^\infty \frac{(x-t)^z e^{(A-C)z}}{\Gamma(z+1)}\, dz - f'_x(0) \int_0^\infty \frac{x^z e^{(A-C)z}}{\Gamma(z+1)}\, dz,$$

⊙ Reference: S. G. Samko, A. A. Kilbas, and O. I. Marichev (1993).

4. $\displaystyle\int_a^x (A \ln x + B \ln t) y(t)\, dt = f(x).$

This is a special case of equation 1.9.4 with $g(x) = \ln x$. For $B = -A$, see equation 1.4.1.
    Solution:

$$y(x) = \frac{\operatorname{sign}(\ln x)}{A + B} \frac{d}{dx}\left\{ |\ln x|^{-\frac{A}{A+B}} \int_a^x |\ln t|^{-\frac{B}{A+B}} f'_t(t)\, dt \right\}.$$

5. $\displaystyle\int_a^x (A \ln x + B \ln t + C) y(t)\, dt = f(x).$

This is a special case of equation 1.9.5 with $g(x) = x$.

6. $\displaystyle\int_a^x \left[\ln^2(\lambda x) - \ln^2(\lambda t)\right] y(t)\, dt = f(x), \qquad f(a) = f'_x(a) = 0.$

Solution: $\displaystyle y(x) = \frac{d}{dx}\left[\frac{x f'_x(x)}{2\ln(\lambda x)}\right].$

7. $\displaystyle\int_a^x \left[A\ln^2(\lambda x) + B\ln^2(\lambda t)\right] y(t)\, dt = f(x).$

For $B = -A$, see equation 1.4.7. This is a special case of equation 1.9.4 with $g(x) = \ln^2(\lambda x)$.
    Solution:

$$y(x) = \frac{1}{A+B}\frac{d}{dx}\left\{|\ln(\lambda x)|^{-\frac{2A}{A+B}}\int_a^x |\ln(\lambda t)|^{-\frac{2B}{A+B}} f'_t(t)\, dt\right\}.$$

8. $\displaystyle\int_a^x \left[A\ln^2(\lambda x) + B\ln^2(\mu t) + C\right] y(t)\, dt = f(x).$

This is a special case of equation 1.9.6 with $g(x) = \ln^2(\lambda x)$ and $h(t) = \ln^2(\mu t) + C$.

9. $\displaystyle\int_a^x \left[\ln(x/t)\right]^n y(t)\, dt = f(x), \qquad n = 1, 2, \ldots$

The right-hand side of the equation is assumed to satisfy the conditions $f(a) = f'_x(a) = \cdots = f_x^{(n)}(a) = 0$.
    Solution: $\displaystyle y(x) = \frac{1}{n!\, x}\left(x\frac{d}{dx}\right)^{n+1} f(x).$

10. $\displaystyle\int_a^x \left(\ln^2 x - \ln^2 t\right)^n y(t)\, dt = f(x), \qquad n = 1, 2, \ldots$

The right-hand side of the equation is assumed to satisfy the conditions $f(a) = f'_x(a) = \cdots = f_x^{(n)}(a) = 0$.
    Solution: $\displaystyle y(x) = \frac{\ln x}{2^n n!\, x}\left(\frac{x}{\ln x}\frac{d}{dx}\right)^{n+1} f(x).$

11. $\displaystyle\int_a^x \ln\left(\frac{x+b}{t+b}\right) y(t)\, dt = f(x).$

This is a special case of equation 1.9.2 with $g(x) = \ln(x+b)$.
    Solution: $y(x) = (x+b)f''_{xx}(x) + f'_x(x).$

12. $\displaystyle\int_a^x \sqrt{\ln(x/t)}\, y(t)\, dt = f(x).$

Solution:

$$y(x) = \frac{2}{\pi x}\left(x\frac{d}{dx}\right)^2 \int_a^x \frac{f(t)\, dt}{t\sqrt{\ln(x/t)}}.$$

13. $\displaystyle\int_a^x \frac{y(t)\, dt}{\sqrt{\ln(x/t)}} = f(x).$

Solution:

$$y(x) = \frac{1}{\pi}\frac{d}{dx}\int_a^x \frac{f(t)\, dt}{t\sqrt{\ln(x/t)}}.$$

14. $\displaystyle\int_a^x \left[\ln^\mu(\lambda x) - \ln^\mu(\lambda t)\right] y(t)\, dt = f(x).$

This is a special case of equation 1.9.2 with $g(x) = \ln^\mu(\lambda x)$.

Solution: $\displaystyle y(x) = \frac{1}{\mu}\frac{d}{dx}\left[x \ln^{1-\mu}(\lambda x) f_x'(x)\right].$

15. $\displaystyle\int_a^x \left[A \ln^\beta(\lambda x) + B \ln^\gamma(\mu t) + C\right] y(t)\, dt = f(x).$

This is a special case of equation 1.9.6 with $g(x) = A \ln^\beta(\lambda x)$ and $h(t) = B \ln^\gamma(\mu t) + C$.

16. $\displaystyle\int_a^x [\ln(x/t)]^\lambda y(t)\, dt = f(x), \qquad 0 < \lambda < 1.$

Solution:

$$y(x) = \frac{k}{x}\left(x\frac{d}{dx}\right)^2 \int_a^x \frac{f(t)\,dt}{t[\ln(x/t)]^\lambda}, \qquad k = \frac{\sin(\pi\lambda)}{\pi\lambda}.$$

17. $\displaystyle\int_a^x \frac{y(t)\,dt}{[\ln(x/t)]^\lambda} = f(x), \qquad 0 < \lambda < 1.$

This is a special case of equation 1.9.42 with $g(x) = \ln x$ and $h(x) \equiv 1$.

Solution:

$$y(x) = \frac{\sin(\pi\lambda)}{\pi}\frac{d}{dx}\int_a^x \frac{f(t)\,dt}{t[\ln(x/t)]^{1-\lambda}}.$$

1.4-2. Kernels Containing Power-Law and Logarithmic Functions

18. $\displaystyle\int_a^x (x-t)\left[\ln(x-t) + A\right] y(t)\, dt = f(x).$

Solution:

$$y(x) = -\frac{d^2}{dx^2}\int_a^x \nu_A(x-t) f(t)\, dt, \qquad \nu_A(x) = \frac{d}{dx}\int_0^\infty \frac{x^z e^{(A-C)z}}{\Gamma(z+1)}\, dz,$$

where $C = 0.5772\ldots$ is the Euler constant and $\Gamma(z)$ is the gamma function.

⊙ Reference: S. G. Samko, A. A. Kilbas, and O. I. Marichev (1993).

19. $\displaystyle\int_a^x \frac{\ln(x-t) + A}{(x-t)^\lambda} y(t)\, dt = f(x), \qquad 0 < \lambda < 1.$

Solution:

$$y(x) = -\frac{\sin(\pi\lambda)}{\pi}\frac{d}{dx}\int_a^x \frac{F(t)\,dt}{(x-t)^{1-\lambda}}, \qquad F(x) = \int_a^x \nu_h(x-t) f(t)\, dt,$$

$$\nu_h(x) = \frac{d}{dx}\int_0^\infty \frac{x^z e^{hz}}{\Gamma(z+1)}\, dz, \qquad h = A + \psi(1-\lambda),$$

where $\Gamma(z)$ is the gamma function and $\psi(z) = \left[\Gamma(z)\right]_z'$ is the logarithmic derivative of the gamma function.

⊙ Reference: S. G. Samko, A. A. Kilbas, and O. I. Marichev (1993).

20. $\displaystyle\int_a^x (t^\beta \ln^\lambda x - x^\beta \ln^\lambda t)y(t)\,dt = f(x).$

This is a special case of equation 1.9.11 with $g(x) = \ln^\lambda x$ and $h(t) = t^\beta$.

21. $\displaystyle\int_a^x (At^\beta \ln^\lambda x + Bx^\mu \ln^\gamma t)y(t)\,dt = f(x).$

This is a special case of equation 1.9.15 with $g_1(x) = A\ln^\lambda x$, $h_1(t) = t^\beta$, $g_2(x) = Bx^\mu$, and $h_2(t) = \ln^\gamma t$.

22. $\displaystyle\int_a^x \ln\left(\frac{x^\mu + b}{ct^\lambda + s}\right)y(t)\,dt = f(x).$

This is a special case of equation 1.9.6 with $g(x) = \ln(x^\mu + b)$ and $h(t) = -\ln(ct^\lambda + s)$.

1.5. Equations Whose Kernels Contain Trigonometric Functions

1.5-1. Kernels Containing Cosine

1. $\displaystyle\int_a^x \cos[\lambda(x - t)]y(t)\,dt = f(x).$

Solution: $y(x) = f_x'(x) + \lambda^2 \displaystyle\int_a^x f(x)\,dx.$

2. $\displaystyle\int_a^x \big\{\cos[\lambda(x - t)] - 1\big\}y(t)\,dt = f(x), \qquad f(a) = f_x'(a) = f_{xx}''(x) = 0.$

Solution: $y(x) = -\dfrac{1}{\lambda^2} f_{xxx}'''(x) - f_x'(x).$

3. $\displaystyle\int_a^x \big\{\cos[\lambda(x - t)] + b\big\}y(t)\,dt = f(x).$

For $b = 0$, see equation 1.5.1. For $b = -1$, see equation 1.5.2. For $\lambda = 0$, see equation 1.1.1. Differentiating the equation with respect to x, we arrive at an equation of the form 2.5.16:

$$y(x) - \frac{\lambda}{b + 1}\int_a^x \sin[\lambda(x - t)]y(t)\,dt = \frac{f_x'(x)}{b + 1}.$$

1°. Solution with $b(b + 1) > 0$:

$$y(x) = \frac{f_x'(x)}{b + 1} + \frac{\lambda^2}{k(b + 1)^2}\int_a^x \sin[k(x - t)]f_t'(t)\,dt, \qquad \text{where} \quad k = \lambda\sqrt{\frac{b}{b + 1}}.$$

2°. Solution with $b(b + 1) < 0$:

$$y(x) = \frac{f_x'(x)}{b + 1} + \frac{\lambda^2}{k(b + 1)^2}\int_a^x \sinh[k(x - t)]f_t'(t)\,dt, \qquad \text{where} \quad k = \lambda\sqrt{\frac{-b}{b + 1}}.$$

4. $\displaystyle\int_a^x \cos(\lambda x + \beta t) y(t)\, dt = f(x).$

Differentiating the equation with respect to x twice yields

$$\cos[(\lambda + \beta)x]y(x) - \lambda \int_a^x \sin(\lambda x + \beta t)y(t)\, dt = f_x'(x), \tag{1}$$

$$\left\{\cos[(\lambda + \beta)x]y(x)\right\}_x' - \lambda \sin[(\lambda + \beta)x]y(x) - \lambda^2 \int_a^x \cos(\lambda x + \beta t)y(t)\, dt = f_{xx}''(x). \tag{2}$$

Eliminating the integral term from (2) with the aid of the original equation, we arrive at the first-order linear ordinary differential equation

$$w_x' - \lambda \tan[(\lambda + \beta)x]w = f_{xx}''(x) + \lambda^2 f(x), \qquad w = \cos[(\lambda + \beta)x]y(x). \tag{3}$$

Setting $x = a$ in (1) yields the initial condition $w(a) = f_x'(a)$. On solving equation (3) under this condition, after some transformations we obtain the solution of the original integral equation in the form

$$y(x) = \frac{1}{\cos[(\lambda + \beta)x]} f_x'(x) + \frac{\lambda \sin[(\lambda + \beta)x]}{\cos^2[(\lambda + \beta)x]} f(x)$$
$$- \frac{\lambda\beta}{\cos^{k+1}[(\lambda + \beta)x]} \int_a^x f(t) \cos^{k-2}[(\lambda + \beta)t]\, dt, \qquad k = \frac{\lambda}{\lambda + \beta}.$$

5. $\displaystyle\int_a^x \left[\cos(\lambda x) - \cos(\lambda t)\right] y(t)\, dt = f(x).$

This is a special case of equation 1.9.2 with $g(x) = \cos(\lambda x)$.

Solution: $\displaystyle y(x) = -\frac{1}{\lambda} \frac{d}{dx}\left[\frac{f_x'(x)}{\sin(\lambda x)}\right].$

6. $\displaystyle\int_a^x \left[A \cos(\lambda x) + B \cos(\lambda t)\right] y(t)\, dt = f(x).$

This is a special case of equation 1.9.4 with $g(x) = \cos(\lambda x)$. For $B = -A$, see equation 1.5.5.

Solution with $B \neq -A$:

$$y(x) = \frac{\operatorname{sign}\cos(\lambda x)}{A + B} \frac{d}{dx}\left\{ |\cos(\lambda x)|^{-\frac{A}{A+B}} \int_a^x |\cos(\lambda t)|^{-\frac{B}{A+B}} f_t'(t)\, dt \right\}.$$

7. $\displaystyle\int_a^x \left[A \cos(\lambda x) + B \cos(\mu t) + C\right] y(t)\, dt = f(x).$

This is a special case of equation 1.9.6 with $g(x) = A \cos(\lambda x)$ and $h(t) = B \cos(\mu t) + C$.

8. $\displaystyle\int_a^x \left\{ A_1 \cos[\lambda_1(x - t)] + A_2 \cos[\lambda_2(x - t)]\right\} y(t)\, dt = f(x).$

The equation is equivalent to the equation

$$\int_a^x \left\{ B_1 \sin[\lambda_1(x - t)] + B_2 \sin[\lambda_2(x - t)]\right\} y(t)\, dt = F(x),$$

$$B_1 = \frac{A_1}{\lambda_1}, \qquad B_2 = \frac{A_2}{\lambda_2}, \qquad F(x) = \int_a^x f(t)\, dt.$$

which has the form 1.5.41. (Differentiation of this equation yields the original integral equation.)

9. $\displaystyle\int_a^x \cos^2[\lambda(x-t)]y(t)\,dt = f(x).$

Differentiating yields an equation of the form 2.5.16:

$$y(x) - \lambda \int_a^x \sin[2\lambda(x-t)]y(t)\,dt = f'_x(x).$$

Solution:

$$y(x) = f'_x(x) + \frac{2\lambda^2}{k}\int_a^x \sin[k(x-t)]f'_t(t)\,dt, \qquad \text{where} \quad k = \lambda\sqrt{2}.$$

10. $\displaystyle\int_a^x \left[\cos^2(\lambda x) - \cos^2(\lambda t)\right]y(t)\,dt = f(x), \qquad f(a) = f'_x(a) = 0.$

Solution: $y(x) = -\dfrac{1}{\lambda}\dfrac{d}{dx}\left[\dfrac{f'_x(x)}{\sin(2\lambda x)}\right].$

11. $\displaystyle\int_a^x \left[A\cos^2(\lambda x) + B\cos^2(\lambda t)\right]y(t)\,dt = f(x).$

For $B = -A$, see equation 1.5.10. This is a special case of equation 1.9.4 with $g(x) = \cos^2(\lambda x)$.
Solution:

$$y(x) = \frac{1}{A+B}\frac{d}{dx}\left\{\left[\cos(\lambda x)\right]^{-\frac{2A}{A+B}}\int_a^x \left[\cos(\lambda t)\right]^{-\frac{2B}{A+B}}f'_t(t)\,dt\right\}.$$

12. $\displaystyle\int_a^x \left[A\cos^2(\lambda x) + B\cos^2(\mu t) + C\right]y(t)\,dt = f(x).$

This is a special case of equation 1.9.6 with $g(x) = A\cos^2(\lambda x)$ and $h(t) = B\cos^2(\mu t) + C$.

13. $\displaystyle\int_a^x \cos[\lambda(x-t)]\cos[\lambda(x+t)]y(t)\,dt = f(x).$

Using the trigonometric formula

$$\cos(\alpha - \beta)\cos(\alpha + \beta) = \tfrac{1}{2}\left[\cos(2\alpha) + \cos(2\beta)\right], \quad \alpha = \lambda x, \quad \beta = \lambda t,$$

we reduce the original equation to an equation of the form 1.5.6 with $A = B = 1$:

$$\int_a^x \left[\cos(2\lambda x) + \cos(2\lambda t)\right]y(t)\,dt = 2f(x).$$

Solution with $\cos(2\lambda x) > 0$:

$$y(x) = \frac{d}{dx}\left[\frac{1}{\sqrt{\cos(2\lambda x)}}\int_a^x \frac{f'_t(t)\,dt}{\sqrt{\cos(2\lambda t)}}\right].$$

14. $\displaystyle\int_a^x \left[A\cos(\lambda x)\cos(\mu t) + B\cos(\beta x)\cos(\gamma t)\right]y(t)\,dt = f(x).$

This is a special case of equation 1.9.15 with $g_1(x) = A\cos(\lambda x)$, $h_1(t) = \cos(\mu t)$, $g_2(x) = B\cos(\beta x)$, and $h_2(t) = \cos(\gamma t)$.

15. $\displaystyle\int_a^x \cos^3[\lambda(x-t)]y(t)\,dt = f(x).$

Using the formula $\cos^3\beta = \frac{1}{4}\cos 3\beta + \frac{3}{4}\cos\beta$, we arrive at an equation of the form 1.5.8:

$$\int_a^x \left\{\tfrac{1}{4}\cos[3\lambda(x-t)] + \tfrac{3}{4}\cos[\lambda(x-t)]\right\}y(t)\,dt = f(x).$$

16. $\displaystyle\int_a^x \left[\cos^3(\lambda x) - \cos^3(\lambda t)\right]y(t)\,dt = f(x),\qquad f(a) = f'_x(a) = 0.$

Solution: $\displaystyle y(x) = -\frac{1}{3\lambda}\frac{d}{dx}\left[\frac{f'_x(x)}{\sin(\lambda x)\cos^2(\lambda x)}\right].$

17. $\displaystyle\int_a^x \left[A\cos^3(\lambda x) + B\cos^3(\lambda t)\right]y(t)\,dt = f(x).$

For $B = -A$, see equation 1.3.16. This is a special case of equation 1.9.4 with $g(x) = \cos^3(\lambda x)$.
Solution:

$$y(x) = \frac{1}{A+B}\frac{d}{dx}\left\{\left[\cos(\lambda x)\right]^{-\frac{3A}{A+B}}\int_a^x \left[\cos(\lambda t)\right]^{-\frac{3B}{A+B}}f'_t(t)\,dt\right\}.$$

18. $\displaystyle\int_a^x \left[\cos^2(\lambda x)\cos(\mu t) + \cos(\beta x)\cos^2(\gamma t)\right]y(t)\,dt = f(x).$

This is a special case of equation 1.9.15 with $g_1(x) = \cos^2(\lambda x)$, $h_1(t) = \cos(\mu t)$, $g_2(x) = \cos(\beta x)$, and $h_2(t) = \cos^2(\gamma t)$.

19. $\displaystyle\int_a^x \cos^4[\lambda(x-t)]y(t)\,dt = f(x).$

Let us transform the kernel of the integral equation using the trigonometric formula $\cos^4\beta = \frac{1}{8}\cos 4\beta + \frac{1}{2}\cos 2\beta + \frac{3}{8}$, where $\beta = \lambda(x-t)$, and differentiate the resulting equation with respect to x. Then we arrive at an equation of the form 2.5.18:

$$y(x) - \lambda\int_a^x \left\{\tfrac{1}{2}\sin[4\lambda(x-t)] + \sin[2\lambda(x-t)]\right\}y(t)\,dt = f'_x(x).$$

20. $\displaystyle\int_a^x \left[\cos(\lambda x) - \cos(\lambda t)\right]^n y(t)\,dt = f(x),\qquad n = 1, 2, \ldots$

The right-hand side of the equation is assumed to satisfy the conditions $f(a) = f'_x(a) = \cdots = f_x^{(n)}(a) = 0$.

Solution: $\displaystyle y(x) = \frac{(-1)^n}{\lambda^n n!}\sin(\lambda x)\left[\frac{1}{\sin(\lambda x)}\frac{d}{dx}\right]^{n+1}f(x).$

21. $\displaystyle\int_a^x \sqrt{\cos t - \cos x}\,y(t)\,dt = f(x).$

This is a special case of equation 1.9.38 with $g(x) = 1 - \cos x$.
Solution:

$$y(x) = \frac{2}{\pi}\sin x\left(\frac{1}{\sin x}\frac{d}{dx}\right)^2\int_a^x \frac{\sin t\,f(t)\,dt}{\sqrt{\cos t - \cos x}}.$$

22. $\displaystyle\int_a^x \frac{y(t)\,dt}{\sqrt{\cos t - \cos x}} = f(x).$

Solution:
$$y(x) = \frac{1}{\pi}\frac{d}{dx}\int_a^x \frac{\sin t\, f(t)\,dt}{\sqrt{\cos t - \cos x}}.$$

23. $\displaystyle\int_a^x (\cos t - \cos x)^\lambda y(t)\,dt = f(x), \qquad 0 < \lambda < 1.$

Solution:
$$y(x) = k\sin x\left(\frac{1}{\sin x}\frac{d}{dx}\right)^2 \int_a^x \frac{\sin t\, f(t)\,dt}{(\cos t - \cos x)^\lambda}, \qquad k = \frac{\sin(\pi\lambda)}{\pi\lambda}.$$

24. $\displaystyle\int_a^x (\cos^\mu x - \cos^\mu t)y(t)\,dt = f(x).$

This is a special case of equation 1.9.2 with $g(x) = \cos^\mu x$.

Solution: $\displaystyle y(x) = -\frac{1}{\mu}\frac{d}{dx}\left[\frac{f'_x(x)}{\sin x\,\cos^{\mu-1} x}\right].$

25. $\displaystyle\int_a^x \left(A\cos^\mu x + B\cos^\mu t\right)y(t)\,dt = f(x).$

For $B = -A$, see equation 1.5.24. This is a special case of equation 1.9.4 with $g(x) = \cos^\mu x$.

Solution:
$$y(x) = \frac{1}{A+B}\frac{d}{dx}\left\{|\cos x|^{-\frac{A\mu}{A+B}}\int_a^x |\cos t|^{-\frac{B\mu}{A+B}} f'_t(t)\,dt\right\}.$$

26. $\displaystyle\int_a^x \frac{y(t)\,dt}{(\cos t - \cos x)^\lambda} = f(x), \qquad 0 < \lambda < 1.$

Solution:
$$y(x) = \frac{\sin(\pi\lambda)}{\pi}\frac{d}{dx}\int_a^x \frac{\sin t\, f(t)\,dt}{(\cos t - \cos x)^{1-\lambda}}.$$

27. $\displaystyle\int_a^x (x-t)\cos[\lambda(x-t)]y(t)\,dt = f(x), \qquad f(a) = f'_x(a) = 0.$

Differentiating the equation twice yields
$$y(x) - 2\lambda\int_a^x \sin[\lambda(x-t)]y(t)\,dt - \lambda^2\int_a^x (x-t)\cos[\lambda(x-t)]y(t)\,dt = f''_{xx}(x).$$

Eliminating the third term on the left-hand side with the aid of the original equation, we arrive at an equation of the form 2.5.16:
$$y(x) - 2\lambda\int_a^x \sin[\lambda(x-t)]y(t)\,dt = f''_{xx}(x) + \lambda^2 f(x).$$

28. $\displaystyle\int_a^x \frac{\cos\left(\lambda\sqrt{x-t}\right)}{\sqrt{x-t}}\,y(t)\,dt = f(x).$

Solution:
$$y(x) = \frac{1}{\pi}\frac{d}{dx}\int_a^x \frac{\cosh\left(\lambda\sqrt{x-t}\right)}{\sqrt{x-t}}f(t)\,dt.$$

29. $\displaystyle\int_0^x \frac{\cos\left(\lambda\sqrt{x^2 - t^2}\right)}{\sqrt{x^2 - t^2}}\, y(t)\, dt = f(x).$

Solution:

$$y(x) = \frac{2}{\pi}\frac{d}{dx}\int_0^x t\,\frac{\cosh\left(\lambda\sqrt{x^2 - t^2}\right)}{\sqrt{x^2 - t^2}}f(t)\, dt.$$

30. $\displaystyle\int_x^\infty \frac{\cos\left(\lambda\sqrt{t^2 - x^2}\right)}{\sqrt{t^2 - x^2}}\, y(t)\, dt = f(x).$

Solution:

$$y(x) = -\frac{2}{\pi}\frac{d}{dx}\int_x^\infty t\,\frac{\cosh\left(\lambda\sqrt{t^2 - x^2}\right)}{\sqrt{t^2 - x^2}}f(t)\, dt.$$

31. $\displaystyle\int_a^x \left[Ax^\beta + B\cos^\gamma(\lambda t) + C\right]y(t)\, dt = f(x).$

This is a special case of equation 1.9.6 with $g(x) = Ax^\beta$ and $h(t) = B\cos^\gamma(\lambda t) + C$.

32. $\displaystyle\int_a^x \left[A\cos^\gamma(\lambda x) + Bt^\beta + C\right]y(t)\, dt = f(x).$

This is a special case of equation 1.9.6 with $g(x) = A\cos^\gamma(\lambda x)$ and $h(t) = Bt^\beta + C$.

33. $\displaystyle\int_a^x \left(Ax^\lambda\cos^\mu t + Bt^\beta\cos^\gamma x\right)y(t)\, dt = f(x).$

This is a special case of equation 1.9.15 with $g_1(x) = Ax^\lambda$, $h_1(t) = \cos^\mu t$, $g_2(x) = B\cos^\gamma x$, and $h_2(t) = t^\beta$.

1.5-2. Kernels Containing Sine

34. $\displaystyle\int_a^x \sin[\lambda(x - t)]y(t)\, dt = f(x), \qquad f(a) = f_x'(a) = 0.$

Solution: $y(x) = \dfrac{1}{\lambda}f_{xx}''(x) + \lambda f(x).$

35. $\displaystyle\int_a^x \left\{\sin[\lambda(x - t)] + b\right\}y(t)\, dt = f(x).$

Differentiating the equation with respect to x yields an equation of the form 2.5.3:

$$y(x) + \frac{\lambda}{b}\int_a^x \cos[\lambda(x - t)]y(t)\, dt = \frac{1}{b}f_x'(x).$$

36. $\displaystyle\int_a^x \sin(\lambda x + \beta t)y(t)\, dt = f(x).$

For $\beta = -\lambda$, see equation 1.5.34. Assume that $\beta \neq -\lambda$.

Differentiating the equation with respect to x twice yields

$$\sin[(\lambda + \beta)x]y(x) + \lambda\int_a^x \cos(\lambda x + \beta t)y(t)\, dt = f_x'(x), \tag{1}$$

$$\left\{\sin[(\lambda + \beta)x]y(x)\right\}_x' + \lambda\cos[(\lambda + \beta)x]y(x) - \lambda^2\int_a^x \sin(\lambda x + \beta t)y(t)\, dt = f_{xx}''(x). \tag{2}$$

Eliminating the integral term from (2) with the aid of the original equation, we arrive at the first-order linear ordinary differential equation

$$w'_x + \lambda \cot[(\lambda + \beta)x]w = f''_{xx}(x) + \lambda^2 f(x), \qquad w = \sin[(\lambda + \beta)x]y(x). \qquad (3)$$

Setting $x = a$ in (1) yields the initial condition $w(a) = f'_x(a)$. On solving equation (3) under this condition, after some transformation we obtain the solution of the original integral equation in the form

$$y(x) = \frac{1}{\sin[(\lambda + \beta)x]} f'_x(x) - \frac{\lambda \cos[(\lambda + \beta)x]}{\sin^2[(\lambda + \beta)x]} f(x)$$
$$- \frac{\lambda \beta}{\sin^{k+1}[(\lambda + \beta)x]} \int_a^x f(t) \sin^{k-2}[(\lambda + \beta)t] \, dt, \qquad k = \frac{\lambda}{\lambda + \beta}.$$

37. $\displaystyle \int_a^x \big[\sin(\lambda x) - \sin(\lambda t)\big] y(t) \, dt = f(x).$

This is a special case of equation 1.9.2 with $g(x) = \sin(\lambda x)$.

Solution: $\displaystyle y(x) = \frac{1}{\lambda} \frac{d}{dx} \left[\frac{f'_x(x)}{\cos(\lambda x)} \right].$

38. $\displaystyle \int_a^x \big[A \sin(\lambda x) + B \sin(\lambda t)\big] y(t) \, dt = f(x).$

This is a special case of equation 1.9.4 with $g(x) = \sin(\lambda x)$. For $B = -A$, see equation 1.5.37.

Solution with $B \neq -A$:

$$y(x) = \frac{\text{sign} \sin(\lambda x)}{A + B} \frac{d}{dx} \left\{ |\sin(\lambda x)|^{-\frac{A}{A+B}} \int_a^x |\sin(\lambda t)|^{-\frac{B}{A+B}} f'_t(t) \, dt \right\}.$$

39. $\displaystyle \int_a^x \big[A \sin(\lambda x) + B \sin(\mu t) + C\big] y(t) \, dt = f(x).$

This is a special case of equation 1.9.6 with $g(x) = A \sin(\lambda x)$ and $h(t) = B \sin(\mu t) + C$.

40. $\displaystyle \int_a^x \big\{\mu \sin[\lambda(x - t)] - \lambda \sin[\mu(x - t)]\big\} y(t) \, dt = f(x).$

It is assumed that $f(a) = f'_x(a) = f''_{xx}(a) = f'''_{xxx}(a) = 0$.

Solution:

$$y(x) = \frac{f''''_{xxxx} + (\lambda^2 + \mu^2)f''_{xx} + \lambda^2\mu^2 f}{\lambda\mu^3 - \lambda^3\mu}, \qquad f = f(x).$$

41. $\displaystyle \int_a^x \big\{A_1 \sin[\lambda_1(x - t)] + A_2 \sin[\lambda_2(x - t)]\big\} y(t) \, dt = f(x), \quad f(a) = f'_x(a) = 0.$

This equation can be solved in the same manner as equation 1.3.41, i.e., by reducing it to a second-order linear ordinary differential equation with constant coefficients.

Let

$$\Delta = -\lambda_1 \lambda_2 \frac{A_1 \lambda_2 + A_2 \lambda_1}{A_1 \lambda_1 + A_2 \lambda_2}.$$

1°. Solution for $\Delta > 0$:

$$(A_1 \lambda_1 + A_2 \lambda_2)y(x) = f''_{xx}(x) + Bf(x) + C \int_a^x \sinh[k(x - t)]f(t) \, dt,$$

$$k = \sqrt{\Delta}, \quad B = \Delta + \lambda_1^2 + \lambda_2^2, \quad C = \frac{1}{\sqrt{\Delta}}[\Delta^2 + (\lambda_1^2 + \lambda_2^2)\Delta + \lambda_1^2\lambda_2^2].$$

$2°$. Solution for $\Delta < 0$:

$$(A_1\lambda_1 + A_2\lambda_2)y(x) = f''_{xx}(x) + Bf(x) + C\int_a^x \sin[k(x-t)]f(t)\,dt,$$

$$k = \sqrt{-\Delta}, \quad B = \Delta + \lambda_1^2 + \lambda_2^2, \quad C = \frac{1}{\sqrt{-\Delta}}\left[\Delta^2 + (\lambda_1^2 + \lambda_2^2)\Delta + \lambda_1^2\lambda_2^2\right].$$

$3°$. Solution for $\Delta = 0$:

$$(A_1\lambda_1 + A_2\lambda_2)y(x) = f''_{xx}(x) + (\lambda_1^2 + \lambda_2^2)f(x) + \lambda_1^2\lambda_2^2\int_a^x (x-t)f(t)\,dt.$$

$4°$. Solution for $\Delta = \infty$:

$$y(x) = -\frac{f''''_{xxxx} + (\lambda_1^2 + \lambda_2^2)f''_{xx} + \lambda_1^2\lambda_2^2 f}{A_1\lambda_1^3 + A_2\lambda_2^3}, \qquad f = f(x).$$

In the last case, the relation $A_1\lambda_1 + A_2\lambda_2 = 0$ holds and the right-hand side of the integral equation is assumed to satisfy the conditions $f(a) = f'_x(a) = f''_{xx}(a) = f'''_{xxx}(a) = 0$.

Remark. The solution can be obtained from the solution of equation 1.3.41 in which the change of variables $\lambda_k \to i\lambda_k$, $A_k \to -iA_k$, $i^2 = -1$ $(k = 1, 2)$, should be made.

42. $\displaystyle\int_a^x \left\{ A\sin[\lambda(x-t)] + B\sin[\mu(x-t)] + C\sin[\beta(x-t)]\right\}y(t)\,dt = f(x).$

It is assumed that $f(a) = f'_x(a) = 0$. Differentiating the integral equation twice yields

$$(A\lambda + B\mu + C\beta)y(x) - \int_a^x \left\{ A\lambda^2 \sin[\lambda(x-t)] + B\mu^2 \sin[\mu(x-t)]\right\}y(t)\,dt$$

$$-C\beta^2\int_a^x \sin[\beta(x-t)]y(t)\,dt = f''_{xx}(x).$$

Eliminating the last integral with the aid of the original equation, we arrive at an equation of the form 2.5.18:

$$(A\lambda + B\mu + C\beta)y(x) + \int_a^x \left\{ A(\beta^2 - \lambda^2)\sin[\lambda(x-t)]\right.$$

$$\left. + B(\beta^2 - \mu^2)\sin[\mu(x-t)]\right\}y(t)\,dt = f''_{xx}(x) + \beta^2 f(x).$$

In the special case $A\lambda + B\mu + C\beta = 0$, this is an equation of the form 1.5.41.

43. $\displaystyle\int_a^x \sin^2[\lambda(x-t)]y(t)\,dt = f(x), \qquad f(a) = f'_x(a) = f''_{xx}(a) = 0.$

Differentiation yields an equation of the form 1.5.34:

$$\int_a^x \sin[2\lambda(x-t)]y(t)\,dt = \frac{1}{\lambda}f'_x(x).$$

Solution: $y(x) = \frac{1}{2}\lambda^{-2}f'''_{xxx}(x) + 2f'_x(x).$

44. $\displaystyle\int_a^x \left[\sin^2(\lambda x) - \sin^2(\lambda t)\right]y(t)\,dt = f(x), \qquad f(a) = f'_x(a) = 0.$

Solution: $y(x) = \dfrac{1}{\lambda}\dfrac{d}{dx}\left[\dfrac{f'_x(x)}{\sin(2\lambda x)}\right].$

45. $\int_a^x \left[A\sin^2(\lambda x) + B\sin^2(\lambda t)\right] y(t)\,dt = f(x).$

For $B = -A$, see equation 1.5.44. This is a special case of equation 1.9.4 with $g(x) = \sin^2(\lambda x)$.

Solution:

$$y(x) = \frac{1}{A+B}\frac{d}{dx}\left\{ |\sin(\lambda x)|^{-\frac{2A}{A+B}} \int_a^x |\sin(\lambda t)|^{-\frac{2B}{A+B}} f_t'(t)\,dt \right\}.$$

46. $\int_a^x \left[A\sin^2(\lambda x) + B\sin^2(\mu t) + C\right] y(t)\,dt = f(x).$

This is a special case of equation 1.9.6 with $g(x) = A\sin^2(\lambda x)$ and $h(t) = B\sin^2(\mu t) + C$.

47. $\int_a^x \sin[\lambda(x-t)]\sin[\lambda(x+t)]y(t)\,dt = f(x), \qquad f(a) = f_x'(a) = 0.$

Using the trigonometric formula

$$\sin(\alpha - \beta)\sin(\alpha + \beta) = \tfrac{1}{2}\left[\cos(2\beta) - \cos(2\alpha)\right], \qquad \alpha = \lambda x, \qquad \beta = \lambda t,$$

we reduce the original equation to an equation of the form 1.5.5 with $A = B = 1$:

$$\int_a^x \left[\cos(2\lambda x) - \cos(2\lambda t)\right] y(t)\,dt = -2f(x).$$

Solution: $y(x) = \dfrac{1}{\lambda}\dfrac{d}{dx}\left[\dfrac{f_x'(x)}{\sin(2\lambda x)}\right].$

48. $\int_a^x \left[\sin(\lambda x)\sin(\mu t) + \sin(\beta x)\sin(\gamma t)\right] y(t)\,dt = f(x).$

This is a special case of equation 1.9.15 with $g_1(x) = \sin(\lambda x)$, $h_1(t) = \sin(\mu t)$, $g_2(x) = \sin(\beta x)$, and $h_2(t) = \sin(\gamma t)$.

49. $\int_a^x \sin^3[\lambda(x-t)]y(t)\,dt = f(x).$

It is assumed that $f(a) = f_x'(a) = f_{xx}''(a) = f_{xxx}'''(a) = 0$.

Using the formula $\sin^3 \beta = -\tfrac{1}{4}\sin 3\beta + \tfrac{3}{4}\sin \beta$, we arrive at an equation of the form 1.5.41:

$$\int_a^x \left\{-\tfrac{1}{4}\sin[3\lambda(x-t)] + \tfrac{3}{4}\sin[\lambda(x-t)]\right\}y(t)\,dt = f(x).$$

50. $\int_a^x \left[\sin^3(\lambda x) - \sin^3(\lambda t)\right] y(t)\,dt = f(x), \qquad f(a) = f_x'(a) = 0.$

This is a special case of equation 1.9.2 with $g(x) = \sin^3(\lambda x)$.

51. $\int_a^x \left[A\sin^3(\lambda x) + B\sin^3(\lambda t)\right] y(t)\,dt = f(x).$

This is a special case of equation 1.9.4 with $g(x) = \sin^3(\lambda x)$.

Solution:

$$y(x) = \frac{\operatorname{sign}\sin(\lambda x)}{A+B}\frac{d}{dx}\left\{ |\sin(\lambda x)|^{-\frac{3A}{A+B}} \int_a^x |\sin(\lambda t)|^{-\frac{3B}{A+B}} f_t'(t)\,dt \right\}.$$

52. $\displaystyle\int_a^x \left[\sin^2(\lambda x)\sin(\mu t) + \sin(\beta x)\sin^2(\gamma t)\right] y(t)\, dt = f(x).$

This is a special case of equation 1.9.15 with $g_1(x) = \sin^2(\lambda x)$, $h_1(t) = \sin(\mu t)$, $g_2(x) = \sin(\beta x)$, and $h_2(t) = \sin^2(\gamma t)$.

53. $\displaystyle\int_a^x \sin^4[\lambda(x - t)] y(t)\, dt = f(x).$

It is assumed that $f(a) = f'_x(a) = \cdots = f''''_{xxxx}(a) = 0$.

Let us transform the kernel of the integral equation using the trigonometric formula $\sin^4\beta = \frac{1}{8}\cos 4\beta - \frac{1}{2}\cos 2\beta + \frac{3}{8}$, where $\beta = \lambda(x - t)$, and differentiate the resulting equation with respect to x. Then we obtain an equation of the form 1.5.41:

$$\lambda\int_a^x \left\{-\tfrac{1}{2}\sin[4\lambda(x - t)] + \sin[2\lambda(x - t)]\right\} y(t)\, dt = f'_x(x).$$

54. $\displaystyle\int_a^x \sin^n[\lambda(x - t)] y(t)\, dt = f(x), \qquad n = 2, 3, \ldots$

It is assumed that $f(a) = f'_x(a) = \cdots = f_x^{(n)}(a) = 0$.

Let us differentiate the equation with respect to x twice and transform the kernel of the resulting integral equation using the formula $\cos^2\beta = 1 - \sin^2\beta$, where $\beta = \lambda(x - t)$. We have

$$-\lambda^2 n^2\int_a^x \sin^n[\lambda(x - t)] y(t)\, dt + \lambda^2 n(n - 1)\int_a^x \sin^{n-2}[\lambda(x - t)] y(t)\, dt = f''_{xx}(x).$$

Eliminating the first term on the left-hand side with the aid of the original equation, we obtain

$$\int_a^x \sin^{n-2}[\lambda(x - t)] y(t)\, dt = \frac{1}{\lambda^2 n(n - 1)}\left[f''_{xx}(x) + \lambda^2 n^2 f(x)\right].$$

This equation has the same form as the original equation, but the degree characterizing the kernel has been reduced by two.

By applying this technique sufficiently many times, we finally arrive at simple integral equations of the form 1.1.1 (for even n) or 1.5.34 (for odd n).

55. $\displaystyle\int_a^x \sin\left(\lambda\sqrt{x - t}\right) y(t)\, dt = f(x).$

Solution:

$$y(x) = \frac{2}{\pi\lambda}\frac{d^2}{dx^2}\int_a^x \frac{\cosh\left(\lambda\sqrt{x - t}\right)}{\sqrt{x - t}}\, f(t)\, dt.$$

56. $\displaystyle\int_a^x \sqrt{\sin x - \sin t}\; y(t)\, dt = f(x).$

Solution:

$$y(x) = \frac{2}{\pi}\cos x\left(\frac{1}{\cos x}\frac{d}{dx}\right)^2\int_a^x \frac{\cos t\, f(t)\, dt}{\sqrt{\sin x - \sin t}}.$$

57. $\displaystyle\int_a^x \frac{y(t)\, dt}{\sqrt{\sin x - \sin t}} = f(x).$

Solution:

$$y(x) = \frac{1}{\pi}\frac{d}{dx}\int_a^x \frac{\cos t\, f(t)\, dt}{\sqrt{\sin x - \sin t}}.$$

58. $\displaystyle\int_a^x (\sin x - \sin t)^\lambda y(t)\,dt = f(x), \qquad 0 < \lambda < 1.$

Solution:

$$y(x) = k\cos x \left(\frac{1}{\cos x}\frac{d}{dx}\right)^2 \int_a^x \frac{\cos t\, f(t)\,dt}{(\sin x - \sin t)^\lambda}, \qquad k = \frac{\sin(\pi\lambda)}{\pi\lambda}.$$

59. $\displaystyle\int_a^x (\sin^\mu x - \sin^\mu t)y(t)\,dt = f(x).$

This is a special case of equation 1.9.2 with $g(x) = \sin^\mu x$.

Solution: $\displaystyle y(x) = \frac{1}{\mu}\frac{d}{dx}\left[\frac{f_x'(x)}{\cos x\,\sin^{\mu-1} x}\right].$

60. $\displaystyle\int_a^x \left\{A|\sin(\lambda x)|^\mu + B|\sin(\lambda t)|^\mu\right\}y(t)\,dt = f(x).$

This is a special case of equation 1.9.4 with $g(x) = |\sin(\lambda x)|^\mu$.

Solution:

$$y(x) = \frac{1}{A+B}\frac{d}{dx}\left\{|\sin(\lambda x)|^{-\frac{A\mu}{A+B}}\int_a^x |\sin(\lambda t)|^{-\frac{B\mu}{A+B}} f_t'(t)\,dt\right\}.$$

61. $\displaystyle\int_a^x \frac{y(t)\,dt}{[\sin(\lambda x) - \sin(\lambda t)]^\mu} = f(x), \qquad 0 < \mu < 1.$

This is a special case of equation 1.9.42 with $g(x) = \sin(\lambda x)$ and $h(x) \equiv 1$.

Solution:

$$y(x) = \frac{\lambda\sin(\pi\mu)}{\pi}\frac{d}{dx}\int_a^x \frac{\cos(\lambda t)f(t)\,dt}{[\sin(\lambda x) - \sin(\lambda t)]^{1-\mu}}.$$

62. $\displaystyle\int_a^x (x-t)\sin[\lambda(x-t)]y(t)\,dt = f(x), \qquad f(a) = f_x'(a) = f_{xx}''(a) = 0.$

Double differentiation yields

$$2\lambda\int_a^x \cos[\lambda(x-t)]y(t)\,dt - \lambda^2\int_a^x (x-t)\sin[\lambda(x-t)]y(t)\,dt = f_{xx}''(x).$$

Eliminating the second integral on the left-hand side of this equation with the aid of the original equation, we arrive at an equation of the form 1.5.1:

$$\int_a^x \cos[\lambda(x-t)]y(t)\,dt = \frac{1}{2\lambda}\left[f_{xx}''(x) + \lambda^2 f(x)\right].$$

Solution:

$$y(x) = \frac{1}{2\lambda}f_{xxx}'''(x) + \lambda f_x'(x) + \frac{1}{2}\lambda^3\int_a^x f(t)\,dt.$$

63. $\displaystyle\int_a^x \left[Ax^\beta + B\sin^\gamma(\lambda t) + C\right]y(t)\,dt = f(x).$

This is a special case of equation 1.9.6 with $g(x) = Ax^\beta$ and $h(t) = B\sin^\gamma(\lambda t) + C$.

64. $\displaystyle\int_a^x \big[A\sin^\gamma(\lambda x) + Bt^\beta + C\big]y(t)\,dt = f(x).$

This is a special case of equation 1.9.6 with $g(x) = A\sin^\gamma(\lambda x)$ and $h(t) = Bt^\beta + C$.

65. $\displaystyle\int_a^x \big(Ax^\lambda \sin^\mu t + Bt^\beta \sin^\gamma x\big)y(t)\,dt = f(x).$

This is a special case of equation 1.9.15 with $g_1(x) = Ax^\lambda$, $h_1(t) = \sin^\mu t$, $g_2(x) = B\sin^\gamma x$, and $h_2(t) = t^\beta$.

1.5-3. Kernels Containing Tangent

66. $\displaystyle\int_a^x \big[\tan(\lambda x) - \tan(\lambda t)\big]y(t)\,dt = f(x).$

This is a special case of equation 1.9.2 with $g(x) = \tan(\lambda x)$.

Solution: $y(x) = \dfrac{1}{\lambda}\dfrac{d}{dx}\big[\cos^2(\lambda x)f_x'(x)\big].$

67. $\displaystyle\int_a^x \big[A\tan(\lambda x) + B\tan(\lambda t)\big]y(t)\,dt = f(x).$

For $B = -A$, see equation 1.5.66. This is a special case of equation 1.9.4 with $g(x) = \tan(\lambda x)$.

Solution: $y(x) = \dfrac{1}{A+B}\dfrac{d}{dx}\left\{\big[\tan(\lambda x)\big]^{-\frac{A}{A+B}}\displaystyle\int_a^x \big[\tan(\lambda t)\big]^{-\frac{B}{A+B}} f_t'(t)\,dt\right\}.$

68. $\displaystyle\int_a^x \big[A\tan(\lambda x) + B\tan(\mu t) + C\big]y(t)\,dt = f(x).$

This is a special case of equation 1.9.6 with $g(x) = A\tan(\lambda x)$ and $h(t) = B\tan(\mu t) + C$.

69. $\displaystyle\int_a^x \big[\tan^2(\lambda x) - \tan^2(\lambda t)\big]y(t)\,dt = f(x).$

This is a special case of equation 1.9.2 with $g(x) = \tan^2(\lambda x)$.

Solution: $y(x) = \dfrac{d}{dx}\left[\dfrac{\cos^3(\lambda x)f_x'(x)}{2\lambda\sin(\lambda x)}\right].$

70. $\displaystyle\int_a^x \big[A\tan^2(\lambda x) + B\tan^2(\lambda t)\big]y(t)\,dt = f(x).$

For $B = -A$, see equation 1.5.69. This is a special case of equation 1.9.4 with $g(x) = \tan^2(\lambda x)$.

Solution: $y(x) = \dfrac{1}{A+B}\dfrac{d}{dx}\left\{\big|\tan(\lambda x)\big|^{-\frac{2A}{A+B}}\displaystyle\int_a^x \big|\tan(\lambda t)\big|^{-\frac{2B}{A+B}} f_t'(t)\,dt\right\}.$

71. $\displaystyle\int_a^x \big[A\tan^2(\lambda x) + B\tan^2(\mu t) + C\big]y(t)\,dt = f(x).$

This is a special case of equation 1.9.6 with $g(x) = A\tan^2(\lambda x)$ and $h(t) = B\tan^2(\mu t) + C$.

72. $\displaystyle\int_a^x \big[\tan(\lambda x) - \tan(\lambda t)\big]^n y(t)\,dt = f(x),\qquad n = 1, 2, \dots$

The right-hand side of the equation is assumed to satisfy the conditions $f(a) = f_x'(a) = \cdots = f_x^{(n)}(a) = 0$.

Solution: $y(x) = \dfrac{1}{\lambda^n n!\cos^2(\lambda x)}\left[\cos^2(\lambda x)\dfrac{d}{dx}\right]^{n+1} f(x).$

73. $\displaystyle\int_a^x \sqrt{\tan x - \tan t}\, y(t)\, dt = f(x).$

Solution:
$$y(x) = \frac{2}{\pi \cos^2 x} \left(\cos^2 x \frac{d}{dx}\right)^2 \int_a^x \frac{f(t)\, dt}{\cos^2 t \sqrt{\tan x - \tan t}}.$$

74. $\displaystyle\int_a^x \frac{y(t)\, dt}{\sqrt{\tan x - \tan t}} = f(x).$

Solution:
$$y(x) = \frac{1}{\pi} \frac{d}{dx} \int_a^x \frac{f(t)\, dt}{\cos^2 t \sqrt{\tan x - \tan t}}.$$

75. $\displaystyle\int_a^x (\tan x - \tan t)^\lambda y(t)\, dt = f(x), \qquad 0 < \lambda < 1.$

Solution:
$$y(x) = \frac{\sin(\pi\lambda)}{\pi\lambda \cos^2 x} \left(\cos^2 x \frac{d}{dx}\right)^2 \int_a^x \frac{f(t)\, dt}{\cos^2 t (\tan x - \tan t)^\lambda}.$$

76. $\displaystyle\int_a^x (\tan^\mu x - \tan^\mu t) y(t)\, dt = f(x).$

This is a special case of equation 1.9.2 with $g(x) = \tan^\mu x$.

Solution: $y(x) = \dfrac{1}{\mu} \dfrac{d}{dx} \left[\dfrac{\cos^{\mu+1} x f_x'(x)}{\sin^{\mu-1} x}\right].$

77. $\displaystyle\int_a^x \left(A \tan^\mu x + B \tan^\mu t\right) y(t)\, dt = f(x).$

For $B = -A$, see equation 1.5.76. This is a special case of equation 1.9.4 with $g(x) = \tan^\mu x$.
Solution:
$$y(x) = \frac{1}{A+B} \frac{d}{dx} \left\{ [\tan(\lambda x)]^{-\frac{A\mu}{A+B}} \int_a^x [\tan(\lambda t)]^{-\frac{B\mu}{A+B}} f_t'(t)\, dt \right\}.$$

78. $\displaystyle\int_a^x \frac{y(t)\, dt}{[\tan(\lambda x) - \tan(\lambda t)]^\mu} = f(x), \qquad 0 < \mu < 1.$

This is a special case of equation 1.9.42 with $g(x) = \tan(\lambda x)$ and $h(x) \equiv 1$.
Solution:
$$y(x) = \frac{\lambda \sin(\pi\mu)}{\pi} \frac{d}{dx} \int_a^x \frac{f(t)\, dt}{\cos^2(\lambda t)[\tan(\lambda x) - \tan(\lambda t)]^{1-\mu}}.$$

79. $\displaystyle\int_a^x \left[Ax^\beta + B \tan^\gamma(\lambda t) + C\right] y(t)\, dt = f(x).$

This is a special case of equation 1.9.6 with $g(x) = Ax^\beta$ and $h(t) = B\tan^\gamma(\lambda t) + C$.

80. $\displaystyle\int_a^x \left[A \tan^\gamma(\lambda x) + Bt^\beta + C\right] y(t)\, dt = f(x).$

This is a special case of equation 1.9.6 with $g(x) = A\tan^\gamma(\lambda x)$ and $h(t) = Bt^\beta + C$.

81. $\displaystyle\int_a^x \left(Ax^\lambda \tan^\mu t + Bt^\beta \tan^\gamma x\right) y(t)\, dt = f(x).$

This is a special case of equation 1.9.15 with $g_1(x) = Ax^\lambda$, $h_1(t) = \tan^\mu t$, $g_2(x) = B\tan^\gamma x$, and $h_2(t) = t^\beta$.

1.5-4. Kernels Containing Cotangent

82. $\displaystyle\int_a^x \left[\cot(\lambda x) - \cot(\lambda t)\right] y(t)\, dt = f(x).$

This is a special case of equation 1.9.2 with $g(x) = \cot(\lambda x)$.

Solution: $\displaystyle y(x) = -\frac{1}{\lambda}\frac{d}{dx}\left[\sin^2(\lambda x) f_x'(x)\right].$

83. $\displaystyle\int_a^x \left[A\cot(\lambda x) + B\cot(\lambda t)\right] y(t)\, dt = f(x).$

For $B = -A$, see equation 1.5.82. This is a special case of equation 1.9.4 with $g(x) = \cot(\lambda x)$.

Solution: $\displaystyle y(x) = \frac{1}{A+B}\frac{d}{dx}\left\{ \left[\tan(\lambda x)\right]^{\frac{A}{A+B}} \int_a^x \left[\tan(\lambda t)\right]^{\frac{B}{A+B}} f_t'(t)\, dt \right\}.$

84. $\displaystyle\int_a^x \left[A\cot(\lambda x) + B\cot(\mu t) + C\right] y(t)\, dt = f(x).$

This is a special case of equation 1.9.6 with $g(x) = A\cot(\lambda x)$ and $h(t) = B\cot(\mu t) + C$.

85. $\displaystyle\int_a^x \left[\cot^2(\lambda x) - \cot^2(\lambda t)\right] y(t)\, dt = f(x).$

This is a special case of equation 1.9.2 with $g(x) = \cot^2(\lambda x)$.

Solution: $\displaystyle y(x) = -\frac{d}{dx}\left[\frac{\sin^3(\lambda x) f_x'(x)}{2\lambda\cos(\lambda x)}\right].$

86. $\displaystyle\int_a^x \left[A\cot^2(\lambda x) + B\cot^2(\lambda t)\right] y(t)\, dt = f(x).$

For $B = -A$, see equation 1.5.85. This is a special case of equation 1.9.4 with $g(x) = \cot^2(\lambda x)$.

Solution: $\displaystyle y(x) = \frac{1}{A+B}\frac{d}{dx}\left\{ \left|\tan(\lambda x)\right|^{\frac{2A}{A+B}} \int_a^x \left|\tan(\lambda t)\right|^{\frac{2B}{A+B}} f_t'(t)\, dt \right\}.$

87. $\displaystyle\int_a^x \left[A\cot^2(\lambda x) + B\cot^2(\mu t) + C\right] y(t)\, dt = f(x).$

This is a special case of equation 1.9.6 with $g(x) = A\cot^2(\lambda x)$ and $h(t) = B\cot^2(\mu t) + C$.

88. $\displaystyle\int_a^x \left[\cot(\lambda x) - \cot(\lambda t)\right]^n y(t)\, dt = f(x), \qquad n = 1, 2, \ldots$

The right-hand side of the equation is assumed to satisfy the conditions $f(a) = f_x'(a) = \cdots = f_x^{(n)}(a) = 0$.

Solution: $\displaystyle y(x) = \frac{(-1)^n}{\lambda^n n!\sin^2(\lambda x)}\left[\sin^2(\lambda x)\frac{d}{dx}\right]^{n+1} f(x).$

89. $\displaystyle\int_a^x \left(\cot^\mu x - \cot^\mu t\right) y(t)\, dt = f(x).$

This is a special case of equation 1.9.2 with $g(x) = \cot^\mu x$.

Solution: $\displaystyle y(x) = -\frac{1}{\mu}\frac{d}{dx}\left[\frac{\sin^{\mu+1} x\, f_x'(x)}{\cos^{\mu-1} x}\right].$

90. $\displaystyle\int_a^x \left(A \cot^\mu x + B \cot^\mu t\right) y(t)\, dt = f(x).$

For $B = -A$, see equation 1.5.89. This is a special case of equation 1.9.4 with $g(x) = \cot^\mu x$.
Solution:

$$y(x) = \frac{1}{A+B} \frac{d}{dx} \left\{ |\tan x|^{\frac{A\mu}{A+B}} \int_a^x |\tan t|^{\frac{B\mu}{A+B}} f_t'(t)\, dt \right\}.$$

91. $\displaystyle\int_a^x \left[A x^\beta + B \cot^\gamma(\lambda t) + C\right] y(t)\, dt = f(x).$

This is a special case of equation 1.9.6 with $g(x) = A x^\beta$ and $h(t) = B \cot^\gamma(\lambda t) + C$.

92. $\displaystyle\int_a^x \left[A \cot^\gamma(\lambda x) + B t^\beta + C\right] y(t)\, dt = f(x).$

This is a special case of equation 1.9.6 with $g(x) = A \cot^\gamma(\lambda x)$ and $h(t) = B t^\beta + C$.

93. $\displaystyle\int_a^x \left(A x^\lambda \cot^\mu t + B t^\beta \cot^\gamma x\right) y(t)\, dt = f(x).$

This is a special case of equation 1.9.15 with $g_1(x) = A x^\lambda$, $h_1(t) = \cot^\mu t$, $g_2(x) = B \cot^\gamma x$, and $h_2(t) = t^\beta$.

1.5-5. Kernels Containing Combinations of Trigonometric Functions

94. $\displaystyle\int_a^x \left\{\cos[\lambda(x - t)] + A \sin[\mu(x - t)]\right\} y(t)\, dt = f(x).$

Differentiating the equation with respect to x followed by eliminating the integral with the cosine yields an equation of the form 2.3.16:

$$y(x) - (\lambda + A^2 \mu) \int_a^x \sin[\mu(x - t)]\, y(t)\, dt = f_x'(x) - A\mu f(x).$$

95. $\displaystyle\int_a^x \left[A \cos(\lambda x) + B \sin(\mu t) + C\right] y(t)\, dt = f(x).$

This is a special case of equation 1.9.6 with $g(x) = A \cos(\lambda x)$ and $h(t) = B \sin(\mu t) + C$.

96. $\displaystyle\int_a^x \left[A \sin(\lambda x) + B \cos(\mu t) + C\right] y(t)\, dt = f(x).$

This is a special case of equation 1.9.6 with $g(x) = A \sin(\lambda x)$ and $h(t) = B \cos(\mu t) + C$.

97. $\displaystyle\int_a^x \left[A \cos^2(\lambda x) + B \sin^2(\mu t)\right] y(t)\, dt = f(x).$

This is a special case of equation 1.9.6 with $g(x) = A \cos^2(\lambda x)$ and $h(t) = B \sin^2(\mu t)$.

98. $\displaystyle\int_a^x \sin[\lambda(x-t)]\cos[\lambda(x+t)]y(t)\,dt = f(x), \qquad f(a) = f'_x(a) = 0.$

Using the trigonometric formula

$$\sin(\alpha - \beta)\cos(\alpha + \beta) = \tfrac{1}{2}\left[\sin(2\alpha) - \sin(2\beta)\right], \quad \alpha = \lambda x, \quad \beta = \lambda t,$$

we reduce the original equation to an equation of the form 1.5.37:

$$\int_a^x \left[\sin(2\lambda x) - \sin(2\lambda t)\right]y(t)\,dt = 2f(x).$$

Solution: $\displaystyle y(x) = \frac{1}{\lambda}\frac{d}{dx}\left[\frac{f'_x(x)}{\cos(2\lambda x)}\right].$

99. $\displaystyle\int_a^x \cos[\lambda(x-t)]\sin[\lambda(x+t)]y(t)\,dt = f(x).$

Using the trigonometric formula

$$\cos(\alpha - \beta)\sin(\alpha + \beta) = \tfrac{1}{2}\left[\sin(2\alpha) + \sin(2\beta)\right], \quad \alpha = \lambda x, \quad \beta = \lambda t,$$

we reduce the original equation to an equation of the form 1.5.38 with $A = B = 1$:

$$\int_a^x \left[\sin(2\lambda x) + \sin(2\lambda t)\right]y(t)\,dt = 2f(x).$$

Solution with $\sin(2\lambda x) > 0$:

$$y(x) = \frac{d}{dx}\left[\frac{1}{\sqrt{\sin(2\lambda x)}}\int_a^x \frac{f'_t(t)\,dt}{\sqrt{\sin(2\lambda t)}}\right].$$

100. $\displaystyle\int_a^x \left[A\cos(\lambda x)\sin(\mu t) + B\cos(\beta x)\sin(\gamma t)\right]y(t)\,dt = f(x).$

This is a special case of equation 1.9.15 with $g_1(x) = A\cos(\lambda x)$, $h_1(t) = \sin(\mu t)$, $g_2(x) = B\cos(\beta x)$, and $h_2(t) = \sin(\gamma t)$.

101. $\displaystyle\int_a^x \left[A\sin(\lambda x)\cos(\mu t) + B\sin(\beta x)\cos(\gamma t)\right]y(t)\,dt = f(x).$

This is a special case of equation 1.9.15 with $g_1(x) = A\sin(\lambda x)$, $h_1(t) = \cos(\mu t)$, $g_2(x) = B\sin(\beta x)$, and $h_2(t) = \cos(\gamma t)$.

102. $\displaystyle\int_a^x \left[A\cos(\lambda x)\cos(\mu t) + B\sin(\beta x)\sin(\gamma t)\right]y(t)\,dt = f(x).$

This is a special case of equation 1.9.15 with $g_1(x) = A\cos(\lambda x)$, $h_1(t) = \cos(\mu t)$, $g_2(x) = B\sin(\beta x)$, and $h_2(t) = \sin(\gamma t)$.

103. $\displaystyle\int_a^x \left[A\cos^\beta(\lambda x) + B\sin^\gamma(\mu t)\right]y(t)\,dt = f(x).$

This is a special case of equation 1.9.6 with $g(x) = A\cos^\beta(\lambda x)$ and $h(t) = B\sin^\gamma(\mu t)$.

104. $\displaystyle\int_a^x \left[A \sin^\beta(\lambda x) + B \cos^\gamma(\mu t) \right] y(t)\, dt = f(x).$

This is a special case of equation 1.9.6 with $g(x) = A \sin^\beta(\lambda x)$ and $h(t) = B \cos^\gamma(\mu t)$.

105. $\displaystyle\int_a^x \left(A x^\lambda \cos^\mu t + B t^\beta \sin^\gamma x \right) y(t)\, dt = f(x).$

This is a special case of equation 1.9.15 with $g_1(x) = A x^\lambda$, $h_1(t) = \cos^\mu t$, $g_2(x) = B \sin^\gamma x$, and $h_2(t) = t^\beta$.

106. $\displaystyle\int_a^x \left(A x^\lambda \sin^\mu t + B t^\beta \cos^\gamma x \right) y(t)\, dt = f(x).$

This is a special case of equation 1.9.15 with $g_1(x) = A x^\lambda$, $h_1(t) = \sin^\mu t$, $g_2(x) = B \cos^\gamma x$, and $h_2(t) = t^\beta$.

107. $\displaystyle\int_a^x \left\{ (x - t)\sin[\lambda(x - t)] - \lambda(x - t)^2 \cos[\lambda(x - t)] \right\} y(t)\, dt = f(x).$

Solution:

$$y(x) = \int_a^x g(t)\, dt,$$

where

$$g(t) = \sqrt{\frac{\pi}{2\lambda}} \frac{1}{64\lambda^5} \left(\frac{d^2}{dt^2} + \lambda^2 \right)^6 \int_a^t (t - \tau)^{5/2} J_{5/2}[\lambda(t - \tau)]\, f(\tau)\, d\tau.$$

108. $\displaystyle\int_a^x \left\{ \frac{\sin[\lambda(x - t)]}{x - t} - \lambda \cos[\lambda(x - t)] \right\} y(t)\, dt = f(x).$

Solution:

$$y(x) = \frac{1}{2\lambda^4} \left(\frac{d^2}{dx^2} + \lambda^2 \right)^3 \int_a^x \sin[\lambda(x - t)] f(t)\, dt.$$

109. $\displaystyle\int_a^x \left[\sin\left(\lambda \sqrt{x - t} \right) - \lambda \sqrt{x - t} \cos\left(\lambda \sqrt{x - t} \right) \right] y(t)\, dt = f(x), \quad f(a) = f_x'(a) = 0.$

Solution:

$$y(x) = \frac{4}{\pi \lambda^3} \frac{d^3}{dx^3} \int_a^x \frac{\cosh\left(\lambda \sqrt{x - t} \right)}{\sqrt{x - t}} f(t)\, dt.$$

110. $\displaystyle\int_a^x \left[A \tan(\lambda x) + B \cot(\mu t) + C \right] y(t)\, dt = f(x).$

This is a special case of equation 1.9.6 with $g(x) = A \tan(\lambda x)$ and $h(t) = B \cot(\mu t) + C$.

111. $\displaystyle\int_a^x \left[A \tan^2(\lambda x) + B \cot^2(\mu t) \right] y(t)\, dt = f(x).$

This is a special case of equation 1.9.6 with $g(x) = A \tan^2(\lambda x)$ and $h(t) = B \cot^2(\mu t)$.

112. $\displaystyle\int_a^x \left[\tan(\lambda x) \cot(\mu t) + \tan(\beta x) \cot(\gamma t) \right] y(t)\, dt = f(x).$

This is a special case of equation 1.9.15 with $g_1(x) = \tan(\lambda x)$, $h_1(t) = \cot(\mu t)$, $g_2(x) = \tan(\beta x)$, and $h_2(t) = \cot(\gamma t)$.

113. $\displaystyle\int_a^x \big[\cot(\lambda x)\tan(\mu t) + \cot(\beta x)\tan(\gamma t)\big]y(t)\,dt = f(x).$

This is a special case of equation 1.9.15 with $g_1(x) = \cot(\lambda x)$, $h_1(t) = \tan(\mu t)$, $g_2(x) = \cot(\beta x)$, and $h_2(t) = \tan(\gamma t)$.

114. $\displaystyle\int_a^x \big[\tan(\lambda x)\tan(\mu t) + \cot(\beta x)\cot(\gamma t)\big]y(t)\,dt = f(x).$

This is a special case of equation 1.9.15 with $g_1(x) = \tan(\lambda x)$, $h_1(t) = \tan(\mu t)$, $g_2(x) = \cot(\beta x)$, and $h_2(t) = \cot(\gamma t)$.

115. $\displaystyle\int_a^x \big[A\tan^\beta(\lambda x) + B\cot^\gamma(\mu t)\big]y(t)\,dt = f(x).$

This is a special case of equation 1.9.6 with $g(x) = A\tan^\beta(\lambda x)$ and $h(t) = B\cot^\gamma(\mu t)$.

116. $\displaystyle\int_a^x \big[A\cot^\beta(\lambda x) + B\tan^\gamma(\mu t)\big]y(t)\,dt = f(x).$

This is a special case of equation 1.9.6 with $g(x) = A\cot^\beta(\lambda x)$ and $h(t) = B\tan^\gamma(\mu t)$.

117. $\displaystyle\int_a^x \big(Ax^\lambda\tan^\mu t + Bt^\beta\cot^\gamma x\big)y(t)\,dt = f(x).$

This is a special case of equation 1.9.15 with $g_1(x) = Ax^\lambda$, $h_1(t) = \tan^\mu t$, $g_2(x) = B\cot^\gamma x$, and $h_2(t) = t^\beta$.

118. $\displaystyle\int_a^x \big(Ax^\lambda\cot^\mu t + Bt^\beta\tan^\gamma x\big)y(t)\,dt = f(x).$

This is a special case of equation 1.9.15 with $g_1(x) = Ax^\lambda$, $h_1(t) = \cot^\mu t$, $g_2(x) = B\tan^\gamma x$, and $h_2(t) = t^\beta$.

1.6. Equations Whose Kernels Contain Inverse Trigonometric Functions

1.6-1. Kernels Containing Arccosine

1. $\displaystyle\int_a^x \big[\arccos(\lambda x) - \arccos(\lambda t)\big]y(t)\,dt = f(x).$

This is a special case of equation 1.9.2 with $g(x) = \arccos(\lambda x)$.

Solution: $\displaystyle y(x) = -\frac{1}{\lambda}\frac{d}{dx}\Big[\sqrt{1 - \lambda^2 x^2}\, f'_x(x)\Big].$

2. $\displaystyle\int_a^x \big[A\arccos(\lambda x) + B\arccos(\lambda t)\big]y(t)\,dt = f(x).$

For $B = -A$, see equation 1.6.1. This is a special case of equation 1.9.4 with $g(x) = \arccos(\lambda x)$.
Solution:

$$y(x) = \frac{1}{A+B}\frac{d}{dx}\left\{\big[\arccos(\lambda x)\big]^{-\frac{A}{A+B}}\int_a^x \big[\arccos(\lambda t)\big]^{-\frac{B}{A+B}} f'_t(t)\,dt\right\}.$$

3. $\displaystyle\int_a^x \left[A\arccos(\lambda x) + B\arccos(\mu t) + C\right] y(t)\,dt = f(x).$

This is a special case of equation 1.9.6 with $g(x) = \arccos(\lambda x)$ and $h(t) = B\arccos(\mu t) + C$.

4. $\displaystyle\int_a^x \left[\arccos(\lambda x) - \arccos(\lambda t)\right]^n y(t)\,dt = f(x), \qquad n = 1, 2, \ldots$

The right-hand side of the equation is assumed to satisfy the conditions $f(a) = f'_x(a) = \cdots = f_x^{(n)}(a) = 0$.

Solution:
$$y(x) = \frac{(-1)^n}{\lambda^n n!\,\sqrt{1 - \lambda^2 x^2}} \left(\sqrt{1 - \lambda^2 x^2}\,\frac{d}{dx}\right)^{n+1} f(x).$$

5. $\displaystyle\int_a^x \sqrt{\arccos(\lambda t) - \arccos(\lambda x)}\,y(t)\,dt = f(x).$

This is a special case of equation 1.9.38 with $g(x) = 1 - \arccos(\lambda x)$.

Solution:
$$y(x) = \frac{2}{\pi}\varphi(x)\left(\frac{1}{\varphi(x)}\frac{d}{dx}\right)^2 \int_a^x \frac{\varphi(t)f(t)\,dt}{\sqrt{\arccos(\lambda t) - \arccos(\lambda x)}}, \qquad \varphi(x) = \frac{1}{\sqrt{1 - \lambda^2 x^2}}.$$

6. $\displaystyle\int_a^x \frac{y(t)\,dt}{\sqrt{\arccos(\lambda t) - \arccos(\lambda x)}} = f(x).$

Solution:
$$y(x) = \frac{\lambda}{\pi}\frac{d}{dx}\int_a^x \frac{\varphi(t)f(t)\,dt}{\sqrt{\arccos(\lambda t) - \arccos(\lambda x)}}, \qquad \varphi(x) = \frac{1}{\sqrt{1 - \lambda^2 x^2}}.$$

7. $\displaystyle\int_a^x \left[\arccos(\lambda t) - \arccos(\lambda x)\right]^\mu y(t)\,dt = f(x), \qquad 0 < \mu < 1.$

Solution:
$$y(x) = k\varphi(x)\left(\frac{1}{\varphi(x)}\frac{d}{dx}\right)^2 \int_a^x \frac{\varphi(t)f(t)\,dt}{[\arccos(\lambda t) - \arccos(\lambda x)]^\mu},$$
$$\varphi(x) = \frac{1}{\sqrt{1 - \lambda^2 x^2}}, \qquad k = \frac{\sin(\pi\mu)}{\pi\mu}.$$

8. $\displaystyle\int_a^x \left[\arccos^\mu(\lambda x) - \arccos^\mu(\lambda t)\right] y(t)\,dt = f(x).$

This is a special case of equation 1.9.2 with $g(x) = \arccos^\mu(\lambda x)$.

Solution: $\displaystyle y(x) = -\frac{1}{\lambda\mu}\frac{d}{dx}\left[\frac{f'_x(x)\sqrt{1 - \lambda^2 x^2}}{\arccos^{\mu-1}(\lambda x)}\right].$

9. $\displaystyle\int_a^x \frac{y(t)\,dt}{\left[\arccos(\lambda t) - \arccos(\lambda x)\right]^\mu} = f(x), \qquad 0 < \mu < 1.$

Solution:
$$y(x) = \frac{\lambda\sin(\pi\mu)}{\pi}\frac{d}{dx}\int_a^x \frac{\varphi(t)f(t)\,dt}{[\arccos(\lambda t) - \arccos(\lambda x)]^{1-\mu}}, \qquad \varphi(x) = \frac{1}{\sqrt{1 - \lambda^2 x^2}}.$$

10. $\displaystyle\int_a^x \left[A\arccos^\beta(\lambda x) + B\arccos^\gamma(\mu t) + C\right] y(t)\,dt = f(x).$

This is a special case of equation 1.9.6 with $g(x) = A\arccos^\beta(\lambda x)$ and $h(t) = B\arccos^\gamma(\mu t) + C$.

1.6-2. Kernels Containing Arcsine

11. $\displaystyle\int_a^x \big[\arcsin(\lambda x) - \arcsin(\lambda t)\big] y(t)\, dt = f(x).$

This is a special case of equation 1.9.2 with $g(x) = \arcsin(\lambda x)$.

Solution: $y(x) = \dfrac{1}{\lambda}\dfrac{d}{dx}\Big[\sqrt{1 - \lambda^2 x^2}\, f_x'(x)\Big].$

12. $\displaystyle\int_a^x \big[A \arcsin(\lambda x) + B \arcsin(\lambda t)\big] y(t)\, dt = f(x).$

For $B = -A$, see equation 1.6.11. This is a special case of equation 1.9.4 with $g(x) = \arcsin(\lambda x)$.

Solution:

$$y(x) = \frac{\operatorname{sign} x}{A + B}\frac{d}{dx}\left\{ \big|\arcsin(\lambda x)\big|^{-\frac{A}{A+B}} \int_a^x \big|\arcsin(\lambda t)\big|^{-\frac{B}{A+B}} f_t'(t)\, dt\right\}.$$

13. $\displaystyle\int_a^x \big[A \arcsin(\lambda x) + B \arcsin(\mu t) + C\big] y(t)\, dt = f(x).$

This is a special case of equation 1.9.6 with $g(x) = A \arcsin(\lambda x)$ and $h(t) = B \arcsin(\mu t) + C$.

14. $\displaystyle\int_a^x \big[\arcsin(\lambda x) - \arcsin(\lambda t)\big]^n y(t)\, dt = f(x), \qquad n = 1, 2, \ldots$

The right-hand side of the equation is assumed to satisfy the conditions $f(a) = f_x'(a) = \cdots = f_x^{(n)}(a) = 0$.

Solution:

$$y(x) = \frac{1}{\lambda^n n!\, \sqrt{1 - \lambda^2 x^2}}\left(\sqrt{1 - \lambda^2 x^2}\,\frac{d}{dx}\right)^{n+1} f(x).$$

15. $\displaystyle\int_a^x \sqrt{\arcsin(\lambda x) - \arcsin(\lambda t)}\, y(t)\, dt = f(x).$

Solution:

$$y(x) = \frac{2}{\pi}\varphi(x)\left(\frac{1}{\varphi(x)}\frac{d}{dx}\right)^2 \int_a^x \frac{\varphi(t) f(t)\, dt}{\sqrt{\arcsin(\lambda x) - \arcsin(\lambda t)}}, \qquad \varphi(x) = \frac{1}{\sqrt{1 - \lambda^2 x^2}}.$$

16. $\displaystyle\int_a^x \frac{y(t)\, dt}{\sqrt{\arcsin(\lambda x) - \arcsin(\lambda t)}} = f(x).$

Solution:

$$y(x) = \frac{\lambda}{\pi}\frac{d}{dx}\int_a^x \frac{\varphi(t) f(t)\, dt}{\sqrt{\arcsin(\lambda x) - \arcsin(\lambda t)}}, \qquad \varphi(x) = \frac{1}{\sqrt{1 - \lambda^2 x^2}}.$$

17. $\displaystyle\int_a^x \big[\arcsin(\lambda x) - \arcsin(\lambda t)\big]^\mu y(t)\, dt = f(x), \qquad 0 < \mu < 1.$

Solution:

$$y(x) = k\varphi(x)\left(\frac{1}{\varphi(x)}\frac{d}{dx}\right)^2 \int_a^x \frac{\varphi(t) f(t)\, dt}{[\arcsin(\lambda x) - \arcsin(\lambda t)]^\mu},$$

$$\varphi(x) = \frac{1}{\sqrt{1 - \lambda^2 x^2}}, \qquad k = \frac{\sin(\pi\mu)}{\pi\mu}.$$

18. $\displaystyle\int_a^x \left[\arcsin^\mu(\lambda x) - \arcsin^\mu(\lambda t)\right] y(t)\, dt = f(x).$

This is a special case of equation 1.9.2 with $g(x) = \arcsin^\mu(\lambda x)$.

Solution: $\displaystyle y(x) = \frac{1}{\lambda\mu}\frac{d}{dx}\left[\frac{f_x'(x)\sqrt{1-\lambda^2 x^2}}{\arcsin^{\mu-1}(\lambda x)}\right].$

19. $\displaystyle\int_a^x \frac{y(t)\, dt}{\left[\arcsin(\lambda x) - \arcsin(\lambda t)\right]^\mu} = f(x), \qquad 0 < \mu < 1.$

Solution:

$$y(x) = \frac{\lambda \sin(\pi\mu)}{\pi}\frac{d}{dx}\int_a^x \frac{\varphi(t) f(t)\, dt}{\left[\arcsin(\lambda x) - \arcsin(\lambda t)\right]^{1-\mu}}, \qquad \varphi(x) = \frac{1}{\sqrt{1-\lambda^2 x^2}}.$$

20. $\displaystyle\int_a^x \left[A\arcsin^\beta(\lambda x) + B\arcsin^\gamma(\mu t) + C\right] y(t)\, dt = f(x).$

This is a special case of equation 1.9.6 with $g(x) = A\arcsin^\beta(\lambda x)$ and $h(t) = B\arcsin^\gamma(\mu t) + C$.

1.6-3. Kernels Containing Arctangent

21. $\displaystyle\int_a^x \left[\arctan(\lambda x) - \arctan(\lambda t)\right] y(t)\, dt = f(x).$

This is a special case of equation 1.9.2 with $g(x) = \arctan(\lambda x)$.

Solution: $\displaystyle y(x) = \frac{1}{\lambda}\frac{d}{dx}\left[(1 + \lambda^2 x^2)\, f_x'(x)\right].$

22. $\displaystyle\int_a^x \left[A\arctan(\lambda x) + B\arctan(\lambda t)\right] y(t)\, dt = f(x).$

For $B = -A$, see equation 1.6.21. This is a special case of equation 1.9.4 with $g(x) = \arctan(\lambda x)$.

Solution:

$$y(x) = \frac{\operatorname{sign} x}{A + B}\frac{d}{dx}\left\{\left|\arctan(\lambda x)\right|^{-\frac{A}{A+B}}\int_a^x \left|\arctan(\lambda t)\right|^{-\frac{B}{A+B}} f_t'(t)\, dt\right\}.$$

23. $\displaystyle\int_a^x \left[A\arctan(\lambda x) + B\arctan(\mu t) + C\right] y(t)\, dt = f(x).$

This is a special case of equation 1.9.6 with $g(x) = A\arctan(\lambda x)$ and $h(t) = B\arctan(\mu t) + C$.

24. $\displaystyle\int_a^x \left[\arctan(\lambda x) - \arctan(\lambda t)\right]^n y(t)\, dt = f(x), \qquad n = 1, 2, \ldots$

The right-hand side of the equation is assumed to satisfy the conditions $f(a) = f_x'(a) = \cdots = f_x^{(n)}(a) = 0$.

Solution:

$$y(x) = \frac{1}{\lambda^n n!\,(1 + \lambda^2 x^2)}\left((1 + \lambda^2 x^2)\frac{d}{dx}\right)^{n+1} f(x).$$

25. $\displaystyle\int_a^x \sqrt{\arctan(\lambda x) - \arctan(\lambda t)}\, y(t)\, dt = f(x).$

Solution:

$$y(x) = \frac{2}{\pi}\varphi(x)\left(\frac{1}{\varphi(x)}\frac{d}{dx}\right)^2 \int_a^x \frac{\varphi(t)f(t)\, dt}{\sqrt{\arctan(\lambda x) - \arctan(\lambda t)}}, \qquad \varphi(x) = \frac{1}{1 + \lambda^2 x^2}.$$

26. $\displaystyle\int_a^x \frac{y(t)\, dt}{\sqrt{\arctan(\lambda x) - \arctan(\lambda t)}} = f(x).$

Solution:

$$y(x) = \frac{\lambda}{\pi}\frac{d}{dx}\int_a^x \frac{\varphi(t)f(t)\, dt}{\sqrt{\arctan(\lambda x) - \arctan(\lambda t)}}, \qquad \varphi(x) = \frac{1}{1 + \lambda^2 x^2}.$$

27. $\displaystyle\int_a^x \sqrt{t}\, \arctan\left(\sqrt{\frac{x-t}{t}}\,\right) y(t)\, dt = f(x).$

The equation can be rewritten in terms of the Gaussian hypergeometric function in the form

$$\int_a^x (x-t)^{\gamma-1} F\left(\alpha, \beta, \gamma; 1 - \frac{x}{t}\right) y(t)\, dt = f(x), \qquad \text{where} \quad \alpha = \tfrac{1}{2}, \quad \beta = 1, \quad \gamma = \tfrac{3}{2}.$$

See 1.8.86 for the solution of this equation.

28. $\displaystyle\int_a^x \big[\arctan(\lambda x) - \arctan(\lambda t)\big]^\mu y(t)\, dt = f(x), \qquad 0 < \mu < 1.$

Solution:

$$y(x) = k\varphi(x)\left(\frac{1}{\varphi(x)}\frac{d}{dx}\right)^2 \int_a^x \frac{\varphi(t)f(t)\, dt}{[\arctan(\lambda x) - \arctan(\lambda t)]^\mu},$$

$$\varphi(x) = \frac{1}{1 + \lambda^2 x^2}, \qquad k = \frac{\sin(\pi\mu)}{\pi\mu}.$$

29. $\displaystyle\int_a^x \big[\arctan^\mu(\lambda x) - \arctan^\mu(\lambda t)\big] y(t)\, dt = f(x).$

This is a special case of equation 1.9.2 with $g(x) = \arctan^\mu(\lambda x)$.

Solution: $\displaystyle y(x) = \frac{1}{\lambda\mu}\frac{d}{dx}\left[\frac{(1 + \lambda^2 x^2)f'_x(x)}{\arctan^{\mu-1}(\lambda x)}\right].$

30. $\displaystyle\int_a^x \frac{y(t)\, dt}{\big[\arctan(\lambda x) - \arctan(\lambda t)\big]^\mu} = f(x), \qquad 0 < \mu < 1.$

Solution:

$$y(x) = \frac{\lambda \sin(\pi\mu)}{\pi}\frac{d}{dx}\int_a^x \frac{\varphi(t)f(t)\, dt}{[\arctan(\lambda x) - \arctan(\lambda t)]^{1-\mu}}, \qquad \varphi(x) = \frac{1}{1 + \lambda^2 x^2}.$$

31. $\displaystyle\int_a^x \big[A\arctan^\beta(\lambda x) + B\arctan^\gamma(\mu t) + C\big] y(t)\, dt = f(x).$

This is a special case of equation 1.9.6 with $g(x) = A\arctan^\beta(\lambda x)$ and $h(t) = B\arctan^\gamma(\mu t) + C$.

1.6-4. Kernels Containing Arccotangent

32. $\displaystyle\int_a^x \big[\mathbf{arccot}(\lambda x) - \mathbf{arccot}(\lambda t)\big] y(t)\, dt = f(x).$

This is a special case of equation 1.9.2 with $g(x) = \mathrm{arccot}(\lambda x)$.

Solution: $\displaystyle y(x) = -\frac{1}{\lambda}\frac{d}{dx}\big[(1 + \lambda^2 x^2)\, f'_x(x)\big].$

33. $\displaystyle\int_a^x \big[A\,\mathbf{arccot}(\lambda x) + B\,\mathbf{arccot}(\lambda t)\big] y(t)\, dt = f(x).$

For $B = -A$, see equation 1.6.32. This is a special case of equation 1.9.4 with $g(x) = \mathrm{arccot}(\lambda x)$.
Solution:

$$y(x) = \frac{1}{A+B}\frac{d}{dx}\left\{ [\mathrm{arccot}(\lambda x)]^{-\frac{A}{A+B}} \int_a^x [\mathrm{arccot}(\lambda t)]^{-\frac{B}{A+B}} f'_t(t)\, dt \right\}.$$

34. $\displaystyle\int_a^x \big[A\,\mathbf{arccot}(\lambda x) + B\,\mathbf{arccot}(\mu t) + C\big] y(t)\, dt = f(x).$

This is a special case of equation 1.9.6 with $g(x) = A\,\mathrm{arccot}(\lambda x)$ and $h(t) = B\,\mathrm{arccot}(\mu t) + C$.

35. $\displaystyle\int_a^x \big[\mathbf{arccot}(\lambda x) - \mathbf{arccot}(\lambda t)\big]^n y(t)\, dt = f(x), \qquad n = 1, 2, \dots$

The right-hand side of the equation is assumed to satisfy the conditions $f(a) = f'_x(a) = \cdots = f_x^{(n)}(a) = 0$.
Solution:

$$y(x) = \frac{(-1)^n}{\lambda^n n!\,(1 + \lambda^2 x^2)}\left((1 + \lambda^2 x^2)\frac{d}{dx}\right)^{n+1} f(x).$$

36. $\displaystyle\int_a^x \sqrt{\mathbf{arccot}(\lambda t) - \mathbf{arccot}(\lambda x)}\; y(t)\, dt = f(x).$

Solution:

$$y(x) = \frac{2}{\pi}\varphi(x)\left(\frac{1}{\varphi(x)}\frac{d}{dx}\right)^2 \int_a^x \frac{\varphi(t)f(t)\, dt}{\sqrt{\mathrm{arccot}(\lambda t) - \mathrm{arccot}(\lambda x)}}, \qquad \varphi(x) = \frac{1}{1 + \lambda^2 x^2}.$$

37. $\displaystyle\int_a^x \frac{y(t)\, dt}{\sqrt{\mathbf{arccot}(\lambda t) - \mathbf{arccot}(\lambda x)}} = f(x).$

Solution:

$$y(x) = \frac{\lambda}{\pi}\frac{d}{dx}\int_a^x \frac{\varphi(t)f(t)\, dt}{\sqrt{\mathrm{arccot}(\lambda t) - \mathrm{arccot}(\lambda x)}}, \qquad \varphi(x) = \frac{1}{1 + \lambda^2 x^2}.$$

38. $\displaystyle\int_a^x \big[\mathbf{arccot}(\lambda t) - \mathbf{arccot}(\lambda x)\big]^\mu y(t)\, dt = f(x), \qquad 0 < \mu < 1.$

Solution:

$$y(x) = k\varphi(x)\left(\frac{1}{\varphi(x)}\frac{d}{dx}\right)^2 \int_a^x \frac{\varphi(t)f(t)\, dt}{[\mathrm{arccot}(\lambda t) - \mathrm{arccot}(\lambda x)]^\mu},$$

$$\varphi(x) = \frac{1}{1 + \lambda^2 x^2}, \qquad k = \frac{\sin(\pi\mu)}{\pi\mu}.$$

39. $\displaystyle\int_a^x \left[\operatorname{arccot}^\mu(\lambda x) - \operatorname{arccot}^\mu(\lambda t)\right] y(t)\, dt = f(x).$

This is a special case of equation 1.9.2 with $g(x) = \operatorname{arccot}^\mu(\lambda x)$.

Solution: $\displaystyle y(x) = -\frac{1}{\lambda\mu}\frac{d}{dx}\left[\frac{(1 + \lambda^2 x^2) f'_x(x)}{\operatorname{arccot}^{\mu-1}(\lambda x)}\right].$

40. $\displaystyle\int_a^x \frac{y(t)\, dt}{\left[\operatorname{arccot}(\lambda t) - \operatorname{arccot}(\lambda x)\right]^\mu} = f(x), \qquad 0 < \mu < 1.$

Solution:

$$y(x) = \frac{\lambda\sin(\pi\mu)}{\pi}\frac{d}{dx}\int_a^x \frac{\varphi(t) f(t)\, dt}{[\operatorname{arccot}(\lambda t) - \operatorname{arccot}(\lambda x)]^{1-\mu}}, \qquad \varphi(x) = \frac{1}{1 + \lambda^2 x^2}.$$

41. $\displaystyle\int_a^x \left[A\,\operatorname{arccot}^\beta(\lambda x) + B\,\operatorname{arccot}^\gamma(\mu t) + C\right] y(t)\, dt = f(x).$

This is a special case of equation 1.9.6 with $g(x) = A\,\operatorname{arccot}^\beta(\lambda x)$ and $h(t) = B\,\operatorname{arccot}^\gamma(\mu t) + C$.

1.7. Equations Whose Kernels Contain Combinations of Elementary Functions

1.7-1. Kernels Containing Exponential and Hyperbolic Functions

1. $\displaystyle\int_a^x e^{\mu(x-t)}\left\{A_1\cosh[\lambda_1(x-t)] + A_2\cosh[\lambda_2(x-t)]\right\} y(t)\, dt = f(x).$

The substitution $w(x) = e^{-\mu x} y(x)$ leads to an equation of the form 1.3.8:

$$\int_a^x \left\{A_1\cosh[\lambda_1(x-t)] + A_2\cosh[\lambda_2(x-t)]\right\} w(t)\, dt = e^{-\mu x} f(x).$$

2. $\displaystyle\int_a^x e^{\mu(x-t)}\cosh^2[\lambda(x-t)] y(t)\, dt = f(x).$

Solution:

$$y(x) = \varphi(x) - \frac{2\lambda^2}{k}\int_a^x e^{\mu(x-t)}\sinh[k(x-t)]\varphi(t)\, dt, \qquad k = \lambda\sqrt{2}, \quad \varphi(x) = f'_x(x) - \mu f(x).$$

3. $\displaystyle\int_a^x e^{\mu(x-t)}\cosh^3[\lambda(x-t)] y(t)\, dt = f(x).$

The substitution $w(x) = e^{-\mu x} y(x)$ leads to an equation of the form 1.3.15:

$$\int_a^x \cosh^3[\lambda(x-t)] w(t)\, dt = e^{-\mu x} f(x).$$

4. $\displaystyle\int_a^x e^{\mu(x-t)}\cosh^4[\lambda(x-t)] y(t)\, dt = f(x).$

The substitution $w(x) = e^{-\mu x} y(x)$ leads to an equation of the form 1.3.19:

$$\int_a^x \cosh^4[\lambda(x-t)] w(t)\, dt = e^{-\mu x} f(x).$$

5. $\displaystyle\int_a^x e^{\mu(x-t)}\big[\cosh(\lambda x) - \cosh(\lambda t)\big]^n y(t)\, dt = f(x), \qquad n = 1, 2, \dots$

Solution:

$$y(x) = \frac{1}{\lambda^n n!}e^{\mu x}\sinh(\lambda x)\left[\frac{1}{\sinh(\lambda x)}\frac{d}{dx}\right]^{n+1} F_\mu(x), \qquad F_\mu(x) = e^{-\mu x}f(x).$$

6. $\displaystyle\int_a^x e^{\mu(x-t)}\sqrt{\cosh x - \cosh t}\; y(t)\, dt = f(x), \qquad f(a) = 0.$

Solution:

$$y(x) = \frac{2}{\pi}e^{\mu x}\sinh x\left(\frac{1}{\sinh x}\frac{d}{dx}\right)^2 \int_a^x \frac{e^{-\mu t}\sinh t\, f(t)\, dt}{\sqrt{\cosh x - \cosh t}}.$$

7. $\displaystyle\int_a^x \frac{e^{\mu(x-t)}y(t)\, dt}{\sqrt{\cosh x - \cosh t}} = f(x).$

Solution:

$$y(x) = \frac{1}{\pi}e^{\mu x}\frac{d}{dx}\int_a^x \frac{e^{-\mu t}\sinh t\, f(t)\, dt}{\sqrt{\cosh x - \cosh t}}.$$

8. $\displaystyle\int_a^x e^{\mu(x-t)}(\cosh x - \cosh t)^\lambda y(t)\, dt = f(x), \qquad 0 < \lambda < 1.$

The substitution $w(x) = e^{-\mu x}y(x)$ leads to an equation of the form 1.3.23:

$$\int_a^x (\cosh x - \cosh t)^\lambda w(t)\, dt = e^{-\mu x}f(x).$$

9. $\displaystyle\int_a^x \big[Ae^{\mu(x-t)} + B\cosh^\lambda x\big]y(t)\, dt = f(x).$

This is a special case of equation 1.9.15 with $g_1(x) = Ae^{\mu x}$, $h_1(t) = e^{-\mu t}$, $g_2(x) = B\cosh^\lambda x$, and $h_2(t) = 1$.

10. $\displaystyle\int_a^x \big[Ae^{\mu(x-t)} + B\cosh^\lambda t\big]y(t)\, dt = f(x).$

This is a special case of equation 1.9.15 with $g_1(x) = Ae^{\mu x}$, $h_1(t) = e^{-\mu t}$, $g_2(x) = B$, and $h_2(t) = \cosh^\lambda t$.

11. $\displaystyle\int_a^x e^{\mu(x-t)}(\cosh^\lambda x - \cosh^\lambda t)y(t)\, dt = f(x).$

The substitution $w(x) = e^{-\mu x}y(x)$ leads to an equation of the form 1.3.24:

$$\int_a^x (\cosh^\lambda x - \cosh^\lambda t)w(t)\, dt = e^{-\mu x}f(x).$$

12. $\displaystyle\int_a^x e^{\mu(x-t)}\big(A\cosh^\lambda x + B\cosh^\lambda t\big)y(t)\, dt = f(x).$

The substitution $w(x) = e^{-\mu x}y(x)$ leads to an equation of the form 1.3.25:

$$\int_a^x \big(A\cosh^\lambda x + B\cosh^\lambda t\big)w(t)\, dt = e^{-\mu x}f(x).$$

13. $\displaystyle\int_a^x \frac{e^{\mu(x-t)}y(t)\,dt}{(\cosh x - \cosh t)^\lambda} = f(x), \qquad 0 < \lambda < 1.$

Solution:

$$y(x) = \frac{\sin(\pi\lambda)}{\pi} e^{\mu x} \frac{d}{dx} \int_a^x \frac{e^{-\mu t} \sinh t\, f(t)\,dt}{(\cosh x - \cosh t)^{1-\lambda}}.$$

14. $\displaystyle\int_a^x e^{\mu(x-t)}\{A_1 \sinh[\lambda_1(x-t)] + A_2 \sinh[\lambda_2(x-t)]\}y(t)\,dt = f(x).$

The substitution $w(x) = e^{-\mu x}y(x)$ leads to an equation of the form 1.3.41:

$$\int_a^x \{A_1 \sinh[\lambda_1(x-t)] + A_2 \sinh[\lambda_2(x-t)]\}w(t)\,dt = e^{-\mu x} f(x).$$

15. $\displaystyle\int_a^x e^{\mu(x-t)} \sinh^2[\lambda(x-t)]y(t)\,dt = f(x).$

The substitution $w(x) = e^{-\mu x}y(x)$ leads to an equation of the form 1.3.43:

$$\int_a^x \sinh^2[\lambda(x-t)]w(t)\,dt = e^{-\mu x} f(x).$$

16. $\displaystyle\int_a^x e^{\mu(x-t)} \sinh^3[\lambda(x-t)]y(t)\,dt = f(x).$

The substitution $w(x) = e^{-\mu x}y(x)$ leads to an equation of the form 1.3.49:

$$\int_a^x \sinh^3[\lambda(x-t)]w(t)\,dt = e^{-\mu x} f(x).$$

17. $\displaystyle\int_a^x e^{\mu(x-t)} \sinh^n[\lambda(x-t)]y(t)\,dt = f(x), \qquad n = 2, 3, \dots$

The substitution $w(x) = e^{-\mu x}y(x)$ leads to an equation of the form 1.3.54:

$$\int_a^x \sinh^n[\lambda(x-t)]w(t)\,dt = e^{-\mu x} f(x).$$

18. $\displaystyle\int_a^x e^{\mu(x-t)} \sinh\left(k\sqrt{x-t}\right)y(t)\,dt = f(x).$

Solution:

$$y(x) = \frac{2}{\pi k} e^{\mu x} \frac{d^2}{dx^2} \int_a^x \frac{e^{-\mu t} \cos\left(k\sqrt{x-t}\right)}{\sqrt{x-t}} f(t)\,dt.$$

19. $\displaystyle\int_a^x e^{\mu(x-t)} \sqrt{\sinh x - \sinh t}\, y(t)\,dt = f(x).$

Solution:

$$y(x) = \frac{2}{\pi} e^{\mu x} \cosh x \left(\frac{1}{\cosh x} \frac{d}{dx}\right)^2 \int_a^x \frac{e^{-\mu t} \cosh t\, f(t)\,dt}{\sqrt{\sinh x - \sinh t}}.$$

20. $\displaystyle\int_a^x \frac{e^{\mu(x-t)}y(t)\,dt}{\sqrt{\sinh x - \sinh t}} = f(x).$

Solution:

$$y(x) = \frac{1}{\pi} e^{\mu x} \frac{d}{dx} \int_a^x \frac{e^{-\mu t} \cosh t\, f(t)\,dt}{\sqrt{\sinh x - \sinh t}}.$$

21. $\displaystyle\int_a^x e^{\mu(x-t)}(\sinh x - \sinh t)^\lambda y(t)\,dt = f(x), \qquad 0 < \lambda < 1.$

The substitution $w(x) = e^{-\mu x} y(x)$ leads to an equation of the form 1.3.58:

$$\int_a^x (\sinh x - \sinh t)^\lambda w(t)\,dt = e^{-\mu x} f(x).$$

22. $\displaystyle\int_a^x e^{\mu(x-t)}(\sinh^\lambda x - \sinh^\lambda t) y(t)\,dt = f(x).$

The substitution $w(x) = e^{-\mu x} y(x)$ leads to an equation of the form 1.3.59:

$$\int_a^x (\sinh^\lambda x - \sinh^\lambda t) w(t)\,dt = e^{-\mu x} f(x).$$

23. $\displaystyle\int_a^x e^{\mu(x-t)}\big(A\sinh^\lambda x + B\sinh^\lambda t\big) y(t)\,dt = f(x).$

The substitution $w(x) = e^{-\mu x} y(x)$ leads to an equation of the form 1.3.60:

$$\int_a^x \big(A\sinh^\lambda x + B\sinh^\lambda t\big) w(t)\,dt = e^{-\mu x} f(x).$$

24. $\displaystyle\int_a^x \big[Ae^{\mu(x-t)} + B\sinh^\lambda x\big] y(t)\,dt = f(x).$

This is a special case of equation 1.9.15 with $g_1(x) = Ae^{\mu x}$, $h_1(t) = e^{-\mu t}$, $g_2(x) = B\sinh^\lambda x$, and $h_2(t) = 1$.

25. $\displaystyle\int_a^x \big[Ae^{\mu(x-t)} + B\sinh^\lambda t\big] y(t)\,dt = f(x).$

This is a special case of equation 1.9.15 with $g_1(x) = Ae^{\mu x}$, $h_1(t) = e^{-\mu t}$, $g_2(x) = B$, and $h_2(t) = \sinh^\lambda t$.

26. $\displaystyle\int_a^x \frac{e^{\mu(x-t)} y(t)\,dt}{(\sinh x - \sinh t)^\lambda} = f(x), \qquad 0 < \lambda < 1.$

Solution:

$$y(x) = \frac{\sin(\pi\lambda)}{\pi} e^{\mu x} \frac{d}{dx} \int_a^x \frac{e^{-\mu t}\cosh t\, f(t)\,dt}{(\sinh x - \sinh t)^{1-\lambda}}.$$

27. $\displaystyle\int_a^x e^{\mu(x-t)}\big(A\tanh^\lambda x + B\tanh^\lambda t\big) y(t)\,dt = f(x).$

The substitution $w(x) = e^{-\mu x} y(x)$ leads to an equation of the form 1.3.77:

$$\int_a^x \big(A\tanh^\lambda x + B\tanh^\lambda t\big) w(t)\,dt = e^{-\mu x} f(x).$$

28. $\displaystyle\int_a^x e^{\mu(x-t)}\big(A\tanh^\lambda x + B\tanh^\beta t + C\big) y(t)\,dt = f(x).$

The substitution $w(x) = e^{-\mu x} y(x)$ leads to an equation of the form 1.9.6 with $g(x) = A\tanh^\lambda x$, $g(t) = B\tanh^\beta t + C$:

$$\int_a^x \big(A\tanh^\lambda x + B\tanh^\beta t + C\big) w(t)\,dt = e^{-\mu x} f(x).$$

29. $\displaystyle\int_a^x \left[Ae^{\mu(x-t)} + B\tanh^\lambda x\right]y(t)\,dt = f(x).$

This is a special case of equation 1.9.15 with $g_1(x) = Ae^{\mu x}$, $h_1(t) = e^{-\mu t}$, $g_2(x) = B\tanh^\lambda x$, and $h_2(t) = 1$.

30. $\displaystyle\int_a^x \left[Ae^{\mu(x-t)} + B\tanh^\lambda t\right]y(t)\,dt = f(x).$

This is a special case of equation 1.9.15 with $g_1(x) = Ae^{\mu x}$, $h_1(t) = e^{-\mu t}$, $g_2(x) = B$, and $h_2(t) = \tanh^\lambda t$.

31. $\displaystyle\int_a^x e^{\mu(x-t)}\left(A\coth^\lambda x + B\coth^\lambda t\right)y(t)\,dt = f(x).$

The substitution $w(x) = e^{-\mu x}y(x)$ leads to an equation of the form 1.3.90:

$$\int_a^x \left(A\coth^\lambda x + B\coth^\lambda t\right)w(t)\,dt = e^{-\mu x}f(x).$$

32. $\displaystyle\int_a^x e^{\mu(x-t)}\left(A\coth^\lambda x + B\coth^\beta t + C\right)y(t)\,dt = f(x).$

The substitution $w(x) = e^{-\mu x}y(x)$ leads to an equation of the form 1.9.6 with $g(x) = A\coth^\lambda x$, $h(t) = B\coth^\beta t + C$:

$$\int_a^x \left(A\coth^\lambda x + B\coth^\beta t + C\right)w(t)\,dt = e^{-\mu x}f(x).$$

33. $\displaystyle\int_a^x \left[Ae^{\mu(x-t)} + B\coth^\lambda x\right]y(t)\,dt = f(x).$

This is a special case of equation 1.9.15 with $g_1(x) = Ae^{\mu x}$, $h_1(t) = e^{-\mu t}$, $g_2(x) = B\coth^\lambda x$, and $h_2(t) = 1$.

34. $\displaystyle\int_a^x \left[Ae^{\mu(x-t)} + B\coth^\lambda t\right]y(t)\,dt = f(x).$

This is a special case of equation 1.9.15 with $g_1(x) = Ae^{\mu x}$, $h_1(t) = e^{-\mu t}$, $g_2(x) = B$, and $h_2(t) = \coth^\lambda t$.

1.7-2. Kernels Containing Exponential and Logarithmic Functions

35. $\displaystyle\int_a^x e^{\lambda(x-t)}(\ln x - \ln t)y(t)\,dt = f(x).$

Solution:
$$y(x) = e^{\lambda x}\left[x\varphi''_{xx}(x) + \varphi'_x(x)\right], \qquad \varphi(x) = e^{-\lambda x}f(x).$$

36. $\displaystyle\int_0^x e^{\lambda(x-t)}\ln(x-t)y(t)\,dt = f(x).$

The substitution $w(x) = e^{-\lambda x}y(x)$ leads to an equation of the form 1.4.2:

$$\int_0^x \ln(x-t)w(t)\,dt = e^{-\lambda x}f(x).$$

37. $\displaystyle\int_a^x e^{\lambda(x-t)}(A \ln x + B \ln t)y(t)\,dt = f(x).$

The substitution $w(x) = e^{-\lambda x}y(x)$ leads to an equation of the form 1.4.4:

$$\int_a^x (A \ln x + B \ln t)w(t)\,dt = e^{-\lambda x}f(x).$$

38. $\displaystyle\int_a^x e^{\mu(x-t)}\big[A \ln^2(\lambda x) + B \ln^2(\lambda t)\big]y(t)\,dt = f(x).$

The substitution $w(x) = e^{-\lambda x}y(x)$ leads to an equation of the form 1.4.7:

$$\int_a^x \big[A \ln^2(\lambda x) + B \ln^2(\lambda t)\big]w(t)\,dt = e^{-\lambda x}f(x).$$

39. $\displaystyle\int_a^x e^{\lambda(x-t)}\big[\ln(x/t)\big]^n y(t)\,dt = f(x), \qquad n = 1, 2, \ldots$

Solution:

$$y(x) = \frac{1}{n!\,x}e^{\lambda x}\left(x\frac{d}{dx}\right)^{n+1}F_\lambda(x), \qquad F_\lambda(x) = e^{-\lambda x}f(x).$$

40. $\displaystyle\int_a^x e^{\lambda(x-t)}\sqrt{\ln(x/t)}\,y(t)\,dt = f(x).$

Solution:

$$y(x) = \frac{2e^{\lambda x}}{\pi x}\left(x\frac{d}{dx}\right)^2\int_a^x \frac{e^{-\lambda t}f(t)\,dt}{t\sqrt{\ln(x/t)}}.$$

41. $\displaystyle\int_a^x \frac{e^{\lambda(x-t)}}{\sqrt{\ln(x/t)}}y(t)\,dt = f(x).$

Solution:

$$y(x) = \frac{1}{\pi}e^{\lambda x}\frac{d}{dx}\int_a^x \frac{e^{-\lambda t}f(t)\,dt}{t\sqrt{\ln(x/t)}}.$$

42. $\displaystyle\int_a^x \big[Ae^{\mu(x-t)} + B \ln^\nu(\lambda x)\big]y(t)\,dt = f(x).$

This is a special case of equation 1.9.15 with $g_1(x) = Ae^{\mu x}$, $h_1(t) = e^{-\mu t}$, $g_2(x) = B \ln^\nu(\lambda x)$, and $h_2(t) = 1$.

43. $\displaystyle\int_a^x \big[Ae^{\mu(x-t)} + B \ln^\nu(\lambda t)\big]y(t)\,dt = f(x).$

This is a special case of equation 1.9.15 with $g_1(x) = Ae^{\mu x}$, $h_1(t) = e^{-\mu t}$, $g_2(x) = B$, and $h_2(t) = \ln^\nu(\lambda t)$.

44. $\displaystyle\int_a^x e^{\mu(x-t)}[\ln(x/t)]^\lambda y(t)\,dt = f(x), \qquad 0 < \lambda < 1.$

The substitution $w(x) = e^{-\mu x}y(x)$ leads to an equation of the form 1.4.16:

$$\int_a^x [\ln(x/t)]^\lambda w(t)\,dt = e^{-\mu x}f(x).$$

45. $\displaystyle\int_a^x \frac{e^{\mu(x-t)}}{[\ln(x/t)]^\lambda} y(t)\, dt = f(x), \qquad 0 < \lambda < 1.$

Solution:

$$y(x) = \frac{\sin(\pi\lambda)}{\pi} e^{\mu x} \frac{d}{dx} \int_a^x \frac{f(t)\, dt}{te^{\mu t}[\ln(x/t)]^{1-\lambda}}.$$

1.7-3. Kernels Containing Exponential and Trigonometric Functions

46. $\displaystyle\int_a^x e^{\mu(x-t)} \cos[\lambda(x-t)] y(t)\, dt = f(x).$

Solution: $\displaystyle y(x) = f'_x(x) - \mu f(x) + \lambda^2 \int_a^x e^{\mu(x-t)} f(t)\, dt.$

47. $\displaystyle\int_a^x e^{\mu(x-t)} \big\{ A_1 \cos[\lambda_1(x-t)] + A_2 \cos[\lambda_2(x-t)] \big\} y(t)\, dt = f(x).$

The substitution $w(x) = e^{-\mu x} y(x)$ leads to an equation of the form 1.5.8:

$$\int_a^x \big\{ A_1 \cos[\lambda_1(x-t)] + A_2 \cos[\lambda_2(x-t)] \big\} w(t)\, dt = e^{-\mu x} f(x).$$

48. $\displaystyle\int_a^x e^{\mu(x-t)} \cos^2[\lambda(x-t)] y(t)\, dt = f(x).$

The substitution $w(x) = e^{-\mu x} y(x)$ leads to an equation of the form 1.5.9.

Solution:

$$y(x) = \varphi(x) + \frac{2\lambda^2}{k} \int_a^x e^{\mu(x-t)} \sin[k(x-t)]\varphi(t)\, dt, \qquad k = \lambda\sqrt{2}, \quad \varphi(x) = f'_x(x) - \mu f(x).$$

49. $\displaystyle\int_a^x e^{\mu(x-t)} \cos^3[\lambda(x-t)] y(t)\, dt = f(x).$

The substitution $w(x) = e^{-\mu x} y(x)$ leads to an equation of the form 1.5.15:

$$\int_a^x \cos^3[\lambda(x-t)] w(t)\, dt = e^{-\mu x} f(x).$$

50. $\displaystyle\int_a^x e^{\mu(x-t)} \cos^4[\lambda(x-t)] y(t)\, dt = f(x).$

The substitution $w(x) = e^{-\mu x} y(x)$ leads to an equation of the form 1.5.19:

$$\int_a^x \cos^4[\lambda(x-t)] w(t)\, dt = e^{-\mu x} f(x).$$

51. $\displaystyle\int_a^x e^{\mu(x-t)} \big[\cos(\lambda x) - \cos(\lambda t)\big]^n y(t)\, dt = f(x), \qquad n = 1, 2, \dots$

The right-hand side of the equation is assumed to satisfy the conditions $f(a) = f'_x(a) = \cdots = f_x^{(n)}(a) = 0$.

Solution:

$$y(x) = \frac{(-1)^n}{\lambda^n n!} e^{\mu x} \sin(\lambda x) \left[\frac{1}{\sin(\lambda x)} \frac{d}{dx} \right]^{n+1} F_\mu(x), \qquad F_\mu(x) = e^{-\mu x} f(x).$$

52. $\displaystyle\int_a^x e^{\mu(x-t)}\sqrt{\cos t - \cos x}\, y(t)\, dt = f(x).$

Solution:

$$y(x) = \frac{2}{\pi} e^{\mu x} \sin x \left(\frac{1}{\sin x}\frac{d}{dx}\right)^2 \int_a^x \frac{e^{-\mu t} \sin t\, f(t)\, dt}{\sqrt{\cos t - \cos x}}.$$

53. $\displaystyle\int_a^x \frac{e^{\mu(x-t)} y(t)\, dt}{\sqrt{\cos t - \cos x}} = f(x).$

Solution:

$$y(x) = \frac{1}{\pi} e^{\mu x} \frac{d}{dx} \int_a^x \frac{e^{-\mu t} \sin t\, f(t)\, dt}{\sqrt{\cos t - \cos x}}.$$

54. $\displaystyle\int_a^x e^{\mu(x-t)}(\cos t - \cos x)^\lambda y(t)\, dt = f(x), \qquad 0 < \lambda < 1.$

Solution:

$$y(x) = k e^{\mu x} \sin x \left(\frac{1}{\sin x}\frac{d}{dx}\right)^2 \int_a^x \frac{e^{-\mu t} \sin t\, f(t)\, dt}{(\cos t - \cos x)^\lambda}, \qquad k = \frac{\sin(\pi\lambda)}{\pi\lambda}.$$

55. $\displaystyle\int_a^x e^{\mu(x-t)}(\cos^\lambda x - \cos^\lambda t) y(t)\, dt = f(x).$

The substitution $w(x) = e^{-\mu x} y(x)$ leads to an equation of the form 1.5.24:

$$\int_a^x (\cos^\lambda x - \cos^\lambda t) w(t)\, dt = e^{-\mu x} f(x).$$

56. $\displaystyle\int_a^x e^{\mu(x-t)}\left(A\cos^\lambda x + B\cos^\lambda t\right) y(t)\, dt = f(x).$

The substitution $w(x) = e^{-\mu x} y(x)$ leads to an equation of the form 1.5.25:

$$\int_a^x \left(A\cos^\lambda x + B\cos^\lambda t\right) w(t)\, dt = e^{-\mu x} f(x).$$

57. $\displaystyle\int_a^x \frac{e^{\mu(x-t)} y(t)\, dt}{(\cos t - \cos x)^\lambda} = f(x), \qquad 0 < \lambda < 1.$

The substitution $w(x) = e^{-\mu x} y(x)$ leads to an equation of the form 1.5.26:

$$\int_a^x \frac{w(t)\, dt}{(\cos t - \cos x)^\lambda} = e^{-\mu x} f(x).$$

58. $\displaystyle\int_a^x \left[A e^{\mu(x-t)} + B\cos^\nu(\lambda x)\right] y(t)\, dt = f(x).$

This is a special case of equation 1.9.15 with $g_1(x) = A e^{\mu x}$, $h_1(t) = e^{-\mu t}$, $g_2(x) = B\cos^\nu(\lambda x)$, and $h_2(t) = 1$.

59. $\displaystyle\int_a^x \left[A e^{\mu(x-t)} + B\cos^\nu(\lambda t)\right] y(t)\, dt = f(x).$

This is a special case of equation 1.9.15 with $g_1(x) = A e^{\mu x}$, $h_1(t) = e^{-\mu t}$, $g_2(x) = B$, and $h_2(t) = \cos^\nu(\lambda t)$.

60. $\displaystyle\int_a^x e^{\mu(x-t)} \sin[\lambda(x-t)]y(t)\,dt = f(x), \qquad f(a) = f'_x(a) = 0.$

Solution: $y(x) = \frac{1}{\lambda}\left[f''_{xx}(x) - 2\mu f'_x(x) + (\lambda^2 + \mu^2)f(x)\right].$

61. $\displaystyle\int_a^x e^{\mu(x-t)}\left\{A_1 \sin[\lambda_1(x-t)] + A_2 \sin[\lambda_2(x-t)]\right\}y(t)\,dt = f(x).$

The substitution $w(x) = e^{-\mu x}y(x)$ leads to an equation of the form 1.5.41:

$$\int_a^x \left\{A_1 \sin[\lambda_1(x-t)] + A_2 \sin[\lambda_2(x-t)]\right\}w(t)\,dt = e^{-\mu x}f(x).$$

62. $\displaystyle\int_a^x e^{\mu(x-t)} \sin^2[\lambda(x-t)]y(t)\,dt = f(x).$

The substitution $w(x) = e^{-\mu x}y(x)$ leads to an equation of the form 1.5.43:

$$\int_a^x \sin^2[\lambda(x-t)]w(t)\,dt = e^{-\mu x}f(x).$$

63. $\displaystyle\int_a^x e^{\mu(x-t)} \sin^3[\lambda(x-t)]y(t)\,dt = f(x).$

The substitution $w(x) = e^{-\mu x}y(x)$ leads to an equation of the form 1.5.49:

$$\int_a^x \sin^3[\lambda(x-t)]w(t)\,dt = e^{-\mu x}f(x).$$

64. $\displaystyle\int_a^x e^{\mu(x-t)} \sin^n[\lambda(x-t)]y(t)\,dt = f(x), \qquad n = 2, 3, \ldots$

The substitution $w(x) = e^{-\mu x}y(x)$ leads to an equation of the form 1.5.54:

$$\int_a^x \sin^n[\lambda(x-t)]w(t)\,dt = e^{-\mu x}f(x).$$

65. $\displaystyle\int_a^x e^{\mu(x-t)} \sin\left(k\sqrt{x-t}\right)y(t)\,dt = f(x).$

Solution:

$$y(x) = \frac{2}{\pi k}e^{\mu x}\frac{d^2}{dx^2}\int_a^x \frac{e^{-\mu t}\cosh\left(k\sqrt{x-t}\right)}{\sqrt{x-t}}f(t)\,dt.$$

66. $\displaystyle\int_a^x e^{\mu(x-t)}\sqrt{\sin x - \sin t}\; y(t)\,dt = f(x).$

Solution:

$$y(x) = \frac{2}{\pi}e^{\mu x}\cos x\left(\frac{1}{\cos x}\frac{d}{dx}\right)^2 \int_a^x \frac{e^{-\mu t}\cos t\, f(t)\,dt}{\sqrt{\sin x - \sin t}}.$$

67. $\displaystyle\int_a^x \frac{e^{\mu(x-t)}y(t)\,dt}{\sqrt{\sin x - \sin t}} = f(x).$

Solution:

$$y(x) = \frac{1}{\pi}e^{\mu x}\frac{d}{dx}\int_a^x \frac{e^{-\mu t}\cos t\, f(t)\,dt}{\sqrt{\sin x - \sin t}}.$$

68. $\displaystyle\int_a^x e^{\mu(x-t)}(\sin x - \sin t)^\lambda y(t)\, dt = f(x), \qquad 0 < \lambda < 1.$

Solution:

$$y(x) = ke^{\mu x}\cos x\left(\frac{1}{\cos x}\frac{d}{dx}\right)^2\int_a^x \frac{e^{-\mu t}\cos t\, f(t)\, dt}{(\sin x - \sin t)^\lambda}, \qquad k = \frac{\sin(\pi\lambda)}{\pi\lambda}.$$

69. $\displaystyle\int_a^x e^{\mu(x-t)}(\sin^\lambda x - \sin^\lambda t)y(t)\, dt = f(x).$

The substitution $w(x) = e^{-\mu x}y(x)$ leads to an equation of the form 1.5.59:

$$\int_a^x (\sin^\lambda x - \sin^\lambda t)w(t)\, dt = e^{-\mu x}f(x).$$

70. $\displaystyle\int_a^x e^{\mu(x-t)}\left(A\sin^\lambda x + B\sin^\lambda t\right)y(t)\, dt = f(x).$

The substitution $w(x) = e^{-\mu x}y(x)$ leads to an equation of the form 1.5.60:

$$\int_a^x \left(A\sin^\lambda x + B\sin^\lambda t\right)w(t)\, dt = e^{-\mu x}f(x).$$

71. $\displaystyle\int_a^x \frac{e^{\mu(x-t)}y(t)\, dt}{(\sin x - \sin t)^\lambda} = f(x), \qquad 0 < \lambda < 1.$

The substitution $w(x) = e^{-\mu x}y(x)$ leads to an equation of the form 1.5.61:

$$\int_a^x \frac{w(t)\, dt}{(\sin x - \sin t)^\lambda} = e^{-\mu x}f(x).$$

72. $\displaystyle\int_a^x \left[Ae^{\mu(x-t)} + B\sin^\nu(\lambda x)\right]y(t)\, dt = f(x).$

This is a special case of equation 1.9.15 with $g_1(x) = Ae^{\mu x}$, $h_1(t) = e^{-\mu t}$, $g_2(x) = B\sin^\nu(\lambda x)$, and $h_2(t) = 1$.

73. $\displaystyle\int_a^x \left[Ae^{\mu(x-t)} + B\sin^\nu(\lambda t)\right]y(t)\, dt = f(x).$

This is a special case of equation 1.9.15 with $g_1(x) = Ae^{\mu x}$, $h_1(t) = e^{-\mu t}$, $g_2(x) = B$, and $h_2(t) = \sin^\nu(\lambda t)$.

74. $\displaystyle\int_a^x e^{\mu(x-t)}\left(A\tan^\lambda x + B\tan^\lambda t\right)y(t)\, dt = f(x).$

The substitution $w(x) = e^{-\mu x}y(x)$ leads to an equation of the form 1.5.77:

$$\int_a^x \left(A\tan^\lambda x + B\tan^\lambda t\right)w(t)\, dt = e^{-\mu x}f(x).$$

75. $\displaystyle\int_a^x e^{\mu(x-t)}\left(A\tan^\lambda x + B\tan^\beta t + C\right)y(t)\, dt = f(x).$

The substitution $w(x) = e^{-\mu x}y(x)$ leads to an equation of the form 1.9.6:

$$\int_a^x \left(A\tan^\lambda x + B\tan^\beta t + C\right)w(t)\, dt = e^{-\mu x}f(x).$$

76. $\displaystyle\int_a^x \left[Ae^{\mu(x-t)} + B\tan^\nu(\lambda x)\right] y(t)\,dt = f(x).$

This is a special case of equation 1.9.15 with $g_1(x) = Ae^{\mu x}$, $h_1(t) = e^{-\mu t}$, $g_2(x) = B\tan^\nu(\lambda x)$, and $h_2(t) = 1$.

77. $\displaystyle\int_a^x \left[Ae^{\mu(x-t)} + B\tan^\nu(\lambda t)\right] y(t)\,dt = f(x).$

This is a special case of equation 1.9.15 with $g_1(x) = Ae^{\mu x}$, $h_1(t) = e^{-\mu t}$, $g_2(x) = B$, and $h_2(t) = \tan^\nu(\lambda t)$.

78. $\displaystyle\int_a^x e^{\mu(x-t)}\left(A\cot^\lambda x + B\cot^\lambda t\right) y(t)\,dt = f(x).$

The substitution $w(x) = e^{-\mu x}y(x)$ leads to an equation of the form 1.5.90:

$$\int_a^x \left(A\cot^\lambda x + B\cot^\lambda t\right) w(t)\,dt = e^{-\mu x}f(x).$$

79. $\displaystyle\int_a^x e^{\mu(x-t)}\left(A\cot^\lambda x + B\cot^\beta t + C\right) y(t)\,dt = f(x).$

The substitution $w(x) = e^{-\mu x}y(x)$ leads to an equation of the form 1.9.6:

$$\int_a^x \left(A\cot^\lambda x + B\cot^\beta t + C\right) w(t)\,dt = e^{-\mu x}f(x).$$

80. $\displaystyle\int_a^x \left[Ae^{\mu(x-t)} + B\cot^\nu(\lambda x)\right] y(t)\,dt = f(x).$

This is a special case of equation 1.9.15 with $g_1(x) = Ae^{\mu x}$, $h_1(t) = e^{-\mu t}$, $g_2(x) = B\cot^\nu(\lambda x)$, and $h_2(t) = 1$.

81. $\displaystyle\int_a^x \left[Ae^{\mu(x-t)} + B\cot^\nu(\lambda t)\right] y(t)\,dt = f(x).$

This is a special case of equation 1.9.15 with $g_1(x) = Ae^{\mu x}$, $h_1(t) = e^{-\mu t}$, $g_2(x) = B$, and $h_2(t) = \cot^\nu(\lambda t)$.

1.7-4. Kernels Containing Hyperbolic and Logarithmic Functions

82. $\displaystyle\int_a^x \left[A\cosh^\beta(\lambda x) + B\ln^\gamma(\mu t) + C\right] y(t)\,dt = f(x).$

This is a special case of equation 1.9.6 with $g(x) = A\cosh^\beta(\lambda x)$ and $h(t) = B\ln^\gamma(\mu t) + C$.

83. $\displaystyle\int_a^x \left[A\cosh^\beta(\lambda t) + B\ln^\gamma(\mu x) + C\right] y(t)\,dt = f(x).$

This is a special case of equation 1.9.6 with $g(x) = B\ln^\gamma(\mu x) + C$ and $h(t) = A\cosh^\beta(\lambda t)$.

84. $\displaystyle\int_a^x \left[A\sinh^\beta(\lambda x) + B\ln^\gamma(\mu t) + C\right] y(t)\,dt = f(x).$

This is a special case of equation 1.9.6 with $g(x) = A\sinh^\beta(\lambda x)$ and $h(t) = B\ln^\gamma(\mu t) + C$.

85. $\displaystyle\int_a^x \left[A\,\sinh^\beta(\lambda t) + B\,\ln^\gamma(\mu x) + C\right] y(t)\,dt = f(x).$

This is a special case of equation 1.9.6 with $g(x) = B\,\ln^\gamma(\mu x)$ and $h(t) = A\,\sinh^\beta(\lambda t) + C$.

86. $\displaystyle\int_a^x \left[A\,\tanh^\beta(\lambda x) + B\,\ln^\gamma(\mu t) + C\right] y(t)\,dt = f(x).$

This is a special case of equation 1.9.6 with $g(x) = A\,\tanh^\beta(\lambda x)$ and $h(t) = B\,\ln^\gamma(\mu t) + C$.

87. $\displaystyle\int_a^x \left[A\,\tanh^\beta(\lambda t) + B\,\ln^\gamma(\mu x) + C\right] y(t)\,dt = f(x).$

This is a special case of equation 1.9.6 with $g(x) = B\,\ln^\gamma(\mu x)$ and $h(t) = A\,\tanh^\beta(\lambda t) + C$.

88. $\displaystyle\int_a^x \left[A\,\coth^\beta(\lambda x) + B\,\ln^\gamma(\mu t) + C\right] y(t)\,dt = f(x).$

This is a special case of equation 1.9.6 with $g(x) = A\,\coth^\beta(\lambda x)$ and $h(t) = B\,\ln^\gamma(\mu t) + C$.

89. $\displaystyle\int_a^x \left[A\,\coth^\beta(\lambda t) + B\,\ln^\gamma(\mu x) + C\right] y(t)\,dt = f(x).$

This is a special case of equation 1.9.6 with $g(x) = B\,\ln^\gamma(\mu x)$ and $h(t) = A\,\coth^\beta(\lambda t) + C$.

1.7-5. Kernels Containing Hyperbolic and Trigonometric Functions

90. $\displaystyle\int_a^x \left[A\,\cosh^\beta(\lambda x) + B\,\cos^\gamma(\mu t) + C\right] y(t)\,dt = f(x).$

This is a special case of equation 1.9.6 with $g(x) = A\,\cosh^\beta(\lambda x)$ and $h(t) = B\,\cos^\gamma(\mu t) + C$.

91. $\displaystyle\int_a^x \left[A\,\cosh^\beta(\lambda t) + B\,\sin^\gamma(\mu x) + C\right] y(t)\,dt = f(x).$

This is a special case of equation 1.9.6 with $g(x) = B\,\sin^\gamma(\mu x) + C$ and $h(t) = A\,\cosh^\beta(\lambda t)$.

92. $\displaystyle\int_a^x \left[A\,\cosh^\beta(\lambda x) + B\,\tan^\gamma(\mu t) + C\right] y(t)\,dt = f(x).$

This is a special case of equation 1.9.6 with $g(x) = A\,\cosh^\beta(\lambda x)$ and $h(t) = B\,\tan^\gamma(\mu t) + C$.

93. $\displaystyle\int_a^x \left[A\,\sinh^\beta(\lambda x) + B\,\cos^\gamma(\mu t) + C\right] y(t)\,dt = f(x).$

This is a special case of equation 1.9.6 with $g(x) = A\,\sinh^\beta(\lambda x)$ and $h(t) = B\,\cos^\gamma(\mu t) + C$.

94. $\displaystyle\int_a^x \left[A\,\sinh^\beta(\lambda t) + B\,\sin^\gamma(\mu x) + C\right] y(t)\,dt = f(x).$

This is a special case of equation 1.9.6 with $g(x) = B\,\sin^\gamma(\mu x)$ and $h(t) = A\,\sinh^\beta(\lambda t) + C$.

95. $\displaystyle\int_a^x \left[A\,\sinh^\beta(\lambda x) + B\,\tan^\gamma(\mu t) + C\right] y(t)\,dt = f(x).$

This is a special case of equation 1.9.6 with $g(x) = A\,\sinh^\beta(\lambda x)$ and $h(t) = B\,\tan^\gamma(\mu t) + C$.

96. $\displaystyle\int_a^x \left[A\tanh^\beta(\lambda x) + B\cos^\gamma(\mu t) + C\right]y(t)\,dt = f(x).$

This is a special case of equation 1.9.6 with $g(x) = A\tanh^\beta(\lambda x)$ and $h(t) = B\cos^\gamma(\mu t) + C$.

97. $\displaystyle\int_a^x \left[A\tanh^\beta(\lambda x) + B\sin^\gamma(\mu t) + C\right]y(t)\,dt = f(x).$

This is a special case of equation 1.9.6 with $g(x) = A\tanh^\beta(\lambda x)$ and $h(t) = B\sin^\gamma(\mu t) + C$.

1.7-6. Kernels Containing Logarithmic and Trigonometric Functions

98. $\displaystyle\int_a^x \left[A\cos^\beta(\lambda x) + B\ln^\gamma(\mu t) + C\right]y(t)\,dt = f(x).$

This is a special case of equation 1.9.6 with $g(x) = A\cos^\beta(\lambda x)$ and $h(t) = B\ln^\gamma(\mu t) + C$.

99. $\displaystyle\int_a^x \left[A\cos^\beta(\lambda t) + B\ln^\gamma(\mu x) + C\right]y(t)\,dt = f(x).$

This is a special case of equation 1.9.6 with $g(x) = B\ln^\gamma(\mu x) + C$ and $h(t) = A\cos^\beta(\lambda t)$.

100. $\displaystyle\int_a^x \left[A\sin^\beta(\lambda x) + B\ln^\gamma(\mu t) + C\right]y(t)\,dt = f(x).$

This is a special case of equation 1.9.6 with $g(x) = A\sin^\beta(\lambda x)$ and $h(t) = B\ln^\gamma(\mu t) + C$.

101. $\displaystyle\int_a^x \left[A\sin^\beta(\lambda t) + B\ln^\gamma(\mu x) + C\right]y(t)\,dt = f(x).$

This is a special case of equation 1.9.6 with $g(x) = B\ln^\gamma(\mu x)$ and $h(t) = A\sin^\beta(\lambda t) + C$.

1.8. Equations Whose Kernels Contain Special Functions

1.8-1. Kernels Containing Bessel Functions

1. $\displaystyle\int_a^x J_0[\lambda(x-t)]y(t)\,dt = f(x).$

Solution:

$$y(x) = \frac{1}{\lambda}\left(\frac{d^2}{dx^2} + \lambda^2\right)^2 \int_a^x (x-t)\,J_1[\lambda(x-t)]\,f(t)\,dt.$$

Example. In the special case $\lambda = 1$ and $f(x) = A\sin x$, the solution has the form $y(x) = AJ_0(x)$.

2. $\displaystyle\int_a^x [J_0(\lambda x) - J_0(\lambda t)]y(t)\,dt = f(x).$

Solution: $\displaystyle y(x) = -\frac{d}{dx}\left[\frac{f_x'(x)}{\lambda J_1(\lambda x)}\right].$

3. $\int_a^x [AJ_0(\lambda x) + BJ_0(\lambda t)]y(t)\,dt = f(x).$

For $B = -A$, see equation 1.8.2. We consider the interval $[a, x]$ in which $J_0(\lambda x)$ does not change its sign.

Solution with $B \neq -A$:

$$y(x) = \pm \frac{1}{A + B} \frac{d}{dx} \left\{ |J_0(\lambda x)|^{-\frac{A}{A+B}} \int_a^x |J_0(\lambda t)|^{-\frac{B}{A+B}} f_t'(t)\,dt \right\}.$$

Here the sign of $J_0(\lambda x)$ should be taken.

4. $\int_a^x (x - t) J_0[\lambda(x - t)]y(t)\,dt = f(x).$

Solution:

$$y(x) = \int_a^x g(t)\,dt,$$

where

$$g(t) = \frac{1}{\lambda} \left(\frac{d^2}{dt^2} + \lambda^2 \right)^3 \int_a^t (t - \tau) J_1[\lambda(t - \tau)]\, f(\tau)\,d\tau.$$

5. $\int_a^x (x - t)J_1[\lambda(x - t)]y(t)\,dt = f(x).$

Solution:

$$y(x) = \frac{1}{3\lambda^3} \left(\frac{d^2}{dx^2} + \lambda^2 \right)^4 \int_a^x (x - t)^2\, J_2[\lambda(x - t)]\, f(t)\,dt.$$

6. $\int_a^x (x - t)^2 J_1[\lambda(x - t)]y(t)\,dt = f(x).$

Solution:

$$y(x) = \int_a^x g(t)\,dt,$$

where

$$g(t) = \frac{1}{9\lambda^3} \left(\frac{d^2}{dt^2} + \lambda^2 \right)^5 \int_a^t (t - \tau)^2\, J_2[\lambda(t - \tau)]\, f(\tau)\,d\tau.$$

7. $\int_a^x (x - t)^n J_n[\lambda(x - t)]y(t)\,dt = f(x), \qquad n = 0, 1, 2, \ldots$

Solution:

$$y(x) = A\left(\frac{d^2}{dx^2} + \lambda^2 \right)^{2n+2} \int_a^x (x - t)^{n+1} J_{n+1}[\lambda(x - t)]\, f(t)\,dt,$$

$$A = \left(\frac{2}{\lambda} \right)^{2n+1} \frac{n!\,(n + 1)!}{(2n)!\,(2n + 2)!}.$$

If the right-hand side of the equation is differentiable sufficiently many times and the conditions $f(a) = f_x'(a) = \cdots = f_x^{(2n+1)}(a) = 0$ are satisfied, then the solution of the integral equation can be written in the form

$$y(x) = A \int_a^x (x - t)^{2n+1} J_{2n+1}[\lambda(x - t)]F(t)\,dt, \qquad F(t) = \left(\frac{d^2}{dt^2} + \lambda^2 \right)^{2n+2} f(t)\,dt.$$

8. $\displaystyle\int_a^x (x-t)^{n+1} J_n[\lambda(x-t)] y(t)\, dt = f(x), \qquad n = 0, 1, 2, \ldots$

Solution:

$$y(x) = \int_a^x g(t)\, dt,$$

where

$$g(t) = A \left(\frac{d^2}{dt^2} + \lambda^2 \right)^{2n+3} \int_a^t (t-\tau)^{n+1} J_{n+1}[\lambda(t-\tau)]\, f(\tau)\, d\tau,$$

$$A = \left(\frac{2}{\lambda} \right)^{2n+1} \frac{n!\,(n+1)!}{(2n+1)!\,(2n+2)!}.$$

If the right-hand side of the equation is differentiable sufficiently many times and the conditions $f(a) = f_x'(a) = \cdots = f_x^{(2n+2)}(a) = 0$ are satisfied, then the function $g(t)$ defining the solution can be written in the form

$$g(t) = A \int_a^t (t-\tau)^{n-\nu-2} J_{n-\nu-2}[\lambda(t-\tau)] F(\tau)\, d\tau, \quad F(\tau) = \left(\frac{d^2}{d\tau^2} + \lambda^2 \right)^{2n+2} f(\tau).$$

9. $\displaystyle\int_a^x (x-t)^{1/2} J_{1/2}[\lambda(x-t)] y(t)\, dt = f(x).$

Solution:

$$y(x) = \frac{\pi}{4\lambda^2} \left(\frac{d^2}{dx^2} + \lambda^2 \right)^3 \int_a^x (x-t)^{3/2}\, J_{3/2}[\lambda(x-t)]\, f(t)\, dt.$$

10. $\displaystyle\int_a^x (x-t)^{3/2} J_{1/2}[\lambda(x-t)] y(t)\, dt = f(x).$

Solution:

$$y(x) = \int_a^x g(t)\, dt,$$

where

$$g(t) = \frac{\pi}{8\lambda^2} \left(\frac{d^2}{dt^2} + \lambda^2 \right)^4 \int_a^t (t-\tau)^{3/2}\, J_{3/2}[\lambda(t-\tau)]\, f(\tau)\, d\tau.$$

11. $\displaystyle\int_a^x (x-t)^{3/2} J_{3/2}[\lambda(x-t)] y(t)\, dt = f(x).$

Solution:

$$y(x) = \frac{\sqrt{\pi}}{2^{3/2}\lambda^{5/2}} \left(\frac{d^2}{dx^2} + \lambda^2 \right)^3 \int_a^x \sin[\lambda(x-t)]\, f(t)\, dt.$$

12. $\displaystyle\int_a^x (x-t)^{5/2} J_{3/2}[\lambda(x-t)] y(t)\, dt = f(x).$

Solution:

$$y(x) = \int_a^x g(t)\, dt,$$

where

$$g(t) = \frac{\pi}{128\lambda^4} \left(\frac{d^2}{dt^2} + \lambda^2 \right)^6 \int_a^t (t-\tau)^{5/2}\, J_{5/2}[\lambda(t-\tau)]\, f(\tau)\, d\tau.$$

13. $\displaystyle\int_a^x (x-t)^{\frac{2n-1}{2}} J_{\frac{2n-1}{2}}[\lambda(x-t)] y(t)\, dt = f(x), \qquad n = 2, 3, \ldots$

Solution:

$$y(x) = \frac{\sqrt{\pi}}{\sqrt{2}\,\lambda^{\frac{2n+1}{2}}(2n-2)!!} \left(\frac{d^2}{dx^2} + \lambda^2\right)^n \int_a^x \sin[\lambda(x-t)]\, f(t)\, dt.$$

14. $\displaystyle\int_a^x [J_\nu(\lambda x) - J_\nu(\lambda t)] y(t)\, dt = f(x).$

This is a special case of equation 1.9.2 with $g(x) = J_\nu(\lambda x)$.

Solution: $\displaystyle y(x) = \frac{d}{dx}\left[\frac{x f'_x(x)}{\nu J_\nu(\lambda x) - \lambda x J_{\nu+1}(\lambda x)}\right].$

15. $\displaystyle\int_a^x [A J_\nu(\lambda x) + B J_\nu(\lambda t)] y(t)\, dt = f(x).$

For $B = -A$, see equation 1.8.14. We consider the interval $[a, x]$ in which $J_\nu(\lambda x)$ does not change its sign.

Solution with $B \neq -A$:

$$y(x) = \pm\frac{1}{A+B}\frac{d}{dx}\left\{ \left|J_\nu(\lambda x)\right|^{-\frac{A}{A+B}} \int_a^x \left|J_\nu(\lambda t)\right|^{-\frac{B}{A+B}} f'_t(t)\, dt \right\}.$$

Here the sign of $J_\nu(\lambda x)$ should be taken.

16. $\displaystyle\int_a^x [A J_\nu(\lambda x) + B J_\mu(\beta t)] y(t)\, dt = f(x).$

This is a special case of equation 1.9.6 with $g(x) = A J_\nu(\lambda x)$ and $h(t) = B J_\mu(\beta t)$.

17. $\displaystyle\int_a^x (x-t)^\nu J_\nu[\lambda(x-t)] y(t)\, dt = f(x).$

Solution:

$$y(x) = A\left(\frac{d^2}{dx^2} + \lambda^2\right)^n \int_a^x (x-t)^{n-\nu-1} J_{n-\nu-1}[\lambda(x-t)]\, f(t)\, dt,$$

$$A = \left(\frac{2}{\lambda}\right)^{n-1} \frac{\Gamma(\nu+1)\,\Gamma(n-\nu)}{\Gamma(2\nu+1)\,\Gamma(2n-2\nu-1)},$$

where $-\frac{1}{2} < \nu < \frac{n-1}{2}$ and $n = 1, 2, \ldots$

If the right-hand side of the equation is differentiable sufficiently many times and the conditions $f(a) = f'_x(a) = \cdots = f_x^{(n-1)}(a) = 0$ are satisfied, then the solution of the integral equation can be written in the form

$$y(x) = A\int_a^x (x-t)^{n-\nu-1} J_{n-\nu-1}[\lambda(x-t)] F(t)\, dt, \qquad F(t) = \left(\frac{d^2}{dt^2} + \lambda^2\right)^n f(t).$$

⊙ Reference: S. G. Samko, A. A. Kilbas, and O. I. Marichev (1993).

18. $\int_a^x (x-t)^{\nu+1} J_\nu[\lambda(x-t)]y(t)\,dt = f(x).$

Solution:

$$y(x) = \int_a^x g(t)\,dt,$$

where

$$g(t) = A\left(\frac{d^2}{dt^2} + \lambda^2\right)^n \int_a^t (t-\tau)^{n-\nu-2} J_{n-\nu-2}[\lambda(t-\tau)]\,f(\tau)\,d\tau,$$

$$A = \left(\frac{2}{\lambda}\right)^{n-2} \frac{\Gamma(\nu+1)\,\Gamma(n-\nu-1)}{\Gamma(2\nu+2)\,\Gamma(2n-2\nu-3)},$$

where $-1 < \nu < \frac{n}{2} - 1$ and $n = 1, 2, \ldots$

If the right-hand side of the equation is differentiable sufficiently many times and the conditions $f(a) = f'_x(a) = \cdots = f_x^{(n-1)}(a) = 0$ are satisfied, then the function $g(t)$ defining the solution can be written in the form

$$g(t) = A\int_a^t (t-\tau)^{n-\nu-2}\,J_{n-\nu-2}[\lambda(t-\tau)]F(\tau)\,d\tau, \quad F(\tau) = \left(\frac{d^2}{d\tau^2} + \lambda^2\right)^n f(\tau).$$

⊙ Reference: S. G. Samko, A. A. Kilbas, and O. I. Marichev (1993).

19. $\int_a^x J_0\left(\lambda\sqrt{x-t}\,\right)y(t)\,dt = f(x).$

Solution:

$$y(x) = \frac{d^2}{dx^2}\int_a^x I_0\left(\lambda\sqrt{x-t}\,\right)f(t)\,dt.$$

20. $\int_a^x \left[AJ_\nu\left(\lambda\sqrt{x}\,\right) + BJ_\nu\left(\lambda\sqrt{t}\,\right)\right]y(t)\,dt = f(x).$

We consider the interval $[a, x]$ in which $J_\nu\left(\lambda\sqrt{x}\,\right)$ does not change its sign.
Solution with $B \neq -A$:

$$y(x) = \pm\frac{1}{A+B}\frac{d}{dx}\left\{\left|J_\nu\left(\lambda\sqrt{x}\,\right)\right|^{-\frac{A}{A+B}} \int_a^x \left|J_\nu\left(\lambda\sqrt{t}\,\right)\right|^{-\frac{B}{A+B}} f'_t(t)\,dt\right\}.$$

Here the sign $J_\nu\left(\lambda\sqrt{x}\,\right)$ should be taken.

21. $\int_a^x \left[AJ_\nu\left(\lambda\sqrt{x}\,\right) + BJ_\mu\left(\beta\sqrt{t}\,\right)\right]y(t)\,dt = f(x).$

This is a special case of equation 1.9.6 with $g(x) = AJ_\nu\left(\lambda\sqrt{x}\,\right)$ and $h(t) = BJ_\mu\left(\beta\sqrt{t}\,\right)$.

22. $\int_a^x \sqrt{x-t}\,J_1\left(\lambda\sqrt{x-t}\,\right)y(t)\,dt = f(x).$

Solution:

$$y(x) = \frac{2}{\lambda}\frac{d^3}{dx^3}\int_a^x I_0\left(\lambda\sqrt{x-t}\,\right)f(t)\,dt.$$

23. $\int_a^x (x-t)^{1/4} J_{1/2}(\lambda\sqrt{x-t}) y(t)\, dt = f(x).$

Solution:

$$y(x) = \sqrt{\frac{2}{\pi\lambda}} \frac{d^2}{dx^2} \int_a^x \frac{\cosh(\lambda\sqrt{x-t})}{\sqrt{x-t}} f(t)\, dt.$$

24. $\int_a^x (x-t)^{3/4} J_{3/2}(\lambda\sqrt{x-t}) y(t)\, dt = f(x).$

Solution:

$$y(x) = \frac{2^{3/2}}{\sqrt{\pi}\,\lambda^{3/2}} \frac{d^3}{dx^3} \int_a^x \frac{\cosh(\lambda\sqrt{x-t})}{\sqrt{x-t}} f(t)\, dt.$$

25. $\int_a^x (x-t)^{n/2} J_n(\lambda\sqrt{x-t}) y(t)\, dt = f(x), \qquad n = 0, 1, 2, \dots$

Solution:

$$y(x) = \left(\frac{2}{\lambda}\right)^n \frac{d^{n+2}}{dx^{n+2}} \int_a^x I_0(\lambda\sqrt{x-t}) f(t)\, dt.$$

26. $\int_a^x (x-t)^{\frac{2n-3}{4}} J_{\frac{2n-3}{2}}(\lambda\sqrt{x-t}) y(t)\, dt = f(x), \qquad n = 1, 2, \dots$

Solution:

$$y(x) = \frac{1}{\sqrt{\pi}} \left(\frac{2}{\lambda}\right)^{\frac{2n-3}{2}} \frac{d^n}{dx^n} \int_a^x \frac{\cosh(\lambda\sqrt{x-t})}{\sqrt{x-t}} f(t)\, dt.$$

27. $\int_a^x (x-t)^{-1/4} J_{-1/2}(\lambda\sqrt{x-t}) y(t)\, dt = f(x).$

Solution:

$$y(x) = \sqrt{\frac{\lambda}{2\pi}} \frac{d}{dx} \int_a^x \frac{\cosh(\lambda\sqrt{x-t})}{\sqrt{x-t}} f(t)\, dt.$$

28. $\int_a^x (x-t)^{\nu/2} J_\nu(\lambda\sqrt{x-t}) y(t)\, dt = f(x).$

Solution:

$$y(x) = \left(\frac{2}{\lambda}\right)^{n-2} \frac{d^n}{dx^n} \int_a^x (x-t)^{\frac{n-\nu-2}{2}} I_{n-\nu-2}(\lambda\sqrt{x-t}) f(t)\, dt,$$

where $-1 < \nu < n-1, n = 1, 2, \dots$

If the right-hand side of the equation is differentiable sufficiently many times and the conditions $f(a) = f'_x(a) = \cdots = f_x^{(n-1)}(a) = 0$ are satisfied, then the solution of the integral equation can be written in the form

$$y(x) = \left(\frac{2}{\lambda}\right)^{n-2} \int_a^x (x-t)^{\frac{n-\nu-2}{2}} I_{n-\nu-2}(\lambda\sqrt{x-t}) f_t^{(n)}(t)\, dt.$$

29. $\int_0^x (x^2 - t^2)^{-1/4} J_{-1/2}(\lambda\sqrt{x^2 - t^2}) y(t)\, dt = f(x).$

Solution:

$$y(x) = \sqrt{\frac{2\lambda}{\pi}} \frac{d}{dx} \int_0^x t\, \frac{\cosh(\lambda\sqrt{x^2 - t^2})}{\sqrt{x^2 - t^2}} f(t)\, dt.$$

⊙ Reference: S. G. Samko, A. A. Kilbas, and O. I. Marichev (1993).

30. $\int_x^\infty (t^2 - x^2)^{-1/4} J_{-1/2}(\lambda\sqrt{t^2 - x^2})y(t)\,dt = f(x).$

Solution:

$$y(x) = -\sqrt{\frac{2\lambda}{\pi}}\frac{d}{dx}\int_x^\infty t\,\frac{\cosh(\lambda\sqrt{t^2 - x^2})}{\sqrt{t^2 - x^2}}f(t)\,dt.$$

31. $\int_0^x (x^2 - t^2)^{\nu/2} J_\nu(\lambda\sqrt{x^2 - t^2})y(t)\,dt = f(x), \qquad -1 < \nu < 0.$

Solution:

$$y(x) = \lambda\frac{d}{dx}\int_0^x t\,(x^2 - t^2)^{-(\nu+1)/2} I_{\nu-1}(\lambda\sqrt{x^2 - t^2})f(t)\,dt.$$

⊙ Reference: S. G. Samko, A. A. Kilbas, and O. I. Marichev (1993).

32. $\int_x^\infty (t^2 - x^2)^{\nu/2} J_\nu(\lambda\sqrt{t^2 - x^2})y(t)\,dt = f(x), \qquad -1 < \nu < 0.$

Solution:

$$y(x) = -\lambda\frac{d}{dx}\int_x^\infty t\,(t^2 - x^2)^{-(\nu+1)/2} I_{\nu-1}(\lambda\sqrt{t^2 - x^2})f(t)\,dt.$$

⊙ Reference: S. G. Samko, A. A. Kilbas, and O. I. Marichev (1993).

33. $\int_a^x [At^k J_\nu(\lambda x) + Bx^m J_\mu(\lambda t)]y(t)\,dt = f(x).$

This is a special case of equation 1.9.15 with $g_1(x) = AJ_\nu(\lambda x)$, $h_1(t) = t^k$, $g_2(x) = Bx^m$, and $h_2(t) = J_\mu(\lambda t)$.

34. $\int_a^x [AJ_\nu^2(\lambda x) + BJ_\nu^2(\lambda t)]y(t)\,dt = f(x).$

Solution with $B \neq -A$:

$$y(x) = \frac{1}{A + B}\frac{d}{dx}\left\{|J_\nu(\lambda x)|^{-\frac{2A}{A+B}}\int_a^x |J_\nu(\lambda t)|^{-\frac{2B}{A+B}} f_t'(t)\,dt\right\}.$$

35. $\int_a^x \left[AJ_\nu^k(\lambda x) + BJ_\mu^m(\beta t)\right]y(t)\,dt = f(x).$

This is a special case of equation 1.9.6 with $g(x) = AJ_\nu^k(\lambda x)$ and $h(t) = BJ_\mu^m(\beta t)$.

36. $\int_a^x [Y_0(\lambda x) - Y_0(\lambda t)]y(t)\,dt = f(x).$

Solution: $y(x) = -\dfrac{d}{dx}\left[\dfrac{f_x'(x)}{\lambda Y_1(\lambda x)}\right].$

37. $\int_a^x [Y_\nu(\lambda x) - Y_\nu(\lambda t)]y(t)\,dt = f(x).$

Solution: $y(x) = \dfrac{d}{dx}\left[\dfrac{xf_x'(x)}{\nu Y_\nu(\lambda x) - \lambda x Y_{\nu+1}(\lambda x)}\right].$

38. $\displaystyle\int_a^x [AY_\nu(\lambda x) + BY_\nu(\lambda t)]y(t)\,dt = f(x).$

For $B = -A$, see equation 1.8.37. We consider the interval $[a, x]$ in which $Y_\nu(\lambda x)$ does not change its sign.

Solution with $B \neq -A$:

$$y(x) = \pm\frac{1}{A+B}\frac{d}{dx}\left\{\left|Y_\nu(\lambda x)\right|^{-\frac{A}{A+B}}\int_a^x \left|Y_\nu(\lambda t)\right|^{-\frac{B}{A+B}} f_t'(t)\,dt\right\}.$$

Here the sign of $Y_\nu(\lambda x)$ should be taken.

39. $\displaystyle\int_a^x [At^k Y_\nu(\lambda x) + Bx^m Y_\mu(\lambda t)]y(t)\,dt = f(x).$

This is a special case of equation 1.9.15 with $g_1(x) = AY_\nu(\lambda x)$, $h_1(t) = t^k$, $g_2(x) = Bx^m$, and $h_2(t) = Y_\mu(\lambda t)$.

40. $\displaystyle\int_a^x [AJ_\nu(\lambda x)Y_\mu(\beta t) + BJ_\nu(\lambda t)Y_\mu(\beta x)]y(t)\,dt = f(x).$

This is a special case of equation 1.9.15 with $g_1(x) = AY_\nu(\lambda x)$, $h_1(t) = Y_\mu(\beta t)$, $g_2(x) = BY_\mu(\beta x)$, and $h_2(t) = J_\nu(\lambda t)$.

1.8-2. Kernels Containing Modified Bessel Functions

41. $\displaystyle\int_a^x I_0[\lambda(x - t)]y(t)\,dt = f(x).$

Solution:

$$y(x) = \frac{1}{\lambda}\left(\frac{d^2}{dx^2} - \lambda^2\right)^2 \int_a^x (x - t)\,I_1[\lambda(x - t)]\,f(t)\,dt.$$

42. $\displaystyle\int_a^x [I_0(\lambda x) - I_0(\lambda t)]y(t)\,dt = f(x), \qquad f(a) = f_x'(a) = 0.$

Solution: $y(x) = \dfrac{d}{dx}\left[\dfrac{f_x'(x)}{\lambda I_1(\lambda x)}\right].$

43. $\displaystyle\int_a^x [AI_0(\lambda x) + BI_0(\lambda t)]y(t)\,dt = f(x).$

For $B = -A$, see equation 1.8.42. Solution with $B \neq -A$:

$$y(x) = \pm\frac{1}{A+B}\frac{d}{dx}\left\{\left|I_0(\lambda x)\right|^{-\frac{A}{A+B}}\int_a^x \left|I_0(\lambda t)\right|^{-\frac{B}{A+B}} f_t'(t)\,dt\right\}.$$

Here the sign of $I_\nu(\lambda x)$ should be taken.

44. $\displaystyle\int_a^x (x - t)I_0[\lambda(x - t)]y(t)\,dt = f(x).$

Solution:

$$y(x) = \int_a^x g(t)\,dt,$$

where

$$g(t) = \frac{1}{\lambda}\left(\frac{d^2}{dt^2} - \lambda^2\right)^3 \int_a^t (t - \tau)\,I_1[\lambda(t - \tau)]\,f(\tau)\,d\tau.$$

45. $\displaystyle\int_a^x (x-t) I_1[\lambda(x-t)] y(t)\, dt = f(x).$

Solution:

$$y(x) = \frac{1}{3\lambda^3} \left(\frac{d^2}{dx^2} - \lambda^2 \right)^4 \int_a^x (x-t)^2\, I_2[\lambda(x-t)]\, f(t)\, dt.$$

46. $\displaystyle\int_a^x (x-t)^2\, I_1[\lambda(x-t)] y(t)\, dt = f(x).$

Solution:

$$y(x) = \int_a^x g(t)\, dt,$$

where

$$g(t) = \frac{1}{9\lambda^3} \left(\frac{d^2}{dt^2} - \lambda^2 \right)^5 \int_a^t (t-\tau)^2\, I_2[\lambda(t-\tau)]\, f(\tau)\, d\tau.$$

47. $\displaystyle\int_a^x (x-t)^n\, I_n[\lambda(x-t)] y(t)\, dt = f(x), \qquad n = 0, 1, 2, \ldots$

Solution:

$$y(x) = A \left(\frac{d^2}{dx^2} - \lambda^2 \right)^{2n+2} \int_a^x (x-t)^{n+1}\, I_{n+1}[\lambda(x-t)]\, f(t)\, dt,$$

$$A = \left(\frac{2}{\lambda} \right)^{2n+1} \frac{n!\,(n+1)!}{(2n)!\,(2n+2)!}.$$

If the right-hand side of the equation is differentiable sufficiently many times and the conditions $f(a) = f_x'(a) = \cdots = f_x^{(2n+1)}(a) = 0$ are satisfied, then the solution of the integral equation can be written in the form

$$y(x) = A \int_a^x (x-t)^{2n+1} I_{2n+1}[\lambda(x-t)] F(t)\, dt, \qquad F(t) = \left(\frac{d^2}{dt^2} - \lambda^2 \right)^{2n+2} f(t).$$

48. $\displaystyle\int_a^x (x-t)^{n+1} I_n[\lambda(x-t)] y(t)\, dt = f(x), \qquad n = 0, 1, 2, \ldots$

Solution:

$$y(x) = \int_a^x g(t)\, dt,$$

where

$$g(t) = A \left(\frac{d^2}{dt^2} - \lambda^2 \right)^{2n+3} \int_a^t (t-\tau)^{n+1}\, I_{n+1}[\lambda(t-\tau)]\, f(\tau)\, d\tau,$$

$$A = \left(\frac{2}{\lambda} \right)^{2n+1} \frac{n!\,(n+1)!}{(2n+1)!\,(2n+2)!}.$$

If the right-hand side of the equation is differentiable sufficiently many times and the conditions $f(a) = f_x'(a) = \cdots = f_x^{(2n+2)}(a) = 0$ are satisfied, then the function $g(t)$ defining the solution can be written in the form

$$g(t) = A \int_a^t (t-\tau)^{n-\nu-2}\, I_{n-\nu-2}[\lambda(t-\tau)] F(\tau)\, d\tau, \qquad F(\tau) = \left(\frac{d^2}{d\tau^2} - \lambda^2 \right)^{2n+2} f(\tau).$$

49. $\int_a^x (x-t)^{1/2} I_{1/2}[\lambda(x-t)] y(t)\, dt = f(x).$

Solution:

$$y(x) = \frac{\pi}{4\lambda^2} \left(\frac{d^2}{dx^2} - \lambda^2 \right)^3 \int_a^x (x-t)^{3/2} I_{3/2}[\lambda(x-t)]\, f(t)\, dt.$$

50. $\int_a^x (x-t)^{3/2} I_{1/2}[\lambda(x-t)] y(t)\, dt = f(x).$

Solution:

$$y(x) = \int_a^x g(t)\, dt,$$

where

$$g(t) = \frac{\pi}{8\lambda^2} \left(\frac{d^2}{dt^2} - \lambda^2 \right)^4 \int_a^t (t-\tau)^{3/2} I_{3/2}[\lambda(t-\tau)]\, f(\tau)\, d\tau.$$

51. $\int_a^x (x-t)^{3/2} I_{3/2}[\lambda(x-t)] y(t)\, dt = f(x).$

Solution:

$$y(x) = \frac{\sqrt{\pi}}{2^{3/2} \lambda^{5/2}} \left(\frac{d^2}{dx^2} - \lambda^2 \right)^3 \int_a^x \sinh[\lambda(x-t)]\, f(t)\, dt.$$

52. $\int_a^x (x-t)^{5/2} I_{3/2}[\lambda(x-t)] y(t)\, dt = f(x).$

Solution:

$$y(x) = \int_a^x g(t)\, dt,$$

where

$$g(t) = \frac{\pi}{128\lambda^4} \left(\frac{d^2}{dt^2} - \lambda^2 \right)^6 \int_a^t (t-\tau)^{5/2} I_{5/2}[\lambda(t-\tau)]\, f(\tau)\, d\tau.$$

53. $\int_a^x (x-t)^{\frac{2n-1}{2}} I_{\frac{2n-1}{2}}[\lambda(x-t)] y(t)\, dt = f(x), \qquad n = 2, 3, \ldots$

Solution:

$$y(x) = \frac{\sqrt{\pi}}{\sqrt{2}\, \lambda^{\frac{2n+1}{2}} (2n-2)!!} \left(\frac{d^2}{dx^2} - \lambda^2 \right)^n \int_a^x \sinh[\lambda(x-t)]\, f(t)\, dt.$$

54. $\int_a^x [I_\nu(\lambda x) - I_\nu(\lambda t)] y(t)\, dt = f(x).$

This is a special case of equation 1.9.2 with $g(x) = I_\nu(\lambda x)$.

55. $\int_a^x [A I_\nu(\lambda x) + B I_\nu(\lambda t)] y(t)\, dt = f(x).$

Solution with $B \neq -A$:

$$y(x) = \frac{1}{A+B} \frac{d}{dx} \left\{ [I_\nu(\lambda x)]^{-\frac{A}{A+B}} \int_a^x [I_\nu(\lambda t)]^{-\frac{B}{A+B}} f_t'(t)\, dt \right\}.$$

56. $\int_a^x [AI_\nu(\lambda x) + BI_\mu(\beta t)] y(t)\, dt = f(x).$

This is a special case of equation 1.9.6 with $g(x) = AI_\nu(\lambda x)$ and $h(t) = BI_\mu(\beta t)$.

57. $\int_a^x (x - t)^\nu I_\nu[\lambda(x - t)] y(t)\, dt = f(x).$

Solution:

$$y(x) = A\left(\frac{d^2}{dx^2} - \lambda^2\right)^n \int_a^x (x - t)^{n-\nu-1} I_{n-\nu-1}[\lambda(x - t)]\, f(t)\, dt,$$

$$A = \left(\frac{2}{\lambda}\right)^{n-1} \frac{\Gamma(\nu + 1)\,\Gamma(n - \nu)}{\Gamma(2\nu + 1)\,\Gamma(2n - 2\nu - 1)},$$

where $-\frac{1}{2} < \nu < \frac{n-1}{2}$ and $n = 1, 2, \ldots$

If the right-hand side of the equation is differentiable sufficiently many times and the conditions $f(a) = f'_x(a) = \cdots = f_x^{(n-1)}(a) = 0$ are satisfied, then the solution of the integral equation can be written in the form

$$y(x) = A \int_a^x (x - t)^{n-\nu-1} I_{n-\nu-1}[\lambda(x - t)] F(t)\, dt, \quad F(t) = \left(\frac{d^2}{dt^2} - \lambda^2\right)^n f(t).$$

⊙ Reference: S. G. Samko, A. A. Kilbas, and O. I. Marichev (1993).

58. $\int_a^x (x - t)^{\nu+1} I_\nu[\lambda(x - t)] y(t)\, dt = f(x).$

Solution:

$$y(x) = \int_a^x g(t)\, dt,$$

where

$$g(t) = A\left(\frac{d^2}{dt^2} - \lambda^2\right)^n \int_a^t (t - \tau)^{n-\nu-2} I_{n-\nu-2}[\lambda(t - \tau)]\, f(\tau)\, d\tau,$$

$$A = \left(\frac{2}{\lambda}\right)^{n-2} \frac{\Gamma(\nu + 1)\,\Gamma(n - \nu - 1)}{\Gamma(2\nu + 2)\,\Gamma(2n - 2\nu - 3)},$$

where $-1 < \nu < \frac{n}{2} - 1$ and $n = 1, 2, \ldots$

If the right-hand side of the equation is differentiable sufficiently many times and the conditions $f(a) = f'_x(a) = \cdots = f_x^{(n-1)}(a) = 0$ are satisfied, then the function $g(t)$ defining the solution can be written in the form

$$g(t) = A \int_a^t (t - \tau)^{n-\nu-2} I_{n-\nu-2}[\lambda(t - \tau)] F(\tau)\, d\tau, \quad F(\tau) = \left(\frac{d^2}{d\tau^2} - \lambda^2\right)^n f(\tau).$$

⊙ Reference: S. G. Samko, A. A. Kilbas, and O. I. Marichev (1993).

59. $\int_a^x I_0\left(\lambda\sqrt{x - t}\right) y(t)\, dt = f(x).$

Solution:

$$y(x) = \frac{d^2}{dx^2} \int_a^x J_0\left(\lambda\sqrt{x - t}\right) f(t)\, dt.$$

60. $\int_a^x \left[AI_\nu\left(\lambda\sqrt{x}\right) + BI_\nu\left(\lambda\sqrt{t}\right) \right] y(t)\, dt = f(x).$

Solution with $B \neq -A$:

$$y(x) = \frac{1}{A+B} \frac{d}{dx} \left\{ \left[I_\nu\left(\lambda\sqrt{x}\right) \right]^{-\frac{A}{A+B}} \int_a^x \left[I_\nu\left(\lambda\sqrt{t}\right) \right]^{-\frac{B}{A+B}} f'_t(t)\, dt \right\}.$$

61. $\int_a^x \left[AI_\nu\left(\lambda\sqrt{x}\right) + BI_\mu\left(\beta\sqrt{t}\right) \right] y(t)\, dt = f(x).$

This is a special case of equation 1.9.6 with $g(x) = AI_\nu\left(\lambda\sqrt{x}\right)$ and $h(t) = BI_\mu\left(\beta\sqrt{t}\right)$.

62. $\int_a^x \sqrt{x-t}\, I_1\left(\lambda\sqrt{x-t}\right) y(t)\, dt = f(x).$

Solution:

$$y(x) = \frac{2}{\lambda} \frac{d^3}{dx^3} \int_a^x J_0\left(\lambda\sqrt{x-t}\right) f(t)\, dt.$$

63. $\int_a^x (x-t)^{1/4} I_{1/2}\left(\lambda\sqrt{x-t}\right) y(t)\, dt = f(x).$

Solution:

$$y(x) = \sqrt{\frac{2}{\pi\lambda}} \frac{d^2}{dx^2} \int_a^x \frac{\cos\left(\lambda\sqrt{x-t}\right)}{\sqrt{x-t}}\, f(t)\, dt.$$

64. $\int_a^x (x-t)^{3/4} I_{3/2}\left(\lambda\sqrt{x-t}\right) y(t)\, dt = f(x).$

Solution:

$$y(x) = \frac{2^{3/2}}{\sqrt{\pi}\, \lambda^{3/2}} \frac{d^3}{dx^3} \int_a^x \frac{\cos\left(\lambda\sqrt{x-t}\right)}{\sqrt{x-t}}\, f(t)\, dt.$$

65. $\int_a^x (x-t)^{n/2} I_n\left(\lambda\sqrt{x-t}\right) y(t)\, dt = f(x), \qquad n = 0, 1, 2, \ldots$

Solution:

$$y(x) = \left(\frac{2}{\lambda}\right)^n \frac{d^{n+2}}{dx^{n+2}} \int_a^x J_0\left(\lambda\sqrt{x-t}\right) f(t)\, dt.$$

66. $\int_a^x (x-t)^{\frac{2n-3}{4}} I_{\frac{2n-3}{2}}\left(\lambda\sqrt{x-t}\right) y(t)\, dt = f(x), \qquad n = 1, 2, \ldots$

Solution:

$$y(x) = \frac{1}{\sqrt{\pi}} \left(\frac{2}{\lambda}\right)^{\frac{2n-3}{2}} \frac{d^n}{dx^n} \int_a^x \frac{\cos\left(\lambda\sqrt{x-t}\right)}{\sqrt{x-t}}\, f(t)\, dt.$$

67. $\int_a^x (x-t)^{-1/4} I_{-1/2}\left(\lambda\sqrt{x-t}\right) y(t)\, dt = f(x).$

Solution:

$$y(x) = \sqrt{\frac{\lambda}{2\pi}} \frac{d}{dx} \int_a^x \frac{\cos\left(\lambda\sqrt{x-t}\right)}{\sqrt{x-t}}\, f(t)\, dt.$$

68. $\displaystyle\int_a^x (x-t)^{\nu/2} I_\nu\left(\lambda\sqrt{x-t}\right) y(t)\, dt = f(x).$

Solution:

$$y(x) = \left(\frac{2}{\lambda}\right)^{n-2} \frac{d^n}{dx^n} \int_a^x (x-t)^{\frac{n-\nu-2}{2}} J_{n-\nu-2}\left(\lambda\sqrt{x-t}\right) f(t)\, dt,$$

where $-1 < \nu < n-1$, $n = 1, 2, \ldots$

If the right-hand side of the equation is differentiable sufficiently many times and the conditions $f(a) = f_x'(a) = \cdots = f_x^{(n-1)}(a) = 0$ are satisfied, then the solution of the integral equation can be written in the form

$$y(x) = \left(\frac{2}{\lambda}\right)^{n-2} \int_a^x (x-t)^{\frac{n-\nu-2}{2}} J_{n-\nu-2}\left(\lambda\sqrt{x-t}\right) f_t^{(n)}(t)\, dt.$$

⊙ Reference: S. G. Samko, A. A. Kilbas, and O. I. Marichev (1993).

69. $\displaystyle\int_0^x (x^2 - t^2)^{-1/4} I_{-1/2}\left(\lambda\sqrt{x^2 - t^2}\right) y(t)\, dt = f(x).$

Solution:

$$y(x) = \sqrt{\frac{2\lambda}{\pi}} \frac{d}{dx} \int_0^x t\, \frac{\cos\left(\lambda\sqrt{x^2 - t^2}\right)}{\sqrt{x^2 - t^2}} f(t)\, dt.$$

70. $\displaystyle\int_x^\infty (t^2 - x^2)^{-1/4} I_{-1/2}\left(\lambda\sqrt{t^2 - x^2}\right) y(t)\, dt = f(x).$

Solution:

$$y(x) = -\sqrt{\frac{2\lambda}{\pi}} \frac{d}{dx} \int_x^\infty t\, \frac{\cos\left(\lambda\sqrt{t^2 - x^2}\right)}{\sqrt{t^2 - x^2}} f(t)\, dt.$$

71. $\displaystyle\int_0^x (x^2 - t^2)^{\nu/2} I_\nu\left(\lambda\sqrt{x^2 - t^2}\right) y(t)\, dt = f(x), \qquad -1 < \nu < 0.$

Solution:

$$y(x) = \lambda \frac{d}{dx} \int_0^x t\left(x^2 - t^2\right)^{-(\nu+1)/2} J_{-\nu-1}\left(\lambda\sqrt{x^2 - t^2}\right) f(t)\, dt.$$

⊙ Reference: S. G. Samko, A. A. Kilbas, and O. I. Marichev (1993).

72. $\displaystyle\int_x^\infty (t^2 - x^2)^{\nu/2} I_\nu\left(\lambda\sqrt{t^2 - x^2}\right) y(t)\, dt = f(x), \qquad -1 < \nu < 0.$

Solution:

$$y(x) = -\lambda \frac{d}{dx} \int_x^\infty t\, (t^2 - x^2)^{-(\nu+1)/2} J_{-\nu-1}\left(\lambda\sqrt{t^2 - x^2}\right) f(t)\, dt.$$

⊙ Reference: S. G. Samko, A. A. Kilbas, and O. I. Marichev (1993).

73. $\displaystyle\int_a^x [At^k I_\nu(\lambda x) + Bx^s I_\mu(\lambda t)] y(t)\, dt = f(x).$

This is a special case of equation 1.9.15 with $g_1(x) = A I_\nu(\lambda x)$, $h_1(t) = t^k$, $g_2(x) = Bx^s$, and $h_2(t) = I_\mu(\lambda t)$.

74. $\int_a^x [AI_\nu^2(\lambda x) + BI_\nu^2(\lambda t)]y(t)\,dt = f(x).$

Solution with $B \ne -A$:

$$y(x) = \frac{1}{A+B}\frac{d}{dx}\left\{|I_\nu(\lambda x)|^{-\frac{2A}{A+B}}\int_a^x |I_\nu(\lambda t)|^{-\frac{2B}{A+B}} f_t'(t)\,dt\right\}.$$

75. $\int_a^x [AI_\nu^k(\lambda x) + BI_\mu^s(\beta t)]y(t)\,dt = f(x).$

This is a special case of equation 1.9.6 with $g(x) = AI_\nu^k(\lambda x)$ and $h(t) = BI_\mu^s(\beta t)$.

76. $\int_a^x [K_0(\lambda x) - K_0(\lambda t)]y(t)\,dt = f(x).$

Solution: $y(x) = -\dfrac{d}{dx}\left[\dfrac{f_x'(x)}{\lambda K_1(\lambda x)}\right].$

77. $\int_a^x [K_\nu(\lambda x) - K_\nu(\lambda t)]y(t)\,dt = f(x).$

This is a special case of equation 1.9.2 with $g(x) = K_\nu(\lambda x)$.

78. $\int_a^x [AK_\nu(\lambda x) + BK_\nu(\lambda t)]y(t)\,dt = f(x).$

Solution with $B \ne -A$:

$$y(x) = \frac{1}{A+B}\frac{d}{dx}\left\{[K_\nu(\lambda x)]^{-\frac{A}{A+B}}\int_a^x [K_\nu(\lambda t)]^{-\frac{B}{A+B}} f_t'(t)\,dt\right\}.$$

79. $\int_a^x [At^k K_\nu(\lambda x) + Bx^s K_\mu(\lambda t)]y(t)\,dt = f(x).$

This is a special case of equation 1.9.15 with $g_1(x) = AK_\nu(\lambda x)$, $h_1(t) = t^k$, $g_2(x) = Bx^s$, and $h_2(t) = K_\mu(\lambda t)$.

80. $\int_a^x [AI_\nu(\lambda x)K_\mu(\beta t) + BI_\nu(\lambda t)K_\mu(\beta x)]y(t)\,dt = f(x).$

This is a special case of equation 1.9.15 with $g_1(x) = AI_\nu(\lambda x)$, $h_1(t) = K_\mu(\beta t)$, $g_2(x) = BK_\mu(\beta x)$, and $h_2(t) = I_\nu(\lambda t)$.

1.8-3. Kernels Containing Associated Legendre Functions

81. $\int_a^x (x^2 - t^2)^{-\mu/2} P_\nu^\mu\left(\dfrac{x}{t}\right)y(t)\,dt = f(x), \qquad 0 < a < \infty.$

Here $P_\nu^\mu(x)$ is the associated Legendre function (see Supplement 10).
  Solution:

$$y(x) = x^{n+\mu-1}\frac{d^n}{dx^n}\left[x^{1-\mu}\int_a^x (x^2 - t^2)^{\frac{n+\mu-2}{2}} t^{-n} P_\nu^{2-n-\mu}\left(\frac{t}{x}\right)f(t)\,dt\right],$$

where $\mu < 1$, $\nu \ge -\frac{1}{2}$, and $n = 1, 2, \ldots$

⊙ Reference: S. G. Samko, A. A. Kilbas, and O. I. Marichev (1993).

82. $\displaystyle\int_a^x (x^2 - t^2)^{-\mu/2} P_\nu^\mu\left(\frac{t}{x}\right) y(t)\, dt = f(x), \qquad 0 < a < \infty.$

Here $P_\nu^\mu(x)$ is the associated Legendre function (see Supplement 10).
    Solution:

$$y(x) = \frac{d^n}{dx^n} \int_a^x (x^2 - t^2)^{\frac{n+\mu-2}{2}} P_\nu^{2-n-\mu}\left(\frac{x}{t}\right) f(t)\, dt,$$

where $\mu < 1$, $\nu \ge -\frac{1}{2}$, and $n = 1, 2, \dots$

⊙ Reference: S. G. Samko, A. A. Kilbas, and O. I. Marichev (1993).

83. $\displaystyle\int_x^b (t^2 - x^2)^{-\mu/2} P_\nu^\mu\left(\frac{x}{t}\right) y(t)\, dt = f(x), \qquad 0 < b < \infty.$

Here $P_\nu^\mu(x)$ is the associated Legendre function (see Supplement 10).
    Solution:

$$y(x) = (-1)^n x^{n+\mu-1} \frac{d^n}{dx^n}\left[x^{1-\mu} \int_x^b (t^2 - x^2)^{\frac{n+\mu-2}{2}} t^{-n} P_\nu^{2-n-\mu}\left(\frac{t}{x}\right) f(t)\, dt\right],$$

where $\mu < 1$, $\nu \ge -\frac{1}{2}$, and $n = 1, 2, \dots$

⊙ Reference: S. G. Samko, A. A. Kilbas, and O. I. Marichev (1993).

84. $\displaystyle\int_x^b (t^2 - x^2)^{-\mu/2} P_\nu^\mu\left(\frac{t}{x}\right) y(t)\, dt = f(x), \qquad 0 < b < \infty.$

Here $P_\nu^\mu(x)$ is the associated Legendre function (see Supplement 10).
    Solution:

$$y(x) = (-1)^n \frac{d^n}{dx^n} \int_x^b (t^2 - x^2)^{\frac{n+\mu-2}{2}} P_\nu^{2-n-\mu}\left(\frac{x}{t}\right) f(t)\, dt,$$

where $\mu < 1$, $\nu \ge -\frac{1}{2}$, and $n = 1, 2, \dots$

⊙ Reference: S. G. Samko, A. A. Kilbas, and O. I. Marichev (1993).

1.8-4. Kernels Containing Hypergeometric Functions

85. $\displaystyle\int_s^x (x - t)^{b-1} \Phi\big(a, b; \lambda(x - t)\big) y(t)\, dt = f(x).$

Here $\Phi(a, b; z)$ is the degenerate hypergeometric function (see Supplement 10).
    Solution:

$$y(x) = \frac{d^n}{dx^n} \int_s^x \frac{(x - t)^{n-b-1}}{\Gamma(b)\Gamma(n - b)} \Phi\big(-a, n - b;\ \lambda(x - t)\big) f(t)\, dt,$$

where $0 < b < n$ and $n = 1, 2, \dots$

    If the right-hand side of the equation is differentiable sufficiently many times and the conditions $f(s) = f'_x(s) = \cdots = f_x^{(n-1)}(s) = 0$ are satisfied, then the solution of the integral equation can be written in the form

$$y(x) = \int_s^x \frac{(x - t)^{n-b-1}}{\Gamma(b)\Gamma(n - b)} \Phi\big(-a, n - b;\ \lambda(x - t)\big) f_t^{(n)}(t)\, dt.$$

⊙ Reference: S. G. Samko, A. A. Kilbas, and O. I. Marichev (1993).

86. $\displaystyle\int_s^x (x-t)^{c-1} F\left(a,b,c;\ 1-\frac{x}{t}\right) y(t)\,dt = f(x).$

Here $\Phi(a,b,c;z)$ is the Gaussian hypergeometric function (see Supplement 10).

Solution:

$$y(x) = x^{-a}\frac{d^n}{dx^n}\left\{x^a \int_s^x \frac{(x-t)^{n-c-1}}{\Gamma(c)\Gamma(n-c)} F\left(-a,\ n-b,\ n-c;\ 1-\frac{t}{x}\right) f(t)\,dt\right\},$$

where $0 < c < n$ and $n = 1, 2, \ldots$

If the right-hand side of the equation is differentiable sufficiently many times and the conditions $f(s) = f'_x(s) = \cdots = f_x^{(n-1)}(s) = 0$ are satisfied, then the solution of the integral equation can be written in the form

$$y(x) = \int_s^x \frac{(x-t)^{n-c-1}}{\Gamma(c)\Gamma(n-c)} F\left(-a,\ -b,\ n-c;\ 1-\frac{t}{x}\right) f_t^{(n)}(t)\,dt.$$

⊙ Reference: S. G. Samko, A. A. Kilbas, and O. I. Marichev (1993).

1.9. Equations Whose Kernels Contain Arbitrary Functions

1.9-1. Equations With Degenerate Kernel: $K(x,t) = g_1(x)h_1(t) + g_2(x)h_2(t)$

1. $\displaystyle\int_a^x g(x)h(t)y(t)\,dt = f(x).$

Solution: $\displaystyle y = \frac{1}{h(x)}\frac{d}{dx}\left[\frac{f(x)}{g(x)}\right] = \frac{1}{g(x)h(x)}f'_x(x) - \frac{g'_x(x)}{g^2(x)h(x)}f(x).$

2. $\displaystyle\int_a^x [g(x) - g(t)]y(t)\,dt = f(x).$

It is assumed that $f(a) = f'_x(a) = 0$ and $f'_x/g'_x \neq \text{const}.$

Solution: $\displaystyle y(x) = \frac{d}{dx}\left[\frac{f'_x(x)}{g'_x(x)}\right].$

3. $\displaystyle\int_a^x [g(x) - g(t) + b]y(t)\,dt = f(x).$

Differentiation with respect to x yields an equation of the form 2.9.2:

$$y(x) + \frac{1}{b}g'_x(x)\int_a^x y(t)\,dt = \frac{1}{b}f'_x(x).$$

Solution:

$$y(x) = \frac{1}{b}f'_x(x) - \frac{1}{b^2}g'_x(x)\int_a^x \exp\left[\frac{g(t)-g(x)}{b}\right]f'_t(t)\,dt.$$

4. $\displaystyle\int_a^x [Ag(x) + Bg(t)]y(t)\,dt = f(x).$

For $B = -A$, see equation 1.9.2.

Solution with $B \neq -A$:

$$y(x) = \frac{\text{sign}\,g(x)}{A+B}\frac{d}{dx}\left\{|g(x)|^{-\frac{A}{A+B}}\int_a^x |g(t)|^{-\frac{B}{A+B}} f'_t(t)\,dt\right\}.$$

5. $\int_a^x [Ag(x) + Bg(t) + C]y(t)\, dt = f(x).$

For $B = -A$, see equation 1.9.3. Assume that $B \neq -A$ and $(A + B)g(x) + C > 0$.
   Solution:

$$y(x) = \frac{d}{dx}\left\{ \left|(A + B)g(x) + C\right|^{-\frac{A}{A+B}} \int_a^x \left|(A + B)g(t) + C\right|^{-\frac{B}{A+B}} f_t'(t)\, dt\right\}.$$

6. $\int_a^x [g(x) + h(t)]y(t)\, dt = f(x).$

Solution:

$$y(x) = \frac{d}{dx}\left[\frac{\Phi(x)}{g(x) + h(x)} \int_a^x \frac{f_t'(t)\, dt}{\Phi(t)}\right], \qquad \Phi(x) = \exp\left[\int_a^x \frac{h_t'(t)\, dt}{g(t) + h(t)}\right].$$

7. $\int_a^x \left[g(x) + (x - t)h(x)\right]y(t)\, dt = f(x).$

This is a special case of equation 1.9.15 with $g_1(x) = g(x) + xh(x)$, $h_1(t) = 1$, $g_2(x) = h(x)$, and $h_2(t) = -t$.
   Solution:

$$y(x) = \frac{d}{dx}\left\{ \Phi(x)\frac{h(x)}{g(x)} \int_a^x \left[\frac{f(t)}{h(t)}\right]_t' \frac{dt}{\Phi(t)}\right\}, \qquad \Phi(t) = \exp\left[-\int_a^x \frac{h(t)}{g(t)}\, dt\right].$$

8. $\int_a^x \left[g(t) + (x - t)h(t)\right]y(t)\, dt = f(x).$

This is a special case of equation 1.9.15 with $g_1(x) = x$, $h_1(t) = h(t)$, $g_2(x) = 1$, and $h_2(t) = g(t) - th(t)$.

9. $\int_a^x \left[g(x) + (Ax^\lambda + Bt^\mu)h(x)\right]y(t)\, dt = f(x).$

This is a special case of equation 1.9.15 with $g_1(x) = g(x) + Ax^\lambda h(x)$, $h_1(t) = 1$, $g_2(x) = h(x)$, and $h_2(t) = Bt^\mu$.

10. $\int_a^x \left[g(t) + (Ax^\lambda + Bt^\mu)h(t)\right]y(t)\, dt = f(x).$

This is a special case of equation 1.9.15 with $g_1(x) = Ax^\lambda$, $h_1(t) = h(t)$, $g_2(x) = 1$, and $h_2(t) = g(t) + Bt^\mu h(t)$.

11. $\int_a^x [g(x)h(t) - h(x)g(t)]y(t)\, dt = f(x), \qquad f(a) = f_x'(a) = 0.$

For $g = $ const or $h = $ const, see equation 1.9.2.
   Solution:

$$y(x) = \frac{1}{h}\frac{d}{dx}\left[\frac{(f/h)_x'}{(g/h)_x'}\right], \qquad \text{where} \quad f = f(x), \quad g = g(x), \quad h = h(x).$$

Here $Af + Bg + Ch \not\equiv 0$, with A, B, and C being some constants.

12. $\displaystyle\int_a^x [Ag(x)h(t) + Bg(t)h(x)]y(t)\,dt = f(x).$

For $B = -A$, see equation 1.9.11.
    Solution with $B \neq -A$:

$$y(x) = \frac{1}{(A+B)h(x)} \frac{d}{dx}\left\{ \left[\frac{h(x)}{g(x)}\right]^{\frac{A}{A+B}} \int_a^x \left[\frac{h(t)}{g(t)}\right]^{\frac{B}{A+B}} \frac{d}{dt}\left[\frac{f(t)}{h(t)}\right] dt \right\}.$$

13. $\displaystyle\int_a^x \{1 + [g(t) - g(x)]h(x)\}y(t)\,dt = f(x).$

This is a special case of equation 1.9.15 with $g_1(x) = 1 - g(x)h(x)$, $h_1(t) = 1$, $g_2(x) = h(x)$, and $h_2(t) = g(t)$.
    Solution:

$$y(x) = \frac{d}{dx}\left\{ h(x)\Phi(x) \int_a^x \left[\frac{f(t)}{h(t)}\right]_t' \frac{dt}{\Phi(t)} \right\}, \qquad \Phi(x) = \exp\left[\int_a^x g_t'(t)h(t)\,dt\right].$$

14. $\displaystyle\int_a^x \left\{e^{-\lambda(x-t)} + \left[e^{\lambda x}g(t) - e^{\lambda t}g(x)\right]h(x)\right\}y(t)\,dt = f(x).$

This is a special case of equation 1.9.15 with $g_1(x) = e^{\lambda x}h(x)$, $h_1(t) = g(t)$, $g_2(x) = e^{-\lambda x} - g(x)h(x)$, and $h_2(t) = e^{\lambda t}$.

15. $\displaystyle\int_a^x [g_1(x)h_1(t) + g_2(x)h_2(t)]y(t)\,dt = f(x).$

For $g_2/g_1 = \text{const}$ or $h_2/h_1 = \text{const}$, see equation 1.9.1.

$1°$. Solution with $g_1(x)h_1(x) + g_2(x)h_2(x) \not\equiv 0$ and $f(x) \not\equiv \text{const}\, g_2(x)$:

$$y(x) = \frac{1}{h_1(x)} \frac{d}{dx}\left\{ \frac{g_2(x)h_1(x)\Phi(x)}{g_1(x)h_1(x) + g_2(x)h_2(x)} \int_a^x \left[\frac{f(t)}{g_2(t)}\right]_t' \frac{dt}{\Phi(t)} \right\}, \qquad (1)$$

where

$$\Phi(x) = \exp\left\{ \int_a^x \left[\frac{h_2(t)}{h_1(t)}\right]_t' \frac{g_2(t)h_1(t)\,dt}{g_1(t)h_1(t) + g_2(t)h_2(t)} \right\}. \qquad (2)$$

If $f(x) \equiv \text{const}\, g_2(x)$, the solution is given by formulas (1) and (2) in which the subscript 1 must be changed by 2 and vice versa.

$2°$. Solution with $g_1(x)h_1(x) + g_2(x)h_2(x) \equiv 0$:

$$y(x) = \frac{1}{h_1} \frac{d}{dx}\left[\frac{(f/g_2)_x'}{(g_1/g_2)_x'}\right] = -\frac{1}{h_1} \frac{d}{dx}\left[\frac{(f/g_2)_x'}{(h_2/h_1)_x'}\right],$$

where $f = f(x)$, $g_2 = g_2(x)$, $h_1 = h_1(x)$, and $h_2 = h_2(x)$.

1.9-2. Equations With Difference Kernel: $K(x,t) = K(x-t)$

16. $\displaystyle\int_a^x K(x-t)y(t)\,dt = f(x).$

$1°$. Let $K(0) = 1$ and $f(a) = 0$. Differentiating the equation with respect to x yields a Volterra equation of the second kind:

$$y(x) + \int_a^x K'_x(x-t)y(t)\,dt = f'_x(x).$$

The solution of this equation can be represented in the form

$$y(x) = f'_x(x) + \int_a^x R(x-t)f'_t(t)\,dt. \tag{1}$$

Here the resolvent $R(x)$ is related to the kernel $K(x)$ of the original equation by

$$R(x) = \mathfrak{L}^{-1}\left[\frac{1}{p\tilde{K}(p)} - 1\right], \qquad \tilde{K}(p) = \mathfrak{L}\left[K(x)\right],$$

where $\mathfrak{L}$ and $\mathfrak{L}^{-1}$ are the operators of the direct and inverse Laplace transforms, respectively.

$$\tilde{K}(p) = \mathfrak{L}\left[K(x)\right] = \int_0^\infty e^{-px}K(x)\,dx, \qquad R(x) = \mathfrak{L}^{-1}\left[\tilde{R}(p)\right] = \frac{1}{2\pi i}\int_{c-i\infty}^{c+i\infty} e^{px}\tilde{R}(p)\,dp.$$

$2°$. Let $K(x)$ have an integrable power-law singularity at $x = 0$. Denote by $w = w(x)$ the solution of the simpler auxiliary equation (compared with the original equation) with $a = 0$ and constant right-hand side $f \equiv 1$,

$$\int_0^x K(x-t)w(t)\,dt = 1. \tag{2}$$

Then the solution of the original integral equation with arbitrary right-hand side is expressed in terms of w as follows:

$$y(x) = \frac{d}{dx}\int_a^x w(x-t)f(t)\,dt = f(a)w(x-a) + \int_a^x w(x-t)f'_t(t)\,dt.$$

17. $\displaystyle\int_{-\infty}^x K(x-t)y(t)\,dt = Ax^n, \qquad n = 0,1,2,\dots$

This is a special case of equation 1.9.19 with $\lambda = 0$.

$1°$. Solution with $n = 0$:

$$y(x) = \frac{A}{B}, \qquad B = \int_0^\infty K(z)\,dz.$$

$2°$. Solution with $n = 1$:

$$y(x) = \frac{A}{B}x + \frac{AC}{B^2}, \qquad B = \int_0^\infty K(z)\,dz, \quad C = \int_0^\infty zK(z)\,dz.$$

$3°$. Solution with $n = 2$:

$$y_2(x) = \frac{A}{B}x^2 + 2\frac{AC}{B^2}x + 2\frac{AC^2}{B^3} - \frac{AD}{B^2},$$

$$B = \int_0^\infty K(z)\,dz, \quad C = \int_0^\infty zK(z)\,dz, \quad D = \int_0^\infty z^2 K(z)\,dz.$$

$4°$. Solution with $n = 3,4,\dots$ is given by:

$$y_n(x) = A\left\{\frac{\partial^n}{\partial\lambda^n}\left[\frac{e^{\lambda x}}{B(\lambda)}\right]\right\}_{\lambda=0}, \qquad B(\lambda) = \int_0^\infty K(z)e^{-\lambda z}\,dz.$$

18. $\displaystyle\int_{-\infty}^{x} K(x-t)y(t)\,dt = Ae^{\lambda x}$.

Solution:

$$y(x) = \frac{A}{B}e^{\lambda x}, \qquad B = \int_{0}^{\infty} K(z)e^{-\lambda z}\,dz = \mathfrak{L}\{K(z), \lambda\}.$$

19. $\displaystyle\int_{-\infty}^{x} K(x-t)y(t)\,dt = Ax^n e^{\lambda x}, \qquad n = 1, 2, \ldots$

$1°$. Solution with $n = 1$:

$$y_1(x) = \frac{A}{B}xe^{\lambda x} + \frac{AC}{B^2}e^{\lambda x},$$

$$B = \int_{0}^{\infty} K(z)e^{-\lambda z}\,dz, \quad C = \int_{0}^{\infty} zK(z)e^{-\lambda z}\,dz.$$

It is convenient to calculate the coefficients B and C using tables of Laplace transforms according to the formulas $B = \mathfrak{L}\{K(z), \lambda\}$ and $C = \mathfrak{L}\{zK(z), \lambda\}$.

$2°$. Solution with $n = 2$:

$$y_2(x) = \frac{A}{B}x^2 e^{\lambda x} + 2\frac{AC}{B^2}xe^{\lambda x} + \left(2\frac{AC^2}{B^3} - \frac{AD}{B^2}\right)e^{\lambda x},$$

$$B = \int_{0}^{\infty} K(z)e^{-\lambda z}\,dz, \quad C = \int_{0}^{\infty} zK(z)e^{-\lambda z}\,dz, \quad D = \int_{0}^{\infty} z^2 K(z)e^{-\lambda z}\,dz.$$

$3°$. Solution with $n = 3, 4, \ldots$ is given by:

$$y_n(x) = \frac{\partial}{\partial\lambda}y_{n-1}(x) = A\frac{\partial^n}{\partial\lambda^n}\left[\frac{e^{\lambda x}}{B(\lambda)}\right], \qquad B(\lambda) = \int_{0}^{\infty} K(z)e^{-\lambda z}\,dz.$$

20. $\displaystyle\int_{-\infty}^{x} K(x-t)y(t)\,dt = A\cosh(\lambda x)$.

Solution:

$$y(x) = \frac{A}{2B_-}e^{\lambda x} + \frac{A}{2B_+}e^{-\lambda x} = \frac{1}{2}\left(\frac{A}{B_-} + \frac{A}{B_+}\right)\cosh(\lambda x) + \frac{1}{2}\left(\frac{A}{B_-} - \frac{A}{B_+}\right)\sinh(\lambda x),$$

$$B_- = \int_{0}^{\infty} K(z)e^{-\lambda z}\,dz, \quad B_+ = \int_{0}^{\infty} K(z)e^{\lambda z}\,dz.$$

21. $\displaystyle\int_{-\infty}^{x} K(x-t)y(t)\,dt = A\sinh(\lambda x)$.

Solution:

$$y(x) = \frac{A}{2B_-}e^{\lambda x} - \frac{A}{2B_+}e^{-\lambda x} = \frac{1}{2}\left(\frac{A}{B_-} - \frac{A}{B_+}\right)\cosh(\lambda x) + \frac{1}{2}\left(\frac{A}{B_-} + \frac{A}{B_+}\right)\sinh(\lambda x),$$

$$B_- = \int_{0}^{\infty} K(z)e^{-\lambda z}\,dz, \quad B_+ = \int_{0}^{\infty} K(z)e^{\lambda z}\,dz.$$

22. $\displaystyle\int_{-\infty}^{x} K(x-t)y(t)\,dt = A\cos(\lambda x)$.

Solution:

$$y(x) = \frac{A}{B_c^2 + B_s^2}\left[B_c\cos(\lambda x) - B_s\sin(\lambda x)\right],$$

$$B_c = \int_{0}^{\infty} K(z)\cos(\lambda z)\,dz, \quad B_s = \int_{0}^{\infty} K(z)\sin(\lambda z)\,dz.$$

23. $\displaystyle\int_{-\infty}^{x} K(x-t)y(t)\,dt = A\sin(\lambda x).$

Solution:

$$y(x) = \frac{A}{B_c^2 + B_s^2}\left[B_c\sin(\lambda x) + B_s\cos(\lambda x)\right],$$

$$B_c = \int_0^\infty K(z)\cos(\lambda z)\,dz, \quad B_s = \int_0^\infty K(z)\sin(\lambda z)\,dz.$$

24. $\displaystyle\int_{-\infty}^{x} K(x-t)y(t)\,dt = Ae^{\mu x}\cos(\lambda x).$

Solution:

$$y(x) = \frac{A}{B_c^2 + B_s^2}e^{\mu x}\left[B_c\cos(\lambda x) - B_s\sin(\lambda x)\right],$$

$$B_c = \int_0^\infty K(z)e^{-\mu z}\cos(\lambda z)\,dz, \quad B_s = \int_0^\infty K(z)e^{-\mu z}\sin(\lambda z)\,dz.$$

25. $\displaystyle\int_{-\infty}^{x} K(x-t)y(t)\,dt = Ae^{\mu x}\sin(\lambda x).$

Solution:

$$y(x) = \frac{A}{B_c^2 + B_s^2}e^{\mu x}\left[B_c\sin(\lambda x) + B_s\cos(\lambda x)\right],$$

$$B_c = \int_0^\infty K(z)e^{-\mu z}\cos(\lambda z)\,dz, \quad B_s = \int_0^\infty K(z)e^{-\mu z}\sin(\lambda z)\,dz.$$

26. $\displaystyle\int_{-\infty}^{x} K(x-t)y(t)\,dt = f(x).$

$1°$. For a polynomial right-hand side of the equation, $f(x) = \sum_{k=0}^{n} A_k x^k$, the solution has the form

$$y(x) = \sum_{k=0}^{n} B_k x^k,$$

where the constants B_k are found by the method of undetermined coefficients. The solution can also be obtained by the formula given in 1.9.17 (item $4°$).

$2°$. For $f(x) = e^{\lambda x}\sum_{k=0}^{n} A_k x^k$, the solution has the form

$$y(x) = e^{\lambda x}\sum_{k=0}^{n} B_k x^k,$$

where the constants B_k are found by the method of undetermined coefficients. The solution can also be obtained by the formula given in 1.9.19 (item $3°$).

$3°$. For $f(x) = \sum_{k=0}^{n} A_k \exp(\lambda_k x)$, the solution has the form

$$y(x) = \sum_{k=0}^{n} \frac{A_k}{B_k} \exp(\lambda_k x), \qquad B_k = \int_0^\infty K(z) \exp(-\lambda_k z)\, dz.$$

$4°$. For $f(x) = \cos(\lambda x) \sum_{k=0}^{n} A_k x^k$, the solution has the form

$$y(x) = \cos(\lambda x) \sum_{k=0}^{n} B_k x^k + \sin(\lambda x) \sum_{k=0}^{n} C_k x^k,$$

where the constants B_k and C_k are found by the method of undetermined coefficients.

$5°$. For $f(x) = \sin(\lambda x) \sum_{k=0}^{n} A_k x^k$, the solution has the form

$$y(x) = \cos(\lambda x) \sum_{k=0}^{n} B_k x^k + \sin(\lambda x) \sum_{k=0}^{n} C_k x^k,$$

where the constants B_k and C_k are found by the method of undetermined coefficients.

$6°$. For $f(x) = \sum_{k=0}^{n} A_k \cos(\lambda_k x)$, the solution has the form

$$y(x) = \sum_{k=0}^{n} \frac{A_k}{B_{ck}^2 + B_{sk}^2} \left[B_{ck} \cos(\lambda_k x) - B_{sk} \sin(\lambda_k x) \right],$$

$$B_{ck} = \int_0^\infty K(z) \cos(\lambda_k z)\, dz, \quad B_{sk} = \int_0^\infty K(z) \sin(\lambda_k z)\, dz.$$

$7°$. For $f(x) = \sum_{k=0}^{n} A_k \sin(\lambda_k x)$, the solution has the form

$$y(x) = \sum_{k=0}^{n} \frac{A_k}{B_{ck}^2 + B_{sk}^2} \left[B_{ck} \sin(\lambda_k x) + B_{sk} \cos(\lambda_k x) \right],$$

$$B_{ck} = \int_0^\infty K(z) \cos(\lambda_k z)\, dz, \quad B_{sk} = \int_0^\infty K(z) \sin(\lambda_k z)\, dz.$$

27. $\displaystyle \int_x^\infty K(x - t) y(t)\, dt = A x^n, \qquad n = 0, 1, 2, \ldots$

This is a special case of equation 1.9.29 with $\lambda = 0$.

$1°$. Solution with $n = 0$:

$$y(x) = \frac{A}{B}, \qquad B = \int_0^\infty K(-z)\, dz.$$

$2°$. Solution with $n = 1$:

$$y(x) = \frac{A}{B} x - \frac{AC}{B^2}, \qquad B = \int_0^\infty K(-z)\, dz, \quad C = \int_0^\infty z K(-z)\, dz.$$

$3°$. Solution with $n = 2$:

$$y_2(x) = \frac{A}{B}x^2 - 2\frac{AC}{B^2}x + 2\frac{AC^2}{B^3} - \frac{AD}{B^2},$$

$$B = \int_0^\infty K(-z)\,dz, \quad C = \int_0^\infty zK(-z)\,dz, \quad D = \int_0^\infty z^2K(-z)\,dz.$$

$4°$. Solution with $n = 3, 4, \ldots$ is given by

$$y_n(x) = A\left\{\frac{\partial^n}{\partial\lambda^n}\left[\frac{e^{\lambda x}}{B(\lambda)}\right]\right\}_{\lambda=0}, \qquad B(\lambda) = \int_0^\infty K(-z)e^{\lambda z}\,dz.$$

28. $\displaystyle\int_x^\infty K(x-t)y(t)\,dt = Ae^{\lambda x}.$

Solution:

$$y(x) = \frac{A}{B}e^{\lambda x}, \qquad B = \int_0^\infty K(-z)e^{\lambda z}\,dz.$$

The expression for B is the Laplace transform of the function $K(-z)$ with parameter $p = -\lambda$ and can be calculated with the aid of tables of Laplace transforms given (e.g., see Supplement 4).

29. $\displaystyle\int_x^\infty K(x-t)y(t)\,dt = Ax^n e^{\lambda x}, \qquad n = 1, 2, \ldots$

$1°$. Solution with $n = 1$:

$$y_1(x) = \frac{A}{B}xe^{\lambda x} - \frac{AC}{B^2}e^{\lambda x},$$

$$B = \int_0^\infty K(-z)e^{\lambda z}\,dz, \quad C = \int_0^\infty zK(-z)e^{\lambda z}\,dz.$$

It is convenient to calculate the coefficients B and C using tables of Laplace transforms with parameter $p = -\lambda$.

$2°$. Solution with $n = 2$:

$$y_2(x) = \frac{A}{B}x^2 e^{\lambda x} - 2\frac{AC}{B^2}xe^{\lambda x} + \left(2\frac{AC^2}{B^3} - \frac{AD}{B^2}\right)e^{\lambda x},$$

$$B = \int_0^\infty K(-z)e^{\lambda z}\,dz, \quad C = \int_0^\infty zK(-z)e^{\lambda z}\,dz, \quad D = \int_0^\infty z^2K(-z)e^{\lambda z}\,dz.$$

$3°$. Solution with $n = 3, 4, \ldots$ is given by:

$$y_n(x) = \frac{\partial}{\partial\lambda}y_{n-1}(x) = A\frac{\partial^n}{\partial\lambda^n}\left[\frac{e^{\lambda x}}{B(\lambda)}\right], \qquad B(\lambda) = \int_0^\infty K(-z)e^{\lambda z}\,dz.$$

30. $\displaystyle\int_x^\infty K(x-t)y(t)\,dt = A\cosh(\lambda x).$

Solution:

$$y(x) = \frac{A}{2B_+}e^{\lambda x} + \frac{A}{2B_-}e^{-\lambda x} = \frac{1}{2}\left(\frac{A}{B_+} + \frac{A}{B_-}\right)\cosh(\lambda x) + \frac{1}{2}\left(\frac{A}{B_+} - \frac{A}{B_-}\right)\sinh(\lambda x),$$

$$B_+ = \int_0^\infty K(-z)e^{\lambda z}\,dz, \quad B_- = \int_0^\infty K(-z)e^{-\lambda z}\,dz.$$

31. $\displaystyle\int_x^\infty K(x-t)y(t)\,dt = A\sinh(\lambda x).$

Solution:

$$y(x) = \frac{A}{2B_+}e^{\lambda x} - \frac{A}{2B_-}e^{-\lambda x} = \frac{1}{2}\left(\frac{A}{B_+} - \frac{A}{B_-}\right)\cosh(\lambda x) + \frac{1}{2}\left(\frac{A}{B_+} + \frac{A}{B_-}\right)\sinh(\lambda x),$$

$$B_+ = \int_0^\infty K(-z)e^{\lambda z}\,dz, \quad B_- = \int_0^\infty K(-z)e^{-\lambda z}\,dz.$$

32. $\displaystyle\int_x^\infty K(x-t)y(t)\,dt = A\cos(\lambda x).$

Solution:

$$y(x) = \frac{A}{B_c^2 + B_s^2}\left[B_c\cos(\lambda x) + B_s\sin(\lambda x)\right],$$

$$B_c = \int_0^\infty K(-z)\cos(\lambda z)\,dz, \quad B_s = \int_0^\infty K(-z)\sin(\lambda z)\,dz.$$

33. $\displaystyle\int_x^\infty K(x-t)y(t)\,dt = A\sin(\lambda x).$

Solution:

$$y(x) = \frac{A}{B_c^2 + B_s^2}\left[B_c\sin(\lambda x) - B_s\cos(\lambda x)\right],$$

$$B_c = \int_0^\infty K(-z)\cos(\lambda z)\,dz, \quad B_s = \int_0^\infty K(-z)\sin(\lambda z)\,dz.$$

34. $\displaystyle\int_x^\infty K(x-t)y(t)\,dt = Ae^{\mu x}\cos(\lambda x).$

Solution:

$$y(x) = \frac{A}{B_c^2 + B_s^2}e^{\mu x}\left[B_c\cos(\lambda x) + B_s\sin(\lambda x)\right],$$

$$B_c = \int_0^\infty K(-z)e^{\mu z}\cos(\lambda z)\,dz, \quad B_s = \int_0^\infty K(-z)e^{\mu z}\sin(\lambda z)\,dz.$$

35. $\displaystyle\int_x^\infty K(x-t)y(t)\,dt = Ae^{\mu x}\sin(\lambda x).$

Solution:

$$y(x) = \frac{A}{B_c^2 + B_s^2}e^{\mu x}\left[B_c\sin(\lambda x) - B_s\cos(\lambda x)\right],$$

$$B_c = \int_0^\infty K(-z)e^{\mu z}\cos(\lambda z)\,dz, \quad B_s = \int_0^\infty K(-z)e^{\mu z}\sin(\lambda z)\,dz.$$

36. $\displaystyle\int_x^\infty K(x-t)y(t)\,dt = f(x).$

$1°$. For a polynomial right-hand side of the equation, $f(x) = \sum_{k=0}^n A_k x^k$, the solution has the form

$$y(x) = \sum_{k=0}^n B_k x^k,$$

where the constants B_k are found by the method of undetermined coefficients. The solution can also be obtained by the formula given in 1.9.27 (item $4°$).

2°. For $f(x) = e^{\lambda x} \sum_{k=0}^{n} A_k x^k$, the solution has the form

$$y(x) = e^{\lambda x} \sum_{k=0}^{n} B_k x^k,$$

where the constants B_k are found by the method of undetermined coefficients. The solution can also be obtained by the formula given in 1.9.29 (item 3°).

3°. For $f(x) = \sum_{k=0}^{n} A_k \exp(\lambda_k x)$, the solution has the form

$$y(x) = \sum_{k=0}^{n} \frac{A_k}{B_k} \exp(\lambda_k x), \qquad B_k = \int_0^\infty K(-z) \exp(\lambda_k z)\, dz.$$

4°. For $f(x) = \cos(\lambda x) \sum_{k=0}^{n} A_k x^k$, the solution has the form

$$y(x) = \cos(\lambda x) \sum_{k=0}^{n} B_k x^k + \sin(\lambda x) \sum_{k=0}^{n} C_k x^k,$$

where the constants B_k and C_k are found by the method of undetermined coefficients.

5°. For $f(x) = \sin(\lambda x) \sum_{k=0}^{n} A_k x^k$, the solution has the form

$$y(x) = \cos(\lambda x) \sum_{k=0}^{n} B_k x^k + \sin(\lambda x) \sum_{k=0}^{n} C_k x^k,$$

where the constants B_k and C_k are found by the method of undetermined coefficients.

6°. For $f(x) = \sum_{k=0}^{n} A_k \cos(\lambda_k x)$, the solution has the form

$$y(x) = \sum_{k=0}^{n} \frac{A_k}{B_{ck}^2 + B_{sk}^2} \left[B_{ck} \cos(\lambda_k x) + B_{sk} \sin(\lambda_k x) \right],$$

$$B_{ck} = \int_0^\infty K(-z) \cos(\lambda_k z)\, dz, \quad B_{sk} = \int_0^\infty K(-z) \sin(\lambda_k z)\, dz.$$

7°. For $f(x) = \sum_{k=0}^{n} A_k \sin(\lambda_k x)$, the solution has the form

$$y(x) = \sum_{k=0}^{n} \frac{A_k}{B_{ck}^2 + B_{sk}^2} \left[B_{ck} \sin(\lambda_k x) - B_{sk} \cos(\lambda_k x) \right],$$

$$B_{ck} = \int_0^\infty K(-z) \cos(\lambda_k z)\, dz, \quad B_{sk} = \int_0^\infty K(-z) \sin(\lambda_k z)\, dz.$$

8°. For arbitrary right-hand side $f = f(x)$, the solution of the integral equation can be calculated by the formula

$$y(x) = \frac{1}{2\pi i} \int_{c-i\infty}^{c+i\infty} \frac{\tilde{f}(p)}{\tilde{k}(-p)} e^{px}\, dp,$$

$$\tilde{f}(p) = \int_0^\infty f(x) e^{-px}\, dx, \qquad \tilde{k}(-p) = \int_0^\infty K(-z) e^{pz}\, dz.$$

To calculate $\tilde{f}(p)$ and $\tilde{k}(-p)$, it is convenient to use tables of Laplace transforms, and to determine $y(x)$, tables of inverse Laplace transforms.

1.9-3. Other Equations

37. $\displaystyle\int_a^x \big[g(x) - g(t)\big]^n y(t)\,dt = f(x), \qquad n = 1, 2, \ldots$

The right-hand side of the equation is assumed to satisfy the conditions $f(a) = f_x'(a) = \cdots = f_x^{(n)}(a) = 0$.

Solution: $\displaystyle y(x) = \frac{1}{n!} g_x'(x)\left(\frac{1}{g_x'(x)}\frac{d}{dx}\right)^{n+1} f(x).$

38. $\displaystyle\int_a^x \sqrt{g(x) - g(t)}\, y(t)\,dt = f(x), \qquad f(a) = 0.$

Solution:

$$y(x) = \frac{2}{\pi} g_x'(x)\left(\frac{1}{g_x'(x)}\frac{d}{dx}\right)^2 \int_a^x \frac{f(t)g_t'(t)\,dt}{\sqrt{g(x) - g(t)}}.$$

39. $\displaystyle\int_a^x \frac{y(t)\,dt}{\sqrt{g(x) - g(t)}} = f(x), \qquad g_x' > 0.$

Solution:

$$y(x) = \frac{1}{\pi}\frac{d}{dx}\int_a^x \frac{f(t)g_t'(t)\,dt}{\sqrt{g(x) - g(t)}}.$$

40. $\displaystyle\int_a^x \frac{e^{\lambda(x-t)}y(t)\,dt}{\sqrt{g(x) - g(t)}} = f(x), \qquad g_x' > 0.$

Solution:

$$y(x) = \frac{1}{\pi}e^{\lambda x}\frac{d}{dx}\int_a^x \frac{e^{-\lambda t}f(t)g_t'(t)}{\sqrt{g(x) - g(t)}}\,dt.$$

41. $\displaystyle\int_a^x \big[g(x) - g(t)\big]^\lambda y(t)\,dt = f(x), \qquad f(a) = 0, \quad 0 < \lambda < 1.$

Solution:

$$y(x) = kg_x'(x)\left(\frac{1}{g_x'(x)}\frac{d}{dx}\right)^2 \int_a^x \frac{g_t'(t)f(t)\,dt}{[g(x) - g(t)]^\lambda}, \qquad k = \frac{\sin(\pi\lambda)}{\pi\lambda}.$$

42. $\displaystyle\int_a^x \frac{h(t)y(t)\,dt}{[g(x) - g(t)]^\lambda} = f(x), \qquad g_x' > 0, \quad 0 < \lambda < 1.$

Solution:

$$y(x) = \frac{\sin(\pi\lambda)}{\pi h(x)}\frac{d}{dx}\int_a^x \frac{f(t)g_t'(t)\,dt}{[g(x) - g(t)]^{1-\lambda}}.$$

43. $\displaystyle\int_0^x K\!\left(\frac{t}{x}\right) y(t)\,dt = Ax^\lambda + Bx^\mu.$

Solution:

$$y(x) = \frac{A}{I_\lambda}x^{\lambda-1} + \frac{B}{I_\mu}x^{\mu-1}, \qquad I_\lambda = \int_0^1 K(z)z^{\lambda-1}\,dz, \quad I_\mu = \int_0^1 K(z)z^{\mu-1}\,dz.$$

44. $\displaystyle\int_0^x K\!\left(\frac{t}{x}\right) y(t)\,dt = P_n(x), \qquad P_n(x) = x^\lambda \sum_{m=0}^n A_m x^m.$

Solution:

$$y(x) = x^\lambda \sum_{m=0}^n \frac{A_m}{I_m} x^{m-1}, \qquad I_m = \int_0^1 K(z) z^{\lambda+m-1}\, dz.$$

The integral I_0 is supposed to converge.

45. $\displaystyle\int_a^x \left\{ g_1(x)\big[h_1(t) - h_1(x)\big] + g_2(x)\big[h_2(t) - h_2(x)\big] \right\} y(t)\,dt = f(x).$

This is a special case of equation 1.9.50 with $g_3(x) = -g_1(x)h_1(x) - g_2(x)h_2(x)$ and $h_3(t) = 1$.

The substitution $Y(x) = \displaystyle\int_a^x y(t)\,dt$ followed by integration by parts leads to an integral equation of the form 1.9.15:

$$\int_a^x \left\{ g_1(x)\big[h_1(t)\big]_t' + g_2(x)\big[h_2(t)\big]_t' \right\} Y(t)\,dt = -f(x).$$

46. $\displaystyle\int_a^x \left\{ g_1(x)\big[h_1(t) - e^{\lambda(x-t)}h_1(x)\big] + g_2(x)\big[h_2(t) - e^{\lambda(x-t)}h_2(x)\big] \right\} y(t)\,dt = f(x).$

This is a special case of equation 1.9.50 with $g_3(x) = -e^{\lambda x}\big[g_1(x)h_1(x) + g_2(x)h_2(x)\big]$, and $h_3(t) = e^{-\lambda t}$.

The substitution $Y(x) = \displaystyle\int_a^x e^{-\lambda t} y(t)\,dt$ followed by integration by parts leads to an integral equation of the form 1.9.15:

$$\int_a^x \left\{ g_1(x)\big[e^{\lambda t}h_1(t)\big]_t' + g_2(x)\big[e^{\lambda t}h_2(t)\big]_t' \right\} Y(t)\,dt = -f(x).$$

47. $\displaystyle\int_a^x \big[Ag^\lambda(x)g^\mu(t) + Bg^{\lambda+\beta}(x)g^{\mu-\beta}(t) - (A+B)g^{\lambda+\gamma}(x)g^{\mu-\gamma}(t) \big] y(t)\,dt = f(x).$

This is a special case of equation 1.9.50 with $g_1(x) = Ag^\lambda(x)$, $h_1(t) = g^\mu(t)$, $g_2(x) = Bg^{\lambda+\beta}(x)$, $h_2(t) = g^{\mu-\beta}(t)$, $g_3(x) = -(A+B)g^{\lambda+\gamma}(x)$, and $h_3(t) = g^{\mu-\gamma}(t)$.

48. $\displaystyle\int_a^x \big[Ag^\lambda(x)h(x)g^\mu(t) + Bg^{\lambda+\beta}(x)h(x)g^{\mu-\beta}(t)$
$$- (A+B)g^{\lambda+\gamma}(x)g^{\mu-\gamma}(t)h(t) \big] y(t)\,dt = f(x).$$

This is a special case of equation 1.9.50 with $g_1(x) = Ag^\lambda(x)h(x)$, $h_1(t) = g^\mu(t)$, $g_2(x) = Bg^{\lambda+\beta}(x)h(x)$, $h_2(t) = g^{\mu-\beta}(t)$, $g_3(x) = -(A+B)g^{\lambda+\gamma}(x)$, and $h_3(t) = g^{\mu-\gamma}(t)h(t)$.

49. $\displaystyle\int_a^x \big[Ag^\lambda(x)h(x)g^\mu(t) + Bg^{\lambda+\beta}(x)h(t)g^{\mu-\beta}(t)$
$$- (A+B)g^{\lambda+\gamma}(x)g^{\mu-\gamma}(t)h(t) \big] y(t)\,dt = f(x).$$

This is a special case of equation 1.9.50 with $g_1(x) = Ag^\lambda(x)h(x)$, $h_1(t) = g^\mu(t)$, $g_2(x) = Bg^{\lambda+\beta}(x)$, $h_2(t) = g^{\mu-\beta}(t)h(t)$, $g_3(x) = -(A+B)g^{\lambda+\gamma}(x)$, and $h_3(t) = g^{\mu-\gamma}(t)h(t)$.

50. $\displaystyle\int_a^x \big[g_1(x)h_1(t) + g_2(x)h_2(t) + g_3(x)h_3(t)\big]y(t)\,dt = f(x),$

$$\text{where} \quad g_1(x)h_1(x) + g_2(x)h_2(x) + g_3(x)h_3(x) \equiv 0.$$

The substitution $Y(x) = \displaystyle\int_a^x h_3(t)y(t)\,dt$ followed by integration by parts leads to an integral equation of the form 1.9.15:

$$\int_a^x \left\{ g_1(x)\left[\frac{h_1(t)}{h_3(t)}\right]_t' + g_2(x)\left[\frac{h_2(t)}{h_3(t)}\right]_t' \right\} Y(t)\,dt = -f(x).$$

51. $\displaystyle\int_{-\infty}^x Q(x-t)e^{\alpha t}y(\xi)\,dt = Ae^{px}, \qquad \xi = e^{\beta t}g(x-t).$

Solution:

$$y(\xi) = \frac{A}{q}\,\xi^{\frac{p-\alpha}{\beta}}, \qquad q = \int_0^\infty Q(z)[g(z)]^{\frac{p-\alpha}{\beta}} e^{-pz}\,dz.$$

1.10. Some Formulas and Transformations

1. Let the solution of the integral equation

$$\int_a^x K(x,t)y(t)\,dt = f(x) \tag{1}$$

have the form

$$y(x) = \mathcal{F}\big[f(x)\big], \tag{2}$$

where $\mathcal{F}$ is some linear integro-differential operator. Then the solution of the more complicated integral equation

$$\int_a^x K(x,t)g(x)h(t)y(t)\,dt = f(x) \tag{3}$$

has the form

$$y(x) = \frac{1}{h(x)}\,\mathcal{F}\left[\frac{f(x)}{g(x)}\right]. \tag{4}$$

Below are formulas for the solutions of integral equations of the form (3) for some specific functions $g(x)$ and $h(t)$. In all cases, it is assumed that the solution of equation (1) is known and is determined by formula (2).

(a) The solution of the equation

$$\int_a^x K(x,t)(x/t)^\lambda y(t)\,dt = f(x)$$

has the form

$$y(x) = x^\lambda \mathcal{F}\big[x^{-\lambda} f(x)\big].$$

(b) The solution of the equation

$$\int_a^x K(x,t)e^{\lambda(x-t)}y(t)\,dt = f(x)$$

has the form

$$y(x) = e^{\lambda x} \mathcal{F}\big[e^{-\lambda x} f(x)\big].$$

2. Let the solution of the integral equation (1) have the form

$$y(x) = L_1\left(x, \frac{d}{dx}\right) f(x) + L_2\left(x, \frac{d}{dx}\right) \int_a^x R(x, t) f(t)\, dt, \tag{5}$$

where L_1 and L_2 are some linear differential operators.

The solution of the more complicated integral equation

$$\int_a^x K\big(\varphi(x),\, \varphi(t)\big) y(t)\, dt = f(x), \tag{6}$$

where $\varphi(x)$ is an arbitrary monotone function (differentiable sufficiently many times, $\varphi_x' > 0$), is determined by the formula

$$
\begin{aligned}
y(x) = {}& \varphi_x'(x) L_1\left(\varphi(x),\, \frac{1}{\varphi_x'(x)}\frac{d}{dx}\right) f(x) \\
& + \varphi_x'(x) L_2\left(\varphi(x),\, \frac{1}{\varphi_x'(x)}\frac{d}{dx}\right) \int_a^x R\big(\varphi(x),\, \varphi(t)\big) \varphi_t'(t) f(t)\, dt.
\end{aligned}
\tag{7}
$$

Below are formulas for the solutions of integral equations of the form (6) for some specific functions $\varphi(x)$. In all cases, it is assumed that the solution of equation (1) is known and is determined by formula (5).

(a) For $\varphi(x) = x^\lambda$,

$$y(x) = \lambda x^{\lambda-1} L_1\left(x^\lambda,\, \frac{1}{\lambda x^{\lambda-1}}\frac{d}{dx}\right) f(x) + \lambda^2 x^{\lambda-1} L_2\left(x^\lambda,\, \frac{1}{\lambda x^{\lambda-1}}\frac{d}{dx}\right) \int_a^x R(x^\lambda,\, t^\lambda) t^{\lambda-1} f(t)\, dt.$$

(b) For $\varphi(x) = e^{\lambda x}$,

$$y(x) = \lambda e^{\lambda x} L_1\left(e^{\lambda x},\, \frac{1}{\lambda e^{\lambda x}}\frac{d}{dx}\right) f(x) + \lambda^2 e^{\lambda x} L_2\left(e^{\lambda x},\, \frac{1}{\lambda e^{\lambda x}}\frac{d}{dx}\right) \int_a^x R(e^{\lambda x},\, e^{\lambda t}) e^{\lambda t} f(t)\, dt.$$

(c) For $\varphi(x) = \ln(\lambda x)$,

$$y(x) = \frac{1}{x} L_1\left(\ln(\lambda x),\, x\frac{d}{dx}\right) f(x) + \frac{1}{x} L_2\left(\ln(\lambda x),\, x\frac{d}{dx}\right) \int_a^x \frac{1}{t} R\big(\ln(\lambda x),\, \ln(\lambda t)\big) f(t)\, dt.$$

(d) For $\varphi(x) = \cos(\lambda x)$,

$$
\begin{aligned}
y(x) = {}& -\lambda \sin(\lambda x) L_1\left(\cos(\lambda x),\, \frac{-1}{\lambda \sin(\lambda x)}\frac{d}{dx}\right) f(x) \\
& + \lambda^2 \sin(\lambda x) L_2\left(\cos(\lambda x),\, \frac{-1}{\lambda \sin(\lambda x)}\frac{d}{dx}\right) \int_a^x R\big(\cos(\lambda x),\, \cos(\lambda t)\big) \sin(\lambda t) f(t)\, dt.
\end{aligned}
$$

(e) For $\varphi(x) = \sin(\lambda x)$,

$$
\begin{aligned}
y(x) = {}& \lambda \cos(\lambda x) L_1\left(\sin(\lambda x),\, \frac{1}{\lambda \cos(\lambda x)}\frac{d}{dx}\right) f(x) \\
& + \lambda^2 \cos(\lambda x) L_2\left(\sin(\lambda x),\, \frac{1}{\lambda \cos(\lambda x)}\frac{d}{dx}\right) \int_a^x R\big(\sin(\lambda x),\, \sin(\lambda t)\big) \cos(\lambda t) f(t)\, dt.
\end{aligned}
$$

Chapter 2

Linear Equations of the Second Kind With Variable Limit of Integration

▶ *Notation:* $f = f(x)$, $g = g(x)$, $h = h(x)$, $K = K(x)$, and $M = M(x)$ are arbitrary functions (these may be composite functions of the argument depending on two variables x and t); A, B, C, D, a, b, c, α, β, γ, λ, and μ are free parameters; and m and n are nonnegative integers.

2.1. Equations Whose Kernels Contain Power-Law Functions

2.1-1. Kernels Linear in the Arguments x and t

1. $\quad y(x) - \lambda \int_a^x y(t)\, dt = f(x).$

Solution:
$$y(x) = f(x) + \lambda \int_a^x e^{\lambda(x-t)} f(t)\, dt.$$

2. $\quad y(x) + \lambda x \int_a^x y(t)\, dt = f(x).$

Solution:
$$y(x) = f(x) - \lambda \int_a^x x \exp\left[\tfrac{1}{2}\lambda(t^2 - x^2)\right] f(t)\, dt.$$

3. $\quad y(x) + \lambda \int_a^x t y(t)\, dt = f(x).$

Solution:
$$y(x) = f(x) - \lambda \int_a^x t \exp\left[\tfrac{1}{2}\lambda(t^2 - x^2)\right] f(t)\, dt.$$

4. $\quad y(x) + \lambda \int_a^x (x - t) y(t)\, dt = f(x).$

This is a special case of equation 2.1.34 with $n = 1$.

1°. Solution with $\lambda > 0$:
$$y(x) = f(x) - k \int_a^x \sin[k(x - t)] f(t)\, dt, \qquad k = \sqrt{\lambda}.$$

$2°$. Solution with $\lambda < 0$:

$$y(x) = f(x) + k \int_a^x \sinh[k(x-t)]f(t)\,dt, \qquad k = \sqrt{-\lambda}.$$

5. $y(x) + \displaystyle\int_a^x \big[A + B(x-t)\big]y(t)\,dt = f(x).$

$1°$. Solution with $A^2 > 4B$:

$$y(x) = f(x) - \int_a^x R(x-t)f(t)\,dt,$$

$$R(x) = \exp\left(-\tfrac{1}{2}Ax\right)\left[A\cosh(\beta x) + \frac{2B - A^2}{2\beta}\sinh(\beta x)\right], \qquad \beta = \sqrt{\tfrac{1}{4}A^2 - B}.$$

$2°$. Solution with $A^2 < 4B$:

$$y(x) = f(x) - \int_a^x R(x-t)f(t)\,dt,$$

$$R(x) = \exp\left(-\tfrac{1}{2}Ax\right)\left[A\cos(\beta x) + \frac{2B - A^2}{2\beta}\sin(\beta x)\right], \qquad \beta = \sqrt{B - \tfrac{1}{4}A^2}.$$

$3°$. Solution with $A^2 = 4B$:

$$y(x) = f(x) - \int_a^x R(x-t)f(t)\,dt, \qquad R(x) = \exp\left(-\tfrac{1}{2}Ax\right)\left(A - \tfrac{1}{4}A^2 x\right).$$

6. $y(x) - \displaystyle\int_a^x \big(Ax + Bt + C\big)y(t)\,dt = f(x).$

For $B = -A$ see equation 2.1.5. This is a special case of equation 2.9.6 with $g(x) = -Ax$ and $h(t) = -Bt - C$.

By differentiation followed by the substitution $Y(x) = \displaystyle\int_a^x y(t)\,dt$, the original equation can be reduced to the second-order linear ordinary differential equation

$$Y''_{xx} - \big[(A+B)x + C\big]Y'_x - AY = f'_x(x) \tag{1}$$

under the initial conditions

$$Y(a) = 0, \qquad Y'_x(a) = f(a). \tag{2}$$

A fundamental system of solutions of the homogeneous equation (1) with $f \equiv 0$ has the form

$$Y_1(x) = \Phi\left(\alpha, \tfrac{1}{2}; kz^2\right), \qquad Y_2(x) = \Psi\left(\alpha, \tfrac{1}{2}; kz^2\right),$$

$$\alpha = \frac{A}{2(A+B)}, \qquad k = \frac{A+B}{2}, \qquad z = x + \frac{C}{A+B},$$

where $\Phi(\alpha, \beta; x)$ and $\Psi(\alpha, \beta; x)$ are degenerate hypergeometric functions.

Solving the homogeneous equation (1) under conditions (2) for an arbitrary function $f = f(x)$ and taking into account the relation $y(x) = Y'_x(x)$, we thus obtain the solution of the integral equation in the form

$$y(x) = f(x) - \int_a^x R(x,t)f(t)\,dt,$$

$$R(x,t) = \frac{\partial^2}{\partial x \partial t}\left[\frac{Y_1(x)Y_2(t) - Y_2(x)Y_1(t)}{W(t)}\right], \qquad W(t) = \frac{2\sqrt{\pi k}}{\Gamma(\alpha)}\exp\left[k\left(t + \frac{C}{A+B}\right)^2\right].$$

2.1-2. Kernels Quadratic in the Arguments x and t

7. $\quad y(x) + A \displaystyle\int_a^x x^2 y(t)\,dt = f(x).$

This is a special case of equation 2.1.50 with $\lambda = 2$ and $\mu = 0$.
  Solution:
$$y(x) = f(x) - A \int_a^x x^2 \exp\left[\tfrac{1}{3}A(t^3 - x^3)\right] f(t)\,dt.$$

8. $\quad y(x) + A \displaystyle\int_a^x xt\, y(t)\,dt = f(x).$

This is a special case of equation 2.1.50 with $\lambda = 1$ and $\mu = 1$.
  Solution:
$$y(x) = f(x) - A \int_a^x xt \exp\left[\tfrac{1}{3}A(t^3 - x^3)\right] f(t)\,dt.$$

9. $\quad y(x) + A \displaystyle\int_a^x t^2 y(t)\,dt = f(x).$

This is a special case of equation 2.1.50 with $\lambda = 0$ and $\mu = 2$.
  Solution:
$$y(x) = f(x) - A \int_a^x t^2 \exp\left[\tfrac{1}{3}A(t^3 - x^3)\right] f(t)\,dt.$$

10. $\quad y(x) + \lambda \displaystyle\int_a^x (x - t)^2 y(t)\,dt = f(x).$

This is a special case of equation 2.1.34 with $n = 2$.
  Solution:
$$y(x) = f(x) - \int_a^x R(x - t) f(t)\,dt,$$
$$R(x) = \tfrac{2}{3}ke^{-2kx} - \tfrac{2}{3}ke^{kx}\left[\cos\left(\sqrt{3}\,kx\right) - \sqrt{3}\sin\left(\sqrt{3}\,kx\right)\right], \qquad k = \left(\tfrac{1}{4}\lambda\right)^{1/3}.$$

11. $\quad y(x) + A \displaystyle\int_a^x (x^2 - t^2) y(t)\,dt = f(x).$

This is a special case of equation 2.9.5 with $g(x) = Ax^2$.
  Solution:
$$y(x) = f(x) + \frac{1}{W} \int_a^x \left[u_1'(x)u_2'(t) - u_2'(x)u_1'(t)\right] f(t)\,dt,$$

where the primes denote differentiation with respect to the argument specified in the parentheses; $u_1(x)$, $u_2(x)$ is a fundamental system of solutions of the second-order linear homogeneous ordinary differential equation $u_{xx}'' + 2Axu = 0$; and the functions $u_1(x)$ and $u_2(x)$ are expressed in terms of Bessel functions or modified Bessel functions, depending on the sign of the parameter A:
  For $A > 0$,
$$W = 3/\pi, \quad u_1(x) = \sqrt{x}\, J_{1/3}\left(\sqrt{\tfrac{8}{9}A}\, x^{3/2}\right), \quad u_2(x) = \sqrt{x}\, Y_{1/3}\left(\sqrt{\tfrac{8}{9}A}\, x^{3/2}\right).$$
  For $A < 0$,
$$W = -\tfrac{3}{2}, \quad u_1(x) = \sqrt{x}\, I_{1/3}\left(\sqrt{\tfrac{8}{9}|A|}\, x^{3/2}\right), \quad u_2(x) = \sqrt{x}\, K_{1/3}\left(\sqrt{\tfrac{8}{9}|A|}\, x^{3/2}\right).$$

12. $y(x) + A \int_a^x (xt - t^2) y(t)\, dt = f(x).$

This is a special case of equation 2.9.4 with $g(t) = At$. Solution:

$$y(x) = f(x) + \frac{A}{W} \int_a^x t[y_1(x)y_2(t) - y_2(x)y_1(t)] f(t)\, dt,$$

where $y_1(x), y_2(x)$ is a fundamental system of solutions of the second-order linear homogeneous ordinary differential equation $y''_{xx} + Axy = 0$; the functions $y_1(x)$ and $y_2(x)$ are expressed in terms of Bessel functions or modified Bessel functions, depending on the sign of the parameter A:

For $A > 0$,

$$W = 3/\pi, \quad y_1(x) = \sqrt{x}\, J_{1/3}\left(\tfrac{2}{3}\sqrt{A}\, x^{3/2}\right), \quad y_2(x) = \sqrt{x}\, Y_{1/3}\left(\tfrac{2}{3}\sqrt{A}\, x^{3/2}\right).$$

For $A < 0$,

$$W = -\tfrac{3}{2}, \quad y_1(x) = \sqrt{x}\, I_{1/3}\left(\tfrac{2}{3}\sqrt{|A|}\, x^{3/2}\right), \quad y_2(x) = \sqrt{x}\, K_{1/3}\left(\tfrac{2}{3}\sqrt{|A|}\, x^{3/2}\right).$$

13. $y(x) + A \int_a^x (x^2 - xt) y(t)\, dt = f(x).$

This is a special case of equation 2.9.3 with $g(x) = Ax$. Solution:

$$y(x) = f(x) + \frac{A}{W} \int_a^x x[y_1(x)y_2(t) - y_2(x)y_1(t)] f(t)\, dt,$$

where $y_1(x), y_2(x)$ is a fundamental system of solutions of the second-order linear homogeneous ordinary differential equation $y''_{xx} + Axy = 0$; the functions $y_1(x)$ and $y_2(x)$ are expressed in terms of Bessel functions or modified Bessel functions, depending on the sign of the parameter A:

For $A > 0$,

$$W = 3/\pi, \quad y_1(x) = \sqrt{x}\, J_{1/3}\left(\tfrac{2}{3}\sqrt{A}\, x^{3/2}\right), \quad y_2(x) = \sqrt{x}\, Y_{1/3}\left(\tfrac{2}{3}\sqrt{A}\, x^{3/2}\right).$$

For $A < 0$,

$$W = -\tfrac{3}{2}, \quad y_1(x) = \sqrt{x}\, I_{1/3}\left(\tfrac{2}{3}\sqrt{|A|}\, x^{3/2}\right), \quad y_2(x) = \sqrt{x}\, K_{1/3}\left(\tfrac{2}{3}\sqrt{|A|}\, x^{3/2}\right).$$

14. $y(x) + A \int_a^x (t^2 - 3x^2) y(t)\, dt = f(x).$

This is a special case of equation 2.1.55 with $\lambda = 1$ and $\mu = 2$.

15. $y(x) + A \int_a^x (2xt - 3x^2) y(t)\, dt = f(x).$

This is a special case of equation 2.1.55 with $\lambda = 2$ and $\mu = 1$.

16. $y(x) - \int_a^x (ABxt - ABx^2 + Ax + B) y(t)\, dt = f(x).$

This is a special case of equation 2.9.16 with $g(x) = Ax$ and $h(x) = B$.
   Solution:

$$y(x) = f(x) + \int_a^x R(x, t) f(t)\, dt,$$

$$R(x, t) = (Ax + B) \exp\left[\tfrac{1}{2} A(x^2 - t^2)\right] + B^2 \int_t^x \exp\left[\tfrac{1}{2} A(s^2 - t^2) + B(x - s)\right] ds.$$

17. $y(x) + \displaystyle\int_a^x \left(Ax^2 - At^2 + Bx - Ct + D\right) y(t)\, dt = f(x).$

This is a special case of equation 2.9.6 with $g(x) = Ax^2 + Bx + D$ and $h(t) = -At^2 - Ct$.
   Solution:

$$y(x) = f(x) + \int_a^x \frac{\partial^2}{\partial x \partial t}\left[\frac{Y_1(x)Y_2(t) - Y_2(x)Y_1(t)}{W(t)}\right] f(t)\, dt.$$

Here $Y_1(x), Y_2(x)$ is a fundamental system of solutions of the second-order homogeneous ordinary differential equation $Y''_{xx} + \left[(B-C)x + D\right]Y'_x + (2Ax + B)Y = 0$ (see A. D. Polyanin and V. F. Zaitsev (1996) for details about this equation):

$$Y_1(x) = \exp(-kx)\Phi\left(\alpha, \tfrac{1}{2}; \tfrac{1}{2}(C-B)z^2\right), \quad Y_2(x) = \exp(-kx)\Psi\left(\alpha, \tfrac{1}{2}; \tfrac{1}{2}(C-B)z^2\right),$$

$$W(x) = -\frac{\sqrt{2\pi(C-B)}}{\Gamma(\alpha)}\exp\left[\tfrac{1}{2}(C-B)z^2 - 2kx\right], \quad k = \frac{2A}{B-C},$$

$$\alpha = -\frac{4A^2 + 2AD(C-B) + B(C-B)^2}{2(C-B)^3}, \quad z = x - \frac{4A + (C-B)D}{(C-B)^2},$$

where $\Phi\left(\alpha, \beta; x\right)$ and $\Psi\left(\alpha, \beta; x\right)$ are degenerate hypergeometric functions and $\Gamma(\alpha)$ is the gamma function.

18. $y(x) - \displaystyle\int_a^x \left[Ax + B + (Cx + D)(x - t)\right] y(t)\, dt = f(x).$

This is a special case of equation 2.9.11 with $g(x) = Ax + B$ and $h(x) = Cx + D$.
   Solution with $A \neq 0$:

$$y(x) = f(x) + \int_a^x \left[Y''_2(x)Y_1(t) - Y''_1(x)Y_2(t)\right] \frac{f(t)}{W(t)}\, dt.$$

Here $Y_1(x), Y_2(x)$ is a fundamental system of solutions of the second-order homogeneous ordinary differential equation $Y''_{xx} - (Ax + B)Y'_x - (Cx + D)Y = 0$ (see A. D. Polyanin and V. F. Zaitsev (1996) for details about this equation):

$$Y_1(x) = \exp(-kx)\Phi\left(\alpha, \tfrac{1}{2}; \tfrac{1}{2}Az^2\right), \quad Y_2(x) = \exp(-kx)\Psi\left(\alpha, \tfrac{1}{2}; \tfrac{1}{2}Az^2\right),$$

$$W(x) = -\sqrt{2\pi A}\left[\Gamma(\alpha)\right]^{-1}\exp\left(\tfrac{1}{2}Az^2 - 2kx\right), \quad k = C/A,$$

$$\alpha = \tfrac{1}{2}(A^2 D - ABC - C^2)A^{-3}, \quad z = x + (AB + 2C)A^{-2},$$

where $\Phi\left(\alpha, \beta; x\right)$ and $\Psi\left(\alpha, \beta; x\right)$ are degenerate hypergeometric functions, $\Gamma(\alpha)$ is the gamma function.

19. $y(x) + \displaystyle\int_a^x \left[At + B + (Ct + D)(t - x)\right] y(t)\, dt = f(x).$

This is a special case of equation 2.9.12 with $g(t) = -At - B$ and $h(t) = -Ct - D$.
   Solution with $A \neq 0$:

$$y(x) = f(x) - \int_a^x \left[Y_1(x)Y''_2(t) - Y''_1(t)Y_2(x)\right] \frac{f(t)}{W(x)}\, dt.$$

Here $Y_1(x), Y_2(x)$ is a fundamental system of solutions of the second-order homogeneous ordinary differential equation $Y''_{xx} - (Ax + B)Y'_x - (Cx + D)Y = 0$ (see A. D. Polyanin and V. F. Zaitsev (1996) for details about this equation):

$$Y_1(x) = \exp(-kx)\Phi\left(\alpha, \tfrac{1}{2}; \tfrac{1}{2}Az^2\right), \quad Y_2(x) = \exp(-kx)\Psi\left(\alpha, \tfrac{1}{2}; \tfrac{1}{2}Az^2\right),$$

$$W(x) = -\sqrt{2\pi A}\left[\Gamma(\alpha)\right]^{-1}\exp\left(\tfrac{1}{2}Az^2 - 2kx\right), \quad k = C/A,$$

$$\alpha = \tfrac{1}{2}(A^2 D - ABC - C^2)A^{-3}, \quad z = x + (AB + 2C)A^{-2},$$

where $\Phi\left(\alpha, \beta; x\right)$ and $\Psi\left(\alpha, \beta; x\right)$ are degenerate hypergeometric functions and $\Gamma(\alpha)$ is the gamma function.

2.1-3. Kernels Cubic in the Arguments x and t

20. $y(x) + A \displaystyle\int_a^x x^3 y(t)\, dt = f(x).$

Solution:
$$y(x) = f(x) - A \int_a^x x^3 \exp\left[\tfrac{1}{4}A(t^4 - x^4)\right] f(t)\, dt.$$

21. $y(x) + A \displaystyle\int_a^x x^2 t y(t)\, dt = f(x).$

Solution:
$$y(x) = f(x) - A \int_a^x x^2 t \exp\left[\tfrac{1}{4}A(t^4 - x^4)\right] f(t)\, dt.$$

22. $y(x) + A \displaystyle\int_a^x x t^2 y(t)\, dt = f(x).$

Solution:
$$y(x) = f(x) - A \int_a^x x t^2 \exp\left[\tfrac{1}{4}A(t^4 - x^4)\right] f(t)\, dt.$$

23. $y(x) + A \displaystyle\int_a^x t^3 y(t)\, dt = f(x).$

Solution:
$$y(x) = f(x) - A \int_a^x t^3 \exp\left[\tfrac{1}{4}A(t^4 - x^4)\right] f(t)\, dt.$$

24. $y(x) + \lambda \displaystyle\int_a^x (x - t)^3 y(t)\, dt = f(x).$

This is a special case of equation 2.1.34 with $n = 3$.

Solution:
$$y(x) = f(x) - \int_a^x R(x - t) f(t)\, dt,$$

where

$$R(x) = \begin{cases} k\left[\cosh(kx)\sin(kx) - \sinh(kx)\cos(kx)\right], & k = \left(\tfrac{3}{2}\lambda\right)^{1/4} & \text{for } \lambda > 0, \\ \tfrac{1}{2}s\left[\sin(sx) - \sinh(sx)\right], & s = (-6\lambda)^{1/4} & \text{for } \lambda < 0. \end{cases}$$

25. $y(x) + A \displaystyle\int_a^x (x^3 - t^3) y(t)\, dt = f(x).$

This is a special case of equation 2.1.52 with $\lambda = 3$.

26. $y(x) - A \displaystyle\int_a^x \left(4x^3 - t^3\right) y(t)\, dt = f(x).$

This is a special case of equation 2.1.55 with $\lambda = 1$ and $\mu = 3$.

27. $y(x) + A \displaystyle\int_a^x (x t^2 - t^3) y(t)\, dt = f(x).$

This is a special case of equation 2.1.49 with $\lambda = 2$.

28. $y(x) + A \int_a^x (x^2 t - t^3) y(t)\, dt = f(x).$

The transformation $z = x^2$, $\tau = t^2$, $y(x) = w(z)$ leads to an equation of the form 2.1.4:

$$w(z) + \tfrac{1}{2} A \int_{a^2}^z (z - \tau) w(\tau)\, d\tau = F(z), \qquad F(z) = f(x).$$

29. $y(x) + \int_a^x (Ax^2 t + Bt^3) y(t)\, dt = f(x).$

The transformation $z = x^2$, $\tau = t^2$, $y(x) = w(z)$ leads to an equation of the form 2.1.6:

$$w(z) + \int_{a^2}^z \left(\tfrac{1}{2} Az + \tfrac{1}{2} B\tau \right) w(\tau)\, d\tau = F(z), \qquad F(z) = f(x).$$

30. $y(x) + B \int_a^x (2x^3 - xt^2) y(t)\, dt = f(x).$

This is a special case of equation 2.1.55 with $\lambda = 2$, $\mu = 2$, and $B = -2A$.

31. $y(x) - A \int_a^x (4x^3 - 3x^2 t) y(t)\, dt = f(x).$

This is a special case of equation 2.1.55 with $\lambda = 3$ and $\mu = 1$.

32. $y(x) + \int_a^x (ABx^3 - ABx^2 t - Ax^2 - B) y(t)\, dt = f(x).$

This is a special case of equation 2.9.7 with $g(x) = Ax^2$ and $\lambda = B$.
 Solution:

$$y(x) = f(x) + \int_a^x R(x - t) f(t)\, dt,$$

$$R(x, t) = (Ax^2 + B) \exp\left[\tfrac{1}{3} A(x^3 - t^3) \right] + B^2 \int_t^x \exp\left[\tfrac{1}{3} A(s^3 - t^3) + B(x - s) \right] ds.$$

33. $y(x) + \int_a^x (ABxt^2 - ABt^3 + At^2 + B) y(t)\, dt = f(x).$

This is a special case of equation 2.9.8 with $g(t) = At^2$ and $\lambda = B$.
 Solution:

$$y(x) = f(x) + \int_a^x R(x - t) f(t)\, dt,$$

$$R(x, t) = -(At^2 + B) \exp\left[\tfrac{1}{3} A(t^3 - x^3) \right] + B^2 \int_t^x \exp\left[\tfrac{1}{3} A(s^3 - x^3) + B(t - s) \right] ds.$$

2.1-4. Kernels Containing Higher-Order Polynomials in x and t

34. $y(x) + A \int_a^x (x - t)^n y(t)\, dt = f(x), \qquad n = 1, 2, \ldots$

1°. Differentiating the equation $n + 1$ times with respect to x yields an $(n + 1)$st-order linear ordinary differential equation with constant coefficients for $y = y(x)$:

$$y_x^{(n+1)} + An!\, y = f_x^{(n+1)}(x).$$

This equation under the initial conditions $y(a) = f(a)$, $y_x'(a) = f_x'(a)$, $\ldots$, $y_x^{(n)}(a) = f_x^{(n)}(a)$ determines the solution of the original integral equation.

2°. Solution:

$$y(x) = f(x) + \int_a^x R(x-t)f(t)\,dt,$$

$$R(x) = \frac{1}{n+1} \sum_{k=0}^n \exp(\sigma_k x)\big[\sigma_k \cos(\beta_k x) - \beta_k \sin(\beta_k x)\big],$$

where the coefficients σ_k and β_k are given by

$$\sigma_k = |An!|^{\frac{1}{n+1}} \cos\left(\frac{2\pi k}{n+1}\right), \qquad \beta_k = |An!|^{\frac{1}{n+1}} \sin\left(\frac{2\pi k}{n+1}\right) \qquad \text{for} \quad A < 0,$$

$$\sigma_k = |An!|^{\frac{1}{n+1}} \cos\left(\frac{2\pi k + \pi}{n+1}\right), \quad \beta_k = |An!|^{\frac{1}{n+1}} \sin\left(\frac{2\pi k + \pi}{n+1}\right) \qquad \text{for} \quad A > 0.$$

35. $y(x) + A \displaystyle\int_x^\infty (t-x)^n y(t)\,dt = f(x), \qquad n = 1, 2, \ldots$

The Picard–Goursat equation. This is a special case of equation 2.9.62 with $K(z) = A(-z)^n$.

1°. A solution of the homogeneous equation ($f \equiv 0$) is

$$y(x) = Ce^{-\lambda x}, \qquad \lambda = \left(-An!\right)^{\frac{1}{n+1}},$$

where C is an arbitrary constant and $A < 0$. This is a unique solution for $n = 0, 1, 2, 3$.

The general solution of the homogeneous equation for any sign of A has the form

$$y(x) = \sum_{k=1}^s C_k \exp(-\lambda_k x). \tag{1}$$

Here C_k are arbitrary constants and λ_k are the roots of the algebraic equation $\lambda^{n+1} + An! = 0$ that satisfy the condition $\operatorname{Re} \lambda_k > 0$. The number of terms in (1) is determined by the inequality $s \le 2\left[\frac{n}{4}\right] + 1$, where $[a]$ stands for the integral part of a number a. For more details about the solution of the homogeneous Picard–Goursat equation, see Subsection 9.11-1 (Example 1).

2°. For $f(x) = \displaystyle\sum_{k=1}^m a_k \exp(-\beta_k x)$, where $\beta_k > 0$, a solution of the equation has the form

$$y(x) = \sum_{k=1}^m \frac{a_k \beta_k^{n+1}}{\beta_k^{n+1} + An!} \exp(-\beta_k x), \tag{2}$$

where $\beta_k^{n+1} + An! \ne 0$. For $A > 0$, this formula can also be used for arbitrary $f(x)$ expandable into a convergent exponential series (which corresponds to $m = \infty$).

3°. For $f(x) = e^{-\beta x} \displaystyle\sum_{k=1}^m a_k x^k$, where $\beta > 0$, a solution of the equation has the form

$$y(x) = e^{-\beta x} \sum_{k=0}^m B_k x^k, \tag{3}$$

where the constants B_k are found by the method of undetermined coefficients. The solution can also be constructed using the formulas given in item 3°, equation 2.9.55.

4°. For $f(x) = \cos(\beta x) \sum\limits_{k=1}^{m} a_k \exp(-\mu_k x)$, a solution of the equation has the form

$$y(x) = \cos(\beta x) \sum_{k=1}^{m} B_k \exp(-\mu_k x) + \sin(\beta x) \sum_{k=1}^{m} C_k \exp(-\mu_k x), \qquad (4)$$

where the constants B_k and C_k are found by the method of undetermined coefficients. The solution can also be constructed using the formulas given in 2.9.60.

5°. For $f(x) = \sin(\beta x) \sum\limits_{k=1}^{m} a_k \exp(-\mu_k x)$, a solution of the equation has the form

$$y(x) = \cos(\beta x) \sum_{k=1}^{m} B_k \exp(-\mu_k x) + \sin(\beta x) \sum_{k=1}^{m} C_k \exp(-\mu_k x), \qquad (5)$$

where the constants B_k and C_k are found by the method of undetermined coefficients. The solution can also be constructed using the formulas given in 2.9.61.

6°. To obtain the general solution in item 2°–5°, the solution (1) of the homogeneous equation must be added to each right-hand side of (2)–(5).

36. $\quad y(x) + A \displaystyle\int_a^x (x-t) t^n y(t)\, dt = f(x), \qquad n = 1, 2, \ldots$

This is a special case of equation 2.1.49 with $\lambda = n$.

37. $\quad y(x) + A \displaystyle\int_a^x (x^n - t^n) y(t)\, dt = f(x), \qquad n = 1, 2, \ldots$

This is a special case of equation 2.1.52 with $\lambda = n$.

38. $\quad y(x) + \displaystyle\int_a^x \left(ABx^{n+1} - ABx^n t - Ax^n - B \right) y(t)\, dt = f(x), \qquad n = 1, 2, \ldots$

This is a special case of equation 2.9.7 with $g(x) = Ax^n$ and $\lambda = B$.
  Solution:

$$y(x) = f(x) + \int_a^x R(x-t) f(t)\, dt,$$

$$R(x,t) = (Ax^n + B) \exp\left[\frac{A}{n+1} \left(x^{n+1} - t^{n+1} \right) \right] + B^2 \int_t^x \exp\left[\frac{A}{n+1} \left(s^{n+1} - t^{n+1} \right) + B(x-s) \right] ds.$$

39. $\quad y(x) + \displaystyle\int_a^x \left(ABxt^n - ABt^{n+1} + At^n + B \right) y(t)\, dt = f(x), \qquad n = 1, 2, \ldots$

This is a special case of equation 2.9.8 with $g(t) = At^n$ and $\lambda = B$.
  Solution:

$$y(x) = f(x) + \int_a^x R(x-t) f(t)\, dt,$$

$$R(x,t) = -(At^n + B) \exp\left[\frac{A}{n+1} \left(t^{n+1} - x^{n+1} \right) \right] + B^2 \int_t^x \exp\left[\frac{A}{n+1} \left(s^{n+1} - x^{n+1} \right) + B(t-s) \right] ds.$$

| **2.1-5. Kernels Containing Rational Functions** |

40. $y(x) + x^{-3} \int_a^x t[2Ax + (1 - A)t] y(t)\, dt = f(x).$

This equation can be obtained by differentiating the equation

$$\int_a^x [Ax^2 t + (1 - A)xt^2] y(t)\, dt = F(x), \qquad F(x) = \int_a^x t^3 f(t)\, dt,$$

which has the form 1.1.17:
Solution:

$$y(x) = \frac{1}{x} \frac{d}{dx}\left[x^{-A} \int_a^x t^{A-1} \varphi'_t(t)\, dt \right], \qquad \varphi(x) = \frac{1}{x} \int_a^x t^3 f(t)\, dt.$$

41. $y(x) - \lambda \int_0^x \dfrac{y(t)\, dt}{x + t} = f(x).$

Dixon's equation. This is a special case of equation 2.1.62 with $a = b = 1$ and $\mu = 0$.

1°. The solution of the homogeneous equation ($f \equiv 0$) is

$$y(x) = Cx^\beta \qquad (\beta > -1,\ \lambda > 0). \tag{1}$$

Here C is an arbitrary constant, and $\beta = \beta(\lambda)$ is determined by the transcendental equation

$$\lambda I(\beta) = 1, \qquad \text{where} \quad I(\beta) = \int_0^1 \frac{z^\beta\, dz}{1 + z}. \tag{2}$$

2°. For a polynomial right-hand side,

$$f(x) = \sum_{n=0}^N A_n x^n$$

the solution bounded at zero is given by

$$y(x) = \begin{cases} \displaystyle\sum_{n=0}^N \frac{A_n}{1 - (\lambda/\lambda_n)} x^n & \text{for } \lambda < \lambda_0, \\[4mm] \displaystyle\sum_{n=0}^N \frac{A_n}{1 - (\lambda/\lambda_n)} x^n + Cx^\beta & \text{for } \lambda > \lambda_0 \text{ and } \lambda \neq \lambda_n, \end{cases}$$

$$\lambda_n = \frac{1}{I(n)}, \qquad I(n) = (-1)^n \left[\ln 2 + \sum_{m=1}^n \frac{(-1)^m}{m} \right],$$

where C is an arbitrary constant, and $\beta = \beta(\lambda)$ is determined by the transcendental equation (2).
For special $\lambda = \lambda_n$ ($n = 1, 2, \dots$), the solution differs in one term and has the form

$$y(x) = \sum_{m=0}^{n-1} \frac{A_m}{1 - (\lambda_n/\lambda_m)} x^m + \sum_{m=n+1}^N \frac{A_m}{1 - (\lambda_n/\lambda_m)} x^m - A_n \frac{\bar\lambda_n}{\lambda_n} x^n \ln x + Cx^n,$$

where $\bar\lambda_n = (-1)^{n+1} \left[\dfrac{\pi^2}{12} + \displaystyle\sum_{k=1}^n \frac{(-1)^k}{k^2} \right]^{-1}.$

Remark. For arbitrary $f(x)$, expandable into power series, the formulas of item 2° can be used, in which one should set $N = \infty$. In this case, the radius of convergence of the solution $y(x)$ is equal to the radius of convergence of $f(x)$.

3°. For logarithmic-polynomial right-hand side,

$$f(x) = \ln x \left(\sum_{n=0}^{N} A_n x^n \right),$$

the solution with logarithmic singularity at zero is given by

$$y(x) = \begin{cases} \ln x \displaystyle\sum_{n=0}^{N} \frac{A_n}{1 - (\lambda/\lambda_n)} x^n + \displaystyle\sum_{n=0}^{N} \frac{A_n D_n \lambda}{[1 - (\lambda/\lambda_n)]^2} x^n & \text{for } \lambda < \lambda_0, \\[4mm] \ln x \displaystyle\sum_{n=0}^{N} \frac{A_n}{1 - (\lambda/\lambda_n)} x^n + \displaystyle\sum_{n=0}^{N} \frac{A_n D_n \lambda}{[1 - (\lambda/\lambda_n)]^2} x^n + C x^\beta & \text{for } \lambda > \lambda_0 \text{ and } \lambda \neq \lambda_n, \end{cases}$$

$$\lambda_n = \frac{1}{I(n)}, \quad I(n) = (-1)^n \left[\ln 2 + \sum_{k=1}^{n} \frac{(-1)^k}{k} \right], \quad D_n = (-1)^{n+1} \left[\frac{\pi^2}{12} + \sum_{k=1}^{n} \frac{(-1)^k}{k^2} \right].$$

4°. For arbitrary $f(x)$, the transformation

$$x = \tfrac{1}{2} e^{2z}, \quad t = \tfrac{1}{2} e^{2\tau}, \quad y(x) = e^{-z} w(z), \quad f(x) = e^{-z} g(z)$$

leads to an integral equation with difference kernel of the form 2.9.51:

$$w(z) - \lambda \int_{-\infty}^{z} \frac{w(\tau)\, d\tau}{\cosh(z - \tau)} = g(z).$$

42. $\quad y(x) - \lambda \displaystyle\int_{a}^{x} \frac{x + b}{t + b} y(t)\, dt = f(x).$

This is a special case of equation 2.9.1 with $g(x) = x + b$.

Solution:

$$y(x) = f(x) + \lambda \int_{a}^{x} \frac{x + b}{t + b} e^{\lambda(x - t)} f(t)\, dt.$$

43. $\quad y(x) = \dfrac{2}{(1 - \lambda^2) x^2} \displaystyle\int_{\lambda x}^{x} \frac{t}{1 + t} y(t)\, dt.$

This equation is encountered in nuclear physics and describes deceleration of neutrons in matter.

1°. Solution with $\lambda = 0$:

$$y(x) = \frac{C}{(1 + x)^2},$$

where C is an arbitrary constant.

2°. For $\lambda \neq 0$, the solution can be found in the series form

$$y(x) = \sum_{n=0}^{\infty} A_n x^n.$$

$\odot$ Reference: I. Sneddon (1951).

2.1-6. Kernels Containing Square Roots and Fractional Powers

44. $\quad y(x) + A \int_a^x (x - t)\sqrt{t}\, y(t)\, dt = f(x).$

This is a special case of equation 2.1.49 with $\lambda = \frac{1}{2}$.

45. $\quad y(x) + A \int_a^x \left(\sqrt{x} - \sqrt{t}\,\right) y(t)\, dt = f(x).$

This is a special case of equation 2.1.52 with $\lambda = \frac{1}{2}$.

46. $\quad y(x) + \lambda \int_a^x \dfrac{y(t)\, dt}{\sqrt{x - t}} = f(x).$

Abel's equation of the second kind. This equation is encountered in problems of heat and mass transfer.

Solution:

$$y(x) = F(x) + \pi\lambda^2 \int_a^x \exp[\pi\lambda^2(x - t)] F(t)\, dt,$$

where

$$F(x) = f(x) - \lambda \int_a^x \frac{f(t)\, dt}{\sqrt{x - t}}.$$

⊙ References: H. Brakhage, K. Nickel, and P. Rieder (1965), Yu. I. Babenko (1986).

47. $\quad y(x) - \lambda \int_0^x \dfrac{y(t)\, dt}{\sqrt{ax^2 + bt^2}} = f(x), \qquad a > 0, \quad b > 0.$

$1°.$ The solution of the homogeneous equation ($f \equiv 0$) is

$$y(x) = Cx^\beta \qquad (\beta > -1, \ \lambda > 0). \tag{1}$$

Here C is an arbitrary constant, and $\beta = \beta(\lambda)$ is determined by the transcendental equation

$$\lambda I(\beta) = 1, \qquad \text{where} \quad I(\beta) = \int_0^1 \frac{z^\beta\, dz}{\sqrt{a + bz^2}}. \tag{2}$$

$2°.$ For a polynomial right-hand side,

$$f(x) = \sum_{n=0}^N A_n x^n$$

the solution bounded at zero is given by

$$y(x) = \begin{cases} \displaystyle\sum_{n=0}^N \frac{A_n}{1 - (\lambda/\lambda_n)} x^n & \text{for } \lambda < \lambda_0, \\[3mm] \displaystyle\sum_{n=0}^N \frac{A_n}{1 - (\lambda/\lambda_n)} x^n + Cx^\beta & \text{for } \lambda > \lambda_0 \text{ and } \lambda \neq \lambda_n, \end{cases}$$

$$\lambda_0 = \frac{\sqrt{b}}{\operatorname{Arsinh}\left(\sqrt{b/a}\,\right)}, \qquad \lambda_n = \frac{1}{I(n)}, \qquad I(n) = \int_0^1 \frac{z^n\, dz}{\sqrt{a + bz^2}}.$$

Here C is an arbitrary constant, and $\beta = \beta(\lambda)$ is determined by the transcendental equation (2).

$3°$. For special $\lambda = \lambda_n$, $(n = 1, 2, \ldots)$, the solution differs in one term and has the form

$$y(x) = \sum_{m=0}^{n-1} \frac{A_m}{1 - (\lambda_n/\lambda_m)} x^m + \sum_{m=n+1}^{N} \frac{A_m}{1 - (\lambda_n/\lambda_m)} x^m - A_n \frac{\bar{\lambda}_n}{\lambda_n} x^n \ln x + Cx^n,$$

where $\bar{\lambda}_n = \left[\int_0^1 \frac{z^n \ln z \, dz}{\sqrt{a + bz^2}} \right]^{-1}$.

$4°$. For arbitrary $f(x)$, expandable into power series, the formulas of item $2°$ can be used, in which one should set $N = \infty$. In this case, the radius of convergence of the solution $y(x)$ is equal to the radius of convergence of $f(x)$.

48. $\quad y(x) + \lambda \int_a^x \frac{y(t) \, dt}{(x - t)^{3/4}} = f(x).$

This equation admits solution by quadratures (see equation 2.1.60 and Section 9.4-2).

2.1-7. Kernels Containing Arbitrary Powers

49. $\quad y(x) + A \int_a^x (x - t)t^\lambda y(t) \, dt = f(x).$

This is a special case of equation 2.9.4 with $g(t) = At^\lambda$.
  Solution:

$$y(x) = f(x) + \frac{A}{W} \int_a^x \left[y_1(x)y_2(t) - y_2(x)y_1(t) \right] t^\lambda f(t) \, dt,$$

where $y_1(x), y_2(x)$ is a fundamental system of solutions of the second-order linear homogeneous ordinary differential equation $y''_{xx} + Ax^\lambda y = 0$; the functions $y_1(x)$ and $y_2(x)$ are expressed in terms of Bessel functions or modified Bessel functions, depending on the sign of A:
  For $A > 0$,

$$W = \frac{2q}{\pi}, \quad y_1(x) = \sqrt{x}\, J_{\frac{1}{2q}}\left(\frac{\sqrt{A}}{q} x^q \right), \quad y_2(x) = \sqrt{x}\, Y_{\frac{1}{2q}}\left(\frac{\sqrt{A}}{q} x^q \right), \quad q = \frac{\lambda + 2}{2},$$

For $A < 0$,

$$W = -q, \quad y_1(x) = \sqrt{x}\, I_{\frac{1}{2q}}\left(\frac{\sqrt{|A|}}{q} x^q \right), \quad y_2(x) = \sqrt{x}\, K_{\frac{1}{2q}}\left(\frac{\sqrt{|A|}}{q} x^q \right), \quad q = \frac{\lambda + 2}{2}.$$

50. $\quad y(x) + A \int_a^x x^\lambda t^\mu y(t) \, dt = f(x).$

This is a special case of equation 2.9.2 with $g(x) = -Ax^\lambda$ and $h(t) = t^\mu$ (λ and μ are arbitrary numbers).
  Solution:

$$y(x) = f(x) - \int_a^x R(x, t)f(t) \, dt,$$

$$R(x, t) = \begin{cases} Ax^\lambda t^\mu \exp\left[\dfrac{A}{\lambda + \mu + 1}\left(t^{\lambda + \mu + 1} - x^{\lambda + \mu + 1} \right) \right] & \text{for } \lambda + \mu + 1 \neq 0, \\ Ax^{\lambda - A} t^{\mu + A} & \text{for } \lambda + \mu + 1 = 0. \end{cases}$$

51. $y(x) + A \displaystyle\int_a^x (x - t)x^\lambda t^\mu y(t)\, dt = f(x).$

The substitution $u(x) = x^{-\lambda} y(x)$ leads to an equation of the form 2.1.49:

$$u(x) + A \int_a^x (x - t)t^{\lambda+\mu} u(t)\, dt = f(x)x^{-\lambda}.$$

52. $y(x) + A \displaystyle\int_a^x (x^\lambda - t^\lambda)y(t)\, dt = f(x).$

This is a special case of equation 2.9.5 with $g(x) = Ax^\lambda$.

    Solution:

$$y(x) = f(x) + \frac{1}{W} \int_a^x \left[u_1'(x)u_2'(t) - u_2'(x)u_1'(t) \right] f(t)\, dt,$$

where the primes denote differentiation with respect to the argument specified in the parentheses, and $u_1(x)$, $u_2(x)$ is a fundamental system of solutions of the second-order linear homogeneous ordinary differential equation $u_{xx}'' + A\lambda x^{\lambda-1} u = 0$; the functions $u_1(x)$ and $u_2(x)$ are expressed in terms of Bessel functions or modified Bessel functions, depending on the sign of A:

    For $A\lambda > 0$,

$$W = \frac{2q}{\pi}, \quad u_1(x) = \sqrt{x}\, J_{\frac{1}{2q}}\left(\frac{\sqrt{A\lambda}}{q} x^q \right), \quad u_2(x) = \sqrt{x}\, Y_{\frac{1}{2q}}\left(\frac{\sqrt{A\lambda}}{q} x^q \right), \quad q = \frac{\lambda + 1}{2},$$

    For $A\lambda < 0$,

$$W = -q, \quad u_1(x) = \sqrt{x}\, I_{\frac{1}{2q}}\left(\frac{\sqrt{|A\lambda|}}{q} x^q \right), \quad u_2(x) = \sqrt{x}\, K_{\frac{1}{2q}}\left(\frac{\sqrt{|A\lambda|}}{q} x^q \right), \quad q = \frac{\lambda + 1}{2}.$$

53. $y(x) - \displaystyle\int_a^x \left(Ax^\lambda t^{\lambda-1} + Bt^{2\lambda-1} \right) y(t)\, dt = f(x).$

The transformation

$$z = x^\lambda, \quad \tau = t^\lambda, \quad y(x) = Y(z)$$

leads to an equation of the form 2.1.6:

$$Y(z) - \int_b^z \left(\frac{A}{\lambda} z + \frac{B}{\lambda} \tau \right) Y(\tau)\, d\tau = F(z), \qquad F(z) = f(x),\ b = a^\lambda.$$

54. $y(x) - \displaystyle\int_a^x \left(Ax^{\lambda+\mu} t^{\lambda-\mu-1} + Bx^\mu t^{2\lambda-\mu-1} \right) y(t)\, dt = f(x).$

The substitution $y(x) = x^\mu w(x)$ leads to an equation of the form 2.1.53:

$$w(x) - \int_a^x \left(Ax^\lambda t^{\lambda-1} + Bt^{2\lambda-1} \right) w(t)\, dt = x^{-\mu} f(x).$$

55. $y(x) + A \displaystyle\int_a^x \left[\lambda x^{\lambda-1} t^\mu - (\lambda + \mu)x^{\lambda+\mu-1} \right] y(t)\, dt = f(x).$

This equation can be obtained by differentiating equation 1.1.51:

$$\int_a^x \left[1 + A(x^\lambda t^\mu - x^{\lambda+\mu}) \right] y(t)\, dt = F(x), \qquad F(x) = \int_a^x f(x)\, dx.$$

    Solution:

$$y(x) = \frac{d}{dx}\left\{ \frac{x^\lambda}{\Phi(x)} \int_a^x \left[t^{-\lambda} F(t) \right]_t' \Phi(t)\, dt \right\}, \qquad \Phi(x) = \exp\left(-\frac{A\mu}{\mu + \lambda} x^{\mu+\lambda} \right).$$

56. $y(x) + \int_a^x \left(ABx^{\lambda+1} - ABx^\lambda t - Ax^\lambda - B \right) y(t)\, dt = f(x).$

This is a special case of equation 2.9.7.

Solution:

$$y(x) = f(x) + \int_a^x R(x-t)f(t)\, dt,$$

$$R(x, t) = (Ax^\lambda + B) \exp\left[\frac{A}{\lambda+1} \left(x^{\lambda+1} - t^{\lambda+1} \right) \right] + B^2 \int_t^x \exp\left[\frac{A}{\lambda+1} \left(s^{\lambda+1} - t^{\lambda+1} \right) + B(x-s) \right] ds.$$

57. $y(x) + \int_a^x \left(ABxt^\lambda - ABt^{\lambda+1} + At^\lambda + B \right) y(t)\, dt = f(x).$

This is a special case of equation 2.9.8.

Solution:

$$y(x) = f(x) + \int_a^x R(x-t)f(t)\, dt,$$

$$R(x, t) = -(At^\lambda + B) \exp\left[\frac{A}{\lambda+1} \left(t^{\lambda+1} - x^{\lambda+1} \right) \right] + B^2 \int_t^x \exp\left[\frac{A}{\lambda+1} \left(s^{\lambda+1} - x^{\lambda+1} \right) + B(t-s) \right] ds.$$

58. $y(x) - \lambda \int_a^x \left(\dfrac{x+b}{t+b} \right)^\mu y(t)\, dt = f(x).$

This is a special case of equation 2.9.1 with $g(x) = (x+b)^\mu$.

Solution:

$$y(x) = f(x) + \lambda \int_a^x \left(\frac{x+b}{t+b} \right)^\mu e^{\lambda(x-t)} f(t)\, dt.$$

59. $y(x) - \lambda \int_a^x \dfrac{x^\mu + b}{t^\mu + b}\, y(t)\, dt = f(x).$

This is a special case of equation 2.9.1 with $g(x) = x^\mu + b$.

Solution:

$$y(x) = f(x) + \lambda \int_a^x \frac{x^\mu + b}{t^\mu + b} e^{\lambda(x-t)} f(t)\, dt.$$

60. $y(x) - \lambda \int_0^x \dfrac{y(t)\, dt}{(x-t)^\alpha} = f(x), \qquad 0 < \alpha < 1.$

Generalized Abel equation of the second kind.

$1°$. Assume that the number α can be represented in the form

$$\alpha = 1 - \frac{m}{n}, \qquad \text{where} \quad m = 1, 2, \dots, \quad n = 2, 3, \dots \quad (m < n).$$

In this case, the solution of the generalized Abel equation of the second kind can be written in closed form (in quadratures):

$$y(x) = f(x) + \int_0^x R(x-t)f(t)\, dt,$$

where

$$R(x) = \sum_{\nu=1}^{n-1} \frac{\lambda^\nu \Gamma^\nu(m/n)}{\Gamma(\nu m/n)} x^{(\nu m/n)-1} + \frac{b}{m} \sum_{\mu=0}^{m-1} \varepsilon_\mu \exp\left(\varepsilon_\mu bx\right)$$

$$+ \frac{b}{m} \sum_{\nu=1}^{n-1} \frac{\lambda^\nu \Gamma^\nu(m/n)}{\Gamma(\nu m/n)} \left[\sum_{\mu=0}^{m-1} \varepsilon_\mu \exp\left(\varepsilon_\mu bx\right) \int_0^x t^{(\nu m/n)-1} \exp\left(-\varepsilon_\mu bt\right) dt \right],$$

$$b = \lambda^{n/m} \Gamma^{n/m}(m/n), \quad \varepsilon_\mu = \exp\left(\frac{2\pi\mu i}{m}\right), \quad i^2 = -1, \quad \mu = 0, 1, \dots, m-1.$$

$2°$. Solution with any α from $0 < \alpha < 1$:

$$y(x) = f(x) + \int_0^x R(x - t) f(t)\, dt, \qquad \text{where} \quad R(x) = \sum_{n=1}^\infty \frac{\left[\lambda\Gamma(1-\alpha)x^{1-\alpha}\right]^n}{x\Gamma\left[n(1-\alpha)\right]}.$$

⊙ References: H. Brakhage, K. Nickel, and P. Rieder (1965), V. I. Smirnov (1974).

61. $y(x) - \dfrac{\lambda}{x^\alpha} \displaystyle\int_0^x \frac{y(t)\, dt}{(x-t)^{1-\alpha}} = f(x), \qquad 0 < \alpha \le 1.$

$1°$. The solution of the homogeneous equation ($f \equiv 0$) is

$$y(x) = Cx^\beta \qquad (\beta > -1,\ \lambda > 0). \tag{1}$$

Here C is an arbitrary constant, and $\beta = \beta(\lambda)$ is determined by the transcendental equation

$$\lambda B(\alpha, \beta + 1) = 1, \tag{2}$$

where $B(p, q) = \int_0^1 z^{p-1}(1-z)^{q-1}\, dz$ is the beta function.

$2°$. For a polynomial right-hand side,

$$f(x) = \sum_{n=0}^N A_n x^n$$

the solution bounded at zero is given by

$$y(x) = \begin{cases} \displaystyle\sum_{n=0}^N \frac{A_n}{1 - (\lambda/\lambda_n)} x^n & \text{for } \lambda < \alpha, \\[4mm] \displaystyle\sum_{n=0}^N \frac{A_n}{1 - (\lambda/\lambda_n)} x^n + Cx^\beta & \text{for } \lambda > \alpha \text{ and } \lambda \ne \lambda_n, \end{cases}$$

$$\lambda_n = \frac{(\alpha)_{n+1}}{n!}, \qquad (\alpha)_{n+1} = \alpha(\alpha+1)\dots(\alpha+n).$$

Here C is an arbitrary constant, and $\beta = \beta(\lambda)$ is determined by the transcendental equation (2).
For special $\lambda = \lambda_n$ ($n = 1, 2, \dots$), the solution differs in one term and has the form

$$y(x) = \sum_{m=0}^{n-1} \frac{A_m}{1 - (\lambda_n/\lambda_m)} x^m + \sum_{m=n+1}^N \frac{A_m}{1 - (\lambda_n/\lambda_m)} x^m - A_n \frac{\bar{\lambda}_n}{\lambda_n} x^n \ln x + Cx^n,$$

where $\bar{\lambda}_n = \left[\displaystyle\int_0^1 (1-z)^{\alpha-1} z^n \ln z\, dz \right]^{-1}.$

$3°$. For arbitrary $f(x)$, expandable into power series, the formulas of item $2°$ can be used, in which one should set $N = \infty$. In this case, the radius of convergence of the solution $y(x)$ is equal to the radius of convergence of $f(x)$.

$4°$. For

$$f(x) = \ln(kx) \sum_{n=0}^{N} A_n x^n,$$

a solution has the form

$$y(x) = \ln(kx) \sum_{n=0}^{N} B_n x^n + \sum_{n=0}^{N} D_n x^n,$$

where the constants B_n and D_n are found by the method of undetermined coefficients. To obtain the general solution we must add the solution (1) of the homogeneous equation.

In Mikhailov (1966), solvability conditions for the integral equation in question were investigated for various classes of $f(x)$.

62. $y(x) - \dfrac{\lambda}{x^\mu} \displaystyle\int_0^x \dfrac{y(t)\,dt}{(ax + bt)^{1-\mu}} = f(x).$

Here $a > 0$, $b > 0$, and μ is an arbitrary number.

$1°$. The solution of the homogeneous equation ($f \equiv 0$) is

$$y(x) = Cx^\beta \qquad (\beta > -1, \ \lambda > 0). \tag{1}$$

Here C is an arbitrary constant, and $\beta = \beta(\lambda)$ is determined by the transcendental equation

$$\lambda I(\beta) = 1, \qquad \text{where} \quad I(\beta) = \int_0^1 z^\beta (a + bz)^{\mu-1} \, dz. \tag{2}$$

$2°$. For a polynomial right-hand side,

$$f(x) = \sum_{n=0}^{N} A_n x^n$$

the solution bounded at zero is given by

$$y(x) = \begin{cases} \displaystyle\sum_{n=0}^{N} \dfrac{A_n}{1 - (\lambda/\lambda_n)} x^n & \text{for } \lambda < \lambda_0, \\[4mm] \displaystyle\sum_{n=0}^{N} \dfrac{A_n}{1 - (\lambda/\lambda_n)} x^n + Cx^\beta & \text{for } \lambda > \lambda_0 \text{ and } \lambda \neq \lambda_n, \end{cases}$$

$$\lambda_n = \frac{1}{I(n)}, \qquad I(n) = \int_0^1 z^n (a + bz)^{\mu-1} \, dz.$$

Here C is an arbitrary constant, and $\beta = \beta(\lambda)$ is determined by the transcendental equation (2).

$3°$. For special $\lambda = \lambda_n$ ($n = 1, 2, \dots$), the solution differs in one term and has the form

$$y(x) = \sum_{m=0}^{n-1} \frac{A_m}{1 - (\lambda_n/\lambda_m)} x^m + \sum_{m=n+1}^{N} \frac{A_m}{1 - (\lambda_n/\lambda_m)} x^m - A_n \frac{\bar{\lambda}_n}{\lambda_n} x^n \ln x + Cx^n,$$

where $\bar{\lambda}_n = \left[\displaystyle\int_0^1 z^n (a + bz)^{\mu-1} \ln z \, dz \right]^{-1}.$

$4°$. For arbitrary $f(x)$ expandable into power series, the formulas of item $2°$ can be used, in which one should set $N = \infty$. In this case, the radius of convergence of the solution $y(x)$ is equal to the radius of convergence of $f(x)$.

2.2. Equations Whose Kernels Contain Exponential Functions

1. $\quad y(x) + A\displaystyle\int_a^x e^{\lambda(x-t)}y(t)\,dt = f(x).$

Solution:

$$y(x) = f(x) - A\int_a^x e^{(\lambda-A)(x-t)}f(t)\,dt.$$

2. $\quad y(x) + A\displaystyle\int_a^x e^{\lambda x+\beta t}y(t)\,dt = f(x).$

For $\beta = -\lambda$, see equation 2.2.1. This is a special case of equation 2.9.2 with $g(x) = -Ae^{\lambda x}$ and $h(t) = e^{\beta t}$.

Solution:

$$y(x) = f(x) - \int_a^x R(x,t)f(t)\,dt, \quad R(x,t) = Ae^{\lambda x+\beta t}\exp\left\{\frac{A}{\lambda+\beta}\left[e^{(\lambda+\beta)t} - e^{(\lambda+\beta)x}\right]\right\}.$$

3. $\quad y(x) + A\displaystyle\int_a^x \left[e^{\lambda(x-t)} - 1\right]y(t)\,dt = f(x).$

$1°$. Solution with $D \equiv \lambda(\lambda - 4A) > 0$:

$$y(x) = f(x) - \frac{2A\lambda}{\sqrt{D}}\int_a^x R(x-t)f(t)\,dt, \qquad R(x) = \exp\left(\tfrac{1}{2}\lambda x\right)\sinh\left(\tfrac{1}{2}\sqrt{D}\,x\right).$$

$2°$. Solution with $D \equiv \lambda(\lambda - 4A) < 0$:

$$y(x) = f(x) - \frac{2A\lambda}{\sqrt{|D|}}\int_a^x R(x-t)f(t)\,dt, \qquad R(x) = \exp\left(\tfrac{1}{2}\lambda x\right)\sin\left(\tfrac{1}{2}\sqrt{|D|}\,x\right).$$

$3°$. Solution with $\lambda = 4A$:

$$y(x) = f(x) - 4A^2\int_a^x (x-t)\exp\left[2A(x-t)\right]f(t)\,dt.$$

4. $\quad y(x) + \displaystyle\int_a^x \left[Ae^{\lambda(x-t)} + B\right]y(t)\,dt = f(x).$

This is a special case of equation 2.2.10 with $A_1 = A$, $A_2 = B$, $\lambda_1 = \lambda$, and $\lambda_2 = 0$.

$1°$. The structure of the solution depends on the sign of the discriminant

$$D \equiv (A - B - \lambda)^2 + 4AB \tag{1}$$

of the square equation

$$\mu^2 + (A + B - \lambda)\mu - B\lambda = 0. \tag{2}$$

$2°$. If $D > 0$, then equation (2) has the real different roots

$$\mu_1 = \tfrac{1}{2}(\lambda - A - B) + \tfrac{1}{2}\sqrt{D}, \qquad \mu_2 = \tfrac{1}{2}(\lambda - A - B) - \tfrac{1}{2}\sqrt{D}.$$

In this case, the original integral equation has the solution

$$y(x) = f(x) + \int_a^x \left[E_1 e^{\mu_1(x-t)} + E_2 e^{\mu_2(x-t)} \right] f(t)\, dt,$$

where

$$E_1 = A\frac{\mu_1}{\mu_2 - \mu_1} + B\frac{\mu_1 - \lambda}{\mu_2 - \mu_1}, \qquad E_2 = A\frac{\mu_2}{\mu_1 - \mu_2} + B\frac{\mu_2 - \lambda}{\mu_1 - \mu_2}.$$

3°. If $D < 0$, then equation (2) has the complex conjugate roots

$$\mu_1 = \sigma + i\beta, \quad \mu_2 = \sigma - i\beta, \qquad \sigma = \tfrac{1}{2}(\lambda - A - B), \quad \beta = \tfrac{1}{2}\sqrt{-D}.$$

In this case, the original integral equation has the solution

$$y(x) = f(x) + \int_a^x \left\{ E_1 e^{\sigma(x-t)} \cos[\beta(x-t)] + E_2 e^{\sigma(x-t)} \sin[\beta(x-t)] \right\} f(t)\, dt,$$

where

$$E_1 = -A - B, \qquad E_2 = \frac{1}{\beta}(-A\sigma - B\sigma + B\lambda).$$

5. $\quad y(x) + A \displaystyle\int_a^x (e^{\lambda x} - e^{\lambda t}) y(t)\, dt = f(x).$

This is a special case of equation 2.9.5 with $g(x) = Ae^{\lambda x}$.

Solution:

$$y(x) = f(x) + \frac{1}{W} \int_a^x \left[u_1'(x) u_2'(t) - u_2'(x) u_1'(t) \right] f(t)\, dt,$$

where the primes denote differentiation with respect to the argument specified in the parentheses, and $u_1(x), u_2(x)$ is a fundamental system of solutions of the second-order linear homogeneous ordinary differential equation $u_{xx}'' + A\lambda e^{\lambda x} u = 0$; the functions $u_1(x)$ and $u_2(x)$ are expressed in terms of Bessel functions or modified Bessel functions, depending on the sign of A:

For $A\lambda > 0$,

$$W = \frac{\lambda}{\pi}, \quad u_1(x) = J_0\left(\frac{2\sqrt{A\lambda}}{\lambda} e^{\lambda x/2} \right), \quad u_2(x) = Y_0\left(\frac{2\sqrt{A\lambda}}{\lambda} e^{\lambda x/2} \right),$$

For $A\lambda < 0$,

$$W = -\frac{\lambda}{2}, \quad u_1(x) = I_0\left(\frac{2\sqrt{|A\lambda|}}{\lambda} e^{\lambda x/2} \right), \quad u_2(x) = K_0\left(\frac{2\sqrt{|A\lambda|}}{\lambda} e^{\lambda x/2} \right).$$

6. $\quad y(x) + \displaystyle\int_a^x \left(Ae^{\lambda x} + Be^{\lambda t} \right) y(t)\, dt = f(x).$

For $B = -A$, see equation 2.2.5. This is a special case of equation 2.9.6 with $g(x) = Ae^{\lambda x}$ and $h(t) = Be^{\lambda t}$.

Differentiating the original integral equation followed by substituting $Y(x) = \displaystyle\int_a^x y(t)\, dt$ yields the second-order linear ordinary differential equation

$$Y_{xx}'' + (A + B)e^{\lambda x} Y_x' + A\lambda e^{\lambda x} Y = f_x'(x) \tag{1}$$

under the initial conditions

$$Y(a) = 0, \quad Y'_x(a) = f(a). \tag{2}$$

A fundamental system of solutions of the homogeneous equation (1) with $f \equiv 0$ has the form

$$Y_1(x) = \Phi\left(\frac{A}{m}, 1; -\frac{m}{\lambda}e^{\lambda x}\right), \quad Y_2(x) = \Psi\left(\frac{A}{m}, 1; -\frac{m}{\lambda}e^{\lambda x}\right), \quad m = A + B,$$

where $\Phi(\alpha, \beta; x)$ and $\Psi(\alpha, \beta; x)$ are degenerate hypergeometric functions.

Solving the homogeneous equation (1) under conditions (2) for an arbitrary function $f = f(x)$ and taking into account the relation $y(x) = Y'_x(x)$, we thus obtain the solution of the integral equation in the form

$$y(x) = f(x) - \int_a^x R(x,t)f(t)\, dt,$$

$$R(x,t) = \frac{\Gamma(A/m)}{\lambda}\frac{\partial^2}{\partial x \partial t}\left\{\exp\left(\frac{m}{\lambda}e^{\lambda t}\right)\left[Y_1(x)Y_2(t) - Y_2(x)Y_1(t)\right]\right\}.$$

7. $\quad y(x) + A\displaystyle\int_a^x \left[e^{\lambda(x+t)} - e^{2\lambda t}\right]y(t)\, dt = f(x).$

The transformation $z = e^{\lambda x}$, $\tau = e^{\lambda t}$ leads to an equation of the form 2.1.4.

$1°$. Solution with $A\lambda > 0$:

$$y(x) = f(x) - \lambda k\int_a^x e^{\lambda t}\sin\left[k(e^{\lambda x} - e^{\lambda t})\right]f(t)\, dt, \qquad k = \sqrt{A/\lambda}.$$

$2°$. Solution with $A\lambda < 0$:

$$y(x) = f(x) + \lambda k\int_a^x e^{\lambda t}\sinh\left[k(e^{\lambda x} - e^{\lambda t})\right]f(t)\, dt, \qquad k = \sqrt{|A/\lambda|}.$$

8. $\quad y(x) + A\displaystyle\int_a^x \left[e^{\lambda x+\mu t} - e^{(\lambda+\mu)t}\right]y(t)\, dt = f(x).$

The transformation $z = e^{\mu x}$, $\tau = e^{\mu t}$, $Y(z) = y(x)$ leads to an equation of the form 2.1.52:

$$Y(z) + \frac{A}{\mu}\int_b^z (z^k - \tau^k)Y(\tau)\, d\tau = F(z), \qquad F(z) = f(x),$$

where $k = \lambda/\mu$, $b = e^{\mu a}$.

9. $\quad y(x) + A\displaystyle\int_a^x \left[\lambda e^{\lambda x+\mu t} - (\lambda + \mu)e^{(\lambda+\mu)x}\right]y(t)\, dt = f(x).$

This equation can be obtained by differentiating an equation of the form 1.2.22:

$$\int_a^x \left[1 + Ae^{\lambda x}(e^{\mu t} - e^{\mu x})\right]y(t)\, dt = F(x), \qquad F(x) = \int_a^x f(t)\, dt.$$

Solution:

$$y(x) = \frac{d}{dx}\left\{e^{\lambda x}\Phi(x)\int_a^x \left[\frac{F(t)}{e^{\lambda t}}\right]'_t\frac{dt}{\Phi(t)}\right\}, \qquad \Phi(x) = \exp\left[\frac{A\mu}{\lambda+\mu}e^{(\lambda+\mu)x}\right].$$

10. $\quad y(x) + \displaystyle\int_a^x \left[A_1 e^{\lambda_1(x-t)} + A_2 e^{\lambda_2(x-t)} \right] y(t)\, dt = f(x).$

1°. Introduce the notation

$$I_1 = \int_a^x e^{\lambda_1(x-t)} y(t)\, dt, \qquad I_2 = \int_a^x e^{\lambda_2(x-t)} y(t)\, dt.$$

Differentiating the integral equation twice yields (the first line is the original equation)

$$y + A_1 I_1 + A_2 I_2 = f, \qquad f = f(x), \tag{1}$$
$$y'_x + (A_1 + A_2)y + A_1\lambda_1 I_1 + A_2\lambda_2 I_2 = f'_x, \tag{2}$$
$$y''_{xx} + (A_1 + A_2)y'_x + (A_1\lambda_1 + A_2\lambda_2)y + A_1\lambda_1^2 I_1 + A_2\lambda_2^2 I_2 = f''_{xx}. \tag{3}$$

Eliminating I_1 and I_2, we arrive at the second-order linear ordinary differential equation with constant coefficients

$$y''_{xx} + (A_1 + A_2 - \lambda_1 - \lambda_2)y'_x + (\lambda_1\lambda_2 - A_1\lambda_2 - A_2\lambda_1)y = f''_{xx} - (\lambda_1 + \lambda_2)f'_x + \lambda_1\lambda_2 f. \tag{4}$$

Substituting $x = a$ into (1) and (2) yields the initial conditions

$$y(a) = f(a), \qquad y'_x(a) = f'_x(a) - (A_1 + A_2)f(a). \tag{5}$$

Solving the differential equation (4) under conditions (5), we can find the solution of the integral equation.

2°. Consider the characteristic equation

$$\mu^2 + (A_1 + A_2 - \lambda_1 - \lambda_2)\mu + \lambda_1\lambda_2 - A_1\lambda_2 - A_2\lambda_1 = 0 \tag{6}$$

which corresponds to the homogeneous differential equation (4) (with $f(x) \equiv 0$). The structure of the solution of the integral equation depends on the sign of the discriminant

$$D \equiv (A_1 - A_2 - \lambda_1 + \lambda_2)^2 + 4A_1 A_2$$

of the quadratic equation (6).

If $D > 0$, the quadratic equation (6) has the real different roots

$$\mu_1 = \tfrac{1}{2}(\lambda_1 + \lambda_2 - A_1 - A_2) + \tfrac{1}{2}\sqrt{D}, \quad \mu_2 = \tfrac{1}{2}(\lambda_1 + \lambda_2 - A_1 - A_2) - \tfrac{1}{2}\sqrt{D}.$$

In this case, the solution of the original integral equation has the form

$$y(x) = f(x) + \int_a^x \left[B_1 e^{\mu_1(x-t)} + B_2 e^{\mu_2(x-t)} \right] f(t)\, dt,$$

where

$$B_1 = A_1 \frac{\mu_1 - \lambda_2}{\mu_2 - \mu_1} + A_2 \frac{\mu_1 - \lambda_1}{\mu_2 - \mu_1}, \qquad B_2 = A_1 \frac{\mu_2 - \lambda_2}{\mu_1 - \mu_2} + A_2 \frac{\mu_2 - \lambda_1}{\mu_1 - \mu_2}.$$

If $D < 0$, the quadratic equation (6) has the complex conjugate roots

$$\mu_1 = \sigma + i\beta, \quad \mu_2 = \sigma - i\beta, \qquad \sigma = \tfrac{1}{2}(\lambda_1 + \lambda_2 - A_1 - A_2), \quad \beta = \tfrac{1}{2}\sqrt{-D}.$$

In this case, the solution of the original integral equation has the form

$$y(x) = f(x) + \int_a^x \left\{ B_1 e^{\sigma(x-t)} \cos[\beta(x-t)] + B_2 e^{\sigma(x-t)} \sin[\beta(x-t)] \right\} f(t)\, dt.$$

where

$$B_1 = -A_1 - A_2, \qquad B_2 = \frac{1}{\beta}\left[A_1(\lambda_2 - \sigma) + A_2(\lambda_1 - \sigma) \right].$$

11. $y(x) + \int_a^x \left[Ae^{\lambda(x+t)} - Ae^{2\lambda t} + Be^{\lambda t} \right] y(t)\, dt = f(x).$

The transformation $z = e^{\lambda x}$, $\tau = e^{\lambda t}$, $Y(z) = y(x)$ leads to an equation of the form 2.1.5:

$$Y(z) + \int_b^z \left[B_1(z - \tau) + A_1 \right] Y(\tau)\, d\tau = F(z), \qquad F(z) = f(x),$$

where $A_1 = B/\lambda$, $B_1 = A/\lambda$, $b = e^{\lambda a}$.

12. $y(x) + \int_a^x \left[Ae^{\lambda(x+t)} + Be^{2\lambda t} + Ce^{\lambda t} \right] y(t)\, dt = f(x).$

The transformation $z = e^{\lambda x}$, $\tau = e^{\lambda t}$, $Y(z) = y(x)$ leads to an equation of the form 2.1.6:

$$Y(z) - \int_b^z (A_1 z + B_1 \tau + C_1) Y(\tau)\, d\tau = F(z), \qquad F(z) = f(x),$$

where $A_1 = -A/\lambda$, $B_1 = -B/\lambda$, $C_1 = -C/\lambda$, $b = e^{\lambda a}$.

13. $y(x) + \int_a^x \left[\lambda e^{\lambda(x-t)} + A\left(\mu e^{\mu x + \lambda t} - \lambda e^{\lambda x + \mu t} \right) \right] y(t)\, dt = f(x).$

This is a special case of equation 2.9.23 with $h(t) = A$.
 Solution:

$$y(x) = \frac{1}{e^{\lambda x}} \frac{d}{dx} \left\{ \Phi(x) \int_a^x \left[\frac{F(t)}{e^{\lambda t}} \right]'_t \frac{e^{2\lambda t}}{\Phi(t)}\, dt \right\},$$

$$\Phi(x) = \exp\left[A \frac{\lambda - \mu}{\lambda + \mu} e^{(\lambda + \mu)x} \right], \qquad F(x) = \int_a^x f(t)\, dt.$$

14. $y(x) - \int_a^x \left[\lambda e^{-\lambda(x-t)} + A\left(\mu e^{\lambda x + \mu t} - \lambda e^{\mu x + \lambda t} \right) \right] y(t)\, dt = f(x).$

This is a special case of equation 2.9.24 with $h(x) = A$.
 Assume that $f(a) = 0$. Solution:

$$y(x) = \int_a^x w(t)\, dt, \qquad w(x) = e^{-\lambda x} \frac{d}{dx} \left\{ \frac{e^{2\lambda x}}{\Phi(x)} \int_a^x \left[\frac{f(t)}{e^{\lambda t}} \right]'_t \Phi(t)\, dt \right\},$$

$$\Phi(x) = \exp\left[A \frac{\lambda - \mu}{\lambda + \mu} e^{(\lambda + \mu)x} \right].$$

15. $y(x) + \int_a^x \left[\lambda e^{\lambda(x-t)} + Ae^{\beta t}\left(\mu e^{\mu x + \lambda t} - \lambda e^{\lambda x + \mu t} \right) \right] y(t)\, dt = f(x).$

This is a special case of equation 2.9.23 with $h(t) = Ae^{\beta t}$.
 Solution:

$$y(x) = e^{-(\lambda + \beta)x} \frac{d}{dx} \left\{ \Phi(x) \int_a^x \left[\frac{F(t)}{e^{\lambda t}} \right]'_t \frac{e^{(2\lambda + \beta)t}}{\Phi(t)}\, dt \right\},$$

$$\Phi(x) = \exp\left[A \frac{\lambda - \mu}{\lambda + \mu + \beta} e^{(\lambda + \mu + \beta)x} \right], \qquad F(x) = \int_a^x f(t)\, dt.$$

16. $y(x) - \int_a^x \left[\lambda e^{-\lambda(x-t)} + A e^{\beta x} \left(\mu e^{\lambda x + \mu t} - \lambda e^{\mu x + \lambda t} \right) \right] y(t)\, dt = f(x).$

This is a special case of equation 2.9.24 with $h(x) = A e^{\beta x}$.
Assume that $f(a) = 0$. Solution:

$$y(x) = \int_a^x w(t)\, dt, \qquad w(x) = e^{-\lambda x} \frac{d}{dx} \left\{ \frac{e^{(2\lambda+\beta)x}}{\Phi(x)} \int_a^x \left[\frac{f(t)}{e^{(\lambda+\beta)t}} \right]' \Phi(t)\, dt \right\},$$

$$\Phi(x) = \exp\left[A \frac{\lambda - \mu}{\lambda + \mu + \beta} e^{(\lambda+\mu+\beta)x} \right].$$

17. $y(x) + \int_a^x \left[A B e^{(\lambda+1)x+t} - A B e^{\lambda x + 2t} - A e^{\lambda x + t} - B e^t \right] y(t)\, dt = f(x).$

The transformation $z = e^x$, $\tau = e^t$, $Y(z) = y(x)$ leads to an equation of the form 2.1.56:

$$Y(z) + \int_b^z \left(A B z^{\lambda+1} - A B z^\lambda \tau - A z^\lambda - B \right) Y(\tau)\, d\tau = F(z),$$

where $F(z) = f(x)$ and $b = e^a$.

18. $y(x) + \int_a^x \left[A B e^{x+\lambda t} - A B e^{(\lambda+1)t} + A e^{\lambda t} + B e^t \right] y(t)\, dt = f(x).$

The transformation $z = e^x$, $\tau = e^t$, $Y(z) = y(x)$ leads to an equation of the form 2.1.57 (in which λ is substituted by $\lambda - 1$):

$$Y(z) + \int_b^z \left(A B z \tau^{\lambda-1} - A B \tau^\lambda + A \tau^{\lambda-1} + B \right) Y(\tau)\, d\tau = F(z),$$

where $F(z) = f(x)$ and $b = e^a$.

19. $y(x) + \int_a^x \left[\sum_{k=1}^n A_k e^{\lambda_k(x-t)} \right] y(t)\, dt = f(x).$

$1°$. This integral equation can be reduced to an nth-order linear nonhomogeneous ordinary differential equation with constant coefficients. Set

$$I_k(x) = \int_a^x e^{\lambda_k(x-t)} y(t)\, dt. \tag{1}$$

Differentiating (1) with respect to x yields

$$I_k' = y(x) + \lambda_k \int_a^x e^{\lambda_k(x-t)} y(t)\, dt, \tag{2}$$

where the prime stands for differentiation with respect to x. From the comparison of (1) with (2) we see that

$$I_k' = y(x) + \lambda_k I_k, \qquad I_k = I_k(x). \tag{3}$$

The integral equation can be written in terms of $I_k(x)$ as follows:

$$y(x) + \sum_{k=1}^n A_k I_k = f(x). \tag{4}$$

Differentiating (4) with respect to x and taking account of (3), we obtain

$$y_x'(x) + \sigma_n y(x) + \sum_{k=1}^{n} A_k \lambda_k I_k = f_x'(x), \qquad \sigma_n = \sum_{k=1}^{n} A_k. \tag{5}$$

Eliminating the integral I_n from (4) and (5), we find that

$$y_x'(x) + (\sigma_n - \lambda_n)y(x) + \sum_{k=1}^{n-1} A_k(\lambda_k - \lambda_n)I_k = f_x'(x) - \lambda_n f(x). \tag{6}$$

Differentiating (6) with respect to x and eliminating I_{n-1} from the resulting equation with the aid of (6), we obtain a similar equation whose left-hand side is a second-order linear differential operator (acting on y) with constant coefficients plus the sum $\sum_{k=1}^{n-2} A_k^1 I_k$. If we proceed with successively eliminating I_{n-2}, I_{n-3}, ..., I_1 with the aid of differentiation and formula (3), then we will finally arrive at an nth-order linear nonhomogeneous ordinary differential equation with constant coefficients.

The initial conditions for $y(x)$ can be obtained by setting $x = a$ in the integral equation and all its derivative equations.

$2°$. The solution of the equation can be represented in the form

$$y(x) = f(x) + \int_a^x \left[\sum_{k=1}^{n} B_k e^{\mu_k (x-t)} \right] f(t) \, dt. \tag{7}$$

The unknown constants μ_k are the roots of the algebraic equation

$$\sum_{k=1}^{n} \frac{A_k}{z - \lambda_k} + 1 = 0, \tag{8}$$

which is reduced (by separating the numerator) to the problem of finding the roots of an nth-order characteristic polynomial.

After the μ_k have been calculated, the coefficients B_k can be found from the following linear system of algebraic equations:

$$\sum_{k=1}^{n} \frac{B_k}{\lambda_m - \mu_k} + 1 = 0, \qquad m = 1, \ldots, n. \tag{9}$$

Another way of determining the B_k is presented in item $3°$ below.

If all the roots μ_k of equation (8) are real and different, then the solution of the original integral equation can be calculated by formula (7).

To a pair of complex conjugate roots $\mu_{k,k+1} = \alpha \pm i\beta$ of the characteristic polynomial (8) there corresponds a pair of complex conjugate coefficients $B_{k,k+1}$ in equation (9). In this case, the corresponding terms $B_k e^{\mu_k(x-t)} + B_{k+1} e^{\mu_{k+1}(x-t)}$ in solution (7) can be written in the form $\overline{B}_k e^{\alpha(x-t)} [\cos \beta(x-t)] + \overline{B}_{k+1} e^{\alpha(x-t)} [\sin \beta(x-t)]$, where $\overline{B}_k$ and $\overline{B}_{k+1}$ are real coefficients.

$3°$. For $a = 0$, the solution of the original integral equation is given by

$$y(x) = f(x) - \int_0^x R(x-t)f(t) \, dt, \qquad R(x) = \mathfrak{L}^{-1}\left[\overline{R}(p) \right], \tag{10}$$

where $\mathcal{L}^{-1}\left[\overline{R}(p)\right]$ is the inverse Laplace transform of the function

$$\overline{R}(p) = \frac{\overline{K}(p)}{1 + \overline{K}(p)}, \qquad \overline{K}(p) = \sum_{k=1}^{n} \frac{A_k}{p - \lambda_k}. \tag{11}$$

The transform $\overline{R}(p)$ of the resolvent $R(x)$ can be represented as a regular fractional function:

$$\overline{R}(p) = \frac{Q(p)}{P(p)}, \qquad P(p) = (p - \mu_1)(p - \mu_2)\ldots(p - \mu_n),$$

where $Q(p)$ is a polynomial in p of degree $< n$. The roots μ_k of the polynomial $P(p)$ coincide with the roots of equation (8). If all μ_k are real and different, then the resolvent can be determined by the formula

$$R(x) = \sum_{k=1}^{n} B_k e^{\mu_k x}, \qquad B_k = \frac{Q(\mu_k)}{P'(\mu_k)},$$

where the prime stands for differentiation.

2.2-2. Kernels Containing Power-Law and Exponential Functions

20. $y(x) + A \displaystyle\int_a^x x e^{\lambda(x-t)} y(t)\, dt = f(x).$

Solution:

$$y(x) = f(x) - A \int_a^x x \exp\left[\tfrac{1}{2} A(t^2 - x^2) + \lambda(x - t)\right] f(t)\, dt.$$

21. $y(x) + A \displaystyle\int_a^x t e^{\lambda(x-t)} y(t)\, dt = f(x).$

Solution:

$$y(x) = f(x) - A \int_a^x t \exp\left[\tfrac{1}{2} A(t^2 - x^2) + \lambda(x - t)\right] f(t)\, dt.$$

22. $y(x) + A \displaystyle\int_a^x (x - t) e^{\lambda t} y(t)\, dt = f(x).$

This is a special case of equation 2.9.4 with $g(t) = Ae^{\lambda t}$.
   Solution:

$$y(x) = f(x) + \frac{A}{W} \int_a^x \left[u_1(x)u_2(t) - u_2(x)u_1(t)\right] e^{\lambda t} f(t)\, dt,$$

where $u_1(x), u_2(x)$ is a fundamental system of solutions of the second-order linear homogeneous ordinary differential equation $u''_{xx} + Ae^{\lambda x} u = 0$; the functions $u_1(x)$ and $u_2(x)$ are expressed in terms of Bessel functions or modified Bessel functions, depending on sign A:

$$W = \frac{\lambda}{\pi}, \quad u_1(x) = J_0\left(\frac{2\sqrt{A}}{\lambda} e^{\lambda x/2}\right), \quad u_2(x) = Y_0\left(\frac{2\sqrt{A}}{\lambda} e^{\lambda x/2}\right) \qquad \text{for } A > 0,$$

$$W = -\frac{\lambda}{2}, \quad u_1(x) = I_0\left(\frac{2\sqrt{|A|}}{\lambda} e^{\lambda x/2}\right), \quad u_2(x) = K_0\left(\frac{2\sqrt{|A|}}{\lambda} e^{\lambda x/2}\right) \qquad \text{for } A < 0.$$

23. $y(x) + A \int_a^x (x - t)e^{\lambda(x-t)}y(t)\, dt = f(x).$

1°. Solution with $A > 0$:

$$y(x) = f(x) - k \int_a^x e^{\lambda(x-t)} \sin[k(x - t)]f(t)\, dt, \qquad k = \sqrt{A}.$$

2°. Solution with $A < 0$:

$$y(x) = f(x) + k \int_a^x e^{\lambda(x-t)} \sinh[k(x - t)]f(t)\, dt, \qquad k = \sqrt{-A}.$$

24. $y(x) + A \int_a^x (x - t)e^{\lambda x + \mu t}y(t)\, dt = f(x).$

The substitution $u(x) = e^{-\lambda x}y(x)$ leads to an equation of the form 2.2.22:

$$u(x) + A \int_a^x (x - t)e^{(\lambda+\mu)t}u(t)\, dt = f(x)e^{-\lambda x}.$$

25. $y(x) - \int_a^x (Ax + Bt + C)e^{\lambda(x-t)}y(t)\, dt = f(x).$

The substitution $u(x) = e^{-\lambda x}y(x)$ leads to an equation of the form 2.1.6:

$$u(x) - A \int_a^x (Ax + Bt + C)u(t)\, dt = f(x)e^{-\lambda x}.$$

26. $y(x) + A \int_a^x x^2 e^{\lambda(x-t)}y(t)\, dt = f(x).$

Solution:

$$y(x) = f(x) - A \int_a^x x^2 \exp\left[\tfrac{1}{3}A(t^3 - x^3) + \lambda(x - t)\right] f(t)\, dt.$$

27. $y(x) + A \int_a^x xt e^{\lambda(x-t)}y(t)\, dt = f(x).$

Solution:

$$y(x) = f(x) - A \int_a^x xt \exp\left[\tfrac{1}{3}A(t^3 - x^3) + \lambda(x - t)\right] f(t)\, dt.$$

28. $y(x) + A \int_a^x t^2 e^{\lambda(x-t)}y(t)\, dt = f(x).$

Solution:

$$y(x) = f(x) - A \int_a^x t^2 \exp\left[\tfrac{1}{3}A(t^3 - x^3) + \lambda(x - t)\right] f(t)\, dt.$$

29. $y(x) + A \int_a^x (x - t)^2 e^{\lambda(x-t)}y(t)\, dt = f(x).$

Solution:

$$y(x) = f(x) - \int_a^x R(x - t)f(t)\, dt,$$

$$R(x) = \tfrac{2}{3}ke^{(\lambda-2k)x} - \tfrac{2}{3}ke^{(\lambda+k)x}\left[\cos(\sqrt{3}\, kx) - \sqrt{3}\sin(\sqrt{3}\, kx)\right], \qquad k = \left(\tfrac{1}{4}A\right)^{1/3}.$$

30. $y(x) + A \int_0^x (x^2 - t^2) e^{\lambda(x-t)} y(t) \, dt = f(x).$

The substitution $u(x) = e^{-\lambda x} y(x)$ leads to an equation of the form 2.1.11:

$$u(x) + A \int_0^x (x^2 - t^2) u(t) \, dt = f(x) e^{-\lambda x}.$$

31. $y(x) + A \int_a^x (x - t)^n e^{\lambda(x-t)} y(t) \, dt = f(x), \qquad n = 1, 2, \ldots$

Solution:

$$y(x) = f(x) + \int_a^x R(x - t) f(t) \, dt,$$

$$R(x) = \frac{1}{n+1} e^{\lambda x} \sum_{k=0}^n \exp(\sigma_k x) \big[\sigma_k \cos(\beta_k x) - \beta_k \sin(\beta_k x) \big],$$

where

$$\sigma_k = |An!|^{\frac{1}{n+1}} \cos\left(\frac{2\pi k}{n+1}\right), \qquad \beta_k = |An!|^{\frac{1}{n+1}} \sin\left(\frac{2\pi k}{n+1}\right) \qquad \text{for} \quad A < 0,$$

$$\sigma_k = |An!|^{\frac{1}{n+1}} \cos\left(\frac{2\pi k + \pi}{n+1}\right), \qquad \beta_k = |An!|^{\frac{1}{n+1}} \sin\left(\frac{2\pi k + \pi}{n+1}\right) \qquad \text{for} \quad A > 0.$$

32. $y(x) + b \int_a^x \frac{\exp[\lambda(x-t)]}{\sqrt{x-t}} y(t) \, dt = f(x).$

Solution:

$$y(x) = e^{\lambda x} \left\{ F(x) + \pi b^2 \int_a^x \exp[\pi b^2 (x-t)] F(t) \, dt \right\},$$

where

$$F(x) = e^{-\lambda x} f(x) - b \int_a^x \frac{e^{-\lambda t} f(t)}{\sqrt{x-t}} \, dt.$$

33. $y(x) + A \int_a^x (x - t) t^k e^{\lambda(x-t)} y(t) \, dt = f(x).$

The substitution $u(x) = e^{-\lambda x} y(x)$ leads to an equation of the form 2.1.49:

$$u(x) + A \int_a^x (x - t) t^k u(t) \, dt = f(x) e^{-\lambda x}.$$

34. $y(x) + A \int_a^x (x^k - t^k) e^{\lambda(x-t)} y(t) \, dt = f(x).$

The substitution $u(x) = e^{-\lambda x} y(x)$ leads to an equation of the form 2.1.52:

$$u(x) + A \int_a^x (x^k - t^k) u(t) \, dt = f(x) e^{-\lambda x}.$$

35. $y(x) - \lambda \int_0^x \frac{e^{\mu(x-t)}}{(x-t)^\alpha} y(t) \, dt = f(x), \qquad 0 < \alpha < 1.$

Solution:

$$y(x) = f(x) + \int_0^x R(x - t) f(t) \, dt, \qquad \text{where} \quad R(x) = e^{\mu x} \sum_{n=1}^\infty \frac{\left[\lambda \Gamma(1 - \alpha) x^{1-\alpha} \right]^n}{x \Gamma[n(1-\alpha)]}.$$

36. $\quad y(x) + A \int_a^x \exp\left[\lambda(x^2 - t^2)\right] y(t)\, dt = f(x).$

Solution:

$$y(x) = f(x) - A \int_a^x \exp\left[\lambda(x^2 - t^2) - A(x - t)\right] f(t)\, dt.$$

37. $\quad y(x) + A \int_a^x \exp\left(\lambda x^2 + \beta t^2\right) y(t)\, dt = f(x).$

In the case $\beta = -\lambda$, see equation 2.2.36. This is a special case of equation 2.9.2 with $g(x) = -A \exp\left(\lambda x^2\right)$ and $h(t) = \exp\left(\beta t^2\right)$.

38. $\quad y(x) + A \int_x^\infty \exp\left(-\lambda\sqrt{t - x}\,\right) y(t)\, dt = f(x).$

This is a special case of equation 2.9.62 with $K(x) = A \exp\left(-\lambda\sqrt{-x}\,\right)$.

39. $\quad y(x) + A \int_a^x \exp\left[\lambda(x^\mu - t^\mu)\right] y(t)\, dt = f(x), \qquad \mu > 0.$

This is a special case of equation 2.9.2 with $g(x) = -A \exp\left(\lambda x^\mu\right)$ and $h(t) = \exp\left(-\lambda t^\mu\right)$.
Solution:

$$y(x) = f(x) - A \int_a^x \exp\left[\lambda(x^\mu - t^\mu) - A(x - t)\right] f(t)\, dt.$$

40. $\quad y(x) + k \int_0^x \frac{1}{x} \exp\left(-\lambda\frac{t}{x}\right) y(t)\, dt = g(x).$

This is a special case of equation 2.9.71 with $f(z) = ke^{-\lambda z}$.

For a polynomial right-hand side, $g(x) = \sum_{n=0}^N A_n x^n$, a solution is given by

$$y(x) = \sum_{n=0}^N \frac{A_n}{1 + kB_n} x^n, \qquad B_n = \frac{n!}{\lambda^{n+1}} - e^{-\lambda} \sum_{k=0}^n \frac{n!}{k!} \frac{1}{\lambda^{n-k+1}}.$$

2.3. Equations Whose Kernels Contain Hyperbolic Functions

2.3-1. Kernels Containing Hyperbolic Cosine

1. $\quad y(x) - A \int_a^x \cosh(\lambda x) y(t)\, dt = f(x).$

This is a special case of equation 2.9.2 with $g(x) = A \cosh(\lambda x)$ and $h(t) = 1$.
Solution:

$$y(x) = f(x) + A \int_a^x \cosh(\lambda x) \exp\left\{\frac{A}{\lambda}\left[\sinh(\lambda x) - \sinh(\lambda t)\right]\right\} f(t)\, dt.$$

2. $\quad y(x) - A \displaystyle\int_a^x \cosh(\lambda t) y(t)\, dt = f(x).$

This is a special case of equation 2.9.2 with $g(x) = A$ and $h(t) = \cosh(\lambda t)$.
  Solution:

$$y(x) = f(x) + A \int_a^x \cosh(\lambda t) \exp\left\{ \frac{A}{\lambda} \left[\sinh(\lambda x) - \sinh(\lambda t) \right] \right\} f(t)\, dt.$$

3. $\quad y(x) + A \displaystyle\int_a^x \cosh[\lambda(x - t)] y(t)\, dt = f(x).$

This is a special case of equation 2.9.28 with $g(t) = A$. Therefore, solving the original integral equation is reduced to solving the second-order linear nonhomogeneous ordinary differential equation with constant coefficients

$$y''_{xx} + Ay'_x - \lambda^2 y = f''_{xx} - \lambda^2 f, \qquad f = f(x),$$

under the initial conditions

$$y(a) = f(a), \quad y'_x(a) = f'_x(a) - Af(a).$$

  Solution:

$$y(x) = f(x) + \int_a^x R(x - t) f(t)\, dt,$$

$$R(x) = \exp\left(-\tfrac{1}{2}Ax\right) \left[\frac{A^2}{2k} \sinh(kx) - A \cosh(kx) \right], \quad k = \sqrt{\lambda^2 + \tfrac{1}{4}A^2}.$$

4. $\quad y(x) + \displaystyle\int_a^x \left\{ \sum_{k=1}^n A_k \cosh[\lambda_k(x - t)] \right\} y(t)\, dt = f(x).$

This equation can be reduced to an equation of the form 2.2.19 by using the identity $\cosh z \equiv \tfrac{1}{2}\left(e^z + e^{-z}\right)$. Therefore, the integral equation in question can be reduced to a linear nonhomogeneous ordinary differential equation of order $2n$ with constant coefficients.

5. $\quad y(x) - A \displaystyle\int_a^x \frac{\cosh(\lambda x)}{\cosh(\lambda t)} y(t)\, dt = f(x).$

  Solution:
$$y(x) = f(x) + A \int_a^x e^{A(x-t)} \frac{\cosh(\lambda x)}{\cosh(\lambda t)} f(t)\, dt.$$

6. $\quad y(x) - A \displaystyle\int_a^x \frac{\cosh(\lambda t)}{\cosh(\lambda x)} y(t)\, dt = f(x).$

  Solution:
$$y(x) = f(x) + A \int_a^x e^{A(x-t)} \frac{\cosh(\lambda t)}{\cosh(\lambda x)} f(t)\, dt.$$

7. $\quad y(x) - A \displaystyle\int_a^x \cosh^k(\lambda x) \cosh^m(\mu t) y(t)\, dt = f(x).$

This is a special case of equation 2.9.2 with $g(x) = A \cosh^k(\lambda x)$ and $h(t) = \cosh^m(\mu t)$.

8. $\quad y(x) + A \displaystyle\int_a^x t \cosh[\lambda(x-t)] y(t)\,dt = f(x).$

This is a special case of equation 2.9.28 with $g(t) = At$.

9. $\quad y(x) + A \displaystyle\int_a^x t^k \cosh^m(\lambda x) y(t)\,dt = f(x).$

This is a special case of equation 2.9.2 with $g(x) = -A \cosh^m(\lambda x)$ and $h(t) = t^k$.

10. $\quad y(x) + A \displaystyle\int_a^x x^k \cosh^m(\lambda t) y(t)\,dt = f(x).$

This is a special case of equation 2.9.2 with $g(x) = -Ax^k$ and $h(t) = \cosh^m(\lambda t)$.

11. $\quad y(x) - \displaystyle\int_a^x \Big[A \cosh(kx) + B - AB(x-t)\cosh(kx) \Big] y(t)\,dt = f(x).$

This is a special case of equation 2.9.7 with $\lambda = B$ and $g(x) = A \cosh(kx)$.
Solution:

$$y(x) = f(x) + \int_a^x R(x,t) f(t)\,dt,$$

$$R(x,t) = [A\cosh(kx) + B]\frac{G(x)}{G(t)} + \frac{B^2}{G(t)} \int_t^x e^{B(x-s)} G(s)\,ds, \quad G(x) = \exp\left[\frac{A}{k}\sinh(kx)\right].$$

12. $\quad y(x) + \displaystyle\int_a^x \Big[A\cosh(kt) + B + AB(x-t)\cosh(kt) \Big] y(t)\,dt = f(x).$

This is a special case of equation 2.9.8 with $\lambda = B$ and $g(t) = A\cosh(kt)$.
Solution:

$$y(x) = f(x) + \int_a^x R(x,t) f(t)\,dt,$$

$$R(x,t) = -[A\cosh(kt) + B]\frac{G(t)}{G(x)} + \frac{B^2}{G(x)} \int_t^x e^{B(t-s)} G(s)\,ds, \quad G(x) = \exp\left[\frac{A}{k}\sinh(kx)\right].$$

13. $\quad y(x) + A \displaystyle\int_x^\infty \cosh\!\left(\lambda\sqrt{t-x}\,\right) y(t)\,dt = f(x).$

This is a special case of equation 2.9.62 with $K(x) = A\cosh\!\left(\lambda\sqrt{-x}\right)$.

2.3-2. Kernels Containing Hyperbolic Sine

14. $\quad y(x) - A \displaystyle\int_a^x \sinh(\lambda x) y(t)\,dt = f(x).$

This is a special case of equation 2.9.2 with $g(x) = A\sinh(\lambda x)$ and $h(t) = 1$.
Solution:

$$y(x) = f(x) + A \int_a^x \sinh(\lambda x) \exp\left\{ \frac{A}{\lambda}\big[\cosh(\lambda x) - \cosh(\lambda t)\big] \right\} f(t)\,dt.$$

15. $y(x) - A \int_a^x \sinh(\lambda t) y(t)\, dt = f(x).$

This is a special case of equation 2.9.2 with $g(x) = A$ and $h(t) = \sinh(\lambda t)$.

Solution:

$$y(x) = f(x) + A \int_a^x \sinh(\lambda t) \exp\left\{ \frac{A}{\lambda} \left[\cosh(\lambda x) - \cosh(\lambda t) \right] \right\} f(t)\, dt.$$

16. $y(x) + A \int_a^x \sinh[\lambda(x - t)] y(t)\, dt = f(x).$

This is a special case of equation 2.9.30 with $g(x) = A$.

$1°$. Solution with $\lambda(A - \lambda) > 0$:

$$y(x) = f(x) - \frac{A\lambda}{k} \int_a^x \sin[k(x - t)] f(t)\, dt, \qquad \text{where} \quad k = \sqrt{\lambda(A - \lambda)}.$$

$2°$. Solution with $\lambda(A - \lambda) < 0$:

$$y(x) = f(x) - \frac{A\lambda}{k} \int_a^x \sinh[k(x - t)] f(t)\, dt, \qquad \text{where} \quad k = \sqrt{\lambda(\lambda - A)}.$$

$3°$. Solution with $A = \lambda$:

$$y(x) = f(x) - \lambda^2 \int_a^x (x - t) f(t)\, dt.$$

17. $y(x) + A \int_a^x \sinh^3[\lambda(x - t)] y(t)\, dt = f(x).$

Using the formula $\sinh^3 \beta = \frac{1}{4} \sinh 3\beta - \frac{3}{4} \sinh \beta$, we arrive at an equation of the form 2.3.18:

$$y(x) + \int_a^x \left\{ \tfrac{1}{4} A \sinh\left[3\lambda(x - t)\right] - \tfrac{3}{4} A \sinh[\lambda(x - t)] \right\} y(t)\, dt = f(x).$$

18. $y(x) + \int_a^x \left\{ A_1 \sinh[\lambda_1(x - t)] + A_2 \sinh[\lambda_2(x - t)] \right\} y(t)\, dt = f(x).$

$1°$. Introduce the notation

$$I_1 = \int_a^x \sinh[\lambda_1(x - t)] y(t)\, dt, \quad I_2 = \int_a^x \sinh[\lambda_2(x - t)] y(t)\, dt,$$

$$J_1 = \int_a^x \cosh[\lambda_1(x - t)] y(t)\, dt, \quad J_2 = \int_a^x \cosh[\lambda_2(x - t)] y(t)\, dt.$$

Successively differentiating the integral equation four times yields (the first line is the original equation)

$$y + A_1 I_1 + A_2 I_2 = f, \qquad f = f(x), \tag{1}$$

$$y'_x + A_1 \lambda_1 J_1 + A_2 \lambda_2 J_2 = f'_x, \tag{2}$$

$$y''_{xx} + (A_1\lambda_1 + A_2\lambda_2) y + A_1\lambda_1^2 I_1 + A_2\lambda_2^2 I_2 = f''_{xx}, \tag{3}$$

$$y'''_{xxx} + (A_1\lambda_1 + A_2\lambda_2) y'_x + A_1\lambda_1^3 J_1 + A_2\lambda_2^3 J_2 = f'''_{xxx}, \tag{4}$$

$$y''''_{xxxx} + (A_1\lambda_1 + A_2\lambda_2) y''_{xx} + (A_1\lambda_1^3 + A_2\lambda_2^3) y + A_1\lambda_1^4 I_1 + A_2\lambda_2^4 I_2 = f''''_{xxxx}. \tag{5}$$

Eliminating I_1 and I_2 from (1), (3), and (5), we arrive at a fourth-order linear ordinary differential equation with constant coefficients:

$$y''''_{xxxx} - (\lambda_1^2 + \lambda_2^2 - A_1\lambda_1 - A_2\lambda_2)y''_{xx} + (\lambda_1^2\lambda_2^2 - A_1\lambda_1\lambda_2^2 - A_2\lambda_1^2\lambda_2)y =$$
$$f''''_{xxxx} - (\lambda_1^2 + \lambda_2^2)f''_{xx} + \lambda_1^2\lambda_2^2 f. \tag{6}$$

The initial conditions can be obtained by setting $x = a$ in (1)–(4):

$$y(a) = f(a), \quad y'_x(a) = f'_x(a),$$
$$y''_{xx}(a) = f''_{xx}(a) - (A_1\lambda_1 + A_2\lambda_2)f(a), \tag{7}$$
$$y'''_{xxx}(a) = f'''_{xxx}(a) - (A_1\lambda_1 + A_2\lambda_2)f'_x(a).$$

On solving the differential equation (6) under conditions (7), we thus find the solution of the integral equation.

$2°$. Consider the characteristic equation

$$z^2 - (\lambda_1^2 + \lambda_2^2 - A_1\lambda_1 - A_2\lambda_2)z + \lambda_1^2\lambda_2^2 - A_1\lambda_1\lambda_2^2 - A_2\lambda_1^2\lambda_2 = 0, \tag{8}$$

whose roots, z_1 and z_2, determine the solution structure of the integral equation.

Assume that the discriminant of equation (8) is positive:

$$D \equiv (A_1\lambda_1 - A_2\lambda_2 - \lambda_1^2 + \lambda_2^2)^2 + 4A_1A_2\lambda_1\lambda_2 > 0.$$

In this case, the quadratic equation (8) has the real (different) roots

$$z_1 = \tfrac{1}{2}(\lambda_1^2 + \lambda_2^2 - A_1\lambda_1 - A_2\lambda_2) + \tfrac{1}{2}\sqrt{D}, \quad z_2 = \tfrac{1}{2}(\lambda_1^2 + \lambda_2^2 - A_1\lambda_1 - A_2\lambda_2) - \tfrac{1}{2}\sqrt{D}.$$

Depending on the signs of z_1 and z_2 the following three cases are possible.

Case 1. If $z_1 > 0$ and $z_2 > 0$, then the solution of the integral equation has the form $(i = 1, 2)$:

$$y(x) = f(x) + \int_a^x \{B_1\sinh[\mu_1(x-t)] + B_2\sinh[\mu_2(x-t)]\}f(t)\,dt, \qquad \mu_i = \sqrt{z_i},$$

where

$$B_1 = A_1\frac{\lambda_1(\mu_1^2 - \lambda_2^2)}{\mu_1(\mu_2^2 - \mu_1^2)} + A_2\frac{\lambda_2(\mu_1^2 - \lambda_1^2)}{\mu_1(\mu_2^2 - \mu_1^2)}, \quad B_2 = A_1\frac{\lambda_1(\mu_2^2 - \lambda_2^2)}{\mu_2(\mu_1^2 - \mu_2^2)} + A_2\frac{\lambda_2(\mu_2^2 - \lambda_1^2)}{\mu_2(\mu_1^2 - \mu_2^2)}.$$

Case 2. If $z_1 < 0$ and $z_2 < 0$, then the solution of the integral equation has the form

$$y(x) = f(x) + \int_a^x \{B_1\sin[\mu_1(x-t)] + B_2\sin[\mu_2(x-t)]\}f(t)\,dt, \qquad \mu_i = \sqrt{|z_i|},$$

where the coefficients B_1 and B_2 are found by solving the following system of linear algebraic equations:

$$\frac{B_1\mu_1}{\lambda_1^2 + \mu_1^2} + \frac{B_2\mu_2}{\lambda_1^2 + \mu_2^2} + 1 = 0, \qquad \frac{B_1\mu_1}{\lambda_2^2 + \mu_1^2} + \frac{B_2\mu_2}{\lambda_2^2 + \mu_2^2} + 1 = 0.$$

Case 3. If $z_1 > 0$ and $z_2 < 0$, then the solution of the integral equation has the form

$$y(x) = f(x) + \int_a^x \{B_1\sinh[\mu_1(x-t)] + B_2\sin[\mu_2(x-t)]\}f(t)\,dt, \qquad \mu_i = \sqrt{|z_i|},$$

where B_1 and B_2 are determined from the following system of linear algebraic equations:

$$\frac{B_1\mu_1}{\lambda_1^2 - \mu_1^2} + \frac{B_2\mu_2}{\lambda_1^2 + \mu_2^2} + 1 = 0, \qquad \frac{B_1\mu_1}{\lambda_2^2 - \mu_1^2} + \frac{B_2\mu_2}{\lambda_2^2 + \mu_2^2} + 1 = 0.$$

19. $y(x) + \int_a^x \left\{ \sum_{k=1}^{n} A_k \sinh[\lambda_k(x-t)] \right\} y(t)\, dt = f(x).$

$1°$. This equation can be reduced to an equation of the form 2.2.19 with the aid of the formula $\sinh z = \frac{1}{2}\left(e^z - e^{-z}\right)$. Therefore, the original integral equation can be reduced to a linear nonhomogeneous ordinary differential equation of order $2n$ with constant coefficients.

$2°$. Let us find the roots z_k of the algebraic equation

$$\sum_{k=1}^{n} \frac{\lambda_k A_k}{z - \lambda_k^2} + 1 = 0. \tag{1}$$

By reducing it to a common denominator, we arrive at the problem of determining the roots of an nth-degree characteristic polynomial.

Assume that all z_k are real, different, and nonzero. Let us divide the roots into two groups

$$z_1 > 0, \quad z_2 > 0, \quad \ldots, \quad z_s > 0 \quad \text{(positive roots)};$$
$$z_{s+1} < 0, \quad z_{s+2} < 0, \quad \ldots, \quad z_n < 0 \quad \text{(negative roots)}.$$

Then the solution of the integral equation can be written in the form

$$y(x) = f(x) + \int_a^x \left\{ \sum_{k=1}^{s} B_k \sinh\left[\mu_k(x-t)\right] + \sum_{k=s+1}^{n} C_k \sin\left[\mu_k(x-t)\right] \right\} f(t)\, dt, \quad \mu_k = \sqrt{|z_k|}. \tag{2}$$

The coefficients B_k and C_k are determined from the following system of linear algebraic equations:

$$\sum_{k=0}^{s} \frac{B_k \mu_k}{\lambda_m^2 - \mu_k^2} + \sum_{k=s+1}^{n} \frac{C_k \mu_k}{\lambda_m^2 + \mu_k^2} + 1 = 0, \quad \mu_k = \sqrt{|z_k|}, \quad m = 1, \ldots, n. \tag{3}$$

In the case of a nonzero root $z_s = 0$, we can introduce the new constant $D = B_s \mu_s$ and proceed to the limit $\mu_s \to 0$. As a result, the term $D(x-t)$ appears in solution (2) instead of $B_s \sinh\left[\mu_s(x-t)\right]$ and the corresponding terms $D\lambda_m^{-2}$ appear in system (3).

20. $y(x) - A \int_a^x \frac{\sinh(\lambda x)}{\sinh(\lambda t)} y(t)\, dt = f(x).$

Solution:
$$y(x) = f(x) + A \int_a^x e^{A(x-t)} \frac{\sinh(\lambda x)}{\sinh(\lambda t)} f(t)\, dt.$$

21. $y(x) - A \int_a^x \frac{\sinh(\lambda t)}{\sinh(\lambda x)} y(t)\, dt = f(x).$

Solution:
$$y(x) = f(x) + A \int_a^x e^{A(x-t)} \frac{\sinh(\lambda t)}{\sinh(\lambda x)} f(t)\, dt.$$

22. $y(x) - A \int_a^x \sinh^k(\lambda x) \sinh^m(\mu t) y(t)\, dt = f(x).$

This is a special case of equation 2.9.2 with $g(x) = A \sinh^k(\lambda x)$ and $h(t) = \sinh^m(\mu t)$.

23. $y(x) + A \displaystyle\int_a^x t \sinh[\lambda(x - t)] y(t)\, dt = f(x).$

This is a special case of equation 2.9.30 with $g(t) = At$.

Solution:

$$y(x) = f(x) + \frac{A\lambda}{W} \int_a^x t \big[u_1(x)u_2(t) - u_2(x)u_1(t)\big] f(t)\, dt,$$

where $u_1(x), u_2(x)$ is a fundamental system of solutions of the second-order linear ordinary differential equation $u''_{xx} + \lambda(Ax - \lambda)u = 0$, and W is the Wronskian.

The functions $u_1(x)$ and $u_2(x)$ are expressed in terms of Bessel functions or modified Bessel functions, depending on the sign of $A\lambda$, as follows:

if $A\lambda > 0$, then

$$u_1(x) = \xi^{1/2} J_{1/3}\big(\tfrac{2}{3}\sqrt{A\lambda}\,\xi^{3/2}\big), \quad u_2(x) = \xi^{1/2} Y_{1/3}\big(\tfrac{2}{3}\sqrt{A\lambda}\,\xi^{3/2}\big),$$
$$W = 3/\pi, \quad \xi = x - (\lambda/A);$$

if $A\lambda < 0$, then

$$u_1(x) = \xi^{1/2} I_{1/3}\big(\tfrac{2}{3}\sqrt{-A\lambda}\,\xi^{3/2}\big), \quad u_2(x) = \xi^{1/2} K_{1/3}\big(\tfrac{2}{3}\sqrt{-A\lambda}\,\xi^{3/2}\big),$$
$$W = -\tfrac{3}{2}, \quad \xi = x - (\lambda/A).$$

24. $y(x) + A \displaystyle\int_a^x x \sinh[\lambda(x - t)] y(t)\, dt = f(x).$

This is a special case of equation 2.9.31 with $g(x) = Ax$ and $h(t) = 1$.

Solution:

$$y(x) = f(x) + \frac{A\lambda}{W} \int_a^x x \big[u_1(x)u_2(t) - u_2(x)u_1(t)\big] f(t)\, dt,$$

where $u_1(x), u_2(x)$ is a fundamental system of solutions of the second-order linear ordinary differential equation $u''_{xx} + \lambda(Ax - \lambda)u = 0$, and W is the Wronskian.

The functions $u_1(x)$, $u_2(x)$, and W are specified in 2.3.23.

25. $y(x) + A \displaystyle\int_a^x t^k \sinh^m(\lambda x) y(t)\, dt = f(x).$

This is a special case of equation 2.9.2 with $g(x) = -A \sinh^m(\lambda x)$ and $h(t) = t^k$.

26. $y(x) + A \displaystyle\int_a^x x^k \sinh^m(\lambda t) y(t)\, dt = f(x).$

This is a special case of equation 2.9.2 with $g(x) = -Ax^k$ and $h(t) = \sinh^m(\lambda t)$.

27. $y(x) - \displaystyle\int_a^x \big[A \sinh(kx) + B - AB(x - t)\sinh(kx)\big] y(t)\, dt = f(x).$

This is a special case of equation 2.9.7 with $\lambda = B$ and $g(x) = A \sinh(kx)$.

Solution:

$$y(x) = f(x) + \int_a^x R(x, t) f(t)\, dt,$$

$$R(x, t) = [A \sinh(kx) + B]\frac{G(x)}{G(t)} + \frac{B^2}{G(t)} \int_t^x e^{B(x-s)} G(s)\, ds, \quad G(x) = \exp\left[\frac{A}{k}\cosh(kx)\right].$$

28. $y(x) + \int_a^x \left[A \sinh(kt) + B + AB(x - t) \sinh(kt) \right] y(t)\, dt = f(x).$

This is a special case of equation 2.9.8 with $\lambda = B$ and $g(t) = A \sinh(kt)$.
Solution:

$$y(x) = f(x) + \int_a^x R(x, t) f(t)\, dt,$$

$$R(x, t) = -[\sinh(kt) + B] \frac{G(t)}{G(x)} + \frac{B^2}{G(x)} \int_t^x e^{B(t-s)} G(s)\, ds, \quad G(x) = \exp\left[\frac{A}{k} \cosh(kx) \right].$$

29. $y(x) + A \int_x^\infty \sinh\left(\lambda \sqrt{t - x} \right) y(t)\, dt = f(x).$

This is a special case of equation 2.9.62 with $K(x) = A \sinh\left(\lambda \sqrt{-x} \right)$.

2.3-3. Kernels Containing Hyperbolic Tangent

30. $y(x) - A \int_a^x \tanh(\lambda x) y(t)\, dt = f(x).$

This is a special case of equation 2.9.2 with $g(x) = A \tanh(\lambda x)$ and $h(t) = 1$.
Solution:

$$y(x) = f(x) + A \int_a^x \tanh(\lambda x) \left[\frac{\cosh(\lambda x)}{\cosh(\lambda t)} \right]^{A/\lambda} f(t)\, dt.$$

31. $y(x) - A \int_a^x \tanh(\lambda t) y(t)\, dt = f(x).$

This is a special case of equation 2.9.2 with $g(x) = A$ and $h(t) = \tanh(\lambda t)$.
Solution:

$$y(x) = f(x) + A \int_a^x \tanh(\lambda t) \left[\frac{\cosh(\lambda x)}{\cosh(\lambda t)} \right]^{A/\lambda} f(t)\, dt.$$

32. $y(x) + A \int_a^x \left[\tanh(\lambda x) - \tanh(\lambda t) \right] y(t)\, dt = f(x).$

This is a special case of equation 2.9.5 with $g(x) = A \tanh(\lambda x)$.
Solution:

$$y(x) = f(x) + \frac{1}{W} \int_a^x \left[Y_1'(x) Y_2'(t) - Y_2'(x) Y_1'(t) \right] f(t)\, dt,$$

where $Y_1(x), Y_2(x)$ is a fundamental system of solutions of the second-order linear ordinary differential equation $\cosh^2(\lambda x) Y_{xx}'' + A \lambda Y = 0$, W is the Wronskian, and the primes stand for the differentiation with respect to the argument specified in the parentheses.

As shown in A. D. Polyanin and V. F. Zaitsev (1996), the functions $Y_1(x)$ and $Y_2(x)$ can be represented in the form

$$Y_1(x) = F\left(\alpha, \beta, 1; \frac{e^{\lambda x}}{1 + e^{\lambda x}} \right), \quad Y_2(x) = Y_1(x) \int_a^x \frac{d\xi}{Y_1^2(\xi)}, \quad W = 1,$$

where $F(\alpha, \beta, \gamma; z)$ is the hypergeometric function, in which α and β are determined from the algebraic system $\alpha + \beta = 1$, $\alpha\beta = -A/\lambda$.

33. $y(x) - A \displaystyle\int_a^x \frac{\tanh(\lambda x)}{\tanh(\lambda t)} y(t)\, dt = f(x).$

Solution:
$$y(x) = f(x) + A \int_a^x e^{A(x-t)} \frac{\tanh(\lambda x)}{\tanh(\lambda t)} f(t)\, dt.$$

34. $y(x) - A \displaystyle\int_a^x \frac{\tanh(\lambda t)}{\tanh(\lambda x)} y(t)\, dt = f(x).$

Solution:
$$y(x) = f(x) + A \int_a^x e^{A(x-t)} \frac{\tanh(\lambda t)}{\tanh(\lambda x)} f(t)\, dt.$$

35. $y(x) - A \displaystyle\int_a^x \tanh^k(\lambda x) \tanh^m(\mu t) y(t)\, dt = f(x).$

This is a special case of equation 2.9.2 with $g(x) = A \tanh^k(\lambda x)$ and $h(t) = \tanh^m(\mu t)$.

36. $y(x) + A \displaystyle\int_a^x t^k \tanh^m(\lambda x) y(t)\, dt = f(x).$

This is a special case of equation 2.9.2 with $g(x) = -A \tanh^m(\lambda x)$ and $h(t) = t^k$.

37. $y(x) + A \displaystyle\int_a^x x^k \tanh^m(\lambda t) y(t)\, dt = f(x).$

This is a special case of equation 2.9.2 with $g(x) = -Ax^k$ and $h(t) = \tanh^m(\lambda t)$.

38. $y(x) + A \displaystyle\int_x^\infty \tanh[\lambda(t - x)] y(t)\, dt = f(x).$

This is a special case of equation 2.9.62 with $K(z) = A \tanh(-\lambda z)$.

39. $y(x) + A \displaystyle\int_x^\infty \tanh\left(\lambda\sqrt{t - x}\right) y(t)\, dt = f(x).$

This is a special case of equation 2.9.62 with $K(z) = A \tanh\left(\lambda\sqrt{-z}\right)$.

40. $y(x) - \displaystyle\int_a^x \left[A \tanh(kx) + B - AB(x - t) \tanh(kx)\right] y(t)\, dt = f(x).$

This is a special case of equation 2.9.7 with $\lambda = B$ and $g(x) = A \tanh(kx)$.

41. $y(x) + \displaystyle\int_a^x \left[A \tanh(kt) + B + AB(x - t) \tanh(kt)\right] y(t)\, dt = f(x).$

This is a special case of equation 2.9.8 with $\lambda = B$ and $g(t) = A \tanh(kt)$.

2.3-4. Kernels Containing Hyperbolic Cotangent

42. $y(x) - A \displaystyle\int_a^x \coth(\lambda x) y(t)\, dt = f(x).$

This is a special case of equation 2.9.2 with $g(x) = A \coth(\lambda x)$ and $h(t) = 1$.

Solution:
$$y(x) = f(x) + A \int_a^x \coth(\lambda x) \left[\frac{\sinh(\lambda x)}{\sinh(\lambda t)}\right]^{A/\lambda} f(t)\, dt.$$

43. $y(x) - A \int_a^x \coth(\lambda t) y(t)\, dt = f(x).$

This is a special case of equation 2.9.2 with $g(x) = A$ and $h(t) = \coth(\lambda t)$.

Solution:
$$y(x) = f(x) + A \int_a^x \coth(\lambda t) \left[\frac{\sinh(\lambda x)}{\sinh(\lambda t)}\right]^{A/\lambda} f(t)\, dt.$$

44. $y(x) - A \int_a^x \dfrac{\coth(\lambda t)}{\coth(\lambda x)} y(t)\, dt = f(x).$

Solution:
$$y(x) = f(x) + A \int_a^x e^{A(x-t)} \frac{\coth(\lambda t)}{\coth(\lambda x)} f(t)\, dt.$$

45. $y(x) - A \int_a^x \dfrac{\coth(\lambda x)}{\coth(\lambda t)} y(t)\, dt = f(x).$

Solution:
$$y(x) = f(x) + A \int_a^x e^{A(x-t)} \frac{\coth(\lambda x)}{\coth(\lambda t)} f(t)\, dt.$$

46. $y(x) - A \int_a^x \coth^k(\lambda x) \coth^m(\mu t) y(t)\, dt = f(x).$

This is a special case of equation 2.9.2 with $g(x) = A \coth^k(\lambda x)$ and $h(t) = \coth^m(\mu t)$.

47. $y(x) + A \int_a^x t^k \coth^m(\lambda x) y(t)\, dt = f(x).$

This is a special case of equation 2.9.2 with $g(x) = -A \coth^m(\lambda x)$ and $h(t) = t^k$.

48. $y(x) + A \int_a^x x^k \coth^m(\lambda t) y(t)\, dt = f(x).$

This is a special case of equation 2.9.2 with $g(x) = -Ax^k$ and $h(t) = \coth^m(\lambda t)$.

49. $y(x) + A \int_x^\infty \coth[\lambda(t - x)] y(t)\, dt = f(x).$

This is a special case of equation 2.9.62 with $K(z) = A \coth(-\lambda z)$.

50. $y(x) + A \int_x^\infty \coth\left(\lambda \sqrt{t - x}\right) y(t)\, dt = f(x).$

This is a special case of equation 2.9.62 with $K(z) = A \coth\left(\lambda \sqrt{-z}\right)$.

51. $y(x) - \int_a^x \left[A \coth(kx) + B - AB(x - t) \coth(kx)\right] y(t)\, dt = f(x).$

This is a special case of equation 2.9.7 with $\lambda = B$ and $g(x) = A \coth(kx)$.

52. $y(x) + \int_a^x \left[A \coth(kt) + B + AB(x - t) \coth(kt)\right] y(t)\, dt = f(x).$

This is a special case of equation 2.9.8 with $\lambda = B$ and $g(t) = A \coth(kt)$.

2.3-5. Kernels Containing Combinations of Hyperbolic Functions

53. $y(x) - A \displaystyle\int_a^x \cosh^k(\lambda x) \sinh^m(\mu t) y(t)\, dt = f(x).$

This is a special case of equation 2.9.2 with $g(x) = A \cosh^k(\lambda x)$ and $h(t) = \sinh^m(\mu t)$.

54. $y(x) - \displaystyle\int_a^x \left\{ A + B \cosh(\lambda x) + B(x - t)[\lambda \sinh(\lambda x) - A \cosh(\lambda x)] \right\} y(t)\, dt = f(x).$

This is a special case of equation 2.9.32 with $b = B$ and $g(x) = A$.

55. $y(x) - \displaystyle\int_a^x \left\{ A + B \sinh(\lambda x) + B(x - t)[\lambda \cosh(\lambda x) - A \sinh(\lambda x)] \right\} y(t)\, dt = f(x).$

This is a special case of equation 2.9.33 with $b = B$ and $g(x) = A$.

56. $y(x) - A \displaystyle\int_a^x \tanh^k(\lambda x) \coth^m(\mu t) y(t)\, dt = f(x).$

This is a special case of equation 2.9.2 with $g(x) = A \tanh^k(\lambda x)$ and $h(t) = \coth^m(\mu t)$.

2.4. Equations Whose Kernels Contain Logarithmic Functions

2.4-1. Kernels Containing Logarithmic Functions

1. $y(x) - A \displaystyle\int_a^x \ln(\lambda x) y(t)\, dt = f(x).$

This is a special case of equation 2.9.2 with $g(x) = A \ln(\lambda x)$ and $h(t) = 1$.

Solution:
$$y(x) = f(x) + A \int_a^x \ln(\lambda x) e^{-A(x-t)} \frac{(\lambda x)^{Ax}}{(\lambda t)^{At}} f(t)\, dt.$$

2. $y(x) - A \displaystyle\int_a^x \ln(\lambda t) y(t)\, dt = f(x).$

This is a special case of equation 2.9.2 with $g(x) = A$ and $h(t) = \ln(\lambda t)$.

Solution:
$$y(x) = f(x) + A \int_a^x \ln(\lambda t) e^{-A(x-t)} \frac{(\lambda x)^{Ax}}{(\lambda t)^{At}} f(t)\, dt.$$

3. $y(x) + A \displaystyle\int_a^x (\ln x - \ln t) y(t)\, dt = f(x).$

This is a special case of equation 2.9.5 with $g(x) = A \ln x$.

Solution:
$$y(x) = f(x) + \frac{1}{W} \int_a^x \left[u_1'(x) u_2'(t) - u_2'(x) u_1'(t) \right] f(t)\, dt,$$

where the primes denote differentiation with respect to the argument specified in the parentheses; and $u_1(x), u_2(x)$ is a fundamental system of solutions of the second-order linear homogeneous ordinary differential equation $u_{xx}'' + A x^{-1} u = 0$, with $u_1(x)$ and $u_2(x)$ expressed in terms of Bessel functions or modified Bessel functions, depending on the sign of A:

$$W = \tfrac{1}{\pi}, \quad u_1(x) = \sqrt{x}\, J_1\big(2\sqrt{Ax}\,\big), \quad u_2(x) = \sqrt{x}\, Y_1\big(2\sqrt{Ax}\,\big) \qquad \text{for } A > 0,$$

$$W = -\tfrac{1}{2}, \quad u_1(x) = \sqrt{x}\, I_1\big(2\sqrt{-Ax}\,\big), \quad u_2(x) = \sqrt{x}\, K_1\big(2\sqrt{-Ax}\,\big) \qquad \text{for } A < 0.$$

4. $y(x) - A \int_a^x \dfrac{\ln(\lambda x)}{\ln(\lambda t)} y(t)\, dt = f(x).$

Solution:

$$y(x) = f(x) + A \int_a^x e^{A(x-t)} \frac{\ln(\lambda x)}{\ln(\lambda t)} f(t)\, dt.$$

5. $y(x) - A \int_a^x \dfrac{\ln(\lambda t)}{\ln(\lambda x)} y(t)\, dt = f(x).$

Solution:

$$y(x) = f(x) + A \int_a^x e^{A(x-t)} \frac{\ln(\lambda t)}{\ln(\lambda x)} f(t)\, dt.$$

6. $y(x) - A \int_a^x \ln^k(\lambda x) \ln^m(\mu t) y(t)\, dt = f(x).$

This is a special case of equation 2.9.2 with $g(x) = A \ln^k(\lambda x)$ and $h(t) = \ln^m(\mu t)$.

7. $y(x) + a \int_x^\infty \ln(t - x) y(t)\, dt = f(x).$

This is a special case of equation 2.9.62 with $K(x) = a \ln(-x)$.

For $f(x) = \sum_{k=1}^m A_k \exp(-\lambda_k x)$, where $\lambda_k > 0$, a solution of the equation has the form

$$y(x) = \sum_{k=1}^m \frac{A_k}{B_k} \exp(-\lambda_k x), \qquad B_k = 1 - \frac{a}{\lambda_k}(\ln \lambda_k + C),$$

where $C = 0.5772\ldots$ is the Euler constant.

8. $y(x) + a \int_x^\infty \ln^2(t - x) y(t)\, dt = f(x).$

This is a special case of equation 2.9.62 with $K(x) = a \ln^2(-x)$.

For $f(x) = \sum_{k=1}^m A_k \exp(-\lambda_k x)$, where $\lambda_k > 0$, a solution of the equation has the form

$$y(x) = \sum_{k=1}^m \frac{A_k}{B_k} \exp(-\lambda_k x), \qquad B_k = 1 + \frac{a}{\lambda_k}\left[\tfrac{1}{6}\pi^2 + (\ln \lambda_k + C)^2\right],$$

where $C = 0.5772\ldots$ is the Euler constant.

2.4-2. Kernels Containing Power-Law and Logarithmic Functions

9. $y(x) - A \int_a^x x^k \ln^m(\lambda t) y(t)\, dt = f(x).$

This is a special case of equation 2.9.2 with $g(x) = Ax^k$ and $h(t) = \ln^m(\lambda t)$.

10. $y(x) - A \int_a^x t^k \ln^m(\lambda x) y(t)\, dt = f(x).$

This is a special case of equation 2.9.2 with $g(x) = A \ln^m(\lambda x)$ and $h(t) = t^k$.

11. $y(x) - \int_a^x \left[A \ln(kx) + B - AB(x - t) \ln(kx)\right] y(t)\, dt = f(x).$

This is a special case of equation 2.9.7 with $\lambda = B$ and $g(x) = A \ln(kx)$.

12. $y(x) + \int_a^x \left[A \ln(kt) + B + AB(x - t) \ln(kt)\right] y(t)\, dt = f(x).$

This is a special case of equation 2.9.8 with $\lambda = B$ and $g(t) = A \ln(kt)$.

13. $y(x) + a \int_x^\infty (t - x)^n \ln(t - x) y(t)\, dt = f(x), \qquad n = 1, 2, \ldots$

For $f(x) = \sum_{k=1}^m A_k \exp(-\lambda_k x)$, where $\lambda_k > 0$, a solution of the equation has the form

$$y(x) = \sum_{k=1}^m \frac{A_k}{B_k} \exp(-\lambda_k x), \qquad B_k = 1 + \frac{an!}{\lambda_k^{n+1}} \left(1 + \tfrac{1}{2} + \tfrac{1}{3} + \cdots + \tfrac{1}{n} - \ln \lambda_k - C\right),$$

where $C = 0.5772 \ldots$ is the Euler constant.

14. $y(x) + a \int_x^\infty \frac{\ln(t - x)}{\sqrt{t - x}} y(t)\, dt = f(x).$

This is a special case of equation 2.9.62 with $K(-x) = ax^{-1/2} \ln x$.

For $f(x) = \sum_{k=1}^m A_k \exp(-\lambda_k x)$, where $\lambda_k > 0$, a solution of the equation has the form

$$y(x) = \sum_{k=1}^m \frac{A_k}{B_k} \exp(-\lambda_k x), \qquad B_k = 1 - a\sqrt{\frac{\pi}{\lambda_k}} \left[\ln(4\lambda_k) + C\right],$$

where $C = 0.5772 \ldots$ is the Euler constant.

2.5. Equations Whose Kernels Contain Trigonometric Functions

2.5-1. Kernels Containing Cosine

1. $y(x) - A \int_a^x \cos(\lambda x) y(t)\, dt = f(x).$

This is a special case of equation 2.9.2 with $g(x) = A \cos(\lambda x)$ and $h(t) = 1$.
    Solution:

$$y(x) = f(x) + A \int_a^x \cos(\lambda x) \exp\left\{\frac{A}{\lambda}\left[\sin(\lambda x) - \sin(\lambda t)\right]\right\} f(t)\, dt.$$

2. $y(x) - A \int_a^x \cos(\lambda t) y(t)\, dt = f(x).$

This is a special case of equation 2.9.2 with $g(x) = A$ and $h(t) = \cos(\lambda t)$.
    Solution:

$$y(x) = f(x) + A \int_a^x \cos(\lambda t) \exp\left\{\frac{A}{\lambda}\left[\sin(\lambda x) - \sin(\lambda t)\right]\right\} f(t)\, dt.$$

3. $y(x) + A \displaystyle\int_a^x \cos[\lambda(x-t)]y(t)\, dt = f(x).$

This is a special case of equation 2.9.34 with $g(t) = A$. Therefore, solving this integral equation is reduced to solving the following second-order linear nonhomogeneous ordinary differential equation with constant coefficients:

$$y''_{xx} + Ay'_x + \lambda^2 y = f''_{xx} + \lambda^2 f, \qquad f = f(x),$$

with the initial conditions

$$y(a) = f(a), \quad y'_x(a) = f'_x(a) - Af(a).$$

1°. Solution with $|A| > 2|\lambda|$:

$$y(x) = f(x) + \int_a^x R(x-t)f(t)\, dt,$$

$$R(x) = \exp\left(-\tfrac{1}{2}Ax\right)\left[\frac{A^2}{2k}\sinh(kx) - A\cosh(kx)\right], \quad k = \sqrt{\tfrac{1}{4}A^2 - \lambda^2}.$$

2°. Solution with $|A| < 2|\lambda|$:

$$y(x) = f(x) + \int_a^x R(x-t)f(t)\, dt,$$

$$R(x) = \exp\left(-\tfrac{1}{2}Ax\right)\left[\frac{A^2}{2k}\sin(kx) - A\cos(kx)\right], \quad k = \sqrt{\lambda^2 - \tfrac{1}{4}A^2}.$$

3°. Solution with $\lambda = \pm\tfrac{1}{2}A$:

$$y(x) = f(x) + \int_a^x R(x-t)f(t)\, dt, \qquad R(x) = \exp\left(-\tfrac{1}{2}Ax\right)\left(\tfrac{1}{2}A^2 x - A\right).$$

4. $y(x) + \displaystyle\int_a^x \left\{ \sum_{k=1}^{n} A_k \cos[\lambda_k(x-t)] \right\} y(t)\, dt = f(x).$

This integral equation is reduced to a linear nonhomogeneous ordinary differential equation of order $2n$ with constant coefficients. Set

$$I_k(x) = \int_a^x \cos[\lambda_k(x-t)]y(t)\, dt. \tag{1}$$

Differentiating (1) with respect to x twice yields

$$\begin{aligned} I'_k &= y(x) - \lambda_k \int_a^x \sin[\lambda_k(x-t)]y(t)\, dt, \\ I''_k &= y'_x(x) - \lambda_k^2 \int_a^x \cos[\lambda_k(x-t)]y(t)\, dt, \end{aligned} \tag{2}$$

where the primes stand for differentiation with respect to x. Comparing (1) and (2), we see that

$$I''_k = y'_x(x) - \lambda_k^2 I_k, \qquad I_k = I_k(x). \tag{3}$$

With the aid of (1), the integral equation can be rewritten in the form

$$y(x) + \sum_{k=1}^{n} A_k I_k = f(x). \tag{4}$$

Differentiating (4) with respect to x twice taking into account (3) yields

$$y''_{xx}(x) + \sigma_n y'_x(x) - \sum_{k=1}^{n} A_k \lambda_k^2 I_k = f''_{xx}(x), \qquad \sigma_n = \sum_{k=1}^{n} A_k. \tag{5}$$

Eliminating the integral I_n from (4) and (5), we obtain

$$y''_{xx}(x) + \sigma_n y'_x(x) + \lambda_n^2 y(x) + \sum_{k=1}^{n-1} A_k(\lambda_n^2 - \lambda_k^2) I_k = f''_{xx}(x) + \lambda_n^2 f(x). \tag{6}$$

Differentiating (6) with respect to x twice followed by eliminating I_{n-1} from the resulting expression with the aid of (6) yields a similar equation whose left-hand side is a fourth-order differential operator (acting on y) with constant coefficients plus the sum $\sum_{k=1}^{n-2} B_k I_k$. Successively eliminating the terms $I_{n-2}, I_{n-3}, \ldots$ using double differentiation and formula (3), we finally arrive at a linear nonhomogeneous ordinary differential equation of order $2n$ with constant coefficients.

The initial conditions for $y(x)$ can be obtained by setting $x = a$ in the integral equation and all its derivative equations.

5. $y(x) - A \displaystyle\int_a^x \frac{\cos(\lambda x)}{\cos(\lambda t)} y(t)\, dt = f(x).$

Solution:

$$y(x) = f(x) + A \int_a^x e^{A(x-t)} \frac{\cos(\lambda x)}{\cos(\lambda t)} f(t)\, dt.$$

6. $y(x) - A \displaystyle\int_a^x \frac{\cos(\lambda t)}{\cos(\lambda x)} y(t)\, dt = f(x).$

Solution:

$$y(x) = f(x) + A \int_a^x e^{A(x-t)} \frac{\cos(\lambda t)}{\cos(\lambda x)} f(t)\, dt.$$

7. $y(x) - A \displaystyle\int_a^x \cos^k(\lambda x) \cos^m(\mu t) y(t)\, dt = f(x).$

This is a special case of equation 2.9.2 with $g(x) = A \cos^k(\lambda x)$ and $h(t) = \cos^m(\mu t)$.

8. $y(x) + A \displaystyle\int_a^x t \cos[\lambda(x - t)] y(t)\, dt = f(x).$

This is a special case of equation 2.9.34 with $g(t) = At$.

9. $y(x) + A \displaystyle\int_a^x t^k \cos^m(\lambda x) y(t)\, dt = f(x).$

This is a special case of equation 2.9.2 with $g(x) = -A \cos^m(\lambda x)$ and $h(t) = t^k$.

10. $y(x) + A \int_a^x x^k \cos^m(\lambda t) y(t)\, dt = f(x).$

This is a special case of equation 2.9.2 with $g(x) = -Ax^k$ and $h(t) = \cos^m(\lambda t)$.

11. $y(x) - \int_a^x \left[A\cos(kx) + B - AB(x-t)\cos(kx) \right] y(t)\, dt = f(x).$

This is a special case of equation 2.9.7 with $\lambda = B$ and $g(x) = A\cos(kx)$.
Solution:

$$y(x) = f(x) + \int_a^x R(x,t) f(t)\, dt,$$

$$R(x,t) = [A\cos(kx) + B]\frac{G(x)}{G(t)} + \frac{B^2}{G(t)} \int_t^x e^{B(x-s)} G(s)\, ds, \quad G(x) = \exp\left[\frac{A}{k}\sin(kx) \right].$$

12. $y(x) + \int_a^x \left[A\cos(kt) + B + AB(x-t)\cos(kt) \right] y(t)\, dt = f(x).$

This is a special case of equation 2.9.8 with $\lambda = B$ and $g(t) = A\cos(kt)$.
Solution:

$$y(x) = f(x) + \int_a^x R(x,t) f(t)\, dt,$$

$$R(x,t) = -[A\cos(kt) + B]\frac{G(t)}{G(x)} + \frac{B^2}{G(x)} \int_t^x e^{B(t-s)} G(s)\, ds, \quad G(x) = \exp\left[\frac{A}{k}\sin(kx) \right].$$

13. $y(x) + A \int_x^\infty \cos\left(\lambda\sqrt{t-x} \right) y(t)\, dt = f(x).$

This is a special case of equation 2.9.62 with $K(x) = A\cos\left(\lambda\sqrt{-x} \right)$.

2.5-2. Kernels Containing Sine

14. $y(x) - A \int_a^x \sin(\lambda x) y(t)\, dt = f(x).$

This is a special case of equation 2.9.2 with $g(x) = A\sin(\lambda x)$ and $h(t) = 1$.
Solution:

$$y(x) = f(x) + A \int_a^x \sin(\lambda x) \exp\left\{ \frac{A}{\lambda}\left[\cos(\lambda t) - \cos(\lambda x) \right] \right\} f(t)\, dt.$$

15. $y(x) - A \int_a^x \sin(\lambda t) y(t)\, dt = f(x).$

This is a special case of equation 2.9.2 with $g(x) = A$ and $h(t) = \sin(\lambda t)$.
Solution:

$$y(x) = f(x) + A \int_a^x \sin(\lambda t) \exp\left\{ \frac{A}{\lambda}\left[\cos(\lambda t) - \cos(\lambda x) \right] \right\} f(t)\, dt.$$

16. $y(x) + A \displaystyle\int_a^x \sin[\lambda(x - t)]y(t)\, dt = f(x).$

This is a special case of equation 2.9.36 with $g(t) = A$.

1°. Solution with $\lambda(A + \lambda) > 0$:

$$y(x) = f(x) - \frac{A\lambda}{k} \int_a^x \sin[k(x - t)]f(t)\, dt, \qquad \text{where} \quad k = \sqrt{\lambda(A + \lambda)}.$$

2°. Solution with $\lambda(A + \lambda) < 0$:

$$y(x) = f(x) - \frac{A\lambda}{k} \int_a^x \sinh[k(x - t)]f(t)\, dt, \qquad \text{where} \quad k = \sqrt{-\lambda(\lambda + A)}.$$

3°. Solution with $A = -\lambda$:

$$y(x) = f(x) + \lambda^2 \int_a^x (x - t)f(t)\, dt.$$

17. $y(x) + A \displaystyle\int_a^x \sin^3[\lambda(x - t)]y(t)\, dt = f(x).$

Using the formula $\sin^3 \beta = -\frac{1}{4} \sin 3\beta + \frac{3}{4} \sin \beta$, we arrive at an equation of the form 2.5.18:

$$y(x) + \int_a^x \left\{ -\tfrac{1}{4} A \sin[3\lambda(x - t)] + \tfrac{3}{4} A \sin[\lambda(x - t)] \right\} y(t)\, dt = f(x).$$

18. $y(x) + \displaystyle\int_a^x \left\{ A_1 \sin[\lambda_1(x - t)] + A_2 \sin[\lambda_2(x - t)] \right\} y(t)\, dt = f(x).$

This equation can be solved by the same method as equation 2.3.18, by reducing it to a fourth-order linear ordinary differential equation with constant coefficients.

Consider the characteristic equation

$$z^2 + (\lambda_1^2 + \lambda_2^2 + A_1\lambda_1 + A_2\lambda_2)z + \lambda_1^2\lambda_2^2 + A_1\lambda_1\lambda_2^2 + A_2\lambda_1^2\lambda_2 = 0, \qquad (1)$$

whose roots, z_1 and z_2, determine the solution structure of the integral equation.

Assume that the discriminant of equation (1) is positive:

$$D \equiv (A_1\lambda_1 - A_2\lambda_2 + \lambda_1^2 - \lambda_2^2)^2 + 4A_1A_2\lambda_1\lambda_2 > 0.$$

In this case, the quadratic equation (1) has the real (different) roots

$$z_1 = -\tfrac{1}{2}(\lambda_1^2 + \lambda_2^2 + A_1\lambda_1 + A_2\lambda_2) + \tfrac{1}{2}\sqrt{D}, \quad z_2 = -\tfrac{1}{2}(\lambda_1^2 + \lambda_2^2 + A_1\lambda_1 + A_2\lambda_2) - \tfrac{1}{2}\sqrt{D}.$$

Depending on the signs of z_1 and z_2 the following three cases are possible.

Case 1. If $z_1 > 0$ and $z_2 > 0$, then the solution of the integral equation has the form $(i = 1, 2)$:

$$y(x) = f(x) + \int_a^x \left\{ B_1 \sinh[\mu_1(x - t)] + B_2 \sinh\left[\mu_2(x - t)\right] \right\} f(t)\, dt, \qquad \mu_i = \sqrt{z_i},$$

where the coefficients B_1 and B_2 are determined from the following system of linear algebraic equations:

$$\frac{B_1\mu_1}{\lambda_1^2 + \mu_1^2} + \frac{B_2\mu_2}{\lambda_1^2 + \mu_2^2} - 1 = 0, \qquad \frac{B_1\mu_1}{\lambda_2^2 + \mu_1^2} + \frac{B_2\mu_2}{\lambda_2^2 + \mu_2^2} - 1 = 0.$$

Case 2. If $z_1 < 0$ and $z_2 < 0$, then the solution of the integral equation has the form

$$y(x) = f(x) + \int_a^x \{B_1 \sin[\mu_1(x-t)] + B_2 \sin[\mu_2(x-t)]\} f(t)\, dt, \qquad \mu_i = \sqrt{|z_i|},$$

where B_1 and B_2 are determined from the system

$$\frac{B_1\mu_1}{\lambda_1^2 - \mu_1^2} + \frac{B_2\mu_2}{\lambda_1^2 - \mu_2^2} - 1 = 0, \qquad \frac{B_1\mu_1}{\lambda_2^2 - \mu_1^2} + \frac{B_2\mu_2}{\lambda_2^2 - \mu_2^2} - 1 = 0.$$

Case 3. If $z_1 > 0$ and $z_2 < 0$, then the solution of the integral equation has the form

$$y(x) = f(x) + \int_a^x \{B_1 \sinh[\mu_1(x-t)] + B_2 \sin[\mu_2(x-t)]\} f(t)\, dt, \qquad \mu_i = \sqrt{|z_i|},$$

where B_1 and B_2 are determined from the system

$$\frac{B_1\mu_1}{\lambda_1^2 + \mu_1^2} + \frac{B_2\mu_2}{\lambda_1^2 - \mu_2^2} - 1 = 0, \qquad \frac{B_1\mu_1}{\lambda_2^2 + \mu_1^2} + \frac{B_2\mu_2}{\lambda_2^2 - \mu_2^2} - 1 = 0.$$

Remark. The solution of the original integral equation can be obtained from the solution of equation 2.3.18 by performing the following change of parameters:

$$\lambda_k \to i\lambda_k, \quad \mu_k \to i\mu_k, \quad A_k \to -iA_k, \quad B_k \to -iB_k, \quad i^2 = -1 \quad (k = 1, 2).$$

19. $$y(x) + \int_a^x \left\{ \sum_{k=1}^n A_k \sin[\lambda_k(x-t)] \right\} y(t)\, dt = f(x).$$

$1°.$ This integral equation can be reduced to a linear nonhomogeneous ordinary differential equation of order $2n$ with constant coefficients. Set

$$I_k(x) = \int_a^x \sin[\lambda_k(x-t)] y(t)\, dt. \tag{1}$$

Differentiating (1) with respect to x twice yields

$$I_k' = \lambda_k \int_a^x \cos[\lambda_k(x-t)] y(t)\, dt, \qquad I_k'' = \lambda_k y(x) - \lambda_k^2 \int_a^x \sin[\lambda_k(x-t)] y(t)\, dt, \tag{2}$$

where the primes stand for differentiation with respect to x. Comparing (1) and (2), we see that

$$I_k'' = \lambda_k y(x) - \lambda_k^2 I_k, \qquad I_k = I_k(x). \tag{3}$$

With aid of (1), the integral equation can be rewritten in the form

$$y(x) + \sum_{k=1}^n A_k I_k = f(x). \tag{4}$$

Differentiating (4) with respect to x twice taking into account (3) yields

$$y_{xx}''(x) + \sigma_n y(x) - \sum_{k=1}^n A_k \lambda_k^2 I_k = f_{xx}''(x), \qquad \sigma_n = \sum_{k=1}^n A_k \lambda_k. \tag{5}$$

Eliminating the integral I_n from (4) and (5), we obtain

$$y''_{xx}(x) + (\sigma_n + \lambda_n^2)y(x) + \sum_{k=1}^{n-1} A_k(\lambda_n^2 - \lambda_k^2)I_k = f''_{xx}(x) + \lambda_n^2 f(x). \tag{6}$$

Differentiating (6) with respect to x twice followed by eliminating I_{n-1} from the resulting expression with the aid of (6) yields a similar equation whose left-hand side is a fourth-order differential operator (acting on y) with constant coefficients plus the sum $\sum_{k=1}^{n-2} B_k I_k$.
Successively eliminating the terms $I_{n-2}, I_{n-3}, \ldots$ using double differentiation and formula (3), we finally arrive at a linear nonhomogeneous ordinary differential equation of order $2n$ with constant coefficients.

The initial conditions for $y(x)$ can be obtained by setting $x = a$ in the integral equation and all its derivative equations.

$2°$. Let us find the roots z_k of the algebraic equation

$$\sum_{k=1}^{n} \frac{\lambda_k A_k}{z + \lambda_k^2} + 1 = 0. \tag{7}$$

By reducing it to a common denominator, we arrive at the problem of determining the roots of an nth-degree characteristic polynomial.

Assume that all z_k are real, different, and nonzero. Let us divide the roots into two groups

$$z_1 > 0, \quad z_2 > 0, \quad \ldots, \quad z_s > 0 \quad \text{(positive roots)};$$
$$z_{s+1} < 0, \quad z_{s+2} < 0, \quad \ldots, \quad z_n < 0 \quad \text{(negative roots)}.$$

Then the solution of the integral equation can be written in the form

$$y(x) = f(x) + \int_a^x \left\{ \sum_{k=1}^{s} B_k \sinh\left[\mu_k(x-t)\right] + \sum_{k=s+1}^{n} C_k \sin\left[\mu_k(x-t)\right] \right\} f(t)\, dt, \quad \mu_k = \sqrt{|z_k|}. \tag{8}$$

The coefficients B_k and C_k are determined from the following system of linear algebraic equations:

$$\sum_{k=0}^{s} \frac{B_k \mu_k}{\lambda_m^2 + \mu_k^2} + \sum_{k=s+1}^{n} \frac{C_k \mu_k}{\lambda_m^2 - \mu_k^2} - 1 = 0, \quad \mu_k = \sqrt{|z_k|} \quad m = 1, 2, \ldots, n. \tag{9}$$

In the case of a nonzero root $z_s = 0$, we can introduce the new constant $D = B_s \mu_s$ and proceed to the limit $\mu_s \to 0$. As a result, the term $D(x - t)$ appears in solution (8) instead of $B_s \sinh\left[\mu_s(x - t)\right]$ and the corresponding terms $D\lambda_m^{-2}$ appear in system (9).

20. $\quad y(x) - A \displaystyle\int_a^x \frac{\sin(\lambda x)}{\sin(\lambda t)} y(t)\, dt = f(x).$

Solution:
$$y(x) = f(x) + A \int_a^x e^{A(x-t)} \frac{\sin(\lambda x)}{\sin(\lambda t)} f(t)\, dt.$$

21. $\quad y(x) - A \displaystyle\int_a^x \frac{\sin(\lambda t)}{\sin(\lambda x)} y(t)\, dt = f(x).$

Solution:
$$y(x) = f(x) + A \int_a^x e^{A(x-t)} \frac{\sin(\lambda t)}{\sin(\lambda x)} f(t)\, dt.$$

22. $y(x) - A \displaystyle\int_a^x \sin^k(\lambda x) \sin^m(\mu t) y(t)\, dt = f(x).$

This is a special case of equation 2.9.2 with $g(x) = A \sin^k(\lambda x)$ and $h(t) = \sin^m(\mu t)$.

23. $y(x) + A \displaystyle\int_a^x t \sin[\lambda(x - t)] y(t)\, dt = f(x).$

This is a special case of equation 2.9.36 with $g(t) = At$.

    Solution:

$$y(x) = f(x) + \frac{A\lambda}{W} \int_a^x t \big[u_1(x)u_2(t) - u_2(x)u_1(t)\big] f(t)\, dt,$$

where $u_1(x), u_2(x)$ is a fundamental system of solutions of the second-order linear ordinary differential equation $u''_{xx} + \lambda(Ax + \lambda)u = 0$, and W is the Wronskian.

    Depending on the sign of $A\lambda$, the functions $u_1(x)$ and $u_2(x)$ are expressed in terms of Bessel functions or modified Bessel functions as follows:

    if $A\lambda > 0$, then

$$u_1(x) = \xi^{1/2} J_{1/3}\big(\tfrac{2}{3}\sqrt{A\lambda}\,\xi^{3/2}\big), \quad u_2(x) = \xi^{1/2} Y_{1/3}\big(\tfrac{2}{3}\sqrt{A\lambda}\,\xi^{3/2}\big),$$
$$W = 3/\pi, \quad \xi = x + (\lambda/A);$$

    if $A\lambda < 0$, then

$$u_1(x) = \xi^{1/2} I_{1/3}\big(\tfrac{2}{3}\sqrt{-A\lambda}\,\xi^{3/2}\big), \quad u_2(x) = \xi^{1/2} K_{1/3}\big(\tfrac{2}{3}\sqrt{-A\lambda}\,\xi^{3/2}\big),$$
$$W = -\tfrac{3}{2}, \quad \xi = x + (\lambda/A).$$

24. $y(x) + A \displaystyle\int_a^x x \sin[\lambda(x - t)] y(t)\, dt = f(x).$

This is a special case of equation 2.9.37 with $g(x) = Ax$ and $h(t) = 1$.

    Solution:

$$y(x) = f(x) + \frac{A\lambda}{W} \int_a^x x \big[u_1(x)u_2(t) - u_2(x)u_1(t)\big] f(t)\, dt,$$

where $u_1(x), u_2(x)$ is a fundamental system of solutions of the second-order linear ordinary differential equation $u''_{xx} + \lambda(Ax + \lambda)u = 0$, and W is the Wronskian.

    The functions $u_1(x)$, $u_2(x)$, and W are specified in 2.5.23.

25. $y(x) + A \displaystyle\int_a^x t^k \sin^m(\lambda x) y(t)\, dt = f(x).$

This is a special case of equation 2.9.2 with $g(x) = -A \sin^m(\lambda x)$ and $h(t) = t^k$.

26. $y(x) + A \displaystyle\int_a^x x^k \sin^m(\lambda t) y(t)\, dt = f(x).$

This is a special case of equation 2.9.2 with $g(x) = -A x^k$ and $h(t) = \sin^m(\lambda t)$.

27. $y(x) - \displaystyle\int_a^x \big[A \sin(kx) + B - AB(x - t)\sin(kx)\big] y(t)\, dt = f(x).$

This is a special case of equation 2.9.7 with $\lambda = B$ and $g(x) = A \sin(kx)$.

    Solution:

$$y(x) = f(x) + \int_a^x R(x, t) f(t)\, dt,$$

$$R(x, t) = [A \sin(kx) + B]\frac{G(x)}{G(t)} + \frac{B^2}{G(t)} \int_t^x e^{B(x-s)} G(s)\, ds, \quad G(x) = \exp\left[-\frac{A}{k}\cos(kx)\right].$$

28. $y(x) + \int_a^x \left[A \sin(kt) + B + AB(x - t) \sin(kt) \right] y(t)\, dt = f(x).$

This is a special case of equation 2.9.8 with $\lambda = B$ and $g(t) = A \sin(kt)$.
  Solution:

$$y(x) = f(x) + \int_a^x R(x, t) f(t)\, dt,$$

$$R(x, t) = -[A \sin(kt) + B] \frac{G(t)}{G(x)} + \frac{B^2}{G(x)} \int_t^x e^{B(t-s)} G(s)\, ds, \quad G(x) = \exp\left[-\frac{A}{k} \cos(kx) \right].$$

29. $y(x) + A \int_x^\infty \sin\left(\lambda \sqrt{t - x} \right) y(t)\, dt = f(x).$

This is a special case of equation 2.9.62 with $K(x) = A \sin\left(\lambda \sqrt{-x} \right)$.

2.5-3. Kernels Containing Tangent

30. $y(x) - A \int_a^x \tan(\lambda x) y(t)\, dt = f(x).$

This is a special case of equation 2.9.2 with $g(x) = A \tan(\lambda x)$ and $h(t) = 1$.
  Solution:

$$y(x) = f(x) + A \int_a^x \tan(\lambda x) \left| \frac{\cos(\lambda t)}{\cos(\lambda x)} \right|^{A/\lambda} f(t)\, dt.$$

31. $y(x) - A \int_a^x \tan(\lambda t) y(t)\, dt = f(x).$

This is a special case of equation 2.9.2 with $g(x) = A$ and $h(t) = \tan(\lambda t)$.
  Solution:

$$y(x) = f(x) + A \int_a^x \tanh(\lambda t) \left| \frac{\cos(\lambda t)}{\cos(\lambda x)} \right|^{A/\lambda} f(t)\, dt.$$

32. $y(x) + A \int_a^x \left[\tan(\lambda x) - \tan(\lambda t) \right] y(t)\, dt = f(x).$

This is a special case of equation 2.9.5 with $g(x) = A \tan(\lambda x)$.
  Solution:

$$y(x) = f(x) + \frac{1}{W} \int_a^x \left[Y_1'(x) Y_2'(t) - Y_2'(x) Y_1'(t) \right] f(t)\, dt,$$

where $Y_1(x), Y_2(x)$ is a fundamental system of solutions of the second-order linear ordinary differential equation $\cos^2(\lambda x) Y_{xx}'' + A\lambda Y = 0$, W is the Wronskian, and the primes stand for the differentiation with respect to the argument specified in the parentheses.
  As shown in A. D. Polyanin and V. F. Zaitsev (1995, 1996), the functions $Y_1(x)$ and $Y_2(x)$ can be expressed via the hypergeometric function.

33. $y(x) - A \int_a^x \frac{\tan(\lambda x)}{\tan(\lambda t)} y(t)\, dt = f(x).$

Solution:

$$y(x) = f(x) + A \int_a^x e^{A(x-t)} \frac{\tan(\lambda x)}{\tan(\lambda t)} f(t)\, dt.$$

34. $y(x) - A \displaystyle\int_a^x \frac{\tan(\lambda t)}{\tan(\lambda x)} y(t)\, dt = f(x).$

Solution:
$$y(x) = f(x) + A \int_a^x e^{A(x-t)} \frac{\tan(\lambda t)}{\tan(\lambda x)} f(t)\, dt.$$

35. $y(x) - A \displaystyle\int_a^x \tan^k(\lambda x) \tan^m(\mu t) y(t)\, dt = f(x).$

This is a special case of equation 2.9.2 with $g(x) = A \tan^k(\lambda x)$ and $h(t) = \tan^m(\mu t)$.

36. $y(x) + A \displaystyle\int_a^x t^k \tan^m(\lambda x) y(t)\, dt = f(x).$

This is a special case of equation 2.9.2 with $g(x) = -A \tan^m(\lambda x)$ and $h(t) = t^k$.

37. $y(x) + A \displaystyle\int_a^x x^k \tan^m(\lambda t) y(t)\, dt = f(x).$

This is a special case of equation 2.9.2 with $g(x) = -Ax^k$ and $h(t) = \tan^m(\lambda t)$.

38. $y(x) - \displaystyle\int_a^x \big[A \tan(kx) + B - AB(x - t)\tan(kx) \big] y(t)\, dt = f(x).$

This is a special case of equation 2.9.7 with $\lambda = B$ and $g(x) = A\tan(kx)$.

39. $y(x) + \displaystyle\int_a^x \big[A \tan(kt) + B + AB(x - t)\tan(kt) \big] y(t)\, dt = f(x).$

This is a special case of equation 2.9.8 with $\lambda = B$ and $g(t) = A\tan(kt)$.

2.5-4. Kernels Containing Cotangent

40. $y(x) - A \displaystyle\int_a^x \cot(\lambda x) y(t)\, dt = f(x).$

This is a special case of equation 2.9.2 with $g(x) = A\cot(\lambda x)$ and $h(t) = 1$.
Solution:
$$y(x) = f(x) + A \int_a^x \cot(\lambda x) \left| \frac{\sin(\lambda x)}{\sin(\lambda t)} \right|^{A/\lambda} f(t)\, dt.$$

41. $y(x) - A \displaystyle\int_a^x \cot(\lambda t) y(t)\, dt = f(x).$

This is a special case of equation 2.9.2 with $g(x) = A$ and $h(t) = \cot(\lambda t)$.
Solution:
$$y(x) = f(x) + A \int_a^x \coth(\lambda t) \left| \frac{\sin(\lambda x)}{\sin(\lambda t)} \right|^{A/\lambda} f(t)\, dt.$$

42. $y(x) - A \displaystyle\int_a^x \frac{\cot(\lambda x)}{\cot(\lambda t)} y(t)\, dt = f(x).$

Solution:
$$y(x) = f(x) + A \int_a^x e^{A(x-t)} \frac{\cot(\lambda x)}{\cot(\lambda t)} f(t)\, dt.$$

43. $y(x) - A \displaystyle\int_a^x \frac{\cot(\lambda t)}{\cot(\lambda x)} y(t)\, dt = f(x).$

Solution:

$$y(x) = f(x) + A \int_a^x e^{A(x-t)} \frac{\cot(\lambda t)}{\cot(\lambda x)} f(t)\, dt.$$

44. $y(x) + A \displaystyle\int_a^x t^k \cot^m(\lambda x) y(t)\, dt = f(x).$

This is a special case of equation 2.9.2 with $g(x) = -A \cot^m(\lambda x)$ and $h(t) = t^k$.

45. $y(x) + A \displaystyle\int_a^x x^k \cot^m(\lambda t) y(t)\, dt = f(x).$

This is a special case of equation 2.9.2 with $g(x) = -Ax^k$ and $h(t) = \cot^m(\lambda t)$.

46. $y(x) - \displaystyle\int_a^x \big[A \cot(kx) + B - AB(x - t) \cot(kx) \big] y(t)\, dt = f(x).$

This is a special case of equation 2.9.7 with $\lambda = B$ and $g(x) = A \cot(kx)$.

47. $y(x) + \displaystyle\int_a^x \big[A \cot(kt) + B + AB(x - t) \cot(kt) \big] y(t)\, dt = f(x).$

This is a special case of equation 2.9.8 with $\lambda = B$ and $g(t) = A \cot(kt)$.

2.5-5. Kernels Containing Combinations of Trigonometric Functions

48. $y(x) - A \displaystyle\int_a^x \cos^k(\lambda x) \sin^m(\mu t) y(t)\, dt = f(x).$

This is a special case of equation 2.9.2 with $g(x) = A \cos^k(\lambda x)$ and $h(t) = \sin^m(\mu t)$.

49. $y(x) - \displaystyle\int_a^x \big\{ A + B \cos(\lambda x) - B(x - t)[\lambda \sin(\lambda x) + A \cos(\lambda x)] \big\} y(t)\, dt = f(x).$

This is a special case of equation 2.9.38 with $b = B$ and $g(x) = A$.

50. $y(x) - \displaystyle\int_a^x \big\{ A + B \sin(\lambda x) + B(x - t)[\lambda \cos(\lambda x) - A \sin(\lambda x)] \big\} y(t)\, dt = f(x).$

This is a special case of equation 2.9.39 with $b = B$ and $g(x) = A$.

51. $y(x) - A \displaystyle\int_a^x \tan^k(\lambda x) \cot^m(\mu t) y(t)\, dt = f(x).$

This is a special case of equation 2.9.2 with $g(x) = A \tan^k(\lambda x)$ and $h(t) = \cot^m(\mu t)$.

2.6. Equations Whose Kernels Contain Inverse Trigonometric Functions

2.6-1. Kernels Containing Arccosine

1. $y(x) - A \displaystyle\int_a^x \arccos(\lambda x) y(t)\, dt = f(x).$

This is a special case of equation 2.9.2 with $g(x) = A \arccos(\lambda x)$ and $h(t) = 1$.

2. $y(x) - A \int_a^x \arccos(\lambda t) y(t)\, dt = f(x).$

This is a special case of equation 2.9.2 with $g(x) = A$ and $h(t) = \arccos(\lambda t)$.

3. $y(x) - A \int_a^x \dfrac{\arccos(\lambda x)}{\arccos(\lambda t)} y(t)\, dt = f(x).$

Solution:
$$y(x) = f(x) + A \int_a^x e^{A(x-t)} \frac{\arccos(\lambda x)}{\arccos(\lambda t)} f(t)\, dt.$$

4. $y(x) - A \int_a^x \dfrac{\arccos(\lambda t)}{\arccos(\lambda x)} y(t)\, dt = f(x).$

Solution:
$$y(x) = f(x) + A \int_a^x e^{A(x-t)} \frac{\arccos(\lambda t)}{\arccos(\lambda x)} f(t)\, dt.$$

5. $y(x) - \int_a^x \big[A \arccos(kx) + B - AB(x-t) \arccos(kx) \big] y(t)\, dt = f(x).$

This is a special case of equation 2.9.7 with $\lambda = B$ and $g(x) = A \arccos(kx)$.

6. $y(x) + \int_a^x \big[A \arccos(kt) + B + AB(x-t) \arccos(kt) \big] y(t)\, dt = f(x).$

This is a special case of equation 2.9.8 with $\lambda = B$ and $g(t) = A \arccos(kt)$.

2.6-2. Kernels Containing Arcsine

7. $y(x) - A \int_a^x \arcsin(\lambda x) y(t)\, dt = f(x).$

This is a special case of equation 2.9.2 with $g(x) = A \arcsin(\lambda x)$ and $h(t) = 1$.

8. $y(x) - A \int_a^x \arcsin(\lambda t) y(t)\, dt = f(x).$

This is a special case of equation 2.9.2 with $g(x) = A$ and $h(t) = \arcsin(\lambda t)$.

9. $y(x) - A \int_a^x \dfrac{\arcsin(\lambda x)}{\arcsin(\lambda t)} y(t)\, dt = f(x).$

Solution:
$$y(x) = f(x) + A \int_a^x e^{A(x-t)} \frac{\arcsin(\lambda x)}{\arcsin(\lambda t)} f(t)\, dt.$$

10. $y(x) - A \int_a^x \dfrac{\arcsin(\lambda t)}{\arcsin(\lambda x)} y(t)\, dt = f(x).$

Solution:
$$y(x) = f(x) + A \int_a^x e^{A(x-t)} \frac{\arcsin(\lambda t)}{\arcsin(\lambda x)} f(t)\, dt.$$

11. $y(x) - \int_a^x \left[A \arcsin(kx) + B - AB(x-t) \arcsin(kx) \right] y(t)\, dt = f(x).$

This is a special case of equation 2.9.7 with $\lambda = B$ and $g(x) = A \arcsin(kx)$.

12. $y(x) + \int_a^x \left[A \arcsin(kt) + B + AB(x-t) \arcsin(kt) \right] y(t)\, dt = f(x).$

This is a special case of equation 2.9.8 with $\lambda = B$ and $g(t) = A \arcsin(kt)$.

2.6-3. Kernels Containing Arctangent

13. $y(x) - A \int_a^x \arctan(\lambda x) y(t)\, dt = f(x).$

This is a special case of equation 2.9.2 with $g(x) = A \arctan(\lambda x)$ and $h(t) = 1$.

14. $y(x) - A \int_a^x \arctan(\lambda t) y(t)\, dt = f(x).$

This is a special case of equation 2.9.2 with $g(x) = A$ and $h(t) = \arctan(\lambda t)$.

15. $y(x) - A \int_a^x \dfrac{\arctan(\lambda x)}{\arctan(\lambda t)} y(t)\, dt = f(x).$

Solution:

$$y(x) = f(x) + A \int_a^x e^{A(x-t)} \frac{\arctan(\lambda x)}{\arctan(\lambda t)} f(t)\, dt.$$

16. $y(x) - A \int_a^x \dfrac{\arctan(\lambda t)}{\arctan(\lambda x)} y(t)\, dt = f(x).$

Solution:

$$y(x) = f(x) + A \int_a^x e^{A(x-t)} \frac{\arctan(\lambda t)}{\arctan(\lambda x)} f(t)\, dt.$$

17. $y(x) + A \int_x^\infty \arctan[\lambda(t-x)] y(t)\, dt = f(x).$

This is a special case of equation 2.9.62 with $K(x) = A \arctan(-\lambda x)$.

18. $y(x) - \int_a^x \left[A \arctan(kx) + B - AB(x-t) \arctan(kx) \right] y(t)\, dt = f(x).$

This is a special case of equation 2.9.7 with $\lambda = B$ and $g(x) = A \arctan(kx)$.

19. $y(x) + \int_a^x \left[A \arctan(kt) + B + AB(x-t) \arctan(kt) \right] y(t)\, dt = f(x).$

This is a special case of equation 2.9.8 with $\lambda = B$ and $g(t) = A \arctan(kt)$.

2.6-4. Kernels Containing Arccotangent

20. $y(x) - A \int_a^x \mathrm{arccot}(\lambda x) y(t)\, dt = f(x).$

This is a special case of equation 2.9.2 with $g(x) = A \,\mathrm{arccot}(\lambda x)$ and $h(t) = 1$.

21. $\quad y(x) - A \displaystyle\int_a^x \text{arccot}(\lambda t) y(t)\, dt = f(x).$

This is a special case of equation 2.9.2 with $g(x) = A$ and $h(t) = \text{arccot}(\lambda t)$.

22. $\quad y(x) - A \displaystyle\int_a^x \dfrac{\text{arccot}(\lambda x)}{\text{arccot}(\lambda t)} y(t)\, dt = f(x).$

Solution:

$$y(x) = f(x) + A \int_a^x e^{A(x-t)} \frac{\text{arccot}(\lambda x)}{\text{arccot}(\lambda t)} f(t)\, dt.$$

23. $\quad y(x) - A \displaystyle\int_a^x \dfrac{\text{arccot}(\lambda t)}{\text{arccot}(\lambda x)} y(t)\, dt = f(x).$

Solution:

$$y(x) = f(x) + A \int_a^x e^{A(x-t)} \frac{\text{arccot}(\lambda t)}{\text{arccot}(\lambda x)} f(t)\, dt.$$

24. $\quad y(x) + A \displaystyle\int_x^\infty \text{arccot}[\lambda(t-x)] y(t)\, dt = f(x).$

This is a special case of equation 2.9.62 with $K(x) = A\,\text{arccot}(-\lambda x)$.

25. $\quad y(x) - \displaystyle\int_a^x \Big[A\,\text{arccot}(kx) + B - AB(x-t)\,\text{arccot}(kx)\Big] y(t)\, dt = f(x).$

This is a special case of equation 2.9.7 with $\lambda = B$ and $g(x) = A\,\text{arccot}(kx)$.

26. $\quad y(x) + \displaystyle\int_a^x \Big[A\,\text{arccot}(kt) + B + AB(x-t)\,\text{arccot}(kt)\Big] y(t)\, dt = f(x).$

This is a special case of equation 2.9.8 with $\lambda = B$ and $g(t) = A\,\text{arccot}(kt)$.

2.7. Equations Whose Kernels Contain Combinations of Elementary Functions

2.7-1. Kernels Containing Exponential and Hyperbolic Functions

1. $\quad y(x) + A \displaystyle\int_a^x e^{\mu(x-t)} \cosh[\lambda(x-t)] y(t)\, dt = f(x).$

Solution:

$$y(x) = f(x) + \int_a^x R(x-t) f(t)\, dt,$$

$$R(x) = \exp\big[(\mu - \tfrac{1}{2}A)x\big]\left[\frac{A^2}{2k}\sinh(kx) - A\cosh(kx)\right], \quad k = \sqrt{\lambda^2 + \tfrac{1}{4}A^2}.$$

2. $\quad y(x) + A \displaystyle\int_a^x e^{\mu(x-t)} \sinh[\lambda(x-t)]y(t)\,dt = f(x).$

1°. Solution with $\lambda(A - \lambda) > 0$:

$$y(x) = f(x) - \frac{A\lambda}{k}\int_a^x e^{\mu(x-t)}\sin[k(x-t)]f(t)\,dt, \qquad \text{where} \quad k = \sqrt{\lambda(A-\lambda)}.$$

2°. Solution with $\lambda(A - \lambda) < 0$:

$$y(x) = f(x) - \frac{A\lambda}{k}\int_a^x e^{\mu(x-t)}\sinh[k(x-t)]f(t)\,dt, \qquad \text{where} \quad k = \sqrt{\lambda(\lambda-A)}.$$

3°. Solution with $A = \lambda$:

$$y(x) = f(x) - \lambda^2\int_a^x (x-t)e^{\mu(x-t)}f(t)\,dt.$$

3. $\quad y(x) + \displaystyle\int_a^x e^{\mu(x-t)}\big\{A_1\sinh[\lambda_1(x-t)] + A_2\sinh[\lambda_2(x-t)]\big\}y(t)\,dt = f(x).$

The substitution $w(x) = e^{-\mu x}y(x)$ leads to an equation of the form 2.3.18:

$$w(x) + \int_a^x \big\{A_1\sinh[\lambda_1(x-t)] + A_2\sinh[\lambda_2(x-t)]\big\}w(t)\,dt = e^{-\mu x}f(x).$$

4. $\quad y(x) + A\displaystyle\int_a^x te^{\mu(x-t)}\sinh[\lambda(x-t)]y(t)\,dt = f(x).$

The substitution $w(x) = e^{-\mu x}y(x)$ leads to an equation of the form 2.3.23:

$$w(x) + A\int_a^x t\sinh[\lambda(x-t)]w(t)\,dt = e^{-\mu x}f(x).$$

2.7-2. Kernels Containing Exponential and Logarithmic Functions

5. $\quad y(x) - A\displaystyle\int_a^x e^{\mu t}\ln(\lambda x)y(t)\,dt = f(x).$

This is a special case of equation 2.9.2 with $g(x) = A\ln(\lambda x)$ and $h(t) = e^{\mu t}$.

6. $\quad y(x) - A\displaystyle\int_a^x e^{\mu x}\ln(\lambda t)y(t)\,dt = f(x).$

This is a special case of equation 2.9.2 with $g(x) = Ae^{\mu x}$ and $h(t) = \ln(\lambda t)$.

7. $\quad y(x) - A\displaystyle\int_a^x e^{\mu(x-t)}\ln(\lambda x)y(t)\,dt = f(x).$

Solution:

$$y(x) = f(x) + A\int_a^x e^{(\mu-A)(x-t)}\ln(\lambda x)\frac{(\lambda x)^{Ax}}{(\lambda t)^{At}}f(t)\,dt.$$

8. $\quad y(x) - A \displaystyle\int_a^x e^{\mu(x-t)} \ln(\lambda t) y(t)\, dt = f(x).$

Solution:

$$y(x) = f(x) + A \int_a^x e^{(\mu - A)(x-t)} \ln(\lambda t) \frac{(\lambda x)^{Ax}}{(\lambda t)^{At}} f(t)\, dt.$$

9. $\quad y(x) + A \displaystyle\int_a^x e^{\mu(x-t)} (\ln x - \ln t) y(t)\, dt = f(x).$

Solution:

$$y(x) = f(x) + \frac{1}{W} \int_a^x e^{\mu(x-t)} \big[u_1'(x) u_2'(t) - u_2'(x) u_1'(t) \big] f(t)\, dt,$$

where the primes stand for the differentiation with respect to the argument specified in the parentheses, and $u_1(x)$, $u_2(x)$ is a fundamental system of solutions of the second-order linear homogeneous ordinary differential equation $u''_{xx} + Ax^{-1}u = 0$, with $u_1(x)$ and $u_2(x)$ expressed in terms of Bessel functions or modified Bessel functions, depending on the sign of A:

$$W = \tfrac{1}{\pi}, \quad u_1(x) = \sqrt{x}\, J_1\big(2\sqrt{Ax}\big), \quad u_2(x) = \sqrt{x}\, Y_1\big(2\sqrt{Ax}\big) \qquad \text{for} \quad A > 0,$$

$$W = -\tfrac{1}{2}, \quad u_1(x) = \sqrt{x}\, I_1\big(2\sqrt{-Ax}\big), \quad u_2(x) = \sqrt{x}\, K_1\big(2\sqrt{-Ax}\big) \qquad \text{for} \quad A < 0.$$

10. $\quad y(x) + a \displaystyle\int_x^\infty e^{\lambda(x-t)} \ln(t - x) y(t)\, dt = f(x).$

This is a special case of equation 2.9.62 with $K(x) = ae^{\lambda x} \ln(-x)$.

3.7-3. Kernels Containing Exponential and Trigonometric Functions

11. $\quad y(x) - A \displaystyle\int_a^x e^{\mu t} \cos(\lambda x) y(t)\, dt = f(x).$

This is a special case of equation 2.9.2 with $g(x) = A \cos(\lambda x)$ and $h(t) = e^{\mu t}$.

12. $\quad y(x) - A \displaystyle\int_a^x e^{\mu x} \cos(\lambda t) y(t)\, dt = f(x).$

This is a special case of equation 2.9.2 with $g(x) = Ae^{\mu x}$ and $h(t) = \cos(\lambda t)$.

13. $\quad y(x) + A \displaystyle\int_a^x e^{\mu(x-t)} \cos[\lambda(x - t)] y(t)\, dt = f(x).$

$1°$. Solution with $|A| > 2|\lambda|$:

$$y(x) = f(x) + \int_a^x R(x - t) f(t)\, dt,$$

$$R(x) = \exp\big[(\mu - \tfrac{1}{2}A)x\big] \left[\frac{A^2}{2k} \sinh(kx) - A \cosh(kx) \right], \quad k = \sqrt{\tfrac{1}{4}A^2 - \lambda^2}.$$

$2°$. Solution with $|A| < 2|\lambda|$:

$$y(x) = f(x) + \int_a^x R(x - t) f(t)\, dt,$$

$$R(x) = \exp\big[(\mu - \tfrac{1}{2}A)x\big] \left[\frac{A^2}{2k} \sin(kx) - A \cos(kx) \right], \quad k = \sqrt{\lambda^2 - \tfrac{1}{4}A^2}.$$

$3°$. Solution with $\lambda = \pm\tfrac{1}{2}A$:

$$y(x) = f(x) + \int_a^x R(x - t) f(t)\, dt, \qquad R(x) = \big(\tfrac{1}{2}A^2 x - A\big) \exp\big[(\mu - \tfrac{1}{2}A)x\big].$$

14. $y(x) - \int_a^x e^{\mu(x-t)} \big[A \cos(kx) + B - AB(x-t) \cos(kx) \big] y(t)\, dt = f(x).$

Solution:

$$y(x) = f(x) + \int_a^x e^{\mu(x-t)} M(x,t) f(t)\, dt,$$

$$M(x,t) = [A \cos(kx) + B] \frac{G(x)}{G(t)} + \frac{B^2}{G(t)} \int_t^x e^{B(x-s)} G(s)\, ds, \quad G(x) = \exp\left[\frac{A}{k} \sin(kx)\right].$$

15. $y(x) + \int_a^x e^{\mu(x-t)} \big[A \cos(kt) + B + AB(x-t) \cos(kt) \big] y(t)\, dt = f(x).$

Solution:

$$y(x) = f(x) + \int_a^x e^{\mu(x-t)} M(x,t) f(t)\, dt,$$

$$M(x,t) = -[A \cos(kt) + B] \frac{G(t)}{G(x)} + \frac{B^2}{G(x)} \int_t^x e^{B(t-s)} G(s)\, ds, \quad G(x) = \exp\left[\frac{A}{k} \sin(kx)\right].$$

16. $y(x) - A \int_a^x e^{\mu t} \sin(\lambda x) y(t)\, dt = f(x).$

This is a special case of equation 2.9.2 with $g(x) = A \sin(\lambda x)$ and $h(t) = e^{\mu t}$.

17. $y(x) - A \int_a^x e^{\mu x} \sin(\lambda t) y(t)\, dt = f(x).$

This is a special case of equation 2.9.2 with $g(x) = A e^{\mu x}$ and $h(t) = \sin(\lambda t)$.

18. $y(x) + A \int_a^x e^{\mu(x-t)} \sin[\lambda(x-t)] y(t)\, dt = f(x).$

1°. Solution with $\lambda(A + \lambda) > 0$:

$$y(x) = f(x) - \frac{A\lambda}{k} \int_a^x e^{\mu(x-t)} \sin[k(x-t)] f(t)\, dt, \qquad \text{where} \quad k = \sqrt{\lambda(A+\lambda)}.$$

2°. Solution with $\lambda(A + \lambda) < 0$:

$$y(x) = f(x) - \frac{A\lambda}{k} \int_a^x e^{\mu(x-t)} \sinh[k(x-t)] f(t)\, dt, \qquad \text{where} \quad k = \sqrt{-\lambda(\lambda+A)}.$$

3°. Solution with $A = -\lambda$:

$$y(x) = f(x) + \lambda^2 \int_a^x (x-t) e^{\mu(x-t)} f(t)\, dt.$$

19. $y(x) + A \int_a^x e^{\mu(x-t)} \sin^3[\lambda(x-t)] y(t)\, dt = f(x).$

The substitution $w(x) = e^{-\mu x} y(x)$ leads to an equation of the form 2.5.17:

$$w(x) + A \int_a^x \sin^3[\lambda(x-t)] w(t)\, dt = e^{-\mu x} f(x).$$

20. $y(x) + \int_a^x e^{\mu(x-t)}\{A_1\sin[\lambda_1(x-t)] + A_2\sin[\lambda_2(x-t)]\}y(t)\,dt = f(x).$

The substitution $w(x) = e^{-\mu x}y(x)$ leads to an equation of the form 2.5.18:

$$w(x) + \int_a^x \{A_1\sin[\lambda_1(x-t)] + A_2\sin[\lambda_2(x-t)]\}w(t)\,dt = e^{-\mu x}f(x).$$

21. $y(x) + \int_a^x e^{\mu(x-t)}\left\{\sum_{k=1}^n A_k\sin[\lambda_k(x-t)]\right\}y(t)\,dt = f(x).$

The substitution $w(x) = e^{-\mu x}y(x)$ leads to an equation of the form 2.5.19:

$$w(x) + \int_a^x \left\{\sum_{k=1}^n A_k\sin[\lambda_k(x-t)]\right\}w(t)\,dt = e^{-\mu x}f(x).$$

22. $y(x) + A\int_a^x te^{\mu(x-t)}\sin[\lambda(x-t)]y(t)\,dt = f(x).$

Solution:

$$y(x) = f(x) + \frac{A\lambda}{W}\int_a^x te^{\mu(x-t)}\big[u_1(x)u_2(t) - u_2(x)u_1(t)\big]f(t)\,dt,$$

where $u_1(x), u_2(x)$ is a fundamental system of solutions of the second-order linear ordinary differential equation $u_{xx}'' + \lambda(Ax + \lambda)u = 0$, and W is the Wronskian.

Depending on the sign of $A\lambda$, the functions $u_1(x)$ and $u_2(x)$ are expressed in terms of Bessel functions or modified Bessel functions as follows:

if $A\lambda > 0$, then

$$u_1(x) = \xi^{1/2}J_{1/3}\big(\tfrac{2}{3}\sqrt{A\lambda}\,\xi^{3/2}\big), \quad u_2(x) = \xi^{1/2}Y_{1/3}\big(\tfrac{2}{3}\sqrt{A\lambda}\,\xi^{3/2}\big),$$
$$W = 3/\pi, \quad \xi = x + (\lambda/A);$$

if $A\lambda < 0$, then

$$u_1(x) = \xi^{1/2}I_{1/3}\big(\tfrac{2}{3}\sqrt{-A\lambda}\,\xi^{3/2}\big), \quad u_2(x) = \xi^{1/2}K_{1/3}\big(\tfrac{2}{3}\sqrt{-A\lambda}\,\xi^{3/2}\big),$$
$$W = -\tfrac{3}{2}, \quad \xi = x + (\lambda/A).$$

23. $y(x) + A\int_a^x xe^{\mu(x-t)}\sin[\lambda(x-t)]y(t)\,dt = f(x).$

Solution:

$$y(x) = f(x) + \frac{A\lambda}{W}\int_a^x xe^{\mu(x-t)}\big[u_1(x)u_2(t) - u_2(x)u_1(t)\big]f(t)\,dt,$$

where $u_1(x), u_2(x)$ is a fundamental system of solutions of the second-order linear ordinary differential equation $u_{xx}'' + \lambda(Ax + \lambda)u = 0$, and W is the Wronskian.

The functions $u_1(x), u_2(x)$, and W are specified in 2.7.22.

24. $y(x) + A\int_x^\infty e^{\mu(t-x)}\sin\big(\lambda\sqrt{t-x}\,\big)y(t)\,dt = f(x).$

This is a special case of equation 2.9.62 with $K(x) = Ae^{-\mu x}\sin\big(\lambda\sqrt{-x}\,\big).$

25. $y(x) - \displaystyle\int_a^x e^{\mu(x-t)} \big[A\sin(kx) + B - AB(x-t)\sin(kx)\big] y(t)\, dt = f(x).$

Solution:

$$y(x) = f(x) + \int_a^x e^{\mu(x-t)} M(x,t) f(t)\, dt,$$

$$M(x,t) = [A\sin(kx) + B]\frac{G(x)}{G(t)} + \frac{B^2}{G(t)}\int_t^x e^{B(x-s)}G(s)\, ds, \quad G(x) = \exp\!\left[-\frac{A}{k}\cos(kx)\right].$$

26. $y(x) + \displaystyle\int_a^x e^{\mu(x-t)} \big[A\sin(kt) + B + AB(x-t)\sin(kt)\big] y(t)\, dt = f(x).$

Solution:

$$y(x) = f(x) + \int_a^x e^{\mu(x-t)} M(x,t) f(t)\, dt,$$

$$M(x,t) = -[A\sin(kt) + B]\frac{G(t)}{G(x)} + \frac{B^2}{G(x)}\int_t^x e^{B(t-s)}G(s)\, ds, \quad G(x) = \exp\!\left[-\frac{A}{k}\cos(kx)\right].$$

27. $y(x) - A\displaystyle\int_a^x e^{\mu t}\tan(\lambda x) y(t)\, dt = f(x).$

This is a special case of equation 2.9.2 with $g(x) = A\tan(\lambda x)$ and $h(t) = e^{\mu t}$.

28. $y(x) - A\displaystyle\int_a^x e^{\mu x}\tan(\lambda t) y(t)\, dt = f(x).$

This is a special case of equation 2.9.2 with $g(x) = Ae^{\mu x}$ and $h(t) = \tan(\lambda t)$.

29. $y(x) + A\displaystyle\int_a^x e^{\mu(x-t)} \big[\tan(\lambda x) - \tan(\lambda t)\big] y(t)\, dt = f(x).$

The substitution $w(x) = e^{-\mu x} y(x)$ leads to an equation of the form 2.5.32:

$$w(x) + A\int_a^x \big[\tan(\lambda x) - \tan(\lambda t)\big] w(t)\, dt = e^{-\mu x} f(x).$$

30. $y(x) - \displaystyle\int_a^x e^{\mu(x-t)} \big[A\tan(kx) + B - AB(x-t)\tan(kx)\big] y(t)\, dt = f(x).$

The substitution $w(x) = e^{-\mu x} y(x)$ leads to an equation of the form 2.9.7 with $\lambda = B$ and $g(x) = A\tan(kx)$:

$$w(x) - \int_a^x \big[A\tan(kx) + B - AB(x-t)\tan(kx)\big] w(t)\, dt = e^{-\mu x} f(x).$$

31. $y(x) + \displaystyle\int_a^x e^{\mu(x-t)} \big[A\tan(kt) + B + AB(x-t)\tan(kt)\big] y(t)\, dt = f(x).$

The substitution $w(x) = e^{-\mu x} y(x)$ leads to an equation of the form 2.9.8 with $\lambda = B$ and $g(t) = A\tan(kt)$:

$$w(x) + \int_a^x \big[A\tan(kt) + B + AB(x-t)\tan(kt)\big] w(t)\, dt = e^{-\mu x} f(x).$$

32. $y(x) - A \int_a^x e^{\mu t} \cot(\lambda x) y(t)\, dt = f(x).$

This is a special case of equation 2.9.2 with $g(x) = A \cot(\lambda x)$ and $h(t) = e^{\mu t}$.

33. $y(x) - A \int_a^x e^{\mu x} \cot(\lambda t) y(t)\, dt = f(x).$

This is a special case of equation 2.9.2 with $g(x) = A e^{\mu x}$ and $h(t) = \cot(\lambda t)$.

2.7-4. Kernels Containing Hyperbolic and Logarithmic Functions

34. $y(x) - A \int_a^x \cosh^k(\lambda x) \ln^m(\mu t) y(t)\, dt = f(x).$

This is a special case of equation 2.9.2 with $g(x) = A \cosh^k(\lambda x)$ and $h(t) = \ln^m(\mu t)$.

35. $y(x) - A \int_a^x \cosh^k(\lambda t) \ln^m(\mu x) y(t)\, dt = f(x).$

This is a special case of equation 2.9.2 with $g(x) = A \ln^m(\mu x)$ and $h(t) = \cosh^k(\lambda t)$.

36. $y(x) - A \int_a^x \sinh^k(\lambda x) \ln^m(\mu t) y(t)\, dt = f(x).$

This is a special case of equation 2.9.2 with $g(x) = A \sinh^k(\lambda x)$ and $h(t) = \ln^m(\mu t)$.

37. $y(x) - A \int_a^x \sinh^k(\lambda t) \ln^m(\mu x) y(t)\, dt = f(x).$

This is a special case of equation 2.9.2 with $g(x) = A \ln^m(\mu x)$ and $h(t) = \sinh^k(\lambda t)$.

38. $y(x) - A \int_a^x \tanh^k(\lambda x) \ln^m(\mu t) y(t)\, dt = f(x).$

This is a special case of equation 2.9.2 with $g(x) = A \tanh^k(\lambda x)$ and $h(t) = \ln^m(\mu t)$.

39. $y(x) - A \int_a^x \tanh^k(\lambda t) \ln^m(\mu x) y(t)\, dt = f(x).$

This is a special case of equation 2.9.2 with $g(x) = A \ln^m(\mu x)$ and $h(t) = \tanh^k(\lambda t)$.

40. $y(x) - A \int_a^x \coth^k(\lambda x) \ln^m(\mu t) y(t)\, dt = f(x).$

This is a special case of equation 2.9.2 with $g(x) = A \coth^k(\lambda x)$ and $h(t) = \ln^m(\mu t)$.

41. $y(x) - A \int_a^x \coth^k(\lambda t) \ln^m(\mu x) y(t)\, dt = f(x).$

This is a special case of equation 2.9.2 with $g(x) = A \ln^m(\mu x)$ and $h(t) = \coth^k(\lambda t)$.

2.7-5. Kernels Containing Hyperbolic and Trigonometric Functions

42. $y(x) - A \displaystyle\int_a^x \cosh^k(\lambda x) \cos^m(\mu t) y(t)\, dt = f(x).$

This is a special case of equation 2.9.2 with $g(x) = A \cosh^k(\lambda x)$ and $h(t) = \cos^m(\mu t)$.

43. $y(x) - A \displaystyle\int_a^x \cosh^k(\lambda t) \cos^m(\mu x) y(t)\, dt = f(x).$

This is a special case of equation 2.9.2 with $g(x) = A \cos^m(\mu x)$ and $h(t) = \cosh^k(\lambda t)$.

44. $y(x) - A \displaystyle\int_a^x \cosh^k(\lambda x) \sin^m(\mu t) y(t)\, dt = f(x).$

This is a special case of equation 2.9.2 with $g(x) = A \cosh^k(\lambda x)$ and $h(t) = \sin^m(\mu t)$.

45. $y(x) - A \displaystyle\int_a^x \cosh^k(\lambda t) \sin^m(\mu x) y(t)\, dt = f(x).$

This is a special case of equation 2.9.2 with $g(x) = A \sin^m(\mu x)$ and $h(t) = \cosh^k(\lambda t)$.

46. $y(x) - A \displaystyle\int_a^x \sinh^k(\lambda x) \cos^m(\mu t) y(t)\, dt = f(x).$

This is a special case of equation 2.9.2 with $g(x) = A \sinh^k(\lambda x)$ and $h(t) = \cos^m(\mu t)$.

47. $y(x) - A \displaystyle\int_a^x \sinh^k(\lambda t) \cos^m(\mu x) y(t)\, dt = f(x).$

This is a special case of equation 2.9.2 with $g(x) = A \cos^m(\mu x)$ and $h(t) = \sinh^k(\lambda t)$.

48. $y(x) - A \displaystyle\int_a^x \sinh^k(\lambda x) \sin^m(\mu t) y(t)\, dt = f(x).$

This is a special case of equation 2.9.2 with $g(x) = A \sinh^k(\lambda x)$ and $h(t) = \sin^m(\mu t)$.

49. $y(x) - A \displaystyle\int_a^x \sinh^k(\lambda t) \sin^m(\mu x) y(t)\, dt = f(x).$

This is a special case of equation 2.9.2 with $g(x) = A \sin^m(\mu x)$ and $h(t) = \sinh^k(\lambda t)$.

50. $y(x) - A \displaystyle\int_a^x \tanh^k(\lambda x) \cos^m(\mu t) y(t)\, dt = f(x).$

This is a special case of equation 2.9.2 with $g(x) = A \tanh^k(\lambda x)$ and $h(t) = \cos^m(\mu t)$.

51. $y(x) - A \displaystyle\int_a^x \tanh^k(\lambda t) \cos^m(\mu x) y(t)\, dt = f(x).$

This is a special case of equation 2.9.2 with $g(x) = A \cos^m(\mu x)$ and $h(t) = \tanh^k(\lambda t)$.

52. $y(x) - A \displaystyle\int_a^x \tanh^k(\lambda x) \sin^m(\mu t) y(t)\, dt = f(x).$

This is a special case of equation 2.9.2 with $g(x) = A \tanh^k(\lambda x)$ and $h(t) = \sin^m(\mu t)$.

53. $y(x) - A \displaystyle\int_a^x \tanh^k(\lambda t) \sin^m(\mu x) y(t)\, dt = f(x).$

This is a special case of equation 2.9.2 with $g(x) = A \sin^m(\mu x)$ and $h(t) = \tanh^k(\lambda t)$.

2.7-6. Kernels Containing Logarithmic and Trigonometric Functions

54. $y(x) - A \int_a^x \cos^k(\lambda x) \ln^m(\mu t) y(t)\, dt = f(x).$

This is a special case of equation 2.9.2 with $g(x) = A \cos^k(\lambda x)$ and $h(t) = \ln^m(\mu t)$.

55. $y(x) - A \int_a^x \cos^k(\lambda t) \ln^m(\mu x) y(t)\, dt = f(x).$

This is a special case of equation 2.9.2 with $g(x) = A \ln^m(\mu x)$ and $h(t) = \cos^k(\lambda t)$.

56. $y(x) - A \int_a^x \sin^k(\lambda x) \ln^m(\mu t) y(t)\, dt = f(x).$

This is a special case of equation 2.9.2 with $g(x) = A \sin^k(\lambda x)$ and $h(t) = \ln^m(\mu t)$.

57. $y(x) - A \int_a^x \sin^k(\lambda t) \ln^m(\mu x) y(t)\, dt = f(x).$

This is a special case of equation 2.9.2 with $g(x) = A \ln^m(\mu x)$ and $h(t) = \sin^k(\lambda t)$.

58. $y(x) - A \int_a^x \tan^k(\lambda x) \ln^m(\mu t) y(t)\, dt = f(x).$

This is a special case of equation 2.9.2 with $g(x) = A \tan^k(\lambda x)$ and $h(t) = \ln^m(\mu t)$.

59. $y(x) - A \int_a^x \tan^k(\lambda t) \ln^m(\mu x) y(t)\, dt = f(x).$

This is a special case of equation 2.9.2 with $g(x) = A \ln^m(\mu x)$ and $h(t) = \tan^k(\lambda t)$.

60. $y(x) - A \int_a^x \cot^k(\lambda x) \ln^m(\mu t) y(t)\, dt = f(x).$

This is a special case of equation 2.9.2 with $g(x) = A \cot^k(\lambda x)$ and $h(t) = \ln^m(\mu t)$.

61. $y(x) - A \int_a^x \cot^k(\lambda t) \ln^m(\mu x) y(t)\, dt = f(x).$

This is a special case of equation 2.9.2 with $g(x) = A \ln^m(\mu x)$ and $h(t) = \cot^k(\lambda t)$.

2.8. Equations Whose Kernels Contain Special Functions

2.8-1. Kernels Containing Bessel Functions

1. $y(x) - \lambda \int_0^x J_0(x - t) y(t)\, dt = f(x).$

Solution:

$$y(x) = f(x) + \int_0^x R(x - t) f(t)\, dt,$$

where

$$R(x) = \lambda \cos\left(\sqrt{1-\lambda^2}\, x\right) + \frac{\lambda^2}{\sqrt{1-\lambda^2}} \sin\left(\sqrt{1-\lambda^2}\, x\right) + \frac{\lambda}{\sqrt{1-\lambda^2}} \int_0^x \sin\left[\sqrt{1-\lambda^2}\,(x-t)\right] \frac{J_1(t)}{t}\, dt.$$

⊙ Reference: V. I. Smirnov (1974).

2. $y(x) - A \displaystyle\int_a^x J_\nu(\lambda x) y(t)\, dt = f(x).$

This is a special case of equation 2.9.2 with $g(x) = A J_\nu(\lambda x)$ and $h(t) = 1$.

3. $y(x) - A \displaystyle\int_a^x J_\nu(\lambda t) y(t)\, dt = f(x).$

This is a special case of equation 2.9.2 with $g(x) = A$ and $h(t) = J_\nu(\lambda t)$.

4. $y(x) - A \displaystyle\int_a^x \frac{J_\nu(\lambda x)}{J_\nu(\lambda t)} y(t)\, dt = f(x).$

Solution:
$$y(x) = f(x) + A \int_a^x e^{A(x-t)} \frac{J_\nu(\lambda x)}{J_\nu(\lambda t)} f(t)\, dt.$$

5. $y(x) - A \displaystyle\int_a^x \frac{J_\nu(\lambda t)}{J_\nu(\lambda x)} y(t)\, dt = f(x).$

Solution:
$$y(x) = f(x) + A \int_a^x e^{A(x-t)} \frac{J_\nu(\lambda t)}{J_\nu(\lambda x)} f(t)\, dt.$$

6. $y(x) + A \displaystyle\int_x^\infty J_\nu[\lambda(t - x)] y(t)\, dt = f(x).$

This is a special case of equation 2.9.62 with $K(x) = A J_\nu(-\lambda x)$.

7. $y(x) - \displaystyle\int_a^x \left[A J_\nu(kx) + B - AB(x - t) J_\nu(kx) \right] y(t)\, dt = f(x).$

This is a special case of equation 2.9.7 with $\lambda = B$ and $g(x) = A J_\nu(kx)$.

8. $y(x) + \displaystyle\int_a^x \left[A J_\nu(kt) + B + AB(x - t) J_\nu(kt) \right] y(t)\, dt = f(x).$

This is a special case of equation 2.9.8 with $\lambda = B$ and $g(t) = A J_\nu(kt)$.

9. $y(x) - \lambda \displaystyle\int_0^x e^{\mu(x-t)} J_0(x - t) y(t)\, dt = f(x).$

Solution:
$$y(x) = f(x) + \int_0^x R(x - t) f(t)\, dt,$$

where
$$R(x) = e^{\mu x} \left\{ \lambda \cos\left(\sqrt{1 - \lambda^2}\, x \right) + \frac{\lambda^2}{\sqrt{1 - \lambda^2}} \sin\left(\sqrt{1 - \lambda^2}\, x \right) + \right.$$
$$\left. \frac{\lambda}{\sqrt{1 - \lambda^2}} \int_0^x \sin\left[\sqrt{1 - \lambda^2}\, (x - t) \right] \frac{J_1(t)}{t}\, dt \right\}.$$

10. $y(x) - A \displaystyle\int_a^x Y_\nu(\lambda x) y(t)\, dt = f(x).$

This is a special case of equation 2.9.2 with $g(x) = A Y_\nu(\lambda x)$ and $h(t) = 1$.

11. $\quad y(x) - A \displaystyle\int_a^x Y_\nu(\lambda t) y(t)\, dt = f(x).$

This is a special case of equation 2.9.2 with $g(x) = A$ and $h(t) = Y_\nu(\lambda t)$.

12. $\quad y(x) - A \displaystyle\int_a^x \dfrac{Y_\nu(\lambda x)}{Y_\nu(\lambda t)} y(t)\, dt = f(x).$

Solution:
$$y(x) = f(x) + A \int_a^x e^{A(x-t)} \frac{Y_\nu(\lambda x)}{Y_\nu(\lambda t)} f(t)\, dt.$$

13. $\quad y(x) - A \displaystyle\int_a^x \dfrac{Y_\nu(\lambda t)}{Y_\nu(\lambda x)} y(t)\, dt = f(x).$

Solution:
$$y(x) = f(x) + A \int_a^x e^{A(x-t)} \frac{Y_\nu(\lambda t)}{Y_\nu(\lambda x)} f(t)\, dt.$$

14. $\quad y(x) + A \displaystyle\int_x^\infty Y_\nu[\lambda(t - x)] y(t)\, dt = f(x).$

This is a special case of equation 2.9.62 with $K(x) = A Y_\nu(-\lambda x)$.

15. $\quad y(x) - \displaystyle\int_a^x \big[A Y_\nu(kx) + B - AB(x - t) Y_\nu(kx) \big] y(t)\, dt = f(x).$

This is a special case of equation 2.9.7 with $\lambda = B$ and $g(x) = A Y_\nu(kx)$.

16. $\quad y(x) + \displaystyle\int_a^x \big[A Y_\nu(kt) + B + AB(x - t) Y_\nu(kt) \big] y(t)\, dt = f(x).$

This is a special case of equation 2.9.8 with $\lambda = B$ and $g(t) = A Y_\nu(kt)$.

2.8-2. Kernels Containing Modified Bessel Functions

17. $\quad y(x) - A \displaystyle\int_a^x I_\nu(\lambda x) y(t)\, dt = f(x).$

This is a special case of equation 2.9.2 with $g(x) = A I_\nu(\lambda x)$ and $h(t) = 1$.

18. $\quad y(x) - A \displaystyle\int_a^x I_\nu(\lambda t) y(t)\, dt = f(x).$

This is a special case of equation 2.9.2 with $g(x) = A$ and $h(t) = I_\nu(\lambda t)$.

19. $\quad y(x) - A \displaystyle\int_a^x \dfrac{I_\nu(\lambda x)}{I_\nu(\lambda t)} y(t)\, dt = f(x).$

Solution:
$$y(x) = f(x) + A \int_a^x e^{A(x-t)} \frac{I_\nu(\lambda x)}{I_\nu(\lambda t)} f(t)\, dt.$$

20. $\quad y(x) - A \displaystyle\int_a^x \dfrac{I_\nu(\lambda t)}{I_\nu(\lambda x)} y(t)\, dt = f(x).$

Solution:
$$y(x) = f(x) + A \int_a^x e^{A(x-t)} \frac{I_\nu(\lambda t)}{I_\nu(\lambda x)} f(t)\, dt.$$

21. $y(x) + A \displaystyle\int_x^\infty I_\nu[\lambda(t-x)]y(t)\,dt = f(x).$

This is a special case of equation 2.9.62 with $K(x) = AI_\nu(-\lambda x)$.

22. $y(x) - \displaystyle\int_a^x \Big[AI_\nu(kx) + B - AB(x-t)I_\nu(kx)\Big]y(t)\,dt = f(x).$

This is a special case of equation 2.9.7 with $\lambda = B$ and $g(x) = AI_\nu(kx)$.

23. $y(x) + \displaystyle\int_a^x \Big[AI_\nu(kt) + B + AB(x-t)I_\nu(kt)\Big]y(t)\,dt = f(x).$

This is a special case of equation 2.9.8 with $\lambda = B$ and $g(t) = AI_\nu(kt)$.

24. $y(x) - A \displaystyle\int_a^x K_\nu(\lambda x)y(t)\,dt = f(x).$

This is a special case of equation 2.9.2 with $g(x) = AK_\nu(\lambda x)$ and $h(t) = 1$.

25. $y(x) - A \displaystyle\int_a^x K_\nu(\lambda t)y(t)\,dt = f(x).$

This is a special case of equation 2.9.2 with $g(x) = A$ and $h(t) = K_\nu(\lambda t)$.

26. $y(x) - A \displaystyle\int_a^x \dfrac{K_\nu(\lambda x)}{K_\nu(\lambda t)} y(t)\,dt = f(x).$

Solution:
$$y(x) = f(x) + A \int_a^x e^{A(x-t)}\frac{K_\nu(\lambda x)}{K_\nu(\lambda t)} f(t)\,dt.$$

27. $y(x) - A \displaystyle\int_a^x \dfrac{K_\nu(\lambda t)}{K_\nu(\lambda x)} y(t)\,dt = f(x).$

Solution:
$$y(x) = f(x) + A \int_a^x e^{A(x-t)}\frac{K_\nu(\lambda t)}{K_\nu(\lambda x)} f(t)\,dt.$$

28. $y(x) + A \displaystyle\int_x^\infty K_\nu[\lambda(t-x)]y(t)\,dt = f(x).$

This is a special case of equation 2.9.62 with $K(x) = AK_\nu(-\lambda x)$.

29. $y(x) - \displaystyle\int_a^x \Big[AK_\nu(kx) + B - AB(x-t)K_\nu(kx)\Big]y(t)\,dt = f(x).$

This is a special case of equation 2.9.7 with $\lambda = B$ and $g(x) = AK_\nu(kx)$.

30. $y(x) + \displaystyle\int_a^x \Big[AK_\nu(kt) + B + AB(x-t)K_\nu(kt)\Big]y(t)\,dt = f(x).$

This is a special case of equation 2.9.8 with $\lambda = B$ and $g(t) = AK_\nu(kt)$.

2.9. Equations Whose Kernels Contain Arbitrary Functions

2.9-1. Equations With Degenerate Kernel: $K(x,t) = g_1(x)h_1(t) + \cdots + g_n(x)h_n(t)$

1. $y(x) - \lambda \displaystyle\int_a^x \frac{g(x)}{g(t)} y(t)\, dt = f(x).$

   Solution:
   $$y(x) = f(x) + \lambda \int_a^x e^{\lambda(x-t)} \frac{g(x)}{g(t)} f(t)\, dt.$$

2. $y(x) - \displaystyle\int_a^x g(x)h(t)y(t)\, dt = f(x).$

   Solution:
   $$y(x) = f(x) + \int_a^x R(x,t)f(t)\, dt, \qquad \text{where} \quad R(x,t) = g(x)h(t)\exp\left[\int_t^x g(s)h(s)\, ds\right].$$

3. $y(x) + \displaystyle\int_a^x (x-t)g(x)y(t)\, dt = f(x).$

   This is a special case of equation 2.9.11.

   1°. Solution:
   $$y(x) = f(x) + \frac{1}{W} \int_a^x \left[Y_1(x)Y_2(t) - Y_2(x)Y_1(t)\right] g(x)f(t)\, dt, \tag{1}$$

   where $Y_1 = Y_1(x)$ and $Y_2 = Y_2(x)$ are two linearly independent solutions $(Y_1/Y_2 \not\equiv \text{const})$ of the second-order linear homogeneous differential equation $Y''_{xx} + g(x)Y = 0$. In this case, the Wronskian is a constant: $W = Y_1(Y_2)'_x - Y_2(Y_1)'_x \equiv \text{const}.$

   2°. Given only one nontrivial solution $Y_1 = Y_1(x)$ of the linear homogeneous differential equation $Y''_{xx} + g(x)Y = 0$, one can obtain the solution of the integral equation by formula (1) with
   $$W = 1, \qquad Y_2(x) = Y_1(x) \int_b^x \frac{d\xi}{Y_1^2(\xi)},$$

   where b is an arbitrary number.

4. $y(x) + \displaystyle\int_a^x (x-t)g(t)y(t)\, dt = f(x).$

   This is a special case of equation 2.9.12.

   1°. Solution:
   $$y(x) = f(x) + \frac{1}{W} \int_a^x \left[Y_1(x)Y_2(t) - Y_2(x)Y_1(t)\right] g(t)f(t)\, dt, \tag{1}$$

   where $Y_1 = Y_1(x)$ and $Y_2 = Y_2(x)$ are two linearly independent solutions $(Y_1/Y_2 \not\equiv \text{const})$ of the second-order linear homogeneous differential equation $Y''_{xx} + g(x)Y = 0$. In this case, the Wronskian is a constant: $W = Y_1(Y_2)'_x - Y_2(Y_1)'_x \equiv \text{const}.$

   2°. Given only one nontrivial solution $Y_1 = Y_1(x)$ of the linear homogeneous differential equation $Y''_{xx} + g(x)Y = 0$, one can obtain the solution of the integral equation by formula (1) with
   $$W = 1, \qquad Y_2(x) = Y_1(x) \int_b^x \frac{d\xi}{Y_1^2(\xi)},$$

   where b is an arbitrary number.

5. $y(x) + \displaystyle\int_a^x \big[g(x) - g(t)\big] y(t)\, dt = f(x).$

1°. Differentiating the equation with respect to x yields

$$y'_x(x) + g'_x(x) \int_a^x y(t)\, dt = f'_x(x). \tag{1}$$

Introducing the new variable $Y(x) = \displaystyle\int_a^x y(t)\, dt$, we obtain the second-order linear ordinary differential equation

$$Y''_{xx} + g'_x(x)Y = f'_x(x), \tag{2}$$

which must be supplemented by the initial conditions

$$Y(a) = 0, \quad Y'_x(a) = f(a). \tag{3}$$

Conditions (3) follow from the original equation and the definition of $Y(x)$.

For exact solutions of second-order linear ordinary differential equations (2) with various $f(x)$, see E. Kamke (1977), G. M. Murphy (1960), and A. D. Polyanin and V. F. Zaitsev (1995, 1996).

2°. Let $Y_1 = Y_1(x)$ and $Y_2 = Y_2(x)$ be two linearly independent solutions ($Y_1/Y_2 \not\equiv$ const) of the second-order linear homogeneous differential equation $Y''_{xx} + g'_x(x)Y = 0$, which follows from (2) for $f(x) \equiv 0$. In this case, the Wronskian is a constant:

$$W = Y_1(Y_2)'_x - Y_2(Y_1)'_x \equiv \text{const}.$$

Solving the nonhomogeneous equation (2) under the initial conditions (3) with arbitrary $f = f(x)$ and taking into account $y(x) = Y'_x(x)$, we obtain the solution of the original integral equation in the form

$$y(x) = f(x) + \frac{1}{W} \int_a^x \big[Y'_1(x)Y'_2(t) - Y'_2(x)Y'_1(t)\big] f(t)\, dt, \tag{4}$$

where the primes stand for the differentiation with respect to the argument specified in the parentheses.

3°. Given only one nontrivial solution $Y_1 = Y_1(x)$ of the linear homogeneous differential equation $Y''_{xx} + g'_x(x)Y = 0$, one can obtain the solution of the nonhomogeneous equation (2) under the initial conditions (3) by formula (4) with

$$W = 1, \qquad Y_2(x) = Y_1(x) \int_b^x \frac{d\xi}{Y_1^2(\xi)},$$

where b is an arbitrary number.

6. $y(x) + \displaystyle\int_a^x \big[g(x) + h(t)\big] y(t)\, dt = f(x).$

1°. Differentiating the equation with respect to x yields

$$y'_x(x) + \big[g(x) + h(x)\big] y(x) + g'_x(x) \int_a^x y(t)\, dt = f'_x(x).$$

Introducing the new variable $Y(x) = \int_a^x y(t)\, dt$, we obtain the second-order linear ordinary differential equation

$$Y''_{xx} + \left[g(x) + h(x)\right]Y'_x + g'_x(x)Y = f'_x(x), \tag{1}$$

which must be supplemented by the initial conditions

$$Y(a) = 0, \quad Y'_x(a) = f(a). \tag{2}$$

Conditions (3) follow from the original equation and the definition of $Y(x)$.

For exact solutions of second-order linear ordinary differential equations (1) with various $f(x)$, see E. Kamke (1977), G. M. Murphy (1960), and A. D. Polyanin and V. F. Zaitsev (1995, 1996).

$2°$. Let $Y_1 = Y_1(x)$ and $Y_2 = Y_2(x)$ be two linearly independent solutions ($Y_1/Y_2 \not\equiv \mathrm{const}$) of the second-order linear homogeneous differential equation $Y''_{xx} + \left[g(x) + h(x)\right]Y'_x + g'_x(x)Y = 0$, which follows from (1) for $f(x) \equiv 0$.

Solving the nonhomogeneous equation (1) under the initial conditions (2) with arbitrary $f = f(x)$ and taking into account $y(x) = Y'_x(x)$, we obtain the solution of the original integral equation in the form

$$y(x) = f(x) + \int_a^x R(x,t)f(t)\, dt,$$

$$R(x,t) = \frac{\partial^2}{\partial x \partial t}\left[\frac{Y_1(x)Y_2(t) - Y_2(x)Y_1(t)}{W(t)}\right], \quad W(x) = Y_1(x)Y'_2(x) - Y_2(x)Y'_1(x),$$

where $W(x)$ is the Wronskian and the primes stand for the differentiation with respect to the argument specified in the parentheses.

7. $\quad y(x) - \int_a^x \left[g(x) + \lambda - \lambda(x - t)g(x)\right]y(t)\, dt = f(x).$

This is a special case of equation 2.9.16 with $h(x) = \lambda$.

    Solution:

$$y(x) = f(x) + \int_a^x R(x,t)f(t)\, dt,$$

$$R(x,t) = [g(x) + \lambda]\frac{G(x)}{G(t)} + \frac{\lambda^2}{G(t)}\int_t^x e^{\lambda(x-s)}G(s)\, ds, \quad G(x) = \exp\left[\int_a^x g(s)\, ds\right].$$

8. $\quad y(x) + \int_a^x \left[g(t) + \lambda + \lambda(x - t)g(t)\right]y(t)\, dt = f(x).$

    Solution:

$$y(x) = f(x) + \int_a^x R(x,t)f(t)\, dt,$$

$$R(x,t) = -[g(t) + \lambda]\frac{G(t)}{G(x)} + \frac{\lambda^2}{G(x)}\int_t^x e^{\lambda(t-s)}G(s)\, ds, \quad G(x) = \exp\left[\int_a^x g(s)\, ds\right].$$

9. $\quad y(x) - \int_a^x \left[g_1(x) + g_2(x)t\right]y(t)\, dt = f(x).$

This equation can be rewritten in the form of equation 2.9.11 with $g_1(x) = g(x) + xh(x)$ and $g_2(x) = -h(x)$.

10. $\quad y(x) - \displaystyle\int_a^x \left[g_1(t) + g_2(t)x\right] y(t)\,dt = f(x).$

This equation can be rewritten in the form of equation 2.9.12 with $g_1(t) = g(t) + th(t)$ and $g_2(t) = -h(t)$.

11. $\quad y(x) - \displaystyle\int_a^x \left[g(x) + h(x)(x - t)\right] y(t)\,dt = f(x).$

1°. The solution of the integral equation can be represented in the form $y(x) = Y''_{xx}$, where $Y = Y(x)$ is the solution of the second-order linear nonhomogeneous ordinary differential equation

$$Y''_{xx} - g(x)Y'_x - h(x)Y = f(x), \tag{1}$$

under the initial conditions

$$Y(a) = Y'_x(a) = 0. \tag{2}$$

2°. Let $Y_1 = Y_1(x)$ and $Y_2 = Y_2(x)$ be two nontrivial linearly independent solutions of the second-order linear homogeneous differential equation $Y''_{xx} - g(x)Y'_x - h(x)Y = 0$, which follows from (1) for $f(x) \equiv 0$. Then the solution of the nonhomogeneous differential equation (1) under conditions (2) is given by

$$Y(x) = \int_a^x \left[Y_2(x)Y_1(t) - Y_1(x)Y_2(t)\right] \frac{f(t)}{W(t)}\,dt, \qquad W(t) = Y_1(t)Y'_2(t) - Y_2(t)Y'_1(t), \tag{3}$$

where $W(t)$ is the Wronskian and the primes denote the derivatives.

Substituting (3) into (1), we obtain the solution of the original integral equation in the form

$$y(x) = f(x) + \int_a^x R(x,t)f(t)\,dt, \qquad R(x,t) = \frac{1}{W(t)}[Y''_2(x)Y_1(t) - Y''_1(x)Y_2(t)]. \tag{4}$$

3°. Let $Y_1 = Y_1(x)$ be a nontrivial particular solution of the homogeneous differential equation (1) (with $f \equiv 0$) satisfying the initial condition $Y_1(a) \neq 0$. Then the function

$$Y_2(x) = Y_1(x) \int_a^x \frac{W(t)}{[Y_1(t)]^2}\,dt, \qquad W(x) = \exp\left[\int_a^x g(s)\,ds\right] \tag{5}$$

is another nontrivial solution of the homogeneous equation. Substituting (5) into (4) yields the solution of the original integral equation in the form

$$y(x) = f(x) + \int_a^x R(x,t)f(t)\,dt,$$

$$R(x,t) = g(x)\frac{W(x)}{Y_1(x)}\frac{Y_1(t)}{W(t)} + [g(x)Y'_1(x) + h(x)Y_1(x)]\frac{Y_1(t)}{W(t)}\int_t^x \frac{W(s)}{[Y_1(s)]^2}\,ds,$$

where $W(x) = \exp\left[\displaystyle\int_a^x g(s)\,ds\right].$

12. $\quad y(x) - \displaystyle\int_a^x \left[g(t) + h(t)(t - x)\right] y(t)\,dt = f(x).$

Solution:

$$y(x) = f(x) + \int_a^x R(x,t)f(t)\,dt,$$

$$R(x,t) = g(t)\frac{Y(x)W(x)}{Y(t)W(t)} + Y(x)W(x)[g(t)Y'_t(t) + h(t)Y(t)]\int_x^t \frac{ds}{W(s)[Y(s)]^2},$$

$$W(t) = \exp\left[\int_b^t g(t)\,dt\right],$$

where $Y = Y(x)$ is an arbitrary nontrivial solution of the second-order homogeneous differential equation

$$Y''_{xx} + g(x)Y'_x + h(x)Y = 0$$

satisfying the condition $Y(a) \neq 0$.

13. $\quad y(x) + \displaystyle\int_a^x (x - t)g(x)h(t)y(t)\, dt = f(x).$

The substitution $y(x) = g(x)u(x)$ leads to an equation of the form 2.9.4:

$$u(x) + \int_a^x (x - t)g(t)h(t)u(t)\, dt = f(x)/g(x).$$

14. $\quad y(x) - \displaystyle\int_a^x \{g(x) + \lambda x^n + \lambda(x - t)x^{n-1}[n - xg(x)]\}y(t)\, dt = f(x).$

This is a special case of equation 2.9.16 with $h(x) = \lambda x^n$.
    Solution:

$$y(x) = f(x) + \int_a^x R(x, t)f(t)\, dt,$$

$$R(x, t) = [g(x) + \lambda x^n]\frac{G(x)}{G(t)} + \lambda(\lambda x^{2n} + nx^{n-1})\frac{H(x)}{G(t)}\int_t^x \frac{G(s)}{H(s)}\, ds,$$

where $\ G(x) = \exp\left[\displaystyle\int_a^x g(s)\, ds\right]$ and $H(x) = \exp\left(\dfrac{\lambda}{n+1}x^{n+1}\right).$

15. $\quad y(x) - \displaystyle\int_a^x \{g(x) + \lambda + (x - t)[g'_x(x) - \lambda g(x)]\}y(t)\, dt = f(x).$

This is a special case of equation 2.9.16.
    Solution:

$$y(x) = f(x) + \int_a^x R(x, t)f(t)\, dt,$$

$$R(x, t) = [g(x) + \lambda]e^{\lambda(x-t)} + \{[g(x)]^2 + g'_x(x)\}G(x)\int_t^x \frac{e^{\lambda(s-t)}}{G(s)}\, ds,$$

where $\ G(x) = \exp\left[\displaystyle\int_a^x g(s)\, ds\right].$

16. $\quad y(x) - \displaystyle\int_a^x \{g(x) + h(x) + (x - t)[h'_x(x) - g(x)h(x)]\}y(t)\, dt = f(x).$

Solution:

$$y(x) = f(x) + \int_a^x R(x, t)f(t)\, dt,$$

$$R(x, t) = [g(x) + h(x)]\frac{G(x)}{G(t)} + \{[h(x)]^2 + h'_x(x)\}\frac{H(x)}{G(t)}\int_t^x \frac{G(s)}{H(s)}\, ds,$$

where $\ G(x) = \exp\left[\displaystyle\int_a^x g(s)\, ds\right]$ and $H(x) = \exp\left[\displaystyle\int_a^x h(s)\, ds\right].$

17. $y(x) + \displaystyle\int_a^x \left\{ \dfrac{\varphi'_x(x)}{\varphi(t)} + \left[\varphi(t)g'_x(x) - \varphi'_x(x)g(t)\right]h(t) \right\} y(t)\, dt = f(x).$

1°. This equation is equivalent to the equation

$$\int_a^x \left\{ \frac{\varphi(x)}{\varphi(t)} + \left[\varphi(t)g(x) - \varphi(x)g(t)\right]h(t) \right\} y(t)\, dt = F(x), \quad F(x) = \int_a^x f(x)\, dx, \quad (1)$$

obtained by differentiating the original equation with respect to x. Equation (1) is a special case of equation 1.9.15 with

$$g_1(x) = g(x), \quad h_1(t) = \varphi(t)h(t), \quad g_2(x) = \varphi(x), \quad h_2(t) = \frac{1}{\varphi(t)} - g(t)h(t).$$

2°. Solution:

$$y(x) = \frac{1}{\varphi(x)h(x)} \frac{d}{dx} \left\{ \Xi(x) \int_a^x \left[\frac{F(t)}{\varphi(t)}\right]'_t \frac{\varphi^2(t)h(t)}{\Xi(t)}\, dt \right\},$$

$$F(x) = \int_a^x f(x)\, dx, \quad \Xi(x) = \exp\left\{ -\int_a^x \left[\frac{g(t)}{\varphi(t)}\right]'_t \varphi^2(t)h(t)\, dt \right\}.$$

18. $y(x) - \displaystyle\int_a^x \left\{ \dfrac{\varphi'_t(t)}{\varphi(x)} + \left[\varphi(x)g'_t(t) - \varphi'_t(t)g(x)\right]h(x) \right\} y(t)\, dt = f(x).$

1°. Let $f(a) = 0$. The change

$$y(x) = \int_a^x w(t)\, dt \tag{1}$$

followed by the integration by parts leads to the equation

$$\int_a^x \left\{ \frac{\varphi(t)}{\varphi(x)} + \left[\varphi(x)g(t) - \varphi(t)g(x)\right]h(x) \right\} w(t)\, dt = f(x), \tag{2}$$

which is a special case of equation 1.9.15 with

$$g_1(x) = \frac{1}{\varphi(x)} - g(x)h(x), \quad h_1(t) = \varphi(t), \quad g_2(x) = \varphi(x)h(x), \quad h_2(t) = g(t).$$

The solution of equation (2) is given by

$$y(x) = \frac{1}{\varphi(x)} \frac{d}{dx} \left\{ \varphi^2(x)h(x)\Phi(x) \int_a^x \left[\frac{f(t)}{\varphi(t)h(t)}\right]'_t \frac{dt}{\Phi(t)} \right\},$$

$$\Phi(x) = \exp\left\{ \int_a^x \left[\frac{g(t)}{\varphi(t)}\right]'_t \varphi^2(t)h(t)\, dt \right\}.$$

2°. Let $f(a) \neq 0$. The substitution $y(x) = \bar{y}(x) + f(a)$ leads to the integral equation $\bar{y}(x)$ with the right-hand side $\bar{f}(x)$ satisfying the condition $\bar{f}(a) = 0$. Thus we obtain case 1°.

19. $y(x) - \int_a^x \left[\sum_{k=1}^n g_k(x)(x-t)^{k-1} \right] y(t)\, dt = f(x).$

The solution can be represented in the form

$$y(x) = f(x) + \int_a^x R(x,t) f(t)\, dt. \tag{1}$$

Here the resolvent $R(x,t)$ is given by

$$R(x,t) = w_x^{(n)}, \qquad w_x^{(n)} = \frac{d^n w}{dx^n}, \tag{2}$$

where w is the solution of the nth-order linear homogeneous ordinary differential equation

$$w_x^{(n)} - g_1(x)w_x^{(n-1)} - g_2(x)w_x^{(n-2)} - 2g_3(x)w_x^{(n-3)} - \cdots - (n-1)!\, g_n(x)w = 0 \tag{3}$$

satisfying the following initial conditions at $x = t$:

$$w\big|_{x=t} = w_x'\big|_{x=t} = \cdots = w_x^{(n-2)}\big|_{x=t} = 0, \quad w_x^{(n-1)}\big|_{x=t} = 1. \tag{4}$$

Note that the differential equation (3) implicitly depends on t via the initial conditions (4).

⊙ References: E. Goursat (1923), A. F. Verlan' and V. S. Sizikov (1987).

20. $y(x) - \int_a^x \left[\sum_{k=1}^n g_k(t)(t-x)^{k-1} \right] y(t)\, dt = f(x).$

The solution can be represented in the form

$$y(x) = f(x) + \int_a^x R(x,t) f(t)\, dt. \tag{1}$$

Here the resolvent $R(x,t)$ is given by

$$R(x,t) = -u_t^{(n)}, \qquad u_t^{(n)} = \frac{d^n u}{dt^n}, \tag{2}$$

where u is the solution of the nth-order linear homogeneous ordinary differential equation

$$u_t^{(n)} + g_1(t)u_t^{(n-1)} + g_2(t)u_t^{(n-2)} + 2g_3(t)u_t^{(n-3)} + \ldots + (n-1)!\, g_n(t)u = 0, \tag{3}$$

satisfying the following initial conditions at $t = x$:

$$u\big|_{t=x} = u_t'\big|_{t=x} = \cdots = u_t^{(n-2)}\big|_{t=x} = 0, \quad u_t^{(n-1)}\big|_{t=x} = 1. \tag{4}$$

Note that the differential equation (3) implicitly depends on x via the initial conditions (4).

⊙ References: E. Goursat (1923), A. F. Verlan' and V. S. Sizikov (1987).

21. $y(x) + \int_a^x \left(e^{\lambda x + \mu t} - e^{\mu x + \lambda t} \right) g(t) y(t)\, dt = f(x).$

Let us differentiate the equation twice and then eliminate the integral terms from the resulting relations and the original equation. As a result, we arrive at the second-order linear ordinary differential equation

$$y_{xx}'' - (\lambda + \mu)y_x' + \big[(\lambda - \mu)e^{(\lambda+\mu)x} g(x) + \lambda\mu\big] y = f_{xx}''(x) - (\lambda + \mu)f_x'(x) + \lambda\mu f(x),$$

which must be supplemented by the initial conditions $y(a) = f(a)$, $y_x'(a) = f_x'(a)$.

22. $y(x) + \displaystyle\int_a^x \left[e^{\lambda x} g(t) + e^{\mu x} h(t) \right] y(t)\, dt = f(x).$

Let us differentiate the equation twice and then eliminate the integral terms from the resulting relations and the original equation. As a result, we arrive at the second-order linear ordinary differential equation

$$y''_{xx} + \left[e^{\lambda x} g(x) + e^{\mu x} h(x) - \lambda - \mu \right] y'_x + \left[e^{\lambda x} g'_x(x) + e^{\mu x} h'_x(x) \right.$$
$$\left. + (\lambda - \mu) e^{\lambda x} g(x) + (\mu - \lambda) e^{\mu x} h(x) + \lambda\mu \right] y = f''_{xx}(x) - (\lambda + \mu) f'_x(x) + \lambda\mu f(x),$$

which must be supplemented by the initial conditions

$$y(a) = f(a), \qquad y'_x(a) = f'_x(a) - \left[e^{\lambda a} g(a) + e^{\mu a} h(a) \right] f(a).$$

Example. The Arutyunyan equation

$$y(x) - \int_a^x \varphi(t) \frac{\partial}{\partial t} \left\{ \frac{1}{\varphi(t)} + \psi(t) \left[1 - e^{-\lambda(x-t)} \right] \right\} y(t)\, dt = f(x),$$

can be reduced to the above equation. The former is encountered in the theory of viscoelasticity for aging solids. The solution of the Arutyunyan equation is given by

$$y(x) = f(x) - \int_a^x \frac{1}{\varphi(t)} \frac{\partial}{\partial t} \left[\varphi(t) - \lambda\psi(t)\varphi^2(t) e^{\eta(t)} \int_t^x e^{-\eta(s)}\, ds \right] f(t)\, dt,$$

where

$$\eta(x) = \int_a^x \left\{ \lambda \left[1 + \psi(t)\varphi(t) \right] - \frac{\varphi'(t)}{\varphi(t)} \right\} dt.$$

⊙ Reference: N. Kh. Arutyunyan (1966).

23. $y(x) + \displaystyle\int_a^x \left[\lambda e^{\lambda(x-t)} + \left(\mu e^{\mu x + \lambda t} - \lambda e^{\lambda x + \mu t} \right) h(t) \right] y(t)\, dt = f(x).$

This is a special case of equation 2.9.17 with $\varphi(x) = e^{\lambda x}$ and $g(x) = e^{\mu x}$.
   Solution:

$$y(x) = \frac{1}{e^{\lambda x} h(x)} \frac{d}{dx} \left\{ \Phi(x) \int_a^x \left[\frac{F(t)}{e^{\lambda t}} \right]'_t \frac{e^{2\lambda t} h(t)}{\Phi(t)}\, dt \right\},$$
$$F(x) = \int_a^x f(t)\, dt, \quad \Phi(x) = \exp\left[(\lambda - \mu) \int_a^x e^{(\lambda + \mu)t} h(t)\, dt \right].$$

24. $y(x) - \displaystyle\int_a^x \left[\lambda e^{-\lambda(x-t)} + \left(\mu e^{\lambda x + \mu t} - \lambda e^{\mu x + \lambda t} \right) h(x) \right] y(t)\, dt = f(x).$

This is a special case of equation 2.9.18 with $\varphi(x) = e^{\lambda x}$ and $g(x) = e^{\mu x}$.
   Assume that $f(a) = 0$. Solution:

$$y(x) = \int_a^x w(t)\, dt, \quad w(x) = e^{-\lambda x} \frac{d}{dx} \left\{ \frac{e^{2\lambda x} h(x)}{\Phi(x)} \int_a^x \left[\frac{f(t)}{e^{\lambda t} h(t)} \right]'_t \Phi(t)\, dt \right\},$$
$$\Phi(x) = \exp\left[(\lambda - \mu) \int_a^x e^{(\lambda + \mu)t} h(t)\, dt \right].$$

25. $y(x) - \int_a^x \{g(x) + be^{\lambda x} + b(x-t)e^{\lambda x}[\lambda - g(x)]\}y(t)\,dt = f(x).$

This is a special case of equation 2.9.16 with $h(x) = be^{\lambda x}$.

Solution:

$$y(x) = f(x) + \int_a^x R(x,t)f(t)\,dt,$$

$$R(x,t) = [g(x) + be^{\lambda x}]\frac{G(x)}{G(t)} + (b^2 e^{2\lambda x} + b\lambda e^{\lambda x})\frac{H(x)}{G(t)}\int_t^x \frac{G(s)}{H(s)}\,ds,$$

where $G(x) = \exp\left[\int_a^x g(s)\,ds\right]$ and $H(x) = \exp\left(\dfrac{b}{\lambda}e^{\lambda x}\right).$

26. $y(x) + \int_a^x \{\lambda e^{\lambda(x-t)} + [e^{\lambda t}g_x'(x) - \lambda e^{\lambda x}g(t)]h(t)\}y(t)\,dt = f(x).$

This is a special case of equation 2.9.17 with $\varphi(x) = e^{\lambda x}$.

27. $y(x) - \int_a^x \{\lambda e^{-\lambda(x-t)} + [e^{\lambda x}g_t'(t) - \lambda e^{\lambda t}g(x)]h(x)\}y(t)\,dt = f(x).$

This is a special case of equation 2.9.18 with $\varphi(x) = e^{\lambda x}$.

28. $y(x) + \int_a^x \cosh[\lambda(x-t)]g(t)y(t)\,dt = f(x).$

Differentiating the equation with respect to x twice yields

$$y_x'(x) + g(x)y(x) + \lambda\int_a^x \sinh[\lambda(x-t)]g(t)y(t)\,dt = f_x'(x), \tag{1}$$

$$y_{xx}''(x) + [g(x)y(x)]_x' + \lambda^2\int_a^x \cosh[\lambda(x-t)]g(t)y(t)\,dt = f_{xx}''(x). \tag{2}$$

Eliminating the integral term from (2) with the aid of the original equation, we arrive at the second-order linear ordinary differential equation

$$y_{xx}'' + [g(x)y]_x' - \lambda^2 y = f_{xx}''(x) - \lambda^2 f(x). \tag{3}$$

By setting $x = a$ in the original equation and (1), we obtain the initial conditions for $y = y(x)$:

$$y(a) = f(a), \qquad y_x'(a) = f_x'(a) - f(a)g(a). \tag{4}$$

Equation (3) under conditions (4) determines the solution of the original integral equation.

29. $y(x) + \int_a^x \cosh[\lambda(x-t)]g(x)h(t)y(t)\,dt = f(x).$

The substitution $y(x) = g(x)u(x)$ leads to an equation of the form 2.9.28:

$$u(x) + \int_a^x \cosh[\lambda(x-t)]g(t)h(t)u(t)\,dt = f(x)/g(x).$$

30. $\quad y(x) + \displaystyle\int_a^x \sinh[\lambda(x - t)]g(t)y(t)\,dt = f(x).$

1°. Differentiating the equation with respect to x twice yields

$$y'_x(x) + \lambda \int_a^x \cosh[\lambda(x - t)]g(t)y(t)\,dt = f'_x(x), \tag{1}$$

$$y''_{xx}(x) + \lambda g(x)y(x) + \lambda^2 \int_a^x \sinh[\lambda(x - t)]g(t)y(t)\,dt = f''_{xx}(x). \tag{2}$$

Eliminating the integral term from (2) with the aid of the original equation, we arrive at the second-order linear ordinary differential equation

$$y''_{xx} + \lambda\big[g(x) - \lambda\big]y = f''_{xx}(x) - \lambda^2 f(x). \tag{3}$$

By setting $x = a$ in the original equation and (1), we obtain the initial conditions for $y = y(x)$:

$$y(a) = f(a), \qquad y'_x(a) = f'_x(a). \tag{4}$$

For exact solutions of second-order linear ordinary differential equations (3) with various $g(x)$, see E. Kamke (1977), G. M. Murphy (1960), and A. D. Polyanin and V. F. Zaitsev (1995, 1996).

2°. Let $y_1 = y_1(x)$ and $y_2 = y_2(x)$ be two linearly independent solutions ($y_1/y_2 \not\equiv \text{const}$) of the homogeneous differential equation $y''_{xx} + \lambda\big[g(x) - \lambda\big]y = 0$, which follows from (3) for $f(x) \equiv 0$. In this case, the Wronskian is a constant:

$$W = y_1(y_2)'_x - y_2(y_1)'_x \equiv \text{const}.$$

The solution of the nonhomogeneous equation (3) under conditions (4) with arbitrary $f = f(x)$ has the form

$$y(x) = f(x) + \frac{\lambda}{W} \int_a^x \big[y_1(x)y_2(t) - y_2(x)y_1(t)\big]g(t)f(t)\,dt \tag{5}$$

and determines the solution of the original integral equation.

3°. Given only one nontrivial solution $y_1 = y_1(x)$ of the linear homogeneous differential equation $y''_{xx} + \lambda\big[g(x) - \lambda\big]y = 0$, one can obtain the solution of the nonhomogeneous equation (3) under the initial conditions (4) by formula (5) with

$$W = 1, \qquad y_2(x) = y_1(x) \int_b^x \frac{d\xi}{y_1^2(\xi)},$$

where b is an arbitrary number.

31. $\quad y(x) + \displaystyle\int_a^x \sinh[\lambda(x - t)]g(x)h(t)y(t)\,dt = f(x).$

The substitution $y(x) = g(x)u(x)$ leads to an equation of the form 2.9.30:

$$u(x) + \int_a^x \sinh[\lambda(x - t)]g(t)h(t)u(t)\,dt = f(x)/g(x).$$

32. $y(x) - \int_a^x \{g(x) + b\cosh(\lambda x) + b(x - t)[\lambda\sinh(\lambda x) - \cosh(\lambda x)g(x)]\}y(t)\,dt = f(x).$

This is a special case of equation 2.9.16 with $h(x) = b\cosh(\lambda x)$.

Solution:

$$y(x) = f(x) + \int_a^x R(x,t)f(t)\,dt,$$

$$R(x,t) = [g(x) + b\cosh(\lambda x)]\frac{G(x)}{G(t)} + [b^2\cosh^2(\lambda x) + b\lambda\sinh(\lambda x)]\frac{H(x)}{G(t)}\int_t^x \frac{G(s)}{H(s)}\,ds,$$

where $G(x) = \exp\left[\int_a^x g(s)\,ds\right]$ and $H(x) = \exp\left[\frac{b}{\lambda}\sinh(\lambda x)\right]$.

33. $y(x) - \int_a^x \{g(x) + b\sinh(\lambda x) + b(x - t)[\lambda\cosh(\lambda x) - \sinh(\lambda x)g(x)]\}y(t)\,dt = f(x).$

This is a special case of equation 2.9.16 with $h(x) = b\sinh(\lambda x)$.

Solution:

$$y(x) = f(x) + \int_a^x R(x,t)f(t)\,dt,$$

$$R(x,t) = [g(x) + b\sinh(\lambda x)]\frac{G(x)}{G(t)} + [b^2\sinh^2(\lambda x) + b\lambda\cosh(\lambda x)]\frac{H(x)}{G(t)}\int_t^x \frac{G(s)}{H(s)}\,ds,$$

where $G(x) = \exp\left[\int_a^x g(s)\,ds\right]$ and $H(x) = \exp\left[\frac{b}{\lambda}\cosh(\lambda x)\right]$.

34. $y(x) + \int_a^x \cos[\lambda(x - t)]g(t)y(t)\,dt = f(x).$

Differentiating the equation with respect to x twice yields

$$y_x'(x) + g(x)y(x) - \lambda\int_a^x \sin[\lambda(x - t)]g(t)y(t)\,dt = f_x'(x), \tag{1}$$

$$y_{xx}''(x) + [g(x)y(x)]_x' - \lambda^2\int_a^x \cos[\lambda(x - t)]g(t)y(t)\,dt = f_{xx}''(x). \tag{2}$$

Eliminating the integral term from (2) with the aid of the original equation, we arrive at the second-order linear ordinary differential equation

$$y_{xx}'' + [g(x)y]_x' + \lambda^2 y = f_{xx}''(x) + \lambda^2 f(x). \tag{3}$$

By setting $x = a$ in the original equation and (1), we obtain the initial conditions for $y = y(x)$:

$$y(a) = f(a), \qquad y_x'(a) = f_x'(a) - f(a)g(a). \tag{4}$$

35. $y(x) + \int_a^x \cos[\lambda(x - t)]g(x)h(t)y(t)\,dt = f(x).$

The substitution $y(x) = g(x)u(x)$ leads to an equation of the form 2.9.34:

$$u(x) + \int_a^x \cos[\lambda(x - t)]g(t)h(t)u(t)\,dt = f(x)/g(x).$$

36. $y(x) + \int_a^x \sin[\lambda(x - t)]g(t)y(t)\, dt = f(x).$

1°. Differentiating the equation with respect to x twice yields

$$y_x'(x) + \lambda \int_a^x \cos[\lambda(x - t)]g(t)y(t)\, dt = f_x'(x), \tag{1}$$

$$y_{xx}''(x) + \lambda g(x)y(x) - \lambda^2 \int_a^x \sin[\lambda(x - t)]g(t)y(t)\, dt = f_{xx}''(x). \tag{2}$$

Eliminating the integral term from (2) with the aid of the original equation, we arrive at the second-order linear ordinary differential equation

$$y_{xx}'' + \lambda\big[g(x) + \lambda\big]y = f_{xx}''(x) + \lambda^2 f(x). \tag{3}$$

By setting $x = a$ in the original equation and (1), we obtain the initial conditions for $y = y(x)$:

$$y(a) = f(a), \qquad y_x'(a) = f_x'(a). \tag{4}$$

For exact solutions of second-order linear ordinary differential equations (3) with various $f(x)$, see E. Kamke (1977) and A. D. Polyanin and V. F. Zaitsev (1995, 1996).

2°. Let $y_1 = y_1(x)$ and $y_2 = y_2(x)$ be two linearly independent solutions ($y_1/y_2 \not\equiv \text{const}$) of the homogeneous differential equation $y_{xx}'' + \lambda\big[g(x) - \lambda\big]y = 0$, which follows from (3) for $f(x) \equiv 0$. In this case, the Wronskian is a constant:

$$W = y_1(y_2)_x' - y_2(y_1)_x' \equiv \text{const}.$$

The solution of the nonhomogeneous equation (3) under conditions (4) with arbitrary $f = f(x)$ has the form

$$y(x) = f(x) + \frac{\lambda}{W} \int_a^x \big[y_1(x)y_2(t) - y_2(x)y_1(t)\big]g(t)f(t)\, dt \tag{5}$$

and determines the solution of the original integral equation.

3°. Given only one nontrivial solution $y_1 = y_1(x)$ of the linear homogeneous differential equation $y_{xx}'' + \lambda\big[g(x) + \lambda\big]y = 0$, one can obtain the solution of the nonhomogeneous equation (3) under the initial conditions (4) by formula (5) with

$$W = 1, \qquad y_2(x) = y_1(x) \int_b^x \frac{d\xi}{y_1^2(\xi)},$$

where b is an arbitrary number.

37. $y(x) + \int_a^x \sin[\lambda(x - t)]g(x)h(t)y(t)\, dt = f(x).$

The substitution $y(x) = g(x)u(x)$ leads to an equation of the form 2.9.36:

$$u(x) + \int_a^x \sin[\lambda(x - t)]g(t)h(t)u(t)\, dt = f(x)/g(x).$$

38. $y(x) - \int_a^x \{g(x) + b\cos(\lambda x) - b(x-t)[\lambda\sin(\lambda x) + \cos(\lambda x)g(x)]\}y(t)\,dt = f(x).$

This is a special case of equation 2.9.16 with $h(x) = b\cos(\lambda x)$.

Solution:

$$y(x) = f(x) + \int_a^x R(x,t)f(t)\,dt,$$

$$R(x,t) = [g(x) + b\cos(\lambda x)]\frac{G(x)}{G(t)} + [b^2\cos^2(\lambda x) - b\lambda\sin(\lambda x)]\frac{H(x)}{G(t)}\int_t^x \frac{G(s)}{H(s)}\,ds,$$

where $G(x) = \exp\left[\int_a^x g(s)\,ds\right]$ and $H(x) = \exp\left[\frac{b}{\lambda}\sin(\lambda x)\right].$

39. $y(x) - \int_a^x \{g(x) + b\sin(\lambda x) + b(x-t)[\lambda\cos(\lambda x) - \sin(\lambda x)g(x)]\}y(t)\,dt = f(x).$

This is a special case of equation 2.9.16 with $h(x) = b\sin(\lambda x)$.

Solution:

$$y(x) = f(x) + \int_a^x R(x,t)f(t)\,dt,$$

$$R(x,t) = [g(x) + b\sin(\lambda x)]\frac{G(x)}{G(t)} + [b^2\sin^2(\lambda x) + b\lambda\cos(\lambda x)]\frac{H(x)}{G(t)}\int_t^x \frac{G(s)}{H(s)}\,ds,$$

where $G(x) = \exp\left[\int_a^x g(s)\,ds\right]$ and $H(x) = \exp\left[-\frac{b}{\lambda}\cos(\lambda x)\right].$

2.9-2. Equations With Difference Kernel: $K(x,t) = K(x-t)$

40. $y(x) + \int_a^x K(x-t)y(t)\,dt = f(x).$

Renewal equation.

$1°$. To solve this integral equation, direct and inverse Laplace transforms are used. The solution can be represented in the form

$$y(x) = f(x) - \int_a^x R(x-t)f(t)\,dt. \tag{1}$$

Here the resolvent $R(x)$ is expressed via the kernel $K(x)$ of the original equation as follows:

$$R(x) = \frac{1}{2\pi i}\int_{c-i\infty}^{c+i\infty} \tilde{R}(p)e^{px}\,dp,$$

$$\tilde{R}(p) = \frac{\tilde{K}(p)}{1 + \tilde{K}(p)}, \qquad \tilde{K}(p) = \int_0^\infty K(x)e^{-px}\,dx.$$

⊙ References: R. Bellman and K. L. Cooke (1963), M. L. Krasnov, A. I. Kisilev, and G. I. Makarenko (1971), V. I. Smirnov (1974).

2°. Let $w = w(x)$ be the solution of the simpler auxiliary equation with $a = 0$ and $f \equiv 1$:

$$w(x) + \int_0^x K(x - t)w(t)\, dt = 1. \tag{2}$$

Then the solution of the original integral equation with arbitrary $f = f(x)$ is expressed via the solution of the auxiliary equation (2) as

$$y(x) = \frac{d}{dx} \int_a^x w(x - t)f(t)\, dt = f(a)w(x - a) + \int_a^x w(x - t)f'_t(t)\, dt.$$

⊙ Reference: R. Bellman and K. L. Cooke (1963).

41. $\quad y(x) + \displaystyle\int_{-\infty}^x K(x - t)y(t)\, dt = 0.$

Eigenfunctions of this integral equation are determined by the roots of the following transcendental (algebraic) equation for the parameter λ:

$$\int_0^\infty K(z)e^{-\lambda z}\, dz = -1. \tag{1}$$

The left-hand side of this equation is the Laplace transform of the kernel of the integral equation.

1°. For a real simple root λ_k of equation (1) there is a corresponding eigenfunction

$$y_k(x) = \exp(\lambda_k x).$$

2°. For a real root λ_k of multiplicity r there are corresponding r eigenfunctions

$$y_{k1}(x) = \exp(\lambda_k x), \quad y_{k2}(x) = x \exp(\lambda_k x), \quad \ldots, \quad y_{kr}(x) = x^{r-1}\exp(\lambda_k x).$$

3°. For a complex simple root $\lambda_k = \alpha_k + i\beta_k$ of equation (1) there is a corresponding eigenfunction pair

$$y_k^{(1)}(x) = \exp(\alpha_k x)\cos(\beta_k x), \quad y_k^{(2)}(x) = \exp(\alpha_k x)\sin(\beta_k x).$$

4°. For a complex root $\lambda_k = \alpha_k + i\beta_k$ of multiplicity r there are corresponding r eigenfunction pairs

$$\begin{aligned}
y_{k1}^{(1)}(x) &= \exp(\alpha_k x)\cos(\beta_k x), & y_{k1}^{(2)}(x) &= \exp(\alpha_k x)\sin(\beta_k x), \\
y_{k2}^{(1)}(x) &= x\exp(\alpha_k x)\cos(\beta_k x), & y_{k2}^{(2)}(x) &= x\exp(\alpha_k x)\sin(\beta_k x), \\
&\cdots\cdots\cdots\cdots\cdots\cdots & &\cdots\cdots\cdots\cdots\cdots\cdots \\
y_{kr}^{(1)}(x) &= x^{r-1}\exp(\alpha_k x)\cos(\beta_k x), & y_{kr}^{(2)}(x) &= x^{r-1}\exp(\alpha_k x)\sin(\beta_k x).
\end{aligned}$$

The general solution is the combination (with arbitrary constants) of the eigenfunctions of the homogeneous integral equation.

▶ *For equations 2.9.42–2.9.51, only particular solutions are given. To obtain the general solution, one must add the general solution of the corresponding homogeneous equation 2.9.41 to the particular solution.*

42. $\quad y(x) + \displaystyle\int_{-\infty}^{x} K(x - t)y(t)\, dt = Ax^n, \qquad n = 0, 1, 2, \ldots$

This is a special case of equation 2.9.44 with $\lambda = 0$.

1°. A solution with $n = 0$:
$$y(x) = \frac{A}{B}, \qquad B = 1 + \int_0^\infty K(z)\, dz.$$

2°. A solution with $n = 1$:
$$y(x) = \frac{A}{B}x + \frac{AC}{B^2}, \qquad B = 1 + \int_0^\infty K(z)\, dz, \quad C = \int_0^\infty zK(z)\, dz.$$

3°. A solution with $n = 2$:
$$y_2(x) = \frac{A}{B}x^2 + 2\frac{AC}{B^2}x + 2\frac{AC^2}{B^3} - \frac{AD}{B^2},$$
$$B = 1 + \int_0^\infty K(z)\, dz, \quad C = \int_0^\infty zK(z)\, dz, \quad D = \int_0^\infty z^2 K(z)\, dz.$$

4°. A solution with $n = 3, 4, \ldots$ is given by:
$$y_n(x) = A\left\{ \frac{\partial^n}{\partial\lambda^n}\left[\frac{e^{\lambda x}}{B(\lambda)} \right] \right\}_{\lambda=0}, \qquad B(\lambda) = 1 + \int_0^\infty K(z)e^{-\lambda z}\, dz.$$

43. $\quad y(x) + \displaystyle\int_{-\infty}^{x} K(x - t)y(t)\, dt = Ae^{\lambda x}.$

A solution:
$$y(x) = \frac{A}{B}e^{\lambda x}, \qquad B = 1 + \int_0^\infty K(z)e^{-\lambda z}\, dz.$$

The integral term in the expression for B is the Laplace transform of $K(z)$, which may be calculated using tables of Laplace transforms (e.g., see Supplement 4).

44. $\quad y(x) + \displaystyle\int_{-\infty}^{x} K(x - t)y(t)\, dt = Ax^n e^{\lambda x}, \qquad n = 1, 2, \ldots$

1°. A solution with $n = 1$:
$$y_1(x) = \frac{A}{B}xe^{\lambda x} + \frac{AC}{B^2}e^{\lambda x},$$
$$B = 1 + \int_0^\infty K(z)e^{-\lambda z}\, dz, \quad C = \int_0^\infty zK(z)e^{-\lambda z}\, dz.$$

It is convenient to calculate B and C using tables of Laplace transforms.

2°. A solution with $n = 2$:
$$y_2(x) = \frac{A}{B}x^2 e^{\lambda x} + 2\frac{AC}{B^2}xe^{\lambda x} + \left(2\frac{AC^2}{B^3} - \frac{AD}{B^2} \right)e^{\lambda x},$$
$$B = 1 + \int_0^\infty K(z)e^{-\lambda z}\, dz, \quad C = \int_0^\infty zK(z)e^{-\lambda z}\, dz, \quad D = \int_0^\infty z^2 K(z)e^{-\lambda z}\, dz.$$

3°. A solution with $n = 3, 4, \ldots$ is given by:
$$y_n(x) = \frac{\partial}{\partial\lambda}y_{n-1}(x) = A\frac{\partial^n}{\partial\lambda^n}\left[\frac{e^{\lambda x}}{B(\lambda)} \right], \qquad B(\lambda) = 1 + \int_0^\infty K(z)e^{-\lambda z}\, dz.$$

45. $y(x) + \displaystyle\int_{-\infty}^{x} K(x - t)y(t)\, dt = A \cosh(\lambda x).$

A solution:

$$y(x) = \frac{A}{2B_-} e^{\lambda x} + \frac{A}{2B_+} e^{-\lambda x} = \frac{1}{2}\left(\frac{A}{B_-} + \frac{A}{B_+}\right)\cosh(\lambda x) + \frac{1}{2}\left(\frac{A}{B_-} - \frac{A}{B_+}\right)\sinh(\lambda x),$$

$$B_- = 1 + \int_0^\infty K(z)e^{-\lambda z}\, dz, \quad B_+ = 1 + \int_0^\infty K(z)e^{\lambda z}\, dz.$$

46. $y(x) + \displaystyle\int_{-\infty}^{x} K(x - t)y(t)\, dt = A \sinh(\lambda x).$

A solution:

$$y(x) = \frac{A}{2B_-} e^{\lambda x} - \frac{A}{2B_+} e^{-\lambda x} = \frac{1}{2}\left(\frac{A}{B_-} - \frac{A}{B_+}\right)\cosh(\lambda x) + \frac{1}{2}\left(\frac{A}{B_-} + \frac{A}{B_+}\right)\sinh(\lambda x),$$

$$B_- = 1 + \int_0^\infty K(z)e^{-\lambda z}\, dz, \quad B_+ = 1 + \int_0^\infty K(z)e^{\lambda z}\, dz.$$

47. $y(x) + \displaystyle\int_{-\infty}^{x} K(x - t)y(t)\, dt = A \cos(\lambda x).$

A solution:

$$y(x) = \frac{A}{B_c^2 + B_s^2}\left[B_c \cos(\lambda x) - B_s \sin(\lambda x)\right],$$

$$B_c = 1 + \int_0^\infty K(z)\cos(\lambda z)\, dz, \quad B_s = \int_0^\infty K(z)\sin(\lambda z)\, dz.$$

48. $y(x) + \displaystyle\int_{-\infty}^{x} K(x - t)y(t)\, dt = A \sin(\lambda x).$

A solution:

$$y(x) = \frac{A}{B_c^2 + B_s^2}\left[B_c \sin(\lambda x) + B_s \cos(\lambda x)\right],$$

$$B_c = 1 + \int_0^\infty K(z)\cos(\lambda z)\, dz, \quad B_s = \int_0^\infty K(z)\sin(\lambda z)\, dz.$$

49. $y(x) + \displaystyle\int_{-\infty}^{x} K(x - t)y(t)\, dt = A e^{\mu x} \cos(\lambda x).$

A solution:

$$y(x) = \frac{A}{B_c^2 + B_s^2} e^{\mu x}\left[B_c \cos(\lambda x) - B_s \sin(\lambda x)\right],$$

$$B_c = 1 + \int_0^\infty K(z)e^{-\mu z}\cos(\lambda z)\, dz, \quad B_s = \int_0^\infty K(z)e^{-\mu z}\sin(\lambda z)\, dz.$$

50. $y(x) + \displaystyle\int_{-\infty}^{x} K(x - t)y(t)\, dt = A e^{\mu x} \sin(\lambda x).$

A solution:

$$y(x) = \frac{A}{B_c^2 + B_s^2} e^{\mu x}\left[B_c \sin(\lambda x) + B_s \cos(\lambda x)\right],$$

$$B_c = 1 + \int_0^\infty K(z)e^{-\mu z}\cos(\lambda z)\, dz, \quad B_s = \int_0^\infty K(z)e^{-\mu z}\sin(\lambda z)\, dz.$$

51. $y(x) + \displaystyle\int_{-\infty}^{x} K(x - t)y(t)\, dt = f(x).$

1°. For a polynomial right-hand side, $f(x) = \displaystyle\sum_{k=0}^{n} A_k x^k$, a solution has the form

$$y(x) = \sum_{k=0}^{n} B_k x^k,$$

where the constants B_k are found by the method of undetermined coefficients. One can also make use of the formula given in item 4° of equation 2.9.42 to construct the solution.

2°. For $f(x) = e^{\lambda x} \displaystyle\sum_{k=0}^{n} A_k x^k$, a solution of the equation has the form

$$y(x) = e^{\lambda x} \sum_{k=0}^{n} B_k x^k,$$

where the B_k are found by the method of undetermined coefficients. One can also make use of the formula given in item 3° of equation 2.9.44 to construct the solution.

3°. For $f(x) = \displaystyle\sum_{k=0}^{n} A_k \exp(\lambda_k x)$, a solution of the equation has the form

$$y(x) = \sum_{k=0}^{n} \frac{A_k}{B_k} \exp(\lambda_k x), \qquad B_k = 1 + \int_0^{\infty} K(z) \exp(-\lambda_k z)\, dz.$$

4°. For $f(x) = \cos(\lambda x) \displaystyle\sum_{k=0}^{n} A_k x^k$, a solution of the equation has the form

$$y(x) = \cos(\lambda x) \sum_{k=0}^{n} B_k x^k + \sin(\lambda x) \sum_{k=0}^{n} C_k x^k,$$

where the constants B_k and C_k are found by the method of undetermined coefficients.

5°. For $f(x) = \sin(\lambda x) \displaystyle\sum_{k=0}^{n} A_k x^k$, a solution of the equation has the form

$$y(x) = \cos(\lambda x) \sum_{k=0}^{n} B_k x^k + \sin(\lambda x) \sum_{k=0}^{n} C_k x^k,$$

where the constants B_k and C_k are found by the method of undetermined coefficients.

6°. For $f(x) = \displaystyle\sum_{k=0}^{n} A_k \cos(\lambda_k x)$, the solution of a equation has the form

$$y(x) = \sum_{k=0}^{n} \frac{A_k}{B_{ck}^2 + B_{sk}^2} \left[B_{ck} \cos(\lambda_k x) - B_{sk} \sin(\lambda_k x) \right],$$

$$B_{ck} = 1 + \int_0^{\infty} K(z) \cos(\lambda_k z)\, dz, \qquad B_{sk} = \int_0^{\infty} K(z) \sin(\lambda_k z)\, dz.$$

$7°$. For $f(x) = \sum_{k=0}^{n} A_k \sin(\lambda_k x)$, a solution of the equation has the form

$$y(x) = \sum_{k=0}^{n} \frac{A_k}{B_{ck}^2 + B_{sk}^2} \left[B_{ck} \sin(\lambda_k x) + B_{sk} \cos(\lambda_k x) \right],$$

$$B_{ck} = 1 + \int_0^{\infty} K(z) \cos(\lambda_k z) \, dz, \quad B_{sk} = \int_0^{\infty} K(z) \sin(\lambda_k z) \, dz.$$

$8°$. For $f(x) = \cos(\lambda x) \sum_{k=0}^{n} A_k \exp(\mu_k x)$, a solution of the equation has the form

$$y(x) = \cos(\lambda x) \sum_{k=0}^{n} \frac{A_k B_{ck}}{B_{ck}^2 + B_{sk}^2} \exp(\mu_k x) - \sin(\lambda x) \sum_{k=0}^{n} \frac{A_k B_{sk}}{B_{ck}^2 + B_{sk}^2} \exp(\mu_k x),$$

$$B_{ck} = 1 + \int_0^{\infty} K(z) \exp(-\mu_k z) \cos(\lambda z) \, dz, \quad B_{sk} = \int_0^{\infty} K(z) \exp(-\mu_k z) \sin(\lambda z) \, dz.$$

$9°$. For $f(x) = \sin(\lambda x) \sum_{k=0}^{n} A_k \exp(\mu_k x)$, a solution of the equation has the form

$$y(x) = \sin(\lambda x) \sum_{k=0}^{n} \frac{A_k B_{ck}}{B_{ck}^2 + B_{sk}^2} \exp(\mu_k x) + \cos(\lambda x) \sum_{k=0}^{n} \frac{A_k B_{sk}}{B_{ck}^2 + B_{sk}^2} \exp(\mu_k x),$$

$$B_{ck} = 1 + \int_0^{\infty} K(z) \exp(-\mu_k z) \cos(\lambda z) \, dz, \quad B_{sk} = \int_0^{\infty} K(z) \exp(-\mu_k z) \sin(\lambda z) \, dz.$$

52. $y(x) + \int_x^{\infty} K(x - t) y(t) \, dt = 0.$

Eigenfunctions of this integral equation are determined by the roots of the following transcendental (algebraic) equation for the parameter λ:

$$\int_0^{\infty} K(-z) e^{\lambda z} \, dz = -1. \tag{1}$$

The left-hand side of this equation is the Laplace transform of the function $K(-z)$ with parameter $-\lambda$.

$1°$. For a real simple root λ_k of equation (1) there is a corresponding eigenfunction

$$y_k(x) = \exp(\lambda_k x).$$

$2°$. For a real root λ_k of multiplicity r there are corresponding r eigenfunctions

$$y_{k1}(x) = \exp(\lambda_k x), \quad y_{k2}(x) = x \exp(\lambda_k x), \quad \ldots, \quad y_{kr}(x) = x^{r-1} \exp(\lambda_k x).$$

$3°$. For a complex simple root $\lambda_k = \alpha_k + i\beta_k$ of equation (1) there is a corresponding eigenfunction pair

$$y_k^{(1)}(x) = \exp(\alpha_k x) \cos(\beta_k x), \quad y_k^{(2)}(x) = \exp(\alpha_k x) \sin(\beta_k x).$$

$4°$. For a complex root $\lambda_k = \alpha_k + i\beta_k$ of multiplicity r there are corresponding r eigenfunction pairs

$$y_{k1}^{(1)}(x) = \exp(\alpha_k x) \cos(\beta_k x), \quad\quad y_{k1}^{(2)}(x) = \exp(\alpha_k x) \sin(\beta_k x),$$

$$y_{k2}^{(1)}(x) = x \exp(\alpha_k x) \cos(\beta_k x), \quad\quad y_{k2}^{(2)}(x) = x \exp(\alpha_k x) \sin(\beta_k x),$$

$$\cdots\cdots\cdots\cdots\cdots\cdots\cdots\cdots\cdots\cdots\cdots\cdots\cdots\cdots\cdots\cdots$$

$$y_{kr}^{(1)}(x) = x^{r-1} \exp(\alpha_k x) \cos(\beta_k x), \quad\quad y_{kr}^{(2)}(x) = x^{r-1} \exp(\alpha_k x) \sin(\beta_k x).$$

The general solution is the combination (with arbitrary constants) of the eigenfunctions of the homogeneous integral equation.

▶ *For equations 2.9.53–2.9.62, only particular solutions are given. To obtain the general solution, one must add the general solution of the corresponding homogeneous equation 2.9.52 to the particular solution.*

53. $y(x) + \displaystyle\int_x^\infty K(x-t)y(t)\,dt = Ax^n, \qquad n = 0, 1, 2, \ldots$

This is a special case of equation 2.9.55 with $\lambda = 0$.

1°. A solution with $n = 0$:

$$y(x) = \frac{A}{B}, \qquad B = 1 + \int_0^\infty K(-z)\,dz.$$

2°. A solution with $n = 1$:

$$y(x) = \frac{A}{B}x - \frac{AC}{B^2}, \qquad B = 1 + \int_0^\infty K(-z)\,dz, \quad C = \int_0^\infty zK(-z)\,dz.$$

3°. A solution with $n = 2$:

$$y_2(x) = \frac{A}{B}x^2 - 2\frac{AC}{B^2}x + 2\frac{AC^2}{B^3} - \frac{AD}{B^2},$$

$$B = 1 + \int_0^\infty K(-z)\,dz, \quad C = \int_0^\infty zK(-z)\,dz, \quad D = \int_0^\infty z^2 K(-z)\,dz.$$

4°. A solution with $n = 3, 4, \ldots$ is given by:

$$y_n(x) = A\left\{\frac{\partial^n}{\partial\lambda^n}\left[\frac{e^{\lambda x}}{B(\lambda)}\right]\right\}_{\lambda=0}, \qquad B(\lambda) = 1 + \int_0^\infty K(-z)e^{\lambda z}\,dz.$$

54. $y(x) + \displaystyle\int_x^\infty K(x-t)y(t)\,dt = Ae^{\lambda x}.$

A solution:

$$y(x) = \frac{A}{B}e^{\lambda x}, \qquad B = 1 + \int_0^\infty K(-z)e^{\lambda z}\,dz = 1 + \mathfrak{L}\{K(-z), -\lambda\}.$$

The integral term in the expression for B is the Laplace transform of $K(-z)$ with parameter $-\lambda$, which may be calculated using tables of Laplace transforms (e.g., see H. Bateman and A. Erdélyi (vol. 1, 1954) and V. A. Ditkin and A. P. Prudnikov (1965)).

55. $y(x) + \displaystyle\int_x^\infty K(x-t)y(t)\,dt = Ax^n e^{\lambda x}, \qquad n = 1, 2, \ldots$

1°. A solution with $n = 1$:

$$y_1(x) = \frac{A}{B}xe^{\lambda x} - \frac{AC}{B^2}e^{\lambda x},$$

$$B = 1 + \int_0^\infty K(-z)e^{\lambda z}\,dz, \quad C = \int_0^\infty zK(-z)e^{\lambda z}\,dz.$$

It is convenient to calculate B and C using tables of Laplace transforms (with parameter $-\lambda$).

2°. A solution with $n = 2$:

$$y_2(x) = \frac{A}{B}x^2 e^{\lambda x} - 2\frac{AC}{B^2}xe^{\lambda x} + \left(2\frac{AC^2}{B^3} - \frac{AD}{B^2}\right)e^{\lambda x},$$

$$B = 1 + \int_0^\infty K(-z)e^{\lambda z}\,dz, \quad C = \int_0^\infty zK(-z)e^{\lambda z}\,dz, \quad D = \int_0^\infty z^2 K(-z)e^{\lambda z}\,dz.$$

3°. A solution with $n = 3, 4, \ldots$ is given by

$$y_n(x) = \frac{\partial}{\partial\lambda}y_{n-1}(x) = A\frac{\partial^n}{\partial\lambda^n}\left[\frac{e^{\lambda x}}{B(\lambda)}\right], \qquad B(\lambda) = 1 + \int_0^\infty K(-z)e^{\lambda z}\,dz.$$

56. $y(x) + \displaystyle\int_x^\infty K(x-t)y(t)\,dt = A\cosh(\lambda x).$

A solution:

$$y(x) = \frac{A}{2B_+}e^{\lambda x} + \frac{A}{2B_-}e^{-\lambda x} = \frac{1}{2}\left(\frac{A}{B_+} + \frac{A}{B_-}\right)\cosh(\lambda x) + \frac{1}{2}\left(\frac{A}{B_+} - \frac{A}{B_-}\right)\sinh(\lambda x),$$

$$B_+ = 1 + \int_0^\infty K(-z)e^{\lambda z}\,dz, \quad B_- = 1 + \int_0^\infty K(-z)e^{-\lambda z}\,dz.$$

57. $y(x) + \displaystyle\int_x^\infty K(x-t)y(t)\,dt = A\sinh(\lambda x).$

A solution:

$$y(x) = \frac{A}{2B_+}e^{\lambda x} - \frac{A}{2B_-}e^{-\lambda x} = \frac{1}{2}\left(\frac{A}{B_+} - \frac{A}{B_-}\right)\cosh(\lambda x) + \frac{1}{2}\left(\frac{A}{B_+} + \frac{A}{B_-}\right)\sinh(\lambda x),$$

$$B_+ = 1 + \int_0^\infty K(-z)e^{\lambda z}\,dz, \quad B_- = 1 + \int_0^\infty K(-z)e^{-\lambda z}\,dz.$$

58. $y(x) + \displaystyle\int_x^\infty K(x-t)y(t)\,dt = A\cos(\lambda x).$

A solution:

$$y(x) = \frac{A}{B_c^2 + B_s^2}\left[B_c\cos(\lambda x) + B_s\sin(\lambda x)\right],$$

$$B_c = 1 + \int_0^\infty K(-z)\cos(\lambda z)\,dz, \quad B_s = \int_0^\infty K(-z)\sin(\lambda z)\,dz.$$

59. $y(x) + \displaystyle\int_x^\infty K(x-t)y(t)\,dt = A\sin(\lambda x).$

A solution:

$$y(x) = \frac{A}{B_c^2 + B_s^2}\left[B_c\sin(\lambda x) - B_s\cos(\lambda x)\right],$$

$$B_c = 1 + \int_0^\infty K(-z)\cos(\lambda z)\,dz, \quad B_s = \int_0^\infty K(-z)\sin(\lambda z)\,dz.$$

60. $y(x) + \displaystyle\int_x^\infty K(x-t)y(t)\,dt = Ae^{\mu x}\cos(\lambda x).$

A solution:

$$y(x) = \frac{A}{B_c^2 + B_s^2}e^{\mu x}\left[B_c\cos(\lambda x) + B_s\sin(\lambda x)\right],$$

$$B_c = 1 + \int_0^\infty K(-z)e^{\mu z}\cos(\lambda z)\,dz, \quad B_s = \int_0^\infty K(-z)e^{\mu z}\sin(\lambda z)\,dz.$$

61. $y(x) + \displaystyle\int_x^\infty K(x-t)y(t)\,dt = Ae^{\mu x}\sin(\lambda x).$

A solution:

$$y(x) = \frac{A}{B_c^2 + B_s^2}e^{\mu x}\left[B_c\sin(\lambda x) - B_s\cos(\lambda x)\right],$$

$$B_c = 1 + \int_0^\infty K(-z)e^{\mu z}\cos(\lambda z)\,dz, \quad B_s = \int_0^\infty K(-z)e^{\mu z}\sin(\lambda z)\,dz.$$

62. $y(x) + \displaystyle\int_x^\infty K(x-t)y(t)\,dt = f(x).$

$1°$. For a polynomial right-hand side, $f(x) = \sum\limits_{k=0}^{n} A_k x^k$, a solution has the form

$$y(x) = \sum_{k=0}^{n} B_k x^k,$$

where the constants B_k are found by the method of undetermined coefficients. One can also make use of the formula given in item $4°$ of equation 2.9.53 to construct the solution.

$2°$. For $f(x) = e^{\lambda x} \sum\limits_{k=0}^{n} A_k x^k$, a solution of the equation has the form

$$y(x) = e^{\lambda x} \sum_{k=0}^{n} B_k x^k,$$

where the constants B_k are found by the method of undetermined coefficients. One can also make use of the formula given in item $3°$ of equation 2.9.55 to construct the solution.

$3°$. For $f(x) = \sum\limits_{k=0}^{n} A_k \exp(\lambda_k x)$, a solution of the equation has the form

$$y(x) = \sum_{k=0}^{n} \frac{A_k}{B_k} \exp(\lambda_k x), \qquad B_k = 1 + \int_0^\infty K(-z)\exp(\lambda_k z)\,dz.$$

$4°$. For $f(x) = \cos(\lambda x) \sum\limits_{k=0}^{n} A_k x^k$ a solution of the equation has the form

$$y(x) = \cos(\lambda x) \sum_{k=0}^{n} B_k x^k + \sin(\lambda x) \sum_{k=0}^{n} C_k x^k,$$

where the constants B_k and C_k are found by the method of undetermined coefficients.

$5°$. For $f(x) = \sin(\lambda x) \sum\limits_{k=0}^{n} A_k x^k$, a solution of the equation has the form

$$y(x) = \cos(\lambda x) \sum_{k=0}^{n} B_k x^k + \sin(\lambda x) \sum_{k=0}^{n} C_k x^k,$$

where the B_k and C_k are found by the method of undetermined coefficients.

$6°$. For $f(x) = \sum\limits_{k=0}^{n} A_k \cos(\lambda_k x)$, a solution of the equation has the form

$$y(x) = \sum_{k=0}^{n} \frac{A_k}{B_{ck}^2 + B_{sk}^2} \big[B_{ck} \cos(\lambda_k x) + B_{sk} \sin(\lambda_k x) \big],$$

$$B_{ck} = 1 + \int_0^\infty K(-z)\cos(\lambda_k z)\,dz, \qquad B_{sk} = \int_0^\infty K(-z)\sin(\lambda_k z)\,dz.$$

$7°$. For $f(x) = \sum\limits_{k=0}^{n} A_k \sin(\lambda_k x)$, a solution of the equation has the form

$$y(x) = \sum_{k=0}^{n} \frac{A_k}{B_{ck}^2 + B_{sk}^2} \left[B_{ck} \sin(\lambda_k x) - B_{sk} \cos(\lambda_k x) \right],$$

$$B_{ck} = 1 + \int_0^\infty K(-z) \cos(\lambda_k z) \, dz, \quad B_{sk} = \int_0^\infty K(-z) \sin(\lambda_k z) \, dz.$$

$8°$. For $f(x) = \cos(\lambda x) \sum\limits_{k=0}^{n} A_k \exp(\mu_k x)$, a solution of the equation has the form

$$y(x) = \cos(\lambda x) \sum_{k=0}^{n} \frac{A_k B_{ck}}{B_{ck}^2 + B_{sk}^2} \exp(\mu_k x) + \sin(\lambda x) \sum_{k=0}^{n} \frac{A_k B_{sk}}{B_{ck}^2 + B_{sk}^2} \exp(\mu_k x),$$

$$B_{ck} = 1 + \int_0^\infty K(-z) \exp(\mu_k z) \cos(\lambda z) \, dz, \quad B_{sk} = \int_0^\infty K(-z) \exp(\mu_k z) \sin(\lambda z) \, dz.$$

$9°$. For $f(x) = \sin(\lambda x) \sum\limits_{k=0}^{n} A_k \exp(\mu_k x)$, a solution of the equation has the form

$$y(x) = \sin(\lambda x) \sum_{k=0}^{n} \frac{A_k B_{ck}}{B_{ck}^2 + B_{sk}^2} \exp(\mu_k x) - \cos(\lambda x) \sum_{k=0}^{n} \frac{A_k B_{sk}}{B_{ck}^2 + B_{sk}^2} \exp(\mu_k x),$$

$$B_{ck} = 1 + \int_0^\infty K(-z) \exp(\mu_k z) \cos(\lambda z) \, dz, \quad B_{sk} = \int_0^\infty K(-z) \exp(\mu_k z) \sin(\lambda z) \, dz.$$

$10°$. In the general case of arbitrary right-hand side $f = f(x)$, the solution of the integral equation can be represented in the form

$$y(x) = \frac{1}{2\pi i} \int_{c-i\infty}^{c+i\infty} \frac{\tilde{f}(p)}{1 + \tilde{k}(-p)} e^{px} \, dp,$$

$$\tilde{f}(p) = \int_0^\infty f(x) e^{-px} \, dx, \quad \tilde{k}(-p) = \int_0^\infty K(-z) e^{pz} \, dz.$$

To calculate $\tilde{f}(p)$ and $\tilde{k}(-p)$, it is convenient to use tables of Laplace transforms, and to determine $y(x)$, tables of inverse Laplace transforms.

2.9-3. Other Equations

63. $y(x) + \int_0^x \dfrac{1}{x} f\left(\dfrac{t}{x}\right) y(t) \, dt = 0.$

Eigenfunctions of this integral equation are determined by the roots of the following transcendental (algebraic) equation for the parameter λ:

$$\int_0^1 f(z) z^\lambda \, dz = -1. \tag{1}$$

$1°$. For a real simple root λ_k of equation (1) there is a corresponding eigenfunction

$$y_k(x) = x^{\lambda_k}.$$

2°. For a real root λ_k of multiplicity r there are corresponding r eigenfunctions

$$y_{k1}(x) = x^{\lambda_k}, \quad y_{k2}(x) = x^{\lambda_k} \ln x, \quad \ldots, \quad y_{kr}(x) = x^{\lambda_k} \ln^{r-1} x.$$

3°. For a complex simple root $\lambda_k = \alpha_k + i\beta_k$ of equation (1) there is a corresponding eigenfunction pair

$$y_k^{(1)}(x) = x^{\alpha_k} \cos(\beta_k \ln x), \quad y_k^{(2)}(x) = x^{\alpha_k} \sin(\beta_k \ln x).$$

4°. For a complex root $\lambda_k = \alpha_k + i\beta_k$ of multiplicity r there are corresponding r eigenfunction pairs

$$y_{k1}^{(1)}(x) = x^{\alpha_k} \cos(\beta_k \ln x), \qquad y_{k1}^{(2)}(x) = x^{\alpha_k} \sin(\beta_k \ln x),$$
$$y_{k2}^{(1)}(x) = x^{\alpha_k} \ln x \cos(\beta_k \ln x), \qquad y_{k2}^{(2)}(x) = x^{\alpha_k} \ln x \sin(\beta_k \ln x),$$
$$\ldots\ldots\ldots\ldots\ldots\ldots\ldots\ldots\ldots \qquad \ldots\ldots\ldots\ldots\ldots\ldots\ldots\ldots\ldots$$
$$y_{kr}^{(1)}(x) = x^{\alpha_k} \ln^{r-1} x \cos(\beta_k \ln x), \quad y_{kr}^{(2)}(x) = x^{\alpha_k} \ln^{r-1} x \sin(\beta_k \ln x).$$

The general solution is the combination (with arbitrary constants) of the eigenfunctions of the homogeneous integral equation.

▶ *For equations 2.9.64–2.9.71, only particular solutions are given. To obtain the general solution, one must add the general solution of the corresponding homogeneous equation 2.9.63 to the particular solution.*

64. $y(x) + \displaystyle\int_0^x \frac{1}{x} f\left(\frac{t}{x}\right) y(t)\, dt = Ax + B.$

A solution:

$$y(x) = \frac{A}{1 + I_1} x + \frac{B}{1 + I_0}, \qquad I_0 = \int_0^1 f(t)\, dt, \quad I_1 = \int_0^1 t f(t)\, dt.$$

65. $y(x) + \displaystyle\int_0^x \frac{1}{x} f\left(\frac{t}{x}\right) y(t)\, dt = Ax^\beta.$

A solution:

$$y(x) = \frac{A}{B} x^\beta, \qquad B = 1 + \int_0^1 f(t) t^\beta\, dt.$$

66. $y(x) + \displaystyle\int_0^x \frac{1}{x} f\left(\frac{t}{x}\right) y(t)\, dt = A \ln x + B.$

A solution:

$$y(x) = p \ln x + q,$$

where

$$p = \frac{A}{1 + I_0}, \quad q = \frac{B}{1 + I_0} - \frac{AI_l}{(1 + I_0)^2}, \qquad I_0 = \int_0^1 f(t)\, dt, \quad I_l = \int_0^1 f(t) \ln t\, dt.$$

67. $y(x) + \displaystyle\int_0^x \frac{1}{x} f\left(\frac{t}{x}\right) y(t)\, dt = Ax^\beta \ln x.$

A solution:

$$y(x) = px^\beta \ln x + qx^\beta,$$

where

$$p = \frac{A}{1 + I_1}, \quad q = -\frac{AI_2}{(1 + I_1)^2}, \qquad I_1 = \int_0^1 f(t) t^\beta\, dt, \quad I_2 = \int_0^1 f(t) t^\beta \ln t\, dt.$$

68. $y(x) + \int_0^x \frac{1}{x} f\left(\frac{t}{x}\right) y(t)\, dt = A\cos(\ln x).$

A solution:

$$y(x) = \frac{AI_c}{I_c^2 + I_s^2} \cos(\ln x) + \frac{AI_s}{I_c^2 + I_s^2} \sin(\ln x),$$

$$I_c = 1 + \int_0^1 f(t)\cos(\ln t)\, dt, \quad I_s = \int_0^1 f(t)\sin(\ln t)\, dt.$$

69. $y(x) + \int_0^x \frac{1}{x} f\left(\frac{t}{x}\right) y(t)\, dt = A\sin(\ln x).$

A solution:

$$y(x) = -\frac{AI_s}{I_c^2 + I_s^2} \cos(\ln x) + \frac{AI_c}{I_c^2 + I_s^2} \sin(\ln x),$$

$$I_c = 1 + \int_0^1 f(t)\cos(\ln t)\, dt, \quad I_s = \int_0^1 f(t)\sin(\ln t)\, dt.$$

70. $y(x) + \int_0^x \frac{1}{x} f\left(\frac{t}{x}\right) y(t)\, dt = Ax^\beta \cos(\ln x) + Bx^\beta \sin(\ln x).$

A solution:

$$y(x) = px^\beta \cos(\ln x) + qx^\beta \sin(\ln x),$$

where

$$p = \frac{AI_c - BI_s}{I_c^2 + I_s^2}, \quad q = \frac{AI_s + BI_c}{I_c^2 + I_s^2},$$

$$I_c = 1 + \int_0^1 f(t)t^\beta \cos(\ln t)\, dt, \quad I_s = \int_0^1 f(t)t^\beta \sin(\ln t)\, dt.$$

71. $y(x) + \int_0^x \frac{1}{x} f\left(\frac{t}{x}\right) y(t)\, dt = g(x).$

$1°$. For a polynomial right-hand side,

$$g(x) = \sum_{n=0}^N A_n x^n$$

a solution bounded at zero is given by

$$y(x) = \sum_{n=0}^N \frac{A_n}{1 + f_n} x^n, \quad f_n = \int_0^1 f(z) z^n\, dz.$$

Here its is assumed that $f_0 < \infty$ and $f_n \neq -1$ $(n = 0, 1, 2, \dots)$.

If for some n the relation $f_n = -1$ holds, then a solution differs from the above case in one term and has the form

$$y(x) = \sum_{m=0}^{n-1} \frac{A_m}{1 + f_m} x^m + \sum_{m=n+1}^N \frac{A_m}{1 + f_m} x^m + \frac{A_n}{\bar{f}_n} x^n \ln x, \quad \bar{f}_n = \int_0^1 f(z) z^n \ln z\, dz.$$

For arbitrary $g(x)$ expandable into power series, the formulas of item $1°$ can be used, in which one should set $N = \infty$. In this case, the convergence radius of the obtained solution $y(x)$ is equal to that of the function $g(x)$.

2°. For $g(x) = \ln x \sum_{k=0}^{n} A_k x^k$, a solution has the form

$$y(x) = \ln x \sum_{k=0}^{n} B_k x^k + \sum_{k=0}^{n} C_k x^k,$$

where the constants B_k and C_k are found by the method of undetermined coefficients.

3°. For $g(x) = \sum_{k=0}^{n} A_k (\ln x)^k$, a solution of the equation has the form

$$y(x) = \sum_{k=0}^{n} B_k (\ln x)^k,$$

where the B_k are found by the method of undetermined coefficients.

4°. For $g(x) = \sum_{k=1}^{n} A_k \cos(\lambda_k \ln x)$, a solution of the equation has the form

$$y(x) = \sum_{k=1}^{n} B_k \cos(\lambda_k \ln x) + \sum_{k=1}^{n} C_k \sin(\lambda_k \ln x),$$

where the B_k and C_k are found by the method of undetermined coefficients.

5°. For $g(x) = \sum_{k=1}^{n} A_k \sin(\lambda_k \ln x)$ a solution of the equation has the form

$$y(x) = \sum_{k=1}^{n} B_k \cos(\lambda_k \ln x) + \sum_{k=1}^{n} C_k \sin(\lambda_k \ln x),$$

where the B_k and C_k are found by the method of undetermined coefficients.

6°. For arbitrary right-hand side $g(x)$, the transformation

$$x = e^{-z}, \quad t = e^{-\tau}, \quad y(x) = e^z w(z), \quad f(\xi) = F(\ln \xi), \quad g(x) = e^z G(z)$$

leads to an equation with difference kernel of the form 2.9.62:

$$w(z) + \int_z^{\infty} F(z - \tau) w(\tau) \, d\tau = G(z).$$

7°. For arbitrary right-hand side $g(x)$, the solution of the integral equation can be expressed via the inverse Mellin transform (see Section 7.3-1).

2.10. Some Formulas and Transformations

Let the solution of the integral equation

$$y(x) + \int_a^x K(x, t) y(t) \, dt = f(x) \tag{1}$$

have the form

$$y(x) = f(x) + \int_a^x R(x, t) f(t) \, dt. \tag{2}$$

Then the solution of the more complicated integral equation

$$y(x) + \int_a^x K(x,t)\frac{g(x)}{g(t)}y(t)\,dt = f(x) \tag{3}$$

has the form

$$y(x) = f(x) + \int_a^x R(x,t)\frac{g(x)}{g(t)}f(t)\,dt. \tag{4}$$

Below are formulas for the solutions of integral equations of the form (3) for some specific functions $g(x)$. In all cases, it is assumed that the solution of equation (1) is known and is given by (2).

1°. The solution of the equation

$$y(x) + \int_a^x K(x,t)(x/t)^\lambda y(t)\,dt = f(x)$$

has the form

$$y(x) = f(x) + \int_a^x R(x,t)(x/t)^\lambda f(t)\,dt.$$

2°. The solution of the equation

$$y(x) + \int_a^x K(x,t)e^{\lambda(x-t)}y(t)\,dt = f(x)$$

has the form

$$y(x) = f(x) + \int_a^x R(x,t)e^{\lambda(x-t)}f(t)\,dt.$$

Chapter 3

Linear Equation of the First Kind With Constant Limits of Integration

▶ *Notation:* $f = f(x)$, $g = g(x)$, $h = h(x)$, $K = K(x)$, *and* $M = M(x)$ *are arbitrary functions (these may be composite functions of the argument depending on two variables x and t); A, B, C, a, b, c, k, α, β, γ, λ, and μ are free parameters; and n is a nonnegative integer.*

3.1. Equations Whose Kernels Contain Power-Law Functions

3.1-1. Kernels Linear in the Arguments x and t

1. $$\int_0^1 |x - t|\, y(t)\, dt = f(x).$$

   1°. Let us remove the modulus in the integrand:

   $$\int_0^x (x - t)y(t)\, dt + \int_x^1 (t - x)y(t)\, dt = f(x). \tag{1}$$

   Differentiating (1) with respect to x yields

   $$\int_0^x y(t)\, dt - \int_x^1 y(t)\, dt = f_x'(x). \tag{2}$$

   Differentiating (2) yields the solution

   $$y(x) = \tfrac{1}{2} f_{xx}''(x). \tag{3}$$

   2°. Let us demonstrate that the right-hand side $f(x)$ of the integral equation must satisfy certain relations. By setting $x = 0$ and $x = 1$ in (1), we obtain two corollaries $\int_0^1 ty(t)\, dt = f(0)$ and $\int_0^1 (1 - t)y(t)\, dt = f(1)$, which can be rewritten in the form

   $$\int_0^1 ty(t)\, dt = f(0), \qquad \int_0^1 y(t)\, dt = f(0) + f(1). \tag{4}$$

In Section 3.1, we mean that kernels of the integral equations discussed may contain power-law functions or modulus of power-law functions.

Substitute $y(x)$ of (3) into (4). Integration by parts yields $f'_x(1) = f(1) + f(0)$ and $f'_x(1) - f'_x(0) = 2f(1) + 2f(0)$. Hence, we obtain the desired constraints for $f(x)$:

$$f'_x(1) = f(0) + f(1), \qquad f'_x(0) + f'_x(1) = 0. \tag{5}$$

Conditions (5) make it possible to find the admissible general form of the right-hand side of the integral equation:

$$f(x) = F(x) + Ax + B,$$
$$A = -\tfrac{1}{2}\left[F'_x(1) + F'_x(0)\right], \qquad B = \tfrac{1}{2}\left[F'_x(1) - F(1) - F(0)\right],$$

where $F(x)$ is an arbitrary bounded twice differentiable function with bounded first derivative.

2. $\displaystyle\int_a^b |x - t|\, y(t)\, dt = f(x), \qquad 0 \le a < b < \infty.$

This is a special case of equation 3.8.3 with $g(x) = x$.

    Solution:

$$y(x) = \tfrac{1}{2} f''_{xx}(x).$$

The right-hand side $f(x)$ of the integral equation must satisfy certain relations. The general form of $f(x)$ is as follows:

$$f(x) = F(x) + Ax + B,$$
$$A = -\tfrac{1}{2}\left[F'_x(a) + F'_x(b)\right], \qquad B = \tfrac{1}{2}\left[aF'_x(a) + bF'_x(b) - F(a) - F(b)\right],$$

where $F(x)$ is an arbitrary bounded twice differentiable function (with bounded first derivative).

3. $\displaystyle\int_0^a |\lambda x - t|\, y(t)\, dt = f(x), \qquad \lambda > 0.$

Here $0 \le x \le a$ and $0 \le t \le a$.

$1°$. Let us remove the modulus in the integrand:

$$\int_0^{\lambda x} (\lambda x - t) y(t)\, dt + \int_{\lambda x}^a (t - \lambda x) y(t)\, dt = f(x). \tag{1}$$

Differentiating (1) with respect to x, we find that

$$\lambda \int_0^{\lambda x} y(t)\, dt - \lambda \int_{\lambda x}^a y(t)\, dt = f'_x(x). \tag{2}$$

Differentiating (2) yields $2\lambda^2 y(\lambda x) = f''_{xx}(x)$. Hence, we obtain the solution

$$y(x) = \frac{1}{2\lambda^2} f''_{xx}\left(\frac{x}{\lambda}\right). \tag{3}$$

$2°$. Let us demonstrate that the right-hand side $f(x)$ of the integral equation must satisfy certain relations. By setting $x = 0$ in (1) and (2), we obtain two corollaries

$$\int_0^a t y(t)\, dt = f(0), \qquad \lambda \int_0^a y(t)\, dt = -f'_x(0), \tag{4}$$

Substitute $y(x)$ from (3) into (4). Integrating by parts yields the desired constraints for $f(x)$:

$$(a/\lambda) f'_x(a/\lambda) = f(0) + f(a/\lambda), \qquad f'_x(0) + f'_x(a/\lambda) = 0. \tag{5}$$

Conditions (5) make it possible to establish the admissible general form of the right-hand side of the integral equation:

$$f(x) = F(z) + Az + B, \qquad z = \lambda x;$$
$$A = -\tfrac{1}{2}\left[F'_z(a) + F'_z(0)\right], \qquad B = \tfrac{1}{2}\left[aF'_z(a) - F(a) - F(0)\right],$$

where $F(x)$ is an arbitrary bounded twice differentiable function (with bounded first derivative).

4. $\displaystyle\int_0^a |x - \lambda t|\, y(t)\, dt = f(x), \qquad \lambda > 0.$

Here $0 \le x \le a$ and $0 \le t \le a$.

Solution:
$$y(x) = \tfrac{1}{2}\lambda f''_{xx}(\lambda x).$$

The right-hand side $f(x)$ of the integral equation must satisfy the relations
$$a\lambda f'_x(a\lambda) = f(0) + f(a\lambda), \qquad f'_x(0) + f'_x(a\lambda) = 0.$$

Hence, it follows the general form of the right-hand side:
$$f(x) = F(x) + Ax + B, \qquad A = -\tfrac{1}{2}\big[F'_x(\lambda a) + F'_x(0)\big], \qquad B = \tfrac{1}{2}\big[a\lambda F'_x(a\lambda) - F(\lambda a) - F(0)\big],$$

where $F(x)$ is an arbitrary bounded twice differentiable function (with bounded first derivative).

3.1-2. Kernels Quadratic in the Arguments x and t

5. $\displaystyle\int_0^a |Ax + Bx^2 - t|\, y(t)\, dt = f(x), \qquad A > 0, \quad B > 0.$

This is a special case of equation 3.8.5 with $g(x) = Ax + Bx^2$.

6. $\displaystyle\int_0^a |x - At - Bt^2|\, y(t)\, dt = f(x), \qquad A > 0, \quad B > 0.$

This is a special case of equation 3.8.6 with $g(x) = At + Bt^2$.

7. $\displaystyle\int_a^b |xt - t^2|\, y(t)\, dt = f(x) \qquad 0 \le a < b < \infty.$

The substitution $w(t) = ty(t)$ leads to an equation of the form 1.3.2:
$$\int_a^b |x - t| w(t)\, dt = f(x).$$

8. $\displaystyle\int_a^b |x^2 - t^2|\, y(t)\, dt = f(x).$

This is a special case of equation 3.8.3 with $g(x) = x^2$.

Solution: $y(x) = \dfrac{d}{dx}\left[\dfrac{f'_x(x)}{4x}\right]$. The right-hand side $f(x)$ of the equation must satisfy certain constraints, given in 3.8.3.

9. $\displaystyle\int_0^a |x^2 - \beta t^2|\, y(t)\, dt = f(x), \qquad \beta > 0.$

This is a special case of equation 3.8.4 with $g(x) = x^2$ and $\beta = \lambda^2$.

10. $\displaystyle\int_0^a |Ax + Bx^2 - A\lambda t - B\lambda^2 t^2|\, y(t)\, dt = f(x), \qquad \lambda > 0.$

This is a special case of equation 3.8.4 with $g(x) = Ax + Bx^2$.

3.1-3. Kernels Containing Integer Powers of x and t or Rational Functions

11. $\displaystyle\int_a^b |x - t|^3 y(t)\, dt = f(x).$

Let us remove the modulus in the integrand:

$$\int_a^x (x - t)^3 y(t)\, dt + \int_x^b (t - x)^3 y(t)\, dt = f(x). \tag{1}$$

Differentiating (1) twice yields

$$6\int_a^x (x - t) y(t)\, dt + 6\int_x^b (t - x) y(t)\, dt = f''_{xx}(x).$$

This equation can be rewritten in the form 3.1.2:

$$\int_a^b |x - t|\, y(t)\, dt = \tfrac{1}{6} f''_{xx}(x). \tag{2}$$

Therefore the solution of the integral equation is given by

$$y(x) = \tfrac{1}{12} y''''_{xxxx}(x). \tag{3}$$

The right-hand side $f(x)$ of the equation must satisfy certain conditions. To obtain these conditions, one must substitute solution (3) into (1) with $x = a$ and $x = b$ and into (2) with $x = a$ and $x = b$, and then integrate the four resulting relations by parts.

12. $\displaystyle\int_a^b |x^3 - t^3|\, y(t)\, dt = f(x).$

This is a special case of equation 3.8.3 with $g(x) = x^3$.

13. $\displaystyle\int_a^b |xt^2 - t^3|\, y(t)\, dt = f(x) \qquad 0 \le a < b < \infty.$

The substitution $w(t) = t^2 y(t)$ leads to an equation of the form 3.1.2:

$$\int_a^b |x - t| w(t)\, dt = f(x).$$

14. $\displaystyle\int_a^b |x^2 t - t^3|\, y(t)\, dt = f(x).$

The substitution $w(t) = |t|\, y(t)$ leads to an equation of the form 3.1.8:

$$\int_a^b |x^2 - t^2| w(t)\, dt = f(x).$$

15. $\displaystyle\int_0^a |x^3 - \beta t^3|\, y(t)\, dt = f(x), \qquad \beta > 0.$

This is a special case of equation 3.8.4 with $g(x) = x^3$ and $\beta = \lambda^3$.

16. $\displaystyle\int_a^b |x - t|^{2n+1} y(t)\, dt = f(x), \qquad n = 0, 1, 2, \ldots$

Solution:

$$y(x) = \frac{1}{2(2n + 1)!}\, f_x^{(2n+2)}(x). \tag{1}$$

The right-hand side $f(x)$ of the equation must satisfy certain conditions. To obtain these conditions, one must substitute solution (1) into the relations

$$\int_a^b (t - a)^{2n+1} y(t)\, dt = f(a), \qquad \int_a^b (t - a)^{2n-k} y(t)\, dt = \frac{(-1)^{k+1}}{A_k}\, f_x^{(k+1)}(a),$$
$$A_k = (2n + 1)(2n)\ldots(2n + 1 - k); \qquad k = 0, 1, \ldots, 2n,$$

and then integrate the resulting equations by parts.

17. $\displaystyle\int_0^\infty \frac{y(t)\, dt}{x + t} = f(x).$

The left-hand side of this equation is the Stieltjes transform.

$1°.$ By setting

$$x = e^z, \quad t = e^\tau, \quad y(t) = e^{-\tau/2} w(\tau), \quad f(x) = e^{-z/2} g(z),$$

we obtain an integral equation with difference kernel of the form 3.8.15:

$$\int_{-\infty}^\infty \frac{w(\tau)\, d\tau}{2\cosh\left[\frac{1}{2}(z - \tau)\right]} = g(z),$$

whose solution is given by

$$w(z) = \frac{1}{\sqrt{2\pi^3}} \int_{-\infty}^\infty \cosh(\pi u)\, \tilde{g}(u) e^{iux}\, du, \qquad \tilde{g}(u) = \frac{1}{\sqrt{2\pi}} \int_{-\infty}^\infty g(z) e^{-iuz}\, dz.$$

⊙ Reference: P. P. Zabreyko, A. I. Koshelev, et al. (1975).

$2°.$ Under some assumptions, the solution of the original equation can be represented in the form

$$y(x) = \lim_{n\to\infty} \frac{(-1)^n}{(n - 1)!\,(n + 1)!} \left[x^{2n+1} f_x^{(n)}(x)\right]_x^{(n+1)}, \tag{1}$$

which is the real inversion of the Stieltjes transform.

An alternative form of the solution is

$$y(x) = \lim_{n\to\infty} \frac{(-1)^n}{2\pi} \left(\frac{e}{n}\right)^{2n} \left[x^{2n} f_x^{(n)}(x)\right]_x^{(n)}. \tag{2}$$

To obtain an approximate solution of the integral equation, one restricts oneself to a specific value of n in (1) or (2) instead of taking the limit.

⊙ Reference: I. I. Hirschman and D. V. Widder (1955).

3.1-4. Kernels Containing Square Roots

18. $\displaystyle\int_0^a \left| \sqrt{x} - \sqrt{t} \, \right| y(t)\, dt = f(x), \qquad 0 < a < \infty.$

This is a special case of equation 3.8.3 with $g(x) = \sqrt{x}$.
    Solution:

$$y(x) = \frac{d}{dx}\left[\sqrt{x}\, f'_x(x) \right].$$

    The right-hand side $f(x)$ of the equation must satisfy certain conditions. The general form of the right-hand side is

$$f(x) = F(x) + Ax + B, \qquad A = -F'_x(a), \quad B = \tfrac{1}{2}\left[aF'_x(a) - F(a) - F(0) \right],$$

where $F(x)$ is an arbitrary bounded twice differentiable function (with bounded first derivative).

19. $\displaystyle\int_0^a \left| \sqrt{x} - \beta\sqrt{t} \, \right| y(t)\, dt = f(x), \qquad \beta > 0.$

This is a special case of equation 3.8.4 with $g(x) = \sqrt{x}$ and $\beta = \sqrt{\lambda}$.

20. $\displaystyle\int_0^a \left| \sqrt{x} - t \right| y(t)\, dt = f(x).$

This is a special case of equation 3.8.5 with $g(x) = \sqrt{x}$ (see item 3° of 3.8.5).

21. $\displaystyle\int_0^a \left| x - \sqrt{t} \, \right| y(t)\, dt = f(x).$

This is a special case of equation 3.8.6 with $g(t) = \sqrt{t}$ (see item 3° of 3.8.6).

22. $\displaystyle\int_0^a \frac{y(t)}{\sqrt{|x - t|}}\, dt = f(x), \qquad 0 < a \le \infty.$

This is a special case of equation 3.1.29 with $k = \tfrac{1}{2}$.
    Solution:

$$y(x) = -\frac{A}{x^{1/4}}\frac{d}{dx}\left[\int_x^a \frac{dt}{(t-x)^{1/4}} \int_0^t \frac{f(s)\, ds}{s^{1/4}(t-s)^{1/4}} \right], \qquad A = \frac{1}{\sqrt{8\pi}\,\Gamma^2(3/4)}.$$

23. $\displaystyle\int_{-\infty}^{\infty} \frac{y(t)}{\sqrt{|x - t|}}\, dt = f(x).$

This is a special case of equation 3.1.34 with $\lambda = \tfrac{1}{2}$.
    Solution:

$$y(x) = \frac{1}{4\pi}\int_{-\infty}^{\infty} \frac{f(x) - f(t)}{|x - t|^{3/2}}\, dt.$$

3.1-5. Kernels Containing Arbitrary Powers

24. $\displaystyle\int_0^a |x^k - t^k|\, y(t)\, dt = f(x), \qquad 0 < k < 1, \quad 0 < a < \infty.$

$1°.$ Let us remove the modulus in the integrand:

$$\int_0^x (x^k - t^k) y(t)\, dt + \int_x^a (t^k - x^k) y(t)\, dt = f(x). \tag{1}$$

Differentiating (1) with respect to x yields

$$k x^{k-1} \int_0^x y(t)\, dt - k x^{k-1} \int_x^a y(t)\, dt = f'_x(x). \tag{2}$$

Let us divide both sides of (2) by $k x^{k-1}$ and differentiate the resulting equation. As a result, we obtain the solution

$$y(x) = \frac{1}{2k} \frac{d}{dx}\left[x^{1-k} f'_x(x)\right]. \tag{3}$$

$2°.$ Let us demonstrate that the right-hand side $f(x)$ of the integral equation must satisfy certain relations. By setting $x = 0$ and $x = a$, in (1), we obtain two corollaries $\displaystyle\int_0^a t^k y(t)\, dt = f(0)$ and $\displaystyle\int_0^a (a^k - t^k) y(t)\, dt = f(a)$, which can be rewritten in the form

$$\int_0^a t^k y(t)\, dt = f(0), \qquad a^k \int_0^a y(t)\, dt = f(0) + f(a). \tag{4}$$

Substitute $y(x)$ of (3) into (4). Integrating by parts yields the relations $a f'_x(a) = k f(a) + k f(0)$ and $a f'_x(a) = 2k f(a) + 2k f(0)$. Hence, the desired constraints for $f(x)$ have the form

$$f(0) + f(a) = 0, \qquad f'_x(a) = 0. \tag{5}$$

Conditions (5) make it possible to find the admissible general form of the right-hand side of the integral equation:

$$f(x) = F(x) + Ax + B, \qquad A = -F'_x(a), \quad B = \tfrac{1}{2}\left[a F'_x(a) - F(a) - F(0)\right],$$

where $F(x)$ is an arbitrary bounded twice differentiable function with bounded first derivative. The first derivative may be unbounded at $x = 0$, in which case the conditions $\left[x^{1-k} F'_x\right]_{x=0} = 0$ must hold.

25. $\displaystyle\int_0^a |x^k - \beta t^k|\, y(t)\, dt = f(x), \qquad 0 < k < 1, \quad \beta > 0.$

This is a special case of equation 3.8.4 with $g(x) = x^k$ and $\beta = \lambda^k$.

26. $\displaystyle\int_0^a |x^k t^m - t^{k+m}|\, y(t)\, dt = f(x), \qquad 0 < k < 1, \quad 0 < a < \infty.$

The substitution $w(t) = t^m y(t)$ leads to an equation of the form 3.1.24:

$$\int_0^a |x^k - t^k|\, w(t)\, dt = f(x).$$

27. $\displaystyle\int_0^1 |x^k - t^m|\, y(t)\, dt = f(x), \qquad k > 0, \quad m > 0.$

The transformation

$$z = x^k, \quad \tau = t^m, \quad w(\tau) = \tau^{\frac{1-m}{m}} y(t)$$

leads to an equation of the form 3.1.1:

$$\int_0^1 |z - \tau|\, w(\tau)\, d\tau = F(z), \qquad F(z) = m f(z^{1/k}).$$

28. $\displaystyle\int_a^b |x - t|^{1+\lambda}\, y(t)\, dt = f(x), \qquad 0 \le \lambda < 1.$

For $\lambda = 0$, see equation 3.1.2. Assume that $0 < \lambda < 1$.

1°. Let us remove the modulus in the integrand:

$$\int_a^x (x - t)^{1+\lambda} y(t)\, dt + \int_x^b (t - x)^{1+\lambda} y(t)\, dt = f(x). \tag{1}$$

Let us differentiate (1) with respect to x twice and then divide both the sides by $\lambda(\lambda + 1)$. As a result, we obtain

$$\int_a^x (x - t)^{\lambda-1} y(t)\, dt + \int_x^b (t - x)^{\lambda-1} y(t)\, dt = \frac{1}{\lambda(\lambda + 1)} f''_{xx}(x). \tag{2}$$

Rewrite equation (2) in the form

$$\int_a^b \frac{y(t)\, dt}{|x - t|^k} = \frac{1}{\lambda(\lambda + 1)} f''_{xx}(x), \qquad k = 1 - \lambda. \tag{3}$$

See 3.1.29 and 3.1.30 for the solutions of equation (3) for various a and b.

2°. The right-hand side $f(x)$ of the integral equation must satisfy certain relations. By setting $x = a$ and $x = b$ in (1), we obtain two corollaries

$$\int_a^b (t - a)^{1+\lambda} y(t)\, dt = f(a), \qquad \int_a^b (b - t)^{1+\lambda} y(t)\, dt = f(b). \tag{4}$$

On substituting the solution $y(x)$ of (3) into (4) and then integrating by parts, we obtain the desired constraints for $f(x)$.

29. $\displaystyle\int_0^a \frac{y(t)}{|x - t|^k}\, dt = f(x), \qquad 0 < k < 1, \quad 0 < a \le \infty.$

1°. Solution:

$$y(x) = -A x^{\frac{k-1}{2}} \frac{d}{dx} \left[\int_x^a \frac{t^{\frac{1-2k}{2}}\, dt}{(t - x)^{\frac{1-k}{2}}} \int_0^t \frac{f(s)\, ds}{s^{\frac{1-k}{2}} (t - s)^{\frac{1-k}{2}}} \right],$$

$$A = \frac{1}{2\pi} \cos\left(\frac{\pi k}{2}\right) \Gamma(k) \left[\Gamma\left(\frac{1+k}{2}\right) \right]^{-2},$$

where $\Gamma(k)$ is the gamma function.

2°. The transformation $x = z^2$, $t = \xi^2$, $w(\xi) = 2\xi y(t)$ leads to an equation of the form 3.1.31:

$$\int_0^{\sqrt{a}} \frac{w(\xi)}{|z^2 - \xi^2|^k}\, d\xi = f(z^2).$$

30. $\displaystyle\int_a^b \frac{y(t)}{|x-t|^k}\,dt = f(x), \qquad 0 < k < 1.$

It is assumed that $|a| + |b| < \infty$. Solution:

$$y(x) = \frac{1}{2\pi}\cot(\tfrac{1}{2}\pi k)\frac{d}{dx}\int_a^x \frac{f(t)\,dt}{(x-t)^{1-k}} - \frac{1}{\pi^2}\cos^2(\tfrac{1}{2}\pi k)\int_a^x \frac{Z(t)F(t)}{(x-t)^{1-k}}\,dt,$$

where

$$Z(t) = (t-a)^{\frac{1+k}{2}}(b-t)^{\frac{1-k}{2}}, \qquad F(t) = \frac{d}{dt}\left[\int_a^t \frac{d\tau}{(t-\tau)^k}\int_\tau^b \frac{f(s)\,ds}{Z(s)(s-\tau)^{1-k}}\right].$$

⊙ Reference: F. D. Gakhov (1977).

31. $\displaystyle\int_0^a \frac{y(t)}{|x^2-t^2|^k}\,dt = f(x), \qquad 0 < k < 1, \quad 0 < a \le \infty.$

Solution:

$$y(x) = -\frac{2\Gamma(k)\cos\left(\tfrac{1}{2}\pi k\right)}{\pi\left[\Gamma\left(\frac{1+k}{2}\right)\right]^2}x^{k-1}\frac{d}{dx}\int_x^a \frac{t^{2-2k}F(t)\,dt}{(t^2-x^2)^{\frac{1-k}{2}}}, \qquad F(t) = \int_0^t \frac{s^k f(s)\,ds}{(t^2-s^2)^{\frac{1-k}{2}}}.$$

⊙ Reference: P. P. Zabreyko, A. I. Koshelev, et al. (1975).

32. $\displaystyle\int_a^b \frac{y(t)}{|x^\lambda-t^\lambda|^k}\,dt = f(x), \qquad 0 < k < 1, \quad \lambda > 0.$

1°. The transformation

$$z = x^\lambda, \quad \tau = t^\lambda, \quad w(\tau) = \tau^{\frac{1-\lambda}{\lambda}}y(t)$$

leads to an equation of the form 3.8.30:

$$\int_A^B \frac{w(\tau)}{|z-\tau|^k}\,d\tau = F(z),$$

where $A = a^\lambda$, $B = b^\lambda$, $F(z) = \lambda f(z^{1/\lambda})$.

2°. Solution with $a = 0$:

$$y(x) = -Ax^{\frac{\lambda(k-1)}{2}}\frac{d}{dx}\left[\int_x^b \frac{t^{\frac{\lambda(3-2k)-2}{2}}\,dt}{(t^\lambda-x^\lambda)^{\frac{1-k}{2}}}\int_0^t \frac{s^{\frac{\lambda(k+1)-2}{2}}f(s)\,ds}{(t^\lambda-s^\lambda)^{\frac{1-k}{2}}}\right],$$

$$A = \frac{\lambda^2}{2\pi}\cos\left(\frac{\pi k}{2}\right)\Gamma(k)\left[\Gamma\left(\frac{1+k}{2}\right)\right]^{-2},$$

where $\Gamma(k)$ is the gamma function.

33. $\displaystyle\int_0^1 \frac{y(t)}{|x^\lambda-t^m|^k}\,dt = f(x), \qquad 0 < k < 1, \quad \lambda > 0, \quad m > 0.$

The transformation

$$z = x^\lambda, \quad \tau = t^m, \quad w(\tau) = \tau^{\frac{1-m}{m}}y(t)$$

leads to an equation of the form 3.8.30:

$$\int_0^1 \frac{w(\tau)}{|z-\tau|^k}\,d\tau = F(z), \qquad F(z) = mf(z^{1/\lambda}).$$

34. $\displaystyle\int_{-\infty}^{\infty} \frac{y(t)}{|x-t|^{1-\lambda}}\, dt = f(x), \qquad 0 < \lambda < 1.$

Solution:

$$y(x) = \frac{\lambda}{2\pi} \tan\left(\frac{\pi\lambda}{2}\right) \int_{-\infty}^{\infty} \frac{f(x)-f(t)}{|x-t|^{1+\lambda}}\, dt.$$

It assumed that the condition $\displaystyle\int_{-\infty}^{\infty} |f(x)|^p dx < \infty$ is satisfied for some p, $1 < p < 1/\lambda$.

⊙ Reference: S. G. Samko, A. A. Kilbas, and A. A. Marichev (1993).

35. $\displaystyle\int_{-\infty}^{\infty} \frac{y(t)}{|x^3-t|^{1-\lambda}}\, dt = f(x), \qquad 0 < \lambda < 1.$

The substitution $z = x^3$ leads to an equation of the form 3.1.34:

$$\int_{-\infty}^{\infty} \frac{y(t)}{|z-t|^{1-\lambda}}\, dt = f\left(z^{1/3}\right).$$

36. $\displaystyle\int_{-\infty}^{\infty} \frac{y(t)}{|x^3-t^3|^{1-\lambda}}\, dt = f(x), \qquad 0 < \lambda < 1.$

The transformation

$$z = x^3, \qquad \tau = t^3, \qquad w(\tau) = \tau^{-2/3} y(t)$$

leads to an equation of the form 3.1.34:

$$\int_{-\infty}^{\infty} \frac{w(\tau)}{|z-\tau|^{1-\lambda}}\, d\tau = F(z), \qquad F(z) = 3f\left(z^{1/3}\right).$$

37. $\displaystyle\int_{-\infty}^{\infty} \frac{\text{sign}(x-t)}{|x-t|^{1-\lambda}} y(t)\, dt = f(x), \qquad 0 < \lambda < 1.$

Solution:

$$y(x) = \frac{\lambda}{2\pi} \cot\left(\frac{\pi\lambda}{2}\right) \int_{-\infty}^{\infty} \frac{f(x)-f(t)}{|x-t|^{1+\lambda}} \text{sign}(x-t)\, dt.$$

⊙ Reference: S. G. Samko, A. A. Kilbas, and A. A. Marichev (1993).

38. $\displaystyle\int_{-\infty}^{\infty} \frac{a + b\,\text{sign}(x-t)}{|x-t|^{1-\lambda}} y(t)\, dt = f(x), \qquad 0 < \lambda < 1.$

Solution:

$$y(x) = \frac{\lambda \sin(\pi\lambda)}{4\pi\left[a^2 \cos^2\left(\frac{1}{2}\pi\lambda\right) + b^2 \sin^2\left(\frac{1}{2}\pi\lambda\right)\right]} \int_{-\infty}^{\infty} \frac{a + b\,\text{sign}(x-t)}{|x-t|^{1+\lambda}} \left[f(x)-f(t)\right] dt.$$

⊙ Reference: S. G. Samko, A. A. Kilbas, and A. A. Marichev (1993).

39. $\displaystyle\int_{0}^{\infty} \frac{y(t)\, dt}{(ax + bt)^k} = f(x), \qquad a > 0, \quad b > 0, \quad k > 0.$

By setting

$$x = \frac{1}{2a} e^{2z}, \qquad t = \frac{1}{2b} e^{2\tau}, \qquad y(t) = b e^{(k-2)\tau} w(\tau), \qquad f(x) = e^{-kz} g(z).$$

we obtain an integral equation with the difference kernel of the form 3.8.15:

$$\int_{-\infty}^{\infty} \frac{w(\tau)\, d\tau}{\cosh^k(z-\tau)} = g(z).$$

40. $\int_0^\infty t^{z-1} y(t)\, dt = f(z).$

The left-hand side of this equation is the Mellin transform of $y(t)$ (z is treated as a complex variable).

Solution:

$$y(t) = \frac{1}{2\pi i} \int_{c-i\infty}^{c+i\infty} t^{-z} f(z)\, dz, \qquad i^2 = -1.$$

For specific $f(z)$, one can use tables of Mellin and Laplace integral transforms to calculate the integral.

⊙ References: H. Bateman and A. Erdélyi (vol. 2, 1954), V. A. Ditkin and A. P. Prudnikov (1965).

3.1-6. Equation Containing the Unknown Function of a Complicated Argument

41. $\int_0^1 y(xt)\, dt = f(x).$

Solution:

$$y(x) = x f'_x(x) + f(x).$$

The function $f(x)$ is assumed to satisfy the condition $\big[x f(x)\big]_{x=0} = 0$.

42. $\int_0^1 t^\lambda y(xt)\, dt = f(x).$

The substitution $\xi = xt$ leads to equation $\int_0^x \xi^\lambda y(\xi)\, d\xi = x^{\lambda+1} f(x)$. Differentiating with respect to x yields the solution

$$y(x) = x f'_x(x) + (\lambda + 1) f(x).$$

The function $f(x)$ is assumed to satisfy the condition $\big[x^{\lambda+1} f(x)\big]_{x=0} = 0$.

43. $\int_0^1 (A x^k + B t^m) y(xt)\, dt = f(x).$

The substitution $\xi = xt$ leads to an equation of the form 1.1.50:

$$\int_0^x \big(A x^{k+m} + B \xi^m\big) y(\xi)\, d\xi = x^{m+1} f(x).$$

44. $\int_0^1 \dfrac{y(xt)\, dt}{\sqrt{1-t}} = f(x).$

The substitution $\xi = xt$ leads to Abel's equation 1.1.36:

$$\int_0^x \frac{y(\xi)\, d\xi}{\sqrt{x-\xi}} = \sqrt{x}\, f(x).$$

45. $\int_0^1 \dfrac{y(xt)\, dt}{(1-t)^\lambda} = f(x), \qquad 0 < \lambda < 1.$

The substitution $\xi = xt$ leads to the generalized Abel equation 1.1.46:

$$\int_0^x \frac{y(\xi)\, d\xi}{(x-\xi)^\lambda} = x^{1-\lambda} f(x).$$

46. $\displaystyle\int_0^1 \frac{t^\mu y(xt)}{(1-t)^\lambda}\,dt = f(x), \qquad 0 < \lambda < 1.$

The transformation $\xi = xt$, $w(\xi) = \xi^\mu y(\xi)$ leads to the generalized Abel equation 1.1.46:

$$\int_0^x \frac{w(\xi)\,d\xi}{(x-\xi)^\lambda} = x^{1+\mu-\lambda} f(x).$$

47. $\displaystyle\int_0^\infty \frac{y(x+t) - y(x-t)}{t}\,dt = f(x).$

Solution:

$$y(x) = -\frac{1}{\pi^2}\int_0^\infty \frac{f(x+t) - f(x-t)}{t}\,dt.$$

⊙ Reference: V. A. Ditkin and A. P. Prudnikov (1965).

3.1-7. Singular Equations

In this subsection, all singular integrals are understood in the sense of the Cauchy principal value.

48. $\displaystyle\int_{-\infty}^\infty \frac{y(t)\,dt}{t-x} = f(x).$

Solution:

$$y(x) = -\frac{1}{\pi^2}\int_{-\infty}^\infty \frac{f(t)\,dt}{t-x}.$$

The integral equation and its solution form a Hilbert transform pair (in the asymmetric form).

⊙ Reference: V. A. Ditkin and A. P. Prudnikov (1965).

49. $\displaystyle\int_a^b \frac{y(t)\,dt}{t-x} = f(x).$

This equation is encountered in hydrodynamics in solving the problem on the flow of an ideal inviscid fluid around a thin profile ($a \leq x \leq b$). It is assumed that $|a| + |b| < \infty$.

1°. The solution bounded at the endpoints is

$$y(x) = -\frac{1}{\pi^2}\sqrt{(x-a)(b-x)}\int_a^b \frac{f(t)}{\sqrt{(t-a)(b-t)}}\,\frac{dt}{t-x},$$

provided that

$$\int_a^b \frac{f(t)\,dt}{\sqrt{(t-a)(b-t)}} = 0.$$

2°. The solution bounded at the endpoint $x = a$ and unbounded at the endpoint $x = b$ is

$$y(x) = -\frac{1}{\pi^2}\sqrt{\frac{x-a}{b-x}}\int_a^b \sqrt{\frac{b-t}{t-a}}\,\frac{f(t)}{t-x}\,dt.$$

3°. The solution unbounded at the endpoints is

$$y(x) = -\frac{1}{\pi^2\sqrt{(x-a)(b-x)}}\left[\int_a^b \frac{\sqrt{(t-a)(b-t)}}{t-x}f(t)\,dt + C\right],$$

where C is an arbitrary constant. The formula $\displaystyle\int_a^b y(t)\,dt = C/\pi$ holds.

Solutions that have a singularity point $x = s$ inside the interval $[a, b]$ can be found in Subsection 12.4-3.

⊙ Reference: F. D. Gakhov (1977).

3.2. Equations Whose Kernels Contain Exponential Functions

3.2-1. Kernels Containing Exponential Functions

1. $\displaystyle\int_a^b e^{\lambda|x-t|} y(t)\,dt = f(x), \qquad -\infty < a < b < \infty.$

1°. Let us remove the modulus in the integrand:

$$\int_a^x e^{\lambda(x-t)} y(t)\,dt + \int_x^b e^{\lambda(t-x)} y(t)\,dt = f(x). \tag{1}$$

Differentiating (1) with respect to x twice yields

$$2\lambda y(x) + \lambda^2 \int_a^x e^{\lambda(x-t)} y(t)\,dt + \lambda^2 \int_x^b e^{\lambda(t-x)} y(t)\,dt = f''_{xx}(x). \tag{2}$$

By eliminating the integral terms from (1) and (2), we obtain the solution

$$y(x) = \frac{1}{2\lambda}\left[f''_{xx}(x) - \lambda^2 f(x) \right]. \tag{3}$$

2°. The right-hand side $f(x)$ of the integral equation must satisfy certain relations. By setting $x = a$ and $x = b$ in (1), we obtain two corollaries

$$\int_a^b e^{\lambda t} y(t)\,dt = e^{\lambda a} f(a), \qquad \int_a^b e^{-\lambda t} y(t)\,dt = e^{-\lambda b} f(b). \tag{4}$$

On substituting the solution $y(x)$ of (3) into (4) and then integrating by parts, we see that

$$e^{\lambda b} f'_x(b) - e^{\lambda a} f'_x(a) = \lambda e^{\lambda a} f(a) + \lambda e^{\lambda b} f(b),$$
$$e^{-\lambda b} f'_x(b) - e^{-\lambda a} f'_x(a) = \lambda e^{-\lambda a} f(a) + \lambda e^{-\lambda b} f(b).$$

Hence, we obtain the desired constraints for $f(x)$:

$$f'_x(a) + \lambda f(a) = 0, \qquad f'_x(b) - \lambda f(b) = 0. \tag{5}$$

The general form of the right-hand side satisfying conditions (5) is given by

$$f(x) = F(x) + Ax + B,$$

$$A = \frac{1}{b\lambda - a\lambda - 2}\left[F'_x(a) + F'_x(b) + \lambda F(a) - \lambda F(b) \right], \qquad B = -\frac{1}{\lambda}\left[F'_x(a) + \lambda F(a) + Aa\lambda + A \right],$$

where $F(x)$ is an arbitrary bounded, twice differentiable function.

2. $\displaystyle\int_a^b \left(Ae^{\lambda|x-t|} + Be^{\mu|x-t|} \right) y(t)\,dt = f(x), \qquad -\infty < a < b < \infty.$

Let us remove the modulus in the integrand and differentiate the resulting equation with respect to x twice to obtain

$$2(A\lambda + B\mu)y(x) + \int_a^b \left(A\lambda^2 e^{\lambda|x-t|} + B\mu^2 e^{\mu|x-t|} \right) y(t)\,dt = f''_{xx}(x). \tag{1}$$

Eliminating the integral term with $e^{\mu|x-t|}$ from (1) with the aid of the original integral equation, we find that

$$2(A\lambda + B\mu)y(x) + A(\lambda^2 - \mu^2) \int_a^b e^{\lambda|x-t|} y(t)\,dt = f''_{xx}(x) - \mu^2 f(x). \tag{2}$$

For $A\lambda + B\mu = 0$, this is an equation of the form 3.2.1, and for $A\lambda + B\mu \neq 0$, this is an equation of the form 4.2.15.

The right-hand side $f(x)$ must satisfy certain relations, which can be obtained by setting $x = a$ and $x = b$ in the original equation (a similar procedure is used in 3.2.1).

3. $\displaystyle\int_a^b |e^{\lambda x} - e^{\lambda t}|\, y(t)\, dt = f(x), \qquad \lambda > 0.$

This is a special case of equation 3.8.3 with $g(x) = e^{\lambda x}$.
Solution:

$$y(x) = \frac{1}{2\lambda}\frac{d}{dx}\left[e^{-\lambda x} f'_x(x)\right].$$

The right-hand side $f(x)$ of the integral equation must satisfy certain relations (see item $2°$ of equation 3.8.3).

4. $\displaystyle\int_0^a |e^{\beta x} - e^{\mu t}|\, y(t)\, dt = f(x), \qquad \beta > 0, \quad \mu > 0.$

This is a special case of equation 3.8.4 with $g(x) = e^{\beta x}$ and $\lambda = \mu/\beta$.

5. $\displaystyle\int_a^b \left[\sum_{k=1}^n A_k \exp\left(\lambda_k |x - t|\right)\right] y(t)\, dt = f(x), \qquad -\infty < a < b < \infty.$

$1°$. Let us remove the modulus in the kth summand of the integrand:

$$I_k(x) = \int_a^b \exp\left(\lambda_k |x-t|\right) y(t)\, dt = \int_a^x \exp[\lambda_k(x-t)] y(t)\, dt + \int_x^b \exp[\lambda_k(t-x)] y(t)\, dt. \quad (1)$$

Differentiating (1) with respect to x twice yields

$$I'_k = \lambda_k \int_a^x \exp[\lambda_k(x-t)] y(t)\, dt - \lambda_k \int_x^b \exp[\lambda_k(t-x)] y(t)\, dt,$$

$$I''_k = 2\lambda_k y(x) + \lambda_k^2 \int_a^x \exp[\lambda_k(x-t)] y(t)\, dt + \lambda_k^2 \int_x^b \exp[\lambda_k(t-x)] y(t)\, dt, \qquad (2)$$

where the primes denote the derivatives with respect to x. By comparing formulas (1) and (2), we find the relation between I''_k and I_k:

$$I''_k = 2\lambda_k y(x) + \lambda_k^2 I_k, \qquad I_k = I_k(x). \qquad (3)$$

$2°$. With the aid of (1), the integral equation can be rewritten in the form

$$\sum_{k=1}^n A_k I_k = f(x). \qquad (4)$$

Differentiating (4) with respect to x twice and taking into account (3), we obtain

$$\sigma_1 y(x) + \sum_{k=1}^n A_k \lambda_k^2 I_k = f''_{xx}(x), \qquad \sigma_1 = 2\sum_{k=1}^n A_k \lambda_k. \qquad (5)$$

Eliminating the integral I_n from (4) and (5) yields

$$\sigma_1 y(x) + \sum_{k=1}^{n-1} A_k(\lambda_k^2 - \lambda_n^2) I_k = f''_{xx}(x) - \lambda_n^2 f(x). \qquad (6)$$

Differentiating (6) with respect to x twice and eliminating I_{n-1} from the resulting equation with the aid of (6), we obtain a similar equation whose right-hand side is a second-order linear differential operator (acting on y) with constant coefficients plus the sum $\sum_{k=1}^{n-2} B_k I_k$. If we successively eliminate $I_{n-2}, I_{n-3}, \ldots, I_1$ with the aid of double differentiation, then we finally arrive at a linear nonhomogeneous ordinary differential equation of order $2(n-1)$ with constant coefficients.

$3°$. The right-hand side $f(x)$ must satisfy certain conditions. To find these conditions, one must set $x = a$ in the integral equation and its derivatives. (Alternatively, these conditions can be found by setting $x = a$ and $x = b$ in the integral equation and all its derivatives obtained by means of double differentiation.)

6.
$$\int_a^b \frac{y(t)\,dt}{|e^{\lambda x} - e^{\lambda t}|^k} = f(x), \qquad 0 < k < 1.$$

The transformation $z = e^{\lambda x}$, $\tau = e^{\lambda t}$, $w(\tau) = e^{-\lambda t}y(t)$ leads to an equation of the form 3.1.30:

$$\int_A^B \frac{w(\tau)}{|z - \tau|^k}\,d\tau = F(z),$$

where $A = e^{\lambda a}$, $B = e^{\lambda b}$, $F(z) = \lambda f\left(\frac{1}{\lambda}\ln z\right)$.

7.
$$\int_0^\infty \frac{y(t)\,dt}{(e^{\lambda x} + e^{\lambda t})^k} = f(x), \qquad \lambda > 0, \quad k > 0.$$

This equation can be rewritten as an equation with difference kernel in the form 3.8.16:

$$\int_0^\infty \frac{w(t)\,dt}{\cosh^k\left[\frac{1}{2}\lambda(x - t)\right]} = g(x),$$

where $w(t) = 2^{-k}\exp\left(-\frac{1}{2}\lambda k t\right)y(t)$ and $g(x) = \exp\left(\frac{1}{2}\lambda k x\right)f(x)$.

8.
$$\int_0^\infty e^{-zt}y(t)\,dt = f(z).$$

The left-hand side of the equation is the Laplace transform of $y(t)$ (z is treated as a complex variable).

$1°$. Solution:

$$y(t) = \frac{1}{2\pi i}\int_{c-i\infty}^{c+i\infty} e^{zt}f(z)\,dz, \qquad i^2 = -1.$$

For specific functions $f(z)$, one may use tables of inverse Laplace transforms to calculate the integral (e.g., see Supplement 5).

$2°$. For real $z = x$, under some assumptions the solution of the original equation can be represented in the form

$$y(x) = \lim_{n\to\infty} \frac{(-1)^n}{n!}\left(\frac{n}{x}\right)^{n+1} f_x^{(n)}\left(\frac{n}{x}\right),$$

which is the real inversion of the Laplace transform. To calculate the solution approximately, one should restrict oneself to a specific value of n in this formula instead of taking the limit.

⊙ References: H. Bateman and A. Erdélyi (vol. 1, 1954), I. I. Hirschman and D. V. Widder (1955), V. A. Ditkin and A. P. Prudnikov (1965).

3.2-2. Kernels Containing Power-Law and Exponential Functions

9.
$$\int_0^a \left|ke^{\lambda x} - k - t\right| y(t)\,dt = f(x).$$

This is a special case of equation 3.8.5 with $g(x) = ke^{\lambda x} - k$.

10. $\displaystyle\int_0^a \left|x - ke^{\lambda t} - k\right| y(t)\, dt = f(x).$

This is a special case of equation 3.8.6 with $g(t) = ke^{\lambda t} - k$.

11. $\displaystyle\int_a^b \left|\exp(\lambda x^2) - \exp(\lambda t^2)\right| y(t)\, dt = f(x), \qquad \lambda > 0.$

This is a special case of equation 3.8.3 with $g(x) = \exp(\lambda x^2)$.

    Solution:
$$y(x) = \frac{1}{4\lambda}\frac{d}{dx}\left[\frac{1}{x}\exp(-\lambda x^2)f_x'(x)\right].$$

    The right-hand side $f(x)$ of the integral equation must satisfy certain relations (see item $2°$ of equation 3.8.3).

12. $\displaystyle\frac{1}{\sqrt{\pi x}}\int_0^\infty \exp\left(-\frac{t^2}{4x}\right) y(t)\, dt = f(x).$

Applying the Laplace transformation to the equation, we obtain
$$\frac{\tilde{y}(\sqrt{p})}{\sqrt{p}} = \tilde{f}(p), \qquad \tilde{f}(p) = \int_0^\infty e^{-pt}f(t)\, dt.$$

Substituting p by p^2 and solving for the transform $\tilde{y}$, we find that $\tilde{y}(p) = p\tilde{f}(p^2)$. The inverse Laplace transform provides the solution of the original integral equation:
$$y(t) = \mathcal{L}^{-1}\{p\tilde{f}(p^2)\}, \qquad \mathcal{L}^{-1}\{g(p)\} \equiv \frac{1}{2\pi i}\int_{c-i\infty}^{c+i\infty} e^{pt}g(p)\, dp.$$

13. $\displaystyle\int_0^\infty \exp[-g(x)t^2]y(t)\, dt = f(x).$

Assume that $g(0) = \infty$, $g(\infty) = 0$, and $g_x' < 0$.

    The substitution $z = \dfrac{1}{4g(x)}$ leads to equation 3.2.12:
$$\frac{1}{\sqrt{\pi z}}\int_0^\infty \exp\left(-\frac{t^2}{4z}\right) y(t)\, dt = F(z),$$

where the function $F(z)$ is determined by the relations $F = \dfrac{2}{\sqrt{\pi}}f(x)\sqrt{g(x)}$ and $z = \dfrac{1}{4g(x)}$ by means of eliminating x.

3.3. Equations Whose Kernels Contain Hyperbolic Functions

3.3-1. Kernels Containing Hyperbolic Cosine

1. $\displaystyle\int_a^b \left|\cosh(\lambda x) - \cosh(\lambda t)\right| y(t)\, dt = f(x).$

This is a special case of equation 3.8.3 with $g(x) = \cosh(\lambda x)$.

    Solution:
$$y(x) = \frac{1}{2\lambda}\frac{d}{dx}\left[\frac{f_x'(x)}{\sinh(\lambda x)}\right].$$

    The right-hand side $f(x)$ of the integral equation must satisfy certain relations (see item $2°$ of equation 3.8.3).

2. $\displaystyle\int_0^a \big|\cosh(\beta x) - \cosh(\mu t)\big|\, y(t)\, dt = f(x), \qquad \beta > 0, \quad \mu > 0.$

This is a special case of equation 3.8.4 with $g(x) = \cosh(\beta x)$ and $\lambda = \mu/\beta$.

3. $\displaystyle\int_a^b \big|\cosh^k x - \cosh^k t\big|\, y(t)\, dt = f(x), \qquad 0 < k < 1.$

This is a special case of equation 3.8.3 with $g(x) = \cosh^k x$.

Solution:
$$y(x) = \frac{1}{2k}\frac{d}{dx}\left[\frac{f'_x(x)}{\sinh x\,\cosh^{k-1} x}\right].$$

The right-hand side $f(x)$ of the integral equation must satisfy certain relations (see item 2° of equation 3.8.3).

4. $\displaystyle\int_a^b \frac{y(t)}{|\cosh(\lambda x) - \cosh(\lambda t)|^k}\, dt = f(x), \qquad 0 < k < 1.$

This is a special case of equation 3.8.7 with $g(x) = \cosh(\lambda x) + \beta$, where β is an arbitrary number.

3.3-2. Kernels Containing Hyperbolic Sine

5. $\displaystyle\int_a^b \sinh\big(\lambda|x - t|\big)\, y(t)\, dt = f(x), \qquad -\infty < a < b < \infty.$

1°. Let us remove the modulus in the integrand:

$$\int_a^x \sinh[\lambda(x - t)]y(t)\, dt + \int_x^b \sinh[\lambda(t - x)]y(t)\, dt = f(x). \tag{1}$$

Differentiating (1) with respect to x twice yields

$$2\lambda y(x) + \lambda^2\int_a^x \sinh[\lambda(x - t)]y(t)\, dt + \lambda^2\int_x^b \sinh[\lambda(t - x)]y(t)\, dt = f''_{xx}(x). \tag{2}$$

Eliminating the integral terms from (1) and (2), we obtain the solution

$$y(x) = \frac{1}{2\lambda}\big[f''_{xx}(x) - \lambda^2 f(x)\big]. \tag{3}$$

2°. The right-hand side $f(x)$ of the integral equation must satisfy certain relations. By setting $x = a$ and $x = b$ in (1), we obtain two corollaries

$$\int_a^b \sinh[\lambda(t - a)]y(t)\, dt = f(a), \qquad \int_a^b \sinh[\lambda(b - t)]y(t)\, dt = f(b). \tag{4}$$

Substituting solution (3) into (4) and integrating by parts yields the desired conditions for $f(x)$:

$$\begin{aligned}
\sinh[\lambda(b - a)]f'_x(b) - \lambda\cosh[\lambda(b - a)]f(b) &= \lambda f(a), \\
\sinh[\lambda(b - a)]f'_x(a) + \lambda\cosh[\lambda(b - a)]f(a) &= -\lambda f(b).
\end{aligned} \tag{5}$$

The general form of the right-hand side is given by

$$f(x) = F(x) + Ax + B, \tag{6}$$

where $F(x)$ is an arbitrary bounded twice differentiable function, and the coefficients A and B are expressed in terms of $F(a)$, $F(b)$, $F'_x(a)$, and $F'_x(b)$ and can be determined by substituting formula (6) into conditions (5).

6. $\displaystyle\int_a^b \Big\{ A \sinh(\lambda|x-t|) + B \sinh(\mu|x-t|) \Big\} y(t)\, dt = f(x), \qquad -\infty < a < b < \infty.$

Let us remove the modulus in the integrand and differentiate the equation with respect to x twice to obtain

$$2(A\lambda + B\mu)y(x) + \int_a^b \big\{ A\lambda^2 \sinh(\lambda|x-t|) + B\mu^2 \sinh(\mu|x-t|) \big\} y(t)\, dt = f''_{xx}(x), \qquad (1)$$

Eliminating the integral term with $\sinh(\mu|x-t|)$ from (1) yields

$$2(A\lambda + B\mu)y(x) + A(\lambda^2 - \mu^2) \int_a^b \sinh(\lambda|x-t|) y(t)\, dt = f''_{xx}(x) - \mu^2 f(x). \qquad (2)$$

For $A\lambda + B\mu = 0$, this is an equation of the form 3.3.5, and for $A\lambda + B\mu \neq 0$, this is an equation of the form 4.3.26.

The right-hand side $f(x)$ must satisfy certain relations, which can be obtained by setting $x = a$ and $x = b$ in the original equation (a similar procedure is used in 3.3.5).

7. $\displaystyle\int_a^b \big| \sinh(\lambda x) - \sinh(\lambda t) \big|\, y(t)\, dt = f(x).$

This is a special case of equation 3.8.3 with $g(x) = \sinh(\lambda x)$.

Solution:

$$y(x) = \frac{1}{2\lambda} \frac{d}{dx} \left[\frac{f'_x(x)}{\cosh(\lambda x)} \right].$$

The right-hand side $f(x)$ of the integral equation must satisfy certain relations (see item 2° of equation 3.8.3).

8. $\displaystyle\int_0^a \big| \sinh(\beta x) - \sinh(\mu t) \big|\, y(t)\, dt = f(x), \qquad \beta > 0, \quad \mu > 0.$

This is a special case of equation 3.8.4 with $g(x) = \sinh(\beta x)$ and $\lambda = \mu/\beta$.

9. $\displaystyle\int_a^b \sinh^3(\lambda|x-t|) y(t)\, dt = f(x).$

Using the formula $\sinh^3 \beta = \frac{1}{4} \sinh 3\beta - \frac{3}{4} \sinh \beta$, we arrive at an equation of the form 3.3.6:

$$\int_a^b \left[\tfrac{1}{4} A \sinh(3\lambda|x-t|) - \tfrac{3}{4} A \sinh(\lambda|x-t|) \right] y(t)\, dt = f(x).$$

10. $\displaystyle\int_a^b \left[\sum_{k=1}^n A_k \sinh(\lambda_k|x-t|) \right] y(t)\, dt = f(x), \qquad -\infty < a < b < \infty.$

1°. Let us remove the modulus in the kth summand of the integrand:

$$I_k(x) = \int_a^b \sinh(\lambda_k|x-t|) y(t)\, dt = \int_a^x \sinh[\lambda_k(x-t)]y(t)\, dt + \int_x^b \sinh[\lambda_k(t-x)]y(t)\, dt. \quad (1)$$

Differentiating (1) with respect to x twice yields

$$I'_k = \lambda_k \int_a^x \cosh[\lambda_k(x-t)]y(t)\, dt - \lambda_k \int_x^b \cosh[\lambda_k(t-x)]y(t)\, dt,$$

$$I''_k = 2\lambda_k y(x) + \lambda_k^2 \int_a^x \sinh[\lambda_k(x-t)]y(t)\, dt + \lambda_k^2 \int_x^b \sinh[\lambda_k(t-x)]y(t)\, dt,$$

$$(2)$$

where the primes denote the derivatives with respect to x. By comparing formulas (1) and (2), we find the relation between I_k'' and I_k:

$$I_k'' = 2\lambda_k y(x) + \lambda_k^2 I_k, \qquad I_k = I_k(x). \tag{3}$$

$2°$. With the aid of (1), the integral equation can be rewritten in the form

$$\sum_{k=1}^{n} A_k I_k = f(x). \tag{4}$$

Differentiating (4) with respect to x twice and taking into account (3), we find that

$$\sigma_1 y(x) + \sum_{k=1}^{n} A_k \lambda_k^2 I_k = f_{xx}''(x), \qquad \sigma_1 = 2\sum_{k=1}^{n} A_k \lambda_k. \tag{5}$$

Eliminating the integral I_n from (4) and (5) yields

$$\sigma_1 y(x) + \sum_{k=1}^{n-1} A_k (\lambda_k^2 - \lambda_n^2) I_k = f_{xx}''(x) - \lambda_n^2 f(x). \tag{6}$$

Differentiating (6) with respect to x twice and eliminating I_{n-1} from the resulting equation with the aid of (6), we obtain a similar equation whose right-hand side is a second-order linear differential operator (acting on y) with constant coefficients plus the sum $\sum_{k=1}^{n-2} B_k I_k$. If we successively eliminate I_{n-2}, I_{n-3}, ..., with the aid of double differentiation, then we finally arrive at a linear nonhomogeneous ordinary differential equation of order $2(n-1)$ with constant coefficients.

$3°$. The right-hand side $f(x)$ must satisfy certain conditions. To find these conditions, one should set $x = a$ in the integral equation and its derivatives. (Alternatively, these conditions can be found by setting $x = a$ and $x = b$ in the integral equation and all its derivatives obtained by means of double differentiation.)

11. $\displaystyle\int_0^b \left| \sinh^k x - \sinh^k t \right| y(t)\, dt = f(x), \qquad 0 < k < 1.$

This is a special case of equation 3.8.3 with $g(x) = \sinh^k x$.

Solution:

$$y(x) = \frac{1}{2k} \frac{d}{dx} \left[\frac{f_x'(x)}{\cosh x \, \sinh^{k-1} x} \right].$$

The right-hand side $f(x)$ must satisfy certain conditions. As follows from item $3°$ of equation 3.8.3, the admissible general form of the right-hand side is given by

$$f(x) = F(x) + Ax + B, \qquad A = -F_x'(b), \qquad B = \tfrac{1}{2}\left[bF_x'(b) - F(0) - F(b) \right],$$

where $F(x)$ is an arbitrary bounded twice differentiable function (with bounded first derivative).

12. $\displaystyle\int_a^b \frac{y(t)}{|\sinh(\lambda x) - \sinh(\lambda t)|^k}\, dt = f(x), \qquad 0 < k < 1.$

This is a special case of equation 3.8.7 with $g(x) = \sinh(\lambda x) + \beta$, where β is an arbitrary number.

13. $\displaystyle\int_0^a \big|k\sinh(\lambda x) - t\big|\, y(t)\, dt = f(x).$

This is a special case of equation 3.8.5 with $g(x) = k\sinh(\lambda x)$.

14. $\displaystyle\int_0^a \big|x - k\sinh(\lambda t)\big|\, y(t)\, dt = f(x).$

This is a special case of equation 3.8.6 with $g(x) = k\sinh(\lambda t)$.

3.3-3. Kernels Containing Hyperbolic Tangent

15. $\displaystyle\int_a^b \big|\tanh(\lambda x) - \tanh(\lambda t)\big|\, y(t)\, dt = f(x).$

This is a special case of equation 3.8.3 with $g(x) = \tanh(\lambda x)$.
   Solution:
$$y(x) = \frac{1}{2\lambda}\frac{d}{dx}\Big[\cosh^2(\lambda x)f_x'(x)\Big].$$

The right-hand side $f(x)$ of the integral equation must satisfy certain relations (see item 2° of equation 3.8.3).

16. $\displaystyle\int_0^a \big|\tanh(\beta x) - \tanh(\mu t)\big|\, y(t)\, dt = f(x), \qquad \beta > 0, \quad \mu > 0.$

This is a special case of equation 3.8.4 with $g(x) = \tanh(\beta x)$ and $\lambda = \mu/\beta$.

17. $\displaystyle\int_0^b \big|\tanh^k x - \tanh^k t\big|\, y(t)\, dt = f(x), \qquad 0 < k < 1.$

This is a special case of equation 3.8.3 with $g(x) = \tanh^k x$.
   Solution:
$$y(x) = \frac{1}{2k}\frac{d}{dx}\Big[\cosh^2 x \,\coth^{k-1} x\, f_x'(x)\Big].$$

The right-hand side $f(x)$ must satisfy certain conditions. As follows from item 3° of equation 3.8.3, the admissible general form of the right-hand side is given by

$$f(x) = F(x) + Ax + B, \qquad A = -F_x'(b), \quad B = \tfrac{1}{2}\big[bF_x'(b) - F(0) - F(b)\big],$$

where $F(x)$ is an arbitrary bounded twice differentiable function (with bounded first derivative).

18. $\displaystyle\int_a^b \frac{y(t)}{|\tanh(\lambda x) - \tanh(\lambda t)|^k}\, dt = f(x), \qquad 0 < k < 1.$

This is a special case of equation 3.8.7 with $g(x) = \tanh(\lambda x) + \beta$, where β is an arbitrary number.

19. $\displaystyle\int_0^a \big|k\tanh(\lambda x) - t\big|\, y(t)\, dt = f(x).$

This is a special case of equation 3.8.5 with $g(x) = k\tanh(\lambda x)$.

20. $\displaystyle\int_0^a \big|x - k\tanh(\lambda t)\big|\, y(t)\, dt = f(x).$

This is a special case of equation 3.8.6 with $g(x) = k\tanh(\lambda t)$.

3.3-4. Kernels Containing Hyperbolic Cotangent

21. $\displaystyle\int_a^b \big| \coth(\lambda x) - \coth(\lambda t) \big|\, y(t)\, dt = f(x).$

This is a special case of equation 3.8.3 with $g(x) = \coth(\lambda x)$.

22. $\displaystyle\int_0^b \big| \coth^k x - \coth^k t \big|\, y(t)\, dt = f(x), \qquad 0 < k < 1.$

This is a special case of equation 3.8.3 with $g(x) = \coth^k x$.

3.4. Equations Whose Kernels Contain Logarithmic Functions

3.4-1. Kernels Containing Logarithmic Functions

1. $\displaystyle\int_a^b \big| \ln(x/t) \big|\, y(t)\, dt = f(x).$

This is a special case of equation 3.8.3 with $g(x) = \ln x$.
Solution:

$$y(x) = \frac{1}{2} \frac{d}{dx} \big[x f'_x(x) \big].$$

The right-hand side $f(x)$ of the integral equation must satisfy certain relations (see item 2° of equation 3.8.3).

2. $\displaystyle\int_a^b \ln|x - t|\, y(t)\, dt = f(x).$

Carleman's equation.

1°. Solution with $b - a \neq 4$:

$$y(x) = \frac{1}{\pi^2 \sqrt{(x-a)(b-x)}} \left[\int_a^b \frac{\sqrt{(t-a)(b-t)}\, f'_t(t)\, dt}{t-x} + \frac{1}{\pi \ln\left[\frac{1}{4}(b-a)\right]} \int_a^b \frac{f(t)\, dt}{\sqrt{(t-a)(b-t)}} \right].$$

2°. If $b - a = 4$, then for the equation to be solvable, the condition

$$\int_a^b f(t)(t-a)^{-1/2}(b-t)^{-1/2}\, dt = 0$$

must be satisfied. In this case, the solution has the form

$$y(x) = \frac{1}{\pi^2 \sqrt{(x-a)(b-x)}} \left[\int_a^b \frac{\sqrt{(t-a)(b-t)}\, f'_t(t)\, dt}{t-x} + C \right],$$

where C is an arbitrary constant.

⊙ Reference: F. D. Gakhov (1977).

3. $\displaystyle\int_a^b \left(\ln|x-t|+\beta\right)y(t)\,dt = f(x).$

By setting

$$x = e^{-\beta}z, \quad t = e^{-\beta}\tau, \quad y(t) = Y(\tau), \quad f(x) = e^{-\beta}g(z),$$

we arrive at an equation of the form 3.4.2:

$$\int_A^B \ln|z-\tau|\,Y(\tau)\,d\tau = g(z), \qquad A = ae^\beta,\ B = be^\beta.$$

4. $\displaystyle\int_{-a}^a \left(\ln\frac{A}{|x-t|}\right)y(t)\,dt = f(x), \qquad -a \le x \le a.$

This is a special case of equation 3.4.3 with $b = -a$. Solution with $0 < a < 2A$:

$$y(x) = \frac{1}{2M'(a)}\left[\frac{d}{da}\int_{-a}^a w(t,a)f(t)\,dt\right]w(x,a)$$
$$-\frac{1}{2}\int_{|x|}^a w(x,\xi)\frac{d}{d\xi}\left[\frac{1}{M'(\xi)}\frac{d}{d\xi}\int_{-\xi}^\xi w(t,\xi)f(t)\,dt\right]d\xi$$
$$-\frac{1}{2}\frac{d}{dx}\int_{|x|}^a \frac{w(x,\xi)}{M'(\xi)}\left[\int_{-\xi}^\xi w(t,\xi)\,df(t)\right]d\xi,$$

where

$$M(\xi) = \left(\ln\frac{2A}{\xi}\right)^{-1}, \qquad w(x,\xi) = \frac{M(\xi)}{\pi\sqrt{\xi^2-x^2}},$$

and the prime stands for the derivative.

  ⊙ Reference: I. C. Gohberg and M. G. Krein (1967).

5. $\displaystyle\int_0^a \ln\left|\frac{x+t}{x-t}\right| y(t)\,dt = f(x).$

Solution:

$$y(x) = -\frac{2}{\pi^2}\frac{d}{dx}\int_x^a \frac{F(t)\,dt}{\sqrt{t^2-x^2}}, \qquad F(t) = \frac{d}{dt}\int_0^t \frac{sf(s)\,ds}{\sqrt{t^2-s^2}}.$$

  ⊙ Reference: P. P. Zabreyko, A. I. Koshelev, et al. (1975).

6. $\displaystyle\int_a^b \left|\ln\frac{1+\lambda x}{1+\lambda t}\right| y(t)\,dt = f(x).$

This is a special case of equation 3.8.3 with $g(x) = \ln(1+\lambda x)$.

    Solution:

$$y(x) = \frac{1}{2\lambda}\frac{d}{dx}\left[(1+\lambda x)f_x'(x)\right].$$

    The right-hand side $f(x)$ of the integral equation must satisfy certain relations (see item 2° of equation 3.8.3).

7. $\displaystyle\int_a^b \left|\ln^\beta x - \ln^\beta t\right| y(t)\,dt = f(x), \qquad 0 < \beta < 1.$

This is a special case of equation 3.8.3 with $g(x) = \ln^\beta x$.

8. $\displaystyle\int_a^b \frac{y(t)}{|\ln(x/t)|^\beta}\,dt = f(x), \qquad 0 < \beta < 1.$

This is a special case of equation 3.8.7 with $g(x) = \ln x + A$, where A is an arbitrary number.

3.4-2. Kernels Containing Power-Law and Logarithmic Functions

9. $\displaystyle\int_0^a \big|k\ln(1+\lambda x) - t\big|\, y(t)\, dt = f(x).$

This is a special case of equation 3.8.5 with $g(x) = k\ln(1+\lambda x)$.

10. $\displaystyle\int_0^a \big|x - k\ln(1+\lambda t)\big|\, y(t)\, dt = f(x).$

This is a special case of equation 3.8.6 with $g(x) = k\ln(1+\lambda t)$.

11. $\displaystyle\int_0^\infty \frac{1}{t}\ln\left|\frac{x+t}{x-t}\right| y(t)\, dt = f(x).$

Solution:

$$y(x) = \frac{x}{\pi^2}\frac{d}{dx}\int_0^\infty \frac{df(t)}{dt}\ln\left|1 - \frac{x^2}{t^2}\right| dt.$$

⊙ Reference: P. P. Zabreyko, A. I. Koshelev, et al. (1975).

12. $\displaystyle\int_0^\infty \frac{\ln x - \ln t}{x - t}\, y(t)\, dt = f(x).$

The left-hand side of this equation is the iterated Stieltjes transform.

Under some assumptions, the solution of the integral equation can be represented in the form

$$y(x) = \frac{1}{4\pi^2}\lim_{n\to\infty}\left(\frac{e}{n}\right)^{4n} D^n x^{2n} D^{2n} x^{2n} D^n f(x), \qquad D = \frac{d}{dx}.$$

To calculate the solution approximately, one should restrict oneself to a specific value of n in this formula instead of taking the limit.

⊙ Reference: I. I. Hirschman and D. V. Widder (1955).

13. $\displaystyle\int_a^b \ln\big|x^\beta - t^\beta\big|\, y(t)\, dt = f(x), \qquad \beta > 0.$

The transformation

$$z = x^\beta, \quad \tau = t^\beta, \quad w(\tau) = t^{1-\beta} y(t)$$

leads to Carleman's equation 3.4.2:

$$\int_A^B \ln|z - \tau|\, w(\tau)\, d\tau = F(z), \qquad A = a^\beta, \quad B = b^\beta,$$

where $F(z) = \beta f\big(z^{1/\beta}\big)$.

14. $\displaystyle\int_0^1 \ln\big|x^\beta - t^\mu\big|\, y(t)\, dt = f(x), \qquad \beta > 0,\ \mu > 0.$

The transformation

$$z = x^\beta, \quad \tau = t^\mu, \quad w(\tau) = t^{1-\mu} y(t)$$

leads to an equation of the form 3.4.2:

$$\int_0^1 \ln|z - \tau|\, w(\tau)\, d\tau = F(z), \qquad F(z) = \mu f\big(z^{1/\beta}\big).$$

3.4-3. An Equation Containing the Unknown Function of a Complicated Argument

15. $\displaystyle\int_0^1 (A \ln t + B) y(xt)\, dt = f(x).$

The substitution $\xi = xt$ leads to an equation of the form 1.9.3 with $g(x) = -A \ln x$:

$$\int_0^x (A \ln \xi - A \ln x + B) y(\xi)\, d\xi = x f(x).$$

3.5. Equations Whose Kernels Contain Trigonometric Functions

3.5-1. Kernels Containing Cosine

1. $\displaystyle\int_0^\infty \cos(xt) y(t)\, dt = f(x).$

Solution: $y(x) = \dfrac{2}{\pi} \displaystyle\int_0^\infty \cos(xt) f(t)\, dt.$

Up to constant factors, the function $f(x)$ and the solution $y(t)$ are the Fourier cosine transform pair.

⊙ References: H. Bateman and A. Erdélyi (vol. 1, 1954), V. A. Ditkin and A. P. Prudnikov (1965).

2. $\displaystyle\int_a^b \big|\cos(\lambda x) - \cos(\lambda t)\big| y(t)\, dt = f(x).$

This is a special case of equation 3.8.3 with $g(x) = \cos(\lambda x)$.
Solution:

$$y(x) = -\frac{1}{2\lambda} \frac{d}{dx}\left[\frac{f'_x(x)}{\sin(\lambda x)}\right].$$

The right-hand side $f(x)$ of the integral equation must satisfy certain relations (see item 2° of equation 3.8.3).

3. $\displaystyle\int_0^a \big|\cos(\beta x) - \cos(\mu t)\big| y(t)\, dt = f(x), \qquad \beta > 0, \quad \mu > 0.$

This is a special case of equation 3.8.4 with $g(x) = \cos(\beta x)$ and $\lambda = \mu/\beta$.

4. $\displaystyle\int_a^b \big|\cos^k x - \cos^k t\big| y(t)\, dt = f(x), \qquad 0 < k < 1.$

This is a special case of equation 3.8.3 with $g(x) = \cos^k x$.
Solution:

$$y(x) = -\frac{1}{2k} \frac{d}{dx}\left[\frac{f'_x(x)}{\sin x \, \cos^{k-1} x}\right].$$

The right-hand side $f(x)$ of the integral equation must satisfy certain relations (see item 2° of equation 3.8.3).

5. $\displaystyle\int_a^b \frac{y(t)}{|\cos(\lambda x) - \cos(\lambda t)|^k}\, dt = f(x), \qquad 0 < k < 1.$

This is a special case of equation 3.8.7 with $g(x) = \cos(\lambda x) + \beta$, where β is an arbitrary number.

3.5-2. Kernels Containing Sine

6.
$$\int_0^\infty \sin(xt)y(t)\,dt = f(x).$$

Solution: $y(x) = \dfrac{2}{\pi} \displaystyle\int_0^\infty \sin(xt)f(t)\,dt.$

Up to constant factors, the function $f(x)$ and the solution $y(t)$ are the Fourier sine transform pair.

⊙ References: H. Bateman and A. Erdélyi (vol. 1, 1954), V. A. Ditkin and A. P. Prudnikov (1965).

7.
$$\int_a^b \sin\left(\lambda|x - t|\right)y(t)\,dt = f(x), \qquad -\infty < a < b < \infty.$$

1°. Let us remove the modulus in the integrand:

$$\int_a^x \sin[\lambda(x - t)]y(t)\,dt + \int_x^b \sin[\lambda(t - x)]y(t)\,dt = f(x). \tag{1}$$

Differentiating (1) with respect to x twice yields

$$2\lambda y(x) - \lambda^2 \int_a^x \sin[\lambda(x - t)]y(t)\,dt - \lambda^2 \int_x^b \sin[\lambda(t - x)]y(t)\,dt = f''_{xx}(x). \tag{2}$$

Eliminating the integral terms from (1) and (2), we obtain the solution

$$y(x) = \frac{1}{2\lambda}\left[f''_{xx}(x) + \lambda^2 f(x)\right]. \tag{3}$$

2°. The right-hand side $f(x)$ of the integral equation must satisfy certain relations. By setting $x = a$ and $x = b$ in (1), we obtain two corollaries

$$\int_a^b \sin[\lambda(t - a)]y(t)\,dt = f(a), \qquad \int_a^b \sin[\lambda(b - t)]y(t)\,dt = f(b). \tag{4}$$

Substituting solution (3) into (4) followed by integrating by parts yields the desired conditions for $f(x)$:

$$\begin{aligned}
\sin[\lambda(b - a)]f'_x(b) - \lambda \cos[\lambda(b - a)]f(b) &= \lambda f(a), \\
\sin[\lambda(b - a)]f'_x(a) + \lambda \cos[\lambda(b - a)]f(a) &= -\lambda f(b).
\end{aligned} \tag{5}$$

The general form of the right-hand side of the integral equation is given by

$$f(x) = F(x) + Ax + B, \tag{6}$$

where $F(x)$ is an arbitrary bounded twice differentiable function, and the coefficients A and B are expressed in terms of $F(a)$, $F(b)$, $F'_x(a)$, and $F'_x(b)$ and can be determined by substituting formula (6) into conditions (5).

8. $\displaystyle\int_a^b \left\{ A \sin\left(\lambda |x - t|\right) + B \sin\left(\mu |x - t|\right) \right\} y(t)\, dt = f(x), \qquad -\infty < a < b < \infty.$

Let us remove the modulus in the integrand and differentiate the equation with respect to x twice to obtain

$$2(A\lambda + B\mu)y(x) - \int_a^b \left\{ A\lambda^2 \sin\left(\lambda |x - t|\right) + B\mu^2 \sin\left(\mu |x - t|\right) \right\} y(t)\, dt = f''_{xx}(x). \qquad (1)$$

Eliminating the integral term with $\sin\left(\mu |x - t|\right)$ from (1) with the aid of the original equation, we find that

$$2(A\lambda + B\mu)y(x) + A(\mu^2 - \lambda^2) \int_a^b \sin\left(\lambda |x - t|\right) y(t)\, dt = f''_{xx}(x) + \mu^2 f(x). \qquad (2)$$

For $A\lambda + B\mu = 0$, this is an equation of the form 3.5.7 and for $A\lambda + B\mu \neq 0$, this is an equation of the form 4.5.29.

The right-hand side $f(x)$ must satisfy certain relations, which can be obtained by setting $x = a$ and $x = b$ in the original equation (a similar procedure is used in 3.5.7).

9. $\displaystyle\int_a^b \left| \sin(\lambda x) - \sin(\lambda t) \right| y(t)\, dt = f(x).$

This is a special case of equation 3.8.3 with $g(x) = \sin(\lambda x)$.

    Solution:

$$y(x) = \frac{1}{2\lambda} \frac{d}{dx} \left[\frac{f'_x(x)}{\cos(\lambda x)} \right].$$

The right-hand side $f(x)$ of the integral equation must satisfy certain relations (see item 2° of equation 3.8.3).

10. $\displaystyle\int_0^a \left| \sin(\beta x) - \sin(\mu t) \right| y(t)\, dt = f(x), \qquad \beta > 0, \quad \mu > 0.$

This is a special case of equation 3.8.4 with $g(x) = \sin(\beta x)$ and $\lambda = \mu/\beta$.

11. $\displaystyle\int_a^b \sin^3\left(\lambda |x - t|\right) y(t)\, dt = f(x).$

Using the formula $\sin^3 \beta = -\frac{1}{4} \sin 3\beta + \frac{3}{4} \sin \beta$, we arrive at an equation of the form 3.5.8:

$$\int_a^b \left[-\tfrac{1}{4} A \sin\left(3\lambda |x - t|\right) + \tfrac{3}{4} A \sin\left(\lambda |x - t|\right) \right] y(t)\, dt = f(x).$$

12. $\displaystyle\int_a^b \left[\sum_{k=1}^n A_k \sin\left(\lambda_k |x - t|\right) \right] y(t)\, dt = f(x), \qquad -\infty < a < b < \infty.$

1°. Let us remove the modulus in the kth summand of the integrand:

$$I_k(x) = \int_a^b \sin\left(\lambda_k |x - t|\right) y(t)\, dt = \int_a^x \sin[\lambda_k(x - t)] y(t)\, dt + \int_x^b \sin[\lambda_k(t - x)] y(t)\, dt. \qquad (1)$$

Differentiating (1) with respect to x yields

$$I'_k = \lambda_k \int_a^x \cos[\lambda_k(x - t)] y(t)\, dt - \lambda_k \int_x^b \cos[\lambda_k(t - x)] y(t)\, dt,$$

$$I''_k = 2\lambda_k y(x) - \lambda_k^2 \int_a^x \sin[\lambda_k(x - t)] y(t)\, dt - \lambda_k^2 \int_x^b \sin[\lambda_k(t - x)] y(t)\, dt,$$

$$(2)$$

where the primes denote the derivatives with respect to x. By comparing formulas (1) and (2), we find the relation between I_k'' and I_k:

$$I_k'' = 2\lambda_k y(x) - \lambda_k^2 I_k, \qquad I_k = I_k(x). \tag{3}$$

$2°$. With the aid of (1), the integral equation can be rewritten in the form

$$\sum_{k=1}^{n} A_k I_k = f(x). \tag{4}$$

Differentiating (4) with respect to x twice and taking into account (3), we find that

$$\sigma_1 y(x) - \sum_{k=1}^{n} A_k \lambda_k^2 I_k = f_{xx}''(x), \qquad \sigma_1 = 2 \sum_{k=1}^{n} A_k \lambda_k. \tag{5}$$

Eliminating the integral I_n from (4) and (5) yields

$$\sigma_1 y(x) + \sum_{k=1}^{n-1} A_k (\lambda_n^2 - \lambda_k^2) I_k = f_{xx}''(x) + \lambda_n^2 f(x). \tag{6}$$

Differentiating (6) with respect to x twice and eliminating I_{n-1} from the resulting equation with the aid of (6), we obtain a similar equation whose left-hand side is a second-order linear differential operator (acting on y) with constant coefficients plus the sum $\sum_{k=1}^{n-2} B_k I_k$. If we successively eliminate I_{n-2}, I_{n-3}, ..., with the aid of double differentiation, then we finally arrive at a linear nonhomogeneous ordinary differential equation of order $2(n-1)$ with constant coefficients.

$3°$. The right-hand side $f(x)$ must satisfy certain conditions. To find these conditions, one should set $x = a$ in the integral equation and its derivatives. (Alternatively, these conditions can be found by setting $x = a$ and $x = b$ in the integral equation and all its derivatives obtained by means of double differentiation.)

13. $\displaystyle\int_0^b \left|\sin^k x - \sin^k t\right| y(t)\, dt = f(x), \qquad 0 < k < 1.$

This is a special case of equation 3.8.3 with $g(x) = \sin^k x$.

Solution:

$$y(x) = \frac{1}{2k} \frac{d}{dx} \left[\frac{f_x'(x)}{\cos x \, \sin^{k-1} x} \right].$$

The right-hand side $f(x)$ must satisfy certain conditions. As follows from item $3°$ of equation 3.8.3, the admissible general form of the right-hand side is given by

$$f(x) = F(x) + Ax + B, \qquad A = -F_x'(b), \quad B = \tfrac{1}{2}\left[bF_x'(b) - F(0) - F(b)\right],$$

where $F(x)$ is an arbitrary bounded twice differentiable function (with bounded first derivative).

14. $\displaystyle\int_a^b \frac{y(t)}{|\sin(\lambda x) - \sin(\lambda t)|^k}\, dt = f(x), \qquad 0 < k < 1.$

This is a special case of equation 3.8.7 with $g(x) = \sin(\lambda x) + \beta$, where β is an arbitrary number.

15. $\displaystyle\int_0^a \left|k \sin(\lambda x) - t\right| y(t)\, dt = f(x).$

This is a special case of equation 3.8.5 with $g(x) = k \sin(\lambda x)$.

16. $\displaystyle\int_0^a \left|x - k \sin(\lambda t)\right| y(t)\, dt = f(x).$

This is a special case of equation 3.8.6 with $g(x) = k \sin(\lambda t)$.

3.5-3. Kernels Containing Tangent

17. $\displaystyle \int_a^b \left| \tan(\lambda x) - \tan(\lambda t) \right| y(t)\, dt = f(x).$

This is a special case of equation 3.8.3 with $g(x) = \tan(\lambda x)$.

    Solution:

$$ y(x) = \frac{1}{2\lambda} \frac{d}{dx} \left[\cos^2(\lambda x) f_x'(x) \right]. $$

The right-hand side $f(x)$ of the integral equation must satisfy certain relations (see item 2° of equation 3.8.3).

18. $\displaystyle \int_0^a \left| \tan(\beta x) - \tan(\mu t) \right| y(t)\, dt = f(x), \qquad \beta > 0, \quad \mu > 0.$

This is a special case of equation 3.8.4 with $g(x) = \tan(\beta x)$ and $\lambda = \mu/\beta$.

19. $\displaystyle \int_0^b \left| \tan^k x - \tan^k t \right| y(t)\, dt = f(x), \qquad 0 < k < 1.$

This is a special case of equation 3.8.3 with $g(x) = \tan^k x$.

    Solution:

$$ y(x) = \frac{1}{2k} \frac{d}{dx} \left[\cos^2 x\, \cot^{k-1} x f_x'(x) \right]. $$

The right-hand side $f(x)$ must satisfy certain conditions. As follows from item 3° of equation 3.8.3, the admissible general form of the right-hand side is given by

$$ f(x) = F(x) + Ax + B, \qquad A = -F_x'(b), \quad B = \tfrac{1}{2}\left[bF_x'(b) - F(0) - F(b) \right], $$

where $F(x)$ is an arbitrary bounded twice differentiable function (with bounded first derivative).

20. $\displaystyle \int_a^b \frac{y(t)}{\left| \tan(\lambda x) - \tan(\lambda t) \right|^k}\, dt = f(x), \qquad 0 < k < 1.$

This is a special case of equation 3.8.7 with $g(x) = \tan(\lambda x) + \beta$, where β is an arbitrary number.

21. $\displaystyle \int_0^a \left| k \tan(\lambda x) - t \right| y(t)\, dt = f(x).$

This is a special case of equation 3.8.5 with $g(x) = k \tan(\lambda x)$.

22. $\displaystyle \int_0^a \left| x - k \tan(\lambda t) \right| y(t)\, dt = f(x).$

This is a special case of equation 3.8.6 with $g(x) = k \tan(\lambda t)$.

3.5-4. Kernels Containing Cotangent

23. $\displaystyle \int_a^b \left| \cot(\lambda x) - \cot(\lambda t) \right| y(t)\, dt = f(x).$

This is a special case of equation 3.8.3 with $g(x) = \cot(\lambda x)$.

24. $\displaystyle \int_a^b \left| \cot^k x - \cot^k t \right| y(t)\, dt = f(x), \qquad 0 < k < 1.$

This is a special case of equation 3.8.3 with $g(x) = \cot^k x$.

3.5-5. Kernels Containing a Combination of Trigonometric Functions

25. $\displaystyle\int_{-\infty}^{\infty} \big[\cos(xt) + \sin(xt)\big] y(t)\, dt = f(x).$

Solution:

$$y(x) = \frac{1}{2\pi} \int_{-\infty}^{\infty} \big[\cos(xt) + \sin(xt)\big] f(t)\, dt.$$

Up to constant factors, the function $f(x)$ and the solution $y(t)$ are the Hartley transform pair.

⊙ Reference: D. Zwillinger (1989).

26. $\displaystyle\int_{0}^{\infty} \big[\sin(xt) - xt \cos(xt)\big] y(t)\, dt = f(x).$

This equation can be reduced to a special case of equation 3.7.1 with $\nu = \frac{3}{2}$.

Solution:

$$y(x) = \frac{2}{\pi} \int_{0}^{\infty} \frac{\sin(xt) - xt \cos(xt)}{x^2 t^2} f(t)\, dt.$$

3.5-6. Equations Containing the Unknown Function of a Complicated Argument

27. $\displaystyle\int_{0}^{\pi/2} y(\xi)\, dt = f(x), \qquad \xi = x \sin t.$

Schlömilch equation.

Solution:

$$y(x) = \frac{2}{\pi}\left[f(0) + x \int_{0}^{\pi/2} f_{\xi}'(\xi)\, dt \right], \qquad \xi = x \sin t.$$

⊙ References: E. T. Whittaker and G. N. Watson (1958), F. D. Gakhov (1977).

28. $\displaystyle\int_{0}^{\pi/2} y(\xi)\, dt = f(x), \qquad \xi = x \sin^k t.$

Generalized Schlömilch equation.

This is a special case of equation 3.5.29 for $n = 0$ and $m = 0$.

Solution:

$$y(x) = \frac{2k}{\pi} x^{\frac{k-1}{k}} \frac{d}{dx}\left[x^{\frac{1}{k}} \int_{0}^{x} \sin t\, f(\xi)\, dt \right], \qquad \xi = x \sin^k t.$$

29. $\displaystyle\int_{0}^{\pi/2} \sin^\lambda t\, y(\xi)\, dt = f(x), \qquad \xi = x \sin^k t.$

This is a special case of equation 3.5.29 for $m = 0$.

Solution:

$$y(x) = \frac{2k}{\pi} x^{\frac{k-\lambda-1}{k}} \frac{d}{dx}\left[x^{\frac{\lambda+1}{k}} \int_{0}^{x} \sin^{\lambda+1} t\, f(\xi)\, dt \right], \qquad \xi = x \sin^k t.$$

30. $\displaystyle\int_0^{\pi/2} \sin^\lambda t \cos^m t \, y(\xi)\, dt = f(x), \qquad \xi = x\sin^k t.$

$1°$. Let $\lambda > -1$, $m > -1$, and $k > 0$. The transformation

$$z = x^{\frac{2}{k}}, \quad \zeta = z\sin^2 t, \quad w(\zeta) = \zeta^{\frac{\lambda-1}{2}} y\!\left(\zeta^{\frac{k}{2}}\right)$$

leads to an equation of the form 1.1.43:

$$\int_0^z (z-\zeta)^{\frac{m-1}{2}} w(\zeta)\, d\zeta = F(z), \qquad F(z) = 2z^{\frac{\lambda+m}{2}} f\!\left(z^{\frac{k}{2}}\right).$$

$2°$. Solution with $-1 < m < 1$:

$$y(x) = \frac{2k}{\pi} \sin\!\left[\frac{\pi(1-m)}{2}\right] x^{\frac{k-\lambda-1}{k}} \frac{d}{dx}\left[x^{\frac{\lambda+1}{k}} \int_0^{\pi/2} \sin^{\lambda+1} t \tan^m t \, f(\xi)\, dt\right],$$

where $\xi = x\sin^k t$.

3.5-7. A Singular Equation

31. $\displaystyle\int_0^{2\pi} \cot\!\left(\frac{t-x}{2}\right) y(t)\, dt = f(x), \qquad 0 \le x \le 2\pi.$

Here the integral is understood in the sense of the Cauchy principal value and the right-hand side is assumed to satisfy the condition $\displaystyle\int_0^{2\pi} f(t)\, dt = 0$.

    Solution:

$$y(x) = -\frac{1}{4\pi^2} \int_0^{2\pi} \cot\!\left(\frac{t-x}{2}\right) f(t)\, dt + C,$$

where C is an arbitrary constant.

    It follows from the solution that $\displaystyle\int_0^{2\pi} y(t)\, dt = 2\pi C$.

    The equation and its solution form a Hilbert transform pair (in the asymmetric form).

   ⊙   Reference: F. D. Gakhov (1977).

3.6. Equations Whose Kernels Contain Combinations of Elementary Functions

3.6-1. Kernels Containing Hyperbolic and Logarithmic Functions

1. $\displaystyle\int_a^b \ln\bigl|\cosh(\lambda x) - \cosh(\lambda t)\bigr| \, y(t)\, dt = f(x).$

This is a special case of equation 1.8.9 with $g(x) = \cosh(\lambda x)$.

2. $\displaystyle\int_a^b \ln\bigl|\sinh(\lambda x) - \sinh(\lambda t)\bigr| \, y(t)\, dt = f(x).$

This is a special case of equation 1.8.9 with $g(x) = \sinh(\lambda x)$.

3. $\displaystyle\int_{-a}^{a} \ln\left[\frac{\sinh\left(\frac{1}{2}A\right)}{2\sinh\left(\frac{1}{2}|x-t|\right)} \right] y(t)\,dt = f(x), \qquad -a \le x \le a.$

Solution with $0 < a < A$:

$$y(x) = \frac{1}{2M'(a)} \left[\frac{d}{da} \int_{-a}^{a} w(t,a)f(t)\,dt \right] w(x,a)$$

$$- \frac{1}{2} \int_{|x|}^{a} w(x,\xi) \frac{d}{d\xi} \left[\frac{1}{M'(\xi)} \frac{d}{d\xi} \int_{-\xi}^{\xi} w(t,\xi)f(t)\,dt \right] d\xi$$

$$- \frac{1}{2} \frac{d}{dx} \int_{|x|}^{a} \frac{w(x,\xi)}{M'(\xi)} \left[\int_{-\xi}^{\xi} w(t,\xi)\,df(t) \right] d\xi,$$

where the prime stands for the derivative with respect to the argument and

$$M(\xi) = \left[\ln\left(\frac{\sinh\left(\frac{1}{2}A\right)}{\sinh\left(\frac{1}{2}\xi\right)} \right) \right]^{-1}, \qquad w(x,\xi) = \frac{\cosh\left(\frac{1}{2}x\right)M(\xi)}{\pi\sqrt{2\cosh\xi - 2\cosh x}}.$$

⊙ Reference: I. C. Gohberg and M. G. Krein (1967).

4. $\displaystyle\int_{a}^{b} \ln\left|\tanh(\lambda x) - \tanh(\lambda t)\right| y(t)\,dt = f(x).$

This is a special case of equation 1.8.9 with $g(x) = \tanh(\lambda x)$.

5. $\displaystyle\int_{-a}^{a} \ln\left[\coth\left(\frac{1}{4}|x-t|\right)\right] y(t)\,dt = f(x), \qquad -a \le x \le a.$

Solution:

$$y(x) = \frac{1}{2M'(a)} \left[\frac{d}{da} \int_{-a}^{a} w(t,a)f(t)\,dt \right] w(x,a)$$

$$- \frac{1}{2} \int_{|x|}^{a} w(x,\xi) \frac{d}{d\xi} \left[\frac{1}{M'(\xi)} \frac{d}{d\xi} \int_{-\xi}^{\xi} w(t,\xi)f(t)\,dt \right] d\xi$$

$$- \frac{1}{2} \frac{d}{dx} \int_{|x|}^{a} \frac{w(x,\xi)}{M'(\xi)} \left[\int_{-\xi}^{\xi} w(t,\xi)\,df(t) \right] d\xi,$$

where the prime stands for the derivative with respect to the argument and

$$M(\xi) = \frac{P_{-1/2}(\cosh\xi)}{Q_{-1/2}(\cosh\xi)}, \qquad w(x,\xi) = \frac{1}{\pi Q_{-1/2}(\cosh\xi)\sqrt{2\cosh\xi - 2\cosh x}},$$

and $P_{-1/2}(\cosh\xi)$ and $Q_{-1/2}(\cosh\xi)$ are the Legendre functions of the first and second kind, respectively.

⊙ Reference: I. C. Gohberg and M. G. Krein (1967).

3.6-2. Kernels Containing Logarithmic and Trigonometric Functions

6. $\displaystyle\int_{a}^{b} \ln\left|\cos(\lambda x) - \cos(\lambda t)\right| y(t)\,dt = f(x).$

This is a special case of equation 1.8.9 with $g(x) = \cos(\lambda x)$.

7. $\displaystyle\int_a^b \ln\bigl|\sin(\lambda x) - \sin(\lambda t)\bigr|\, y(t)\, dt = f(x).$

This is a special case of equation 1.8.9 with $g(x) = \sin(\lambda x)$.

8. $\displaystyle\int_{-a}^a \ln\left[\frac{\sin\bigl(\tfrac{1}{2}A\bigr)}{2\sin\bigl(\tfrac{1}{2}|x - t|\bigr)}\right] y(t)\, dt = f(x), \qquad -a \le x \le a.$

Solution with $0 < a < A$:

$$y(x) = \frac{1}{2M'(a)}\left[\frac{d}{da}\int_{-a}^a w(t,a)f(t)\, dt\right] w(x,a)$$

$$-\frac{1}{2}\int_{|x|}^a w(x,\xi)\frac{d}{d\xi}\left[\frac{1}{M'(\xi)}\frac{d}{d\xi}\int_{-\xi}^\xi w(t,\xi)f(t)\, dt\right] d\xi$$

$$-\frac{1}{2}\frac{d}{dx}\int_{|x|}^a \frac{w(x,\xi)}{M'(\xi)}\left[\int_{-\xi}^\xi w(t,\xi)\, df(t)\right] d\xi,$$

where the prime stands for the derivative with respect to the argument and

$$M(\xi) = \left[\ln\left(\frac{\sin\bigl(\tfrac{1}{2}A\bigr)}{\sin\bigl(\tfrac{1}{2}\xi\bigr)}\right)\right]^{-1}, \qquad w(x,\xi) = \frac{\cos\bigl(\tfrac{1}{2}\xi\bigr)M(\xi)}{\pi\sqrt{2\cos x - 2\cos\xi}}.$$

⊙ Reference: I. C. Gohberg and M. G. Krein (1967).

3.7. Equations Whose Kernels Contain Special Functions

3.7-1. Kernels Containing Bessel Functions

1. $\displaystyle\int_0^\infty tJ_\nu(xt)y(t)\, dt = f(x), \qquad \nu > -\tfrac{1}{2}.$

Here J_ν is the Bessel function of the first kind.

Solution:

$$y(x) = \int_0^\infty tJ_\nu(xt)f(t)\, dt.$$

The function $f(x)$ and the solution $y(t)$ are the Hankel transform pair.

⊙ Reference: V. A. Ditkin and A. P. Prudnikov (1965).

2. $\displaystyle\int_a^b \bigl|J_\nu(\lambda x) - J_\nu(\lambda t)\bigr|\, y(t)\, dt = f(x).$

This is a special case of equation 3.8.3 with $g(x) = J_\nu(\lambda x)$, where J_ν is the Bessel function of the first kind.

3. $\displaystyle\int_a^b \bigl|Y_\nu(\lambda x) - Y_\nu(\lambda t)\bigr|\, y(t)\, dt = f(x).$

This is a special case of equation 3.8.3 with $g(x) = Y_\nu(\lambda x)$, where Y_ν is the Bessel function of the second kind.

3.7-2. Kernels Containing Modified Bessel Functions

4. $\int_a^b \big| I_\nu(\lambda x) - I_\nu(\lambda t)\big| y(t)\, dt = f(x).$

This is a special case of equation 3.8.3 with $g(x) = I_\nu(\lambda x)$, where I_ν is the modified Bessel function of the first kind.

5. $\int_a^b \big| K_\nu(\lambda x) - K_\nu(\lambda t)\big| y(t)\, dt = f(x).$

This is a special case of equation 3.8.3 with $g(x) = K_\nu(\lambda x)$, where K_ν is the modified Bessel function of the second kind (the Macdonald function).

6. $\int_0^\infty \sqrt{zt}\, K_\nu(zt) y(t)\, dt = f(z).$

Here K_ν is the modified Bessel function of the second kind.

   Up to a constant factor, the left-hand side of this equation is the Meijer transform of $y(t)$ (z is treated as a complex variable).

   Solution:

$$y(t) = \frac{1}{\pi i} \int_{c-i\infty}^{c+i\infty} \sqrt{zt}\, I_\nu(zt) f(z)\, dz.$$

For specific $f(z)$, one may use tables of Meijer integral transforms to calculate the integral.

⊙ Reference: V. A. Ditkin and A. P. Prudnikov (1965).

7. $\int_{-\infty}^\infty K_0\big(|x - t|\big) y(t)\, dt = f(x).$

Here K_0 is the modified Bessel function of the second kind.

   Solution:

$$y(x) = -\frac{1}{\pi^2}\left(\frac{d^2}{dx^2} - 1\right) \int_{-\infty}^\infty K_0\big(|x - t|\big) f(t)\, dt.$$

⊙ Reference: D. Naylor (1986).

3.7-3. Other Kernels

8. $\int_0^a K\left(\frac{2\sqrt{xt}}{x + t}\right) \frac{y(t)\, dt}{x + t} = f(x).$

Here $K(z) = \int_0^1 \dfrac{dt}{\sqrt{(1 - t^2)(1 - z^2 t^2)}}$ is the complete elliptic integral of the first kind.

   Solution:

$$y(x) = -\frac{4}{\pi^2} \frac{d}{dx} \int_x^a \frac{t F(t)\, dt}{\sqrt{t^2 - x^2}}, \qquad F(t) = \frac{d}{dt} \int_0^t \frac{s f(s)\, ds}{\sqrt{t^2 - s^2}}.$$

⊙ Reference: P. P. Zabreyko, A. I. Koshelev, et al. (1975).

9. $\displaystyle \int_0^a F\left(\frac{\beta}{2}, \frac{\beta+1}{2}, \mu; \frac{4x^2t^2}{(x^2+t^2)^2}\right) \frac{y(t)\,dt}{(x^2+t^2)^\beta} = f(x).$

Here $0 < a \le \infty$, $0 < \beta < \mu < \beta + 1$, and $F(a, b, c; z)$ is the hypergeometric function.

Solution:

$$y(x) = \frac{x^{2\mu-2}}{\Gamma(1+\beta-\mu)} \frac{d}{dx} \int_x^a \frac{tg(t)\,dt}{(t^2-x^2)^{\mu-\beta}},$$

$$g(t) = \frac{2\,\Gamma(\beta)\sin[(\beta-\mu)\pi]}{\pi\Gamma(\mu)} t^{1-2\beta} \frac{d}{dt} \int_0^t \frac{s^{2\mu-1}f(s)\,ds}{(t^2-s^2)^{\mu-\beta}}.$$

If $a = \infty$ and $f(x)$ is a differentiable function, then the solution can be represented in the form

$$y(x) = A \frac{d}{dt} \int_0^\infty \frac{(xt)^{2\mu} f_t'(t)}{(x^2+t^2)^{2\mu-\beta}} F\left(\mu-\frac{\beta}{2}, \mu+\frac{1-\beta}{2}, \mu+1; \frac{4x^2t^2}{(x^2+t^2)^2}\right) dt,$$

where $\displaystyle A = \frac{\Gamma(\beta)\,\Gamma(2\mu-\beta)\sin[(\beta-\mu)\pi]}{\pi\Gamma(\mu)\,\Gamma(1+\mu)}.$

⊙   Reference: P. P. Zabreyko, A. I. Koshelev, et al. (1975).

3.8. Equations Whose Kernels Contain Arbitrary Functions

3.8-1. Equations With Degenerate Kernel

1. $\displaystyle \int_a^b \left[g_1(x)h_1(t) + g_2(x)h_2(t)\right] y(t)\,dt = f(x).$

This integral equation has solutions only if its right-hand side is representable in the form

$$f(x) = A_1 g_1(x) + A_2 g_2(x), \qquad A_1 = \text{const}, \; A_2 = \text{const}. \tag{1}$$

In this case, any function $y = y(x)$ satisfying the normalization type conditions

$$\int_a^b h_1(t)y(t)\,dt = A_1, \qquad \int_a^b h_2(t)y(t)\,dt = A_2 \tag{2}$$

is a solution of the integral equation. Otherwise, the equation has no solutions.

2. $\displaystyle \int_a^b \left[\sum_{k=0}^n g_k(x)h_k(t)\right] y(t)\,dt = f(x).$

This integral equation has solutions only if its right-hand side is representable in the form

$$f(x) = \sum_{k=0}^n A_k g_k(x), \tag{1}$$

where the A_k are some constants. In this case, any function $y = y(x)$ satisfying the normalization type conditions

$$\int_a^b h_k(t)y(t)\,dt = A_k \qquad (k = 1, \ldots, n) \tag{2}$$

is a solution of the integral equation. Otherwise, the equation has no solutions.

3.8-2. Equations Containing Modulus

3. $\displaystyle\int_a^b |g(x) - g(t)|\, y(t)\, dt = f(x).$

Let $a \leq x \leq b$ and $a \leq t \leq b$; it is assumed in items 1° and 2° that $0 < g_x'(x) < \infty$.

1°. Let us remove the modulus in the integrand:

$$\int_a^x \big[g(x) - g(t)\big] y(t)\, dt + \int_x^b \big[g(t) - g(x)\big] y(t)\, dt = f(x). \tag{1}$$

Differentiating (1) with respect to x yields

$$g_x'(x) \int_a^x y(t)\, dt - g_x'(x) \int_x^b y(t)\, dt = f_x'(x). \tag{2}$$

Divide both sides of (2) by $g_x'(x)$ and differentiate the resulting equation to obtain the solution

$$y(x) = \frac{1}{2} \frac{d}{dx}\left[\frac{f_x'(x)}{g_x'(x)}\right]. \tag{3}$$

2°. Let us demonstrate that the right-hand side $f(x)$ of the integral equation must satisfy certain relations. By setting $x = a$ and $x = b$, in (1), we obtain two corollaries

$$\int_a^b \big[g(t) - g(a)\big] y(t)\, dt = f(a), \qquad \int_a^b \big[g(b) - g(t)\big] y(t)\, dt = f(b). \tag{4}$$

Substitute $y(x)$ of (3) into (4). Integrating by parts yields the desired constraints for $f(x)$:

$$\begin{aligned}
\big[g(b) - g(a)\big] \frac{f_x'(b)}{g_x'(b)} &= f(a) + f(b), \\
\big[g(a) - g(b)\big] \frac{f_x'(a)}{g_x'(a)} &= f(a) + f(b).
\end{aligned} \tag{5}$$

Let us point out a useful property of these constraints: $f_x'(b)g_x'(a) + f_x'(a)g_x'(b) = 0$.

Conditions (5) make it possible to find the admissible general form of the right-hand side of the integral equation:

$$f(x) = F(x) + Ax + B, \tag{6}$$

where $F(x)$ is an arbitrary bounded twice differentiable function (with bounded first derivative), and the coefficients A and B are given by

$$A = -\frac{g_x'(a)F_x'(b) + g_x'(b)F_x'(a)}{g_x'(a) + g_x'(b)},$$

$$B = -\tfrac{1}{2}A(a + b) - \tfrac{1}{2}\big[F(a) + F(b)\big] - \frac{g(b) - g(a)}{2g_x'(a)}\big[A + F_x'(a)\big].$$

3°. If $g(x)$ is representable in the form $g(x) = O(x - a)^k$ with $0 < k < 1$ in the vicinity of the point $x = a$ (in particular, the derivative g_x' is unbounded as $x \to a$), then the solution of the integral equation is given by formula (3) as well. In this case, the right-hand side of the integral equation must satisfy the conditions

$$f(a) + f(b) = 0, \qquad f_x'(b) = 0. \tag{7}$$

As before, the right-hand side of the integral equation is given by (6), with

$$A = -F'_x(b), \qquad B = \tfrac{1}{2}\big[(a+b)F'_x(b) - F(a) - F(b)\big].$$

$4°$. For $g'_x(a) = 0$, the right-hand side of the integral equation must satisfy the conditions

$$f'_x(a) = 0, \qquad \big[g(b) - g(a)\big]f'_x(b) = \big[f(a) + f(b)\big]g'_x(b).$$

As before, the right-hand side of the integral equation is given by (6), with

$$A = -F'_x(a), \qquad B = \tfrac{1}{2}\big[(a+b)F'_x(a) - F(a) - F(b)\big] + \frac{g(b) - g(a)}{2g'_x(b)}\big[F'_x(b) - F'_x(a)\big].$$

4. $\displaystyle \int_0^a \big|g(x) - g(\lambda t)\big|\, y(t)\, dt = f(x), \qquad \lambda > 0.$

Assume that $0 \le x \le a$, $0 \le t \le a$ and $0 < g'_x(x) < \infty$.

$1°$. Let us remove the modulus in the integrand:

$$\int_0^{x/\lambda} \big[g(x) - g(\lambda t)\big]y(t)\, dt + \int_{x/\lambda}^a \big[g(\lambda t) - g(x)\big]y(t)\, dt = f(x). \tag{1}$$

Differentiating (1) with respect to x yields

$$g'_x(x) \int_0^{x/\lambda} y(t)\, dt - g'_x(x) \int_{x/\lambda}^a y(t)\, dt = f'_x(x). \tag{2}$$

Let us divide both sides of (2) by $g'_x(x)$ and differentiate the resulting equation to obtain $y(x/\lambda) = \tfrac{1}{2}\lambda\big[f'_x(x)/g'_x(x)\big]'_x$. Substituting x by λx yields the solution

$$y(x) = \frac{\lambda}{2}\frac{d}{dz}\left[\frac{f'_z(z)}{g'_z(z)}\right], \qquad z = \lambda x. \tag{3}$$

$2°$. Let us demonstrate that the right-hand side $f(x)$ of the integral equation must satisfy certain relations. By setting $x = 0$ in (1) and (2), we obtain two corollaries

$$\int_0^a \big[g(\lambda t) - g(0)\big]y(t)\, dt = f(0), \qquad g'_x(0) \int_0^a y(t)\, dt = -f'_x(0). \tag{4}$$

Substitute $y(x)$ of (3) into (4). Integrating by parts yields the desired constraints for $f(x)$:

$$\begin{aligned} &f'_x(0)g'_x(\lambda a) + f'_x(\lambda a)g'_x(0) = 0, \\ &\big[g(\lambda a) - g(0)\big]\frac{f'_x(\lambda a)}{g'_x(\lambda a)} = f(0) + f(\lambda a). \end{aligned} \tag{5}$$

Conditions (5) make it possible to find the admissible general form of the right-hand side of the integral equation:

$$f(x) = F(x) + Ax + B, \tag{6}$$

where $F(x)$ is an arbitrary bounded twice differentiable function (with bounded first derivative), and the coefficients A and B are given by

$$A = -\frac{g'_x(0)F'_x(\lambda a) + g'_x(\lambda a)F'_x(0)}{g'_x(0) + g'_x(\lambda a)},$$

$$B = -\tfrac{1}{2}Aa\lambda - \tfrac{1}{2}\big[F(0) + F(\lambda a)\big] - \frac{g(\lambda a) - g(0)}{2g'_x(0)}\big[A + F'_x(0)\big].$$

3°. If $g(x)$ is representable in the form $g(x) = O(x)^k$ with $0 < k < 1$ in the vicinity of the point $x = 0$ (in particular, the derivative g'_x is unbounded as $x \to 0$), then the solution of the integral equation is given by formula (3) as well. In this case, the right-hand side of the integral equation must satisfy the conditions

$$f(0) + f(\lambda a) = 0, \qquad f'_x(\lambda a) = 0. \tag{7}$$

As before, the right-hand side of the integral equation is given by (6), with

$$A = -F'_x(\lambda a), \qquad B = \tfrac{1}{2}\left[a\lambda F'_x(\lambda a) - F(0) - F(\lambda a)\right].$$

5. $\displaystyle\int_0^a |g(x) - t|\, y(t)\, dt = f(x).$

Assume that $0 \le x \le a$, $0 \le t \le a$; $g(0) = 0$, and $0 < g'_x(x) < \infty$.

1°. Let us remove the modulus in the integrand:

$$\int_0^{g(x)} \left[g(x) - t\right] y(t)\, dt + \int_{g(x)}^a \left[t - g(x)\right] y(t)\, dt = f(x). \tag{1}$$

Differentiating (1) with respect to x yields

$$g'_x(x) \int_0^{g(x)} y(t)\, dt - g'_x(x) \int_{g(x)}^a y(t)\, dt = f'_x(x). \tag{2}$$

Let us divide both sides of (2) by $g'_x(x)$ and differentiate the resulting equation to obtain $2g'_x(x)y\big(g(x)\big) = \left[f'_x(x)/g'_x(x)\right]'_x$. Hence, we find the solution:

$$y(x) = \frac{1}{2g'_z(z)} \frac{d}{dz}\left[\frac{f'_z(z)}{g'_z(z)}\right], \qquad z = g^{-1}(x), \tag{3}$$

where g^{-1} is the inverse of g.

2°. Let us demonstrate that the right-hand side $f(x)$ of the integral equation must satisfy certain relations. By setting $x = 0$ in (1) and (2), we obtain two corollaries

$$\int_0^a ty(t)\, dt = f(0), \qquad g'_x(0) \int_0^a y(t)\, dt = -f'_x(0). \tag{4}$$

Substitute $y(x)$ of (3) into (4). Integrating by parts yields the desired constraints for $f(x)$:

$$f'_x(0)g'_x(x_a) + f'_x(x_a)g'_x(0) = 0, \quad x_a = g^{-1}(a);$$
$$g(x_a)\frac{f'_x(x_a)}{g'_x(x_a)} = f(0) + f(x_a). \tag{5}$$

Conditions (5) make it possible to find the admissible general form of the right-hand side of the integral equation in question:

$$f(x) = F(x) + Ax + B, \tag{6}$$

where $F(x)$ is an arbitrary bounded twice differentiable function (with bounded first derivative), and the coefficients A and B are given by

$$A = -\frac{g_x'(0)F_x'(x_a) + g_x'(x_a)F_x'(0)}{g_x'(0) + g_x'(x_a)}, \qquad x_a = g^{-1}(a),$$

$$B = -\tfrac{1}{2}Ax_a - \tfrac{1}{2}\big[F(0) + F(x_a)\big] - \frac{g(x_a)}{2g_x'(0)}\big[A + F_x'(0)\big].$$

3°. If $g(x)$ is representable in the vicinity of the point $x = 0$ in the form $g(x) = O(x)^k$ with $0 < k < 1$ (i.e., the derivative g_x' is unbounded as $x \to 0$), then the solution of the integral equation is given by formula (3) as well. In this case, the right-hand side of the integral equation must satisfy the conditions

$$f(0) + f(x_a) = 0, \qquad f_x'(x_a) = 0. \tag{7}$$

As before, the right-hand side of the integral equation is given by (6), with

$$A = -F_x'(x_a), \qquad B = \tfrac{1}{2}\big[x_a F_x'(x_a) - F(0) - F(x_a)\big].$$

6. $\displaystyle\int_0^a \big|x - g(t)\big|\, y(t)\, dt = f(x).$

Assume that $0 \le x \le a$, $0 \le t \le a$; $g(0) = 0$, and $0 < g_x'(x) < \infty$.

1°. Let us remove the modulus in the integrand:

$$\int_0^{g^{-1}(x)} \big[x - g(t)\big] y(t)\, dt + \int_{g^{-1}(x)}^a \big[g(t) - x\big] y(t)\, dt = f(x), \tag{1}$$

where g^{-1} is the inverse of g. Differentiating (1) with respect to x yields

$$\int_0^{g^{-1}(x)} y(t)\, dt - \int_{g^{-1}(x)}^a y(t)\, dt = f_x'(x). \tag{2}$$

Differentiating the resulting equation yields $2y\big(g^{-1}(x)\big) = g_x'(x)f_{xx}''(x)$. Hence, we obtain the solution

$$y(x) = \tfrac{1}{2}g_z'(z)f_{zz}''(z), \qquad z = g(x). \tag{3}$$

2°. Let us demonstrate that the right-hand side $f(x)$ of the integral equation must satisfy certain relations. By setting $x = 0$ in (1) and (2), we obtain two corollaries

$$\int_0^a g(t)y(t)\, dt = f(0), \qquad \int_0^a y(t)\, dt = -f_x'(0). \tag{4}$$

Substitute $y(x)$ of (3) into (4). Integrating by parts yields the desired constraints for $f(x)$:

$$x_a f_x'(x_a) = f(0) + f(x_a), \qquad f_x'(0) + f_x'(x_a) = 0, \qquad x_a = g(a). \tag{5}$$

Conditions (5) make it possible to find the admissible general form of the right-hand side of the integral equation:

$$f(x) = F(x) + Ax + B,$$

$$A = -\tfrac{1}{2}\big[F_x'(0) + F_x'(x_a)\big], \qquad B = \tfrac{1}{2}\big[x_a F_x'(0) - F(x_a) - F(0)\big], \qquad x_a = g(a),$$

where $F(x)$ is an arbitrary bounded twice differentiable function (with bounded first derivative).

7. $\displaystyle\int_a^b \frac{y(t)}{|g(x) - g(t)|^k}\, dt = f(x), \qquad 0 < k < 1.$

Let $g_x' \neq 0$. The transformation

$$z = g(x), \quad \tau = g(t), \quad w(\tau) = \frac{1}{g_t'(t)} y(t)$$

leads to an equation of the form 3.1.30:

$$\int_A^B \frac{w(\tau)}{|z - \tau|^k}\, d\tau = F(z), \qquad A = g(a), \quad B = g(b),$$

where $F = F(z)$ is the function which is obtained from $z = g(x)$ and $F = f(x)$ by eliminating x.

8. $\displaystyle\int_0^1 \frac{y(t)}{|g(x) - h(t)|^k}\, dt = f(x), \qquad 0 < k < 1.$

Let $g(0) = 0$, $g(1) = 1$, $g_x' > 0$; $h(0) = 0$, $h(1) = 1$, and $h_t' > 0$.
    The transformation

$$z = g(x), \quad \tau = h(t), \quad w(\tau) = \frac{1}{h_t'(t)} y(t)$$

leads to an equation of the form 3.1.29:

$$\int_0^1 \frac{w(\tau)}{|z - \tau|^k}\, d\tau = F(z),$$

where $F = F(z)$ is the function which is obtained from $z = g(x)$ and $F = f(x)$ by eliminating x.

9. $\displaystyle\int_a^b y(t) \ln|g(x) - g(t)|\, dt = f(x).$

Let $g_x' \neq 0$. The transformation

$$z = g(x), \quad \tau = g(t), \quad w(\tau) = \frac{1}{g_t'(t)} y(t)$$

leads to Carleman's equation 3.4.2:

$$\int_A^B \ln|z - \tau| w(\tau)\, d\tau = F(z), \qquad A = g(a), \quad B = g(b),$$

where $F = F(z)$ is the function which is obtained from $z = g(x)$ and $F = f(x)$ by eliminating x.

10. $\displaystyle\int_0^1 y(t) \ln|g(x) - h(t)|\, dt = f(x).$

Let $g(0) = 0$, $g(1) = 1$, $g_x' > 0$; $h(0) = 0$, $h(1) = 1$, and $h_t' > 0$.
    The transformation

$$z = g(x), \quad \tau = h(t), \quad w(\tau) = \frac{1}{h_t'(t)} y(t)$$

leads to an equation of the form 3.4.2:

$$\int_0^1 \ln|z - \tau| w(\tau)\, d\tau = F(z),$$

where $F = F(z)$ is the function which is obtained from $z = g(x)$ and $F = f(x)$ by eliminating x.

3.8-3. Equations With Difference Kernel: $K(x, t) = K(x - t)$

11. $\displaystyle\int_{-\infty}^{\infty} K(x - t)y(t)\, dt = Ax^n, \qquad n = 0, 1, 2, \ldots$

$1°$. Solution with $n = 0$:

$$y(x) = \frac{A}{B}, \qquad B = \int_{-\infty}^{\infty} K(x)\, dx.$$

$2°$. Solution with $n = 1$:

$$y(x) = \frac{A}{B}x + \frac{AC}{B^2}, \qquad B = \int_{-\infty}^{\infty} K(x)\, dx, \quad C = \int_{-\infty}^{\infty} xK(x).$$

$3°$. Solution with $n \geq 2$:

$$y(x) = \left\{ \frac{d^n}{d\lambda^n}\left[\frac{Ae^{\lambda x}}{B(\lambda)} \right] \right\}_{\lambda = 0}, \qquad B(\lambda) = \int_{-\infty}^{\infty} K(x)e^{-\lambda x}\, dx.$$

12. $\displaystyle\int_{-\infty}^{\infty} K(x - t)y(t)\, dt = Ae^{\lambda x}.$

Solution:

$$y(x) = \frac{A}{B}e^{\lambda x}, \qquad B = \int_{-\infty}^{\infty} K(x)e^{-\lambda x}\, dx.$$

13. $\displaystyle\int_{-\infty}^{\infty} K(x - t)y(t)\, dt = Ax^n e^{\lambda x}, \qquad n = 1, 2, \ldots$

$1°$. Solution with $n = 1$:

$$y(x) = \frac{A}{B}xe^{\lambda x} + \frac{AC}{B^2}e^{\lambda x},$$

$$B = \int_{-\infty}^{\infty} K(x)e^{-\lambda x}\, dx, \quad C = \int_{-\infty}^{\infty} xK(x)e^{-\lambda x}\, dx.$$

$2°$. Solution with $n \geq 2$:

$$y(x) = \frac{d^n}{d\lambda^n}\left[\frac{Ae^{\lambda x}}{B(\lambda)} \right], \qquad B(\lambda) = \int_{-\infty}^{\infty} K(x)e^{-\lambda x}\, dx.$$

14. $\displaystyle\int_{-\infty}^{\infty} K(x - t)y(t)\, dt = A\cos(\lambda x) + B\sin(\lambda x).$

Solution:

$$y(x) = \frac{AI_c + BI_s}{I_c^2 + I_s^2}\cos(\lambda x) + \frac{BI_c - AI_s}{I_c^2 + I_s^2}\sin(\lambda x),$$

$$I_c = \int_{-\infty}^{\infty} K(z)\cos(\lambda z)\, dz, \quad I_s = \int_{-\infty}^{\infty} K(z)\sin(\lambda z)\, dz.$$

15. $\displaystyle\int_{-\infty}^{\infty} K(x-t)y(t)\,dt = f(x).$

The Fourier transform is used to solve this equation.

1°. Solution:

$$y(x) = \frac{1}{2\pi}\int_{-\infty}^{\infty}\frac{\tilde{f}(u)}{\tilde{K}(u)}e^{iux}\,du,$$

$$\tilde{f}(u) = \frac{1}{\sqrt{2\pi}}\int_{-\infty}^{\infty}f(x)e^{-iux}\,dx, \quad \tilde{K}(u) = \frac{1}{\sqrt{2\pi}}\int_{-\infty}^{\infty}K(x)e^{-iux}\,dx.$$

The following statement is valid. Let $f(x) \in L_2(-\infty, \infty)$ and $K(x) \in L_1(-\infty, \infty)$. Then for a solution $y(x) \in L_2(-\infty, \infty)$ of the integral equation to exist, it is necessary and sufficient that $\tilde{f}(u)/\tilde{K}(u) \in L_2(-\infty, \infty)$.

2°. Let the function $P(s)$ defined by the formula

$$\frac{1}{P(s)} = \int_{-\infty}^{\infty} e^{-st}K(t)\,dt$$

be a polynomial of degree n with real roots of the form

$$P(s) = \left(1 - \frac{s}{a_1}\right)\left(1 - \frac{s}{a_2}\right)\cdots\left(1 - \frac{s}{a_n}\right).$$

Then the solution of the integral equation is given by

$$y(x) = P(D)f(x), \quad D = \frac{d}{dx}.$$

⊙ References: I. I. Hirschman and D. V. Widder (1955), V. A. Ditkin and A. P. Prudnikov (1965).

16. $\displaystyle\int_{0}^{\infty} K(x-t)y(t)\,dt = f(x).$

The Wiener–Hopf equation of the first kind. This equation is discussed in Subsection 10.5-1 in detail.

3.8-4. Other Equations of the Form $\int_a^b K(x,t)y(t)\,dt = F(x)$

17. $\displaystyle\int_{-\infty}^{\infty} K(ax-t)y(t)\,dt = Ae^{\lambda x}.$

Solution:

$$y(x) = \frac{A}{B}\exp\left(\frac{\lambda}{a}x\right), \quad B = \int_{-\infty}^{\infty} K(z)\exp\left(-\frac{\lambda}{a}z\right)dz.$$

18. $\displaystyle\int_{-\infty}^{\infty} K(ax-t)y(t)\,dt = f(x).$

The substitution $z = ax$ leads to an equation of the form 3.8.15:

$$\int_{-\infty}^{\infty} K(z-t)y(t)\,dt = f(z/a).$$

19. $\displaystyle\int_{-\infty}^{\infty} K(ax+t)y(t)\,dt = Ae^{\lambda x}.$

Solution:
$$y(x) = \frac{A}{B}\exp\left(-\frac{\lambda}{a}x\right), \qquad B = \int_{-\infty}^{\infty} K(z)\exp\left(-\frac{\lambda}{a}z\right)dz.$$

20. $\displaystyle\int_{-\infty}^{\infty} K(ax+t)y(t)\,dt = f(x).$

The transformation $\tau = -t$, $z = ax$, $y(t) = Y(\tau)$ leads to an equation of the form 3.8.15:
$$\int_{-\infty}^{\infty} K(z-\tau)Y(\tau)\,dt = f(z/a).$$

21. $\displaystyle\int_{-\infty}^{\infty} [e^{\beta t}K(ax+t) + e^{\mu t}M(ax-t)]y(t)\,dt = Ae^{\lambda x}.$

Solution:
$$y(x) = A\,\frac{I_k(q)e^{px} - I_m(p)e^{qx}}{I_k(p)I_k(q) - I_m(p)I_m(q)}, \qquad p = -\frac{\lambda}{a} - \beta, \quad q = \frac{\lambda}{a} - \mu,$$

where
$$I_k(q) = \int_{-\infty}^{\infty} K(z)e^{(\beta+q)z}\,dz, \quad I_m(q) = \int_{-\infty}^{\infty} M(z)e^{-(\mu+q)z}\,dz.$$

22. $\displaystyle\int_{0}^{\infty} g(xt)y(t)\,dt = f(x).$

By setting
$$x = e^z, \quad t = e^{-\tau}, \quad y(t) = e^\tau w(\tau), \quad g(\xi) = G(\ln\xi), \quad f(\xi) = F(\ln\xi),$$

we arrive at an integral equation with difference kernel of the form 3.8.15:
$$\int_{-\infty}^{\infty} G(z-\tau)w(\tau)\,d\tau = F(z).$$

23. $\displaystyle\int_{0}^{\infty} g\left(\frac{x}{t}\right)y(t)\,dt = f(x).$

By setting
$$x = e^z, \quad t = e^\tau, \quad y(t) = e^{-\tau} w(\tau), \quad g(\xi) = G(\ln\xi), \quad f(\xi) = F(\ln\xi),$$

we arrive at an integral equation with difference kernel of the form 3.8.15:
$$\int_{-\infty}^{\infty} G(z-\tau)w(\tau)\,d\tau = F(z).$$

24. $\displaystyle\int_{0}^{\infty} g\left(x^\beta t^\lambda\right)y(t)\,dt = f(x), \qquad \beta > 0, \quad \lambda > 0.$

By setting
$$x = e^{z/\beta}, \quad t = e^{-\tau/\lambda}, \quad y(t) = e^{\tau/\lambda} w(\tau), \quad g(\xi) = G(\ln\xi), \quad f(\xi) = \tfrac{1}{\lambda}F(\beta\ln\xi),$$

we arrive at an integral equation with difference kernel of the form 3.8.15:
$$\int_{-\infty}^{\infty} G(z-\tau)w(\tau)\,d\tau = F(z).$$

25. $\displaystyle\int_0^\infty g\left(\frac{x^\beta}{t^\lambda}\right) y(t)\, dt = f(x), \qquad \beta > 0, \quad \lambda > 0.$

By setting

$$x = e^{z/\beta}, \quad t = e^{\tau/\lambda}, \quad y(t) = e^{-\tau/\lambda} w(\tau), \quad g(\xi) = G(\ln \xi), \quad f(\xi) = \tfrac{1}{\lambda} F(\beta \ln \xi),$$

we arrive at an integral equation with difference kernel of the form 3.8.15:

$$\int_{-\infty}^\infty G(z - \tau) w(\tau)\, d\tau = F(z).$$

3.8-5. Equations of the Form $\int_a^b K(x,t)y(\cdots)\,dt = F(x)$

26. $\displaystyle\int_a^b f(t)y(xt)\, dt = Ax + B.$

Solution:

$$y(x) = \frac{A}{I_1} x + \frac{B}{I_0}, \qquad I_0 = \int_a^b f(t)\, dt, \quad I_1 = \int_a^b t f(t)\, dt.$$

27. $\displaystyle\int_a^b f(t)y(xt)\, dt = Ax^\beta.$

Solution:

$$y(x) = \frac{A}{B} x^\beta, \qquad B = \int_a^b f(t) t^\beta\, dt.$$

28. $\displaystyle\int_a^b f(t)y(xt)\, dt = A \ln x + B.$

Solution:

$$y(x) = p \ln x + q,$$

where

$$p = \frac{A}{I_0}, \quad q = \frac{B}{I_0} - \frac{A I_l}{I_0^2}, \qquad I_0 = \int_a^b f(t)\, dt, \quad I_l = \int_a^b f(t) \ln t\, dt.$$

29. $\displaystyle\int_a^b f(t)y(xt)\, dt = Ax^\beta \ln x.$

Solution:

$$y(x) = p x^\beta \ln x + q x^\beta,$$

where

$$p = \frac{A}{I_1}, \quad q = -\frac{A I_2}{I_1^2}, \qquad I_1 = \int_a^b f(t) t^\beta dt, \quad I_2 = \int_a^b f(t) t^\beta \ln t\, dt.$$

30. $\displaystyle\int_a^b f(t)y(xt)\, dt = A \cos(\ln x).$

Solution:

$$y(x) = \frac{A I_c}{I_c^2 + I_s^2} \cos(\ln x) + \frac{A I_s}{I_c^2 + I_s^2} \sin(\ln x),$$

$$I_c = \int_a^b f(t) \cos(\ln t)\, dt, \quad I_s = \int_a^b f(t) \sin(\ln t)\, dt.$$

31. $\displaystyle\int_a^b f(t)y(xt)\,dt = A\sin(\ln x).$

Solution:

$$y(x) = -\frac{AI_s}{I_c^2 + I_s^2}\cos(\ln x) + \frac{AI_c}{I_c^2 + I_s^2}\sin(\ln x),$$

$$I_c = \int_a^b f(t)\cos(\ln t)\,dt, \quad I_s = \int_a^b f(t)\sin(\ln t)\,dt.$$

32. $\displaystyle\int_a^b f(t)y(xt)\,dt = Ax^\beta\cos(\ln x) + Bx^\beta\sin(\ln x).$

Solution:

$$y(x) = px^\beta\cos(\ln x) + qx^\beta\sin(\ln x),$$

where

$$p = \frac{AI_c - BI_s}{I_c^2 + I_s^2}, \quad q = \frac{AI_s + BI_c}{I_c^2 + I_s^2},$$

$$I_c = \int_a^b f(t)t^\beta\cos(\ln t)\,dt, \quad I_s = \int_a^b f(t)t^\beta\sin(\ln t)\,dt.$$

33. $\displaystyle\int_a^b f(t)y(x - t)\,dt = Ax + B.$

Solution:

$$y(x) = px + q,$$

where

$$p = \frac{A}{I_0}, \quad q = \frac{AI_1}{I_0^2} + \frac{B}{I_0}, \quad I_0 = \int_a^b f(t)\,dt, \quad I_1 = \int_a^b t f(t)\,dt.$$

34. $\displaystyle\int_a^b f(t)y(x - t)\,dt = Ae^{\lambda x}.$

Solution:

$$y(x) = \frac{A}{B}e^{\lambda x}, \quad B = \int_a^b f(t)\exp(-\lambda t)\,dt.$$

35. $\displaystyle\int_a^b f(t)y(x - t)\,dt = A\cos(\lambda x).$

Solution:

$$y(x) = -\frac{AI_s}{I_c^2 + I_s^2}\sin(\lambda x) + \frac{AI_c}{I_c^2 + I_s^2}\cos(\lambda x),$$

$$I_c = \int_a^b f(t)\cos(\lambda t)\,dt, \quad I_s = \int_a^b f(t)\sin(\lambda t)\,dt.$$

36. $\displaystyle\int_a^b f(t)y(x - t)\,dt = A\sin(\lambda x).$

Solution:

$$y(x) = \frac{AI_c}{I_c^2 + I_s^2}\sin(\lambda x) + \frac{AI_s}{I_c^2 + I_s^2}\cos(\lambda x),$$

$$I_c = \int_a^b f(t)\cos(\lambda t)\,dt, \quad I_s = \int_a^b f(t)\sin(\lambda t)\,dt.$$

37. $\displaystyle\int_a^b f(t)y(x-t)\,dt = e^{\mu x}(A\sin\lambda x + B\cos\lambda x).$

Solution:

$$y(x) = e^{\mu x}(p\sin\lambda x + q\cos\lambda x),$$

where

$$p = \frac{AI_c - BI_s}{I_c^2 + I_s^2}, \qquad q = \frac{AI_s + BI_c}{I_c^2 + I_s^2},$$

$$I_c = \int_a^b f(t)e^{-\mu t}\cos(\lambda t)\,dt, \qquad I_s = \int_a^b f(t)e^{-\mu t}\sin(\lambda t)\,dt.$$

38. $\displaystyle\int_a^b f(t)y(x-t)\,dt = g(x).$

1°. For $g(x) = \sum\limits_{k=1}^{n} A_k\exp(\lambda_k x)$, the solution of the equation has the form

$$y(x) = \sum_{k=1}^{n}\frac{A_k}{B_k}\exp(\lambda_k x), \qquad B_k = \int_a^b f(t)\exp(-\lambda_k t)\,dt.$$

2°. For a polynomial right-hand side, $g(x) = \sum\limits_{k=0}^{n} A_k x^k$, the solution has the form

$$y(x) = \sum_{k=0}^{n} B_k x^k,$$

where the constants B_k are found by the method of undetermined coefficients.

3°. For $g(x) = e^{\lambda x}\sum\limits_{k=0}^{n} A_k x^k$, the solution has the form

$$y(x) = e^{\lambda x}\sum_{k=0}^{n} B_k x^k,$$

where the constants B_k are found by the method of undetermined coefficients.

4°. For $g(x) = \sum\limits_{k=1}^{n} A_k\cos(\lambda_k x)$, the solution has the form

$$y(x) = \sum_{k=1}^{n} B_k\cos(\lambda_k x) + \sum_{k=1}^{n} C_k\sin(\lambda_k x),$$

where the constants B_k and C_k are found by the method of undetermined coefficients.

5°. For $g(x) = \sum\limits_{k=1}^{n} A_k\sin(\lambda_k x)$, the solution has the form

$$y(x) = \sum_{k=1}^{n} B_k\cos(\lambda_k x) + \sum_{k=1}^{n} C_k\sin(\lambda_k x),$$

where the constants B_k and C_k are found by the method of undetermined coefficients.

$6°$. For $g(x) = \cos(\lambda x) \sum_{k=0}^{n} A_k x^k$, the solution has the form

$$y(x) = \cos(\lambda x) \sum_{k=0}^{n} B_k x^k + \sin(\lambda x) \sum_{k=0}^{n} C_k x^k,$$

where the constants B_k and C_k are found by the method of undetermined coefficients.

$7°$. For $g(x) = \sin(\lambda x) \sum_{k=0}^{n} A_k x^k$, the solution has the form

$$y(x) = \cos(\lambda x) \sum_{k=0}^{n} B_k x^k + \sin(\lambda x) \sum_{k=0}^{n} C_k x^k,$$

where the constants B_k and C_k are found by the method of undetermined coefficients.

$8°$. For $g(x) = e^{\mu x} \sum_{k=1}^{n} A_k \cos(\lambda_k x)$, the solution has the form

$$y(x) = e^{\mu x} \sum_{k=1}^{n} B_k \cos(\lambda_k x) + e^{\mu x} \sum_{k=1}^{n} C_k \sin(\lambda_k x),$$

where the constants B_k and C_k are found by the method of undetermined coefficients.

$9°$. For $g(x) = e^{\mu x} \sum_{k=1}^{n} A_k \sin(\lambda_k x)$, the solution has the form

$$y(x) = e^{\mu x} \sum_{k=1}^{n} B_k \cos(\lambda_k x) + e^{\mu x} \sum_{k=1}^{n} C_k \sin(\lambda_k x),$$

where the constants B_k and C_k are found by the method of undetermined coefficients.

$10°$. For $g(x) = \cos(\lambda x) \sum_{k=1}^{n} A_k \exp(\mu_k x)$, the solution has the form

$$y(x) = \cos(\lambda x) \sum_{k=1}^{n} B_k \exp(\mu_k x) + \sin(\lambda x) \sum_{k=1}^{n} C_k \exp(\mu_k x),$$

where the constants B_k and C_k are found by the method of undetermined coefficients.

$11°$. For $g(x) = \sin(\lambda x) \sum_{k=1}^{n} A_k \exp(\mu_k x)$, the solution has the form

$$y(x) = \cos(\lambda x) \sum_{k=1}^{n} B_k \exp(\mu_k x) + \sin(\lambda x) \sum_{k=1}^{n} C_k \exp(\mu_k x),$$

where the constants B_k and C_k are found by the method of undetermined coefficients.

39. $\displaystyle\int_a^b f(t) y(x + \beta t)\, dt = Ax + B.$

Solution:

$$y(x) = px + q,$$

where

$$p = \frac{A}{I_0}, \quad q = \frac{B}{I_0} - \frac{A I_1 \beta}{I_0^2}, \qquad I_0 = \int_a^b f(t)\, dt, \quad I_1 = \int_a^b t f(t)\, dt.$$

40. $\displaystyle\int_a^b f(t)y(x + \beta t)\, dt = Ae^{\lambda x}.$

Solution:

$$y(x) = \frac{A}{B}e^{\lambda x}, \qquad B = \int_a^b f(t)\exp(\lambda\beta t)\, dt.$$

41. $\displaystyle\int_a^b f(t)y(x + \beta t)\, dt = A\sin\lambda x + B\cos\lambda x.$

Solution:

$$y(x) = p\sin\lambda x + q\cos\lambda x,$$

where

$$p = \frac{AI_c + BI_s}{I_c^2 + I_s^2}, \qquad q = \frac{BI_c - AI_s}{I_c^2 + I_s^2},$$

$$I_c = \int_a^b f(t)\cos(\lambda\beta t)\, dt, \qquad I_s = \int_a^b f(t)\sin(\lambda\beta t)\, dt.$$

42. $\displaystyle\int_0^1 y(\xi)\, dt = f(x), \qquad \xi = g(x)t.$

Assume that $g(0) = 0$, $g(1) = 1$, and $g'_x \geq 0$.

$1°$. The substitution $z = g(x)$ leads to an equation of the form 3.1.41: $\displaystyle\int_0^1 y(zt)\, dt = F(z)$, where the function $F(z)$ is obtained from $z = g(x)$ and $F = f(x)$ by eliminating x.

$2°$. Solution $y = y(z)$ in the parametric form:

$$y(z) = \frac{g(x)}{g'_x(x)}f'_x(x) + f(x), \qquad z = g(x).$$

43. $\displaystyle\int_0^1 t^\lambda y(\xi)\, dt = f(x), \qquad \xi = g(x)t.$

Assume that $g(0) = 0$, $g(1) = 1$, and $g'_x \geq 0$.

$1°$. The substitution $z = g(x)$ leads to an equation of the form 3.1.42: $\displaystyle\int_0^1 t^\lambda y(zt)\, dt = F(z)$, where the function $F(z)$ is obtained from $z = g(x)$ and $F = f(x)$ by eliminating x.

$2°$. Solution $y = y(z)$ in the parametric form:

$$y(z) = \frac{g(x)}{g'_x(x)}f'_x(x) + (\lambda + 1)f(x), \qquad z = g(x).$$

44. $\displaystyle\int_a^b f(t)y(\xi)\, dt = Ax^\beta, \qquad \xi = x\varphi(t).$

Solution:

$$y(x) = \frac{A}{B}x^\beta, \qquad B = \int_a^b f(t)\big[\varphi(t)\big]^\beta\, dt. \tag{1}$$

45. $\displaystyle\int_a^b f(t)y(\xi)\,dt = g(x), \qquad \xi = x\varphi(t).$

1°. For $g(x) = \displaystyle\sum_{k=0}^n A_k x^k$, the solution of the equation has the form

$$y(x) = \sum_{k=0}^n \frac{A_k}{B_k} x^k, \qquad B_k = \int_a^b f(t)\big[\varphi(t)\big]^k\,dt.$$

2°. For $g(x) = \displaystyle\sum_{k=0}^n A_k x^{\lambda_k}$, the solution has the form

$$y(x) = \sum_{k=0}^n \frac{A_k}{B_k} x^{\lambda_k}, \qquad B_k = \int_a^b f(t)\big[\varphi(t)\big]^{\lambda_k}\,dt.$$

3°. For $g(x) = \ln x \displaystyle\sum_{k=0}^n A_k x^k$, the solution has the form

$$y(x) = \ln x \sum_{k=0}^n B_k x^k + \sum_{k=0}^n C_k x^k,$$

where the constants B_k and C_k are found by the method of undetermined coefficients.

4°. For $g(x) = \displaystyle\sum_{k=0}^n A_k \big(\ln x\big)^k$, the solution has the form

$$y(x) = \sum_{k=0}^n B_k \big(\ln x\big)^k,$$

where the constants B_k are found by the method of undetermined coefficients.

5°. For $g(x) = \displaystyle\sum_{k=1}^n A_k \cos(\lambda_k \ln x)$, the solution has the form

$$y(x) = \sum_{k=1}^n B_k \cos(\lambda_k \ln x) + \sum_{k=1}^n C_k \sin(\lambda_k \ln x),$$

where the constants B_k and C_k are found by the method of undetermined coefficients.

6°. For $g(x) = \displaystyle\sum_{k=1}^n A_k \sin(\lambda_k \ln x)$, the solution has the form

$$y(x) = \sum_{k=1}^n B_k \cos(\lambda_k \ln x) + \sum_{k=1}^n C_k \sin(\lambda_k \ln x),$$

where the constants B_k and C_k are found by the method of undetermined coefficients.

46. $\displaystyle\int_a^b f(t)y(\xi)\,dt = g(x), \qquad \xi = x + \varphi(t).$

1°. For $g(x) = \displaystyle\sum_{k=1}^{n} A_k \exp(\lambda_k x)$, the solution of the equation has the form

$$y(x) = \sum_{k=1}^{n} \frac{A_k}{B_k} \exp(\lambda_k x), \qquad B_k = \int_a^b f(t) \exp\left[\lambda_k \varphi(t)\right] dt.$$

2°. For a polynomial right-hand side, $g(x) = \displaystyle\sum_{k=0}^{n} A_k x^k$, the solution has the form

$$y(x) = \sum_{k=0}^{n} B_k x^k,$$

where the constants B_k are found by the method of undetermined coefficients.

3°. For $g(x) = e^{\lambda x} \displaystyle\sum_{k=0}^{n} A_k x^k$, the solution has the form

$$y(x) = e^{\lambda x} \sum_{k=0}^{n} B_k x^k,$$

where the constants B_k are found by the method of undetermined coefficients.

4°. For $g(x) = \displaystyle\sum_{k=1}^{n} A_k \cos(\lambda_k x)$ the solution has the form

$$y(x) = \sum_{k=1}^{n} B_k \cos(\lambda_k x) + \sum_{k=1}^{n} C_k \sin(\lambda_k x),$$

where the constants B_k and C_k are found by the method of undetermined coefficients.

5°. For $g(x) = \displaystyle\sum_{k=1}^{n} A_k \sin(\lambda_k x)$, the solution has the form

$$y(x) = \sum_{k=1}^{n} B_k \cos(\lambda_k x) + \sum_{k=1}^{n} C_k \sin(\lambda_k x),$$

where the constants B_k and C_k are found by the method of undetermined coefficients.

6°. For $g(x) = \cos(\lambda x) \displaystyle\sum_{k=0}^{n} A_k x^k$, the solution has the form

$$y(x) = \cos(\lambda x) \sum_{k=0}^{n} B_k x^k + \sin(\lambda x) \sum_{k=0}^{n} C_k x^k,$$

where the constants B_k and C_k are found by the method of undetermined coefficients.

7°. For $g(x) = \sin(\lambda x) \displaystyle\sum_{k=0}^{n} A_k x^k$, the solution has the form

$$y(x) = \cos(\lambda x) \sum_{k=0}^{n} B_k x^k + \sin(\lambda x) \sum_{k=0}^{n} C_k x^k,$$

where the constants B_k and C_k are found by the method of undetermined coefficients.

8°. For $g(x) = e^{\mu x} \sum\limits_{k=1}^{n} A_k \cos(\lambda_k x)$, the solution has the form

$$y(x) = e^{\mu x} \sum_{k=1}^{n} B_k \cos(\lambda_k x) + e^{\mu x} \sum_{k=1}^{n} C_k \sin(\lambda_k x),$$

where the constants B_k and C_k are found by the method of undetermined coefficients.

9°. For $g(x) = e^{\mu x} \sum\limits_{k=1}^{n} A_k \sin(\lambda_k x)$, the solution has the form

$$y(x) = e^{\mu x} \sum_{k=1}^{n} B_k \cos(\lambda_k x) + e^{\mu x} \sum_{k=1}^{n} C_k \sin(\lambda_k x),$$

where the constants B_k and C_k are found by the method of undetermined coefficients.

10°. For $g(x) = \cos(\lambda x) \sum\limits_{k=1}^{n} A_k \exp(\mu_k x)$, the solution has the form

$$y(x) = \cos(\lambda x) \sum_{k=1}^{n} B_k \exp(\mu_k x) + \sin(\lambda x) \sum_{k=1}^{n} C_k \exp(\mu_k x),$$

where the constants B_k and C_k are found by the method of undetermined coefficients.

11°. For $g(x) = \sin(\lambda x) \sum\limits_{k=1}^{n} A_k \exp(\mu_k x)$, the solution has the form

$$y(x) = \cos(\lambda x) \sum_{k=1}^{n} B_k \exp(\mu_k x) + \sin(\lambda x) \sum_{k=1}^{n} C_k \exp(\mu_k x),$$

where the constants B_k and C_k are found by the method of undetermined coefficients.

Chapter 4

Linear Equations of the Second Kind With Constant Limits of Integration

▶ *Notation:* $f = f(x)$, $g = g(x)$, $h = h(x)$, $v = v(x)$, $w = w(x)$, $K = K(x)$ *are arbitrary functions;* A, B, C, D, E, a, b, c, l, α, β, γ, δ, μ, *and* ν *are arbitrary parameters;* n *is a nonnegative integer; and* i *is the imaginary unit.*

▶ **Preliminary remarks.** A number λ is called a *characteristic value* of the integral equation

$$y(x) - \lambda \int_a^b K(x,t)y(t)\,dt = f(x)$$

if there exist nontrivial solutions of the corresponding homogeneous equation (with $f(x) \equiv 0$). The nontrivial solutions themselves are called the *eigenfunctions* of the integral equation corresponding to the characteristic value λ. If λ is a characteristic value, the number $1/\lambda$ is called an *eigenvalue* of the integral equation. A value of the parameter λ is said to be *regular* if for this value the homogeneous equation has only the trivial solution. Sometimes the characteristic values and the eigenfunctions of a Fredholm integral equation are called the *characteristic values* and the *eigenfunctions of the kernel* $K(x,t)$. In the above equation, it is usually assumed that $a \leq x \leq b$.

4.1. Equations Whose Kernels Contain Power-Law Functions

4.1-1. Kernels Linear in the Arguments x and t

1. $\quad y(x) - \lambda \int_a^b (x - t)y(t)\,dt = f(x).$

Solution:
$$y(x) = f(x) + \lambda(A_1 x + A_2),$$

where

$$A_1 = \frac{12f_1 + 6\lambda\,(f_1\Delta_2 - 2f_2\Delta_1)}{\lambda^2\Delta_1^4 + 12}, \qquad A_2 = \frac{-12f_2 + 2\lambda\,(3f_2\Delta_2 - 2f_1\Delta_3)}{\lambda^2\Delta_1^4 + 12},$$

$$f_1 = \int_a^b f(x)\,dx, \qquad f_2 = \int_a^b x f(x)\,dx, \qquad \Delta_n = b^n - a^n.$$

2. $y(x) - \lambda \displaystyle\int_a^b (x + t) y(t)\, dt = f(x).$

The characteristic values of the equation:

$$\lambda_1 = \frac{6(b + a) + 4\sqrt{3(a^2 + ab + b^2)}}{(a - b)^3}, \qquad \lambda_2 = \frac{6(b + a) - 4\sqrt{3(a^2 + ab + b^2)}}{(a - b)^3}.$$

1°. Solution with $\lambda \neq \lambda_{1,2}$:

$$y(x) = f(x) + \lambda(A_1 x + A_2),$$

where

$$A_1 = \frac{12 f_1 - 6\lambda(f_1 \Delta_2 - 2 f_2 \Delta_1)}{12 - 12\lambda\Delta_2 - \lambda^2 \Delta_1^4}, \qquad A_2 = \frac{12 f_2 - 2\lambda(3 f_2 \Delta_2 - 2 f_1 \Delta_3)}{12 - 12\lambda\Delta_2 - \lambda^2 \Delta_1^4},$$

$$f_1 = \int_a^b f(x)\, dx, \quad f_2 = \int_a^b x f(x)\, dx, \quad \Delta_n = b^n - a^n.$$

2°. Solution with $\lambda = \lambda_1 \neq \lambda_2$ and $f_1 = f_2 = 0$:

$$y(x) = f(x) + C y_1(x),$$

where C is an arbitrary constant and $y_1(x)$ is an eigenfunction of the equation corresponding to the characteristic value λ_1:

$$y_1(x) = x + \frac{1}{\lambda_1(b - a)} - \frac{b + a}{2}.$$

3°. Solution with $\lambda = \lambda_2 \neq \lambda_1$ and $f_1 = f_2 = 0$ is given by the formulas of item 2° in which one must replace λ_1 and $y_1(x)$ by λ_2 and $y_2(x)$, respectively.

4°. The equation has no multiple characteristic values.

3. $y(x) - \lambda \displaystyle\int_a^b (Ax + Bt) y(t)\, dt = f(x).$

The characteristic values of the equation:

$$\lambda_{1,2} = \frac{3(A + B)(b + a) \pm \sqrt{9(A - B)^2(b + a)^2 + 48AB(a^2 + ab + b^2)}}{AB(a - b)^3}.$$

1°. Solution with $\lambda \neq \lambda_{1,2}$:

$$y(x) = f(x) + \lambda(A_1 x + A_2),$$

where the constants A_1 and A_2 are given by

$$A_1 = \frac{12 A f_1 - 6 A B \lambda(f_1 \Delta_2 - 2 f_2 \Delta_1)}{12 - 6(A + B)\lambda\Delta_2 - A B \lambda^2 \Delta_1^4}, \qquad A_2 = \frac{12 B f_2 - 2 A B \lambda(3 f_2 \Delta_2 - 2 f_1 \Delta_3)}{12 - 6(A + B)\lambda\Delta_2 - A B \lambda^2 \Delta_1^4},$$

$$f_1 = \int_a^b f(x)\, dx, \quad f_2 = \int_a^b x f(x)\, dx, \quad \Delta_n = b^n - a^n.$$

2°. Solution with $\lambda = \lambda_1 \neq \lambda_2$ and $f_1 = f_2 = 0$:

$$y(x) = f(x) + C y_1(x),$$

where C is an arbitrary constant and $y_1(x)$ is an eigenfunction of the equation corresponding to the characteristic value λ_1:

$$y_1(x) = x + \frac{1}{\lambda_1 A(b - a)} - \frac{b + a}{2}.$$

3°. Solution with $\lambda = \lambda_2 \neq \lambda_1$ and $f_1 = f_2 = 0$ is given by the formulas of item 2° in which one must replace λ_1 and $y_1(x)$ by λ_2 and $y_2(x)$, respectively.

4°. Solution with $\lambda = \lambda_{1,2} = \lambda_*$ and $f_1 = f_2 = 0$, where the characteristic value $\lambda_* = \dfrac{4}{(A+B)(b^2 - a^2)}$ is double:

$$y(x) = f(x) + Cy_*(x),$$

where C is an arbitrary constant and $y_*(x)$ is an eigenfunction of the equation corresponding to λ_*:

$$y_*(x) = x - \frac{(A-B)(b+a)}{4A}.$$

4. $\quad y(x) - \lambda \displaystyle\int_a^b [A + B(x-t)]y(t)\, dt = f(x).$

This is a special case of equation 4.9.8 with $h(t) = 1$.
　　Solution:

$$y(x) = f(x) + \lambda(A_1 + A_2 x),$$

where A_1 and A_2 are the constants determined by the formulas presented in 4.9.8.

5. $\quad y(x) - \lambda \displaystyle\int_a^b (Ax + Bt + C)y(t)\, dt = f(x).$

This is a special case of equation 4.9.7 with $g(x) = x$ and $h(t) = 1$.
　　Solution:

$$y(x) = f(x) + \lambda(A_1 x + A_2),$$

where A_1 and A_2 are the constants determined by the formulas presented in 4.9.7.

6. $\quad y(x) + A \displaystyle\int_a^b |x - t|\, y(t)\, dt = f(x).$

This is a special case of equation 4.9.36 with $g(t) = A$.

1°. The function $y = y(x)$ obeys the following second-order linear nonhomogeneous ordinary differential equation with constant coefficients:

$$y''_{xx} + 2Ay = f''_{xx}(x). \tag{1}$$

The boundary conditions for (1) have the form (see 4.9.36)

$$y'_x(a) + y'_x(b) = f'_x(a) + f'_x(b),$$
$$y(a) + y(b) + (b-a)y'_x(a) = f(a) + f(b) + (b-a)f'_x(a). \tag{2}$$

　　Equation (1) under the boundary conditions (2) determines the solution of the original integral equation.

2°. For $A < 0$, the general solution of equation (1) is given by

$$y(x) = C_1 \cosh(kx) + C_2 \sinh(kx) + f(x) + k \int_a^x \sinh[k(x-t)]f(t)\, dt, \quad k = \sqrt{-2A}, \tag{3}$$

where C_1 and C_2 are arbitrary constants.
　　For $A > 0$, the general solution of equation (1) is given by

$$y(x) = C_1 \cos(kx) + C_2 \sin(kx) + f(x) - k \int_a^x \sin[k(x-t)]f(t)\, dt, \quad k = \sqrt{2A}. \tag{4}$$

The constants C_1 and C_2 in solutions (3) and (4) are determined by conditions (2).

3°. In the special case $a = 0$ and $A > 0$, the solution of the integral equation is given by formula (4) with

$$C_1 = k\frac{I_s(1 + \cos\lambda) - I_c(\lambda + \sin\lambda)}{2 + 2\cos\lambda + \lambda\sin\lambda}, \qquad C_2 = k\frac{I_s\sin\lambda + I_c(1 + \cos\lambda)}{2 + 2\cos\lambda + \lambda\sin\lambda},$$

$$k = \sqrt{2A}, \quad \lambda = bk, \quad I_s = \int_0^b \sin[k(b-t)]f(t)\,dt, \quad I_c = \int_0^b \cos[k(b-t)]f(t)\,dt.$$

4.1-2. Kernels Quadratic in the Arguments x and t

7. $y(x) - \lambda \displaystyle\int_a^b (x^2 + t^2)y(t)\,dt = f(x).$

The characteristic values of the equation:

$$\lambda_1 = \frac{1}{\frac{1}{3}(b^3 - a^3) + \sqrt{\frac{1}{5}(b^5 - a^5)(b - a)}}, \qquad \lambda_2 = \frac{1}{\frac{1}{3}(b^3 - a^3) - \sqrt{\frac{1}{5}(b^5 - a^5)(b - a)}}.$$

1°. Solution with $\lambda \neq \lambda_{1,2}$:

$$y(x) = f(x) + \lambda(A_1 x^2 + A_2),$$

where the constants A_1 and A_2 are given by

$$A_1 = \frac{f_1 - \lambda\left(\frac{1}{3}f_1\Delta_3 - f_2\Delta_1\right)}{\lambda^2\left(\frac{1}{9}\Delta_3^2 - \frac{1}{5}\Delta_1\Delta_5\right) - \frac{2}{3}\lambda\Delta_3 + 1}, \qquad A_2 = \frac{f_2 - \lambda\left(\frac{1}{3}f_2\Delta_3 - \frac{1}{5}f_1\Delta_5\right)}{\lambda^2\left(\frac{1}{9}\Delta_3^2 - \frac{1}{5}\Delta_1\Delta_5\right) - \frac{2}{3}\lambda\Delta_3 + 1},$$

$$f_1 = \int_a^b f(x)\,dx, \quad f_2 = \int_a^b x^2 f(x)\,dx, \quad \Delta_n = b^n - a^n.$$

2°. Solution with $\lambda = \lambda_1 \neq \lambda_2$ and $f_1 = f_2 = 0$:

$$y(x) = f(x) + Cy_1(x), \qquad y_1(x) = x^2 + \sqrt{\frac{b^5 - a^5}{5(b - a)}},$$

where C is an arbitrary constant and $y_1(x)$ is an eigenfunction of the equation corresponding to the characteristic value λ_1.

3°. Solution with $\lambda = \lambda_2 \neq \lambda_1$ and $f_1 = f_2 = 0$:

$$y(x) = f(x) + Cy_2(x), \qquad y_2(x) = x^2 - \sqrt{\frac{b^5 - a^5}{5(b - a)}}.$$

where C is an arbitrary constant and $y_2(x)$ is an eigenfunction of the equation corresponding to the characteristic value λ_2.

4°. The equation has no multiple characteristic values.

8. $\quad y(x) - \lambda \int_a^b (x^2 - t^2) y(t)\, dt = f(x).$

The characteristic values of the equation:

$$\lambda_{1,2} = \pm \frac{1}{\sqrt{\frac{1}{9}(b^3 - a^3)^2 - \frac{1}{5}(b^5 - a^5)(b - a)}}.$$

$1°$. Solution with $\lambda \neq \lambda_{1,2}$:

$$y(x) = f(x) + \lambda(A_1 x^2 + A_2),$$

where the constants A_1 and A_2 are given by

$$A_1 = \frac{f_1 + \lambda\left(\frac{1}{3}f_1\Delta_3 - f_2\Delta_1\right)}{\lambda^2\left(\frac{1}{5}\Delta_1\Delta_5 - \frac{1}{9}\Delta_3^2\right) + 1}, \qquad A_2 = \frac{-f_2 + \lambda\left(\frac{1}{3}f_2\Delta_3 - \frac{1}{5}f_1\Delta_5\right)}{\lambda^2\left(\frac{1}{5}\Delta_1\Delta_5 - \frac{1}{9}\Delta_3^2\right) + 1},$$

$$f_1 = \int_a^b f(x)\, dx, \quad f_2 = \int_a^b x^2 f(x)\, dx, \quad \Delta_n = b^n - a^n.$$

$2°$. Solution with $\lambda = \lambda_1 \neq \lambda_2$ and $f_1 = f_2 = 0$:

$$y(x) = f(x) + C y_1(x), \qquad y_1(x) = x^2 + \frac{3 - \lambda_1(b^3 - a^3)}{3\lambda_1(b - a)},$$

where C is an arbitrary constant and $y_1(x)$ is an eigenfunction of the equation corresponding to the characteristic value λ_1.

$3°$. The solution with $\lambda = \lambda_2 \neq \lambda_1$ and $f_1 = f_2 = 0$ is given by the formulas of item $2°$ in which one must replace λ_1 and $y_1(x)$ by λ_2 and $y_2(x)$, respectively.

$4°$. The equation has no multiple characteristic values.

9. $\quad y(x) - \lambda \int_a^b (Ax^2 + Bt^2) y(t)\, dt = f(x).$

The characteristic values of the equation:

$$\lambda_{1,2} = \frac{\frac{1}{3}(A + B)\Delta_3 \pm \sqrt{\frac{1}{9}(A - B)^2\Delta_3^2 + \frac{4}{5}AB\Delta_1\Delta_5}}{2AB\left(\frac{1}{9}\Delta_3^2 - \frac{1}{5}\Delta_1\Delta_5\right)}, \qquad \Delta_n = b^n - a^n.$$

$1°$. Solution with $\lambda \neq \lambda_{1,2}$:

$$y(x) = f(x) + \lambda(A_1 x^2 + A_2),$$

where the constants A_1 and A_2 are given by

$$A_1 = \frac{Af_1 - AB\lambda\left(\frac{1}{3}f_1\Delta_3 - f_2\Delta_1\right)}{AB\lambda^2\left(\frac{1}{9}\Delta_3^2 - \frac{1}{5}\Delta_1\Delta_5\right) - \frac{1}{3}(A + B)\lambda\Delta_3 + 1},$$

$$A_2 = \frac{Bf_2 - AB\lambda\left(\frac{1}{3}f_2\Delta_3 - \frac{1}{5}f_1\Delta_5\right)}{AB\lambda^2\left(\frac{1}{9}\Delta_3^2 - \frac{1}{5}\Delta_1\Delta_5\right) - \frac{1}{3}(A + B)\lambda\Delta_3 + 1},$$

$$f_1 = \int_a^b f(x)\, dx, \quad f_2 = \int_a^b x^2 f(x)\, dx.$$

$2°$. Solution with $\lambda = \lambda_1 \neq \lambda_2$ and $f_1 = f_2 = 0$:

$$y(x) = f(x) + C y_1(x), \qquad y_1(x) = x^2 + \frac{3 - \lambda_1 A(b^3 - a^3)}{3\lambda_1 A(b - a)},$$

where C is an arbitrary constant and $y_1(x)$ is an eigenfunction of the equation corresponding to the characteristic value λ_1.

3°. The solution with $\lambda = \lambda_2 \neq \lambda_1$ and $f_1 = f_2 = 0$ is given by the formulas of item 2° in which one must replace λ_1 and $y_1(x)$ by λ_2 and $y_2(x)$, respectively.

4°. Solution with $\lambda = \lambda_{1,2} = \lambda_*$ and $f_1 = f_2 = 0$, where $\lambda_* = \dfrac{6}{(A+B)(b^3 - a^3)}$ is the double characteristic value:

$$y(x) = f(x) + C_1 y_*(x),$$

where C_1 is an arbitrary constant and $y_*(x)$ is an eigenfunction of the equation corresponding to λ_*:

$$y_*(x) = x^2 - \frac{(A-B)(b^3 - a^3)}{6A(b-a)}.$$

10. $y(x) - \lambda \displaystyle\int_a^b (xt - t^2) y(t)\, dt = f(x).$

This is a special case of equation 4.9.8 with $A = 0$, $B = 1$, and $h(t) = t$.
     Solution:

$$y(x) = f(x) + \lambda(A_1 + A_2 x),$$

where A_1 and A_2 are the constants determined by the formulas presented in 4.9.8.

11. $y(x) - \lambda \displaystyle\int_a^b (x^2 - xt) y(t)\, dt = f(x).$

This is a special case of equation 4.9.10 with $A = 0$, $B = 1$, and $h(x) = x$.
     Solution:

$$y(x) = f(x) + \lambda(E_1 x^2 + E_2 x),$$

where E_1 and E_2 are the constants determined by the formulas presented in 4.9.10.

12. $y(x) - \lambda \displaystyle\int_a^b (Bxt + Ct^2) y(t)\, dt = f(x).$

This is a special case of equation 4.9.9 with $A = 0$ and $h(t) = t$.
     Solution:

$$y(x) = f(x) + \lambda(A_1 + A_2 x),$$

where A_1 and A_2 are the constants determined by the formulas presented in 4.9.9.

13. $y(x) - \lambda \displaystyle\int_a^b (Bx^2 + Cxt) y(t)\, dt = f(x).$

This is a special case of equation 4.9.11 with $A = 0$ and $h(x) = x$.
     Solution:

$$y(x) = f(x) + \lambda(A_1 x^2 + A_2 x),$$

where A_1 and A_2 are the constants determined by the formulas presented in 4.9.11.

14. $y(x) - \lambda \displaystyle\int_a^b (Axt + Bx^2 + Cx + D) y(t)\, dt = f(x).$

This is a special case of equation 4.9.18 with $g_1(x) = Bx^2 + Cx + D$, $h_1(t) = 1$, $g_2(x) = x$, and $h_2(t) = At$.
     Solution:

$$y(x) = f(x) + \lambda[A_1(Bx^2 + Cx + D) + A_2 x],$$

where A_1 and A_2 are the constants determined by the formulas presented in 4.9.18.

15. $y(x) - \lambda \int_a^b (Ax^2 + Bt^2 + Cx + Dt + E)y(t)\, dt = f(x).$

This is a special case of equation 4.9.18 with $g_1(x) = Ax^2 + Cx$, $h_1(t) = 1$, $g_2(x) = 1$, and $h_2(t) = Bt^2 + Dt + E$.
     Solution:
$$y(x) = f(x) + \lambda[A_1(Ax^2 + Cx) + A_2],$$

where A_1 and A_2 are the constants determined by the formulas presented in 4.9.18.

16. $y(x) - \lambda \int_a^b [Ax + B + (Cx + D)(x - t)]y(t)\, dt = f(x).$

This is a special case of equation 4.9.18 with $g_1(x) = Cx^2 + (A + D)x + B$, $h_1(t) = 1$, $g_2(x) = Cx + D$, and $h_2(t) = -t$.
     Solution:
$$y(x) = f(x) + \lambda[A_1(Cx^2 + Ax + Dx + B) + A_2(Cx + D)],$$

where A_1 and A_2 are the constants determined by the formulas presented in 4.9.18.

17. $y(x) - \lambda \int_a^b [At + B + (Ct + D)(t - x)]y(t)\, dt = f(x).$

This is a special case of equation 4.9.18 with $g_1(x) = 1$, $h_1(t) = Ct^2 + (A + D)t + B$, $g_2(x) = x$, and $h_2(t) = -(Ct + D)$.
     Solution:
$$y(x) = f(x) + \lambda(A_1 + A_2 x),$$

where A_1 and A_2 are the constants determined by the formulas presented in 4.9.18.

18. $y(x) - \lambda \int_a^b (x - t)^2 y(t)\, dt = f(x).$

This is a special case of equation 4.9.19 with $g(x) = x$, $h(t) = -t$, and $m = 2$.

19. $y(x) - \lambda \int_a^b (Ax + Bt)^2 y(t)\, dt = f(x).$

This is a special case of equation 4.9.19 with $g(x) = Ax$, $h(t) = Bt$, and $m = 2$.

4.1-3. Kernels Cubic in the Arguments x and t

20. $y(x) - \lambda \int_a^b (x^3 + t^3)y(t)\, dt = f(x).$

The characteristic values of the equation:

$$\lambda_1 = \frac{1}{\frac{1}{4}(b^4 - a^4) + \sqrt{\frac{1}{7}(b^7 - a^7)(b - a)}}, \qquad \lambda_2 = \frac{1}{\frac{1}{4}(b^4 - a^4) - \sqrt{\frac{1}{7}(b^7 - a^7)(b - a)}}.$$

$1°$. Solution with $\lambda \neq \lambda_{1,2}$:

$$y(x) = f(x) + \lambda(A_1 x^3 + A_2),$$

where the constants A_1 and A_2 are given by

$$A_1 = \frac{f_1 - \lambda \left(\frac{1}{4} f_1 \Delta_4 - f_2 \Delta_1 \right)}{\lambda^2 \left(\frac{1}{16} \Delta_4^2 - \frac{1}{7} \Delta_1 \Delta_7 \right) - \frac{1}{2} \lambda \Delta_4 + 1}, \quad A_2 = \frac{f_2 - \lambda \left(\frac{1}{4} f_2 \Delta_4 - \frac{1}{7} f_1 \Delta_7 \right)}{\lambda^2 \left(\frac{1}{16} \Delta_4^2 - \frac{1}{7} \Delta_1 \Delta_7 \right) - \frac{1}{2} \lambda \Delta_4 + 1},$$

$$f_1 = \int_a^b f(x)\, dx, \quad f_2 = \int_a^b x^3 f(x)\, dx, \quad \Delta_n = b^n - a^n.$$

2°. Solution with $\lambda = \lambda_1 \neq \lambda_2$ and $f_1 = f_2 = 0$:

$$y(x) = f(x) + C y_1(x), \qquad y_1(x) = x^3 + \sqrt{\frac{b^7 - a^7}{7(b-a)}},$$

where C is an arbitrary constant and $y_1(x)$ is an eigenfunction of the equation corresponding to the characteristic value λ_1.

3°. Solution with $\lambda = \lambda_2 \neq \lambda_1$ and $f_1 = f_2 = 0$:

$$y(x) = f(x) + C y_2(x), \qquad y_2(x) = x^3 - \sqrt{\frac{b^7 - a^7}{7(b-a)}},$$

where C is an arbitrary constant and $y_2(x)$ is an eigenfunction of the equation corresponding to the characteristic value λ_2.

4°. The equation has no multiple characteristic values.

21. $y(x) - \lambda \int_a^b (x^3 - t^3) y(t)\, dt = f(x).$

The characteristic values of the equation:

$$\lambda_{1,2} = \pm \frac{1}{\sqrt{\frac{1}{4}(a^4 - b^4)^2 - \frac{1}{7}(a^7 - b^7)(b - a)}}.$$

1°. Solution with $\lambda \neq \lambda_{1,2}$:

$$y(x) = f(x) + \lambda (A_1 x^3 + A_2),$$

where the constants A_1 and A_2 are given by

$$A_1 = \frac{f_1 + \lambda \left(\frac{1}{4} f_1 \Delta_4 - f_2 \Delta_1 \right)}{\lambda^2 \left(\frac{1}{7} \Delta_1 \Delta_7 - \frac{1}{16} \Delta_4^2 \right) + 1}, \quad A_2 = \frac{-f_2 + \lambda \left(\frac{1}{4} f_2 \Delta_4 - \frac{1}{7} f_1 \Delta_7 \right)}{\lambda^2 \left(\frac{1}{7} \Delta_1 \Delta_7 - \frac{1}{16} \Delta_4^2 \right) + 1},$$

$$f_1 = \int_a^b f(x)\, dx, \quad f_2 = \int_a^b x^3 f(x)\, dx, \quad \Delta_n = b^n - a^n.$$

2°. Solution with $\lambda = \lambda_1 \neq \lambda_2$ and $f_1 = f_2 = 0$:

$$y(x) = f(x) + C y_1(x), \qquad y_1(x) = x^3 + \frac{4 - \lambda_1 (b^4 - a^4)}{4 \lambda_1 (b - a)},$$

where C is an arbitrary constant and $y_1(x)$ is an eigenfunction of the equation corresponding to the characteristic value λ_1.

3°. The solution with $\lambda = \lambda_2 \neq \lambda_1$ and $f_1 = f_2 = 0$ is given by the formulas of item 2° in which one must replace λ_1 and $y_1(x)$ by λ_2 and $y_2(x)$, respectively.

4°. The equation has no multiple characteristic values.

22. $y(x) - \lambda \int_a^b (Ax^3 + Bt^3) y(t)\, dt = f(x).$

The characteristic values of the equation:

$$\lambda_{1,2} = \frac{\frac{1}{4}(A+B)\Delta_4 \pm \sqrt{\frac{1}{16}(A-B)^2\Delta_4^2 + \frac{4}{7}AB\Delta_1\Delta_7}}{2AB\left(\frac{1}{16}\Delta_4^2 - \frac{1}{7}\Delta_1\Delta_7\right)}, \qquad \Delta_n = b^n - a^n.$$

$1°$. Solution with $\lambda \neq \lambda_{1,2}$:

$$y(x) = f(x) + \lambda(A_1 x^3 + A_2),$$

where the constants A_1 and A_2 are given by

$$A_1 = \frac{Af_1 - AB\lambda\left(\frac{1}{4}f_1\Delta_4 - f_2\Delta_1\right)}{AB\lambda^2\left(\frac{1}{16}\Delta_4^2 - \frac{1}{7}\Delta_1\Delta_7\right) - \frac{1}{4}\lambda(A+B)\Delta_4 + 1},$$

$$A_2 = \frac{Bf_2 - AB\lambda\left(\frac{1}{4}f_2\Delta_4 - \frac{1}{7}f_1\Delta_7\right)}{AB\lambda^2\left(\frac{1}{16}\Delta_4^2 - \frac{1}{7}\Delta_1\Delta_7\right) - \frac{1}{4}\lambda(A+B)\Delta_4 + 1},$$

$$f_1 = \int_a^b f(x)\, dx, \quad f_2 = \int_a^b x^3 f(x)\, dx.$$

$2°$. Solution with $\lambda = \lambda_1 \neq \lambda_2$ and $f_1 = f_2 = 0$:

$$y(x) = f(x) + Cy_1(x), \qquad y_1(x) = x^3 + \frac{4 - \lambda_1 A(b^4 - a^4)}{4\lambda_1 A(b-a)},$$

where C is an arbitrary constant and $y_1(x)$ is an eigenfunction of the equation corresponding to the characteristic value λ_1.

$3°$. The solution with $\lambda = \lambda_2 \neq \lambda_1$ and $f_1 = f_2 = 0$ is given by the formulas of item $2°$ in which one must replace λ_1 and $y_1(x)$ by λ_2 and $y_2(x)$, respectively.

$4°$. Solution with $\lambda = \lambda_{1,2} = \lambda_*$ and $f_1 = f_2 = 0$, where $\lambda_* = \dfrac{8}{(A+B)(b^4 - a^4)}$ is the double characteristic value:

$$y(x) = f(x) + Cy_*(x), \qquad y_*(x) = x^3 - \frac{(A-B)(b^4 - a^4)}{8A(b-a)},$$

where C is an arbitrary constant and $y_*(x)$ is an eigenfunction of the equation corresponding to λ_*.

23. $y(x) - \lambda \int_a^b (xt^2 - t^3) y(t)\, dt = f(x).$

This is a special case of equation 4.9.8 with $A = 0$, $B = 1$, and $h(t) = t^2$.
Solution:

$$y(x) = f(x) + \lambda(A_1 + A_2 x),$$

where A_1 and A_2 are the constants determined by the formulas presented in 4.9.8.

24. $y(x) - \lambda \int_a^b (Bxt^2 + Ct^3)y(t)\, dt = f(x).$

This is a special case of equation 4.9.9 with $A = 0$ and $h(t) = t^2$.
    Solution:
$$y(x) = f(x) + \lambda(A_1 + A_2 x),$$

where A_1 and A_2 are the constants determined by the formulas presented in 4.9.9.

25. $y(x) - \lambda \int_a^b (Ax^2 t + Bxt^2)y(t)\, dt = f(x).$

This is a special case of equation 4.9.17 with $g(x) = x^2$ and $h(x) = x$.
    Solution:
$$y(x) = f(x) + \lambda(A_1 x^2 + A_2 x),$$

where A_1 and A_2 are the constants determined by the formulas presented in 4.9.17.

26. $y(x) - \lambda \int_a^b (Ax^3 + Bxt^2)y(t)\, dt = f(x).$

This is a special case of equation 4.9.18 with $g_1(x) = x^3$, $h_1(t) = A$, $g_2(x) = x$, and $h_2(t) = Bt^2$.
    Solution:
$$y(x) = f(x) + \lambda(A_1 x^3 + A_2 x),$$

where A_1 and A_2 are the constants determined by the formulas presented in 4.9.18.

27. $y(x) - \lambda \int_a^b (Ax^3 + Bx^2 t + Cx^2 + D)y(t)\, dt = f(x).$

This is a special case of equation 4.9.18 with $g_1(x) = Ax^3 + Cx^2 + D$, $h_1(t) = 1$, $g_2(x) = x^2$, and $h_2(t) = Bt$.
    Solution:
$$y(x) = f(x) + \lambda[A_1(Ax^3 + Cx^2 + D) + A_2 x^2],$$

where A_1 and A_2 are the constants determined by the formulas presented in 4.9.18.

28. $y(x) - \lambda \int_a^b (Axt^2 + Bt^3 + Ct^2 + D)y(t)\, dt = f(x).$

This is a special case of equation 4.9.18 with $g_1(x) = x$, $h_1(t) = At^2$, $g_2(x) = 1$, and $h_2(t) = Bt^3 + Ct^2 + D$.
    Solution:
$$y(x) = f(x) + \lambda(A_1 x + A_2),$$

where A_1 and A_2 are the constants determined by the formulas presented in 4.9.18.

29. $y(x) - \lambda \int_a^b (x - t)^3 y(t)\, dt = f(x).$

This is a special case of equation 4.9.19 with $g(x) = x$, $h(t) = -t$, and $m = 3$.

30. $y(x) - \lambda \int_a^b (Ax + Bt)^3 y(t)\, dt = f(x).$

This is a special case of equation 4.9.19 with $g(x) = Ax$, $h(t) = Bt$, and $m = 3$.

4.1-4. Kernels Containing Higher-Order Polynomials in x and t

31. $y(x) - \lambda \int_a^b (x^n + t^n) y(t)\, dt = f(x), \qquad n = 1, 2, \ldots$

The characteristic values of the equation:

$$\lambda_{1,2} = \frac{1}{\Delta_n \pm \sqrt{\Delta_0 \Delta_{2n}}}, \qquad \text{where} \quad \Delta_n = \frac{1}{n+1}(b^{n+1} - a^{n+1}).$$

1°. Solution with $\lambda \neq \lambda_{1,2}$:

$$y(x) = f(x) + \lambda(A_1 x^n + A_2),$$

where the constants A_1 and A_2 are given by

$$A_1 = \frac{f_1 - \lambda(f_1 \Delta_n - f_2 \Delta_0)}{\lambda^2(\Delta_n^2 - \Delta_0 \Delta_{2n}) - 2\lambda \Delta_n + 1}, \quad A_2 = \frac{f_2 - \lambda(f_2 \Delta_n - f_1 \Delta_{2n})}{\lambda^2(\Delta_n^2 - \Delta_0 \Delta_{2n}) - 2\lambda \Delta_n + 1},$$

$$f_1 = \int_a^b f(x)\, dx, \quad f_2 = \int_a^b x^n f(x)\, dx, \quad \Delta_n = \frac{1}{n+1}(b^{n+1} - a^{n+1}).$$

2°. Solution with $\lambda = \lambda_1 \neq \lambda_2$ and $f_1 = f_2 = 0$:

$$y(x) = f(x) + C y_1(x), \qquad y_1(x) = x^n + \sqrt{\Delta_{2n}/\Delta_0},$$

where C is an arbitrary constant and $y_1(x)$ is an eigenfunction of the equation corresponding to the characteristic value λ_1.

3°. Solution with $\lambda = \lambda_2 \neq \lambda_1$ and $f_1 = f_2 = 0$:

$$y(x) = f(x) + C y_2(x), \qquad y_2(x) = x^n - \sqrt{\Delta_{2n}/\Delta_0},$$

where C is an arbitrary constant and $y_2(x)$ is an eigenfunction of the equation corresponding to the characteristic value λ_2.

4°. The equation has no multiple characteristic values.

32. $y(x) - \lambda \int_a^b (x^n - t^n) y(t)\, dt = f(x), \qquad n = 1, 2, \ldots$

The characteristic values of the equation:

$$\lambda_{1,2} = \pm \left[\frac{1}{(n+1)^2}(b^{n+1} - a^{n+1})^2 - \frac{1}{2n+1}(b^{2n+1} - a^{2n+1})(b - a) \right]^{-1/2}.$$

1°. Solution with $\lambda \neq \lambda_{1,2}$:

$$y(x) = f(x) + \lambda(A_1 x^n + A_2),$$

where the constants A_1 and A_2 are given by

$$A_1 = \frac{f_1 + \lambda(f_1 \Delta_n - f_2 \Delta_0)}{\lambda^2(\Delta_0 \Delta_{2n} - \Delta_n^2) + 1}, \quad A_2 = \frac{-f_2 + \lambda(f_2 \Delta_n - f_1 \Delta_{2n})}{\lambda^2(\Delta_0 \Delta_{2n} - \Delta_n^2) + 1},$$

$$f_1 = \int_a^b f(x)\, dx, \quad f_2 = \int_a^b x^n f(x)\, dx, \quad \Delta_n = \frac{1}{n+1}(b^{n+1} - a^{n+1}).$$

2°. Solution with $\lambda = \lambda_1 \neq \lambda_2$ and $f_1 = f_2 = 0$:

$$y(x) = f(x) + C y_1(x), \qquad y_1(x) = x^n + \frac{1 - \lambda_1 \Delta_n}{\lambda_1 \Delta_0},$$

where C is an arbitrary constant and $y_1(x)$ is an eigenfunction of the equation corresponding to the characteristic value λ_1.

3°. The solution with $\lambda = \lambda_2 \neq \lambda_1$ and $f_1 = f_2 = 0$ is given by the formulas of item 2° in which one must replace λ_1 and $y_1(x)$ by λ_2 and $y_2(x)$, respectively.

4°. The equation has no multiple characteristic values.

33. $\quad y(x) - \lambda \displaystyle\int_a^b (Ax^n + Bt^n)y(t)\,dt = f(x), \qquad n = 1, 2, \ldots$

The characteristic values of the equation:

$$\lambda_{1,2} = \frac{(A+B)\Delta_n \pm \sqrt{(A-B)^2\Delta_n^2 + 4AB\Delta_0\Delta_{2n}}}{2AB(\Delta_n^2 - \Delta_0\Delta_{2n})}, \qquad \Delta_n = \frac{1}{n+1}(b^{n+1} - a^{n+1}).$$

1°. Solution with $\lambda \neq \lambda_{1,2}$:

$$y(x) = f(x) + \lambda(A_1 x^n + A_2),$$

where the constants A_1 and A_2 are given by

$$A_1 = \frac{Af_1 - AB\lambda(f_1\Delta_n - f_2\Delta_0)}{AB\lambda^2(\Delta_n^2 - \Delta_0\Delta_{2n}) - (A+B)\lambda\Delta_n + 1},$$

$$A_2 = \frac{Bf_2 - AB\lambda(f_2\Delta_n - f_1\Delta_{2n})}{AB\lambda^2(\Delta_n^2 - \Delta_0\Delta_{2n}) - (A+B)\lambda\Delta_n + 1},$$

$$f_1 = \int_a^b f(x)\,dx, \quad f_2 = \int_a^b x^n f(x)\,dx.$$

2°. Solution with $\lambda = \lambda_1 \neq \lambda_2$ and $f_1 = f_2 = 0$:

$$y(x) = f(x) + Cy_1(x), \qquad y_1(x) = x^n + \frac{1 - A\lambda_1\Delta_n}{A\lambda_1\Delta_0},$$

where C is an arbitrary constant and $y_1(x)$ is an eigenfunction of the equation corresponding to the characteristic value λ_1.

3°. The solution with $\lambda = \lambda_2 \neq \lambda_1$ and $f_1 = f_2 = 0$ is given by the formulas of item 2° in which one must replace λ_1 and $y_1(x)$ by λ_2 and $y_2(x)$, respectively.

4°. Solution with $\lambda = \lambda_{1,2} = \lambda_*$ and $f_1 = f_2 = 0$, where the characteristic value $\lambda_* = \dfrac{2}{(A+B)\Delta_n}$ is double:

$$y(x) = f(x) + Cy_*(x), \qquad y_*(x) = x^n - \frac{(A-B)\Delta_n}{2A\Delta_0}.$$

Here C is an arbitrary constant and $y_*(x)$ is an eigenfunction of the equation corresponding to λ_*.

34. $\quad y(x) - \lambda \displaystyle\int_a^b (x-t)t^m y(t)\,dt = f(x), \qquad m = 1, 2, \ldots$

This is a special case of equation 4.9.8 with $A = 0$, $B = 1$, and $h(t) = t^m$.
    Solution:

$$y(x) = f(x) + \lambda(A_1 + A_2 x),$$

where A_1 and A_2 are the constants determined by the formulas presented in 4.9.8.

35. $\quad y(x) - \lambda \displaystyle\int_a^b (x-t)x^m y(t)\,dt = f(x), \qquad m = 1, 2, \ldots$

This is a special case of equation 4.9.10 with $A = 0$, $B = 1$, and $h(x) = x^m$.
    Solution:

$$y(x) = f(x) + \lambda(A_1 x^{m+1} + A_2 x^m),$$

where A_1 and A_2 are the constants determined by the formulas presented in 4.9.10.

36. $\quad y(x) - \lambda \displaystyle\int_a^b (Ax^{m+1} + Bx^m t + Cx^m + D)y(t)\, dt = f(x), \qquad m = 1, 2, \ldots$

This is a special case of equation 4.9.18 with $g_1(x) = Ax^{m+1} + Cx^m + D$, $h_1(t) = 1$, $g_2(x) = x^m$, and $h_2(t) = Bt$.
$\quad$ Solution:
$$y(x) = f(x) + \lambda[A_1(Ax^{m+1} + Cx^m + D) + A_2 x^m],$$

where A_1 and A_2 are the constants determined by the formulas presented in 4.9.18.

37. $\quad y(x) - \lambda \displaystyle\int_a^b (Axt^m + Bt^{m+1} + Ct^m + D)y(t)\, dt = f(x), \qquad m = 1, 2, \ldots$

This is a special case of equation 4.9.18 with $g_1(x) = x$, $h_1(t) = At^m$, $g_2(x) = 1$, and $h_2(t) = Bt^{m+1} + Ct^m + D$.
$\quad$ Solution:
$$y(x) = f(x) + \lambda(A_1 x + A_2),$$

where A_1 and A_2 are the constants determined by the formulas presented in 4.9.18.

38. $\quad y(x) - \lambda \displaystyle\int_a^b (Ax^n t^n + Bx^m t^m)y(t)\, dt = f(x), \qquad n, m = 1, 2, \ldots, \quad n \neq m.$

This is a special case of equation 4.9.14 with $g(x) = x^n$ and $h(t) = t^m$.
$\quad$ Solution:
$$y(x) = f(x) + \lambda(A_1 x^n + A_2 x^m),$$

where A_1 and A_2 are the constants determined by the formulas presented in 4.9.14.

39. $\quad y(x) - \lambda \displaystyle\int_a^b (Ax^n t^m + Bx^m t^n)y(t)\, dt = f(x), \qquad n, m = 1, 2, \ldots, \quad n \neq m.$

This is a special case of equation 4.9.17 with $g(x) = x^n$ and $h(t) = t^m$.
$\quad$ Solution:
$$y(x) = f(x) + \lambda(A_1 x^n + A_2 x^m),$$

where A_1 and A_2 are the constants determined by the formulas presented in 4.9.17.

40. $\quad y(x) - \lambda \displaystyle\int_a^b (x - t)^m y(t)\, dt = f(x), \qquad m = 1, 2, \ldots$

This is a special case of equation 4.9.19 with $g(x) = x$ and $h(t) = -t$.

41. $\quad y(x) - \lambda \displaystyle\int_a^b (Ax + Bt)^m y(t)\, dt = f(x), \qquad m = 1, 2, \ldots$

This is a special case of equation 4.9.19 with $g(x) = Ax$ and $h(t) = Bt$.

42. $\quad y(x) + A \displaystyle\int_a^b |x - t| t^k y(t)\, dt = f(x).$

This is a special case of equation 4.9.36 with $g(t) = At^k$. Solving the integral equation is reduced to solving the ordinary differential equation $y''_{xx} + 2Ax^k y = f''_{xx}(x)$, the general solution of which can be expressed via Bessel functions or modified Bessel functions (the boundary conditions are given in 4.9.36).

43. $\quad y(x) + A \displaystyle\int_a^b |x - t|^{2n+1} y(t)\, dt = f(x), \qquad n = 0, 1, 2, \ldots$

Let us remove the modulus in the integrand:

$$y(x) + A \int_a^x (x - t)^{2n+1} y(t)\, dt + A \int_x^b (t - x)^{2n+1} y(t)\, dt = f(x). \tag{1}$$

The k-fold differentiation of (1) with respect to x yields

$$y_x^{(k)}(x) + A B_k \int_a^x (x - t)^{2n+1-k} y(t)\, dt + (-1)^k A B_k \int_x^b (t - x)^{2n+1-k} y(t)\, dt = f_x^{(k)}(x), \tag{2}$$

$$B_k = (2n + 1)(2n)\ldots(2n + 2 - k), \qquad k = 1, 2, \ldots, 2n + 1.$$

Differentiating (2) with $k = 2n + 1$, we arrive at the following linear nonhomogeneous differential equation with constant coefficients for $y = y(x)$:

$$y_x^{(2n+2)} + 2(2n + 1)!\, A y = f_x^{(2n+2)}(x). \tag{3}$$

Equation (3) must satisfy the initial conditions which can be obtained by setting $x = a$ in (1) and (2):

$$y(a) + A \int_a^b (t - a)^{2n+1} y(t)\, dt = f(a),$$

$$y_x^{(k)}(a) + (-1)^k A B_k \int_a^b (t - a)^{2n+1-k} y(t)\, dt = f_x^{(k)}(a), \quad k = 1, 2, \ldots, 2n + 1. \tag{4}$$

These conditions can be reduced to a more habitual form containing no integrals. To this end, y must be expressed from equation (3) in terms of $y_x^{(2n+2)}$ and $f_x^{(2n+2)}$ and substituted into (4), and then one must integrate the resulting expressions by parts (sufficiently many times).

4.1-5. Kernels Containing Rational Functions

44. $\quad y(x) - \lambda \displaystyle\int_a^b \left(\dfrac{1}{x} + \dfrac{1}{t} \right) y(t)\, dt = f(x).$

This is a special case of equation 4.9.2 with $g(x) = 1/x$.
    Solution:
$$y(x) = f(x) + \lambda \left(\frac{A_1}{x} + A_2 \right),$$

where A_1 and A_2 are the constants determined by the formulas presented in 4.9.2.

45. $\quad y(x) - \lambda \displaystyle\int_a^b \left(\dfrac{1}{x} - \dfrac{1}{t} \right) y(t)\, dt = f(x).$

This is a special case of equation 4.9.3 with $g(x) = 1/x$.
    Solution:
$$y(x) = f(x) + \lambda \left(\frac{A_1}{x} + A_2 \right),$$

where A_1 and A_2 are the constants determined by the formulas presented in 4.9.3.

46. $y(x) - \lambda \int_a^b \left(\dfrac{A}{x} + \dfrac{B}{t} \right) y(t)\, dt = f(x).$

This is a special case of equation 4.9.4 with $g(x) = 1/x$.

Solution:
$$y(x) = f(x) + \lambda \left(\dfrac{A_1}{x} + A_2 \right),$$

where A_1 and A_2 are the constants determined by the formulas presented in 4.9.4.

47. $y(x) - \lambda \int_a^b \left(\dfrac{A}{x + \alpha} + \dfrac{B}{t + \beta} \right) y(t)\, dt = f(x).$

This is a special case of equation 4.9.5 with $g(x) = \dfrac{A}{x + \alpha}$ and $h(t) = \dfrac{B}{t + \beta}$.

Solution:
$$y(x) = f(x) + \lambda \left(A_1 \dfrac{A}{x + \alpha} + A_2 \right),$$

where A_1 and A_2 are the constants determined by the formulas presented in 4.9.5.

48. $y(x) - \lambda \int_a^b \left(\dfrac{x}{t} - \dfrac{t}{x} \right) y(t)\, dt = f(x).$

This is a special case of equation 4.9.16 with $g(x) = x$ and $h(t) = 1/t$.

Solution:
$$y(x) = f(x) + \lambda \left(A_1 x + \dfrac{A_2}{x} \right),$$

where A_1 and A_2 are the constants determined by the formulas presented in 4.9.16.

49. $y(x) - \lambda \int_a^b \left(\dfrac{Ax}{t} + \dfrac{Bt}{x} \right) y(t)\, dt = f(x).$

This is a special case of equation 4.9.17 with $g(x) = x$ and $h(t) = 1/t$.

Solution:
$$y(x) = f(x) + \lambda \left(A_1 x + \dfrac{A_2}{x} \right),$$

where A_1 and A_2 are the constants determined by the formulas presented in 4.9.17.

50. $y(x) - \lambda \int_a^b \left(A \dfrac{x + \alpha}{t + \beta} + B \dfrac{t + \alpha}{x + \beta} \right) y(t)\, dt = f(x).$

This is a special case of equation 4.9.17 with $g(x) = x + \alpha$ and $h(t) = \dfrac{1}{t + \beta}$.

Solution:
$$y(x) = f(x) + \lambda \left[A_1 (x + \alpha) + \dfrac{A_2}{x + \beta} \right],$$

where A_1 and A_2 are the constants determined by the formulas presented in 4.9.17.

51. $y(x) - \lambda \int_a^b \left[A \dfrac{(x + \alpha)^n}{(t + \beta)^m} + B \dfrac{(t + \alpha)^n}{(x + \beta)^m} \right] y(t)\, dt = f(x), \qquad n, m = 0, 1, 2, \ldots$

This is a special case of equation 4.9.17 with $g(x) = (x + \alpha)^n$ and $h(t) = (t + \beta)^{-m}$.

Solution:
$$y(x) = f(x) + \lambda \left[A_1 (x + \alpha)^n + \dfrac{A_2}{(x + \beta)^m} \right],$$

where A_1 and A_2 are the constants determined by the formulas presented in 4.9.17.

52. $\quad y(x) - \lambda \displaystyle\int_1^\infty \frac{y(t)}{x+t}\, dt = f(x), \qquad 1 \le x < \infty, \quad -\infty < \pi\lambda < 1.$

Solution:

$$y(x) = \int_0^\infty \frac{\tau \sinh(\pi\tau)\, F(\tau)}{\cosh(\pi\tau) - \pi\lambda} P_{-\frac{1}{2}+i\tau}(x)\, d\tau,$$

$$F(\tau) = \int_1^\infty f(x) P_{-\frac{1}{2}+i\tau}(x)\, dx,$$

where $P_\nu(x) = F\left(-\nu,\, \nu+1,\, 1;\, \frac{1}{2}(1-x)\right)$ is the Legendre spherical function of the first kind, for which the integral representation

$$P_{-\frac{1}{2}+i\tau}(\cosh\alpha) = \frac{2}{\pi} \int_0^\alpha \frac{\cos(\tau s)\, ds}{\sqrt{2(\cosh\alpha - \cosh s)}} \qquad (\alpha \ge 0)$$

can be used.

⊙   Reference: V. A. Ditkin and A. P. Prudnikov (1965).

53. $\quad (x^2 + b^2)y(x) = \dfrac{\lambda}{\pi} \displaystyle\int_{-\infty}^\infty \frac{a^3 y(t)}{a^2 + (x-t)^2}\, dt.$

This equation is encountered in atomic and nuclear physics.

We seek the solution in the form

$$y(x) = \sum_{m=0}^\infty \frac{A_m x}{x^2 + (am+b)^2}. \tag{1}$$

The coefficients A_m obey the equations

$$m A_m \left(\frac{m+2b}{a}\right) + \lambda A_{m-1} = 0, \quad \sum_{m=0}^\infty A_m = 0. \tag{2}$$

Using the first equation of (2) to express all A_m via A_0 (A_0 can be chosen arbitrarily), substituting the result into the second equation of (2), and dividing by A_0, we obtain

$$1 + \sum_{m=1}^\infty \frac{(-\lambda)^m}{m!} \frac{1}{(1+2b/a)(2+2b/a)\ldots(m+2b/a)} = 0. \tag{3}$$

It follows from the definitions of the Bessel functions of the first kind that equation (3) can be rewritten in the form

$$\lambda^{-b/a} J_{2b/a}\left(2\sqrt{\lambda}\right) = 0. \tag{4}$$

In this sort of problem, a and λ are usually assumed to be given and b, which is proportional to the system energy, to be unknown. The quantity b can be determined by tables of zeros of Bessel functions. In some cases, b and a are given and λ is unknown.

⊙   Reference: I. Sneddon (1951).

4.1-6. Kernels Containing Arbitrary Powers

54. $\quad y(x) - \lambda \int_a^b (x - t)t^\mu y(t)\, dt = f(x).$

This is a special case of equation 4.9.8 with $A = 0$, $B = 1$, and $h(t) = t^\mu$.
  Solution:
$$y(x) = f(x) + \lambda(A_1 + A_2 x),$$

where A_1 and A_2 are the constants determined by the formulas presented in 4.9.8.

55. $\quad y(x) - \lambda \int_a^b (x - t)x^\nu y(t)\, dt = f(x).$

This is a special case of equation 4.9.10 with $A = 0$, $B = 1$, and $h(x) = x^\nu$.
  Solution:
$$y(x) = f(x) + \lambda(E_1 x^{\nu+1} + E_2 x^\nu),$$

where E_1 and E_2 are the constants determined by the formulas presented in 4.9.10.

56. $\quad y(x) - \lambda \int_a^b (x^\mu - t^\mu)y(t)\, dt = f(x).$

This is a special case of equation 4.9.3 with $g(x) = x^\mu$.
  Solution:
$$y(x) = f(x) + \lambda(A_1 x^\mu + A_2),$$

where A_1 and A_2 are the constants determined by the formulas presented in 4.9.3.

57. $\quad y(x) - \lambda \int_a^b (Ax^\nu + Bt^\nu)t^\mu y(t)\, dt = f(x).$

This is a special case of equation 4.9.6 with $g(x) = x^\nu$ and $h(t) = t^\mu$.
  Solution:
$$y(x) = f(x) + \lambda(A_1 x^\nu + A_2),$$

where A_1 and A_2 are the constants determined by the formulas presented in 4.9.6.

58. $\quad y(x) - \lambda \int_a^b (Dx^\nu + Et^\mu)x^\gamma y(t)\, dt = f(x).$

This is a special case of equation 4.9.18 with $g_1(x) = x^{\nu+\gamma}$, $h_1(t) = D$, $g_2(x) = x^\gamma$, and $h_2(t) = Et^\mu$.
  Solution:
$$y(x) = f(x) + \lambda(A_1 x^{\nu+\gamma} + A_2 x^\gamma),$$

where A_1 and A_2 are the constants determined by the formulas presented in 4.9.18.

59. $\quad y(x) - \lambda \int_a^b (Ax^\nu t^\mu + Bx^\gamma t^\delta)y(t)\, dt = f(x).$

This is a special case of equation 4.9.18 with $g_1(x) = x^\nu$, $h_1(t) = At^\mu$, $g_2(x) = x^\gamma$, and $h_2(t) = Bt^\delta$.
  Solution:
$$y(x) = f(x) + \lambda(A_1 x^\nu + A_2 x^\gamma),$$

where A_1 and A_2 are the constants determined by the formulas presented in 4.9.18.

60. $y(x) - \lambda \int_a^b (A + Bxt^\mu + Ct^{\mu+1})y(t)\, dt = f(x).$

This is a special case of equation 4.9.9 with $h(t) = t^\mu$.
  Solution:

$$y(x) = f(x) + \lambda(A_1 + A_2 x),$$

where A_1 and A_2 are the constants determined by the formulas presented in 4.9.9.

61. $y(x) - \lambda \int_a^b (At^\alpha + Bx^\beta t^\mu + Ct^{\mu+\gamma})y(t)\, dt = f(x).$

This is a special case of equation 4.9.18 with $g_1(x) = 1$, $h_1(t) = At^\alpha + Ct^{\mu+\gamma}$, $g_2(x) = x^\beta$, and $h_2(t) = Bt^\mu$.
  Solution:

$$y(x) = f(x) + \lambda(A_1 + A_2 x^\beta),$$

where A_1 and A_2 are the constants determined by the formulas presented in 4.9.18.

62. $y(x) - \lambda \int_a^b (Ax^\alpha t^\gamma + Bx^\beta t^\gamma + Cx^\mu t^\nu)y(t)\, dt = f(x).$

This is a special case of equation 4.9.18 with $g_1(x) = Ax^\alpha + Bx^\beta$, $h_1(t) = t^\gamma$, $g_2(x) = x^\mu$, and $h_2(t) = Ct^\nu$.
  Solution:

$$y(x) = f(x) + \lambda[A_1(Ax^\alpha + Bx^\beta) + A_2 x^\mu],$$

where A_1 and A_2 are the constants determined by the formulas presented in 4.9.18.

63. $y(x) - \lambda \int_a^b \left[A\dfrac{(x + p_1)^\beta}{(t + q_1)^\gamma} + B\dfrac{(x + p_2)^\mu}{(t + q_2)^\delta} \right] y(t)\, dt = f(x).$

This is a special case of equation 4.9.18 with $g_1(x) = (x + p_1)^\beta$, $h_1(t) = A(t + q_1)^{-\gamma}$, $g_2(x) = (x + p_2)^\mu$, and $h_2(t) = B(t + q_2)^{-\delta}$.
  Solution:

$$y(x) = f(x) + \lambda\left[A_1(x + p_1)^\beta + A_2(x + p_2)^\mu\right],$$

where A_1 and A_2 are the constants determined by the formulas presented in 4.9.18.

64. $y(x) - \lambda \int_a^b \left(A\dfrac{x^\mu + a}{t^\nu + b} + B\dfrac{x^\gamma + c}{t^\delta + d} \right) y(t)\, dt = f(x).$

This is a special case of equation 4.9.18 with $g_1(x) = x^\mu + a$, $h_1(t) = \dfrac{A}{t^\nu + b}$, $g_2(x) = x^\gamma + c$, and $h_2(t) = \dfrac{B}{t^\delta + d}$.
  Solution:

$$y(x) = f(x) + \lambda[A_1(x^\mu + a) + A_2(x^\gamma + c)],$$

where A_1 and A_2 are the constants determined by the formulas presented in 4.9.18.

4.1-7. Singular Equations

In this subsection, all singular integrals are understood in the sense of the Cauchy principal value.

65. $\quad Ay(x) + \dfrac{B}{\pi} \displaystyle\int_{-1}^{1} \dfrac{y(t)\,dt}{t-x} = f(x), \qquad -1 < x < 1.$

Without loss of generality we may assume that $A^2 + B^2 = 1$.

1°. The solution bounded at the endpoints:

$$y(x) = Af(x) - \frac{B}{\pi} \int_{-1}^{1} \frac{g(x)}{g(t)} \frac{f(t)\,dt}{t-x}, \qquad g(x) = (1+x)^\alpha (1-x)^{1-\alpha}, \tag{1}$$

where α is the solution of the trigonometric equation

$$A + B\cot(\pi\alpha) = 0 \tag{2}$$

on the interval $0 < \alpha < 1$. This solution $y(x)$ exists if and only if $\displaystyle\int_{-1}^{1} \frac{f(t)}{g(t)}\,dt = 0$.

2°. The solution bounded at the endpoint $x = 1$ and unbounded at the endpoint $x = -1$:

$$y(x) = Af(x) - \frac{B}{\pi} \int_{-1}^{1} \frac{g(x)}{g(t)} \frac{f(t)\,dt}{t-x}, \qquad g(x) = (1+x)^\alpha (1-x)^{-\alpha}, \tag{3}$$

where α is the solution of the trigonometric equation (2) on the interval $-1 < \alpha < 0$.

3°. The solution unbounded at the endpoints:

$$y(x) = Af(x) - \frac{B}{\pi} \int_{-1}^{1} \frac{g(x)}{g(t)} \frac{f(t)\,dt}{t-x} + Cg(x), \qquad g(x) = (1+x)^\alpha (1-x)^{-1-\alpha}, \tag{4}$$

where C is an arbitrary constant and α is the solution of the trigonometric equation (2) on the interval $-1 < \alpha < 0$.

⊙ Reference: I. K. Lifanov (1996).

66. $\quad y(x) - \lambda \displaystyle\int_{0}^{1} \left(\dfrac{1}{t-x} - \dfrac{1}{x+t-2xt} \right) y(t)\,dt = f(x), \quad 0 < x < 1.$

Tricomi's equation.
  Solution:

$$y(x) = \frac{1}{1 + \lambda^2 \pi^2} \left[f(x) + \int_{0}^{1} \frac{t^\alpha (1-x)^\alpha}{x^\alpha (1-t)^\alpha} \left(\frac{1}{t-x} - \frac{1}{x+t-2xt} \right) f(t)\,dt \right] + \frac{C(1-x)^\beta}{x^{1+\beta}},$$

$$\alpha = \frac{2}{\pi} \arctan(\lambda\pi) \ (-1 < \alpha < 1), \quad \tan\frac{\beta\pi}{2} = \lambda\pi \ (-2 < \beta < 0),$$

where C is an arbitrary constant.

⊙ References: P. P. Zabreyko, A. I. Koshelev, et al. (1975), F. G. Tricomi (1985).

4.2. Equations Whose Kernels Contain Exponential Functions

4.2-1. Kernels Containing Exponential Functions

1. $y(x) - \lambda \displaystyle\int_a^b (e^{\beta x} + e^{\beta t})y(t)\,dt = f(x).$

The characteristic values of the equation:

$$\lambda_{1,2} = \frac{\beta}{e^{\beta b} - e^{\beta a} \pm \sqrt{\frac{1}{2}\beta(b-a)(e^{2\beta b} - e^{2\beta a})}}.$$

1°. Solution with $\lambda \neq \lambda_{1,2}$:

$$y(x) = f(x) + \lambda(A_1 e^{\beta x} + A_2),$$

where the constants A_1 and A_2 are given by

$$A_1 = \frac{f_1 - \lambda\big[f_1\Delta_\beta - (b-a)f_2\big]}{\lambda^2\big[\Delta_\beta^2 - (b-a)\Delta_{2\beta}\big] - 2\lambda\Delta_\beta + 1}, \quad A_2 = \frac{f_2 - \lambda(f_2\Delta_\beta - f_1\Delta_{2\beta})}{\lambda^2\big[\Delta_\beta^2 - (b-a)\Delta_{2\beta}\big] - 2\lambda\Delta_\beta + 1},$$

$$f_1 = \int_a^b f(x)\,dx, \quad f_2 = \int_a^b f(x)e^{\beta x}\,dx, \quad \Delta_\beta = \frac{1}{\beta}(e^{\beta b} - e^{\beta a}).$$

2°. Solution with $\lambda = \lambda_1 \neq \lambda_2$ and $f_1 = f_2 = 0$:

$$y(x) = f(x) + Cy_1(x), \quad y_1(x) = e^{\beta x} + \sqrt{\frac{e^{2\beta b} - e^{2\beta a}}{2\beta(b-a)}},$$

where C is an arbitrary constant and $y_1(x)$ is an eigenfunction of the equation corresponding to the characteristic value λ_1.

3°. Solution with $\lambda = \lambda_2 \neq \lambda_1$ and $f_1 = f_2 = 0$:

$$y(x) = f(x) + Cy_2(x), \quad y_2(x) = e^{\beta x} - \sqrt{\frac{e^{2\beta b} - e^{2\beta a}}{2\beta(b-a)}},$$

where C is an arbitrary constant and $y_2(x)$ is an eigenfunction of the equation corresponding to the characteristic value λ_2.

4°. The equation has no multiple characteristic values.

2. $y(x) - \lambda \displaystyle\int_a^b (e^{\beta x} - e^{\beta t})y(t)\,dt = f(x).$

The characteristic values of the equation:

$$\lambda_{1,2} = \pm\frac{\beta}{\sqrt{(e^{\beta b} - e^{\beta a})^2 - \frac{1}{2}\beta(b-a)(e^{2\beta b} - e^{2\beta a})}}.$$

1°. Solution with $\lambda \neq \lambda_{1,2}$:

$$y(x) = f(x) + \lambda(A_1 e^{\beta x} + A_2),$$

where the constants A_1 and A_2 are given by

$$A_1 = \frac{f_1 + \lambda[f_1 \Delta_\beta - (b-a)f_2]}{\lambda^2[(b-a)\Delta_{2\beta} - \Delta_\beta^2] + 1}, \quad A_2 = \frac{-f_2 + \lambda(f_2 \Delta_\beta - f_1 \Delta_{2\beta})}{\lambda^2[(b-a)\Delta_{2\beta} - \Delta_\beta^2] + 1},$$

$$f_1 = \int_a^b f(x)\,dx, \quad f_2 = \int_a^b f(x)e^{\beta x}\,dx, \quad \Delta_\beta = \frac{1}{\beta}(e^{\beta b} - e^{\beta a}).$$

2°. Solution with $\lambda = \lambda_1 \neq \lambda_2$ and $f_1 = f_2 = 0$:

$$y(x) = f(x) + Cy_1(x), \qquad y_1(x) = e^{\beta x} + \frac{1 - \lambda_1 \Delta_\beta}{\lambda_1(b-a)},$$

where C is an arbitrary constant and $y_1(x)$ is an eigenfunction of the equation corresponding to the characteristic value λ_1.

3°. The solution with $\lambda = \lambda_2 \neq \lambda_1$ and $f_1 = f_2 = 0$ is given by the formulas of item 2° in which one must replace λ_1 and $y_1(x)$ by λ_2 and $y_2(x)$, respectively.

4°. The equation has no multiple characteristic values.

3. $\quad y(x) - \lambda \int_a^b (Ae^{\beta x} + Be^{\beta t})y(t)\,dt = f(x).$

The characteristic values of the equation:

$$\lambda_{1,2} = \frac{(A+B)\Delta_\beta \pm \sqrt{(A-B)^2\Delta_\beta^2 + 4AB(b-a)\Delta_{2\beta}}}{2AB[\Delta_\beta^2 - (b-a)\Delta_{2\beta}]}, \quad \Delta_\beta = \frac{1}{\beta}(e^{\beta b} - e^{\beta a}).$$

1°. Solution with $\lambda \neq \lambda_{1,2}$:

$$y(x) = f(x) + \lambda(A_1 e^{\beta x} + A_2),$$

where the constants A_1 and A_2 are given by

$$A_1 = \frac{Af_1 - AB\lambda[f_1 \Delta_\beta - (b-a)f_2]}{AB\lambda^2[\Delta_\beta^2 - (b-a)\Delta_{2\beta}] - (A+B)\lambda\Delta_\beta + 1},$$

$$A_2 = \frac{Bf_2 - AB\lambda(f_2 \Delta_\beta - f_1 \Delta_{2\beta})}{AB\lambda^2[\Delta_\beta^2 - (b-a)\Delta_{2\beta}] - (A+B)\lambda\Delta_\beta + 1},$$

$$f_1 = \int_a^b f(x)\,dx, \quad f_2 = \int_a^b f(x)e^{\beta x}\,dx.$$

2°. Solution with $\lambda = \lambda_1 \neq \lambda_2$ and $f_1 = f_2 = 0$:

$$y(x) = f(x) + Cy_1(x), \qquad y_1(x) = e^{\beta x} + \frac{1 - A\lambda_1 \Delta_\beta}{A(b-a)\lambda_1},$$

where C is an arbitrary constant and $y_1(x)$ is an eigenfunction of the equation corresponding to the characteristic value λ_1.

3°. The solution with $\lambda = \lambda_2 \neq \lambda_1$ and $f_1 = f_2 = 0$ is given by the formulas of item 2° in which one must replace λ_1 and $y_1(x)$ by λ_2 and $y_2(x)$, respectively.

4°. Solution with $\lambda = \lambda_{1,2} = \lambda_*$ and $f_1 = f_2 = 0$, where the characteristic value $\lambda_* = \dfrac{2}{(A+B)\Delta_\beta}$ is double:

$$y(x) = f(x) + Cy_*(x), \qquad y_*(x) = e^{\beta x} - \frac{(A-B)\Delta_\beta}{2A(b-a)},$$

where C is an arbitrary constant and $y_*(x)$ is an eigenfunction of the equation corresponding to λ_*.

4. $y(x) - \lambda \int_a^b \left[A e^{\beta(x-t)} + B \right] y(t)\, dt = f(x).$

This is a special case of equation 4.9.18 with $g_1(x) = e^{\beta x}$, $h_1(t) = A e^{-\beta t}$, $g_2(x) = 1$, and $h_2(t) = B$.

Solution:
$$y(x) = f(x) + \lambda(A_1 e^{\beta x} + A_2),$$

where A_1 and A_2 are the constants determined by the formulas presented in 4.9.18.

5. $y(x) - \lambda \int_a^b \left[A e^{\beta x + \mu t} + B e^{(\beta + \mu)t} \right] y(t)\, dt = f(x).$

This is a special case of equation 4.9.6 with $g(x) = e^{\beta x}$ and $h(t) = e^{\mu t}$.

Solution:
$$y(x) = f(x) + \lambda(A_1 e^{\beta x} + A_2),$$

where A_1 and A_2 are the constants determined by the formulas presented in 4.9.6.

6. $y(x) - \lambda \int_a^b \left[A e^{\alpha(x+t)} + B e^{\beta(x+t)} \right] y(t)\, dt = f(x).$

This is a special case of equation 4.9.14 with $g(x) = e^{\alpha x}$ and $h(t) = e^{\beta t}$.

Solution:
$$y(x) = f(x) + \lambda(A_1 e^{\alpha x} + A_2 e^{\beta x}),$$

where A_1 and A_2 are the constants determined by the formulas presented in 4.9.14.

7. $y(x) - \lambda \int_a^b \left(A e^{\alpha x + \beta t} + B e^{\beta x + \alpha t} \right) y(t)\, dt = f(x).$

This is a special case of equation 4.9.17 with $g(x) = e^{\alpha x}$ and $h(t) = e^{\beta t}$.

Solution:
$$y(x) = f(x) + \lambda(A_1 e^{\alpha x} + A_2 e^{\beta x}),$$

where A_1 and A_2 are the constants determined by the formulas presented in 4.9.17.

8. $y(x) - \lambda \int_a^b \left[D e^{(\gamma + \mu)x} + E e^{\nu t + \mu x} \right] y(t)\, dt = f(x).$

This is a special case of equation 4.9.18 with $g_1(x) = e^{(\gamma + \mu)x}$, $h_1(t) = D$, $g_2(x) = e^{\mu x}$, and $h_2(t) = E e^{\nu t}$.

Solution:
$$y(x) = f(x) + \lambda[A_1 e^{(\gamma + \mu)x} + A_2 e^{\mu x}],$$

where A_1 and A_2 are the constants determined by the formulas presented in 4.9.18.

9. $y(x) - \lambda \int_a^b \left(A e^{\alpha x + \beta t} + B e^{\gamma x + \delta t} \right) y(t)\, dt = f(x).$

This is a special case of equation 4.9.18 with $g_1(x) = e^{\alpha x}$, $h_1(t) = A e^{\beta t}$, $g_2(x) = e^{\gamma x}$, and $h_2(t) = B e^{\delta t}$.

Solution:
$$y(x) = f(x) + \lambda(A_1 e^{\alpha x} + A_2 e^{\gamma x}),$$

where A_1 and A_2 are the constants determined by the formulas presented in 4.9.18.

10. $y(x) - \lambda \int_a^b \left[\sum_{k=1}^n A_k e^{\gamma_k(x-t)} \right] y(t)\, dt = f(x).$

This is a special case of equation 4.9.20 with $g_k(x) = e^{\gamma_k x}$ and $h_k(t) = A_k e^{-\gamma_k t}$.

11. $y(x) - \dfrac{1}{2} \int_0^\infty e^{-|x-t|} y(t)\, dt = A e^{\mu x},\qquad 0 < \mu < 1.$

Solution:
$$y(x) = C(1 + x) + A\mu^{-2}\left[(\mu^2 - 1)e^{\mu x} - \mu + 1\right],$$

where C is an arbitrary constant.

⊙ Reference: P. P. Zabreyko, A. I. Koshelev, et al. (1975).

12. $y(x) + \lambda \int_0^\infty e^{-|x-t|} y(t)\, dt = f(x).$

Solution:
$$y(x) = f(x) - \frac{\lambda}{\sqrt{1+2\lambda}} \int_0^\infty \exp\left(-\sqrt{1+2\lambda}\,|x-t|\right) f(t)\, dt$$
$$+ \left(1 - \frac{\lambda+1}{\sqrt{1+2\lambda}}\right) \int_0^\infty \exp\left[-\sqrt{1+2\lambda}\,(x+t)\right] f(t)\, dt,$$

where $\lambda > -\frac{1}{2}$.

⊙ Reference: F. D. Gakhov and Yu. I. Cherskii (1978).

13. $y(x) - \lambda \int_{-\infty}^\infty e^{-|x-t|} y(t)\, dt = 0,\qquad \lambda > 0.$

The Lalesco–Picard equation.
Solution:
$$y(x) = \begin{cases} C_1 \exp\left(x\sqrt{1-2\lambda}\right) + C_2 \exp\left(-x\sqrt{1-2\lambda}\right) & \text{for } 0 < \lambda < \frac{1}{2}, \\ C_1 + C_2 x & \text{for } \lambda = \frac{1}{2}, \\ C_1 \cos\left(x\sqrt{2\lambda-1}\right) + C_2 \sin\left(x\sqrt{2\lambda-1}\right) & \text{for } \lambda > \frac{1}{2}, \end{cases}$$

where C_1 and C_2 are arbitrary constants.

⊙ Reference: M. L. Krasnov, A. I. Kisilev, and G. I. Makarenko (1971).

14. $y(x) + \lambda \int_{-\infty}^\infty e^{-|x-t|} y(t)\, dt = f(x).$

$1°$. Solution with $\lambda > -\frac{1}{2}$:

$$y(x) = f(x) - \frac{\lambda}{\sqrt{1+2\lambda}} \int_{-\infty}^\infty \exp\left(-\sqrt{1+2\lambda}\,|x-t|\right) f(t)\, dt.$$

$2°$. If $\lambda \le -\frac{1}{2}$, for the equation to be solvable the conditions

$$\int_{-\infty}^\infty f(x) \cos(ax)\, dx = 0, \qquad \int_{-\infty}^\infty f(x) \sin(ax)\, dx = 0,$$

where $a = \sqrt{-1 - 2\lambda}$, must be satisfied. In this case, the solution has the form

$$y(x) = f(x) - \frac{a^2 + 1}{2a} \int_0^\infty \sin(at) f(x + t)\, dt, \qquad (-\infty < x < \infty).$$

In the class of solutions not belonging to $L_2(-\infty, \infty)$, the homogeneous equation (with $f(x) \equiv 0$) has a nontrivial solution. In this case, the general solution of the corresponding nonhomogeneous equation with $\lambda \leq -\frac{1}{2}$ has the form

$$y(x) = C_1 \sin(ax) + C_2 \cos(ax) + f(x) - \frac{a^2 + 1}{4a} \int_{-\infty}^\infty \sin(a|x - t|) f(t)\, dt.$$

⊙ Reference: F. D. Gakhov and Yu. I. Cherskii (1978).

15. $\quad y(x) + A \displaystyle\int_a^b e^{\lambda|x-t|} y(t)\, dt = f(x).$

This is a special case of equation 4.9.37 with $g(t) = A$.

$1°$. The function $y = y(x)$ obeys the following second-order linear nonhomogeneous ordinary differential equation with constant coefficients:

$$y''_{xx} + \lambda(2A - \lambda)y = f''_{xx}(x) - \lambda^2 f(x). \tag{1}$$

The boundary conditions for (1) have the form (see 4.9.37)

$$\begin{aligned} y'_x(a) + \lambda y(a) &= f'_x(a) + \lambda f(a), \\ y'_x(b) - \lambda y(b) &= f'_x(b) - \lambda f(b). \end{aligned} \tag{2}$$

Equation (1) under the boundary conditions (2) determines the solution of the original integral equation.

$2°$. For $\lambda(2A - \lambda) < 0$, the general solution of equation (1) is given by

$$y(x) = C_1 \cosh(kx) + C_2 \sinh(kx) + f(x) - \frac{2A\lambda}{k} \int_a^x \sinh[k(x - t)]\, f(t)\, dt,$$
$$k = \sqrt{\lambda(\lambda - 2A)}, \tag{3}$$

where C_1 and C_2 are arbitrary constants.

For $\lambda(2A - \lambda) > 0$, the general solution of equation (1) is given by

$$y(x) = C_1 \cos(kx) + C_2 \sin(kx) + f(x) - \frac{2A\lambda}{k} \int_a^x \sin[k(x - t)]\, f(t)\, dt,$$
$$k = \sqrt{\lambda(2A - \lambda)}. \tag{4}$$

For $\lambda = 2A$, the general solution of equation (1) is given by

$$y(x) = C_1 + C_2 x + f(x) - 4A^2 \int_a^x (x - t) f(t)\, dt. \tag{5}$$

The constants C_1 and C_2 in solutions (3)–(5) are determined by conditions (2).

$3°$. In the special case $a = 0$ and $\lambda(2A - \lambda) > 0$, the solution of the integral equation is given by formula (4) with

$$C_1 = \frac{A(kI_c - \lambda I_s)}{(\lambda - A)\sin\mu - k\cos\mu}, \qquad C_2 = -\frac{\lambda}{k} \frac{A(kI_c - \lambda I_s)}{(\lambda - A)\sin\mu - k\cos\mu},$$

$$k = \sqrt{\lambda(2A - \lambda)}, \quad \mu = bk, \quad I_s = \int_0^b \sin[k(b - t)] f(t)\, dt, \quad I_c = \int_0^b \cos[k(b - t)] f(t)\, dt.$$

16. $y(x) + \int_a^b \left[\sum_{k=1}^n A_k \exp(\lambda_k |x - t|) \right] y(t)\, dt = f(x), \qquad -\infty < a < b < \infty.$

$1°$. Let us remove the modulus in the kth summand of the integrand:

$$I_k(x) = \int_a^b \exp(\lambda_k |x - t|) y(t)\, dt = \int_a^x \exp[\lambda_k (x - t)] y(t)\, dt + \int_x^b \exp[\lambda_k (t - x)] y(t)\, dt. \quad (1)$$

Differentiating (1) with respect to x twice yields

$$I_k' = \lambda_k \int_a^x \exp[\lambda_k (x - t)] y(t)\, dt - \lambda_k \int_x^b \exp[\lambda_k (t - x)] y(t)\, dt,$$

$$I_k'' = 2\lambda_k y(x) + \lambda_k^2 \int_a^x \exp[\lambda_k (x - t)] y(t)\, dt + \lambda_k^2 \int_x^b \exp[\lambda_k (t - x)] y(t)\, dt, \qquad (2)$$

where the primes denote the derivatives with respect to x. By comparing formulas (1) and (2), we find the relation between I_k'' and I_k:

$$I_k'' = 2\lambda_k y(x) + \lambda_k^2 I_k, \qquad I_k = I_k(x). \qquad (3)$$

$2°$. With the aid of (1), the integral equation can be rewritten in the form

$$y(x) + \sum_{k=1}^n A_k I_k = f(x). \qquad (4)$$

Differentiating (4) with respect to x twice and taking into account (3), we find that

$$y_{xx}''(x) + \sigma_n y(x) + \sum_{k=1}^n A_k \lambda_k^2 I_k = f_{xx}''(x), \qquad \sigma_n = 2 \sum_{k=1}^n A_k \lambda_k. \qquad (5)$$

Eliminating the integral I_n from (4) and (5) yields

$$y_{xx}''(x) + (\sigma_n - \lambda_n^2) y(x) + \sum_{k=1}^{n-1} A_k (\lambda_k^2 - \lambda_n^2) I_k = f_{xx}''(x) - \lambda_n^2 f(x). \qquad (6)$$

Differentiating (6) with respect to x twice and eliminating I_{n-1} from the resulting equation with the aid of (6), we obtain a similar equation whose left-hand side is a second-order linear differential operator (acting on y) with constant coefficients plus the sum $\sum_{k=1}^{n-2} B_k I_k$. If we successively eliminate I_{n-2}, I_{n-3}, ..., with the aid of double differentiation, then we finally arrive at a linear nonhomogeneous ordinary differential equation of order $2n$ with constant coefficients.

$3°$. The boundary conditions for $y(x)$ can be found by setting $x = a$ in the integral equation and all its derivatives. (Alternatively, these conditions can be found by setting $x = a$ and $x = b$ in the integral equation and all its derivatives obtained by means of double differentiation.)

4.2-2. Kernels Containing Power-Law and Exponential Functions

17. $y(x) - \lambda \int_a^b (x - t) e^{\gamma t} y(t)\, dt = f(x).$

This is a special case of equation 4.9.8 with $A = 0$, $B = 1$, and $h(t) = e^{\gamma t}$.

18. $y(x) - \lambda \int_a^b (x-t) e^{\gamma x} y(t)\, dt = f(x).$

This is a special case of equation 4.9.10 with $A = 0$, $B = 1$, and $h(x) = e^{\gamma x}$.

19. $y(x) - \lambda \int_a^b (x-t) e^{\gamma x + \mu t} y(t)\, dt = f(x).$

This is a special case of equation 4.9.18 with $g_1(x) = x e^{\gamma x}$, $h_1(t) = e^{\mu t}$, $g_2(x) = e^{\gamma x}$, and $h_2(t) = -t e^{\mu t}$.

20. $y(x) - \lambda \int_a^b [A + (Bx + Ct) e^{\gamma x}] y(t)\, dt = f(x).$

This is a special case of equation 4.9.11 with $h(x) = e^{\gamma x}$.

21. $y(x) - \lambda \int_0^b (x^2 + t^2) e^{\gamma(x+t)} y(t)\, dt = f(x).$

This is a special case of equation 4.9.15 with $g(x) = x^2 e^{\gamma x}$ and $h(t) = e^{\gamma t}$.

22. $y(x) - \lambda \int_0^b (x^2 - t^2) e^{\gamma(x-t)} y(t)\, dt = f(x).$

This is a special case of equation 4.9.18 with $g_1(x) = x^2 e^{\gamma x}$, $h_1(t) = e^{-\gamma t}$, $g_2(x) = e^{\gamma x}$, and $h_2(t) = -t^2 e^{-\gamma t}$.

23. $y(x) - \lambda \int_0^b (A x^n + B t^n) e^{\alpha x + \beta t} y(t)\, dt = f(x), \qquad n = 1, 2, \ldots$

This is a special case of equation 4.9.18 with $g_1(x) = x^n e^{\alpha x}$, $h_1(t) = A e^{\beta t}$, $g_2(x) = e^{\alpha x}$, and $h_2(t) = B t^n e^{\beta t}$.

24. $y(x) - \lambda \int_a^b \left[\sum_{k=1}^n A_k t^{\nu_k} e^{\alpha_k x + \beta_k t} \right] y(t)\, dt = f(x), \qquad n = 1, 2, \ldots$

This is a special case of equation 4.9.20 with $g_k(x) = e^{\alpha_k x}$ and $h_k(t) = A_k t^{\nu_k} e^{\beta_k t}$.

25. $y(x) - \lambda \int_a^b \left[\sum_{k=1}^n A_k x^{\nu_k} e^{\alpha_k x + \beta_k t} \right] y(t)\, dt = f(x), \qquad n = 1, 2, \ldots$

This is a special case of equation 4.9.20 with $g_k(x) = A_k x^{\nu_k} e^{\alpha_k x}$ and $h_k(t) = e^{\beta_k t}$.

26. $y(x) - \lambda \int_a^b (x-t)^n e^{\gamma(x-t)} y(t)\, dt = f(x), \qquad n = 1, 2, \ldots$

This is a special case of equation 4.9.20.

27. $y(x) - \lambda \int_a^b (x-t)^n e^{\alpha x + \beta t} y(t)\, dt = f(x), \qquad n = 1, 2, \ldots$

This is a special case of equation 4.9.20.

28. $y(x) - \lambda \int_a^b (Ax + Bt)^n e^{\alpha x + \beta t} y(t)\, dt = f(x), \qquad n = 1, 2, \ldots$

This is a special case of equation 4.9.20.

29. $\quad y(x) + A \displaystyle\int_a^b te^{\lambda|x-t|} y(t)\, dt = f(x).$

This is a special case of equation 4.9.37 with $g(t) = At$. The solution of the integral equation can be written via the Bessel functions (or modified Bessel functions) of order 1/3.

30. $\quad y(x) + \displaystyle\int_0^\infty (a + b|x - t|)\exp(-|x - t|)y(t)\, dt = f(x).$

Let the biquadratic polynomial $P(k) = k^4 + 2(a - b + 1)k^2 + 2a + 2b + 1$ have no real roots and let $k = \alpha + i\beta$ be a root of the equation $P(k) = 0$ such that $\alpha > 0$ and $\beta > 0$. In this case, the solution has the form

$$y(x) = f(x) + \rho \int_0^\infty \exp(-\beta|x - t|)\cos(\theta + \alpha|x - t|)f(t)\, dt$$

$$+ \frac{[\alpha + (\beta - 1)^2]^2}{4\alpha^2\beta} \int_0^\infty \exp[-\beta(x + t)]\cos[\alpha(x - t)]f(t)\, dt$$

$$+ \frac{R}{4\alpha^2} \int_0^\infty \exp[-\beta(x + t)]\cos[\psi + \alpha(x + t)]f(t)\, dt,$$

where the parameters ρ, θ, R, and ψ are determined from the system of algebraic equations obtained by separating real and imaginary parts in the relations

$$\rho e^{i\theta} = \frac{\mu}{\beta - i\alpha}, \qquad Re^{i\psi} = \frac{(\beta - 1 - i\alpha)^4}{8\alpha^2(\beta - i\alpha)}.$$

⊙ Reference: F. D. Gakhov and Yu. I. Cherskii (1978).

4.3. Equations Whose Kernels Contain Hyperbolic Functions

4.3-1. Kernels Containing Hyperbolic Cosine

1. $\quad y(x) - \lambda \displaystyle\int_a^b \cosh(\beta x)y(t)\, dt = f(x).$

This is a special case of equation 4.9.1 with $g(x) = \cosh(\beta x)$ and $h(t) = 1$.

2. $\quad y(x) - \lambda \displaystyle\int_a^b \cosh(\beta t)y(t)\, dt = f(x).$

This is a special case of equation 4.9.1 with $g(x) = 1$ and $h(t) = \cosh(\beta t)$.

3. $\quad y(x) - \lambda \displaystyle\int_a^b \cosh[\beta(x - t)]y(t)\, dt = f(x).$

This is a special case of equation 4.9.13 with $g(x) = \cosh(\beta x)$ and $h(t) = \sinh(\beta t)$.
    Solution:

$$y(x) = f(x) + \lambda\big[A_1 \cosh(\beta x) + A_2 \sinh(\beta x)\big],$$

where A_1 and A_2 are the constants determined by the formulas presented in 4.9.13.

4. $y(x) - \lambda \int_a^b \cosh[\beta(x + t)]y(t)\, dt = f(x).$

This is a special case of equation 4.9.12 with $g(x) = \cosh(\beta x)$ and $h(t) = \sinh(\beta t)$.
Solution:
$$y(x) = f(x) + \lambda\big[A_1 \cosh(\beta x) + A_2 \sinh(\beta x)\big],$$
where A_1 and A_2 are the constants determined by the formulas presented in 4.9.12.

5. $y(x) - \lambda \int_a^b \left\{ \sum_{k=1}^{n} A_k \cosh[\beta_k(x - t)] \right\} y(t)\, dt = f(x), \qquad n = 1, 2, \dots$

This is a special case of equation 4.9.20.

6. $y(x) - \lambda \int_a^b \dfrac{\cosh(\beta x)}{\cosh(\beta t)} y(t)\, dt = f(x).$

This is a special case of equation 4.9.1 with $g(x) = \cosh(\beta x)$ and $h(t) = \dfrac{1}{\cosh(\beta t)}$.

7. $y(x) - \lambda \int_a^b \dfrac{\cosh(\beta t)}{\cosh(\beta x)} y(t)\, dt = f(x).$

This is a special case of equation 4.9.1 with $g(x) = \dfrac{1}{\cosh(\beta x)}$ and $h(t) = \cosh(\beta t)$.

8. $y(x) - \lambda \int_a^b \cosh^k(\beta x) \cosh^m(\mu t) y(t)\, dt = f(x).$

This is a special case of equation 4.9.1 with $g(x) = \cosh^k(\beta x)$ and $h(t) = \cosh^m(\mu t)$.

9. $y(x) - \lambda \int_a^b t^k \cosh^m(\beta x) y(t)\, dt = f(x).$

This is a special case of equation 4.9.1 with $g(x) = \cosh^m(\beta x)$ and $h(t) = t^k$.

10. $y(x) - \lambda \int_a^b x^k \cosh^m(\beta t) y(t)\, dt = f(x).$

This is a special case of equation 4.9.1 with $g(x) = x^k$ and $h(t) = \cosh^m(\beta t)$.

11. $y(x) - \lambda \int_a^b [A + B(x - t) \cosh(\beta x)] y(t)\, dt = f(x).$

This is a special case of equation 4.9.10 with $h(x) = \cosh(\beta x)$.

12. $y(x) - \lambda \int_a^b [A + B(x - t) \cosh(\beta t)] y(t)\, dt = f(x).$

This is a special case of equation 4.9.8 with $h(t) = \cosh(\beta t)$.

13. $y(x) + \lambda \int_{-\infty}^{\infty} \dfrac{y(t)\, dt}{\cosh[b(x - t)]} = f(x).$

Solution with $b > \pi|\lambda|$:

$$y(x) = f(x) - \frac{2\lambda b}{\sqrt{b^2 - \pi^2\lambda^2}} \int_{-\infty}^{\infty} \frac{\sinh[2k(x - t)]}{\sinh[2b(x - t)]} f(t)\, dt, \qquad k = \frac{b}{\pi} \arccos\left(\frac{\pi\lambda}{b}\right).$$

⊙ Reference: F. D. Gakhov and Yu. I. Cherskii (1978).

4.3-2. Kernels Containing Hyperbolic Sine

14. $\quad y(x) - \lambda \displaystyle\int_a^b \sinh(\beta x) y(t)\, dt = f(x).$

This is a special case of equation 4.9.1 with $g(x) = \sinh(\beta x)$ and $h(t) = 1$.

15. $\quad y(x) - \lambda \displaystyle\int_a^b \sinh(\beta t) y(t)\, dt = f(x).$

This is a special case of equation 4.9.1 with $g(x) = 1$ and $h(t) = \sinh(\beta t)$.

16. $\quad y(x) - \lambda \displaystyle\int_a^b \sinh[\beta(x - t)] y(t)\, dt = f(x).$

This is a special case of equation 4.9.16 with $g(x) = \sinh(\beta x)$ and $h(t) = \cosh(\beta t)$.
Solution:
$$y(x) = f(x) + \lambda\big[A_1 \sinh(\beta x) + A_2 \cosh(\beta x)\big],$$
where A_1 and A_2 are the constants determined by the formulas presented in 4.9.16.

17. $\quad y(x) - \lambda \displaystyle\int_a^b \sinh[\beta(x + t)] y(t)\, dt = f(x).$

This is a special case of equation 4.9.15 with $g(x) = \sinh(\beta x)$ and $h(t) = \cosh(\beta t)$.
Solution:
$$y(x) = f(x) + \lambda\big[A_1 \sinh(\beta x) + A_2 \cosh(\beta x)\big],$$
where A_1 and A_2 are the constants determined by the formulas presented in 4.9.15.

18. $\quad y(x) - \lambda \displaystyle\int_a^b \left\{ \sum_{k=1}^n A_k \sinh[\beta_k(x - t)] \right\} y(t)\, dt = f(x), \qquad n = 1, 2, \ldots$

This is a special case of equation 4.9.20.

19. $\quad y(x) - \lambda \displaystyle\int_a^b \dfrac{\sinh(\beta x)}{\sinh(\beta t)} y(t)\, dt = f(x).$

This is a special case of equation 4.9.1 with $g(x) = \sinh(\beta x)$ and $h(t) = \dfrac{1}{\sinh(\beta t)}$.

20. $\quad y(x) - \lambda \displaystyle\int_a^b \dfrac{\sinh(\beta t)}{\sinh(\beta x)} y(t)\, dt = f(x).$

This is a special case of equation 4.9.1 with $g(x) = \dfrac{1}{\sinh(\beta x)}$ and $h(t) = \sinh(\beta t)$.

21. $\quad y(x) - \lambda \displaystyle\int_a^b \sinh^k(\beta x) \sinh^m(\mu t) y(t)\, dt = f(x).$

This is a special case of equation 4.9.1 with $g(x) = \sinh^k(\beta x)$ and $h(t) = \sinh^m(\mu t)$.

22. $\quad y(x) - \lambda \displaystyle\int_a^b t^k \sinh^m(\beta x) y(t)\, dt = f(x).$

This is a special case of equation 4.9.1 with $g(x) = \sinh^m(\beta x)$ and $h(t) = t^k$.

23. $y(x) - \lambda \int_a^b x^k \sinh^m(\beta t) y(t)\, dt = f(x).$

This is a special case of equation 4.9.1 with $g(x) = x^k$ and $h(t) = \sinh^m(\beta t)$.

24. $y(x) - \lambda \int_a^b [A + B(x - t)\sinh(\beta t)] y(t)\, dt = f(x).$

This is a special case of equation 4.9.8 with $h(t) = \sinh(\beta t)$.

25. $y(x) - \lambda \int_a^b [A + B(x - t)\sinh(\beta x)] y(t)\, dt = f(x).$

This is a special case of equation 4.9.10 with $h(x) = \sinh(\beta x)$.

26. $y(x) + A \int_a^b \sinh(\lambda|x - t|) y(t)\, dt = f(x).$

This is a special case of equation 4.9.38 with $g(t) = A$.

$1°$. The function $y = y(x)$ obeys the following second-order linear nonhomogeneous ordinary differential equation with constant coefficients:

$$y''_{xx} + \lambda(2A - \lambda)y = f''_{xx}(x) - \lambda^2 f(x). \tag{1}$$

The boundary conditions for (1) have the form (see 4.9.38)

$$\sinh[\lambda(b - a)]\varphi'_x(b) - \lambda \cosh[\lambda(b - a)]\varphi(b) = \lambda\varphi(a),$$
$$\sinh[\lambda(b - a)]\varphi'_x(a) + \lambda \cosh[\lambda(b - a)]\varphi(a) = -\lambda\varphi(b), \qquad \varphi(x) = y(x) - f(x). \tag{2}$$

Equation (1) under the boundary conditions (2) determines the solution of the original integral equation.

$2°$. For $\lambda(2A - \lambda) = -k^2 < 0$, the general solution of equation (1) is given by

$$y(x) = C_1 \cosh(kx) + C_2 \sinh(kx) + f(x) - \frac{2A\lambda}{k} \int_a^x \sinh[k(x - t)]f(t)\, dt, \tag{3}$$

where C_1 and C_2 are arbitrary constants.

For $\lambda(2A - \lambda) = k^2 > 0$, the general solution of equation (1) is given by

$$y(x) = C_1 \cos(kx) + C_2 \sin(kx) + f(x) - \frac{2A\lambda}{k} \int_a^x \sin[k(x - t)]f(t)\, dt. \tag{4}$$

For $\lambda = 2A$, the general solution of equation (1) is given by

$$y(x) = C_1 + C_2 x + f(x) - 4A^2 \int_a^x (x - t)f(t)\, dt. \tag{5}$$

The constants C_1 and C_2 in solutions (3)–(5) are determined by conditions (2).

27. $y(x) + A \int_a^b t \sinh(\lambda|x - t|) y(t)\, dt = f(x).$

This is a special case of equation 4.9.38 with $g(t) = At$. The solution of the integral equation can be written via the Bessel functions (or modified Bessel functions) of order 1/3.

28. $\quad y(x) + A \displaystyle\int_a^b \sinh^3(\lambda|x-t|)y(t)\,dt = f(x).$

Using the formula $\sinh^3 \beta = \frac{1}{4}\sinh 3\beta - \frac{3}{4}\sinh\beta$, we arrive at an equation of the form 4.3.29 with $n = 2$:

$$y(x) + \int_a^b \left[\tfrac{1}{4}A\sinh(3\lambda|x-t|) - \tfrac{3}{4}A\sinh(\lambda|x-t|)\right]y(t)\,dt = f(x).$$

29. $\quad y(x) + \displaystyle\int_a^b \left[\sum_{k=1}^n A_k \sinh(\lambda_k|x-t|)\right] y(t)\,dt = f(x), \qquad -\infty < a < b < \infty.$

$1°$. Let us remove the modulus in the kth summand of the integrand:

$$I_k(x) = \int_a^b \sinh(\lambda_k|x-t|)y(t)\,dt = \int_a^x \sinh[\lambda_k(x-t)]y(t)\,dt + \int_x^b \sinh[\lambda_k(t-x)]y(t)\,dt. \quad (1)$$

Differentiating (1) with respect to x twice yields

$$I_k' = \lambda_k \int_a^x \cosh[\lambda_k(x-t)]y(t)\,dt - \lambda_k \int_x^b \cosh[\lambda_k(t-x)]y(t)\,dt,$$

$$I_k'' = 2\lambda_k y(x) + \lambda_k^2 \int_a^x \sinh[\lambda_k(x-t)]y(t)\,dt + \lambda_k^2 \int_x^b \sinh[\lambda_k(t-x)]y(t)\,dt, \qquad (2)$$

where the primes denote the derivatives with respect to x. By comparing formulas (1) and (2), we find the relation between I_k'' and I_k:

$$I_k'' = 2\lambda_k y(x) + \lambda_k^2 I_k, \qquad I_k = I_k(x). \qquad (3)$$

$2°$. With the aid of (1), the integral equation can be rewritten in the form

$$y(x) + \sum_{k=1}^n A_k I_k = f(x). \qquad (4)$$

Differentiating (4) with respect to x twice and taking into account (3), we find that

$$y_{xx}''(x) + \sigma_n y(x) + \sum_{k=1}^n A_k \lambda_k^2 I_k = f_{xx}''(x), \qquad \sigma_n = 2\sum_{k=1}^n A_k \lambda_k. \qquad (5)$$

Eliminating the integral I_n from (4) and (5) yields

$$y_{xx}''(x) + (\sigma_n - \lambda_n^2)y(x) + \sum_{k=1}^{n-1} A_k(\lambda_k^2 - \lambda_n^2)I_k = f_{xx}''(x) - \lambda_n^2 f(x). \qquad (6)$$

Differentiating (6) with respect to x twice and eliminating I_{n-1} from the resulting equation with the aid of (6), we obtain a similar equation whose left-hand side is a second-order linear differential operator (acting on y) with constant coefficients plus the sum $\sum_{k=1}^{n-2} B_k I_k$. If we successively eliminate $I_{n-2}, I_{n-3}, \ldots$, with the aid of double differentiation, then we finally arrive at a linear nonhomogeneous ordinary differential equation of order $2n$ with constant coefficients.

$3°$. The boundary conditions for $y(x)$ can be found by setting $x = a$ in the integral equation and its derivatives. (Alternatively, these conditions can be found by setting $x = a$ and $x = b$ in the integral equation and all its derivatives obtained by means of double differentiation.)

4.3-3. Kernels Containing Hyperbolic Tangent

30. $y(x) - \lambda \displaystyle\int_a^b \tanh(\beta x) y(t)\, dt = f(x).$

This is a special case of equation 4.9.1 with $g(x) = \tanh(\beta x)$ and $h(t) = 1$.

31. $y(x) - \lambda \displaystyle\int_a^b \tanh(\beta t) y(t)\, dt = f(x).$

This is a special case of equation 4.9.1 with $g(x) = 1$ and $h(t) = \tanh(\beta t)$.

32. $y(x) - \lambda \displaystyle\int_a^b [A \tanh(\beta x) + B \tanh(\beta t)] y(t)\, dt = f(x).$

This is a special case of equation 4.9.4 with $g(x) = \tanh(\beta x)$.

33. $y(x) - \lambda \displaystyle\int_a^b \dfrac{\tanh(\beta x)}{\tanh(\beta t)} y(t)\, dt = f(x).$

This is a special case of equation 4.9.1 with $g(x) = \tanh(\beta x)$ and $h(t) = \dfrac{1}{\tanh(\beta t)}$.

34. $y(x) - \lambda \displaystyle\int_a^b \dfrac{\tanh(\beta t)}{\tanh(\beta x)} y(t)\, dt = f(x).$

This is a special case of equation 4.9.1 with $g(x) = \dfrac{1}{\tanh(\beta x)}$ and $h(t) = \tanh(\beta t)$.

35. $y(x) - \lambda \displaystyle\int_a^b \tanh^k(\beta x) \tanh^m(\mu t) y(t)\, dt = f(x).$

This is a special case of equation 4.9.1 with $g(x) = \tanh^k(\beta x)$ and $h(t) = \tanh^m(\mu t)$.

36. $y(x) - \lambda \displaystyle\int_a^b t^k \tanh^m(\beta x) y(t)\, dt = f(x).$

This is a special case of equation 4.9.1 with $g(x) = \tanh^m(\beta x)$ and $h(t) = t^k$.

37. $y(x) - \lambda \displaystyle\int_a^b x^k \tanh^m(\beta t) y(t)\, dt = f(x).$

This is a special case of equation 4.9.1 with $g(x) = x^k$ and $h(t) = \tanh^m(\beta t)$.

38. $y(x) - \lambda \displaystyle\int_a^b [A + B(x - t) \tanh(\beta t)] y(t)\, dt = f(x).$

This is a special case of equation 4.9.8 with $h(t) = \tanh(\beta t)$.

39. $y(x) - \lambda \displaystyle\int_a^b [A + B(x - t) \tanh(\beta x)] y(t)\, dt = f(x).$

This is a special case of equation 4.9.10 with $h(x) = \tanh(\beta x)$.

4.3-4. Kernels Containing Hyperbolic Cotangent

40. $y(x) - \lambda \int_a^b \coth(\beta x) y(t)\, dt = f(x).$

This is a special case of equation 4.9.1 with $g(x) = \coth(\beta x)$ and $h(t) = 1$.

41. $y(x) - \lambda \int_a^b \coth(\beta t) y(t)\, dt = f(x).$

This is a special case of equation 4.9.1 with $g(x) = 1$ and $h(t) = \coth(\beta t)$.

42. $y(x) - \lambda \int_a^b [A \coth(\beta x) + B \coth(\beta t)] y(t)\, dt = f(x).$

This is a special case of equation 4.9.4 with $g(x) = \coth(\beta x)$.

43. $y(x) - \lambda \int_a^b \dfrac{\coth(\beta x)}{\coth(\beta t)} y(t)\, dt = f(x).$

This is a special case of equation 4.9.1 with $g(x) = \coth(\beta x)$ and $h(t) = \dfrac{1}{\coth(\beta t)}$.

44. $y(x) - \lambda \int_a^b \dfrac{\coth(\beta t)}{\coth(\beta x)} y(t)\, dt = f(x).$

This is a special case of equation 4.9.1 with $g(x) = \dfrac{1}{\coth(\beta x)}$ and $h(t) = \coth(\beta t)$.

45. $y(x) - \lambda \int_a^b \coth^k(\beta x) \coth^m(\mu t) y(t)\, dt = f(x).$

This is a special case of equation 4.9.1 with $g(x) = \coth^k(\beta x)$ and $h(t) = \coth^m(\mu t)$.

46. $y(x) - \lambda \int_a^b t^k \coth^m(\beta x) y(t)\, dt = f(x).$

This is a special case of equation 4.9.1 with $g(x) = \coth^m(\beta x)$ and $h(t) = t^k$.

47. $y(x) - \lambda \int_a^b x^k \coth^m(\beta t) y(t)\, dt = f(x).$

This is a special case of equation 4.9.1 with $g(x) = x^k$ and $h(t) = \coth^m(\beta t)$.

48. $y(x) - \lambda \int_a^b [A + B(x - t) \coth(\beta t)] y(t)\, dt = f(x).$

This is a special case of equation 4.9.8 with $h(t) = \coth(\beta t)$.

49. $y(x) - \lambda \int_a^b [A + B(x - t) \coth(\beta x)] y(t)\, dt = f(x).$

This is a special case of equation 4.9.10 with $h(x) = \coth(\beta x)$.

4.3-5. Kernels Containing Combination of Hyperbolic Functions

50. $y(x) - \lambda \int_a^b \cosh^k(\beta x) \sinh^m(\mu t) y(t) \, dt = f(x).$

This is a special case of equation 4.9.1 with $g(x) = \cosh^k(\beta x)$ and $h(t) = \sinh^m(\mu t)$.

51. $y(x) - \lambda \int_a^b [A \sinh(\alpha x) \cosh(\beta t) + B \sinh(\gamma x) \cosh(\delta t)] y(t) \, dt = f(x).$

This is a special case of equation 4.9.18 with $g_1(x) = \sinh(\alpha x)$, $h_1(t) = A \cosh(\beta t)$, $g_2(x) = \sinh(\gamma x)$, and $h_2(t) = B \cosh(\delta t)$.

52. $y(x) - \lambda \int_a^b \tanh^k(\gamma x) \coth^m(\mu t) y(t) \, dt = f(x).$

This is a special case of equation 4.9.1 with $g(x) = \tanh^k(\gamma x)$ and $h(t) = \coth^m(\mu t)$.

53. $y(x) - \lambda \int_a^b [A \tanh(\alpha x) \coth(\beta t) + B \tanh(\gamma x) \coth(\delta t)] y(t) \, dt = f(x).$

This is a special case of equation 4.9.18 with $g_1(x) = \tanh(\alpha x)$, $h_1(t) = A \coth(\beta t)$, $g_2(x) = \tanh(\gamma x)$, and $h_2(t) = B \coth(\delta t)$.

4.4. Equations Whose Kernels Contain Logarithmic Functions

4.4-1. Kernels Containing Logarithmic Functions

1. $y(x) - \lambda \int_a^b \ln(\gamma x) y(t) \, dt = f(x).$

This is a special case of equation 4.9.1 with $g(x) = \ln(\gamma x)$ and $h(t) = 1$.

2. $y(x) - \lambda \int_a^b \ln(\gamma t) y(t) \, dt = f(x).$

This is a special case of equation 4.9.1 with $g(x) = 1$ and $h(t) = \ln(\gamma t)$.

3. $y(x) - \lambda \int_a^b (\ln x - \ln t) y(t) \, dt = f(x).$

This is a special case of equation 4.9.3 with $g(x) = \ln x$.

4. $y(x) - \lambda \int_a^b \frac{\ln(\gamma x)}{\ln(\gamma t)} y(t) \, dt = f(x).$

This is a special case of equation 4.9.1 with $g(x) = \ln(\gamma x)$ and $h(t) = \dfrac{1}{\ln(\gamma t)}$.

5. $y(x) - \lambda \int_a^b \frac{\ln(\gamma t)}{\ln(\gamma x)} y(t) \, dt = f(x).$

This is a special case of equation 4.9.1 with $g(x) = \dfrac{1}{\ln(\gamma x)}$ and $h(t) = \ln(\gamma t)$.

6. $y(x) - \lambda \int_a^b \ln^k(\gamma x) \ln^m(\mu t) y(t)\, dt = f(x).$

This is a special case of equation 4.9.1 with $g(x) = \ln^k(\gamma x)$ and $h(t) = \ln^m(\mu t)$.

4.4-2. Kernels Containing Power-Law and Logarithmic Functions

7. $y(x) - \lambda \int_a^b t^k \ln^m(\gamma x) y(t)\, dt = f(x).$

This is a special case of equation 4.9.1 with $g(x) = \ln^m(\gamma x)$ and $h(t) = t^k$.

8. $y(x) - \lambda \int_a^b x^k \ln^m(\gamma t) y(t)\, dt = f(x).$

This is a special case of equation 4.9.1 with $g(x) = x^k$ and $h(t) = \ln^m(\gamma t)$.

9. $y(x) - \lambda \int_a^b [A + B(x - t) \ln(\gamma t)] y(t)\, dt = f(x).$

This is a special case of equation 4.9.8 with $h(t) = \ln(\gamma t)$.

10. $y(x) - \lambda \int_a^b [A + B(x - t) \ln(\gamma x)] y(t)\, dt = f(x).$

This is a special case of equation 4.9.10 with $h(x) = \ln(\gamma x)$.

11. $y(x) - \lambda \int_a^b [A + (Bx + Ct) \ln(\gamma t)] y(t)\, dt = f(x).$

This is a special case of equation 4.9.9 with $h(t) = \ln(\gamma t)$.

12. $y(x) - \lambda \int_a^b [A + (Bx + Ct) \ln(\gamma x)] y(t)\, dt = f(x).$

This is a special case of equation 4.9.11 with $h(x) = \ln(\gamma x)$.

13. $y(x) - \lambda \int_a^b [At^n \ln^m(\beta x) + Bx^k \ln^l(\gamma t)] y(t)\, dt = f(x).$

This is a special case of equation 4.9.18 with $g_1(x) = \ln^m(\beta x)$, $h_1(t) = At^n$, $g_2(x) = x^k$, and $h_2(t) = B \ln^l(\gamma t)$.

4.5. Equations Whose Kernels Contain Trigonometric Functions

4.5-1. Kernels Containing Cosine

1. $y(x) - \lambda \int_a^b \cos(\beta x) y(t)\, dt = f(x).$

This is a special case of equation 4.9.1 with $g(x) = \cos(\beta x)$ and $h(t) = 1$.

2. $y(x) - \lambda \int_a^b \cos(\beta t) y(t)\, dt = f(x).$

This is a special case of equation 4.9.1 with $g(x) = 1$ and $h(t) = \cos(\beta t)$.

3. $y(x) - \lambda \int_a^b \cos[\beta(x - t)] y(t)\, dt = f(x).$

This is a special case of equation 4.9.12 with $g(x) = \cos(\beta x)$ and $h(t) = \sin(\beta t)$.
    Solution:

$$y(x) = f(x) + \lambda \big[A_1 \cos(\beta x) + A_2 \sin(\beta x) \big],$$

where A_1 and A_2 are the constants determined by the formulas presented in 4.9.12.

4. $y(x) - \lambda \int_a^b \cos[\beta(x + t)] y(t)\, dt = f(x).$

This is a special case of equation 4.9.13 with $g(x) = \cos(\beta x)$ and $h(t) = \sin(\beta t)$.
    Solution:

$$y(x) = f(x) + \lambda \big[A_1 \cos(\beta x) + A_2 \sin(\beta x) \big],$$

where A_1 and A_2 are the constants determined by the formulas presented in 4.9.13.

5. $y(x) - \lambda \int_0^\infty \cos(xt) y(t)\, dt = 0.$

Characteristic values: $\lambda = \pm\sqrt{2/\pi}$. For the characteristic values, the integral equation has infinitely many linearly independent eigenfunctions.
    Eigenfunctions for $\lambda = +\sqrt{2/\pi}$ have the form

$$y_+(x) = f(x) + \sqrt{\frac{2}{\pi}} \int_0^\infty f(t) \cos(xt)\, dt, \tag{1}$$

where $f = f(x)$ is any continuous function absolutely integrable on the interval $[0, \infty)$.
    Eigenfunctions for $\lambda = -\sqrt{2/\pi}$ have the form

$$y_-(x) = f(x) - \sqrt{\frac{2}{\pi}} \int_0^\infty f(t) \cos(xt)\, dt, \tag{2}$$

where $f = f(x)$ is any continuous function absolutely integrable on the interval $[0, \infty)$.
    In particular, from (1) and (2) with $f(x) = e^{-ax}$ we obtain

$$y_+(x) = e^{-ax} + \sqrt{\frac{2}{\pi}} \frac{a}{a^2 + x^2} \qquad \text{for} \quad \lambda = +\sqrt{\frac{2}{\pi}},$$

$$y_-(x) = e^{-ax} - \sqrt{\frac{2}{\pi}} \frac{a}{a^2 + x^2} \qquad \text{for} \quad \lambda = -\sqrt{\frac{2}{\pi}},$$

where a is any positive number.

⊙ Reference: M. L. Krasnov, A. I. Kisilev, and G. I. Makarenko (1971).

6. $y(x) - \lambda \displaystyle\int_0^\infty \cos(xt) y(t)\, dt = f(x).$

Solution:

$$y(x) = \frac{f(x)}{1 - \frac{\pi}{2}\lambda^2} + \frac{\lambda}{1 - \frac{\pi}{2}\lambda^2} \int_0^\infty \cos(xt) f(t)\, dt,$$

where $\lambda \neq \pm\sqrt{2/\pi}$.

⊙ Reference: M. L. Krasnov, A. I. Kisilev, and G. I. Makarenko (1971).

7. $y(x) - \lambda \displaystyle\int_a^b \left\{ \sum_{k=1}^{n} A_k \cos[\beta_k(x - t)] \right\} y(t)\, dt = f(x), \qquad n = 1, 2, \ldots$

This equation can be reduced to a special case of equation 4.9.20; the formula $\cos[\beta(x-t)] = \cos(\beta x)\cos(\beta t) + \sin(\beta x)\sin(\beta t)$ must be used.

8. $y(x) - \lambda \displaystyle\int_a^b \frac{\cos(\beta x)}{\cos(\beta t)} y(t)\, dt = f(x).$

This is a special case of equation 4.9.1 with $g(x) = \cos(\beta x)$ and $h(t) = \dfrac{1}{\cos(\beta t)}$.

9. $y(x) - \lambda \displaystyle\int_a^b \frac{\cos(\beta t)}{\cos(\beta x)} y(t)\, dt = f(x).$

This is a special case of equation 4.9.1 with $g(x) = \dfrac{1}{\cos(\beta x)}$ and $h(t) = \cos(\beta t)$.

10. $y(x) - \lambda \displaystyle\int_a^b \cos^k(\beta x) \cos^m(\mu t) y(t)\, dt = f(x).$

This is a special case of equation 4.9.1 with $g(x) = \cos^k(\beta x)$ and $h(t) = \cos^m(\mu t)$.

11. $y(x) - \lambda \displaystyle\int_a^b t^k \cos^m(\beta x) y(t)\, dt = f(x).$

This is a special case of equation 4.9.1 with $g(x) = \cos^m(\beta x)$ and $h(t) = t^k$.

12. $y(x) - \lambda \displaystyle\int_a^b x^k \cos^m(\beta t) y(t)\, dt = f(x).$

This is a special case of equation 4.9.1 with $g(x) = x^k$ and $h(t) = \cos^m(\beta t)$.

13. $y(x) - \lambda \displaystyle\int_a^b [A + B(x - t) \cos(\beta x)] y(t)\, dt = f(x).$

This is a special case of equation 4.9.10 with $h(x) = \cos(\beta x)$.

14. $y(x) - \lambda \displaystyle\int_a^b [A + B(x - t) \cos(\beta t)] y(t)\, dt = f(x).$

This is a special case of equation 4.9.8 with $h(t) = \cos(\beta t)$.

4.5-2. Kernels Containing Sine

15. $y(x) - \lambda \int_a^b \sin(\beta x) y(t)\, dt = f(x).$

This is a special case of equation 4.9.1 with $g(x) = \sin(\beta x)$ and $h(t) = 1$.

16. $y(x) - \lambda \int_a^b \sin(\beta t) y(t)\, dt = f(x).$

This is a special case of equation 4.9.1 with $g(x) = 1$ and $h(t) = \sin(\beta t)$.

17. $y(x) - \lambda \int_a^b \sin[\beta(x - t)] y(t)\, dt = f(x).$

This is a special case of equation 4.9.16 with $g(x) = \sin(\beta x)$ and $h(t) = \cos(\beta t)$.
     Solution:

$$y(x) = f(x) + \lambda \big[A_1 \sin(\beta x) + A_2 \cos(\beta x) \big],$$

where A_1 and A_2 are the constants determined by the formulas presented in 4.9.16.

18. $y(x) - \lambda \int_a^b \sin[\beta(x + t)] y(t)\, dt = f(x).$

This is a special case of equation 4.9.15 with $g(x) = \sin(\beta x)$ and $h(t) = \cos(\beta t)$.
     Solution:

$$y(x) = f(x) + \lambda \big[A_1 \sin(\beta x) + A_2 \cos(\beta x) \big],$$

where A_1 and A_2 are the constants determined by the formulas presented in 4.9.15.

19. $y(x) - \lambda \int_0^\infty \sin(xt) y(t)\, dt = 0.$

Characteristic values: $\lambda = \pm\sqrt{2/\pi}$. For the characteristic values, the integral equation has infinitely many linearly independent eigenfunctions.
     Eigenfunctions for $\lambda = +\sqrt{2/\pi}$ have the form

$$y_+(x) = f(x) + \sqrt{\frac{2}{\pi}} \int_0^\infty f(t) \sin(xt)\, dt,$$

where $f = f(x)$ is any continuous function absolutely integrable on the interval $[0, \infty)$.
     Eigenfunctions for $\lambda = -\sqrt{2/\pi}$ have the form

$$y_-(x) = f(x) - \sqrt{\frac{2}{\pi}} \int_0^\infty f(t) \sin(xt)\, dt,$$

where $f = f(x)$ is any continuous function absolutely integrable on the interval $[0, \infty)$.

   ⊙   Reference: M. L. Krasnov, A. I. Kisilev, and G. I. Makarenko (1971).

20. $y(x) - \lambda \int_0^\infty \sin(xt) y(t)\, dt = f(x).$

Solution:

$$y(x) = \frac{f(x)}{1 - \frac{\pi}{2}\lambda^2} + \frac{\lambda}{1 - \frac{\pi}{2}\lambda^2} \int_0^\infty \sin(xt) f(t)\, dt,$$

where $\lambda \neq \pm\sqrt{2/\pi}$.

   ⊙   References: M. L. Krasnov, A. I. Kisilev, and G. I. Makarenko (1971), F. D. Gakhov and Yu. I. Cherskii (1978).

21. $y(x) - \lambda \int_a^b \left\{ \sum_{k=1}^n A_k \sin[\beta_k(x-t)] \right\} y(t)\, dt = f(x), \qquad n = 1, 2, \dots$

This equation can be reduced to a special case of equation 4.9.20; the formula $\sin[\beta(x-t)] = \sin(\beta x)\cos(\beta t) - \sin(\beta t)\cos(\beta x)$ must be used.

22. $y(x) - \lambda \int_a^b \dfrac{\sin(\beta x)}{\sin(\beta t)} y(t)\, dt = f(x).$

This is a special case of equation 4.9.1 with $g(x) = \sin(\beta x)$ and $h(t) = \dfrac{1}{\sin(\beta t)}$.

23. $y(x) - \lambda \int_a^b \dfrac{\sin(\beta t)}{\sin(\beta x)} y(t)\, dt = f(x).$

This is a special case of equation 4.9.1 with $g(x) = \dfrac{1}{\sin(\beta x)}$ and $h(t) = \sin(\beta t)$.

24. $y(x) - \lambda \int_a^b \sin^k(\beta x) \sin^m(\mu t) y(t)\, dt = f(x).$

This is a special case of equation 4.9.1 with $g(x) = \sin^k(\beta x)$ and $h(t) = \sin^m(\mu t)$.

25. $y(x) - \lambda \int_a^b t^k \sin^m(\beta x) y(t)\, dt = f(x).$

This is a special case of equation 4.9.1 with $g(x) = \sin^m(\beta x)$ and $h(t) = t^k$.

26. $y(x) - \lambda \int_a^b x^k \sin^m(\beta t) y(t)\, dt = f(x).$

This is a special case of equation 4.9.1 with $g(x) = x^k$ and $h(t) = \sin^m(\beta t)$.

27. $y(x) - \lambda \int_a^b [A + B(x-t)\sin(\beta t)] y(t)\, dt = f(x).$

This is a special case of equation 4.9.8 with $h(t) = \sin(\beta t)$.

28. $y(x) - \lambda \int_a^b [A + B(x-t)\sin(\beta x)] y(t)\, dt = f(x).$

This is a special case of equation 4.9.10 with $h(x) = \sin(\beta x)$.

29. $y(x) + A \int_a^b \sin(\lambda|x-t|) y(t)\, dt = f(x).$

This is a special case of equation 4.9.39 with $g(t) = A$.

$1°$. The function $y = y(x)$ obeys the following second-order linear nonhomogeneous ordinary differential equation with constant coefficients:

$$y''_{xx} + \lambda(2A + \lambda)y = f''_{xx}(x) + \lambda^2 f(x). \qquad (1)$$

The boundary conditions for (1) have the form (see 4.9.39)

$$\sin[\lambda(b-a)]\varphi'_x(b) - \lambda\cos[\lambda(b-a)]\varphi(b) = \lambda\varphi(a),$$
$$\sin[\lambda(b-a)]\varphi'_x(a) + \lambda\cos[\lambda(b-a)]\varphi(a) = -\lambda\varphi(b), \qquad \varphi(x) = y(x) - f(x). \qquad (2)$$

Equation (1) under the boundary conditions (2) determines the solution of the original integral equation.

$2°$. For $\lambda(2A + \lambda) = -k^2 < 0$, the general solution of equation (1) is given by

$$y(x) = C_1 \cosh(kx) + C_2 \sinh(kx) + f(x) - \frac{2A\lambda}{k} \int_a^x \sinh[k(x - t)]\, f(t)\, dt, \qquad (3)$$

where C_1 and C_2 are arbitrary constants.

For $\lambda(2A + \lambda) = k^2 > 0$, the general solution of equation (1) is given by

$$y(x) = C_1 \cos(kx) + C_2 \sin(kx) + f(x) - \frac{2A\lambda}{k} \int_a^x \sin[k(x - t)]\, f(t)\, dt. \qquad (4)$$

For $\lambda = 2A$, the general solution of equation (1) is given by

$$y(x) = C_1 + C_2 x + f(x) + 4A^2 \int_a^x (x - t) f(t)\, dt. \qquad (5)$$

The constants C_1 and C_2 in solutions (3)–(5) are determined by conditions (2).

30. $\quad y(x) + A \displaystyle\int_a^b t \sin(\lambda|x - t|) y(t)\, dt = f(x).$

This is a special case of equation 4.9.39 with $g(t) = At$. The solution of the integral equation can be written via the Bessel functions (or modified Bessel functions) of order 1/3.

31. $\quad y(x) + A \displaystyle\int_a^b \sin^3(\lambda|x - t|) y(t)\, dt = f(x).$

Using the formula $\sin^3 \beta = -\frac{1}{4} \sin 3\beta + \frac{3}{4} \sin \beta$, we arrive at an equation of the form 4.5.32 with $n = 2$:

$$y(x) + \int_a^b \left[-\tfrac{1}{4} A \sin(3\lambda|x - t|) + \tfrac{3}{4} A \sin(\lambda|x - t|)\right] y(t)\, dt = f(x).$$

32. $\quad y(x) + \displaystyle\int_a^b \left[\sum_{k=1}^n A_k \sin(\lambda_k|x - t|)\right] y(t)\, dt = f(x), \qquad -\infty < a < b < \infty.$

$1°$. Let us remove the modulus in the kth summand of the integrand:

$$I_k(x) = \int_a^b \sin(\lambda_k|x - t|) y(t)\, dt = \int_a^x \sin[\lambda_k(x - t)] y(t)\, dt + \int_x^b \sin[\lambda_k(t - x)] y(t)\, dt. \quad (1)$$

Differentiating (1) with respect to x twice yields

$$I_k' = \lambda_k \int_a^x \cos[\lambda_k(x - t)] y(t)\, dt - \lambda_k \int_x^b \cos[\lambda_k(t - x)] y(t)\, dt,$$

$$I_k'' = 2\lambda_k y(x) - \lambda_k^2 \int_a^x \sin[\lambda_k(x - t)] y(t)\, dt - \lambda_k^2 \int_x^b \sin[\lambda_k(t - x)] y(t)\, dt, \qquad (2)$$

where the primes denote the derivatives with respect to x. By comparing formulas (1) and (2), we find the relation between I_k'' and I_k:

$$I_k'' = 2\lambda_k y(x) - \lambda_k^2 I_k, \qquad I_k = I_k(x). \qquad (3)$$

$2°$. With the aid of (1), the integral equation can be rewritten in the form

$$y(x) + \sum_{k=1}^{n} A_k I_k = f(x).$$ (4)

Differentiating (4) with respect to x twice and taking into account (3), we find that

$$y''_{xx}(x) + \sigma_n y(x) - \sum_{k=1}^{n} A_k \lambda_k^2 I_k = f''_{xx}(x), \qquad \sigma_n = 2 \sum_{k=1}^{n} A_k \lambda_k.$$ (5)

Eliminating the integral I_n from (4) and (5) yields

$$y''_{xx}(x) + (\sigma_n + \lambda_n^2)y(x) + \sum_{k=1}^{n-1} A_k(\lambda_n^2 - \lambda_k^2)I_k = f''_{xx}(x) + \lambda_n^2 f(x).$$ (6)

Differentiating (6) with respect to x twice and eliminating I_{n-1} from the resulting equation with the aid of (6), we obtain a similar equation whose left-hand side is a second-order linear differential operator (acting on y) with constant coefficients plus the sum $\sum_{k=1}^{n-2} B_k I_k$. If we successively eliminate $I_{n-2}, I_{n-3}, \ldots$, with the aid of double differentiation, then we finally arrive at a linear nonhomogeneous ordinary differential equation of order $2n$ with constant coefficients.

$3°$. The boundary conditions for $y(x)$ can be found by setting $x = a$ in the integral equation and all its derivatives. (Alternatively, these conditions can be found by setting $x = a$ and $x = b$ in the integral equation and all its derivatives obtained by means of double differentiation.)

33. $\quad y(x) - \lambda \int_{-\infty}^{\infty} \dfrac{\sin(x-t)}{x-t} y(t)\, dt = f(x).$

Solution:

$$y(x) = f(x) + \frac{\lambda}{\sqrt{2\pi} - \pi\lambda} \int_{-\infty}^{\infty} \frac{\sin(x-t)}{x-t} f(t)\, dt, \qquad \lambda \ne \sqrt{\frac{2}{\pi}}.$$

⊙ Reference: F. D. Gakhov and Yu. I. Cherskii (1978).

4.5-3. Kernels Containing Tangent

34. $\quad y(x) - \lambda \int_{a}^{b} \tan(\beta x) y(t)\, dt = f(x).$

This is a special case of equation 4.9.1 with $g(x) = \tan(\beta x)$ and $h(t) = 1$.

35. $\quad y(x) - \lambda \int_{a}^{b} \tan(\beta t) y(t)\, dt = f(x).$

This is a special case of equation 4.9.1 with $g(x) = 1$ and $h(t) = \tan(\beta t)$.

36. $\quad y(x) - \lambda \int_{a}^{b} [A\tan(\beta x) + B\tan(\beta t)] y(t)\, dt = f(x).$

This is a special case of equation 4.9.4 with $g(x) = \tan(\beta x)$.

37. $y(x) - \lambda \int_a^b \dfrac{\tan(\beta x)}{\tan(\beta t)} y(t) \, dt = f(x).$

This is a special case of equation 4.9.1 with $g(x) = \tan(\beta x)$ and $h(t) = \dfrac{1}{\tan(\beta t)}.$

38. $y(x) - \lambda \int_a^b \dfrac{\tan(\beta t)}{\tan(\beta x)} y(t) \, dt = f(x).$

This is a special case of equation 4.9.1 with $g(x) = \dfrac{1}{\tan(\beta x)}$ and $h(t) = \tan(\beta t).$

39. $y(x) - \lambda \int_a^b \tan^k(\beta x) \tan^m(\mu t) y(t) \, dt = f(x).$

This is a special case of equation 4.9.1 with $g(x) = \tan^k(\beta x)$ and $h(t) = \tan^m(\mu t).$

40. $y(x) - \lambda \int_a^b t^k \tan^m(\beta x) y(t) \, dt = f(x).$

This is a special case of equation 4.9.1 with $g(x) = \tan^m(\beta x)$ and $h(t) = t^k.$

41. $y(x) - \lambda \int_a^b x^k \tan^m(\beta t) y(t) \, dt = f(x).$

This is a special case of equation 4.9.1 with $g(x) = x^k$ and $h(t) = \tan^m(\beta t).$

42. $y(x) - \lambda \int_a^b [A + B(x - t) \tan(\beta t)] y(t) \, dt = f(x).$

This is a special case of equation 4.9.8 with $h(t) = \tan(\beta t).$

43. $y(x) - \lambda \int_a^b [A + B(x - t) \tan(\beta x)] y(t) \, dt = f(x).$

This is a special case of equation 4.9.10 with $h(x) = \tan(\beta x).$

4.5-4. Kernels Containing Cotangent

44. $y(x) - \lambda \int_a^b \cot(\beta x) y(t) \, dt = f(x).$

This is a special case of equation 4.9.1 with $g(x) = \cot(\beta x)$ and $h(t) = 1.$

45. $y(x) - \lambda \int_a^b \cot(\beta t) y(t) \, dt = f(x).$

This is a special case of equation 4.9.1 with $g(x) = 1$ and $h(t) = \cot(\beta t).$

46. $y(x) - \lambda \int_a^b [A \cot(\beta x) + B \cot(\beta t)] y(t) \, dt = f(x).$

This is a special case of equation 4.9.4 with $g(x) = \cot(\beta x).$

47. $\quad y(x) - \lambda \displaystyle\int_a^b \frac{\cot(\beta x)}{\cot(\beta t)} y(t)\,dt = f(x).$

This is a special case of equation 4.9.1 with $g(x) = \cot(\beta x)$ and $h(t) = \dfrac{1}{\cot(\beta t)}$.

48. $\quad y(x) - \lambda \displaystyle\int_a^b \frac{\cot(\beta t)}{\cot(\beta x)} y(t)\,dt = f(x).$

This is a special case of equation 4.9.1 with $g(x) = \dfrac{1}{\cot(\beta x)}$ and $h(t) = \cot(\beta t)$.

49. $\quad y(x) - \lambda \displaystyle\int_a^b \cot^k(\beta x) \cot^m(\mu t) y(t)\,dt = f(x).$

This is a special case of equation 4.9.1 with $g(x) = \cot^k(\beta x)$ and $h(t) = \cot^m(\mu t)$.

50. $\quad y(x) - \lambda \displaystyle\int_a^b t^k \cot^m(\beta x) y(t)\,dt = f(x).$

This is a special case of equation 4.9.1 with $g(x) = \cot^m(\beta x)$ and $h(t) = t^k$.

51. $\quad y(x) - \lambda \displaystyle\int_a^b x^k \cot^m(\beta t) y(t)\,dt = f(x).$

This is a special case of equation 4.9.1 with $g(x) = x^k$ and $h(t) = \cot^m(\beta t)$.

52. $\quad y(x) - \lambda \displaystyle\int_a^b [A + B(x-t)\cot(\beta t)] y(t)\,dt = f(x).$

This is a special case of equation 4.9.8 with $h(t) = \cot(\beta t)$.

53. $\quad y(x) - \lambda \displaystyle\int_a^b [A + B(x-t)\cot(\beta x)] y(t)\,dt = f(x).$

This is a special case of equation 4.9.10 with $h(x) = \cot(\beta x)$.

4.5-5. Kernels Containing Combinations of Trigonometric Functions

54. $\quad y(x) - \lambda \displaystyle\int_a^b \cos^k(\beta x) \sin^m(\mu t) y(t)\,dt = f(x).$

This is a special case of equation 4.9.1 with $g(x) = \cos^k(\beta x)$ and $h(t) = \sin^m(\mu t)$.

55. $\quad y(x) - \lambda \displaystyle\int_a^b [A \sin(\alpha x)\cos(\beta t) + B \sin(\gamma x)\cos(\delta t)] y(t)\,dt = f(x).$

This is a special case of equation 4.9.18 with $g_1(x) = \sin(\alpha x)$, $h_1(t) = A\cos(\beta t)$, $g_2(x) = \sin(\gamma x)$, and $h_2(t) = B\cos(\delta t)$.

56. $\quad y(x) - \lambda \displaystyle\int_a^b \tan^k(\gamma x) \cot^m(\mu t) y(t)\,dt = f(x).$

This is a special case of equation 4.9.1 with $g(x) = \tan^k(\gamma x)$ and $h(t) = \cot^m(\mu t)$.

57. $\quad y(x) - \lambda \displaystyle\int_a^b [A \tan(\alpha x)\cot(\beta t) + B \tan(\gamma x)\cot(\delta t)] y(t)\,dt = f(x).$

This is a special case of equation 4.9.18 with $g_1(x) = \tan(\alpha x)$, $h_1(t) = A\cot(\beta t)$, $g_2(x) = \tan(\gamma x)$, and $h_2(t) = B\cot(\delta t)$.

4.5-6. A Singular Equation

58. $\quad Ay(x) - \dfrac{B}{2\pi} \displaystyle\int_0^{2\pi} \cot\left(\dfrac{t-x}{2}\right) y(t)\, dt = f(x), \qquad 0 \le x \le 2\pi.$

Here the integral is understood in the sense of the Cauchy principal value. Without loss of generality we may assume that $A^2 + B^2 = 1$.

Solution:

$$y(x) = Af(x) + \frac{B}{2\pi}\int_0^{2\pi}\cot\left(\frac{t-x}{2}\right)f(t)\,dt + \frac{B^2}{2\pi A}\int_0^{2\pi}f(t)\,dt.$$

⊙ Reference: I. K. Lifanov (1996).

4.6. Equations Whose Kernels Contain Inverse Trigonometric Functions

4.6-1. Kernels Containing Arccosine

1. $\quad y(x) - \lambda\displaystyle\int_a^b \arccos(\beta x) y(t)\, dt = f(x).$

This is a special case of equation 4.9.1 with $g(x) = \arccos(\beta x)$ and $h(t) = 1$.

2. $\quad y(x) - \lambda\displaystyle\int_a^b \arccos(\beta t) y(t)\, dt = f(x).$

This is a special case of equation 4.9.1 with $g(x) = 1$ and $h(t) = \arccos(\beta t)$.

3. $\quad y(x) - \lambda\displaystyle\int_a^b \dfrac{\arccos(\beta x)}{\arccos(\beta t)} y(t)\, dt = f(x).$

This is a special case of equation 4.9.1 with $g(x) = \arccos(\beta x)$ and $h(t) = \dfrac{1}{\arccos(\beta t)}$.

4. $\quad y(x) - \lambda\displaystyle\int_a^b \dfrac{\arccos(\beta t)}{\arccos(\beta x)} y(t)\, dt = f(x).$

This is a special case of equation 4.9.1 with $g(x) = \dfrac{1}{\arccos(\beta x)}$ and $h(t) = \arccos(\beta t)$.

5. $\quad y(x) - \lambda\displaystyle\int_a^b \arccos^k(\beta x)\arccos^m(\mu t) y(t)\, dt = f(x).$

This is a special case of equation 4.9.1 with $g(x) = \arccos^k(\beta x)$ and $h(t) = \arccos^m(\mu t)$.

6. $\quad y(x) - \lambda\displaystyle\int_a^b t^k \arccos^m(\beta x) y(t)\, dt = f(x).$

This is a special case of equation 4.9.1 with $g(x) = \arccos^m(\beta x)$ and $h(t) = t^k$.

7. $\quad y(x) - \lambda\displaystyle\int_a^b x^k \arccos^m(\beta t) y(t)\, dt = f(x).$

This is a special case of equation 4.9.1 with $g(x) = x^k$ and $h(t) = \arccos^m(\beta t)$.

8. $y(x) - \lambda \int_a^b [A + B(x - t) \arccos(\beta x)] y(t)\, dt = f(x).$

This is a special case of equation 4.9.10 with $h(x) = \arccos(\beta x)$.

9. $y(x) - \lambda \int_a^b [A + B(x - t) \arccos(\beta t)] y(t)\, dt = f(x).$

This is a special case of equation 4.9.8 with $h(t) = \arccos(\beta t)$.

4.6-2. Kernels Containing Arcsine

10. $y(x) - \lambda \int_a^b \arcsin(\beta x) y(t)\, dt = f(x).$

This is a special case of equation 4.9.1 with $g(x) = \arcsin(\beta x)$ and $h(t) = 1$.

11. $y(x) - \lambda \int_a^b \arcsin(\beta t) y(t)\, dt = f(x).$

This is a special case of equation 4.9.1 with $g(x) = 1$ and $h(t) = \arcsin(\beta t)$.

12. $y(x) - \lambda \int_a^b \dfrac{\arcsin(\beta x)}{\arcsin(\beta t)} y(t)\, dt = f(x).$

This is a special case of equation 4.9.1 with $g(x) = \arcsin(\beta x)$ and $h(t) = \dfrac{1}{\arcsin(\beta t)}$.

13. $y(x) - \lambda \int_a^b \dfrac{\arcsin(\beta t)}{\arcsin(\beta x)} y(t)\, dt = f(x).$

This is a special case of equation 4.9.1 with $g(x) = \dfrac{1}{\arcsin(\beta x)}$ and $h(t) = \arcsin(\beta t)$.

14. $y(x) - \lambda \int_a^b \arcsin^k(\beta x) \arcsin^m(\mu t) y(t)\, dt = f(x).$

This is a special case of equation 4.9.1 with $g(x) = \arcsin^k(\beta x)$ and $h(t) = \arcsin^m(\mu t)$.

15. $y(x) - \lambda \int_a^b t^k \arcsin^m(\beta x) y(t)\, dt = f(x).$

This is a special case of equation 4.9.1 with $g(x) = \arcsin^m(\beta x)$ and $h(t) = t^k$.

16. $y(x) - \lambda \int_a^b x^k \arcsin^m(\beta t) y(t)\, dt = f(x).$

This is a special case of equation 4.9.1 with $g(x) = x^k$ and $h(t) = \arcsin^m(\beta t)$.

17. $y(x) - \lambda \int_a^b [A + B(x - t) \arcsin(\beta t)] y(t)\, dt = f(x).$

This is a special case of equation 4.9.8 with $h(t) = \arcsin(\beta t)$.

18. $y(x) - \lambda \int_a^b [A + B(x - t) \arcsin(\beta x)] y(t)\, dt = f(x).$

This is a special case of equation 4.9.10 with $h(x) = \arcsin(\beta x)$.

4.6-3. Kernels Containing Arctangent

19. $y(x) - \lambda \int_a^b \arctan(\beta x) y(t)\, dt = f(x).$

This is a special case of equation 4.9.1 with $g(x) = \arctan(\beta x)$ and $h(t) = 1$.

20. $y(x) - \lambda \int_a^b \arctan(\beta t) y(t)\, dt = f(x).$

This is a special case of equation 4.9.1 with $g(x) = 1$ and $h(t) = \arctan(\beta t)$.

21. $y(x) - \lambda \int_a^b [A \arctan(\beta x) + B \arctan(\beta t)] y(t)\, dt = f(x).$

This is a special case of equation 4.9.4 with $g(x) = \arctan(\beta x)$.

22. $y(x) - \lambda \int_a^b \dfrac{\arctan(\beta x)}{\arctan(\beta t)} y(t)\, dt = f(x).$

This is a special case of equation 4.9.1 with $g(x) = \arctan(\beta x)$ and $h(t) = \dfrac{1}{\arctan(\beta t)}$.

23. $y(x) - \lambda \int_a^b \dfrac{\arctan(\beta t)}{\arctan(\beta x)} y(t)\, dt = f(x).$

This is a special case of equation 4.9.1 with $g(x) = \dfrac{1}{\arctan(\beta x)}$ and $h(t) = \arctan(\beta t)$.

24. $y(x) - \lambda \int_a^b \arctan^k(\beta x) \arctan^m(\mu t) y(t)\, dt = f(x).$

This is a special case of equation 4.9.1 with $g(x) = \arctan^k(\beta x)$ and $h(t) = \arctan^m(\mu t)$.

25. $y(x) - \lambda \int_a^b t^k \arctan^m(\beta x) y(t)\, dt = f(x).$

This is a special case of equation 4.9.1 with $g(x) = \arctan^m(\beta x)$ and $h(t) = t^k$.

26. $y(x) - \lambda \int_a^b x^k \arctan^m(\beta t) y(t)\, dt = f(x).$

This is a special case of equation 4.9.1 with $g(x) = x^k$ and $h(t) = \arctan^m(\beta t)$.

27. $y(x) - \lambda \int_a^b [A + B(x - t) \arctan(\beta t)] y(t)\, dt = f(x).$

This is a special case of equation 4.9.8 with $h(t) = \arctan(\beta t)$.

28. $y(x) - \lambda \int_a^b [A + B(x - t) \arctan(\beta x)] y(t)\, dt = f(x).$

This is a special case of equation 4.9.10 with $h(x) = \arctan(\beta x)$.

4.6-4. Kernels Containing Arccotangent

29. $y(x) - \lambda \int_a^b \operatorname{arccot}(\beta x) y(t)\, dt = f(x).$

This is a special case of equation 4.9.1 with $g(x) = \operatorname{arccot}(\beta x)$ and $h(t) = 1$.

30. $y(x) - \lambda \int_a^b \operatorname{arccot}(\beta t) y(t)\, dt = f(x).$

This is a special case of equation 4.9.1 with $g(x) = 1$ and $h(t) = \operatorname{arccot}(\beta t)$.

31. $y(x) - \lambda \int_a^b [A \operatorname{arccot}(\beta x) + B \operatorname{arccot}(\beta t)] y(t)\, dt = f(x).$

This is a special case of equation 4.9.4 with $g(x) = \operatorname{arccot}(\beta x)$.

32. $y(x) - \lambda \int_a^b \dfrac{\operatorname{arccot}(\beta x)}{\operatorname{arccot}(\beta t)} y(t)\, dt = f(x).$

This is a special case of equation 4.9.1 with $g(x) = \operatorname{arccot}(\beta x)$ and $h(t) = \dfrac{1}{\operatorname{arccot}(\beta t)}$.

33. $y(x) - \lambda \int_a^b \dfrac{\operatorname{arccot}(\beta t)}{\operatorname{arccot}(\beta x)} y(t)\, dt = f(x).$

This is a special case of equation 4.9.1 with $g(x) = \dfrac{1}{\operatorname{arccot}(\beta x)}$ and $h(t) = \operatorname{arccot}(\beta t)$.

34. $y(x) - \lambda \int_a^b \operatorname{arccot}^k(\beta x) \operatorname{arccot}^m(\mu t) y(t)\, dt = f(x).$

This is a special case of equation 4.9.1 with $g(x) = \operatorname{arccot}^k(\beta x)$ and $h(t) = \operatorname{arccot}^m(\mu t)$.

35. $y(x) - \lambda \int_a^b t^k \operatorname{arccot}^m(\beta x) y(t)\, dt = f(x).$

This is a special case of equation 4.9.1 with $g(x) = \operatorname{arccot}^m(\beta x)$ and $h(t) = t^k$.

36. $y(x) - \lambda \int_a^b x^k \operatorname{arccot}^m(\beta t) y(t)\, dt = f(x).$

This is a special case of equation 4.9.1 with $g(x) = x^k$ and $h(t) = \operatorname{arccot}^m(\beta t)$.

37. $y(x) - \lambda \int_a^b [A + B(x - t) \operatorname{arccot}(\beta t)] y(t)\, dt = f(x).$

This is a special case of equation 4.9.8 with $h(t) = \operatorname{arccot}(\beta t)$.

38. $y(x) - \lambda \int_a^b [A + B(x - t) \operatorname{arccot}(\beta x)] y(t)\, dt = f(x).$

This is a special case of equation 4.9.10 with $h(x) = \operatorname{arccot}(\beta x)$.

4.7. Equations Whose Kernels Contain Combinations of Elementary Functions

4.7-1. Kernels Containing Exponential and Hyperbolic Functions

1. $y(x) - \lambda \int_a^b e^{\mu(x-t)} \cosh[\beta(x-t)]y(t)\,dt = f(x).$

This is a special case of equation 4.9.18 with $g_1(x) = e^{\mu x}\cosh(\beta x)$, $h_1(t) = e^{-\mu t}\cosh(\beta t)$, $g_2(x) = e^{\mu x}\sinh(\beta x)$, and $h_2(t) = -e^{-\mu t}\sinh(\beta t)$.

2. $y(x) - \lambda \int_a^b e^{\mu(x-t)} \sinh[\beta(x-t)]y(t)\,dt = f(x).$

This is a special case of equation 4.9.18 with $g_1(x) = e^{\mu x}\sinh(\beta x)$, $h_1(t) = e^{-\mu t}\cosh(\beta t)$, $g_2(x) = e^{\mu x}\cosh(\beta x)$, and $h_2(t) = -e^{-\mu t}\sinh(\beta t)$.

3. $y(x) - \lambda \int_a^b t e^{\mu(x-t)} \sinh[\beta(x-t)]y(t)\,dt = f(x).$

This is a special case of equation 4.9.18 with $g_1(x) = e^{\mu x}\sinh(\beta x)$, $h_1(t) = t e^{-\mu t}\cosh(\beta t)$, $g_2(x) = e^{\mu x}\cosh(\beta x)$, and $h_2(t) = -t e^{-\mu t}\sinh(\beta t)$.

4.7-2. Kernels Containing Exponential and Logarithmic Functions

4. $y(x) - \lambda \int_a^b e^{\mu t} \ln(\beta x)y(t)\,dt = f(x).$

This is a special case of equation 4.9.1 with $g(x) = \ln(\beta x)$ and $h(t) = e^{\mu t}$.

5. $y(x) - \lambda \int_a^b e^{\mu x} \ln(\beta t)y(t)\,dt = f(x).$

This is a special case of equation 4.9.1 with $g(x) = e^{\mu x}$ and $h(t) = \ln(\beta t)$.

6. $y(x) - \lambda \int_a^b e^{\mu(x-t)} \ln(\beta x)y(t)\,dt = f(x).$

This is a special case of equation 4.9.1 with $g(x) = e^{\mu x}\ln(\beta x)$ and $h(t) = e^{-\mu t}$.

7. $y(x) - \lambda \int_a^b e^{\mu(x-t)} \ln(\beta t)y(t)\,dt = f(x).$

This is a special case of equation 4.9.1 with $g(x) = e^{\mu x}$ and $h(t) = e^{-\mu t}\ln(\beta t)$.

8. $y(x) - \lambda \int_a^b e^{\mu(x-t)}(\ln x - \ln t)y(t)\,dt = f(x).$

This is a special case of equation 4.9.18 with $g_1(x) = e^{\mu x}\ln x$, $h_1(t) = e^{-\mu t}$, $g_2(x) = e^{\mu x}$, and $h_2(t) = -e^{-\mu t}\ln t$.

9. $y(x) + \dfrac{b^2 - a^2}{2a} \displaystyle\int_0^\infty \dfrac{1}{t} \exp\left(-a\left|\ln\dfrac{x}{t}\right|\right) y(t)\, dt = f(x).$

Solution with $a > 0$, $b > 0$, and $x > 0$:

$$y(x) = f(x) + \frac{a^2 - b^2}{2b} \int_0^\infty \frac{1}{t} \exp\left(-b\left|\ln\frac{x}{t}\right|\right) f(t)\, dt.$$

⊙ Reference: F. D. Gakhov and Yu. I. Cherskii (1978).

4.7-3. Kernels Containing Exponential and Trigonometric Functions

10. $y(x) - \lambda \displaystyle\int_a^b e^{\mu t} \cos(\beta x) y(t)\, dt = f(x).$

This is a special case of equation 4.9.1 with $g(x) = \cos(\beta x)$ and $h(t) = e^{\mu t}$.

11. $y(x) - \lambda \displaystyle\int_a^b e^{\mu x} \cos(\beta t) y(t)\, dt = f(x).$

This is a special case of equation 4.9.1 with $g(x) = e^{\mu x}$ and $h(t) = \cos(\beta t)$.

12. $y(x) - \lambda \displaystyle\int_0^\infty e^{\mu(x-t)} \cos(xt) y(t)\, dt = f(x).$

Solution:

$$y(x) = \frac{f(x)}{1 - \frac{\pi}{2}\lambda^2} + \frac{\lambda}{1 - \frac{\pi}{2}\lambda^2} \int_0^\infty e^{\mu(x-t)} \cos(xt) f(t)\, dt, \qquad \lambda \neq \pm\sqrt{2/\pi}.$$

13. $y(x) - \lambda \displaystyle\int_a^b e^{\mu(x-t)} \cos[\beta(x - t)] y(t)\, dt = f(x).$

This is a special case of equation 4.9.18 with $g_1(x) = e^{\mu x} \cos(\beta x)$, $h_1(t) = e^{-\mu t} \cos(\beta t)$, $g_2(x) = e^{\mu x} \sin(\beta x)$, and $h_2(t) = e^{-\mu t} \sin(\beta t)$.

14. $y(x) - \lambda \displaystyle\int_a^b e^{\mu t} \sin(\beta x) y(t)\, dt = f(x).$

This is a special case of equation 4.9.1 with $g(x) = \sin(\beta x)$ and $h(t) = e^{\mu t}$.

15. $y(x) - \lambda \displaystyle\int_a^b e^{\mu x} \sin(\beta t) y(t)\, dt = f(x).$

This is a special case of equation 4.9.1 with $g(x) = e^{\mu x}$ and $h(t) = \sin(\beta t)$.

16. $y(x) - \lambda \displaystyle\int_0^\infty e^{\mu(x-t)} \sin(xt) y(t)\, dt = f(x).$

Solution:

$$y(x) = \frac{f(x)}{1 - \frac{\pi}{2}\lambda^2} + \frac{\lambda}{1 - \frac{\pi}{2}\lambda^2} \int_0^\infty e^{\mu(x-t)} \sin(xt) f(t)\, dt, \qquad \lambda \neq \pm\sqrt{2/\pi}.$$

17. $y(x) - \lambda \int_a^b e^{\mu(x-t)} \sin[\beta(x-t)]y(t)\, dt = f(x).$

This is a special case of equation 4.9.18 with $g_1(x) = e^{\mu x} \sin(\beta x)$, $h_1(t) = e^{-\mu t} \cos(\beta t)$, $g_2(x) = e^{\mu x} \cos(\beta x)$, and $h_2(t) = -e^{-\mu t} \sin(\beta t)$.

18. $y(x) - \lambda \int_a^b e^{\mu(x-t)} \left\{ \sum_{k=1}^n A_k \sin[\beta_k(x-t)] \right\} y(t)\, dt = f(x), \qquad n = 1, 2, \ldots$

This is a special case of equation 4.9.20.

19. $y(x) - \lambda \int_a^b t e^{\mu(x-t)} \sin[\beta(x-t)]y(t)\, dt = f(x).$

This is a special case of equation 4.9.18 with $g_1(x) = e^{\mu x} \sin(\beta x)$, $h_1(t) = t e^{-\mu t} \cos(\beta t)$, $g_2(x) = e^{\mu x} \cos(\beta x)$, and $h_2(t) = -t e^{-\mu t} \sin(\beta t)$.

20. $y(x) - \lambda \int_a^b x e^{\mu(x-t)} \sin[\beta(x-t)]y(t)\, dt = f(x).$

This is a special case of equation 4.9.18 with $g_1(x) = x e^{\mu x} \sin(\beta x)$, $h_1(t) = e^{-\mu t} \cos(\beta t)$, $g_2(x) = x e^{\mu x} \cos(\beta x)$, and $h_2(t) = -e^{-\mu t} \sin(\beta t)$.

21. $y(x) - \lambda \int_a^b e^{\mu t} \tan(\beta x) y(t)\, dt = f(x).$

This is a special case of equation 4.9.1 with $g(x) = \tan(\beta x)$ and $h(t) = e^{\mu t}$.

22. $y(x) - \lambda \int_a^b e^{\mu x} \tan(\beta t) y(t)\, dt = f(x).$

This is a special case of equation 4.9.1 with $g(x) = e^{\mu x}$ and $h(t) = \tan(\beta t)$.

23. $y(x) - \lambda \int_a^b e^{\mu(x-t)} [\tan(\beta x) - \tan(\beta t)] y(t)\, dt = f(x).$

This is a special case of equation 4.9.18 with $g_1(x) = e^{\mu x} \tan(\beta x)$, $h_1(t) = e^{-\mu t}$, $g_2(x) = e^{\mu x}$, and $h_2(t) = -e^{-\mu t} \tan(\beta t)$.

24. $y(x) - \lambda \int_a^b e^{\mu t} \cot(\beta x) y(t)\, dt = f(x).$

This is a special case of equation 4.9.1 with $g(x) = \cot(\beta x)$ and $h(t) = e^{\mu t}$.

25. $y(x) - \lambda \int_a^b e^{\mu x} \cot(\beta t) y(t)\, dt = f(x).$

This is a special case of equation 4.9.1 with $g(x) = e^{\mu x}$ and $h(t) = \cot(\beta t)$.

4.7-4. Kernels Containing Hyperbolic and Logarithmic Functions

26. $y(x) - \lambda \int_a^b \cosh^k(\beta x) \ln^m(\mu t) y(t)\, dt = f(x).$

This is a special case of equation 4.9.1 with $g(x) = \cosh^k(\beta x)$ and $h(t) = \ln^m(\mu t)$.

27. $y(x) - \lambda \int_a^b \cosh^k(\beta t) \ln^m(\mu x) y(t)\, dt = f(x).$

This is a special case of equation 4.9.1 with $g(x) = \ln^m(\mu x)$ and $h(t) = \cosh^k(\beta t)$.

28. $y(x) - \lambda \int_a^b \sinh^k(\beta x) \ln^m(\mu t) y(t)\, dt = f(x).$

This is a special case of equation 4.9.1 with $g(x) = \sinh^k(\beta x)$ and $h(t) = \ln^m(\mu t)$.

29. $y(x) - \lambda \int_a^b \sinh^k(\beta t) \ln^m(\mu x) y(t)\, dt = f(x).$

This is a special case of equation 4.9.1 with $g(x) = \ln^m(\mu x)$ and $h(t) = \sinh^k(\beta t)$.

30. $y(x) - \lambda \int_a^b \tanh^k(\beta x) \ln^m(\mu t) y(t)\, dt = f(x).$

This is a special case of equation 4.9.1 with $g(x) = \tanh^k(\beta x)$ and $h(t) = \ln^m(\mu t)$.

31. $y(x) - \lambda \int_a^b \tanh^k(\beta t) \ln^m(\mu x) y(t)\, dt = f(x).$

This is a special case of equation 4.9.1 with $g(x) = \ln^m(\mu x)$ and $h(t) = \tanh^k(\beta t)$.

32. $y(x) - \lambda \int_a^b \coth^k(\beta x) \ln^m(\mu t) y(t)\, dt = f(x).$

This is a special case of equation 4.9.1 with $g(x) = \coth^k(\beta x)$ and $h(t) = \ln^m(\mu t)$.

33. $y(x) - \lambda \int_a^b \coth^k(\beta t) \ln^m(\mu x) y(t)\, dt = f(x).$

This is a special case of equation 4.9.1 with $g(x) = \ln^m(\mu x)$ and $h(t) = \coth^k(\beta t)$.

4.7-5. Kernels Containing Hyperbolic and Trigonometric Functions

34. $y(x) - \lambda \int_a^b \cosh^k(\beta x) \cos^m(\mu t) y(t)\, dt = f(x).$

This is a special case of equation 4.9.1 with $g(x) = \cosh^k(\beta x)$ and $h(t) = \cos^m(\mu t)$.

35. $y(x) - \lambda \int_a^b \cosh^k(\beta t) \cos^m(\mu x) y(t)\, dt = f(x).$

This is a special case of equation 4.9.1 with $g(x) = \cos^m(\mu x)$ and $h(t) = \cosh^k(\beta t)$.

36. $y(x) - \lambda \int_a^b \cosh^k(\beta x) \sin^m(\mu t) y(t)\, dt = f(x).$

This is a special case of equation 4.9.1 with $g(x) = \cosh^k(\beta x)$ and $h(t) = \sin^m(\mu t)$.

37. $y(x) - \lambda \int_a^b \cosh^k(\beta t) \sin^m(\mu x) y(t)\, dt = f(x).$

This is a special case of equation 4.9.1 with $g(x) = \sin^m(\mu x)$ and $h(t) = \cosh^k(\beta t)$.

38. $y(x) - \lambda \int_a^b \sinh^k(\beta x) \cos^m(\mu t) y(t)\, dt = f(x).$

This is a special case of equation 4.9.1 with $g(x) = \sinh^k(\beta x)$ and $h(t) = \cos^m(\mu t)$.

39. $y(x) - \lambda \int_a^b \sinh^k(\beta t) \cos^m(\mu x) y(t)\, dt = f(x).$

This is a special case of equation 4.9.1 with $g(x) = \cos^m(\mu x)$ and $h(t) = \sinh^k(\beta t)$.

40. $y(x) - \lambda \int_a^b \sinh^k(\beta x) \sin^m(\mu t) y(t)\, dt = f(x).$

This is a special case of equation 4.9.1 with $g(x) = \sinh^k(\beta x)$ and $h(t) = \sin^m(\mu t)$.

41. $y(x) - \lambda \int_a^b \sinh^k(\beta t) \sin^m(\mu x) y(t)\, dt = f(x).$

This is a special case of equation 4.9.1 with $g(x) = \sin^m(\mu x)$ and $h(t) = \sinh^k(\beta t)$.

42. $y(x) - \lambda \int_a^b \tanh^k(\beta x) \cos^m(\mu t) y(t)\, dt = f(x).$

This is a special case of equation 4.9.1 with $g(x) = \tanh^k(\beta x)$ and $h(t) = \cos^m(\mu t)$.

43. $y(x) - \lambda \int_a^b \tanh^k(\beta t) \cos^m(\mu x) y(t)\, dt = f(x).$

This is a special case of equation 4.9.1 with $g(x) = \cos^m(\mu x)$ and $h(t) = \tanh^k(\beta t)$.

44. $y(x) - \lambda \int_a^b \tanh^k(\beta x) \sin^m(\mu t) y(t)\, dt = f(x).$

This is a special case of equation 4.9.1 with $g(x) = \tanh^k(\beta x)$ and $h(t) = \sin^m(\mu t)$.

45. $y(x) - \lambda \int_a^b \tanh^k(\beta t) \sin^m(\mu x) y(t)\, dt = f(x).$

This is a special case of equation 4.9.1 with $g(x) = \sin^m(\mu x)$ and $h(t) = \tanh^k(\beta t)$.

4.7-6. Kernels Containing Logarithmic and Trigonometric Functions

46. $y(x) - \lambda \int_a^b \cos^k(\beta x) \ln^m(\mu t) y(t)\, dt = f(x).$

This is a special case of equation 4.9.1 with $g(x) = \cos^k(\beta x)$ and $h(t) = \ln^m(\mu t)$.

47.　$y(x) - \lambda \displaystyle\int_a^b \cos^k(\beta t) \ln^m(\mu x) y(t)\, dt = f(x).$

This is a special case of equation 4.9.1 with $g(x) = \ln^m(\mu x)$ and $h(t) = \cos^k(\beta t)$.

48.　$y(x) - \lambda \displaystyle\int_a^b \sin^k(\beta x) \ln^m(\mu t) y(t)\, dt = f(x).$

This is a special case of equation 4.9.1 with $g(x) = \sin^k(\beta x)$ and $h(t) = \ln^m(\mu t)$.

49.　$y(x) - \lambda \displaystyle\int_a^b \sin^k(\beta t) \ln^m(\mu x) y(t)\, dt = f(x).$

This is a special case of equation 4.9.1 with $g(x) = \ln^m(\mu x)$ and $h(t) = \sin^k(\beta t)$.

50.　$y(x) - \lambda \displaystyle\int_a^b \tan^k(\beta x) \ln^m(\mu t) y(t)\, dt = f(x).$

This is a special case of equation 4.9.1 with $g(x) = \tan^k(\beta x)$ and $h(t) = \ln^m(\mu t)$.

51.　$y(x) - \lambda \displaystyle\int_a^b \tan^k(\beta t) \ln^m(\mu x) y(t)\, dt = f(x).$

This is a special case of equation 4.9.1 with $g(x) = \ln^m(\mu x)$ and $h(t) = \tan^k(\beta t)$.

52.　$y(x) - \lambda \displaystyle\int_a^b \cot^k(\beta x) \ln^m(\mu t) y(t)\, dt = f(x).$

This is a special case of equation 4.9.1 with $g(x) = \cot^k(\beta x)$ and $h(t) = \ln^m(\mu t)$.

53.　$y(x) - \lambda \displaystyle\int_a^b \cot^k(\beta t) \ln^m(\mu x) y(t)\, dt = f(x).$

This is a special case of equation 4.9.1 with $g(x) = \ln^m(\mu x)$ and $h(t) = \cot^k(\beta t)$.

4.8. Equations Whose Kernels Contain Special Functions

4.8-1. Kernels Containing Bessel Functions

1.　$y(x) - \lambda \displaystyle\int_a^b J_\nu(\beta x) y(t)\, dt = f(x).$

This is a special case of equation 4.9.1 with $g(x) = J_\nu(\beta x)$ and $h(t) = 1$.

2.　$y(x) - \lambda \displaystyle\int_a^b J_\nu(\beta t) y(t)\, dt = f(x).$

This is a special case of equation 4.9.1 with $g(x) = 1$ and $h(t) = J_\nu(\beta t)$.

3.　$y(x) + \lambda \displaystyle\int_0^\infty t J_\nu(xt) y(t)\, dt = f(x), \qquad \nu > -\tfrac{1}{2}.$

Solution:
$$y(x) = \frac{f(x)}{1 - \lambda^2} - \frac{\lambda}{1 - \lambda^2} \int_0^\infty t J_\nu(xt) f(t)\, dt, \qquad \lambda \neq \pm 1.$$

4. $\quad y(x) + \lambda \displaystyle\int_0^\infty J_\nu\left(2\sqrt{xt}\,\right) y(t)\,dt = f(x).$

By setting $x = \frac{1}{2}z^2$, $t = \frac{1}{2}\tau^2$, $y(x) = Y(z)$, and $f(x) = F(z)$, we arrive at an equation of the form 4.8.3:

$$Y(z) + \lambda \int_0^\infty \tau J_\nu(z\tau) Y(\tau)\,d\tau = F(z).$$

5. $\quad y(x) - \lambda \displaystyle\int_a^b [A + B(x - t)J_\nu(\beta t)]y(t)\,dt = f(x).$

This is a special case of equation 4.9.8 with $h(t) = J_\nu(\beta t)$.

6. $\quad y(x) - \lambda \displaystyle\int_a^b [A + B(x - t)J_\nu(\beta x)]y(t)\,dt = f(x).$

This is a special case of equation 4.9.10 with $h(x) = J_\nu(\beta x)$.

7. $\quad y(x) - \lambda \displaystyle\int_a^b [AJ_\mu(\alpha x) + BJ_\nu(\beta t)]y(t)\,dt = f(x).$

This is a special case of equation 4.9.5 with $g(x) = AJ_\mu(\alpha x)$ and $h(t) = BJ_\nu(\beta t)$.

8. $\quad y(x) - \lambda \displaystyle\int_a^b [AJ_\mu(x)J_\nu(t) + BJ_\nu(x)J_\mu(t)]y(t)\,dt = f(x).$

This is a special case of equation 4.9.17 with $g(x) = J_\mu(x)$ and $h(t) = J_\nu(t)$.

9. $\quad y(x) - \lambda \displaystyle\int_a^b Y_\nu(\beta x)y(t)\,dt = f(x).$

This is a special case of equation 4.9.1 with $g(x) = Y_\nu(\beta x)$ and $h(t) = 1$.

10. $\quad y(x) - \lambda \displaystyle\int_a^b Y_\nu(\beta t)y(t)\,dt = f(x).$

This is a special case of equation 4.9.1 with $g(x) = 1$ and $h(t) = Y_\nu(\beta t)$.

11. $\quad y(x) - \lambda \displaystyle\int_a^b [A + B(x - t)Y_\nu(\beta t)]y(t)\,dt = f(x).$

This is a special case of equation 4.9.8 with $h(t) = Y_\nu(\beta t)$.

12. $\quad y(x) - \lambda \displaystyle\int_a^b [A + B(x - t)Y_\nu(\beta x)]y(t)\,dt = f(x).$

This is a special case of equation 4.9.10 with $h(x) = Y_\nu(\beta x)$.

13. $\quad y(x) - \lambda \displaystyle\int_a^b [AY_\mu(\alpha x) + BY_\nu(\beta t)]y(t)\,dt = f(x).$

This is a special case of equation 4.9.5 with $g(x) = AY_\mu(\alpha x)$ and $h(t) = BY_\nu(\beta t)$.

14. $\quad y(x) - \lambda \displaystyle\int_a^b [AY_\mu(x)Y_\mu(t) + BY_\nu(x)Y_\nu(t)]y(t)\,dt = f(x).$

This is a special case of equation 4.9.14 with $g(x) = Y_\mu(x)$ and $h(t) = Y_\nu(t)$.

15. $\quad y(x) - \lambda \displaystyle\int_a^b [AY_\mu(x)Y_\nu(t) + BY_\nu(x)Y_\mu(t)]y(t)\,dt = f(x).$

This is a special case of equation 4.9.17 with $g(x) = Y_\mu(x)$ and $h(t) = Y_\nu(t)$.

4.8-2. Kernels Containing Modified Bessel Functions

16. $y(x) - \lambda \int_a^b I_\nu(\beta x) y(t)\, dt = f(x).$

This is a special case of equation 4.9.1 with $g(x) = I_\nu(\beta x)$ and $h(t) = 1$.

17. $y(x) - \lambda \int_a^b I_\nu(\beta t) y(t)\, dt = f(x).$

This is a special case of equation 4.9.1 with $g(x) = 1$ and $h(t) = I_\nu(\beta t)$.

18. $y(x) - \lambda \int_a^b [A + B(x - t) I_\nu(\beta t)] y(t)\, dt = f(x).$

This is a special case of equation 4.9.8 with $h(t) = I_\nu(\beta t)$.

19. $y(x) - \lambda \int_a^b [A + B(x - t) I_\nu(\beta x)] y(t)\, dt = f(x).$

This is a special case of equation 4.9.10 with $h(x) = I_\nu(\beta x)$.

20. $y(x) - \lambda \int_a^b [A I_\mu(\alpha x) + B I_\nu(\beta t)] y(t)\, dt = f(x).$

This is a special case of equation 4.9.5 with $g(x) = A I_\mu(\alpha x)$ and $h(t) = B I_\nu(\beta t)$.

21. $y(x) - \lambda \int_a^b [A I_\mu(x) I_\mu(t) + B I_\nu(x) I_\nu(t)] y(t)\, dt = f(x).$

This is a special case of equation 4.9.14 with $g(x) = I_\mu(x)$ and $h(t) = I_\nu(t)$.

22. $y(x) - \lambda \int_a^b [A I_\mu(x) I_\nu(t) + B I_\nu(x) I_\mu(t)] y(t)\, dt = f(x).$

This is a special case of equation 4.9.17 with $g(x) = I_\mu(x)$ and $h(t) = I_\nu(t)$.

23. $y(x) - \lambda \int_a^b K_\nu(\beta x) y(t)\, dt = f(x).$

This is a special case of equation 4.9.1 with $g(x) = K_\nu(\beta x)$ and $h(t) = 1$.

24. $y(x) - \lambda \int_a^b K_\nu(\beta t) y(t)\, dt = f(x).$

This is a special case of equation 4.9.1 with $g(x) = 1$ and $h(t) = K_\nu(\beta t)$.

25. $y(x) - \lambda \int_a^b [A + B(x - t) K_\nu(\beta t)] y(t)\, dt = f(x).$

This is a special case of equation 4.9.8 with $h(t) = K_\nu(\beta t)$.

26. $y(x) - \lambda \int_a^b [A + B(x - t) K_\nu(\beta x)] y(t)\, dt = f(x).$

This is a special case of equation 4.9.10 with $h(x) = K_\nu(\beta x)$.

27. $y(x) - \lambda \int_a^b [AK_\mu(\alpha x) + BK_\nu(\beta t)]y(t)\,dt = f(x).$

This is a special case of equation 4.9.5 with $g(x) = AK_\mu(\alpha x)$ and $h(t) = BK_\nu(\beta t)$.

28. $y(x) - \lambda \int_a^b [AK_\mu(x)K_\mu(t) + BK_\nu(x)K_\nu(t)]y(t)\,dt = f(x).$

This is a special case of equation 4.9.14 with $g(x) = K_\mu(x)$ and $h(t) = K_\nu(t)$.

29. $y(x) - \lambda \int_a^b [AK_\mu(x)K_\nu(t) + BK_\nu(x)K_\mu(t)]y(t)\,dt = f(x).$

This is a special case of equation 4.9.17 with $g(x) = K_\mu(x)$ and $h(t) = K_\nu(t)$.

4.9. Equations Whose Kernels Contain Arbitrary Functions

4.9-1. Equations With Degenerate Kernel: $K(x,t) = g_1(x)h_1(t) + \cdots + g_n(x)h_n(t)$
---

1. $y(x) - \lambda \int_a^b g(x)h(t)y(t)\,dt = f(x).$

$1°.$ Assume that $\lambda \neq \left(\int_a^b g(t)h(t)\,dt \right)^{-1}.$
Solution:

$$y(x) = f(x) + \lambda k g(x), \qquad \text{where} \quad k = \left(1 - \lambda \int_a^b g(t)h(t)\,dt \right)^{-1} \int_a^b h(t)f(t)\,dt.$$

$2°.$ Assume that $\lambda = \left(\int_a^b g(t)h(t)\,dt \right)^{-1}.$

For $\int_a^b h(t)f(t)\,dt = 0$, the solution has the form

$$y = f(x) + Cg(x),$$

where C is an arbitrary constant.

For $\int_a^b h(t)f(t)\,dt \neq 0$, there is no solution.

The limits of integration may take the values $a = -\infty$ and/or $b = \infty$, provided that the corresponding improper integral converges.

2. $y(x) - \lambda \int_a^b [g(x) + g(t)]y(t)\,dt = f(x).$

The characteristic values of the equation:

$$\lambda_1 = \frac{1}{g_1 + \sqrt{(b-a)g_2}}, \qquad \lambda_2 = \frac{1}{g_1 - \sqrt{(b-a)g_2}},$$

where

$$g_1 = \int_a^b g(x)\,dx, \qquad g_2 = \int_a^b g^2(x)\,dx.$$

1°. Solution with $\lambda \neq \lambda_{1,2}$:

$$y(x) = f(x) + \lambda[A_1 g(x) + A_2],$$

where the constants A_1 and A_2 are given by

$$A_1 = \frac{f_1 - \lambda[f_1 g_1 - (b-a)f_2]}{[g_1^2 - (b-a)g_2]\lambda^2 - 2g_1\lambda + 1}, \quad A_2 = \frac{f_2 - \lambda(f_2 g_1 - f_1 g_2)}{[g_1^2 - (b-a)g_2]\lambda^2 - 2g_1\lambda + 1},$$

$$f_1 = \int_a^b f(x)\,dx, \quad f_2 = \int_a^b f(x)g(x)\,dx.$$

2°. Solution with $\lambda = \lambda_1 \neq \lambda_2$ and $f_1 = f_2 = 0$:

$$y(x) = f(x) + C y_1(x), \qquad y_1(x) = g(x) + \sqrt{\frac{g_2}{b-a}},$$

where C is an arbitrary constant and $y_1(x)$ is an eigenfunction of the equation corresponding to the characteristic value λ_1.

3°. Solution with $\lambda = \lambda_2 \neq \lambda_1$ and $f_1 = f_2 = 0$:

$$y(x) = f(x) + C y_2(x), \quad y_2(x) = g(x) - \sqrt{\frac{g_2}{b-a}},$$

where C is an arbitrary constant and $y_2(x)$ is an eigenfunction of the equation corresponding to the characteristic value λ_2.

4°. The equation has no multiple characteristic values.

3. $\quad y(x) - \lambda \int_a^b [g(x) - g(t)]y(t)\,dt = f(x).$

The characteristic values of the equation:

$$\lambda_1 = \frac{1}{\sqrt{g_1^2 - (b-a)g_2}}, \quad \lambda_2 = -\frac{1}{\sqrt{g_1^2 - (b-a)g_2}},$$

where

$$g_1 = \int_a^b g(x)\,dx, \quad g_2 = \int_a^b g^2(x)\,dx.$$

1°. Solution with $\lambda \neq \lambda_{1,2}$:

$$y(x) = f(x) + \lambda[A_1 g(x) + A_2],$$

where the constants A_1 and A_2 are given by

$$A_1 = \frac{f_1 + \lambda[f_1 g_1 - (b-a)f_2]}{[(b-a)g_2 - g_1^2]\lambda^2 + 1}, \quad A_2 = \frac{-f_2 + \lambda(f_2 g_1 - f_1 g_2)}{[(b-a)g_2 - g_1^2]\lambda^2 + 1},$$

$$f_1 = \int_a^b f(x)\,dx, \quad f_2 = \int_a^b f(x)g(x)\,dx.$$

2°. Solution with $\lambda = \lambda_1 \neq \lambda_2$ and $f_1 = f_2 = 0$:

$$y(x) = f(x) + C y_1(x), \qquad y_1(x) = g(x) + \frac{1 - \lambda_1 g_1}{\lambda_1(b-a)},$$

where C is an arbitrary constant and $y_1(x)$ is an eigenfunction of the equation corresponding to the characteristic value λ_1.

3°. The solution with $\lambda = \lambda_2 \neq \lambda_1$ and $f_1 = f_2 = 0$ is given by the formulas of item 2° in which one must replace λ_1 and $y_1(x)$ by λ_2 and $y_2(x)$, respectively.

4°. The equation has no multiple characteristic values.

4. $y(x) - \lambda \int_a^b [Ag(x) + Bg(t)]y(t)\, dt = f(x).$

The characteristic values of the equation:

$$\lambda_{1,2} = \frac{(A+B)g_1 \pm \sqrt{(A-B)^2 g_1^2 + 4AB(b-a)g_2}}{2AB[g_1^2 - (b-a)g_2]},$$

where

$$g_1 = \int_a^b g(x)\, dx, \quad g_2 = \int_a^b g^2(x)\, dx.$$

1°. Solution with $\lambda \neq \lambda_{1,2}$:

$$y(x) = f(x) + \lambda[A_1 g(x) + A_2],$$

where the constants A_1 and A_2 are given by

$$A_1 = \frac{Af_1 - \lambda AB[f_1 g_1 - (b-a)f_2]}{AB[g_1^2 - (b-a)g_2]\lambda^2 - (A+B)g_1\lambda + 1}, \quad A_2 = \frac{Bf_2 - \lambda AB(f_2 g_1 - f_1 g_2)}{AB[g_1^2 - (b-a)g_2]\lambda^2 - (A+B)g_1\lambda + 1},$$

$$f_1 = \int_a^b f(x)\, dx, \quad f_2 = \int_a^b f(x)g(x)\, dx.$$

2°. Solution with $\lambda = \lambda_1 \neq \lambda_2$ and $f_1 = f_2 = 0$:

$$y(x) = f(x) + Cy_1(x), \qquad y_1(x) = g(x) + \frac{1 - \lambda_1 A g_1}{\lambda_1 A(b-a)},$$

where C is an arbitrary constant and $y_1(x)$ is an eigenfunction of the equation corresponding to the characteristic value λ_1.

3°. The solution with $\lambda = \lambda_2 \neq \lambda_1$ and $f_1 = f_2 = 0$ is given by the formulas of item 2° in which one must replace λ_1 and $y_1(x)$ by λ_2 and $y_2(x)$, respectively.

4°. Solution with $\lambda = \lambda_{1,2} = \lambda_*$ and $f_1 = f_2 = 0$, where the characteristic value $\lambda_* = \dfrac{2}{(A+B)g_1}$ is double:

$$y(x) = f(x) + Cy_*(x), \qquad y_*(x) = g(x) - \frac{(A-B)g_1}{2A(b-a)}.$$

Here C is an arbitrary constant and $y_*(x)$ is an eigenfunction of the equation corresponding to λ_*.

5. $y(x) - \lambda \int_a^b [g(x) + h(t)]y(t)\, dt = f(x).$

The characteristic values of the equation:

$$\lambda_{1,2} = \frac{s_1 + s_3 \pm \sqrt{(s_1 - s_3)^2 + 4(b-a)s_2}}{2[s_1 s_3 - (b-a)s_2]},$$

where

$$s_1 = \int_a^b g(x)\, dx, \quad s_2 = \int_a^b g(x)h(x)\, dx, \quad s_3 = \int_a^b h(x)\, dx.$$

$1°$. Solution with $\lambda \neq \lambda_{1,2}$:

$$y(x) = f(x) + \lambda[A_1 g(x) + A_2],$$

where the constants A_1 and A_2 are given by

$$A_1 = \frac{f_1 - \lambda[f_1 s_3 - (b-a)f_2]}{[s_1 s_3 - (b-a)s_2]\lambda^2 - (s_1 + s_3)\lambda + 1}, \quad A_2 = \frac{f_2 - \lambda(f_2 s_1 - f_1 s_2)}{[s_1 s_3 - (b-a)s_2]\lambda^2 - (s_1 + s_3)\lambda + 1},$$

$$f_1 = \int_a^b f(x)\,dx, \quad f_2 = \int_a^b f(x)h(x)\,dx.$$

$2°$. Solution with $\lambda = \lambda_1 \neq \lambda_2$ and $f_1 = f_2 = 0$:

$$y(x) = f(x) + C y_1(x), \qquad y_1(x) = g(x) + \frac{1 - \lambda_1 s_1}{\lambda_1(b-a)},$$

where C is an arbitrary constant and $y_1(x)$ is an eigenfunction of the equation corresponding to the characteristic value λ_1.

$3°$. The solution with $\lambda = \lambda_2 \neq \lambda_1$ and $f_1 = f_2 = 0$ is given by the formulas of item $2°$ in which one must replace λ_1 and $y_1(x)$ by λ_2 and $y_2(x)$, respectively.

$4°$. Solution with $\lambda = \lambda_{1,2} = \lambda_*$ and $f_1 = f_2 = 0$, where the characteristic value $\lambda_* = \dfrac{2}{s_1 + s_3}$ is double:

$$y(x) = f(x) + C y_*(x), \qquad y_*(x) = g(x) - \frac{s_1 - s_3}{2(b-a)}.$$

Here C is an arbitrary constant and $y_*(x)$ is an eigenfunction of the equation corresponding to λ_*.

6. $\quad y(x) - \lambda \displaystyle\int_a^b [Ag(x) + Bg(t)]h(t)\,y(t)\,dt = f(x).$

The characteristic values of the equation:

$$\lambda_{1,2} = \frac{(A+B)s_1 \pm \sqrt{(A-B)^2 s_1^2 + 4AB s_0 s_2}}{2AB(s_1^2 - s_0 s_2)},$$

where

$$s_0 = \int_a^b h(x)\,dx, \quad s_1 = \int_a^b g(x)h(x)\,dx, \quad s_2 = \int_a^b g^2(x)h(x)\,dx.$$

$1°$. Solution with $\lambda \neq \lambda_{1,2}$:

$$y(x) = f(x) + \lambda[A_1 g(x) + A_2],$$

where the constants A_1 and A_2 are given by

$$A_1 = \frac{Af_1 - AB\lambda(f_1 s_1 - f_2 s_0)}{AB(s_1^2 - s_0 s_2)\lambda^2 - (A+B)s_1\lambda + 1}, \quad A_2 = \frac{Bf_2 - AB\lambda(f_2 s_1 - f_1 s_2)}{AB(s_1^2 - s_0 s_2)\lambda^2 - (A+B)s_1\lambda + 1},$$

$$f_1 = \int_a^b f(x)h(x)\,dx, \quad f_2 = \int_a^b f(x)g(x)h(x)\,dx.$$

$2°$. Solution with $\lambda = \lambda_1 \neq \lambda_2$ and $f_1 = f_2 = 0$:

$$y(x) = f(x) + C y_1(x), \qquad y_1(x) = g(x) + \frac{1 - \lambda_1 A s_1}{\lambda_1 A s_0},$$

where C is an arbitrary constant and $y_1(x)$ is an eigenfunction of the equation corresponding to the characteristic value λ_1.

3°. The solution with $\lambda = \lambda_2 \neq \lambda_1$ and $f_1 = f_2 = 0$ is given by the formulas of item 2° in which one must replace λ_1 and $y_1(x)$ by λ_2 and $y_2(x)$, respectively.

4°. Solution with $\lambda = \lambda_{1,2} = \lambda_*$ and $f_1 = f_2 = 0$, where the characteristic value $\lambda_* = \dfrac{2}{(A+B)s_1}$ is double:

$$y(x) = f(x) + Cy_*(x),$$

where C is an arbitrary constant and $y_*(x)$ is an eigenfunction of the equation corresponding to λ_*. Two cases are possible.

(a) If $A \neq B$, then $4ABs_0s_2 = -(A-B)^2 s_1^2$ and

$$y_*(x) = g(x) - \frac{(A-B)s_1}{2As_0}.$$

(b) If $A = B$, then, in view of $4ABs_0s_2 = -(A-B)^2 s_1^2 = 0$, we have

$$y_*(x) = \begin{cases} g(x) & \text{for } s_0 \neq 0 \text{ and } s_2 = 0, \\ 1 & \text{for } s_0 = 0 \text{ and } s_2 \neq 0, \\ C_1 g(x) + C_2 & \text{for } s_0 = s_2 = 0, \end{cases}$$

where C_1 and C_2 are arbitrary constants.

7. $y(x) - \lambda \displaystyle\int_a^b [Ag(x) + Bg(t) + C]h(t)\, y(t)\, dt = f(x).$

The characteristic values of the equation:

$$\lambda_{1,2} = \frac{(A+B)s_1 + Cs_0 \pm \sqrt{(A-B)^2 s_1^2 + 2(A+B)Cs_1 s_0 + C^2 s_0^2 + 4ABs_0s_2}}{2AB(s_1^2 - s_0 s_2)},$$

where

$$s_0 = \int_a^b h(x)\, dx, \quad s_1 = \int_a^b g(x)h(x)\, dx, \quad s_2 = \int_a^b g^2(x)h(x)\, dx.$$

1°. Solution with $\lambda \neq \lambda_{1,2}$:

$$y(x) = f(x) + \lambda[A_1 g(x) + A_2],$$

where the constants A_1 and A_2 are given by

$$A_1 = \frac{Af_1 - AB\lambda(f_1 s_1 - f_2 s_0)}{AB(s_1^2 - s_0 s_2)\lambda^2 - [(A+B)s_1 + Cs_0]\lambda + 1},$$

$$A_2 = \frac{C_1 f_1 + Bf_2 - AB\lambda(f_2 s_1 - f_1 s_2)}{AB(s_1^2 - s_0 s_2)\lambda^2 - [(A+B)s_1 + Cs_0]\lambda + 1},$$

$$f_1 = \int_a^b f(x)h(x)\, dx, \quad f_2 = \int_a^b f(x)g(x)h(x)\, dx.$$

2°. Solution with $\lambda = \lambda_1 \neq \lambda_2$ and $f_1 = f_2 = 0$:

$$y(x) = f(x) + \widetilde{C} y_1(x), \qquad y_1(x) = g(x) + \frac{1 - \lambda_1 A s_1}{\lambda_1 A s_0},$$

where $\widetilde{C}$ is an arbitrary constant and $y_1(x)$ is an eigenfunction of the equation corresponding to the characteristic value λ_1.

$3°$. The solution with $\lambda = \lambda_2 \neq \lambda_1$ and $f_1 = f_2 = 0$ is given by the formulas of item $2°$ in which one must replace λ_1 and $y_1(x)$ by λ_2 and $y_2(x)$, respectively.

$4°$. Solution with $\lambda = \lambda_{1,2} = \lambda_*$ and $f_1 = f_2 = 0$, where the characteristic value $\lambda_* = \dfrac{2}{(A + B)s_1 + Cs_0}$ is double:

$$y(x) = f(x) + \widetilde{C}y_*(x),$$

where $\widetilde{C}$ is an arbitrary constant and $y_*(x)$ is an eigenfunction of the equation corresponding to λ_*. Two cases are possible.

(a) If $As_1 - (Bs_1 + Cs_0) \neq 0$, then $4As_0(Bs_2 + Cs_1) = -[(A - B)s_1 - Cs_0]^2$ and

$$y_*(x) = g(x) - \frac{(A - B)s_1 - Cs_0}{2As_0}.$$

(b) If $(A - B)s_1 = Cs_0$, then, in view of $4As_0(Bs_2 + Cs_1) = -[(A - B)s_1 - Cs_0]^2 = 0$, we have

$$y_*(x) = \begin{cases} g(x) & \text{for } s_0 \neq 0 \text{ and } Bs_2 = -Cs_1, \\ 1 & \text{for } s_0 = 0 \text{ and } Bs_2 \neq -Cs_1, \\ \widetilde{C}_1 g(x) + \widetilde{C}_2 & \text{for } s_0 = 0 \text{ and } Bs_2 = -Cs_1, \end{cases}$$

where $\widetilde{C}_1$ and $\widetilde{C}_2$ are arbitrary constants.

8. $\displaystyle y(x) - \lambda \int_a^b [A + B(x - t)h(t)]y(t)\,dt = f(x).$

The characteristic values of the equation:

$$\lambda_{1,2} = \frac{A(b - a) \pm \sqrt{[A(b - a) - 2Bh_1]^2 + 2Bh_0[A(b^2 - a^2) - 2Bh_2]}}{B\{A(b - a)[2h_1 - (b + a)h_0] - 2B(h_1^2 - h_0h_2)\}},$$

where

$$h_0 = \int_a^b h(x)\,dx, \quad h_1 = \int_a^b xh(x)\,dx, \quad h_2 = \int_a^b x^2 h(x)\,dx.$$

$1°$. Solution with $\lambda \neq \lambda_{1,2}$:

$$y(x) = f(x) + \lambda(A_1 + A_2 x),$$

where the constants A_1 and A_2 are given by

$$A_1 = \frac{f_1 - \lambda\left[B(f_1 h_1 + f_2 h_2) - \frac{1}{2}Af_2(b^2 - a^2)\right]}{B\left\{A(b - a)\left[h_1 - \frac{1}{2}(b + a)h_0\right] - B(h_1^2 - h_0 h_2)\right\}\lambda^2 + A(b - a)\lambda + 1},$$

$$A_2 = \frac{f_2 - \lambda[A(b - a)f_2 - B(f_1 h_0 + f_2 h_1)]}{B\left\{A(b - a)\left[h_1 - \frac{1}{2}(b + a)h_0\right] - B(h_1^2 - h_0 h_2)\right\}\lambda^2 + A(b - a)\lambda + 1},$$

$$f_1 = A\int_a^b f(x)\,dx - B\int_a^b xf(x)h(x)\,dx, \quad f_2 = B\int_a^b f(x)h(x)\,dx.$$

$2°$. Solution with $\lambda = \lambda_1 \neq \lambda_2$ and $f_1 = f_2 = 0$:

$$y(x) = f(x) + Cy_1(x), \quad y_1(x) = 1 + \frac{2 - 2\lambda_1[A(b - a) - Bh_1]}{\lambda_1[A(b^2 - a^2) - 2Bh_2]}\,x,$$

where C is an arbitrary constant, and $y_1(x)$ is an eigenfunction of the equation corresponding to the characteristic value λ_1.

3°. The solution with $\lambda = \lambda_2 \neq \lambda_1$ and $f_1 = f_2 = 0$ is given by the formulas of item 2° in which one must replace λ_1 and $y_1(x)$ by λ_2 and $y_2(x)$, respectively.

4°. Solution with $\lambda = \lambda_{1,2} = \lambda_*$ and $f_1 = f_2 = 0$, where the characteristic value $\lambda_* = \dfrac{2}{A(b-a)}$ ($A \neq 0$) is double:

$$y(x) = f(x) + Cy_*(x),$$

where C is an arbitrary constant, and $y_*(x)$ is an eigenfunction of the equation corresponding to λ_*. Two cases are possible.

(a) If $A(b-a) - 2Bh_1 \neq 0$, then

$$y_*(x) = 1 - \frac{A(b-a) - 2Bh_1}{A(b^2 - a^2) - 2Bh_2}\, x.$$

(b) If $A(b-a) - 2Bh_1 = 0$, then, in view of $h_0[A(b^2 - a^2) - 2Bh_2] = 0$, we have

$$y_*(x) = \begin{cases} 1 & \text{for } h_0 \neq 0 \text{ and } A(b^2 - a^2) = 2Bh_2, \\ x & \text{for } h_0 = 0 \text{ and } A(b^2 - a^2) \neq 2Bh_2, \\ C_1 + C_2 x & \text{for } h_0 = 0 \text{ and } A(b^2 - a^2) = 2Bh_2, \end{cases}$$

where C_1 and C_2 are arbitrary constants.

9. $$y(x) - \lambda \int_a^b [A + (Bx + Ct)h(t)]y(t)\, dt = f(x).$$

The characteristic values of the equation:

$$\lambda_{1,2} = \frac{A(b-a) + (C+B)h_1 \pm \sqrt{D}}{B\{A(b-a)[2h_1 - (b+a)h_0] + 2C(h_1^2 - h_0 h_2)\}},$$

$$D = [A(b-a) + (C-B)h_1]^2 + 2Bh_0[A(b^2 - a^2) + 2Ch_2],$$

where

$$h_0 = \int_a^b h(x)\, dx, \qquad h_1 = \int_a^b xh(x)\, dx, \qquad h_2 = \int_a^b x^2 h(x)\, dx.$$

1°. Solution with $\lambda \neq \lambda_{1,2}$:

$$y(x) = f(x) + \lambda(A_1 + A_2 x),$$

where the constants A_1 and A_2 are given by

$$A_1 = \Delta^{-1}\left\{ f_1 - \lambda\left[Bf_1 h_1 - Cf_2 h_2 - \tfrac{1}{2}A(b^2 - a^2)f_2\right]\right\},$$

$$A_2 = \Delta^{-1}\left\{ f_2 - \lambda\left[A(b-a)f_2 - Bf_1 h_0 + Cf_2 h_1\right]\right\},$$

$$\Delta = B\left\{A(b-a)\left[h_1 - \tfrac{1}{2}(b+a)h_0\right] + C(h_1^2 - h_0 h_2)\right\}\lambda^2 + [A(b-a) + (B+C)h_1]\lambda + 1,$$

$$f_1 = A\int_a^b f(x)\, dx + C\int_a^b xf(x)h(x)\, dx, \qquad f_2 = B\int_a^b f(x)h(x)\, dx.$$

2°. Solution with $\lambda = \lambda_1 \neq \lambda_2$ and $f_1 = f_2 = 0$:

$$y(x) = f(x) + \widetilde{C}y_1(x), \qquad y_1(x) = 1 + \frac{2 - 2\lambda_1[A(b-a) + Ch_1]}{\lambda_1[A(b^2 - a^2) + 2Ch_2]}\, x,$$

where $\widetilde{C}$ is an arbitrary constant and $y_1(x)$ is an eigenfunction of the equation corresponding to the characteristic value λ_1.

3°. The solution with $\lambda = \lambda_2 \neq \lambda_1$ and $f_1 = f_2 = 0$ is given by the formulas of item 2° in which one must replace λ_1 and $y_1(x)$ by λ_2 and $y_2(x)$, respectively.

4°. Solution with $\lambda = \lambda_{1,2} = \lambda_*$ and $f_1 = f_2 = 0$, where the characteristic value $\lambda_* = \dfrac{2}{A(b-a) + (B+C)h_1}$ is double:

$$y(x) = f(x) + \widetilde{C}y_*(x),$$

where $\widetilde{C}$ is an arbitrary constant and $y_*(x)$ is an eigenfunction of the equation corresponding to λ_*. Two cases are possible.

(a) If $A(b-a) + (C-B)h_1 \neq 0$, then

$$y_*(x) = 1 - \frac{A(b-a) + (C-B)h_1}{A(b^2 - a^2) + 2Ch_2} x.$$

(b) If $A(b-a) + (C-B)h_1 = 0$, then, in view of $h_0[A(b^2-a^2) + 2Ch_2] = 0$, we have

$$y_*(x) = \begin{cases} 1 & \text{for } h_0 \neq 0 \text{ and } A(b^2 - a^2) = -2Ch_2, \\ x & \text{for } h_0 = 0 \text{ and } A(b^2 - a^2) \neq -2Ch_2, \\ \widetilde{C}_1 + \widetilde{C}_2 x & \text{for } h_0 = 0 \text{ and } A(b^2 - a^2) = -2Ch_2, \end{cases}$$

where $\widetilde{C}_1$ and $\widetilde{C}_2$ are arbitrary constants.

10. $y(x) - \lambda \displaystyle\int_a^b [A + B(x-t)h(x)]y(t)\,dt = f(x).$

The characteristic values of the equation:

$$\lambda_{1,2} = \frac{A(b-a) \pm \sqrt{[A(b-a) + 2Bh_1]^2 - 2Bh_0[A(b^2-a^2) + 2Bh_2]}}{B\{h_0[A(b^2-a^2) + 2Bh_2] - 2h_1[A(b-a) + Bh_1]\}},$$

where

$$h_0 = \int_a^b h(x)\,dx, \quad h_1 = \int_a^b xh(x)\,dx, \quad h_2 = \int_a^b x^2 h(x)\,dx.$$

1°. Solution with $\lambda \neq \lambda_{1,2}$:

$$y(x) = f(x) + \lambda\big[AE_1 + (BE_1 x + E_2)h(x)\big],$$

where the constants E_1 and E_2 are given by

$$E_1 = \Delta^{-1}\big[f_1 + \lambda B(f_1 h_1 - f_2 h_0)\big],$$
$$E_2 = B\Delta^{-1}\big\{-f_2 + \lambda f_2\big[A(b-a) + Bh_1\big] + \lambda f_1\big[\tfrac{1}{2}A(b^2 - a^2) + Bh_2\big]\big\},$$
$$\Delta = B\big\{h_0\big[\tfrac{1}{2}A(b^2 - a^2) + Bh_2\big] - h_1[A(b-a) + Bh_1]\big\}\lambda^2 - A(b-a)\lambda + 1,$$
$$f_1 = \int_a^b f(x)\,dx, \quad f_2 = \int_a^b xf(x)\,dx.$$

2°. Solution with $\lambda = \lambda_1 \neq \lambda_2$ and $f_1 = f_2 = 0$:

$$y(x) = f(x) + Cy_1(x), \qquad y_1(x) = A + Bxh(x) + \frac{1 - \lambda_1[A(b-a) + Bh_1]}{\lambda_1 h_0}h(x),$$

where C is an arbitrary constant and $y_1(x)$ is an eigenfunction of the equation corresponding to the characteristic value λ_1.

3°. The solution with $\lambda = \lambda_2 \neq \lambda_1$ and $f_1 = f_2 = 0$ is given by the formulas of item 2° in which one must replace λ_1 and $y_1(x)$ by λ_2 and $y_2(x)$, respectively.

4°. Solution with $\lambda = \lambda_{1,2} = \lambda_*$ and $f_1 = f_2 = 0$, where the characteristic value $\lambda_* = \dfrac{2}{A(b-a)}$ ($A \neq 0$) is double:

$$y(x) = f(x) + Cy_*(x),$$

where C is an arbitrary constant and $y_*(x)$ is an eigenfunction of the equation corresponding to λ_*. Two cases are possible.

(a) If $A(b-a) \neq -2Bh_1$, then

$$y_*(x) = A + Bxh(x) - \frac{A(b-a) + 2Bh_1}{2h_0}h(x).$$

(b) If $A(b-a) = -2Bh_1$, then, in view of $h_0[A(b^2 - a^2) + 2Bh_2] = 0$, we have

$$y_*(x) = \begin{cases} A + Bxh(x) & \text{for } h_0 \neq 0 \text{ and } A(b^2 - a^2) = -2Bh_2, \\ h(x) & \text{for } h_0 = 0 \text{ and } A(b^2 - a^2) \neq -2Bh_2, \\ C_1[A + Bxh(x)] + C_2h(x) & \text{for } h_0 = 0 \text{ and } A(b^2 - a^2) = -2Bh_2, \end{cases}$$

where C_1 and C_2 are arbitrary constants.

11. $y(x) - \lambda \displaystyle\int_a^b [A + (Bx + Ct)h(x)]y(t)\,dt = f(x).$

The characteristic values of the equation:

$$\lambda_{1,2} = \frac{A(b-a) + (B+C)h_1 \pm \sqrt{D}}{C\{2h_1[A(b-a) + Bh_1] - h_0[A(b^2 - a^2) + 2Bh_2]\}},$$
$$D = [A(b-a) + (B-C)h_1]^2 + 2Ch_0[A(b^2 - a^2) + 2Bh_2],$$

where

$$h_0 = \int_a^b h(x)\,dx, \quad h_1 = \int_a^b xh(x)\,dx, \quad h_2 = \int_a^b x^2 h(x)\,dx.$$

1°. Solution with $\lambda \neq \lambda_{1,2}$:

$$y(x) = f(x) + \lambda\big[AE_1 + (BE_1x + E_2)h(x)\big],$$

where the constants E_1 and E_2 are given by

$E_1 = \Delta^{-1}[f_1 - \lambda C(f_1 h_1 - f_2 h_0)],$
$E_2 = C\Delta^{-1}\big\{f_2 - \lambda f_2[A(b-a) + Bh_1] - \lambda f_1\big[\tfrac{1}{2}A(b^2 - a^2) + Bh_2\big]\big\},$
$\Delta = C\big\{h_1[A(b-a) + Bh_1] - h_0\big[\tfrac{1}{2}A(b^2 - a^2) + Bh_2\big]\big\}\lambda^2 - [A(b-a) + (B+C)h_1]\lambda + 1,$
$f_1 = \displaystyle\int_a^b f(x)\,dx, \quad f_2 = \int_a^b xf(x)\,dx.$

2°. Solution with $\lambda = \lambda_1 \neq \lambda_2$ and $f_1 = f_2 = 0$:

$$y(x) = f(x) + \widetilde{C}y_1(x), \qquad y_1(x) = A + Bxh(x) + \frac{1 - \lambda_1[A(b-a) + Bh_1]}{\lambda_1 h_0}h(x),$$

where $\widetilde{C}$ is an arbitrary constant and $y_1(x)$ is an eigenfunction of the equation corresponding to the characteristic value λ_1.

3°. The solution with $\lambda = \lambda_2 \neq \lambda_1$ and $f_1 = f_2 = 0$ is given by the formulas of item 2° in which one must replace λ_1 and $y_1(x)$ by λ_2 and $y_2(x)$, respectively.

4°. Solution with $\lambda = \lambda_{1,2} = \lambda_*$ and $f_1 = f_2 = 0$, where the characteristic value $\lambda_* = \dfrac{2}{A(b-a) + (B+C)h_1}$ $(A \neq 0)$ is double:

$$y(x) = f(x) + \widetilde{C} y_*(x),$$

where $\widetilde{C}$ is an arbitrary constant and $y_*(x)$ is an eigenfunction of the equation corresponding to λ_*. Two cases are possible.

(a) If $A(b-a) + Bh_1 \neq Ch_1$, then

$$y_*(x) = A + Bxh(x) - \frac{A(b-a) + (B-C)h_1}{2h_0} h(x).$$

(b) If $A(b-a) + Bh_1 = Ch_1$, then, in view of $h_0[A(b^2 - a^2) + 2Bh_2] = 0$, we have

$$y_*(x) = \begin{cases} A + Bxh(x) & \text{for } h_0 \neq 0 \text{ and } A(b^2 - a^2) = -2Bh_2, \\ h(x) & \text{for } h_0 = 0 \text{ and } A(b^2 - a^2) \neq -2Bh_2, \\ \widetilde{C}_1[A + Bxh(x)] + \widetilde{C}_2 h(x) & \text{for } h_0 = 0 \text{ and } A(b^2 - a^2) = -2Bh_2, \end{cases}$$

where $\widetilde{C}_1$ and $\widetilde{C}_2$ are arbitrary constants.

12. $\quad y(x) - \lambda \displaystyle\int_a^b [g(x)g(t) + h(x)h(t)]y(t)\,dt = f(x).$

The characteristic values of the equation:

$$\lambda_1 = \frac{s_1 + s_3 + \sqrt{(s_1 - s_3)^2 + 4s_2^2}}{2(s_1 s_3 - s_2^2)}, \qquad \lambda_2 = \frac{s_1 + s_3 - \sqrt{(s_1 - s_3)^2 + 4s_2^2}}{2(s_1 s_3 - s_2^2)},$$

where

$$s_1 = \int_a^b g^2(x)\,dx, \qquad s_2 = \int_a^b g(x)h(x)\,dx, \qquad s_3 = \int_a^b h^2(x)\,dx.$$

1°. Solution with $\lambda \neq \lambda_{1,2}$:

$$y(x) = f(x) + \lambda[A_1 g(x) + A_2 h(x)],$$

where the constants A_1 and A_2 are given by

$$A_1 = \frac{f_1 - \lambda(f_1 s_3 - f_2 s_2)}{(s_1 s_3 - s_2^2)\lambda^2 - (s_1 + s_3)\lambda + 1}, \qquad A_2 = \frac{f_2 - \lambda(f_2 s_1 - f_1 s_2)}{(s_1 s_3 - s_2^2)\lambda^2 - (s_1 + s_3)\lambda + 1},$$

$$f_1 = \int_a^b f(x)g(x)\,dx, \qquad f_2 = \int_a^b f(x)h(x)\,dx.$$

2°. Solution with $\lambda = \lambda_1 \neq \lambda_2$ and $f_1 = f_2 = 0$:

$$y(x) = f(x) + C y_1(x), \qquad y_1(x) = g(x) + \frac{1 - \lambda_1 s_1}{\lambda_1 s_2} h(x),$$

where C is an arbitrary constant and $y_1(x)$ is an eigenfunction of the equation corresponding to the characteristic value λ_1.

3°. The solution with $\lambda = \lambda_2 \neq \lambda_1$ and $f_1 = f_2 = 0$ is given by the formulas of item 2° in which one must replace λ_1 and $y_1(x)$ by λ_2 and $y_2(x)$, respectively.

4°. Solution with $\lambda = \lambda_{1,2} = \lambda_*$ and $f_1 = f_2 = 0$, where the characteristic value $\lambda_* = 1/s_1$ is double:

$$y(x) = f(x) + \widetilde{C}_1 g(x) + \widetilde{C}_2 h(x),$$

where $\widetilde{C}_1$ and $\widetilde{C}_2$ are arbitrary constants.

13. $y(x) - \lambda \int_a^b [g(x)g(t) - h(x)h(t)]y(t)\, dt = f(x).$

The characteristic values of the equation:

$$\lambda_1 = \frac{s_1 - s_3 + \sqrt{(s_1 + s_3)^2 - 4s_2^2}}{2(s_2^2 - s_1 s_3)}, \qquad \lambda_2 = \frac{s_1 - s_3 - \sqrt{(s_1 + s_3)^2 - 4s_2^2}}{2(s_2^2 - s_1 s_3)},$$

where

$$s_1 = \int_a^b g^2(x)\, dx, \quad s_2 = \int_a^b g(x)h(x)\, dx, \quad s_3 = \int_a^b h^2(x)\, dx.$$

$1°$. Solution with $\lambda \neq \lambda_{1,2}$:

$$y(x) = f(x) + \lambda[A_1 g(x) + A_2 h(x)],$$

where the constants A_1 and A_2 are given by

$$A_1 = \frac{f_1 + \lambda(f_1 s_3 - f_2 s_2)}{(s_2^2 - s_1 s_3)\lambda^2 - (s_1 - s_3)\lambda + 1}, \qquad A_2 = \frac{-f_2 + \lambda(f_2 s_1 - f_1 s_2)}{(s_2^2 - s_1 s_3)\lambda^2 - (s_1 - s_3)\lambda + 1},$$

$$f_1 = \int_a^b f(x)g(x)\, dx, \quad f_2 = \int_a^b f(x)h(x)\, dx.$$

$2°$. Solution with $\lambda = \lambda_1 \neq \lambda_2$ and $f_1 = f_2 = 0$:

$$y(x) = f(x) + C y_1(x), \qquad y_1(x) = g(x) + \frac{1 - \lambda_1 s_1}{\lambda_1 s_2} h(x),$$

where C is an arbitrary constant and $y_1(x)$ is an eigenfunction of the equation corresponding to the characteristic value λ_1.

$3°$. The solution with $\lambda = \lambda_2 \neq \lambda_1$ and $f_1 = f_2 = 0$ is given by the formulas of item $2°$ in which one must replace λ_1 and $y_1(x)$ by λ_2 and $y_2(x)$, respectively.

$4°$. Solution with $\lambda = \lambda_{1,2} = \lambda_*$ and $f_1 = f_2 = 0$, where the characteristic value $\lambda_* = \dfrac{2}{s_1 - s_3}$ is double:

$$y(x) = f(x) + C y_*(x), \qquad y_*(x) = g(x) - \frac{s_1 + s_3}{2s_2} h(x),$$

where C is an arbitrary constant and $y_*(x)$ is an eigenfunction of the equation corresponding to λ_*.

14. $y(x) - \lambda \int_a^b [Ag(x)g(t) + Bh(x)h(t)]y(t)\, dt = f(x).$

The characteristic values of the equation:

$$\lambda_{1,2} = \frac{As_1 + Bs_3 \pm \sqrt{(As_1 - Bs_3)^2 + 4ABs_2^2}}{2AB(s_1 s_3 - s_2^2)},$$

where

$$s_1 = \int_a^b g^2(x)\, dx, \quad s_2 = \int_a^b g(x)h(x)\, dx, \quad s_3 = \int_a^b h^2(x)\, dx.$$

$1°$. Solution with $\lambda \neq \lambda_{1,2}$:

$$y(x) = f(x) + \lambda[A_1 g(x) + A_2 h(x)],$$

where the constants A_1 and A_2 are given by

$$A_1 = \frac{Af_1 - \lambda AB(f_1s_3 - f_2s_2)}{AB(s_1s_3 - s_2^2)\lambda^2 - (As_1 + Bs_3)\lambda + 1}, \quad A_2 = \frac{Bf_2 - \lambda AB(f_2s_1 - f_1s_2)}{AB(s_1s_3 - s_2^2)\lambda^2 - (As_1 + Bs_3)\lambda + 1},$$

$$f_1 = \int_a^b f(x)g(x)\,dx, \quad f_2 = \int_a^b f(x)h(x)\,dx.$$

2°. Solution with $\lambda = \lambda_1 \neq \lambda_2$ and $f_1 = f_2 = 0$:

$$y(x) = f(x) + Cy_1(x), \qquad y_1(x) = g(x) + \frac{1 - \lambda_1 As_1}{\lambda_1 As_2} h(x),$$

where C is an arbitrary constant and $y_1(x)$ is an eigenfunction of the equation corresponding to the characteristic value λ_1.

3°. The solution with $\lambda = \lambda_2 \neq \lambda_1$ and $f_1 = f_2 = 0$ is given by the formulas of item 2° in which one must replace λ_1 and $y_1(x)$ by λ_2 and $y_2(x)$, respectively.

4°. Solution with $\lambda = \lambda_{1,2} = \lambda_*$ and $f_1 = f_2 = 0$, where the characteristic value $\lambda_* = \dfrac{2}{As_1 + Bs_3}$ is double:

$$y(x) = f(x) + Cy_*(x),$$

where C is an arbitrary constant and $y_*(x)$ is an eigenfunction of the equation corresponding to λ_*. Two cases are possible.

(a) If $As_1 \neq Bs_3$, then $4ABs_2^2 = -(As_1 - Bs_3)^2$, $AB < 0$, and

$$y_*(x) = g(x) - \frac{As_1 - Bs_3}{2As_2} h(x).$$

(b) If $As_1 = Bs_3$, then, in view of $s_2 = 0$, we have

$$y_*(x) = C_1 g(x) + C_2 h(x),$$

where C_1 and C_2 are arbitrary constants.

15. $\quad y(x) - \lambda \displaystyle\int_a^b [g(x)h(t) + h(x)g(t)]y(t)\,dt = f(x).$

The characteristic values of the equation:

$$\lambda_1 = \frac{1}{s_1 + \sqrt{s_2s_3}}, \quad \lambda_2 = \frac{1}{s_1 - \sqrt{s_2s_3}},$$

where

$$s_1 = \int_a^b h(x)g(x)\,dx, \quad s_2 = \int_a^b h^2(x)\,dx, \quad s_3 = \int_a^b g^2(x)\,dx.$$

1°. Solution with $\lambda \neq \lambda_{1,2}$:

$$y(x) = f(x) + \lambda[A_1 g(x) + A_2 h(x)],$$

where the constants A_1 and A_2 are given by

$$A_1 = \frac{f_1 - \lambda(f_1s_1 - f_2s_2)}{(s_1^2 - s_2s_3)\lambda^2 - 2s_1\lambda + 1}, \quad A_2 = \frac{f_2 - \lambda(f_2s_1 - f_1s_3)}{(s_1^2 - s_2s_3)\lambda^2 - 2s_1\lambda + 1},$$

$$f_1 = \int_a^b f(x)h(x)\,dx, \quad f_2 = \int_a^b f(x)g(x)\,dx.$$

2°. Solution with $\lambda = \lambda_1 \neq \lambda_2$ and $f_1 = f_2 = 0$:

$$y(x) = f(x) + Cy_1(x), \qquad y_1(x) = g(x) + \sqrt{\frac{s_3}{s_2}}\, h(x),$$

where C is an arbitrary constant and $y_1(x)$ is an eigenfunction of the equation corresponding to the characteristic value λ_1.

3°. Solution with $\lambda = \lambda_2 \ne \lambda_1$ and $f_1 = f_2 = 0$:

$$y(x) = f(x) + Cy_2(x), \qquad y_2(x) = g(x) - \sqrt{\frac{s_3}{s_2}}\, h(x),$$

where C is an arbitrary constant and $y_2(x)$ is an eigenfunction of the equation corresponding to the characteristic value λ_2.

4°. The equation has no multiple characteristic values.

16. $\quad y(x) - \lambda \displaystyle\int_a^b [g(x)h(t) - h(x)g(t)]y(t)\,dt = f(x).$

The characteristic values of the equation:

$$\lambda_1 = \frac{1}{\sqrt{s_1^2 - s_2 s_3}}, \qquad \lambda_2 = -\frac{1}{\sqrt{s_1^2 - s_2 s_3}},$$

where

$$s_1 = \int_a^b h(x)g(x)\,dx, \qquad s_2 = \int_a^b h^2(x)\,dx, \qquad s_3 = \int_a^b g^2(x)\,dx.$$

1°. Solution with $\lambda \ne \lambda_{1,2}$:

$$y(x) = f(x) + \lambda[A_1 g(x) + A_2 h(x)],$$

where the constants A_1 and A_2 are given by

$$A_1 = \frac{f_1 + \lambda(f_1 s_1 - f_2 s_2)}{(s_2 s_3 - s_1^2)\lambda^2 + 1}, \qquad A_2 = \frac{-f_2 + \lambda(f_2 s_1 - f_1 s_3)}{(s_2 s_3 - s_1^2)\lambda^2 + 1},$$

$$f_1 = \int_a^b f(x)h(x)\,dx, \qquad f_2 = \int_a^b f(x)g(x)\,dx.$$

2°. Solution with $\lambda = \lambda_1 \ne \lambda_2$ and $f_1 = f_2 = 0$:

$$y(x) = f(x) + Cy_1(x), \qquad y_1(x) = g(x) + \frac{\sqrt{s_1^2 - s_2 s_3} - s_1}{s_2}\, h(x),$$

where C is an arbitrary constant and $y_1(x)$ is an eigenfunction of the equation corresponding to the characteristic value λ_1.

3°. Solution with $\lambda = \lambda_2 \ne \lambda_1$ and $f_1 = f_2 = 0$:

$$y(x) = f(x) + Cy_2(x), \qquad y_2(x) = g(x) - \frac{\sqrt{s_1^2 - s_2 s_3} + s_1}{s_2}\, h(x),$$

where C is an arbitrary constant and $y_2(x)$ is an eigenfunction of the equation corresponding to the characteristic value λ_2.

4°. The equation has no multiple characteristic values.

17. $y(x) - \lambda \int_a^b [Ag(x)h(t) + Bh(x)g(t)]y(t)\, dt = f(x).$

The characteristic values of the equation:

$$\lambda_{1,2} = \frac{(A+B)s_1 \pm \sqrt{(A-B)^2 s_1^2 + 4ABs_2s_3}}{2AB(s_1^2 - s_2s_3)},$$

where

$$s_1 = \int_a^b h(x)g(x)\, dx, \quad s_2 = \int_a^b h^2(x)\, dx, \quad s_3 = \int_a^b g^2(x)\, dx.$$

$1°.$ Solution with $\lambda \neq \lambda_{1,2}$:

$$y(x) = f(x) + \lambda[A_1 g(x) + A_2 h(x)],$$

where the constants A_1 and A_2 are given by

$$A_1 = \frac{Af_1 - \lambda AB(f_1 s_1 - f_2 s_2)}{AB(s_1^2 - s_2 s_3)\lambda^2 - (A+B)s_1\lambda + 1}, \quad A_2 = \frac{Bf_2 - \lambda AB(f_2 s_1 - f_1 s_3)}{AB(s_1^2 - s_2 s_3)\lambda^2 - (A+B)s_1\lambda + 1},$$

$$f_1 = \int_a^b f(x)h(x)\, dx, \quad f_2 = \int_a^b f(x)g(x)\, dx.$$

$2°.$ Solution with $\lambda = \lambda_1 \neq \lambda_2$ and $f_1 = f_2 = 0$:

$$y(x) = f(x) + Cy_1(x), \qquad y_1(x) = g(x) + \frac{1 - \lambda_1 As_1}{\lambda_1 As_2}h(x),$$

where C is an arbitrary constant and $y_1(x)$ is an eigenfunction of the equation corresponding to the characteristic value λ_1.

$3°.$ The solution with $\lambda = \lambda_2 \neq \lambda_1$ and $f_1 = f_2 = 0$ is given by the formulas of item $2°$ in which one must replace λ_1 and $y_1(x)$ by λ_2 and $y_2(x)$, respectively.

$4°.$ Solution with $\lambda = \lambda_{1,2} = \lambda_*$ and $f_1 = f_2 = 0$, where the characteristic value $\lambda_* = \dfrac{2}{(A+B)s_1}$ is double:

$$y(x) = f(x) + Cy_*(x), \qquad y_*(x) = g(x) - \frac{(A-B)s_1}{2As_2}h(x).$$

Here C is an arbitrary constant and $y_*(x)$ is an eigenfunction of the equation corresponding to λ_*.

18. $y(x) - \lambda \int_a^b [g_1(x)h_1(t) + g_2(x)h_2(t)]y(t)\, dt = f(x).$

The characteristic values of the equation λ_1 and λ_2 are given by

$$\lambda_{1,2} = \frac{s_{11} + s_{22} \pm \sqrt{(s_{11} - s_{22})^2 + 4s_{12}s_{21}}}{2(s_{11}s_{22} - s_{12}s_{21})},$$

provided that the integrals

$$s_{11} = \int_a^b h_1(x)g_1(x)\, dx, \quad s_{12} = \int_a^b h_1(x)g_2(x)\, dx, \quad s_{21} = \int_a^b h_2(x)g_1(x)\, dx, \quad s_{22} = \int_a^b h_2(x)g_2(x)\, dx$$

are convergent.

1°. Solution with $\lambda \neq \lambda_{1,2}$:

$$y(x) = f(x) + \lambda[A_1 g_1(x) + A_2 g_2(x)],$$

where the constants A_1 and A_2 are given by

$$A_1 = \frac{f_1 - \lambda(f_1 s_{22} - f_2 s_{12})}{(s_{11} s_{22} - s_{12} s_{21})\lambda^2 - (s_{11} + s_{22})\lambda + 1}, \quad A_2 = \frac{f_2 - \lambda(f_2 s_{11} - f_1 s_{21})}{(s_{11} s_{22} - s_{12} s_{21})\lambda^2 - (s_{11} + s_{22})\lambda + 1},$$

$$f_1 = \int_a^b f(x) h_1(x)\, dx, \quad f_2 = \int_a^b f(x) h_2(x)\, dx.$$

2°. Solution with $\lambda = \lambda_1 \neq \lambda_2$ and $f_1 = f_2 = 0$:

$$y(x) = f(x) + C y_1(x),$$

where C is an arbitrary constant and $y_1(x)$ is an eigenfunction of the equation corresponding to the characteristic value λ_1:

$$y_1(x) = g_1(x) + \frac{1 - \lambda_1 s_{11}}{\lambda_1 s_{12}} g_2(x) = g_1(x) + \frac{\lambda_1 s_{21}}{1 - \lambda_1 s_{22}} g_2(x).$$

3°. The solution with $\lambda = \lambda_2 \neq \lambda_1$ and $f_1 = f_2 = 0$ is given by the formulas of item 2° in which one must replace λ_1 and $y_1(x)$ by λ_2 and $y_2(x)$, respectively.

4°. Solution with $\lambda = \lambda_{1,2} = \lambda_*$ and $f_1 = f_2 = 0$, where the characteristic value $\lambda_* = \dfrac{2}{s_{11} + s_{22}}$ is double (there is no double characteristic value provided that $s_{11} = -s_{22}$):

$$y(x) = f(x) + C y_*(x),$$

where C is an arbitrary constant and $y_*(x)$ is an eigenfunction of the equation corresponding to λ_*. Two cases are possible.

(a) If $s_{11} \neq s_{22}$, then $s_{12} = -\frac{1}{4}(s_{11} - s_{22})^2/s_{21}$, $s_{21} \neq 0$, and

$$y_*(x) = g_1(x) - \frac{s_{11} - s_{22}}{2 s_{12}} g_2(x).$$

Note that in this case, s_{12} and s_{21} have opposite signs.

(b) If $s_{11} = s_{22}$, then, in view of $s_{12} s_{21} = -\frac{1}{4}(s_{11} - s_{22})^2 = 0$, we have

$$y_*(x) = \begin{cases} g_1(x) & \text{for } s_{12} \neq 0 \text{ and } s_{21} = 0, \\ g_2(x) & \text{for } s_{12} = 0 \text{ and } s_{21} \neq 0, \\ C_1 g_1(x) + C_2 g_2(x) & \text{for } s_{12} = 0 \text{ and } s_{21} = 0, \end{cases}$$

where C_1 and C_2 are arbitrary constants.

19. $$y(x) - \lambda \int_a^b [g(x) + h(t)]^m\, y(t)\, dt = f(x), \qquad m = 1, 2, \ldots$$

This is a special case of equation 4.9.20, with $g_k(x) = g^k(x)$, $h_k(t) = C_m^k h^{m-k}(t)$, and $k = 1, \ldots, m$.

Solution:

$$y(x) = f(x) + \lambda \sum_{k=0}^m A_k g^k(x),$$

where the A_k are constants that can be determined from 4.9.20.

20. $y(x) - \lambda \int_a^b \left[\sum_{k=1}^n g_k(x) h_k(t) \right] y(t)\, dt = f(x), \quad n = 2, 3, \ldots$

The characteristic values of the integral equation (counting the multiplicity, we have exactly n of them) are the roots of the algebraic equation

$$\Delta(\lambda) = 0,$$

where

$$\Delta(\lambda) = \begin{vmatrix} 1 - \lambda s_{11} & -\lambda s_{12} & \cdots & -\lambda s_{1n} \\ -\lambda s_{21} & 1 - \lambda s_{22} & \cdots & -\lambda s_{2n} \\ \vdots & \vdots & \ddots & \vdots \\ -\lambda s_{n1} & -\lambda s_{n2} & \cdots & 1 - \lambda s_{nn} \end{vmatrix} = (-\lambda)^n \begin{vmatrix} s_{11} - \lambda^{-1} & s_{12} & \cdots & s_{1n} \\ s_{21} & s_{22} - \lambda^{-1} & \cdots & s_{2n} \\ \vdots & \vdots & \ddots & \vdots \\ s_{n1} & s_{n2} & \cdots & s_{nn} - \lambda^{-1} \end{vmatrix},$$

and the integrals

$$s_{mk} = \int_a^b h_m(x) g_k(x)\, dx; \qquad m, k = 1, \ldots, n,$$

are assumed to be convergent.

Solution with regular λ:

$$y(x) = f(x) + \lambda \sum_{k=1}^n A_k g_k(x),$$

where the constants A_k form the solution of the following system of algebraic equations:

$$A_m - \lambda \sum_{k=1}^n s_{mk} A_k = f_m, \qquad f_m = \int_a^b f(x) h_m(x)\, dx, \quad m = 1, \ldots, n.$$

The A_k can be calculated by Cramer's rule:

$$A_k = \Delta_k(\lambda) / \Delta(\lambda),$$

where

$$\Delta_k(\lambda) = \begin{vmatrix} 1 - \lambda s_{11} & \cdots & -\lambda s_{1k-1} & f_1 & -\lambda s_{1k+1} & \cdots & -\lambda s_{1n} \\ -\lambda s_{21} & \cdots & -\lambda s_{2k-1} & f_2 & -\lambda s_{2k+1} & \cdots & -\lambda s_{2n} \\ \cdots\cdots\cdots\cdots\cdots\cdots\cdots\cdots\cdots\cdots\cdots\cdots\cdots\cdots\cdots\cdots\cdots\cdots\cdots\cdots\cdots \\ -\lambda s_{n1} & \cdots & -\lambda s_{nk-1} & f_n & -\lambda s_{nk+1} & \cdots & 1 - \lambda s_{nn} \end{vmatrix}.$$

For solutions of the equation in the case in which λ is a characteristic value, see Subsection 11.2-2.

⊙ Reference: S. G. Mikhlin (1960).

4.9-2. Equations With Difference Kernel: $K(x, t) = K(x - t)$

21. $y(x) = \lambda \int_{-\pi}^{\pi} K(x - t) y(t)\, dt, \qquad K(x) = K(-x).$

Characteristic values:

$$\lambda_n = \frac{1}{\pi a_n}, \qquad a_n = \frac{1}{\pi} \int_{-\pi}^{\pi} K(x) \cos(nx)\, dx \quad (n = 0, 1, 2, \ldots).$$

The corresponding eigenfunctions are

$$y_0(x) = 1, \quad y_n^{(1)}(x) = \cos(nx), \quad y_n^{(2)}(x) = \sin(nx) \quad (n = 1, 2, \ldots).$$

For each value λ_n with $n \neq 0$, there are two corresponding linearly independent eigenfunctions $y_n^{(1)}(x)$ and $y_n^{(2)}(x)$.

⊙ Reference: M. L. Krasnov, A. I. Kisilev, and G. I. Makarenko (1971).

22. $y(x) + \displaystyle\int_{-\infty}^{\infty} K(x - t)y(t)\,dt = Ae^{\lambda x}.$

Solution:

$$y(x) = \frac{A}{1 + q} e^{\lambda x}, \qquad q = \int_{-\infty}^{\infty} K(x)e^{-\lambda x}\,dx.$$

23. $y(x) + \displaystyle\int_{-\infty}^{\infty} K(x - t)y(t)\,dt = A\cos(\lambda x) + B\sin(\lambda x).$

Solution:

$$y(x) = \frac{AI_c + BI_s}{I_c^2 + I_s^2}\cos(\lambda x) + \frac{BI_c - AI_s}{I_c^2 + I_s^2}\sin(\lambda x),$$

$$I_c = 1 + \int_{-\infty}^{\infty} K(z)\cos(\lambda z)\,dz, \quad I_s = \int_{-\infty}^{\infty} K(z)\sin(\lambda z)\,dz.$$

24. $y(x) - \displaystyle\int_{-\infty}^{\infty} K(x - t)y(t)\,dt = f(x).$

Here $-\infty < x < \infty$, $f(x) \in L_1(-\infty, \infty)$, and $K(x) \in L_1(-\infty, \infty)$.

For the integral equation to be solvable (in L_1), it is necessary and sufficient that

$$1 - \sqrt{2\pi}\,\widetilde{K}(u) \neq 0, \qquad -\infty < u < \infty, \tag{1}$$

where $\widetilde{K}(u) = \frac{1}{\sqrt{2\pi}}\displaystyle\int_{-\infty}^{\infty} K(x)e^{-iux}\,dx$ is the Fourier transform of $K(x)$. In this case, the equation has a unique solution, which is given by

$$y(x) = f(x) + \int_{-\infty}^{\infty} R(x - t)f(t)\,dt,$$

$$R(x) = \frac{1}{\sqrt{2\pi}}\int_{-\infty}^{\infty} \widetilde{R}(u)e^{iux}\,du, \qquad \widetilde{R}(u) = \frac{\widetilde{K}(u)}{1 - \sqrt{2\pi}\,\widetilde{K}(u)}.$$

⊙ Reference: V. A. Ditkin and A. P. Prudnikov (1965).

25. $y(x) - \displaystyle\int_{0}^{\infty} K(x - t)y(t)\,dt = f(x).$

The Wiener–Hopf equation of the second kind.*

Here $0 \leq x < \infty$, $K(x) \in L_1(-\infty, \infty)$, $f(x) \in L_1(0, \infty)$, and $y(x) \in L_1(0, \infty)$.

For the integral equation to be solvable, it is necessary and sufficient that

$$\Omega(u) = 1 - \check{K}(u) \neq 0, \qquad -\infty < u < \infty, \tag{1}$$

where $\check{K}(u) = \displaystyle\int_{-\infty}^{\infty} K(x)e^{iux}\,dx$ is the Fourier transform (in the asymmetric form) of $K(x)$. In this case, the index of the equation can be introduced,

$$\nu = -\text{ind}\,\Omega(u) = -\frac{1}{2\pi}\big[\arg\Omega(u)\big]_{-\infty}^{\infty}.$$

1°. Solution with $\nu = 0$:

$$y(x) = f(x) + \int_{0}^{\infty} R(x, t)f(t)\,dt,$$

---

* A comprehensive discussion of this equation is given in Subsection 11.9-1, Section 11.10, and Section 11.11.

where

$$R(x,t) = R_+(x-t) + R_-(t-x) + \int_0^\infty R_+(x-s)R_-(t-s)\,ds,$$

and the functions $R_+(x)$ and $R_-(x)$ satisfy the conditions $R_+(x) = 0$ and $R_-(x) = 0$ for $x < 0$ and are uniquely defined by their Fourier transforms as follows:

$$1 + \int_0^\infty R_\pm(t)e^{\pm iut}\,dt = \exp\left[-\frac{1}{2}\ln\Omega(u) \mp \frac{1}{2\pi i}\int_{-\infty}^\infty \frac{\ln\Omega(t)}{t-u}\,dt\right].$$

Alternatively, $R_+(x)$ and $R_-(x)$ can be obtained by constructing the solutions of the equations

$$R_+(x) + \int_0^\infty K(x-t)R_+(t)\,dt = K(x), \qquad 0 \le x \le \infty,$$

$$R_-(x) + \int_0^\infty K(t-x)R_-(t)\,dt = K(-x), \qquad 0 \le x \le \infty.$$

2°. Solution with $\nu > 0$:

$$y(x) = f(x) + \sum_{m=1}^\nu C_m x^{m-1} e^{-x} + \int_0^\infty R^\circ(x,t)\left[f(t) + \sum_{m=1}^\nu C_m t^{m-1} e^{-t}\right] dt,$$

where the C_m are arbitrary constants,

$$R^\circ(x,t) = R_+^{(0)}(x-t) + R_-^{(1)}(t-x) + \int_0^\infty R_+^{(0)}(x-s)R_-^{(1)}(t-s)\,ds,$$

and the functions $R_+^{(0)}(x)$ and $R_-^{(1)}(x)$ are uniquely defined by their Fourier transforms:

$$1 + \int_0^\infty R_\pm^{(1)}(t)e^{\pm iut}\,dt = \left(\frac{u-i}{u+i}\right)^\nu \left[1 + \int_0^\infty R_\pm^{(0)}(t)e^{\pm iut}\,dt\right],$$

$$1 + \int_0^\infty R_\pm^{(0)}(t)e^{\pm iut}\,dt = \exp\left[-\frac{1}{2}\ln\Omega^\circ(u) \mp \frac{1}{2\pi i}\int_{-\infty}^\infty \frac{\ln\Omega^\circ(t)}{t-u}\,dt\right],$$

$$\Omega^\circ(u)(u+i)^\nu = \Omega(u)(u-i)^\nu.$$

3°. For $\nu < 0$, the solution exists only if the conditions

$$\int_0^\infty f(x)\psi_m(x)\,dx = 0, \qquad m = 1, 2, \ldots, -\nu,$$

are satisfied. Here $\psi_1(x), \ldots, \psi_\nu(x)$ is the system of linearly independent solutions of the transposed homogeneous equation

$$\psi(x) - \int_0^\infty K(t-x)\psi(t)\,dt = 0.$$

Then

$$y(x) = f(x) + \int_0^\infty R^*(x,t)f(t)\,dt,$$

where

$$R^*(x,t) = R_+^{(1)}(x-t) + R_-^{(0)}(t-x) + \int_0^\infty R_+^{(1)}(x-s)R_-^{(0)}(t-s)\,ds,$$

and the functions $R_+^{(1)}(x)$ and $R_-^{(0)}(x)$ are uniquely defined in item 2° by their Fourier transforms.

⊙ References: V. I. Smirnov (1974), F. D. Gakhov and Yu. I. Cherskii (1978), I. M. Vinogradov (1979).

> **4.9-3. Other Equations of the Form** $y(x) + \int_a^b K(x,t)y(t)\,dt = F(x)$

26. $\quad y(x) - \displaystyle\int_{-\infty}^{\infty} K(x+t)y(t)\,dt = f(x).$

The Fourier transform is used to solving this equation.

Solution:

$$y(x) = \frac{1}{\sqrt{2\pi}} \int_{-\infty}^{\infty} \frac{\tilde{f}(u) + \sqrt{2\pi}\,\tilde{f}(-u)\tilde{K}(u)}{1 - \sqrt{2\pi}\,\tilde{K}(u)\tilde{K}(-u)} e^{iux}\,du,$$

where

$$\tilde{f}(u) = \frac{1}{\sqrt{2\pi}} \int_{-\infty}^{\infty} f(x)e^{-iux}\,dx, \quad \tilde{K}(u) = \frac{1}{\sqrt{2\pi}} \int_{-\infty}^{\infty} K(x)e^{-iux}\,dx.$$

⊙ Reference: V. A. Ditkin and A. P. Prudnikov (1965).

27. $\quad y(x) + \displaystyle\int_{-\infty}^{\infty} e^{\beta t} K(x+t)y(t)\,dt = Ae^{\lambda x}.$

Solution:

$$y(x) = \frac{e^{\lambda x} - k(\lambda)e^{-(\beta+\lambda)x}}{1 - k(\lambda)k(-\beta-\lambda)}, \qquad k(\lambda) = \int_{-\infty}^{\infty} K(x)e^{(\lambda+\beta)x}\,dx.$$

28. $\quad y(x) + \displaystyle\int_{-\infty}^{\infty} [e^{\beta t} K(x+t) + M(x-t)]y(t)\,dt = Ae^{\lambda x}.$

Solution:

$$y(x) = A \frac{I_k(\lambda)e^{px} - [1 + I_m(p)]e^{\lambda x}}{I_k(\lambda)I_k(p) - [1 + I_m(\lambda)][1 + I_m(p)]}, \qquad p = -\lambda - \beta,$$

where

$$I_k(\lambda) = \int_{-\infty}^{\infty} K(z)e^{(\beta+\lambda)z}\,dz, \quad I_m(\lambda) = \int_{-\infty}^{\infty} M(z)e^{-\lambda z}\,dz.$$

29. $\quad y(x) - \displaystyle\int_{0}^{\infty} K(xt)y(t)\,dt = f(x).$

The solution can be obtained with the aid of the inverse Mellin transform:

$$y(x) = \frac{1}{2\pi i} \int_{c-i\infty}^{c+i\infty} \frac{\tilde{f}(s) + \tilde{K}(s)\tilde{f}(1-s)}{1 - \tilde{K}(s)\tilde{K}(1-s)} x^{-s}\,ds,$$

where $\tilde{f}$ and $\tilde{K}$ stand for the Mellin transforms of the right-hand side and of the kernel of the integral equation,

$$\tilde{f}(s) = \int_0^{\infty} f(x)x^{s-1}\,dx, \quad \tilde{K}(s) = \int_0^{\infty} K(x)x^{s-1}\,dx.$$

⊙ Reference: M. L. Krasnov, A. I. Kisilev, and G. I. Makarenko (1971).

30. $\quad y(x) - \displaystyle\int_{0}^{\infty} K(xt)t^{\beta}y(t)\,dt = Ax^{\lambda}.$

Solution:

$$y(x) = A\frac{x^{\lambda} + I_{\beta+\lambda}x^{-\beta-\lambda-1}}{1 - I_{\beta+\lambda}I_{-\lambda-1}}, \qquad I_{\mu} = \int_0^{\infty} K(\xi)\xi^{\mu}\,d\xi.$$

It is assumed that all improper integrals are convergent.

31. $y(x) - \displaystyle\int_0^\infty K(xt)t^\beta y(t)\, dt = f(x).$

The solution can be obtained with the aid of the inverse Mellin transform as follows:

$$y(x) = \frac{1}{2\pi i} \int_{c-i\infty}^{c+i\infty} \frac{\widetilde{f}(s) + \widetilde{K}(s)\widetilde{f}(1+\beta-s)}{1 - \widetilde{K}(s)\widetilde{K}(1+\beta-s)} x^{-s}\, ds,$$

where $\widetilde{f}$ and $\widetilde{K}$ stand for the Mellin transforms of the right-hand side and of the kernel of the integral equation,

$$\widetilde{f}(s) = \int_0^\infty f(x)x^{s-1}\, dx, \quad \widetilde{K}(s) = \int_0^\infty K(x)x^{s-1}\, dx.$$

32. $y(x) - \displaystyle\int_0^\infty g(xt)x^\lambda t^\mu y(t)\, dt = f(x).$

This equation can be rewritten in the form of equation 4.9.31 by setting $K(z) = z^\lambda g(z)$ and $\beta = \mu - \lambda$.

33. $y(x) - \displaystyle\int_0^\infty \frac{1}{t} K\left(\frac{x}{t}\right) y(t)\, dt = 0.$

Eigenfunctions of this integral equation are determined by the roots of the following transcendental (algebraic) equation for the parameter λ:

$$\int_0^\infty K\left(\frac{1}{z}\right) z^{\lambda-1}\, dz = 1. \tag{1}$$

$1°$. For a real simple root λ_n of equation (1), there is a corresponding eigenfunction

$$y_n(x) = x^{\lambda_n}.$$

$2°$. For a real root λ_n of multiplicity r, there are corresponding r eigenfunctions

$$y_{n1}(x) = x^{\lambda_n}, \quad y_{n2}(x) = x^{\lambda_n} \ln x, \quad \ldots, \quad y_{nr}(x) = x^{\lambda_n} \ln^{r-1} x.$$

$3°$. For a complex simple root $\lambda_n = \alpha_n + i\beta_n$ of equation (1), there is a corresponding pair of eigenfunctions

$$y_n^{(1)}(x) = x^{\alpha_n} \cos(\beta_n \ln x), \quad y_n^{(2)}(x) = x^{\alpha_n} \sin(\beta_n \ln x).$$

$4°$. For a complex root $\lambda_n = \alpha_n + i\beta_n$ of multiplicity r, there are corresponding r eigenfunction pairs

$$\begin{aligned}
y_{n1}^{(1)}(x) &= x^{\alpha_n} \cos(\beta_n \ln x), & y_{n1}^{(2)}(x) &= x^{\alpha_n} \sin(\beta_n \ln x), \\
y_{n2}^{(1)}(x) &= x^{\alpha_n} \ln x \cos(\beta_n \ln x), & y_{n2}^{(2)}(x) &= x^{\alpha_n} \ln x \sin(\beta_n \ln x), \\
& \cdots\cdots\cdots\cdots\cdots\cdots & & \cdots\cdots\cdots\cdots\cdots\cdots\cdots \\
y_{nr}^{(1)}(x) &= x^{\alpha_n} \ln^{r-1} x \cos(\beta_n \ln x), & y_{nr}^{(2)}(x) &= x^{\alpha_n} \ln^{r-1} x \sin(\beta_n \ln x).
\end{aligned}$$

The general solution is the linear combination (with arbitrary constants) of the eigenfunctions of the homogeneous integral equation.

34. $y(x) - \int_0^\infty \frac{1}{t} K\left(\frac{x}{t}\right) y(t)\, dt = Ax^b.$

A solution:

$$y(x) = \frac{A}{B} x^b, \qquad B = 1 - \int_0^\infty K\left(\frac{1}{\xi}\right) \xi^{b-1}\, d\xi.$$

It is assumed that the improper integral is convergent and $B \neq 0$. The general solution of the integral equations is the sum of the above solution and the solution of the homogeneous equation 4.9.33.

35. $y(x) - \int_0^\infty \frac{1}{t} K\left(\frac{x}{t}\right) y(t)\, dt = f(x).$

The solution can be obtained with the aid of the inverse Mellin transform:

$$y(x) = \frac{1}{2\pi i} \int_{c-i\infty}^{c+i\infty} \frac{\widetilde{f}(s)}{1 - \widetilde{K}(s)} x^{-s}\, ds,$$

where $\widetilde{f}$ and $\widetilde{K}$ stand for the Mellin transforms of the right-hand side and the kernel of the integral equation,

$$\widetilde{f}(s) = \int_0^\infty f(x) x^{s-1}\, dx, \qquad \widetilde{K}(s) = \int_0^\infty K(x) x^{s-1}\, dx.$$

Example. For $f(x) = Ae^{-\lambda x}$ and $K(x) = \frac{1}{2} e^{-x}$, the solution of the integral equation has the form

$$y(x) = \begin{cases} \dfrac{4A}{(3 - 2C)(\lambda x)^3} & \text{for } \lambda x > 1, \\[2ex] -2A \displaystyle\sum_{k=1}^\infty \dfrac{1}{(\lambda x)^{s_k}\, \psi(s_k)} & \text{for } \lambda x < 1. \end{cases}$$

Here $C = 0.5772\ldots$ is the Euler constant, $\psi(z) = [\ln \Gamma(z)]'_z$ is the logarithmic derivative of the gamma function, and the s_k are the negative roots of the transcendental equation $\Gamma(s_k) = 2$, where $\Gamma(z)$ is the gamma function.

⊙ Reference: M. L. Krasnov, A. I. Kisilev, and G. I. Makarenko (1971).

36. $y(x) + \displaystyle\int_a^b |x - t| g(t) y(t)\, dt = f(x), \qquad a \le x \le b.$

1°. Let us remove the modulus in the integrand,

$$y(x) + \int_a^x (x - t) g(t) y(t)\, dt + \int_x^b (t - x) g(t) y(t)\, dt = f(x). \tag{1}$$

Differentiating (1) with respect to x yields

$$y_x'(x) + \int_a^x g(t) y(t)\, dt - \int_x^b g(t) y(t)\, dt = f_x'(x). \tag{2}$$

Differentiating (2), we arrive at a second-order ordinary differential equation for $y = y(x)$,

$$y_{xx}'' + 2g(x)y = f_{xx}''(x). \tag{3}$$

2°. Let us derive the boundary conditions for equation (3). We assume that the limits of integration satisfy the conditions $-\infty < a < b < \infty$. By setting $x = a$ and $x = b$ in (1), we obtain two consequences

$$y(a) + \int_a^b (t-a)g(t)y(t)\,dt = f(a),$$

$$y(b) + \int_a^b (b-t)g(t)y(t)\,dt = f(b).$$

(4)

Let us express $g(x)y$ from (3) via y''_{xx} and f''_{xx} and substitute the result into (4). Integrating by parts yields the desired boundary conditions for $y(x)$,

$$y(a) + y(b) + (b-a)[f'_x(b) - y'_x(b)] = f(a) + f(b),$$
$$y(a) + y(b) + (a-b)[f'_x(a) - y'_x(a)] = f(a) + f(b).$$

(5)

Note a useful consequence of (5),

$$y'_x(a) + y'_x(b) = f'_x(a) + f'_x(b),$$

(6)

which can be used together with one of conditions (5).

Equation (3) under the boundary conditions (5) determines the solution of the original integral equation. Conditions (5) make it possible to calculate the constants of integration that occur in the solution of the differential equation (3).

37. $\quad y(x) + \displaystyle\int_a^b e^{\lambda|x-t|}g(t)y(t)\,dt = f(x), \qquad a \le x \le b.$

1°. Let us remove the modulus in the integrand:

$$y(x) + \int_a^x e^{\lambda(x-t)}g(t)y(t)\,dt + \int_x^b e^{\lambda(t-x)}g(t)y(t)\,dt = f(x).$$

(1)

Differentiating (1) with respect to x twice yields

$$y''_{xx}(x) + 2\lambda g(x)y(x) + \lambda^2 \int_a^x e^{\lambda(x-t)}g(t)y(t)\,dt + \lambda^2 \int_x^b e^{\lambda(t-x)}g(t)y(t)\,dt = f''_{xx}(x).$$

(2)

Eliminating the integral terms from (1) and (2), we arrive at a second-order ordinary differential equation for $y = y(x)$,

$$y''_{xx} + 2\lambda g(x)y - \lambda^2 y = f''_{xx}(x) - \lambda^2 f(x).$$

(3)

2°. Let us derive the boundary conditions for equation (3). We assume that the limits of integration satisfy the conditions $-\infty < a < b < \infty$. By setting $x = a$ and $x = b$ in (1), we obtain two consequences

$$y(a) + e^{-\lambda a} \int_a^b e^{\lambda t}g(t)y(t)\,dt = f(a),$$

$$y(b) + e^{\lambda b} \int_a^b e^{-\lambda t}g(t)y(t)\,dt = f(b).$$

(4)

Let us express $g(x)y$ from (3) via y''_{xx} and f''_{xx} and substitute the result into (4). Integrating by parts yields the conditions

$$e^{\lambda b}\varphi'_x(b) - e^{\lambda a}\varphi'_x(a) = \lambda e^{\lambda a}\varphi(a) + \lambda e^{\lambda b}\varphi(b),$$
$$e^{-\lambda b}\varphi'_x(b) - e^{-\lambda a}\varphi'_x(a) = \lambda e^{-\lambda a}\varphi(a) + \lambda e^{-\lambda b}\varphi(b), \qquad \varphi(x) = y(x) - f(x).$$

Finally, after some manipulations, we arrive at the desired boundary conditions for $y(x)$:

$$\varphi'_x(a) + \lambda\varphi(a) = 0, \quad \varphi'_x(b) - \lambda\varphi(b) = 0; \qquad \varphi(x) = y(x) - f(x).$$

(5)

Equation (3) under the boundary conditions (5) determines the solution of the original integral equation. Conditions (5) make it possible to calculate the constants of integration that occur in solving the differential equation (3).

38. $y(x) + \displaystyle\int_a^b \sinh(\lambda|x - t|)g(t)y(t)\, dt = f(x),$ **$a \le x \le b.$**

1°. Let us remove the modulus in the integrand:

$$y(x) + \int_a^x \sinh[\lambda(x - t)]g(t)y(t)\, dt + \int_x^b \sinh[\lambda(t - x)]g(t)y(t)\, dt = f(x). \tag{1}$$

Differentiating (1) with respect to x twice yields

$$y_{xx}''(x) + 2\lambda g(x)y(x) + \lambda^2 \int_a^x \sinh[\lambda(x - t)]g(t)y(t)\, dt$$

$$+ \lambda^2 \int_x^b \sinh[\lambda(t - x)]g(t)y(t)\, dt = f_{xx}''(x). \tag{2}$$

Eliminating the integral terms from (1) and (2), we arrive at a second-order ordinary differential equation for $y = y(x)$,

$$y_{xx}'' + 2\lambda g(x)y - \lambda^2 y = f_{xx}''(x) - \lambda^2 f(x). \tag{3}$$

2°. Let us derive the boundary conditions for equation (3). We assume that the limits of integration satisfy the conditions $-\infty < a < b < \infty$. By setting $x = a$ and $x = b$ in (1), we obtain two corollaries

$$y(a) + \int_a^b \sinh[\lambda(t - a)]g(t)y(t)\, dt = f(a),$$

$$y(b) + \int_a^b \sinh[\lambda(b - t)]g(t)y(t)\, dt = f(b). \tag{4}$$

Let us express $g(x)y$ from (3) via y_{xx}'' and f_{xx}'' and substitute the result into (4). Integrating by parts yields the desired boundary conditions for $y(x)$,

$$\sinh[\lambda(b - a)]\varphi_x'(b) - \lambda \cosh[\lambda(b - a)]\varphi(b) = \lambda\varphi(a),$$
$$\sinh[\lambda(b - a)]\varphi_x'(a) + \lambda \cosh[\lambda(b - a)]\varphi(a) = -\lambda\varphi(b); \quad \varphi(x) = y(x) - f(x). \tag{5}$$

Equation (3) under the boundary conditions (5) determines the solution of the original integral equation. Conditions (5) make it possible to calculate the constants of integration that occur in solving the differential equation (3).

39. $y(x) + \displaystyle\int_a^b \sin(\lambda|x - t|)g(t)y(t)\, dt = f(x),$ **$a \le x \le b.$**

1°. Let us remove the modulus in the integrand:

$$y(x) + \int_a^x \sin[\lambda(x - t)]g(t)y(t)\, dt + \int_x^b \sin[\lambda(t - x)]g(t)y(t)\, dt = f(x). \tag{1}$$

Differentiating (1) with respect to x twice yields

$$y_{xx}''(x) + 2\lambda g(x)y(x) - \lambda^2 \int_a^x \sin[\lambda(x - t)]g(t)y(t)\, dt$$

$$- \lambda^2 \int_x^b \sin[\lambda(t - x)]g(t)y(t)\, dt = f_{xx}''(x). \tag{2}$$

Eliminating the integral terms from (1) and (2), we arrive at a second-order ordinary differential equation for $y = y(x)$,

$$y''_{xx} + 2\lambda g(x)y + \lambda^2 y = f''_{xx}(x) + \lambda^2 f(x). \tag{3}$$

2°. Let us derive the boundary conditions for equation (3). We assume that the limits of integration satisfy the conditions $-\infty < a < b < \infty$. By setting $x = a$ and $x = b$ in (1), we obtain two consequences

$$y(a) + \int_a^b \sin[\lambda(t-a)]g(t)y(t)\,dt = f(a),$$

$$y(b) + \int_a^b \sin[\lambda(b-t)]g(t)y(t)\,dt = f(b). \tag{4}$$

Let us express $g(x)y$ from (3) via y''_{xx} and f''_{xx} and substitute the result into (4). Integrating by parts yields the desired boundary conditions for $y(x)$,

$$\sin[\lambda(b-a)]\varphi'_x(b) - \lambda\cos[\lambda(b-a)]\varphi(b) = \lambda\varphi(a),$$

$$\sin[\lambda(b-a)]\varphi'_x(a) + \lambda\cos[\lambda(b-a)]\varphi(a) = -\lambda\varphi(b); \quad \varphi(x) = y(x) - f(x). \tag{5}$$

Equation (3) under the boundary conditions (5) determines the solution of the original integral equation. Conditions (5) make it possible to calculate the constants of integration that occur in solving the differential equation (3).

4.9-4. Equations of the Form $y(x) + \int_a^b K(x,t)y(\cdots)\,dt = F(x)$

40. $y(x) + \displaystyle\int_a^b f(t)y(x-t)\,dt = 0.$

Eigenfunctions of this integral equation* are determined by the roots of the following characteristic (transcendental or algebraic) equation for μ:

$$\int_a^b f(t)\exp(-\mu t)\,dt = -1. \tag{1}$$

1°. For a real (simple) root μ_k of equation (1), there is a corresponding eigenfunction

$$y_k(x) = \exp(\mu_k x).$$

2°. For a real root μ_k of multiplicity r, there are corresponding r eigenfunctions

$$y_{k1}(x) = \exp(\mu_k x), \quad y_{k2}(x) = x\exp(\mu_k x), \quad \dots, \quad y_{kr}(x) = x^{r-1}\exp(\mu_k x).$$

3°. For a complex (simple) root $\mu_k = \alpha_k + i\beta_k$ of equation (1), there is a corresponding pair of eigenfunctions

$$y_k^{(1)}(x) = \exp(\alpha_k x)\cos(\beta_k x), \quad y_k^{(2)}(x) = \exp(\alpha_k x)\sin(\beta_k x).$$

4°. For a complex root $\mu_k = \alpha_k + i\beta_k$ of multiplicity r, there are corresponding r pairs of eigenfunctions

$$y_{k1}^{(1)}(x) = \exp(\alpha_k x)\cos(\beta_k x), \qquad y_{k1}^{(2)}(x) = \exp(\alpha_k x)\sin(\beta_k x),$$

$$y_{k2}^{(1)}(x) = x\exp(\alpha_k x)\cos(\beta_k x), \qquad y_{k2}^{(2)}(x) = x\exp(\alpha_k x)\sin(\beta_k x),$$

$$\dots\dots\dots\dots\dots\dots\dots\dots\dots\dots\dots\dots\dots\dots\dots\dots\dots\dots$$

$$y_{kr}^{(1)}(x) = x^{r-1}\exp(\alpha_k x)\cos(\beta_k x), \qquad y_{kr}^{(2)}(x) = x^{r-1}\exp(\alpha_k x)\sin(\beta_k x).$$

The general solution is the linear combination (with arbitrary constants) of the eigenfunctions of the homogeneous integral equation.

---

* In the equations below that contain $y(x-t)$ in the integrand, the arguments can have, for example, the domain (a) $-\infty < x < \infty$, $-\infty < t < \infty$ for $a = -\infty$ and $b = \infty$ or (b) $a \le t \le b$, $-\infty \le x < \infty$, for a and b such that $-\infty < a < b < \infty$. Case (b) is a special case of (a) if $f(t)$ is nonzero only on the interval $a \le t \le b$.

▶ *For equations 4.9.41–4.9.46, only particular solutions are given. To obtain the general solution, one must add the particular solution to the general solution of the corresponding homogeneous equation 4.9.40.*

41. $y(x) + \displaystyle\int_a^b f(t)y(x - t)\,dt = Ax + B.$

A solution:

$$y(x) = px + q,$$

where the coefficients p and q are given by

$$p = \frac{A}{1 + I_0}, \quad q = \frac{AI_1}{(1 + I_0)^2} + \frac{B}{1 + I_0}, \quad I_0 = \int_a^b f(t)\,dt, \quad I_1 = \int_a^b tf(t)\,dt.$$

42. $y(x) + \displaystyle\int_a^b f(t)y(x - t)\,dt = Ae^{\lambda x}.$

A solution:

$$y(x) = \frac{A}{B}e^{\lambda x}, \quad B = 1 + \int_a^b f(t)\exp(-\lambda t)\,dt.$$

The general solution of the integral equation is the sum of the specified particular solution and the general solution of the homogeneous equation 4.9.40.

43. $y(x) + \displaystyle\int_a^b f(t)y(x - t)\,dt = A\sin(\lambda x).$

A solution:

$$y(x) = \frac{AI_c}{I_c^2 + I_s^2}\sin(\lambda x) + \frac{AI_s}{I_c^2 + I_s^2}\cos(\lambda x),$$

where the coefficients I_c and I_s are given by

$$I_c = 1 + \int_a^b f(t)\cos(\lambda t)\,dt, \quad I_s = \int_a^b f(t)\sin(\lambda t)\,dt.$$

44. $y(x) + \displaystyle\int_a^b f(t)y(x - t)\,dt = A\cos(\lambda x).$

A solution:

$$y(x) = -\frac{AI_s}{I_c^2 + I_s^2}\sin(\lambda x) + \frac{AI_c}{I_c^2 + I_s^2}\cos(\lambda x),$$

where the coefficients I_c and I_s are given by

$$I_c = 1 + \int_a^b f(t)\cos(\lambda t)\,dt, \quad I_s = \int_a^b f(t)\sin(\lambda t)\,dt.$$

45. $y(x) + \displaystyle\int_a^b f(t)y(x - t)\,dt = e^{\mu x}(A\sin \lambda x + B\cos \lambda x).$

A solution:

$$y(x) = e^{\mu x}(p\sin \lambda x + q\cos \lambda x),$$

where the coefficients p and q are given by

$$p = \frac{AI_c - BI_s}{I_c^2 + I_s^2}, \quad q = \frac{AI_s + BI_c}{I_c^2 + I_s^2},$$

$$I_c = 1 + \int_a^b f(t)e^{-\mu t}\cos(\lambda t)\,dt, \quad I_s = \int_a^b f(t)e^{-\mu t}\sin(\lambda t)\,dt.$$

46. $y(x) + \displaystyle\int_a^b f(t)y(x-t)\,dt = g(x).$

1°. For $g(x) = \sum_{k=1}^n A_k \exp(\lambda_k x)$, the equation has a solution

$$y(x) = \sum_{k=1}^n \frac{A_k}{B_k} \exp(\lambda_k x), \qquad B_k = 1 + \int_a^b f(t)\exp(-\lambda_k t)\,dt.$$

2°. For polynomial right-hand side of the equation, $g(x) = \sum_{k=0}^n A_k x^k$, a solution has the form

$$y(x) = \sum_{k=0}^n B_k x^k,$$

where the constants B_k can be found by the method of undetermined coefficients.

3°. For $g(x) = e^{\lambda x} \sum_{k=0}^n A_k x^k$, a solution of the equation has the form

$$y(x) = e^{\lambda x} \sum_{k=0}^n B_k x^k,$$

where the constants B_k can be found by the method of undetermined coefficients.

4°. For $g(x) = \sum_{k=1}^n A_k \cos(\lambda_k x)$, a solution of the equation has the form

$$y(x) = \sum_{k=1}^n B_k \cos(\lambda_k x) + \sum_{k=1}^n C_k \sin(\lambda_k x),$$

where the constants B_k and C_k can be found by the method of undetermined coefficients.

5°. For $g(x) = \sum_{k=1}^n A_k \sin(\lambda_k x)$, a solution of the equation has the form

$$y(x) = \sum_{k=1}^n B_k \cos(\lambda_k x) + \sum_{k=1}^n C_k \sin(\lambda_k x),$$

where the constants B_k and C_k can be found by the method of undetermined coefficients.

6°. For $g(x) = \cos(\lambda x) \sum_{k=0}^n A_k x^k$, a solution of the equation has the form

$$y(x) = \cos(\lambda x) \sum_{k=0}^n B_k x^k + \sin(\lambda x) \sum_{k=0}^n C_k x^k,$$

where the constants B_k and C_k can be found by the method of undetermined coefficients.

7°. For $g(x) = \sin(\lambda x) \sum_{k=0}^n A_k x^k$, a solution of the equation has the form

$$y(x) = \cos(\lambda x) \sum_{k=0}^n B_k x^k + \sin(\lambda x) \sum_{k=0}^n C_k x^k,$$

where the constants B_k and C_k can be found by the method of undetermined coefficients.

8°. For $g(x) = e^{\mu x} \sum\limits_{k=1}^{n} A_k \cos(\lambda_k x)$, a solution of the equation has the form

$$y(x) = e^{\mu x} \sum_{k=1}^{n} B_k \cos(\lambda_k x) + e^{\mu x} \sum_{k=1}^{n} C_k \sin(\lambda_k x),$$

where the constants B_k and C_k can be found by the method of undetermined coefficients.

9°. For $g(x) = e^{\mu x} \sum\limits_{k=1}^{n} A_k \sin(\lambda_k x)$, a solution of the equation has the form

$$y(x) = e^{\mu x} \sum_{k=1}^{n} B_k \cos(\lambda_k x) + e^{\mu x} \sum_{k=1}^{n} C_k \sin(\lambda_k x),$$

where the constants B_k and C_k can be found by the method of undetermined coefficients.

10°. For $g(x) = \cos(\lambda x) \sum\limits_{k=1}^{n} A_k \exp(\mu_k x)$, a solution of the equation has the form

$$y(x) = \cos(\lambda x) \sum_{k=1}^{n} B_k \exp(\mu_k x) + \sin(\lambda x) \sum_{k=1}^{n} C_k \exp(\mu_k x),$$

where the constants B_k and C_k can be found by the method of undetermined coefficients.

11°. For $g(x) = \sin(\lambda x) \sum\limits_{k=1}^{n} A_k \exp(\mu_k x)$, a solution of the equation has the form

$$y(x) = \cos(\lambda x) \sum_{k=1}^{n} B_k \exp(\mu_k x) + \sin(\lambda x) \sum_{k=1}^{n} C_k \exp(\mu_k x),$$

where the constants B_k and C_k can be found by the method of undetermined coefficients.

47. $y(x) + \int_a^b f(t)y(x + \beta t)\, dt = Ax + B.$

A solution:*

$$y(x) = px + q,$$

where

$$p = \frac{A}{1 + I_0}, \quad q = \frac{B}{1 + I_0} - \frac{A I_1 \beta}{(1 + I_0)^2}, \quad I_0 = \int_a^b f(t)\, dt, \quad I_1 = \int_a^b t f(t)\, dt.$$

48. $y(x) + \int_a^b f(t)y(x + \beta t)\, dt = A e^{\lambda x}.$

A solution:

$$y(x) = \frac{A}{B} e^{\lambda x}, \quad B = 1 + \int_a^b f(t) \exp(\lambda \beta t)\, dt.$$

---

* In the equations below that contain $y(x + \beta t)$, $\beta > 0$, in the integrand, the arguments can have, for example, the domain (a) $0 \le x < \infty$, $0 \le t < \infty$ for $a = 0$ and $b = \infty$ or (b) $a \le t \le b$, $0 \le x < \infty$ for a and b such that $0 \le a < b < \infty$. Case (b) is a special case of (a) if $f(t)$ is nonzero only on the interval $a \le t \le b$.

49. $\quad y(x) + \displaystyle\int_a^b f(t)y(x + \beta t)\, dt = A \sin \lambda x + B \cos \lambda x.$

A solution:

$$y(x) = p \sin \lambda x + q \cos \lambda x,$$

where the coefficients p and q are given by

$$p = \frac{AI_c + BI_s}{I_c^2 + I_s^2}, \qquad q = \frac{BI_c - AI_s}{I_c^2 + I_s^2},$$

$$I_c = 1 + \int_a^b f(t) \cos(\lambda\beta t)\, dt, \qquad I_s = \int_a^b f(t) \sin(\lambda\beta t)\, dt.$$

50. $\quad y(x) + \displaystyle\int_a^b f(t)y(x + \beta t)\, dt = g(x).$

$1°$. For $g(x) = \displaystyle\sum_{k=1}^n A_k \exp(\lambda_k x)$, a solution of the equation has the form

$$y(x) = \sum_{k=1}^n \frac{A_k}{B_k} \exp(\lambda_k x), \qquad B_k = 1 + \int_a^b f(t) \exp(\beta\lambda_k t)\, dt.$$

$2°$. For polynomial right-hand side of the equation, $g(x) = \displaystyle\sum_{k=0}^n A_k x^k$, a solution has the form

$$y(x) = \sum_{k=0}^n B_k x^k,$$

where the constants B_k can be found by the method of undetermined coefficients.

$3°$. For $g(x) = e^{\lambda x} \displaystyle\sum_{k=0}^n A_k x^k$, a solution of the equation has the form

$$y(x) = e^{\lambda x} \sum_{k=0}^n B_k x^k,$$

where the constants B_k can be found by the method of undetermined coefficients.

$4°$. For $g(x) = \displaystyle\sum_{k=1}^n A_k \cos(\lambda_k x)$, a solution of the equation has the form

$$y(x) = \sum_{k=1}^n B_k \cos(\lambda_k x) + \sum_{k=1}^n C_k \sin(\lambda_k x),$$

where the constants B_k and C_k can be found by the method of undetermined coefficients.

$5°$. For $g(x) = \displaystyle\sum_{k=1}^n A_k \sin(\lambda_k x)$, a solution of the equation has the form

$$y(x) = \sum_{k=1}^n B_k \cos(\lambda_k x) + \sum_{k=1}^n C_k \sin(\lambda_k x),$$

where the constants B_k and C_k can be found by the method of undetermined coefficients.

6°. For $g(x) = \cos(\lambda x) \sum\limits_{k=0}^{n} A_k x^k$, a solution of the equation has the form

$$y(x) = \cos(\lambda x) \sum_{k=0}^{n} B_k x^k + \sin(\lambda x) \sum_{k=0}^{n} C_k x^k,$$

where the constants B_k and C_k can be found by the method of undetermined coefficients.

7°. For $g(x) = \sin(\lambda x) \sum\limits_{k=0}^{n} A_k x^k$, a solution of the equation has the form

$$y(x) = \cos(\lambda x) \sum_{k=0}^{n} B_k x^k + \sin(\lambda x) \sum_{k=0}^{n} C_k x^k,$$

where the constants B_k and C_k can be found by the method of undetermined coefficients.

8°. For $g(x) = e^{\mu x} \sum\limits_{k=1}^{n} A_k \cos(\lambda_k x)$, a solution of the equation has the form

$$y(x) = e^{\mu x} \sum_{k=1}^{n} B_k \cos(\lambda_k x) + e^{\mu x} \sum_{k=1}^{n} C_k \sin(\lambda_k x),$$

where the constants B_k and C_k can be found by the method of undetermined coefficients.

9°. For $g(x) = e^{\mu x} \sum\limits_{k=1}^{n} A_k \sin(\lambda_k x)$, a solution of the equation has the form

$$y(x) = e^{\mu x} \sum_{k=1}^{n} B_k \cos(\lambda_k x) + e^{\mu x} \sum_{k=1}^{n} C_k \sin(\lambda_k x),$$

where the constants B_k and C_k can be found by the method of undetermined coefficients.

10°. For $g(x) = \cos(\lambda x) \sum\limits_{k=1}^{n} A_k \exp(\mu_k x)$, a solution of the equation has the form

$$y(x) = \cos(\lambda x) \sum_{k=1}^{n} B_k \exp(\mu_k x) + \sin(\lambda x) \sum_{k=1}^{n} C_k \exp(\mu_k x),$$

where the constants B_k and C_k can be found by the method of undetermined coefficients.

11°. For $g(x) = \sin(\lambda x) \sum\limits_{k=1}^{n} A_k \exp(\mu_k x)$, a solution of the equation has the form

$$y(x) = \cos(\lambda x) \sum_{k=1}^{n} B_k \exp(\mu_k x) + \sin(\lambda x) \sum_{k=1}^{n} C_k \exp(\mu_k x),$$

where the constants B_k and C_k can be found by the method of undetermined coefficients.

51. $y(x) + \displaystyle\int_a^b f(t)y(xt)\,dt = 0.$

Eigenfunctions of this integral equation* are determined by the roots of the following transcendental (or algebraic) equation for λ:

$$\int_a^b f(t)t^\lambda\,dt = -1. \tag{1}$$

$1°$. For a real (simple) root λ_k of equation (1), there is a corresponding eigenfunction

$$y_k(x) = x^{\lambda_k}.$$

$2°$. For a real root λ_k of multiplicity r, there are corresponding r eigenfunctions

$$y_{k1}(x) = x^{\lambda_k}, \quad y_{k2}(x) = x^{\lambda_k}\ln x, \quad \ldots, \quad y_{kr}(x) = x^{\lambda_k}\ln^{r-1} x.$$

$3°$. For a complex (simple) root $\lambda_k = \alpha_k + i\beta_k$ of equation (1), there is a corresponding pair of eigenfunctions

$$y_k^{(1)}(x) = x^{\alpha_k}\cos(\beta_k\ln x), \quad y_k^{(2)}(x) = x^{\alpha_k}\sin(\beta_k\ln x).$$

$4°$. For a complex root $\lambda_k = \alpha_k + i\beta_k$ of multiplicity r, there are corresponding r pairs of eigenfunctions

$$
\begin{aligned}
&y_{k1}^{(1)}(x) = x^{\alpha_k}\cos(\beta_k\ln x), &\quad &y_{k1}^{(2)}(x) = x^{\alpha_k}\sin(\beta_k\ln x),\\
&y_{k2}^{(1)}(x) = x^{\alpha_k}\ln x\cos(\beta_k\ln x), &\quad &y_{k2}^{(2)}(x) = x^{\alpha_k}\ln x\sin(\beta_k\ln x),\\
&\quad\cdots\cdots\cdots\cdots\cdots\cdots &\quad &\quad\cdots\cdots\cdots\cdots\cdots\cdots\\
&y_{kr}^{(1)}(x) = x^{\alpha_k}\ln^{r-1} x\cos(\beta_k\ln x), &\quad &y_{kr}^{(2)}(x) = x^{\alpha_k}\ln^{r-1} x\sin(\beta_k\ln x).
\end{aligned}
$$

The general solution is the linear combination (with arbitrary constants) of the eigenfunctions of the homogeneous integral equation.

▶ *For equations 4.9.52–4.9.58, only particular solutions are given. To obtain the general solution, one must add the particular solution to the general solution of the corresponding homogeneous equation 4.9.51.*

52. $y(x) + \displaystyle\int_a^b f(t)y(xt)\,dt = Ax + B.$

A solution:

$$y(x) = \frac{A}{1+I_1}x + \frac{B}{1+I_0}, \qquad I_0 = \int_a^b f(t)\,dt, \quad I_1 = \int_a^b tf(t)\,dt.$$

53. $y(x) + \displaystyle\int_a^b f(t)y(xt)\,dt = Ax^\beta.$

A solution:

$$y(x) = \frac{A}{B}x^\beta, \qquad B = 1 + \int_a^b f(t)t^\beta\,dt.$$

---

* In the equations below that contain $y(xt)$ in the integrand, the arguments can have, for example, the domain (a) $0 \le x \le 1$, $0 \le t \le 1$ for $a = 0$ and $b = 1$, (b) $1 \le x < \infty$, $1 \le t < \infty$ for $a = 1$ and $b = \infty$, (c) $0 \le x < \infty$, $0 \le t < \infty$ for $a = 0$ and $b = \infty$, or (d) $a \le t \le b$, $0 \le x < \infty$ for a and b such that $0 \le a < b \le \infty$. Case (d) is a special case of (c) if $f(t)$ is nonzero only on the interval $a \le t \le b$.

54. $y(x) + \displaystyle\int_a^b f(t)y(xt)\,dt = A \ln x + B.$

A solution:

$$y(x) = p \ln x + q,$$

where

$$p = \frac{A}{1 + I_0}, \quad q = \frac{B}{1 + I_0} - \frac{AI_l}{(1 + I_0)^2}, \quad I_0 = \int_a^b f(t)\,dt, \quad I_l = \int_a^b f(t) \ln t\,dt.$$

55. $y(x) + \displaystyle\int_a^b f(t)y(xt)\,dt = Ax^\beta \ln x.$

A solution:

$$y(x) = px^\beta \ln x + qx^\beta,$$

where

$$p = \frac{A}{1 + I_1}, \quad q = -\frac{AI_2}{(1 + I_1)^2}, \quad I_1 = \int_a^b f(t)t^\beta\,dt, \quad I_2 = \int_a^b f(t)t^\beta \ln t\,dt.$$

56. $y(x) + \displaystyle\int_a^b f(t)y(xt)\,dt = A \cos(\ln x).$

A solution:

$$y(x) = \frac{AI_c}{I_c^2 + I_s^2} \cos(\ln x) + \frac{AI_s}{I_c^2 + I_s^2} \sin(\ln x),$$

$$I_c = 1 + \int_a^b f(t) \cos(\ln t)\,dt, \quad I_s = \int_a^b f(t) \sin(\ln t)\,dt.$$

57. $y(x) + \displaystyle\int_a^b f(t)y(xt)\,dt = A \sin(\ln x).$

A solution:

$$y(x) = -\frac{AI_s}{I_c^2 + I_s^2} \cos(\ln x) + \frac{AI_c}{I_c^2 + I_s^2} \sin(\ln x),$$

$$I_c = 1 + \int_a^b f(t) \cos(\ln t)\,dt, \quad I_s = \int_a^b f(t) \sin(\ln t)\,dt.$$

58. $y(x) + \displaystyle\int_a^b f(t)y(xt)\,dt = Ax^\beta \cos(\lambda \ln x) + Bx^\beta \sin(\lambda \ln x).$

A solution:

$$y(x) = px^\beta \cos(\lambda \ln x) + qx^\beta \sin(\lambda \ln x),$$

where

$$p = \frac{AI_c - BI_s}{I_c^2 + I_s^2}, \quad q = \frac{AI_s + BI_c}{I_c^2 + I_s^2},$$

$$I_c = 1 + \int_a^b f(t)t^\beta \cos(\lambda \ln t)\,dt, \quad I_s = \int_a^b f(t)t^\beta \sin(\lambda \ln t)\,dt.$$

59. $y(x) + \displaystyle\int_a^b f(t)y(\xi)\,dt = 0, \qquad \xi = x\varphi(t).$

Eigenfunctions of this integral equation are determined by the roots of the following transcendental (or algebraic) equation for λ:

$$\int_a^b f(t)[\varphi(t)]^\lambda\,dt = -1. \tag{1}$$

1°. For a real (simple) root λ_k of equation (1), there is a corresponding eigenfunction

$$y_k(x) = x^{\lambda_k}.$$

2°. For a real root λ_k of multiplicity r, there are corresponding r eigenfunctions

$$y_{k1}(x) = x^{\lambda_k}, \quad y_{k2}(x) = x^{\lambda_k}\ln x, \quad \ldots, \quad y_{kr}(x) = x^{\lambda_k}\ln^{r-1} x.$$

3°. For a complex (simple) root $\lambda_k = \alpha_k + i\beta_k$ of equation (1), there is a corresponding pair of eigenfunctions

$$y_k^{(1)}(x) = x^{\alpha_k}\cos(\beta_k\ln x), \quad y_k^{(2)}(x) = x^{\alpha_k}\sin(\beta_k\ln x).$$

4°. For a complex root $\lambda_k = \alpha_k + i\beta_k$ of multiplicity r, there are corresponding r pairs of eigenfunctions

$$y_{k1}^{(1)}(x) = x^{\alpha_k}\cos(\beta_k\ln x), \qquad y_{k1}^{(2)}(x) = x^{\alpha_k}\sin(\beta_k\ln x),$$
$$y_{k2}^{(1)}(x) = x^{\alpha_k}\ln x\cos(\beta_k\ln x), \qquad y_{k2}^{(2)}(x) = x^{\alpha_k}\ln x\sin(\beta_k\ln x),$$
$$\cdots\cdots\cdots\cdots\cdots\cdots\cdots\cdots \qquad \cdots\cdots\cdots\cdots\cdots\cdots\cdots\cdots$$
$$y_{kr}^{(1)}(x) = x^{\alpha_k}\ln^{r-1} x\cos(\beta_k\ln x), \qquad y_{kr}^{(2)}(x) = x^{\alpha_k}\ln^{r-1} x\sin(\beta_k\ln x).$$

The general solution is the linear combination (with arbitrary constants) of the eigenfunctions of the homogeneous integral equation.

60. $y(x) + \displaystyle\int_a^b f(t)y(\xi)\,dt = Ax^\beta, \qquad \xi = x\varphi(t).$

A solution:

$$y(x) = \frac{A}{B}x^\beta, \qquad B = 1 + \int_a^b f(t)[\varphi(t)]^\beta\,dt.$$

It is assumed that $B \neq 0$. A linear combination of eigenfunctions of the corresponding homogeneous equation (see 4.9.59) can be added to this solution.

61. $y(x) + \displaystyle\int_a^b f(t)y(\xi)\,dt = g(x), \qquad \xi = x\varphi(t).$

1°. For $g(x) = \displaystyle\sum_{k=0}^n A_k x^k$, a solution of the equation has the form

$$y(x) = \sum_{k=0}^n \frac{A_k}{B_k}x^k, \qquad B_k = 1 + \int_a^b f(t)[\varphi(t)]^k\,dt. \tag{1}$$

2°. For $g(x) = \ln x\displaystyle\sum_{k=0}^n A_k x^k$, a solution has the form

$$y(x) = \ln x\sum_{k=0}^n B_k x^k + \sum_{k=0}^n C_k x^k, \tag{2}$$

where the constants B_k and C_k can be found by the method of undetermined coefficients.

$3°$. For $g(x) = \sum\limits_{k=0}^{n} A_k (\ln x)^k$, a solution of the equation has the form

$$y(x) = \sum_{k=0}^{n} B_k (\ln x)^k, \tag{3}$$

where the constants B_k can be found by the method of undetermined coefficients.

$4°$. For $g(x) = \sum\limits_{k=1}^{n} A_k \cos(\lambda_k \ln x)$, a solution of the equation has the form

$$y(x) = \sum_{k=1}^{n} B_k \cos(\lambda_k \ln x) + \sum_{k=1}^{n} C_k \sin(\lambda_k \ln x), \tag{4}$$

where the constants B_k and C_k can be found by the method of undetermined coefficients.

$5°$. For $g(x) = \sum\limits_{k=1}^{n} A_k \sin(\lambda_k \ln x)$, a solution of the equation has the form

$$y(x) = \sum_{k=1}^{n} B_k \cos(\lambda_k \ln x) + \sum_{k=1}^{n} C_k \sin(\lambda_k \ln x), \tag{5}$$

where the constants B_k and C_k can be found by the method of undetermined coefficients.

Remark. A linear combination of eigenfunctions of the corresponding homogeneous equation (see 4.9.59) can be added to solutions (1)–(5).

4.10. Some Formulas and Transformations

Let the solution of the integral equation

$$y(x) + \int_a^b K(x,t) y(t)\, dt = f(x) \tag{1}$$

have the form

$$y(x) = f(x) + \int_a^b R(x,t) f(t)\, dt. \tag{2}$$

Then the solution of the more complicated integral equation

$$y(x) + \int_a^b K(x,t) \frac{g(x)}{g(t)} y(t)\, dt = f(x) \tag{3}$$

has the form

$$y(x) = f(x) + \int_a^b R(x,t) \frac{g(x)}{g(t)} f(t)\, dt. \tag{4}$$

Below are formulas for the solutions of integral equations of the form (3) for some specific functions $g(x)$. In all cases, it is assumed that the solution of equation (1) is known and is given by (2).

1°. The solution of the equation

$$y(x) + \int_a^b K(x,t)(x/t)^\lambda y(t)\, dt = f(x)$$

has the form

$$y(x) = f(x) + \int_a^b R(x,t)(x/t)^\lambda f(t)\, dt.$$

2°. The solution of the equation

$$y(x) + \int_a^b K(x,t)e^{\lambda(x-t)}y(t)\, dt = f(x)$$

has the form

$$y(x) = f(x) + \int_a^b R(x,t)e^{\lambda(x-t)}f(t)\, dt.$$

Chapter 5

Nonlinear Equations With Variable Limit of Integration

▶ *Notation: f, g, h, and φ are arbitrary functions of an argument specified in the parentheses (the argument can depend on t, x, and y); A, B, C, a, b, c, k, β, λ, and μ are arbitrary parameters.*

5.1. Equations With Quadratic Nonlinearity That Contain Arbitrary Parameters

5.1-1. Equations of the Form $\int_0^x y(t)y(x-t)\,dt = F(x)$

1. $$\int_0^x y(t)y(x-t)\,dt = Ax + B, \qquad A, B > 0.$$

    Solutions:
    $$y(x) = \pm\sqrt{B}\left[\frac{1}{\sqrt{\pi x}}\exp\left(-\frac{A}{B}x\right) + \sqrt{\frac{A}{B}}\,\mathrm{erf}\left(\sqrt{\frac{A}{B}}\,x\right)\right],$$

    where $\mathrm{erf}\,z = \dfrac{2}{\sqrt{\pi}}\displaystyle\int_0^z \exp(-t^2)\,dt$ is the error function.

2. $$\int_0^x y(t)y(x-t)\,dt = A^2 x^\lambda.$$

    Solutions:
    $$y(x) = \pm A\frac{\sqrt{\Gamma(\lambda+1)}}{\Gamma\left(\frac{\lambda+1}{2}\right)}x^{\frac{\lambda-1}{2}},$$

    where $\Gamma(z)$ is the gamma function.

3. $$\int_0^x y(t)y(x-t)\,dt = Ax^{\lambda-1} + Bx^\lambda, \qquad \lambda > 0.$$

    Solutions:
    $$y(x) = \pm\frac{\sqrt{A\Gamma(\lambda)}}{\Gamma(\lambda/2)}x^{\frac{\lambda-2}{2}}\exp\left(-\lambda\frac{B}{A}x\right)\Phi\left(\frac{\lambda+1}{2}, \frac{\lambda}{2}; \lambda\frac{B}{A}x\right),$$

    where $\Phi(a, c; x)$ is the degenerate hypergeometric function (Kummer's function).

4. $$\int_0^x y(t)y(x-t)\,dt = A^2 e^{\lambda x}.$$

    Solutions: $y(x) = \pm\dfrac{A}{\sqrt{\pi x}}e^{\lambda x}.$

5. $\displaystyle\int_0^x y(t)y(x-t)\,dt = (Ax+B)e^{\lambda x}, \qquad A, B > 0.$

Solutions:

$$y(x) = \pm\sqrt{B}\,e^{\lambda x}\left[\frac{1}{\sqrt{\pi x}}\exp\left(-\frac{A}{B}x\right) + \sqrt{\frac{A}{B}}\,\mathrm{erf}\left(\sqrt{\frac{A}{B}\,x}\right)\right],$$

where $\mathrm{erf}\,z = \dfrac{2}{\sqrt{\pi}}\displaystyle\int_0^z \exp(-t^2)\,dt$ is the error function.

6. $\displaystyle\int_0^x y(t)y(x-t)\,dt = A^2 x^\mu e^{\lambda x}.$

Solutions:

$$y(x) = \pm\frac{A\sqrt{\Gamma(\mu+1)}}{\Gamma\left(\frac{\mu+1}{2}\right)}x^{\frac{\mu-1}{2}}e^{\lambda x}.$$

7. $\displaystyle\int_0^x y(t)y(x-t)\,dt = \left(Ax^{\mu-1} + Bx^\mu\right)e^{\lambda x}.$

Solutions:

$$y(x) = \pm\frac{\sqrt{A\Gamma(\mu)}}{\Gamma(\mu/2)}x^{\frac{\mu-2}{2}}\exp\left[\left(\lambda - \mu\frac{B}{A}\right)x\right]\Phi\left(\frac{\mu+1}{2},\ \frac{\mu}{2};\ \mu\frac{B}{A}\,x\right),$$

where $\Phi(a,c;x)$ is the degenerate hypergeometric function (Kummer's function).

8. $\displaystyle\int_0^x y(t)y(x-t)\,dt = A^2\cosh(\lambda x).$

Solutions: $y(x) = \pm\dfrac{A}{\sqrt{\pi}}\dfrac{d}{dx}\displaystyle\int_0^x \dfrac{I_0(\lambda t)\,dt}{\sqrt{x-t}}$, where I_0 is the modified Bessel function.

9. $\displaystyle\int_0^x y(t)y(x-t)\,dt = A\sinh(\lambda x).$

Solutions: $y = \pm\sqrt{A\lambda}\,I_0(\lambda x)$, where I_0 is the modified Bessel function.

10. $\displaystyle\int_0^x y(t)y(x-t)\,dt = A\sinh(\lambda\sqrt{x}\,).$

Solutions: $y = \pm\sqrt{A}\,\pi^{1/4}2^{-7/8}\lambda^{3/4}x^{-1/8}I_{-1/4}\left(\lambda\sqrt{\frac{1}{2}x}\right)$, where $I_{-1/4}$ is the modified Bessel function.

11. $\displaystyle\int_0^x y(t)y(x-t)\,dt = A^2\cos(\lambda x).$

Solutions: $y(x) = \pm\dfrac{A}{\sqrt{\pi}}\dfrac{d}{dx}\displaystyle\int_0^x \dfrac{J_0(\lambda t)\,dt}{\sqrt{x-t}}$, where J_0 is the Bessel function.

12. $\displaystyle\int_0^x y(t)y(x-t)\,dt = A\sin(\lambda x).$

Solutions: $y = \pm\sqrt{A\lambda}\,J_0(\lambda x)$, where J_0 is the Bessel function.

13. $\int_0^x y(t)y(x-t)\,dt = A\sin(\lambda\sqrt{x}\,).$

Solutions: $y = \pm\sqrt{A}\,\pi^{1/4}2^{-7/8}\lambda^{3/4}x^{-1/8}J_{-1/4}\left(\lambda\sqrt{\tfrac{1}{2}x}\right)$, where $J_{-1/4}$ is the Bessel function.

14. $\int_0^x y(t)y(x-t)\,dt = A^2e^{\mu x}\cosh(\lambda x).$

Solutions: $y(x) = \pm\dfrac{A}{\sqrt{\pi}}e^{\mu x}\dfrac{d}{dx}\displaystyle\int_0^x \dfrac{I_0(\lambda t)\,dt}{\sqrt{x-t}}$, where I_0 is the modified Bessel function.

15. $\int_0^x y(t)y(x-t)\,dt = Ae^{\mu x}\sinh(\lambda x).$

Solutions: $y = \pm\sqrt{A\lambda}\,e^{\mu x}I_0(\lambda x)$, where I_0 is the modified Bessel function.

16. $\int_0^x y(t)y(x-t)\,dt = A^2e^{\mu x}\cos(\lambda x).$

Solutions: $y(x) = \pm\dfrac{A}{\sqrt{\pi}}e^{\mu x}\dfrac{d}{dx}\displaystyle\int_0^x \dfrac{J_0(\lambda t)\,dt}{\sqrt{x-t}}$, where J_0 is the Bessel function.

17. $\int_0^x y(t)y(x-t)\,dt = Ae^{\mu x}\sin(\lambda x).$

Solutions: $y = \pm\sqrt{A\lambda}\,e^{\mu x}J_0(\lambda x)$, where J_0 is the Bessel function.

5.1-2. Equations of the Form $\int_0^x K(x,t)y(t)y(x-t)\,dt = F(x)$

18. $\int_0^x t^k y(t)y(x-t)\,dt = Ax^\lambda, \qquad A > 0.$

Solutions:
$$y(x) = \pm\left[\dfrac{A\Gamma(\lambda+1)}{\Gamma\left(\frac{\lambda+1+k}{2}\right)\Gamma\left(\frac{\lambda+1-k}{2}\right)}\right]^{1/2}x^{\frac{\lambda-k-1}{2}},$$

where $\Gamma(z)$ is the gamma function.

19. $\int_0^x t^k y(t)y(x-t)\,dt = Ae^{\lambda x}.$

Solutions:
$$y(x) = \pm\left[\dfrac{A}{\Gamma\left(\frac{k+1}{2}\right)\Gamma\left(\frac{1-k}{2}\right)}\right]^{1/2}x^{-\frac{k+1}{2}}e^{\lambda x},$$

where $\Gamma(z)$ is the gamma function.

20. $\int_0^x t^k y(t)y(x-t)\,dt = Ax^\mu e^{\lambda x}.$

Solutions:
$$y(x) = \pm\left[\dfrac{A\Gamma(\mu+1)}{\Gamma\left(\frac{\mu+k+1}{2}\right)\Gamma\left(\frac{\mu-k+1}{2}\right)}\right]^{1/2}x^{\frac{\mu-k-1}{2}}e^{\lambda x},$$

where $\Gamma(z)$ is the gamma function.

21. $\int_0^x \dfrac{y(t)y(x-t)}{ax+bt}\,dt = Ax^\lambda.$

Solutions:

$$y(x) = \pm\sqrt{\dfrac{A}{I}}\,x^{\lambda/2}, \qquad I = \int_0^1 z^{\lambda/2}(1-z)^{\lambda/2}\dfrac{dz}{a+bz}.$$

22. $\int_0^x \dfrac{y(t)y(x-t)}{ax+bt}\,dt = Ae^{\lambda x}.$

Solutions:

$$y(x) = \pm\sqrt{\dfrac{A}{I}}\,e^{\lambda x}, \qquad I = \dfrac{1}{b}\ln\left(1+\dfrac{b}{a}\right).$$

23. $\int_0^x \dfrac{y(t)y(x-t)}{ax+bt}\,dt = Ax^\mu e^{\lambda x}.$

Solutions:

$$y(x) = \pm\sqrt{\dfrac{A}{I}}\,x^{\mu/2}e^{\lambda x}, \qquad I = \int_0^1 z^{\mu/2}(1-z)^{\mu/2}\dfrac{dz}{a+bz}.$$

24. $\int_0^x \dfrac{y(t)y(x-t)}{\sqrt{ax^2+bt^2}}\,dt = Ax^\lambda.$

Solutions:

$$y(x) = \pm\sqrt{\dfrac{A}{I}}\,x^{\lambda/2}, \qquad I = \int_0^1 z^{\lambda/2}(1-z)^{\lambda/2}\dfrac{dz}{\sqrt{a+bz^2}}.$$

25. $\int_0^x \dfrac{y(t)y(x-t)}{\sqrt{ax^2+bt^2}}\,dt = Ae^{\lambda x}.$

Solutions:

$$y(x) = \pm\sqrt{\dfrac{A}{I}}\,e^{\lambda x}, \qquad I = \int_0^1 \dfrac{dz}{\sqrt{a+bz^2}}.$$

26. $\int_0^x \dfrac{y(t)y(x-t)}{\sqrt{ax^2+bt^2}}\,dt = Ax^\mu e^{\lambda x}.$

Solutions:

$$y(x) = \pm\sqrt{\dfrac{A}{I}}\,x^{\mu/2}e^{\lambda x}, \qquad I = \int_0^1 z^{\mu/2}(1-z)^{\mu/2}\dfrac{dz}{\sqrt{a+bz^2}}.$$

5.1-3. Equations of the Form $\int_0^x G(\cdots)\,dt = F(x)$

27. $\int_0^x y(t)y(ax+bt)\,dt = Ax^\lambda.$

Solutions:

$$y(x) = \pm\sqrt{\dfrac{A}{I}}\,x^{\frac{\lambda-1}{2}}, \qquad I = \int_0^1 z^{\frac{\lambda-1}{2}}(a+bz)^{\frac{\lambda-1}{2}}\,dz.$$

28. $\displaystyle\int_0^x y(t)y(ax - t)\, dt = Ae^{\lambda x}, \qquad a \geq 1.$

Solutions:

$$y(x) = \pm\sqrt{\frac{A}{I}}\,\frac{\exp(\lambda x/a)}{\sqrt{x}}, \qquad I = \int_0^1 \frac{dz}{\sqrt{z(a-z)}}.$$

29. $\displaystyle\int_0^x y(t)y(ax - t)\, dt = Ax^\mu e^{\lambda x}, \qquad a \geq 1.$

Solutions:

$$y(x) = \pm\sqrt{\frac{A}{I}}\, x^{\frac{\mu-1}{2}}\exp(\lambda x/a), \qquad I = \int_0^1 z^{\frac{\mu-1}{2}}(a-z)^{\frac{\mu-1}{2}}\,dz.$$

5.1-4. Equations of the Form $y(x) + \int_a^x K(x,t)y^2(t)\, dt = F(x)$

---

30. $\displaystyle y(x) + A\int_a^x y^2(t)\, dt = Bx + C.$

By differentiation, this integral equation can be reduced to a separable ordinary differential equation.

1°. Solution with $AB > 0$:

$$y(x) = k\frac{(k + y_a)\exp[2Ak(x-a)] + y_a - k}{(k + y_a)\exp[2Ak(x-a)] - y_a + k}, \qquad k = \sqrt{\frac{B}{A}}, \quad y_a = aB + C.$$

2°. Solution with $AB < 0$:

$$y(x) = k\tan\left[Ak(a-x) + \arctan\frac{y_a}{k}\right], \qquad k = \sqrt{-\frac{B}{A}}, \quad y_a = aB + C.$$

3°. Solution with $B = 0$:

$$y(x) = \frac{C}{AC(x-a) + 1}.$$

31. $\displaystyle y(x) + k\int_a^x (x - t)y^2(t)\, dt = Ax^2 + Bx + C.$

This is a special case of equation 5.8.5 with $f(y) = ky^2$.

Solution in an implicit form:

$$\int_{y_0}^y \left[4Au - 2kF(u) + B^2 - 4AC\right]^{-1/2}du = \pm(x-a),$$

$$F(u) = \tfrac{1}{3}\left(u^3 - y_0^3\right), \quad y_0 = Aa^2 + Ba + C.$$

32. $\displaystyle y(x) + A\int_a^x t^\lambda y^2(t)\, dt = Bx^{\lambda+1} + C.$

This is a special case of equation 5.8.6 with $f(y) = Ay^2$. By differentiation, this integral equation can be reduced to a separable ordinary differential equation.

Solution in an implicit form:

$$(\lambda + 1)\int_{y_a}^y \frac{du}{Au^2 - B(\lambda+1)} + x^{\lambda+1} - a^{\lambda+1} = 0, \qquad y_a = Ba^{\lambda+1} + C.$$

33. $\quad y(x) + A \displaystyle\int_0^x x^{-\lambda-1} y^2(t)\, dt = Bx^\lambda, \qquad \lambda > -\frac{1}{2}.$

Solutions: $y_1(x) = \beta_1 x^\lambda$ and $y_2(x) = \beta_2 x^\lambda$, where $\beta_{1,2}$ are the roots of the quadratic equation $A\beta^2 + (2\lambda + 1)\beta - B(2\lambda + 1) = 0$.

34. $\quad y(x) + \displaystyle\int_0^x \frac{y^2(t)\, dt}{ax + bt} = A.$

Solutions: $y_1(x) = \lambda_1$ and $y_2(x) = \lambda_2$, where $\lambda_{1,2}$ are the roots of the quadratic equation $\ln\left(1 + \dfrac{b}{a}\right)\lambda^2 + b\lambda - Ab = 0$.

35. $\quad y(x) + A \displaystyle\int_0^x \frac{y^2(t)\, dt}{x^2 + t^2} = Bx.$

Solutions: $y_1(x) = \lambda_1 x$ and $y_2(x) = \lambda_2 x$, where $\lambda_{1,2}$ are the roots of the quadratic equation $\left(1 - \frac{1}{4}\pi\right) A\lambda^2 + \lambda - B = 0$.

36. $\quad y(x) + \displaystyle\int_0^x \frac{y^2(t)\, dt}{\sqrt{ax^2 + bt^2}} = A.$

Solutions: $y_1(x) = \lambda_1$ and $y_2(x) = \lambda_2$, where $\lambda_{1,2}$ are the roots of the quadratic equation

$$I\lambda^2 + \lambda - A = 0, \qquad I = \int_0^1 \frac{dz}{\sqrt{a + bz^2}}.$$

37. $\quad y(x) + A \displaystyle\int_0^x \left(ax^n + bt^n\right)^{-\frac{\lambda+1}{n}} y^2(t)\, dt = Bx^\lambda.$

Solutions: $y_1(x) = \beta_1 x^\lambda$ and $y_2(x) = \beta_2 x^\lambda$, where $\beta_{1,2}$ are the roots of the quadratic equation

$$AI\beta^2 + \beta - B = 0, \qquad I = \int_0^1 z^{2\lambda}\left(a + bz^n\right)^{-\frac{\lambda+1}{n}}\, dz.$$

38. $\quad y(x) + A \displaystyle\int_a^x e^{\lambda t} y^2(t)\, dt = Be^{\lambda x} + C.$

This is a special case of equation 5.8.11 with $f(y) = Ay^2$. By differentiation, this integral equation can be reduced to a separable ordinary differential equation.

Solution in an implicit form:

$$\lambda \int_{y_0}^y \frac{du}{Au^2 - B\lambda} + e^{\lambda x} - e^{\lambda a} = 0, \qquad y_0 = Be^{\lambda a} + C.$$

39. $\quad y(x) + A \displaystyle\int_a^x e^{\lambda(x-t)} y^2(t)\, dt = B.$

This is a special case of equation 5.8.12. By differentiation, this integral equation can be reduced to the separable ordinary differential equation

$$y_x' + Ay^2 - \lambda y + \lambda B = 0, \qquad y(a) = B.$$

Solution in an implicit form:

$$\int_B^y \frac{du}{Au^2 - \lambda u + \lambda B} + x - a = 0.$$

40. $y(x) + k \int_a^x e^{\lambda(x-t)} y^2(t)\, dt = Ae^{\lambda x} + B.$

Solution in an implicit form:

$$\int_{y_0}^y \frac{du}{\lambda u - ku^2 - \lambda B} = x - a, \qquad y_0 = Ae^{\lambda a} + B.$$

41. $y(x) + k \int_a^x \sinh[\lambda(x-t)] y^2(t)\, dt = Ae^{\lambda x} + Be^{-\lambda x} + C.$

This is a special case of equation 5.8.14 with $f(y) = ky^2$.
Solution in an implicit form:

$$\int_{y_0}^y \left[\lambda^2 u^2 - 2\lambda^2 Cu - 2k\lambda F(u) + \lambda^2(C^2 - 4AB) \right]^{-1/2} du = \pm(x-a),$$

$$F(u) = \tfrac{1}{3}\left(u^3 - y_0^3\right), \quad y_0 = Ae^{\lambda a} + Be^{-\lambda a} + C.$$

42. $y(x) + k \int_a^x \sinh[\lambda(x-t)] y^2(t)\, dt = A\cosh(\lambda x) + B.$

This is a special case of equation 5.8.15 with $f(y) = ky^2$.
Solution in an implicit form:

$$\int_{y_0}^y \left[\lambda^2 u^2 - 2\lambda^2 Bu - 2k\lambda F(u) + \lambda^2(B^2 - A^2) \right]^{-1/2} du = \pm(x-a),$$

$$F(u) = \tfrac{1}{3}\left(u^3 - y_0^3\right), \quad y_0 = A\cosh(\lambda a) + B.$$

43. $y(x) + k \int_a^x \sinh[\lambda(x-t)] y^2(t)\, dt = A\sinh(\lambda x) + B.$

This is a special case of equation 5.8.16 with $f(y) = ky^2$.
Solution in an implicit form:

$$\int_{y_0}^y \left[\lambda^2 u^2 - 2\lambda^2 Bu - 2k\lambda F(u) + \lambda^2(A^2 + B^2) \right]^{-1/2} du = \pm(x-a),$$

$$F(u) = \tfrac{1}{3}\left(u^3 - y_0^3\right), \quad y_0 = A\sinh(\lambda a) + B.$$

44. $y(x) + k \int_a^x \sin[\lambda(x-t)] y^2(t)\, dt = A\sin(\lambda x) + B\cos(\lambda x) + C.$

This is a special case of equation 5.8.17 with $f(y) = ky^2$.
Solution in an implicit form:

$$\int_{y_0}^y \left[\lambda^2 D - \lambda^2 u^2 + 2\lambda^2 Cu - 2k\lambda F(u) \right]^{-1/2} du = \pm(x-a),$$

$$y_0 = A\sin(\lambda a) + B\cos(\lambda a) + C, \quad D = A^2 + B^2 - C^2, \quad F(u) = \tfrac{1}{3}\left(u^3 - y_0^3\right).$$

5.1-5. Equations of the Form $y(x) + \int_a^x K(x,t) y(t) y(x-t)\, dt = F(x)$

45. $y(x) + A \int_0^x y(t) y(x-t)\, dt = AB^2 x + B.$

A solution: $y(x) = B.$

46. $y(x) + A \displaystyle\int_0^x y(t)y(x-t)\,dt = (AB^2x + B)e^{\lambda x}.$

A solution: $y(x) = Be^{\lambda x}.$

47. $y(x) + \dfrac{\lambda}{2\beta} \displaystyle\int_0^x y(t)y(x-t)\,dt = \tfrac{1}{2}\beta \sinh(\lambda x).$

A solution: $y(x) = \beta I_1(\lambda x)$, where $I_1(x)$ is the modified Bessel function.

48. $y(x) - \dfrac{\lambda}{2\beta} \displaystyle\int_0^x y(t)y(x-t)\,dt = \tfrac{1}{2}\beta \sin(\lambda x).$

A solution: $y(x) = \beta J_1(\lambda x)$, where $J_1(x)$ is the Bessel function.

49. $y(x) + A \displaystyle\int_0^x x^{-\lambda-1}y(t)y(x-t)\,dt = Bx^\lambda.$

Solutions: $y_1(x) = \beta_1 x^\lambda$ and $y_2(x) = \beta_2 x^\lambda$, where $\beta_{1,2}$ are the roots of the quadratic equation

$$AI\beta^2 + \beta - B = 0, \qquad I = \int_0^1 z^\lambda (1-z)^\lambda \, dz = \frac{\Gamma^2(\lambda+1)}{\Gamma(2\lambda+2)}.$$

5.2. Equations With Quadratic Nonlinearity That Contain Arbitrary Functions

5.2-1. Equations of the Form $\int_a^x G(\cdots)\,dt = F(x)$

1. $\displaystyle\int_a^x K(x,t)y^2(t)\,dt = f(x).$

The substitution $w(x) = y^2(x)$ leads to the linear equation

$$\int_a^x K(x,t)w(t)\,dt = f(x).$$

2. $\displaystyle\int_a^x K(t)y(x)y(t)\,dt = f(x).$

Solutions:

$$y(x) = \pm f(x)\left[2\int_a^x K(t)f(t)\,dt\right]^{-1/2}.$$

3. $\displaystyle\int_0^x f\!\left(\frac{t}{x}\right)y(t)y(x-t)\,dt = Ax^\lambda.$

Solutions:

$$y(x) = \pm\sqrt{\frac{A}{I}}\,x^{\frac{\lambda-1}{2}}, \qquad I = \int_0^1 f(z)z^{\frac{\lambda-1}{2}}(1-z)^{\frac{\lambda-1}{2}}\,dz.$$

4. $\displaystyle\int_0^x f\!\left(\frac{t}{x}\right)y(t)y(x-t)\,dt = Ae^{\lambda x}.$

Solutions:

$$y(x) = \pm\sqrt{\frac{A}{I}}\,\frac{e^{\lambda x}}{\sqrt{x}}, \qquad I = \int_0^1 \frac{f(z)\,dz}{\sqrt{z(1-z)}}.$$

5. $\displaystyle\int_0^x f\left(\frac{t}{x}\right) y(t)y(x-t)\,dt = Ax^\mu e^{\lambda x}.$

Solutions:

$$y(x) = \pm\sqrt{\frac{A}{I}}\, x^{\frac{\mu-1}{2}} e^{\lambda x}, \qquad I = \int_0^1 f(z) z^{\frac{\mu-1}{2}} (1-z)^{\frac{\mu-1}{2}}\,dz.$$

6. $\displaystyle\int_0^x f\left(\frac{t}{x}\right) y(t)y(ax+bt)\,dt = Ax^\lambda.$

Solutions:

$$y(x) = \pm\sqrt{\frac{A}{I}}\, x^{\frac{\lambda-1}{2}}, \qquad I = \int_0^1 f(z) z^{\frac{\lambda-1}{2}} (a+bz)^{\frac{\lambda-1}{2}}\,dz.$$

7. $\displaystyle\int_0^x f\left(\frac{t}{x}\right) y(t)y(ax-t)\,dt = Ae^{\lambda x}, \qquad a \ge 1.$

Solutions:

$$y(x) = \pm\sqrt{\frac{A}{I}}\, \frac{\exp(\lambda x/a)}{\sqrt{x}}, \qquad I = \int_0^1 \frac{f(z)\,dz}{\sqrt{z(a-z)}}.$$

8. $\displaystyle\int_0^x f\left(\frac{t}{x}\right) y(t)y(ax-t)\,dt = Ax^\mu e^{\lambda x}, \qquad a \ge 1.$

Solutions:

$$y(x) = \pm\sqrt{\frac{A}{I}}\, x^{\frac{\mu-1}{2}} \exp(\lambda x/a), \qquad I = \int_0^1 f(z) z^{\frac{\mu-1}{2}} (a-z)^{\frac{\mu-1}{2}}\,dz.$$

5.2-2. Equations of the Form $y(x) + \int_a^x K(x,t)y^2(t)\,dt = F(x)$

9. $\displaystyle y(x) + \int_a^x f(t)y^2(t)\,dt = A.$

Solution:

$$y(x) = A\left[1 + A\int_a^x f(t)\,dt\right]^{-1}.$$

10. $\displaystyle y(x) + \int_a^x e^{\lambda(x-t)}g(t)y^2(t)\,dt = f(x).$

Differentiating the equation with respect to x yields

$$y_x' + g(x)y^2 + \lambda \int_a^x e^{\lambda(x-t)}g(t)y^2(t)\,dt = f_x'(x). \tag{1}$$

Eliminating the integral term from (1) with the aid of the original equation, we arrive at a Riccati ordinary differential equation,

$$y_x' + g(x)y^2 - \lambda y + \lambda f(x) - f_x'(x) = 0, \tag{2}$$

under the initial condition $y(a) = f(a)$. Equation (2) can be reduced to a second-order linear ordinary differential equation. For the exact solutions of equation (2) with various specific functions f and g, see, for example, E. Kamke (1977) and A. D. Polyanin and V. F. Zaitsev (1995).

11. $\quad y(x) + \displaystyle\int_a^x g(x)h(t)y^2(t)\,dt = f(x).$

Differentiating the equation with respect to x yields

$$y'_x + g(x)h(x)y^2 + g'_x(x)\int_a^x h(t)y^2(t)\,dt = f'_x(x). \tag{1}$$

Eliminating the integral term from (1) with the aid of the original equation, we arrive at a Riccati ordinary differential equation,

$$y'_x + g(x)h(x)y^2 - \frac{g'_x(x)}{g(x)}y = f'_x(x) - \frac{g'_x(x)}{g(x)}f(x), \tag{2}$$

under the initial condition $y(a) = f(a)$. Equation (2) can be reduced to a second-order linear ordinary differential equation. For the exact solutions of equation (2) with various specific functions f, g, and h, see, for example, E. Kamke (1977) and A. D. Polyanin and V. F. Zaitsev (1995).

12. $\quad y(x) + \displaystyle\int_0^x x^{-\lambda-1} f\left(\frac{t}{x}\right) y^2(t)\,dt = Ax^\lambda.$

Solutions: $y_1(x) = \beta_1 x^\lambda$ and $y_2(x) = \beta_2 x^\lambda$, where $\beta_{1,2}$ are the roots of the quadratic equation

$$I\beta^2 + \beta - A = 0, \qquad I = \int_0^1 f(z)z^{2\lambda}\,dz.$$

13. $\quad y(x) - \displaystyle\int_{-\infty}^x e^{\lambda t + \beta x} f(x-t)y^2(t)\,dt = 0.$

This is a special case of equation 5.3.19 with $k = 2$.

14. $\quad y(x) - \displaystyle\int_x^\infty e^{\lambda t + \beta x} f(x-t)y^2(t)\,dt = 0.$

A solution:

$$y(x) = \frac{1}{A}e^{-(\lambda+\beta)x}, \qquad A = \int_0^\infty e^{-(\lambda+2\beta)z} f(-z)\,dz.$$

5.2-3. Equations of the Form $y(x) + \int_a^x G(\cdots)\,dt = F(x)$

15. $\quad y(x) + \displaystyle\int_0^x \frac{1}{x} f\left(\frac{t}{x}\right) y(t)y(x-t)\,dt = Ae^{\lambda x}.$

Solutions:

$$y_1(x) = B_1 e^{\lambda x}, \qquad y_2(x) = B_2 e^{\lambda x},$$

where B_1 and B_2 are the roots of the quadratic equation

$$IB^2 + B - A = 0, \qquad I = \int_0^1 f(z)\,dz.$$

16. $y(x) + A \int_0^x x^{-\lambda-1} f\left(\dfrac{t}{x}\right) y(t) y(x - t)\, dt = Bx^\lambda.$

    Solutions: $y_1(x) = \beta_1 x^\lambda$ and $y_2(x) = \beta_2 x^\lambda$, where $\beta_{1,2}$ are the roots of the quadratic equation

$$AI\beta^2 + \beta - B = 0, \qquad I = \int_0^1 f(z) z^\lambda (1 - z)^\lambda\, dz.$$

17. $y(x) + \int_x^\infty f(t - x) y(t - x) y(t)\, dt = ae^{-\lambda x}.$

    Solutions: $y(x) = b_k e^{-\lambda x}$, where b_k ($k = 1, 2$) are the roots of the quadratic equation

$$b^2 I + b - a = 0, \qquad I = \int_0^\infty f(z) e^{-2\lambda z}\, dz.$$

To calculate the integral I, it is convenient to use tables of Laplace transforms (with parameter $p = 2\lambda$).

5.3. Equations With Power-Law Nonlinearity

5.3-1. Equations Containing Arbitrary Parameters

1. $y(x) + A \int_a^x t^\lambda y^k(t)\, dt = Bx^{\lambda+1} + C.$

    By differentiation, this integral equation can be reduced to a separable ordinary differential equation.
    Solution in an implicit form:

$$(\lambda + 1) \int_{y_0}^y \frac{du}{Au^k - B(\lambda + 1)} + x^{\lambda+1} - a^{\lambda+1} = 0, \qquad y_0 = Ba^{\lambda+1} + C.$$

2. $y(x) + \int_0^x \dfrac{y^k(t)}{ax + bt}\, dt = A.$

    A solution: $y(x) = \lambda$, where λ is a root of the algebraic (or transcendental) equation

$$\ln\left(1 + \frac{b}{a}\right)\lambda^k + b\lambda - Ab = 0.$$

3. $y(x) + Ax \int_0^x \dfrac{y^k(t)\, dt}{x^2 + t^2} = B.$

    A solution: $y(x) = \lambda$, where λ is a root of the algebraic (or transcendental) equation

$$\lambda + \tfrac{1}{4} A\pi \lambda^k = B.$$

4. $y(x) + \int_0^x \dfrac{y^k(t)\, dt}{\sqrt{ax^2 + bt^2}} = A.$

    A solution: $y(x) = \lambda$, where λ is a root of the algebraic (or transcendental) equation

$$I\lambda^k + \lambda - A = 0, \qquad I = \int_0^1 \frac{dz}{\sqrt{a + bz^2}}.$$

5. $y(x) + A \displaystyle\int_a^x \left(ax^n + bt^n\right)^{\frac{\lambda - k\lambda - 1}{n}} y^k(t)\, dt = Bx^\lambda.$

A solution: $y = \beta x^\lambda$, where β is a root of the algebraic (or transcendental) equation

$$ AI\beta^k + \beta - B = 0, \qquad I = \int_0^1 z^{k\lambda}\left(a + bz^n\right)^{\frac{\lambda - k\lambda - 1}{n}}\, dz. $$

6. $y(x) + A \displaystyle\int_a^x e^{\lambda t} y^\mu(t)\, dt = Be^{\lambda x} + C.$

By differentiation, this integral equation can be reduced to a separable ordinary differential equation.

Solution in an implicit form:

$$ \lambda \int_{y_0}^y \frac{du}{Au^\mu - B\lambda} + e^{\lambda x} - e^{\lambda a} = 0, \qquad y_0 = Be^{\lambda a} + C. $$

7. $y(x) + k \displaystyle\int_a^x e^{\lambda(x-t)} y^\mu(t)\, dt = Ae^{\lambda x} + B.$

Solution in an implicit form:

$$ \int_{y_0}^y \frac{dt}{\lambda t - kt^\mu - \lambda B} = x - a, \qquad y_0 = Ae^{\lambda a} + B. $$

8. $y(x) + k \displaystyle\int_a^x \sinh[\lambda(x - t)] y^\mu(t)\, dt = Ae^{\lambda x} + Be^{-\lambda x} + C.$

This is a special case of equation 5.8.14 with $f(y) = ky^\mu$.

Solution in an implicit form:

$$ \int_{y_0}^y \left[\lambda^2 u^2 - 2\lambda^2 Cu - 2k\lambda F(u) + \lambda^2(C^2 - 4AB)\right]^{-1/2} du = \pm(x - a), $$

$$ F(u) = \frac{1}{\mu + 1}\left(u^{\mu+1} - y_0^{\mu+1}\right), \qquad y_0 = Ae^{\lambda a} + Be^{-\lambda a} + C. $$

9. $y(x) + k \displaystyle\int_a^x \sinh[\lambda(x - t)] y^\mu(t)\, dt = A\cosh(\lambda x) + B.$

This is a special case of equation 5.8.15 with $f(y) = ky^\mu$.

Solution in an implicit form:

$$ \int_{y_0}^y \left[\lambda^2 u^2 - 2\lambda^2 Bu - 2k\lambda F(u) + \lambda^2(B^2 - A^2)\right]^{-1/2} du = \pm(x - a), $$

$$ F(u) = \frac{1}{\mu + 1}\left(u^{\mu+1} - y_0^{\mu+1}\right), \qquad y_0 = A\cosh(\lambda a) + B. $$

10. $y(x) + k \displaystyle\int_a^x \sinh[\lambda(x - t)] y^\mu(t)\, dt = A\sinh(\lambda x) + B.$

This is a special case of equation 5.8.16 with $f(y) = ky^\mu$.

Solution in an implicit form:

$$ \int_{y_0}^y \left[\lambda^2 u^2 - 2\lambda^2 Bu - 2k\lambda F(u) + \lambda^2(A^2 + B^2)\right]^{-1/2} du = \pm(x - a), $$

$$ F(u) = \frac{1}{\mu + 1}\left(u^{\mu+1} - y_0^{\mu+1}\right), \qquad y_0 = A\sinh(\lambda a) + B. $$

11. $y(x) + k \int_a^x \sin[\lambda(x - t)] y^\mu(t)\, dt = A \sin(\lambda x) + B \cos(\lambda x) + C.$

This is a special case of equation 5.8.17 with $f(y) = ky^\mu$.

Solution in an implicit form:

$$\int_{y_0}^{y} \left[\lambda^2 D - \lambda^2 u^2 + 2\lambda^2 C u - 2k\lambda F(u) \right]^{-1/2} du = \pm (x - a),$$

$$y_0 = A \sin(\lambda a) + B \cos(\lambda a) + C, \quad D = A^2 + B^2 - C^2, \quad F(u) = \frac{1}{\mu + 1} \left(u^{\mu+1} - y_0^{\mu+1} \right).$$

5.3-2. Equations Containing Arbitrary Functions

12. $\int_a^x K(x, t) y^\lambda(t)\, dt = f(x).$

The substitution $w(x) = y^\lambda(x)$ leads to the linear equation

$$\int_a^x K(x, t) w(t)\, dt = f(x).$$

13. $y(x) + \int_a^x f(t) y^k(t)\, dt = A.$

Solution:

$$y(x) = \left[A^{1-k} + (k - 1) \int_a^x f(t)\, dt \right]^{\frac{1}{1-k}}.$$

14. $y(x) - \int_a^x f(x) g(t) y^k(t)\, dt = 0.$

1°. Differentiating the equation with respect to x and eliminating the integral term (using the original equation), we obtain the Bernoulli ordinary differential equation

$$y'_x - f(x) g(x) y^k - \frac{f'_x(x)}{f(x)} y = 0, \qquad y(a) = 0.$$

2°. Solution with $k < 1$:

$$y(x) = f(x) \left[(1 - k) \int_a^x f^k(t) g(t)\, dt \right]^{\frac{1}{1-k}}.$$

Additionally, for $k > 0$, there is the trivial solution $y(x) \equiv 0$.

15. $y(x) + \int_0^x x^{\lambda - k\lambda - 1} f\!\left(\frac{t}{x} \right) y^k(t)\, dt = A x^\lambda.$

A solution: $y(x) = \beta x^\lambda$, where β is a root of the algebraic equation

$$I\beta^k + \beta - A = 0, \qquad I = \int_0^1 f(z) z^{k\lambda}\, dz.$$

16. $\quad y(x) + \displaystyle\int_0^x f\left(\dfrac{t}{x}\right) \sqrt{y(t)}\, dt = Ax^2.$

Solutions: $y_k(x) = B_k^2 x^2$, where B_k $(k = 1, 2)$ are the roots of the quadratic equations

$$B^2 \pm IB - A = 0, \qquad I = \int_0^1 z f(z)\, dz.$$

17. $\quad y(x) - \displaystyle\int_0^x t^a f\left(\dfrac{t}{x}\right) y^k(t)\, dt = 0, \qquad k \neq 1.$

A solution:

$$y(x) = A x^{\frac{1+a}{1-k}}, \qquad A^{1-k} = \int_0^1 z^{\frac{a+k}{1-k}} f(z)\, dz.$$

18. $\quad y(x) - \displaystyle\int_x^\infty e^{\lambda t + \beta x} f(x - t) y^k(t)\, dt = 0, \qquad k \neq 1.$

A solution:

$$y(x) = A \exp\left(\dfrac{\lambda + \beta}{1 - k} x\right), \qquad A^{1-k} = \int_0^\infty \exp\left(\dfrac{\lambda + \beta k}{1 - k} z\right) f(-z)\, dz.$$

19. $\quad y(x) - \displaystyle\int_{-\infty}^x e^{\lambda t + \beta x} f(x - t) y^k(t)\, dt = 0, \qquad k \neq 1.$

A solution:

$$y(x) = A \exp\left(\dfrac{\lambda + \beta}{1 - k} x\right), \qquad A^{1-k} = \int_0^\infty \exp\left(-\dfrac{\lambda + \beta k}{1 - k} z\right) f(z)\, dz.$$

5.4. Equations With Exponential Nonlinearity

5.4-1. Equations Containing Arbitrary Parameters

1. $\quad y(x) + A \displaystyle\int_a^x \exp[\lambda y(t)]\, dt = B.$

Solution:

$$y(x) = -\frac{1}{\lambda} \ln\left[A\lambda(x - a) + e^{-B\lambda}\right].$$

2. $\quad y(x) + A \displaystyle\int_a^x \exp[\lambda y(t)]\, dt = Bx + C.$

For $B = 0$, see equation 5.4.1.
    Solution with $B \neq 0$:

$$y(x) = -\frac{1}{\lambda} \ln\left[\frac{A}{B} + \left(e^{-\lambda y_0} - \frac{A}{B}\right) e^{\lambda B(a - x)}\right], \qquad y_0 = aB + C.$$

3. $y(x) + k \int_a^x (x - t) \exp[\lambda y(t)]\, dt = Ax^2 + Bx + C.$

1°. This is a special case of equation 5.8.5 with $f(y) = ke^{\lambda y}$. The solution of this integral equation is determined by the solution of the second-order autonomous ordinary differential equation

$$y''_{xx} + ke^{\lambda y} - 2A = 0$$

under the initial conditions

$$y(a) = Aa^2 + Ba + C, \qquad y'_x(a) = 2Aa + B.$$

2°. Solution in an implicit form:

$$\int_{y_0}^y \left[4Au - 2F(u) + B^2 - 4AC\right]^{-1/2} du = \pm(x - a),$$

$$F(u) = \frac{k}{\lambda}\left(e^{\lambda u} - e^{\lambda y_0}\right), \quad y_0 = Aa^2 + Ba + C.$$

4. $y(x) + A \int_a^x t^\lambda \exp[\beta y(t)]\, dt = Bx^{\lambda+1} + C.$

By differentiation, this integral equation can be reduced to a separable ordinary differential equation.

   Solution in an implicit form:

$$(\lambda + 1) \int_{y_0}^y \frac{du}{Ae^{\beta u} - B(\lambda + 1)} + x^{\lambda+1} - a^{\lambda+1} = 0, \qquad y_0 = Ba^{\lambda+1} + C.$$

5. $y(x) + \int_0^x \frac{\exp[\lambda y(t)]}{ax + bt}\, dt = A.$

A solution: $y(x) = \beta$, where β is a root of the transcendental equation

$$\ln\left(1 + \frac{b}{a}\right)e^{\lambda\beta} + b\beta - Ab = 0.$$

6. $y(x) + \int_0^x \frac{\exp[\lambda y(t)]}{\sqrt{ax^2 + bt^2}}\, dt = A.$

A solution: $y(x) = \beta$, where β is a root of the transcendental equation

$$ke^{\lambda\beta} + \beta - A = 0, \qquad k = \int_0^1 \frac{dz}{\sqrt{a + bz^2}}.$$

7. $y(x) + A \int_a^x \exp\left[\lambda t + \beta y(t)\right] dt = Be^{\lambda x} + C.$

By differentiation, this integral equation can be reduced to a separable ordinary differential equation.

   Solution in an implicit form:

$$\lambda \int_{y_0}^y \frac{du}{Ae^{\beta u} - B\lambda} + e^{\lambda x} - e^{\lambda a} = 0, \qquad y_0 = Be^{\lambda a} + C.$$

8. $y(x) + k \displaystyle\int_a^x \exp\left[\lambda(x-t) + \beta y(t)\right] dt = A.$

Solution in an implicit form:

$$\int_A^y \frac{dt}{\lambda t - ke^{\beta t} - \lambda A} = x - a.$$

9. $y(x) + k \displaystyle\int_a^x \exp\left[\lambda(x-t) + \beta y(t)\right] dt = Ae^{\lambda x} + B.$

Solution in an implicit form:

$$\int_{y_0}^y \frac{dt}{\lambda t - ke^{\beta t} - \lambda B} = x - a, \qquad y_0 = Ae^{\lambda a} + B.$$

10. $y(x) + k \displaystyle\int_a^x \sinh[\lambda(x-t)] \exp[\beta y(t)]\, dt = Ae^{\lambda x} + Be^{-\lambda x} + C.$

This is a special case of equation 5.8.14 with $f(y) = ke^{\beta y}$.

11. $y(x) + k \displaystyle\int_a^x \sinh[\lambda(x-t)] \exp[\beta y(t)]\, dt = A\cosh(\lambda x) + B.$

This is a special case of equation 5.8.15 with $f(y) = ke^{\beta y}$.

12. $y(x) + k \displaystyle\int_a^x \sinh[\lambda(x-t)] \exp[\beta y(t)]\, dt = A\sinh(\lambda x) + B.$

This is a special case of equation 5.8.16 with $f(y) = ke^{\beta y}$.

13. $y(x) + k \displaystyle\int_a^x \sin[\lambda(x-t)] \exp[\beta y(t)]\, dt = A\sin(\lambda x) + B\cos(\lambda x) + C.$

This is a special case of equation 5.8.17 with $f(y) = ke^{\beta y}$.

5.4-2. Equations Containing Arbitrary Functions

14. $y(x) + \displaystyle\int_a^x f(t) \exp[\lambda y(t)]\, dt = A.$

Solution:

$$y(x) = -\frac{1}{\lambda} \ln\left[\lambda \int_a^x f(t)\, dt + e^{-A\lambda}\right].$$

15. $y(x) + \displaystyle\int_a^x g(t) \exp[\lambda y(t)]\, dt = f(x).$

$1°$. By differentiation, this integral equation can be reduced to the first-order ordinary differential equation

$$y_x' + g(x)e^{\lambda y} = f_x'(x) \tag{1}$$

under the initial condition $y(a) = f(a)$. The substitution $w = e^{-\lambda y}$ reduces (1) to the linear equation

$$w_x' + \lambda f_x'(x)w - \lambda g(x) = 0, \qquad w(a) = \exp[-\lambda f(a)].$$

$2°$. Solution:

$$y(x) = f(x) - \frac{1}{\lambda} \ln\left\{1 + \lambda \int_a^x g(t)\exp[\lambda f(t)]\, dt\right\}.$$

16. $y(x) + \dfrac{1}{x} \displaystyle\int_0^x f\left(\dfrac{t}{x}\right) \exp[\lambda y(t)]\, dt = A.$

A solution: $y(x) = \beta$, where β is a root of the transcendental equation

$$\beta + Ie^{\lambda\beta} - A = 0, \qquad I = \int_0^1 f(z)\, dz.$$

5.5. Equations With Hyperbolic Nonlinearity

5.5-1. Integrands With Nonlinearity of the Form $\cosh[\beta y(t)]$

1. $y(x) + k \displaystyle\int_a^x \cosh[\beta y(t)]\, dt = A.$

This is a special case of equation 5.8.3 with $f(y) = k\cosh(\beta y)$.

2. $y(x) + k \displaystyle\int_a^x \cosh[\beta y(t)]\, dt = Ax + B.$

This is a special case of equation 5.8.4 with $f(y) = k\cosh(\beta y)$.

3. $y(x) + k \displaystyle\int_a^x (x - t)\cosh[\beta y(t)]\, dt = Ax^2 + Bx + C.$

This is a special case of equation 5.8.5 with $f(y) = k\cosh(\beta y)$.

4. $y(x) + k \displaystyle\int_a^x t^\lambda \cosh[\beta y(t)]\, dt = Bx^{\lambda+1} + C.$

This is a special case of equation 5.8.6 with $f(y) = k\cosh(\beta y)$.

5. $y(x) + \displaystyle\int_a^x g(t)\cosh[\beta y(t)]\, dt = A.$

This is a special case of equation 5.8.7 with $f(y) = \cosh(\beta y)$.

6. $y(x) + \displaystyle\int_0^x \dfrac{\cosh[\beta y(t)]}{ax + bt}\, dt = A.$

This is a special case of equation 5.8.8 with $f(y) = \cosh(\beta y)$.

7. $y(x) + \displaystyle\int_0^x \dfrac{\cosh[\beta y(t)]}{\sqrt{ax^2 + bt^2}}\, dt = A.$

This is a special case of equation 5.8.9 with $f(y) = \cosh(\beta y)$.

8. $y(x) + k \displaystyle\int_a^x e^{\lambda t} \cosh[\beta y(t)]\, dt = Be^{\lambda x} + C.$

This is a special case of equation 5.8.11 with $f(y) = k\cosh(\beta y)$.

9. $y(x) + k \displaystyle\int_a^x e^{\lambda(x-t)} \cosh[\beta y(t)]\, dt = A.$

This is a special case of equation 5.8.12 with $f(y) = k\cosh(\beta y)$.

10. $y(x) + k \int_a^x e^{\lambda(x-t)} \cosh[\beta y(t)] \, dt = Ae^{\lambda x} + B.$

This is a special case of equation 5.8.13 with $f(y) = k \cosh(\beta y)$.

11. $y(x) + k \int_a^x \sinh[\lambda(x-t)] \cosh[\beta y(t)] \, dt = Ae^{\lambda x} + Be^{-\lambda x} + C.$

This is a special case of equation 5.8.14 with $f(y) = k \cosh(\beta y)$.

12. $y(x) + k \int_a^x \sinh[\lambda(x-t)] \cosh[\beta y(t)] \, dt = A \cosh(\lambda x) + B.$

This is a special case of equation 5.8.15 with $f(y) = k \cosh(\beta y)$.

13. $y(x) + k \int_a^x \sinh[\lambda(x-t)] \cosh[\beta y(t)] \, dt = A \sinh(\lambda x) + B.$

This is a special case of equation 5.8.16 with $f(y) = k \cosh(\beta y)$.

14. $y(x) + k \int_a^x \sin[\lambda(x-t)] \cosh[\beta y(t)] \, dt = A \sin(\lambda x) + B \cos(\lambda x) + C.$

This is a special case of equation 5.8.17 with $f(y) = k \cosh(\beta y)$.

5.5-2. Integrands With Nonlinearity of the Form $\sinh[\beta y(t)]$

15. $y(x) + k \int_a^x \sinh[\beta y(t)] \, dt = A.$

This is a special case of equation 5.8.3 with $f(y) = k \sinh(\beta y)$.

16. $y(x) + k \int_a^x \sinh[\beta y(t)] \, dt = Ax + B.$

This is a special case of equation 5.8.4 with $f(y) = k \sinh(\beta y)$.

17. $y(x) + k \int_a^x (x-t) \sinh[\beta y(t)] \, dt = Ax^2 + Bx + C.$

This is a special case of equation 5.8.5 with $f(y) = k \sinh(\beta y)$.

18. $y(x) + k \int_a^x t^\lambda \sinh[\beta y(t)] \, dt = Bx^{\lambda+1} + C.$

This is a special case of equation 5.8.6 with $f(y) = k \sinh(\beta y)$.

19. $y(x) + \int_a^x g(t) \sinh[\beta y(t)] \, dt = A.$

This is a special case of equation 5.8.7 with $f(y) = \sinh(\beta y)$.

20. $y(x) + \int_0^x \dfrac{\sinh[\beta y(t)]}{ax + bt} \, dt = A.$

This is a special case of equation 5.8.8 with $f(y) = \sinh(\beta y)$.

21. $y(x) + \displaystyle\int_0^x \frac{\sinh[\beta y(t)]}{\sqrt{ax^2 + bt^2}}\, dt = A.$

This is a special case of equation 5.8.9 with $f(y) = \sinh(\beta y)$.

22. $y(x) + k \displaystyle\int_a^x e^{\lambda t} \sinh[\beta y(t)]\, dt = Be^{\lambda x} + C.$

This is a special case of equation 5.8.11 with $f(y) = k \sinh(\beta y)$.

23. $y(x) + k \displaystyle\int_a^x e^{\lambda(x-t)} \sinh[\beta y(t)]\, dt = A.$

This is a special case of equation 5.8.12 with $f(y) = k \sinh(\beta y)$.

24. $y(x) + k \displaystyle\int_a^x e^{\lambda(x-t)} \sinh[\beta y(t)]\, dt = Ae^{\lambda x} + B.$

This is a special case of equation 5.8.13 with $f(y) = k \sinh(\beta y)$.

25. $y(x) + k \displaystyle\int_a^x \sinh[\lambda(x - t)] \sinh[\beta y(t)]\, dt = Ae^{\lambda x} + Be^{-\lambda x} + C.$

This is a special case of equation 5.8.14 with $f(y) = k \sinh(\beta y)$.

26. $y(x) + k \displaystyle\int_a^x \sinh[\lambda(x - t)] \sinh[\beta y(t)]\, dt = A \cosh(\lambda x) + B.$

This is a special case of equation 5.8.15 with $f(y) = k \sinh(\beta y)$.

27. $y(x) + k \displaystyle\int_a^x \sinh[\lambda(x - t)] \sinh[\beta y(t)]\, dt = A \sinh(\lambda x) + B.$

This is a special case of equation 5.8.16 with $f(y) = k \sinh(\beta y)$.

28. $y(x) + k \displaystyle\int_a^x \sin[\lambda(x - t)] \sinh[\beta y(t)]\, dt = A \sin(\lambda x) + B \cos(\lambda x) + C.$

This is a special case of equation 5.8.17 with $f(y) = k \sinh(\beta y)$.

5.5-3. Integrands With Nonlinearity of the Form tanh[$\beta y(t)$]

29. $y(x) + k \displaystyle\int_a^x \tanh[\beta y(t)]\, dt = A.$

This is a special case of equation 5.8.3 with $f(y) = k \tanh(\beta y)$.

30. $y(x) + k \displaystyle\int_a^x \tanh[\beta y(t)]\, dt = Ax + B.$

This is a special case of equation 5.8.4 with $f(y) = k \tanh(\beta y)$.

31. $y(x) + k \displaystyle\int_a^x (x - t) \tanh[\beta y(t)]\, dt = Ax^2 + Bx + C.$

This is a special case of equation 5.8.5 with $f(y) = k \tanh(\beta y)$.

32. $y(x) + k \int_a^x t^\lambda \tanh[\beta y(t)] \, dt = Bx^{\lambda+1} + C.$

This is a special case of equation 5.8.6 with $f(y) = k \tanh(\beta y)$.

33. $y(x) + \int_a^x g(t) \tanh[\beta y(t)] \, dt = A.$

This is a special case of equation 5.8.7 with $f(y) = \tanh(\beta y)$.

34. $y(x) + \int_0^x \dfrac{\tanh[\beta y(t)]}{ax + bt} \, dt = A.$

This is a special case of equation 5.8.8 with $f(y) = \tanh(\beta y)$.

35. $y(x) + \int_0^x \dfrac{\tanh[\beta y(t)]}{\sqrt{ax^2 + bt^2}} \, dt = A.$

This is a special case of equation 5.8.9 with $f(y) = \tanh(\beta y)$.

36. $y(x) + k \int_a^x e^{\lambda t} \tanh[\beta y(t)] \, dt = Be^{\lambda x} + C.$

This is a special case of equation 5.8.11 with $f(y) = k \tanh(\beta y)$.

37. $y(x) + k \int_a^x e^{\lambda(x-t)} \tanh[\beta y(t)] \, dt = A.$

This is a special case of equation 5.8.12 with $f(y) = k \tanh(\beta y)$.

38. $y(x) + k \int_a^x e^{\lambda(x-t)} \tanh[\beta y(t)] \, dt = Ae^{\lambda x} + B.$

This is a special case of equation 5.8.13 with $f(y) = k \tanh(\beta y)$.

39. $y(x) + k \int_a^x \sinh[\lambda(x - t)] \tanh[\beta y(t)] \, dt = Ae^{\lambda x} + Be^{-\lambda x} + C.$

This is a special case of equation 5.8.14 with $f(y) = k \tanh(\beta y)$.

40. $y(x) + k \int_a^x \sinh[\lambda(x - t)] \tanh[\beta y(t)] \, dt = A \cosh(\lambda x) + B.$

This is a special case of equation 5.8.15 with $f(y) = k \tanh(\beta y)$.

41. $y(x) + k \int_a^x \sinh[\lambda(x - t)] \tanh[\beta y(t)] \, dt = A \sinh(\lambda x) + B.$

This is a special case of equation 5.8.16 with $f(y) = k \tanh(\beta y)$.

42. $y(x) + k \int_a^x \sin[\lambda(x - t)] \tanh[\beta y(t)] \, dt = A \sin(\lambda x) + B \cos(\lambda x) + C.$

This is a special case of equation 5.8.17 with $f(y) = k \tanh(\beta y)$.

5.5-4. Integrands With Nonlinearity of the Form $\coth[\beta y(t)]$

43. $y(x) + k \displaystyle\int_a^x \coth[\beta y(t)]\, dt = A.$

This is a special case of equation 5.8.3 with $f(y) = k\coth(\beta y)$.

44. $y(x) + k \displaystyle\int_a^x \coth[\beta y(t)]\, dt = Ax + B.$

This is a special case of equation 5.8.4 with $f(y) = k\coth(\beta y)$.

45. $y(x) + k \displaystyle\int_a^x (x - t)\coth[\beta y(t)]\, dt = Ax^2 + Bx + C.$

This is a special case of equation 5.8.5 with $f(y) = k\coth(\beta y)$.

46. $y(x) + k \displaystyle\int_a^x t^\lambda \coth[\beta y(t)]\, dt = Bx^{\lambda+1} + C.$

This is a special case of equation 5.8.6 with $f(y) = k\coth(\beta y)$.

47. $y(x) + \displaystyle\int_a^x g(t)\coth[\beta y(t)]\, dt = A.$

This is a special case of equation 5.8.7 with $f(y) = \coth(\beta y)$.

48. $y(x) + \displaystyle\int_0^x \frac{\coth[\beta y(t)]}{ax + bt}\, dt = A.$

This is a special case of equation 5.8.8 with $f(y) = \coth(\beta y)$.

49. $y(x) + \displaystyle\int_0^x \frac{\coth[\beta y(t)]}{\sqrt{ax^2 + bt^2}}\, dt = A.$

This is a special case of equation 5.8.9 with $f(y) = \coth(\beta y)$.

50. $y(x) + k \displaystyle\int_a^x e^{\lambda t}\coth[\beta y(t)]\, dt = Be^{\lambda x} + C.$

This is a special case of equation 5.8.11 with $f(y) = k\coth(\beta y)$.

51. $y(x) + k \displaystyle\int_a^x e^{\lambda(x-t)}\coth[\beta y(t)]\, dt = A.$

This is a special case of equation 5.8.12 with $f(y) = k\coth(\beta y)$.

52. $y(x) + k \displaystyle\int_a^x e^{\lambda(x-t)}\coth[\beta y(t)]\, dt = Ae^{\lambda x} + B.$

This is a special case of equation 5.8.13 with $f(y) = k\coth(\beta y)$.

53. $y(x) + k \displaystyle\int_a^x \sinh[\lambda(x - t)]\coth[\beta y(t)]\, dt = Ae^{\lambda x} + Be^{-\lambda x} + C.$

This is a special case of equation 5.8.14 with $f(y) = k\coth(\beta y)$.

54. $\quad y(x) + k \displaystyle\int_a^x \sinh[\lambda(x - t)] \coth[\beta y(t)] \, dt = A \cosh(\lambda x) + B.$

This is a special case of equation 5.8.15 with $f(y) = k \coth(\beta y)$.

55. $\quad y(x) + k \displaystyle\int_a^x \sinh[\lambda(x - t)] \coth[\beta y(t)] \, dt = A \sinh(\lambda x) + B.$

This is a special case of equation 5.8.16 with $f(y) = k \coth(\beta y)$.

56. $\quad y(x) + k \displaystyle\int_a^x \sin[\lambda(x - t)] \coth[\beta y(t)] \, dt = A \sin(\lambda x) + B \cos(\lambda x) + C.$

This is a special case of equation 5.8.17 with $f(y) = k \coth(\beta y)$.

5.6. Equations With Logarithmic Nonlinearity

5.6-1. Integrands Containing Power-Law Functions of x and t

1. $\quad y(x) + k \displaystyle\int_a^x \ln[\lambda y(t)] \, dt = A.$

This is a special case of equation 5.8.3 with $f(y) = k \ln(\lambda y)$.

2. $\quad y(x) + k \displaystyle\int_a^x \ln[\lambda y(t)] \, dt = Ax + B.$

This is a special case of equation 5.8.4 with $f(y) = k \ln(\lambda y)$.

3. $\quad y(x) + k \displaystyle\int_a^x (x - t) \ln[\lambda y(t)] \, dt = Ax^2 + Bx + C.$

This is a special case of equation 5.8.5 with $f(y) = k \ln(\lambda y)$.

4. $\quad y(x) + k \displaystyle\int_a^x t^\lambda \ln[\mu y(t)] \, dt = Bx^{\lambda+1} + C.$

This is a special case of equation 5.8.6 with $f(y) = k \ln(\mu y)$.

5. $\quad y(x) + \displaystyle\int_0^x \frac{\ln[\lambda y(t)]}{ax + bt} \, dt = A.$

This is a special case of equation 5.8.8 with $f(y) = \ln(\lambda y)$.

6. $\quad y(x) + \displaystyle\int_0^x \frac{\ln[\lambda y(t)]}{\sqrt{ax^2 + bt^2}} \, dt = A.$

This is a special case of equation 5.8.9 with $f(y) = \ln(\lambda y)$.

5.6-2. Integrands Containing Exponential Functions of x and t

7. $\quad y(x) + k \displaystyle\int_a^x e^{\lambda t} \ln[\mu y(t)] \, dt = Be^{\lambda x} + C.$

This is a special case of equation 5.8.11 with $f(y) = k \ln(\mu y)$.

8. $y(x) + k \int_a^x e^{\lambda(x-t)} \ln[\mu y(t)]\, dt = A.$

   This is a special case of equation 5.8.12 with $f(y) = k \ln(\mu y)$.

9. $y(x) + k \int_a^x e^{\lambda(x-t)} \ln[\mu y(t)]\, dt = Ae^{\lambda x} + B.$

   This is a special case of equation 5.8.13 with $f(y) = k \ln(\mu y)$.

5.6-3. Other Integrands

10. $y(x) + \int_a^x g(t) \ln[\lambda y(t)]\, dt = A.$

    This is a special case of equation 5.8.7 with $f(y) = \ln(\lambda y)$.

11. $y(x) + k \int_a^x \sinh[\lambda(x-t)] \ln[\mu y(t)]\, dt = Ae^{\lambda x} + Be^{-\lambda x} + C.$

    This is a special case of equation 5.8.14 with $f(y) = k \ln(\mu y)$.

12. $y(x) + k \int_a^x \sinh[\lambda(x-t)] \ln[\mu y(t)]\, dt = A \cosh(\lambda x) + B.$

    This is a special case of equation 5.8.15 with $f(y) = k \ln(\mu y)$.

13. $y(x) + k \int_a^x \sinh[\lambda(x-t)] \ln[\mu y(t)]\, dt = A \sinh(\lambda x) + B.$

    This is a special case of equation 5.8.16 with $f(y) = k \ln(\mu y)$.

14. $y(x) + k \int_a^x \sin[\lambda(x-t)] \ln[\mu y(t)]\, dt = A \sin(\lambda x) + B \cos(\lambda x) + C.$

    This is a special case of equation 5.8.17 with $f(y) = k \ln(\mu y)$.

5.7. Equations With Trigonometric Nonlinearity

5.7-1. Integrands With Nonlinearity of the Form $\cos[\beta y(t)]$

1. $y(x) + k \int_a^x \cos[\beta y(t)]\, dt = A.$

   This is a special case of equation 5.8.3 with $f(y) = k \cos(\beta y)$.

2. $y(x) + k \int_a^x \cos[\beta y(t)]\, dt = Ax + B.$

   This is a special case of equation 5.8.4 with $f(y) = k \cos(\beta y)$.

3. $y(x) + k \int_a^x (x - t) \cos[\beta y(t)]\, dt = Ax^2 + Bx + C.$

   This is a special case of equation 5.8.5 with $f(y) = k \cos(\beta y)$.

4. $\quad y(x) + k \displaystyle\int_a^x t^\lambda \cos[\beta y(t)]\, dt = Bx^{\lambda+1} + C.$

This is a special case of equation 5.8.6 with $f(y) = k\cos(\beta y)$.

5. $\quad y(x) + \displaystyle\int_a^x g(t) \cos[\beta y(t)]\, dt = A.$

This is a special case of equation 5.8.7 with $f(y) = \cos(\beta y)$.

6. $\quad y(x) + \displaystyle\int_0^x \dfrac{\cos[\beta y(t)]}{ax + bt}\, dt = A.$

This is a special case of equation 5.8.8 with $f(y) = \cos(\beta y)$.

7. $\quad y(x) + \displaystyle\int_0^x \dfrac{\cos[\beta y(t)]}{\sqrt{ax^2 + bt^2}}\, dt = A.$

This is a special case of equation 5.8.9 with $f(y) = \cos(\beta y)$.

8. $\quad y(x) + k \displaystyle\int_a^x e^{\lambda t} \cos[\beta y(t)]\, dt = Be^{\lambda x} + C.$

This is a special case of equation 5.8.11 with $f(y) = k\cos(\beta y)$.

9. $\quad y(x) + k \displaystyle\int_a^x e^{\lambda(x-t)} \cos[\beta y(t)]\, dt = A.$

This is a special case of equation 5.8.12 with $f(y) = k\cos(\beta y)$.

10. $\quad y(x) + k \displaystyle\int_a^x e^{\lambda(x-t)} \cos[\beta y(t)]\, dt = Ae^{\lambda x} + B.$

This is a special case of equation 5.8.13 with $f(y) = k\cos(\beta y)$.

11. $\quad y(x) + k \displaystyle\int_a^x \sinh[\lambda(x-t)] \cos[\beta y(t)]\, dt = Ae^{\lambda x} + Be^{-\lambda x} + C.$

This is a special case of equation 5.8.14 with $f(y) = k\cos(\beta y)$.

12. $\quad y(x) + k \displaystyle\int_a^x \sinh[\lambda(x-t)] \cos[\beta y(t)]\, dt = A\cosh(\lambda x) + B.$

This is a special case of equation 5.8.15 with $f(y) = k\cos(\beta y)$.

13. $\quad y(x) + k \displaystyle\int_a^x \sinh[\lambda(x-t)] \cos[\beta y(t)]\, dt = A\sinh(\lambda x) + B.$

This is a special case of equation 5.8.16 with $f(y) = k\cos(\beta y)$.

14. $\quad y(x) + k \displaystyle\int_a^x \sin[\lambda(x-t)] \cos[\beta y(t)]\, dt = A\sin(\lambda x) + B\cos(\lambda x) + C.$

This is a special case of equation 5.8.17 with $f(y) = k\cos(\beta y)$.

5.7-2. Integrands With Nonlinearity of the Form $\sin[\beta y(t)]$

15. $y(x) + k \displaystyle\int_a^x \sin[\beta y(t)]\,dt = A.$

This is a special case of equation 5.8.3 with $f(y) = k\sin(\beta y)$.

16. $y(x) + k \displaystyle\int_a^x \sin[\beta y(t)]\,dt = Ax + B.$

This is a special case of equation 5.8.4 with $f(y) = k\sin(\beta y)$.

17. $y(x) + k \displaystyle\int_a^x (x - t)\sin[\beta y(t)]\,dt = Ax^2 + Bx + C.$

This is a special case of equation 5.8.5 with $f(y) = k\sin(\beta y)$.

18. $y(x) + k \displaystyle\int_a^x t^\lambda \sin[\beta y(t)]\,dt = Bx^{\lambda+1} + C.$

This is a special case of equation 5.8.6 with $f(y) = k\sin(\beta y)$.

19. $y(x) + \displaystyle\int_a^x g(t)\sin[\beta y(t)]\,dt = A.$

This is a special case of equation 5.8.7 with $f(y) = \sin(\beta y)$.

20. $y(x) + \displaystyle\int_0^x \frac{\sin[\beta y(t)]}{ax + bt}\,dt = A.$

This is a special case of equation 5.8.8 with $f(y) = \sin(\beta y)$.

21. $y(x) + \displaystyle\int_0^x \frac{\sin[\beta y(t)]}{\sqrt{ax^2 + bt^2}}\,dt = A.$

This is a special case of equation 5.8.9 with $f(y) = \sin(\beta y)$.

22. $y(x) + k \displaystyle\int_a^x e^{\lambda t}\sin[\beta y(t)]\,dt = Be^{\lambda x} + C.$

This is a special case of equation 5.8.11 with $f(y) = k\sin(\beta y)$.

23. $y(x) + k \displaystyle\int_a^x e^{\lambda(x-t)}\sin[\beta y(t)]\,dt = A.$

This is a special case of equation 5.8.12 with $f(y) = k\sin(\beta y)$.

24. $y(x) + k \displaystyle\int_a^x e^{\lambda(x-t)}\sin[\beta y(t)]\,dt = Ae^{\lambda x} + B.$

This is a special case of equation 5.8.13 with $f(y) = k\sin(\beta y)$.

25. $y(x) + k \displaystyle\int_a^x \sinh[\lambda(x - t)]\sin[\beta y(t)]\,dt = Ae^{\lambda x} + Be^{-\lambda x} + C.$

This is a special case of equation 5.8.14 with $f(y) = k\sin(\beta y)$.

26. $y(x) + k \int_a^x \sinh[\lambda(x - t)] \sin[\beta y(t)] \, dt = A \cosh(\lambda x) + B.$

This is a special case of equation 5.8.15 with $f(y) = k \sin(\beta y)$.

27. $y(x) + k \int_a^x \sinh[\lambda(x - t)] \sin[\beta y(t)] \, dt = A \sinh(\lambda x) + B.$

This is a special case of equation 5.8.16 with $f(y) = k \sin(\beta y)$.

28. $y(x) + k \int_a^x \sin[\lambda(x - t)] \sin[\beta y(t)] \, dt = A \sin(\lambda x) + B \cos(\lambda x) + C.$

This is a special case of equation 5.8.17 with $f(y) = k \sin(\beta y)$.

5.7-3. Integrands With Nonlinearity of the Form $\tan[\beta y(t)]$

29. $y(x) + k \int_a^x \tan[\beta y(t)] \, dt = A.$

This is a special case of equation 5.8.3 with $f(y) = k \tan(\beta y)$.

30. $y(x) + k \int_a^x \tan[\beta y(t)] \, dt = Ax + B.$

This is a special case of equation 5.8.4 with $f(y) = k \tan(\beta y)$.

31. $y(x) + k \int_a^x (x - t) \tan[\beta y(t)] \, dt = Ax^2 + Bx + C.$

This is a special case of equation 5.8.5 with $f(y) = k \tan(\beta y)$.

32. $y(x) + k \int_a^x t^\lambda \tan[\beta y(t)] \, dt = Bx^{\lambda+1} + C.$

This is a special case of equation 5.8.6 with $f(y) = k \tan(\beta y)$.

33. $y(x) + \int_a^x g(t) \tan[\beta y(t)] \, dt = A.$

This is a special case of equation 5.8.7 with $f(y) = \tan(\beta y)$.

34. $y(x) + \int_0^x \dfrac{\tan[\beta y(t)]}{ax + bt} \, dt = A.$

This is a special case of equation 5.8.8 with $f(y) = \tan(\beta y)$.

35. $y(x) + \int_0^x \dfrac{\tan[\beta y(t)]}{\sqrt{ax^2 + bt^2}} \, dt = A.$

This is a special case of equation 5.8.9 with $f(y) = \tan(\beta y)$.

36. $y(x) + k \int_a^x e^{\lambda t} \tan[\beta y(t)] \, dt = Be^{\lambda x} + C.$

This is a special case of equation 5.8.11 with $f(y) = k \tan(\beta y)$.

37. $y(x) + k \int_a^x e^{\lambda(x-t)} \tan[\beta y(t)]\, dt = A.$

This is a special case of equation 5.8.12 with $f(y) = k \tan(\beta y)$.

38. $y(x) + k \int_a^x e^{\lambda(x-t)} \tan[\beta y(t)]\, dt = Ae^{\lambda x} + B.$

This is a special case of equation 5.8.13 with $f(y) = k \tan(\beta y)$.

39. $y(x) + k \int_a^x \sinh[\lambda(x-t)] \tan[\beta y(t)]\, dt = Ae^{\lambda x} + Be^{-\lambda x} + C.$

This is a special case of equation 5.8.14 with $f(y) = k \tan(\beta y)$.

40. $y(x) + k \int_a^x \sinh[\lambda(x-t)] \tan[\beta y(t)]\, dt = A \cosh(\lambda x) + B.$

This is a special case of equation 5.8.15 with $f(y) = k \tan(\beta y)$.

41. $y(x) + k \int_a^x \sinh[\lambda(x-t)] \tan[\beta y(t)]\, dt = A \sinh(\lambda x) + B.$

This is a special case of equation 5.8.16 with $f(y) = k \tan(\beta y)$.

42. $y(x) + k \int_a^x \sin[\lambda(x-t)] \tan[\beta y(t)]\, dt = A \sin(\lambda x) + B \cos(\lambda x) + C.$

This is a special case of equation 5.8.17 with $f(y) = k \tan(\beta y)$.

5.7-4. Integrands With Nonlinearity of the Form $\cot[\beta y(t)]$

43. $y(x) + k \int_a^x \cot[\beta y(t)]\, dt = A.$

This is a special case of equation 5.8.3 with $f(y) = k \cot(\beta y)$.

44. $y(x) + k \int_a^x \cot[\beta y(t)]\, dt = Ax + B.$

This is a special case of equation 5.8.4 with $f(y) = k \cot(\beta y)$.

45. $y(x) + k \int_a^x (x-t) \cot[\beta y(t)]\, dt = Ax^2 + Bx + C.$

This is a special case of equation 5.8.5 with $f(y) = k \cot(\beta y)$.

46. $y(x) + k \int_a^x t^\lambda \cot[\beta y(t)]\, dt = Bx^{\lambda+1} + C.$

This is a special case of equation 5.8.6 with $f(y) = k \cot(\beta y)$.

47. $y(x) + \int_a^x g(t) \cot[\beta y(t)]\, dt = A.$

This is a special case of equation 5.8.7 with $f(y) = \cot(\beta y)$.

48. $\quad y(x) + \displaystyle\int_0^x \frac{\cot[\beta y(t)]}{ax + bt}\, dt = A.$

This is a special case of equation 5.8.8 with $f(y) = \cot(\beta y)$.

49. $\quad y(x) + \displaystyle\int_0^x \frac{\cot[\beta y(t)]}{\sqrt{ax^2 + bt^2}}\, dt = A.$

This is a special case of equation 5.8.9 with $f(y) = \cot(\beta y)$.

50. $\quad y(x) + k\displaystyle\int_a^x e^{\lambda t} \cot[\beta y(t)]\, dt = Be^{\lambda x} + C.$

This is a special case of equation 5.8.11 with $f(y) = k\cot(\beta y)$.

51. $\quad y(x) + k\displaystyle\int_a^x e^{\lambda(x-t)} \cot[\beta y(t)]\, dt = A.$

This is a special case of equation 5.8.12 with $f(y) = k\cot(\beta y)$.

52. $\quad y(x) + k\displaystyle\int_a^x e^{\lambda(x-t)} \cot[\beta y(t)]\, dt = Ae^{\lambda x} + B.$

This is a special case of equation 5.8.13 with $f(y) = k\cot(\beta y)$.

53. $\quad y(x) + k\displaystyle\int_a^x \sinh[\lambda(x - t)] \cot[\beta y(t)]\, dt = Ae^{\lambda x} + Be^{-\lambda x} + C.$

This is a special case of equation 5.8.14 with $f(y) = k\cot(\beta y)$.

54. $\quad y(x) + k\displaystyle\int_a^x \sinh[\lambda(x - t)] \cot[\beta y(t)]\, dt = A\cosh(\lambda x) + B.$

This is a special case of equation 5.8.15 with $f(y) = k\cot(\beta y)$.

55. $\quad y(x) + k\displaystyle\int_a^x \sinh[\lambda(x - t)] \cot[\beta y(t)]\, dt = A\sinh(\lambda x) + B.$

This is a special case of equation 5.8.16 with $f(y) = k\cot(\beta y)$.

56. $\quad y(x) + k\displaystyle\int_a^x \sin[\lambda(x - t)] \cot[\beta y(t)]\, dt = A\sin(\lambda x) + B\cos(\lambda x) + C.$

This is a special case of equation 5.8.17 with $f(y) = k\cot(\beta y)$.

5.8. Equations With Nonlinearity of General Form

5.8-1. Equations of the Form $\int_a^x G(\cdots)\, dt = F(x)$

1. $\quad \displaystyle\int_a^x K(x, t)\varphi\big(y(t)\big)\, dt = f(x).$

The substitution $w(x) = \varphi\big(y(x)\big)$ leads to the linear equation

$$\int_a^x K(x, t)w(t)\, dt = f(x).$$

2. $\displaystyle\int_a^x K(x,t)\varphi\big(t,y(t)\big)\,dt = f(x).$

The substitution $w(x) = \varphi\big(x, y(x)\big)$ leads to the linear equation

$$\int_a^x K(x,t)w(t)\,dt = f(x).$$

5.8-2. Equations of the Form $y(x) + \int_a^x K(x,t)G\big(y(t)\big)\,dt = F(x)$

3. $\displaystyle y(x) + \int_a^x f\big(y(t)\big)\,dt = A.$

Solution in an implicit form:
$$\int_A^y \frac{du}{f(u)} + x - a = 0.$$

4. $\displaystyle y(x) + \int_a^x f\big(y(t)\big)\,dt = Ax + B.$

Solution in an implicit form:
$$\int_{y_0}^y \frac{du}{A - f(u)} = x - a, \qquad y_0 = Aa + B.$$

5. $\displaystyle y(x) + \int_a^x (x - t)f\big(y(t)\big)\,dt = Ax^2 + Bx + C.$

1°. This is a special case of equation 5.8.19. The solution of this integral equation is determined by the solution of the second-order autonomous ordinary differential equation

$$y''_{xx} + f(y) - 2A = 0$$

under the initial conditions

$$y(a) = Aa^2 + Ba + C, \qquad y'_x(a) = 2Aa + B.$$

2°. Solutions in an implicit form:

$$\int_{y_0}^y \big[4Au - 2F(u) + B^2 - 4AC\big]^{-1/2}\,du = \pm(x - a),$$

$$F(u) = \int_{y_0}^u f(t)\,dt, \quad y_0 = Aa^2 + Ba + C.$$

6. $\displaystyle y(x) + \int_a^x t^\lambda f\big(y(t)\big)\,dt = Bx^{\lambda+1} + C.$

By differentiation, this integral equation can be reduced to a separable ordinary differential equation.
Solution in an implicit form:

$$(\lambda + 1)\int_{y_a}^y \frac{du}{f(u) - B(\lambda + 1)} + x^{\lambda+1} - a^{\lambda+1} = 0, \qquad y_a = Ba^{\lambda+1} + C.$$

7. $\quad y(x) + \displaystyle\int_a^x g(t) f(y(t))\, dt = A.$

Solution in an implicit form:

$$\int_A^y \frac{du}{f(u)} + \int_a^x g(t)\, dt = 0.$$

8. $\quad y(x) + \displaystyle\int_0^x \frac{f(y(t))}{ax + bt}\, dt = A.$

A solution: $y(x) = \lambda$, where λ is a root of the algebraic (or transcendental) equation

$$\ln\left(1 + \frac{b}{a}\right) f(\lambda) + b\lambda - Ab = 0.$$

9. $\quad y(x) + \displaystyle\int_0^x \frac{f(y(t))}{\sqrt{ax^2 + bt^2}}\, dt = A.$

A solution: $y(x) = \lambda$, where λ is a root of the algebraic (or transcendental) equation

$$k f(\lambda) + \lambda - A = 0, \qquad k = \int_0^1 \frac{dz}{\sqrt{a + bz^2}}.$$

10. $\quad y(x) + x \displaystyle\int_0^x f(y(t)) \frac{dt}{x^2 + t^2} = A.$

A solution: $y(x) = \lambda$, where λ is a root of the algebraic (or transcendental) equation

$$\lambda + \tfrac{1}{4}\pi f(\lambda) = A.$$

11. $\quad y(x) + \displaystyle\int_a^x e^{\lambda t} f(y(t))\, dt = B e^{\lambda x} + C.$

By differentiation, this integral equation can be reduced to a separable ordinary differential equation.

Solution in an implicit form:

$$\lambda \int_{y_0}^y \frac{du}{f(u) - B\lambda} + e^{\lambda x} - e^{\lambda a} = 0, \qquad y_0 = B e^{\lambda a} + C.$$

12. $\quad y(x) + \displaystyle\int_a^x e^{\lambda(x-t)} f(y(t))\, dt = A.$

Solution in an implicit form:

$$\int_A^y \frac{du}{\lambda u - f(u) - \lambda A} = x - a.$$

13. $\quad y(x) + \displaystyle\int_a^x e^{\lambda(x-t)} f(y(t))\, dt = A e^{\lambda x} + B.$

Solution in an implicit form:

$$\int_{y_0}^y \frac{du}{\lambda u - f(u) - \lambda B} = x - a, \qquad y_0 = A e^{\lambda a} + B.$$

14. $y(x) + \displaystyle\int_a^x \sinh[\lambda(x-t)]f\big(y(t)\big)\,dt = Ae^{\lambda x} + Be^{-\lambda x} + C.$

1°. This is a special case of equation 5.8.23. The solution of this integral equation is determined by the solution of the second-order autonomous ordinary differential equation

$$y''_{xx} + \lambda f(y) - \lambda^2 y + \lambda^2 C = 0$$

under the initial conditions

$$y(a) = Ae^{\lambda a} + Be^{-\lambda a} + C, \qquad y'_x(a) = A\lambda e^{\lambda a} - B\lambda e^{-\lambda a}.$$

2°. Solution in an implicit form:

$$\int_{y_0}^y \big[\lambda^2 u^2 - 2\lambda^2 Cu - 2\lambda F(u) + \lambda^2(C^2 - 4AB)\big]^{-1/2}\,du = \pm(x-a),$$

$$F(u) = \int_{y_0}^u f(t)\,dt, \quad y_0 = Ae^{\lambda a} + Be^{-\lambda a} + C.$$

15. $y(x) + \displaystyle\int_a^x \sinh[\lambda(x-t)]f\big(y(t)\big)\,dt = A\cosh(\lambda x) + B.$

This is a special case of equation 5.8.14.

Solution in an implicit form:

$$\int_{y_0}^y \big[\lambda^2 u^2 - 2\lambda^2 Bu - 2\lambda F(u) + \lambda^2(B^2 - A^2)\big]^{-1/2}\,du = \pm(x-a),$$

$$F(u) = \int_{y_0}^u f(t)\,dt, \quad y_0 = A\cosh(\lambda a) + B.$$

16. $y(x) + \displaystyle\int_a^x \sinh[\lambda(x-t)]f\big(y(t)\big)\,dt = A\sinh(\lambda x) + B.$

This is a special case of equation 5.8.23.

Solution in an implicit form:

$$\int_{y_0}^y \big[\lambda^2 u^2 - 2\lambda^2 Bu - 2\lambda F(u) + \lambda^2(A^2 + B^2)\big]^{-1/2}\,du = \pm(x-a),$$

$$F(u) = \int_{y_0}^u f(t)\,dt, \quad y_0 = A\sinh(\lambda a) + B.$$

17. $y(x) + \displaystyle\int_a^x \sin[\lambda(x-t)]f\big(y(t)\big)\,dt = A\sin(\lambda x) + B\cos(\lambda x) + C.$

1°. This is a special case of equation 5.8.25. The solution of this integral equation is determined by the solution of the second-order autonomous ordinary differential equation

$$y''_{xx} + \lambda f(y) + \lambda^2 y - \lambda^2 C = 0$$

under the initial conditions

$$y(a) = A\sin(\lambda a) + B\cos(\lambda a) + C, \qquad y'_x(a) = A\lambda\cos(\lambda a) - B\lambda\sin(\lambda a).$$

2°. Solution in an implicit form:

$$\int_{y_0}^y \big[\lambda^2 D - \lambda^2 u^2 + 2\lambda^2 Cu - 2\lambda F(u)\big]^{-1/2}\,du = \pm(x-a),$$

$$y_0 = A\sin(\lambda a) + B\cos(\lambda a) + C, \quad D = A^2 + B^2 - C^2, \quad F(u) = \int_{y_0}^u f(t)\,dt.$$

5.8-3. Equations of the Form $y(x) + \int_a^x K(x,t)G\big(t, y(t)\big)\, dt = F(x)$

18. $y(x) + \displaystyle\int_a^x f\big(t, y(t)\big)\, dt = g(x).$

The solution of this integral equation is determined by the solution of the first-order ordinary differential equation

$$y'_x + f(x, y) - g'_x(x) = 0$$

under the initial condition $y(a) = g(a)$. For the exact solutions of the first-order differential equations with various $f(x, y)$ and $g(x)$, see E. Kamke (1977), A. D. Polyanin and V. F. Zaitsev (1995), and V. F. Zaitsev and A. D. Polyanin (1994).

19. $y(x) + \displaystyle\int_a^x (x - t) f\big(t, y(t)\big)\, dt = g(x).$

Differentiating the equation with respect to x yields

$$y'_x + \int_a^x f\big(t, y(t)\big)\, dt = g'_x(x). \tag{1}$$

In turn, differentiating this equation with respect to x yields the second-order nonlinear ordinary differential equation

$$y''_{xx} + f(x, y) - g''_{xx}(x) = 0. \tag{2}$$

By setting $x = a$ in the original equation and equation (1), we obtain the initial conditions for $y = y(x)$:

$$y(a) = g(a), \qquad y'_x(a) = g'_x(a). \tag{3}$$

Equation (2) under conditions (3) defines the solution of the original integral equation. For the exact solutions of the second-order differential equation (2) with various $f(x, y)$ and $g(x)$, see A. D. Polyanin and V. F. Zaitsev (1995), and V. F. Zaitsev and A. D. Polyanin (1994).

20. $y(x) + \displaystyle\int_a^x (x - t)^n f\big(t, y(t)\big)\, dt = g(x), \qquad n = 1, 2, \dots$

Differentiating the equation $n+1$ times with respect to x, we obtain an $(n+1)$st-order nonlinear ordinary differential equation for $y = y(x)$:

$$y_x^{(n+1)} + n!\, f(x, y) - g_x^{(n+1)}(x) = 0.$$

This equation under the initial conditions

$$y(a) = g(a), \quad y'_x(a) = g'_x(a), \quad \dots, \quad y_x^{(n)}(a) = g_x^{(n)}(a),$$

defines the solution of the original integral equation.

21. $y(x) + \displaystyle\int_a^x e^{\lambda(x-t)} f\big(t, y(t)\big)\, dt = g(x).$

Differentiating the equation with respect to x yields

$$y'_x + f\big(x, y(x)\big) + \lambda \int_a^x e^{\lambda(x-t)} f\big(t, y(t)\big)\, dt = g'_x(x).$$

Eliminating the integral term herefrom with the aid of the original equation, we obtain the first-order nonlinear ordinary differential equation

$$y'_x + f(x, y) - \lambda y + \lambda g(x) - g'_x(x) = 0.$$

The unknown function $y = y(x)$ must satisfy the initial condition $y(a) = g(a)$. For the exact solutions of the first-order differential equations with various $f(x, y)$ and $g(x)$, see E. Kamke (1977), A. D. Polyanin and V. F. Zaitsev (1995), and V. F. Zaitsev and A. D. Polyanin (1994).

22. $y(x) + \displaystyle\int_a^x \cosh[\lambda(x-t)]f\big(t, y(t)\big)\, dt = g(x).$

Differentiating the equation with respect to x twice yields

$$y_x'(x) + f\big(x, y(x)\big) + \lambda \int_a^x \sinh[\lambda(x-t)]f\big(t, y(t)\big)\, dt = g_x'(x), \tag{1}$$

$$y_{xx}''(x) + \big[f\big(x, y(x)\big)\big]_x' + \lambda^2 \int_a^x \cosh[\lambda(x-t)]f\big(t, y(t)\big)\, dt = g_{xx}''(x). \tag{2}$$

Eliminating the integral term from (2) with the aid of the original equation, we arrive at the second-order nonlinear ordinary differential equation

$$y_{xx}'' + \big[f(x, y)\big]_x' - \lambda^2 y + \lambda^2 g(x) - g_{xx}''(x) = 0. \tag{3}$$

By setting $x = a$ in the original equation and in (1), we obtain the initial conditions for $y = y(x)$:

$$y(a) = g(a), \qquad y_x'(a) = g_x'(a) - f\big(a, g(a)\big). \tag{4}$$

Equation (3) under conditions (4) defines the solution of the original integral equation.

23. $y(x) + \displaystyle\int_a^x \sinh[\lambda(x-t)]f\big(t, y(t)\big)\, dt = g(x).$

Differentiating the equation with respect to x twice yields

$$y_x'(x) + \lambda \int_a^x \cosh[\lambda(x-t)]f\big(t, y(t)\big)\, dt = g_x'(x), \tag{1}$$

$$y_{xx}''(x) + \lambda f\big(x, y(x)\big) + \lambda^2 \int_a^x \sinh[\lambda(x-t)]f\big(t, y(t)\big)\, dt = g_{xx}''(x). \tag{2}$$

Eliminating the integral term from (2) with the aid of the original equation, we arrive at the second-order nonlinear ordinary differential equation

$$y_{xx}'' + \lambda f(x, y) - \lambda^2 y + \lambda^2 g(x) - g_{xx}''(x) = 0. \tag{3}$$

By setting $x = a$ in the original equation and in (1), we obtain the initial conditions for $y = y(x)$:

$$y(a) = g(a), \qquad y_x'(a) = g_x'(a). \tag{4}$$

Equation (3) under conditions (4) defines the solution of the original integral equation. For the exact solutions of the second-order differential equation (3) with various $f(x, y)$ and $g(x)$, see A. D. Polyanin and V. F. Zaitsev (1995), and V. F. Zaitsev and A. D. Polyanin (1994).

24. $y(x) + \displaystyle\int_a^x \cos[\lambda(x-t)]f\big(t, y(t)\big)\, dt = g(x).$

Differentiating the equation with respect to x twice yields

$$y_x'(x) + f\big(x, y(x)\big) - \lambda \int_a^x \sin[\lambda(x-t)]f\big(t, y(t)\big)\, dt = g_x'(x), \tag{1}$$

$$y_{xx}''(x) + \big[f\big(x, y(x)\big)\big]_x' - \lambda^2 \int_a^x \cos[\lambda(x-t)]f\big(t, y(t)\big)\, dt = g_{xx}''(x). \tag{2}$$

Eliminating the integral term from (2) with the aid of the original equation, we arrive at the second-order nonlinear ordinary differential equation

$$y_{xx}'' + \big[f(x, y)\big]_x' + \lambda^2 y - \lambda^2 g(x) - g_{xx}''(x) = 0. \tag{3}$$

By setting $x = a$ in the original equation and in (1), we obtain the initial conditions for $y = y(x)$:

$$y(a) = g(a), \qquad y_x'(a) = g_x'(a) - f\big(a, g(a)\big). \tag{4}$$

Equation (3) under conditions (4) defines the solution of the original integral equation.

25. $y(x) + \displaystyle\int_a^x \sin[\lambda(x-t)] f\big(t, y(t)\big)\, dt = g(x).$

Differentiating the equation with respect to x twice yields

$$y_x'(x) + \lambda \int_a^x \cos[\lambda(x-t)] f\big(t, y(t)\big)\, dt = g_x'(x), \tag{1}$$

$$y_{xx}''(x) + \lambda f\big(x, y(x)\big) - \lambda^2 \int_a^x \sin[\lambda(x-t)] f\big(t, y(t)\big)\, dt = g_{xx}''(x). \tag{2}$$

Eliminating the integral term from (2) with the aid of the original equation, we arrive at the second-order nonlinear ordinary differential equation

$$y_{xx}'' + \lambda f(x, y) + \lambda^2 y - \lambda^2 g(x) - g_{xx}''(x) = 0. \tag{3}$$

By setting $x = a$ in the original equation and in (1), we obtain the initial conditions for $y = y(x)$:

$$y(a) = g(a), \qquad y_x'(a) = g_x'(a). \tag{4}$$

Equation (3) under conditions (4) defines the solution of the original integral equation. For the exact solutions of the second-order differential equation (3) with various $f(x, y)$ and $g(x)$, see A. D. Polyanin and V. F. Zaitsev (1995), and V. F. Zaitsev and A. D. Polyanin (1994).

5.8-4. Other Equations

26. $y(x) + \dfrac{1}{x} \displaystyle\int_0^x f\left(\dfrac{t}{x}, y(t)\right) dt = A.$

A solution: $y(x) = \lambda$, where λ is a root of the algebraic (or transcendental) equation

$$\lambda + F(\lambda) - A = 0, \qquad F(\lambda) = \int_0^1 f(z, \lambda)\, dz.$$

27. $y(x) + \displaystyle\int_0^x f\left(\dfrac{t}{x}, \dfrac{y(t)}{t}\right) dt = Ax.$

A solution: $y(x) = \lambda x$, where λ is a root of the algebraic (or transcendental) equation

$$\lambda + F(\lambda) - A = 0, \qquad F(\lambda) = \int_0^1 f(z, \lambda)\, dz.$$

28. $y(x) + \displaystyle\int_x^\infty f\big(t - x,\, y(t - x)\big) y(t)\, dt = a e^{-\lambda x}.$

Solutions: $y(x) = b_k e^{-\lambda x}$, where b_k are roots of the algebraic (or transcendental) equation

$$b + bI(b) = a, \qquad I(b) = \int_0^\infty f(z, b e^{-\lambda z}) e^{-\lambda z}\, dz.$$

Chapter 6

Nonlinear Equations
With Constant Limits of Integration

▶ *Notation:* f, g, h, and φ *are arbitrary functions of an argument specified in the parentheses (the argument can depend on t, x, and y); and A, B, C, a, b, c, s, β, γ, λ, and μ are arbitrary parameters; and k, m, and n are nonnegative integers.*

6.1. Equations With Quadratic Nonlinearity That Contain Arbitrary Parameters

6.1-1. Equations of the Form $\int_a^b K(t)y(x)y(t)\,dt = F(x)$

1. $$\int_0^1 y(x)y(t)\,dt = Ax^\lambda, \qquad A > 0, \quad \lambda > -1.$$

    This is a special case of equation 6.2.1 with $f(x) = Ax^\lambda$, $g(t) = 1$, $a = 0$, and $b = 1$.
    Solutions: $y(x) = \pm\sqrt{A(\lambda + 1)}\,x^\lambda$.

2. $$\int_0^1 y(x)y(t)\,dt = Ae^{\beta x}, \qquad A > 0.$$

    This is a special case of equation 6.2.1 with $f(x) = Ae^{\beta x}$, $g(t) = 1$, $a = 0$, and $b = 1$.
    Solutions: $y(x) = \pm\sqrt{\dfrac{A\beta}{e^\beta - 1}}\,e^{\beta x}$.

3. $$\int_0^1 y(x)y(t)\,dt = A\cosh(\beta x), \qquad A > 0.$$

    This is a special case of equation 6.2.1 with $f(x) = A\cosh(\beta x)$, $g(t) = 1$, $a = 0$, and $b = 1$.
    Solutions: $y(x) = \pm\sqrt{\dfrac{A\beta}{\sinh\beta}}\,\cosh(\beta x)$.

4. $$\int_0^1 y(x)y(t)\,dt = A\sinh(\beta x), \qquad A\beta > 0.$$

    This is a special case of equation 6.2.1 with $f(x) = A\sinh(\beta x)$, $g(t) = 1$, $a = 0$, and $b = 1$.
    Solutions: $y(x) = \pm\sqrt{\dfrac{A\beta}{\cosh\beta - 1}}\,\sinh(\beta x)$.

5. $\displaystyle\int_0^1 y(x)y(t)\,dt = A\tanh(\beta x), \qquad A\beta > 0.$

This is a special case of equation 6.2.1 with $f(x) = A\tanh(\beta x)$, $g(t) = 1$, $a = 0$, and $b = 1$.

Solutions: $\displaystyle y(x) = \pm\sqrt{\frac{A\beta}{\ln\cosh\beta}}\,\tanh(\beta x).$

6. $\displaystyle\int_0^1 y(x)y(t)\,dt = A\ln(\beta x), \qquad A(\ln\beta - 1) > 0.$

This is a special case of equation 6.2.1 with $f(x) = A\ln(\beta x)$, $g(t) = 1$, $a = 0$, and $b = 1$.

Solutions: $\displaystyle y(x) = \pm\sqrt{\frac{A}{\ln\beta - 1}}\,\ln(\beta x).$

7. $\displaystyle\int_0^1 y(x)y(t)\,dt = A\cos(\beta x), \qquad A > 0.$

This is a special case of equation 6.2.1 with $f(x) = A\cos(\beta x)$, $g(t) = 1$, $a = 0$, and $b = 1$.

Solutions: $\displaystyle y(x) = \pm\sqrt{\frac{A\beta}{\sin\beta}}\,\cos(\beta x).$

8. $\displaystyle\int_0^1 y(x)y(t)\,dt = A\sin(\beta x), \qquad A\beta > 0.$

This is a special case of equation 6.2.1 with $f(x) = A\sin(\beta x)$, $g(t) = 1$, $a = 0$, and $b = 1$.

Solutions: $\displaystyle y(x) = \pm\sqrt{\frac{A\beta}{1 - \cos\beta}}\,\sin(\beta x).$

9. $\displaystyle\int_0^1 y(x)y(t)\,dt = A\tan(\beta x), \qquad A\beta > 0.$

This is a special case of equation 6.2.1 with $f(x) = A\tan(\beta x)$, $g(t) = 1$, $a = 0$, and $b = 1$.

Solutions: $\displaystyle y(x) = \pm\sqrt{\frac{-A\beta}{\ln|\cos\beta|}}\,\tan(\beta x).$

10. $\displaystyle\int_0^1 t^\mu y(x)y(t)\,dt = Ax^\lambda, \qquad A > 0, \quad \mu + \lambda > -1.$

This is a special case of equation 6.2.1 with $f(x) = Ax^\lambda$, $g(t) = t^\mu$, $a = 0$, and $b = 1$.
Solutions: $y(x) = \pm\sqrt{A(\mu + \lambda + 1)}\,x^\lambda.$

11. $\displaystyle\int_0^1 e^{\mu t} y(x)y(t)\,dt = Ae^{\beta x}, \qquad A > 0.$

This is a special case of equation 6.2.1 with $f(x) = Ae^{\beta x}$, $g(t) = e^{\mu t}$, $a = 0$, and $b = 1$.

Solutions: $\displaystyle y(x) = \pm\sqrt{\frac{A(\mu + \beta)}{e^{\mu+\beta} - 1}}\,e^{\beta x}.$

6.1-2. Equations of the Form $\int_a^b G(\cdots)\,dt = F(x)$

12. $\displaystyle\int_0^1 y(t)y(xt)\,dt = A, \qquad 0 \le x \le 1.$

This is a special case of equation 6.2.2 with $f(t) = 1$, $a = 0$, and $b = 1$.

1°. Solutions:

$$y_1(x) = \sqrt{A}, \qquad\qquad y_2(x) = -\sqrt{A},$$
$$y_3(x) = \sqrt{A}\,(3x - 2), \qquad\qquad y_4(x) = -\sqrt{A}\,(3x - 2),$$
$$y_5(x) = \sqrt{A}\,(10x^2 - 12x + 3), \qquad y_6(x) = -\sqrt{A}\,(10x^2 - 12x + 3).$$

2°. The integral equation has some other solutions; for example,

$$y_7(x) = \frac{\sqrt{A}}{C}\left[(2C + 1)x^C - C - 1\right], \qquad y_8(x) = -\frac{\sqrt{A}}{C}\left[(2C + 1)x^C - C - 1\right],$$
$$y_9(x) = \sqrt{A}\,(\ln x + 1), \qquad\qquad y_{10}(x) = -\sqrt{A}\,(\ln x + 1),$$

where C is an arbitrary constant.

3°. See 6.2.2 for some other solutions.

13. $\displaystyle\int_0^1 y(t)y(xt^\beta)\,dt = A, \qquad \beta > 0.$

1°. Solutions:

$$y_1(x) = \sqrt{A}, \qquad\qquad y_2(x) = -\sqrt{A},$$
$$y_3(x) = \sqrt{B}\left[(\beta + 2)x - \beta - 1\right], \qquad y_4(x) = -\sqrt{B}\left[(\beta + 2)x - \beta - 1\right],$$

where $B = \sqrt{\dfrac{2A}{\beta(\beta + 1)}}$.

2°. The integral equation has some other (more complicated solutions) of the polynomial form $y(x) = \sum\limits_{k=0}^{n} B_k x^k$, where the constants B_k can be found from the corresponding system of algebraic equations.

14. $\displaystyle\int_1^\infty y(t)y(xt)\,dt = Ax^{-\lambda}, \qquad \lambda > 0, \quad 1 \le x < \infty.$

This is a special case of equation 6.2.3 with $f(t) = 1$, $a = 1$, and $b = \infty$.

1°. Solutions:

$$y_1(x) = Bx^{-\lambda}, \qquad\qquad y_2(x) = -Bx^{-\lambda}, \qquad\qquad \lambda > \tfrac{1}{2};$$
$$y_3(x) = B\left[(2\lambda - 3)x - 2\lambda + 2\right]x^{-\lambda}, \quad y_4(x) = -B\left[(2\lambda - 3)x - 2\lambda + 2\right]x^{-\lambda}, \qquad \lambda > \tfrac{3}{2};$$

where $B = \sqrt{A(2\lambda - 1)}$.

2°. For sufficiently large λ, the integral equation has some other (more complicated) solutions of the polynomial form $y(x) = \sum\limits_{k=0}^{n} B_k x^k$, where the constants B_k can be found from the corresponding system of algebraic equations. See 6.2.2 for some other solutions.

15. $\displaystyle\int_0^\infty e^{-\lambda t} y(t) y(xt)\, dt = A, \qquad \lambda > 0, \quad 0 \le x < \infty.$

This is a special case of equation 6.2.2 with $f(t) = e^{-\lambda t}$, $a = 0$, and $b = \infty$.

1°. Solutions:

$$y_1(x) = \sqrt{A\lambda}, \qquad\qquad y_2(x) = -\sqrt{A\lambda},$$
$$y_3(x) = \sqrt{\tfrac{1}{2}A\lambda}\,(\lambda x - 2), \quad y_4(x) = -\sqrt{\tfrac{1}{2}A\lambda}\,(\lambda x - 2).$$

2°. The integral equation has some other (more complicated) solutions of the polynomial form $y(x) = \sum_{k=0}^{n} B_k x^k$, where the constants B_k can be found from the corresponding system of algebraic equations. See 6.2.2 for some other solutions.

16. $\displaystyle\int_0^1 y(t) y(x + \lambda t)\, dt = A, \qquad 0 \le x < \infty.$

This is a special case of equation 6.2.7 with $f(t) \equiv 1$, $a = 0$, and $b = 1$.
    Solutions:

$$y_1(x) = \sqrt{A}, \qquad\qquad y_2(x) = -\sqrt{A},$$
$$y_3(x) = \sqrt{3A/\lambda}\,(1 - 2x), \quad y_4(x) = -\sqrt{3A/\lambda}\,(1 - 2x).$$

17. $\displaystyle\int_0^\infty y(t) y(x + \lambda t)\, dt = A e^{-\beta x}, \qquad A, \lambda, \beta > 0, \quad 0 \le x < \infty.$

This is a special case of equation 6.2.9 with $f(t) \equiv 1$, $a = 0$, and $b = \infty$.
    Solutions:

$$y_1(x) = \sqrt{A\beta(\lambda + 1)}\,e^{-\beta x}, \qquad y_2(x) = -\sqrt{A\beta(\lambda + 1)}\,e^{-\beta x},$$
$$y_3(x) = B\big[\beta(\lambda + 1)x - 1\big]e^{-\beta x}, \quad y_4(x) = -B\big[\beta(\lambda + 1)x - 1\big]e^{-\beta x},$$

where $B = \sqrt{A\beta(\lambda + 1)/\lambda}$.

18. $\displaystyle\int_0^1 y(t) y(x - t)\, dt = A, \qquad -\infty < x < \infty.$

This is a special case of equation 6.2.10 with $f(t) \equiv 1$, $a = 0$, and $b = 1$.

1°. Solutions with $A > 0$:

$$y_1(x) = \sqrt{A}, \qquad\qquad y_2(x) = -\sqrt{A},$$
$$y_3(x) = \sqrt{5A}(6x^2 - 6x + 1), \quad y_4(x) = -\sqrt{5A}(6x^2 - 6x + 1).$$

2°. Solutions with $A < 0$:

$$y_1(x) = \sqrt{-3A}\,(1 - 2x), \quad y_2(x) = -\sqrt{-3A}\,(1 - 2x).$$

    The integral equation has some other (more complicated) solutions of the polynomial form $y(x) = \sum_{k=0}^{n} B_k x^k$, where the constants B_k can be found from the corresponding system of algebraic equations.

19. $\displaystyle\int_0^\infty e^{-\lambda t} y\!\left(\frac{x}{t}\right) y(t)\, dt = A x^b, \qquad \lambda > 0.$

Solutions: $y(x) = \pm\sqrt{A\lambda}\,x^b$.

6.1-3. Equations of the Form $y(x) + \int_a^b K(x,t)y^2(t)\,dt = F(x)$

20. $y(x) + A \displaystyle\int_a^b x^\lambda y^2(t)\,dt = 0.$

Solutions:
$$y_1(x) = 0, \qquad y_2(x) = -\frac{2\lambda + 1}{A(b^{2\lambda+1} - a^{2\lambda+1})}x^\lambda.$$

21. $y(x) + A \displaystyle\int_a^b x^\lambda t^\mu y^2(t)\,dt = 0.$

Solutions:
$$y_1(x) = 0, \qquad y_2(x) = -\frac{2\lambda + \mu + 1}{A(b^{2\lambda+\mu+1} - a^{2\lambda+\mu+1})}x^\lambda.$$

22. $y(x) + A \displaystyle\int_a^b e^{-\lambda x} y^2(t)\,dt = 0.$

Solutions:
$$y_1(x) = 0, \qquad y_2(x) = \frac{2\lambda}{A(e^{-2\lambda b} - e^{-2\lambda a})}e^{-\lambda x}.$$

23. $y(x) + A \displaystyle\int_a^b e^{-\lambda x - \mu t} y^2(t)\,dt = 0.$

Solutions:
$$y_1(x) = 0, \qquad y_2(x) = \frac{2\lambda + \mu}{A[e^{-(2\lambda+\mu)b} - e^{-(2\lambda+\mu)a}]}e^{-\lambda x}.$$

24. $y(x) + A \displaystyle\int_a^b x^\lambda e^{-\mu t} y^2(t)\,dt = 0.$

This is a special case of equation 6.2.20 with $f(x) = Ax^\lambda$ and $g(t) = e^{-\mu t}$.

25. $y(x) + A \displaystyle\int_a^b e^{-\mu x} t^\lambda y^2(t)\,dt = 0.$

This is a special case of equation 6.2.20 with $f(x) = Ae^{-\mu x}$ and $g(t) = t^\lambda$.

26. $y(x) + A \displaystyle\int_0^1 y^2(t)\,dt = Bx^\mu, \qquad \mu > -1.$

This is a special case of equation 6.2.22 with $g(t) = A$, $f(x) = Bx^\mu$, $a = 0$, and $b = 1$.
    A solution: $y(x) = Bx^\mu + \lambda$, where λ is determined by the quadratic equation

$$\lambda^2 + \frac{1}{A}\left(1 + \frac{2AB}{\mu+1}\right)\lambda + \frac{B^2}{2\mu+1} = 0.$$

27. $y(x) + A \displaystyle\int_a^b t^\beta y^2(t)\,dt = Bx^\mu.$

This is a special case of equation 6.2.22 with $g(t) = At^\beta$ and $f(x) = Bx^\mu$.

28. $y(x) + A \int_a^b e^{\beta t} y^2(t)\, dt = B e^{\mu x}.$

This is a special case of equation 6.2.22 with $g(t) = A e^{\beta t}$ and $f(x) = B e^{\mu x}$.

29. $y(x) + A \int_a^b x^\beta y^2(t)\, dt = B x^\mu.$

This is a special case of equation 6.2.23 with $g(x) = A x^\beta$ and $f(x) = B x^\mu$.

30. $y(x) + A \int_a^b e^{\beta x} y^2(t)\, dt = B e^{\mu x}.$

This is a special case of equation 6.2.23 with $g(x) = A e^{\beta x}$ and $f(x) = B e^{\mu x}$.

6.1-4. Equations of the Form $y(x) + \int_a^b K(x,t) y(x) y(t)\, dt = F(x)$

31. $y(x) + A \int_a^b t^\beta y(x) y(t)\, dt = B x^\mu.$

This is a special case of equation 6.2.25 with $g(t) = A t^\beta$ and $f(x) = B x^\mu$.

32. $y(x) + A \int_a^b e^{\beta t} y(x) y(t)\, dt = B e^{\mu x}.$

This is a special case of equation 6.2.25 with $g(t) = A e^{\beta t}$ and $f(x) = B e^{\mu x}$.

33. $y(x) + A \int_a^b x^\beta y(x) y(t)\, dt = B x^\mu.$

This is a special case of equation 6.2.26 with $g(x) = A x^\beta$ and $f(x) = B x^\mu$.

34. $y(x) + A \int_a^b e^{\beta x} y(x) y(t)\, dt = B e^{\mu x}.$

This is a special case of equation 6.2.26 with $g(x) = A e^{\beta x}$ and $f(x) = B e^{\mu x}$.

6.1-5. Equations of the Form $y(x) + \int_a^b G(\cdots)\, dt = F(x)$

35. $y(x) + A \int_0^1 y(t) y(xt)\, dt = 0.$

This is a special case of equation 6.2.30 with $f(t) = A$, $a = 0$, and $b = 1$.

1°. Solutions:

$$y_1(x) = -\frac{1}{A}(2C + 1)x^C, \quad y_2(x) = \frac{(I_1 - I_0)x + I_1 - I_2}{I_0 I_2 - I_1^2} x^C,$$

$$I_m = \frac{A}{2C + m + 1}, \quad m = 0, 1, 2,$$

where C is an arbitrary nonnegative constant.

There are more complicated solutions of the form $y(x) = x^C \sum_{k=0}^{n} B_k x^k$, where C is an arbitrary constant and the coefficients B_k can be found from the corresponding system of algebraic equations.

$2°$. A solution:

$$y_3(x) = \frac{(I_1 - I_0)x^\beta + I_1 - I_2}{I_0 I_2 - I_1^2} x^C, \qquad I_m = \frac{A}{2C + m\beta + 1}, \qquad m = 0, 1, 2,$$

where C and β are arbitrary constants.

There are more complicated solutions of the form $y(x) = x^C \sum_{k=0}^{n} D_k x^{k\beta}$, where C and β are arbitrary constants and the coefficients D_k can be found from the corresponding system of algebraic equations.

$3°$. A solution:

$$y_4(x) = \frac{x^C (J_1 \ln x - J_2)}{J_0 J_2 - J_1^2}, \qquad J_m = \int_0^1 t^{2C} (\ln t)^m dt, \qquad m = 0, 1, 2,$$

where C is an arbitrary constant.

There are more complicated solutions of the form $y(x) = x^C \sum_{k=0}^{n} E_k (\ln x)^k$, where C is an arbitrary constant and the coefficients E_k can be found from the corresponding system of algebraic equations.

36. $y(x) + A \int_1^\infty y(t)y(xt)\, dt = 0.$

This is a special case of equation 6.2.30 with $f(t) = A$, $a = 1$, and $b = \infty$.

37. $y(x) + \lambda \int_1^\infty y(t)y(xt)\, dt = Ax^\beta.$

This is a special case of equation 6.2.31 with $f(t) = \lambda$, $a = 0$, and $b = 1$.

38. $y(x) + A \int_0^1 y(t)y(x + \lambda t)\, dt = 0.$

This is a special case of equation 6.2.35 with $f(t) \equiv A$, $a = 0$, and $b = 1$.

$1°$. A solution:
$$y(x) = \frac{C(\lambda + 1)}{A[1 - e^{C(\lambda+1)}]} e^{Cx},$$

where C is an arbitrary constant.

$2°$. There are more complicated solutions of the form $y(x) = e^{Cx} \sum_{m=0}^{n} B_m x^m$, where C is an arbitrary constant and the coefficients B_m can be found from the corresponding system of algebraic equations.

39. $y(x) + A \int_0^\infty y(t)y(x + \lambda t)\, dt = 0, \qquad \lambda > 0, \quad 0 \le x < \infty.$

This is a special case of equation 6.2.35 with $f(t) \equiv A$, $a = 0$, and $b = \infty$.

A solution:
$$y(x) = -\frac{C(\lambda + 1)}{A} e^{-Cx},$$

where C is an arbitrary positive constant.

40. $\quad y(x) + A \displaystyle\int_0^\infty e^{-\lambda t} y\left(\dfrac{x}{t}\right) y(t)\, dt = 0, \qquad \lambda > 0.$

A solution: $y(x) = -\dfrac{\lambda}{A} x^C$, where C is an arbitrary constant.

41. $\quad y(x) + A \displaystyle\int_0^\infty e^{-\lambda t} y\left(\dfrac{x}{t}\right) y(t)\, dt = Bx^b, \qquad \lambda > 0.$

Solutions:

$$y_1(x) = \beta_1 x^b, \qquad y_2(x) = \beta_2 x^b,$$

where β_1 and β_2 are the roots of the quadratic equation $A\beta^2 + \lambda\beta - B\lambda = 0$.

6.2. Equations With Quadratic Nonlinearity That Contain Arbitrary Functions

---
6.2-1. Equations of the Form $\int_a^b G(\cdots)\, dt = F(x)$
---

1. $\quad \displaystyle\int_a^b g(t)y(x)y(t)\, dt = f(x).$

Solutions:

$$y(x) = \pm\lambda f(x), \qquad \lambda = \left[\int_a^b f(t)g(t)\, dt\right]^{-1/2}.$$

2. $\quad \displaystyle\int_a^b f(t)y(t)y(xt)\, dt = A.$

1°. Solutions*

$$y_1(x) = \sqrt{A/I_0}, \qquad y_2(x) = -\sqrt{A/I_0},$$
$$y_3(x) = q(I_1 x - I_2), \quad y_4(x) = -q(I_1 x - I_2),$$

where

$$I_m = \int_a^b t^m f(t)\, dt, \quad q = \left(\frac{A}{I_0 I_2^2 - I_1^2 I_2}\right)^{1/2}, \quad m = 0, 1, 2.$$

The integral equation has some other (more complicated) solutions of the polynomial form $y(x) = \displaystyle\sum_{k=0}^n B_k x^k$, where the constants B_k can be found from the corresponding system of algebraic equations.

2°. Solutions:

$$y_5(x) = q(I_1 x^C - I_2), \qquad y_6(x) = -q(I_1 x^C - I_2),$$
$$q = \left(\frac{A}{I_0 I_2^2 - I_1^2 I_2}\right)^{1/2}, \quad I_m = \int_a^b t^{mC} f(t)\, dt, \qquad m = 0, 1, 2,$$

where C is an arbitrary constant.

The equation has more complicated solutions of the form $y(x) = \displaystyle\sum_{k=0}^n B_k x^{kC}$, where C is an arbitrary constant and the coefficients B_k can be found from the corresponding system of algebraic equations.

---

* The arguments of the equations containing $y(xt)$ in the integrand can vary, for example, within the following intervals: (a) $0 \le t \le 1, 0 \le x \le 1$ for $a = 0$ and $b = 1$; (b) $1 \le t < \infty, 1 \le x < \infty$ for $a = 1$ and $b = \infty$; (c) $0 \le t < \infty, 0 \le x < \infty$ for $a = 0$ and $b = \infty$; or (d) $a \le t \le b, 0 \le x < \infty$ for arbitrary a and b such that $0 \le a < b \le \infty$. Case (d) is a special case of (c) if $f(t)$ is nonzero only on the interval $a \le t \le b$.

$3°$. Solutions:

$$y_7(x) = p(J_0 \ln x - J_1), \qquad y_8(x) = -p(J_0 \ln x - J_1),$$

$$p = \left(\frac{A}{J_0^2 J_2 - J_0 J_1^2} \right)^{1/2}, \qquad J_m = \int_a^b (\ln t)^m f(t)\, dt.$$

The equation has more complicated solutions of the form $y(x) = \sum_{k=0}^{n} E_k (\ln x)^k$, where the constants E_k can be found from the corresponding system of algebraic equations.

3. $\displaystyle \int_a^b f(t)y(t)y(xt)\, dt = Ax^\beta.$

$1°$. Solutions:
$$y_1(x) = \sqrt{A/I_0}\, x^\beta, \qquad y_2(x) = -\sqrt{A/I_0}\, x^\beta,$$
$$y_3(x) = q(I_1 x - I_2)\, x^\beta, \qquad y_4(x) = -q(I_1 x - I_2)\, x^\beta,$$

where
$$I_m = \int_a^b t^{2\beta + m} f(t)\, dt, \qquad q = \sqrt{\frac{A}{I_2(I_0 I_2 - I_1^2)}}, \qquad m = 0, 1, 2.$$

$2°$. The substitution $y(x) = x^\beta w(x)$ leads to an equation of the form 6.2.2:

$$\int_a^b g(t)w(t)w(xt)\, dt = A, \qquad g(x) = f(x)x^{2\beta}.$$

Therefore, the integral equation in question has more complicated solutions.

4. $\displaystyle \int_a^b f(t)y(t)y(xt)\, dt = A \ln x + B.$

This equation has solutions of the form $y(x) = p \ln x + q$. The constants p and q are determined from the following system of two second-order algebraic equations:

$$I_1 p^2 + I_0 pq = A, \qquad I_2 p^2 + 2I_1 pq + I_0 q^2 = B,$$

where
$$I_m = \int_a^b f(t)(\ln t)^m\, dt, \qquad m = 0, 1, 2.$$

5. $\displaystyle \int_a^b f(t)y(t)y(xt)\, dt = Ax^\lambda \ln x + Bx^\lambda.$

The substitution $y(x) = x^\lambda w(x)$ leads to an equation of the form 6.2.4:

$$\int_a^b g(t)w(t)w(xt)\, dt = A \ln x + B, \qquad g(t) = f(t)t^{2\lambda}.$$

6. $\displaystyle \int_0^\infty f(t)y(t)y\left(\frac{x}{t}\right) dt = Ax^\lambda.$

Solutions:
$$y_1(x) = \sqrt{\frac{A}{I}}\, x^\lambda, \qquad y_2(x) = -\sqrt{\frac{A}{I}}\, x^\lambda, \qquad I = \int_0^\infty f(t)\, dt.$$

7. $\displaystyle\int_a^b f(t)y(t)y(x + \lambda t)\, dt = A, \qquad \lambda > 0.$

1°. Solutions*

$$y_1(x) = \sqrt{A/I_0}, \qquad y_2(x) = -\sqrt{A/I_0},$$
$$y_3(x) = q(I_0 x - I_1), \qquad y_4(x) = -q(I_0 x - I_1),$$

where

$$I_m = \int_a^b t^m f(t)\, dt, \quad q = \sqrt{\frac{A}{\lambda(I_0^2 I_2 - I_0 I_1^2)}}, \qquad m = 0, 1, 2.$$

2°. The integral equation has some other (more complicated) solutions of the polynomial form $y(x) = \sum_{k=0}^{n} B_k x^k$, where the constants B_k can be found from the corresponding system of algebraic equations.

8. $\displaystyle\int_a^b f(t)y(t)y(x + \lambda t)\, dt = Ax + B, \qquad \lambda > 0.$

A solution: $y(x) = \beta x + \mu$, where the constants β and μ are determined from the following system of two second-order algebraic equations:

$$I_0\beta\mu + I_1\beta^2 = A, \quad I_0\mu^2 + (\lambda + 1)I_1\beta\mu + \lambda I_2\beta^2 = B, \qquad I_m = \int_a^b t^m f(t)\, dt. \quad (1)$$

Multiplying the first equation by B and the second by $-A$ and adding the resulting equations, we obtain the quadratic equation

$$AI_0 z^2 + \big[(\lambda + 1)AI_1 - BI_0\big]z + \lambda AI_2 - BI_1 = 0, \qquad z = \mu/\beta. \quad (2)$$

In general, to each root of equation (2) two solutions of system (1) correspond. Therefore, the original integral equation can have at most four solutions of this form. If the discriminant of equation (2) is negative, then the integral equation has no such solutions.

     The integral equation has some other (more complicated) solutions of the polynomial form $y(x) = \sum_{k=0}^{n} B_k x^k$, where the constants β_k can be found from the corresponding system of algebraic equations.

9. $\displaystyle\int_a^b f(t)y(t)y(x + \lambda t)\, dt = Ae^{-\beta x}, \qquad \lambda > 0.$

1°. Solutions:

$$y_1(x) = \sqrt{A/I_0}\, e^{-\beta x}, \qquad y_2(x) = -\sqrt{A/I_0}\, e^{-\beta x},$$
$$y_3(x) = q(I_0 x - I_1)e^{-\beta x}, \qquad y_4(x) = -q(I_0 x - I_1)e^{-\beta x},$$

where

$$I_m = \int_a^b t^m e^{-\beta(\lambda+1)t} f(t)\, dt, \quad q = \sqrt{\frac{A}{\lambda(I_0^2 I_2 - I_0 I_1^2)}}, \qquad m = 0, 1, 2.$$

2°. The equation has more complicated solutions of the form $y(x) = e^{-\beta x}\sum_{k=0}^{n} B_k x^k$, where the constants B_k can be found from the corresponding system of algebraic equations.

---

   * The arguments of the equations containing $y(x+\lambda t)$ in the integrand can vary within the following intervals: (a) $0 \le t < \infty$, $0 \le x < \infty$ for $a = 0$ and $b = \infty$ or (b) $a \le t \le b$, $0 \le x < \infty$ for arbitrary a and b such that $0 \le a < b < \infty$. Case (b) is a special case of (a) if $f(t)$ is nonzero only on the interval $a \le t \le b$.

3°. The substitution $y(x) = e^{-\beta x} w(x)$ leads to an equation of the form 6.2.7:

$$\int_a^b e^{-\beta(\lambda+1)t} f(t) w(t) w(x + \lambda t)\, dt = A.$$

10. $\displaystyle\int_a^b f(t) y(t) y(x - t)\, dt = A.$

1°. Solutions*

$$y_1(x) = \sqrt{A/I_0}, \qquad y_2(x) = -\sqrt{A/I_0},$$
$$y_3(x) = q(I_0 x - I_1), \qquad y_4(x) = -q(I_0 x - I_1),$$

where

$$I_m = \int_a^b t^m f(t)\, dt, \quad q = \sqrt{\frac{A}{I_0 I_1^2 - I_0^2 I_2}}, \quad m = 0, 1, 2.$$

2°. The integral equation has some other (more complicated) solutions of the polynomial form $y(x) = \sum_{k=0}^{n} \lambda_k x^k$, where the constants λ_k can be found from the corresponding system of algebraic equations. For $n = 3$, such a solution is presented in 6.1.18.

11. $\displaystyle\int_a^b f(t) y(t) y(x - t)\, dt = Ax + B.$

A solution: $y(x) = \lambda x + \mu$, where the constants λ and μ are determined from the following system of two second-order algebraic equations:

$$I_0 \lambda \mu + I_1 \lambda^2 = A, \quad I_0 \mu^2 - I_2 \lambda^2 = B, \quad I_m = \int_a^b t^m f(t)\, dt, \quad m = 0, 1, 2. \tag{1}$$

Multiplying the first equation by B and the second by $-A$ and adding the results, we obtain the quadratic equation

$$A I_0 z^2 - B I_0 z - A I_2 - B I_1 = 0, \quad z = \mu/\lambda. \tag{2}$$

In general, to each root of equation (2) two solutions of system (1) correspond. Therefore, the original integral equation can have at most four solutions of this form. If the discriminant of equation (2) is negative, then the integral equation has no such solutions.

The integral equation has some other (more complicated) solutions of the polynomial form $y(x) = \sum_{k=0}^{n} \lambda_k x^k$, where the constants λ_k can be found from the corresponding system of algebraic equations.

12. $\displaystyle\int_a^b f(t) y(t) y(x - t)\, dt = \sum_{k=0}^{n} A_k x^k.$

This equation has solutions of the form

$$y(x) = \sum_{k=0}^{n} \lambda_k x^k, \tag{1}$$

where the constants λ_k are determined from the system of algebraic equations obtained by substituting solution (1) into the original integral equation and matching the coefficients of like powers of x.

---

* The arguments of the equations containing $y(x-t)$ in the integrand can vary within the following intervals: (a) $-\infty < t < \infty$, $-\infty < x < \infty$ for $a = -\infty$ and $b = \infty$ or (b) $a \le t \le b$, $-\infty \le x < \infty$, for arbitrary a and b such that $-\infty < a < b < \infty$. Case (b) is a special case of (a) if $f(t)$ is nonzero only on the interval $a \le t \le b$.

13. $\displaystyle\int_a^b f(t)y(x-t)y(t)\,dt = Ae^{\lambda x}.$

Solutions:

$$y_1(x) = \sqrt{A/I_0}\,e^{\lambda x}, \qquad y_2(x) = -\sqrt{A/I_0}\,e^{\lambda x},$$
$$y_3(x) = q(I_0 x - I_1)e^{\lambda x}, \qquad y_4(x) = -q(I_0 x - I_1)e^{\lambda x},$$

where

$$I_m = \int_a^b t^m f(t)\,dt, \quad q = \sqrt{\frac{A}{I_0 I_1^2 - I_0^2 I_2}}, \quad m = 0, 1, 2.$$

The integral equation has more complicated solutions of the form $y(x) = e^{\lambda x}\sum_{k=0}^{n} B_k x^k$, where the constants B_k can be found from the corresponding system of algebraic equations.

14. $\displaystyle\int_a^b f(t)y(t)y(x-t)\,dt = A\sinh\lambda x.$

A solution:

$$y(x) = p\sinh\lambda x + q\cosh\lambda x. \tag{1}$$

Here p and q are roots of the algebraic system

$$I_0 pq + I_{cs}(p^2 - q^2) = A, \qquad I_{cc}q^2 - I_{ss}p^2 = 0, \tag{2}$$

where the notation

$$I_0 = \int_a^b f(t)\,dt, \qquad I_{cs} = \int_a^b f(t)\cosh(\lambda t)\sinh(\lambda t)\,dt,$$
$$I_{cc} = \int_a^b f(t)\cosh^2(\lambda t)\,dt, \qquad I_{ss} = \int_a^b f(t)\sinh^2(\lambda t)\,dt$$

is used. Different solutions of system (2) generate different solutions (1) of the integral equation.

It follows from the second equation of (2) that $q = \pm\sqrt{I_{ss}/I_{cc}}\,p$. Using this expression to eliminate q from the first equation of (2), we obtain the following four solutions:

$$y_{1,2}(x) = p\big(\sinh\lambda x \pm k\cosh\lambda x\big), \quad y_{3,4}(x) = -p\big(\sinh\lambda x \pm k\cosh\lambda x\big),$$
$$k = \sqrt{\frac{I_{ss}}{I_{cc}}}, \quad p = \sqrt{\frac{A}{(1-k^2)I_{cs} \pm kI_0}}.$$

15. $\displaystyle\int_a^b f(t)y(t)y(x-t)\,dt = A\cosh\lambda x.$

A solution:

$$y(x) = p\sinh\lambda x + q\cosh\lambda x. \tag{1}$$

Here p and q are roots of the algebraic system

$$I_0 pq + I_{cs}(p^2 - q^2) = 0, \qquad I_{cc}q^2 - I_{ss}p^2 = A, \tag{2}$$

where we use the notation introduced in 6.2.14. Different solutions of system (2) generate different solutions (1) of the integral equation.

16. $\displaystyle\int_a^b f(t)y(t)y(x-t)\,dt = A\sin\lambda x.$

A solution:

$$y(x) = p\sin\lambda x + q\cos\lambda x. \tag{1}$$

Here p and q are roots of the algebraic system

$$I_0 pq + I_{cs}(p^2 + q^2) = A, \qquad I_{cc}q^2 - I_{ss}p^2 = 0, \tag{2}$$

where

$$I_0 = \int_a^b f(t)\,dt, \qquad I_{cs} = \int_a^b f(t)\cos(\lambda t)\sin(\lambda t)\,dt,$$

$$I_{cc} = \int_a^b f(t)\cos^2(\lambda t)\,dt, \qquad I_{ss} = \int_a^b f(t)\sin^2(\lambda t)\,dt.$$

It follows from the second equation of (2) that $q = \pm\sqrt{I_{ss}/I_{cc}}\,p$. Using this expression to eliminate q from the first equation of (2), we obtain the following four solutions:

$$y_{1,2}(x) = p\big(\sin\lambda x \pm k\cos\lambda x\big), \qquad y_{3,4}(x) = -p\big(\sin\lambda x \pm k\cos\lambda x\big),$$

$$k = \sqrt{\frac{I_{ss}}{I_{cc}}}, \qquad p = \sqrt{\frac{A}{(1+k^2)I_{cs}\pm kI_0}}.$$

17. $\displaystyle\int_a^b f(t)y(t)y(x-t)\,dt = A\cos\lambda x.$

A solution:

$$y(x) = p\sin\lambda x + q\cos\lambda x. \tag{1}$$

Here p and q are roots of the algebraic system

$$I_0 pq + I_{cs}(p^2 + q^2) = 0, \qquad I_{cc}q^2 - I_{ss}p^2 = A, \tag{2}$$

where we use the notation introduced in 6.2.16. Different solutions of system (2) generate different solutions (1) of the integral equation.

18. $\displaystyle\int_0^1 y(t)y(\xi)\,dt = A, \qquad \xi = f(x)t.$

1°. Solutions:

$$y_1(t) = \sqrt{A}, \qquad\qquad y_2(t) = -\sqrt{A},$$

$$y_3(t) = \sqrt{A}\,(3t-2), \qquad\qquad y_4(t) = -\sqrt{A}\,(3t-2),$$

$$y_5(t) = \sqrt{A}\,(10t^2 - 12t + 3), \qquad y_6(t) = -\sqrt{A}\,(10t^2 - 12t + 3).$$

2°. The integral equation has some other (more complicated) solutions of the polynomial form $y(t) = \sum_{k=0}^n B_k t^k$, where the constants B_k can be found from the corresponding system of algebraic equations.

3°. The substitution $z = f(x)$ leads to an equation of the form 6.1.12.

6.2-2. Equations of the Form $y(x) + \int_a^b K(x,t)y^2(t)\,dt = F(x)$

19. $\quad y(x) + \displaystyle\int_a^b f(x)y^2(t)\,dt = 0.$

Solutions: $y_1(x) = 0$ and $y_2(x) = \lambda f(x)$, where $\lambda = -\left[\displaystyle\int_a^b f^2(t)\,dt\right]^{-1}$.

20. $\quad y(x) + \displaystyle\int_a^b f(x)g(t)y^2(t)\,dt = 0.$

This is a special case of equation 6.8.29.

Solutions: $y_1(x) = 0$ and $y_2(x) = \lambda f(x)$, where $\lambda = -\left[\displaystyle\int_a^b f^2(t)g(t)\,dt\right]^{-1}$.

21. $\quad y(x) + A\displaystyle\int_a^b y^2(t)\,dt = f(x).$

This is a special case of equation 6.8.27.

A solution: $y(x) = f(x) + \lambda$, where λ is determined by the quadratic equation

$$A(b-a)\lambda^2 + (1+2AI_1)\lambda + AI_2 = 0, \quad \text{where} \quad I_1 = \int_a^b f(t)\,dt, \quad I_2 = \int_a^b f^2(t)\,dt.$$

22. $\quad y(x) + \displaystyle\int_a^b g(t)y^2(t)\,dt = f(x).$

This is a special case of equation 6.8.29.

A solution: $y(x) = f(x) + \lambda$, where λ is determined by the quadratic equation

$$I_0\lambda^2 + (1+2I_1)\lambda + I_2 = 0, \quad \text{where} \quad I_m = \int_a^b f^m(t)g(t)\,dt, \quad m = 0,1,2.$$

23. $\quad y(x) + \displaystyle\int_a^b g(x)y^2(t)\,dt = f(x).$

Solution: $y(x) = \lambda g(x) + f(x)$, where λ is determined by the quadratic equation

$$I_{gg}\lambda^2 + (1+2I_{fg})\lambda + I_{ff} = 0,$$

$$I_{gg} = \int_a^b g^2(t)\,dt, \quad I_{fg} = \int_a^b f(t)g(t)\,dt, \quad I_{ff} = \int_a^b f^2(t)\,dt.$$

24. $\quad y(x) + \displaystyle\int_a^b \left[g_1(x)h_1(t) + g_2(x)h_2(t)\right]y^2(t)\,dt = f(x).$

A solution: $y(x) = \lambda_1 g_1(x) + \lambda_2 g_2(x) + f(x)$, where the constants λ_1 and λ_2 can be found from a system of two second-order algebraic equations (this system can be obtained from the more general system presented in 6.8.42).

6.2-3. Equations of the Form $y(x) + \int_a^b \sum K_{nm}(x,t)y^n(x)y^m(t)\,dt = F(x)$, $n + m \leq 2$

25. $y(x) + \int_a^b g(t)y(x)y(t)\,dt = f(x)$.

Solutions:
$$y_1(x) = \lambda_1 f(x), \qquad y_2(x) = \lambda_2 f(x),$$

where λ_1 and λ_2 are the roots of the quadratic equation
$$I\lambda^2 + \lambda - 1 = 0, \qquad I = \int_a^b f(t)g(t)\,dt.$$

26. $y(x) + \int_a^b g(x)y(x)y(t)\,dt = f(x)$.

A solution:
$$y(x) = \frac{f(x)}{1 + \lambda g(x)},$$

where λ is a root of the algebraic (or transcendental) equation
$$\lambda - \int_a^b \frac{f(t)\,dt}{1 + \lambda g(t)} = 0.$$

Different roots generate different solutions of the integral equation.

27. $y(x) + \int_a^b \left[g_1(t)y^2(x) + g_2(x)y(t) \right] dt = f(x)$.

Solution in an implicit form:
$$y(x) + Iy^2(x) + \lambda g_2(x) - f(x) = 0, \qquad I = \int_a^b g_1(t)\,dt, \tag{1}$$

where λ is determined by the algebraic equation
$$\lambda = \int_a^b y(t)\,dt. \tag{2}$$

Here the function $y(x) = y(x, \lambda)$ obtained by solving the quadratic equation (1) must be substituted in the integrand of (2).

28. $y(x) + \int_a^b \left[g_1(t)y^2(x) + g_2(x)y^2(t) \right] dt = f(x)$.

Solution in an implicit form:
$$y(x) + Iy^2(x) + \lambda g_2(x) - f(x) = 0, \qquad I = \int_a^b g_1(t)\,dt, \tag{1}$$

where λ is determined by the algebraic equation
$$\lambda = \int_a^b y^2(t)\,dt. \tag{2}$$

Here the function $y(x) = y(x, \lambda)$ obtained by solving the quadratic equation (1) must be substituted into the integrand of (2).

29. $y(x) + \int_a^b \left[g_{11}(x)h_{11}(t)y^2(x) + g_{12}(x)h_{12}(t)y(x)y(t) + g_{22}(x)h_{22}(t)y^2(t) \right.$
$$\left. + g_1(x)h_1(t)y(x) + g_2(x)h_2(t)y(t) \right] dt = f(x).$$

This is a special case of equation 6.8.44.

6.2-4. Equations of the Form $y(x) + \int_a^b G(\cdots)\, dt = F(x)$

---

30. $y(x) + \displaystyle\int_a^b f(t)y(t)y(xt)\, dt = 0.$

1°. Solutions:

$$y_1(x) = -\frac{1}{I_0}x^C, \quad y_2(x) = \frac{(I_1 - I_0)x + I_1 - I_2}{I_0 I_2 - I_1^2}x^C,$$

$$I_m = \int_a^b f(t)t^{2C+m}\, dt, \quad m = 0, 1, 2,$$

where C is an arbitrary constant.

There are more complicated solutions of the form $y(x) = x^C \sum_{k=0}^n B_k x^k$, where C is an arbitrary constant and the coefficients B_k can be found from the corresponding system of algebraic equations.

2°. A solution:

$$y_3(x) = \frac{(I_1 - I_0)x^\beta + I_1 - I_2}{I_0 I_2 - I_1^2}x^C,$$

$$I_m = \int_a^b f(t)t^{2C+m\beta}\, dt, \quad m = 0, 1, 2,$$

where C and β are arbitrary constants.

There are more complicated solutions of the form $y(x) = x^C \sum_{k=0}^n D_k x^{k\beta}$, where C and β are arbitrary constants and the coefficients D_k can be found from the corresponding system of algebraic equations.

3°. A solution:

$$y_4(x) = \frac{x^C(J_1 \ln x - J_2)}{J_0 J_2 - J_1^2},$$

$$J_m = \int_a^b f(t)t^{2C}(\ln t)^m dt, \quad m = 0, 1, 2,$$

where C is an arbitrary constant.

There are more complicated solutions of the form $y(x) = x^C \sum_{k=0}^n E_k(\ln x)^k$, where C is an arbitrary constant and the coefficients E_k can be found from the corresponding system of algebraic equations.

4°. The equation also has the trivial solution $y(x) \equiv 0$.

5°. The substitution $y(x) = x^\beta w(x)$ leads to an equation of the same form,

$$w(x) + \int_a^b g(t)w(t)w(xt)\, dt = 0, \qquad g(x) = f(x)x^{2\beta}.$$

31. $y(x) + \displaystyle\int_a^b f(t)y(t)y(xt)\,dt = Ax^\beta.$

1°. Solutions:
$$y_1(x) = k_1 x^\beta, \quad y_2(x) = k_2 x^\beta,$$

where k_1 and k_2 are the roots of the quadratic equation
$$Ik^2 + k - A = 0, \qquad I = \int_a^b f(t)t^{2\beta}\,dt.$$

2°. Solutions:
$$y(x) = x^\beta(\lambda x + \mu),$$

where λ and μ are determined from the following system of two algebraic equations (this system can be reduced to a quadratic equation):
$$I_2\lambda + I_1\mu + 1 = 0, \quad I_1\lambda\mu + I_0\mu^2 + \mu - A = 0$$

where $I_m = \displaystyle\int_a^b f(t)t^{2\beta+m}\,dt, \ m = 0, 1, 2.$

3°. There are more complicated solutions of the form $y(x) = x^\beta \displaystyle\sum_{m=0}^n B_m x^m$, where the B_m can be found from the corresponding system of algebraic equations.

32. $y(x) + \displaystyle\int_a^b f(t)y(t)y(xt)\,dt = A\ln x + B.$

This equation has solutions of the form $y(x) = p\ln x + q$, where the constants p and q can be found from a system of two second-order algebraic equations.

33. $y(x) + \displaystyle\int_0^\infty f(t)y(t)y\left(\dfrac{x}{t}\right) dt = 0.$

1°. A solution:
$$y(x) = -kx^C, \qquad k = \left[\int_0^\infty f(t)\,dt\right]^{-1},$$

where C is an arbitrary constant.

2°. The equation has the trivial solution $y(x) \equiv 0$.

3°. The substitution $y(x) = x^\beta w(x)$ leads to an equation of the same form,
$$w(x) + \int_0^\infty f(t)w(t)w\left(\dfrac{x}{t}\right) dt = 0.$$

34. $y(x) + \displaystyle\int_0^\infty f(t)y\left(\dfrac{x}{t}\right)y(t)\,dt = Ax^b.$

Solutions:
$$y_1(x) = \lambda_1 x^b, \qquad y_2(x) = \lambda_2 x^b,$$

where λ_1 and λ_2 are the roots of the quadratic equation
$$I\lambda^2 + \lambda - A = 0, \qquad I = \int_0^\infty f(t)\,dt.$$

35. $y(x) + \displaystyle\int_a^b f(t)y(t)y(x + \lambda t)\, dt = 0, \qquad \lambda > 0.$

1°. Solutions:

$$y_1(x) = -\frac{1}{I_0} \exp(-Cx), \quad y_2(x) = \frac{I_2 - I_1 x}{I_1^2 - I_0 I_2} \exp(-Cx),$$

$$I_m = \int_a^b t^m \exp\big[-C(\lambda + 1)t\big] f(t)\, dt, \qquad m = 0, 1, 2,$$

where C is an arbitrary constant.

2°. There are more complicated solutions of the form $y(x) = \exp(-Cx) \sum_{k=0}^n A_k x^k$, where C is an arbitrary constant and the coefficients A_k can be found from the corresponding system of algebraic equations.

3°. The equation also has the trivial solution $y(x) \equiv 0$.

4°. The substitution $y(x) = e^{\beta x} w(x)$ leads to a similar equation:

$$w(x) + \int_a^b g(t)w(t)w(x + \lambda t)\, dt = 0, \qquad g(t) = e^{\beta(\lambda+1)t} f(t).$$

36. $y(x) + \displaystyle\int_a^b f(t)y(x + \lambda t)y(t)\, dt = Ae^{-\mu x}, \qquad \lambda > 0.$

1°. Solutions:

$$y_1(x) = k_1 e^{-\mu x}, \qquad y_2(x) = k_2 e^{-\mu x},$$

where k_1 and k_2 are the roots of the quadratic equation

$$Ik^2 + k - A = 0, \qquad I = \int_a^b e^{-\mu(\lambda+1)t} f(t)\, dt.$$

2°. There are more complicated solutions of the form $y(x) = e^{-\mu x} \sum_{m=0}^n B_m x^m$, where the B_m can be found from the corresponding system of algebraic equations.

3°. The substitution $y(x) = e^{\beta x} w(x)$ leads to an equation of the same form,

$$w(x) + \int_a^b g(t)w(t)w(x - t)\, dt = Ae^{(\lambda-\beta)x}, \qquad g(t) = f(t)e^{\beta(\lambda+1)t}.$$

37. $y(x) + \displaystyle\int_a^b f(t)y(t)y(x - t)\, dt = 0.$

1°. Solutions:

$$y_1(x) = -\frac{1}{I_0} \exp(Cx), \quad y_2(x) = \frac{I_2 - I_1 x}{I_1^2 - I_0 I_2} \exp(Cx), \quad I_m = \int_a^b t^m f(t)\, dt,$$

where C is an arbitrary constant and $m = 0, 1, 2$.

2°. There are more complicated solutions of the form $y(x) = \exp(Cx) \sum_{k=0}^n A_k x^k$, where C is an arbitrary constant and the coefficients A_k can be found from the corresponding system of algebraic equations.

3°. The equation also has the trivial solution $y(x) \equiv 0$.

4°. The substitution $y(x) = \exp(Cx)w(x)$ leads to an equation of the same form:

$$w(x) + \int_a^b f(t)w(t)w(x - t)\, dt = 0.$$

38. $y(x) + \int_a^b f(t)y(x-t)y(t)\,dt = Ae^{\lambda x}.$

1°. Solutions:

$$y_1(x) = k_1 e^{\lambda x}, \qquad y_2(x) = k_2 e^{\lambda x},$$

where k_1 and k_2 are the roots of the quadratic equation

$$Ik^2 + k - A = 0, \qquad I = \int_a^b f(t)\,dt.$$

2°. The substitution $y(x) = e^{\beta x} w(x)$ leads to an equation of the same form,

$$w(x) + \int_a^b f(t)w(t)w(x-t)\,dt = Ae^{(\lambda-\beta)x}.$$

39. $y(x) + \int_a^b f(t)y(t)y(x-t)\,dt = A \sinh \lambda x.$

A solution:

$$y(x) = p \sinh \lambda x + q \cosh \lambda x. \tag{1}$$

Here p and q are roots of the algebraic system

$$p + I_0 pq + I_{cs}(p^2 - q^2) = A, \qquad q + I_{cc}q^2 - I_{ss}p^2 = 0, \tag{2}$$

where

$$I_0 = \int_a^b f(t)\,dt, \qquad I_{cs} = \int_a^b f(t)\cosh(\lambda t)\sinh(\lambda t)\,dt,$$

$$I_{cc} = \int_a^b f(t)\cosh^2(\lambda t)\,dt, \qquad I_{ss} = \int_a^b f(t)\sinh^2(\lambda t)\,dt.$$

Different solutions of system (2) generate different solutions (1) of the integral equation.

40. $y(x) + \int_a^b f(t)y(t)y(x-t)\,dt = A \cosh \lambda x.$

A solution:

$$y(x) = p \sinh \lambda x + q \cosh \lambda x. \tag{1}$$

Here p and q are roots of the algebraic system

$$p + I_0 pq + I_{cs}(p^2 - q^2) = 0, \qquad q + I_{cc}q^2 - I_{ss}p^2 = A, \tag{2}$$

where we use the notation introduced in 6.2.39. Different solutions of system (2) generate different solutions (1) of the integral equation.

41. $y(x) + \int_a^b f(t)y(t)y(x-t)\,dt = A \sin \lambda x.$

A solution:

$$y(x) = p \sin \lambda x + q \cos \lambda x. \tag{1}$$

Here p and q are roots of the algebraic system

$$p + I_0 pq + I_{cs}(p^2 + q^2) = A, \qquad q + I_{cc}q^2 - I_{ss}p^2 = 0, \tag{2}$$

where

$$I_0 = \int_a^b f(t)\,dt, \qquad I_{cs} = \int_a^b f(t)\cos(\lambda t)\sin(\lambda t)\,dt,$$

$$I_{cc} = \int_a^b f(t)\cos^2(\lambda t)\,dt, \qquad I_{ss} = \int_a^b f(t)\sin^2(\lambda t)\,dt.$$

Different solutions of system (2) generate different solutions (1) of the integral equation.

42. $y(x) + \displaystyle\int_a^b f(t)y(t)y(x-t)\,dt = A\cos\lambda x.$

A solution:

$$y(x) = p\sin\lambda x + q\cos\lambda x. \tag{1}$$

Here p and q are roots of the algebraic system

$$p + I_0 pq + I_{cs}(p^2 + q^2) = 0, \qquad q + I_{cc}q^2 - I_{ss}p^2 = A, \tag{2}$$

where we use the notation introduced in 6.2.41. Different solutions of system (2) generate different solutions (1) of the integral equation.

6.3. Equations With Power-Law Nonlinearity

6.3-1. Equations of the Form $\int_a^b G(\cdots)\,dt = F(x)$

1. $\displaystyle\int_a^b t^\lambda y^\mu(x)y^\beta(t)\,dt = f(x).$

A solution:

$$y(x) = A\big[f(x)\big]^{\frac{1}{\mu}}, \qquad A = \left\{\int_a^b t^\lambda \big[f(t)\big]^{\frac{\beta}{\mu}}\,dt\right\}^{-\frac{1}{\mu+\beta}}.$$

2. $\displaystyle\int_a^b e^{\lambda t}y^\mu(x)y^\beta(t)\,dt = f(x).$

A solution:

$$y(x) = A\big[f(x)\big]^{\frac{1}{\mu}}, \qquad A = \left\{\int_a^b e^{\lambda t}\big[f(t)\big]^{\frac{\beta}{\mu}}\,dt\right\}^{-\frac{1}{\mu+\beta}}.$$

3. $\displaystyle\int_0^\infty f(x^a t)t^b y\big(x^k t\big)\big[y(t)\big]^s\,dt = Ax^c.$

A solution:

$$y(x) = \left(\frac{A}{I}\right)^{\frac{1}{s+1}} x^\lambda, \qquad \lambda = \frac{a+c+ab}{k-a-as},$$

$$I = \int_0^\infty f(t)t^\beta\,dt, \qquad \beta = \frac{a+c+as+bk+cs}{k-a-as}.$$

6.3-2. Equations of the Form $y(x) + \int_a^b K(x,t)y^\beta(t)\,dt = F(x)$

4. $y(x) + A\displaystyle\int_a^b t^\lambda y^\beta(t)\,dt = g(x).$

This is a special case of equation 6.8.27 with $f(t,y) = At^\lambda y^\beta$.

5. $y(x) + A\displaystyle\int_a^b e^{\mu t}y^\beta(t)\,dt = g(x).$

This is a special case of equation 6.8.27 with $f(t,y) = Ae^{\mu t}y^\beta$.

6. $y(x) + A \displaystyle\int_a^b e^{\lambda(x-t)} y^\beta(t)\, dt = g(x).$

    This is a special case of equation 6.8.28 with $f(t, y) = Ay^\beta$.

7. $y(x) - \displaystyle\int_a^b g(x) y^\beta(t)\, dt = 0.$

    A solution:

    $$y(x) = \lambda g(x), \qquad \lambda = \left[\int_a^b g^\beta(t)\, dt \right]^{\frac{1}{1-\beta}}.$$

    For $\beta > 0$, the equation also has the trivial solution $y(x) \equiv 0$.

8. $y(x) - \displaystyle\int_a^b g(x) y^\beta(t)\, dt = h(x).$

    This is a special case of equation 6.8.29 with $f(t, y) = -y^\beta$.

9. $y(x) + A \displaystyle\int_a^b \cosh(\lambda x + \mu t) y^\beta(t)\, dt = h(x).$

    This is a special case of equation 6.8.31 with $f(t, y) = Ay^\beta$.

10. $y(x) + A \displaystyle\int_a^b \sinh(\lambda x + \mu t) y^\beta(t)\, dt = h(x).$

    This is a special case of equation 6.8.32 with $f(t, y) = Ay^\beta$.

11. $y(x) + A \displaystyle\int_a^b \cos(\lambda x + \mu t) y^\beta(t)\, dt = h(x).$

    This is a special case of equation 6.8.33 with $f(t, y) = Ay^\beta$.

12. $y(x) + A \displaystyle\int_a^b \sin(\lambda x + \mu t) y^\beta(t)\, dt = h(x).$

    This is a special case of equation 6.8.34 with $f(t, y) = Ay^\beta$.

13. $y(x) + \displaystyle\int_0^\infty f\!\left(\frac{t}{x}\right) \sqrt{y(t)}\, dt = Ax^2.$

    Solutions: $y_k(x) = \beta_k^2 x^2$, where β_k $(k = 1, 2)$ are the roots of the quadratic equations

    $$\beta^2 \pm I\beta - A = 0, \qquad I = \int_0^\infty z f(z)\, dz.$$

14. $y(x) - \displaystyle\int_0^\infty t^\lambda f\!\left(\frac{t}{x}\right) [y(t)]^\beta\, dt = 0, \qquad \beta \neq 1.$

    A solution:

    $$y(x) = Ax^{\frac{1+\lambda}{1-\beta}}, \qquad A^{1-\beta} = \int_0^\infty z^{\frac{\lambda+\beta}{1-\beta}} f(z)\, dz.$$

15. $y(x) - \displaystyle\int_{-\infty}^\infty e^{\lambda t} f(ax + bt) [y(t)]^\beta\, dt = 0, \qquad b \neq 0, \ a\beta \neq -b.$

    A solution:

    $$y(x) = A \exp\!\left(-\frac{a\lambda}{a\beta + b}\, x\right), \qquad A^{1-\beta} = \int_{-\infty}^\infty \exp\!\left(\frac{\lambda b}{a\beta + b}\, z\right) f(bz)\, dz.$$

> **6.3-3. Equations of the Form $y(x) + \int_a^b G(\cdots)\, dt = F(x)$**

16. $y(x) + A \displaystyle\int_a^b y^\beta(x) y^\mu(t)\, dt = f(x).$

Solution in an implicit form:

$$y(x) + A\lambda y^\beta(x) - f(x) = 0, \tag{1}$$

where λ is determined by the algebraic (or transcendental) equation

$$\lambda = \int_a^b y^\mu(t)\, dt. \tag{2}$$

Here the function $y(x) = y(x, \lambda)$ obtained by solving the quadratic equation (1) must be substituted in the integrand of (2).

17. $y(x) + \displaystyle\int_a^b g(t) y(x) y^\mu(t)\, dt = f(x).$

A solution: $y(x) = \lambda f(x)$, where λ is determined from the algebraic (or transcendental) equation

$$I\lambda^{\mu+1} + \lambda - 1 = 0, \qquad I = \int_a^b g(t) f^\mu(t)\, dt.$$

18. $y(x) + \displaystyle\int_a^b g(x) y(x) y^\mu(t)\, dt = f(x).$

A solution:

$$y(x) = \frac{f(x)}{1 + \lambda g(x)},$$

where λ is a root of the algebraic (or transcendental) equation

$$\lambda - \int_a^b \frac{f^\mu(t)\, dt}{[1 + \lambda g(t)]^\mu} = 0.$$

Different roots generate different solutions of the integral equation.

19. $y(x) + \displaystyle\int_a^b \left[g_1(t) y^2(x) + g_2(x) y^\mu(t) \right] dt = f(x).$

Solution in an implicit form:

$$y(x) + I y^2(x) + \lambda g_2(x) - f(x) = 0, \qquad I = \int_a^b g_1(t)\, dt, \tag{1}$$

where λ is determined by the algebraic (or transcendental) equation

$$\lambda = \int_a^b y^\mu(t)\, dt. \tag{2}$$

Here the function $y(x) = y(x, \lambda)$ obtained by solving the quadratic equation (1) must be substituted in the integrand of (2).

20. $y(x) + \displaystyle\int_a^b \left[g_1(x) h_1(t) y^k(x) y^s(t) + g_2(x) h_2(t) y^p(x) y^q(t) \right] dt = f(x).$

This is a special case of equation 6.8.44.

21. $y(x) + A \displaystyle\int_a^b y(xt) y^\beta(t)\, dt = 0.$

This is a special case of equation 6.8.45 with $f(t, y) = Ay^\beta$.

6.4. Equations With Exponential Nonlinearity

6.4-1. Integrands With Nonlinearity of the Form exp[$\beta y(t)$]

1. $y(x) + A \displaystyle\int_a^b \exp[\beta y(t)]\, dt = g(x).$

This is a special case of equation 6.8.27 with $f(t, y) = A \exp(\beta y)$.

2. $y(x) + A \displaystyle\int_a^b t^\mu \exp[\beta y(t)]\, dt = g(x).$

This is a special case of equation 6.8.27 with $f(t, y) = A t^\mu \exp(\beta y)$.

3. $y(x) + A \displaystyle\int_a^b \exp\big[\mu t + \beta y(t)\big]\, dt = g(x).$

This is a special case of equation 6.8.27 with $f(t, y) = A \exp(\mu t) \exp(\beta y)$.

4. $y(x) + A \displaystyle\int_a^b \exp\big[\lambda(x - t) + \beta y(t)\big]\, dt = g(x).$

This is a special case of equation 6.8.28 with $f(t, y) = A \exp(\beta y)$.

5. $y(x) + \displaystyle\int_a^b g(x) \exp[\beta y(t)]\, dt = h(x).$

This is a special case of equation 6.8.29 with $f(t, y) = \exp(\beta y)$.

6. $y(x) + A \displaystyle\int_a^b \cosh(\lambda x + \mu t) \exp[\beta y(t)]\, dt = h(x).$

This is a special case of equation 6.8.31 with $f(t, y) = A \exp(\beta y)$.

7. $y(x) + A \displaystyle\int_a^b \sinh(\lambda x + \mu t) \exp[\beta y(t)]\, dt = h(x).$

This is a special case of equation 6.8.32 with $f(t, y) = A \exp(\beta y)$.

8. $y(x) + A \displaystyle\int_a^b \cos(\lambda x + \mu t) \exp[\beta y(t)]\, dt = h(x).$

This is a special case of equation 6.8.33 with $f(t, y) = A \exp(\beta y)$.

9. $y(x) + A \displaystyle\int_a^b \sin(\lambda x + \mu t) \exp[\beta y(t)]\, dt = h(x).$

This is a special case of equation 6.8.34 with $f(t, y) = A \exp(\beta y)$.

6.4-2. Other Integrands

10. $y(x) + A \displaystyle\int_a^b \exp\big[\beta y(x) + \gamma y(t)\big]\, dt = h(x).$

This is a special case of equation 6.8.43 with $g(x, y) = A \exp(\beta y)$ and $f(t, y) = \exp(\gamma y)$.

11. $y(x) + A \displaystyle\int_a^b y(xt) \exp[\beta y(t)]\, dt = 0.$

This is a special case of equation 6.8.45 with $f(t, y) = A \exp(\beta y)$.

6.5. Equations With Hyperbolic Nonlinearity

6.5-1. Integrands With Nonlinearity of the Form $\cosh[\beta y(t)]$

1. $y(x) + A \displaystyle\int_a^b \cosh[\beta y(t)]\, dt = g(x).$

This is a special case of equation 6.8.27 with $f(t, y) = A \cosh(\beta y)$.

2. $y(x) + A \displaystyle\int_a^b t^\mu \cosh^k[\beta y(t)]\, dt = g(x).$

This is a special case of equation 6.8.27 with $f(t, y) = At^\mu \cosh^k(\beta y)$.

3. $y(x) + A \displaystyle\int_a^b \cosh(\mu t) \cosh[\beta y(t)]\, dt = g(x).$

This is a special case of equation 6.8.27 with $f(t, y) = A \cosh(\mu t) \cosh(\beta y)$.

4. $y(x) + A \displaystyle\int_a^b e^{\lambda(x-t)} \cosh[\beta y(t)]\, dt = g(x).$

This is a special case of equation 6.8.28 with $f(t, y) = A \cosh(\beta y)$.

5. $y(x) + \displaystyle\int_a^b g(x) \cosh[\beta y(t)]\, dt = h(x).$

This is a special case of equation 6.8.29 with $f(t, y) = \cosh(\beta y)$.

6. $y(x) + A \displaystyle\int_a^b \cosh(\lambda x + \mu t) \cosh[\beta y(t)]\, dt = h(x).$

This is a special case of equation 6.8.31 with $f(t, y) = A \cosh(\beta y)$.

7. $y(x) + A \displaystyle\int_a^b \sinh(\lambda x + \mu t) \cosh[\beta y(t)]\, dt = h(x).$

This is a special case of equation 6.8.32 with $f(t, y) = A \cosh(\beta y)$.

8. $y(x) + A \displaystyle\int_a^b \cos(\lambda x + \mu t) \cosh[\beta y(t)]\, dt = h(x).$

This is a special case of equation 6.8.33 with $f(t, y) = A \cosh(\beta y)$.

9. $y(x) + A \displaystyle\int_a^b \sin(\lambda x + \mu t) \cosh[\beta y(t)]\, dt = h(x).$

This is a special case of equation 6.8.34 with $f(t, y) = A \cosh(\beta y)$.

6.5-2. Integrands With Nonlinearity of the Form $\sinh[\beta y(t)]$

10. $y(x) + A \displaystyle\int_a^b \sinh[\beta y(t)]\, dt = g(x).$

This is a special case of equation 6.8.27 with $f(t, y) = A \sinh(\beta y)$.

11. $y(x) + A \displaystyle\int_a^b t^\mu \sinh^k[\beta y(t)]\, dt = g(x).$

This is a special case of equation 6.8.27 with $f(t, y) = At^\mu \sinh^k(\beta y)$.

12. $y(x) + A \displaystyle\int_a^b \sinh(\mu t) \sinh[\beta y(t)]\, dt = g(x).$

This is a special case of equation 6.8.27 with $f(t, y) = A \sinh(\mu t) \sinh(\beta y)$.

13. $y(x) + A \displaystyle\int_a^b e^{\lambda(x-t)} \sinh[\beta y(t)]\, dt = g(x).$

This is a special case of equation 6.8.28 with $f(t, y) = A \sinh(\beta y)$.

14. $y(x) + \displaystyle\int_a^b g(x) \sinh[\beta y(t)]\, dt = h(x).$

This is a special case of equation 6.8.29 with $f(t, y) = \sinh(\beta y)$.

15. $y(x) + A \displaystyle\int_a^b \cosh(\lambda x + \mu t) \sinh[\beta y(t)]\, dt = h(x).$

This is a special case of equation 6.8.31 with $f(t, y) = A \sinh(\beta y)$.

16. $y(x) + A \displaystyle\int_a^b \sinh(\lambda x + \mu t) \sinh[\beta y(t)]\, dt = h(x).$

This is a special case of equation 6.8.32 with $f(t, y) = A \sinh(\beta y)$.

17. $y(x) + A \displaystyle\int_a^b \cos(\lambda x + \mu t) \sinh[\beta y(t)]\, dt = h(x).$

This is a special case of equation 6.8.33 with $f(t, y) = A \sinh(\beta y)$.

18. $y(x) + A \displaystyle\int_a^b \sin(\lambda x + \mu t) \sinh[\beta y(t)]\, dt = h(x).$

This is a special case of equation 6.8.34 with $f(t, y) = A \sinh(\beta y)$.

6.5-3. Integrands With Nonlinearity of the Form $\tanh[\beta y(t)]$

19. $y(x) + A \displaystyle\int_a^b \tanh[\beta y(t)]\, dt = g(x).$

This is a special case of equation 6.8.27 with $f(t, y) = A \tanh(\beta y)$.

20. $y(x) + A \displaystyle\int_a^b t^\mu \tanh^k[\beta y(t)]\, dt = g(x).$

This is a special case of equation 6.8.27 with $f(t, y) = At^\mu \tanh^k(\beta y)$.

21. $y(x) + A \displaystyle\int_a^b \tanh(\mu t) \tanh[\beta y(t)]\, dt = g(x).$

This is a special case of equation 6.8.27 with $f(t, y) = A \tanh(\mu t) \tanh(\beta y)$.

22. $y(x) + A \int_a^b e^{\lambda(x-t)} \tanh[\beta y(t)] \, dt = g(x).$

This is a special case of equation 6.8.28 with $f(t,y) = A \tanh(\beta y)$.

23. $y(x) + \int_a^b g(x) \tanh[\beta y(t)] \, dt = h(x).$

This is a special case of equation 6.8.29 with $f(t,y) = \tanh(\beta y)$.

24. $y(x) + A \int_a^b \cosh(\lambda x + \mu t) \tanh[\beta y(t)] \, dt = h(x).$

This is a special case of equation 6.8.31 with $f(t,y) = A \tanh(\beta y)$.

25. $y(x) + A \int_a^b \sinh(\lambda x + \mu t) \tanh[\beta y(t)] \, dt = h(x).$

This is a special case of equation 6.8.32 with $f(t,y) = A \tanh(\beta y)$.

26. $y(x) + A \int_a^b \cos(\lambda x + \mu t) \tanh[\beta y(t)] \, dt = h(x).$

This is a special case of equation 6.8.33 with $f(t,y) = A \tanh(\beta y)$.

27. $y(x) + A \int_a^b \sin(\lambda x + \mu t) \tanh[\beta y(t)] \, dt = h(x).$

This is a special case of equation 6.8.34 with $f(t,y) = A \tanh(\beta y)$.

6.5-4. Integrands With Nonlinearity of the Form $\coth[\beta y(t)]$

28. $y(x) + A \int_a^b \coth[\beta y(t)] \, dt = g(x).$

This is a special case of equation 6.8.27 with $f(t,y) = A \coth(\beta y)$.

29. $y(x) + A \int_a^b t^\mu \coth^k[\beta y(t)] \, dt = g(x).$

This is a special case of equation 6.8.27 with $f(t,y) = A t^\mu \coth^k(\beta y)$.

30. $y(x) + A \int_a^b \coth(\mu t) \coth[\beta y(t)] \, dt = g(x).$

This is a special case of equation 6.8.27 with $f(t,y) = A \coth(\mu t) \coth(\beta y)$.

31. $y(x) + A \int_a^b e^{\lambda(x-t)} \coth[\beta y(t)] \, dt = g(x).$

This is a special case of equation 6.8.28 with $f(t,y) = A \coth(\beta y)$.

32. $y(x) + \int_a^b g(x) \coth[\beta y(t)] \, dt = h(x).$

This is a special case of equation 6.8.29 with $f(t,y) = \coth(\beta y)$.

33. $y(x) + A \int_a^b \cosh(\lambda x + \mu t) \coth[\beta y(t)] \, dt = h(x).$

   This is a special case of equation 6.8.31 with $f(t, y) = A \coth(\beta y)$.

34. $y(x) + A \int_a^b \sinh(\lambda x + \mu t) \coth[\beta y(t)] \, dt = h(x).$

   This is a special case of equation 6.8.32 with $f(t, y) = A \coth(\beta y)$.

35. $y(x) + A \int_a^b \cos(\lambda x + \mu t) \coth[\beta y(t)] \, dt = h(x).$

   This is a special case of equation 6.8.33 with $f(t, y) = A \coth(\beta y)$.

36. $y(x) + A \int_a^b \sin(\lambda x + \mu t) \coth[\beta y(t)] \, dt = h(x).$

   This is a special case of equation 6.8.34 with $f(t, y) = A \coth(\beta y)$.

6.5-5. Other Integrands

37. $y(x) + A \int_a^b \cosh[\beta y(x)] \cosh[\gamma y(t)] \, dt = h(x).$

   This is a special case of equation 6.8.43 with $g(x, y) = A \cosh(\beta y)$ and $f(t, y) = \cosh(\gamma y)$.

38. $y(x) + A \int_a^b y(xt) \cosh[\beta y(t)] \, dt = 0.$

   This is a special case of equation 6.8.45 with $f(t, y) = A \cosh(\beta y)$.

39. $y(x) + A \int_a^b \sinh[\beta y(x)] \sinh[\gamma y(t)] \, dt = h(x).$

   This is a special case of equation 6.8.43 with $g(x, y) = A \sinh(\beta y)$ and $f(t, y) = \sinh(\gamma y)$.

40. $y(x) + A \int_a^b y(xt) \sinh[\beta y(t)] \, dt = 0.$

   This is a special case of equation 6.8.45 with $f(t, y) = A \sinh(\beta y)$.

41. $y(x) + A \int_a^b \tanh[\beta y(x)] \tanh[\gamma y(t)] \, dt = h(x).$

   This is a special case of equation 6.8.43 with $g(x, y) = A \tanh(\beta y)$ and $f(t, y) = \tanh(\gamma y)$.

42. $y(x) + A \int_a^b y(xt) \tanh[\beta y(t)] \, dt = 0.$

   This is a special case of equation 6.8.45 with $f(t, y) = A \tanh(\beta y)$.

43. $y(x) + A \int_a^b \coth[\beta y(x)] \coth[\gamma y(t)] \, dt = h(x).$

   This is a special case of equation 6.8.43 with $g(x, y) = A \coth(\beta y)$ and $f(t, y) = \coth(\gamma y)$.

44. $y(x) + A \int_a^b y(xt) \coth[\beta y(t)] \, dt = 0.$

   This is a special case of equation 6.8.45 with $f(t, y) = A \coth(\beta y)$.

6.6. Equations With Logarithmic Nonlinearity

6.6-1. Integrands With Nonlinearity of the Form $\ln[\beta y(t)]$

1. $y(x) + A\displaystyle\int_a^b \ln[\beta y(t)]\,dt = g(x).$

   This is a special case of equation 6.8.27 with $f(t,y) = A\ln(\beta y)$.

2. $y(x) + A\displaystyle\int_a^b t^\mu \ln^k[\beta y(t)]\,dt = g(x).$

   This is a special case of equation 6.8.27 with $f(t,y) = At^\mu \ln^k(\beta y)$.

3. $y(x) + A\displaystyle\int_a^b \ln(\mu t)\ln[\beta y(t)]\,dt = g(x).$

   This is a special case of equation 6.8.27 with $f(t,y) = A\ln(\mu t)\ln(\beta y)$.

4. $y(x) + A\displaystyle\int_a^b e^{\lambda(x-t)}\ln[\beta y(t)]\,dt = g(x).$

   This is a special case of equation 6.8.28 with $f(t,y) = A\ln(\beta y)$.

5. $y(x) + \displaystyle\int_a^b g(x)\ln[\beta y(t)]\,dt = h(x).$

   This is a special case of equation 6.8.29 with $f(t,y) = \ln(\beta y)$.

6. $y(x) + A\displaystyle\int_a^b \cosh(\lambda x + \mu t)\ln[\beta y(t)]\,dt = h(x).$

   This is a special case of equation 6.8.31 with $f(t,y) = A\ln(\beta y)$.

7. $y(x) + A\displaystyle\int_a^b \sinh(\lambda x + \mu t)\ln[\beta y(t)]\,dt = h(x).$

   This is a special case of equation 6.8.32 with $f(t,y) = A\ln(\beta y)$.

8. $y(x) + A\displaystyle\int_a^b \cos(\lambda x + \mu t)\ln[\beta y(t)]\,dt = h(x).$

   This is a special case of equation 6.8.33 with $f(t,y) = A\ln(\beta y)$.

9. $y(x) + A\displaystyle\int_a^b \sin(\lambda x + \mu t)\ln[\beta y(t)]\,dt = h(x).$

   This is a special case of equation 6.8.34 with $f(t,y) = A\ln(\beta y)$.

6.6-2. Other Integrands

10. $y(x) + A\displaystyle\int_a^b \ln[\beta y(x)]\ln[\gamma y(t)]\,dt = h(x).$

    This is a special case of equation 6.8.43 with $g(x,y) = A\ln(\beta y)$ and $f(t,y) = \ln(\gamma y)$.

11. $y(x) + A\displaystyle\int_a^b y(xt)\ln[\beta y(t)]\,dt = 0.$

    This is a special case of equation 6.8.45 with $f(t,y) = A\ln(\beta y)$.

6.7. Equations With Trigonometric Nonlinearity

1. $y(x) + A \displaystyle\int_a^b \cos[\beta y(t)] \, dt = g(x).$

This is a special case of equation 6.8.27 with $f(t, y) = A\cos(\beta y)$.

2. $y(x) + A \displaystyle\int_a^b t^\mu \cos^k[\beta y(t)] \, dt = g(x).$

This is a special case of equation 6.8.27 with $f(t, y) = At^\mu \cos^k(\beta y)$.

3. $y(x) + A \displaystyle\int_a^b \cos(\mu t) \cos[\beta y(t)] \, dt = g(x).$

This is a special case of equation 6.8.27 with $f(t, y) = A\cos(\mu t)\cos(\beta y)$.

4. $y(x) + A \displaystyle\int_a^b e^{\lambda(x-t)} \cos[\beta y(t)] \, dt = g(x).$

This is a special case of equation 6.8.28 with $f(t, y) = A\cos(\beta y)$.

5. $y(x) + \displaystyle\int_a^b g(x) \cos[\beta y(t)] \, dt = h(x).$

This is a special case of equation 6.8.29 with $f(t, y) = \cos(\beta y)$.

6. $y(x) + A \displaystyle\int_a^b \cosh(\lambda x + \mu t) \cos[\beta y(t)] \, dt = h(x).$

This is a special case of equation 6.8.31 with $f(t, y) = A\cos(\beta y)$.

7. $y(x) + A \displaystyle\int_a^b \sinh(\lambda x + \mu t) \cos[\beta y(t)] \, dt = h(x).$

This is a special case of equation 6.8.32 with $f(t, y) = A\cos(\beta y)$.

8. $y(x) + A \displaystyle\int_a^b \cos(\lambda x + \mu t) \cos[\beta y(t)] \, dt = h(x).$

This is a special case of equation 6.8.33 with $f(t, y) = A\cos(\beta y)$.

9. $y(x) + A \displaystyle\int_a^b \sin(\lambda x + \mu t) \cos[\beta y(t)] \, dt = h(x).$

This is a special case of equation 6.8.34 with $f(t, y) = A\cos(\beta y)$.

10. $y(x) + A \displaystyle\int_a^b \sin[\beta y(t)] \, dt = g(x).$

This is a special case of equation 6.8.27 with $f(t, y) = A\sin(\beta y)$.

11. $y(x) + A \displaystyle\int_a^b t^\mu \sin^k[\beta y(t)]\, dt = g(x).$

This is a special case of equation 6.8.27 with $f(t, y) = A t^\mu \sin^k(\beta y)$.

12. $y(x) + A \displaystyle\int_a^b \sin(\mu t) \sin[\beta y(t)]\, dt = g(x).$

This is a special case of equation 6.8.27 with $f(t, y) = A \sin(\mu t) \sin(\beta y)$.

13. $y(x) + A \displaystyle\int_a^b e^{\lambda(x-t)} \sin[\beta y(t)]\, dt = g(x).$

This is a special case of equation 6.8.28 with $f(t, y) = A \sin(\beta y)$.

14. $y(x) + \displaystyle\int_a^b g(x) \sin[\beta y(t)]\, dt = h(x).$

This is a special case of equation 6.8.29 with $f(t, y) = \sin(\beta y)$.

15. $y(x) + A \displaystyle\int_a^b \cosh(\lambda x + \mu t) \sin[\beta y(t)]\, dt = h(x).$

This is a special case of equation 6.8.31 with $f(t, y) = A \sin(\beta y)$.

16. $y(x) + A \displaystyle\int_a^b \sinh(\lambda x + \mu t) \sin[\beta y(t)]\, dt = h(x).$

This is a special case of equation 6.8.32 with $f(t, y) = A \sin(\beta y)$.

17. $y(x) + A \displaystyle\int_a^b \cos(\lambda x + \mu t) \sin[\beta y(t)]\, dt = h(x).$

This is a special case of equation 6.8.33 with $f(t, y) = A \sin(\beta y)$.

18. $y(x) + A \displaystyle\int_a^b \sin(\lambda x + \mu t) \sin[\beta y(t)]\, dt = h(x).$

This is a special case of equation 6.8.34 with $f(t, y) = A \sin(\beta y)$.

6.7-3. Integrands With Nonlinearity of the Form $\tan[\beta y(t)]$

19. $y(x) + A \displaystyle\int_a^b \tan[\beta y(t)]\, dt = g(x).$

This is a special case of equation 6.8.27 with $f(t, y) = A \tan(\beta y)$.

20. $y(x) + A \displaystyle\int_a^b t^\mu \tan^k[\beta y(t)]\, dt = g(x).$

This is a special case of equation 6.8.27 with $f(t, y) = A t^\mu \tan^k(\beta y)$.

21. $y(x) + A \displaystyle\int_a^b \tan(\mu t) \tan[\beta y(t)]\, dt = g(x).$

This is a special case of equation 6.8.27 with $f(t, y) = A \tan(\mu t) \tan(\beta y)$.

22. $y(x) + A \int_a^b e^{\lambda(x-t)} \tan[\beta y(t)] \, dt = g(x).$

This is a special case of equation 6.8.28 with $f(t, y) = A \tan(\beta y)$.

23. $y(x) + \int_a^b g(x) \tan[\beta y(t)] \, dt = h(x).$

This is a special case of equation 6.8.29 with $f(t, y) = \tan(\beta y)$.

24. $y(x) + A \int_a^b \cosh(\lambda x + \mu t) \tan[\beta y(t)] \, dt = h(x).$

This is a special case of equation 6.8.31 with $f(t, y) = A \tan(\beta y)$.

25. $y(x) + A \int_a^b \sinh(\lambda x + \mu t) \tan[\beta y(t)] \, dt = h(x).$

This is a special case of equation 6.8.32 with $f(t, y) = A \tan(\beta y)$.

26. $y(x) + A \int_a^b \cos(\lambda x + \mu t) \tan[\beta y(t)] \, dt = h(x).$

This is a special case of equation 6.8.33 with $f(t, y) = A \tan(\beta y)$.

27. $y(x) + A \int_a^b \sin(\lambda x + \mu t) \tan[\beta y(t)] \, dt = h(x).$

This is a special case of equation 6.8.34 with $f(t, y) = A \tan(\beta y)$.

6.7-4. Integrands With Nonlinearity of the Form $\cot[\beta y(t)]$

28. $y(x) + A \int_a^b \cot[\beta y(t)] \, dt = g(x).$

This is a special case of equation 6.8.27 with $f(t, y) = A \cot(\beta y)$.

29. $y(x) + A \int_a^b t^\mu \cot^k[\beta y(t)] \, dt = g(x).$

This is a special case of equation 6.8.27 with $f(t, y) = A t^\mu \cot^k(\beta y)$.

30. $y(x) + A \int_a^b \cot(\mu t) \cot[\beta y(t)] \, dt = g(x).$

This is a special case of equation 6.8.27 with $f(t, y) = A \cot(\mu t) \cot(\beta y)$.

31. $y(x) + A \int_a^b e^{\lambda(x-t)} \cot[\beta y(t)] \, dt = g(x).$

This is a special case of equation 6.8.28 with $f(t, y) = A \cot(\beta y)$.

32. $y(x) + \int_a^b g(x) \cot[\beta y(t)] \, dt = h(x).$

This is a special case of equation 6.8.29 with $f(t, y) = \cot(\beta y)$.

33. $y(x) + A \int_a^b \cosh(\lambda x + \mu t) \cot[\beta y(t)] \, dt = h(x).$

This is a special case of equation 6.8.31 with $f(t, y) = A \cot(\beta y)$.

34. $y(x) + A \int_a^b \sinh(\lambda x + \mu t) \cot[\beta y(t)] \, dt = h(x).$

This is a special case of equation 6.8.32 with $f(t, y) = A \cot(\beta y)$.

35. $y(x) + A \int_a^b \cos(\lambda x + \mu t) \cot[\beta y(t)] \, dt = h(x).$

This is a special case of equation 6.8.33 with $f(t, y) = A \cot(\beta y)$.

36. $y(x) + A \int_a^b \sin(\lambda x + \mu t) \cot[\beta y(t)] \, dt = h(x).$

This is a special case of equation 6.8.34 with $f(t, y) = A \cot(\beta y)$.

6.7-5. Other Integrands

37. $y(x) + A \int_a^b \cos[\beta y(x)] \cos[\gamma y(t)] \, dt = h(x).$

This is a special case of equation 6.8.43 with $g(x, y) = A \cos(\beta y)$ and $f(t, y) = \cos(\gamma y)$.

38. $y(x) + A \int_a^b y(xt) \cos[\beta y(t)] \, dt = 0.$

This is a special case of equation 6.8.45 with $f(t, y) = A \cos(\beta y)$.

39. $y(x) + A \int_a^b \sin[\beta y(x)] \sin[\gamma y(t)] \, dt = h(x).$

This is a special case of equation 6.8.43 with $g(x, y) = A \sin(\beta y)$ and $f(t, y) = \sin(\gamma y)$.

40. $y(x) + A \int_a^b y(xt) \sin[\beta y(t)] \, dt = 0.$

This is a special case of equation 6.8.45 with $f(t, y) = A \sin(\beta y)$.

41. $y(x) + A \int_a^b \tan[\beta y(x)] \tan[\gamma y(t)] \, dt = h(x).$

This is a special case of equation 6.8.43 with $g(x, y) = A \tan(\beta y)$ and $f(t, y) = \tan(\gamma y)$.

42. $y(x) + A \int_a^b y(xt) \tan[\beta y(t)] \, dt = 0.$

This is a special case of equation 6.8.45 with $f(t, y) = A \tan(\beta y)$.

43. $y(x) + A \int_a^b \cot[\beta y(x)] \cot[\gamma y(t)] \, dt = h(x).$

This is a special case of equation 6.8.43 with $g(x, y) = A \cot(\beta y)$ and $f(t, y) = \cot(\gamma y)$.

44. $y(x) + A \int_a^b y(xt) \cot[\beta y(t)] \, dt = 0.$

This is a special case of equation 6.8.45 with $f(t, y) = A \cot(\beta y)$.

6.8. Equations With Nonlinearity of General Form

6.8-1. Equations of the Form $\int_a^b G(\cdots)\,dt = F(x)$

1. $\displaystyle\int_a^b y(x)f\big(t, y(t)\big)\,dt = g(x).$

A solution: $y(x) = \lambda g(x)$, where λ is determined by the algebraic (or transcendental) equation $\lambda \displaystyle\int_a^b f\big(t, \lambda g(t)\big)\,dt = 1.$

2. $\displaystyle\int_a^b y^k(x)f\big(t, y(t)\big)\,dt = g(x).$

A solution: $y(x) = \lambda[g(x)]^{1/k}$, where λ is determined from the algebraic (or transcendental) equation $\lambda^k \displaystyle\int_a^b f\big(t, \lambda g^{1/k}(t)\big)\,dt = 1.$

3. $\displaystyle\int_a^b \varphi\big(y(x)\big)f\big(t, y(t)\big)\,dt = g(x).$

A solution in an implicit form:

$$\lambda\varphi\big(y(x)\big) - g(x) = 0, \tag{1}$$

where λ is determined by the algebraic (or transcendental) equation

$$\lambda - F(\lambda) = 0, \qquad F(\lambda) = \int_a^b f\big(t, y(t)\big)\,dt. \tag{2}$$

Here the function $y(x) = y(x, \lambda)$ obtained by solving (1) must be substituted into (2).

The number of solutions of the integral equation is determined by the number of the solutions obtained from (1) and (2).

4. $\displaystyle\int_a^b y(xt)f\big(t, y(t)\big)\,dt = A.$

1°. Solutions: $y(x) = \lambda_k$, where λ_k are roots of the algebraic (or transcendental) equation $\lambda \displaystyle\int_a^b f(t, \lambda)\,dt = A.$

2°. Solutions: $y(x) = px + q$, where p and q are roots of the following system of algebraic (or transcendental) equations:

$$\int_a^b tf(t, pt + q)\,dt = 0, \qquad q\int_a^b f(t, pt + q)\,dt = A.$$

In the case $f\big(t, y(t)\big) = \bar{f}(t)y(t)$, see 6.2.2 for solutions of this system.

2°. The integral equation has some other (more complicated) solutions of the polynomial form $y(x) = \displaystyle\sum_{k=0}^{n} B_k x^k$, where the constants B_k can be found from the corresponding system of algebraic (or transcendental) equations.

4°. The integral equation can have logarithmic solutions similar to those presented in item 3° of equation 6.2.2.

5. $\displaystyle\int_a^b y(xt)f\big(t,y(t)\big)\,dt = Ax + B.$

1°. A solution:

$$y(x) = px + q, \tag{1}$$

where p and q are roots of the following system of algebraic (or transcendental) equations:

$$p\int_a^b tf(t, pt + q)\,dt - A = 0, \qquad q\int_a^b f(t, pt + q)\,dt - B = 0. \tag{2}$$

Different solutions of system (2) generate different solutions (1) of the integral equation.

2°. The integral equation has some other (more complicated) solutions of the polynomial form $y(x) = \sum_{k=0}^{n} B_k x^k$, where the constants B_k can be found from the corresponding system of algebraic (or transcendental) equations.

6. $\displaystyle\int_a^b y(xt)f\big(t,y(t)\big)\,dt = Ax^\beta.$

A solution:

$$y(x) = kx^\beta, \tag{1}$$

where k is a root of the algebraic (or transcendental) equation

$$kF(k) - A = 0, \qquad F(k) = \int_a^b t^\beta f\big(t, kt^\beta\big)\,dt. \tag{2}$$

Each root of equation (2) generates a solution of the integral equation which has the form (1).

7. $\displaystyle\int_a^b y(xt)f\big(t,y(t)\big)\,dt = A\ln x + B.$

A solution:

$$y(x) = p\ln x + q, \tag{1}$$

where p and q are roots of the following system of algebraic (or transcendental) equations:

$$p\int_a^b f(t, p\ln t + q)\,dt - A = 0, \qquad \int_a^b (p\ln t + q)f(t, p\ln t + q)\,dt - B = 0. \tag{2}$$

Different solutions of system (2) generate different solutions (1) of the integral equation.

8. $\displaystyle\int_a^b y(xt)f\big(t,y(t)\big)\,dt = Ax^\beta \ln x.$

This equation has solutions of the form $y(x) = px^\beta \ln x + qx^\beta$, where p and q are some constants.

9. $\displaystyle\int_a^b y(xt)f\big(t,y(t)\big)\,dt = A\cos(\beta\ln x).$

This equation has solutions of the form $y(x) = p\cos(\beta\ln x) + q\sin(\beta\ln x)$, where p and q are some constants.

10. $\displaystyle\int_a^b y(xt)f\big(t,y(t)\big)\,dt = A\sin(\beta\ln x).$

This equation has solutions of the form $y(x) = p\cos(\beta\ln x) + q\sin(\beta\ln x)$, where p and q are some constants.

11. $\displaystyle\int_a^b y(xt)f\big(t,y(t)\big)\,dt = Ax^\beta\cos(\beta\ln x) + Bx^\beta\sin(\beta\ln x).$

This equation has solutions of the form $y(x) = px^\beta\cos(\beta\ln x) + qx^\beta\sin(\beta\ln x)$, where p and q are some constants.

12. $\displaystyle\int_a^b y(x+\beta t)f\big(t,y(t)\big)\,dt = Ax + B, \qquad \beta > 0.$

A solution:
$$y(x) = px + q, \tag{1}$$

where p and q are roots of the following system of algebraic (or transcendental) equations:

$$p\int_a^b f(t, pt+q)\,dt - A = 0, \qquad \int_a^b (\beta pt+q)f(t, pt+q)\,dt - B = 0. \tag{2}$$

Different solutions of system (2) generate different solutions (1) of the integral equation.

13. $\displaystyle\int_a^b y(x+\beta t)f\big(t,y(t)\big)\,dt = Ae^{-\lambda x}, \qquad \beta > 0.$

Solutions:
$$y(x) = k_n e^{-\lambda x},$$

where k_n are roots of the algebraic (or transcendental) equation

$$kF(k) - A = 0, \qquad F(k) = \int_a^b f\big(t, ke^{-\lambda t}\big)e^{-\beta\lambda t}\,dt.$$

14. $\displaystyle\int_a^b y(x+\beta t)f\big(t,y(t)\big)\,dt = A\cos\lambda x, \qquad \beta > 0.$

This equation has solutions of the form $y(x) = p\sin\lambda x + q\cos\lambda x$, where p and q are some constants.

15. $\displaystyle\int_a^b y(x+\beta t)f\big(t,y(t)\big)\,dt = A\sin\lambda x, \qquad \beta > 0.$

This equation has solutions of the form $y(x) = p\sin\lambda x + q\cos\lambda x$, where p and q are some constants.

16. $\displaystyle\int_a^b y(x+\beta t)f\big(t,y(t)\big)\,dt = e^{-\mu x}(A\cos\lambda x + B\sin\lambda x), \qquad \beta > 0.$

This equation has solutions of the form $y(x) = e^{-\mu x}(p\sin\lambda x + q\cos\lambda x)$, where p and q are some constants.

17. $\displaystyle\int_a^b y(x-t)f\big(t,y(t)\big)\,dt = Ax + B.$

This equation has solutions of the form $y(x) = px + q$, where p and q are some constants.

18. $\displaystyle\int_a^b y(x-t)f\big(t,y(t)\big)\,dt = Ae^{\lambda x}.$

This equation has solutions of the form $y(x) = pe^{\lambda x}$, where p is some constant.

19. $\displaystyle\int_a^b y(x-t)f\big(t,y(t)\big)\,dt = A\cos\lambda x.$

This equation has solutions of the form $y(x) = p\sin\lambda x + q\cos\lambda x$, where p and q are some constants.

20. $\displaystyle\int_a^b y(x-t)f\big(t,y(t)\big)\,dt = e^{-\mu x}(A\cos\lambda x + B\sin\lambda x).$

This equation has solutions of the form $y(x) = e^{-\mu x}(p\sin\lambda x + q\cos\lambda x)$, where p and q are some constants.

6.8-2. Equations of the Form $y(x) + \int_a^b K(x,t)G\big(y(t)\big)\,dt = F(x)$

21. $\displaystyle y(x) + \int_a^b |x-t|f\big(y(t)\big)\,dt = Ax^2 + Bx + C.$

This is a special case of equation 6.8.35 with $f(t,y) = f(y)$ and $g(x) = Ax^2 + Bx + C$.

The function $y = y(x)$ obeys the second-order autonomous differential equation

$$y''_{xx} + 2f(y) = 2A,$$

whose solution can be represented in an implicit form:

$$\int_{y_a}^{y} \frac{du}{\sqrt{w_a^2 + 4A(u-y_a) - 4F(u,y_a)}} = \pm(x-a), \qquad F(u,v) = \int_v^u f(t)\,dt, \qquad (1)$$

where $y_a = y(a)$ and $w_a = y'_x(a)$ are constants of integration. These constants, as well as the unknowns $y_b = y(b)$ and $w_b = y'_x(b)$, are determined by the algebraic (or transcendental) system

$$\begin{aligned}
&y_a + y_b - (a-b)w_a = (b^2 + 2ab - a^2)A + 2bB + 2C,\\
&w_a + w_b = 2(a+b)A + 2B,\\
&w_b^2 = w_a^2 + 4A(y_b - y_a) - 4F(y_b, y_a),\\
&\int_{y_a}^{y_b} \frac{du}{\sqrt{w_a^2 + 4A(u-y_a) - 4F(u,y_a)}} = \pm(b-a).
\end{aligned} \qquad (2)$$

Here the first equation is obtained from the second condition of (5) in 6.8.35, the second equation is obtained from condition (6) in 6.8.35, and the third and fourth equations are consequences of (1).

Each solution of system (2) generates a solution of the integral equation.

22. $\displaystyle y(x) + \int_a^b e^{\lambda|x-t|}f\big(y(t)\big)\,dt = A + Be^{\lambda x} + Ce^{-\lambda x}.$

This is a special case of equation 6.8.36 with $f(t,y) = f(y)$ and $g(x) = A + Be^{\lambda x} + Ce^{-\lambda x}$.

The function $y = y(x)$ satisfies the second-order autonomous differential equation

$$y''_{xx} + 2\lambda f(y) - \lambda^2 y = -\lambda^2 A, \qquad (1)$$

whose solution can be written in an implicit form:

$$\int_{y_a}^{y} \frac{du}{\sqrt{w_a^2 + \lambda^2(u^2 - y_a^2) - 2A\lambda^2(u - y_a) - 4\lambda F(u, y_a)}} = \pm(x - a), \quad F(u, v) = \int_{v}^{u} f(t)\, dt, \quad (2)$$

where $y_a = y(a)$ and $w_a = y_x'(a)$ are constants of integration. These constants, as well as the unknowns $y_b = y(b)$ and $w_b = y_x'(b)$, are determined by the algebraic (or transcendental) system

$$w_a + \lambda y_a = A\lambda + 2B\lambda e^{\lambda a},$$
$$w_b - \lambda y_b = -A\lambda - 2C\lambda e^{-\lambda b},$$
$$w_b^2 = w_a^2 + \lambda^2(y_b^2 - y_a^2) - 2A\lambda^2(y_b - y_a) - 4\lambda F(y_b, y_a), \quad (3)$$
$$\int_{y_a}^{y_b} \frac{du}{\sqrt{w_a^2 + \lambda^2(u^2 - y_a^2) - 2A\lambda^2(u - y_a) - 4\lambda F(u, y_a)}} = \pm(b - a).$$

Here the first and second equations are obtained from conditions (5) in 6.8.86, and the third and fourth equations are consequences of (2).

Each solution of system (3) generates a solution of the integral equation.

23. $\quad y(x) + \displaystyle\int_a^b e^{\lambda|x-t|} f\big(y(t)\big)\, dt = \beta \cosh(\lambda x).$

This is a special case of equation 6.8.22 with $A = 0$ and $B = C = \frac{1}{2}\beta$.

24. $\quad y(x) + \displaystyle\int_a^b e^{\lambda|x-t|} f\big(y(t)\big)\, dt = \beta \sinh(\lambda x).$

This is a special case of equation 6.8.22 with $A = 0$, $B = \frac{1}{2}\beta$, and $C = -\frac{1}{2}\beta$.

25. $\quad y(x) + \displaystyle\int_a^b \sinh\big(\lambda|x - t|\big) f\big(y(t)\big)\, dt = A + B \cosh(\lambda x) + C \sinh(\lambda x).$

This is a special case of equation 6.8.37 with $f(t, y) = f(y)$ and $g(x) = A + B \cosh(\lambda x) + C \sinh(\lambda x)$.

The function $y = y(x)$ satisfies the second-order autonomous differential equation

$$y_{xx}'' + 2\lambda f(y) - \lambda^2 y = -\lambda^2 A,$$

whose solution can be represented in an implicit form:

$$\int_{y_a}^{y} \frac{du}{\sqrt{w_a^2 + \lambda^2(u^2 - y_a^2) - 2A\lambda^2(u - y_a) - 4\lambda F(u, y_a)}} = \pm(x - a), \quad F(u, v) = \int_{v}^{u} f(t)\, dt,$$

where $y_a = y(a)$ and $w_a = y_x'(a)$ are constants of integration, which can be determined from the boundary conditions (5) in 6.8.37.

26. $\quad y(x) + \displaystyle\int_a^b \sin\big(\lambda|x - t|\big) f\big(y(t)\big)\, dt = A + B \cos(\lambda x) + C \sin(\lambda x).$

This is a special case of equation 6.8.38 with $f(t, y) = f(y)$ and $g(x) = A + B \cos(\lambda x) + C \sin(\lambda x)$.

The function $y = y(x)$ satisfies the second-order autonomous differential equation

$$y_{xx}'' + 2\lambda f(y) + \lambda^2 y = \lambda^2 A,$$

whose solution can be represented in an implicit form:

$$\int_{y_a}^{y} \frac{du}{\sqrt{w_a^2 - \lambda^2(u^2 - y_a^2) + 2A\lambda^2(u - y_a) - 4\lambda F(u, y_a)}} = \pm(x - a), \quad F(u, v) = \int_{v}^{u} f(t)\, dt,$$

where $y_a = y(a)$ and $w_a = y_x'(a)$ are constants of integration, which can be determined from the boundary conditions (5) in 6.8.38.

6.8-3. Equations of the Form $y(x) + \int_a^b K(x,t)G\big(t, y(t)\big)\, dt = F(x)$

27. $y(x) + \displaystyle\int_a^b f\big(t, y(t)\big)\, dt = g(x).$

A solution: $y(x) = g(x) + \lambda$, where λ is determined by the algebraic (or transcendental) equation

$$\lambda + F(\lambda) = 0, \qquad F(\lambda) = \int_a^b f\big(t,\, g(t) + \lambda\big)\, dt.$$

28. $y(x) + \displaystyle\int_a^b e^{\lambda(x-t)} f\big(t, y(t)\big)\, dt = g(x).$

A solution: $y(x) = \beta e^{\lambda x} + g(x)$, where λ is determined by the algebraic (or transcendental) equation

$$\beta + F(\beta) = 0, \qquad F(\beta) = \int_a^b e^{-\lambda t} f\big(t,\, \beta e^{\lambda t} + g(t)\big)\, dt.$$

29. $y(x) + \displaystyle\int_a^b g(x) f\big(t, y(t)\big)\, dt = h(x).$

A solution: $y(x) = \lambda g(x) + h(x)$, where λ is determined by the algebraic (or transcendental) equation

$$\lambda + F(\lambda) = 0, \qquad F(\lambda) = \int_a^b f\big(t,\, \lambda g(t) + h(t)\big)\, dt.$$

30. $y(x) + \displaystyle\int_a^b (Ax + Bt) f\big(t, y(t)\big)\, dt = g(x).$

A solution: $y(x) = g(x) + \lambda x + \mu$, where the constants λ and μ are determined from the algebraic (or transcendental) system

$$\lambda + A \int_a^b f\big(t,\, g(t) + \lambda t + \mu\big)\, dt = 0, \qquad \mu + B \int_a^b t f\big(t,\, g(t) + \lambda t + \mu\big)\, dt = 0.$$

31. $y(x) + \displaystyle\int_a^b \cosh(\lambda x + \mu t) f\big(t, y(t)\big)\, dt = h(x).$

Using the formula $\cosh(\lambda x + \mu t) = \cosh(\lambda x)\cosh(\mu t) + \sinh(\mu t)\sinh(\lambda x)$, we arrive at an equation of the form 6.8.39:

$$y(x) + \int_a^b \big[\cosh(\lambda x) f_1\big(t, y(t)\big) + \sinh(\lambda x) f_2\big(t, y(t)\big)\big]\, dt = h(x),$$
$$f_1\big(t, y(t)\big) = \cosh(\mu t) f\big(t, y(t)\big), \qquad f_2\big(t, y(t)\big) = \sinh(\mu t) f\big(t, y(t)\big).$$

32. $y(x) + \displaystyle\int_a^b \sinh(\lambda x + \mu t) f\big(t, y(t)\big)\, dt = h(x).$

Using the formula $\sinh(\lambda x + \mu t) = \cosh(\lambda x)\sinh(\mu t) + \cosh(\mu t)\sinh(\lambda x)$, we arrive at an equation of the form 6.8.39:

$$y(x) + \int_a^b \big[\cosh(\lambda x) f_1\big(t, y(t)\big) + \sinh(\lambda x) f_2\big(t, y(t)\big)\big]\, dt = h(x),$$
$$f_1\big(t, y(t)\big) = \sinh(\mu t) f\big(t, y(t)\big), \qquad f_2\big(t, y(t)\big) = \cosh(\mu t) f\big(t, y(t)\big).$$

33. $y(x) + \int_a^b \cos(\lambda x + \mu t) f\big(t, y(t)\big)\, dt = h(x).$

Using the formula $\cos(\lambda x + \mu t) = \cos(\lambda x)\cos(\mu t) - \sin(\mu t)\sin(\lambda x)$, we arrive at an equation of the form 6.8.39:

$$y(x) + \int_a^b \big[\cos(\lambda x) f_1\big(t, y(t)\big) + \sin(\lambda x) f_2\big(t, y(t)\big)\big]\, dt = h(x),$$

$$f_1\big(t, y(t)\big) = \cos(\mu t) f\big(t, y(t)\big), \quad f_2\big(t, y(t)\big) = -\sin(\mu t) f\big(t, y(t)\big).$$

34. $y(x) + \int_a^b \sin(\lambda x + \mu t) f\big(t, y(t)\big)\, dt = h(x).$

Using the formula $\sin(\lambda x + \mu t) = \cos(\lambda x)\sin(\mu t) + \cos(\mu t)\sin(\lambda x)$, we arrive at an equation of the form 6.8.39:

$$y(x) + \int_a^b \big[\cos(\lambda x) f_1\big(t, y(t)\big) + \sin(\lambda x) f_2\big(t, y(t)\big)\big]\, dt = h(x),$$

$$f_1\big(t, y(t)\big) = \sin(\mu t) f\big(t, y(t)\big), \quad f_2\big(t, y(t)\big) = \cos(\mu t) f\big(t, y(t)\big).$$

35. $y(x) + \int_a^b |x - t| f\big(t, y(t)\big)\, dt = g(x), \qquad a \le x \le b.$

$1°$. Let us remove the modulus in the integrand:

$$y(x) + \int_a^x (x - t) f\big(t, y(t)\big)\, dt + \int_x^b (t - x) f\big(t, y(t)\big)\, dt = g(x). \tag{1}$$

Differentiating (1) with respect to x yields

$$y_x'(x) + \int_a^x f\big(t, y(t)\big)\, dt - \int_x^b f\big(t, y(t)\big)\, dt = g_x'(x). \tag{2}$$

Differentiating (2), we arrive at a second-order ordinary differential equation for $y = y(x)$:

$$y_{xx}'' + 2f(x, y) = g_{xx}''(x). \tag{3}$$

$2°$. Let us derive the boundary conditions for equation (3). We assume that $-\infty < a < b < \infty$. By setting $x = a$ and $x = b$ in (1), we obtain the relations

$$y(a) + \int_a^b (t - a) f\big(t, y(t)\big)\, dt = g(a),$$
$$y(b) + \int_a^b (b - t) f\big(t, y(t)\big)\, dt = g(b). \tag{4}$$

Let us solve equation (3) for $f(x, y)$ and substitute the result into (4). Integrating by parts yields the desired boundary conditions for $y(x)$:

$$y(a) + y(b) + (b - a)\big[g_x'(b) - y_x'(b)\big] = g(a) + g(b),$$
$$y(a) + y(b) + (a - b)\big[g_x'(a) - y_x'(a)\big] = g(a) + g(b). \tag{5}$$

Let us point out a useful consequence of (5):

$$y_x'(a) + y_x'(b) = g_x'(a) + g_x'(b), \tag{6}$$

which can be used together with one of conditions (5).

Equation (3) under the boundary conditions (5) determines the solution of the original integral equation (there may be several solutions). Conditions (5) make it possible to calculate the constants of integration that occur in solving the differential equation (3).

36.　$y(x) + \displaystyle\int_a^b e^{\lambda|x-t|} f(t, y(t))\, dt = g(x), \qquad a \le x \le b.$

$1°$. Let us remove the modulus in the integrand:

$$y(x) + \int_a^x e^{\lambda(x-t)} f(t, y(t))\, dt + \int_x^b e^{\lambda(t-x)} f(t, y(t))\, dt = g(x). \tag{1}$$

Differentiating (1) with respect to x twice yields

$$y''_{xx}(x) + 2\lambda f(x, y(x)) + \lambda^2 \int_a^x e^{\lambda(x-t)} f(t, y(t))\, dt + \lambda^2 \int_x^b e^{\lambda(t-x)} f(t, y(t))\, dt = g''_{xx}(x). \tag{2}$$

Eliminating the integral terms from (1) and (2), we arrive at a second-order ordinary differential equation for $y = y(x)$:

$$y''_{xx} + 2\lambda f(x, y) - \lambda^2 y = g''_{xx}(x) - \lambda^2 g(x). \tag{3}$$

$2°$. Let us derive the boundary conditions for equation (3). We assume that $-\infty < a < b < \infty$. By setting $x = a$ and $x = b$ in (1), we obtain the relations

$$
\begin{aligned}
y(a) + e^{-\lambda a} \int_a^b e^{\lambda t} f(t, y(t))\, dt &= g(a), \\
y(b) + e^{\lambda b} \int_a^b e^{-\lambda t} f(t, y(t))\, dt &= g(b).
\end{aligned}
\tag{4}
$$

Let us solve equation (3) for $f(x, y)$ and substitute the result into (4). Integrating by parts yields

$$e^{\lambda b} \varphi'_x(b) - e^{\lambda a} \varphi'_x(a) = \lambda e^{\lambda a} \varphi(a) + \lambda e^{\lambda b} \varphi(b), \quad \varphi(x) = y(x) - g(x);$$

$$e^{-\lambda b} \varphi'_x(b) - e^{-\lambda a} \varphi'_x(a) = \lambda e^{-\lambda a} \varphi(a) + \lambda e^{-\lambda b} \varphi(b).$$

Hence, we obtain the boundary conditions for $y(x)$:

$$\varphi'_x(a) + \lambda \varphi(a) = 0, \quad \varphi'_x(b) - \lambda \varphi(b) = 0; \qquad \varphi(x) = y(x) - g(x). \tag{5}$$

Equation (3) under the boundary conditions (5) determines the solution of the original integral equation (there may be several solutions). Conditions (5) make it possible to calculate the constants of integration that occur in solving the differential equation (3).

37.　$y(x) + \displaystyle\int_a^b \sinh(\lambda|x-t|) f(t, y(t))\, dt = g(x), \qquad a \le x \le b.$

$1°$. Let us remove the modulus in the integrand:

$$y(x) + \int_a^x \sinh[\lambda(x-t)] f(t, y(t))\, dt + \int_x^b \sinh[\lambda(t-x)] f(t, y(t))\, dt = g(x). \tag{1}$$

Differentiating (1) with respect to x twice yields

$$y''_{xx}(x) + 2\lambda f(x, y(x)) + \lambda^2 \int_a^x \sinh[\lambda(x-t)] f(t, y(t))\, dt$$

$$+ \lambda^2 \int_x^b \sinh[\lambda(t-x)] f(t, y(t))\, dt = g''_{xx}(x). \tag{2}$$

Eliminating the integral terms from (1) and (2), we arrive at a second-order ordinary differential equation for $y = y(x)$:

$$y''_{xx} + 2\lambda f(x, y) - \lambda^2 y = g''_{xx}(x) - \lambda^2 g(x). \tag{3}$$

2°. Let us derive the boundary conditions for equation (3). We assume that $-\infty < a < b < \infty$. By setting $x = a$ and $x = b$ in (1), we obtain the relations

$$y(a) + \int_a^b \sinh[\lambda(t-a)]f\big(t, y(t)\big)\, dt = g(a),$$
$$y(b) + \int_a^b \sinh[\lambda(b-t)]f\big(t, y(t)\big)\, dt = g(b). \tag{4}$$

Let us solve equation (3) for $f(x, y)$ and substitute the result into (4). Integrating by parts yields

$$\sinh[\lambda(b-a)]\varphi'_x(b) - \lambda\cosh[\lambda(b-a)]\varphi(b) = \lambda\varphi(a), \quad \varphi(x) = y(x) - g(x);$$
$$\sinh[\lambda(b-a)]\varphi'_x(a) + \lambda\cosh[\lambda(b-a)]\varphi(a) = -\lambda\varphi(b). \tag{5}$$

Equation (3) under the boundary conditions (5) determines the solution of the original integral equation (there may be several solutions). Conditions (5) make it possible to calculate the constants of integration that occur in solving the differential equation (3).

38. $\quad y(x) + \displaystyle\int_a^b \sin(\lambda|x-t|)\, f\big(t, y(t)\big)\, dt = g(x), \qquad a \le x \le b.$

1°. Let us remove the modulus in the integrand:

$$y(x) + \int_a^x \sin[\lambda(x-t)]f\big(t, y(t)\big)\, dt + \int_x^b \sin[\lambda(t-x)]f\big(t, y(t)\big)\, dt = g(x). \tag{1}$$

Differentiating (1) with respect to x twice yields

$$y''_{xx}(x) + 2\lambda f\big(x, y(x)\big) - \lambda^2 \int_a^x \sin[\lambda(x-t)]f\big(t, y(t)\big)\, dt$$
$$- \lambda^2 \int_x^b \sin[\lambda(t-x)]f\big(t, y(t)\big)\, dt = g''_{xx}(x). \tag{2}$$

Eliminating the integral terms from (1) and (2), we arrive at a second-order ordinary differential equation for $y = y(x)$:

$$y''_{xx} + 2\lambda f(x, y) + \lambda^2 y = g''_{xx}(x) + \lambda^2 g(x). \tag{3}$$

2°. Let us derive the boundary conditions for equation (3). We assume that $-\infty < a < b < \infty$. By setting $x = a$ and $x = b$ in (1), we obtain the relations

$$y(a) + \int_a^b \sin[\lambda(t-a)]\, f\big(t, y(t)\big)\, dt = g(a),$$
$$y(b) + \int_a^b \sin[\lambda(b-t)]\, f\big(t, y(t)\big)\, dt = g(b). \tag{4}$$

Let us solve equation (3) for $f(x, y)$ and substitute the result into (4). Integrating by parts yields

$$\sin[\lambda(b-a)]\, \varphi'_x(b) - \lambda\cos[\lambda(b-a)]\, \varphi(b) = \lambda\varphi(a), \quad \varphi(x) = y(x) - g(x);$$
$$\sin[\lambda(b-a)]\, \varphi'_x(a) + \lambda\cos[\lambda(b-a)]\, \varphi(a) = -\lambda\varphi(b). \tag{5}$$

Equation (3) under the boundary conditions (5) determines the solution of the original integral equation (there may be several solutions). Conditions (5) make it possible to calculate the constants of integration that occur in solving the differential equation (3).

6.8-4. Equations of the Form $y(x) + \int_a^b G(x, t, y(t))\, dt = F(x)$

39. $y(x) + \displaystyle\int_a^b \left[g_1(x) f_1(t, y(t)) + g_2(x) f_2(t, y(t)) \right] dt = h(x).$

A solution:

$$y(x) = h(x) + \lambda_1 g_1(x) + \lambda_2 g_2(x),$$

where the constants λ_1 and λ_2 are determined from the algebraic (or transcendental) system

$$\lambda_1 + \int_a^b f_1(t, h(t) + \lambda_1 g_1(t) + \lambda_2 g_2(t))\, dt = 0,$$

$$\lambda_2 + \int_a^b f_2(t, h(t) + \lambda_1 g_1(t) + \lambda_2 g_2(t))\, dt = 0.$$

40. $y(x) + \displaystyle\int_a^b \left[\sum_{k=1}^n g_k(x) f_k(t, y(t)) \right] dt = h(x).$

A solution:

$$y(x) = h(x) + \sum_{k=1}^n \lambda_k g_k(x),$$

where the coefficients λ_k are determined from the algebraic (or transcendental) system

$$\lambda_m + \int_a^b f_m\left(t, h(t) + \sum_{k=1}^n \lambda_k g_k(t)\right) dt = 0; \qquad m = 1, \dots, n.$$

Different roots of this system generate different solutions of the integral equation.

⊙   Reference: A. F. Verlan' and V. S. Sizikov (1986).

6.8-5. Equations of the Form $F(x, y(x)) + \int_a^b G(x, t, y(x), y(t))\, dt = 0$

41. $y(x) + \displaystyle\int_a^b y(x) f(t, y(t))\, dt = g(x).$

A solution: $y(x) = \lambda g(x)$, where λ is determined by the algebraic (or transcendental) equation

$$\lambda + \lambda F(\lambda) - 1 = 0, \qquad F(\lambda) = \int_a^b f(t, \lambda g(t))\, dt.$$

42. $y(x) + \displaystyle\int_a^b g(x) y(x) f(t, y(t))\, dt = h(x).$

A solution: $y(x) = \dfrac{h(x)}{1 + \lambda g(x)}$, where λ is determined from the algebraic (or transcendental) equation

$$\lambda - F(\lambda) = 0, \qquad F(\lambda) = \int_a^b f\left(t, \frac{h(t)}{1 + \lambda g(t)}\right) dt.$$

43. $\quad y(x) + \displaystyle\int_a^b g\big(x, y(x)\big) f\big(t, y(t)\big)\, dt = h(x).$

Solution in an implicit form:

$$y(x) + \lambda g\big(x, y(x)\big) - h(x) = 0, \tag{1}$$

where λ is determined from the algebraic (or transcendental) equation

$$\lambda - F(\lambda) = 0, \qquad F(\lambda) = \int_a^b f\big(t, y(t)\big)\, dt. \tag{2}$$

Here the function $y(x) = y(x, \lambda)$ obtained by solving (1) must be substituted into (2).

The number of solutions of the integral equation is determined by the number of the solutions obtained from (1) and (2).

44. $\quad f\big(x, y(x)\big) + \displaystyle\int_a^b \left[\sum_{k=1}^n g_k\big(x, y(x)\big) h_k\big(t, y(t)\big) \right] dt = 0.$

Solution in an implicit form:

$$f\big(x,\, y(x)\big) + \sum_{k=1}^n \lambda_k g_k\big(x, y(x)\big) = 0, \tag{1}$$

where the λ_k are determined from the algebraic (or transcendental) system

$$\lambda_k - H_k(\vec{\lambda}) = 0, \qquad k = 1, \ldots, n;$$
$$H_k(\vec{\lambda}) = \int_a^b h_k\big(t, y(t)\big)\, dt, \qquad \vec{\lambda} = \{\lambda_1, \ldots, \lambda_n\}. \tag{2}$$

Here the function $y(x) = y(x, \vec{\lambda})$ obtained by solving (1) must be substituted into (2).

The number of solutions of the integral equation is determined by the number of the solutions obtained from (1) and (2).

6.8-6. Other Equations

45. $\quad y(x) + \displaystyle\int_a^b y(xt) f\big(t, y(t)\big)\, dt = 0.$

1°. A solution:
$$y(x) = k x^C, \tag{1}$$

where C is an arbitrary constant and the dependence $k = k(C)$ is determined by the algebraic (or transcendental) equation

$$1 + \int_a^b t^C f\big(t, kt^C\big)\, dt = 0. \tag{2}$$

Each root of equation (2) generates a solution of the integral equation which has the form (1).

2°. The integral equation can have some other solutions similar to those indicated in items 1°–3° of equation 6.2.30.

46. $y(x) + \displaystyle\int_a^b y(xt)f\left(t, y(t)\right) dt = Ax + B.$

A solution:

$$y(x) = px + q, \tag{1}$$

where p and q are roots of the following system of algebraic (or transcendental) equations:

$$p + p \int_a^b tf(t, pt + q)\, dt - A = 0,$$
$$q + q \int_a^b f(t, pt + q)\, dt - B = 0. \tag{2}$$

Different solutions of system (2) generate different solutions (1) of the integral equation.

47. $y(x) + \displaystyle\int_a^b y(xt)f\left(t, y(t)\right) dt = Ax^\beta.$

A solution:

$$y(x) = kx^\beta, \tag{1}$$

where k is a root of the algebraic (or transcendental) equation

$$k + kF(k) - A = 0, \qquad F(k) = \int_a^b t^\beta f\left(t, kt^\beta\right) dt. \tag{2}$$

Each root of equation (2) generates a solution of the integral equation which has the form (1).

48. $y(x) + \displaystyle\int_a^b y(xt)f\left(t, y(t)\right) dt = A \ln x + B.$

A solution:

$$y(x) = p \ln x + q, \tag{1}$$

where p and q are roots of the following system of algebraic (or transcendental) equations:

$$p + p \int_a^b f(t,\, p \ln t + q)\, dt - A = 0,$$
$$q + \int_a^b (p \ln t + q) f(t,\, p \ln t + q)\, dt - B = 0. \tag{2}$$

Different solutions of system (2) generate different solutions (1) of the integral equation.

49. $y(x) + \displaystyle\int_a^b y(xt)f\left(t, y(t)\right) dt = Ax^\beta \ln x.$

A solution:

$$y(x) = px^\beta \ln x + qx^\beta, \tag{1}$$

where p and q are roots of the following system of algebraic (or transcendental) equations:

$$p + p \int_a^b t^\beta f(t, pt^\beta \ln t + qt^\beta)\, dt = A,$$
$$q + \int_a^b (pt^\beta \ln t + qt^\beta) f(t,\, pt^\beta \ln t + qt^\beta)\, dt = 0. \tag{2}$$

Different solutions of system (2) generate different solutions (1) of the integral equation.

50. $y(x) + \displaystyle\int_a^b y(xt) f\big(t, y(t)\big)\, dt = A\cos(\ln x).$

A solution:
$$y(x) = p\cos(\ln x) + q\sin(\ln x),$$

where p and q are roots of the following system of algebraic (or transcendental) equations:

$$p + \int_a^b \big[p\cos(\ln t) + q\sin(\ln t)\big] f\big(t,\, p\cos(\ln t) + q\sin(\ln t)\big)\, dt = A,$$

$$q + \int_a^b \big[q\cos(\ln t) - p\sin(\ln t)\big] f\big(t,\, p\cos(\ln t) + q\sin(\ln t)\big)\, dt = 0.$$

51. $y(x) + \displaystyle\int_a^b y(xt) f\big(t, y(t)\big)\, dt = A\sin(\ln x).$

A solution:
$$y(x) = p\cos(\ln x) + q\sin(\ln x),$$

where p and q are roots of the following system of algebraic (or transcendental) equations:

$$p + \int_a^b \big[p\cos(\ln t) + q\sin(\ln t)\big] f\big(t,\, p\cos(\ln t) + q\sin(\ln t)\big)\, dt = 0,$$

$$q + \int_a^b \big[q\cos(\ln t) - p\sin(\ln t)\big] f\big(t,\, p\cos(\ln t) + q\sin(\ln t)\big)\, dt = A.$$

52. $y(x) + \displaystyle\int_a^b y(xt) f\big(t, y(t)\big)\, dt = Ax^\beta \cos(\ln x) + Bx^\beta \sin(\ln x).$

A solution:
$$y(x) = px^\beta \cos(\ln x) + qx^\beta \sin(\ln x), \tag{1}$$

where p and q are roots of the following system of algebraic (or transcendental) equations:

$$p + \int_a^b t^\beta \big[p\cos(\ln t) + q\sin(\ln t)\big] f\big(t,\, pt^\beta \cos(\ln t) + qt^\beta \sin(\ln t)\big)\, dt = A, \tag{2}$$

$$q + \int_a^b t^\beta \big[q\cos(\ln t) - p\sin(\ln t)\big] f\big(t,\, pt^\beta \cos(\ln t) + qt^\beta \sin(\ln t)\big)\, dt = B.$$

Different solutions of system (2) generate different solutions (1) of the integral equation.

53. $y(x) + \displaystyle\int_a^b y\big(xt^\beta\big) f\big(t, y(t)\big)\, dt = g(x), \qquad \beta > 0.$

1°. For $g(x) = \displaystyle\sum_{k=1}^n A_k x^k$, the equation has a solution of the form

$$y(x) = \sum_{k=1}^n B_k x^k,$$

where B_k are roots of the algebraic (or transcendental) equations

$$B_k + B_k F_k(\vec{B}) - A_k = 0, \qquad F_k(\vec{B}) = \int_a^b t^{k\beta} f\bigg(t, \sum_{m=1}^n B_m t^m\bigg)\, dt.$$

Different roots of this system generate different solutions of the integral equation.

$2°$. For $g(x) = \ln x \sum_{k=0}^{n} A_k x^k$, the equation has a solution of the form

$$y(x) = \ln x \sum_{k=0}^{n} B_k x^k + \sum_{k=0}^{n} C_k x^k,$$

where the constants B_k and C_k can be found by the method of undetermined coefficients.

$3°$. For $g(x) = \sum_{k=0}^{n} A_k (\ln x)^k$, the equation has a solution of the form

$$y(x) = \sum_{k=0}^{n} B_k (\ln x)^k,$$

where the constants B_k can be found by the method of undetermined coefficients.

$4°$. For $g(x) = \sum_{k=1}^{n} A_k \cos(\lambda_k \ln x)$, the equation has a solution of the form

$$y(x) = \sum_{k=1}^{n} B_k \cos(\lambda_k \ln x) + \sum_{k=1}^{n} C_k \sin(\lambda_k \ln x),$$

where the constants B_k and C_k can be found by the method of undetermined coefficients.

$5°$. For $g(x) = \sum_{k=1}^{n} A_k \sin(\lambda_k \ln x)$, the equation has a solution of the form

$$y(x) = \sum_{k=1}^{n} B_k \cos(\lambda_k \ln x) + \sum_{k=1}^{n} C_k \sin(\lambda_k \ln x),$$

where the constants B_k and C_k can be found by the method of undetermined coefficients.

54. $y(x) + \int_a^b y(x - t) f(t, y(t))\, dt = 0.$

$1°$. A solution:

$$y(x) = k e^{Cx}, \tag{1}$$

where C is an arbitrary constant and the dependence $k = k(C)$ is determined by the algebraic (or transcendental) equation

$$1 + \int_a^b f(t, k e^{Ct}) e^{-Ct}\, dt = 0. \tag{2}$$

Each root of equation (2) generates a solution of the integral equation which has the form (1).

$2°$. The equation has solutions of the form $y(x) = \sum_{m=0}^{n} E_m x^m$, where the constants E_m can be found by the method of undetermined coefficients.

55. $y(x) + \displaystyle\int_a^b y(x-t)f\big(t,y(t)\big)\,dt = Ax + B.$

A solution:

$$y(x) = px + q, \tag{1}$$

where p and q are roots of the following system of algebraic (or transcendental) equations:

$$p + p\int_a^b f(t, pt + q)\,dt - A = 0,$$
$$q + \int_a^b (q - pt)f(t, pt + q)\,dt - B = 0. \tag{2}$$

Different solutions of system (2) generate different solutions (1) of the integral equation.

56. $y(x) + \displaystyle\int_a^b y(x-t)f\big(t,y(t)\big)\,dt = Ae^{\lambda x}.$

Solutions:

$$y(x) = k_n e^{\lambda x},$$

where k_n are roots of the algebraic (or transcendental) equation

$$k + kF(k) - A = 0, \qquad F(k) = \int_a^b f\big(t, ke^{\lambda t}\big)e^{-\lambda t}\,dt.$$

57. $y(x) + \displaystyle\int_a^b y(x-t)f\big(t,y(t)\big)\,dt = A\sinh\lambda x.$

A solution:

$$y(x) = p\sinh\lambda x + q\cosh\lambda x, \tag{1}$$

where p and q are roots of the following system of algebraic (or transcendental) equations:

$$p + \int_a^b (p\cosh\lambda t - q\sinh\lambda t)f\big(t,\, p\sinh\lambda t + q\cosh\lambda t\big)\,dt = A,$$
$$q + \int_a^b (q\cosh\lambda t - p\sinh\lambda t)f\big(t,\, p\sinh\lambda t + q\cosh\lambda t\big)\,dt = 0. \tag{2}$$

Different solutions of system (2) generate different solutions (1) of the integral equation.

58. $y(x) + \displaystyle\int_a^b y(x-t)f\big(t,y(t)\big)\,dt = A\cosh\lambda x.$

A solution:

$$y(x) = p\sinh\lambda x + q\cosh\lambda x,$$

where p and q are roots of the following system of algebraic (or transcendental) equations:

$$p + \int_a^b (p\cosh\lambda t - q\sinh\lambda t)f\big(t,\, p\sinh\lambda t + q\cosh\lambda t\big)\,dt = 0,$$
$$q + \int_a^b (q\cosh\lambda t - p\sinh\lambda t)f\big(t,\, p\sinh\lambda t + q\cosh\lambda t\big)\,dt = A.$$

59. $y(x) + \displaystyle\int_a^b y(x - t)f\left(t, y(t)\right) dt = A \sin \lambda x.$

A solution:

$$y(x) = p \sin \lambda x + q \cos \lambda x, \tag{1}$$

where p and q are roots of the following system of algebraic (or transcendental) equations:

$$p + \int_a^b (p \cos \lambda t + q \sin \lambda t)f\left(t, p \sin \lambda t + q \cos \lambda t\right) dt = A,$$
$$q + \int_a^b (q \cos \lambda t - p \sin \lambda t)f\left(t, p \sin \lambda t + q \cos \lambda t\right) dt = 0. \tag{2}$$

Different solutions of system (2) generate different solutions (1) of the integral equation.

60. $y(x) + \displaystyle\int_a^b y(x - t)f\left(t, y(t)\right) dt = A \cos \lambda x.$

A solution:

$$y(x) = p \sin \lambda x + q \cos \lambda x,$$

where p and q are roots of the following system of algebraic (or transcendental) equations:

$$p + \int_a^b (p \cos \lambda t + q \sin \lambda t)f\left(t, p \sin \lambda t + q \cos \lambda t\right) dt = 0,$$
$$q + \int_a^b (q \cos \lambda t - p \sin \lambda t)f\left(t, p \sin \lambda t + q \cos \lambda t\right) dt = A.$$

61. $y(x) + \displaystyle\int_a^b y(x - t)f\left(t, y(t)\right) dt = e^{\mu x}(A \sin \lambda x + B \cos \lambda x).$

A solution:

$$y(x) = e^{\mu x}(p \sin \lambda x + q \cos \lambda x), \tag{1}$$

where p and q are roots of the following system of algebraic (or transcendental) equations:

$$p + \int_a^b (p \cos \lambda t + q \sin \lambda t)e^{-\mu t} f\left(t, pe^{\mu t} \sin \lambda t + qe^{\mu t} \cos \lambda t\right) dt = A,$$
$$q + \int_a^b (q \cos \lambda t - p \sin \lambda t)e^{-\mu t} f\left(t, pe^{\mu t} \sin \lambda t + qe^{\mu t} \cos \lambda t\right) dt = B. \tag{2}$$

Different solutions of system (2) generate different solutions (1) of the integral equation.

62. $y(x) + \displaystyle\int_a^b y(x - t)f\left(t, y(t)\right) dt = g(x).$

1°. For $g(x) = \displaystyle\sum_{k=1}^{n} A_k \exp(\lambda_k x)$, the equation has a solution of the form

$$y(x) = \sum_{k=1}^{n} B_k \exp(\lambda_k x),$$

where the constants B_k are determined from the nonlinear algebraic (or transcendental) system

$$B_k + B_k F_k(\vec{B}) - A_k = 0, \qquad k = 1, \ldots, n,$$
$$\vec{B} = \{B_1, \ldots, B_n\}, \qquad F_k(\vec{B}) = \int_a^b f\left(t, \sum_{m=1}^{n} B_m \exp(\lambda_m t)\right) \exp(-\lambda_k t) \, dt.$$

Different solutions of this system generate different solutions of the integral equation.

$2°$. For a polynomial right-hand side, $g(x) = \sum\limits_{k=0}^{n} A_k x^k$, the equation has a solution of the form

$$y(x) = \sum_{k=0}^{n} B_k x^k,$$

where the constants B_k can be found by the method of undetermined coefficients.

$3°$. For $g(x) = e^{\lambda x} \sum\limits_{k=0}^{n} A_k x^k$, the equation has a solution of the form

$$y(x) = e^{\lambda x} \sum_{k=0}^{n} B_k x^k,$$

where the constants B_k can be found by the method of undetermined coefficients.

$4°$. For $g(x) = \sum\limits_{k=1}^{n} A_k \cos(\lambda_k x)$, the equation has a solution of the form

$$y(x) = \sum_{k=1}^{n} B_k \cos(\lambda_k x) + \sum_{k=1}^{n} C_k \sin(\lambda_k x),$$

where the constants B_k and C_k can be found by the method of undetermined coefficients.

$5°$. For $g(x) = \sum\limits_{k=1}^{n} A_k \sin(\lambda_k x)$, the equation has a solution of the form

$$y(x) = \sum_{k=1}^{n} B_k \cos(\lambda_k x) + \sum_{k=1}^{n} C_k \sin(\lambda_k x),$$

where the constants B_k and C_k can be found by the method of undetermined coefficients.

$6°$. For $g(x) = \cos(\lambda x) \sum\limits_{k=0}^{n} A_k x^k$, the equation has a solution of the form

$$y(x) = \cos(\lambda x) \sum_{k=0}^{n} B_k x^k + \sin(\lambda x) \sum_{k=0}^{n} C_k x^k,$$

where the constants B_k and C_k can be found by the method of undetermined coefficients.

$7°$. For $g(x) = \sin(\lambda x) \sum\limits_{k=0}^{n} A_k x^k$, the equation has a solution of the form

$$y(x) = \cos(\lambda x) \sum_{k=0}^{n} B_k x^k + \sin(\lambda x) \sum_{k=0}^{n} C_k x^k,$$

where the constants B_k and C_k can be found by the method of undetermined coefficients.

$8°$. For $g(x) = e^{\mu x} \sum\limits_{k=1}^{n} A_k \cos(\lambda_k x)$, the equation has a solution of the form

$$y(x) = e^{\mu x} \sum_{k=1}^{n} B_k \cos(\lambda_k x) + e^{\mu x} \sum_{k=1}^{n} C_k \sin(\lambda_k x),$$

where the constants B_k and C_k can be found by the method of undetermined coefficients.

$9°$. For $g(x) = e^{\mu x} \sum\limits_{k=1}^{n} A_k \sin(\lambda_k x)$, the equation has a solution of the form

$$y(x) = e^{\mu x} \sum_{k=1}^{n} B_k \cos(\lambda_k x) + e^{\mu x} \sum_{k=1}^{n} C_k \sin(\lambda_k x),$$

where the constants B_k and C_k can be found by the method of undetermined coefficients.

$10°$. For $g(x) = \cos(\lambda x) \sum\limits_{k=1}^{n} A_k \exp(\mu_k x)$, the equation has a solution of the form

$$y(x) = \cos(\lambda x) \sum_{k=1}^{n} B_k \exp(\mu_k x) + \sin(\lambda x) \sum_{k=1}^{n} C_k \exp(\mu_k x),$$

where the constants B_k and C_k can be found by the method of undetermined coefficients.

$11°$. For $g(x) = \sin(\lambda x) \sum\limits_{k=1}^{n} A_k \exp(\mu_k x)$, the equation has a solution of the form

$$y(x) = \cos(\lambda x) \sum_{k=1}^{n} B_k \exp(\mu_k x) + \sin(\lambda x) \sum_{k=1}^{n} C_k \exp(\mu_k x),$$

where the constants B_k and C_k can be found by the method of undetermined coefficients.

63. $y(x) + \int_a^b y(x + \beta t) f\big(t, y(t)\big)\, dt = Ax + B.$

A solution:

$$y(x) = px + q, \tag{1}$$

where p and q are roots of the following system of algebraic (or transcendental) equations:

$$p + p \int_a^b f(t, pt + q)\, dt - A = 0,$$
$$q + \int_a^b (\beta pt + q) f(t, pt + q)\, dt - B = 0. \tag{2}$$

Different solutions of system (2) generate different solutions (1) of the integral equation.

64. $y(x) + \int_a^b y(x + \beta t) f\big(t, y(t)\big)\, dt = Ae^{\lambda x}.$

Solutions:

$$y(x) = k_n e^{\lambda x},$$

where k_n are roots of the algebraic (or transcendental) equation

$$k + kF(k) - A = 0, \qquad F(k) = \int_a^b f\big(t, k e^{\lambda t}\big) e^{\beta \lambda t}\, dt.$$

65. $y(x) + \displaystyle\int_a^b y(x + \beta t)f\big(t, y(t)\big)\, dt = A \sin \lambda x + B \cos \lambda x.$

A solution:

$$y(x) = p \sin \lambda x + q \cos \lambda x, \tag{1}$$

where p and q are roots of the following system of algebraic (or transcendental) equations:

$$\begin{aligned}
p + \int_a^b \big[p\cos(\lambda\beta t) - q\sin(\lambda\beta t)\big] f\big(t,\, p\sin \lambda t + q\cos \lambda t\big)\, dt &= A, \\
q + \int_a^b \big[q\cos(\lambda\beta t) + p\sin(\lambda\beta t)\big] f\big(t,\, p\sin \lambda t + q\cos \lambda t\big)\, dt &= B.
\end{aligned} \tag{2}$$

Different solutions of system (2) generate different solutions (1) of the integral equation.

66. $y(x) + \displaystyle\int_a^b y(x + \beta t)f\big(t, y(t)\big)\, dt = g(x).$

1°. For $g(x) = \sum\limits_{k=1}^n A_k \exp(\lambda_k x)$, the equation has a solution of the form

$$y(x) = \sum_{k=1}^n B_k \exp(\lambda_k x),$$

where the constants B_k are determined from the nonlinear algebraic (or transcendental) system

$$B_k + B_k F_k(\vec{B}) - A_k = 0, \qquad k = 1, \dots, n,$$

$$\vec{B} = \{B_1, \dots, B_n\}, \quad F_k(\vec{B}) = \int_a^b f\left(t, \sum_{m=1}^n B_m \exp(\lambda_m t)\right) \exp(\lambda_k \beta t)\, dt.$$

Different solutions of this system generate different solutions of the integral equation.

2°. For a polynomial right-hand side, $g(x) = \sum\limits_{k=0}^n A_k x^k$, the equation has a solution of the form

$$y(x) = \sum_{k=0}^n B_k x^k,$$

where the constants B_k can be found by the method of undetermined coefficients.

3°. For $g(x) = e^{\lambda x} \sum\limits_{k=0}^n A_k x^k$, the equation has a solution of the form

$$y(x) = e^{\lambda x} \sum_{k=0}^n B_k x^k,$$

where the constants B_k can be found by the method of undetermined coefficients.

4°. For $g(x) = \sum\limits_{k=1}^n A_k \cos(\lambda_k x)$, the equation has a solution of the form

$$y(x) = \sum_{k=1}^n B_k \cos(\lambda_k x) + \sum_{k=1}^n C_k \sin(\lambda_k x),$$

where the constants B_k and C_k can be found by the method of undetermined coefficients.

$5°$. For $g(x) = \sum_{k=1}^{n} A_k \sin(\lambda_k x)$, the equation has a solution of the form

$$y(x) = \sum_{k=1}^{n} B_k \cos(\lambda_k x) + \sum_{k=1}^{n} C_k \sin(\lambda_k x),$$

where the constants B_k and C_k can be found by the method of undetermined coefficients.

$6°$. For $g(x) = \cos(\lambda x) \sum_{k=0}^{n} A_k x^k$, the equation has a solution of the form

$$y(x) = \cos(\lambda x) \sum_{k=0}^{n} B_k x^k + \sin(\lambda x) \sum_{k=0}^{n} C_k x^k,$$

where the constants B_k and C_k can be found by the method of undetermined coefficients.

$7°$. For $g(x) = \sin(\lambda x) \sum_{k=0}^{n} A_k x^k$, the equation has a solution of the form

$$y(x) = \cos(\lambda x) \sum_{k=0}^{n} B_k x^k + \sin(\lambda x) \sum_{k=0}^{n} C_k x^k,$$

where the constants B_k and C_k can be found by the method of undetermined coefficients.

$8°$. For $g(x) = e^{\mu x} \sum_{k=1}^{n} A_k \cos(\lambda_k x)$, the equation has a solution of the form

$$y(x) = e^{\mu x} \sum_{k=1}^{n} B_k \cos(\lambda_k x) + e^{\mu x} \sum_{k=1}^{n} C_k \sin(\lambda_k x),$$

where the constants B_k and C_k can be found by the method of undetermined coefficients.

$9°$. For $g(x) = e^{\mu x} \sum_{k=1}^{n} A_k \sin(\lambda_k x)$, the equation has a solution of the form

$$y(x) = e^{\mu x} \sum_{k=1}^{n} B_k \cos(\lambda_k x) + e^{\mu x} \sum_{k=1}^{n} C_k \sin(\lambda_k x),$$

where the constants B_k and C_k can be found by the method of undetermined coefficients.

$10°$. For $g(x) = \cos(\lambda x) \sum_{k=1}^{n} A_k \exp(\mu_k x)$, the equation has a solution of the form

$$y(x) = \cos(\lambda x) \sum_{k=1}^{n} B_k \exp(\mu_k x) + \sin(\lambda x) \sum_{k=1}^{n} C_k \exp(\mu_k x),$$

where the constants B_k and C_k can be found by the method of undetermined coefficients.

$11°$. For $g(x) = \sin(\lambda x) \sum_{k=1}^{n} A_k \exp(\mu_k x)$, the equation has a solution of the form

$$y(x) = \cos(\lambda x) \sum_{k=1}^{n} B_k \exp(\mu_k x) + \sin(\lambda x) \sum_{k=1}^{n} C_k \exp(\mu_k x),$$

where the constants B_k and C_k can be found by the method of undetermined coefficients.

67. $y(x) + \displaystyle\int_a^b y(\xi)f\big(t, y(t)\big)\,dt = 0, \qquad \xi = x\varphi(t).$

1°. A solution:

$$y(x) = kx^C, \tag{1}$$

where C is an arbitrary constant and the dependence $k = k(C)$ is determined by the algebraic (or transcendental) equation

$$1 + \int_a^b \big[\varphi(t)\big]^C f\big(t, kt^C\big)\,dt = 0. \tag{2}$$

Each root of equation (2) generates a solution of the integral equation which has the form (1).

2°. The equation has solutions of the form $y(x) = \displaystyle\sum_{m=0}^{n} E_m x^m$, where the constants E_m can be found by the method of undetermined coefficients.

68. $y(x) + \displaystyle\int_a^b y(\xi)f\big(t, y(t)\big)\,dt = g(x), \qquad \xi = x\varphi(t).$

1°. For $g(x) = \displaystyle\sum_{k=1}^{n} A_k x^k$, the equation has a solution of the form

$$y(x) = \sum_{k=1}^{n} B_k x^k,$$

where B_k are roots of the algebraic (or transcendental) equations

$$B_k + B_k F_k(\vec{B}) - A_k = 0, \qquad k = 1, \ldots, n,$$

$$\vec{B} = \{B_1, \ldots, B_n\}, \quad F_k(\vec{B}) = \int_a^b \big[\varphi(t)\big]^k f\Big(t, \sum_{m=1}^{n} B_m t^m\Big)\,dt.$$

Different roots generate different solutions of the integral equation.

2°. For solutions with some other functions $g(x)$, see items 2°–5° of equation 6.8.53.

69. $y(x) + \displaystyle\int_a^b y(\xi)f\big(t, y(t)\big)\,dt = 0, \qquad \xi = x + \varphi(t).$

1°. A solution:

$$y(x) = ke^{Cx}, \tag{1}$$

where C is an arbitrary constant and the dependence $k = k(C)$ is determined by the algebraic (or transcendental) equation

$$1 + \int_a^b e^{C\varphi(t)} f\big(t, ke^{Ct}\big)\,dt = 0. \tag{2}$$

Each root of equation (2) generates a solution of the integral equation which has the form (1).

2°. The equation has a solution of the form $y(x) = \displaystyle\sum_{m=0}^{n} E_m x^m$, where the constants E_m can be found by the method of undetermined coefficients.

70. $\quad y(x) + \displaystyle\int_a^b y(\xi) f\big(t, y(t)\big)\, dt = g(x), \qquad \xi = x + \varphi(t).$

1°. For $g(x) = \displaystyle\sum_{k=1}^n A_k \exp(\lambda_k x)$ the equation has a solution of the form

$$y(x) = \sum_{k=1}^n B_k \exp(\lambda_k x),$$

where the constants B_k are determined from the nonlinear algebraic (or transcendental) system

$$B_k + B_k F_k(\vec{B}) - A_k = 0, \qquad k = 1, \ldots, n,$$

$$\vec{B} = \{B_1, \ldots, B_n\}, \quad F_k(\vec{B}) = \int_a^b f\left(t, \sum_{m=1}^n B_m \exp(\lambda_m t)\right) \exp\big[\lambda_k \varphi(t)\big]\, dt.$$

2°. Solutions for some other functions $g(x)$ can be found in items 2°–11° of equation 6.8.66.

Part II

Methods for Solving Integral Equations

Chapter 7

Main Definitions and Formulas. Integral Transforms

7.1. Some Definitions, Remarks, and Formulas

7.1-1. Some Definitions

A function $f(x)$ is said to be *square integrable* on an interval $[a, b]$ if $f^2(x)$ is integrable on $[a, b]$. The set of all square integrable functions is denoted by $L_2(a, b)$ or, briefly, L_2.* Likewise, the set of all integrable functions on $[a, b]$ is denoted by $L_1(a, b)$ or, briefly, L_1.

Let us list the main properties of functions from L_2.

$1°$. The sum of two square integrable functions is a square integrable function.

$2°$. The product of a square integrable function by a constant is a square integrable function.

$3°$. The product of two square integrable functions is an integrable function.

$4°$. If $f(x) \in L_2$ and $g(x) \in L_2$, then the following Cauchy–Schwarz–Bunyakovsky inequality holds:

$$(f, g)^2 \leq \|f\|^2 \|g\|^2,$$

$$(f, g) = \int_a^b f(x) g(x) \, dx, \quad \|f\|^2 = (f, f) = \int_a^b f^2(x) \, dx.$$

The number (f, g) is called the *inner product* of the functions $f(x)$ and $g(x)$ and the number $\|f\|$ is called the L_2-*norm* of $f(x)$.

$5°$. For $f(x) \in L_2$ and $g(x) \in L_2$, the following *triangle inequality* holds:

$$\|f + g\| \leq \|f\| + \|g\|.$$

$6°$. Let functions $f(x)$ and $f_1(x), f_2(x), \ldots, f_n(x), \ldots$ be square integrable on an interval $[a, b]$. If

$$\lim_{n \to \infty} \int_a^b \left[f_n(x) - f(x) \right]^2 dx = 0,$$

then the sequence $f_1(x), f_2(x), \ldots$ is said to be *mean-square convergent* to $f(x)$.

Note that if a sequence of functions $\{f_n(x)\}$ from L_2 converges uniformly to $f(x)$, then $f(x) \in L_2$ and $\{f_n(x)\}$ is mean-square convergent to $f(x)$.

---

* In the most general case the integral is understood as the Lebesgue integral of measurable functions. As usual, two equivalent functions (i.e., equal everywhere, or distinct on a negligible set (of zero measure)) are regarded as one and the same element of L_2.

The notion of an integrable function of several variables is similar. For instance, a function $f(x, t)$ is said to be *square integrable* in a domain $S = \{a \leq x \leq b, \; a \leq t \leq b\}$ if $f(x)$ is measurable and

$$\|f\|^2 \equiv \int_a^b \int_a^b f^2(x, t)\, dx\, dt < \infty.$$

Here $\|f\|$ denotes the norm of the function $f(x, t)$, as above.

7.1-2. The Structure of Solutions to Linear Integral Equations

A linear integral equation with variable integration limit has the form

$$\beta y(x) + \int_a^x K(x, t) y(t)\, dt = f(x), \tag{1}$$

where $y(x)$ is the unknown function.

A linear integral equation with constant integration limits has the form

$$\beta y(x) + \int_a^b K(x, t) y(t)\, dt = f(x). \tag{2}$$

For $\beta = 0$, Eqs. (1) and (2) are called *linear integral equations of the first kind*, and for $\beta \neq 0$, *linear integral equations of the second kind.**

Equations of the form (1) and (2) with specific conditions imposed on the kernels and the right-hand sides form various classes of integral equations (Volterra equations, Fredholm equations, convolution equations, etc.), which are considered in detail in Chapters 8–12.

For brevity, we shall sometimes represent the linear equations (1) and (2) in the operator form

$$\mathbf{L}\,[y] = f(x). \tag{3}$$

A linear operator $\mathbf{L}$ possesses the properties

$$\mathbf{L}\,[y_1 + y_2] = \mathbf{L}\,[y_1] + \mathbf{L}\,[y_2],$$
$$\mathbf{L}\,[\sigma y] = \sigma \mathbf{L}\,[y], \quad \sigma = \text{const}.$$

A linear equation is called *homogeneous* if $f(x) \equiv 0$ and *nonhomogeneous* otherwise.

An arbitrary homogeneous linear integral equation has the trivial solution $y \equiv 0$.

If $y_1 = y_1(x)$ and $y_2 = y_2(x)$ are particular solutions of a linear homogeneous integral equation, then the linear combination $C_1 y_1 + C_2 y_2$ with arbitrary constants C_1 and C_2 is also a solution (in physical problems, this property is called the *linear superposition principle*).

The general solution of a linear nonhomogeneous integral equation (3) is the sum of the general solution $Y = Y(x)$ of the corresponding homogeneous equation $\mathbf{L}\,[Y] = 0$ and an arbitrary particular solution $\bar{y} = \bar{y}(x)$ of the nonhomogeneous equation $\mathbf{L}\,[\bar{y}] = f(x)$, that is,

$$y = Y + \bar{y}. \tag{4}$$

If the homogeneous integral equation has only the trivial solution $Y \equiv 0$, then the solution of the corresponding nonhomogeneous equation is unique (if it exists).

Let $\bar{y}_1$ and $\bar{y}_2$ be solutions of nonhomogeneous linear integral equations with the same left-hand sides and different right-hand sides, $\mathbf{L}\,[\bar{y}_1] = f_1(x)$ and $\mathbf{L}\,[\bar{y}_2] = f_2(x)$. Then the function $\bar{y} = \bar{y}_1 + \bar{y}_2$ is a solution of the equation $\mathbf{L}\,[\bar{y}] = f_1(x) + f_2(x)$.

The transformation

$$x = g(z), \quad t = g(\tau), \quad y(x) = \varphi(z) w(z) + \psi(z), \tag{5}$$

where $g(z)$, $\varphi(z)$, and $\psi(z)$ are arbitrary continuous functions ($g_z' \neq 0$), reduces Eqs. (1) and (2) to linear equations of the same form for the unknown function $w = w(z)$. Such transformations are frequently used for constructing exact solutions of linear integral equations.

---

* In Chapters 1–4, which deal with equations with variable and constant limits of integration, we sometimes consider more general equations in which the integrand contains the unknown function $y(z)$, where $z = z(x, t)$, instead of $y(t)$.

Integral transforms have the form

$$\tilde{f}(\lambda) = \int_a^b \varphi(x, \lambda) f(x) \, dx.$$

The function $\tilde{f}(\lambda)$ is called the *transform* of the function $f(x)$ and $\varphi(x, \lambda)$ is called the *kernel* of the integral transform. The function $f(x)$ is called the *inverse transform* of $\tilde{f}(\lambda)$. The limits of integration a and b are real numbers (usually, $a = 0$, $b = \infty$ or $a = -\infty$, $b = \infty$).

In Subsections 7.2–7.6, the most popular (Laplace, Mellin, Fourier, etc.) integral transforms, applied in this book to the solution of specific integral equations, are described. These subsections also describe the corresponding inversion formulas, which have the form

$$f(x) = \int_{\mathcal{L}} \psi(x, \lambda) \tilde{f}(\lambda) \, d\lambda$$

and make it possible to recover $f(x)$ if $\tilde{f}(\lambda)$ is given. The integration path $\mathcal{L}$ can lie either on the real axis or in the complex plane.

Integral transforms are used in the solution of various differential and integral equations. Figure 1 outlines the overall scheme of solving some special classes of linear integral equations by means of integral transforms (by applying appropriate integral transforms to this sort of integral equations, one obtains first-order linear algebraic equations for $\tilde{f}(\lambda)$).

In many cases, to calculate definite integrals, in particular, to find the inverse Laplace, Mellin, and Fourier transforms, methods of the theory of functions of a complex variable can be applied, including the residue theorem and the Jordan lemma, which are presented below in Subsections 7.1-4 and 7.1-5.

7.1-4. Residues. Calculation Formulas

The *residue* of a function $f(z)$ holomorphic in a deleted neighborhood of a point $z = a$ (thus, a is an isolated singularity of f) of the complex plane z is the number

$$\operatorname*{res}_{z=a} f(z) = \frac{1}{2\pi i} \int_{c_\varepsilon} f(z) \, dz, \quad i^2 = -1,$$

where c_ε is a circle of sufficiently small radius ε described by the equation $|z - a| = \varepsilon$.

If the point $z = a$ is a pole of order n* of the function $f(z)$, then we have

$$\operatorname*{res}_{z=a} f(z) = \frac{1}{(n-1)!} \lim_{z \to a} \frac{d^{n-1}}{dx^{n-1}} \left[(z - a)^n f(z) \right].$$

For a simple pole, which corresponds to $n = 1$, this implies

$$\operatorname*{res}_{z=a} f(z) = \lim_{z \to a} \left[(z - a) f(z) \right].$$

If $f(z) = \dfrac{\varphi(z)}{\psi(z)}$, where $\varphi(a) \neq 0$ and $\psi(z)$ has a simple zero at the point $z = a$, i.e., $\psi(a) = 0$ and $\psi'_z(a) \neq 0$, then

$$\operatorname*{res}_{z=a} f(z) = \frac{\varphi(a)}{\psi'_z(a)}.$$

---

* In a neighborhood of this point we have $f(z) \approx \operatorname{const} (z - a)^{-n}$.

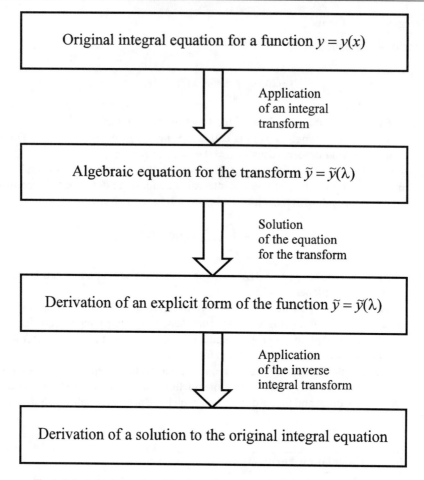

Fig. 1. Principal scheme of applying integral transforms for solving integral equations

If a function $f(z)$ is continuous in the domain $|z| \geq R_0$, $\operatorname{Im} z \geq \alpha$, where α is a chosen real number, and if $\lim\limits_{z \to \infty} f(z) = 0$, then

$$\lim_{R \to \infty} \int_{C_R} e^{i\lambda z} f(z)\, dz = 0$$

for any $\lambda > 0$, where C_R is the arc of the circle $|z| = R$ that lies in this domain.

⊙ References for Section 7.1: A. G. Sveshnikov and A. N. Tikhonov (1970), M. L. Krasnov, A. I. Kiselev, and G. I. Makarenko (1971).

7.2. The Laplace Transform

7.2-1. Definition. The Inversion Formula

The *Laplace transform* of an arbitrary (complex-valued) function $f(x)$ of a real variable x ($x \geq 0$) is defined by

$$\tilde{f}(p) = \int_0^\infty e^{-px} f(x)\, dx, \tag{1}$$

where $p = s + i\sigma$ is a complex variable.

The Laplace transform exists for any continuous or piecewise-continuous function satisfying the condition $|f(x)| < Me^{\sigma_0 x}$ with some $M > 0$ and $\sigma_0 \geq 0$. In the following, σ_0 often means the greatest lower bound of the possible values of σ_0 in this estimate; this value is called the *growth exponent* of the function $f(x)$.

For any $f(x)$, the transform $\tilde{f}(p)$ is defined in the half-plane $\operatorname{Re} p > \sigma_0$ and is analytic there.

For brevity, we shall write formula (1) as follows:

$$\tilde{f}(p) = \mathfrak{L}\left\{f(x)\right\}, \qquad \text{or} \qquad \tilde{f}(p) = \mathfrak{L}\left\{f(x),\, p\right\}.$$

Given the transform $\tilde{f}(p)$, the function can be found by means of the inverse Laplace transform

$$f(x) = \frac{1}{2\pi i} \int_{c-i\infty}^{c+i\infty} \tilde{f}(p)e^{px}\, dp, \qquad i^2 = -1, \tag{2}$$

where the integration path is parallel to the imaginary axis and lies to the right of all singularities of $\tilde{f}(p)$, which corresponds to $c > \sigma_0$.

The integral in (2) is understood in the sense of the Cauchy principal value:

$$\int_{c-i\infty}^{c+i\infty} \tilde{f}(p)e^{px}\, dp = \lim_{\omega \to \infty} \int_{c-i\omega}^{c+i\omega} \tilde{f}(p)e^{px}\, dp.$$

In the domain $x < 0$, formula (2) gives $f(x) \equiv 0$.

Formula (2) holds for continuous functions. If $f(x)$ has a (finite) jump discontinuity at a point $x = x_0 > 0$, then the left-hand side of (2) is equal to $\frac{1}{2}[f(x_0 - 0) + f(x_0 + 0)]$ at this point (for $x_0 = 0$, the first term in the square brackets must be omitted).

For brevity, we write the Laplace inversion formula (2) as follows:

$$f(x) = \mathfrak{L}^{-1}\left\{\tilde{f}(p)\right\}, \qquad \text{or} \qquad f(x) = \mathfrak{L}^{-1}\left\{\tilde{f}(p),\, x\right\}.$$

7.2-2. The Inverse Transforms of Rational Functions

Consider the important case in which the transform is a rational function of the form

$$\tilde{f}(p) = \frac{R(p)}{Q(p)}, \tag{3}$$

where $Q(p)$ and $R(p)$ are polynomials in the variable p and the degree of $Q(p)$ exceeds that of $R(p)$.

Assume that the zeros of the denominator are simple, i.e.,

$$Q(p) \equiv \operatorname{const}(p - \lambda_1)(p - \lambda_2)\ldots(p - \lambda_n).$$

Then the inverse transform can be determined by the formula

$$f(x) = \sum_{k=1}^{n} \frac{R(\lambda_k)}{Q'(\lambda_k)} \exp(\lambda_k x), \tag{4}$$

where the primes denote the derivatives.

If $Q(p)$ has multiple zeros, i.e.,

$$Q(p) \equiv \mathrm{const}\,(p-\lambda_1)^{s_1}(p-\lambda_2)^{s_2}\ldots(p-\lambda_m)^{s_m},$$

then

$$f(x) = \sum_{k=1}^{m} \frac{1}{(s_k-1)!} \lim_{p \to s_k} \frac{d^{s_k-1}}{dp^{s_k-1}}\left[(p-\lambda_k)^{s_k}\,\tilde{f}(p)e^{px}\right].$$

7.2-3. The Convolution Theorem for the Laplace Transform

The *convolution* of two functions $f(x)$ and $g(x)$ is defined as the integral $\int_0^x f(t)g(x-t)\,dt$, and is usually denoted by $f(x) * g(x)$. The *convolution theorem* states that

$$\mathfrak{L}\left\{f(x)*g(x)\right\} = \mathfrak{L}\left\{f(x)\right\}\mathfrak{L}\left\{g(x)\right\},$$

and is frequently applied to solve Volterra equations with kernels depending on the difference of the arguments.

7.2-4. Limit Theorems

Let $0 \le x < \infty$ and $\tilde{f}(p) = \mathfrak{L}\left\{f(x)\right\}$ be the Laplace transform of $f(x)$. If a limit of $f(x)$ as $x \to 0$ exists, then

$$\lim_{x \to 0} f(x) = \lim_{p \to \infty}\left[p\tilde{f}(p)\right].$$

If a limit of $f(x)$ as $x \to \infty$ exists, then

$$\lim_{x \to \infty} f(x) = \lim_{p \to 0}\left[p\tilde{f}(p)\right].$$

7.2-5. Main Properties of the Laplace Transform

The main properties of the correspondence between functions and their Laplace transforms are gathered in Table 1.

    There are tables of direct and inverse Laplace transforms (see Supplements 4 and 5), which are handy in solving linear integral and differential equations.

7.2-6. The Post–Widder Formula

In applications, one can find $f(x)$ if the Laplace transform $\tilde{f}(t)$ on the real semiaxis is known for $t = p \ge 0$. To this end, one uses the Post–Widder formula

$$f(x) = \lim_{n \to \infty}\left[\frac{(-1)^n}{n!}\left(\frac{n}{x}\right)^{n+1}\tilde{f}_t^{(n)}\left(\frac{n}{x}\right)\right]. \tag{5}$$

Approximate inversion formulas are obtained by taking sufficiently large positive integer n in (5) instead of passing to the limit.

⊙ References for Section 7.2: G. Doetsch (1950, 1956, 1958), H. Bateman and A. Erdélyi (1954), I. I. Hirschman and D. V. Widder (1955), V. A. Ditkin and A. P. Prudnikov (1965), J. W. Miles (1971), B. Davis (1978), Yu. A. Brychkov and A. P. Prudnikov (1989), W. H. Beyer (1991).

TABLE 1
Main properties of the Laplace transform

No	Function	Laplace Transform	Operation
1	$af_1(x) + bf_2(x)$	$a\tilde{f}_1(p) + b\tilde{f}_2(p)$	Linearity
2	$f(x/a),\ a > 0$	$a\tilde{f}(ap)$	Scaling
3	$f(x-a),$ $f(\xi) \equiv 0$ for $\xi < 0$	$e^{-ap}\tilde{f}(p)$	Shift of the argument
4	$x^n f(x);\ n = 1, 2, \ldots$	$(-1)^n \tilde{f}_p^{(n)}(p)$	Differentiation of the transform
5	$\dfrac{1}{x} f(x)$	$\displaystyle\int_p^\infty \tilde{f}(q)\,dq$	Integration of the transform
6	$e^{ax} f(x)$	$\tilde{f}(p-a)$	Shift in the complex plane
7	$f'_x(x)$	$p\tilde{f}(p) - f(+0)$	Differentiation
8	$f_x^{(n)}(x)$	$p^n \tilde{f}(p) - \displaystyle\sum_{k=1}^n p^{n-k} f_x^{(k-1)}(+0)$	Differentiation
9	$x^m f_x^{(n)}(x),\ m \geq n$	$\left(-\dfrac{d}{dp}\right)^m \left[p^n \tilde{f}(p)\right]$	Differentiation
10	$\dfrac{d^n}{dx^n}\left[x^m f(x)\right],\ m \geq n$	$(-1)^m p^n \dfrac{d^m}{dp^m} \tilde{f}(p)$	Differentiation
11	$\displaystyle\int_0^x f(t)\,dt$	$\dfrac{\tilde{f}(p)}{p}$	Integration
12	$\displaystyle\int_0^x f_1(t) f_2(x-t)\,dt$	$\tilde{f}_1(p)\tilde{f}_2(p)$	Convolution

7.3. The Mellin Transform

7.3-1. Definition. The Inversion Formula

Suppose that a function $f(x)$ is defined for positive x and satisfies the conditions

$$\int_0^1 |f(x)|x^{\sigma_1-1}\,dx < \infty, \qquad \int_1^\infty |f(x)|x^{\sigma_2-1}\,dx < \infty$$

for some real numbers σ_1 and σ_2, $\sigma_1 < \sigma_2$.

The Mellin transform of $f(x)$ is defined by

$$\hat{f}(s) = \int_0^\infty f(x)x^{s-1}\,dx, \tag{1}$$

where $s = \sigma + i\tau$ is a complex variable ($\sigma_1 < \sigma < \sigma_2$).

For brevity, we rewrite formula (1) as follows:

$$\hat{f}(s) = \mathfrak{M}\{f(x)\}, \qquad \text{or} \qquad \hat{f}(s) = \mathfrak{M}\{f(x), s\}.$$

Given $\hat{f}(s)$, the function can be found by means of the *inverse Mellin transform*

$$f(x) = \frac{1}{2\pi i} \int_{\sigma-i\infty}^{\sigma+i\infty} \hat{f}(s)x^{-s}\,ds, \qquad (\sigma_1 < \sigma < \sigma_2) \tag{2}$$

where the integration path is parallel to the imaginary axis of the complex plane s and the integral is understood in the sense of the Cauchy principal value.

Formula (2) holds for continuous functions. If $f(x)$ has a (finite) jump discontinuity at a point $x = x_0 > 0$, then the left-hand side of (2) is equal to $\frac{1}{2}\left[f(x_0 - 0) + f(x_0 + 0)\right]$ at this point (for $x_0 = 0$, the first term in the square brackets must be omitted).

For brevity, we rewrite formula (2) in the form

$$f(x) = \mathfrak{M}^{-1}\{\hat{f}(s)\}, \qquad \text{or} \qquad f(x) = \mathfrak{M}^{-1}\{\hat{f}(s),\, x\}.$$

7.3-2. Main Properties of the Mellin Transform

The main properties of the correspondence between the functions and their Mellin transforms are gathered in Table 2.

TABLE 2
Main properties of the Mellin transform

No	Function	Mellin Transform	Operation
1	$af_1(x) + bf_2(x)$	$a\hat{f}_1(s) + b\hat{f}_2(s)$	Linearity
2	$f(ax),\ a > 0$	$a^{-s}\hat{f}(s)$	Scaling
3	$x^a f(x)$	$\hat{f}(s + a)$	Shift of the argument of the transform
4	$f(x^2)$	$\frac{1}{2}\hat{f}\left(\frac{1}{2}s\right)$	Squared argument
5	$f(1/x)$	$\hat{f}(-s)$	Inversion of the argument of the transform
6	$x^\lambda f(ax^\beta),\ \ a > 0, \beta \neq 0$	$\frac{1}{\beta}a^{-\frac{s+\lambda}{\beta}}\hat{f}\left(\frac{s+\lambda}{\beta}\right)$	Power law transform
7	$f'_x(x)$	$-(s-1)\hat{f}(s-1)$	Differentiation
8	$xf'_x(x)$	$-s\hat{f}(s)$	Differentiation
9	$f_x^{(n)}(x)$	$(-1)^n\dfrac{\Gamma(s)}{\Gamma(s-n)}\hat{f}(s-n)$	Multiple differentiation
10	$\left(x\dfrac{d}{dx}\right)^n f(x)$	$(-1)^n s^n \hat{f}(s)$	Multiple differentiation
11	$x^\alpha \displaystyle\int_0^\infty t^\beta f_1(xt)f_2(t)\,dt$	$\hat{f}_1(s+\alpha)\hat{f}_2(1-s-\alpha+\beta)$	Complicated integration
12	$x^\alpha \displaystyle\int_0^\infty t^\beta f_1\left(\dfrac{x}{t}\right)f_2(t)\,dt$	$\hat{f}_1(s+\alpha)\hat{f}_2(s+\alpha+\beta+1)$	Complicated integration

7.3-3. The Relation Among the Mellin, Laplace, and Fourier Transforms

There are tables of direct and inverse Mellin transforms (see Supplements 8 and 9), which are useful in solving specific integral and differential equations. The Mellin transform is related to the Laplace and Fourier transforms as follows:

$$\mathfrak{M}\{f(x), s\} = \mathfrak{L}\{f(e^x), -s\} + \mathfrak{L}\{f(e^{-x}), s\} = \mathfrak{F}\{f(e^x), is\},$$

which makes it possible to apply much more common tables of direct and inverse Laplace and Fourier transforms.

⊙ References for Section 7.3: V. A. Ditkin and A. P. Prudnikov (1965), Yu. A. Brychkov and A. P. Prudnikov (1989).

7.4. The Fourier Transform

7.4-1. Definition. The Inversion Formula

The *Fourier transform* is defined as follows:

$$\tilde{f}(u) = \frac{1}{\sqrt{2\pi}} \int_{-\infty}^{\infty} f(x)e^{-iux}\, dx. \tag{1}$$

For brevity, we rewrite formula (1) as follows:

$$\tilde{f}(u) = \mathfrak{F}\{f(x)\}, \qquad \text{or} \qquad \tilde{f}(u) = \mathfrak{F}\{f(x), u\}.$$

Given $\tilde{f}(u)$, the function $f(x)$ can be found by means of the *inverse Fourier transform*

$$f(x) = \frac{1}{\sqrt{2\pi}} \int_{-\infty}^{\infty} \tilde{f}(u)e^{iux}\, du. \tag{2}$$

Formula (2) holds for continuous functions. If $f(x)$ has a (finite) jump discontinuity at a point $x = x_0$, then the left-hand side of (2) is equal to $\frac{1}{2}\left[f(x_0 - 0) + f(x_0 + 0)\right]$ at this point.

For brevity, we rewrite formula (2) as follows:

$$f(x) = \mathfrak{F}^{-1}\{\tilde{f}(u)\}, \qquad \text{or} \qquad f(x) = \mathfrak{F}^{-1}\{\tilde{f}(u),\, x\}.$$

7.4-2. An Asymmetric Form of the Transform

Sometimes it is more convenient to define the Fourier transform by

$$\check{f}(u) = \int_{-\infty}^{\infty} f(x)e^{-iux}\, dx. \tag{3}$$

For brevity, we rewrite formula (3) as follows: $\check{f}(u) = \mathcal{F}\{f(x)\}$ or $\check{f}(u) = \mathcal{F}\{f(x), u\}$.

In this case, the *Fourier inversion formula* reads

$$f(x) = \frac{1}{2\pi} \int_{-\infty}^{\infty} \check{f}(u)e^{iux}\, du, \tag{4}$$

and we use the following symbolic notation for relation (4): $f(x) = \mathcal{F}^{-1}\{\check{f}(u)\}$, or $f(x) = \mathcal{F}^{-1}\{\check{f}(u),\, x\}$.

7.4-3. The Alternative Fourier Transform

Sometimes, for instance, in the theory of boundary value problems, the alternative Fourier transform is used (and called merely the *Fourier transform*) in the form

$$\mathcal{F}(u) = \frac{1}{\sqrt{2\pi}} \int_{-\infty}^{\infty} f(x)e^{iux}\, dx. \tag{5}$$

For brevity, we rewrite formula (5) as follows:

$$\mathcal{F}(u) = \mathbf{F}\{f(x)\}, \qquad \text{or} \qquad \mathcal{F}(u) = \mathbf{F}\{f(x), u\}.$$

For given $\mathcal{F}(u)$, the function $f(x)$ can be found by means of the inverse transform

$$f(x) = \frac{1}{\sqrt{2\pi}} \int_{-\infty}^{\infty} \mathcal{F}(u)e^{-iux}\,du. \tag{6}$$

For brevity, we rewrite formula (6) as follows:

$$f(x) = \mathbf{F}^{-1}\{\mathcal{F}(u)\}, \qquad \text{or} \qquad f(x) = \mathbf{F}^{-1}\{\mathcal{F}(u),\,x\}.$$

The function $\mathcal{F}(u)$ is also called the *Fourier integral* of $f(x)$.

We can introduce an asymmetric form for the alternative Fourier transform similarly to that of the Fourier transform:

$$\check{\mathcal{F}}(u) = \int_{-\infty}^{\infty} f(x)e^{iux}\,dx, \qquad f(x) = \frac{1}{2\pi} \int_{-\infty}^{\infty} \check{\mathcal{F}}(u)e^{-iux}\,du, \tag{7}$$

where the direct and the inverse transforms (7) are briefly denoted by $\check{\mathcal{F}}(u) = \check{\mathbf{F}}\{f(x)\}$ and $f(x) = \check{\mathbf{F}}^{-1}\{\check{\mathcal{F}}(u)\}$, or by $\check{\mathcal{F}}(u) = \check{\mathbf{F}}\{f(x),\,u\}$ and $f(x) = \check{\mathbf{F}}^{-1}\{\check{\mathcal{F}}(u)\,x\}$.

7.4-4. The Convolution Theorem for the Fourier Transform

The *convolution* of two functions $f(x)$ and $g(x)$ is defined as

$$f(x) * g(x) \equiv \frac{1}{\sqrt{2\pi}} \int_{-\infty}^{\infty} f(x-t)g(t)\,dt.$$

By performing substitution $x - t = u$, we see that the convolution is symmetric with respect to the convolved functions: $f(x) * g(x) = g(x) * f(x)$.

The *convolution theorem* states that

$$\mathfrak{F}\{f(x) * g(x)\} = \mathfrak{F}\{f(x)\}\,\mathfrak{F}\{g(x)\}. \tag{8}$$

For the alternative Fourier transform, the convolution theorem reads

$$\mathbf{F}\{f(x) * g(x)\} = \mathbf{F}\{f(x)\}\,\mathbf{F}\{g(x)\}. \tag{9}$$

Formulas (8) and (9) will be used in Chapters 10 and 11 for solving linear integral equations with difference kernel.

⊙ References for Section 7.4: V. A. Ditkin and A. P. Prudnikov (1965), J. W. Miles (1971), B. Davis (1978), Yu. A. Brychkov and A. P. Prudnikov (1989), W. H. Beyer (1991).

7.5. The Fourier Sine and Cosine Transforms

7.5-1. The Fourier Cosine Transform

Let a function $f(x)$ be integrable on the semiaxis $0 \leq x < \infty$. The Fourier cosine transform is defined by

$$\tilde{f}_c(u) = \sqrt{\frac{2}{\pi}} \int_0^{\infty} f(x)\cos(xu)\,dx, \qquad 0 < u < \infty. \tag{1}$$

For given $\tilde{f}_c(u)$, the function can be found by means of the Fourier cosine inversion formula

$$f(x) = \sqrt{\frac{2}{\pi}} \int_0^\infty \tilde{f}_c(u) \cos(xu)\, du, \qquad 0 < x < \infty. \tag{2}$$

The Fourier cosine transform (1) is denoted for brevity by $\tilde{f}_c(u) = \mathfrak{F}_c\{f(x)\}$. It follows from formula (2) that the Fourier cosine transform has the property $\mathfrak{F}_c^2 = 1$. There are tables of the Fourier cosine transform (see Supplement 7) which prove useful in the solution of specific integral equations.

Sometimes the asymmetric form of the Fourier cosine transform is applied, which is given by the pair of formulas

$$\check{f}_c(u) = \int_0^\infty f(x) \cos(xu)\, dx, \qquad f(x) = \frac{2}{\pi} \int_0^\infty \check{f}_c(u) \cos(xu)\, du. \tag{3}$$

The direct and inverse Fourier cosine transforms (3) are denoted by $\check{f}_c(u) = \mathcal{F}_c\{f(x)\}$ and $f(x) = \mathcal{F}_c^{-1}\{\check{f}_c(u)\}$, respectively.

7.5-2. The Fourier Sine Transform

Let a function $f(x)$ be integrable on the semiaxis $0 \le x < \infty$. The Fourier sine transform is defined by

$$\tilde{f}_s(u) = \sqrt{\frac{2}{\pi}} \int_0^\infty f(x) \sin(xu)\, dx, \qquad 0 < u < \infty. \tag{4}$$

For given $\tilde{f}_s(u)$, the function $f(x)$ can be found by means of the inverse Fourier sine transform

$$f(x) = \sqrt{\frac{2}{\pi}} \int_0^\infty \tilde{f}_s(u) \sin(xu)\, du, \qquad 0 < x < \infty. \tag{5}$$

The Fourier sine transform (4) is briefly denoted by $\tilde{f}_s(u) = \mathfrak{F}_s\{f(x)\}$. It follows from formula (5) that the Fourier sine transform has the property $\mathfrak{F}_s^2 = 1$. There are tables of the Fourier sine transform (see Supplement 6), which are useful in solving specific integral equations.

Sometimes it is more convenient to apply the asymmetric form of the Fourier sine transform defined by the following two formulas:

$$\check{f}_s(u) = \int_0^\infty f(x) \sin(xu)\, dx, \qquad f(x) = \frac{2}{\pi} \int_0^\infty \check{f}_s(u) \sin(xu)\, du. \tag{6}$$

The direct and inverse Fourier sine transforms (6) are denoted by $\check{f}_s(u) = \mathcal{F}_s\{f(x)\}$ and $f(x) = \mathcal{F}_s^{-1}\{\check{f}_s(u)\}$, respectively.

⊙ References for Section 7.5: V. A. Ditkin and A. P. Prudnikov (1965), J. W. Miles (1971), Yu. A. Brychkov and A. P. Prudnikov (1989), W. H. Beyer (1991).

7.6. Other Integral Transforms

7.6-1. The Hankel Transform

The Hankel transform is defined as follows:

$$\tilde{f}_\nu(u) = \int_0^\infty x J_\nu(ux) f(x)\, dx, \qquad 0 < u < \infty, \tag{1}$$

where $\nu > -\frac{1}{2}$ and $J_\nu(x)$ is the Bessel function of the first kind of order ν (see Supplement 10).

For given $\tilde{f}_\nu(u)$, the function $f(x)$ can be found by means of the Hankel inversion formula

$$f(x) = \int_0^\infty u J_\nu(ux) \tilde{f}_\nu(u)\, du, \qquad 0 < x < \infty. \tag{2}$$

Note that if $f(x) = O(x^\alpha)$ as $x \to 0$, where $\alpha + \nu + 2 > 0$, and $f(x) = O(x^\beta)$ as $x \to \infty$, where $\beta + \frac{3}{2} < 0$, then the integral (1) is convergent.

The inversion formula (2) holds for continuous functions. If $f(x)$ has a (finite) jump discontinuity at a point $x = x_0$, then the left-hand side of (2) is equal to $\frac{1}{2}[f(x_0 - 0) + f(x_0 + 0)]$ at this point.

For brevity, we denote the Hankel transform (1) by $\tilde{f}_\nu(u) = \mathfrak{H}_\nu\{f(x)\}$. It follows from formula (2) that the Hankel transform has the property $\mathfrak{H}_\nu^2 = 1$.

7.6-2. The Meijer Transform

The Meijer transform is defined as follows:

$$\hat{f}_\mu(s) = \sqrt{\frac{2}{\pi}} \int_0^\infty \sqrt{sx}\, K_\mu(sx) f(x)\, dx, \qquad 0 < s < \infty, \tag{3}$$

where $K_\mu(x)$ is the modified Bessel function of the second kind (the Macdonald function) of order μ (see Supplement 10).

For given $\tilde{f}_\mu(s)$, the function $f(x)$ can be found by means of the *Meijer inversion formula*

$$f(x) = \frac{1}{i\sqrt{2\pi}} \int_{c-i\infty}^{c+i\infty} \sqrt{sx}\, I_\mu(sx) \hat{f}_\mu(s)\, ds, \qquad 0 < x < \infty, \tag{4}$$

where $I_\mu(x)$ is the modified Bessel function of the first kind of order μ (see Supplement 10). For the Meijer transform, a convolution is defined and an operational calculus is developed.

7.6-3. The Kontorovich–Lebedev Transform and Other Transforms

The Kontorovich–Lebedev transform is introduced as follows:

$$F(\tau) = \int_0^\infty K_{i\tau}(x) f(x)\, dx, \qquad 0 < \tau < \infty, \tag{5}$$

where $K_\mu(x)$ is the modified Bessel function of the second kind (the Macdonald function) of order μ (see Supplement 10) and $i = \sqrt{-1}$.

For given $F(\tau)$, the function can be found by means of the *Kontorovich–Lebedev inversion formula*

$$f(x) = \frac{2}{\pi^2 x} \int_0^\infty \tau \sinh(\pi\tau) K_{i\tau}(x) F(\tau)\, d\tau, \qquad 0 < x < \infty. \tag{6}$$

There are also other integral transforms, of which the most important are listed in Table 3 (for the constraints imposed on the functions and parameters occurring in the integrand, see the references given at the end of this section).

TABLE 3
Main integral transforms

Integral Transform	Definition	Inversion Formula
Laplace transform	$\widetilde{f}(p) = \displaystyle\int_0^\infty e^{-px} f(x)\,dx$	$f(x) = \dfrac{1}{2\pi i} \displaystyle\int_{c-i\infty}^{c+i\infty} e^{px} \widetilde{f}(p)\,dp$
Two-sided Laplace transform	$\widetilde{f}_*(p) = \displaystyle\int_{-\infty}^\infty e^{-px} f(x)\,dx$	$f(x) = \dfrac{1}{2\pi i} \displaystyle\int_{c-i\infty}^{c+i\infty} e^{px} \widetilde{f}_*(p)\,dp$
Fourier transform	$\widetilde{f}(u) = \dfrac{1}{\sqrt{2\pi}} \displaystyle\int_{-\infty}^\infty e^{-iux} f(x)\,dx$	$f(x) = \dfrac{1}{\sqrt{2\pi}} \displaystyle\int_{-\infty}^\infty e^{iux} \widetilde{f}(u)\,du$
Fourier sine transform	$\widetilde{f}_s(u) = \sqrt{\dfrac{2}{\pi}} \displaystyle\int_0^\infty \sin(xu) f(x)\,dx$	$f(x) = \sqrt{\dfrac{2}{\pi}} \displaystyle\int_0^\infty \sin(xu) \widetilde{f}_s(u)\,du$
Fourier cosine transform	$\widetilde{f}_c(u) = \sqrt{\dfrac{2}{\pi}} \displaystyle\int_0^\infty \cos(xu) f(x)\,dx$	$f(x) = \sqrt{\dfrac{2}{\pi}} \displaystyle\int_0^\infty \cos(xu) \widetilde{f}_c(u)\,du$
Hartley transform	$\widetilde{f}_h(u) = \dfrac{1}{\sqrt{2\pi}} \displaystyle\int_{-\infty}^\infty (\cos xu + \sin xu) f(x)\,dx$	$f(x) = \dfrac{1}{\sqrt{2\pi}} \displaystyle\int_{-\infty}^\infty (\cos xu + \sin xu) \widetilde{f}_h(u)\,du$
Mellin transform	$\widehat{f}(s) = \displaystyle\int_0^\infty x^{s-1} f(x)\,dx$	$f(x) = \dfrac{1}{2\pi i} \displaystyle\int_{c-i\infty}^{c+i\infty} x^{-s} \widehat{f}(s)\,ds$
Hankel transform	$\widehat{f}_\nu(w) = \displaystyle\int_0^\infty x J_\nu(xw) f(x)\,dx$	$f(x) = \displaystyle\int_0^\infty w J_\nu(xw) \widehat{f}_\nu(w)\,dw$
Y-transform	$F_\nu(u) = \displaystyle\int_0^\infty \sqrt{ux}\, Y_\nu(ux) f(x)\,dx$	$f(x) = \displaystyle\int_0^\infty \sqrt{ux}\, \mathbf{H}_\nu(ux) F_\nu(u)\,du$
Meijer transform (K-transform)	$\widehat{f}(s) = \sqrt{\dfrac{2}{\pi}} \displaystyle\int_0^\infty \sqrt{sx}\, K_\nu(sx) f(x)\,dx$	$f(x) = \dfrac{1}{i\sqrt{2\pi}} \displaystyle\int_{c-i\infty}^{c+i\infty} \sqrt{sx}\, I_\nu(sx) \widehat{f}(s)\,ds$
Bochner transform	$\widetilde{f}(r) = \displaystyle\int_0^\infty J_{n/2-1}(2\pi xr) G(x,r) f(x)\,dx,$ $G(x,r) = 2\pi r(x/r)^{n/2}, \quad n = 1,2,\dots$	$f(x) = \displaystyle\int_0^\infty J_{n/2-1}(2\pi rx) G(r,x) \widetilde{f}(r)\,dr$
Weber transform	$F_a(u) = \displaystyle\int_a^\infty W_\nu(xu, au) x f(x)\,dx,$ $W_\nu(\beta, \mu) \equiv J_\nu(\beta) Y_\nu(\mu) - J_\nu(\mu) Y_\nu(\beta)$	$f(x) = \displaystyle\int_0^\infty \dfrac{W_\nu(xu, au)}{J_\nu^2(au) + Y_\nu^2(au)} u F_a(u)\,du$
Kontorovich–Lebedev transform	$F(\tau) = \displaystyle\int_0^\infty K_{i\tau}(x) f(x)\,dx$	$f(x) = \dfrac{2}{\pi^2 x} \displaystyle\int_0^\infty \tau \sinh(\pi\tau) K_{i\tau}(x) F(\tau)\,d\tau$
Meler–Fock transform	$\widetilde{F}(\tau) = \displaystyle\int_1^\infty P_{-\frac{1}{2}+i\tau}(x) f(x)\,dx$	$f(x) = \displaystyle\int_0^\infty \tau \tanh(\pi\tau) P_{-\frac{1}{2}+i\tau}(x) \widetilde{F}(\tau)\,d\tau$
Hilbert transform*	$\widehat{F}(s) = \dfrac{1}{\pi} \displaystyle\int_{-\infty}^\infty \dfrac{f(x)}{x-s}\,dx$	$f(x) = -\dfrac{1}{\pi} \displaystyle\int_{-\infty}^\infty \dfrac{\widehat{F}(s)}{s-x}\,ds$

Notation: $i = \sqrt{-1}$, $J_\mu(x)$ and $Y_\mu(x)$ are the Bessel functions of the first and the second kind, respectively, $I_\mu(x)$ and $K_\mu(x)$ are the modified Bessel functions of the first and the second kind, respectively, $P_\mu(x)$ is the Legendre spherical function of the second kind, and $\mathbf{H}_\mu(x)$ is the Struve function, $\mathbf{H}_\mu(x) = \displaystyle\sum_{j=0}^\infty \dfrac{(-1)^j (x/2)^{\mu+2j+1}}{\Gamma\left(j+\frac{3}{2}\right) \Gamma\left(\mu+j+\frac{3}{2}\right)}$.

* REMARK. In the direct and inverse Hilbert transforms, the integrals are understood in the sense of the Cauchy principal value.

References for Section 7.6: H. Bateman and A. Erdélyi (1954), V. A. Ditkin and A. P. Prudnikov (1965), J. W. Miles (1971), B. Davis (1978), D. Zwillinger (1989), Yu. A. Brychkov and A. P. Prudnikov (1989), W. H. Beyer (1991).

Chapter 8

Methods for Solving Linear Equations of the Form $\int_a^x K(x,t)y(t)\,dt = f(x)$

8.1. Volterra Equations of the First Kind

8.1-1. Equations of the First Kind. Function and Kernel Classes

In this chapter we present methods for solving Volterra linear equations of the first kind. These equations have the form

$$\int_a^x K(x,t)y(t)\,dt = f(x), \tag{1}$$

where $y(x)$ is the unknown function ($a \leq x \leq b$), $K(x,t)$ is the kernel of the integral equation, and $f(x)$ is a given function, the *right-hand side* of Eq. (1). The functions $y(x)$ and $f(x)$ are usually assumed to be continuous or square integrable on $[a,b]$. The kernel $K(x,t)$ is usually assumed either to be continuous on the square $S = \{a \leq x \leq b,\ a \leq t \leq b\}$ or to satisfy the condition

$$\int_a^b \int_a^b K^2(x,t)\,dx\,dt = B^2 < \infty, \tag{2}$$

where B is a constant, that is, to be square integrable on this square. It is assumed in (2) that $K(x,t) \equiv 0$ for $t > x$.

The kernel $K(x,t)$ is said to be *degenerate* if it can be represented in the form $K(x,t) = g_1(x)h_1(t) + \cdots + g_n(x)h_n(t)$.

The kernel $K(x,t)$ of an integral equation is called *difference kernel* if it depends only on the difference of the arguments, $K(x,t) = K(x-t)$.

Polar kernels

$$K(x,t) = \frac{L(x,t)}{(x-t)^\beta} + M(x,t), \qquad 0 < \beta < 1, \tag{3}$$

and logarithmic kernels (kernels with logarithmic singularity)

$$K(x,t) = L(x,t)\ln(x-t) + M(x,t), \tag{4}$$

where $L(x,t)$ and $M(x,t)$ are continuous on S and $L(x,x) \not\equiv 0$, are often considered as well.

Polar and logarithmic kernels form a class of kernels with weak singularity. Equations containing such kernels are called *equations with weak singularity*.

The following *generalized Abel equation* is a special case of Eq. (1) with the kernel of the form (3):

$$\int_a^x \frac{y(t)}{(x-t)^\beta}\,dt = f(x), \qquad 0 < \beta < 1.$$

In case the functions $K(x,t)$ and $f(x)$ are continuous, the right-hand side of Eq. (1) must satisfy the following conditions:

$1°$. If $K(a, a) \neq 0$, then $f(x)$ must be constrained by $f(a) = 0$.

$2°$. If $K(a, a) = K'_x(a, a) = \cdots = K_x^{(n-1)}(a, a) = 0$, $0 < |K_x^{(n)}(a, a)| < \infty$, then the right-hand side of the equation must satisfy the conditions

$$f(a) = f'_x(a) = \cdots = f_x^{(n)}(a) = 0.$$

$3°$. If $K(a, a) = K'_x(a, a) = \cdots = K_x^{(n-1)}(a, a) = 0$, $K_x^{(n)}(a, a) = \infty$, then the right-hand side of the equation must satisfy the conditions

$$f(a) = f'_x(a) = \cdots = f_x^{(n-1)}(a) = 0.$$

For polar kernels of the form (4) and continuous $f(x)$, no additional conditions are imposed on the right-hand side of the integral equation.

Remark 1. Generally, the case in which the integration limit a is infinite is not excluded.

8.1-2. Existence and Uniqueness of a Solution

Assume that in Eq. (1) the functions $f(x)$ and $K(x, t)$ are continuous together with their first derivatives on $[a, b]$ and on S, respectively. If $K(x, x) \neq 0$ ($x \in [a, b]$) and $f(a) = 0$, then there exists a unique continuous solution $y(x)$ of Eq. (1).

Remark 2. The problem of existence and uniqueness of a solution to a Volterra equation of the first kind is closely related to conditions under which this equation can be reduced to Volterra equations of the second kind (see Section 8.3).

Remark 3. A Volterra equation of the first kind can be treated as a Fredholm equation of the first kind whose kernel $K(x, t)$ vanishes for $t > x$ (see Chapter 10).

⊙ References for Section 8.1: E. Goursat (1923), H. M. Müntz (1934), F. G. Tricomi (1957), V. Volterra (1959), S. G. Mikhlin (1960), M. L. Krasnov, A. I. Kiselev, and G. I. Makarenko (1971), J. A. Cochran (1972), C. Corduneanu (1973), V. I. Smirnov (1974), P. P. Zabreyko, A. I. Koshelev, et al. (1975), A. J. Jerry (1985), A. F. Verlan' and V. S. Sizikov (1986).

8.2. Equations With Degenerate Kernel: $K(x, t) = g_1(x)h_1(t) + \cdots + g_n(x)h_n(t)$

8.2-1. Equations With Kernel of the Form $K(x, t) = g_1(x)h_1(t) + g_2(x)h_2(t)$

Any equation of this type can be rewritten in the form

$$g_1(x) \int_a^x h_1(t)y(t)\,dt + g_2(x) \int_a^x h_2(t)y(t)\,dt = f(x). \tag{1}$$

It is assumed that $g_1(x) \neq \text{const}\, g_2(x)$, $h_1(t) \neq \text{const}\, h_2(t)$, $0 < g_1^2(a) + g_2^2(a) < \infty$, and $f(a) = 0$.

The change of variables

$$u(x) = \int_a^x h_1(t)y(t)\,dt \tag{2}$$

followed by the integration by parts in the second integral in (1) with regard to the relation $u(a) = 0$ yields the following Volterra equation of the second kind:

$$[g_1(x)h_1(x) + g_2(x)h_2(x)]u(x) - g_2(x)h_1(x) \int_a^x \left[\frac{h_2(t)}{h_1(t)}\right]'_t u(t)\,dt = h_1(x)f(x). \tag{3}$$

The substitution

$$w(x) = \int_a^x \left[\frac{h_2(t)}{h_1(t)}\right]'_t u(t)\, dt \tag{4}$$

reduces Eq. (3) to the first-order linear ordinary differential equation

$$[g_1(x)h_1(x) + g_2(x)h_2(x)]w'_x - g_2(x)h_1(x)\left[\frac{h_2(x)}{h_1(x)}\right]'_x w = f(x)h_1(x)\left[\frac{h_2(x)}{h_1(x)}\right]'_x. \tag{5}$$

1°. In the case $g_1(x)h_1(x) + g_2(x)h_2(x) \not\equiv 0$, the solution of equation (5) satisfying the condition $w(a) = 0$ (this condition is a consequence of the substitution (4)) has the form

$$w(x) = \Phi(x)\int_a^x \left[\frac{h_2(t)}{h_1(t)}\right]'_t \frac{f(t)h_1(t)\, dt}{\Phi(t)[g_1(t)h_1(t) + g_2(t)h_2(t)]}, \tag{6}$$

$$\Phi(x) = \exp\left\{\int_a^x \left[\frac{h_2(t)}{h_1(t)}\right]'_t \frac{g_2(t)h_1(t)\, dt}{g_1(t)h_1(t) + g_2(t)h_2(t)}\right\}. \tag{7}$$

Let us differentiate relation (4) and substitute the function (6) into the resulting expression. After integrating by parts with regard to the relations $f(a) = 0$ and $w(a) = 0$, for $f \not\equiv \text{const}\, g_2$ we obtain

$$u(x) = \frac{g_2(x)h_1(x)\Phi(x)}{g_1(x)h_1(x) + g_2(x)h_2(x)}\int_a^x \left[\frac{f(t)}{g_2(t)}\right]'_t \frac{dt}{\Phi(t)}.$$

Using formula (2), we find a solution of the original equation in the form

$$y(x) = \frac{1}{h_1(x)}\frac{d}{dx}\left\{\frac{g_2(x)h_1(x)\Phi(x)}{g_1(x)h_1(x) + g_2(x)h_2(x)}\int_a^x \left[\frac{f(t)}{g_2(t)}\right]'_t \frac{dt}{\Phi(t)}\right\}, \tag{8}$$

where the function $\Phi(x)$ is given by (7).

If $f(x) \equiv \text{const}\, g_2(x)$, the solution is given by formulas (8) and (7) in which the subscript 1 must be changed by 2 and vice versa.

2°. In the case $g_1(x)h_1(x) + g_2(x)h_2(x) \equiv 0$, the solution has the form

$$y(x) = \frac{1}{h_1}\frac{d}{dx}\left[\frac{(f/g_2)'_x}{(g_1/g_2)'_x}\right] = -\frac{1}{h_1}\frac{d}{dx}\left[\frac{(f/g_2)'_x}{(h_2/h_1)'_x}\right].$$

8.2-2. Equations With General Degenerate Kernel

A Volterra equation of the first kind with general degenerate kernel has the form

$$\sum_{m=1}^n g_m(x)\int_a^x h_m(t)y(t)\, dt = f(x). \tag{9}$$

Using the notation

$$w_m(x) = \int_a^x h_m(t)y(t)\, dt, \qquad m = 1, \ldots, n, \tag{10}$$

we can rewrite Eq. (9) as follows:

$$\sum_{m=1}^n g_m(x)w_m(x) = f(x). \tag{11}$$

On differentiating formulas (10) and eliminating $y(x)$ from the resulting equations, we arrive at the following linear differential equations for the functions $w_m = w_m(x)$:

$$h_1(x)w_m' = h_m(x)w_1', \qquad m = 2, \ldots, n, \tag{12}$$

(the prime stands for the derivative with respect to x) with the initial conditions

$$w_m(a) = 0, \qquad m = 1, \ldots, n.$$

Any solution of system (11), (12) determines a solution of the original integral equation (9) by each of the expressions

$$y(x) = \frac{w_m'(x)}{h_m(x)}, \qquad m = 1, \ldots, n,$$

which can be obtained by differentiating formula (10).

System (11), (12) can be reduced to a linear differential equation of order $n - 1$ for any function $w_m(x)$ $(m = 1, \ldots, n)$ by multiple differentiation of Eq. (11) with regard to (12).

⊙ References for Section 8.2: E. Goursat (1923), A. F. Verlan' and V. S. Sizikov (1986).

8.3. Reduction of Volterra Equations of the First Kind to Volterra Equations of the Second Kind

8.3-1. The First Method

Suppose that the kernel and the right-hand side of the equation

$$\int_a^x K(x,t)y(t)\,dt = f(x), \tag{1}$$

have continuous derivatives with respect to x and that the condition $K(x,x) \not\equiv 0$ holds. In this case, after differentiating relation (1) and dividing the resulting expression by $K(x,x)$ we arrive at the following Volterra equation of the second kind:

$$y(x) + \int_a^x \frac{K_x'(x,t)}{K(x,x)}y(t)\,dt = \frac{f_x'(x)}{K(x,x)}. \tag{2}$$

Equations of this type are considered in Chapter 9. If $K(x,x) \equiv 0$, then, on differentiating Eq. (1) with respect to x twice and assuming that $K_x'(x,t)|_{t=x} \not\equiv 0$, we obtain the Volterra equation of the second kind

$$y(x) + \int_a^x \frac{K_{xx}''(x,t)}{K_x'(x,t)|_{t=x}}y(t)\,dt = \frac{f_{xx}''(x)}{K_x'(x,t)|_{t=x}}.$$

If $K_x'(x,x) \equiv 0$, we can again apply differentiation, and so on. If the first $m - 2$ partial derivatives of the kernel with respect to x are identically zero and the $(m-1)$st derivative is nonzero, then the m-fold differentiation of the original equation gives the following Volterra equation of the second kind:

$$y(x) + \int_a^x \frac{K_x^{(m)}(x,t)}{K_x^{(m-1)}(x,t)|_{t=x}}y(t)\,dt = \frac{f_x^{(m)}(x)}{K_x^{(m-1)}(x,t)|_{t=x}}.$$

8.3-2. The Second Method

Let us introduce the new variable

$$Y(x) = \int_a^x y(t)\, dt$$

and integrate the right-hand side of Eq. (1) by parts taking into account the relation $f(a) = 0$. After dividing the resulting expression by $K(x, x)$, we arrive at the Volterra equation of the second kind

$$Y(x) - \int_a^x \frac{K_t'(x, t)}{K(x, x)} Y(t)\, dt = \frac{f(x)}{K(x, x)},$$

for which the condition $K(x, x) \not\equiv 0$ must hold.

⊙ References for Section 8.3: E. Goursat (1923), V. Volterra (1959).

8.4. Equations With Difference Kernel: $K(x, t) = K(x - t)$

8.4-1. A Solution Method Based on the Laplace Transform

Volterra equations of the first kind with kernel depending on the difference of the arguments have the form

$$\int_0^x K(x - t) y(t)\, dt = f(x). \tag{1}$$

To solve these equations, the Laplace transform can be used (see Section 7.2). In what follows we need the transforms of the kernel and the right-hand side; they are given by the formulas

$$\tilde{K}(p) = \int_0^\infty K(x) e^{-px}\, dx, \quad \tilde{f}(p) = \int_0^\infty f(x) e^{-px}\, dx. \tag{2}$$

Applying the Laplace transform $\mathfrak{L}$ to Eq. (1) and taking into account the fact that an integral with kernel depending on the difference of the arguments is transformed to the product by the rule (see Subsection 7.2-3)

$$\mathfrak{L}\left\{ \int_0^x K(x - t) y(t)\, dt \right\} = \tilde{K}(p) \tilde{y}(p),$$

we obtain the following equation for the transform $\tilde{y}(p)$:

$$\tilde{K}(p) \tilde{y}(p) = \tilde{f}(p). \tag{3}$$

The solution of Eq. (3) is given by the formula

$$\tilde{y}(p) = \frac{\tilde{f}(p)}{\tilde{K}(p)}. \tag{4}$$

On applying the Laplace inversion formula (if it is applicable) to (4), we obtain a solution of Eq. (1) in the form

$$y(x) = \frac{1}{2\pi i} \int_{c-i\infty}^{c+i\infty} \frac{\tilde{f}(p)}{\tilde{K}(p)} e^{px}\, dp. \tag{5}$$

When applying formula (5) in practice, the following two technical problems occur:

1°. Finding the transform $\tilde{K}(p) = \int_0^\infty K(x) e^{-px}\, dx$ for a given kernel $K(x)$.

2°. Finding the resolvent (5) whose transform $\tilde{R}(p)$ is given by formula (4).

To calculate the corresponding integrals, tables of direct and inverse Laplace transforms can be applied (see Supplements 4 and 5), and, in many cases, to find the inverse transform, methods of the theory of functions of a complex variable are applied, including the Cauchy residue theorem (see Subsection 7.1-4).

Remark. If the lower limit in the integral of a Volterra equation with difference kernel is a, then this equation can be reduced to Eq. (1) by means of the change of variables $x = \bar{x} - a$, $t = \bar{t} - a$.

8.4-2. The Case in Which the Transform of the Solution is a Rational Function

Consider the important special case in which the transform (4) of the solution is a rational function of the form

$$\tilde{y}(p) = \frac{\tilde{f}(p)}{\tilde{K}(p)} \equiv \frac{R(p)}{Q(p)},$$

where $Q(p)$ and $R(p)$ are polynomials in the variable p and the degree of $Q(p)$ exceeds that of $R(p)$.

If the zeros of the denominator $Q(p)$ are simple, i.e.,

$$Q(p) \equiv \mathrm{const}\,(p - \lambda_1)(p - \lambda_2)\ldots(p - \lambda_n),$$

and $\lambda_i \neq \lambda_j$ for $i \neq j$, then the solution has the form

$$y(x) = \sum_{k=1}^{n} \frac{R(\lambda_k)}{Q'(\lambda_k)} \exp(\lambda_k x),$$

where the prime stands for the derivatives.

Example 1. Consider the Volterra integral equation of the first kind

$$\int_0^x e^{-a(x-t)} y(t)\,dt = A\sinh(bx).$$

We apply the Laplace transform to this equation and obtain (see Supplement 4)

$$\frac{1}{p+a}\tilde{y}(p) = \frac{Ab}{p^2 - b^2}.$$

This implies

$$\tilde{y}(p) = \frac{Ab(p+a)}{p^2 - b^2} = \frac{Ab(p+a)}{(p-b)(p+b)}.$$

We have $Q(p) = (p-b)(p+b)$, $R(p) = Ab(p+a)$, $\lambda_1 = b$, and $\lambda_2 = -b$. Therefore, the solution of the integral equation has the form

$$y(x) = \tfrac{1}{2}A(b+a)e^{bx} + \tfrac{1}{2}A(b-a)e^{-bx} = Aa\sinh(bx) + Ab\cosh(bx).$$

8.4-3. Convolution Representation of a Solution

In solving Volterra integral equations of the first kind with difference kernel $K(x-t)$ by means of the Laplace transform, it is sometimes useful to apply the following approach.

Let us represent the transform (4) of a solution in the form

$$\tilde{y}(p) = \tilde{N}(p)\tilde{M}(p)\tilde{f}(p), \qquad \tilde{N}(p) \equiv \frac{1}{\tilde{K}(p)\tilde{M}(p)}. \tag{6}$$

If we can find a function $\tilde{M}(p)$ for which the inverse transforms

$$\mathcal{L}^{-1}\{\tilde{M}(p)\} = M(x), \qquad \mathcal{L}^{-1}\{\tilde{N}(p)\} = N(x) \tag{7}$$

exist and can be found in a closed form, then the solution can be written as the convolution

$$y(x) = \int_0^x N(x-t)F(t)\,dt, \qquad F(t) = \int_0^t M(t-s)f(s)\,ds. \tag{8}$$

Example 2. Consider the equation

$$\int_0^x \sin\!\left(k\sqrt{x-t}\,\right)y(t)\,dt = f(x), \qquad f(0) = 0. \tag{9}$$

Applying the Laplace transform, we obtain (see Supplement 4)

$$\tilde{y}(p) = \frac{2}{\sqrt{\pi}\,k} p^{3/2} \exp(\alpha/p) \tilde{f}(p), \qquad \alpha = \tfrac{1}{4}k^2. \tag{10}$$

Let us rewrite the right-hand side of (10) in the equivalent form

$$\tilde{y}(p) = \frac{2}{\sqrt{\pi}\,k} p^2 \left[p^{-1/2} \exp(\alpha/p) \right] \tilde{f}(p), \qquad \alpha = \tfrac{1}{4}k^2, \tag{11}$$

where the factor in the square brackets corresponds to $\tilde{M}(p)$ in formula (6) and $\tilde{N}(p) = \text{const}\, p^2$.

By applying the Laplace inversion formula according to the above scheme to formula (11) with regard to the relation (see Supplement 5)

$$\mathcal{L}^{-1}\{p^2 \tilde{\varphi}(p)\} = \frac{d^2}{dx^2} \varphi(x), \qquad \mathcal{L}^{-1}\{p^{-1/2} \exp(\alpha/p)\} = \frac{1}{\sqrt{\pi x}} \cosh(k\sqrt{x}),$$

we find the solution

$$y(x) = \frac{2}{\pi k} \frac{d^2}{dx^2} \int_0^x \frac{\cosh(k\sqrt{x-t})}{\sqrt{x-t}} f(t)\, dt.$$

8.4-4. Application of an Auxiliary Equation

Consider the equation

$$\int_a^x K(x - t) y(t)\, dt = f(x), \tag{12}$$

where the kernel $K(x)$ has an integrable singularity at $x = 0$.

Let $w = w(x)$ be the solution of the simpler auxiliary equation with $f(x) \equiv 1$ and $a = 0$,

$$\int_0^x K(x - t) w(t)\, dt = 1. \tag{13}$$

Then the solution of the original equation (12) with arbitrary right-hand side can be expressed as follows via the solution of the auxiliary equation (13):

$$y(x) = \frac{d}{dx} \int_a^x w(x - t) f(t)\, dt = f(a) w(x - a) + \int_a^x w(x - t) f_t'(t)\, dt. \tag{14}$$

Example 3. Consider the generalized Abel equation

$$\int_a^x \frac{y(t)\, dt}{(x - t)^\mu} = f(x), \qquad 0 < \mu < 1. \tag{15}$$

We seek a solution of the corresponding auxiliary equation

$$\int_0^x \frac{w(t)\, dt}{(x - t)^\mu} = 1, \qquad 0 < \mu < 1, \tag{16}$$

by the method of indeterminate coefficients in the form

$$w(x) = Ax^\beta. \tag{17}$$

Let us substitute (17) into (15) and then perform the change of variable $t = x\xi$ in the integral. Taking into account the relationship

$$B(p, q) = \int_0^1 \xi^{p-1}(1 - \xi)^{1-q}\, d\xi = \frac{\Gamma(p)\Gamma(q)}{\Gamma(p + q)}$$

between the beta and gamma functions, we obtain

$$A \frac{\Gamma(\beta + 1)\Gamma(1 - \mu)}{\Gamma(2 + \beta - \mu)} x^{\beta + 1 - \mu} = 1.$$

From this relation we find the coefficients A and β:

$$\beta = \mu - 1, \qquad A = \frac{1}{\Gamma(\mu)\Gamma(1 - \mu)} = \frac{\sin(\pi\mu)}{\pi}. \tag{18}$$

Formulas (17) and (18) define the solution of the auxiliary equation (16) and make it possible to find the solution of the generalized Abel equation (15) by means of formula (14) as follows:

$$y(x) = \frac{\sin(\pi\mu)}{\pi} \frac{d}{dx} \int_a^x \frac{f(t)\, dt}{(x - t)^{1-\mu}} = \frac{\sin(\pi\mu)}{\pi} \left[\frac{f(a)}{(x - a)^{1-\mu}} + \int_a^x \frac{f_t'(t)\, dt}{(x - t)^{1-\mu}} \right]. \tag{19}$$

8.4-5. Reduction to Ordinary Differential Equations

Consider the special case in which the transform of the kernel of the integral equation (1) can be represented in the form

$$\tilde{K}(p) = \frac{M(p)}{N(p)}, \tag{20}$$

where $M(p)$ and $N(p)$ are some polynomials of degrees m and n, respectively:

$$M(p) = \sum_{k=0}^{m} A_k p^k, \quad N(p) = \sum_{k=0}^{n} B_k p^k. \tag{21}$$

In this case, the solution of the integral equation (1) (if it exists) satisfies the following linear nonhomogeneous ordinary differential equation of order m with constant coefficients:

$$\sum_{k=0}^{m} A_k y_x^{(k)}(x) = \sum_{k=0}^{n} B_k f_x^{(k)}(x). \tag{22}$$

We can rewrite Eq. (22) in the operator form

$$M(D)y(x) = N(D)f(x), \qquad D \equiv \frac{d}{dx}.$$

The initial data for the differential equation (22), as well as the conditions that must be imposed on the right-hand side of the integral equation (1), can be obtained from the relation

$$\sum_{k=0}^{m} A_k \sum_{s=0}^{k-1} p^{k-1-s} y_x^{(s)}(0) - \sum_{k=0}^{n} B_k \sum_{s=0}^{k-1} p^{k-1-s} f_x^{(s)}(0) = 0 \tag{23}$$

by matching the coefficients of like powers of the parameter p.

The proof of this assertion can be given by applying the Laplace transform to the differential equation (22) followed by comparing the resulting expression with Eq. (3) with regard to (20).

8.4-6. Reduction of a Volterra Equation to a Wiener–Hopf Equation

A Volterra equation of the first kind with difference kernel of the form

$$\int_0^x K(x-t)y(t)\, dt = f(x), \qquad 0 < x < \infty, \tag{24}$$

can be reduced to the following Wiener–Hopf equation of the first kind:

$$\int_0^\infty K_+(x-t)y(t)\, dt = f(x), \qquad 0 < x < \infty, \tag{25}$$

where the kernel $K_+(x-t)$ is given by

$$K_+(s) = \begin{cases} K(s) & \text{for } s > 0, \\ 0 & \text{for } s < 0. \end{cases}$$

Methods for solving Eq. (25) are presented in Chapter 10.

⊙ References for Section 8.4: G. Doetsch (1956), V. A. Ditkin and A. P. Prudnikov (1965), M. L. Krasnov, A. I. Kiselev, and G. I. Makarenko (1971), V. I. Smirnov (1974), P. P. Zabreyko, A. I. Koshelev, et al. (1975), F. D. Gakhov and Yu. I. Cherskii (1978).

8.5. Method of Fractional Differentiation

8.5-1. The Definition of Fractional Integrals

A function $f(x)$ is said to be *absolutely continuous* on a closed interval $[a, b]$ if for each $\varepsilon > 0$ there exists a $\delta > 0$ such that for any finite system of disjoint intervals $[a_k, b_k] \subset [a, b]$, $k = 1, \ldots, n$, such that $\sum_{k=1}^{n} (b_k - a_k) < \delta$ the inequality $\sum_{k=1}^{n} |f(b_k) - f(a_k)| < \varepsilon$ holds. The class of all these functions is denoted by AC.

Let AC^n, $n = 1, 2, \ldots$, be the class of functions $f(x)$ that are continuously differentiable on $[a, b]$ up to the order $n - 1$ and for which $f^{(n-1)}(x) \in AC$.

Let $\varphi(x) \in L_1(a, b)$. The integrals

$$\mathbf{I}_{a+}^{\mu} \varphi(x) \equiv \frac{1}{\Gamma(\mu)} \int_a^x \frac{\varphi(t)}{(x - t)^{1-\mu}} \, dt, \qquad x > a, \tag{1}$$

$$\mathbf{I}_{b-}^{\mu} \varphi(x) \equiv \frac{1}{\Gamma(\mu)} \int_x^b \frac{\varphi(t)}{(t - x)^{1-\mu}} \, dt, \qquad x < b, \tag{2}$$

where $\mu > 0$, are called the *integrals of fractional order* μ. Sometimes the integral (1) is called *left-sided* and the integral (2) is called *right-sided*. The operators $\mathbf{I}_{a+}^{\mu}$ and $\mathbf{I}_{b-}^{\mu}$ are called the *operators of fractional integration*.

The integrals (1) and (2) are usually called the *Riemann–Liouville fractional integrals*.

The following formula holds:

$$\int_a^b \varphi(x) \mathbf{I}_{a+}^{\mu} \psi(x) \, dx = \int_a^b \psi(x) \mathbf{I}_{b-}^{\mu} \varphi(x) \, dx, \tag{3}$$

which is sometimes called the *formula of fractional integration by parts*.

Fractional integration has the property

$$\mathbf{I}_{a+}^{\mu} \mathbf{I}_{a+}^{\beta} \varphi(x) = \mathbf{I}_{a+}^{\mu+\beta} \varphi(x), \quad \mathbf{I}_{b-}^{\mu} \mathbf{I}_{b-}^{\beta} \varphi(x) = \mathbf{I}_{b-}^{\mu+\beta} \varphi(x), \qquad \mu > 0, \quad \beta > 0. \tag{4}$$

Property (4) is called the *semigroup property of fractional integration*.

8.5-2. The Definition of Fractional Derivatives

It is natural to introduce fractional differentiation as the operation inverse to fractional integration. For a function $f(x)$ defined on a closed interval $[a, b]$, the expressions

$$\mathbf{D}_{a+}^{\mu} f(x) = \frac{1}{\Gamma(1 - \mu)} \frac{d}{dx} \int_a^x \frac{f(t)}{(x - t)^{\mu}} \, dt, \tag{5}$$

$$\mathbf{D}_{b-}^{\mu} f(x) = -\frac{1}{\Gamma(1 - \mu)} \frac{d}{dx} \int_x^b \frac{f(t)}{(t - x)^{\mu}} \, dt \tag{6}$$

are called the *left* and the *right fractional derivative of order* μ, respectively. It is assumed here that $0 < \mu < 1$.

The fractional derivatives (5) and (6) are usually called the *Riemann–Liouville derivatives*.

Note that the fractional integrals are defined for any order $\mu > 0$, but the fractional derivatives are so far defined only for $0 < \mu < 1$.

If $f(x) \in AC$, then the derivatives $\mathbf{D}_{a+}^{\mu} f(x)$ and $\mathbf{D}_{b-}^{\mu} f(x)$, $0 < \mu < 1$, exist almost everywhere, and we have $\mathbf{D}_{a+}^{\mu} f(x) \in L_r(a,b)$ and $\mathbf{D}_{b-}^{\mu} f(x) \in L_r(a,b)$, $1 \leq r < 1/\mu$. These derivatives have the representations

$$\mathbf{D}_{a+}^{\mu} f(x) = \frac{1}{\Gamma(1-\mu)} \left[\frac{f(a)}{(x-a)^{\mu}} + \int_a^x \frac{f_t'(t)}{(x-t)^{\mu}}\,dt \right], \tag{7}$$

$$\mathbf{D}_{b-}^{\mu} f(x) = \frac{1}{\Gamma(1-\mu)} \left[\frac{f(b)}{(b-x)^{\mu}} - \int_x^b \frac{f_t'(t)}{(t-x)^{\mu}}\,dt \right]. \tag{8}$$

Finally, let us pass to the fractional derivatives of order $\mu \geq 1$. We shall use the following notation: $[\mu]$ stands for the integral part of a real number μ and $\{\mu\}$ is the fractional part of μ, $0 \leq \{\mu\} < 1$, so that

$$\mu = [\mu] + \{\mu\}. \tag{9}$$

If μ is an integer, then by the fractional derivative of order μ we mean the ordinary derivative

$$\mathbf{D}_{a+}^{\mu} = \left(\frac{d}{dx} \right)^{\mu}, \quad \mathbf{D}_{b-}^{\mu} = \left(-\frac{d}{dx} \right)^{\mu}, \qquad \mu = 1, 2, \ldots \tag{10}$$

However, if μ is not integral, then $\mathbf{D}_{a+}^{\mu} f$ and $\mathbf{D}_{b-}^{\mu} f$ are introduced by the formulas

$$\mathbf{D}_{a+}^{\mu} f(x) \equiv \left(\frac{d}{dx} \right)^{[\mu]} \mathbf{D}_{a+}^{\{\mu\}} f(x) = \left(\frac{d}{dx} \right)^{[\mu]+1} \mathbf{I}_{a+}^{1-\{\mu\}} f(x), \tag{11}$$

$$\mathbf{D}_{b-}^{\mu} f(x) \equiv \left(-\frac{d}{dx} \right)^{[\mu]} \mathbf{D}_{b-}^{\{\mu\}} f(x) = \left(-\frac{d}{dx} \right)^{[\mu]+1} \mathbf{I}_{b-}^{1-\{\mu\}} f(x). \tag{12}$$

Thus,

$$\mathbf{D}_{a+}^{\mu} f(x) = \frac{1}{\Gamma(n-\mu)} \left(\frac{d}{dx} \right)^n \int_a^x \frac{f(t)}{(x-t)^{\mu-n+1}}\,dt, \qquad n = [\mu] + 1, \tag{13}$$

$$\mathbf{D}_{b-}^{\mu} f(x) = \frac{(-1)^n}{\Gamma(n-\mu)} \left(\frac{d}{dx} \right)^n \int_x^b \frac{f(t)}{(t-x)^{\mu-n+1}}\,dt, \qquad n = [\mu] + 1. \tag{14}$$

A sufficient condition for the existence of the derivatives (13) and (14) is as follows:

$$\int_a^x \frac{f(t)\,dt}{(x-t)^{\{\mu\}}} \in AC^{[\mu]}.$$

This sufficient condition holds whenever $f(x) \in AC^{[\mu]}$.

Remark. The definitions of the fractional integrals and fractional derivatives can be extended to the case of complex μ (e.g., see S. G. Samko, A. A. Kilbas, and O. I. Marichev (1993)).

Let $\mathbf{I}_{a+}^{\mu}(L_1)$, $\mu > 0$, be the class of functions $f(x)$ that can be represented by the left fractional integral of order μ of an integrable function: $f(x) = \mathbf{I}_{a+}^{\mu} \varphi(x)$, $\varphi(x) \in L_1(a,b)$, $1 \leq p < \infty$.

For the relation $f(x) \in \mathbf{I}_{a+}^{\mu}(L_1)$, $\mu > 0$, to hold, it is necessary and sufficient that

$$f_{n-\mu}(x) \equiv \mathbf{I}_{a+}^{n-\mu} f \in AC^n, \tag{15}$$

where $n = [\mu] + 1$, and*

$$f_{n-\mu}^{(k)}(a) = 0, \qquad k = 0, 1, \ldots, n - 1. \tag{16}$$

Let $\mu > 0$. We say that a function $f(x) \in L_1$ has an *integrable fractional derivative* $\mathbf{D}_{a+}^{\mu} f$ if $\mathbf{I}_{a+}^{n-\mu} f(x) \in AC^n$, where $n = [\mu] + 1$.

In other words, this definition introduces a notion involving only the first of the two conditions (15) and (16) describing the class $\mathbf{I}_{a+}^{\mu}(L_1)$.

Let $\mu > 0$. In this case the relation

$$\mathbf{D}_{a+}^{\mu} \mathbf{I}_{a+}^{\mu} \varphi(x) = \varphi(x) \tag{17}$$

holds for any integrable function $\varphi(x)$, and the relation

$$\mathbf{I}_{a+}^{\mu} \mathbf{D}_{a+}^{\mu} f(x) = f(x) \tag{18}$$

holds for any function $f(x)$ such that

$$f(x) \in \mathbf{I}_{a+}^{\mu}(L_1). \tag{19}$$

If we replace (19) by the condition that the function $f(x) \in L_1(a, b)$ has an integrable derivative $\mathbf{D}_{a+}^{\mu} f(x)$, then relation (18) fails in general and must be replaced by the formula

$$\mathbf{I}_{a+}^{\mu} \mathbf{D}_{a+}^{\mu} f(x) = f(x) - \sum_{k=0}^{n-1} \frac{(x-a)^{\mu-k-1}}{\Gamma(\mu-k)} f_{n-\mu}^{(n-k-1)}(a), \tag{20}$$

where $n = [\mu] + 1$ and $f_{n-\mu}(x) = \mathbf{I}_{a+}^{n-\mu} f(x)$. In particular, for $0 < \mu < 1$ we have

$$\mathbf{I}_{a+}^{\mu} \mathbf{D}_{a+}^{\mu} f(x) = f(x) - \frac{f_{1-\mu}(a)}{\Gamma(\mu)} (x-a)^{\mu-1}. \tag{21}$$

8.5-4. The Solution of the Generalized Abel Equation

Consider the Abel integral equation

$$\int_a^x \frac{y(t)}{(x-t)^{\mu}} \, dt = f(x), \tag{22}$$

where $0 < \mu < 1$. Suppose that $x \in [a, b]$, $f(x) \in AC$, and $y(t) \in L_1$, and apply the technique of fractional differentiation. We divide Eq. (22) by $\Gamma(1 - \mu)$, and, by virtue of (1), rewrite this equation as follows:

$$\mathbf{I}_{a+}^{1-\mu} y(x) = \frac{f(x)}{\Gamma(1 - \mu)}, \qquad x > a. \tag{23}$$

Let us apply the operator of fractional differentiation $\mathbf{D}_{a+}^{1-\mu}$ to (23). Using the properties of the operators of fractional integration and differentiation, we obtain

$$y(x) = \frac{\mathbf{D}_{a+}^{1-\mu} f(x)}{\Gamma(1 - \mu)}, \tag{24}$$

---

* From now on in Section 8.5, by $f^{(n)}(x)$ we mean the nth derivative of $f(x)$ with respect to x and $f^{(n)}(a) \equiv f^{(n)}(x)\big|_{x=a}$.

or, in the detailed notation,

$$y(x) = \frac{1}{\Gamma(\mu)\Gamma(1-\mu)}\left[\frac{f(a)}{(x-a)^{1-\mu}} + \int_a^x \frac{f'_t(t)}{(x-t)^{1-\mu}}\,dt\right]. \tag{25}$$

Taking into account the relation

$$\frac{1}{\Gamma(\mu)\Gamma(1-\mu)} = \frac{\sin(\pi\mu)}{\pi},$$

we now arrive at the solution of the generalized Abel equation in the form

$$y(x) = \frac{\sin(\pi\mu)}{\pi}\left[\frac{f(a)}{(x-a)^{1-\mu}} + \int_a^x \frac{f'_t(t)\,dt}{(x-t)^{1-\mu}}\right], \tag{26}$$

which coincides with that obtained above in Subsection 8.4-4.

⊙ References for Section 8.5: K. B. Oldham and J. Spanier (1974), Yu. I. Babenko (1986), S. G. Samko, A. A. Kilbas, and O. I. Marichev (1993).

8.6. Equations With Weakly Singular Kernel

8.6-1. A Method of Transformation of the Kernel

Consider the Volterra integral equation of the first kind with polar kernel

$$K(x, t) = \frac{L(x, t)}{(x-t)^\alpha}, \qquad 0 < \alpha < 1. \tag{1}$$

The integral equation in question can be represented in the form

$$\int_0^x \frac{L(x, t)}{(x-t)^\alpha}y(t)\,dt = f(x), \tag{2}$$

where we assume that the functions $L(x, t)$ and $\partial L(x, t)/\partial x$ are continuous and bounded. To solve Eq. (2), we multiply it by $dx/(\xi - x)^{1-\alpha}$ and integrate from 0 to ξ, thus obtaining

$$\int_0^\xi \left[\int_0^x \frac{L(x, t)}{(x-t)^\alpha}y(t)\,dt\right]\frac{dx}{(\xi-x)^{1-\alpha}} = \int_0^\xi \frac{f(x)\,dx}{(\xi-x)^{1-\alpha}}.$$

By setting

$$K^*(\xi, t) = \int_t^\xi \frac{L(x, t)\,dx}{(\xi-x)^{1-\alpha}(x-t)^\alpha},$$

$$\varphi(\xi) = \int_0^\xi \frac{f(x)\,dx}{(\xi-x)^{1-\alpha}}, \qquad \varphi(0) = 0,$$

we obtain another integral equation of the first kind with the unknown function $y(t)$:

$$\int_0^\xi K^*(\xi, t)y(t)\,dt = \varphi(\xi), \tag{3}$$

in which the kernel $K^*(\xi, t)$ has no singularities.

It can be shown that any solution of Eq. (3) is a solution of Eq. (2). Thus, after transforming Eq. (2) to the form (3), we can apply any methods available for continuous kernels to the latter equation.

8.6-2. Kernel With Logarithmic Singularity

Consider the equation

$$\int_0^x \ln(x-t)y(t)\,dt = f(x), \qquad f(0) = 0. \tag{4}$$

Let us apply the Laplace transform to solve this equation. Note that

$$\mathcal{L}\left\{x^\nu\right\} = \int_0^\infty e^{-px}x^\nu\,dx = \frac{\Gamma(\nu+1)}{p^{\nu+1}}, \qquad \nu > -1. \tag{5}$$

Let us differentiate relation (5) with respect to ν. We obtain

$$\mathcal{L}\left\{x^\nu \ln x\right\} = \frac{\Gamma(\nu+1)}{p^{\nu+1}}\left[\frac{\Gamma'_\nu(\nu+1)}{\Gamma(\nu+1)} + \ln\frac{1}{p}\right]. \tag{6}$$

For $\nu = 0$, it follows from (6) that

$$\frac{\Gamma'_\nu(1)}{\Gamma(1)} = -\mathcal{C},$$

where $\mathcal{C} = 0.5772\ldots$ is the Euler constant. With regard to the last relation, formula (6) becomes

$$\mathcal{L}\left\{\ln x\right\} = -\frac{\ln p + \mathcal{C}}{p}. \tag{7}$$

Applying the Laplace transform to Eq. (4) and taking into account (7), we obtain

$$-\frac{\ln p + \mathcal{C}}{p}\,\tilde{y}(p) = \tilde{f}(p),$$

and hence

$$\tilde{y}(p) = -\frac{p\tilde{f}(p)}{\ln p + \mathcal{C}}. \tag{8}$$

Now let us express $\tilde{y}(p)$ in the form

$$\tilde{y}(p) = -\frac{p^2\tilde{f}(p) - f'_x(0)}{p(\ln p + \mathcal{C})} - \frac{f'_x(0)}{p(\ln p + \mathcal{C})}. \tag{9}$$

Since $f(0) = 0$, it follows that

$$\mathcal{L}\left\{f''_{xx}(x)\right\} = p^2\tilde{f}(p) - f'_x(0). \tag{10}$$

Let us rewrite formula (5) as

$$\mathcal{L}\left\{\frac{x^\nu}{\Gamma(\nu+1)}\right\} = \frac{1}{p^{\nu+1}} \tag{11}$$

and integrate (11) with respect to ν from 0 to ∞. We obtain

$$\mathcal{L}\left\{\int_0^\infty \frac{x^\nu}{\Gamma(\nu+1)}\,d\nu\right\} = \int_0^\infty \frac{d\nu}{p^{\nu+1}} = \frac{1}{p\ln p}.$$

Applying the scaling formula for the Laplace transform (see Table 1 in Subsection 7.2-5) we see that

$$\mathcal{L}\left\{\int_0^\infty \frac{(x/a)^\nu}{\Gamma(\nu+1)}\,d\nu\right\} = \int_0^\infty \frac{d\nu}{p^{\nu+1}} = \frac{1}{p\ln ap} = \frac{1}{p(\ln p + \ln a)}.$$

We set $a = e^C$ and obtain

$$\mathfrak{L}\left\{\int_0^\infty \frac{x^\nu e^{-C\nu}}{\Gamma(\nu+1)}\,d\nu\right\} = \frac{1}{p(\ln p + C)}. \tag{12}$$

Let us proceed with relation (9). By (12), we have

$$\frac{f_x'(0)}{p(\ln p + C)} = \mathfrak{L}\left\{f_x'(0)\int_0^\infty \frac{x^\nu e^{-C\nu}}{\Gamma(\nu+1)}\,d\nu\right\}. \tag{13}$$

Taking into account (10) and (12), we can regard the first summand on the right-hand side in (9) as a product of transforms. To find this summand itself we apply the convolution theorem:

$$\frac{p^2\tilde{f}(p) - f_x'(0)}{p(\ln p + C)} = \mathfrak{L}\left\{\int_0^x f_{tt}''(t)\int_0^\infty \frac{(x-t)^\nu e^{-C\nu}}{\Gamma(\nu+1)}\,d\nu\,dt\right\}. \tag{14}$$

On the basis of relations (9), (13), and (14) we obtain the solution of the integral equation (4) in the form

$$y(x) = -\int_0^x f_{tt}''(t)\int_0^\infty \frac{(x-t)^\nu e^{-C\nu}}{\Gamma(\nu+1)}\,d\nu\,dt - f_x'(0)\int_0^\infty \frac{x^\nu e^{-C\nu}}{\Gamma(\nu+1)}\,d\nu. \tag{15}$$

⊙ References for Section 8.6: V. Volterra (1959), M. L. Krasnov, A. I. Kiselev, and G. I. Makarenko (1971).

8.7. Method of Quadratures

8.7-1. Quadrature Formulas

The *method of quadratures* is a method for constructing an approximate solution of an integral equation based on the replacement of integrals by finite sums according to some formula. Such formulas are called *quadrature formulas* and, in general, have the form

$$\int_a^b \psi(x)\,dx = \sum_{i=1}^n A_i\psi(x_i) + \varepsilon_n[\psi], \tag{1}$$

where x_i $(i = 1,\ldots,n)$ are the abscissas of the partition points of the integration interval $[a, b]$, or *quadrature (interpolation) nodes*, A_i $(i = 1,\ldots,n)$ are numerical coefficients independent of the choice of the function $\psi(x)$, and $\varepsilon_n[\psi]$ is the remainder (the truncation error) of formula (1). As a rule, $A_i \geq 0$ and $\sum_{i=1}^n A_i = b - a$.

There are quite a few quadrature formulas of the form (1). The following formulas are the simplest and most frequently used in practice.

Rectangle rule:

$$A_1 = A_2 = \cdots = A_{n-1} = h, \quad A_n = 0,$$
$$h = \frac{b-a}{n-1}, \quad x_i = a + h(i-1) \quad (i = 1,\ldots,n). \tag{2}$$

Trapezoidal rule:

$$A_1 = A_n = \tfrac{1}{2}h, \quad A_2 = A_3 = \cdots = A_{n-1} = h,$$
$$h = \frac{b-a}{n-1}, \quad x_i = a + h(i-1) \quad (i = 1,\ldots,n). \tag{3}$$

Simpson's rule (or *prizmoidal formula*):

$$A_1 = A_{2m+1} = \tfrac{1}{3}h, \quad A_2 = \cdots = A_{2m} = \tfrac{4}{3}h, \quad A_3 = \cdots = A_{2m-1} = \tfrac{2}{3}h,$$
$$h = \frac{b-a}{n-1}, \quad x_i = a + h(i-1) \quad (n = 2m+1, \ i = 1,\ldots,n), \tag{4}$$

where m is a positive integer.

In formulas (2)–(4), h is a constant integration step.

The quadrature formulas due to Chebyshev and Gauss with various numbers of interpolation nodes are also widely applied. Let us illustrate these formulas by an example.

Example. For the interval $[-1, 1]$, the parameters in formula (1) acquire the following values:
Chebyshev's formula ($n = 6$):

$$A_1 = A_2 = \cdots = \frac{2}{n} = \frac{1}{3}, \qquad x_1 = -x_6 = -0.8662468181,$$
$$x_2 = -x_5 = -0.4225186538, \qquad x_3 = -x_4 = -0.2666354015. \tag{5}$$

Gauss' formula ($n = 7$):

$$A_1 = A_7 = 0.1294849662, \qquad A_2 = A_6 = 0.2797053915,$$
$$A_3 = A_5 = 0.3818300505, \qquad A_4 = 0.4179591837,$$
$$x_1 = -x_7 = -0.9491079123, \qquad x_2 = -x_6 = -0.7415311856,$$
$$x_3 = -x_5 = -0.4058451514, \qquad x_4 = 0. \tag{6}$$

Note that a vast literature is devoted to quadrature formulas, and the reader can find books of interest (e.g., see G. A. Korn and T. M. Korn (1968), N. S. Bakhvalov (1973), S. M. Nikol'skii (1979)).

8.7-2. The General Scheme of the Method

Let us solve the Volterra integral equation of the first kind

$$\int_a^x K(x, t) y(t)\, dt = f(x), \qquad f(a) = 0, \tag{7}$$

on an interval $a \le x \le b$ by the method of quadratures. The procedure of constructing the solution involves two stages:

$1°$. First, we determine the initial value $y(a)$. To this end, we differentiate Eq. (7) with respect to x, thus obtaining

$$K(x, x) y(x) + \int_a^x K_x'(x, t) y(t)\, dt = f_x'(x).$$

By setting $x = a$, we find that

$$y_1 = y(a) = \frac{f_x'(a)}{K(a, a)} = \frac{f_x'(a)}{K_{11}}.$$

$2°$. Let us choose a constant integration step h and consider the discrete set of points $x_i = a + h(i-1)$, $i = 1, \ldots, n$. For $x = x_i$, Eq. (7) acquires the form

$$\int_a^{x_i} K(x_i, t) y(t)\, dt = f(x_i), \qquad i = 2, \ldots, n, \tag{8}$$

Applying the quadrature formula (1) to the integral in (8) and choosing x_j ($j = 1, \ldots, i$) to be the nodes in t, we arrive at the system of equations

$$\sum_{j=1}^{i} A_{ij} K(x_i, x_j) y(x_j) = f(x_i) + \varepsilon_i[y], \qquad i = 2, \ldots, n, \tag{9}$$

where the A_{ij} are the coefficients of the quadrature formula on the interval $[a, x_i]$ and $\varepsilon_i[y]$ is the truncation error. Assume that the $\varepsilon_i[y]$ are small and neglect them; then we obtain a system of linear algebraic equations in the form

$$\sum_{j=1}^{i} A_{ij} K_{ij} y_j = f_i, \qquad i = 2, \ldots, n, \tag{10}$$

where $K_{ij} = K(x_i, x_j)$ $(j = 1, \ldots, i)$, $f_i = f(x_i)$, and y_j are approximate values of the unknown function at the nodes x_i.

Now system (10) permits one, provided that $A_{ii}K_{ii} \neq 0$ $(i = 2, \ldots, n)$, to successively find the desired approximate values by the formulas

$$y_1 = \frac{f_x'(a)}{K_{11}}, \quad y_2 = \frac{f_2 - A_{21}K_{21}y_1}{A_{22}K_{22}}, \quad \ldots, \quad y_n = \frac{f_n - \sum\limits_{j=1}^{n-1} A_{nj}K_{nj}y_j}{A_{nn}K_{nn}},$$

whose specific form depends on the choice of the quadrature formula.

8.7-3. An Algorithm Based on the Trapezoidal Rule

According to the trapezoidal rule (3), we have

$$A_{i1} = A_{ii} = \tfrac{1}{2}h, \quad A_{i2} = \cdots = A_{i,i-1} = h, \quad i = 2, \ldots, n.$$

The application of the trapezoidal rule in the general scheme leads to the following step algorithm:

$$y_1 = \frac{f_x'(a)}{K_{11}}, \quad f_x'(a) = \frac{-3f_1 + 4f_2 - f_3}{2h},$$

$$y_i = \frac{2}{K_{ii}}\left(\frac{f_i}{h} - \sum_{j=1}^{i-1}\beta_j K_{ij}y_j\right), \quad \beta_j = \begin{cases} \frac{1}{2} & \text{for } j = 1, \\ 1 & \text{for } j > 1, \end{cases} \quad i = 2, \ldots, n,$$

where the notation coincides with that introduced in Subsection 8.7-2. The trapezoidal rule is quite simple and effective and frequently used in practice for solving integral equations with variable limit of integration.

On the basis of Subsections 8.7-1 and 8.7-2, one can write out similar expressions for other quadrature formulas. However, they must be used with care. For example, the application of Simpson's rule must be alternated, for odd nodes, with some other rule, e.g., the rectangle rule or the trapezoidal rule. For equations with variable integration limit, the use of Chebyshev's formula or Gauss' formula also has some difficulties as well.

8.7-4. An Algorithm for an Equation With Degenerate Kernel

A general property of the algorithms of the method of quadratures in the solution of the Volterra equations of the first kind with arbitrary kernel is that the amount of computational work at each step is proportional to the number of the step: all operations of the previous step are repeated with new data and another term in the sum is added.

However, if the kernel in Eq. (7) is degenerate, i.e.,

$$K(x, t) = \sum_{k=1}^m p_k(x)q_k(t), \tag{11}$$

or if the kernel under consideration can be approximated by a degenerate kernel, then an algorithm can be constructed for which the number of operations does not depend on the index of the digitalization node. With regard to (11), Eq. (7) becomes

$$\sum_{k=1}^m p_k(x)\int_a^x q_k(t)y(t)\,dt = f(x). \tag{12}$$

By applying the trapezoidal rule to (12), we obtain recurrent expressions for the solution of the equation (see formulas in Subsection 8.7-3):

$$y(a) = \frac{f'_x(a)}{\displaystyle\sum_{k=1}^{m} p_k(a)q_k(a)}, \qquad y_i = \frac{2}{\displaystyle\sum_{k=1}^{m} p_{ki}q_{ki}} \left[\frac{f_i}{h} - \sum_{k=1}^{m} p_{ki} \sum_{j=1}^{i-1} \beta_j q_{kj} y_j \right],$$

where y_i are approximate values of $y(x)$ at x_i, $f_i = f(x_i)$, $p_{ki} = p_k(x_i)$, and $q_{ki} = q_k(x_i)$.

⊙ References for Section 8.7: G. A. Korn and T. M. Korn (1968), N. S. Bakhvalov (1973), V. I. Krylov, V. V. Bobkov, and P. I. Monastyrnyi (1984), A. F. Verlan' and V. S. Sizikov (1986).

8.8. Equations With Infinite Integration Limit

Integral equations of the first kind with difference kernel in which one of the limits of integration is variable and the other is infinite are of interest. Sometimes the kernels and the functions of these equations do not belong to the classes described in the beginning of the chapter. The investigation of these equations can be performed by the method of model solutions (see Section 9.6) or by the method of reducing to equations of the convolution type. Let us consider these methods for an example of an equation of the first kind with variable lower limit of integration.

8.8-1. An Equation of the First Kind With Variable Lower Limit of Integration

Consider the equation of the first kind with difference kernel

$$\int_x^\infty K(x-t)y(t)\,dt = f(x). \tag{1}$$

Equation (1) cannot be solved by direct application of the Laplace transform, because the convolution theorem cannot be used here. According to the method of model solutions whose detailed exposition can be found in Section 9.6, we consider the auxiliary equation with exponential right-hand side

$$\int_x^\infty K(x-t)y(t)\,dt = e^{px}. \tag{2}$$

The solution of (2) has the form

$$Y(x,p) = \frac{1}{\tilde{K}(-p)} e^{px}, \qquad \tilde{K}(-p) = \int_0^\infty K(-z)e^{pz}\,dz. \tag{3}$$

On the basis of these formulas and formula (11) from Section 9.6, we obtain the solution of Eq. (1) for an arbitrary right-hand side $f(x)$ in the form

$$y(x) = \frac{1}{2\pi i} \int_{c-i\infty}^{c+i\infty} \frac{\tilde{f}(p)}{\tilde{K}(-p)} e^{px}\,dp, \tag{4}$$

where $\tilde{f}(p)$ is the Laplace transform of the function $f(x)$.

Example. Consider the following integral equation of the first kind with variable lower limit of integration:

$$\int_x^\infty e^{a(x-t)}y(t)\,dt = A\sin(bx), \qquad a > 0. \tag{5}$$

According to (3) and (4), we can write out the expressions for $\tilde{f}(p)$ (see Supplement 4) and $\tilde{K}(-p)$,

$$\tilde{f}(p) = \frac{Ab}{p^2 + b^2}, \qquad \tilde{K}(-p) = \int_0^\infty e^{(p-a)z}\,dz = \frac{1}{a-p}, \tag{6}$$

and the solution of Eq. (5) in the form

$$y(x) = \frac{1}{2\pi i}\int_{c-i\infty}^{c+i\infty}\frac{Ab(a-p)}{p^2 + b^2}e^{px}\,dp. \tag{7}$$

Now using the tables of inverse Laplace transforms (see Supplement 5), we obtain the exact solution

$$y(x) = Aa\sin(bx) - Ab\cos(bx), \qquad a > 0, \tag{8}$$

which can readily be verified by substituting (8) into (5) and using the tables of integrals in Supplement 2.

8.8-2. Reduction to a Wiener–Hopf Equation of the First Kind

Equation (1) can be reduced to a first-kind one-sided equation

$$\int_0^\infty K_-(x-t)y(t)\,dt = -f(x), \qquad 0 < x < \infty, \tag{9}$$

where the kernel $K_-(x-t)$ has the following form:

$$K_-(s) = \begin{cases} 0 & \text{for } s > 0, \\ -K(s) & \text{for } s < 0. \end{cases}$$

Methods for studying Eq. (9) are described in Chapter 10.

⊙ References for Section 8.8: F. D. Gakhov and Yu. I. Cherskii (1978), A. D. Polyanin and A. V. Manzhirov (1997).

Chapter 9

Methods for Solving Linear Equations of the Form $y(x) - \int_a^x K(x,t)y(t)\,dt = f(x)$

9.1. Volterra Integral Equations of the Second Kind

9.1-1. Preliminary Remarks. Equations for the Resolvent

In this chapter we present methods for solving Volterra integral equations of the second kind, which have the form

$$y(x) - \int_a^x K(x,t)y(t)\,dt = f(x), \tag{1}$$

where $y(x)$ is the unknown function ($a \le x \le b$), $K(x,t)$ is the kernel of the integral equation, and $f(x)$ is the *right-hand side* of the integral equation. The function classes to which $y(x)$, $f(x)$, and $K(x,t)$ can belong are defined in Subsection 8.1-1. In these function classes, there exists a unique solution of the Volterra integral equation of the second kind.

Equation (1) is said to be *homogeneous* if $f(x) \equiv 0$ and *nonhomogeneous* otherwise.

The kernel $K(x,t)$ is said to be *degenerate* if it can be represented in the form $K(x,t) = g_1(x)h_1(t) + \cdots + g_n(x)h_n(t)$.

The kernel $K(x,t)$ of an integral equation is called *difference kernel* if it depends only on the difference of the arguments, $K(x,t) = K(x-t)$.

Remark 1. A homogeneous Volterra integral equation of the second kind has only the trivial solution.

Remark 2. The existence and uniqueness of the solution of a Volterra integral equation of the second kind hold for a much wider class of kernels and functions.

Remark 3. A Volterra equation of the second kind can be regarded as a Fredholm equation of the second kind whose kernel $K(x,t)$ vanishes for $t > x$ (see Chapter 11).

Remark 4. The case in which $a = -\infty$ and/or $b = \infty$ is not excluded, but in this case the square integrability of the kernel $K(x,t)$ on the square $S = \{a \le x \le b,\ a \le t \le b\}$ is especially significant.

The solution of Eq. (1) can be presented in the form

$$y(x) = f(x) + \int_a^x R(x,t)f(t)\,dt, \tag{2}$$

where the resolvent $R(x,t)$ is independent of $f(x)$ and the lower limit of integration a and is determined by the kernel of the integral equation alone.

The resolvent of the Volterra equation (1) satisfies the following two integral equations:

$$R(x,t) = K(x,t) + \int_t^x K(x,s)R(s,t)\,ds, \tag{3}$$

$$R(x,t) = K(x,t) + \int_t^x K(s,t)R(x,s)\,ds, \tag{4}$$

in which the integration is performed with respect to different pairs of variables of the kernel and the resolvent.

9.1-2. A Relationship Between Solutions of Some Integral Equations

Let us present two useful formulas that express the solution of one integral equation via the solutions of other integral equations.

$1°$. Assume that the Volterra equation of the second kind with kernel $K(x,t)$ has a resolvent $R(x,t)$. Then the Volterra equation of the second kind with kernel $K^*(x,t) = -K(t,x)$ has the resolvent $R^*(x,t) = -R(t,x)$.

$2°$. Assume that two Volterra equations of the second kind with kernels $K_1(x,t)$ and $K_2(x,t)$ are given and that resolvents $R_1(x,t)$ and $R_2(x,t)$ correspond to these equations. In this case the Volterra equation with kernel

$$K(x,t) = K_1(x,t) + K_2(x,t) - \int_t^x K_1(x,s)K_2(s,t)\,ds \tag{5}$$

has the resolvent

$$R(x,t) = R_1(x,t) + R_2(x,t) + \int_t^x R_1(s,t)R_2(x,s)\,ds. \tag{6}$$

Note that in formulas (5) and (6), the integration is performed with respect to different pairs of variables.

⊙ References for Section 9.1: E. Goursat (1923), H. M. Müntz (1934), V. Volterra (1959), S. G. Mikhlin (1960), M. L. Krasnov, A. I. Kiselev, and G. I. Makarenko (1971), J. A. Cochran (1972), V. I. Smirnov (1974), P. P. Zabreyko, A. I. Koshelev, et al. (1975), A. J. Jerry (1985), F. G. Tricomi (1985), A. F. Verlan' and V. S. Sizikov (1986), G. Gripenberg, S.-O. Londen, and O. Staffans (1990), C. Corduneanu (1991), R. Gorenflo and S. Vessella (1991), A. C. Pipkin (1991).

9.2. Equations With Degenerate Kernel: $K(x,t) = g_1(x)h_1(t) + \cdots + g_n(x)h_n(t)$

9.2-1. Equations With Kernel of the Form $K(x,t) = \varphi(x) + \psi(x)(x - t)$

The solution of a Volterra equation (see Subsection 9.1-1) with kernel of this type can be expressed by the formula

$$y = w_{xx}'', \tag{1}$$

where $w = w(x)$ is the solution of the second-order linear nonhomogeneous ordinary differential equation

$$w_{xx}'' - \varphi(x)w_x' - \psi(x)w = f(x), \tag{2}$$

with the initial conditions

$$w(a) = w_x'(a) = 0. \tag{3}$$

Let $w_1 = w_1(x)$ be a nontrivial particular solution of the corresponding homogeneous linear differential equation (2) for $f(x) \equiv 0$. Assume that $w_1(a) \neq 0$. In this case, the other nontrivial particular solution $w_2 = w_2(x)$ of this homogeneous linear differential equation has the form

$$w_2(x) = w_1(x) \int_a^x \frac{\Phi(t)}{[w_1(t)]^2}\,dt, \qquad \Phi(x) = \exp\left[\int_a^x \varphi(s)\,ds\right].$$

The solution of the nonhomogeneous equation (2) with the initial conditions (3) is given by the formula

$$w(x) = w_2(x) \int_a^x \frac{w_1(t)}{\Phi(t)} f(t)\, dt - w_1(x) \int_a^x \frac{w_2(t)}{\Phi(t)} f(t)\, dt. \tag{4}$$

On substituting expression (4) into formula (1) we obtain the solution of the original integral equation in the form

$$y(x) = f(x) + \int_a^x R(x,t) f(t)\, dt,$$

where

$$R(x,t) = [w_2''(x)w_1(t) - w_1''(x)w_2(t)] \frac{1}{\Phi(t)}$$

$$= \varphi(x) \frac{\Phi(x)}{w_1(x)} \frac{w_1(t)}{\Phi(t)} + [\varphi(x)w_1'(x) + \psi(x)w_1(x)] \frac{w_1(t)}{\Phi(t)} \int_t^x \frac{\Phi(s)}{[w_1(s)]^2}\, ds.$$

Here $\Phi(x) = \exp\left[\int_a^x \varphi(s)\, ds\right]$ and the primes stand for x-derivatives.

For a degenerate kernel of the above form, the resolvent can be defined by the formula

$$R(x,t) = u_{xx}'',$$

where the auxiliary function u is the solution of the homogeneous linear second-order ordinary differential equation

$$u_{xx}'' - \varphi(x)u_x' - \psi(x)u = 0 \tag{5}$$

with the following initial conditions at $x = t$:

$$u\big|_{x=t} = 0, \quad u_x'\big|_{x=t} = 1. \tag{6}$$

The parameter t occurs only in the initial conditions (6), and Eq. (5) itself is independent of t.

Remark 1. The kernel of the integral equation in question can be rewritten in the form $K(x,t) = G_1(x) + tG_2(x)$, where $G_1(x) = \varphi(x) + x\psi(x)$ and $G_2(x) = -\varphi(x)$.

9.2-2. Equations With Kernel of the Form $K(x,t) = \varphi(t) + \psi(t)(t - x)$

For a degenerate kernel of the above form, the resolvent is determined by the expression

$$R(x,t) = -v_{tt}'', \tag{7}$$

where the auxiliary function v is the solution of the homogeneous linear second-order ordinary differential equation

$$v_{tt}'' + \varphi(t)v_t' + \psi(t)v = 0 \tag{8}$$

with the following initial conditions at $t = x$:

$$v\big|_{t=x} = 0, \quad v_t'\big|_{t=x} = 1. \tag{9}$$

The point x occurs only in the initial data (9) as a parameter, and Eq. (8) itself is independent of x.

Assume that $v_1 = v_1(t)$ is a nontrivial particular solution of Eq. (8). In this case, the general solution of this differential equation is given by the formula

$$v(t) = C_1 v_1(t) + C_2 v_1(t) \int_a^t \frac{ds}{\Phi(s)[v_1(s)]^2}, \qquad \Phi(t) = \exp\left[\int_a^t \varphi(s)\, ds\right].$$

Taking into account the initial data (9), we find the dependence of the integration constants C_1 and C_2 on the parameter x. As a result, we obtain the solution of problem (8), (9):

$$v = v_1(x)\Phi(x) \int_x^t \frac{ds}{\Phi(s)[v_1(s)]^2}. \tag{10}$$

On substituting the expression (10) into formula (7) and eliminating the second derivative by means of Eq. (8) we find the resolvent:

$$R(x,t) = \varphi(t)\frac{v_1(x)\Phi(x)}{v_1(t)\Phi(t)} + v_1(x)\Phi(x)[\varphi(t)v_t'(t) + \psi(t)v_1(t)] \int_x^t \frac{ds}{\Phi(s)[v_1(s)]^2}.$$

Remark 2. The kernel of the integral equation under consideration can be rewritten in the form $K(x,t) = G_1(t) + xG_2(t)$, where $G_1(t) = \varphi(t) + t\psi(t)$ and $G_2(t) = -\varphi(t)$.

9.2-3. Equations With Kernel of the Form $K(x,t) = \sum_{m=1}^n \varphi_m(x)(x-t)^{m-1}$

To find the resolvent, we introduce an auxiliary function as follows:

$$u(x,t) = \frac{1}{(n-1)!} \int_t^x R(s,t)(x-s)^{n-1}\,ds + \frac{(x-t)^{n-1}}{(n-1)!};$$

at $x = t$, this function vanishes together with the first $n-2$ derivatives with respect to x, and the $(n-1)$st derivative at $x = t$ is equal to 1. Moreover,

$$R(x,t) = u_x^{(n)}(x,t), \qquad u_x^{(n)} = \frac{d^n u(x,t)}{dx^n}. \tag{11}$$

On substituting relation (11) into the resolvent equation (3) of Subsection 9.1-1, we see that

$$u_x^{(n)}(x,t) = K(x,t) + \int_t^x K(x,s)u_s^{(n)}(s,t)\,ds. \tag{12}$$

Integrating by parts the right-hand side in (12), we obtain

$$u_x^{(n)}(x,t) = K(x,t) + \sum_{m=0}^{n-1}(-1)^m K_s^{(m)}(x,s)u_s^{(n-m-1)}(s,t)\Big|_{s=t}^{s=x}. \tag{13}$$

On substituting the expressions for $K(x,t)$ and $u(x,t)$ into (13), we arrive at a linear homogeneous ordinary differential equation of order n for the function $u(x,t)$.

Thus, the resolvent $R(x,t)$ of the Volterra integral equation with degenerate kernel of the above form can be obtained by means of (11), where $u(x,t)$ satisfies the following differential equation and initial conditions:

$$u_x^{(n)} - \varphi_1(x)u_x^{(n-1)} - \varphi_2(x)u_x^{(n-2)} - 2\varphi_3(x)u_x^{(n-3)} - \cdots - (n-1)!\,\varphi_n(x)u = 0,$$
$$u\big|_{x=t} = u_x'\big|_{x=t} = \cdots = u_x^{(n-2)}\big|_{x=t} = 0, \quad u_x^{(n-1)}\big|_{x=t} = 1.$$

The parameter t occurs only in the initial conditions, and the equation itself is independent of t explicitly.

Remark 3. A kernel of the form $K(x,t) = \sum_{m=1}^n \phi_m(x)t^{m-1}$ can be reduced to a kernel of the above type by elementary transformations.

9.2-4. Equations With Kernel of the Form $K(x,t) = \sum_{m=1}^{n} \varphi_m(t)(t-x)^{m-1}$

Let us represent the resolvent of this degenerate kernel in the form

$$R(x,t) = -v_t^{(n)}(x,t), \qquad v_t^{(n)} = \frac{d^n v(x,t)}{dt^n},$$

where the auxiliary function $v(x,t)$ vanishes at $t=x$ together with $n-2$ derivatives with respect to t, and the $(n-1)$st derivative with respect to t at $t=x$ is equal to 1. On substituting the expression for the resolvent into Eq. (3) of Subsection 9.1-1, we obtain

$$v_t^{(n)}(x,t) = \int_t^x K(s,t)v_s^{(n)}(x,s)\,ds - K(x,t).$$

Let us apply integration by parts to the integral on the right-hand side. Taking into account the properties of the auxiliary function $v(x,t)$, we arrive at the following Cauchy problem for an nth-order ordinary differential equation:

$$v_t^{(n)} + \varphi_1(t)v_t^{(n-1)} + \varphi_2(t)v_t^{(n-2)} + 2\varphi_3(t)v_t^{(n-3)} + \cdots + (n-1)!\,\varphi_n(t)v = 0,$$
$$v\big|_{t=x} = v_t'\big|_{t=x} = \cdots = v_t^{(n-2)}\big|_{t=x} = 0, \qquad v_t^{(n-1)}\big|_{t=x} = 1.$$

The parameter x occurs only in the initial conditions, and the equation itself is independent of x explicitly.

Remark 4. A kernel of the form $K(x,t) = \sum_{m=1}^{n} \phi_m(t)x^{m-1}$ can be reduced to a kernel of the above type by elementary transformations.

9.2-5. Equations With Degenerate Kernel of the General Form

In this case, the Volterra equation of the second kind can be represented in the form

$$y(x) - \sum_{m=1}^{n} g_m(x) \int_a^x h_m(t)y(t)\,dt = f(x). \tag{14}$$

Let us introduce the notation

$$w_j(x) = \int_a^x h_j(t)y(t)\,dt, \qquad j = 1, \ldots, n, \tag{15}$$

and rewrite Eq. (14) as follows:

$$y(x) = \sum_{m=1}^{n} g_m(x)w_m(x) + f(x). \tag{16}$$

On differentiating the expressions (15) with regard to formula (16), we arrive at the following system of linear differential equations for the functions $w_j = w_j(x)$:

$$w_j' = h_j(x)\left[\sum_{m=1}^{n} g_m(x)w_m + f(x)\right], \qquad j = 1, \ldots, n,$$

with the initial conditions
$$w_j(a) = 0, \qquad j = 1, \ldots, n.$$

Once the solution of this system is found, the solution of the original integral equation (14) is defined by formula (16) or any of the expressions

$$y(x) = \frac{w_j'(x)}{h_j(x)}, \qquad j = 1, \ldots, n,$$

which can be obtained from formula (15) by differentiation.

⊙ References for Section 9.2: E. Goursat (1923), H. M. Müntz (1934), A. F. Verlan' and V. S. Sizikov (1986), A. D. Polyanin and A. V. Manzhirov (1998).

9.3. Equations With Difference Kernel: $K(x,t) = K(x-t)$

9.3-1. A Solution Method Based on the Laplace Transform

Volterra equations of the second kind with kernel depending on the difference of the arguments have the form

$$y(x) - \int_0^x K(x-t)y(t)\,dt = f(x). \tag{1}$$

Applying the Laplace transform $\mathcal{L}$ to Eq. (1) and taking into account the fact that by the convolution theorem (see Subsection 7.2-3) the integral with kernel depending on the difference of the arguments is transformed into the product $\tilde{K}(p)\tilde{y}(p)$, we arrive at the following equation for the transform of the unknown function:

$$\tilde{y}(p) - \tilde{K}(p)\tilde{y}(p) = \tilde{f}(p). \tag{2}$$

The solution of Eq. (2) is given by the formula

$$\tilde{y}(p) = \frac{\tilde{f}(p)}{1 - \tilde{K}(p)}, \tag{3}$$

which can be written equivalently in the form

$$\tilde{y}(p) = \tilde{f}(p) + \tilde{R}(p)\tilde{f}(p), \qquad \tilde{R}(p) = \frac{\tilde{K}(p)}{1 - \tilde{K}(p)}. \tag{4}$$

On applying the Laplace inversion formula to (4), we obtain the solution of Eq. (1) in the form

$$y(x) = f(x) + \int_0^x R(x-t)f(t)\,dt,$$
$$R(x) = \frac{1}{2\pi i}\int_{c-i\infty}^{c+i\infty} \tilde{R}(p)e^{px}\,dp. \tag{5}$$

When applying formula (5) in practice, the following two technical problems occur:

$1°$. Finding the transform $\tilde{K}(p) = \int_0^\infty K(x)e^{-px}\,dx$ for a given kernel $K(x)$.

$2°$. Finding the resolvent (5) whose transform $\tilde{R}(p)$ is given by formula (4).

To calculate the corresponding integrals, tables of direct and inverse Laplace transforms can be applied (see Supplements 4 and 5), and, in many cases, to find the inverse transform, methods of the theory of functions of a complex variable are applied, including the Cauchy residue theorem (see Subsection 7.1-4).

Remark. If the lower limit of the integral in the Volterra equation with kernel depending on the difference of the arguments is equal to a, then this equation can be reduced to Eq. (1) by the change of variables $x = \bar{x} - a$, $t = \bar{t} - a$.

Figure 2 depicts the principal scheme of solving Volterra integral equations of the second kind with difference kernel by means of the Laplace integral transform.

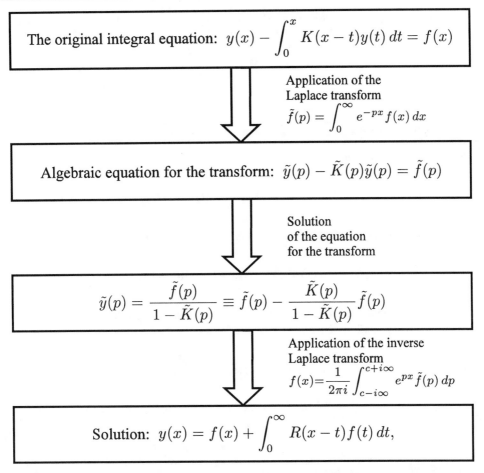

Fig. 2. Scheme of solving Volterra integral equations of the second kind with difference kernel by means of the Laplace integral transform. $R(x)$ is the inverse transform of the function $\tilde{R}(p) = \dfrac{\tilde{K}(p)}{1 - \tilde{K}(p)}$.

Example 1. Consider the equation

$$y(x) + A \int_0^x \sin\left[\lambda(x - t)\right] y(t)\, dt = f(x), \tag{6}$$

which is a special case of Eq. (1) for $K(x) = -A \sin(\lambda x)$.

We first apply the table of Laplace transforms (see Supplement 4) and obtain the transform of the kernel of the integral equation in the form

$$\tilde{K}(p) = -\frac{A\lambda}{p^2 + \lambda^2}.$$

Next, by formula (4) we find the transform of the resolvent:

$$\tilde{R}(p) = -\frac{A\lambda}{p^2 + \lambda(A + \lambda)}.$$

Furthermore, applying the table of inverse Laplace transforms (see Supplement 5) we obtain the resolvent:

$$R(x) = \begin{cases} -\dfrac{A\lambda}{k} \sin(kx) & \text{for } \lambda(A + \lambda) > 0, \\[2mm] -\dfrac{A\lambda}{k} \sinh(kx) & \text{for } \lambda(A + \lambda) < 0, \end{cases} \qquad \text{where} \quad k = |\lambda(A + \lambda)|^{1/2}.$$

Moreover, in the special case $\lambda = -A$, we have $R(x) = A^2 x$. On substituting the expressions for the resolvent into formula (5), we find the solution of the integral equation (6). In particular, for $\lambda(A + \lambda) > 0$, this solution has the form

$$y(x) = f(x) - \frac{A\lambda}{k} \int_0^x \sin\left[k(x - t)\right] f(t)\,dt, \qquad k = \sqrt{\lambda(A + \lambda)}. \tag{7}$$

9.3-2. A Method Based on the Solution of an Auxiliary Equation

Consider the integral equation

$$Ay(x) + B \int_a^x K(x - t)y(t)\,dt = f(x). \tag{8}$$

Let $w = w(x)$ be a solution of the simpler auxiliary equation with $f(x) \equiv 1$ and $a = 0$,

$$Aw(x) + B \int_0^x K(x - t)w(t)\,dt = 1. \tag{9}$$

In this case, the solution of the original equation (8) with an arbitrary right-hand side can be expressed via the solution of the auxiliary equation (9) by the formula

$$y(x) = \frac{d}{dx} \int_a^x w(x - t)f(t)\,dt = f(a)w(x - a) + \int_a^x w(x - t)f_t'(t)\,dt. \tag{10}$$

Let us prove this assertion. We rewrite expression (10) (in which we first redenote the integration parameter t by s) in the form

$$y(x) = \frac{d}{dx}I(x), \qquad I(x) = \int_a^x w(x - s)f(s)\,ds \tag{11}$$

and substitute it into the left-hand side of Eq. (8). After some algebraic manipulations and after changing the order of integration in the double integral with regard to (9), we obtain

$$\frac{d}{dx}AI(x) + B\int_a^x K(x - t)\frac{d}{dt}I(t)\,dt = \frac{d}{dx}AI(x) + \frac{d}{dx}B\int_a^x K(x - t)I(t)\,dt$$

$$= \frac{d}{dx}\left[A\int_a^x w(x - s)f(s)\,ds + B\int_a^x \int_a^t K(x - t)w(t - s)f(s)\,ds\,dt\right]$$

$$= \frac{d}{dx}\left\{\int_a^x f(s)\left[Aw(x - s) + B\int_s^x K(x - t)w(t - s)\,dt\right]ds\right\}$$

$$= \frac{d}{dx}\left\{\int_a^x f(s)\left[Aw(x - s) + B\int_0^{x-s} K(x - s - \lambda)w(\lambda)\,d\lambda\right]ds\right\} = \frac{d}{dx}\int_a^x f(s)\,ds = f(x),$$

which proves the desired assertion.

9.3-3. Reduction to Ordinary Differential Equations

Consider the special case in which the transform of the kernel of the integral equation (1) can be expressed in the form

$$1 - \tilde{K}(p) = \frac{Q(p)}{R(p)}, \tag{12}$$

where $Q(p)$ and $R(p)$ are polynomials of degree n:

$$Q(p) = \sum_{k=0}^n A_k p^k, \qquad R(p) = \sum_{k=0}^n B_k p^k. \tag{13}$$

In this case, the solution of the integral equation (1) satisfies the following linear nonhomogeneous ordinary differential equation of order n with constant coefficients:

$$\sum_{k=0}^{n} A_k y_x^{(k)}(x) = \sum_{k=0}^{n} B_k f_x^{(k)}(x).$$

(14)

Equation (14) can be rewritten in the operator form

$$Q(D)y(x) = R(D)f(x), \qquad D \equiv \frac{d}{dx}.$$

The initial conditions for Eq. (14) can be found from the relation

$$\sum_{k=0}^{n} A_k \sum_{s=0}^{k-1} p^{k-1-s} y_x^{(s)}(0) - \sum_{k=0}^{n} B_k \sum_{s=0}^{k-1} p^{k-1-s} f_x^{(s)}(0) = 0$$

(15)

by matching the coefficients of like powers of the parameter p.

The proof of this assertion can be performed by applying the Laplace transform to the differential equation (14) and by the subsequent comparison of the resulting expression with Eq. (2) with regard to (12).

Another method of reducing an integral equation to an ordinary differential equation is described in Section 9.7.

9.3-4. Reduction to a Wiener–Hopf Equation of the Second Kind

A Volterra equation of the second kind with the difference kernel of the form

$$y(x) + \int_0^x K(x - t)y(t)\, dt = f(x), \qquad 0 < x < \infty,$$

(16)

can be reduced to the Wiener–Hopf equation

$$y(x) + \int_0^\infty K_+(x - t)y(t)\, dt = f(x), \qquad 0 < x < \infty,$$

(17)

where the kernel $K_+(x - t)$ is given by

$$K_+(s) = \begin{cases} K(s) & \text{for } s > 0, \\ 0 & \text{for } s < 0. \end{cases}$$

Methods for studying Eq. (17) are described in Chapter 11, where an example of constructing a solution of a Volterra equation of the second kind with difference kernel by means of constructing a solution of the corresponding Wiener–Hopf equation of the second kind is presented (see Subsection 11.9-3).

9.3-5. Method of Fractional Integration for the Generalized Abel Equation

Consider the generalized Abel equation of the second kind

$$y(x) - \lambda \int_a^x \frac{y(t)}{(x - t)^\mu}\, dt = f(x), \qquad x > a,$$

(18)

where $0 < \mu < 1$. Let us assume that $x \in [a, b]$, $f(x) \in AC$, and $y(t) \in L_1$, and apply the technique of the fractional integration (see Section 8.5). We set

$$\mu = 1 - \beta, \qquad 0 < \beta < 1, \qquad \lambda = \frac{\nu}{\Gamma(\beta)}, \tag{19}$$

and use (8.5.1) to rewrite Eq. (18) in the form

$$\left(1 - \nu \mathbf{I}_{a+}^{\beta}\right)y(x) = f(x), \qquad x > a. \tag{20}$$

Now the solution of the generalized Abel equation of the second kind can be symbolically written as follows:

$$y(x) = \left(1 - \nu \mathbf{I}_{a+}^{\beta}\right)^{-1} f(x), \qquad x > a. \tag{21}$$

On expanding the operator expression in the parentheses in a series in powers of the operator by means of the formula for a geometric progression, we obtain

$$y(x) = \left[1 + \sum_{n=1}^{\infty} \left(\nu \mathbf{I}_{a+}^{\beta}\right)^n\right] f(x), \qquad x > a. \tag{22}$$

Taking into account the relation $(\mathbf{I}_{a+}^{\beta})^n = \mathbf{I}_{a+}^{\beta n}$, we can rewrite formula (22) in the expanded form

$$y(x) = f(x) + \sum_{n=1}^{\infty} \frac{\nu^n}{\Gamma(\beta n)} \int_a^x (x - t)^{\beta n - 1} f(t)\,dt, \qquad x > a. \tag{23}$$

Let us transpose the integration and summation in the expression (23). Note that

$$\sum_{n=1}^{\infty} \frac{\nu^n (x-t)^{\beta n - 1}}{\Gamma(\beta n)} = \frac{d}{dx} \sum_{n=1}^{\infty} \frac{\nu^n (x-t)^{\beta n}}{\Gamma(1 + \beta n)}.$$

In this case, taking into account the change of variables (19), we see that a solution of the generalized Abel equation of the second kind becomes

$$y(x) = f(x) + \int_a^x R(x - t) f(t)\,dt, \qquad x > a, \tag{24}$$

where the resolvent $R(x - t)$ is given by the formula

$$R(x - t) = \frac{d}{dx} \sum_{n=1}^{\infty} \frac{\left[\lambda \Gamma(1 - \mu)(x - t)^{(1-\mu)}\right]^n}{\Gamma[1 + (1 - \mu)n]}. \tag{25}$$

In some cases, the sum of the series in the representation (25) of the resolvent can be found, and a closed-form expression for this sum can be obtained.

Example 2. Consider the Abel equation of the second kind (we set $\mu = \frac{1}{2}$ in Eq. (18))

$$y(x) - \lambda \int_a^x \frac{y(t)}{\sqrt{x - t}}\,dt = f(x), \qquad x > a. \tag{26}$$

By virtue of formula (25), the resolvent for Eq. (26) is given by the expression

$$R(x - t) = \frac{d}{dx} \sum_{n=1}^{\infty} \frac{\left[\lambda \sqrt{\pi(x - t)}\right]^n}{\Gamma\left(1 + \frac{1}{2}n\right)}. \tag{27}$$

We have

$$\sum_{n=1}^{\infty} \frac{x^{n/2}}{\Gamma\left(1 + \frac{1}{2}n\right)} = e^x \operatorname{erf}\sqrt{x}, \qquad \operatorname{erf} x \equiv \frac{2}{\sqrt{\pi}} \int_0^x e^{-t^2}\,dt, \tag{28}$$

where $\operatorname{erf} x$ is the *error function*. By (27) and (28), in this case the expression for the resolvent can be rewritten in the form

$$R(x - t) = \frac{d}{dx} \left\{\exp[\lambda^2 \pi(x - t)] \operatorname{erf}\left[\lambda \sqrt{\pi(x - t)}\right]\right\}. \tag{29}$$

Applying relations (24) and (27), we obtain the solution of the Abel integral equation of the second kind in the form

$$y(x) = f(x) + \frac{d}{dx} \int_a^x \left\{\exp[\lambda^2 \pi(x - t)] \operatorname{erf}\left[\lambda \sqrt{\pi(x - t)}\right]\right\} f(t)\,dt, \qquad x > a. \tag{30}$$

Note that in the case under consideration, the solution is constructed in the closed form.

9.3-6. Systems of Volterra Integral Equations

The Laplace transform can be applied to solve systems of Volterra integral equations of the form

$$y_m(x) - \sum_{k=1}^{n} \int_0^x K_{mk}(x-t)y_k(t)\,dt = f_m(x), \qquad m = 1, \dots, n. \tag{31}$$

Let us apply the Laplace transform to system (31). We obtain the relations

$$\tilde{y}_m(p) - \sum_{k=1}^{n} \tilde{K}_{mk}(p)\tilde{y}_k(p) = \tilde{f}_m(p), \qquad m = 1, \dots, n. \tag{32}$$

On solving this system of linear algebraic equations, we find $\tilde{y}_m(p)$, and the solution of the system under consideration becomes

$$y_m(x) = \frac{1}{2\pi i} \int_{c-i\infty}^{c+i\infty} \tilde{y}_m(p)e^{px}\,dp. \tag{33}$$

The Laplace transform can be applied to construct a solution of systems of Volterra equations of the first kind and of integro-differential equations as well.

⊙ References for Section 9.3: V. A. Ditkin and A. P. Prudnikov (1965), M. L. Krasnov, A. I. Kiselev, and G. I. Makarenko (1971), V. I. Smirnov (1974), K. B. Oldham and J. Spanier (1974), P. P. Zabreyko, A. I. Koshelev, et al. (1975), F. D. Gakhov and Yu. I. Cherskii (1978), Yu. I. Babenko (1986), R. Gorenflo and S. Vessella (1991), S. G. Samko, A. A. Kilbas, and O. I. Marichev (1993).

9.4. Operator Methods for Solving Linear Integral Equations

9.4-1. Application of a Solution of a "Truncated" Equation of the First Kind

Consider the linear equation of the second kind

$$y(x) + \mathbf{L}\,[y] = f(x), \tag{1}$$

where $\mathbf{L}$ is a linear (integral) operator.

Assume that the solution of the auxiliary "truncated" equation of the first kind

$$\mathbf{L}\,[u] = g(x), \tag{2}$$

can be represented in the form

$$u(x) = \mathbf{M}\big[\mathbf{L}[g]\big], \tag{3}$$

where $\mathbf{M}$ is a known linear operator. Formula (3) means that

$$\mathbf{L}^{-1} = \mathbf{ML}.$$

Let us apply the operator $\mathbf{L}^{-1}$ to Eq. (1). The resulting relation has the form

$$\mathbf{M}\big[\mathbf{L}[y]\big] + y(x) = \mathbf{M}\big[\mathbf{L}[f]\big], \tag{4}$$

On eliminating $y(x)$ from (1) and (4) we obtain the equation

$$\mathbf{M}\,[w] - w(x) = F(x), \tag{5}$$

in which the following notation is used:

$$w = \mathbf{L}\,[y], \quad F(x) = \mathbf{M}\big[\mathbf{L}[f]\big] - f(x).$$

In some cases, Eq. (5) is simpler than the original equation (1). For example, this is the case if the operator $\mathbf{M}$ is a constant (see Subsection 11.7-2) or a differential operator:

$$\mathbf{M} = a_n D^n + a_{n-1} D^{n-1} + \cdots + a_1 D + a_0, \quad D \equiv \frac{d}{dx}.$$

In the latter case, Eq. (5) is an ordinary linear differential equation for the function w.

If a solution $w = w(x)$ of Eq. (5) is obtained, then a solution of Eq. (1) is given by the formula $y(x) = \mathbf{M}\big[\mathbf{L}[w]\big]$.

Example 1. Consider the Abel equation of the second kind

$$y(x) + \lambda \int_a^x \frac{y(t)\,dt}{\sqrt{x-t}} = f(x). \tag{6}$$

To solve this equation, we apply a slight modification of the above scheme, which corresponds to the case $\mathbf{M} \equiv \text{const}\,\dfrac{d}{dx}$.

Let us rewrite Eq. (6) as follows:

$$\int_a^x \frac{y(t)\,dt}{\sqrt{x-t}} = \frac{f(x) - y(x)}{\lambda}. \tag{7}$$

Let us assume that the right-hand side of Eq. (7) is known and treat Eq. (7) as an Abel equation of the first kind. Its solution can be written in the following form (see the example in Subsection 8.4-4):

$$y(x) = \frac{1}{\pi}\frac{d}{dx}\int_a^x \frac{f(t) - y(t)}{\lambda\sqrt{x-t}}\,dt$$

or

$$y(x) + \frac{1}{\pi\lambda}\frac{d}{dx}\int_a^x \frac{y(t)\,dt}{\sqrt{x-t}}\,dt = \frac{1}{\pi\lambda}\frac{d}{dx}\int_a^x \frac{f(t)\,dt}{\sqrt{x-t}}. \tag{8}$$

Let us differentiate both sides of Eq. (6) with respect to x, multiply Eq. (8) by $-\pi\lambda^2$, and add the resulting expressions term by term. We eventually arrive at the following first-order linear ordinary differential equation for the function $y = y(x)$:

$$y'_x - \pi\lambda^2 y = F'_x(x), \tag{9}$$

where

$$F(x) = f(x) - \lambda\int_a^x \frac{f(t)\,dt}{\sqrt{x-t}}. \tag{10}$$

We must supplement Eq. (9) with initial condition

$$y(a) = f(a), \tag{11}$$

which is a consequence of (6).

The solution of problem (9)–(11) has the form

$$y(x) = F(x) + \pi\lambda^2\int_a^x \exp[\pi\lambda^2(x-t)]F(t)\,dt, \tag{12}$$

and defines the solution of the Abel equation of the second kind (6).

9.4-2. Application of the Auxiliary Equation of the Second Kind

The solution of the Abel equation of the second kind (6) can also be obtained by another method, presented below.

Consider the linear equation

$$y(x) - \mathbf{L}[y] = f(x), \tag{13}$$

where $\mathbf{L}$ is a linear operator. Assume that the solution of the auxiliary equation

$$w(x) - \mathbf{L}^n[w] = \Phi(x), \qquad \mathbf{L}^n[w] \equiv \mathbf{L}\left[\mathbf{L}^{n-1}[w]\right], \tag{14}$$

which involves the nth power of the operator $\mathbf{L}$, is known and is defined by the formula

$$w(x) = \mathbf{M}[\Phi(x)]. \tag{15}$$

In this case, the solution of the original equation (13) has the form

$$y(x) = \mathbf{M}[\Phi(x)], \qquad \Phi(x) = \mathbf{L}^{n-1}[f] + \mathbf{L}^{n-2}[f] + \cdots + \mathbf{L}[f] + f(x). \tag{16}$$

This assertion can be proved by applying the operator $\mathbf{L}^{n-1} + \mathbf{L}^{n-2} + \cdots + \mathbf{L} + 1$ to Eq. (13), with regard to the operator relation

$$\left(1 - \mathbf{L}\right)\left(\mathbf{L}^{n-1} + \mathbf{L}^{n-2} + \cdots + \mathbf{L} + 1\right) = 1 - \mathbf{L}^n$$

together with formula (16) for $\Phi(x)$. In Eq. (14) we may write $y(x)$ instead of $w(x)$.

Example 2. Let us apply the operator method (for $n = 2$) to solve the generalized Abel equation with exponent 3/4:

$$y(x) - b \int_0^x \frac{y(t)\,dt}{(x-t)^{3/4}} = f(x). \tag{17}$$

We first consider the integral operator with difference kernel

$$\mathbf{L}\,[y(x)] \equiv \int_0^x K(x-t)y(t)\,dt.$$

Let us find $\mathbf{L}^2$:

$$\mathbf{L}^2\,[y] \equiv \mathbf{L}\,\big[\mathbf{L}\,[y]\big] = \int_0^x \int_0^t K(x-t)K(t-s)y(s)\,ds\,dt$$

$$= \int_0^x y(s)\,ds \int_s^x K(x-t)K(t-s)\,dt = \int_0^x K_2(x-s)y(s)\,ds, \tag{18}$$

$$K_2(z) = \int_0^z K(\xi)K(z-\xi)\,d\xi.$$

In the proof of this formula, we have reversed the order of integration and performed the change of variables $\xi = t - s$.

For the power-law kernel

$$K(\xi) = b\xi^\mu,$$

we have

$$K_2(z) = b^2 \frac{\Gamma^2(1+\mu)}{\Gamma(2+2\mu)} z^{1+2\mu}. \tag{19}$$

For Eq. (17) we obtain

$$\mu = -\frac{3}{4}, \quad K_2(z) = A\frac{1}{\sqrt{z}}, \quad A = \frac{b^2}{\sqrt{\pi}}\Gamma^2(\tfrac{1}{4}).$$

Therefore, the auxiliary equation (14) corresponding to $n = 2$ has the form

$$y(x) - A \int_0^x \frac{y(t)\,dt}{\sqrt{x-t}} = \Phi(x), \tag{20}$$

where

$$\Phi(x) = f(x) + b \int_0^x \frac{f(t)\,dt}{(x-t)^{3/4}}.$$

After the substitution $A \to -\lambda$ and $\Phi \to f$, relation (20) coincides with Eq. (6), and the solution of Eq. (20) can be obtained by formula (12).

Remark. It follows from (19) that the solution of the generalized Abel equation with exponent β

$$y(x) + \lambda \int_0^x \frac{y(t)\,dt}{(x-t)^\beta} = f(x)$$

can be reduced to the solution of a similar equation with the different exponent $\beta_1 = 2\beta - 1$. In particular, the Abel equation (6), which corresponds to $\beta = \frac{1}{2}$, is reduced to the solution of an equation with degenerate kernel for $\beta_1 = 0$.

9.4-3. A Method for Solving "Quadratic" Operator Equations

Suppose that the solution of the linear (integral, differential, etc.) equation

$$y(x) - \lambda \mathbf{L}\,[y] = f(x) \tag{21}$$

is known for an arbitrary right-hand side $f(x)$ and for any λ from the interval $(\lambda_{\min}, \lambda_{\max})$. We denote this solution by

$$y = Y(f, \lambda). \tag{22}$$

Let us construct the solution of the more complicated equation

$$y(x) - a\mathbf{L}\,[y] - b\mathbf{L}^2\,[y] = f(x), \tag{23}$$

where a and b are some numbers and $f(x)$ is an arbitrary function. To this end, we represent the left-hand side of Eq. (23) by the product of operators

$$\left(1 - a\mathbf{L} - b\mathbf{L}^2\right)[y] \equiv \left(1 - \lambda_1\mathbf{L}\right)\left(1 - \lambda_2\mathbf{L}\right)[y], \tag{24}$$

where λ_1 and λ_2 are the roots of the quadratic equation

$$\lambda^2 - a\lambda - b = 0. \tag{25}$$

We assume that $\lambda_{\min} < \lambda_1, \lambda_2 < \lambda_{\max}$.

Let us solve the auxiliary equation

$$w(x) - \lambda_2\mathbf{L}[w] = f(x), \tag{26}$$

which is the special case of Eq. (21) for $\lambda = \lambda_2$. The solution of this equation is given by the formula

$$w(x) = Y(f, \lambda_2). \tag{27}$$

Taking into account (24) and (26), we can rewrite Eq. (23) in the form

$$\left(1 - \lambda_1\mathbf{L}\right)\left(1 - \lambda_2\mathbf{L}\right)[y] = \left(1 - \lambda_2\mathbf{L}\right)[w],$$

or, in view of the identity $(1 - \lambda_1\mathbf{L})(1 - \lambda_2\mathbf{L}) \equiv (1 - \lambda_2\mathbf{L})(1 - \lambda_1\mathbf{L})$, in the form

$$\left(1 - \lambda_2\mathbf{L}\right)\left\{\left(1 - \lambda_1\mathbf{L}\right)[y] - w(x)\right\} = 0.$$

This relation holds if the unknown function $y(x)$ satisfies the equation

$$y(x) - \lambda_1\mathbf{L}[y] = w(x). \tag{28}$$

The solution of this equation is given by the formula

$$y(x) = Y(w, \lambda_1), \quad \text{where} \quad w = Y(f, \lambda_2). \tag{29}$$

If the homogeneous equation $y(x) - \lambda_2\mathbf{L}[y] = 0$ has only the trivial* solution $y \equiv 0$, then formula (29) defines the unique solution of the original equation (23).

Example 3. Consider the integral equation

$$y(x) - \int_0^x \left(\frac{A}{\sqrt{x-t}} + B\right)y(t)\,dt = f(x).$$

It follows from the results of Example 2 that this equation can be written in the form of Eq. (23):

$$y(x) - A\mathbf{L}[y] - \frac{1}{\pi}B\mathbf{L}^2[y] = f(x), \qquad \mathbf{L}[y] \equiv \int_0^x \frac{y(t)\,dt}{\sqrt{x-t}}.$$

Therefore, the solution (in the form of antiderivatives) of the integral equation can be given by the formulas

$$y(x) = Y(w, \lambda_1), \quad w = Y(f, \lambda_2),$$

$$Y(f, \lambda) = F(x) + \pi\lambda^2\int_0^x \exp\left[\pi\lambda^2(x-t)\right]F(t)\,dt, \quad F(x) = f(x) + \lambda\int_0^x \frac{f(t)\,dt}{\sqrt{x-t}},$$

where λ_1 and λ_2 are the roots of the quadratic equation $\lambda^2 - A\lambda - \frac{1}{\pi}B = 0$.

This method can also be applied to solve (in the form of antiderivatives) more general equations of the form

$$y(x) - \int_0^x \left[\frac{A}{(x-t)^\beta} + \frac{B}{(x-t)^{2\beta-1}}\right]y(t)\,dt = f(x),$$

where β is a rational number satisfying the condition $0 < \beta < 1$ (see Example 2 and Eq. 2.1.59 from the first part of the book).

---

* If the homogeneous equation $y(x) - \lambda_2\mathbf{L}[y] = 0$ has nontrivial solutions, then the right-hand side of Eq. (28) must contain the function $w(x) + y_0(x)$ instead of $w(x)$, where y_0 is the general solution of the homogeneous equation.

9.4-4. Solution of Operator Equations of Polynomial Form

The method described in Subsection 9.4-3 can be generalized to the case of operator equations of polynomial form. Suppose that the solution of the linear nonhomogeneous equation (21) is given by formula (22) and that the corresponding homogeneous equation has only the trivial solution.

Let us construct the solution of the more complicated equation with polynomial left-hand side with respect to the operator $\mathbf{L}$:

$$y(x) - \sum_{k=1}^{n} A_k \mathbf{L}^k [y] = f(x), \qquad \mathbf{L}^k \equiv \mathbf{L}\big(\mathbf{L}^{k-1}\big), \tag{30}$$

where A_k are some numbers and $f(x)$ is an arbitrary function.

We denote by $\lambda_1, \ldots, \lambda_n$ the roots of the characteristic equation

$$\lambda^n - \sum_{k=1}^{n} A_k \lambda^{n-k} = 0. \tag{31}$$

The left-hand side of Eq. (30) can be expressed in the form of a product of operators:

$$y(x) - \sum_{k=1}^{n} A_k \mathbf{L}^k [y] \equiv \prod_{k=1}^{n} \big(1 - \lambda_k \mathbf{L}\big) [y]. \tag{32}$$

The solution of the auxiliary equation (26), in which we use the substitution $w \to y_{n-1}$ and $\lambda_2 \to \lambda_n$, is given by the formula $y_{n-1}(x) = Y(f, \lambda_n)$. Reasoning similar to that in Subsection 9.4-3 shows that the solution of Eq. (30) is reduced to the solution of the simpler equation

$$\prod_{k=1}^{n-1} \big(1 - \lambda_k \mathbf{L}\big) [y] = y_{n-1}(x), \tag{33}$$

whose degree is less by one than that of the original equation with respect to the operator $\mathbf{L}$. We can show in a similar way that Eq. (33) can be reduced to the solution of the simpler equation

$$\prod_{k=1}^{n-2} \big(1 - \lambda_k \mathbf{L}\big) [y] = y_{n-2}(x), \qquad y_{n-2}(x) = Y(y_{n-1}, \lambda_{n-1}).$$

Successively reducing the order of the equation, we eventually arrive at an equation of the form (28) whose right-hand side contains the function $y_1(x) = Y(y_2, \lambda_2)$. The solution of this equation is given by the formula $y(x) = Y(y_1, \lambda_1)$.

The solution of the original equation (30) is defined recursively by the following formulas:

$$y_{k-1}(x) = Y(y_k, \lambda_k); \quad k = n, \ldots, 1, \qquad \text{where} \quad y_n(x) \equiv f(x), \quad y_0(x) \equiv y(x).$$

Note that here the decreasing sequence $k = n, \ldots, 1$ is used.

9.4-5. A Generalization

Suppose that the left-hand side of a linear (integral) equation

$$y(x) - \mathbf{Q}[y] = f(x) \tag{34}$$

can be represented in the form of a product

$$y(x) - \mathbf{Q}\,[y] \equiv \prod_{k=1}^{n} \left(1 - \mathbf{L}_k\right)[y], \tag{35}$$

where the $\mathbf{L}_k$ are linear operators. Suppose that the solutions of the auxiliary equations

$$y(x) - \mathbf{L}_k\,[y] = f(x), \qquad k = 1, \dots, n \tag{36}$$

are known and are given by the formulas

$$y(x) = Y_k\big[f(x)\big], \qquad k = 1, \dots, n. \tag{37}$$

The solution of the auxiliary equation (36) for $k = n$, in which we apply the substitution $y \to y_{n-1}$, is given by the formula $y_{n-1}(x) = Y_n\big[f(x)\big]$. Reasoning similar to that used in Subsection 9.4-3 shows that the solution of Eq. (34) can be reduced to the solution of the simpler equation

$$\prod_{k=1}^{n-1} \left(1 - \mathbf{L}_k\right)[y] = y_{n-1}(x).$$

Successively reducing the order of the equation, we eventually arrive at an equation of the form (36) for $k = 1$, whose right-hand side contains the function $y_1(x) = Y_2\big[y_2(x)\big]$. The solution of this equation is given by the formula $y(x) = Y_1\big[y_1(x)\big]$.

The solution of the original equation (35) can be defined recursively by the following formulas:

$$y_{k-1}(x) = Y_k\big[y_k(x)\big]; \quad k = n, \dots, 1, \qquad \text{where} \quad y_n(x) \equiv f(x), \quad y_0(x) \equiv y(x).$$

Note that here the decreasing sequence $k = n, \dots, 1$ is used.

⊙ Reference for Section 9.4: A. D. Polyanin and A. V. Manzhirov (1998).

9.5. Construction of Solutions of Integral Equations With Special Right-Hand Side

In this section we describe some approaches to the construction of solutions of integral equations with special right-hand side. These approaches are based on the application of auxiliary solutions that depend on a free parameter.

9.5-1. The General Scheme

Consider a linear equation, which we shall write in the following brief form:

$$\mathbf{L}\,[y] = f_g(x, \lambda), \tag{1}$$

where $\mathbf{L}$ is a linear operator (integral, differential, etc.) that acts with respect to the variable x and is independent of the parameter λ, and $f_g(x, \lambda)$ is a given function that depends on the variable x and the parameter λ.

Suppose that the solution of Eq. (1) is known:

$$y = y(x, \lambda). \tag{2}$$

Let $\mathbf{M}$ be a linear operator (integral, differential, etc.) that acts with respect to the parameter λ and is independent of the variable x. Consider the (usual) case in which $\mathbf{M}$ commutes with $\mathbf{L}$. We apply the operator $\mathbf{M}$ to Eq. (1) and find that the equation

$$\mathbf{L}\,[w] = f_M(x), \qquad f_M(x) = \mathbf{M}\big[f_g(x, \lambda)\big], \tag{3}$$

has the solution

$$w = \mathbf{M}\big[y(x, \lambda)\big]. \tag{4}$$

By choosing the operator $\mathbf{M}$ in a different way, we can obtain solutions for other right-hand sides of Eq. (1). The original function $f_g(x, \lambda)$ is called the *generating function* for the operator $\mathbf{L}$.

Consider a linear equation with exponential right-hand side

$$\mathbf{L}\,[y] = e^{\lambda x}. \tag{5}$$

Suppose that the solution is known and is given by formula (2). In Table 4 we present solutions of the equation $\mathbf{L}\,[y] = f(x)$ with various right-hand sides; these solutions are expressed via the solution of Eq. (5).

Remark 1. When applying the formulas indicated in the table, we need not know the left-hand side of the linear equation (5) (the equation can be integral, differential, etc.) provided that a particular solution of this equation for exponential right-hand side is known. It is only of importance that the left-hand side of the equation is independent of the parameter λ.

Remark 2. When applying formulas indicated in the table, the convergence of the integrals occurring in the resulting solution must be verified.

Example 1. We seek a solution of the equation with exponential right-hand side

$$y(x) + \int_{x}^{\infty} K(x-t)y(t)\,dt = e^{\lambda x} \tag{6}$$

in the form $y(x, \lambda) = ke^{\lambda x}$ by the method of indeterminate coefficients. Then we obtain

$$y(x, \lambda) = \frac{1}{B(\lambda)}e^{\lambda x}, \qquad B(\lambda) = 1 + \int_{0}^{\infty} K(-z)e^{\lambda z}\,dz. \tag{7}$$

It follows from row 3 of Table 4 that the solution of the equation

$$y(x) + \int_{x}^{\infty} K(x-t)y(t)\,dt = Ax \tag{8}$$

has the form

$$y(x) = \frac{A}{D}x - \frac{AC}{D^2},$$

$$D = 1 + \int_{0}^{\infty} K(-z)\,dz, \qquad C = \int_{0}^{\infty} zK(-z)\,dz.$$

For such a solution to exist, it is necessary that the improper integrals of the functions $K(-z)$ and $zK(-z)$ exist. This holds if the function $K(-z)$ decreases more rapidly than z^{-2} as $z \to \infty$. Otherwise a solution can be nonexistent. It is of interest that for functions $K(-z)$ with power-law growth as $z \to \infty$ in the case $\lambda < 0$, the solution of Eq. (6) exists and is given by formula (7), whereas Eq. (8) does not have a solution. Therefore, we must be careful when using formulas from Table 4 and verify the convergence of the integrals occurring in the solution.

It follows from row 15 of Table 4 that the solution of the equation

$$y(x) + \int_{x}^{\infty} K(x-t)y(t)\,dt = A\sin(\lambda x) \tag{9}$$

is given by the formula

$$y(x) = \frac{A}{B_{\mathrm{c}}^2 + B_{\mathrm{s}}^2}\left[B_{\mathrm{c}}\sin(\lambda x) - B_{\mathrm{s}}\cos(\lambda x)\right],$$

$$B_{\mathrm{c}} = 1 + \int_{0}^{\infty} K(-z)\cos(\lambda z)\,dz, \qquad B_{\mathrm{s}} = \int_{0}^{\infty} K(-z)\sin(\lambda z)\,dz.$$

TABLE 4

Solutions of the equation $\mathbf{L}\,[y] = f(x)$ with generating function of the exponential form

No	Right-Hand Side $f(x)$	Solution y	Solution Method
1	$e^{\lambda x}$	$y(x, \lambda)$	Original Equation
2	$A_1 e^{\lambda_1 x} + \cdots + A_n e^{\lambda_n x}$	$A_1 y(x, \lambda_1) + \cdots + A_n y(x, \lambda_n)$	Follows from linearity
3	$Ax + B$	$A\dfrac{\partial}{\partial \lambda}\Big[y(x,\lambda)\Big]_{\lambda=0} + By(x,0)$	Follows from linearity and the results of row No 4
4	$Ax^n,$ $n = 0, 1, 2, \ldots$	$A\left\{\dfrac{\partial^n}{\partial \lambda^n}\Big[y(x,\lambda)\Big]\right\}_{\lambda=0}$	Follows from the results of row No 6 for $\lambda = 0$
5	$\dfrac{A}{x+a},\ a > 0$	$A\displaystyle\int_0^\infty e^{-a\lambda}y(x,-\lambda)\,d\lambda$	Integration with respect to the parameter λ
6	$Ax^n e^{\lambda x},$ $n = 0, 1, 2, \ldots$	$A\dfrac{\partial^n}{\partial \lambda^n}\Big[y(x,\lambda)\Big]$	Differentiation with respect to the parameter λ
7	a^x	$y(x, \ln a)$	Follows from row No 1
8	$A\cosh(\lambda x)$	$\tfrac{1}{2}A[y(x,\lambda) + y(x,-\lambda)]$	Linearity and relations to the exponential
9	$A\sinh(\lambda x)$	$\tfrac{1}{2}A[y(x,\lambda) - y(x,-\lambda)]$	Linearity and relations to the exponential
10	$Ax^m \cosh(\lambda x),$ $m = 1, 3, 5, \ldots$	$\tfrac{1}{2}A\dfrac{\partial^m}{\partial \lambda^m}[y(x,\lambda) - y(x,-\lambda)]$	Differentiation with respect to λ and relation to the exponential
11	$Ax^m \cosh(\lambda x),$ $m = 2, 4, 6, \ldots$	$\tfrac{1}{2}A\dfrac{\partial^m}{\partial \lambda^m}[y(x,\lambda) + y(x,-\lambda)]$	Differentiation with respect to λ and relation to the exponential
12	$Ax^m \sinh(\lambda x),$ $m = 1, 3, 5, \ldots$	$\tfrac{1}{2}A\dfrac{\partial^m}{\partial \lambda^m}[y(x,\lambda) + y(x,-\lambda)]$	Differentiation with respect to λ and relation to the exponential
13	$Ax^m \sinh(\lambda x),$ $m = 2, 4, 6, \ldots$	$\tfrac{1}{2}A\dfrac{\partial^m}{\partial \lambda^m}[y(x,\lambda) - y(x,-\lambda)]$	Differentiation with respect to λ and relation to the exponential
14	$A\cos(\beta x)$	$A\,\mathrm{Re}\big[y(x, i\beta)\big]$	Selection of the real part for $\lambda = i\beta$
15	$A\sin(\beta x)$	$A\,\mathrm{Im}\big[y(x, i\beta)\big]$	Selection of the imaginary part for $\lambda = i\beta$
16	$Ax^n \cos(\beta x),$ $n = 1, 2, 3, \ldots$	$A\,\mathrm{Re}\left\{\dfrac{\partial^n}{\partial \lambda^n}\Big[y(x,\lambda)\Big]\right\}_{\lambda=i\beta}$	Differentiation with respect to λ and selection of the real part for $\lambda = i\beta$
17	$Ax^n \sin(\beta x),$ $n = 1, 2, 3, \ldots$	$A\,\mathrm{Im}\left\{\dfrac{\partial^n}{\partial \lambda^n}\Big[y(x,\lambda)\Big]\right\}_{\lambda=i\beta}$	Differentiation with respect to λ and selection of the imaginary part for $\lambda = i\beta$
18	$Ae^{\mu x}\cos(\beta x)$	$A\,\mathrm{Re}\big[y(x, \mu + i\beta)\big]$	Selection of the real part for $\lambda = \mu + i\beta$
19	$Ae^{\mu x}\sin(\beta x)$	$A\,\mathrm{Im}\big[y(x, \mu + i\beta)\big]$	Selection of the imaginary part for $\lambda = \mu + i\beta$
20	$Ax^n e^{\mu x}\cos(\beta x),$ $n = 1, 2, 3, \ldots$	$A\,\mathrm{Re}\left\{\dfrac{\partial^n}{\partial \lambda^n}\Big[y(x,\lambda)\Big]\right\}_{\lambda=\mu+i\beta}$	Differentiation with respect to λ and selection of the real part for $\lambda = \mu + i\beta$
21	$Ax^n e^{\mu x}\sin(\beta x),$ $n = 1, 2, 3, \ldots$	$A\,\mathrm{Im}\left\{\dfrac{\partial^n}{\partial \lambda^n}\Big[y(x,\lambda)\Big]\right\}_{\lambda=\mu+i\beta}$	Differentiation with respect to λ and selection of the imaginary part for $\lambda = \mu + i\beta$

9.5-3. Power-Law Generating Function

Consider the linear equation with power-law right-hand side

$$\mathbf{L}[y] = x^\lambda. \tag{10}$$

Suppose that the solution is known and is given by formula (2). In Table 5, solutions of the equation $\mathbf{L}[y] = f(x)$ with various right-hand sides are presented which can be expressed via the solution of Eq. (10).

TABLE 5
Solutions of the equation $\mathbf{L}[y] = f(x)$ with generating function of power-law form

No	Right-Hand Side $f(x)$	Solution y	Solution Method
1	x^λ	$y(x, \lambda)$	Original Equation
2	$\sum\limits_{k=0}^{n} A_k x^k$	$\sum\limits_{k=0}^{n} A_k y(x, k)$	Follows from linearity
3	$A \ln x + B$	$A\dfrac{\partial}{\partial\lambda}\big[y(x, \lambda)\big]_{\lambda=0} + B y(x, 0)$	Follows from linearity and from the results of row No 4
4	$A \ln^n x,$ $n = 0, 1, 2, \ldots$	$A\left\{\dfrac{\partial^n}{\partial\lambda^n}\big[y(x, \lambda)\big]\right\}_{\lambda=0}$	Follows from the results of row No 5 for $\lambda = 0$
5	$A x^n x^\lambda,$ $n = 0, 1, 2, \ldots$	$A\dfrac{\partial^n}{\partial\lambda^n}\big[y(x, \lambda)\big]$	Differentiation with respect to the parameter λ
6	$A \cos(\beta \ln x)$	$A \operatorname{Re}\big[y(x, i\beta)\big]$	Selection of the real part for $\lambda = i\beta$
7	$A \sin(\beta \ln x)$	$A \operatorname{Im}\big[y(x, i\beta)\big]$	Selection of the imaginary part for $\lambda = i\beta$
8	$A x^\mu \cos(\beta \ln x)$	$A \operatorname{Re}\big[y(x, \mu + i\beta)\big]$	Selection of the real part for $\lambda = \mu + i\beta$
9	$A x^\mu \sin(\beta \ln x)$	$A \operatorname{Im}\big[y(x, \mu + i\beta)\big]$	Selection of the imaginary part for $\lambda = \mu + i\beta$

Example 2. We seek a solution of the equation with power-law right-hand side

$$y(x) + \int_0^x \frac{1}{x} K\left(\frac{t}{x}\right) y(t)\, dt = x^\lambda$$

in the form $y(x, \lambda) = k x^\lambda$ by the method of indeterminate coefficients. We finally obtain

$$y(x, \lambda) = \frac{1}{1 + B(\lambda)} x^\lambda, \qquad B(\lambda) = \int_0^1 K(t) t^\lambda\, dt.$$

It follows from row 3 of Table 5 that the solution of the equation with logarithmic right-hand side

$$y(x) + \int_0^x \frac{1}{x} K\left(\frac{t}{x}\right) y(t)\, dt = A \ln x$$

has the form

$$y(x) = \frac{A}{1 + I_0} \ln x - \frac{A I_1}{(1 + I_0)^2},$$

$$I_0 = \int_0^1 K(t)\, dt, \qquad I_1 = \int_0^1 K(t) \ln t\, dt.$$

9.5-4. Generating Function Containing Sines and Cosines

Consider the linear equation

$$\mathbf{L}\,[y] = \sin(\lambda x). \tag{11}$$

We assume that the solution of this equation is known and is given by formula (2). In Table 6, solutions of the equation $\mathbf{L}\,[y] = f(x)$ with various right-hand sides are given, which are expressed via the solution of Eq. (11).

Consider the linear equation

$$\mathbf{L}\,[y] = \cos(\lambda x). \tag{12}$$

We assume that the solution of this equation is known and is given by formula (2). In Table 7, solutions of the equation $\mathbf{L}\,[y] = f(x)$ with various right-hand sides are given, which are expressed via the solution of Eq. (12).

TABLE 6

Solutions of the equation $\mathbf{L}\,[y] = f(x)$ with sine-shaped generating function

No	Right-Hand Side $f(x)$	Solution y	Solution Method
1	$\sin(\lambda x)$	$y(x,\lambda)$	Original Equation
2	$\sum\limits_{k=1}^{n} A_k \sin(\lambda_k x)$	$\sum\limits_{k=1}^{n} A_k y(x,\lambda_k)$	Follows from linearity
3	$Ax^m,$ $m = 1, 3, 5, \ldots$	$A(-1)^{\frac{m-1}{2}} \left[\dfrac{\partial^m}{\partial \lambda^m} y(x,\lambda) \right]_{\lambda=0}$	Follows from the results of row 5 for $\lambda = 0$
4	$Ax^m \sin(\lambda x),$ $m = 2, 4, 6, \ldots$	$A(-1)^{\frac{m}{2}} \dfrac{\partial^m}{\partial \lambda^m} y(x,\lambda)$	Differentiation with respect to the parameter λ
5	$Ax^m \cos(\lambda x),$ $m = 1, 3, 5, \ldots$	$A(-1)^{\frac{m-1}{2}} \dfrac{\partial^m}{\partial \lambda^m} y(x,\lambda)$	Differentiation with respect to the parameter λ
6	$\sinh(\beta x)$	$-iy(x,i\beta)$	Relation to the hyperbolic sine, $\lambda = i\beta$
7	$x^m \sinh(\beta x),$ $m = 2, 4, 6, \ldots$	$i(-1)^{\frac{m+2}{2}} \left[\dfrac{\partial^m}{\partial \lambda^m} y(x,\lambda) \right]_{\lambda=i\beta}$	Differentiation with respect to λ and relation to the hyperbolic sine, $\lambda = i\beta$

TABLE 7

Solutions of the equation $\mathbf{L}\,[y] = f(x)$ with cosine-shaped generating function

No	Right-Hand Side $f(x)$	Solution y	Solution Method
1	$\cos(\lambda x)$	$y(x,\lambda)$	Original Equation
2	$\sum\limits_{k=1}^{n} A_k \cos(\lambda_k x)$	$\sum\limits_{k=1}^{n} A_k y(x,\lambda_k)$	Follows from linearity
3	$Ax^m,$ $m = 0, 2, 4, \ldots$	$A(-1)^{\frac{m}{2}} \left[\dfrac{\partial^m}{\partial \lambda^m} y(x,\lambda) \right]_{\lambda=0}$	Follows from the results of row 4 for $\lambda = 0$
4	$Ax^m \cos(\lambda x),$ $m = 2, 4, 6, \ldots$	$A(-1)^{\frac{m}{2}} \dfrac{\partial^m}{\partial \lambda^m} y(x,\lambda)$	Differentiation with respect to the parameter λ
5	$Ax^m \sin(\lambda x),$ $m = 1, 3, 5, \ldots$	$A(-1)^{\frac{m+1}{2}} \dfrac{\partial^m}{\partial \lambda^m} y(x,\lambda)$	Differentiation with respect to the parameter λ
6	$\cosh(\beta x)$	$y(x,i\beta)$	Relation to the hyperbolic cosine, $\lambda = i\beta$
7	$x^m \cosh(\beta x),$ $m = 2, 4, 6, \ldots$	$(-1)^{\frac{m}{2}} \left[\dfrac{\partial^m}{\partial \lambda^m} y(x,\lambda) \right]_{\lambda=i\beta}$	Differentiation with respect to λ and relation to the hyperbolic cosine, $\lambda = i\beta$

9.6. The Method of Model Solutions

9.6-1. Preliminary Remarks*

Consider a linear equation, which we briefly write out in the form

$$\mathbf{L}\,[y(x)] = f(x), \tag{1}$$

where $\mathbf{L}$ is a linear (integral) operator, $y(x)$ is an unknown function, and $f(x)$ is a known function.
   We first define arbitrarily a test solution

$$y_0 = y_0(x, \lambda), \tag{2}$$

which depends on an auxiliary parameter λ (it is assumed that the operator $\mathbf{L}$ is independent of λ and $y_0 \not\equiv \text{const}$). By means of Eq. (1) we define the right-hand side that corresponds to the test solution (2):

$$f_0(x, \lambda) = \mathbf{L}\,[y_0(x, \lambda)].$$

Let us multiply Eq. (1), for $y = y_0$ and $f = f_0$, by some function $\varphi(\lambda)$ and integrate the resulting relation with respect to λ over an interval $[a, b]$. We finally obtain

$$\mathbf{L}\,[y_\varphi(x)] = f_\varphi(x), \tag{3}$$

where

$$y_\varphi(x) = \int_a^b y_0(x, \lambda)\varphi(\lambda)\,d\lambda, \qquad f_\varphi(x) = \int_a^b f_0(x, \lambda)\varphi(\lambda)\,d\lambda. \tag{4}$$

   It follows from formulas (3) and (4) that, for the right-hand side $f = f_\varphi(x)$, the function $y = y_\varphi(x)$ is a solution of the original equation (1). Since the choice of the function $\varphi(\lambda)$ (as well as of the integration interval) is arbitrary, the function $f_\varphi(x)$ can be arbitrary in principle. Here the main problem is how to choose a function $\varphi(\lambda)$ to obtain a given function $f_\varphi(x)$. This problem can be solved if we can find a test solution such that the right-hand side of Eq. (1) is the kernel of a known inverse integral transform (we denote such a test solution by $Y(x, \lambda)$ and call it a *model solution*).

9.6-2. Description of the Method

Indeed, let $\mathfrak{P}$ be an invertible integral transform that takes each function $f(x)$ to the corresponding transform $F(\lambda)$ by the rule

$$F(\lambda) = \mathfrak{P}\{f(x)\}. \tag{5}$$

Assume that the inverse transform $\mathfrak{P}^{-1}$ has the kernel $\psi(x, \lambda)$ and acts as follows:

$$\mathfrak{P}^{-1}\{F(\lambda)\} = f(x), \qquad \mathfrak{P}^{-1}\{F(\lambda)\} \equiv \int_a^b F(\lambda)\psi(x, \lambda)\,d\lambda. \tag{6}$$

The limits of integration a and b and the integration path in (6) may well lie in the complex plane.
   Suppose that we succeeded in finding a model solution $Y(x, \lambda)$ of the auxiliary problem for Eq. (1) whose right-hand side is the kernel of the inverse transform $\mathfrak{P}^{-1}$:

$$\mathbf{L}\,[Y(x, \lambda)] = \psi(x, \lambda). \tag{7}$$

---

* Before reading this section, it is useful to look over Section 9.5.

Let us multiply Eq. (7) by $F(\lambda)$ and integrate with respect to λ within the same limits that stand in the inverse transform (6). Taking into account the fact that the operator $\mathbf{L}$ is independent of λ and applying the relation $\mathfrak{P}^{-1}\{F(\lambda)\} = f(x)$, we obtain

$$\mathbf{L}\left[\int_a^b Y(x,\lambda)F(\lambda)\,d\lambda\right] = f(x).$$

Therefore, the solution of Eq. (1) for an arbitrary function $f(x)$ on the right-hand side is expressed via a solution of the simpler auxiliary equation (7) by the formula

$$y(x) = \int_a^b Y(x,\lambda)F(\lambda)\,d\lambda, \tag{8}$$

where $F(\lambda)$ is the transform (5) of the function $f(x)$.

For the right-hand side of the auxiliary equation (7) we can take, for instance, exponential, power-law, and trigonometric function, which are the kernels of the Laplace, Mellin, and sine and cosine Fourier transforms (up to a constant factor). Sometimes it is rather easy to find a model solution by means of the method of indeterminate coefficients (by prescribing its structure). Afterwards, to construct a solution of the equation with arbitrary right-hand side, we can apply formulas written out below in Subsections 9.6-3–9.6-6.

9.6-3. The Model Solution in the Case of an Exponential Right-Hand Side

Assume that we have found a model solution $Y = Y(x,\lambda)$ that corresponds to the exponential right-hand side:

$$\mathbf{L}\,[Y(x,\lambda)] = e^{\lambda x}. \tag{9}$$

Consider two cases:

1°. *Equations on the semiaxis, $0 \le x < \infty$.* Let $\tilde{f}(p)$ be the Laplace transform of the function $f(x)$:

$$\tilde{f}(p) = \mathfrak{L}\{f(x)\}, \qquad \mathfrak{L}\{f(x)\} \equiv \int_0^\infty f(x)e^{-px}\,dx. \tag{10}$$

The solution of Eq. (1) for an arbitrary right-hand side $f(x)$ can be expressed via the solution of the simpler auxiliary equation with exponential right-hand side (9) for $\lambda = p$ by the formula

$$y(x) = \frac{1}{2\pi i}\int_{c-i\infty}^{c+i\infty} Y(x,p)\tilde{f}(p)\,dp. \tag{11}$$

2°. *Equations on the entire axis, $-\infty < x < \infty$.* Let $\tilde{f}(u)$ the Fourier transform of the function $f(x)$:

$$\tilde{f}(u) = \mathfrak{F}\{f(x)\}, \qquad \mathfrak{F}\{f(x)\} \equiv \frac{1}{\sqrt{2\pi}}\int_{-\infty}^\infty f(x)e^{-iux}\,dx. \tag{12}$$

The solution of Eq. (1) for an arbitrary right-hand side $f(x)$ can be expressed via the solution of the simpler auxiliary equation with exponential right-hand side (9) for $\lambda = iu$ by the formula

$$y(x) = \frac{1}{\sqrt{2\pi}}\int_{-\infty}^\infty Y(x, iu)\tilde{f}(u)\,du. \tag{13}$$

In the calculation of the integrals on the right-hand sides in (11) and (13), methods of the theory of functions of a complex variable are applied, including the Jordan lemma and the Cauchy residue theorem (see Subsections 7.1-4 and 7.1-5).

Remark 1. The structure of a model solution $Y(x, \lambda)$ can differ from that of the kernel of the Laplace or Fourier inversion formula.

Remark 2. When applying the method under consideration, the left-hand side of Eq. (1) need not be known (the equation can be integral, differential, functional, etc.) if a particular solution of this equation is known for the exponential right-hand side. Here only the most general information is important, namely, that the equation is linear, and its left-hand side is independent of the parameter λ.

Remark 3. The above method can be used in the solution of linear integral (differential, integro-differential, and functional) equations with composed argument of the unknown function.

Example 1. Consider the following Volterra equation of the second kind with difference kernel:

$$y(x) + \int_x^\infty K(x - t)y(t)\, dt = f(x). \tag{14}$$

This equation cannot be solved by direct application of the Laplace transform because the convolution theorem cannot be used here.

In accordance with the method of model solutions, we consider the auxiliary equation with exponential right-hand side

$$y(x) + \int_x^\infty K(x - t)y(t)\, dt = e^{px}. \tag{15}$$

Its solution has the form (see Example 1 of Section 9.5)

$$Y(x, p) = \frac{1}{1 + \tilde{K}(-p)} e^{px}, \qquad \tilde{K}(-p) = \int_0^\infty K(-z)e^{pz}\, dz. \tag{16}$$

This, by means of formula (11), yields a solution of Eq. (12) for an arbitrary right-hand side,

$$y(x) = \frac{1}{2\pi i} \int_{c-i\infty}^{c+i\infty} \frac{\tilde{f}(p)}{1 + \tilde{K}(-p)} e^{px}\, dp, \tag{17}$$

where $\tilde{f}(p)$ is the Laplace transform (10) of the function $f(x)$ (see also Section 9.11).

Note that a solution to Eq. (12) was obtained in the book of M. L. Krasnov, A. I. Kiselev, and G. I. Makarenko (1971) in a more complicated way.

9.6-4. The Model Solution in the Case of a Power-Law Right-Hand Side

Suppose that we have succeeded in finding a model solution $Y = Y(x, s)$ that corresponds to a power-law right-hand side of the equation:

$$\mathbf{L}\,[Y(x, s)] = x^{-s}, \qquad \lambda = -s. \tag{18}$$

Let $\hat{f}(s)$ be the Mellin transform of the function $f(x)$:

$$\hat{f}(s) = \mathfrak{M}\{f(x)\}, \qquad \mathfrak{M}\{f(x)\} \equiv \int_0^\infty f(x)x^{s-1}\, dx. \tag{19}$$

The solution of Eq. (1) for an arbitrary right-hand side $f(x)$ can be expressed via the solution of the simpler auxiliary equation with power-law right-hand side (18) by the formula

$$y(x) = \frac{1}{2\pi i} \int_{c-i\infty}^{c+i\infty} Y(x, s)\hat{f}(s)\, ds. \tag{20}$$

In the calculation of the corresponding integrals on the right-hand side of formula (20), one can use tables of inverse Mellin transforms (e.g., see Supplement 9), as well as methods of the theory of functions of a complex variable, including the Jordan lemma and the Cauchy residue theorem (see Subsections 7.1-4 and 7.1-5).

Example 2. Consider the equation

$$y(x) + \int_0^x \frac{1}{x} K\left(\frac{t}{x}\right) y(t)\,dt = f(x). \tag{21}$$

In accordance with the method of model solutions, we consider the following auxiliary equation with power-law right-hand side:

$$y(x) + \int_0^x \frac{1}{x} K\left(\frac{t}{x}\right) y(t)\,dt = x^{-s}. \tag{22}$$

Its solution has the form (see Example 2 for $\lambda = -s$ in Section 9.5)

$$Y(x,s) = \frac{1}{1+B(s)} x^{-s}, \qquad B(s) = \int_0^1 K(t)t^{-s}\,dt. \tag{23}$$

This, by means of formula (20), yields the solution of Eq. (21) for an arbitrary right-hand side:

$$y(x) = \frac{1}{2\pi i} \int_{c-i\infty}^{c+i\infty} \frac{\hat{f}(s)}{1+B(s)} x^{-s}\,ds, \tag{24}$$

where $\hat{f}(s)$ is the Mellin transform (19) of the function $f(x)$.

9.6-5. The Model Solution in the Case of a Sine-Shaped Right-Hand Side

Suppose that we have succeeded in finding a model solution $Y = Y(x,u)$ that corresponds to the sine on the right-hand side:

$$\mathbf{L}\,[Y(x,u)] = \sin(ux), \qquad \lambda = u. \tag{25}$$

Let $\check{f}_s(u)$ be the asymmetric sine Fourier transform of the function $f(x)$:

$$\check{f}_s(u) = \mathcal{F}_s\{f(x)\}, \qquad \mathcal{F}_s\{f(x)\} \equiv \int_0^\infty f(x)\sin(ux)\,dx. \tag{26}$$

The solution of Eq. (1) for an arbitrary right-hand side $f(x)$ can be expressed via the solution of the simpler auxiliary equation with sine-shape right-hand side (25) by the formula

$$y(x) = \frac{2}{\pi} \int_0^\infty Y(x,u)\check{f}_s(u)\,du. \tag{27}$$

9.6-6. The Model Solution in the Case of a Cosine-Shaped Right-Hand Side

Suppose that we have succeeded in finding a model solution $Y = Y(x,u)$ that corresponds to the cosine on the right-hand side:

$$\mathbf{L}\,[Y(x,u)] = \cos(ux), \qquad \lambda = u. \tag{28}$$

Let $\check{f}_c(u)$ be the asymmetric Fourier cosine transform of the function $f(x)$:

$$\check{f}_c(u) = \mathcal{F}_c\{f(x)\}, \qquad \mathcal{F}_c\{f(x)\} \equiv \int_0^\infty f(x)\cos(ux)\,dx. \tag{29}$$

The solution of Eq. (1) for an arbitrary right-hand side $f(x)$ can be expressed via the solution of the simpler auxiliary equation with cosine right-hand side (28) by the formula

$$y(x) = \frac{2}{\pi} \int_0^\infty Y(x,u)\check{f}_c(u)\,du. \tag{30}$$

9.6-7. Some Generalizations

Just as above we assume that $\mathfrak{P}$ is an invertible transform taking each function $f(x)$ to the corresponding transform $F(\lambda)$ by the rule (5) and that the inverse transform is defined by formula (6).

Suppose that we have succeeded in finding a model solution $Y(x, \lambda)$ of the following auxiliary problem for Eq. (1):

$$\mathbf{L}_x\,[Y(x, \lambda)] = \mathbf{H}_\lambda\,[\psi(x, \lambda)]. \tag{31}$$

The right-hand side of Eq. (31) contains an invertible linear operator (which is integral, differential, or functional) that is independent of the variable x and acts with respect to the parameter λ on the kernel $\psi(x, \lambda)$ of the inverse transform, see formula (6). For clarity, the operator on the left-hand side of Eq. (31) is labeled by the subscript x (it acts with respect to the variable x and is independent of λ).

Let us apply the inverse operator $\mathbf{H}_\lambda^{-1}$ to Eq. (31). As a result, we obtain the kernel $\psi(x, \lambda)$ on the right-hand side. On the left-hand side we intertwine the operators by the rule $\mathbf{H}_\lambda^{-1}\,\mathbf{L}_x = \mathbf{L}_x\,\mathbf{H}_\lambda^{-1}$ (this is as a rule possible because the operators act with respect to different variables). Furthermore, let us multiply the resulting relation by $F(\lambda)$ and integrate with respect to λ within the limits that stand in the inverse transform (6). Taking into account the relation $\mathfrak{P}^{-1}\,\{F(\lambda)\} = f(x)$, we finally obtain

$$\mathbf{L}_x\left[\int_a^b F(\lambda)\mathbf{H}_\lambda^{-1}[Y(x, \lambda)]\,d\lambda\right] = f(x). \tag{32}$$

Hence, a solution of Eq. (1) with an arbitrary function $f(x)$ on the right-hand side can be expressed via the solution of the simpler auxiliary equation (31) by the formula

$$y(x) = \int_a^b F(\lambda)\mathbf{H}_\lambda^{-1}[Y(x, \lambda)]\,d\lambda, \tag{33}$$

where $F(\lambda)$ is the transform of the function $f(x)$ obtained by means of the transform $\mathfrak{P}$ (5).

Since the choice of the operator $\mathbf{H}_\lambda$ is arbitrary, this approach extends the abilities of the method of model solutions.

⊙ References for Section 9.6: A. D. Polyanin and A. V. Manzhirov (1997, 1998).

9.7. Method of Differentiation for Integral Equations

In some cases, the differentiation of integral equations (once, twice, and so on) with the subsequent elimination of integral terms by means of the original equation makes it possible to reduce a given equation to an ordinary differential equation. Sometimes by differentiating we can reduce a given equation to a simpler integral equation whose solution is known. Below we list some classes of integral equations that can be reduced to ordinary differential equations with constant coefficients.

9.7-1. Equations With Kernel Containing a Sum of Exponential Functions

Consider the equation

$$y(x) + \int_a^x \left[\sum_{k=1}^n A_k e^{\lambda_k(x-t)}\right] y(t)\,dt = f(x). \tag{1}$$

In the general case, this equation can be reduced to a linear nonhomogeneous ordinary differential equation of nth order with constant coefficients (see equation 2.2.19 of the first part of the book).

In a wide range of the parameters A_k and λ_k, the solution can be represented as follows:

$$y(x) = f(x) + \int_a^x \left[\sum_{k=1}^n B_k e^{\mu_k(x-t)}\right] f(t)\,dt, \tag{2}$$

where the parameters B_k and μ_k of the solution are related to the parameters A_k and λ_k of the equation by algebraic relations.

For the solution of Eq. (1) with $n = 2$, see Section 2.2 of the first part of the book (equation 2.2.10).

9.7-2. Equations With Kernel Containing a Sum of Hyperbolic Functions

By means of the formulas $\cosh \beta = \frac{1}{2}(e^\beta + e^{-\beta})$ and $\sinh \beta = \frac{1}{2}(e^\beta - e^{-\beta})$, any equation with difference kernel of the form

$$y(x) + \int_a^x K(x - t)y(t)\,dt = f(x),$$

$$K(x) = \sum_{k=1}^m A_k \cosh(\lambda_k x) + \sum_{k=1}^s B_k \sinh(\mu_k x), \tag{3}$$

can be represented in the form of Eq. (1) with $n = 2m + 2s$, and hence these equations can be reduced to linear nonhomogeneous ordinary differential equations with constant coefficients.

9.7-3. Equations With Kernel Containing a Sum of Trigonometric Functions

Equations with difference kernel of the form

$$y(x) + \int_a^x K(x - t)y(t)\,dt = f(x), \qquad K(x) = \sum_{k=1}^m A_k \cos(\lambda_k x), \tag{4}$$

$$y(x) + \int_a^x K(x - t)y(t)\,dt = f(x), \qquad K(x) = \sum_{k=1}^m A_k \sin(\lambda_k x), \tag{5}$$

can also be reduced to linear nonhomogeneous ordinary differential equations of order $2m$ with constant coefficients (see equations 2.5.4 and 2.5.15 in the first part of the book).

In a wide range of the parameters A_k and λ_k, the solution of Eq. (5) can be represented in the form

$$y(x) = f(x) + \int_a^x R(x - t)f(t)\,dt, \qquad R(x) = \sum_{k=1}^m B_k \sin(\mu_k x), \tag{6}$$

where the parameters B_k and μ_k of the solution are related to the parameters A_k and λ_k of the equation by algebraic relations.

Equations with difference kernels containing both cosines and sines can also be reduced to linear nonhomogeneous ordinary differential equations with constant coefficients.

9.7-4. Equations Whose Kernels Contain Combinations of Various Functions

Any equation with difference kernel that contains a linear combination of summands of the form

$$(x - t)^m \ (m = 0, 1, 2, \ldots), \quad \exp[\alpha(x - t)],$$
$$\cosh[\beta(x - t)], \quad \sinh[\gamma(x - t)], \quad \cos[\lambda(x - t)], \quad \sin[\mu(x - t)], \tag{7}$$

can also be reduced by differentiation to a linear nonhomogeneous ordinary differential equation with constant coefficients, where exponential, hyperbolic, and trigonometric functions can also be multiplied by $(x - t)^n$ $(n = 1, 2, \ldots)$.

Remark. The method of differentiation can be successfully used to solve more complicated equations with nondifference kernel to which the Laplace transform cannot be applied (see, for instance, Eqs. 2.9.5, 2.9.28, 2.9.30, and 2.9.34 in the first part of the book).

9.8. Reduction of Volterra Equations of the Second Kind to Volterra Equations of the First Kind

The Volterra equation of the second kind

$$y(x) - \int_a^x K(x,t)y(t)\,dt = f(x) \tag{1}$$

can be reduced to a Volterra equation of the first kind in two ways.

9.8-1. The First Method

We integrate Eq. (1) with respect to x from a to x and then reverse the order of integration in the double integral. We finally obtain the Volterra equation of the first kind

$$\int_a^x M(x,t)y(t)\,dt = F(x), \tag{2}$$

where $M(x,t)$ and $F(x)$ are defined as follows:

$$M(x,t) = 1 - \int_t^x K(s,t)\,ds, \qquad F(x) = \int_a^x f(t)\,dt. \tag{3}$$

9.8-2. The Second Method

Assume that the condition $f(a) = 0$ is satisfied. In this case Eq. (1) can be reduced to a Volterra equation of the first kind for the derivative of the unknown function,

$$\int_a^x N(x,t)y_t'(t)\,dt = f(x), \qquad y(a) = 0, \tag{4}$$

where

$$N(x,t) = 1 - \int_t^x K(x,s)\,ds. \tag{5}$$

Indeed, on integrating by parts the right-hand side of formula (4) with regard to formula (5), we arrive at Eq. (1).

Remark. For $f(a) \neq 0$, Eq. (1) implies the relation $y(a) = f(a)$. In this case the substitution $z(x) = y(x) - f(a)$ yields the Volterra equation of the second kind

$$z(x) - \int_a^x K(x,t)z(t)\,dt = \Phi(x),$$

$$\Phi(x) = f(x) - f(a) + f(a)\int_a^x K(x,t)\,dt,$$

whose right-hand side satisfies the condition $\Phi(a) = 0$, and hence this equation can be reduced by the second method to a Volterra equation of the first kind.

⊙ References for Section 9.8: V. Volterra (1959), A. F. Verlan' and V. S. Sizikov (1986).

9.9. The Successive Approximation Method

9.9-1. The General Scheme

$1°$. Consider a Volterra integral equation of the second kind

$$y(x) - \int_a^x K(x,\,t)y(t)\,dt = f(x). \tag{1}$$

Assume that $f(x)$ is continuous on the interval $[a, b]$ and the kernel $K(x,\,t)$ is continuous for $a \le x \le b$ and $a \le t \le x$.

Let us seek the solution by the successive approximation method. To this end, we set

$$y(x) = f(x) + \sum_{n=1}^{\infty} \varphi_n(x), \tag{2}$$

where the $\varphi_n(x)$ are determined by the formulas

$$\varphi_1(x) = \int_a^x K(x,\,t)f(t)\,dt,$$

$$\varphi_2(x) = \int_a^x K(x,\,t)\varphi_1(t)\,dt = \int_a^x K_2(x,\,t)f(t)\,dt,$$

$$\varphi_3(x) = \int_a^x K(x,\,t)\varphi_2(t)\,dt = \int_a^x K_3(x,\,t)f(t)\,dt, \quad \text{etc.}$$

Here

$$K_n(x,\,t) = \int_a^x K(x,\,z)K_{n-1}(z,\,t)\,dz, \tag{3}$$

where $n = 2, 3, \ldots$, and we have the relations $K_1(x,\,t) \equiv K(x,\,t)$ and $K_n(x,\,t) = 0$ for $t > x$. The functions $K_n(x,\,t)$ given by formulas (3) are called *iterated kernels*. These kernels satisfy the relation

$$K_n(x,\,t) = \int_a^x K_m(x,\,s)K_{n-m}(s,\,t)\,ds, \tag{4}$$

where m is an arbitrary positive integer less than n.

$2°$. The successive approximations can be implemented in a more general scheme:

$$y_n(x) = f(x) + \int_a^x K(x,t)y_{n-1}(t)\,dt, \qquad n = 1, 2, \ldots, \tag{5}$$

where the function $y_0(x)$ is continuous on the interval $[a, b]$. The functions $y_1(x)$, $y_2(x)$, $\ldots$ which are obtained from (5) are also continuous on $[a, b]$.

Under the assumptions adopted in item $1°$ for $f(x)$ and $K(x,t)$, the sequence $\{y_n(x)\}$ converges, as $n \to \infty$, to the continuous solution $y(x)$ of the integral equation. A successful choice of the "zeroth" approximation $y_0(x)$ can result in a rapid convergence of the procedure.

Note that in the special case $y_0(x) = f(x)$, this method becomes that described in item $1°$.

Remark 1. If the kernel $K(x,t)$ is square integrable on the square $S = \{a \le x \le b,\ a \le t \le b\}$ and $f(x) \in L_2(a, b)$, then the successive approximations are mean-square convergent to the solution $y(x) \in L_2(a, b)$ of the integral equation (1) for any initial approximation $y_0(x) \in L_2(a, b)$.

9.9-2. A Formula for the Resolvent

The resolvent of the integral equation (1) is determined via the iterated kernels by the formula

$$R(x, t) = \sum_{n=1}^{\infty} K_n(x, t), \tag{6}$$

where the convergent series on the right-hand side is called the *Neumann series* of the kernel $K(x, t)$. Now the solution of the Volterra equation of the second kind (1) can be rewritten in the traditional form

$$y(x) = f(x) + \int_a^x R(x, t)f(t)\,dt. \tag{7}$$

Remark 2. In the case of a kernel with weak singularity, the solution of Eq. (1) can be obtained by the successive approximation method. In this case the kernels $K_n(x, t)$ are continuous starting from some n. For $\alpha < \frac{1}{2}$, even the kernel $K_2(x, t)$ is continuous.

⊙ References for Section 9.9: W. V. Lovitt (1950), V. Volterra (1959), S. G. Mikhlin (1960), M. L. Krasnov, A. I. Kiselev, and G. I. Makarenko (1971), V. I. Smirnov (1974).

9.10. Method of Quadratures

9.10-1. The General Scheme of the Method

Let us consider the linear Volterra integral equation of the second kind

$$y(x) - \int_a^x K(x, t)y(t)\,dt = f(x), \tag{1}$$

on an interval $a \leq x \leq b$. Assume that the kernel and the right-hand side of the equation are continuous functions.

From Eq. (1) we find that $y(a) = f(a)$. Let us choose a constant integration step h and consider the discrete set of points $x_i = a + h(i - 1)$, $i = 1, \ldots, n$. For $x = x_i$, Eq. (1) acquires the form

$$y(x_i) - \int_a^{x_i} K(x_i, t)y(t)\,dt = f(x_i), \qquad i = 1, \ldots, n. \tag{2}$$

Applying the quadrature formula (see Subsection 8.7-1) to the integral in (2) and choosing x_j $(j = 1, \ldots, i)$ to be the nodes in t, we arrive at the system of equations

$$y(x_i) - \sum_{j=1}^{i} A_{ij} K(x_i, x_j)y(x_j) = f(x_i) + \varepsilon_i[y], \qquad i = 2, \ldots, n, \tag{3}$$

where $\varepsilon_i[y]$ is the truncation error and A_{ij} are the coefficients of the quadrature formula on the interval $[a, x_i]$ (see Subsection 8.7-1). Suppose that $\varepsilon_i[y]$ are small and neglect them; then we obtain a system of linear algebraic equations in the form

$$y_1 = f_1, \quad y_i - \sum_{j=1}^{i} A_{ij} K_{ij} y_j = f_i, \qquad i = 2, \ldots, n, \tag{4}$$

where $K_{ij} = K(x_i, x_j)$, $f_i = f(x_i)$, and y_i are approximate values of the unknown function $y(x)$ at the nodes x_i.

From (4) we obtain the recurrent formula

$$y_1 = f_1, \quad y_i = \frac{f_i + \sum\limits_{j=1}^{i-1} A_{ij} K_{ij} y_j}{1 - A_{ii} K_{ii}}, \qquad i = 2, \ldots, n, \tag{5}$$

valid under the condition

$$1 - A_{ii} K_{ii} \neq 0, \tag{6}$$

which can always be ensured by an appropriate choice of the nodes and by guaranteeing that the coefficients A_{ii} are sufficiently small.

9.10-2. Application of the Trapezoidal Rule

According to the trapezoidal rule (see Section 8.7-1), we have

$$A_{i1} = A_{ii} = \tfrac{1}{2}h, \quad A_{i2} = \cdots = A_{i,i-1} = h, \qquad i = 2, \ldots, n.$$

The application of the trapezoidal rule in the general scheme leads to the following step algorithm:

$$y_1 = f_1, \quad y_i = \frac{f_i + h \sum\limits_{j=1}^{i-1} \beta_j K_{ij} y_j}{1 - \tfrac{1}{2} h K_{ii}}, \qquad i = 2, \ldots, n,$$

$$x_i = a + (i-1)h, \quad n = \frac{b-a}{h} + 1, \quad \beta_j = \begin{cases} \tfrac{1}{2} & \text{for } j = 1, \\ 1 & \text{for } j > 1, \end{cases}$$

where the notation coincides with that introduced in Subsection 9.10-1. The trapezoidal rule is quite simple and effective, and frequently used in practice. Some peculiarities of using the quadrature method for solving integral equations with variable limits of integration are indicated in Subsection 8.7-3.

9.10-3. The Case of a Degenerate Kernel

When solving a Volterra integral equation of the second kind with arbitrary kernel, the amount of calculations increases as the index of the integration step increases. However, if the kernel is degenerate, then it is possible to construct algorithms with a constant amount of calculations at each step. Indeed, for a degenerate kernel

$$K(x, t) = \sum_{k=1}^{m} p_k(x) q_k(t),$$

we can rewrite Eq. (1) in the form

$$y(x) = \sum_{k=1}^{m} p_k(x) \int_a^x q_k(t) y(t)\,dt + f(x).$$

The application of the trapezoidal rule makes it possible to obtain the following recurrent expression (see Subsection 9.10-2):

$$y_1 = f_1, \quad y_i = \frac{f_i + h \sum\limits_{k=1}^{m} p_{ki} \sum\limits_{j=1}^{i-1} \beta_j q_{kj} y_j}{1 - \tfrac{1}{2} h \sum\limits_{k=1}^{m} p_{ki} q_{ki}},$$

where y_i are approximate values of the unknown function $y(x)$ at the nodes x_i, $f_i = f(x_i)$, $p_{ki} = p_k(x_i)$, and $q_{ki} = q_k(x_i)$, and this expression shows that the amount of calculations is the same at each step.

⊙ References for Section 9.10: V. I. Krylov, V. V. Bobkov, and P. I. Monastyrnyi (1984), A. F. Verlan' and V. S. Sizikov (1986).

9.11. Equations With Infinite Integration Limit

Integral equations of the second kind with difference kernel and with a variable limit of integration for which the other limit is infinite are also of interest. Kernels and functions in such equations need not belong to the classes described in the beginning of the chapter. In this case their investigation can be performed by the method of model solutions (see Section 9.6) or by the reduction to equations of convolution type. We consider the latter method by an example of an equation of the second kind with variable lower limit.

9.11-1. An Equation of the Second Kind With Variable Lower Integration Limit

Integral equations of the second kind with variable lower limit, in the case of a difference kernel, have the form

$$y(x) + \int_x^\infty K(x-t)y(t)\,dt = f(x), \qquad 0 < x < \infty. \tag{1}$$

This equation substantially differs from Volterra equations of the second kind studied above for which a solution exists and is unique. A solution of the corresponding homogeneous equation

$$y(x) + \int_x^\infty K(x-t)y(t)\,dt = 0 \tag{2}$$

can be nontrivial.

The eigenfunctions of the integral equation (2) are determined by the roots of the following transcendental (or algebraic) equation for the parameter λ:

$$\int_0^\infty K(-z)e^{-\lambda z}\,dz = -1. \tag{3}$$

The left-hand side of this equation is the Laplace transform of the function $K(-z)$ with parameter λ.

To a real simple root λ_k of Eq. (3) there corresponds an eigenfunction

$$y_k(x) = \exp(-\lambda_k x).$$

The general solution is the linear combination (with arbitrary constants) of the eigenfunctions of the homogeneous integral equation (2).

For solutions of Eq. (2) in the case of multiple or complex roots, see equation 52 in Section 2.9 (see also Example 1 below).

The general solution of the integral equation (1) is the sum of the general solution of the homogeneous equation (2) and a particular solution of the nonhomogeneous equation (1).

Example 1. Consider the homogeneous Picard–Goursat equation

$$y(x) + A \int_x^\infty (t-x)^n y(t)\,dt = 0, \qquad n = 0, 1, 2, \ldots, \tag{4}$$

which is a special case of Eq. (1) with $K(z) = A(-z)^n$.

The general solution of the homogeneous equation has the form

$$y(x) = \sum_{k=1}^m C_k \exp(-\lambda_k x), \tag{5}$$

where C_k are arbitrary constants and λ_k are the roots of the algebraic equation

$$\lambda^{n+1} + An! = 0 \tag{6}$$

that satisfy the condition $\operatorname{Re} \lambda_k > 0$ (m is the number of the roots of Eq. (6) that satisfy this condition). Equation (6) is a special case of Eq. (3) with $K(z) = A(-z)^n$. The roots of Eq. (6) such that $\operatorname{Re} \lambda_k \leq 0$ must be dropped out, since for them the integral in (3) is divergent.

Equation (6) has complex roots. Consider two cases that correspond to different signs of A.

$1°$. Let $A < 0$. A solution of the Eq. (4) is

$$y(x) = Ce^{-\lambda x}, \qquad \lambda = (-An!)^{\frac{1}{n+1}}, \tag{7}$$

where C is an arbitrary constant. This solution is unique for $n = 0, 1, 2, 3$.

For $n \geq 4$, taking the real and the imaginary part in (5), one arrives at the general solution of the homogeneous Picard–Goursat equation in the form

$$y(x) = Ce^{-\lambda x} + \sum_{k=1}^{[n/4]} \exp(-\alpha_k x)\left[C_k^{(1)} \cos(\beta_k x) + C_k^{(2)} \sin(\beta_k x)\right], \tag{8}$$

where $C_k^{(1)}$ and $C_k^{(2)}$ are arbitrary constants, $[a]$ stands for the integral part of a number a, λ is defined in (7), and the coefficients α_k and β_k are given by

$$\alpha_k = |An!|^{\frac{1}{n+1}} \cos\left(\frac{2\pi k}{n+1}\right), \qquad \beta_k = |An!|^{\frac{1}{n+1}} \sin\left(\frac{2\pi k}{n+1}\right).$$

Note that Eq. (8) contains an odd number of terms.

$2°$. Let $A > 0$. By taking the real and the imaginary part in (5), one obtains the general solution of the homogeneous Picard–Goursat equation in the form

$$y(x) = \sum_{k=0}^{\left[\frac{n+2}{4}\right]} \exp(-\alpha_k x)\left[C_k^{(1)} \cos(\beta_k x) + C_k^{(2)} \sin(\beta_k x)\right], \tag{9}$$

where $C_k^{(1)}$ and $C_k^{(2)}$ are arbitrary constants, and the coefficients α_k and β_k are given by

$$\alpha_k = (An!)^{\frac{1}{n+1}} \cos\left(\frac{2\pi k + \pi}{n+1}\right), \quad \beta_k = (An!)^{\frac{1}{n+1}} \sin\left(\frac{2\pi k + \pi}{n+1}\right).$$

Note that Eq. (9) contains an even number of terms. In the special cases of $n = 0$ and $n = 1$, Eq. (9) gives the trivial solution $y(x) \equiv 0$.

Example 2. Consider the nonhomogeneous Picard–Goursat equation

$$y(x) + A \int_x^\infty (t - x)^n y(t)\,dt = Be^{-\mu x}, \qquad n = 0, 1, 2, \ldots, \tag{10}$$

which is a special case of Eq. (1) with $K(z) = A(-z)^n$ and $f(x) = Be^{-\mu x}$.

Let $\mu > 0$. Consider two cases.

$1°$. Let $\mu^{n+1} + An! \neq 0$. A particular solution of the nonhomogeneous equation is

$$\bar{y}(x) = De^{-\mu x}, \qquad D = \frac{B\mu^{n+1}}{\mu^{n+1} + An!}. \tag{11}$$

For $A < 0$, the general solution of the nonhomogeneous Picard–Goursat equation is the sum of solutions (8) and (11). For $A > 0$, the general solution of the Eq. (10) is the sum of solutions (9) and (11).

$2°$. Let $\mu^{n+1} + An! = 0$. Since μ is positive, it follows that A must be negative. A particular solution of the nonhomogeneous equation is

$$\bar{y}(x) = Exe^{-\mu x}, \qquad E = \frac{B\mu^{n+2}}{A(n+1)!}. \tag{12}$$

The general solution of the nonhomogeneous Picard–Goursat equation is the sum of solutions (8) and (12).

9.11-2. Reduction to a Wiener–Hopf Equation of the Second Kind

Equation (1) can be reduced to a one-sided equation of the second kind of the form

$$y(x) - \int_0^\infty K_-(x - t)y(t)\,dt = f(x), \qquad 0 < x < \infty, \tag{13}$$

where the kernel $K_-(x - t)$ has the form

$$K_-(s) = \begin{cases} 0 & \text{for } s > 0, \\ -K(s) & \text{for } s < 0. \end{cases}$$

Methods for studying Eq. (13) are described in Chapter 11, where equations of the second kind with constant limits are considered. In the same chapter, in Subsection 11.9-3, an equation of the second kind with difference kernel and variable lower limit is studied by means of reduction to a Wiener–Hopf equation of the second kind.

⊙ Reference for Section 9.11: F. D. Gakhov and Yu. I. Cherskii (1978).

Chapter 10

Methods for Solving Linear Equations of the Form $\int_a^b K(x,t)y(t)\,dt = f(x)$

10.1. Some Definition and Remarks

10.1-1. Fredholm Integral Equations of the First Kind

Linear integral equations of the first kind with constant limits of integration have the form

$$\int_a^b K(x,t)y(t)\,dt = f(x), \tag{1}$$

where $y(x)$ is the unknown function ($a \le x \le b$), $K(x,t)$ is the *kernel* of the integral equation, and $f(x)$ is a given function, which is called the *right-hand side* of Eq. (1). The functions $y(x)$ and $f(x)$ are usually assumed to be continuous or square integrable on $[a,b]$. If the kernel of the integral equation (1) is continuous on the square $S = \{a \le x \le b, a \le t \le b\}$ or at least square integrable on this square, i.e.,

$$\int_a^b \int_a^b K^2(x,t)\,dx\,dt = B^2 < \infty, \tag{2}$$

where B is a constant, then this kernel is called a *Fredholm kernel*. Equations of the form (1) with constant integration limits and Fredholm kernel are called *Fredholm equations of the first kind*.

The kernel $K(x,t)$ of an integral equation is said to be *degenerate* if it can be represented in the form $K(x,t) = g_1(x)h_1(t) + \cdots + g_n(x)h_n(t)$.

The kernel $K(x,t)$ of an integral equation is called a *difference kernel* if it depends only on the difference of the arguments: $K(x,t) = K(x-t)$.

The kernel $K(x,t)$ of an integral equation is said to be *symmetric* if it satisfies the condition $K(x,t) = K(t,x)$.

The integral equation obtained from (1) by replacing the kernel $K(x,t)$ by $K(t,x)$ is said to be *transposed* to (1).

Remark 1. The variables t and x in Eq. (1) may vary within different intervals (e.g., $a \le t \le b$ and $c \le x \le d$).

10.1-2. Integral Equations of the First Kind With Weak Singularity

If the kernel of the integral equation (1) is polar, i.e., if

$$K(x,t) = \frac{L(x,t)}{|x-t|^\alpha} + M(x,t), \qquad 0 < \alpha < 1, \tag{3}$$

or logarithmic, i.e.,

$$K(x,t) = L(x,t)\ln|x-t| + M(x,t), \tag{4}$$

where $L(x,t)$ and $M(x,t)$ are continuous on S and $L(x,x) \not\equiv 0$, then $K(x,t)$ is called a *kernel with weak singularity*, and the equation itself is called an *equation with weak singularity*.

Remark 2. Kernels with logarithmic singularity and polar kernels with $0 < \alpha < \frac{1}{2}$ are Fredholm kernels.

Remark 3. In general, the case in which the limits of integration a and/or b can be infinite is not excluded, but in this case the validity of condition (2) must be verified with special care.

10.1-3. Integral Equations of Convolution Type

The integral equation of the first kind with difference kernel on the entire axis (this equation is sometimes called an *equation of convolution type of the first kind with a single kernel*) has the form

$$\int_{-\infty}^{\infty} K(x-t)y(t)\,dt = f(x), \quad -\infty < x < \infty, \tag{5}$$

where $f(x)$ and $K(x)$ are the right-hand side and the kernel of the integral equation and $y(x)$ is the unknown function (in what follows we use the above notation).

An integral equation of the first kind with difference kernel on the semiaxis has the form

$$\int_0^{\infty} K(x-t)y(t)\,dt = f(x), \quad 0 < x < \infty. \tag{6}$$

Equation (6) is also called a *one-sided equation of the first kind* or a *Wiener–Hopf integral equation of the first kind*.

An integral equation of convolution type with two kernels of the first kind has the form

$$\int_0^{\infty} K_1(x-t)y(t)\,dt + \int_{-\infty}^0 K_2(x-t)y(t)\,dt = f(x), \quad -\infty < x < \infty, \tag{7}$$

where $K_1(x)$ and $K_2(x)$ are the kernels of the integral equation (7).

Recall that a function $g(x)$ satisfies the *Hölder condition on the real axis* if for any real x_1 and x_2 we have the inequality

$$|g(x_2) - g(x_1)| \le A|x_2 - x_1|^\lambda, \quad 0 < \lambda \le 1,$$

and for any x_1 and x_2 sufficiently large in absolute value we have

$$|g(x_2) - g(x_1)| \le A\left|\frac{1}{x_2} - \frac{1}{x_1}\right|^\lambda, \quad 0 < \lambda \le 1,$$

where A and λ are positive (the latter inequality is the Hölder condition in the vicinity of the point at infinity).

Assume that the functions $y(x)$ and $f(x)$ and the kernels $K(x)$, $K_1(x)$, and $K_2(x)$ are such that their Fourier transforms belong to $L_2(-\infty, \infty)$ and, moreover, satisfy the Hölder condition.

For a function $y(x)$ to belong to the above function class it suffices to require $y(x)$ to belong to $L_2(-\infty, \infty)$ and $xy(x)$ to be absolutely integrable on $(-\infty, \infty)$.

10.1-4. Dual Integral Equations of the First Kind

A *dual integral equation of the first kind with difference kernels* (*of convolution type*) has the form

$$\int_{-\infty}^{\infty} K_1(x-t)y(t)\,dt = f(x), \qquad 0 < x < \infty,$$

$$\int_{-\infty}^{\infty} K_2(x-t)y(t)\,dt = f(x), \qquad -\infty < x < 0,$$

(8)

where the notation and the classes of functions and kernels coincide with those introduced above for equations of convolution type.

In the general case, a dual integral equation of the first kind has the form

$$\int_{a}^{\infty} K_1(x,t)y(t)\,dt = f_1(x), \qquad a < x < b,$$

$$\int_{a}^{\infty} K_2(x,t)y(t)\,dt = f_2(x), \qquad b < x < \infty,$$

where $f_1(x)$ and $f_2(x)$ are the right-hand sides, $K_1(x,t)$ and $K_2(x,t)$ are the kernels of Eq. (9), and $y(x)$ is the unknown function. Various forms of this equation are considered in Subsections 10.6-2 and 10.6-3.

The integral equations obtained from (5)–(8) by replacing the kernel $K(x-t)$ with $K(t-x)$ are called *transposed* equations.

Remark 3. Some equations whose kernels contain the product or the ratio of the variables x and t can be reduced to equations of the form (5)–(8).

Remark 4. Equations (5)–(8) of the convolution type are sometimes written in the form in which the integrals are multiplied by the coefficient $1/\sqrt{2\pi}$.

⊙ References for Section 10.1: I. Sneddon (1951), B. Noble (1958), S. G. Mikhlin (1960), I. C. Gohberg and M. G. Krein (1967), L. Ya. Tslaf (1970), M. L. Krasnov, A. I. Kiselev, and G. I. Makarenko (1971), P. P. Zabreyko, A. I. Koshelev, et al. (1975), Ya. S. Uflyand (1977), F. D. Gakhov and Yu. I. Cherskii (1978), A. J. Jerry (1985), A. F. Verlan' and V. S. Sizikov (1986), L. A. Sakhnovich (1996).

10.2. Krein's Method

10.2-1. The Main Equation and the Auxiliary Equation

Here we describe a method for constructing exact closed-form solutions of linear integral equations of the first kind with weak singularity and with arbitrary right-hand side. The method is based on the construction of the auxiliary solution of the simpler equation whose right-hand side is equal to one. The auxiliary solution is then used to construct the solution of the original equation for an arbitrary right-hand side.

Consider the equation

$$\int_{-a}^{a} K(x-t)y(t)\,dt = f(x), \qquad -a \le x \le a.$$

(1)

Suppose that the kernel of the integral equation (1) is polar or logarithmic and that $K(x)$ is an even positive definite function that can be expressed in the form

$$K(x) = \beta|x|^{-\mu} + M(x), \qquad 0 < \mu < 1,$$

$$K(x) = \beta \ln\frac{1}{|x|} + M(x),$$

respectively, where $\beta > 0$, $-2a \le x \le 2a$, and $M(x)$ is a sufficiently smooth function.

Along with (1), we consider the following auxiliary equation containing a parameter ξ ($0 \le \xi \le a$):

$$\int_{-\xi}^{\xi} K(x-t)w(t,\xi)\,dt = 1, \qquad -\xi \le x \le \xi.$$

(2)

10.2-2. Solution of the Main Equation

For any continuous function $f(x)$, the solution of the original equation (1) can be expressed via the solution of the auxiliary equation (2) by the formula

$$
y(x) = \frac{1}{2M'(a)} \left[\frac{d}{da} \int_{-a}^{a} w(t,a) f(t)\,dt \right] w(x,a)
$$
$$
- \frac{1}{2} \int_{|x|}^{a} w(x,\xi) \frac{d}{d\xi} \left[\frac{1}{M'(\xi)} \frac{d}{d\xi} \int_{-\xi}^{\xi} w(t,\xi) f(t)\,dt \right] d\xi \tag{3}
$$
$$
- \frac{1}{2} \frac{d}{dx} \int_{|x|}^{a} \frac{w(x,\xi)}{M'(\xi)} \left[\int_{-\xi}^{\xi} w(t,\xi)\,df(t) \right] d\xi,
$$

where $M(\xi) = \int_0^\xi w(x,\xi)\,dx$, the prime stands for the derivative, and the last inner integral is treated as a Stieltjes integral.

Formula (3) permits one to obtain some exact solutions of integral equations of the form (1) with arbitrary right-hand side, see Section 3.8 of the first part of the book.

Example 1. The solution of the integral equation

$$
\int_{-a}^{a} \ln\left(\frac{A}{|x-t|} \right) y(t)\,dt = f(x),
$$

which arises in elasticity, is given by formula (3), where

$$
M(\xi) = \left(\ln \frac{2A}{\xi} \right)^{-1}, \qquad w(t,\xi) = \frac{M(\xi)}{\pi\sqrt{\xi^2 - t^2}}.
$$

Example 2. Consider the integral equation

$$
\int_{-a}^{a} \frac{y(t)\,dt}{|x-t|^\mu} = f(x), \qquad 0 < \mu < 1,
$$

which arises in the theory of elasticity. The solution is given by formula (3), where

$$
M(\xi) = \frac{2\sqrt{\pi}}{\mu\,\Gamma\left(\frac{\mu}{2}\right)\Gamma\left(\frac{1-\mu}{2}\right)} \xi^\mu, \qquad w(t,\xi) = \frac{1}{\pi} \cos\left(\frac{\pi\mu}{2} \right) (\xi^2 - t^2)^{\frac{\mu-1}{2}}.
$$

⊙ References for Section 10.2: N. Kh. Arutyunyan (1959), I. C. Gohberg and M. G. Krein (1967).

10.3. The Method of Integral Transforms

The method of integral transforms enables one to reduce some integral equations on the entire axis and on the semiaxis to algebraic equations for transforms. These algebraic equations can readily be solved for the transform of the desired function. The solution of the original integral equation is then obtained by applying the inverse integral transform.

10.3-1. Equation With Difference Kernel on the Entire Axis

Consider the integral equation

$$
\int_{-\infty}^{\infty} K(x-t) y(t)\,dt = f(x), \qquad -\infty < x < \infty, \tag{1}
$$

where $f(x)$, $y(x) \in L_2(-\infty, \infty)$ and $K(x) \in L_1(-\infty, \infty)$.

Let us apply the Fourier transform to Eq. (1). In this case, taking into account the convolution theorem (see Subsection 7.4-4), we obtain

$$\sqrt{2\pi}\,\tilde{K}(u)\tilde{y}(u) = \tilde{f}(u). \tag{2}$$

Thus, by means of the Fourier transform we have reduced the solution of the original integral equation (1) to the solution of the algebraic equation (2) for the Fourier transform of the desired solution. The solution of the latter equation has the form

$$\tilde{y}(u) = \frac{1}{\sqrt{2\pi}}\frac{\tilde{f}(u)}{\tilde{K}(u)}, \tag{3}$$

where the function $\tilde{f}(u)/\tilde{K}(u)$ must belong to the space $L_2(-\infty, \infty)$.

Thus, the Fourier transform of the solution of the original integral equation is expressed via the Fourier transforms of known functions, namely, the kernel and the right-hand side of the equation. The solution itself can be expressed via its Fourier transform by means of the Fourier inversion formula:

$$y(x) = \frac{1}{\sqrt{2\pi}}\int_{-\infty}^{\infty}\tilde{y}(u)e^{iux}\,du = \frac{1}{2\pi}\int_{-\infty}^{\infty}\frac{\tilde{f}(u)}{\tilde{K}(u)}\,e^{iux}\,du. \tag{4}$$

10.3-2. Equations With Kernel $K(x,t) = K(x/t)$ on the Semiaxis

The integral equation of the first kind

$$\int_0^\infty K(x/t)y(t)\,dt = f(x), \quad 0 \le x < \infty, \tag{5}$$

can be reduced to the form (1) by the change of variables $x = e^\xi$, $t = e^\tau$, $w(\tau) = ty(t)$. The solution to this equation can also be obtained by straightforward application of the Mellin transform, and this method is applied in a similar situation in the next section.

10.3-3. Equation With Kernel $K(x,t) = K(xt)$ and Some Generalizations

$1°$. We first consider the equation

$$\int_0^\infty K(xt)y(t)\,dt = f(x), \quad 0 \le x < \infty. \tag{6}$$

By changing variables $x = e^\xi$ and $t = e^{-\tau}$ this equation can be reduced to the form (1), but it is more convenient here to apply the Mellin transform (see Section 7.3). On multiplying Eq. (6) by x^{s-1} and integrating with respect to x from 0 to ∞, we obtain

$$\int_0^\infty y(t)\,dt\int_0^\infty K(xt)x^{s-1}\,dx = \int_0^\infty f(x)x^{s-1}\,dx.$$

We make the change of variables $z = xt$ in the inner integral of the double integral. This implies the relation

$$\hat{K}(s)\int_0^\infty y(t)t^{-s}\,dt = \hat{f}(s). \tag{7}$$

Taking into account the formula

$$\int_0^\infty y(t)t^{-s}\,dt = \hat{y}(1-s),$$

we can rewrite Eq. (7) in the form

$$\hat{K}(s)\hat{y}(1-s) = \hat{f}(s). \tag{8}$$

Replacing $1-s$ by s in (8) and solving the resulting relation for $\hat{y}(s)$, we obtain the transform

$$\hat{y}(s) = \frac{\hat{f}(1-s)}{\hat{K}(1-s)} \tag{9}$$

of the desired solution.

Applying the Mellin inversion formula, we obtain the solution of the integral equation (6) in the form

$$y(x) = \frac{1}{2\pi i} \int_{c-i\infty}^{c+i\infty} \frac{\hat{f}(1-s)}{\hat{K}(1-s)} x^{-s}\,ds.$$

$2°$. Now we consider the more complicated equation

$$\int_0^\infty K\big(\varphi(x)\psi(t)\big)g(t)y(t)\,dt = f(x). \tag{10}$$

Assume that the conditions $\varphi(0) = 0$, $\varphi(\infty) = \infty$, $\varphi'_x > 0$, $\psi(0) = 0$, $\psi(\infty) = \infty$, and $\psi'_x > 0$ are satisfied.

The transform

$$z = \varphi(x), \qquad \tau = \psi(t), \qquad y(t) = \frac{g(t)}{\psi'_t(t)} w(\tau)$$

takes (10) to the following equation of the form (6):

$$\int_0^\infty K(z\tau)w(\tau)\,d\tau = F(z),$$

where the function $F(z)$ is defined parametrically by $F = f(x)$, $z = \varphi(x)$. In many cases, on eliminating x from these relations, we obtain the dependence $F = F(z)$ in an explicit form.

⊙ References for Section 10.3: V. A. Ditkin and A. P. Prudnikov (1965), M. L. Krasnov, A. I. Kiselev, and G. I. Makarenko (1971).

10.4. The Riemann Problem for the Real Axis

The Riemann boundary value problem is one of the main tools for constructing solutions of integral equations provided that various integral transforms can be applied to a given equation and the corresponding convolution-type theorems can be applied. This problem is investigated by an example of the Fourier integral transform.

10.4-1. Relationships Between the Fourier Integral and the Cauchy Type Integral

Let $\mathcal{Y}(\tau)$ be a function integrable on a closed or nonclosed contour L on the complex plane of the variable $z = u + iv$ (τ is the complex coordinate of the contour points). Consider the integral of the Cauchy type (see Section 12.2):

$$\frac{1}{2\pi i} \int_L \frac{\mathcal{Y}(\tau)}{\tau - z}\,d\tau.$$

This integral defines a function that is analytic on the complex plane with a cut along the contour L. If L is a closed curve, then the integral is a function that is analytic on each of the connected parts of the plane bounded by L. If the contour L is the real axis, then we have

$$\frac{1}{2\pi i} \int_{-\infty}^{\infty} \frac{\mathcal{Y}(\tau)}{\tau - z}\,d\tau = \begin{cases} \mathcal{Y}^+(z) & \text{if } \operatorname{Im} z > 0, \\ \mathcal{Y}^-(z) & \text{if } \operatorname{Im} z < 0. \end{cases} \tag{1}$$

Moreover, there exist limit values of the functions $\mathcal{Y}^{\pm}(z)$ on the real axis, and these values are related to the density $\mathcal{Y}$ of the integral by the *Sokhotski–Plemelj formulas*

$$\mathcal{Y}^+(u) = \frac{1}{2}\mathcal{Y}(u) + \frac{1}{2\pi i}\int_{-\infty}^{\infty}\frac{\mathcal{Y}(\tau)}{\tau - u}\,d\tau,$$

$$\mathcal{Y}^-(u) = -\frac{1}{2}\mathcal{Y}(u) + \frac{1}{2\pi i}\int_{-\infty}^{\infty}\frac{\mathcal{Y}(\tau)}{\tau - u}\,d\tau, \tag{2}$$

or

$$\mathcal{Y}^+(u) - \mathcal{Y}^-(u) = \mathcal{Y}(u), \quad \mathcal{Y}^+(u) + \mathcal{Y}^-(u) = \frac{1}{\pi i}\int_{-\infty}^{\infty}\frac{\mathcal{Y}(\tau)}{\tau - u}\,d\tau. \tag{3}$$

In the latter formulas, the integral is understood as a singular integral in the sense of the Cauchy principal value.

In the Fourier integral*

$$\mathcal{Y}(u) = \frac{1}{\sqrt{2\pi}}\int_{-\infty}^{\infty} y(x)e^{iux}\,dx,$$

the real parameter u occurs in an analytic function, and therefore we can replace u in this integral by a complex variable z. The function $\mathcal{Y}(z)$ defined by the integral

$$\mathcal{Y}(z) = \frac{1}{\sqrt{2\pi}}\int_{-\infty}^{\infty} y(x)e^{izx}\,dx, \tag{4}$$

is analytic in the part of the complex plane of the variable $z = u + iv$ in which the integral (4) is absolutely convergent. If this is a domain indeed, i.e., if it is not reduced to the real axis, then the integral (4) gives an analytic continuation of the Fourier integral into the complex plane. The integral (4) will also be called the *Fourier integral*.

Let us establish a relationship between this integral and the integral of the Cauchy type with density $\mathcal{Y}(u)$ taken along the entire axis. We have

$$\frac{1}{2\pi i}\int_{-\infty}^{\infty}\frac{\mathcal{Y}(\tau)}{\tau - z}\,d\tau = \frac{1}{\sqrt{2\pi}}\int_{0}^{\infty} y(x)e^{izx}\,dx, \qquad \text{Im } z > 0, \tag{5}$$

$$\frac{1}{2\pi i}\int_{-\infty}^{\infty}\frac{\mathcal{Y}(\tau)}{\tau - z}\,d\tau = -\frac{1}{\sqrt{2\pi}}\int_{-\infty}^{0} y(x)e^{izx}\,dx, \qquad \text{Im } z < 0. \tag{6}$$

10.4-2. One-Sided Fourier Integrals

If $\mathcal{Y}(z) = \mathcal{Y}^+(z)$ is an analytic function in the upper half-plane whose limit value on the real axis is given by the function $\mathcal{Y}(u) = \mathcal{Y}^+(u) \in L_2(-\infty, \infty)$, then the function $\mathcal{Y}^+(z)$ can be expressed by means of the Cauchy integral. Hence, by virtue of (5) we have

$$\mathcal{Y}^+(z) = \frac{1}{\sqrt{2\pi}}\int_{0}^{\infty} y(x)e^{izx}\,dx,$$

and, since the integral defines a continuous function, the limit values on the axis can be obtained from the last relation merely by setting $z = u$:

$$\mathcal{Y}^+(u) = \frac{1}{\sqrt{2\pi}}\int_{0}^{\infty} y(x)e^{iux}\,dx,$$

---

* In Sections 10.4–10.6, the alternative Fourier transform is used (see Subsection 7.4-3).

where, according to (5), $y(x)$ the inverse transform of $\mathcal{Y}(u)$. The right-hand side can be regarded as the Fourier integral of a function that is identically zero for negative x. Hence, by the uniqueness of the representation of the function $\mathcal{Y}^+(u)$ by a Fourier integral, it follows that $y(x) \equiv 0$ on the negative semiaxis.

Conversely, if $y \equiv 0$ for $x < 0$, then the Fourier integral of this function becomes

$$\mathcal{Y}(u) = \frac{1}{\sqrt{2\pi}} \int_0^\infty y(x)e^{iux}\,dx.$$

If we replace the parameter u by a complex number z belonging to the upper half-plane, then the integral will converge even better. This implies the analyticity of the function

$$\mathcal{Y}(z) = \frac{1}{\sqrt{2\pi}} \int_0^\infty y(x)e^{izx}\,dx$$

in the upper half-plane.

The case of the lower half-plane can be treated in a similar way.

The integrals

$$\mathcal{Y}^+(z) = \frac{1}{\sqrt{2\pi}} \int_0^\infty y(x)e^{izx}\,dx, \quad \mathcal{Y}^-(z) = -\frac{1}{\sqrt{2\pi}} \int_{-\infty}^0 y(x)e^{izx}\,dx \tag{7}$$

are called *one-sided Fourier integrals*, namely, the *right* and the *left* Fourier integral, respectively. As well as in formula (1), the symbols $\pm$ over symbols of functions mean that the corresponding function is analytic in the upper or lower half-plane, respectively.

Let us introduce the functions

$$y_+(x) = \begin{cases} y(x) & \text{for } x > 0, \\ 0 & \text{for } x < 0, \end{cases} \qquad y_-(x) = \begin{cases} 0 & \text{for } x > 0, \\ -y(x) & \text{for } x < 0. \end{cases} \tag{8}$$

These functions are said to be *one-sided functions* for $y(x)$, namely, the *right function* and the *left function*, respectively. Obviously, the following relation holds:

$$y(x) = y_+(x) - y_-(x). \tag{9}$$

Applying the well-known function $\operatorname{sign} x$ defined by

$$\operatorname{sign} x = \begin{cases} 1 & \text{for } x > 0, \\ -1 & \text{for } x < 0, \end{cases} \tag{10}$$

we can express $y_\pm$ in terms of y as follows:

$$y_\pm(x) = \tfrac{1}{2}(\pm 1 + \operatorname{sign} x)y(x). \tag{11}$$

The symbols $\pm$ on symbols of one-sided functions will be always subscripts.

The Fourier integrals of the right and left one-sided functions are the boundary values of functions that are analytic on the upper and lower half-planes, respectively.

Let us indicate the following analogs of the Sokhotski–Plemelj formulas (3) in the Fourier integrals:

$$\begin{aligned} \mathcal{Y}(u) &= \frac{1}{\sqrt{2\pi}} \int_{-\infty}^\infty y(x)e^{iux}\,dx \\ &= \frac{1}{\sqrt{2\pi}} \int_0^\infty y(x)e^{iux}\,dx + \frac{1}{\sqrt{2\pi}} \int_{-\infty}^0 y(x)e^{iux}\,dx = \mathcal{Y}^+(u) - \mathcal{Y}^-(u), \\ \frac{1}{\pi i} \int_{-\infty}^\infty \frac{\mathcal{Y}(\tau)}{\tau - u}\,d\tau &= \mathcal{Y}^+(u) + \mathcal{Y}^-(u) \\ &= \frac{1}{\sqrt{2\pi}} \int_0^\infty y(x)e^{iux}\,dx - \frac{1}{\sqrt{2\pi}} \int_{-\infty}^0 y(x)e^{iux}\,dx = \frac{1}{\sqrt{2\pi}} \int_{-\infty}^\infty y(x)\operatorname{sign} x\, e^{iux}\,dx. \end{aligned} \tag{12}$$

Thus, in this setting, the first Sokhotski–Plemelj formula (a representation of an arbitrary function in the form of the difference of boundary values of analytic functions) is an obvious consequence of the decomposition of a Fourier integral into the right and the left integral. The second formula can also be rewritten as follows:

$$\mathbf{F}\{y(x)\operatorname{sign} x\} = \frac{1}{\pi i}\int_{-\infty}^{\infty}\frac{\mathcal{Y}(\tau)}{\tau - u}\,d\tau, \quad \mathbf{F}^{-1}\left\{\frac{1}{\pi i}\int_{-\infty}^{\infty}\frac{\mathcal{Y}(\tau)}{\tau - u}\,d\tau\right\} = y(x)\operatorname{sign} x. \tag{13}$$

10.4-3. The Analytic Continuation Theorem and the Generalized Liouville Theorem

Below is the analytic continuation theorem and the generalized Liouville theorem combined into a single statement, which will be used in Chapters 10 and 11.

Let functions $\mathcal{Y}_1(z)$ and $\mathcal{Y}_2(z)$ be analytic in the upper and lower half-planes, respectively, possibly except for a point $z_* \neq \infty$, at which these functions have a pole. If $\mathcal{Y}_1(z)$ and $\mathcal{Y}_2(z)$ are bounded at infinity, the principal parts of their expansions in a neighborhood of z_* have the form

$$\frac{c_1}{z - z_*} + \frac{c_2}{(z - z_*)^2} + \cdots + \frac{c_m}{(z - z_*)^m} \equiv \frac{\mathcal{P}_{m-1}(z)}{(z - z_*)^m},$$

and if the functions themselves coincide on the real axis, then these functions represent a single rational function on the entire plane:

$$\mathcal{Y}(z) = c_0 + \frac{\mathcal{P}_{m-1}(z)}{(z - z_*)^m},$$

where c_0 is a constant. The pole z_* can belong either to the open half-planes or to the real axis.

Let us also give a more general version of the above statement.

If functions $\mathcal{Y}_1(z)$ and $\mathcal{Y}_2(z)$ are analytic in the upper and lower half-planes, respectively, possibly except for finitely many points $z_0 = \infty$, z_k $(k = 1, \ldots, n)$, at which these functions can have poles, if the principal parts of the expansions of these functions in a neighborhood of a pole have the form

$$c_1^0 z + c_2^0 z^2 + \cdots + c_{m_0}^0 z^{m_0} \equiv \mathcal{P}_0(z) \qquad \text{at the point } z_0,$$

$$\frac{c_1^k}{z - z_k} + \frac{c_2^k}{(z - z_k)^2} + \cdots + \frac{c_{m_k}^k}{(z - z_k)^{m_k}} \equiv \frac{\mathcal{P}_{m_k-1}(z)}{(z - z_k)^{m_k}} \qquad \text{at the points } z_k,$$

and if the functions themselves coincide on the real axis, then these functions represent a single rational function on the entire plane:

$$\mathcal{Y}(z) = C + \mathcal{P}_0(z) + \sum_{k=1}^{n}\frac{\mathcal{P}_{m_k-1}(z)}{(z - z_k)^{m_k}}$$

where C is a constant. The poles z_k can belong either to the open half-planes or to the real axis.

10.4-4. The Riemann Boundary Value Problem

The solution of the Riemann problem in this section differs from the traditional one, because it is expressed not by means of integrals of the Cauchy type (see Subsection 12.3-7) but by means of Fourier integrals. To solve equations of convolution type under consideration, the Fourier integral technique is more convenient.

By the *index* of a continuous complex-valued nonvanishing function $\mathcal{M}(u)$ $(\mathcal{M}(u) = \mathcal{M}_1(u) + i\mathcal{M}_2(u), -\infty < u < \infty, \mathcal{M}(-\infty) = \mathcal{M}(\infty))$ we mean the variation of the argument of this function on the real axis expressed in the number of full rotations:

$$\operatorname{Ind}\mathcal{M}(u) = \frac{1}{2\pi}\big[\arg\mathcal{M}(u)\big]_{-\infty}^{\infty} = \frac{1}{2\pi i}\big[\ln\mathcal{M}(u)\big]_{-\infty}^{\infty} = \frac{1}{2\pi i}\int_{-\infty}^{\infty}d\ln\mathcal{M}(u).$$

If $\mathcal{M}(u)$ is not differentiable but is of bounded variation, then the last integral must be understood as the Stieltjes integral.

If an analytic function $\mathcal{Y}(z)$ has a representation of the form

$$\mathcal{Y}(z) = (z - z_0)^m \mathcal{Y}_1(z)$$

in a neighborhood of some point z_0, where $\mathcal{Y}_1(z)$ is analytic and $\mathcal{Y}_1(z_0) \neq 0$, then the integer m (which can be positive, negative, or zero) is called the *order* of the function $\mathcal{Y}(z)$ at the point z_0. If $m > 0$, then the order of the function is the order of its zero, and if $m < 0$, then the order of the function is minus the order of its pole. If the order of the function at z_0 is zero, then at this point the function takes a finite nonzero value. When considering the point at infinity we must replace the difference $z - z_0$ by $1/z$.

Let us pose the Riemann problem. Let two functions be given on the real axis, namely, $\mathcal{D}(u)$, the *coefficient of the problem*, and $\mathcal{H}(u)$, the *right-hand side*, and let the following normality condition hold: $\mathcal{D}(u) \neq 0$. The functions $\mathcal{H}(u)$ and $\mathcal{D}(u) - 1$ belong to $L_2(-\infty, \infty)$ and simultaneously satisfy the Hölder condition. The problem is to find two functions $\mathcal{Y}^{\pm}(z)$ that are analytic in the upper and the lower half-plane, respectively,* whose limit values on the real axis satisfy the following boundary condition:

$$\mathcal{Y}^+(u) = \mathcal{D}(u)\mathcal{Y}^-(u) + \mathcal{H}(u). \tag{14}$$

It follows from the representation of $\mathcal{D}(u)$ that $\mathcal{D}(\infty) = 1$. The last condition implies no loss of generality of subsequent reasoning because by dividing the boundary condition (14) by $\mathcal{D}(\infty)$ we can always obtain the necessary form of the problem.**

If $\mathcal{D}(u) \equiv 1$, then the Riemann problem is called the *jump problem*. For $\mathcal{H}(u) \equiv 0$, the Riemann problem is said to be *homogeneous*. The index ν of the coefficient $\mathcal{D}(u)$ of the boundary value problem is called the *index of the Riemann problem*.

Consider the *jump problem*, i.e., the problem of finding $\mathcal{Y}^{\pm}(z)$ from the boundary condition

$$\mathcal{Y}^+(u) - \mathcal{Y}^-(u) = \mathcal{H}(u). \tag{15}$$

The solution of this problem is given by the first formula in (12):

$$\mathcal{Y}^+(z) = \frac{1}{\sqrt{2\pi}} \int_0^{\infty} H(x)e^{izx}\,dx, \qquad \mathcal{Y}^-(z) = -\frac{1}{\sqrt{2\pi}} \int_{-\infty}^0 H(x)e^{izx}\,dx, \tag{16}$$

where

$$H(x) = \frac{1}{\sqrt{2\pi}} \int_{-\infty}^{\infty} \mathcal{H}(u)e^{-iux}\,du. \tag{17}$$

Let us construct a particular solution $\mathcal{X}(z)$ of the homogeneous Riemann problem (14), which we need in what follows:

$$\mathcal{X}^+(u) = \mathcal{D}(u)\mathcal{X}^-(u), \qquad \mathcal{D}(\infty) = 1, \tag{18}$$

where $\mathcal{X}(z)$ is assumed to be nonzero on the real axis with the additional condition $\mathcal{X}^{\pm}(\infty) = 1$. Denote by N_+ and N_- the numbers of zeros of the functions $\mathcal{X}^+(z)$ and $\mathcal{X}^-(z)$ in the upper and lower half-planes, respectively. On calculating the index of both sides of the boundary condition (18) and applying the properties of the index, we obtain

$$N_+ + N_- = \operatorname{Ind}\mathcal{D}(u) = \nu. \tag{19}$$

---

* A couple of functions $\mathcal{Y}^{\pm}(z)$ can be treated as a single function $\mathcal{Y}(z)$ piecewise analytic in the entire complex plane. In some cases, we use the latter notation.

** Since the boundary condition is the main analytic expression of the Riemann problem, in references to the corresponding problem we shall often indicate its boundary condition only and write, for instance, "Riemann problem (14)."

We first assume that $\nu = 0$. In this case, $\ln \mathcal{D}(u)$ is a single-valued function. It follows from relation (19) that $N_+ = N_- = 0$, i.e., the solution has no zeros on the entire plane. Therefore, the functions $\ln \mathcal{X}^+(z)$ and $\ln \mathcal{X}^-(z)$ are analytic in the corresponding half-planes, and hence are single-valued together with their boundary values $\ln \mathcal{X}^+(u)$ and $\ln \mathcal{X}^-(u)$. Taking the logarithm of the boundary condition (18), we obtain

$$\ln \mathcal{X}^+(u) - \ln \mathcal{X}^-(u) = \ln \mathcal{D}(u). \tag{20}$$

On choosing a branch of $\ln \mathcal{D}(u)$ such that $\ln \mathcal{D}(\infty) = 0$ (it can be shown that the final result does not depend on the choice of the branch) we arrive at a jump problem. In this case, on the basis of (15)–(17) and (20), the solution of problem (18) can be represented in the form

$$\mathcal{X}^+(z) = e^{\mathcal{G}^+(z)}, \quad \mathcal{X}^-(z) = e^{\mathcal{G}^-(z)},$$

$$\mathcal{G}^+(z) = \frac{1}{\sqrt{2\pi}} \int_0^\infty g(x) e^{izx} \, dx, \quad \mathcal{G}^-(z) = -\frac{1}{\sqrt{2\pi}} \int_{-\infty}^0 g(x) e^{izx} \, dx, \tag{21}$$

$$g(x) = \frac{1}{\sqrt{2\pi}} \int_{-\infty}^\infty \ln \mathcal{D}(u) \, e^{-iux} \, du.$$

Relations (21) imply the following important fact: a function $\mathcal{D}(u)$ of zero index that is nonvanishing on the real axis and satisfies the condition $\mathcal{D}(\infty) = 1$ can be represented as the ratio of functions that are the boundary values of nonzero analytic functions in the upper and the lower half-plane, respectively.

Let us pass to the case in which the index of the homogeneous Riemann problem (18) is arbitrary. By a *canonical function* $\mathcal{X}(z)$ (of the homogeneous Riemann problem) we mean a function that satisfies the boundary condition (18) and the condition $\mathcal{X}^\pm(\infty) = 1$ and has zero order everywhere possibly except for the point $-i$, at which the order of $\mathcal{X}(z)$ is equal to the index ν of the Riemann problem. Such a function can be constructed by reducing the homogeneous Riemann problem to the above case of zero index. Indeed, let us write out the boundary condition of the homogeneous Riemann problem (18) in the form

$$\mathcal{X}^+(u) = \left[\left(\frac{u-i}{u+i} \right)^{-\nu} \mathcal{D}(u) \right] \left[\left(\frac{u-i}{u+i} \right)^\nu \mathcal{X}^-(u) \right]. \tag{22}$$

In this case, the function in the first square brackets has zero index and can be represented as the ratio of the boundary values of functions that are analytic in the upper and the lower half-plane. This, together with the boundary condition (22), gives the following expression for the canonical function:

$$\mathcal{X}^+(z) = e^{\mathcal{G}^+(z)}, \quad \mathcal{X}^-(z) = \left(\frac{z-i}{z+i} \right)^{-\nu} e^{\mathcal{G}^-(z)},$$

$$\mathcal{G}^+(z) = \frac{1}{\sqrt{2\pi}} \int_0^\infty g(x) e^{izx} \, dx, \quad \mathcal{G}^-(z) = -\frac{1}{\sqrt{2\pi}} \int_{-\infty}^0 g(x) e^{izx} \, dx, \tag{23}$$

$$g(x) = \frac{1}{\sqrt{2\pi}} \int_{-\infty}^\infty \ln \left[\left(\frac{u-i}{u+i} \right)^{-\nu} \mathcal{D}(u) \right] e^{-iux} \, du,$$

where, at the point $-i$, $\mathcal{X}^-(z)$ has a zero of order ν for $\nu > 0$ and a pole of order $|\nu|$ for the case $\nu < 0$.

The coefficient $\mathcal{D}(u)$ of the Riemann boundary value problem can be represented as the ratio of the boundary values of the canonical function (see (22) and (23)):

$$\mathcal{D}(u) = \frac{\mathcal{X}^+(u)}{\mathcal{X}^-(u)}. \tag{24}$$

Such a representation of $\mathcal{D}(u)$ in the form of the ratio of boundary values of the canonical function is often called a *factorization*.

Now we consider the homogeneous Riemann problem with the boundary condition

$$\mathcal{Y}^+(u) = \mathcal{D}(u)\mathcal{Y}^-(u), \quad \mathcal{D}(\infty) = 1. \tag{25}$$

On substituting the expression (24) for $\mathcal{D}(u)$ into (25) we reduce the boundary condition to the form

$$\frac{\mathcal{Y}^+(u)}{\mathcal{X}^+(u)} = \frac{\mathcal{Y}^-(u)}{\mathcal{X}^-(u)}. \tag{26}$$

According to formulas (23) for $\mathcal{X}(z)$, the left- and the right-hand sides of Eq. (26) contain the boundary values of functions that are analytic on the upper and lower half-planes, respectively, possibly except for the point $-i$ at which the order is equal to ν. In the chosen function class, each function vanishes at infinity. In this case, it follows from the analytic continuation theorem and the generalized Liouville theorem (see Subsection 10.4-3) that for $\nu > 0$ we have

$$\frac{\mathcal{Y}^+(z)}{\mathcal{X}^+(z)} = \frac{\mathcal{Y}^-(z)}{\mathcal{X}^-(z)} = \frac{\mathcal{P}_{\nu-1}(z)}{(z+i)^\nu}, \tag{27}$$

where $\mathcal{P}_{\nu-1}(z)$ is an arbitrary polynomial of degree $\nu - 1$ (the degree of the numerator is less than that of the denominator because $\mathcal{Y}(\infty) = 0$). Hence,

$$\mathcal{Y}(z) = \mathcal{X}(z)\frac{\mathcal{P}_{\nu-1}(z)}{(z+i)^\nu}. \tag{28}$$

For $\nu \leq 0$, it follows from $\mathcal{Y}(\infty) = 0$ that $\mathcal{Y}(z) \equiv 0$ by the generalized Liouville theorem.

Hence, for $\nu > 0$, the homogeneous Riemann boundary value problem has precisely ν linearly independent solutions of the form

$$\frac{z^{k-1}\mathcal{X}(z)}{(z+i)^\nu}, \quad k = 1, 2, \ldots, \nu,$$

and for $\nu \leq 0$, there are no nontrivial solutions.

The right-hand side of Eq. (28) has exactly ν zeros on the entire plane, including the zero at infinity. These zeros can lie at arbitrary points of the upper and lower half-plane or on the real axis. Denote the number of zeros on the real axis by N_0. In the general case (without the requirement that there are no zeros on the real axis), formula (19) is replaced by the relation

$$N_+ + N_- + N_0 = \operatorname{Ind}\mathcal{D}(u) = \nu. \tag{29}$$

Let us pass to the solution of the nonhomogeneous Riemann problem with the boundary condition (14). We apply relation (24) and reduce the boundary condition to the form

$$\frac{\mathcal{Y}^+(u)}{\mathcal{X}^+(u)} = \frac{\mathcal{Y}^-(u)}{\mathcal{X}^-(u)} + \frac{\mathcal{H}(u)}{\mathcal{X}^+(u)}. \tag{30}$$

Let us express the last summand as the difference of the boundary values of functions that are analytic in the upper and the lower half-plane (see the jump problem), that is,

$$\mathcal{W}^+(u) - \mathcal{W}^-(u) = \frac{\mathcal{H}(u)}{\mathcal{X}^+(u)}, \tag{31}$$

where

$$W^+(z) = \frac{1}{\sqrt{2\pi}} \int_0^\infty w(x)e^{izx}\,dx, \quad W^-(z) = -\frac{1}{\sqrt{2\pi}} \int_{-\infty}^0 w(x)e^{izx}\,dx,$$

$$w(x) = \frac{1}{\sqrt{2\pi}} \int_{-\infty}^\infty \frac{\mathcal{H}(u)}{\mathcal{X}^+(u)} e^{-iux}\,du.$$

(32)

On substituting (31) into (30), we obtain

$$\frac{\mathcal{Y}^+(u)}{\mathcal{X}^+(u)} - W^+(u) = \frac{\mathcal{Y}^-(u)}{\mathcal{X}^-(u)} - W^-(u).$$

(33)

For $\nu > 0$, it follows from the analytic continuation theorem and the generalized Liouville theorem that

$$\frac{\mathcal{Y}^+(z)}{\mathcal{X}^+(z)} - W^+(z) = \frac{\mathcal{Y}^-(z)}{\mathcal{X}^-(z)} - W^-(z) = \frac{\mathcal{P}_{\nu-1}(z)}{(z+i)^\nu}.$$

Hence, for $\nu > 0$ we have

$$\mathcal{Y}(z) = \mathcal{X}(z)\left[W(z) + \frac{\mathcal{P}_{\nu-1}(z)}{(z+i)^\nu} \right].$$

(34)

The right-hand side of formula (34) contains the general solution (28) of the homogeneous problem as a summand, and hence the general solution of the nonhomogeneous problem is obtained.

For $\nu \le 0$ we must set $\mathcal{P}_{\nu-1}(z) \equiv 0$, and the desired solution becomes

$$\mathcal{Y}(z) = \mathcal{X}(z)W(z).$$

(35)

However, formula (35) gives a solution that satisfies all conditions for $\nu = 0$ only. For $\nu < 0$, the function $\mathcal{X}(z)$ has a pole of order $|\nu|$ at the point $-i$. In this case, for the existence of a solution in the chosen class of functions it is necessary that the second factor have a zero of the corresponding order at the point $-i$. On the basis of relations (6) and (32), we represent the function $W^-(z)$ in the form

$$W^-(z) = \frac{1}{2\pi i} \int_{-\infty}^\infty \frac{\mathcal{H}(\tau)}{\mathcal{X}^+(\tau)} \frac{d\tau}{\tau - z}.$$

On expanding the last integral in series in powers of $z + i$ and equating the coefficients of $(z+i)^{k-1}$ ($k = 1, 2, \ldots, |\nu|$) with zero, we obtain the solvability conditions for the problem in the form

$$\int_{-\infty}^\infty \frac{\mathcal{H}(u)}{\mathcal{X}^+(u)} \frac{du}{(u+i)^k} = 0, \qquad k = 1, 2, \ldots, |\nu|.$$

(36)

Figure 3 depicts a scheme of the above method for solving the Riemann problem on the real axis.

Let us state the results concerning the solution of the Riemann problem in the final form. If the index ν of the problem satisfies the condition $\nu > 0$, then the homogeneous and the nonhomogeneous Riemann problems are unconditionally solvable, and their solutions

$$\mathcal{Y}^\pm(z) = \mathcal{X}^\pm(z)\frac{\mathcal{P}_{\nu-1}(z)}{(z+i)^\nu} \qquad \text{(the homogeneous problem)},$$

(37)

$$\mathcal{Y}^\pm(z) = \mathcal{X}^\pm(z)\left[W^\pm(z) + \frac{\mathcal{P}_{\nu-1}(z)}{(z+i)^\nu} \right] \qquad \text{(the nonhomogeneous problem)}$$

(38)

depend on ν arbitrary complex constants, where $\mathcal{P}_{\nu-1}(z)$ is a polynomial of degree $\nu - 1$. If $\nu \le 0$, then the homogeneous problem has only the trivial zero solution, and the nonhomogeneous problem has the unique solution

$$\mathcal{Y}^\pm(z) = \mathcal{X}^\pm(z)W^\pm(z)$$

(39)

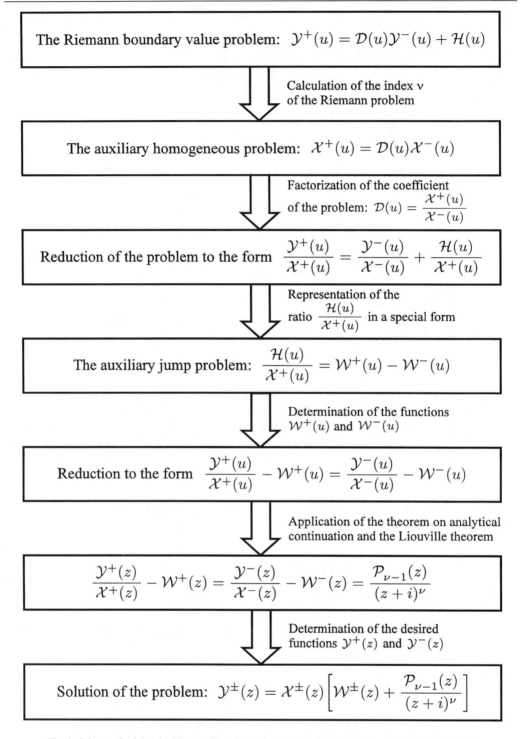

Fig. 3. Scheme of solving the Riemann boundary value problem for the functions $\mathcal{Y}^+(z)$ and $\mathcal{Y}^-(z)$ that are analytic, respectively, in the upper and the lower half-plane of the complex plane $z = u + iv$. It is assumed that $\mathcal{D}(u) \neq 0$ and $\mathcal{P}_{\nu-1}(z) \equiv 0$ for $\nu \le 0$.

provided that $|\nu|$ conditions (36) hold. Here we have

$$g(x) = \frac{1}{\sqrt{2\pi}} \int_{-\infty}^{\infty} \ln\left[\left(\frac{u-i}{u+i}\right)^{-\nu} \mathcal{D}(u)\right] e^{-iux}\, du, \tag{40}$$

$$\mathcal{G}^+(z) = \frac{1}{\sqrt{2\pi}} \int_0^{\infty} g(x)e^{izx}\, dx, \quad \mathcal{G}^-(z) = -\frac{1}{\sqrt{2\pi}} \int_{-\infty}^0 g(x)e^{izx}\, dx, \tag{41}$$

$$\mathcal{X}^+(z) = e^{\mathcal{G}^+(z)}, \quad \mathcal{X}^-(z) = \left(\frac{z-i}{z+i}\right)^{-\nu} e^{\mathcal{G}^-(z)}, \tag{42}$$

$$w(x) = \frac{1}{\sqrt{2\pi}} \int_{-\infty}^{\infty} \frac{\mathcal{H}(u)}{\mathcal{X}^+(u)} e^{-iux}\, du, \tag{43}$$

$$\mathcal{W}^+(z) = \frac{1}{\sqrt{2\pi}} \int_0^{\infty} w(x)e^{izx}\, dx, \quad \mathcal{W}^-(z) = -\frac{1}{\sqrt{2\pi}} \int_{-\infty}^0 w(x)e^{izx}\, dx. \tag{44}$$

The sequence of operations to construct a solution can be described as follows.

$1°$. By virtue of formula (40) we find $g(x)$, and then, with the help of (41), for the given $g(x)$ we find $\mathcal{G}^{\pm}(z)$.

$2°$. By formulas (42) the canonical function $\mathcal{X}^{\pm}(z)$ is determined.

$3°$. By formula (43) we determine $w(x)$, and then apply formula (44) to find $\mathcal{W}^{\pm}(z)$.

After this, solutions of the homogeneous and nonhomogeneous problems can be found by formulas (37)–(39) and (42). For the case $\nu < 0$, it is also necessary to verify the solvability conditions (36).

10.4-5. Problems With Rational Coefficients

The solution of the Riemann problem thus obtained requires evaluation of several Fourier integrals. This can also be readily expressed by means of integrals of the Cauchy type. As a rule, the integrals cannot be evaluated in the closed form and are calculated by various approximate methods. This process is rather cumbersome, and therefore it is of interest to select cases in which the solution can be obtained directly from the boundary condition by applying the method of analytic continuation without using the antiderivatives.

Assume that in the boundary condition (14) we have

$$\mathcal{D}(u) = \frac{\mathcal{R}_+(u)}{\mathcal{Q}_+(u)} \frac{\mathcal{R}_-(u)}{\mathcal{Q}_-(u)}.$$

Here $\mathcal{R}_+(u)$ and $\mathcal{Q}_+(u)$ ($\mathcal{R}_-(u)$ and $\mathcal{Q}_-(u)$) are polynomials whose zeros belong to the upper (lower) half-plane (we must avoid confusing these polynomials with the one-sided functions introduced above, which have similar notation). Denote the degrees of the polynomials $\mathcal{P}_+$, $\mathcal{R}_-$, $\mathcal{Q}_+$, and $\mathcal{Q}_-$ by m_+, m_-, n_+, and n_-, respectively. Since, by the assumption of the problem, the value $\mathcal{D}(\infty)$ can be neither zero nor infinity, it follows that the relation $m_+ + m_- = n_+ + n_-$ holds. The index of the problem can be expressed by the formula

$$\nu = \text{Ind}\,\mathcal{D}(u) = m_+ - n_+ = -(m_- - n_-).$$

On multiplying the boundary condition by $\mathcal{Q}_-(u)/\mathcal{P}_-(u)$ we obtain

$$\frac{\mathcal{Q}_-(u)}{\mathcal{R}_-(u)} \mathcal{Y}^+(u) - \frac{\mathcal{R}_+(u)}{\mathcal{Q}_+(u)} \mathcal{Y}^-(u) = \frac{\mathcal{Q}_-(u)}{\mathcal{R}_-(u)} \mathcal{H}(u).$$

If $\mathcal{H}(u)$ is a rational function as well, then the jump problem can readily be solved:

$$W^+(u) - W^-(u) = \frac{Q_-(u)}{R_-(u)} \mathcal{H}(u). \tag{45}$$

To this end, it suffices to decompose the right-hand side into the sum of partial fractions. Then $W^+(u)$ and $W^-(u)$ are the sums of the partial fractions with poles in the lower and the upper half-planes, respectively. We can directly apply the continuity principle (the analytic continuation theorem) and the generalized Liouville theorem to the resulting relation

$$\frac{Q_-(u)}{R_-(u)} \mathcal{Y}^+(u) - W^+(u) = \frac{R_+(u)}{Q_+(u)} \mathcal{Y}^-(u) - W^-(u).$$

The only exceptional point at which the analytic function, which is the same on the entire complex plane, can have a nonzero order is the point at infinity, at which the order of the function is equal to $\nu - 1 = m_+ - n_+ - 1 = n_- - m_- - 1$.

For $\nu > 0$, the solution can be written in the form

$$\mathcal{Y}^+(z) = \frac{R_-(z)}{Q_-(z)} [W^+(z) + P_{\nu-1}(z)], \quad \mathcal{Y}^-(z) = \frac{Q_+(z)}{R_+(z)} [W^-(z) + P_{\nu-1}(z)].$$

For $\nu \leq 0$ we must set $P_{\nu-1} \equiv 0$; moreover, for $\nu < 0$ we must also write out the solvability conditions that can be obtained by equating with zero the first $|\nu|$ terms of the expansion of the rational function $W(z)$ in a series (in powers of $1/z$) in a neighborhood of the point at infinity.

The solution of the jump problem (45) can be obtained either by applying the method of indeterminate coefficients, as is usually performed in the integration of rational functions, or using the theory of residuals of analytic functions. Let z_k be a pole, of multiplicity m, of the function $[Q_-(z)/R_-(z)]\mathcal{H}(z)$. Then the coefficients of the principal part of the decomposition of this function in a neighborhood of the point z_k, which has the form

$$\frac{c_1^k}{z - z_k} + \cdots + \frac{c_m^k}{(z - z_k)^m},$$

can be found by the formula

$$c_j^k = \frac{1}{(j-1)!} \frac{d^{j-1}}{dz^{j-1}} \left[\frac{Q_-(z)}{R_-(z)} \mathcal{H}(z) \right]_{z=z_k}.$$

The above case is not only of independent interest, as it frequently occurs in practice, but also of importance as a possible way of solving the problem under general assumptions. The approximation of arbitrary coefficients of the class under consideration by rational functions is a widespread method of approximate solution of the Riemann boundary value problem.

10.4-6. Exceptional Cases. The Homogeneous Problem

Assume that the coefficient $\mathcal{D}(u)$ of a Riemann boundary value problem has zeros of orders $\alpha_1, \ldots, \alpha_r$ at points $a_1, \ldots, a_r$, respectively, and poles* of the orders $\beta_1, \ldots, \beta_s$ at points $b_1, \ldots, b_s$ ($\alpha_1, \ldots, \alpha_r$ and $\beta_1, \ldots, \beta_s$ are positive integers). Thus, the coefficient can be represented in the form

$$\mathcal{D}(u) = \frac{\displaystyle\prod_{i=1}^{r} (u - a_i)^{\alpha_i}}{\displaystyle\prod_{j=1}^{s} (u - b_j)^{\beta_j}} \mathcal{D}_1(u), \quad \mathcal{D}_1(u) \neq 0, \quad -\infty < u < \infty, \quad \sum_{i=1}^{r} \alpha_i = m, \quad \sum_{j=1}^{s} \beta_j = n. \tag{46}$$

---

* For the case in which the function $\mathcal{D}(u)$ is not analytic, the term "pole" will be used for points at which the function tends to infinity with integer order.

In turn, we represent the function $\mathcal{D}_1(u)$ (see Subsection 10.4-5) in the form

$$\mathcal{D}_1(u) = \frac{\mathcal{R}_+(u)\mathcal{R}_-(u)}{\mathcal{Q}_+(u)\mathcal{Q}_-(u)}\mathcal{D}_2(u), \tag{47}$$

where, as above, $\mathcal{R}_+(u)$ and $\mathcal{Q}_+(u)$ ($\mathcal{R}_-(u)$ and $\mathcal{Q}_-(u)$) are polynomials of degrees m_+ and n_+ (m_- and n_-) whose zeros belong to the upper (lower) half-plane. The function $\mathcal{D}_2(u)$ satisfies the Hölder condition, has zero index, and nowhere vanishes on the real axis. Moreover, this function can be subjected to some differentiability conditions in neighborhoods of the points a_i and b_j and possibly in a neighborhood of the point at infinity.

The boundary condition of the homogeneous Riemann problem can be rewritten in the form

$$\mathcal{Y}^+(u) = \frac{\displaystyle\prod_{i=1}^{r}(u - a_i)^{\alpha_i}\mathcal{R}_+(u)\mathcal{R}_-(u)}{\displaystyle\prod_{j=1}^{s}(u - b_j)^{\beta_j}\mathcal{Q}_+(u)\mathcal{Q}_-(u)}\mathcal{D}_2(u)\mathcal{Y}^-(u). \tag{48}$$

We seek a solution in the class of functions that are bounded on the real axis and vanish at infinity:

$$\mathcal{Y}(\infty) = 0. \tag{49}$$

The coefficient $\mathcal{D}(u)$ has the order

$$\eta = n + n_+ + n_- - m - m_+ - m_- \tag{50}$$

at infinity. The number

$$\nu = m_+ - n_+ \tag{51}$$

is called the *index* of the problem. Let us introduce the notation

$$h = n_- - m_-. \tag{52}$$

Then the order at infinity is expressed by the formula

$$\eta = h - \nu + n - m. \tag{53}$$

Now let us proceed with the solution of problem (48). Applying general methods, we set

$$\mathcal{D}_2(u) = \frac{e^{\mathcal{G}^+(u)}}{e^{\mathcal{G}^-(u)}}, \quad g(x) = \frac{1}{\sqrt{2\pi}}\int_{-\infty}^{\infty}\ln\mathcal{D}_2(u)e^{-iux}\,du,$$

$$\mathcal{G}^+(z) = \frac{1}{\sqrt{2\pi}}\int_0^{\infty}g(x)e^{izx}\,dx, \quad \mathcal{G}^-(z) = -\frac{1}{\sqrt{2\pi}}\int_{-\infty}^0 g(x)e^{izx}\,dx \tag{54}$$

and rewrite the boundary condition in the form

$$\frac{\mathcal{Q}_-(u)\mathcal{Y}^+(u)}{\displaystyle\prod_{i=1}^{r}(u - a_i)^{\alpha_i}\mathcal{R}_-(u)e^{\mathcal{G}^+(u)}} = \frac{\mathcal{R}_+(u)\mathcal{Y}^-(u)}{\displaystyle\prod_{j=1}^{s}(u - b_j)^{\beta_j}\mathcal{Q}_+(u)e^{\mathcal{G}^-(u)}}. \tag{55}$$

As above, we can apply the analytic continuation and the generalized Liouville theorem and obtain a pole at infinity as the only possible singularity.

Two cases are possible:

$1°$. Let the order η of the coefficient of the boundary value problem at infinity satisfy the condition $\eta \geq 0$, i.e., let $\mathcal{D}(u)$ have a zero of order η at infinity. It follows from (53) that $n - \nu \geq m - h$. On equating the left- and right-hand sides of relation (55) with a polynomial $\mathcal{P}_{\nu-n-1}(z)$, we obtain the solution of the boundary value problem in the form

$$
\begin{aligned}
\mathcal{Y}^+(z) &= \prod_{i=1}^{r}(z - a_i)^{\alpha_i} \frac{\mathcal{R}_-(z)}{\mathcal{Q}_-(z)} e^{\mathcal{G}^+(z)} \mathcal{P}_{\nu-n-1}(z), \\
\mathcal{Y}^-(z) &= \prod_{j=1}^{s}(z - b_j)^{\beta_j} \frac{\mathcal{Q}_+(z)}{\mathcal{R}_+(z)} e^{\mathcal{G}^-(z)} \mathcal{P}_{\nu-n-1}(z).
\end{aligned}
\tag{56}
$$

This problem has $\nu - n$ linearly independent solutions for $\nu - n > 0$ and only the trivial zero solution for $\nu - n \leq 0$.

$2°$. Let $\eta < 0$, i.e., let $\mathcal{D}(u)$ have a pole of order $-\eta$ at infinity. In this case, $m - h > n - \nu$, and we can obtain the general solution from (56) by replacing $\mathcal{P}_{\nu-n-1}(z)$ by $\mathcal{P}_{h-m-1}(z)$ in this expression. In this case, the problem has $h - m$ solutions for $h - m > 0$ and only the trivial zero solution for $h - m \leq 0$.

According to (53), we have

$$
h - m = \nu - n + \eta.
\tag{57}
$$

Thus, in both cases under consideration, the number of linearly independent solutions is equal to the index minus the total number of the poles (including the pole at infinity) of the coefficient $\mathcal{D}(u)$. Hence, we have the following law: the number of linearly independent solutions of a homogeneous Riemann problem is not affected by the number of zeros of the coefficient and is reduced by the total number of its poles.

10.4-7. Exceptional Cases. The Nonhomogeneous Problem

Assume that the right-hand side has the same poles as the coefficient. The boundary condition can be rewritten as follows:

$$
\mathcal{Y}^+(u) = \frac{\displaystyle\prod_{i=1}^{r}(u - a_i)^{\alpha_i}\mathcal{R}_+(u)\mathcal{R}_-(u)}{\displaystyle\prod_{j=1}^{s}(u - b_j)^{\beta_j}\mathcal{Q}_+(u)\mathcal{Q}_-(u)}\mathcal{D}_2(u)\mathcal{Y}^-(u) + \frac{\mathcal{H}_1(u)}{\displaystyle\prod_{j=1}^{s}(u - b_j)^{\beta_j}},
\tag{58}
$$

where $\mathcal{D}_2(u)$ and $\mathcal{H}_1(u)$ satisfy the Hölder condition and some additional differentiability conditions near the points a_i, b_j, and ∞.

$1°$. Assume that the order η at infinity of the coefficient of the boundary value problem satisfies the condition $\eta \geq 0$. Since the first two terms of relation (58) vanish at infinity, it follows that the minimal possible order of $\mathcal{H}_1(u)$ at infinity is equal to $1 - n$. Just as in the homogeneous problem, we replace $\mathcal{D}_2(u)$ by the ratio of two functions (54) and write out the boundary condition in the following form (under the braces, the orders of the functions at infinity are indicated):

$$
\underbrace{\frac{\displaystyle\prod_{j=1}^{s}(u - b_j)^{\beta_j}\mathcal{Q}_-(u)\mathcal{Y}^+(u)}{\mathcal{R}_-(u)e^{\mathcal{G}^+(u)}}}_{1-n-h} = \underbrace{\frac{\displaystyle\prod_{i=1}^{r}(u - a_i)^{\alpha_i}\mathcal{R}_+(u)\mathcal{Y}^-(u)}{\mathcal{Q}_+(u)e^{\mathcal{G}^-(u)}}}_{1-m-\nu} + \underbrace{\frac{\mathcal{H}_1(u)\mathcal{Q}_-(u)}{\mathcal{R}_-(u)e^{\mathcal{G}^+(u)}}}_{1-n-h}
$$

Assume that a polynomial $\mathcal{S}(u)$ of the degree $n + h - 1$ represents the principal part of the decomposition of the last term in a neighborhood of the point at infinity (for the case in which $n + h - 1 \geq 0$):

$$\frac{\mathcal{H}_1(u)\mathcal{Q}_-(u)}{\mathcal{R}_-(u)e^{\mathcal{G}^+(u)}} = \mathcal{S}(u) + \mathcal{W}(u), \qquad \mathcal{W}(\infty) = 0.$$

On replacing the function $\mathcal{W}(u)$ by the difference of boundary values of analytic functions

$$\mathcal{W}(u) = \mathcal{W}^+(u) - \mathcal{W}^-(u), \tag{59}$$

where

$$w(x) = \frac{1}{\sqrt{2\pi}} \int_{-\infty}^{\infty} \mathcal{W}(u)e^{-iux}\, du,$$

$$\mathcal{W}^+(u) = \frac{1}{\sqrt{2\pi}} \int_0^{\infty} w(x)e^{iux}\, dx, \quad \mathcal{W}^-(u) = -\frac{1}{\sqrt{2\pi}} \int_{-\infty}^0 w(x)e^{iux}\, dx, \tag{60}$$

we reduce the boundary condition to the form

$$\frac{\displaystyle\prod_{j=1}^{s}(u - b_j)^{\beta_j}\mathcal{Q}_-(u)\mathcal{Y}^+(u)}{\mathcal{R}_-(u)e^{\mathcal{G}^+(u)}} - \mathcal{S}(u) - \mathcal{W}^+(u) = \frac{\displaystyle\prod_{i=1}^{r}(u - a_i)^{\alpha_i}\mathcal{R}_+(u)\mathcal{Y}^-(u)}{\mathcal{Q}_+(u)e^{\mathcal{G}^-(u)}} - \mathcal{W}^-(u).$$

On applying the analytic continuation theorem and the generalized Liouville theorem and taking into account the fact that the only possible singular point of the function under consideration is the point at infinity, while we have the relation $-n - h \leq -m - \nu$ $(\eta \geq 0)$, we obtain the expressions

$$\mathcal{Y}^+(z) = \frac{\mathcal{R}_-(z)e^{\mathcal{G}^+(z)}}{\displaystyle\prod_{j=1}^{s}(z - b_j)^{\beta_j}\mathcal{Q}_-(z)}[\mathcal{W}^+(z) + \mathcal{S}(z) + \mathcal{P}_{\nu+m-1}(z)],$$

$$\mathcal{Y}^-(z) = \frac{\mathcal{Q}_+(z)e^{\mathcal{G}^-(z)}}{\displaystyle\prod_{i=1}^{r}(z - a_i)^{\alpha_i}\mathcal{R}_+(z)}[\mathcal{W}^-(z) + \mathcal{P}_{\nu+m-1}(z)]. \tag{61}$$

The last formulas define a solution that has pole singularities at the points a_i and b_j. To obtain a bounded solution, we apply the canonical function of the nonhomogeneous problem.

By a *canonical function* $\mathcal{V}(z)$ *of the nonhomogeneous Riemann problem* in the *exceptional case* we mean a piecewise analytic function that satisfies the boundary condition (58), has the zero order on the entire finite part of the complex plane, including the points a_i and b_j, and has the least possible order at infinity.

Let $\mathcal{U}_p(z)$ be the Hermite interpolation polynomial with interpolation nodes of orders α_i and β_j at the points a_i and b_j, respectively. Such a polynomial of degree $p = m + n - 1$ exists and is determined uniquely (see Subsection 12.3-1). The functions $\mathcal{D}_1(u)$ and $\mathcal{H}_1(u)$ must be subjected to the additional condition that in neighborhoods of the points a_i and b_j these functions have derivatives of the orders α_i and β_j, respectively, and these derivatives satisfy the Hölder condition. Then the canonical function of the nonhomogeneous problem can be represented in the form

$$\mathcal{V}^+(z) = \frac{\mathcal{R}_-(z)e^{\mathcal{G}^+(z)}}{\displaystyle\prod_{j=1}^{s}(z - b_j)^{\beta_j}\mathcal{Q}_-(z)}[\mathcal{W}^+(z) + \mathcal{S}(z) - \mathcal{U}_p(z)],$$

$$\mathcal{V}^-(z) = \frac{\mathcal{Q}_+(z)e^{\mathcal{G}^-(z)}}{\displaystyle\prod_{i=1}^{r}(z - a_i)^{\alpha_i}\mathcal{R}_+(z)}[\mathcal{W}^-(z) - \mathcal{U}_p(z)]. \tag{62}$$

Adding $\mathcal{V}(z)$ to the above general solution of the homogeneous problem, we find the general solution of the nonhomogeneous problem under consideration:

$$\mathcal{Y}^+(z) = \mathcal{V}^+(z) + \prod_{i=1}^r (z - a_i)^{\alpha_i} \frac{\mathcal{R}_-(z)}{\mathcal{Q}_-(z)} e^{\mathcal{G}^+(z)} \mathcal{P}_{\nu-n-1}(z),$$

$$\mathcal{Y}^-(z) = \mathcal{V}^-(z) + \prod_{j=1}^s (z - b_j)^{\beta_j} \frac{\mathcal{Q}_+(z)}{\mathcal{R}_+(z)} e^{\mathcal{G}^-(z)} \mathcal{P}_{\nu-n-1}(z).$$

$$(63)$$

For $\nu - n > 0$, the problem has $\nu - n$ linearly independent solutions. In the case $\nu - n \le 0$ we must set $\mathcal{P}_{\nu-n-1}(z) \equiv 0$. For $\nu - n < 0$, the canonical function $\mathcal{V}(z)$ has the order $\nu - n < 0$ at infinity and hence is no longer a solution of the nonhomogeneous problem. However, on subjecting the right-hand side to $n - \nu$ conditions we can increase the order of the function $\mathcal{V}(u)$ at infinity by $n - \nu$ and thus make the canonical function $\mathcal{V}(z)$ be a solution of the nonhomogeneous problem again.

To make the above operations possible, it suffices to require that the functions $u^k \mathcal{H}_1(u)$ and $\mathcal{D}_2(u)$ have derivatives of order $\le n - \nu$ at infinity, and these derivatives satisfy the Hölder condition.

$2°$. Let $\eta < 0$. The least possible order at infinity of $\mathcal{H}_1(u)$ is $h - \nu - m + 1$. In this case, the function $[\mathcal{H}_1(u)\mathcal{Q}_-(u)]/[\mathcal{R}_-(u)e^{\mathcal{G}^+(u)}]$ in the boundary condition (58) has the order $1 - m - \nu$ at infinity. After selecting the principal part of the expansion of $[\mathcal{H}_1(u)\mathcal{Q}_-(u)]/[\mathcal{R}_-(u)e^{\mathcal{G}^+(u)}]$ in a neighborhood of the point at infinity for $m + \nu - 1 > 0$, the boundary condition can be rewritten in the form

$$\frac{\prod_{j=1}^s (u - b_j)^{\beta_j} \mathcal{Q}_-(u) \mathcal{Y}^+(u)}{\mathcal{R}_-(u) e^{\mathcal{G}^+(u)}} - \mathcal{W}^+(u) = \frac{\prod_{i=1}^r (u - a_i)^{\alpha_i} \mathcal{R}_+(u) \mathcal{Y}^-(u)}{\mathcal{Q}_+(u) e^{\mathcal{G}^-(u)}} - \mathcal{W}^-(u) + \mathcal{S}(u).$$

The canonical function of the nonhomogeneous problem can be expressed via the interpolation polynomial as follows:

$$\mathcal{V}_1^+(z) = \frac{\mathcal{R}_-(z) e^{\mathcal{G}^+(z)}}{\prod_{j=1}^s (z - b_j)^{\beta_j} \mathcal{Q}_-(z)} [\mathcal{W}^+(z) - \mathcal{U}_p(z)],$$

$$\mathcal{V}_1^-(z) = \frac{\mathcal{Q}_+(z) e^{\mathcal{G}^-(z)}}{\prod_{i=1}^r (z - a_i)^{\alpha_i} \mathcal{R}_+(z)} [\mathcal{W}^-(z) - \mathcal{S}(z) - \mathcal{U}_p(z)].$$

$$(64)$$

The general solution of problem (58) becomes

$$\mathcal{Y}^+(z) = \mathcal{V}_1^+(z) + \prod_{i=1}^r (z - a_i)^{\alpha_i} \frac{\mathcal{R}_-(z)}{\mathcal{Q}_-(z)} e^{\mathcal{G}^+(z)} \mathcal{P}_{h-m-1}(z),$$

$$\mathcal{Y}^-(z) = \mathcal{V}_1^-(z) + \prod_{j=1}^s (z - b_j)^{\beta_j} \frac{\mathcal{Q}_+(z)}{\mathcal{R}_+(z)} e^{\mathcal{G}^-(z)} \mathcal{P}_{h-m-1}(z).$$

$$(65)$$

For $h - m > 0$, the problem has $h - m$ linearly independent solutions. In the case $h - m \le 0$, we must set the polynomial $\mathcal{P}_{h-m-1}(z)$ to be identically zero and, for the case in which $h - m < 0$, impose $m - h$ conditions of the same type as in the previous case on the right-hand side. Under these conditions, the nonhomogeneous problem (58) has a unique solution.

Remark. In Section 10.5 we consider equations that can be reduced to the problem by applying the convolution theorem for the Fourier transform. Equations to which the convolution theorems for other integral transforms can be applied, for instance, for the Mellin transform, can be investigated in a similar way.

⊙ References for Section 10.4: F. D. Gakhov and Yu. I. Cherskii (1978), S. G. Mikhlin and S. Prössdorf (1986), N. I. Muskhelishvili (1992).

10.5. The Carleman Method for Equations of the Convolution Type of the First Kind

By the *Carleman method* we mean the method of reducing an integral equation to a boundary value problem of the theory of analytic functions, in particular, to the Riemann problem. For equations of convolution type, this reduction can be performed by means of the integral transforms. After solving the boundary value problem, the desired function can be obtained by applying the inverse integral transform.

10.5-1. The Wiener–Hopf Equation of the First Kind

Consider the Wiener–Hopf equation of the first kind

$$\frac{1}{\sqrt{2\pi}} \int_0^\infty K(x-t)y(t)\,dt = f(x), \qquad 0 < x < \infty, \tag{1}$$

which is frequently encountered in applications. Let us extend its domain to the negative semiaxis by introducing one-sided functions,

$$y_+(x) = \begin{cases} y(x) & \text{for } x > 0, \\ 0 & \text{for } x < 0, \end{cases} \qquad f_+(x) = \begin{cases} f(x) & \text{for } x > 0, \\ 0 & \text{for } x < 0, \end{cases} \qquad y_-(x) = 0 \quad \text{for } x > 0.$$

Using these one-sided functions, we can rewrite Eq. (1) in the form

$$\frac{1}{\sqrt{2\pi}} \int_{-\infty}^\infty K(x-t)y_+(t)\,dt = f_+(x) + y_-(x), \qquad -\infty < x < \infty. \tag{2}$$

The auxiliary function $y_-(x)$ is introduced to compensate for the left-hand side of Eq. (2) for $x < 0$. Note that $y_-(x)$ is unknown in the domain $x < 0$ and is to be found in solving the problem.

Let us now apply the alternative Fourier transform to Eq. (2). Then we obtain the boundary value problem

$$\mathcal{Y}^+(u) = \frac{1}{\mathcal{K}(u)}\mathcal{Y}^-(u) + \frac{\mathcal{F}^+(u)}{\mathcal{K}(u)}. \tag{3}$$

If σ is the order of $\mathcal{K}(u)$ at infinity, then the order of the coefficient of the boundary value problem at infinity is $\eta = -\sigma < 0$. The general solution of problem (3) can be obtained on the basis of relations (65) from Subsection 10.4-7 by replacing $\mathcal{P}_{h-m-1}(z)$ with $\mathcal{P}_{\nu-n+\eta-1}(z)$ there. The solution of the original equation (1) can be obtained from the solution of problem (3) by means of the inversion formula

$$y(x) = y_+(x) = \frac{1}{\sqrt{2\pi}} \int_{-\infty}^\infty \mathcal{Y}^+(u)e^{-iux}\,du, \qquad x > 0. \tag{4}$$

Note that in formula (4), only the function $\mathcal{Y}^+(u)$ occurs explicitly, which is related to the function $\mathcal{Y}^-(u)$ by (3).

10.5-2. Integral Equations of the First Kind With Two Kernels

Consider the integral equation of the first kind

$$\frac{1}{\sqrt{2\pi}} \int_0^\infty K_1(x-t)y(t)\,dt + \frac{1}{\sqrt{2\pi}} \int_{-\infty}^0 K_2(x-t)y(t)\,dt = f(x), \qquad -\infty < x < \infty. \tag{5}$$

The Fourier transform of Eq. (5) results in the following boundary value problem:

$$\mathcal{Y}^+(u) = \frac{\mathcal{K}_2(u)}{\mathcal{K}_1(u)}\mathcal{Y}^-(u) + \frac{\mathcal{F}(u)}{\mathcal{K}_1(u)}, \qquad -\infty < u < \infty. \tag{6}$$

The coefficient of this problem is the ratio of functions that vanish at infinity, and hence, in contrast to the preceding case, it can have a zero or a pole of some order at infinity.

Let $\mathcal{K}_1(u) = \mathcal{T}_1(u)/u^\lambda$ and $\mathcal{K}_2(u) = \mathcal{T}_2(u)/u^\mu$, where the functions $\mathcal{T}_1(u)$ and $\mathcal{T}_2(u)$ have zero order at infinity. In the dependence of the sign of the difference $\eta = \mu - \lambda$, two cases can occur. For generality, we assume that there are exceptional points at finite distances as well. Let the functions $\mathcal{K}_1(u)$ and $\mathcal{K}_2(u)$ have the representations

$$\mathcal{K}_1(u) = \prod_{j=1}^{s}(u-b_j)^{\beta_j} \prod_{k=1}^{p}(u-c_k)^{\gamma_k}\mathcal{K}_{11}(u),$$

$$\mathcal{K}_2(u) = \prod_{i=1}^{r}(u-a_i)^{\alpha_i} \prod_{k=1}^{p}(u-c_k)^{\gamma_k}\mathcal{K}_{12}(u).$$

Along with the common zeros at points c_k of multiplicity γ_k, the functions $\mathcal{K}_1(u)$ and $\mathcal{K}_2(u)$ have a common zero of order $\min(\lambda, \mu)$ at infinity.

The coefficient of the Riemann problem can be represented in the form

$$\mathcal{D}(u) = \frac{\displaystyle\prod_{i=1}^{r}(u-a_i)^{\alpha_i}\mathcal{R}_+(u)\mathcal{R}_-(u)}{\displaystyle\prod_{j=1}^{s}(u-b_j)^{\beta_j}\mathcal{Q}_+(u)\mathcal{Q}_-(u)}\mathcal{D}_2(u).$$

It follows from (6) that this problem and the integral equation (5) are solvable if at any point c_k that is a common zero of the functions $\mathcal{K}_1(u)$ and $\mathcal{K}_2(u)$, the function $\mathcal{F}(u)$ has zero of order γ_k, i.e., $\mathcal{F}(u)$ has the form

$$\mathcal{F}(u) = \prod_{k=1}^{p}(u-c_k)^{\gamma_k}\mathcal{F}_1(u).$$

To this end, the following $\gamma_1 + \cdots + \gamma_p = l$ conditions must hold:

$$\left[\mathcal{F}_u^{(j_k)}(u)\right]_{u=c_k} = 0, \qquad j_k = 0, 1, \ldots, \gamma_k - 1, \tag{7}$$

or, which is the same, the conditions

$$\int_{-\infty}^{\infty} f(x)x^{j_k}e^{ic_k x}\,dx = 0. \tag{8}$$

For the case under consideration in which the equation is of the first kind, we must add other d conditions, where

$$d = \min(\lambda, \mu) + 1, \tag{9}$$

that are imposed on the behavior of $\mathcal{F}(u)$ at infinity because the functions $\mathcal{K}_1(u)$ and $\mathcal{K}_2(u)$ have a common zero of order $\min(\lambda, \mu)$ at infinity. Hence, $\mathcal{F}(u)$ must satisfy the conditions (8) and have at least the order d at infinity.

If these conditions are satisfied, then the boundary value problem (6) becomes

$$\mathcal{Y}^+(u) = \frac{\prod\limits_{i=1}^{r}(u-a_i)^{\alpha_i}\mathcal{R}_+(u)\mathcal{R}_-(u)}{\prod\limits_{j=1}^{s}(u-b_j)^{\beta_j}\mathcal{Q}_+(u)\mathcal{Q}_-(u)}\mathcal{D}_2(u)\mathcal{Y}^-(u) + \frac{\mathcal{H}_1(u)}{\prod\limits_{j=1}^{s}(u-b_j)^{\beta_j}}.$$

The solution was given above in Subsection 10.4-7. For the case in which $\eta \geq 0$ ($\mu \geq \lambda$), this solution can be rewritten in the form

$$\mathcal{Y}^+(z) = \mathcal{V}^+(z) + \prod_{i=1}^{r}(z-a_i)^{\alpha_i}\frac{\mathcal{R}_-(z)}{\mathcal{Q}_-(z)}e^{\mathcal{G}^+(z)}\mathcal{P}_{\nu-n+1}(z),$$

$$\mathcal{Y}^-(z) = \mathcal{V}^-(z) + \prod_{j=1}^{s}(z-b_j)^{\beta_j}\frac{\mathcal{Q}_+(z)}{\mathcal{R}_+(z)}e^{\mathcal{G}^-(z)}\mathcal{P}_{\nu-n+1}(z).$$
$$(10)$$

For the case in which $\eta < 0$ ($\mu < \lambda$), this solution becomes

$$\mathcal{Y}^+(z) = \mathcal{V}_1^+(z) + \prod_{i=1}^{r}(z-a_i)^{\alpha_i}\frac{\mathcal{R}_-(z)}{\mathcal{Q}_-(z)}e^{\mathcal{G}^+(z)}\mathcal{P}_{h-m-1}(z),$$

$$\mathcal{Y}^-(z) = \mathcal{V}_1^-(z) + \prod_{j=1}^{s}(z-b_j)^{\beta_j}\frac{\mathcal{Q}_+(z)}{\mathcal{R}_+(z)}e^{\mathcal{G}^-(z)}\mathcal{P}_{h-m-1}(z).$$
$$(11)$$

In both cases, the solution of the original integral equation can be obtained by substituting the expressions (10) and (11) into the formula

$$y(x) = \frac{1}{\sqrt{2\pi}}\int_{-\infty}^{\infty}[\mathcal{Y}^+(u) - \mathcal{Y}^-(u)]e^{-iux}\,du. \qquad (12)$$

Example. Consider the following equation of the first kind:

$$\frac{1}{\sqrt{2\pi}}\int_0^\infty K_1(x-t)y(t)\,dt + \frac{1}{\sqrt{2\pi}}\int_{-\infty}^0 K_2(x-t)y(t)\,dt = f(x),$$

where

$$K_1(x) = \begin{cases} 0 & \text{for } x > 0, \\ \sqrt{2\pi}\,(e^{3x} - e^{2x}) & \text{for } x < 0, \end{cases} \quad K_2(x) = \begin{cases} -\sqrt{2\pi}\,ie^{-2x} & \text{for } x > 0, \\ 0 & \text{for } x < 0, \end{cases} \quad f(x) = \begin{cases} 0 & \text{for } x > 0, \\ \sqrt{2\pi}\,(e^{3x} - e^{2x}) & \text{for } x < 0. \end{cases} \quad (13)$$

Applying the Fourier transform to the functions in (13), we obtain

$$\mathcal{K}_1(u) = \frac{1}{(u-2i)(u-3i)}, \quad \mathcal{K}_2(u) = \frac{1}{u+2i}, \quad \mathcal{F}(u) = \frac{1}{(u-2i)(u-3i)}.$$

Here the boundary value problem (6) becomes

$$\mathcal{Y}^+(u) = \frac{(u-2i)(u-3i)}{u+2i}\mathcal{Y}^-(u) + 1.$$

The coefficient $\mathcal{D}(u)$ has a first-order pole at infinity ($\nu = -1$). In this case

$$m_+ = 2, \quad n_+ = 0, \quad \nu = m_+ - n_+ = 2, \quad \min(\lambda,\mu) = 1, \quad d = 2.$$

The function $\mathcal{F}(u)$ has second-order zero at infinity, and hence the necessary condition for the solvability is satisfied. In the class of functions that vanish at infinity, the homogeneous problem

$$\mathcal{Y}^+(u) = \frac{(u-2i)(u-3i)}{u+2i}\mathcal{Y}^-(u)$$

has the following solution:

$$\mathcal{Y}^+(z) = \frac{C}{z+2i}, \quad \mathcal{Y}^-(z) = \frac{C}{(z-2i)(z-3i)},$$

where C is an arbitrary constant.

The number of linearly independent solutions of problem (13) is less by one than the index, because $\mathcal{D}(u)$ has a first-order pole at infinity.

The solution of the nonhomogeneous problem in the class of functions vanishing at infinity has the form

$$\mathcal{Y}^+(z) = \frac{C}{z+2i}, \quad \mathcal{Y}^-(z) = \frac{C-2i-z}{(z-2i)(z-3i)},$$

$$y(x) = \begin{cases} -\sqrt{2\pi}\,iCe^{-2x} & \text{for } x > 0, \\ \sqrt{2\pi}\,C(e^{2x}-e^{3x}) - 4i\sqrt{2\pi}\,e^{2x} + 5i\sqrt{2\pi}\,e^{3x} & \text{for } x < 0. \end{cases}$$

For the chosen right-hand side, the equation turns out to be solvable. However, if we take, for instance,

$$f(x) = \begin{cases} 0 & \text{for } x > 0, \\ \sqrt{2\pi}\,i(5e^{3x}-4e^{2x}) & \text{for } x < 0, \end{cases} \tag{14}$$

then we have $\mathcal{F}(u) = (u+2i)/[(u-2i)(u-3i)]$. The corresponding Riemann boundary value problem has the form

$$\mathcal{Y}^+(u) = \frac{(u-2i)(u-3i)}{u+2i}\mathcal{Y}^-(u) + u + 2i.$$

In the class of functions bounded at infinity, its solution can be represented in the form

$$\mathcal{Y}^+(z) = \frac{C-z}{z+2i}, \quad \mathcal{Y}^-(z) = \frac{C-z-(z+2i)^2}{(z-2i)(z-3i)}. \tag{15}$$

For no choice of the constant C the solution vanishes at infinity, and hence the equation with the right-hand side defined by (14) has no solutions integrable on the real axis.

⊙ References for Section 10.5: F. D. Gakhov and Yu. I. Cherskii (1978), S. G. Mikhlin and S. Prössdorf (1986), N. I. Muskhelishvili (1992).

10.6. Dual Integral Equations of the First Kind

10.6-1. The Carleman Method for Equations With Difference Kernels

Consider the following dual integral equation of convolution type:

$$\begin{aligned} \frac{1}{\sqrt{2\pi}} \int_{-\infty}^{\infty} K_1(x-t)y(t)\,dt &= f(x), & 0 < x < \infty, \\ \frac{1}{\sqrt{2\pi}} \int_{-\infty}^{\infty} K_2(x-t)y(t)\,dt &= f(x), & -\infty < x < 0, \end{aligned} \tag{1}$$

in which the function $y(x)$ is to be found.

In order to apply the Fourier transform technique (see Subsections 7.4-3, 10.4-1, and 10.4-2), we extend the domain of both conditions in Eq. (1) by formally rewriting them for all real values of x. This can be achieved by introducing new unknown functions into the right-hand sides. These functions must be chosen so that the conditions given on the semiaxis are not violated. Hence, the first condition in (1) must be complemented by a summand that vanishes on the positive semiaxis and the second by a summand that vanishes on the negative semiaxis. Thus, the dual equation can be written in the form

$$\begin{aligned} \frac{1}{\sqrt{2\pi}} \int_{-\infty}^{\infty} K_1(x-t)y(t)\,dt &= f(x) + \xi_-(x), \\ & & -\infty < x < \infty, \\ \frac{1}{\sqrt{2\pi}} \int_{-\infty}^{\infty} K_2(x-t)y(t)\,dt &= f(x) + \xi_+(x), \end{aligned}$$

where the $\xi_\pm(x)$ are some right and left one-sided functions so far unknown.

On applying the Fourier integral transform, we have

$$\mathcal{K}_1(u)\mathcal{Y}(u) = \mathcal{F}(u) + \Xi^-(u), \quad \mathcal{K}_2(u)\mathcal{Y}(u) = \mathcal{F}(u) + \Xi^+(u). \tag{2}$$

Here the three functions $\mathcal{Y}(u)$, $\Xi^+(u)$, and $\Xi^-(u)$ are unknown.

Let us eliminate $\mathcal{Y}(u)$ from relations (2). We obtain the Riemann boundary value problem in the form

$$\Xi^+(u) = \frac{\mathcal{K}_2(u)}{\mathcal{K}_1(u)}\Xi^-(u) + \frac{\mathcal{K}_2(u) - \mathcal{K}_1(u)}{\mathcal{K}_1(u)}\mathcal{F}(u), \quad -\infty < u < \infty.$$

In the present case, the coefficient of the boundary condition is the ratio of functions that vanish at infinity, and hence this coefficient can have a zero or a pole of some order at infinity. The solution of the Riemann boundary value problem can be constructed on the basis of Subsections 10.4-6 and 10.4-7, and the solution of the integral equation (1) can be defined by the formula

$$y(x) = \frac{1}{\sqrt{2\pi}} \int_{-\infty}^{\infty} \frac{\Xi^+(u) + \mathcal{F}(u)}{\mathcal{K}_2(u)} e^{-iux}\, du = \frac{1}{\sqrt{2\pi}} \int_{-\infty}^{\infty} \frac{\Xi^-(u) + \mathcal{F}(u)}{\mathcal{K}_1(u)} e^{-iux}\, du. \tag{3}$$

Example. Let us solve the dual equation (1), where

$$\mathcal{K}_1(x) = \begin{cases} \sqrt{2\pi}\,(e^{3x} - e^{2x}) & \text{for } x < 0, \\ 0 & \text{for } x > 0, \end{cases} \quad \mathcal{K}_2(x) = \begin{cases} 0 & \text{for } x < 0, \\ -\sqrt{2\pi}\,ie^{-2x} & \text{for } x > 0, \end{cases} \quad f(x) = \begin{cases} \frac{1}{4}\sqrt{2\pi}\,e^{2x} & \text{for } x < 0, \\ -\frac{1}{4}\sqrt{2\pi}\,e^{-2x} & \text{for } x > 0. \end{cases}$$

We find the Fourier integrals

$$\mathcal{K}_1(u) = \frac{1}{(u - 2i)(u - 3i)}, \quad \mathcal{K}_2(u) = \frac{1}{u + 2i}, \quad \mathcal{F}(u) = \frac{1}{u^2 + 4}.$$

In this case, the boundary value problem (2) corresponding to this equation becomes

$$\Xi^+(u) = \frac{(u - 2i)(u - 3i)}{u + 2i}\Xi^-(u) + \frac{u - 3i}{(u + 2i)^2} - \frac{1}{u^2 + 4}. \tag{4}$$

The coefficient $\mathcal{D}(u)$ has a first-order pole at infinity (with index $\nu = -1$). The functions $\mathcal{K}_1(u)$ and $\mathcal{K}_2(u)$ have a common zero of the first order at infinity. We find that

$$m_+ = 2, \quad n_+ = 0, \quad \nu = m_+ - n_+ = 2.$$

On representing the boundary condition in the form

$$(u + 2i)\Xi^+(u) - \frac{u - 3i}{u + 2i} = (u - 2i)(u - 3i)\Xi^-(u) - \frac{1}{u - 2i}$$

and applying the analytic continuation and the generalized Liouville theorem, we see that the general solution of problem (4) in the class of functions vanishing at infinity is given by

$$\Xi^+(z) = \frac{1}{z + 2i}\left(\frac{z - 3i}{z + 2i} + C\right), \quad \Xi^-(z) = \frac{1}{(z - 2i)(z - 3i)}\left(\frac{1}{z - 2i} + C\right), \tag{5}$$

where C is an arbitrary constant.

The solution of the integral equation in question is given by the expression

$$y(x) = \frac{1}{\sqrt{2\pi}} \int_{-\infty}^{\infty} \frac{\Xi^+(u) + \mathcal{F}(u)}{\mathcal{K}_2(u)} e^{-iux}\, du.$$

Since the function $\mathcal{K}_2(u)$ has a first-order zero at infinity, it follows that the function $\Xi^+(u) + \mathcal{F}(u)$ must have a zero at infinity whose order is at least two. This condition implies the relation $C = -1$.

For $C = -1$, formulas (5) become

$$\Xi^+(z) = \frac{-5i}{(z + 2i)^2}, \quad \Xi^-(z) = \frac{1 + 2i - z}{(z - 2i)^2(z - 3i)}, \quad y(x) = \begin{cases} i\sqrt{2\pi}\,e^{2x} & \text{for } x < 0, \\ 5\sqrt{2\pi}\,e^{-2x} & \text{for } x > 0. \end{cases}$$

Thus, we have succeeded in satisfying the solvability condition, which follows from the existence of a common zero of the functions $\mathcal{K}_1(u)$ and $\mathcal{K}_2(u)$, by choosing an appropriate constant that enters the general solution, and the integral equation turns out to be unconditionally and uniquely solvable.

10.6-2. Exact Solutions of Some Dual Equations of the First Kind

In applications (for example, in elasticity, thermal conduction, and electrostatics), one encounters dual integral equations of the form

$$\int_0^\infty K_1(x,t)y(t)\,dt = f_1(x) \qquad \text{for} \quad 0 < x < a,$$

$$\int_0^\infty K_2(x,t)y(t)\,dt = f_2(x) \qquad \text{for} \quad a < x < \infty, \tag{6}$$

where $K_1(x,t)$, $K_2(x,t)$, $f_1(x)$, and $f_2(x)$ are known functions and $y(x)$ is the function to be found.

Methods for solving various types of these equations are described, for instance, in the books mentioned in the references at the end of this section. Below we present solutions of some classes of dual integral equations that occur most frequently in applications.

$1°$. Consider the following dual integral equation:

$$\int_0^\infty J_0(xt)y(t)\,dt = f(x) \qquad \text{for} \quad 0 < x < a,$$

$$\int_0^\infty t J_0(xt)y(t)\,dt = 0 \qquad \text{for} \quad a < x < \infty, \tag{7}$$

where $J_0(x)$ is the Bessel function of zero order. We can obtain the solution of Eqs. (7) by applying the Hankel transform. This solution is given by

$$y(x) = \frac{2}{\pi}\int_0^a \cos(xt)\left[\frac{d}{dt}\int_0^t \frac{sf(s)\,ds}{\sqrt{t^2-s^2}}\right]dt. \tag{8}$$

$2°$. The exact solution of the dual integral equation

$$\int_0^\infty t J_0(xt)y(t)\,dt = f(x) \qquad \text{for} \quad 0 < x < a,$$

$$\int_0^\infty J_0(xt)y(t)\,dt = 0 \qquad \text{for} \quad a < x < \infty, \tag{9}$$

where $J_0(x)$ is the Bessel function of zero order, can be constructed by means of the Hankel transform,

$$y(x) = \frac{2}{\pi}\int_0^a \sin(xt)\left[\frac{d}{dt}\int_0^t \frac{sf(s)\,ds}{\sqrt{t^2-s^2}}\right]dt. \tag{10}$$

$3°$. The exact solution of the dual integral equation

$$\int_0^\infty t J_\mu(xt)y(t)\,dt = f(x) \qquad \text{for} \quad 0 < x < a,$$

$$\int_0^\infty J_\mu(xt)y(t)\,dt = 0 \qquad \text{for} \quad a < x < \infty, \tag{11}$$

where $J_\mu(x)$ is the Bessel function of order μ, can be defined by the following expression (here the calculation also involves the Hankel transform):

$$y(x) = \sqrt{\frac{2x}{\pi}}\int_0^a t^{3/2} J_{\mu+\frac{1}{2}}(xt)\left[\int_0^{\pi/2} \sin^{\mu+1}\theta f(t\sin\theta)\,d\theta\right]dt. \tag{12}$$

4°. Consider the dual integral equation

$$\int_0^\infty t^{2\beta} J_\mu(xt)y(t)\, dt = f(x) \qquad \text{for} \quad 0 < x < 1,$$

$$\int_0^\infty J_\mu(xt)y(t)\, dt = 0 \qquad \text{for} \quad 1 < x < \infty, \tag{13}$$

where $J_\mu(x)$ is the Bessel function of order μ.

The solution of Eq. (13) can be obtained by applying the Mellin transform. For $\beta > 0$, this solution is defined by the formulas

$$y(x) = \frac{(2x)^{1-\beta}}{\Gamma(\beta)} \int_0^1 t^{1+\beta} J_{\mu+\beta}(xt)F(t)\, dt, \qquad F(t) = \int_0^1 f(t\zeta)\zeta^{\mu+1}(1-\zeta^2)^{\beta-1}\, d\zeta. \tag{14}$$

For $\beta > -1$, the solution of the dual equation (13) has the form

$$y(x) = \frac{(2x)^{-\beta}}{\Gamma(1+\beta)} \left[x^{1+\beta} J_{\mu+\beta}(x) \int_0^1 t^{\mu+1}(1-t^2)^\beta f(t)\, dt + \int_0^1 t^{\mu+1}(1-t^2)^\beta \Phi(x,t)\, dt \right], \tag{15}$$

$$\Phi(x,t) = \int_0^1 (x\xi)^{2+\beta} J_{\mu+\beta+1}(x\xi)f(\xi t)\, d\xi.$$

Formula (15) holds for $\beta > -1$ and for $-\mu - \frac{1}{2} < 2\beta < \mu + \frac{3}{2}$. It can be shown that for $\beta > 0$ the solution of Eq. (15) can be reduced to the form (14).

5°. The exact solution of the dual integral equation

$$\int_0^\infty t P_{-\frac{1}{2}+it}(\cosh x)y(t)\, dt = f(x) \qquad \text{for} \quad 0 < x < a,$$

$$\int_0^\infty \tanh(\pi t)P_{-\frac{1}{2}+it}(\cosh x)y(t)\, dt = 0 \qquad \text{for} \quad a < x < \infty, \tag{16}$$

where $P_\mu(x)$ is the Legendre spherical function of the first kind (see Supplement 10) and $i^2 = -1$, can be constructed by means of the Meler–Fock integral transform (see Section 7.6) and is given by the formula

$$y(x) = \frac{\sqrt{2}}{\pi} \int_0^a \sin(xt)\left[\int_0^t \frac{f(s)\sinh s}{\sqrt{\cosh t - \cosh s}}\, ds\right] dt. \tag{17}$$

Note that

$$P_{-\frac{1}{2}+it}(\cosh x) = \frac{\sqrt{2}}{\pi} \int_0^x \frac{\cos(ts)}{\sqrt{\cosh x - \cosh s}}\, ds, \qquad x > 0,$$

where the integral on the right-hand side is called the *Meler integral*.

10.6-3. Reduction of Dual Equations to a Fredholm Equation

One of the most effective methods for the approximate solution of dual integral equations of the first kind is the method of reducing these equations to Fredholm integral equations of the second kind (see Chapter 11). In what follows, we present some dual equations encountered in problems of mechanics and physics and related Fredholm equations of the second kind.

1°. The solution of the dual integral equation of the first kind

$$\int_0^\infty g(t)J_0(xt)y(t)\,dt = f(x) \qquad \text{for} \quad 0 < x < a,$$

$$\int_0^\infty tJ_0(xt)y(t)\,dt = 0 \qquad \text{for} \quad a < x < \infty, \tag{18}$$

where $g(x)$ is a given function and $J_0(x)$ is the Bessel function of zero order, has the form

$$y(x) = \int_0^a \varphi(t)\cos(xt)\,dt, \tag{19}$$

where the function $\varphi(x)$ to be found from the following Fredholm equation of the second kind:

$$\varphi(x) - \frac{1}{\pi}\int_0^a K(x,t)\varphi(t)\,dt = \psi(x), \qquad 0 < x < a, \tag{20}$$

where the symmetric kernel $K(x,t)$ and the right-hand side $\psi(x)$ are given by

$$K(x,t) = 2\int_0^\infty [1 - g(s)]\cos(xs)\cos(ts)\,ds, \qquad \psi(x) = \frac{2}{\pi}\frac{d}{dx}\int_0^x \frac{tf(t)}{\sqrt{x^2 - t^2}}\,dt. \tag{21}$$

Methods for the investigation of these equations are presented in Chapter 11.

2°. The solution of the dual integral equation of the first kind

$$\int_0^\infty tg(t)J_0(xt)y(t)\,dt = f(x) \qquad \text{for} \quad 0 < x < a,$$

$$\int_0^\infty J_0(xt)y(t)\,dt = 0 \qquad \text{for} \quad a < x < \infty, \tag{22}$$

where $g(x)$ is a given function and $J_0(x)$ is the Bessel function of zero order, has the form

$$y(x) = \int_0^a \varphi(t)\sin(xt)\,dt, \tag{23}$$

where the function $\varphi(x)$ is to be found from the Fredholm equation (20) of the second kind with

$$K(x,t) = 2\int_0^\infty [1 - g(s)]\sin(xs)\sin(ts)\,ds, \qquad \psi(x) = \frac{2}{\pi}\int_0^x \frac{tf(t)}{\sqrt{x^2 - t^2}}\,dt.$$

Note that the kernel $K(x,t)$ is symmetric.

3°. The solution of the dual integral equation of the first kind

$$\int_0^\infty g(t)J_\mu(xt)y(t)\,dt = f(x) \qquad \text{for} \quad 0 < x < a,$$

$$\int_0^\infty tJ_\mu(xt)y(t)\,dt = 0 \qquad \text{for} \quad a < x < \infty, \tag{24}$$

where $g(x)$ is a given function and $J_\mu(x)$ is the Bessel function of order μ, has the form

$$y(x) = \sqrt{\frac{\pi x}{2}}\int_0^a \sqrt{t}\,J_{\mu-\frac{1}{2}}(xt)\varphi(t)\,dt, \tag{25}$$

where the function $\varphi(x)$ is to be found from the Fredholm equation (20) of the second kind with

$$K(x,t) = \pi\sqrt{xt}\int_0^\infty [1 - g(s)]s\,J_{\mu-\frac{1}{2}}(xs)J_{\mu-\frac{1}{2}}(ts)\,ds,$$

$$\psi(x) = \frac{2}{\pi}\left\{f(0) + \int_0^{\pi/2} \left[\mu(\sin\theta)^{\mu-1}f(x\sin\theta) + x(\sin\theta)^\mu f'(x\sin\theta)\right]d\theta\right\}.$$

Note that $f'(x\sin\theta) = f'_\xi(\xi)\big|_{\xi=x\sin\theta}$, and the kernel $K(x,t)$ is symmetric.

4°. The solution of the integral equation of the first kind

$$\int_0^\infty t g(t) J_\mu(xt) y(t)\, dt = f(x) \qquad \text{for} \quad 0 < x < a,$$

$$\int_0^\infty J_\mu(xt) y(t)\, dt = 0 \qquad \text{for} \quad a < x < \infty, \tag{26}$$

where $g(x)$ is a given function and $J_\mu(x)$ is the Bessel function of order μ, has the form

$$y(x) = \sqrt{\frac{\pi x}{2}} \int_0^a \sqrt{t}\, J_{\mu+\frac{1}{2}}(xt) \varphi(t)\, dt, \tag{27}$$

where the function $\varphi(x)$ is to be found by solving the Fredholm equation (20) of the second kind with

$$K(x,t) = \pi \sqrt{xt} \int_0^\infty [1 - g(s)] s\, J_{\mu+\frac{1}{2}}(xs) J_{\mu+\frac{1}{2}}(ts)\, ds, \qquad \psi(x) = \frac{2x}{\pi} \int_0^{\pi/2} f(x \sin\theta)(\sin\theta)^{\mu+1}\, d\theta,$$

and the kernel $K(x,t)$ is symmetric.

5°. The solution of the dual integral equation of the first kind

$$\int_0^\infty g(t) J_\mu(xt) y(t)\, dt = f(x) \qquad \text{for} \quad 0 < x < a,$$

$$\int_0^\infty J_\mu(xt) y(t)\, dt = 0 \qquad \text{for} \quad a < x < \infty, \tag{28}$$

where $g(x)$ is a given function and $J_\mu(x)$ is the Bessel function of order μ, has the form

$$y(x) = x \sqrt{\frac{\pi x}{2}} \int_0^a \sqrt{t}\, J_{\mu-\frac{1}{2}}(xt) \varphi(t)\, dt, \tag{29}$$

and the function $\varphi(x)$ is to be found from the Fredholm equation (20) of the second kind with

$$K(x,t) = x^\mu \sqrt{2\pi t} \int_x^a \frac{\rho^{1-\mu}}{\sqrt{\rho^2 - x^2}} \int_0^\infty [1 - g(s)] s^{3/2} J_\mu(\rho s) J_{\mu-\frac{1}{2}}(ts)\, ds\, d\rho,$$

$$\psi(x) = \frac{2}{\pi} x^\mu \int_x^a \frac{\rho^{1-\mu}}{\sqrt{\rho^2 - x^2}}\, d\rho.$$

6°. The solution of the dual integral equation of the first kind

$$\int_0^\infty t^{2\beta} g(t) J_\mu(xt) y(t)\, dt = f(x) \qquad \text{for} \quad 0 < x < a,$$

$$\int_0^\infty J_\mu(xt) y(t)\, dt = 0 \qquad \text{for} \quad a < x < \infty, \tag{30}$$

where $0 < \beta < 1$, $g(x)$ is a given function, and $J_\mu(x)$ is the Bessel function of order μ, has the form

$$y(x) = \sqrt{\frac{\pi}{2}} x^{1-\beta} \int_0^a \sqrt{t}\, J_{\mu+\beta}(xt) \varphi(t)\, dt, \tag{31}$$

and the function $\varphi(x)$ is to be found from the Fredholm equation (20) of the second kind with

$$K(x,t) = \pi \sqrt{xt} \int_0^\infty [1 - g(s)] s\, J_{\mu+\beta}(xs) J_{\mu+\beta}(ts)\, ds,$$

$$\psi(x) = \frac{2^{1-\beta}}{\Gamma(\beta)} \sqrt{\frac{2x}{\pi}} x^\beta \int_0^{\pi/2} f(x \sin\theta)(\sin\theta)^{\mu+1}(\cos\theta)^{2\beta-1}\, d\theta,$$

and the kernel $K(x,t)$ is symmetric.

7°. The solution of the dual integral equation of the first kind

$$
\int_0^\infty g(t)P_{-\frac{1}{2}+it}(\cosh x)y(t)\,dt = f(x) \qquad \text{for} \quad 0 < x < a,
$$
$$
\int_0^\infty t\tanh(\pi t)P_{-\frac{1}{2}+it}(\cosh x)y(t)\,dt = 0 \qquad \text{for} \quad a < x < \infty,
$$
(32)

where $P_\mu(x)$ is the Legendre spherical function of the first kind (see Supplement 10), $i^2 = -1$, and $g(x)$ is a given function, is determined by the formula

$$
y(x) = \int_0^a \cos(xt)\varphi(t)\,dt,
$$
(33)

and the function $\varphi(x)$ is to be found from the Fredholm equation (20) of the second kind in which

$$
K(x,t) = \int_0^\infty [1 - g(s)]\{\cos[(x+t)s] + \cos[(x-t)s]\}\,ds,
$$

$$
\psi(x) = \frac{\sqrt{2}}{\pi}\frac{d}{dx}\int_0^x \frac{f(s)\sinh s}{\sqrt{\cosh x - \cosh s}}\,ds.
$$
(34)

On the basis of relations (34), we can readily see that the kernel $K(x,t)$ is symmetric.

8°. The solution of the dual integral equation of the first kind

$$
\int_0^\infty tg(t)P_{-\frac{1}{2}+it}(\cosh x)y(t)\,dt = f(x) \qquad \text{for} \quad 0 < x < a,
$$
$$
\int_0^\infty \tanh(\pi t)P_{-\frac{1}{2}+it}(\cosh x)y(t)\,dt = 0 \qquad \text{for} \quad a < x < \infty,
$$
(35)

where $P_\mu(x)$ is the spherical Legendre function of the first kind (see Supplement 10), $i^2 = -1$, and $g(x)$ is a given function, is determined by the formula

$$
y(x) = \int_0^a \sin(xt)\varphi(t)\,dt,
$$
(36)

and the function $\varphi(x)$ is to be found from the Fredholm equation (20) of the second kind in which

$$
K(x,t) = \int_0^\infty [1 - g(s)]\{\cos[(x-t)s] - \cos[(x+t)s]\}\,ds,
$$

$$
\psi(x) = \frac{\sqrt{2}}{\pi}\int_0^x \frac{f(s)\sinh s}{\sqrt{\cosh x - \cosh s}}\,ds.
$$
(37)

On the basis of relations (37), we can readily see that the kernel $K(x,t)$ is symmetric.

⊙ References for Section 10.6: E. C. Titchmarsh (1948), I. Sneddon (1951), Ya. S. Uflyand (1977), F. D. Gakhov and Yu. I. Cherskii (1978).

10.7. Asymptotic Methods for Solving Equations With Logarithmic Singularity

10.7-1. Preliminary Remarks

Consider the Fredholm integral equation of the first kind of the form

$$\int_{-1}^{1} K\left(\frac{x-t}{\lambda}\right) y(t)\, dt = f(x), \qquad -1 \le x \le 1, \tag{1}$$

with parameter λ $(0 < \lambda < \infty)$.

We assume that the kernel $K = K(x)$ is an even function continuous for $x \ne 0$ which has a logarithmic singularity as $x \to 0$ and exponentially decays as $x \to \infty$. Equations with such a kernel arise in solving various problems of continuum mechanics with mixed boundary conditions.

Let $f(x)$ belong to the space of functions whose first derivatives satisfy the Hölder condition with exponent $\alpha > \frac{1}{2}$ on $[-1, 1]$. In this case, the solution of the integral equation (1) in the class of functions satisfying the Hölder condition exists and is unique for any $\lambda \in (0, \infty)$ and has the structure

$$y(x) = \frac{\omega(x)}{\sqrt{1 - x^2}}, \tag{2}$$

where $\omega(x)$ is a continuous function that does not vanish at $x = \pm 1$.*

It follows from formula (2) that the solution of Eq. (1) is bounded as $x \to \pm 1$. This important circumstance will be taken into account in Subsection 10.7-3 in constructing the asymptotic solution in the case $\lambda \to 0$.

Note that more general equations with difference kernel and arbitrary finite limits of integration can always be reduced to Eq. (1) by a change of variables. The form (1) is taken here for further convenience.

10.7-2. The Solution for Large λ

Let the representation

$$K(x) = \ln|x| \sum_{n=0}^{\infty} a_n |x|^n + \sum_{n=0}^{\infty} b_n |x|^n, \tag{3}$$

where $a_0 \ne 0$, be valid for the kernel of the integral equation (1) as $x \to 0$.

It is obvious from (3) that two different-scale large parameters λ and $\ln \lambda$ occur in Eq. (1) as $\lambda \to \infty$. The latter, "quasiconstant" parameter grows much slower than the former (for instance, for $\lambda = 100$ and $\lambda = 1000$ we have $\ln \lambda \approx 4.6$ and $\ln \lambda \approx 6.9$, respectively).

Let us drop out all terms decaying as $\lambda \to \infty$ in Eq. (1). In view of (3), for the main (zeroth) approximation we have

$$\int_{-1}^{1} \left(a_0 \ln|x - t| - a_0 \ln \lambda + b_0\right) y_0(t)\, dt = f(x), \qquad -1 \le x \le 1. \tag{4}$$

It should be noted that one cannot retain in the integrand only one term proportional to $\ln \lambda$ (since the corresponding "truncated" equation is unsolvable). The constant b_0 must also be included in (4) for the main-approximation equation to be invariant with respect to the scaling parameter λ in Eq. (1).

The exact closed-form solution of Eq. (4) is given in Section 3.4 (see Equations 3 and 4).

---

* The situation $\omega(\pm 1) = 0$ is only possible in exceptional cases for special values of λ.

To construct an asymptotic solution of Eq. (1) as $\lambda \to \infty$, it is convenient to do the following. First, we consider the auxiliary integral equation

$$\int_{-1}^{1} \mathcal{K}(x-t,\beta,\lambda)y(t)\,dt = f(x), \qquad -1 \le x \le 1,$$

$$\mathcal{K}(x,\beta,\lambda) = \left(\ln|x| - \beta\right)\sum_{n=0}^{\infty} \frac{a_n}{\lambda^n}|x|^n + \sum_{n=0}^{\infty}\frac{b_n}{\lambda^n}|x|^n, \tag{5}$$

with two parameters λ and β. We seek its solution in the form of a regular asymptotic expansion in negative powers of λ (for fixed β). That is, we have

$$y(x,\beta,\lambda) = \sum_{n=0}^{N}\lambda^{-n}y_n(x,\beta) + o\left(\lambda^{-N}\right). \tag{6}$$

Substituting (6) into (5) yields a recurrent chain of integral equations of the form (4):

$$\int_{-1}^{1}\left(a_0 \ln|x-t| - a_0\beta + b_0\right)y_n(t,\beta)\,dt = g_n(x,\beta), \qquad -1 \le x \le 1, \tag{7}$$

from which the functions $y_n(x,\beta)$ can be successively calculated. The right-hand sides $g_n(x,\beta)$ depend only on the previously determined functions $y_0, y_1, \ldots, y_{n-1}$.

Note that for $\beta = \ln\lambda$ the auxiliary equation (5) coincides with the original equation (1) into which the expansion (3) is substituted. Therefore, the asymptotic solution of Eq. (1) can be obtained with the aid of (6) and (7) with $\beta = \ln\lambda$.

Some contact problems of elasticity can be reduced to Eq. (1), in which the kernel can be represented in the form (3) with $a_n = 0$ for all $n > 0$ and $b_{2m+1} = 0$ for $m = 0, 1, 2, \ldots$ In this case, one must set $y_n(x,\beta) \equiv 0$ $(n = 1, 3, 5, \ldots)$ in the solution (6). In practice, it usually suffices to retain the terms up to λ^{-4}.

10.7-3. The Solution for Small λ

In analyzing the limit case $\lambda \to 0$, we take into account the singularities of the solution at the endpoints of the interval $-1 \le x \le 1$ (see formula (2)). Consider the following auxiliary system of two integral equations:

$$\int_{-1}^{\infty} K\left(\frac{x-t}{\lambda}\right)y_1(t)\,dt = f_1(x) + \int_{-\infty}^{-1} K\left(\frac{x-t}{\lambda}\right)y_2(t)\,dt, \qquad -1 \le x < \infty,$$

$$\int_{-\infty}^{1} K\left(\frac{x-t}{\lambda}\right)y_2(t)\,dt = f_2(x) + \int_{1}^{\infty} K\left(\frac{x-t}{\lambda}\right)y_1(t)\,dt, \qquad -\infty < x \le 1. \tag{8}$$

The former equation provides for selecting the singularity at $x = -1$ and the latter for selecting the singularity at $x = +1$.

The functions $f_1(x)$ and $f_2(x)$ are such that

$$\begin{aligned}
f_1(x) + f_2(x) &= f(x), & -1 \le x \le 1, \\
f_1(x) &= O\left(e^{-\alpha_1 x}\right) & \text{as} \quad x \to \infty, \\
f_2(x) &= O\left(e^{\alpha_2 x}\right) & \text{as} \quad x \to -\infty,
\end{aligned} \tag{9}$$

where $\alpha_1 > 0$ and $\alpha_2 > 0$.

The first condition in (9) makes it possible to seek the solution of the integral equation (1) as the sum of the solutions of the integral equations (8), that is,

$$y(x) = y_1(x) + y_2(x), \quad -1 \le x \le 1. \tag{10}$$

Note that by virtue of the last two conditions in (9), the relations

$$\begin{aligned} y_1(x) &= O\left(e^{-\beta_1 x}\right) \quad \text{as} \quad x \to \infty, \\ y_2(x) &= O\left(e^{\beta_2 x}\right) \quad \text{as} \quad x \to -\infty, \end{aligned} \tag{11}$$

where $\beta_1 > 0$ and $\beta_2 > 0$, are valid.

Recall that the kernel $K(x)$ is an even function. Therefore, if $f(x)$ in Eq. (1) is an even or odd function, then one must set

$$f_1(x) = \pm f_2(-x), \quad y_1(x) = \pm y_2(-x) \tag{12}$$

in system (8).*

In both cases, system (8) can be reduced by changes of variables to the same integral equation

$$\int_0^\infty K(z - \tau)w(\tau)\,d\tau = F(z) \pm \int_{2/\lambda}^\infty K(2/\lambda - z - \tau)w(\tau)\,d\tau, \quad 0 \le z < \infty, \tag{13}$$

in which the following notation is used:

$$z = \frac{x+1}{\lambda}, \quad \tau = \frac{t+1}{\lambda}, \quad w(\tau) = y(t), \quad F(z) = \frac{1}{\lambda}f_1(x). \tag{14}$$

In view of the properties of the kernel $K(x)$ (see Subsection 10.7-1) and the first relation in (11), the asymptotic estimate

$$I(w) \equiv \int_{2/\lambda}^\infty K(2/\lambda - z - \tau)w(\tau)\,d\tau = O\left(e^{-2\beta_1/\lambda}\right) \tag{15}$$

can be obtained, which is uniform with respect to τ.

According to (15), for small λ the iterative scheme

$$\int_0^\infty K(z - \tau)w_n(\tau)\,d\tau = F(z) \pm I\left(w_{n-1}\right), \quad n = 1, 2, \ldots, \tag{16}$$

can be used to solve the integral equation (13) by the method of successive approximations. In the main approximation, the integral $I(w_0)$ can be omitted on the right-hand side. Equations (16) are Wiener–Hopf integral equations of the first kind, which can be solved in a closed form (see Subsection 10.5-1).

It follows from formulas (10), (12), and (14) that, as $\lambda \to 0$, the leading term of the asymptotic expansion of the solution of the integral equation (1) has the form

$$y(x) = w_1\left(\frac{1+x}{\lambda}\right) \pm w_1\left(\frac{1-x}{\lambda}\right), \tag{17}$$

where $w_1 = w_1(\tau)$ is the solution of Eq. (16) with $n = 1$ and $w_0 \equiv 0$.

For practical purposes, formula (17) is usually sufficient.

---

* In formulas (12), (13), (16), and (17), the plus sign corresponds to even $f(x)$ and the minus sign to odd $f(x)$.

The integral equation (1) whose kernel is given via the Fourier cosine transform,

$$K(x) = \int_0^\infty \frac{L(u)}{u} \cos(ux)\,du, \tag{18}$$

frequently occurs in contact problems of elasticity. The function $L(u)$ in (18) is continuous and positive for $0 < u < \infty$) and satisfies the asymptotic relations

$$
\begin{aligned}
L(u) &= Au + O(u^3) &&\text{as} \quad u \to 0, \\
L(u) &= \sum_{n=0}^{N-1} B_n u^{-n} + O\left(u^{-N}\right) &&\text{as} \quad u \to \infty,
\end{aligned}
\tag{19}
$$

where $A > 0$ and $B_0 > 0$.

Formula (18) implies that the kernel is an even function: $K(x) = K(-x)$.

It is usually assumed that $L(u)u^{-1}$ and $u[L(u)]^{-1}$, treated as functions of the complex variable $w = u + iv$, are regular at the pole $|v| \le \gamma_1$ and the pole $|v| \le \gamma_2$, respectively. It follows in particular that the kernel $K(x)$ decays at least as $\exp(-\gamma_1|t|)$ at infinity.

Formulas (18) and (19) imply that $K(x)$ has a logarithmic singularity at $x = 0$. Moreover, the representation (3) is valid with $a_n = 0$ for $n = 1, 3, 5, \ldots$

Thus, the kernel given by (18) has the same characteristic features as those inherent by assumption in the kernel of the integral equation (1). Therefore, the results of Subsections 10.7-2 and 10.7-3 can be used for the asymptotic analysis of Eq. (1) with kernel (18) as $\lambda \to \infty$ and $\lambda \to 0$.

⊙ References for Section 10.7: I. I. Vorovich, V. M. Aleksandrov, and V. A. Babeshko (1974), V. M. Aleksandrov and E. V. Kovalenko (1986), V. M. Aleksandrov (1993).

10.8. Regularization Methods

Consider the Fredholm equation of the first kind (see also Remark 3, Subsection 11.6-5)

$$\int_a^b K(x,t)y(t)\,dt = f(x), \qquad a \le x \le b, \tag{1}$$

where $f(x) \in L_2(a,b)$ and $y(x) \in L_2(a,b)$. The kernel $K(x,t)$ is square integrable, symmetric, and positive definite (see Subsection 11.6-2), that is, for all $\varphi(x) \in L_2(a,b)$, we have

$$\int_a^b \int_a^b K(x,t)\varphi(x)\varphi(t)\,dx\,dt \ge 0,$$

where the equality is attained only for $\varphi(x) \equiv 0$.

In the above classes of functions and kernels, the problem of finding a solution of Eq. (1) is ill-posed, i.e., unstable with respect to small variations in the right-hand side of the integral equation.

Following the Lavrentiev regularization method, along with Eq. (1) we consider the regularized equation

$$\varepsilon y_\varepsilon(x) + \int_a^b K(x,t)y_\varepsilon(t)\,dt = f(x), \qquad a \le x \le b, \tag{2}$$

where $\varepsilon > 0$ is the regularization parameter. This equation is a Fredholm equation of the second kind, so it can be solved by the methods presented in Chapter 11, whence the solution exists and is unique.

On taking a sufficiently small ε in Eq. (2), we find a solution $y_\varepsilon(x)$ of the equation and substitute this solution into Eq. (1), thus obtaining

$$\int_a^b K(x,t)y_\varepsilon(t)\,dt = f_\varepsilon(x), \qquad a \le x \le b. \tag{3}$$

If the function $f_\varepsilon(x)$ thus obtained differs only slightly from $f(x)$, that is,

$$\|f(x) - f_\varepsilon(x)\| \le \delta, \tag{4}$$

where δ is a prescribed small positive number, then the solution $y_\varepsilon(x)$ is regarded as a sufficiently good approximate solution of Eq. (1).

The parameter δ usually defines the error of the initial data provided that the right-hand side of Eq. (1) is defined or determined by an experiment with some accuracy.

For the case in which, for a given ε, condition (4) fails, we must choose another value of the regularization parameter and repeat the above procedure.

The next subsection describes the regularization method suitable for equations of the first kind with arbitrary square-integrable kernels.

10.8-2. The Tikhonov Regularization Method

Consider the Fredholm integral equation of the first kind

$$\int_a^b K(x,t)y(t)\,dt = f(x), \qquad c \le x \le d. \tag{5}$$

Assume that $K(x,t)$ is any function square-integrable in the domain $\{a \le t \le b,\ c \le x \le d\}$, $f(x) \in L_2(c,d)$, and $y(x) \in L_2(a,b)$. The problem of finding the solution of Eq. (5) is also ill-posed in the above sense.

Following the Tikhonov (zero-order) regularization method, along with (5) we consider the following Fredholm integral equation of the second kind (see Chapter 11):

$$\varepsilon y_\varepsilon(x) + \int_a^b K^*(x,t)y_\varepsilon(t)\,dt = f^*(x), \qquad a \le x \le b, \tag{6}$$

where

$$K^*(x,t) = K^*(t,x) = \int_c^d K(s,x)K(s,t)\,ds, \quad f^*(x) = \int_c^d K(s,x)f(s)\,ds, \tag{7}$$

and the positive number ε is the regularization parameter. Equation (6) is said to be a *regularized integral equation*, and its solution exists and is unique.

Taking a sufficiently small ε in Eq. (6), we find a solution $y_\varepsilon(x)$ of the equation and substitute this solution into Eq. (5), thus obtaining

$$\int_a^b K(x,t)y_\varepsilon(t)\,dt = f_\varepsilon(x), \qquad c \le x \le d. \tag{8}$$

By comparing the right-hand side with the given $f(x)$ using formula (4), we either regard $f_\varepsilon(x)$ as a satisfactory approximate solution obtained in accordance with the above simple algorithm, or continue the procedure for a new value of the regularization parameter.

Presented above are the simplest principles of finding an approximate solution of the Fredholm equation of the first kind. More perfect and complex algorithms can be found in the references cited below.

⊙ References for Section 10.8: M. M. Lavrentiev (1967), A. N. Tikhonov and V. Ya. Arsenin (1979), M. M. Lavrentiev, V. G. Romanov, and S. P. Shishatskii (1980), A. F. Verlan' and V. S. Sizikov (1986).

Chapter 11

Methods for Solving Linear Equations of the Form $y(x) - \int_a^b K(x,t)y(t)\,dt = f(x)$

11.1. Some Definition and Remarks

11.1-1. Fredholm Equations and Equations With Weak Singularity of the Second Kind

Linear integral equations of the second kind with constant limits of integration have the form

$$y(x) - \lambda \int_a^b K(x,t)y(t)\,dt = f(x), \tag{1}$$

where $y(x)$ is the unknown function ($a \le x \le b$), $K(x,t)$ is the *kernel* of the integral equation, and $f(x)$ is a given function, which is called the *right-hand side* of Eq. (1). For convenience of analysis, a number λ is traditionally singled out in Eq. (1), which is called the *parameter of integral equation*. The classes of functions and kernels under consideration were defined above in Subsections 10.1-1 and 10.1-2. Note that equations of the form (1) with constant limits of integration and with Fredholm kernels or kernels with weak singularity are called *Fredholm equations of the second kind* and *equations with weak singularity of the second kind*, respectively.

A number λ is called a *characteristic value* of the integral equation (1) if there exist nontrivial solutions of the corresponding homogeneous equation (with $f(x) \equiv 0$). The nontrivial solutions themselves are called the *eigenfunctions* of the integral equation corresponding to the characteristic value λ. If λ is a characteristic value, the number $1/\lambda$ is called an *eigenvalue* of the integral equation (1). A value of the parameter λ is said to be *regular* if for this value the above homogeneous equation has only the trivial solution. Sometimes the characteristic values and the eigenfunctions of a Fredholm integral equation are called the *characteristic values* and the *eigenfunctions of the kernel* $K(x,t)$.

The kernel $K(x,t)$ of the integral equation (1) is called a *degenerate kernel* if it has the form $K(x,t) = g_1(x)h_1(t) + \cdots + g_n(x)h_n(t)$, a *difference kernel* if it depends on the difference of the arguments ($K(x,t) = K(x-t)$), and a *symmetric kernel* if it satisfies the condition $K(x,t) = K(t,x)$.

The *transposed* integral equation is obtained from (1) by replacing the kernel $K(x,t)$ by $K(t,x)$.

Remark 1. The variables t and x may vary in different ranges (e.g., $a \le t \le b$ and $c \le x \le d$). To be specific, from now on we assume that $c = a$ and $d = b$ (this can be achieved by the linear substitution $x = \alpha\bar{x} + \beta$ with the aid of an appropriate choice of the constants α and β).

Remark 2. In general, the case in which the limits of integration a and/or b can be infinite is not excluded; however, in this case, the validity of the condition that the kernel $K(x,t)$ is square integrable on the square $S = \{a \le x \le b,\ a \le t \le b\}$ is especially significant.

11.1-2. The Structure of the Solution

The solution of Eq. (1) can be presented in the form

$$y(x) = f(x) + \lambda \int_a^b R(x,t;\lambda)f(t)\,dt,$$

where the resolvent $R(x,t;\lambda)$ is independent of $f(x)$ and is determined by the kernel of the integral equation.

The resolvent of the Fredholm equation (1) satisfies the following two integral equations:

$$R(x,t;\lambda) = K(x,t) + \int_a^b K(x,s)R(s,t;\lambda)\,ds,$$

$$R(x,t;\lambda) = K(x,t) + \int_a^b K(s,t)R(x,s;\lambda)\,ds,$$

in which the integration is performed with respect to different pairs of arguments of the kernel and the resolvent.

11.1-3. Integral Equations of Convolution Type of the Second Kind

By the *integral equations of convolution type* (see also Subsection 10.1-3) we mean the integral equations that can be reduced, by applying some integral transform and the convolution theorem for this transform, to an algebraic equation for the transforms or to boundary value problems of the theory of analytic functions. Consider equations of convolution type of the second kind related to the Fourier transform.

An integral equation of the second kind with difference kernel on the entire axis (this equation is sometimes called an *equation of convolution type of the second kind with a single kernel*) has the form

$$y(x) + \int_{-\infty}^{\infty} K(x-t)y(t)\,dt = f(x), \qquad -\infty < x < \infty, \tag{2}$$

where $f(x)$ and $K(x)$ are the right-hand side and the kernel of the integral equation and $y(x)$ is the function to be found.

An integral equation of the second kind with difference kernel on the semiaxis has the form

$$y(x) + \int_0^{\infty} K(x-t)y(t)\,dt = f(x), \qquad 0 < x < \infty. \tag{3}$$

Equation (3) is also called a *one-sided equation of the second kind* or a *Wiener–Hopf integral equation of the second kind*.

An integral equation of convolution type of the second kind with two kernels has the form

$$y(x) + \int_0^{\infty} K_1(x-t)y(t)\,dt + \int_{-\infty}^0 K_2(x-t)y(t)\,dt = f(x), \qquad -\infty < x < \infty, \tag{4}$$

where $K_1(x)$ and $K_2(x)$ are the *kernels* of the integral equation (4). The class of functions and kernels for equations of convolution type was introduced above in Subsection 10.1-3.

11.1-4. Dual Integral Equations of the Second Kind

A *dual integral equation of the second kind* with difference kernels (of convolution type) has the form

$$y(x) + \int_{-\infty}^{\infty} K_1(x - t)y(t)\,dt = f(x), \qquad 0 < x < \infty,$$

$$y(x) + \int_{-\infty}^{\infty} K_2(x - t)y(t)\,dt = f(x), \qquad -\infty < x < 0,$$

(5)

where the notation and the class of the functions and kernels coincide with those introduced for the equations of convolution type in Subsection 10.1-3.

In a sufficiently general case, a dual integral equation of the second kind has the form

$$y(x) + \int_{a}^{\infty} K_1(x, t)y(t)\,dt = f_1(x), \qquad a < x < b,$$

$$y(x) + \int_{a}^{\infty} K_2(x, t)y(t)\,dt = f_2(x), \qquad b < x < \infty,$$

(6)

where $f_1(x)$ and $f_2(x)$ (and $K_1(x, t)$ and $K_2(x, t)$) are the known right-hand sides (and the kernels) of Eq. (6) and $y(x)$ is the function to be found. These equations can be studied by the methods of various integral transforms with reduction to boundary value problems of the theory of analytic functions and also by other methods developed for dual integral equations of the first kind (e.g., see I. Sneddon (1951) and Ya. S. Uflyand (1977)).

The integral equations obtained from (2)–(5) by replacing the kernel $K(x - t)$ by $K(t - x)$ are said to be *transposed* to the original equations.

If the right-hand sides of Eqs. (1)–(6) are identically zero, then these equations are said to be *homogeneous*. For the case in which the right-hand side of an equation of the type (1)–(6) does not vanish on the entire domain, the corresponding equation is said to be *nonhomogeneous*.

Remark 3. Some equations whose kernel contains the product or the ratio of the variables x and t can be reduced to Eqs. (2)–(5).

Remark 4. Sometimes equations of convolution type of the form (2)–(5) are written in the form in which the integrals are multiplied by the coefficient $1/\sqrt{2\pi}$.

Remark 5. The cases in which the class of functions and kernels for equations of convolution type (in particular, for Wiener–Hopf equations) differs from those introduced in Subsections 10.1-3 are always mentioned explicitly (see Sections 11.10 and 11.11).

⊙ References for Section 11.1: E. Goursat (1923), F. Riesz and B. Sz.-Nagy (1955), I. G. Petrovskii (1957), B. Noble (1958), M. G. Krein (1958), S. G. Mikhlin (1960), L. V. Kantorovich and G. P. Akilov (1964), A. N. Kolmogorov and S. V. Fomin (1970), L. Ya. Tslaf (1970), M. L. Krasnov, A. I. Kiselev, and G. I. Makarenko (1971), J. A. Cochran (1972), V. I. Smirnov (1974), P. P. Zabreyko, A. I. Koshelev, et al. (1975), F. D. Gakhov and Yu. I. Cherskii (1978), A. G. Butkovskii (1979), L. M. Delves and J. L. Mohamed (1985), F. G. Tricomi (1985), A. J. Jerry (1985), A. F. Verlan' and V. S. Sizikov (1986), A. Golberg (1990), D. Porter and D. S. G. Stirling (1990), C. Corduneanu (1991), J. Kondo (1991), S. Prössdorf and B. Silbermann (1991), W. Hackbusch (1995), R. P. Kanwal (1997).

11.2. Fredholm Equations of the Second Kind With Degenerate Kernel

11.2-1. The Simplest Degenerate Kernel

Consider Fredholm integral equations of the second kind with the simplest degenerate kernel:

$$y(x) - \lambda \int_{a}^{b} g(x)h(t)y(t)\,dt = f(x), \qquad a \le x \le b.$$

(1)

We seek a solution of Eq. (1) in the form

$$y(x) = f(x) + \lambda A g(x). \tag{2}$$

On substituting the expressions (2) into Eq. (1), after simple algebraic manipulations we obtain

$$A\left[1 - \lambda \int_a^b h(t)g(t)\,dt\right] = \int_a^b f(t)h(t)\,dt. \tag{3}$$

Both integrals occurring in Eq. (3) are supposed to exist. On the basis of (1)–(3) and taking into account the fact that the unique characteristic value λ_1 of Eq. (1) is given by the expression

$$\lambda_1 = \left[\int_a^b h(t)g(t)\,dt\right]^{-1}, \tag{4}$$

we obtain the following results.

1°. If $\lambda \neq \lambda_1$, then for an arbitrary right-hand side there exists a unique solution of Eq. (1), which can be written in the form

$$y(x) = f(x) + \frac{\lambda \lambda_1 f_1}{\lambda_1 - \lambda} g(x), \qquad f_1 = \int_a^b f(t)h(t)\,dt. \tag{5}$$

2°. If $\lambda = \lambda_1$ and $f_1 = 0$, then any solution of Eq. (1) can be represented in the form

$$y = f(x) + C y_1(x), \qquad y_1(x) = g(x), \tag{6}$$

where C is an arbitrary constant and $y_1(x)$ is an eigenfunction that corresponds to the characteristic value λ_1.

3°. If $\lambda = \lambda_1$ and $f_1 \neq 0$, then there are no solutions.

11.2-2. Degenerate Kernel in the General Case

In the general case, a Fredholm integral equation of the second kind with degenerate kernel has the form

$$y(x) - \lambda \int_a^b \left[\sum_{k=1}^n g_k(x)h_k(t)\right] y(t)\,dt = f(x), \qquad n = 2, 3, \dots \tag{7}$$

Let us rewrite Eq. (7) in the form

$$y(x) = f(x) + \lambda \sum_{k=1}^n g_k(x) \int_a^b h_k(t)y(t)\,dt, \qquad n = 2, 3, \dots \tag{8}$$

We assume that Eq. (8) has a solution and introduce the notation

$$A_k = \int_a^b h_k(t)y(t)\,dt. \tag{9}$$

In this case we have

$$y(x) = f(x) + \lambda \sum_{k=1}^n A_k g_k(x), \tag{10}$$

and hence the solution of the integral equation with degenerate kernel is reduced to the definition of the constants A_k.

Let us multiply Eq. (10) by $h_m(x)$ and integrate with respect to x from a to b. We obtain the following system of linear algebraic equations for the coefficients A_k:

$$A_m - \lambda \sum_{k=1}^{n} s_{mk} A_k = f_m, \qquad m = 1, \dots, n, \tag{11}$$

where

$$s_{mk} = \int_a^b h_m(x) g_k(x)\, dx, \quad f_m = \int_a^b f(x) h_m(x)\, dx; \quad m, k = 1, \dots, n. \tag{12}$$

In the calculation of the coefficients s_{mk} and f_m for specific degenerate kernels, the tables of integrals can be applied; see Supplements 2 and 3, as well as I. S. Gradshtein and I. M. Ryzhik (1980), A. P. Prudnikov, Yu. A. Brychkov, and O. I. Marichev (1986).

Once we construct a solution of system (11), we obtain a solution of the integral equation with degenerate kernel (7) as well. The values of the parameter λ at which the determinant of system (11) vanishes are characteristic values of the integral equation (7), and it is clear that there are just n such values counted according to their multiplicities.

Now we can state the main results on the solution of Eq. (7).

1°. If λ is a regular value, then for an arbitrary right-hand side $f(x)$, there exists a unique solution of the Fredholm integral equation with degenerate kernel and this solution can be represented in the form (10), in which the coefficients A_k make up a solution of system (11). The constants A_k can be determined, for instance, by Cramer's rule (see equation 4.9.20, Part I, Chapter 4).

2°. If λ is a characteristic value and $f(x) \equiv 0$, then every solution of the homogeneous equation with degenerate kernel has the form

$$y(x) = \sum_{i=1}^{p} C_i y_i(x), \tag{13}$$

where the C_i are arbitrary constants and the $y_i(x)$ are linearly independent eigenfunctions of the kernel corresponding to the characteristic value λ:

$$y_i(x) = \sum_{k=1}^{n} A_{k(i)} g_k(x). \tag{14}$$

Here the constants $A_{k(i)}$ form p ($p \le n$) linearly independent solutions of the following homogeneous system of algebraic equations:

$$A_{m(i)} - \lambda \sum_{k=1}^{n} s_{mk} A_{k(i)} = 0; \qquad m = 1, \dots, n, \quad i = 1, \dots, p. \tag{15}$$

3°. If λ is a characteristic value and $f(x) \ne 0$, then for the nonhomogeneous integral equation (7) to be solvable, it is necessary and sufficient that the right-hand side $f(x)$ is such that the p conditions

$$\sum_{k=1}^{n} B_{k(i)} f_k = 0, \qquad i = 1, \dots, p, \quad p \le n, \tag{16}$$

are satisfied. Here the constants $B_{k(i)}$ form p linearly independent solutions of the homogeneous system of algebraic equations which is the transpose of system (15). In this case, every solution of Eq. (7) has the form

$$y(x) = y_0(x) + \sum_{i=1}^p C_i y_i(x), \tag{17}$$

where $y_0(x)$ is a particular solution of the nonhomogeneous equation (7) and the sum represents the general solution of the corresponding homogeneous equation (see item 2°). In particular, if $f(x) \neq 0$ but all f_k are zero, we have

$$y(x) = f(x) + \sum_{i=1}^p C_i y_i(x). \tag{18}$$

Remark. When studying Fredholm equations of the second kind with degenerate kernel, it is useful for the reader to be acquainted with equations 4.9.18 and 4.9.20 of the first part of the book.

Example. Let us solve the integral equation

$$y(x) - \lambda \int_{-\pi}^{\pi} (x \cos t + t^2 \sin x + \cos x \sin t)y(t)\,dt = x, \qquad -\pi \leq x \leq \pi. \tag{19}$$

Let us denote

$$A_1 = \int_{-\pi}^{\pi} y(t)\cos t\,dt, \quad A_2 = \int_{-\pi}^{\pi} t^2 y(t)\,dt, \quad A_3 = \int_{-\pi}^{\pi} y(t)\sin t\,dt, \tag{20}$$

where A_1, A_2, and A_3 are unknown constants. Then Eq. (19) can be rewritten in the form

$$y(x) = A_1 \lambda x + A_2 \lambda \sin x + A_3 \lambda \cos x + x. \tag{21}$$

On substituting the expression (21) into relations (20), we obtain

$$A_1 = \int_{-\pi}^{\pi} (A_1 \lambda t + A_2 \lambda \sin t + A_3 \lambda \cos t + t)\cos t\,dt,$$

$$A_2 = \int_{-\pi}^{\pi} (A_1 \lambda t + A_2 \lambda \sin t + A_3 \lambda \cos t + t)t^2\,dt,$$

$$A_3 = \int_{-\pi}^{\pi} (A_1 \lambda t + A_2 \lambda \sin t + A_3 \lambda \cos t + t)\sin t\,dt.$$

On calculating the integrals occurring in these equations, we obtain the following system of algebraic equations for the unknowns A_1, A_2, and A_3:

$$\begin{aligned} A_1 - \lambda\pi A_3 &= 0, \\ A_2 + 4\lambda\pi A_3 &= 0, \\ -2\lambda\pi A_1 - \lambda\pi A_2 + A_3 &= 2\pi. \end{aligned} \tag{22}$$

The determinant of this system is

$$\Delta(\lambda) = \begin{vmatrix} 1 & 0 & -\lambda\pi \\ 0 & 1 & 4\lambda\pi \\ -2\lambda\pi & -\lambda\pi & 1 \end{vmatrix} = 1 + 2\lambda^2\pi^2 \neq 0.$$

Thus, system (22) has the unique solution

$$A_1 = \frac{2\lambda\pi^2}{1 + 2\lambda^2\pi^2}, \quad A_2 = -\frac{8\lambda\pi^2}{1 + 2\lambda^2\pi^2}, \quad A_3 = \frac{2\pi}{1 + 2\lambda^2\pi^2}.$$

On substituting the above values of A_1, A_2, and A_3 into (21), we obtain the solution of the original integral equation:

$$y(x) = \frac{2\lambda\pi}{1 + 2\lambda^2\pi^2}(\lambda\pi x - 4\lambda\pi \sin x + \cos x) + x.$$

⊙ References for Section 11.2: S. G. Mikhlin (1960), M. L. Krasnov, A. I. Kiselev, and G. I. Makarenko (1971), I. S. Gradshteyn and I. M. Ryzhik (1980), A. J. Jerry (1985), A. P. Prudnikov, Yu. A. Brychkov, and O. I. Marichev (1986, 1988).

11.3. Solution as a Power Series in the Parameter. Method of Successive Approximations

11.3-1. Iterated Kernels

Consider the Fredholm integral equation of the second kind:

$$y(x) - \lambda \int_a^b K(x, t)y(t)\, dt = f(x), \qquad a \le x \le b. \tag{1}$$

We seek the solution in the form of a series in powers of the parameter λ:

$$y(x) = f(x) + \sum_{n=1}^{\infty} \lambda^n \psi_n(x). \tag{2}$$

Substitute series (2) into Eq. (1). On matching the coefficients of like powers of λ, we obtain a recurrent system of equations for the functions $\psi_n(x)$. The solution of this system yields

$$\psi_1(x) = \int_a^b K(x, t)f(t)\, dt,$$

$$\psi_2(x) = \int_a^b K(x, t)\psi_1(t)\, dt = \int_a^b K_2(x, t)f(t)\, dt,$$

$$\psi_3(x) = \int_a^b K(x, t)\psi_2(t)\, dt = \int_a^b K_3(x, t)f(t)\, dt, \quad \text{etc.}$$

Here

$$K_n(x, t) = \int_a^b K(x, z)K_{n-1}(z, t)\, dz, \tag{3}$$

where $n = 2, 3, \ldots$, and we have $K_1(x, t) \equiv K(x, t)$. The functions $K_n(x, t)$ defined by formulas (3) are called *iterated kernels*. These kernels satisfy the relation

$$K_n(x, t) = \int_a^b K_m(x, s)K_{n-m}(s, t)\, ds, \tag{4}$$

where m is an arbitrary positive integer less than n.

The iterated kernels $K_n(x, t)$ can be directly expressed via $K(x, t)$ by the formula

$$K_n(x, t) = \underbrace{\int_a^b \int_a^b \cdots \int_a^b}_{n-1} K(x, s_1)K(s_1, s_2) \ldots K(s_{n-1}, t)\, ds_1\, ds_2 \ldots ds_{n-1}.$$

All iterated kernels $K_n(x, t)$, beginning with $K_2(x, t)$, are continuous functions on the square $S = \{a \le x \le b,\ a \le t \le b\}$ if the original kernel $K(x, t)$ is square integrable on S.

If $K(x, t)$ is symmetric, then all iterated kernels $K_n(x, t)$ are also symmetric.

11.3-2. Method of Successive Approximations

The results of Subsection 11.3-1 can also be obtained by means of the method of successive approximations. To this end, one should use the recurrent formula

$$y_n(x) = f(x) + \lambda \int_a^b K(x, t)y_{n-1}(t)\, dt, \qquad n = 1, 2, \ldots,$$

with the zeroth approximation $y_0(x) = f(x)$.

11.3-3. Construction of the Resolvent

The resolvent of the integral equation (1) is defined via the iterated kernels by the formula

$$R(x,t;\lambda) = \sum_{n=1}^{\infty} \lambda^{n-1} K_n(x,t), \qquad (5)$$

where the series on the right-hand side is called the *Neumann series of the kernel* $K(x,t)$. It converges to a unique square integrable solution of Eq. (1) provided that

$$|\lambda| < \frac{1}{B}, \qquad B = \sqrt{\int_a^b \int_a^b K^2(x,t)\,dx\,dt}. \qquad (6)$$

If, in addition, we have

$$\int_a^b K^2(x,t)\,dt \le A, \qquad a \le x \le b,$$

where A is a constant, then the Neumann series converges absolutely and uniformly on $[a,b]$.

A solution of a Fredholm equation of the second kind of the form (1) is expressed by the formula

$$y(x) = f(x) + \lambda \int_a^b R(x,t;\lambda) f(t)\,dt, \qquad a \le x \le b. \qquad (7)$$

Inequality (6) is essential for the convergence of the series (5). However, a solution of Eq. (1) can exist for values $|\lambda| > 1/B$ as well.

Remark 1. A solution of the equation

$$y(x) - \lambda \int_a^b K(x,t)y(t)\,dt = f(x), \qquad a \le x \le b,$$

with weak singularity, where the kernel $K(x,t)$ has the form

$$K(x,t) = \frac{L(x,t)}{|x-t|^\alpha}, \qquad 0 < \alpha < 1,$$

and $L(x,t)$ is a function continuous on the square $S = \{a \le x \le b,\ a \le t \le b\}$, can be obtained by the successive approximation method provided that

$$|\lambda| < \frac{1-\alpha}{2B^*(b-a)^{1-\alpha}}, \qquad B^* = \sup |L(x,t)|.$$

The equation itself can be reduced to a Fredholm equation of the form

$$y(x) - \lambda^n \int_a^b K_n(x,t)y(t)\,dt = F(x), \qquad a \le x \le b,$$

$$F(x) = f(x) + \sum_{p=1}^{n-1} \lambda^p \int_a^b K_p(x,t) f(t)\,dt,$$

where $K_p(x,t)$ $(p = 1, \dots, n)$ is the pth iterated kernel, with $K_n(x,t)$ being a Fredholm kernel for $n > \frac{1}{2}(1-\alpha)^{-1}$ and bounded for $n > (1-\alpha)^{-1}$.

Example 1. Let us solve the integral equation

$$y(x) - \lambda \int_0^1 xt y(t)\, dt = f(x), \qquad 0 \le x \le 1,$$

by the method of successive approximations. Here we have $K(x,t) = xt$, $a = 0$, and $b = 1$. We successively define

$$K_1(x,t) = xt, \quad K_2(x,t) = \int_0^1 (xz)(zt)\, dz = \frac{xt}{3}, \quad K_3(x,t) = \frac{1}{3}\int_0^1 (xz)(zt)\, dz = \frac{xt}{3^2}, \quad \ldots, \quad K_n(x,t) = \frac{xt}{3^{n-1}}.$$

According to formula (5) for the resolvent, we obtain

$$R(x,t;\lambda) = \sum_{n=1}^{\infty} \lambda^{n-1} K_n(x,t) = xt \sum_{n=1}^{\infty} \left(\frac{\lambda}{3}\right)^{n-1} = \frac{3xt}{3 - \lambda},$$

where $|\lambda| < 3$, and it follows from formula (7) that the solution of the integral equation can be rewritten in the form

$$y(x) = f(x) + \lambda \int_0^1 \frac{3xt}{3 - \lambda} f(t)\, dt, \quad 0 \le x \le 1, \quad \lambda \ne 3.$$

In particular, for $f(x) = x$ we obtain

$$y(x) = \frac{3x}{3 - \lambda}, \qquad 0 \le x \le 1, \quad \lambda \ne 3.$$

11.3-4. Orthogonal Kernels

For some Fredholm equations, the Neumann series (5) for the resolvent is convergent for all values of λ. Let us establish this fact.

Assume that two kernels $K(x,t)$ and $L(x,t)$ are given. These kernels are said to be *orthogonal* if the following two conditions hold:

$$\int_a^b K(x,z)L(z,t)\, dz = 0, \qquad \int_a^b L(x,z)K(z,t)\, dz = 0 \tag{8}$$

for all admissible values of x and t.

There exist kernels that are orthogonal to themselves. For these kernels we have $K_2(x,t) \equiv 0$, where $K_2(x,t)$ is the second iterated kernel. It is clear that in this case all the subsequent iterated kernels also vanish, and the resolvent coincides with the kernel $K(x,t)$.

Example 2. Let us find the resolvent of the kernel $K(x,t) = \sin(x - 2t)$, $0 \le x \le 2\pi$, $0 \le t \le 2\pi$. We have

$$\int_0^{2\pi} \sin(x - 2z)\sin(z - 2t)\, dz = \frac{1}{2}\int_0^{2\pi} [\cos(x + 2t - 3z) - \cos(x - 2t - z)]\, dz =$$

$$= \frac{1}{2}\left[-\frac{1}{3}\sin(x + 2t - 3z) + \sin(x - 2t - z)\right]_{z=0}^{z=2\pi} = 0.$$

Thus, in this case the resolvent of the kernel is equal to the kernel itself:

$$R(x,t;\lambda) \equiv \sin(x - 2t),$$

so that the Neumann series (6) consists of a single term and clearly converges for any λ.

Remark 2. If the kernels $M^{(1)}(x,t), \ldots, M^{(n)}(x,t)$ are pairwise orthogonal, then the resolvent corresponding to the sum

$$K(x,t) = \sum_{m=1}^{n} M^{(m)}(x,t)$$

is equal to the sum of the resolvents corresponding to each of the summands.

⊙ References for Section 11.3: S. G. Mikhlin (1960), M. L. Krasnov, A. I. Kiselev, and G. I. Makarenko (1971), J. A. Cochran (1972), V. I. Smirnov (1974), A. J. Jerry (1985).

11.4. Method of Fredholm Determinants

11.4-1. A Formula for the Resolvent

A solution of the Fredholm equation of the second kind

$$y(x) - \lambda \int_a^b K(x,t)y(t)\,dt = f(x), \qquad a \le x \le b, \tag{1}$$

is given by the formula

$$y(x) = f(x) + \lambda \int_a^b R(x,t;\lambda)f(t)\,dt, \qquad a \le x \le b, \tag{2}$$

where the resolvent $R(x,t;\lambda)$ is defined by the relation

$$R(x,t;\lambda) = \frac{D(x,t;\lambda)}{D(\lambda)}, \qquad D(\lambda) \ne 0. \tag{3}$$

Here $D(x,t;\lambda)$ and $D(\lambda)$ are power series in λ,

$$D(x,t;\lambda) = \sum_{n=0}^{\infty} \frac{(-1)^n}{n!} A_n(x,t)\lambda^n, \qquad D(\lambda) = \sum_{n=0}^{\infty} \frac{(-1)^n}{n!} B_n \lambda^n, \tag{4}$$

with coefficients defined by the formulas

$$A_0(x,t) = K(x,t), \quad A_n(x,t) = \underbrace{\int_a^b \cdots \int_a^b}_{n} \begin{vmatrix} K(x,t) & K(x,t_1) & \cdots & K(x,t_n) \\ K(t_1,t) & K(t_1,t_1) & \cdots & K(t_1,t_n) \\ \vdots & \vdots & \ddots & \vdots \\ K(t_n,t) & K(t_n,t_1) & \cdots & K(t_n,t_n) \end{vmatrix} dt_1 \ldots dt_n, \tag{5}$$

$$B_0 = 1, \quad B_n = \underbrace{\int_a^b \cdots \int_a^b}_{n} \begin{vmatrix} K(t_1,t_1) & K(t_1,t_2) & \cdots & K(t_1,t_n) \\ K(t_2,t_1) & K(t_2,t_2) & \cdots & K(t_2,t_n) \\ \vdots & \vdots & \ddots & \vdots \\ K(t_n,t_1) & K(t_n,t_2) & \cdots & K(t_n,t_n) \end{vmatrix} dt_1 \ldots dt_n; \quad n = 0,1,2,\ldots \tag{6}$$

The function $D(x,t;\lambda)$ is called the *Fredholm minor* and $D(\lambda)$ the *Fredholm determinant*. The series (4) converge for all values of λ and hence define entire analytic functions of λ. The resolvent $R(x,t;\lambda)$ is an analytic function of λ everywhere except for the values of λ that are roots of $D(\lambda)$. These roots coincide with the characteristic values of the equation and are poles of the resolvent $R(x,t;\lambda)$.

Example 1. Consider the integral equation

$$y(x) - \lambda \int_0^1 xe^t y(t)\,dt = f(x), \quad 0 \le x \le 1, \quad \lambda \ne 1.$$

We have

$$A_0(x,t) = xe^t, \quad A_1(x,t) = \int_0^1 \begin{vmatrix} xe^t & xe^{t_1} \\ t_1 e^t & t_1 e^{t_1} \end{vmatrix} dt_1 = 0, \quad A_2(x,t) = \int_0^1 \int_0^1 \begin{vmatrix} xe^t & xe^{t_1} & xe^{t_2} \\ t_1 e^t & t_1 e^{t_1} & t_1 e^{t_2} \\ t_2 e^t & t_2 e^{t_1} & t_2 e^{t_2} \end{vmatrix} dt_1\,dt_2 = 0,$$

since the determinants in the integrand are zero. It is clear that the relation $A_n(x,t) = 0$ holds for the subsequent coefficients. Let us find the coefficients B_n:

$$B_1 = \int_0^1 K(t_1,t_1)\,dt_1 = \int_0^1 t_1 e^{t_1}\,dt_1 = 1, \quad B_2 = \int_0^1 \int_0^1 \begin{vmatrix} t_1 e^{t_1} & t_1 e^{t_2} \\ t_2 e^{t_1} & t_2 e^{t_2} \end{vmatrix} dt_1\,dt_2 = 0.$$

It is clear that $B_n = 0$ for all subsequent coefficients as well.

According to formulas (4), we have

$$D(x, t; \lambda) = K(x, t) = xe^t; \qquad D(\lambda) = 1 - \lambda.$$

Thus,

$$R(x, t; \lambda) = \frac{D(x, t; \lambda)}{D(\lambda)} = \frac{xe^t}{1 - \lambda},$$

and the solution of the equation can be represented in the form

$$y(x) = f(x) + \lambda \int_0^1 \frac{xe^t}{1 - \lambda} f(t)\, dt, \quad 0 \le x \le 1, \quad \lambda \ne 1.$$

In particular, for $f(x) = e^{-x}$ we obtain

$$y(x) = e^{-x} + \frac{\lambda}{1 - \lambda} x, \quad 0 \le x \le 1, \quad \lambda \ne 1.$$

11.4-2. Recurrent Relations

In practice, the calculation of the coefficients $A_n(x, t)$ and B_n of the series (4) by means of formulas (5) and (6) is seldom possible. However, formulas (5) and (6) imply the following recurrent relations:

$$A_n(x, t) = B_n K(x, t) - n \int_a^b K(x, s) A_{n-1}(s, t)\, ds, \tag{7}$$

$$B_n = \int_a^b A_{n-1}(s, s)\, ds. \tag{8}$$

Example 2. Let us use formulas (7) and (8) to find the resolvent of the kernel $K(x, t) = x - 2t$, where $0 \le x \le 1$ and $0 \le t \le 1$.

Indeed, we have $B_0 = 1$ and $A_0(x, t) = x - 2t$. Applying formula (8), we see that

$$B_1 = \int_0^1 (-s)\, ds = -\tfrac{1}{2}.$$

Formula (7) implies the relation

$$A_1(x, t) = -\frac{x - 2t}{2} - \int_0^1 (x - 2s)(s - 2t)\, ds = -x - t + 2xt + \tfrac{2}{3}.$$

Furthermore, we have

$$B_2 = \int_0^1 \left(-2s + 2s^2 + \tfrac{2}{3}\right) ds = \tfrac{1}{3},$$

$$A_2(x, t) = \frac{x - 2t}{3} - 2\int_0^1 (x - 2s)\left(-s - t + 2st + \tfrac{2}{3}\right) ds = 0,$$

$$B_3 = B_4 = \cdots = 0, \quad A_3(x, t) = A_4(x, t) = \cdots = 0.$$

Hence,

$$D(\lambda) = 1 + \tfrac{1}{2}\lambda + \tfrac{1}{6}\lambda^2; \quad D(x, t; \lambda) = x - 2t + \lambda\left(x + t - 2xt - \tfrac{2}{3}\right).$$

The resolvent has the form

$$R(x, t; \lambda) = \frac{x - 2t + \lambda\left(x + t - 2xt - \tfrac{2}{3}\right)}{1 + \tfrac{1}{2}\lambda + \tfrac{1}{6}\lambda^2}.$$

⊙ References for Section 11.4: S. G. Mikhlin (1960), M. L. Krasnov, A. I. Kiselev, and G. I. Makarenko (1971), V. I. Smirnov (1974).

11.5. Fredholm Theorems and the Fredholm Alternative

11.5-1. Fredholm Theorems

THEOREM 1. *If λ is a regular value, then both the Fredholm integral equation of the second kind and the transposed equation are solvable for any right-hand side, and both the equations have unique solutions. The corresponding homogeneous equations have only the trivial solutions.*

THEOREM 2. *For the nonhomogeneous integral equation to be solvable, it is necessary and sufficient that the right-hand side $f(x)$ satisfies the conditions*

$$\int_a^b f(x)\psi_k(x)\,dx = 0, \qquad k = 1, \ldots, n,$$

where $\psi_k(x)$ is a complete set of linearly independent solutions of the corresponding transposed homogeneous equation.

THEOREM 3. *If λ is a characteristic value, then both the homogeneous integral equation and the transposed homogeneous equation have nontrivial solutions. The number of linearly independent solutions of the homogeneous integral equation is finite and is equal to the number of linearly independent solutions of the transposed homogeneous equation.*

THEOREM 4. *A Fredholm equation of the second kind has at most countably many characteristic values, whose only possible accumulation point is the point at infinity.*

11.5-2. The Fredholm Alternative

The Fredholm theorems imply the so-called Fredholm alternative, which is most frequently used in the investigation of integral equations.

THE FREDHOLM ALTERNATIVE. *Either the nonhomogeneous equation is solvable for any right-hand side or the corresponding homogeneous equation has nontrivial solutions.*

The first part of the alternative holds if the given value of the parameter is regular and the second if it is characteristic.

Remark. The Fredholm theory is also valid for integral equations of the second kind with weak singularity.

⊙ References for Section 11.5: S. G. Mikhlin (1960), M. L. Krasnov, A. I. Kiselev, and G. I. Makarenko (1971), J. A. Cochran (1972), V. I. Smirnov (1974), A. J. Jerry (1985), D. Porter and D. S. G. Stirling (1990), C. Corduneanu (1991), J. Kondo (1991), W. Hackbusch (1995), R. P. Kanwal (1997).

11.6. Fredholm Integral Equations of the Second Kind With Symmetric Kernel

11.6-1. Characteristic Values and Eigenfunctions

Integral equations whose kernels are *symmetric*, that is, satisfy the condition $K(x,t) = K(t,x)$, are called *symmetric integral equations*.

Each symmetric kernel that is not identically zero has at least one characteristic value.

For any n, the set of characteristic values of the nth iterated kernel coincides with the set of nth powers of the characteristic values of the first kernel.

The eigenfunctions of a symmetric kernel corresponding to distinct characteristic values are orthogonal, i.e., if

$$\varphi_1(x) = \lambda_1 \int_a^b K(x,t)\varphi_1(t)\,dt, \quad \varphi_2(x) = \lambda_2 \int_a^b K(x,t)\varphi_2(t)\,dt, \qquad \lambda_1 \neq \lambda_2,$$

then

$$(\varphi_1, \varphi_2) = 0, \qquad (\varphi, \psi) \equiv \int_a^b \varphi(x)\psi(x)\, dx.$$

The characteristic values of a symmetric kernel are real.

The eigenfunctions can be normalized; namely, we can divide each characteristic function by its norm. If several linearly independent eigenfunctions correspond to the same characteristic value, say, $\varphi_1(x), \ldots, \varphi_n(x)$, then each linear combination of these functions is an eigenfunction as well, and these linear combinations can be chosen so that the corresponding eigenfunctions are orthonormal.

Indeed, the function

$$\psi_1(x) = \frac{\varphi_1(x)}{\|\varphi_1\|}, \qquad \|\varphi_1\| = \sqrt{(\varphi_1, \varphi_1)},$$

has the norm equal to one, i.e., $\|\psi_1\| = 1$. Let us form a linear combination $\alpha\psi_1 + \varphi_2$ and choose α so that

$$(\alpha\psi_1 + \varphi_2, \ \psi_1) = 0,$$

i.e.,

$$\alpha = -\frac{(\varphi_2, \psi_1)}{(\psi_1, \psi_1)} = -(\varphi_2, \psi_1).$$

The function

$$\psi_2(x) = \frac{\alpha\psi_1 + \varphi_2}{\|\alpha\psi_1 + \varphi_2\|}$$

is orthogonal to $\psi_1(x)$ and has the unit norm. Next, we choose a linear combination $\alpha\psi_1 + \beta\psi_2 + \varphi_3$, where the constants α and β can be found from the orthogonality relations

$$(\alpha\psi_1 + \beta\varphi_2 + \varphi_3, \ \psi_1) = 0, \quad (\alpha\psi_1 + \beta\psi_2 + \varphi_3, \ \psi_2) = 0.$$

For the coefficients α and β thus defined, the function

$$\psi_3 = \frac{\alpha\psi_1 + \beta\psi_2 + \varphi_2}{\|\alpha\psi_1 + \beta\varphi_2 + \varphi_3\|}$$

is orthogonal to ψ_1 and ψ_2 and has the unit norm, and so on.

As was noted above, the eigenfunctions corresponding to distinct characteristic values are orthogonal. Hence, the sequence of eigenfunctions of a symmetric kernel can be made orthonormal.

In what follows we assume that the sequence of eigenfunctions of a symmetric kernel is orthonormal.

We also assume that the characteristic values are always numbered in the increasing order of their absolute values. Thus, if

$$\lambda_1, \ \lambda_2, \ \ldots, \ \lambda_n, \ \ldots \tag{1}$$

is the sequence of characteristic values of a symmetric kernel, and if a sequence of eigenfunctions

$$\varphi_1, \ \varphi_2, \ \ldots, \ \varphi_n, \ \ldots \tag{2}$$

corresponds to the sequence (1) so that

$$\varphi_n(x) - \lambda_n \int_a^b K(x,t)\varphi_n(t)\, dt = 0, \tag{3}$$

then

$$\int_a^b \varphi_i(x)\varphi_j(x)\, dx = \begin{cases} 1 & \text{for } i = j, \\ 0 & \text{for } i \neq j, \end{cases} \tag{4}$$

and

$$|\lambda_1| \le |\lambda_2| \le \cdots \le |\lambda_n| \le \cdots . \tag{5}$$

If there are infinitely many characteristic values, then it follows from the fourth Fredholm theorem that their only accumulation point is the point at infinity, and hence $\lambda_n \to \infty$ as $n \to \infty$.

The set of all characteristic values and the corresponding normalized eigenfunctions of a symmetric kernel is called the *system of characteristic values and eigenfunctions* of the kernel. The system of eigenfunctions is said to be *incomplete* if there exists a nonzero square integrable function that is orthogonal to all functions of the system. Otherwise, the system of eigenfunctions is said to be *complete*.

11.6-2. Bilinear Series

Assume that a kernel $K(x,t)$ admits an expansion in a uniformly convergent series with respect to the orthonormal system of its eigenfunctions:

$$K(x,t) = \sum_{k=1}^{\infty} a_k(x)\varphi_k(t) \tag{6}$$

for all x in the case of a continuous kernel or for almost all x in the case of a square integrable kernel.

We have

$$a_k(x) = \int_a^b K(x,t)\varphi_k(t)\,dt = \frac{\varphi_k(x)}{\lambda_k}, \tag{7}$$

and hence

$$K(x,t) = \sum_{k=1}^{\infty} \frac{\varphi_k(x)\varphi_k(t)}{\lambda_k}. \tag{8}$$

Conversely, if the series

$$\sum_{k=1}^{\infty} \frac{\varphi_k(x)\varphi_k(t)}{\lambda_k} \tag{9}$$

is uniformly convergent, then

$$K(x,t) = \sum_{k=1}^{\infty} \frac{\varphi_k(x)\varphi_k(t)}{\lambda_k}.$$

The following assertion holds: the bilinear series (9) converges in mean-square to the kernel $K(x,t)$.

If a symmetric kernel $K(x,t)$ has finitely many characteristic values, then it is degenerate, because in this case we have

$$K(x,t) = \sum_{k=1}^{n} \frac{\varphi_k(x)\varphi_k(t)}{\lambda_k}. \tag{10}$$

A kernel $K(x,t)$ is said to be *positive definite* if for all functions $\varphi(x)$ that are not identically zero we have

$$\int_a^b \int_a^b K(x,t)\varphi(x)\varphi(t)\,dx\,dt > 0,$$

and the above quadratic functional vanishes for $\varphi(x) = 0$ only. Such a kernel has positive characteristic values only. A *negative definite* kernel is defined similarly.

Each symmetric positive definite (or negative definite) continuous kernel can be decomposed in a bilinear series in eigenfunctions that is absolutely and uniformly convergent with respect to the variables x, t.

The assertion remains valid if we assume that the kernel has finitely many negative (positive, respectively) characteristic values.

If a kernel $K(x, t)$ is symmetric, continuous on the square $S = \{a \leq x \leq b,\ a \leq t \leq b\}$, and has uniformly bounded partial derivatives on this square, then this kernel can be expanded in a uniformly convergent bilinear series in eigenfunctions.

11.6-3. The Hilbert–Schmidt Theorem

If a function $f(x)$ can be represented in the form

$$f(x) = \int_a^b K(x, t)g(t)\, dt, \tag{11}$$

where the symmetric kernel $K(x, t)$ is square integrable and $g(t)$ is a square integrable function, then $f(x)$ can be represented by its *Fourier series* with respect to the orthonormal system of eigenfunctions of the kernel $K(x, t)$:

$$f(x) = \sum_{k=1}^{\infty} a_k \varphi_k(x), \tag{12}$$

where

$$a_k = \int_a^b f(x)\varphi_k(x)\, dx, \qquad k = 1, 2, \ldots$$

Moreover, if

$$\int_a^b K^2(x, t)\, dt \leq A < \infty, \tag{13}$$

then the series (12) is absolutely and uniformly convergent for any function $f(x)$ of the form (11).

Remark 1. In the Hilbert–Schmidt theorem, the completeness of the system of eigenfunctions is not assumed.

11.6-4. Bilinear Series of Iterated Kernels

By the definition of the iterated kernels, we have

$$K_m(x, t) = \int_a^b K(x, z)K_{m-1}(z, t)\, dz, \qquad m = 2, 3, \ldots \tag{14}$$

The Fourier coefficients $a_k(t)$ of the kernel $K_m(x, t)$, regarded as a function of the variable x, with respect to the orthonormal system of eigenfunctions of the kernel $K(x, t)$ are equal to

$$a_k(t) = \int_a^b K_m(x, t)\varphi_k(x)\, dx = \frac{\varphi_k(t)}{\lambda_k^m}. \tag{15}$$

On applying the Hilbert–Schmidt theorem to (14), we obtain

$$K_m(x, t) = \sum_{k=1}^{\infty} \frac{\varphi_k(x)\varphi_k(t)}{\lambda_k^m}, \qquad m = 2, 3, \ldots \tag{16}$$

In formula (16), the sum of the series is understood as the limit in mean-square. If in addition to the above assumptions, inequality (13) is satisfied, then the series in (16) is uniformly convergent.

11.6-5. Solution of the Nonhomogeneous Equation

Let us represent an integral equation

$$y(x) - \lambda \int_a^b K(x,t)y(t)\,dt = f(x), \qquad a \le x \le b, \tag{17}$$

where the parameter λ is not a characteristic value, in the form

$$y(x) - f(x) = \lambda \int_a^b K(x,t)y(t)\,dt \tag{18}$$

and apply the Hilbert–Schmidt theorem to the function $y(x) - f(x)$:

$$y(x) - f(x) = \sum_{k=1}^{\infty} A_k \varphi_k(x),$$

$$A_k = \int_a^b [y(x) - f(x)]\varphi_k(x)\,dx = \int_a^b y(x)\varphi_k(x)\,dx - \int_a^b f(x)\varphi_k(x)\,dx = y_k - f_k.$$

Taking into account the expansion (8), we obtain

$$\lambda \int_a^b K(x,t)y(t)\,dt = \lambda \sum_{k=1}^{\infty} \frac{y_k}{\lambda_k}\varphi_k(x),$$

and thus

$$\lambda \frac{y_k}{\lambda_k} = y_k - f_k, \quad y_k = \frac{\lambda_k f_k}{\lambda_k - \lambda}, \quad A_k = \frac{\lambda f_k}{\lambda_k - \lambda}. \tag{19}$$

Hence,

$$y(x) = f(x) + \lambda \sum_{k=1}^{\infty} \frac{f_k}{\lambda_k - \lambda}\varphi_k(x). \tag{20}$$

However, if λ is a characteristic value, i.e.,

$$\lambda = \lambda_p = \lambda_{p+1} = \cdots = \lambda_q, \tag{21}$$

then, for $k \ne p, p+1, \ldots, q$, the terms (20) preserve their form. For $k = p, p+1, \ldots, q$, formula (19) implies the relation $f_k = A_k(\lambda - \lambda_k)/\lambda$, and by (21) we obtain $f_p = f_{p+1} = \cdots = f_q = 0$. The last relation means that

$$\int_a^b f(x)\varphi_k(x)\,dx = 0$$

for $k = p, p+1, \ldots, q$, i.e., the right-hand side of the equation must be orthogonal to the eigenfunctions that correspond to the characteristic value λ.

In this case, the solutions of Eqs. (17) have the form

$$y(x) = f(x) + \lambda \sum_{k=1}^{\infty} \frac{f_k}{\lambda_k - \lambda}\varphi_k(x) + \sum_{k=p}^{q} C_k \varphi_k(x), \tag{22}$$

where the terms in the first of the sums (22) with indices $k = p, p+1, \ldots, q$ must be omitted (for these indices, f_k and $\lambda - \lambda_k$ vanish in this sum simultaneously). The coefficients C_k in the second sum are arbitrary constants.

Remark 2. On the basis of the bilinear expansion (8) and the Hilbert–Schmidt theorem, the solution of the symmetric Fredholm integral equation of the first kind

$$\int_a^b K(x,t)y(t)\,dt = f(x), \qquad a \le x \le b,$$

can be constructed in a similar way in the form

$$y(x) = \sum_{k=1}^{\infty} f_k \lambda_k \varphi_k(x),$$

and the necessary and sufficient condition for the existence and uniqueness of such a solution in $L_2(a,b)$ is the completeness of the system of the eigenfunctions $\varphi_k(x)$ of the kernel $K(x,t)$ together with the convergence of the series $\sum_{k=1}^{\infty} f_k^2 \lambda_k^2$, where the λ_k are the corresponding characteristic values.

It should be noted that the verification of the last condition for specific equations is quite complicated. In the solution of Fredholm equations of the first kind, the methods presented in Chapter 10 are usually applied.

11.6-6. The Fredholm Alternative for Symmetric Equations

The above results can be unified in the following alternative form.

A symmetric integral equation

$$y(x) - \lambda \int_a^b K(x,t)y(t)\,dt = f(x), \qquad a \le x \le b, \tag{23}$$

for a given λ, either has a unique square integrable solution for an arbitrarily given function $f(x) \in L_2(a,b)$, in particular, $y = 0$ for $f = 0$, or the corresponding homogeneous equation has finitely many linearly independent solutions $Y_1(x), \ldots, Y_r(x)$, $r > 0$.

For the second case, the nonhomogeneous equation has a solution if and only if the right-hand side $f(x)$ is orthogonal to all the functions $Y_1(x), \ldots, Y_r(x)$ on the interval $[a,b]$. Here the solution is defined only up to an arbitrary additive linear combination $A_1 Y_1(x) + \cdots + A_r Y_r(x)$.

11.6-7. The Resolvent of a Symmetric Kernel

The solution of a Fredholm equation of the second kind (23) can be written in the form

$$y(x) = f(x) + \lambda \int_a^b R(x,t;\lambda)f(t)\,dt, \tag{24}$$

where the resolvent $R(x,t;\lambda)$ is given by the series

$$R(x,t;\lambda) = \sum_{k=1}^{\infty} \frac{\varphi_k(x)\varphi_k(t)}{\lambda_k - \lambda}. \tag{25}$$

Here the collections $\varphi_k(x)$ and λ_k form the system of eigenfunctions and characteristic values of Eqs. (23). It follows from formula (25) that the resolvent of a symmetric kernel has only simple poles.

11.6-8. Extremal Properties of Characteristic Values and Eigenfunctions

Let us introduce the notation

$$(u, w) = \int_a^b u(x)w(x)\, dx, \quad \|u\|^2 = (u, u),$$

$$(Ku, u) = \int_a^b \int_a^b K(x, t)u(x)u(t)\, dx\, dt,$$

where (u, w) is the *inner product* of functions $u(x)$ and $w(x)$, $\|u\|$ is the *norm* of a function $u(x)$, and (Ku, u) is the *quadratic form* generated by the kernel $K(x, t)$.

Let λ_1 be the characteristic value of the symmetric kernel $K(x, t)$ with minimum absolute value and let $y_1(x)$ be the eigenfunction corresponding to this value. Then

$$\frac{1}{|\lambda_1|} = \max_{y \neq 0} \frac{|(Ky, y)|}{\|y\|^2}; \tag{26}$$

in particular, the maximum is attained, and $y = y_1$ is a maximum point.

Let $\lambda_1, \ldots, \lambda_n$ be the first n characteristic values of a symmetric kernel $K(x, t)$ (in the ascending order of their absolute values) and let $y_1(x), \ldots, y_n(x)$ be orthonormal eigenfunctions corresponding to $\lambda_1, \ldots, \lambda_n$, respectively. Then the formula

$$\frac{1}{|\lambda_{n+1}|} = \max \frac{|(Ky, y)|}{\|y\|^2} \tag{27}$$

is valid for the characteristic value λ_{n+1} following λ_n. The maximum is taken over the set of functions y which are orthogonal to all $y_1, \ldots, y_n$ and are not identically zero, that is, $y \neq 0$

$$(y, y_j) = 0, \qquad j = 1, \ldots, n; \tag{28}$$

in particular, the maximum in (27) is attained, and $y = y_{n+1}$ is a maximum point, where y_{n+1} is any eigenfunction corresponding to the characteristic value λ_{n+1} which is orthogonal to $y_1, \ldots, y_n$.

Remark 3. For a positive definite kernel $K(x, t)$, the symbol of modulus on the right-hand sides of (27) and (28) can be omitted.

11.6-9. Integral Equations Reducible to Symmetric Equations

An equation of the form

$$y(x) - \lambda \int_a^b K(x, t)\rho(t)y(t)\, dt = f(x), \tag{29}$$

where $K(s, t)$ is a symmetric kernel and $\rho(t) > 0$ is a continuous function on $[a, b]$, can be reduced to a symmetric equation. Indeed, on multiplying Eq. (29) by $\sqrt{\rho(x)}$ and introducing the new unknown function $z(x) = \sqrt{\rho(x)}\, y(x)$, we arrive at the integral equation

$$z(x) - \lambda \int_a^b L(x, t)z(t)\, dt = f(x)\sqrt{\rho(x)}, \qquad L(x, t) = K(x, t)\sqrt{\rho(x)\rho(t)}, \tag{30}$$

where $L(x, t)$ is a symmetric kernel.

11.6-10. Skew-Symmetric Integral Equations

By a *skew-symmetric integral equation* we mean an equation whose kernel is skew-symmetric, i.e., an equation of the form

$$y(x) - \lambda \int_a^b K(x, t) y(t)\, dt = f(x) \tag{31}$$

whose kernel $K(x, t)$ has the property

$$K(t, x) = -K(x, t). \tag{32}$$

Equation (31) with the skew-symmetric kernel (32) has at least one characteristic value, and all its characteristic values are purely imaginary.

⊙ References for Section 11.6: E. Goursat (1923), R. Courant and D. Hilbert (1931), S. G. Mikhlin (1960), M. L. Krasnov, A. I. Kiselev, and G. I. Makarenko (1971), J. A. Cochran (1972), V. I. Smirnov (1974), A. J. Jerry (1985), F. G. Tricomi (1985), D. Porter and D. S. G. Stirling (1990), C. Corduneanu (1991), J. Kondo (1991), W. Hackbusch (1995), R. P. Kanwal (1997).

11.7. An Operator Method for Solving Integral Equations of the Second Kind

11.7-1. The Simplest Scheme

Consider a linear equation of the second kind of the special form

$$y(x) - \lambda \mathbf{L}[y] = f(x), \tag{1}$$

where $\mathbf{L}$ is a linear (integral) operator such that $\mathbf{L}^2 = k$, $k = \text{const}$.

Let us apply the operator $\mathbf{L}$ to Eq. (1). We obtain

$$\mathbf{L}[y] - k\lambda y(x) = \mathbf{L}[f(x)]. \tag{2}$$

On eliminating the term $\mathbf{L}[y]$ from (1) and (2), we find the solution

$$y(x) = \frac{1}{1 - k\lambda^2} \big\{ f(x) + \lambda \mathbf{L}[f] \big\}. \tag{3}$$

Remark. In Section 9.4, various generalizations of the above method are described.

11.7-2. Solution of Equations of the Second Kind on the Semiaxis

$1°$. Consider the equation

$$y(x) - \lambda \int_0^\infty \cos(xt) y(t)\, dt = f(x). \tag{4}$$

In this case, the operator $\mathbf{L}$ coincides, up to a constant factor, with the Fourier cosine transform:

$$\mathbf{L}[y] = \int_0^\infty \cos(xt) y(t)\, dt = \sqrt{\frac{\pi}{2}}\, \mathfrak{F}_c[y] \tag{5}$$

and acts by the rule $\mathbf{L}^2 = k$, where $k = \frac{\pi}{2}$ (see Subsection 7.5-1).

We obtain the solution by formula (3) taking into account Eq. (5):

$$y(x) = \frac{2}{2 - \pi\lambda^2}\left[f(x) + \lambda\int_0^\infty \cos(xt)f(t)\,dt\right], \qquad \lambda \neq \pm\sqrt{\frac{2}{\pi}}. \tag{6}$$

2°. Consider the equation

$$y(x) - \lambda\int_0^\infty tJ_\nu(xt)y(t)\,dt = f(x), \tag{7}$$

where $J_\nu(x)$ is the Bessel function, $\operatorname{Re}\nu > -\frac{1}{2}$.

Here the operator **L** coincides, up to a constant factor, with the Hankel transform:

$$\mathbf{L}\,[y] = \int_0^\infty tJ_\nu(xt)y(t)\,dt \tag{8}$$

and acts by the rule $\mathbf{L}^2 = 1$ (see Subsection 7.6-1).

We obtain the solution by formula (3), for $k = 1$, taking into account Eq. (8):

$$y(x) = \frac{1}{1 - \lambda^2}\left[f(x) + \lambda\int_0^\infty tJ_\nu(xt)f(t)\,dt\right], \qquad \lambda \neq \pm 1. \tag{9}$$

⊙ Reference for Section 11.7: A. D. Polyanin and A. V. Manzhirov (1998).

11.8. Methods of Integral Transforms and Model Solutions

11.8-1. Equation With Difference Kernel on the Entire Axis

Consider an integral equation of convolution type of the second kind with one kernel

$$y(x) + \frac{1}{\sqrt{2\pi}}\int_{-\infty}^\infty K(x - t)y(t)\,dt = f(x), \qquad -\infty < x < \infty, \tag{1}$$

where $f(x)$ and $K(x)$ are the known right-hand side and the kernel of the integral equation and $y(x)$ is the unknown function. Let us apply the (alternative) Fourier transform to Eq. (1). In this case, taking into account the convolution theorem (see Subsection 7.4-4), we obtain

$$\mathcal{Y}(u)[1 + \mathcal{K}(u)] = \mathcal{F}(u). \tag{2}$$

Thus, on applying the Fourier transform we reduce the solution of the original integral equation (1) to the solution of the algebraic equation (2) for the transform of the unknown function. The solution of Eq. (2) has the form

$$\mathcal{Y}(u) = \frac{\mathcal{F}(u)}{1 + \mathcal{K}(u)}. \tag{3}$$

Formula (3) gives the transform of the solution of the original integral equation in terms of the transforms of the known functions, namely, the kernel and the right-hand side of the equation. The solution itself can be obtained by applying the Fourier inversion formula:

$$y(x) = \frac{1}{\sqrt{2\pi}}\int_{-\infty}^\infty \mathcal{Y}(u)e^{-iux}\,du = \frac{1}{\sqrt{2\pi}}\int_{-\infty}^\infty \frac{\mathcal{F}(u)}{1 + \mathcal{K}(u)}\,e^{-iux}\,du. \tag{4}$$

In fact, formula (4) solves the problem; however, sometimes it is not convenient because it requires the calculation of the transform $F(u)$ for each right-hand side $f(x)$. In many cases, the representation of the solution of the nonhomogeneous integral equation via the resolvent of the original equation is more convenient. To obtain the desired representation, we note that formula (3) can be transformed to the expression

$$\mathcal{Y}(u) = [1 - \mathcal{R}(u)]\mathcal{F}(u), \qquad \mathcal{R}(u) = \frac{\mathcal{K}(u)}{1 + \mathcal{K}(u)}. \tag{5}$$

On the basis of (5), by applying the Fourier inversion formula and the convolution theorem (for transforms) we obtain

$$y(x) = f(x) - \frac{1}{\sqrt{2\pi}} \int_{-\infty}^{\infty} R(x - t)f(t)\,dt, \tag{6}$$

where the resolvent $R(x - t)$ of the integral equation (1) is given by the relation

$$R(x) = \frac{1}{\sqrt{2\pi}} \int_{-\infty}^{\infty} \frac{\mathcal{K}(u)}{1 + \mathcal{K}(u)} e^{-iux}\,du, \tag{7}$$

Thus, to determine the solution of the original integral equation (1), it suffices to find the function $R(x)$ by formula (7).

The function $R(x)$ is a solution of Eq. (1) for a special form of the function $f(x)$. Indeed, it follows from formulas (3) and (5) that for $\mathcal{Y}(u) = \mathcal{R}(u)$ the function $\mathcal{F}(u)$ is equal to $\mathcal{K}(u)$. This means that, for $f(x) \equiv K(x)$, the function $y(x) \equiv R(x)$ is a solution of Eq. (1), i.e., the resolvent of Eq. (1) satisfies the integral equation

$$R(x) + \frac{1}{\sqrt{2\pi}} \int_{-\infty}^{\infty} K(x - t)R(t)\,dt = K(x), \qquad -\infty < x < \infty. \tag{8}$$

Note that to calculate direct and inverse Fourier transforms, one can use the corresponding tables from Supplements 6 and 7 and the books by H. Bateman and A. Erdélyi (1954) and by V. A. Ditkin and A. P. Prudnikov (1965).

Example. Let us solve the integral equation

$$y(x) - \lambda \int_{-\infty}^{\infty} \exp\big(\alpha|x - t|\big) y(t)\,dt = f(x), \qquad -\infty < x < \infty, \tag{9}$$

which is a special case of Eq. (1) with kernel $K(x - t)$ given by the expression

$$K(x) = -\sqrt{2\pi}\,\lambda e^{-\alpha|x|}, \qquad \alpha > 0. \tag{10}$$

Let us find the function $R(x)$. To this end, we calculate the integral

$$\mathcal{K}(u) = -\int_{-\infty}^{\infty} \lambda e^{-\alpha|x|} e^{iux}\,dx = -\frac{2\alpha\lambda}{u^2 + \alpha^2}. \tag{11}$$

In this case, formula (5) implies

$$\mathcal{R}(u) = \frac{\mathcal{K}(u)}{1 + \mathcal{K}(u)} = -\frac{2\alpha\lambda}{u^2 + \alpha^2 - 2\alpha\lambda}, \tag{12}$$

and hence

$$R(x) = \frac{1}{\sqrt{2\pi}} \int_{-\infty}^{\infty} \mathcal{R}(u) e^{-iux}\,du = -\sqrt{\frac{2}{\pi}} \int_{-\infty}^{\infty} \frac{\alpha\lambda}{u^2 + \alpha^2 - 2\alpha\lambda} e^{-iux}\,du. \tag{13}$$

Assume that $\lambda < \frac{1}{2}\alpha$. In this case the integral (13) makes sense and can be calculated by means of the theory of residues on applying the Jordan lemma (see Subsections 7.1-4 and 7.1-5). After some algebraic manipulations, we obtain

$$R(x) = -\sqrt{2\pi}\,\frac{\alpha\lambda}{\sqrt{\alpha^2 - 2\alpha\lambda}} \exp\big(-|x|\sqrt{\alpha^2 - 2\alpha\lambda}\,\big) \tag{14}$$

and finally, in accordance with (6), we obtain

$$y(x) = f(x) + \frac{\alpha\lambda}{\sqrt{\alpha^2 - 2\alpha\lambda}} \int_{-\infty}^{\infty} \exp\big(-|x - t|\sqrt{\alpha^2 - 2\alpha\lambda}\,\big) f(t)\,dt, \qquad -\infty < x < \infty. \tag{15}$$

11.8-2. An Equation With the Kernel $K(x,t) = t^{-1}Q(x/t)$ on the Semiaxis

Here we consider the following equation on the semiaxis:

$$y(x) - \int_0^\infty \frac{1}{t}Q\left(\frac{x}{t}\right)y(t)\,dt = f(x). \tag{16}$$

To solve this equation we apply the Mellin transform which is defined as follows (see also Section 7.3):

$$\hat{f}(s) = \mathfrak{M}\{f(x),s\} \equiv \int_0^\infty f(x)x^{s-1}\,dx, \tag{17}$$

where $s = \sigma + i\tau$ is a complex variable ($\sigma_1 < \sigma < \sigma_2$) and $\hat{f}(s)$ is the transform of the function $f(x)$. In what follows, we briefly denote the Mellin transform by $\mathfrak{M}\{f(x)\} \equiv \mathfrak{M}\{f(x),s\}$.

For known $\hat{f}(s)$, the original function can be found by means of the Mellin inversion formula

$$f(x) = \mathfrak{M}^{-1}\{\hat{f}(s)\} \equiv \frac{1}{2\pi i}\int_{c-i\infty}^{c+i\infty} \hat{f}(s)x^{-s}\,ds, \qquad \sigma_1 < c < \sigma_2, \tag{18}$$

where the integration path is parallel to the imaginary axis of the complex plane s and the integral is understood in the sense of the Cauchy principal value.

On applying the Mellin transform to Eq. (16) and taking into account the fact that the integral with such a kernel is transformed into the product by the rule (see Subsection 7.3-2)

$$\mathfrak{M}\left\{\int_0^\infty \frac{1}{t}Q\left(\frac{x}{t}\right)y(t)\,dt\right\} = \hat{Q}(s)\hat{y}(s),$$

we obtain the following equation for the transform $\hat{y}(s)$:

$$\hat{y}(s) - \hat{Q}(s)\hat{y}(s) = \hat{f}(s).$$

The solution of this equation is given by the formula

$$\hat{y}(s) = \frac{\hat{f}(s)}{1 - \hat{Q}(s)}. \tag{19}$$

On applying the Mellin inversion formula to Eq. (19) we obtain the solution of the original integral equation

$$y(x) = \frac{1}{2\pi i}\int_{c-i\infty}^{c+i\infty} \frac{\hat{f}(s)}{1 - \hat{Q}(s)}x^{-s}\,ds. \tag{20}$$

This solution can also be represented via the resolvent in the form

$$y(x) = f(x) + \int_0^\infty \frac{1}{t}N\left(\frac{x}{t}\right)f(t)\,dt, \tag{21}$$

where we have used the notation

$$N(x) = \mathfrak{M}^{-1}\{\hat{N}(s)\}, \qquad \hat{N}(s) = \frac{\hat{Q}(s)}{1 - \hat{Q}(s)}. \tag{22}$$

Under the application of this analytical method of solution, the following technical difficulties can occur: (a) in the calculation of the transform for a given kernel $K(x)$ and (b) in the calculation of the solution for the known transform $\hat{y}(s)$. To find the corresponding integrals, tables of direct and inverse Mellin transforms are applied (e.g., see Supplements 8 and 9). In many cases, the relationship between the Mellin transform and the Fourier and Laplace transforms is first used:

$$\mathfrak{M}\{f(x),s\} = \mathfrak{F}\{f(e^x),is\} = \mathfrak{L}\{f(e^x),-s\} + \mathfrak{L}\{f(e^{-x}),s\}, \tag{23}$$

and then tables of direct and inverse Fourier transforms and Laplace transforms are applied (see Supplements 4–7).

Remark 1. The equation

$$y(x) - \int_0^\infty H\left(\frac{x}{t}\right)x^\alpha t^{-\alpha-1}y(t)\,dt = f(x) \tag{24}$$

can be rewritten in the form of Eq. (16) under the notation $K(z) = z^\alpha H(z)$.

11.8-3. Equation With the Kernel $K(x, t) = t^\beta Q(xt)$ on the Semiaxis

Consider the following equation on the semiaxis:

$$y(x) - \int_0^\infty t^\beta Q(xt)y(t)\, dt = f(x). \tag{25}$$

To solve this equation, we apply the Mellin transform. On multiplying Eq. (25) by x^{s-1} and integrating with respect to x from zero to infinity, we obtain

$$\int_0^\infty y(x)x^{s-1}\, dx - \int_0^\infty y(t)t^\beta\, dt \int_0^\infty Q(xt)x^{s-1}\, dx = \int_0^\infty f(x)x^{s-1}\, dx. \tag{26}$$

Let us make the change of variables $z = xt$. We finally obtain

$$\hat{y}(s) - \hat{Q}(s)\int_0^\infty y(t)t^{\beta - s}\, dt = \hat{f}(s). \tag{27}$$

Taking into account the relation

$$\int_0^\infty y(t)t^{\beta - s}\, dt = \hat{y}(1 + \beta - s),$$

we rewrite Eq. (27) in the form

$$\hat{y}(s) - \hat{Q}(s)\hat{y}(1 + \beta - s) = \hat{f}(s). \tag{28}$$

On replacing s by $1 + \beta - s$ in Eq. (28), we obtain

$$\hat{y}(1 + \beta - s) - \hat{Q}(1 + \beta - s)\hat{y}(s) = \hat{f}(1 + \beta - s). \tag{29}$$

Let us eliminate $\hat{y}(1 + \beta - s)$ and solve the resulting equation for $\hat{y}(s)$. We thus find the transform of the solution:

$$\hat{y}(s) = \frac{\hat{f}(s) + \hat{Q}(s)\hat{f}(1 + \beta - s)}{1 - \hat{Q}(s)\hat{Q}(1 + \beta - s)}. \tag{30}$$

On applying the Mellin inversion formula, we obtain the solution of the integral equation (25) in the form

$$y(x) = \frac{1}{2\pi i}\int_{c-i\infty}^{c+i\infty} \frac{\hat{f}(s) + \hat{Q}(s)\hat{f}(1 + \beta - s)}{1 - \hat{Q}(s)\hat{Q}(1 + \beta - s)} x^{-s}\, ds. \tag{31}$$

Remark 2. The equation

$$y(x) - \int_0^\infty H(xt)x^p t^q y(t)\, dt = f(x)$$

can be rewritten in the form of Eq. (25) under the notation $Q(z) = z^p H(z)$, where $\beta = q - p$.

11.8-4. The Method of Model Solutions for Equations on the Entire Axis

Let us illustrate the capability of a generalized modification of the method of model solutions (see Subsection 9.6) by an example of the equation

$$Ay(x) + \int_{-\infty}^{\infty} Q(x+t)e^{\beta t}y(t)\,dt = f(x), \tag{32}$$

where $Q = Q(z)$ and $f(x)$ are arbitrary functions and A and β are arbitrary constants satisfying some constraints.

For clarity, instead of the original equation (32) we write

$$\mathbf{L}\,[y(x)] = f(x). \tag{33}$$

For a test solution, we take the exponential function

$$y_0 = e^{px}. \tag{34}$$

On substituting (34) into the left-hand side of Eq. (33), after some algebraic manipulations we obtain

$$\mathbf{L}\,[e^{px}] = Ae^{px} + q(p)e^{-(p+\beta)x}, \qquad \text{where} \quad q(p) = \int_{-\infty}^{\infty} Q(z)e^{(p+\beta)z}\,dz. \tag{35}$$

The right-hand side of (35) can be regarded as a functional equation for the kernel e^{px} of the inverse Laplace transform. To solve it, we replace p by $-p - \beta$ in Eq. (33). We finally obtain

$$\mathbf{L}\,[e^{-(p+\beta)x}] = Ae^{-(p+\beta)x} + q(-p-\beta)e^{px}. \tag{36}$$

Let us multiply Eq. (35) by A and Eq. (36) by $-q(p)$ and add the resulting relations. This yields

$$\mathbf{L}\,[Ae^{px} - q(p)e^{-(p+\beta)x}] = [A^2 - q(p)q(-p-\beta)]e^{px}. \tag{37}$$

On dividing Eq. (37) by the constant $A^2 - q(p)q(-p-\beta)$, we obtain the original model solution

$$Y(x,p) = \frac{Ae^{px} - q(p)e^{-(p+\beta)x}}{A^2 - q(p)q(-p-\beta)}, \qquad \mathbf{L}\,[Y(x,p)] = e^{px}. \tag{38}$$

Since here $-\infty < x < \infty$, one must set $p = iu$ and use the formulas from Subsection 9.6-3. Then the solution of Eq. (32) for an arbitrary function $f(x)$ can be represented in the form

$$y(x) = \frac{1}{\sqrt{2\pi}} \int_{-\infty}^{\infty} Y(x, iu)\tilde{f}(u)\,du, \qquad \tilde{f}(u) = \int_{-\infty}^{\infty} f(x)e^{-iux}\,dx. \tag{39}$$

⊙ References for Section 11.8: M. L. Krasnov, A. I. Kiselev, and G. I. Makarenko (1971), V. I. Smirnov (1974), P. P. Zabreyko, A. I. Koshelev, et al. (1975), F. D. Gakhov and Yu. I. Cherskii (1978), A. D. Polyanin and A. V. Manzhirov (1997, 1998).

11.9. The Carleman Method for Integral Equations of Convolution Type of the Second Kind

11.9-1. The Wiener–Hopf Equation of the Second Kind

Equations of convolution type of the second kind of the form*

$$y(x) + \frac{1}{\sqrt{2\pi}} \int_0^\infty K(x-t)y(t)\,dt = f(x), \qquad 0 < x < \infty, \tag{1}$$

frequently occur in applications. Here the domain of the kernel $K(x)$ is the entire real axis.

Let us extend the equation domain to the negative semiaxis by introducing one-sided functions,

$$y_+(x) = \begin{cases} y(x) & \text{for } x > 0, \\ 0 & \text{for } x < 0, \end{cases} \qquad f_+(x) = \begin{cases} f(x) & \text{for } x > 0, \\ 0 & \text{for } x < 0, \end{cases} \qquad y_-(x) = 0 \quad \text{for } x > 0.$$

Then we obtain an equation,

$$y_+(x) + \frac{1}{\sqrt{2\pi}} \int_{-\infty}^\infty K(x-t)y_+(t)\,dt = y_-(x) + f_+(x), \qquad -\infty < x < \infty, \tag{2}$$

which coincides with (1) for $x > 0$.

The auxiliary function $y_-(x)$ is introduced to compensate for the left-hand side of Eq. (2) for $x < 0$. Note that $y_-(x)$ is unknown for $x < 0$ and is to be found in solving the problem.

Let us pass to the Fourier integrals in Eq. (2) (see Subsections 7.4-3, 10.4-1, and 10.4-2). We obtain a Riemann problem in the form

$$\mathcal{Y}^+(u) = \frac{\mathcal{Y}^-(u)}{1 + \mathcal{K}(u)} + \frac{\mathcal{F}^+(u)}{1 + \mathcal{K}(u)}, \qquad -\infty < u < \infty. \tag{3}$$

1°. Assume that the normality condition is satisfied, i.e.,

$$1 + \mathcal{K}(u) \neq 0,$$

then we rewrite the Riemann problem in the usual form

$$\mathcal{Y}^+(u) = \mathcal{D}(u)\mathcal{Y}^-(u) + \mathcal{H}(u), \qquad -\infty < u < \infty, \tag{4}$$

where

$$\mathcal{D}(u) = \frac{1}{1 + \mathcal{K}(u)}, \qquad \mathcal{H}(u) = \frac{\mathcal{F}(u)}{1 + \mathcal{K}(u)}. \tag{5}$$

The Riemann problem (4) is equivalent to Eq. (1); in particular, these equations are simultaneously solvable or unsolvable and have an equal number of arbitrary constants in their general solutions. If the *index* ν of the Riemann problem, which is given by the relation

$$\nu = \text{Ind}\,\frac{1}{1 + \mathcal{K}(u)} \tag{6}$$

(which is also sometimes called the *index of the Wiener–Hopf equation of the second kind*), is positive, then the homogeneous equation (1) ($f(x) \equiv 0$) has exactly ν linearly independent solutions,

---

* Prior to reading this section looking through Sections 10.4 and 10.5 is recommended.

and the nonhomogeneous equation is unconditionally solvable and its solution depends on ν arbitrary complex constants.

In the case $\nu \le 0$, the homogeneous equation has no nonzero solutions. For $\nu = 0$, the nonhomogeneous equation is unconditionally solvable, and the solution is unique. If the index ν is negative, then the conditions

$$\int_{-\infty}^{\infty} \frac{\mathcal{F}(u)\,du}{\mathcal{X}^+(u)[1 + \mathcal{K}(u)](u+i)^k} = 0, \qquad k = 1, 2, \ldots, -\nu, \tag{7}$$

are necessary and sufficient for the solvability of the nonhomogeneous equation (see Subsection 10.4-4).

For all cases in which the solution of Eq. (1) exists, it can be found by the formula

$$y(x) = y_+(x) = \frac{1}{\sqrt{2\pi}} \int_{-\infty}^{\infty} \mathcal{Y}^+(u)e^{-iux}\,du, \qquad x > 0, \tag{8}$$

where $\mathcal{Y}^+(u)$ is the solution of the Riemann problem (4) and (5) that is constructed by the scheme of Subsection 10.4-4 (see Fig. 3). The last formula shows that the solution does not depend on $\mathcal{Y}^-(u)$, i.e., is independent of the choice of the extension of the equation to the negative semiaxis.

$2°$. Now let us study the exceptional case of the integral equation (1) in which the normality condition for the Riemann problem (3) (see Subsections 10.4-6 and 10.4-7) is violated. In this case, the coefficient $D(u) = [1 + \mathcal{K}(u)]^{-1}$ has no zeros, and its order at infinity is $\eta = 0$. The general solution to the boundary value problem (3) can be obtained by formulas (63) of Subsection 10.4-7 for $\alpha_i = 0$. The solution of the original integral equation (1) can be determined from the solution of the boundary value problem on applying formula (8).

Figure 4 depicts a scheme of solving the Wiener–Hopf equations (see also Subsection 10.5-1).

Example. Consider the equation

$$y(x) + \int_0^{\infty} (a + b|x - t|)e^{-|x-t|}y(t)\,dt = f(x), \qquad x > 0, \tag{9}$$

where the constants a and b are real, and $b \ne 0$. The kernel $K(x - t)$ of Eq. (1) is given by the expression

$$K(x) = \sqrt{2\pi}\,(a + b|x|)e^{-|x|}.$$

Let us find the transform of the kernel,

$$\mathcal{K}(u) = \int_{-\infty}^{\infty} (a + b|x|)e^{-|x|+iux}\,dx = 2\frac{u^2(a - b) + a + b}{(u^2 + 1)^2}.$$

Hence,

$$1 + \mathcal{K}(u) = \frac{P(u)}{(u^2 + 1)^2}, \qquad P(z) = z^4 + 2(a - b + 1)z^2 + 2a + 2b + 1.$$

On the basis of the normality condition, we assume that the constants a and b are such that the polynomial $P(z)$ has no real roots. Let $\alpha + i\beta$ be a root of the biquadratic equation $P(z) = 0$ such that $\alpha > 0$ and $\beta > 0$. Since the coefficients of the equation are real, it is clear that $(\alpha - i\beta)$, $(-\alpha + i\beta)$, and $(-\alpha - i\beta)$ are the other three roots. Since the function $1 + \mathcal{K}(u)$ is real as well, it follows that it has zero index, and hence Eq. (9) is uniquely solvable.

On factorizing, we obtain the relation $1 + \mathcal{K}(u) = \mathcal{X}^-(u)/\mathcal{X}^+(u)$, where

$$\mathcal{X}^+(u) = \frac{(u+i)^2}{(u + \alpha + i\beta)(u - \alpha + i\beta)}, \qquad \mathcal{X}^-(u) = \frac{(u - \alpha - i\beta)(u + \alpha - i\beta)}{(u - i)^2}.$$

Applying this result, we represent the boundary condition (4), (5) in the form

$$\frac{\mathcal{Y}^+(u)}{\mathcal{X}^+(u)} - \frac{(u - i)^2 \mathcal{F}^+(u)}{(u - \alpha - i\beta)(u + \alpha - i\beta)} = \frac{\mathcal{Y}^-(u)}{\mathcal{X}^-(u)}, \qquad -\infty < u < \infty. \tag{10}$$

It follows from the theorem on the analytic continuation and the generalized Liouville theorem (see Subsection 10.4-3) that both sides of the above relation are equal to

$$\frac{C_1}{u - \alpha - i\beta} + \frac{C_2}{u + \alpha - i\beta},$$

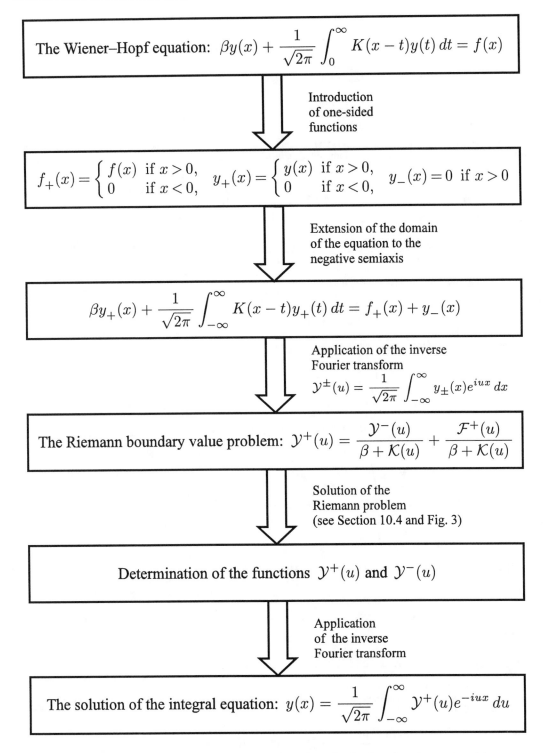

The Wiener–Hopf equation: $\beta y(x) + \dfrac{1}{\sqrt{2\pi}} \displaystyle\int_0^\infty K(x-t)y(t)\,dt = f(x)$

Introduction of one-sided functions

$f_+(x) = \begin{cases} f(x) & \text{if } x > 0, \\ 0 & \text{if } x < 0, \end{cases}$ $y_+(x) = \begin{cases} y(x) & \text{if } x > 0, \\ 0 & \text{if } x < 0, \end{cases}$ $y_-(x) = 0$ if $x > 0$

Extension of the domain of the equation to the negative semiaxis

$\beta y_+(x) + \dfrac{1}{\sqrt{2\pi}} \displaystyle\int_{-\infty}^\infty K(x-t)y_+(t)\,dt = f_+(x) + y_-(x)$

Application of the inverse Fourier transform
$\mathcal{Y}^\pm(u) = \dfrac{1}{\sqrt{2\pi}} \displaystyle\int_{-\infty}^\infty y_\pm(x)e^{iux}\,dx$

The Riemann boundary value problem: $\mathcal{Y}^+(u) = \dfrac{\mathcal{Y}^-(u)}{\beta + \mathcal{K}(u)} + \dfrac{\mathcal{F}^+(u)}{\beta + \mathcal{K}(u)}$

Solution of the Riemann problem (see Section 10.4 and Fig. 3)

Determination of the functions $\mathcal{Y}^+(u)$ and $\mathcal{Y}^-(u)$

Application of the inverse Fourier transform

The solution of the integral equation: $y(x) = \dfrac{1}{\sqrt{2\pi}} \displaystyle\int_{-\infty}^\infty \mathcal{Y}^+(u)e^{-iux}\,du$

Fig. 4. Scheme of solving the Wiener–Hopf integral equations. For $\beta = 0$, we have the equation of the first kind, and for $\beta = 1$, we have the equation of the second kind.

where the constants C_1 and C_2 must be defined. Hence,

$$\mathcal{Y}^+(u) = \mathcal{X}^+(u)\left(\frac{(u-i)^2\mathcal{F}^+(u)}{(u-\alpha-i\beta)(u+\alpha-i\beta)} + \frac{C_1}{u-\alpha-i\beta} + \frac{C_2}{u+\alpha-i\beta}\right). \tag{11}$$

For the poles $(\alpha+i\beta)$ and $(-\alpha+i\beta)$ to be deleted, it is necessary and sufficient that

$$C_1 = -\frac{(\alpha+i\beta-i)^2\mathcal{F}^+(\alpha+i\beta)}{2\alpha}, \quad C_2 = -\frac{(-\alpha+i\beta-i)^2\mathcal{F}^+(-\alpha+i\beta)}{-2\alpha}. \tag{12}$$

Since the problem is more or less cumbersome, we pass from the transform (11) to the corresponding original function in two stages. We first find the inverse transform of the summand

$$\mathcal{Y}_1(u) = \mathcal{X}^+(u)\frac{(u-i)^2\mathcal{F}^+(u)}{(u-\alpha-i\beta)(u+\alpha-i\beta)} = \frac{1}{1+\mathcal{K}(u)}\mathcal{F}^+(u) = \mathcal{F}^+(u) + \mathcal{R}(u)\mathcal{F}^+(u).$$

Here

$$\mathcal{R}(u) = -\frac{2u^2(a-b)+2a+2b}{[u^2-(\alpha+i\beta)^2][u^2-(\alpha-i\beta)^2]} = \frac{\mu}{u^2-(\alpha+i\beta)^2} + \frac{\bar{\mu}}{u^2-(\alpha-i\beta)^2}, \quad \mu = i\frac{(\alpha+i\beta)^2(a-b)+a+b}{2\alpha\beta}.$$

Let us find the inverse transform of the first fraction:

$$\mathbf{F}^{-1}\left\{\frac{\mu}{u^2-(\alpha+i\beta)^2}\right\} = \sqrt{\frac{\pi}{2}}\frac{\mu}{\beta-i\alpha}e^{-(\beta-i\alpha)|x|}.$$

The inverse transform of the second fraction can be found in the form

$$\mathbf{F}^{-1}\left\{\frac{\bar{\mu}}{u^2-(\alpha-i\beta)^2}\right\} = \sqrt{\frac{\pi}{2}}\frac{\bar{\mu}}{\beta+i\alpha}e^{-(\beta+i\alpha)|x|}. \tag{13}$$

Thus,

$$R(x) = \sqrt{\frac{\pi}{2}}\,\rho\left(e^{i\theta+i\alpha|x|} + e^{-i\theta-i\alpha|x|}\right)e^{-\beta|x|} = \sqrt{2\pi}\,\rho e^{-\beta|x|}\cos(\theta+\alpha|x|)$$

and

$$y_1(x) = f(x) + \rho\int_0^\infty e^{-\beta|x-t|}\cos(\theta+\alpha|x-t|)f(t)\,dt, \quad x>0, \quad \rho e^{i\theta} = \frac{\mu}{\beta-i\alpha}. \tag{14}$$

Note that, as a by-product, we have found the resolvent $R(x-t)$ of the following integral equation on the entire axis:

$$y_0(x) + \int_{-\infty}^\infty (a+b|x-t|)e^{-|x-t|}y_0(t)\,dt = f_0(x), \quad -\infty < x < \infty.$$

Now consider the remaining part of the transform (11):

$$\mathcal{Y}_2(u) = \mathcal{X}^+(u)\left(\frac{C_1}{u-\alpha-i\beta} + \frac{C_2}{u+\alpha-i\beta}\right).$$

We can calculate the integrals

$$\mathbf{F}^{-1}\{\mathcal{Y}_2(u)\} = \frac{C_1}{\sqrt{2\pi}}\int_{-\infty}^\infty\frac{(u+i)^2e^{-iux}\,du}{(u+i\beta-\alpha)(u+i\beta+\alpha)(u-\alpha-i\beta)} + \frac{C_2}{\sqrt{2\pi}}\int_{-\infty}^\infty\frac{(u+i)^2e^{-iux}\,du}{(u+i\beta-\alpha)(u+i\beta+\alpha)(u+\alpha-i\beta)}$$

by means of the residue theory (see Subsections 7.1-4 and 7.1-5) and substitute the values (12) into the constants C_1 and C_2. For $x>0$, we obtain

$$y_2(x) = \frac{[\alpha+(\beta-1)^2]^2}{4\alpha^2\beta}\int_0^\infty e^{-\beta(x+t)}\cos[\alpha(x-t)]f(t)\,dt$$

$$+ \frac{\rho_*}{4\alpha^2}\int_0^\infty e^{-\beta(x+t)}\cos[\psi+\alpha(x+t)]f(t)\,dt, \quad \rho_*e^{i\psi} = \frac{(\beta-1-i\alpha)^4}{8\alpha^2(\beta-i\alpha)}. \tag{15}$$

Since $\mathcal{Y}^+(u) = \mathcal{Y}_1(u) + \mathcal{Y}_2(u)$, it follows that the desired solution is the sum of the functions (14) and (15).

11.9-2. An Integral Equation of the Second Kind With Two Kernels

Consider an integral equation of convolution type of the second kind with two kernels of the form

$$y(x) + \frac{1}{\sqrt{2\pi}} \int_0^\infty K_1(x-t)y(t)\,dt + \frac{1}{\sqrt{2\pi}} \int_{-\infty}^0 K_2(x-t)y(t)\,dt = f(x), \qquad -\infty < x < \infty. \quad (16)$$

Note that each of the kernels $K_1(x)$ and $K_2(x)$ is defined on the entire real axis. On representing the desired function as the difference of one-sided functions,

$$y(x) = y_+(x) - y_-(x), \quad (17)$$

we rewrite the equation in the form

$$y_+(x) + \frac{1}{\sqrt{2\pi}} \int_{-\infty}^\infty K_1(x-t)y_+(t)\,dt - y_-(x) - \frac{1}{\sqrt{2\pi}} \int_{-\infty}^\infty K_2(x-t)y_-(t)\,dt = f(x). \quad (18)$$

Applying the Fourier integral transform (see Subsection 7.4-3), we obtain

$$[1 + \mathcal{K}_1(u)]\mathcal{Y}^+(u) - [1 + \mathcal{K}_2(u)]\mathcal{Y}^-(u) = \mathcal{F}(u). \quad (19)$$

This implies the relation

$$\mathcal{Y}^+(u) = \frac{1 + \mathcal{K}_2(u)}{1 + \mathcal{K}_1(u)}\mathcal{Y}^-(u) + \frac{\mathcal{F}(u)}{1 + \mathcal{K}_1(u)}. \quad (20)$$

Here $\mathcal{K}_1(u)$, $\mathcal{K}_2(u)$, and $\mathcal{F}(u)$ stand for the Fourier integrals of known functions. The unknown transforms $\mathcal{Y}^+(u)$ and $\mathcal{Y}^-(u)$ are the boundary values of functions that are analytic on the upper and lower half-planes, respectively. Thus, we have obtained a Riemann boundary value problem.

$1°$. Assume that the normality conditions are satisfied, i.e.,

$$1 + \mathcal{K}_1(u) \neq 0, \quad 1 + \mathcal{K}_2(u) \neq 0,$$

then we can rewrite the Riemann problem in the usual form (see Subsection 10.4-4):

$$\mathcal{Y}^+(u) = \mathcal{D}(u)\mathcal{Y}^-(u) + \mathcal{H}(u), \qquad -\infty < u < \infty, \quad (21)$$

where

$$\mathcal{D}(u) = \frac{1 + \mathcal{K}_2(u)}{1 + \mathcal{K}_1(u)}, \quad \mathcal{H}(u) = \frac{\mathcal{F}(u)}{1 + \mathcal{K}_1(u)}. \quad (22)$$

The Riemann problem (21), (22) is equivalent to Eq. (16): these problems are solvable or unsolvable simultaneously, and have the same number of arbitrary constants in their general solutions.

If the index

$$\nu = \mathrm{Ind}\, \frac{1 + \mathcal{K}_2(u)}{1 + \mathcal{K}_1(u)} \quad (23)$$

is positive, then the homogeneous equation (16) ($f(x) \equiv 0$) has precisely ν linearly independent solutions, and the nonhomogeneous equation is unconditionally solvable; moreover, the solution of this equation depends on ν arbitrary complex constants.

In the case $\nu \leq 0$, the homogeneous equation has no nonzero solutions. The nonhomogeneous equation is unconditionally solvable for $\nu = 0$, and the solution is unique. For the case in which the index ν is negative, the conditions

$$\int_{-\infty}^\infty \frac{\mathcal{F}(u)\,du}{\mathcal{X}^+(u)[1 + \mathcal{K}_1(u)](u+i)^k} = 0, \qquad k = 1, 2, \ldots, -\nu, \quad (24)$$

are necessary and sufficient for the solvability of the nonhomogeneous equation.

In all cases for which the solution of Eq. (16) exists, this solution can be found by the formula

$$y(x) = \frac{1}{\sqrt{2\pi}} \int_{-\infty}^\infty [\mathcal{Y}^+(u) - \mathcal{Y}^-(u)]e^{-iux}\,du, \qquad -\infty < x < \infty, \quad (25)$$

where $\mathcal{Y}^+(u)$, $\mathcal{Y}^-(u)$ is the solution of the Riemann problem (21), (22) constructed with respect to the scheme of Subsection 10.4-4 (see Fig. 3).

Thus, the solution of Eq. (16) is equivalent to the solution of a Riemann boundary value problem and is reduced to the calculation of finitely many Fourier integrals.

2°. Now let us study the exceptional case of an integral equation of the form (16). Assume that the functions $1 + \mathcal{K}_1(u)$ and $1 + \mathcal{K}_2(u)$ can have zeros, and these zeros can be both different and coinciding points of the contour. Let us write out the expansion of these functions on selecting the coinciding zeros:

$$1 + \mathcal{K}_1(u) = \prod_{j=1}^{s} (u - b_j)^{\beta_j} \prod_{k=1}^{p} (u - d_k)^{\gamma_k} \mathcal{K}_{11}(u),$$

$$1 + \mathcal{K}_2(u) = \prod_{i=1}^{r} (u - a_i)^{\alpha_i} \prod_{k=1}^{p} (u - d_k)^{\gamma_k} \mathcal{K}_{12}(u), \qquad \sum_{k=1}^{p} \gamma_k = l. \tag{26}$$

Here $a_i \neq b_j$, but it is possible that some points d_k $(k = 1, \ldots, p)$ coincide with either a_i or b_j. This corresponds to the case in which the functions $1 + \mathcal{K}_1(u)$ and $1 + \mathcal{K}_2(u)$ have a common zero of different multiplicity. We do not select these points especially because their presence does not affect the solvability conditions and the number of solutions of the problem.

It follows from Eq. (19) and from the condition that a solution must be finite on the contour that, for the solvability of the problem, and all the more for the solvability of Eq. (16), it is necessary that the function $\mathcal{F}(u)$ have zero of order γ_k at any point d_k, i.e., $\mathcal{F}(u)$ must have the form

$$\mathcal{F}(u) = \prod_{k=1}^{p} (u - d_k)^{\gamma_k} \mathcal{F}_1(u).$$

To this end, the following $\gamma_1 + \cdots + \gamma_p = l$ conditions must be satisfied:

$$\mathcal{F}_u^{(j_k)}(d_k) = 0, \qquad j_k = 0, 1, \ldots, \gamma_k - 1, \tag{27}$$

or, which is the same,

$$\int_{-\infty}^{\infty} f(x) x^{j_k} e^{id_k x}\, dx = 0. \tag{28}$$

Since the functions $\mathcal{K}_1(u)$ and $\mathcal{K}_2(u)$ vanish at infinity, it follows that the point at infinity is a regular point of $\mathcal{D}(u)$.

Assume that conditions (28) are satisfied. In this case the Riemann boundary value problem (20) can be rewritten in the form (see Subsections 10.4-6 and 10.4-7)

$$\mathcal{Y}^+(u) = \frac{\prod_{i=1}^{r} (u - a_i)^{\alpha_i} \mathcal{R}_+(u) \mathcal{R}_-(u)}{\prod_{j=1}^{s} (u - b_j)^{\beta_j} \mathcal{Q}_+(u) \mathcal{Q}_-(u)} \mathcal{D}_2(u) \mathcal{Y}^-(u) + \frac{\mathcal{H}_1(u)}{\prod_{j=1}^{s} (u - b_j)^{\beta_j}}. \tag{29}$$

On finding its general solution in the exceptional case under consideration, we obtain the general solution of the original equation by means of formula (25).

Let us state the conclusions on the solvability conditions and on the number of solutions of Eq. (16). For the solvability of Eq. (16), it is necessary that the Fourier transform of the right-hand side of the equation satisfies l conditions of the form (27). If these conditions are satisfied, then, for $\nu - n > 0$, problem (20) and the integral equation (16) have exactly $\nu - n$ linearly independent solutions. For $\nu - n \leq 0$, we must take the polynomial $\mathcal{P}_{\nu-n-1}(z)$ to be identically zero, and, for the case in which $\nu - n < 0$, the right-hand side must satisfy another $n - \nu$ conditions. If the latter conditions are satisfied, then the integral equation has a unique solution.

Example. Consider Eq. (16) for which

$$K_1(x) = \begin{cases} -(1+\alpha)\sqrt{2\pi}\,e^{-x} & \text{for } x > 0, \\ 0 & \text{for } x < 0, \end{cases} \quad K_2(x) = \begin{cases} -(1+\beta)\sqrt{2\pi}\,e^{-x} & \text{for } x > 0, \\ 0 & \text{for } x < 0, \end{cases} \quad f(x) = \begin{cases} 0 & \text{for } x > 0, \\ -\sqrt{2\pi}\,e^x & \text{for } x < 0, \end{cases}$$

where α and β are real constants. In this case, $K_1(x-t) = 0$ for $x < t$ and $K_2(x-t) = 0$ for $x < t$. Hence, the equation under consideration has the form

$$y(x) - (1+\alpha)\int_0^x e^{-(x-t)}y(t)\,dt - (1+\beta)\int_{-\infty}^0 e^{-(x-t)}y(t)\,dt = 0, \qquad x > 0,$$

$$y(x) - (1+\beta)\int_{-\infty}^x e^{-(x-t)}y(t)\,dt = -\sqrt{2\pi}\,e^x, \qquad x < 0.$$

Let us calculate the Fourier integrals

$$\mathcal{K}_1(u) = -(1+\alpha)\int_0^\infty e^{-x}e^{iux}\,dx = -\frac{i(1+\alpha)}{u+i}, \qquad \mathcal{K}_2(u) = -\frac{i(1+\beta)}{u+i}, \qquad \mathcal{F}(u) = \frac{i}{u-i}, \qquad \mathcal{D}(u) = \frac{u-i\beta}{u-i\alpha}.$$

The boundary condition can be rewritten in the form

$$\mathcal{Y}^+(u) = \frac{u-i\beta}{u-i\alpha}\mathcal{Y}^-(u) + \frac{i(u+i)}{(u-i)(u-i\alpha)}. \tag{30}$$

The solution of the Riemann problem depends on the signs of α and β.

1°. Let $\alpha > 0$ and $\beta > 0$. In this case we have $\nu = \text{Ind}\,\mathcal{D}(u) = 0$. The left-hand side and the right-hand side of the boundary condition contain functions that have analytic continuations to the upper and the lower half-plane, respectively. On applying the theorem on the analytic continuation directly and the generalized Liouville theorem (Subsection 10.4-3), we see that

$$\mathcal{Y}^+(z) = 0, \qquad \frac{z-i\beta}{z-i\alpha}\mathcal{Y}^-(z) + \frac{i(z+i)}{(z-i)(z-i\alpha)} = 0.$$

Hence,

$$y_+(x) = 0, \qquad y(x) = -y_-(x) = \frac{1}{\sqrt{2\pi}}\int_{-\infty}^\infty \frac{i(u+i)}{(u-i)(u-i\beta)}e^{-iux}\,du.$$

On calculating the last integral, under the assumption that $\beta \neq 1$, by the Cauchy residue theorem (see Subsections 7.1-4 and 7.1-5) we obtain

$$y(x) = \begin{cases} 0 & \text{for } x > 0, \\ -\dfrac{\sqrt{2\pi}}{1-\beta}[2e^x - (1+\beta)e^{\beta x}] & \text{for } x < 0. \end{cases}$$

In the case $\beta = 1$, we have

$$y(x) = \begin{cases} 0 & \text{for } x > 0, \\ -\sqrt{2\pi}\,e^x(1+2x) & \text{for } x < 0. \end{cases}$$

2°. Let $\alpha < 0$ and $\beta < 0$. Here we again have $\nu = 0$, $\mathcal{X}^+(z) = (z-i\beta)(z-i\alpha)^{-1}$, and $\mathcal{X}^-(z) = 1$. On grouping the terms containing the boundary values of functions that are analytic in each of the half-planes and then applying the analytic continuation theorem and the generalized Liouville theorem (Subsection 10.4-3), we see that

$$\frac{\mathcal{Y}^+(z)}{\mathcal{X}^+(z)} + \frac{\beta+1}{i(\beta-1)}\frac{1}{z-i\beta} = \frac{\mathcal{Y}^-(z)}{\mathcal{X}^-(z)} + \frac{2}{i(\beta-1)}\frac{1}{z-i} = 0.$$

Hence,

$$\mathcal{Y}^+(z) = \frac{\beta+1}{\beta-1}\frac{i}{z-i\alpha}, \qquad \mathcal{Y}^-(z) = \frac{2i}{\beta-1}\frac{1}{z-i},$$

$$y(x) = \frac{1}{\sqrt{2\pi}}\int_{-\infty}^\infty [\mathcal{Y}^+(u) - \mathcal{Y}^-(u)]\,e^{-iux}\,du = \begin{cases} \sqrt{2\pi}\,\dfrac{\beta+1}{\beta-1}e^{\alpha x} & \text{for } x > 0, \\ \dfrac{2\sqrt{2\pi}}{\beta-1}e^x & \text{for } x < 0. \end{cases}$$

3°. Let $\alpha < 0$ and $\beta > 0$. In this case we have $\nu = 1$. Let us rewrite the boundary condition (30) in the form

$$\mathcal{Y}^+(u) + \frac{i(1+\alpha)}{1-\alpha}\frac{1}{u-i\alpha} = \frac{u-i\beta}{u-i\alpha}\mathcal{Y}^-(u) - \frac{2i}{1-\alpha}\frac{1}{u-i}.$$

On applying the analytic continuation theorem and the generalized Liouville theorem (Subsection 10.4-3), we see that

$$\mathcal{Y}^+(z) + \frac{i(1+\alpha)}{1-\alpha}\frac{1}{z-i\alpha} = \frac{z-i\beta}{z-i\alpha}\mathcal{Y}^-(z) - \frac{2i}{1-\alpha}\frac{1}{z-i} = \frac{C}{z-i\alpha}.$$

Therefore,

$$\mathcal{Y}^+(z) = \left(C - i\frac{1+\alpha}{1-\alpha}\right)\frac{1}{z - i\alpha}, \qquad \mathcal{Y}^-(z) = \frac{C}{z - i\beta} - \frac{2i}{1-\alpha}\frac{z - i\alpha}{(z-i)(z-i\beta)},$$

where C is an arbitrary constant. Now, by means of the Fourier inversion formula, we obtain the general solution of the integral equation in the form

$$y(x) = \begin{cases} -\sqrt{2\pi}\left(iC + \dfrac{1+\alpha}{1-\alpha}\right)e^{\alpha x} & \text{for } x > 0, \\[2mm] -\sqrt{2\pi}\left[iC + \dfrac{2(\alpha-\beta)}{(1-\alpha)(1-\beta)}\right]e^{\beta x} - \dfrac{2\sqrt{2\pi}}{1-\beta}e^x & \text{for } x < 0. \end{cases}$$

4°. Let $\alpha > 0$ and $\beta < 0$. In this case we have $\nu = -1$. By the Liouville theorem (see Subsection 10.4-3), we obtain

$$\mathcal{Y}^+(z) = \frac{z - i\beta}{z - i\alpha}\mathcal{Y}^-(z) + \frac{i(z+i)}{(z-i)(z-i\alpha)} = 0,$$

and hence

$$\mathcal{Y}^+(z) = 0, \qquad \mathcal{Y}^-(z) = -\frac{i(z+i)}{(z-i)(z-i\beta)}.$$

It can be seen from the expression for $\mathcal{Y}^-(z)$ that the singularity of the function $\mathcal{Y}^-(z)$ at the point $i\beta$ disappears if we set $\beta = -1$. The last condition is exactly the solvability condition of the Riemann problem. In this case we have the unique solution

$$y(x) = \frac{1}{\sqrt{2\pi}}\int_{-\infty}^{\infty}\frac{i}{u-i}e^{-iux}\,du = \begin{cases} 0 & \text{for } x > 0, \\ -\sqrt{2\pi}\,e^x & \text{for } x < 0. \end{cases}$$

Remark 1. Some equations whose kernels contain not the difference but certain other combinations of arguments, namely, the product or, more frequently, the ratio, can be reduced to equations considered in Subsection 11.9-2. For instance, the equation

$$Y(\xi) + \int_0^1 \frac{1}{\tau}N_1\left(\frac{\xi}{\tau}\right)Y(\tau)\,d\tau + \int_1^\infty \frac{1}{\tau}N_2\left(\frac{\xi}{\tau}\right)Y(\tau)\,d\tau = g(\xi), \qquad \xi > 0, \tag{31}$$

becomes a usual equation with two kernels after the following changes of the functions and their arguments: $\xi = e^x$, $\tau = e^t$, $N_1(\xi) = K_1(x)$, $N_2(\xi) = K_2(x)$, $g(\xi) = f(x)$, and $Y(\xi) = y(x)$.

11.9-3. Equations of Convolution Type With Variable Integration Limit

1°. Consider the Volterra integral equation of the second kind

$$y(x) + \frac{1}{\sqrt{2\pi}}\int_0^x K(x-t)y(t)\,dt = f(x), \qquad 0 \leq x < T, \tag{32}$$

where the interval $[0, T)$ can be either finite or infinite. In contrast with Eq. (1), where the kernel is defined on the entire real axis, here the kernel is defined on the positive semiaxis.

Equation (32) can be regarded as a special case of the one-sided equation (1) of Subsection 11.9-1. To see this, we can rewrite Eq. (32) in the form

$$y(x) + \frac{1}{\sqrt{2\pi}}\int_0^\infty K_+(x-t)y(t)\,dt = f(x), \qquad 0 < x < \infty,$$

which can be reduced to the following boundary value problem:

$$\mathcal{Y}^+(u) = \frac{\mathcal{Y}^-(u)}{1 + \mathcal{K}^+(u)} + \frac{\mathcal{F}^+(u)}{1 + \mathcal{K}^+(u)}.$$

Here the coefficient $[1 + \mathcal{K}^+(u)]^{-1}$ of the problem is a function that has an analytic continuation to the upper half-plane, possibly except for finitely many poles that are zeros of the function $1 + \mathcal{K}^+(z)$

(we assume that $1 + \mathcal{K}^+(z) \neq 0$ on the real axis). Therefore, the index ν of the problem is always nonpositive, $\nu \leq 0$. On rewriting the problem in the form $[1 + \mathcal{K}^+(u)]\mathcal{Y}^+(u) = \mathcal{Y}^-(u) + \mathcal{F}^+(u)$, we see that $\mathcal{Y}^-(u) \equiv 0$, which implies

$$\mathcal{Y}^+(u) = \frac{\mathcal{F}^+(u)}{1 + \mathcal{K}^+(u)}. \tag{33}$$

Consider the following cases.

1.1. The function $1 + \mathcal{K}^+(z)$ has no zeros on the upper half-plane (this means that $\nu = 0$). In this case, Eq. (32) has a unique solution for an arbitrary right-hand side $f(x)$, and this solution can be expressed via the resolvent:

$$y(x) = f(x) + \frac{1}{\sqrt{2\pi}} \int_0^x R(x - t) f(t)\, dt, \qquad x > 0, \tag{34}$$

where

$$R(x) = -\frac{1}{\sqrt{2\pi}} \int_{-\infty}^{\infty} \frac{\mathcal{K}^+(u)}{1 + \mathcal{K}^+(u)} e^{-iux}\, du.$$

1.2. The function $1 + \mathcal{K}^+(z)$ has zeros at the points $z = a_1, \ldots, a_m$ of the upper half-plane (in this case we have $\nu < 0$, and ν is equal to the minus total order of the zeros). The following two possibilities can occur.

(a) The function $\mathcal{F}^+(z)$ vanishes at the points $a_1, \ldots, a_m$, and the orders of these zeros are not less than the orders of the corresponding zeros of the function $1 + \mathcal{K}^+(z)$. In this case, the function $\mathcal{F}^+(z)[1 + \mathcal{K}^+(z)]^{-1}$ has no poles again, and thus the equation has the unique solution (34).

The assumption $d^k \mathcal{F}^+(a_j)/dz^k = 0$ on the zeros of the function $\mathcal{F}^+(z)$ is equivalent to the conditions

$$\int_{-\infty}^{\infty} f(t) e^{-ia_j t} t^k\, dt = 0, \qquad k = 0, \ldots, \eta_j - 1, \quad j = 1, \ldots, m, \tag{35}$$

where η_j is the multiplicity of the zero of the function $1 + \mathcal{K}^+(z)$ at the point a_j. In this case, conditions (35) are imposed directly on the right-hand side of the equation.

(b) The function $\mathcal{F}^+(z)$ does not vanish at the points $a_1, \ldots, a_m$ (or vanishes with less multiplicity than $1 + \mathcal{K}^+(z)$). In this case, the function $\mathcal{F}^+(z)[1 + \mathcal{K}^+(z)]^{-1}$ has poles, and therefore the function (33) does not belong to the class under consideration. Equation (32) has no solutions in the chosen class of functions. In this case, conditions (35) fail.

The last result does not contradict the well-known fact that a Volterra equation always has a unique solution. Equation (32) belongs to the class of Volterra type equations, and therefore is also solvable in case (b), but in a broader space of functions with exponential growth.

2°. Another simple special case of Eq. (1) in Subsection 11.9-1 is the following equation with variable lower limit:

$$y(x) + \frac{1}{\sqrt{2\pi}} \int_x^{\infty} K(x - t) y(t)\, dt = f(x), \qquad 0 < x < \infty. \tag{36}$$

This corresponds to the case in which the function $K(x)$ in Eq. (1) is left one-sided: $K(x) = K_-(x)$. Under the assumption $1 + \mathcal{K}^-(u) \neq 0$, the Riemann problem becomes

$$\mathcal{Y}^+(u) = \frac{\mathcal{Y}^-(u)}{1 + \mathcal{K}^-(u)} + \frac{\mathcal{F}^+(u)}{1 + \mathcal{K}^-(u)}. \tag{37}$$

2.1. The function $1 + \mathcal{K}^-(z)$ has no zeros on the lower half-plane. This means that the inverse transform of the function $\mathcal{Y}^-(u)[1 + \mathcal{K}^-(u)]^{-1}$ is left one-sided, and such a function does not influence the relation between the inverse transforms of (37) for $x > 0$. Thus, if we introduce the function

$$\mathcal{R}^-(u) = -\frac{\mathcal{K}^-(u)}{1 + \mathcal{K}^-(u)}$$

(for convenience of the final formula), then by applying the Fourier inversion formula to Eq. (37) and by setting $x > 0$ we obtain the unique solution to Eq. (36),

$$y(x) = f(x) + \frac{1}{\sqrt{2\pi}} \int_x^\infty R_-(x-t)f(t)\,dt, \qquad x > 0.$$

2.2. The function $1 + \mathcal{K}^-(z)$ has zeros in the lower half-plane. Since this function is nonzero both on the entire real axis and at infinity, it follows that the number of zeros is finite. The Riemann problem (37) has a positive index which is just equal to the number of zeros in the lower half-plane (the zeros are counted according to their multiplicities):

$$\nu = \mathrm{Ind}\,\frac{1}{1+\mathcal{K}^-(u)} = -\mathrm{Ind}[1 + \mathcal{K}^-(u)] = \eta_1 + \cdots + \eta_n > 0.$$

Here η_k are the multiplicities of the zeros z_k of the function $1 + \mathcal{K}^-(z)$, $k = 1, \ldots, n$.
Let

$$\frac{C_{1k}}{z - z_k} + \frac{C_{2k}}{(z - z_k)^2} + \cdots + \frac{C_{\eta_k k}}{(z - z_k)^{\eta_k}}$$

be the principal part of the Laurent series expansion of the function $\mathcal{Y}^-(z)[1 + \mathcal{K}^-(z)]^{-1}$ in powers of $(z - z_k)$, $k = 1, \ldots, n$. In this case, Eq. (37) becomes

$$\mathcal{Y}^+(u) = \frac{\mathcal{F}^+(u)}{1 + \mathcal{K}^-(u)} + \sum_{k=1}^n \sum_{j=1}^{\eta_k} \frac{C_{jk}}{(z - z_k)^j} + \cdots, \tag{38}$$

where the dots denote a function whose inverse transform vanishes for $x > 0$. Under the passage to the inverse transforms in Eq. (38), for $x > 0$ we obtain

$$y(x) = f(x) + \frac{1}{\sqrt{2\pi}} \int_x^\infty R_-(x-t)f(t)\,dt + \sum_{k=1}^n P_k(x)e^{-iz_k x}, \qquad x > 0. \tag{39}$$

Here the $P_k(x)$ are polynomials of degree $\eta_k - 1$. We can verify that the function (39) is a solution of Eq. (36) for arbitrary coefficients of the polynomials. Since the number of linearly independent solutions of the homogeneous equation (36) is equal to the index, it follows that the above solution (39) is the general solution of the nonhomogeneous equation.

11.9-4. Dual Equation of Convolution Type of the Second Kind

Consider the dual integral equation of the second kind

$$
\begin{aligned}
y(x) + \frac{1}{\sqrt{2\pi}} \int_{-\infty}^\infty K_1(x-t)y(t)\,dt &= f(x), & 0 < x < \infty, \\
y(x) + \frac{1}{\sqrt{2\pi}} \int_{-\infty}^\infty K_2(x-t)y(t)\,dt &= f(x), & -\infty < x < 0,
\end{aligned}
\tag{40}
$$

in which the function $y(x)$ is to be found.

In order to apply the Fourier transform technique (see Subsections 7.4-3, 10.4-1, and 10.4-2), we extend the domain of both conditions in Eq. (40) by formally rewriting them for all real values of x. This can be achieved by introducing new unknown functions into the right-hand sides. These functions must be chosen so that the conditions given on the semiaxis are not violated. Hence, the first condition in (40) must be complemented by a summand that vanishes on the positive semiaxis and the second by a summand that vanishes on the negative semiaxis. Thus, the dual equation can be written in the form

$$y(x) + \frac{1}{\sqrt{2\pi}} \int_{-\infty}^{\infty} K_1(x-t)y(t)\,dt = f(x) + \xi_-(x),$$
$$\qquad\qquad -\infty < x < \infty, \qquad (41)$$
$$y(x) + \frac{1}{\sqrt{2\pi}} \int_{-\infty}^{\infty} K_2(x-t)y(t)\,dt = f(x) + \xi_+(x),$$

where the $\xi_{\pm}(x)$ are some right and left one-sided functions so far unknown.

On applying the Fourier integral transform, we arrive at the relations

$$[1 + \mathcal{K}_1(u)]\mathcal{Y}(u) = \mathcal{F}(u) + \Xi^-(u), \quad [1 + \mathcal{K}_2(u)]\mathcal{Y}(u) = \mathcal{F}(u) + \Xi^+(u). \qquad (42)$$

Here the three functions $\mathcal{Y}(u)$, $\Xi^+(u)$, and $\Xi^-(u)$ are unknown.

Now on the basis of (42) we can find

$$\mathcal{Y}(u) = \frac{\mathcal{F}(u) + \Xi^-(u)}{1 + \mathcal{K}_1(u)} = \frac{\mathcal{F}(u) + \Xi^+(u)}{1 + \mathcal{K}_2(u)} \qquad (43)$$

and eliminate the function $\mathcal{Y}(u)$ from relations (42) by applying formula (43). We obtain the Riemann boundary value problem in the form

$$\Xi^+(u) = \frac{1 + \mathcal{K}_2(u)}{1 + \mathcal{K}_1(u)}\Xi^-(u) + \frac{\mathcal{K}_2(u) - \mathcal{K}_1(u)}{1 + \mathcal{K}_1(u)}\mathcal{F}(u), \qquad -\infty < u < \infty. \qquad (44)$$

1°. Assume that the normality conditions are satisfied, i.e.,

$$1 + \mathcal{K}_1(u) \neq 0, \quad 1 + \mathcal{K}_2(u) \neq 0;$$

then we can rewrite the Riemann problem (44) in the usual form (see Subsection 10.4-4)

$$\Xi^+(u) = \mathcal{D}(u)\Xi^-(u) + \mathcal{H}(u), \qquad -\infty < u < \infty, \qquad (45)$$

where

$$\mathcal{D}(u) = \frac{1 + \mathcal{K}_2(u)}{1 + \mathcal{K}_1(u)}, \quad \mathcal{H}(u) = \frac{\mathcal{K}_2(u) - \mathcal{K}_1(u)}{1 + \mathcal{K}_1(u)}\mathcal{F}(u). \qquad (46)$$

The Riemann problem (45), (46) is equivalent to Eq. (40); in particular, they are solvable and unsolvable simultaneously and have the same number of arbitrary constants in the general solutions.

If the index

$$\nu = \operatorname{Ind}\frac{1 + \mathcal{K}_2(u)}{1 + \mathcal{K}_1(u)} \qquad (47)$$

is positive, then the homogeneous equation (40) ($f(x) \equiv 0$) has exactly ν linearly independent solutions, and the nonhomogeneous equation is unconditionally solvable and the solution depends on ν arbitrary complex constants.

For the case $\nu \leq 0$, the homogeneous equation has no nonzero solutions. For $\nu = 0$, the nonhomogeneous equation is unconditionally solvable, and a solution is unique. If the index ν is negative, then the conditions

$$\int_{-\infty}^{\infty} \frac{\mathcal{K}_2(u) - \mathcal{K}_1(u)}{\mathcal{X}^+(u)[1 + \mathcal{K}_1(u)]}\mathcal{F}(u)\frac{du}{(u+i)^k} = 0, \qquad k = 1, 2, \ldots, -\nu \qquad (48)$$

are necessary and sufficient for the solvability of the nonhomogeneous equation.

For all cases in which a solution of Eq. (40) exists, it can be found by the formula

$$y(x) = \frac{1}{\sqrt{2\pi}} \int_{-\infty}^{\infty} \frac{\mathcal{F}(u) + \Xi^-(u)}{1 + \mathcal{K}_1(u)}e^{-iux}\,du = \frac{1}{\sqrt{2\pi}} \int_{-\infty}^{\infty} \frac{\mathcal{F}(u) + \Xi^+(u)}{1 + \mathcal{K}_2(u)}e^{-iux}\,du, \qquad (49)$$

where $\Xi^+(u)$, $\Xi^-(u)$ is a solution of the Riemann problem (45), (46) that is constructed by the scheme of Subsection 10.4-4 (see Fig. 3).

$2°$. Let us investigate the exceptional case of the integral equation (40). Assume that the functions $1 + \mathcal{K}_1(u)$ and $1 + \mathcal{K}_2(u)$ can have zeros that can be either different or coinciding points of the contour. Take the expansions of these functions on selecting the coinciding zeros in the form of (26) and further repeat the reasoning performed for the equations of convolution type of the second kind with two kernels. After finding the general solution of the Riemann boundary value problem (44) in this exceptional case (see Subsection 10.4-7), we obtain the general solution of the original equation (40) by formula (49).

The conclusions on the solvability conditions and on the number of solutions of Eq. (40) are similar to those made above for the equations with two kernels in Subsection 11.9-2.

Remark 2. Equations treated in Section 11.9 are sometimes called *characteristic equations of convolution type.*

⊙ Reference for Section 11.9: F. D. Gakhov and Yu. I. Cherskii (1978).

11.10. The Wiener–Hopf Method

11.10-1. Some Remarks

Suppose that the Fourier transform of the function $y(x)$ exists (see Subsection 7.4-3):

$$\mathcal{Y}(z) = \frac{1}{\sqrt{2\pi}} \int_{-\infty}^{\infty} y(x) e^{izx}\,dx. \tag{1}$$

Assume that the parameter z that enters the transform (1) can take complex values as well. Let us study the properties of the function $\mathcal{Y}(z)$ regarded as a function of the complex variable z. To this end, we represent the function $y(x)$ in the form*

$$y(x) = y^+(x) + y^-(x), \tag{2}$$

where the functions $y^+(x)$ and $y^-(x)$ are given by the relations

$$y^+(x) = \begin{cases} y(x) & \text{for } x > 0, \\ 0 & \text{for } x < 0, \end{cases} \qquad y^-(x) = \begin{cases} 0 & \text{for } x > 0, \\ y(x) & \text{for } x < 0. \end{cases} \tag{3}$$

In this case the transform $\mathcal{Y}(z)$ of the function $y(x)$ is clearly equal to the sum of the transforms $\mathcal{Y}_+(z)$ and $\mathcal{Y}_-(z)$ of the functions $y^+(x)$ and $y^-(x)$, respectively. Let us clarify the analytic properties of the function $\mathcal{Y}(z)$ by establishing the analytic properties of the functions $\mathcal{Y}_+(z)$ and $\mathcal{Y}_-(z)$. Consider the function $y^+(x)$ given by relations (3). Its transform is equal to

$$\mathcal{Y}_+(z) = \frac{1}{\sqrt{2\pi}} \int_0^{\infty} y^+(x) e^{izx}\,dx. \tag{4}$$

It can be shown that if the function $y^+(x)$ satisfies the condition

$$|y^+(x)| < M e^{v_- x} \quad \text{as} \quad x \to \infty, \tag{5}$$

where M is a constant, then the function $\mathcal{Y}_+(z)$ given by formula (4) is an analytic function of the complex variable $z = u + iv$ in the domain $\operatorname{Im} z > v_-$, and in this domain we have $\mathcal{Y}_+(z) \to 0$ as $|z| \to \infty$. We can also show that the functions $y^+(x)$ and $\mathcal{Y}_+(z)$ are related as follows:

$$y^+(x) = \frac{1}{\sqrt{2\pi}} \int_{-\infty+iv}^{\infty+iv} \mathcal{Y}_+(z) e^{-izx}\,dz, \tag{6}$$

---

* Do not confuse the functions $y^\pm(x)$ and $\mathcal{Y}_\pm(x)$ introduced in this section with the functions $y_\pm(x)$ and $\mathcal{Y}^\pm(x)$ introduced in Subsection 10.4-2 and used in solving the Riemann boundary value problem on the real axis.

where the integration is performed over any line $\operatorname{Im} z = v > v_-$ in the complex plane z, which is parallel to the real axis.

For $v_- < 0$ (i.e., for functions $y(x)$ with exponential decay at infinity), the real axis belongs to the domain in which the function $\mathcal{Y}_+(z)$ is analytic, and we can integrate over the real axis in formula (6). However, if the only possible values of v_- are positive (for instance, if the function $y^+(x)$ has nontrivial growth at infinity, which does not exceed the exponential growth with linear exponent), then the analyticity domain of the function $\mathcal{Y}_+(z)$ is strictly above the real axis of the complex plane z (and in this case, the integral (4) can be divergent on the real axis). Similarly, if the function $y^-(x)$ in relations (3) satisfies the condition

$$|y^-(x)| < M e^{v_+ x} \quad \text{as} \quad x \to -\infty, \tag{7}$$

then its transform, i.e., the function

$$\mathcal{Y}_-(z) = \frac{1}{\sqrt{2\pi}} \int_{-\infty}^{0} y^-(x) e^{izx} \, dx, \tag{8}$$

is an analytic function of the complex variable z in the domain $\operatorname{Im} z < v_+$. The function $y^-(x)$ can be expressed via $\mathcal{Y}_-(z)$ by means of the relation

$$y^-(x) = \frac{1}{\sqrt{2\pi}} \int_{-\infty+iv}^{\infty+iv} \mathcal{Y}_-(z) e^{-izx} \, dz, \qquad \operatorname{Im} z = v < v_+. \tag{9}$$

For $v_+ > 0$, the analyticity domain of the function $\mathcal{Y}_-(z)$ contains the real axis.

It is clear that for $v_- < v_+$, the function $\mathcal{Y}(z)$ defined by formula (1) is an analytic function of the complex variable z in the strip $v_- < \operatorname{Im} z < v_+$. In this case, the functions $y(x)$ and $\mathcal{Y}(z)$ are related by the Fourier inversion formula

$$y(x) = \frac{1}{\sqrt{2\pi}} \int_{-\infty+iv}^{\infty+iv} \mathcal{Y}(z) e^{-izx} \, dz, \tag{10}$$

where the integration is performed over an arbitrary line in the complex plane z belonging to the strip $v_- < \operatorname{Im} z < v_+$. In particular, for $v_- < 0$ and $v_+ > 0$, the function $\mathcal{Y}(z)$ is analytic in the strip containing the real axis of the complex plane z.

Example 1. For $\alpha > 0$, the function $K(x) = e^{-\alpha|x|}$ has the transform

$$\mathcal{K}(z) = \frac{1}{\sqrt{2\pi}} \frac{2\alpha}{\alpha^2 + z^2},$$

which is an analytic function of the complex variable z in the strip $-\alpha < \operatorname{Im} z < \alpha$, which contains the real axis.

11.10-2. The Homogeneous Wiener–Hopf Equation of the Second Kind

Consider a homogeneous integral Wiener–Hopf equation of the second kind in the form

$$y(x) = \int_{0}^{\infty} K(x-t)y(t) \, dt, \tag{11}$$

whose solution can obviously be determined up to an arbitrary constant factor only. Here the domain of the function $K(x)$ is the entire real axis. This factor can be found from additional conditions of the problem, for instance, from normalization conditions.

We assume that Eq. (11) defines a function $y(x)$ for all values of the variable x, positive and negative. Let us introduce the functions $y^-(x)$ and $y^+(x)$ by formulas (3). Obviously, we have $y(x) = y^+(x) + y^-(x)$, and Eq. (11) can be rewritten in the form

$$y^+(x) = \int_0^\infty K(x-t)y^+(t)\,dt, \qquad x > 0 \tag{12}$$

$$y^-(x) = \int_0^\infty K(x-t)y^+(t)\,dt, \qquad x < 0. \tag{13}$$

That is, the function $y^+(x)$ can be determined by the solution of the integral equation (12) and the function $y^-(x)$ can be expressed via the functions $y^+(x)$ and $K(x)$ by means of formulas (13). In this case, we have the relation

$$y^+(x) + y^-(x) = \int_{-\infty}^\infty K(x-t)y^+(t)\,dt, \tag{14}$$

which is equivalent to the original equation (11).

Let the function $K(x)$ satisfy the condition

$$\begin{aligned} |K(x)| &< Me^{v_- x} \quad \text{as} \quad x \to \infty, \\ |K(x)| &< Me^{v_+ x} \quad \text{as} \quad x \to -\infty, \end{aligned} \tag{15}$$

where $v_- < 0$ and $v_+ > 0$. In this case, the function

$$\mathcal{K}(z) = \frac{1}{\sqrt{2\pi}} \int_{-\infty}^\infty K(x)e^{izx}\,dx, \tag{16}$$

is analytic in the strip $v_- < \operatorname{Im} z < v_+$.

Let us seek the solution of Eq. (11) satisfying the condition

$$|y^+(x)| < M_1 e^{\mu x} \quad \text{as} \quad x \to \infty, \tag{17}$$

where $\mu < v_+$ (such a solution exists). In this case we can readily verify that the integrals on the right-hand sides in (12) and (13) are convergent, and the function $y^-(x)$ satisfies the estimate

$$|y^-(x)| < M_2 e^{v_+ x} \quad \text{as} \quad x \to -\infty. \tag{18}$$

It follows from conditions (17) and (18) that the transforms $\mathcal{Y}_+(z)$ and $\mathcal{Y}_-(z)$ of the functions $y^+(x)$ and $y^-(x)$ are analytic functions of the complex variable z for $\operatorname{Im} z > \mu$ and $\operatorname{Im} z < v_+$, respectively.

Let us pass to the solution of the integral equation (11) or of Eq. (14), which is equivalent to (11). To this end, we apply the (alternative) Fourier transform. By the convolution theorem (see Subsection 7.4-4), it follows from (14) that

$$\mathcal{Y}_+(z) + \mathcal{Y}_-(z) = \sqrt{2\pi}\,\mathcal{K}(z)\mathcal{Y}_+(z),$$

or

$$\mathcal{W}(z)\mathcal{Y}_+(z) + \mathcal{Y}_-(z) = 0, \tag{19}$$

where

$$\mathcal{W}(z) = 1 - \sqrt{2\pi}\,\mathcal{K}(z) \neq 0. \tag{20}$$

Thus, by means of the Fourier transform, we succeeded in the passage from the original integral equation to an algebraic equation for the transforms. However, in this case Eq. (19) involves two unknown functions. In general, a single algebraic equation cannot uniquely determine two unknown functions. The Wiener–Hopf method makes it possible to solve this problem for a certain class of functions. This method is mainly related to the study of the analyticity domains of the functions that enter the equation and to a special representation of this equation. The main idea of the Wiener–Hopf method is as follows.

Let Eq. (19) be representable in the form

$$\mathcal{W}_+(z)\mathcal{Y}_+(z) = -\mathcal{W}_-(z)\mathcal{Y}_-(z), \tag{21}$$

where the left-hand side is analytic in the upper half-plane $\operatorname{Im} z > \mu$ and the right-hand side is analytic in the lower half-plane $\operatorname{Im} z < v_+$, where $\mu < v_+$, so that there exists a common analyticity strip of these functions: $\mu < \operatorname{Im} z < v_+$. Since the analytic continuation is unique, it follows that there exists a unique entire function of the complex variable that coincides with the left-hand side of (21) in the upper half-plane and with the right-hand side of (21) in the lower half-plane, respectively. If, in addition, the functions that enter Eq. (21) have at most power-law growth with respect to z at infinity, then it follows from the generalized Liouville theorem (see Subsection 10.4-3) that the entire function under consideration is a polynomial. In particular, for the case of a function that is bounded at infinity we obtain

$$\mathcal{W}_+(z)\mathcal{Y}_+(z) = -\mathcal{W}_-(z)\mathcal{Y}_-(z) = \text{const}. \tag{22}$$

These relations uniquely determine the functions $\mathcal{Y}_+(z)$ and $\mathcal{Y}_-(z)$.

Thus, let us apply the above scheme to the solution of Eq. (19). It follows from the above reasoning that the analyticity domains of the functions $\mathcal{Y}_+(z)$, $\mathcal{Y}_-(z)$, and $\mathcal{W}(z) = 1 - \sqrt{2\pi}\,\mathcal{K}(z)$, respectively, are the upper half-plane $\operatorname{Im} z > \mu$, the lower half-plane $\operatorname{Im} z < v_+$, and the strip $v_- < \operatorname{Im} z < v_+$. Therefore, this equation holds in the strip* $\mu < \operatorname{Im} z < v_+$, which is the common analyticity domain for all functions that enter the equation. In order to transform Eq. (19) to the form (21), we assume that it is possible to decompose the function $\mathcal{W}(z)$ as follows:

$$\mathcal{W}(z) = \frac{\mathcal{W}_+(z)}{\mathcal{W}_-(z)}, \tag{23}$$

where the functions $\mathcal{W}_+(z)$ and $\mathcal{W}_-(z)$ are analytic for $\operatorname{Im} z > \mu$ and $\operatorname{Im} z < v_+$, respectively. Moreover, we assume that, in the corresponding analyticity domains, these functions grow at infinity no faster than z^n, where n is a positive integer. A representation of an analytic function $\mathcal{W}(z)$ in the form (23) is often called a *factorization* of $\mathcal{W}(z)$.

Thus, as the result of factorization, the original equation is reduced to the form (21). It follows from the above reasoning that this equation determines an entire function of the complex variable z.

Since $\mathcal{Y}_\pm(z) \to 0$ as $|z| \to \infty$ and the growth of the functions $\mathcal{W}_\pm(z)$ does not exceed that of a power function z^n, it follows that the entire function under consideration can be only a polynomial $\mathcal{P}_{n-1}(z)$ of degree at most $n - 1$.

If the growth of the functions $\mathcal{W}_\pm(z)$ at infinity is only linear with respect to the variable z, then it follows from relations (22), by virtue of the Liouville theorem (see Subsection 10.4-3), that the corresponding entire function is a constant C. In this case we obtain the following relations for the unknown functions $\mathcal{Y}_+(z)$ and $\mathcal{Y}_-(z)$:

$$\mathcal{Y}_+(z) = \frac{C}{\mathcal{W}_+(z)}, \quad \mathcal{Y}_-(z) = -\frac{C}{\mathcal{W}_-(z)}, \tag{24}$$

---

* To be definite, we set $\mu > v_-$. Otherwise, the common domain of analyticity is the strip $v_- < \operatorname{Im} z < v_+$.

which define the transform of the solution up to a constant factor, which can be found at least from the normalization conditions. In the general case, the expressions

$$\mathcal{Y}_+(z) = \frac{\mathcal{P}_{n-1}(z)}{\mathcal{W}_+(z)}, \quad \mathcal{Y}_-(z) = -\frac{\mathcal{P}_{n-1}(z)}{\mathcal{W}_-(z)}, \tag{25}$$

define the transform of the desired solution of the integral equation (11) up to indeterminate constants, which can be found from the additional conditions of the problem. The solution itself is defined by means of the Fourier inversion formula (6), (9), and (10).

Example 2. Consider the equation

$$y(x) = \lambda \int_0^\infty e^{-|x-t|} y(t)\,dt, \qquad 0 < \lambda < \infty, \tag{26}$$

whose kernel has the form $K(x) = \lambda e^{-|x|}$.

Let us find the transform of the function $K(x)$:

$$\mathcal{K}(z) = \frac{\lambda}{\sqrt{2\pi}} \int_{-\infty}^\infty K(x) e^{izx}\,dx = \sqrt{\frac{2}{\pi}} \frac{\lambda}{z^2+1}. \tag{27}$$

The function $\mathcal{K}(z)$ is analytic with respect to the complex variable z in the strip $-1 < \operatorname{Im} z < 1$. Let us represent the expression

$$\mathcal{W}(z) = 1 - \sqrt{2\pi}\,\mathcal{K}(z) = \frac{z^2 - 2\lambda + 1}{z^2+1} \tag{28}$$

in the form (23), where

$$\mathcal{W}_+(z) = \frac{z^2 - 2\lambda + 1}{z+i}, \quad \mathcal{W}_-(z) = z - i. \tag{29}$$

The function $\mathcal{W}_+(z)$ in Eq. (29) is analytic with respect to z and nonzero in the domain $\operatorname{Im} z > \operatorname{Im}\sqrt{2\lambda - 1}$. For $0 < \lambda < \frac{1}{2}$, this domain is defined by the condition $\operatorname{Im} z > \sqrt{1 - 2\lambda}$, and $\sqrt{1 - 2\lambda} \leq \mu < 1$. For $\lambda > \frac{1}{2}$, the function $\mathcal{W}_+(z)$ is analytic and nonzero in the domain $\operatorname{Im} z > 0$. It is clear that the function $\mathcal{W}_-(z)$ is a nonzero analytic function in the domain $\operatorname{Im} z < 1$. Therefore, for $0 < \lambda < \frac{1}{2}$ both functions satisfy the required conditions in the domain $\mu < \operatorname{Im} z < 1$.

For $\lambda > \frac{1}{2}$, the strip $0 < \operatorname{Im} z < 1$ is the common domain of analyticity of the functions $\mathcal{W}_+(z)$ and $\mathcal{W}_-(z)$. Thus, we have obtained the desired factorization of the function (28).

Consider the expressions $\mathcal{Y}_\pm(z)\mathcal{W}_\pm(z)$. Since $\mathcal{Y}_\pm(z) \to 0$ as $|z| \to \infty$, and, according to (29), the growth of the functions $\mathcal{W}_\pm(z)$ at infinity is linear with respect to z, it follows that the entire function $\mathcal{P}_{n-1}(z)$ that coincides with $\mathcal{Y}_+(z)\mathcal{W}_+(z)$ for $\operatorname{Im} z > \mu$ and with $\mathcal{Y}_-(z)\mathcal{W}_-(z)$ for $\operatorname{Im} z < 1$ can be a polynomial of zero degree only. Therefore,

$$\mathcal{Y}_+(z)\mathcal{W}_+(z) = C. \tag{30}$$

Hence,

$$\mathcal{Y}_+(z) = C\frac{z+i}{z^2 - 2\lambda + 1}, \tag{31}$$

and it follows from (6) that

$$y^+(x) = \frac{C}{\sqrt{2\pi}} \int_{-\infty+iv}^{\infty+iv} \frac{z+i}{z^2 - 2\lambda + 1} e^{-izx}\,dz, \tag{32}$$

where $\mu < v < 1$.

On closing the integration contour for $x > 0$ by a semicircle in the lower half-plane and estimating the integral over this semicircle by means of the Jordan lemma (see Subsections 7.1-4 and 7.1-5), after some calculations we obtain

$$y^+(x) = C\left[\cos(\sqrt{2\lambda - 1}\,x) + \frac{\sin(\sqrt{2\lambda - 1}\,x)}{\sqrt{2\lambda - 1}}\right], \tag{33}$$

where C is a constant. For $0 < \lambda < \frac{1}{2}$, this solution has exponential growth with respect to x, and for $\frac{1}{2} < \lambda < \infty$, it is bounded at infinity.

11.10-3. The General Scheme of the Method. The Factorization Problem

In the general case, the problem which is solved by the Wiener–Hopf method can be reduced to the following problem. It is required to find functions $\mathcal{Y}_+(z)$ and $\mathcal{Y}_-(z)$ of the complex variable z that are analytic in the half-planes $\operatorname{Im} z > v_-$ and $\operatorname{Im} z < v_+$, respectively ($v_- < v_+$), vanish as $|z| \to \infty$ in their analyticity domains, and satisfy the following functional equation in the strip ($v_- < \operatorname{Im} z < v_+$):

$$A(z)\mathcal{Y}_+(z) + B(z)\mathcal{Y}_-(z) + C(z) = 0. \tag{34}$$

Here $A(z)$, $B(z)$, and $C(z)$ are given functions of the complex variable z that are analytic in the strip $v_- < \operatorname{Im} z < v_+$, and the functions $A(z)$ and $B(z)$ are nonzero in this strip.

The main idea of the solution of this problem is based on the possibility of a factorization of the expression $A(z)/B(z)$, i.e., of a representation in the form

$$\frac{A(z)}{B(z)} = \frac{\mathcal{W}_+(z)}{\mathcal{W}_-(z)}, \tag{35}$$

where the functions $\mathcal{W}_+(z)$ and $\mathcal{W}_-(z)$ are analytic and nonzero in the half-planes $\operatorname{Im} z > v'_-$ and $\operatorname{Im} z < v'_+$, and the strips $v_- < \operatorname{Im} z < v_+$ and $v'_- < \operatorname{Im} z < v'_+$ have a nonempty common part. In this case Eq. (34), with regard to Eq. (35), can be rewritten in the form

$$\mathcal{W}_+(z)\mathcal{Y}_+(z) + \mathcal{W}_-(z)\mathcal{Y}_-(z) + \mathcal{W}_-(z)\frac{C(z)}{B(z)} = 0. \tag{36}$$

If the last summand in Eq. (36) can be represented as the sum

$$\mathcal{W}_-(z)\frac{C(z)}{B(z)} = \mathcal{D}_+(z) + \mathcal{D}_-(z), \tag{37}$$

where the functions $\mathcal{D}_+(z)$ and $\mathcal{D}_-(z)$ are analytic in the half-planes $\operatorname{Im} z > v''_-$ and $\operatorname{Im} z < v''_+$, respectively, and all three strips $v_- < \operatorname{Im} z < v_+$, $v'_- < \operatorname{Im} z < v'_+$, and $v''_- < \operatorname{Im} z < v''_+$ have a nonempty common part, for a strip $v^0_- < \operatorname{Im} z < v^0_+$, then, in this common strip, the following functional equation holds:

$$\mathcal{W}_+(z)\mathcal{Y}_+(z) + \mathcal{D}_+(z) = -\mathcal{W}_-(z)\mathcal{Y}_-(z) - \mathcal{D}_-(z). \tag{38}$$

The left-hand side of Eq. (38) is a function analytic in the half-plane $v^0_- < \operatorname{Im} z$, and the right-hand side is a function analytic in the domain $\operatorname{Im} z < v^0_+$. Since these functions coincide in the strip $v^0_- < \operatorname{Im} z < v^0_+$, it follows that there exists a unique entire function that coincides with the left-hand side and the right-hand side of (38) in their analyticity domains, respectively. If the growth at infinity of all functions that enter the right-hand sides of Eqs. (35) and (37), in their analyticity domains, is at most that of z^n, then it follows from the limit relation $\mathcal{Y}_\pm(z) \to 0$ as $|z| \to \infty$ that this entire function is a polynomial $\mathcal{P}_{n-1}(z)$ of degree at most $n - 1$. Thus, the relations

$$\mathcal{Y}_+(z) = \frac{\mathcal{P}_{n-1}(z) - \mathcal{D}_+(z)}{\mathcal{W}_+(z)}, \qquad \mathcal{Y}_-(z) = \frac{-\mathcal{P}_{n-1}(z) - \mathcal{D}_-(z)}{\mathcal{W}_-(z)} \tag{39}$$

determine the desired functions up to constants. These constants can be found from the additional conditions of the problem.

The application of the Wiener–Hopf method is based on the representations (35) and (37). If a function $\mathcal{G}(z)$ is analytic in the strip $v_- < \operatorname{Im} z < v_+$ and if in this strip the function $\mathcal{G}(z)$ uniformly tends to zero as $|z| \to \infty$, then in this strip the following representation is possible:

$$\mathcal{G}(z) = \mathcal{G}_+(z) + \mathcal{G}_-(z), \tag{40}$$

where the function $\mathcal{G}_+(z)$ is analytic in the half-plane $\operatorname{Im} z > v_-$, the function $\mathcal{G}_-(z)$ is analytic in the half-plane $\operatorname{Im} z < v_+$, and

$$\mathcal{G}_+(z) = \frac{1}{2\pi i} \int_{-\infty+iv_-'}^{\infty+iv_-'} \frac{\mathcal{G}(\tau)}{\tau - z}\,d\tau, \qquad v_- < v_-' < \operatorname{Im} z < v_+, \tag{41}$$

$$\mathcal{G}_-(z) = -\frac{1}{2\pi i} \int_{-\infty+iv_+'}^{\infty+iv_+'} \frac{\mathcal{G}(\tau)}{\tau - z}\,d\tau, \qquad v_- < \operatorname{Im} z < v_+' < v_+. \tag{42}$$

The integrals (41) and (42), being regarded as integrals depending on a parameter, define analytic functions of the complex variable z under the assumption that the point z does not belong to the integration contour.

In particular, $\mathcal{G}_+(z)$ is an analytic function in the half-plane $\operatorname{Im} z > v_-'$ and $\mathcal{G}_-(z)$ in the half-plane $\operatorname{Im} z > v_+'$.

Moreover, if a function $\mathcal{H}(z)$ is analytic and nonzero in the strip $v_- < \operatorname{Im} z < v_+$ and if $\mathcal{H}(z) \to 1$ uniformly in this strip as $|z| \to \infty$, then the following representation holds in the strip:

$$\mathcal{H}(z) = \mathcal{H}_+(z)\mathcal{H}_-(z), \tag{43}$$

$$\mathcal{H}_+(z) = \exp\left[\frac{1}{2\pi i} \int_{-\infty+iv_-'}^{\infty+iv_-'} \frac{\ln \mathcal{H}(\tau)}{\tau - z}\,d\tau\right], \qquad v_- < v_-' < \operatorname{Im} z < v_+, \tag{44}$$

$$\mathcal{H}_-(z) = \exp\left[-\frac{1}{2\pi i} \int_{-\infty+iv_+'}^{\infty+iv_+'} \frac{\ln \mathcal{H}(\tau)}{\tau - z}\,d\tau\right], \qquad v_- < \operatorname{Im} z < v_+' < v_+, \tag{45}$$

where the functions $\mathcal{H}_+(z)$ and $\mathcal{H}_-(z)$ are analytic and nonzero in the half-planes $\operatorname{Im} z > v_-$ and $\operatorname{Im} z < v_+$, respectively. The representation (43) is called a *factorization* of the function $\mathcal{H}(z)$.

11.10-4. The Nonhomogeneous Wiener–Hopf Equation of the Second Kind

Consider the Wiener–Hopf equation of the second kind

$$y(x) - \int_0^\infty K(x-t)y(t)\,dt = f(x), \tag{46}$$

Suppose that the kernel $K(x)$ of the equation and the right-hand side $f(x)$ satisfy conditions (15). Let us seek the solution $y^+(x)$ to Eq. (46) for which condition (17) is satisfied.

In this case, reasoning similar to that in the derivation of the functional equation (19) for a homogeneous integral equation shows that, in the case of Eq. (46), the following functional equation must hold on the strip $\mu < \operatorname{Im} z < v_+$:

$$\mathcal{Y}_+(z) + \mathcal{Y}_-(z) = \sqrt{2\pi}\,\mathcal{K}(z)\mathcal{Y}_+(z) + \mathcal{F}_+(z) + \mathcal{F}_-(z), \tag{47}$$

or

$$\mathcal{W}(z)\mathcal{Y}_+(z) + \mathcal{Y}_-(z) - \mathcal{F}(z) = 0, \tag{48}$$

where $\mathcal{W}(z)$ is subjected to condition (20), as well as in the case of a homogeneous equation.

We now note that Eq. (48) is a special case of Eq. (34). In the strip $v_- < \operatorname{Im} z < v_+$, the function $\mathcal{W}(z)$ is analytic and uniformly tends to 1 as $|z| \to \infty$ because $|\mathcal{K}(z)| \to 0$ as $|z| \to \infty$. In this case, this function has the representation (see (43)–(45))

$$\mathcal{W}(z) = \frac{\mathcal{W}_+(z)}{\mathcal{W}_-(z)}, \tag{49}$$

where the function $\mathcal{W}_+(z)$ is analytic in the upper half-plane $\operatorname{Im} z > v_-$ and $\mathcal{W}_-(z)$ is analytic in the lower half-plane $\operatorname{Im} z < v_+$, and the growth at infinity of the functions $\mathcal{W}_\pm(z)$ does not exceed that of z^n.

On the basis of the representation (49), Eq. (48) becomes

$$\mathcal{W}_+(z)\mathcal{Y}_+(z) + \mathcal{W}_-(z)\mathcal{Y}_-(z) - \mathcal{W}_-(z)\mathcal{F}_-(z) - \mathcal{W}_-(z)\mathcal{F}_+(z) = 0. \tag{50}$$

To reduce Eq. (50) to the form (38), it suffices to decompose the last summand

$$\mathcal{F}_+(z)\mathcal{W}_-(z) = \mathcal{D}_+(z) + \mathcal{D}_-(z) \tag{51}$$

into the sum of functions $\mathcal{D}_+(z)$ and $\mathcal{D}_-(z)$ that are analytic in the half-planes $\operatorname{Im} z > \mu$ and $\operatorname{Im} z < v_+$, respectively.

To establish the possibility of a representation (51), we note that the function $\mathcal{F}_+(z)$ is analytic in the upper half-plane $\operatorname{Im} z > v_-$ and uniformly tends to zero as $|z| \to \infty$. The function $\mathcal{W}_-(z)$ is analytic in the lower half-plane $\operatorname{Im} z < v_+$, and, according to the method of its construction, we can perform the factorization (49) so that the function $\mathcal{W}_-(z)$ remains bounded in the strip $v_- < \operatorname{Im} z < v_+$ as $|z| \to \infty$. Hence (see (40)–(42)), the functions $\mathcal{F}_+(z)\mathcal{W}_-(z)$ in the strip $v_- < \operatorname{Im} z < v_+$ satisfy all conditions that are sufficient for the validity of the representation (51).

The above reasoning makes it possible to take into account the fact that the growth at infinity of the functions $\mathcal{W}_\pm(z)$ does not exceed that of z^n, and thus to present the transform of the solution of the nonhomogeneous integral equation (46) in the form

$$\mathcal{Y}_+(z) = \frac{\mathcal{P}_{n-1}(z) + \mathcal{D}_+(z)}{\mathcal{W}_+(z)}, \quad \mathcal{Y}_-(z) = \frac{-\mathcal{P}_{n-1}(z) + \mathcal{W}_-(z)\mathcal{F}_-(z) + \mathcal{D}_-(z)}{\mathcal{W}_-(z)}. \tag{52}$$

The solution itself can be obtained from (52) by means of the Fourier inversion formula (6), (9), and (10).

11.10-5. The Exceptional Case of a Wiener–Hopf Equation of the Second Kind

Consider the exceptional case of a Wiener–Hopf equation of the second kind in which the function $\mathcal{W}(z) = 1 - \sqrt{2\pi}\,\mathcal{K}(z)$ has finitely many zeros N (counted according to their multiplicities) in the strip $v_- < \operatorname{Im} z < v_+$. In this case, the factorization is also possible. To this end, it suffices to introduce the auxiliary function

$$\mathcal{W}_1(z) = \ln\left[(z^2 + b^2)^{N/2}\mathcal{W}(z)\prod_i (z - z_i)^{-\alpha_i}\right], \tag{53}$$

where α_i is the multiplicity of the zero z_i and a positive constant $b > \{|v_-|, |v_+|\}$ is chosen so that the function in the square brackets has no additional zeros in the strip $v_- < \operatorname{Im} z < v_+$.

However, in the exceptional case, the Wiener–Hopf method gives the answer only if the number of zeros of the function $\mathcal{W}(z)$ is even. This restriction is due to the fact that only for the case in which the number of zeros is even is it possible to achieve the necessary behavior at infinity (for the application of the Wiener–Hopf method) of the function $(z^2 + b^2)^{N/2}$ (see F. D. Gakhov and Yu. I. Cherskii (1978)). The last restriction makes no real obstacle to the broad use of the Wiener–Hopf method in solving applied problems in which the kernel $K(x)$ of the corresponding integral equation is frequently an even function, and thus the reasoning below can be applied completely.

Remark 1. The Wiener–Hopf equation of the second kind for functions vanishing at infinity can be reduced to a Riemann boundary value problem on the real axis (see Subsection 11.9-1). In this case, the assumption that the number of zeros of the function $\mathcal{W}(z)$ is even, as well as the assumption that the kernel $K(x)$ is even in the exceptional case, are unessential.

Remark 2. For functions with nontrivial growth at infinity, the complete solution of Wiener–Hopf equations of the second kind is presented in the cited book by F. D. Gakhov and Yu. I. Cherskii (1978).

Remark 3. The Wiener–Hopf method can be applied to solve Wiener–Hopf integral equations of the first kind under the assumption that the kernels of these equations are even.

⊙ References for Section 11.10: B. Noble (1958), A. G. Sveshnikov and A. N. Tikhonov (1970), V. I. Smirnov (1974), F. D. Gakhov (1977), F. D. Gakhov and Yu. I. Cherskii (1978).

11.11. Krein's Method for Wiener–Hopf Equations

11.11-1. Some Remarks. The Factorization Problem

Consider the Wiener–Hopf equation of the second kind

$$y(x) - \int_0^\infty K(x - t)y(t)\, dt = f(x), \qquad 0 \le x < \infty, \tag{1}$$

where $f(x)$, $y(x) \in L_1(0, \infty)$ and $K(x) \in L_1(-\infty, \infty)$. Let us use the classes of functions that can be represented as Fourier transforms (alternative Fourier transform in the asymmetric form, see Subsection 7.4-3), of functions from $L_1(-\infty, \infty)$, $L_1(0, \infty)$, and $L_1(-\infty, 0)$. For brevity, instead of these symbols we simply write L, L_+, and L_-. Let functions $h(x)$, $h_1(x)$, and $h_2(x)$ belong to L, L_+, and L_-, respectively; in this case, their transforms can be represented in the form

$$\check{\mathcal{H}}(u) = \int_{-\infty}^\infty h(x)e^{iux}\, dx, \quad \check{\mathcal{H}}_1(u) = \int_0^\infty h_1(x)e^{iux}\, dx, \quad \check{\mathcal{H}}_2(u) = \int_{-\infty}^0 h_2(x)e^{iux}\, dx.$$

Let Q, Q_+, and Q_- be the classes of functions representable in the form

$$\check{\mathcal{W}}(u) = 1 + \check{\mathcal{H}}(u), \quad \check{\mathcal{W}}_1(u) = 1 + \check{\mathcal{H}}_1(u), \quad \check{\mathcal{W}}_2(u) = 1 + \check{\mathcal{H}}_2(u), \tag{2}$$

respectively, where the functions from the classes Q_+ and Q_-, treated as functions of the complex variable $z = u + iv$, are analytic for $\text{Im } z > 0$ and $\text{Im } z < 0$, respectively, and are continuous up to the real axis.

Let $T(x)$ belong to L and let $\check{\mathcal{T}}(u)$ be its transform. Assume that

$$1 - \check{\mathcal{T}}(u) \ne 0, \quad \text{Ind}[1 - \check{\mathcal{T}}(u)] = \frac{1}{2\pi}\Big\{\arg[1 - \check{\mathcal{T}}(u)]\Big\}_{-\infty}^\infty = 0, \qquad -\infty < u < \infty. \tag{3}$$

In this case there exists a $q(x) \in L$ such that

$$\ln[1 - \check{\mathcal{T}}(u)] = \int_{-\infty}^\infty q(x)e^{iux}\, dx. \tag{4}$$

This formula readily implies the relation $\ln[1 - \check{\mathcal{T}}(u)] \to 0$ as $u \to \pm\infty$.

In what follows, we apply the *factorization* of functions $\check{\mathcal{M}}(u)$ of the class Q that are continuous on the interval $-\infty \le u \le \infty$. Here the factorization means a representation of the function $\check{\mathcal{M}}(u)$ in the form of a product

$$\check{\mathcal{M}}(u) = \check{\mathcal{M}}_+(u)\left(\frac{u - i}{u + i}\right)^k \check{\mathcal{M}}_-(u), \tag{5}$$

where $\check{\mathcal{M}}_-(z)$ and $\check{\mathcal{M}}_+(z)$ are analytic functions in the corresponding half-planes $\operatorname{Im} z > 0$ and $\operatorname{Im} z < 0$ continuous up to the real axis. Moreover,

$$\check{\mathcal{M}}_+(z) \neq 0 \quad \text{for} \quad \operatorname{Im} z \geq 0 \quad \text{and} \quad \check{\mathcal{M}}_-(z) \neq 0 \quad \text{for} \quad \operatorname{Im} z \leq 0. \tag{6}$$

Relation (5) implies the formula

$$k = \operatorname{Ind} \check{\mathcal{M}}(u).$$

The factorization (5) is said to be *canonical* provided that $k = 0$.

In what follows we consider only functions of the form

$$\check{\mathcal{M}}(u) = 1 - \check{T}(u) \tag{7}$$

such that $\check{\mathcal{M}}(\pm\infty) = 1$. We can also assume that

$$\check{\mathcal{M}}_+(\pm\infty) = \check{\mathcal{M}}_-(\pm\infty) = 1. \tag{8}$$

Let us state the main results concerning the factorization problem.

A function (7) admits a canonical factorization if and only if the following two conditions hold:

$$\check{\mathcal{M}}(u) \neq 0, \qquad \operatorname{Ind} \check{\mathcal{M}}(u) = 0. \tag{9}$$

In this case, the canonical factorization is unique. Moreover, if conditions (9) hold, then there exists a function $M(x)$ in the class L such that

$$\check{\mathcal{M}}(u) = \exp\left[\int_{-\infty}^{\infty} M(x)e^{iux}\,dx\right], \tag{10}$$

$$\check{\mathcal{M}}_+(u) = \exp\left[\int_{0}^{\infty} M(x)e^{iux}\,dx\right], \quad \check{\mathcal{M}}_-(u) = \exp\left[\int_{-\infty}^{0} M(x)e^{iux}\,dx\right]. \tag{11}$$

Hence, we have $\check{\mathcal{M}}(u) \in Q$ and $\check{\mathcal{M}}_\pm(u) \in Q_\pm$. The factors in the canonical factorization are also described by the following formulas:

$$\ln \check{\mathcal{M}}_+(z) = \frac{1}{2\pi i} \int_{-\infty}^{\infty} \frac{\ln \check{\mathcal{M}}(\tau)}{\tau - z}\,d\tau, \qquad \operatorname{Im} z > 0, \tag{12}$$

$$\ln \check{\mathcal{M}}_-(z) = -\frac{1}{2\pi i} \int_{-\infty}^{\infty} \frac{\ln \check{\mathcal{M}}(\tau)}{\tau - z}\,d\tau, \qquad \operatorname{Im} z < 0. \tag{13}$$

In the general case of the factorization, the following assertion holds. A function (7) admits a factorization (5) if and only if the following condition is satisfied:

$$\check{\mathcal{M}}(u) \neq 0, \qquad -\infty < u < \infty. $$

In this case, relation (5) can be rewritten in the form

$$\left(\frac{u-i}{u+i}\right)^{-k} \check{\mathcal{M}}(u) = \check{\mathcal{M}}_-(u)\check{\mathcal{M}}_+(u), \qquad -\infty < u < \infty. $$

The last relation implies the canonical factorization for the function

$$\check{\mathcal{M}}_1(u) = \left(\frac{u-i}{u+i}\right)^{-k} \check{\mathcal{M}}(u). $$

Hence, the factors $\check{\mathcal{M}}_\pm(u)$ satisfy formulas (10)–(13) if we replace $\check{\mathcal{M}}(u)$ in these formulas by $\check{\mathcal{M}}_1(u)$.

Now we return to Eq. (1) for which

$$\check{\mathcal{K}}(u) = \int_{-\infty}^{\infty} K(x)e^{iux}\,dx. \tag{14}$$

11.11-2. The Solution of the Wiener–Hopf Equations of the Second Kind

THEOREM 1. *For Eq. (1) to have a unique solution of the class L_+ for an arbitrary $f(x) \in L_+$, it is necessary and sufficient that the following conditions hold:*

$$1 - \check{K}(u) \neq 0, \qquad -\infty < u < \infty, \tag{15}$$

$$\nu = -\operatorname{Ind}[1 - \check{K}(u)] = 0. \tag{16}$$

THEOREM 2. *If condition (15) holds, then the inequality $\nu > 0$ is necessary and sufficient for the existence of nonzero solutions in the class L_+ of the homogeneous equation*

$$y(x) - \int_0^\infty K(x - t)y(t)\,dt = 0. \tag{17}$$

The set of these solutions has a basis formed by ν functions $\varphi_k(x)$ $(k = 1, \ldots, \nu)$ that tend to zero as $x \to \infty$ and that are related as follows:

$$\varphi_k(x) = \int_0^x \varphi_{k+1}(t)\,dt, \qquad k = 1, 2, \ldots, \nu - 1, \qquad \varphi_\nu(x) = \int_0^x \psi(t)\,dt + C, \tag{18}$$

where C is a nonzero constant and the functions $\varphi_k(t)$ and $\psi(t)$ belong to L_+.

THEOREM 3. *If condition (15) holds and if $\nu > 0$, then for any $f(x) \in L_+$ Eq. (1) has infinitely many solutions in L_+.*

However, if $\nu < 0$, then, for a given $f(x) \in L_+$, Eq. (1) has either no solutions from L_+ or a unique solution. For the latter case to hold, it is necessary and sufficient that the following conditions be satisfied:

$$\int_0^\infty f(x)\psi_k(x)\,dx = 0, \qquad k = 1, 2, \ldots, |\nu|, \tag{19}$$

where $\psi_k(x)$ is a basis of the linear space of all solutions of the transposed homogeneous equation

$$\psi(x) - \int_0^\infty K(t - x)\psi(t)\,dt = 0. \tag{20}$$

1°. If conditions (15) and (16) hold, then there exists a unique factorization

$$[1 - \check{K}(u)]^{-1} = \check{M}_+(u)\check{M}_-(u), \tag{21}$$

and

$$\check{M}_+(u) = 1 + \int_0^\infty R_+(t)e^{iut}\,dt, \qquad \check{M}_-(u) = 1 + \int_0^\infty R_-(t)e^{-iut}\,dt. \tag{22}$$

The resolvent is defined by the formula

$$R(x, t) = R_+(x - t) + R_-(t - x) + \int_0^\infty R_+(x - s)R_-(t - s)\,ds \tag{23}$$

where $0 \leq x < \infty$, $0 \leq t < \infty$, $R_+(x) = 0$, and $R_-(x) = 0$ for $x < 0$, so that, for $f(x)$ from L_+, the solution of the equation is determined by the expression

$$y(x) = f(x) + \int_0^\infty R(x, t)f(t)\,dt. \tag{24}$$

Formula (23) can be rewritten as follows:

$$R(x, t) = R(x - t, 0) + R(0, t - x) + \int_0^\infty R(x - s, 0)R(0, t - s)\, ds. \tag{25}$$

If $K(x - t) = K(t - x)$, then formula (25) becomes

$$R(x, t) = R(|x - t|, 0) + \int_0^{\min(x,t)} R(x - s, 0)R(t - s, 0)\, ds. \tag{26}$$

Note that $R_+(x) = R(x, 0)$ and $R_-(x) = R(0, x)$ are unique solutions, in the class L_+, of the following equations ($0 \le x < \infty$):

$$\begin{aligned}
R_+(x) + \int_0^\infty K(x - t)R_+(t)\, dt &= K(x), \\
R_-(x) + \int_0^\infty K(t - x)R_-(t)\, dt &= K(-x).
\end{aligned} \tag{27}$$

$2°$. Suppose that condition (15) holds, but

$$\nu = -\operatorname{Ind}[1 - \check{K}(u)] > 0.$$

In this case, the function $[1 - \check{K}(u)]^{-1}$ admits the factorization

$$[1 - \check{K}(u)]^{-1} = \check{G}_-(u)\left(\frac{u - i}{u + i}\right)^\nu \check{G}_+(u), \qquad -\infty < u < \infty. \tag{28}$$

For the functions $\check{M}_-(u)$ and $\check{M}_+(u)$ defined by the relations

$$\check{M}_-(u) = \check{G}_-(u) \quad \text{and} \quad \check{M}_+(u) = \left(\frac{u - i}{u + i}\right)^\nu \check{G}_+(u), \tag{29}$$

we have the representation (22) and formula (23) for the resolvent.

Moreover, for $k = 1, \ldots, \nu$, the following representations hold:

$$\frac{i^k \check{M}_+(u)}{(u - i)^k} = \int_0^\infty g_k(x)e^{iux}\, dx, \tag{30}$$

where $g_k(x)$ is the solution of the homogeneous equation (17). The solutions $\varphi_k(x)$ mentioned in Theorem 2 can also naturally be expressed via the functions $g_k(x)$.

$3°$. If $\nu = -\operatorname{Ind}[1 - \check{K}(u)] < 0$, then the transposed equation

$$y(x) - \int_0^\infty K(t - x)y(t)\, dt = f(x) \tag{31}$$

has the index $-\nu > 0$. If formula (28) defines a factorization for Eq. (1), then the transposed equation admits a factorization of the form

$$[1 - \check{K}(u)]^{-1} = \check{M}_-(-u)\check{M}_+(-u),$$

and $\check{M}_-(-u)$ plays the role of $\check{M}_+(u)$, and $\check{M}_+(-u)$ plays the role of $\check{M}_-(u)$.

11.11-3. The Hopf–Fock Formula

Let us give a useful formula that allows one to express the solution of Eq. (1) with an arbitrary right-hand side $f(x)$ via the solution to a simpler auxiliary integral equation with an exponential right-hand side.

Assume that in Eq. (1) we have

$$f(x) = e^{i\zeta x}, \quad \operatorname{Im}\zeta > 0, \quad y(x) = y_\zeta(x), \tag{32}$$

and moreover, conditions (15) and (16) hold. In this case

$$y_\zeta(x) = e^{i\zeta x} + \int_0^\infty R(x,t)e^{i\zeta t}\,dt, \tag{33}$$

where $R(x,t)$ has the form (25). After some manipulations, we can see that

$$y_\zeta(x) = \check{M}_-(-\zeta)\left[1 + \int_0^x R(t,0)e^{-i\zeta t}\,dt\right]e^{i\zeta x}. \tag{34}$$

On setting $x = 0$ in (34), we have

$$y_\zeta(0) = \check{M}_-(-\zeta), \tag{35}$$

and if the function $K(x)$ describing the kernel of the integral equation is even, then

$$y_\zeta(0) = \check{M}_+(\zeta). \tag{36}$$

On the basis of formula (34), we can obtain the solution of Eq. (1) for a general $f(x)$ as well (see also Section 9.6):

$$y(x) = \frac{1}{2\pi}\int_{-\infty}^\infty \check{F}_+(-\zeta)y_\zeta(x)\,d\zeta, \qquad \check{F}_+(u) = \int_0^\infty f(x)e^{iux}\,dx. \tag{37}$$

Remark 1. All results obtained in Section 11.11 concerning Wiener–Hopf equations of the second kind remain valid for continuous, square integrable, and some other classes of functions, which are discussed in detail in the paper by M. G. Krein (1958) and in the book by C. Corduneanu (1973).

Remark 2. The solution of the Wiener–Hopf equation can be also obtained in other classes of functions for the exceptional case in which $1 - \check{K}(u) = 0$ (see Subsections 11.9-1 and 11.10-5).

⊙ References for Section 11.11: V. A. Fock (1942), M. G. Krein (1958), C. Corduneanu (1973), V. I. Smirnov (1974), P. P. Zabreyko, A. I. Koshelev, et al. (1975).

11.12. Methods for Solving Equations With Difference Kernels on a Finite Interval

11.12-1. Krein's Method

Consider a method for constructing exact analytic solutions of linear integral equations with an arbitrary right-hand side. The method is based on the construction of two auxiliary solutions of simpler equations with the right-hand side equal to 1. The auxiliary solutions are used to construct a solution of the original equation for an arbitrary right-hand side.

1°. Let the equation

$$y(x) - \int_a^b K(x,t)y(t)\,dt = f(x), \qquad a \le x \le b, \tag{1}$$

be given. Along with (1), we consider two auxiliary equations depending on a parameter ξ $(a \le \xi \le b)$:

$$w(x,\xi) - \int_a^\xi K(x,t)w(t,\xi)\,dt = 1,$$

$$w^*(x,\xi) - \int_a^\xi K(t,x)w^*(t,\xi)\,dt = 1, \tag{2}$$

where $a \le x \le \xi$. Assume that for any ξ the auxiliary equations (2) have unique continuous solutions $w(x,\xi)$ and $w^*(x,\xi)$, respectively, which satisfy the condition $w(\xi,\xi)w^*(\xi,\xi) \ne 0$ $(a \le \xi \le b)$. In this case, for any continuous function $f(x)$, the unique continuous solution of Eq. (1) can be obtained by the formula

$$y(x) = F(b)w(x,b) - \int_x^b w(x,\xi)F_\xi'(\xi)\,d\xi, \qquad F(\xi) = \frac{1}{m(\xi)}\frac{d}{d\xi}\int_a^\xi w^*(t,\xi)f(t)\,dt, \tag{3}$$

where

$$m(\xi) = w(\xi,\xi)w^*(\xi,\xi).$$

Formula (3) permits one to construct a solution of Eq. (1) with an arbitrary right-hand side $f(x)$ by means of solutions to the two simpler auxiliary equations (2) (depending on the parameter ξ) with a constant right-hand side equal to 1.

2°. Consider now an equation with the kernel depending on the difference of the arguments:

$$y(x) + \int_a^b K(x-t)y(t)\,dt = f(x), \qquad a \le x \le b. \tag{4}$$

It is assumed that $K(x)$ is an even function integrable on $[a-b, b-a]$. Along with (4) we consider the following auxiliary equation depending on a parameter ξ $(a \le \xi \le b)$:

$$w(x,\xi) + \int_a^\xi K(x-t)w(t,\xi)\,dt = 1, \qquad a \le x \le \xi. \tag{5}$$

Assume that for an arbitrary ξ the auxiliary equation (5) has a unique continuous solution $w(x,\xi)$. In this case, for any continuous function $f(x)$, a solution of Eq. (4) can be obtained from formula (3) by setting $w^*(x,t) = w(x,t)$ in this formula.

Now let us indicate another useful formula for equations whose kernel depends on the difference of the arguments:

$$y(x) + \int_{-a}^a K(x-t)y(t)\,dt = f(x), \qquad -a \le x \le a. \tag{6}$$

It is assumed that $K(x)$ is an even function that is integrable on the segment $[-2a, 2a]$. Along with (6) we consider an auxiliary equation depending on a parameter ξ $(0 < \xi \le a)$:

$$w(x,\xi) + \int_{-\xi}^\xi K(x-t)w(t,\xi)\,dt = 1, \qquad -\xi \le x \le \xi. \tag{7}$$

Let the auxiliary equation (7) have a unique continuous solution $w(x, \xi)$ for any ξ. In this case, for an arbitrary continuous function $f(x)$, the solution of Eq. (6) can be obtained by the following formula:

$$y(x) = \frac{1}{2M(a)} \left[\frac{d}{da} \int_{-a}^{a} w(t, a) f(t)\, dt \right] w(x, a)$$
$$- \frac{1}{2} \int_{|x|}^{a} w(x, \xi) \frac{d}{d\xi} \left[\frac{1}{M(\xi)} \frac{d}{d\xi} \int_{-\xi}^{\xi} w(t, \xi) f(t)\, dt \right] d\xi$$
$$- \frac{1}{2} \frac{d}{dx} \int_{|x|}^{a} \frac{w(x, \xi)}{M(\xi)} \left[\int_{-\xi}^{\xi} w(t, \xi)\, df(t) \right] d\xi, \qquad (8)$$

where $M(\xi) = w^2(\xi, \xi)$, and the last inner integral is treated as a Stieltjes integral.

11.12-2. Kernels With Rational Fourier Transforms

Consider an equation of the form

$$y(x) - \int_0^T K(x - t)y(t)\, dt = f(x), \qquad (9)$$

where $0 \le x \le T < \infty$. If the kernel $K(x)$ is integrable on $[-T, T]$, then the Fredholm theory can be applied to this equation.

Since the equation involves the values of the kernel $K(x)$ for the points of $[-T, T]$ only, it follows that we can extend the kernel outside this interval in an arbitrary way. Assume that the kernel is extended to the entire axis so that the extended function is integrable. In the general case, Eq. (9) in the space $L_2(0, T)$ can be reduced to a boundary value problem of the theory of analytic functions (Riemann problem) for two pairs of unknown functions.

If the Fourier transform of the kernel

$$\check{K}(u) = \int_{-\infty}^{\infty} K(x) e^{iux}\, dx$$

is rational, then Eq. (9) can be solved in the closed form. Assume that $1 - \check{K}(u) \ne 0$ $(-\infty < u < \infty)$. In this case, the transform of the solution of the integral equation (9) is given by the formula

$$\check{y}(u) = \frac{1}{1 - \check{K}(u)} \left[\check{F}(u) - \check{W}^+(u) - e^{-iTu} \check{W}^-(u) \right] \qquad (10)$$

in which

$$\check{W}^\pm(u) = \sum_n \sum_{k=1}^{p_n^\pm} \frac{M_{nk}^\pm}{(u - b_n^\pm)^k},$$

where the b_n^+ and the b_n^- are poles of the functions $1 - \check{K}(u)$ that belong to the upper and lower half-planes, respectively, and the $p_n^\pm$ are their multiplicities. The constants $M_{nk}^\pm$ can be determined from the conditions

$$\frac{d^s}{du^s} \left[\check{W}^+(u) + e^{-iTu} \check{W}^-(u) - \check{F}(u) \right]_{u=a_n^+} = 0; \qquad s = 0, 1, \ldots, q_n^+ - 1;$$

$$\frac{d^s}{du^s} \left[\check{W}^+(u) + e^{-iTu} - \check{F}(u) \right]_{u=a_n^-} = 0; \qquad s = 0, 1, \ldots, q_n^- - 1;$$

where a_n^+ and a_n^- are the zeros of the functions $1 - \check{K}(u)$ that belong to the upper and lower half-planes, respectively, and $q_n^\pm$ are their multiplicities. The constants $M_{nk}^\pm$ can also be determined by substituting the solution into the original equation. The solution of the integral equation (9) can be obtained by inverting formula (10).

11.12-3. Reduction to Ordinary Differential Equations

$1°$. Consider the special case in which the Fourier transform of the kernel of the integral equation (9) can be represented in the form

$$\check{\mathcal{K}}(u) = \frac{\check{\mathcal{M}}(u)}{\check{\mathcal{N}}(u)}, \tag{11}$$

where $\check{\mathcal{M}}(u)$ and $\check{\mathcal{N}}(u)$ are some polynomials of degrees m and n, respectively:

$$\check{\mathcal{M}}(u) = \sum_{k=0}^{m} A_k u^k, \quad \check{\mathcal{N}}(u) = \sum_{k=0}^{n} B_k u^k. \tag{12}$$

In this case, the solution of the integral equation (9) (if it exists) satisfies the following linear nonhomogeneous ordinary differential equation of the order m with constant coefficients:

$$\check{\mathcal{M}}\left(i\frac{d}{dx}\right)y(x) = \check{\mathcal{N}}\left(i\frac{d}{dx}\right)f(x), \qquad 0 < x < T. \tag{13}$$

The solution of Eq. (13) contains m arbitrary constants that are defined by substituting the solution into the original equation (9). Here a system of linear algebraic equations is obtained for these constants.

$2°$. Consider the Fredholm equation of the second kind with a difference kernel that contains a sum of the exponential functions:

$$y(x) + \int_a^b \left(\sum_{k=1}^{n} A_k e^{\lambda_k |x-t|} \right) y(t)\, dt = f(x). \tag{14}$$

In the general case, this equation can be reduced to a linear nonhomogeneous ordinary differential equation of order $2n$ with constant coefficients (see equation 4.2.16 in the first part of the book).

For the solution of Eq. (14) with $n = 1$, see equation 4.2.15 in the first part of the book.

$3°$. Equations with a difference kernel that contains a sum of hyperbolic functions,

$$y(x) + \int_a^b K(x-t)y(t)\, dt = f(x), \qquad K(x) = \sum_{k=1}^{n} A_k \sinh(\lambda_k |x|), \tag{15}$$

can be also reduced by differentiation to linear nonhomogeneous ordinary differential equations of order $2n$ with constant coefficients (see equation 4.3.29 in the first part of the book).

For the solution of Eq. (15) with $n = 1$, see equation 4.3.26 in the first part of the book.

$4°$. Equations with a difference kernel containing a sum of trigonometric functions

$$y(x) + \int_a^b K(x-t)y(t)\, dt = f(x), \qquad K(x) = \sum_{k=1}^{n} A_k \sin(\lambda_k |x|), \tag{16}$$

can be also reduced to linear nonhomogeneous ordinary differential equations of order $2n$ with constant coefficients (see equations 4.5.29 and 4.5.32 in the first part of the book).

⊙ References for Section 11.12: W. B. Davenport and W. L. Root (1958), I. C. Gohberg and M. G. Krein (1967), P. P. Zabreyko, A. I. Koshelev, et al. (1975), A. D. Polyanin and A. V. Manzhirov (1998).

11.13. The Method of Approximating a Kernel by a Degenerate One

11.13-1. Approximation of the Kernel

For the approximate solution of the Fredholm integral equation of the second kind

$$y(x) - \int_a^b K(x,t)y(t)\,dt = f(x), \qquad a \le x \le b, \tag{1}$$

where, for simplicity, the functions $f(x)$ and $K(x,t)$ are assumed to be continuous, it is useful to replace the kernel $K(x,t)$ by a close degenerate kernel

$$K_{(n)}(x,t) = \sum_{k=0}^{n} g_k(x)h_k(t). \tag{2}$$

Let us indicate several ways to perform such a change. If the kernel $K(x,t)$ is differentiable with respect to x on $[a,b]$ sufficiently many times, then, for a degenerate kernel $K_{(n)}(x,t)$, we can take a finite segment of the Taylor series:

$$K_{(n)}(x,t) = \sum_{m=0}^{n} \frac{(x-x_0)^m}{m!} K_x^{(m)}(x_0,t), \tag{3}$$

where $x_0 \in [a,b]$. A similar trick can be applied for the case in which $K(x,t)$ is differentiable with respect to t on $[a,b]$ sufficiently many times.

To construct a degenerate kernel, a finite segment of the double Fourier series can be used:

$$K_{(n)}(x,t) = \sum_{p=0}^{n} \sum_{q=0}^{n} a_{pq}(x-x_0)^p(t-t_0)^q, \tag{4}$$

where

$$a_{pq} = \frac{1}{p!\,q!} \frac{\partial^{p+q}}{\partial x^p \partial t^q} K(x,t)\bigg|_{\substack{x=x_0 \\ t=t_0}}, \qquad a \le x_0 \le b, \quad a \le t_0 \le b.$$

A continuous kernel $K(x,t)$ admits an approximation by a trigonometric polynomial of period $2l$, where $l = b - a$.

For instance, we can set

$$K_{(n)}(x,t) = \frac{1}{2}a_0(t) + \sum_{k=1}^{n} a_k(t)\cos\left(\frac{k\pi x}{l}\right), \tag{5}$$

where the $a_k(t)$ $(k = 0, 1, 2, \dots)$ are the Fourier coefficients

$$a_k(t) = \frac{2}{l} \int_a^b K(x,t)\cos\left(\frac{p\pi x}{l}\right) dx. \tag{6}$$

A similar decomposition can be obtained by interchanging the roles of the variables x and t. A finite segment of the double Fourier series can also be applied by setting, for instance,

$$a_k(t) \approx \frac{1}{2}a_{k0} + \sum_{m=1}^{n} a_{km}\cos\left(\frac{m\pi t}{l}\right), \qquad k = 0, 1, \dots, n, \tag{7}$$

and it follows from formulas (5)–(7) that

$$K_{(n)}(x, t) = \frac{1}{4} a_{00} + \frac{1}{2} \sum_{k=1}^{n} a_{k0} \cos\left(\frac{k\pi x}{l}\right) + \frac{1}{2} \sum_{m=1}^{n} a_{0m} \cos\left(\frac{m\pi t}{l}\right)$$
$$+ \sum_{k=1}^{n} \sum_{m=1}^{n} a_{km} \cos\left(\frac{k\pi x}{l}\right) \cos\left(\frac{m\pi t}{l}\right),$$

where

$$a_{km} = \frac{4}{l^2} \int_a^b \int_a^b K(x, t) \cos\left(\frac{k\pi x}{l}\right) \cos\left(\frac{m\pi t}{l}\right) dx \, dt. \tag{8}$$

One can also use other methods of interpolating and approximating the kernel $K(x, t)$.

11.13-2. The Approximate Solution

If $K_{(n)}(x, t)$ is an approximate degenerate kernel for a given exact kernel $K(x, t)$ and if a function $f_n(x)$ is close to $f(x)$, then the solution $y_n(x)$ of the integral equation

$$y_n(x) - \int_a^b K_{(n)}(x, t) y_n(t) \, dt = f_n(x) \tag{9}$$

can be regarded as an approximation to the solution $y(x)$ of Eq. (1).

Assume that the following error estimates hold:

$$\int_a^b |K(x, t) - K_{(n)}(x, t)| \, dt \le \varepsilon, \qquad |f(x) - f_n(x)| \le \delta.$$

Next, let the resolvent $R_n(x, t)$ of Eq. (9) satisfy the relation

$$\int_a^b |R_n(x, t)| \, dt \le M_n$$

for $a \le x \le b$. Finally, assume that the following inequality holds:

$$q = \varepsilon(1 + M_n) < 1.$$

In this case, Eq. (1) has a unique solution $y(x)$ and

$$|y(x) - y_n(x)| \le \varepsilon \frac{N(1 + M_n)^2}{1 - q} + \delta, \qquad N = \max_{a \le x \le b} |f(x)|. \tag{10}$$

Example. Let us find an approximate solution of the equation

$$y(x) - \int_0^{1/2} e^{-x^2 t^2} y(t) \, dt = 1. \tag{11}$$

Applying the expansion in a double Taylor series, we replace the kernel

$$K(x, t) = e^{-x^2 t^2}$$

by the degenerate kernel

$$K_{(2)}(x, t) = 1 - x^2 t^2 + \tfrac{1}{2} x^4 t^4.$$

Hence, instead of Eq. (11) we obtain

$$y_2(x) = 1 + \int_0^{1/2} \left(1 - x^2 t^2 + \tfrac{1}{2} x^4 t^4\right) y_2(t)\, dt. \tag{12}$$

Therefore,

$$y_2(x) = 1 + A_1 + A_2 x^2 + A_3 x^4, \tag{13}$$

where

$$A_1 = \int_0^{1/2} y_2(x)\, dx, \quad A_2 = -\int_0^{1/2} x^2 y_2(x)\, dx, \quad A_3 = \frac{1}{2} \int_0^{1/2} x^4 y_2(x)\, dx. \tag{14}$$

From (13) and (14) we obtain a system of three equations with three unknowns; to the fourth decimal place, the solution is

$$A_1 = 0.9930, \quad A_2 = -0.0833, \quad A_3 = 0.0007.$$

Hence,

$$y(x) \approx y_2(x) = 1.9930 - 0.0833\, x^2 + 0.0007\, x^4, \quad 0 \le x \le \tfrac{1}{2}. \tag{15}$$

An error estimate for the approximate solution (15) can be performed by formula (10).

⊙ References for Section 11.13: L. V. Kantorovich and V. I. Krylov (1958), S. G. Mikhlin (1960), B. P. Demidovich, I. A. Maron, and E. Z. Shuvalova (1963), M. L. Krasnov, A. I. Kiselev, and G. I. Makarenko (1971).

11.14. The Bateman Method

11.14-1. The General Scheme of the Method

In some cases it is useful, instead of replacing a given kernel by a degenerate kernel, to represent the given kernel approximately as the sum of a kernel whose resolvent is known and a degenerate kernel. For the latter, the resolvent can be written out in a closed form.

Consider the Fredholm integral equation of the second kind

$$y(x) - \lambda \int_a^b k(x,t)y(t)\, dt = f(x) \tag{1}$$

with kernel $k(x,t)$ whose resolvent $r(x,t;\lambda)$ is known; thus, the solution of (1) can be represented in the form

$$y(x) = f(x) + \lambda \int_a^b r(x,t;\lambda) f(t)\, dt. \tag{2}$$

Then, for the integral equation with kernel

$$K(x,t) = \frac{1}{\Delta(a_{ij})} \begin{vmatrix} k(x,t) & g_1(x) & \cdots & g_n(x) \\ h_1(t) & a_{11} & \cdots & a_{1n} \\ \vdots & \vdots & \ddots & \vdots \\ h_n(t) & a_{n1} & \cdots & a_{nn} \end{vmatrix}, \quad \Delta(a_{ij}) = \begin{vmatrix} a_{11} & a_{12} & \cdots & a_{1n} \\ a_{21} & a_{22} & \cdots & a_{2n} \\ \vdots & \vdots & \ddots & \vdots \\ a_{n1} & a_{n2} & \cdots & a_{nn} \end{vmatrix}, \tag{3}$$

where $g_k(x)$ and $h_k(t)$ $(k = 1, \ldots, n)$ are arbitrary functions and a_{ij} $(i,j = 1, \ldots, n)$ are arbitrary numbers, the resolvent has the form

$$R(x,t;\lambda) = \frac{1}{\Delta(a_{ij} + \lambda b_{ij})} \begin{vmatrix} r(x,t;\lambda) & \varphi_1(x) & \cdots & \varphi_n(x) \\ \psi_1(t) & a_{11} + \lambda b_{11} & \cdots & a_{1n} + \lambda b_{1n} \\ \vdots & \vdots & \ddots & \vdots \\ \psi_n(t) & a_{n1} + \lambda b_{n1} & \cdots & a_{nn} + \lambda b_{nn} \end{vmatrix}, \tag{4}$$

where

$$\varphi_k(x) = g_k(x) + \lambda \int_a^b r(x,t;\lambda) g_k(t)\, dt, \quad \psi_k(x) = h_k(x) + \lambda \int_a^b r(x,t;\lambda) h_k(t)\, dt,$$

$$b_{ij} = \int_a^b g_j(x) h_i(x)\, dx, \quad k,i,j = 1, \ldots, n. \tag{5}$$

11.14-2. Some Special Cases

Assume that

$$K(x, t) = k(x, t) - \sum_{k=1}^{n} g_k(x) h_k(t), \tag{6}$$

i.e., in formula (3) we have $a_{ij} = 0$ for $i \neq j$ and $a_{ii} = 1$. For this case, the resolvent is equal to

$$R(x, t; \lambda) = \frac{1}{\Delta_*} \begin{vmatrix} r(x, t; \lambda) & \varphi_1(x) & \cdots & \varphi_n(x) \\ \psi_1(t) & 1 + \lambda b_{11} & \cdots & \lambda b_{1n} \\ \vdots & \vdots & \ddots & \vdots \\ \psi_n(t) & \lambda b_{n1} & \cdots & 1 + \lambda b_{nn} \end{vmatrix}, \quad \Delta_* = \begin{vmatrix} 1 + \lambda b_{11} & \lambda b_{12} & \cdots & \lambda b_{1n} \\ \lambda b_{21} & 1 + \lambda b_{22} & \cdots & \lambda b_{2n} \\ \vdots & \vdots & \ddots & \vdots \\ \lambda b_{n1} & \lambda b_{n2} & \cdots & 1 + \lambda b_{nn} \end{vmatrix}. \tag{7}$$

Moreover, assume that $k(x, t) = 0$, i.e., the kernel $K(x, t)$ is degenerate:

$$K(x, t) = -\sum_{k=1}^{n} g_k(x) h_k(t). \tag{8}$$

In this case it is clear that $r(x, t; \lambda) = 0$ and, by virtue of (7),

$$\varphi_k(x) = g_k(x), \quad \psi_k(x) = h_k(x), \quad b_{ij} = \int_a^b g_j(x) h_i(x)\, dx.$$

Therefore, the resolvent becomes

$$R(x, t; \lambda) = \frac{1}{\Delta_*} \begin{vmatrix} 0 & g_1(x) & \cdots & g_n(x) \\ h_1(t) & 1 + \lambda b_{11} & \cdots & \lambda b_{1n} \\ \vdots & \vdots & \ddots & \vdots \\ h_n(t) & \lambda b_{n1} & \cdots & 1 + \lambda b_{nn} \end{vmatrix}. \tag{9}$$

Now we consider an integral equation with some kernel $Q(x, t)$. On the interval (a, b) we arbitrarily choose points $x_1, \ldots, x_n$ and $t_1, \ldots, t_n$, and in relation (3) we set

$$k(x, t) = 0, \quad g_k(x) = Q(x, t_k), \quad h_k(t) = -Q(x_k, t), \quad a_{ij} = Q(x_i, t_j).$$

In this case it is clear that $r(x, t; \lambda) = 0$, and the kernel $K(x, t)$ acquires the form

$$K(x, t) = \frac{1}{D} \begin{vmatrix} 0 & Q(x, t_1) & \cdots & Q(x, t_n) \\ Q(x_1, t) & Q(x_1, t_1) & \cdots & Q(x_1, t_n) \\ \vdots & \vdots & \ddots & \vdots \\ Q(x_n, t) & Q(x_n, t_1) & \cdots & Q(x_n, t_n) \end{vmatrix}, \quad D = \begin{vmatrix} Q(x_1, t_1) & \cdots & Q(x_1, t_n) \\ \vdots & \ddots & \vdots \\ Q(x_n, t_1) & \cdots & Q(x_n, t_n) \end{vmatrix}.$$

It is convenient to rewrite this formula in the form

$$K(x, t) = Q(x, t) - \frac{1}{D} \begin{vmatrix} Q(x, t) & Q(x, t_1) & \cdots & Q(x, t_n) \\ Q(x_1, t) & Q(x_1, t_1) & \cdots & Q(x_1, t_n) \\ \vdots & \vdots & \ddots & \vdots \\ Q(x_n, t) & Q(x_n, t_1) & \cdots & Q(x_n, t_n) \end{vmatrix}. \tag{10}$$

The kernel $K(x, t)$ is degenerate and, moreover, it coincides with the kernel $Q(x, t)$ on the straight lines $x = x_i$, $t = t_j$ ($i, j = 1, \ldots, n$). Indeed, if we set $x = x_i$ or $t = t_j$, then the determinant in the numerator of the second term has two equal rows or columns and hence vanishes, and therefore,

$$K(x_i, t) = Q(x_i, t), \quad K(x, t_j) = Q(x, t_j).$$

This coincidence on $2n$ straight lines permits us to expect that $K(x,t)$ is close to $Q(x,t)$ and the solution of the equation with kernel $K(x,t)$ is close to the solution of the equation with kernel $Q(x,t)$. It should be noted that if $Q(x,t)$ is degenerate, i.e., has the form

$$Q(x,t) = \sum_{k=1}^{n} g_k(x)h_k(t), \tag{11}$$

then the determinant in the numerator is identically zero, and hence in this case we have

$$K(x,t) \equiv Q(x,t). \tag{12}$$

For the kernel $K(x,t)$, the resolvent can be evaluated on the basis of the following relations:

$$r(x,t;\lambda) = 0, \quad \varphi_i(x) = g_i(x) = Q(x,t_i), \quad \psi_j(t) = h_j(t) = -Q(x_j,t),$$
$$b_{ij} = -\int_a^b Q(x,t_j)Q(x_i,x)\,dx = -Q_2(x_i,t_j), \quad i,j = 1,\dots,n, \tag{13}$$

where $Q_2(x,t)$ is the second iterated kernel for $Q(x,t)$:

$$Q_2(x,y) = \int_a^b Q(x,s)Q(s,t)\,ds,$$

and hence

$$R(x,t;\lambda) = \frac{1}{D - \lambda D_2}
\begin{vmatrix}
0 & Q(x,t_1) & \cdots & Q(x,t_n) \\
Q(x_1,t) & Q(x_1,t_1) - \lambda Q_2(x_1,t_1) & \cdots & Q(x_1,t_n) - \lambda Q_2(x_1,t_n) \\
\vdots & \vdots & \ddots & \vdots \\
Q(x_n,t) & Q(x_n,t_1) - \lambda Q_2(x_n,t_1) & \cdots & Q(x_n,t_n) - \lambda Q_2(x_n,t_n)
\end{vmatrix}, \tag{14}$$

where

$$D_2 = \begin{vmatrix}
Q_2(x_1,t_1) & \cdots & Q_2(x_1,t_n) \\
\vdots & \ddots & \vdots \\
Q_2(x_n,t_1) & \cdots & Q_2(x_n,t_n)
\end{vmatrix}.$$

By using the resolvent $R(x,t;\lambda)$, we can obtain an approximate solution of the equation with kernel $Q(x,t)$. In particular, approximate characteristic values $\tilde{\lambda}$ of this kernel can be found by equating the determinant in the denominator of (14) with zero.

Example. Consider the equation

$$y(x) - \lambda \int_0^1 Q(x,t)y(t)\,dt = 0, \quad 0 \le x \le 1, \tag{15}$$
$$Q(x,t) = \begin{cases} x(t-1) & \text{for } x \le t, \\ t(x-1) & \text{for } x \ge t. \end{cases}$$

Let us find its characteristic values. To this end, we apply formula (14), where for the second iterated kernel we have

$$Q_2(x,t) = \int_0^1 Q(x,s)Q(s,t)\,ds = \begin{cases} \frac{1}{6}x(1-t)(2t-x^2-t^2) & \text{for } x \le t, \\ \frac{1}{6}t(1-x)(2x-x^2-t^2) & \text{for } x \ge t. \end{cases}$$

We choose equidistant points x_i and t_j and take $n = 5$. This implies

$$x_1 = t_1 = \tfrac{1}{6}, \quad x_2 = t_2 = \tfrac{2}{6}, \quad x_3 = t_3 = \tfrac{3}{6}, \quad x_4 = t_4 = \tfrac{4}{6}, \quad x_5 = t_5 = \tfrac{5}{6}.$$

Let us equate the determinant in the denominator of (14) with zero. After some algebraic manipulations, we obtain the following equation:

$$130\mu^5 - 441\mu^4 + 488\mu^3 - 206\mu^2 + 30\mu - 1 = 0 \quad (\tilde{\lambda} = 216\mu),$$

which can be rewritten in the form

$$(\mu - 1)(2\mu - 1)(5\mu - 1)(13\mu^2 - 22\mu + 1) = 0. \tag{16}$$

On solving (16), we obtain

$$\bar{\lambda}_1 = 10.02, \quad \bar{\lambda}_2 = 43.2, \quad \bar{\lambda}_3 = 108, \quad \bar{\lambda}_4 = 216, \quad \bar{\lambda}_5 = 355.2.$$

The exact values of the characteristic values of the equation under consideration are known:

$$\lambda_1 = \pi^2 = 9.869\ldots, \quad \lambda_2 = (2\pi)^2 = 39.478\ldots, \quad \lambda_3 = (3\pi)^2 = 88.826\ldots,$$

and hence the calculation error is 2% for the first characteristic value, 9% for the second characteristic value, and 20% for the third characteristic value.

The result can be improved by choosing another collection of points x_i and t_i ($i = 1, \ldots, 5$). However, for this number of ordinates we cannot have very high precision, because the kernel $Q(x, t)$ itself has a singularity, namely, its derivative is discontinuous for $x = t$, and thus the kernels under consideration cannot provide a good approximation of the given kernel.

⊙ References for Section 11.14: H. Bateman (1922), E. Goursat (1923), L. V. Kantorovich and V. I. Krylov (1958).

11.15. The Collocation Method

11.15-1. General Remarks

Let us rewrite the Fredholm integral equation of the second kind in the form

$$\varepsilon[y(x)] \equiv y(x) - \lambda \int_a^b K(x, t) y(t)\, dt - f(x) = 0. \tag{1}$$

Let us seek an approximate solution of Eq. (1) in the special form

$$Y_n(x) = \Phi(x, A_1, \ldots, A_n) \tag{2}$$

with free parameters $A_1, \ldots, A_n$ (undetermined coefficients). On substituting the expression (2) into Eq. (1), we obtain the residual

$$\varepsilon[Y_n(x)] = Y_n(x) - \lambda \int_a^b K(x, t) Y_n(t)\, dt - f(x). \tag{3}$$

If $y(x)$ is an exact solution, then, clearly, the residual $\varepsilon[y(x)]$ is zero. Therefore, one tries to choose the parameters $A_1, \ldots, A_n$ so that, in a sense, the residual $\varepsilon[Y_n(x)]$ is as small as possible. The residual $\varepsilon[Y_n(x)]$ can be minimized in several ways. Usually, to simplify the calculations, a function $Y_n(x)$ linearly depending on the parameters $A_1, \ldots, A_n$ is taken. On finding the parameters $A_1, \ldots, A_n$, we obtain an approximate solution (2). If

$$\lim_{n \to \infty} Y_n(x) = y(x), \tag{4}$$

then, by taking a sufficiently large number of parameters $A_1, \ldots, A_n$, we find that the solution $y(x)$ can be found with an arbitrary prescribed precision.

Now let us go to the description of a concrete method of construction of an approximate solution $Y_n(x)$.

11.15-2. The Approximate Solution

We set

$$Y_n(x) = \varphi_0(x) + \sum_{i=1}^n A_i \varphi_i(x), \tag{5}$$

where $\varphi_0(x)$, $\varphi_1(x)$, ..., $\varphi_n(x)$ are given functions (*coordinate functions*) and A_1, ..., A_n are indeterminate coefficients, and assume that the functions $\varphi_i(x)$ $(i = 1, \ldots, n)$ are linearly independent. Note that, in particular, we can take $\varphi_0(x) = f(x)$ or $\varphi_0(x) \equiv 0$. On substituting the expression (5) into the left-hand side of Eq. (1), we obtain the residual

$$\varepsilon[Y_n(x)] = \varphi_0(x) + \sum_{i=1}^n A_i \varphi_i(x) - f(x) - \lambda \int_a^b K(x,t) \left[\varphi_0(t) + \sum_{i=1}^n A_i \varphi_i(t) \right] dt,$$

or

$$\varepsilon[Y_n(x)] = \psi_0(x, \lambda) + \sum_{i=1}^n A_i \psi_i(x, \lambda), \tag{6}$$

where

$$\psi_0(x, \lambda) = \varphi_0(x) - f(x) - \lambda \int_a^b K(x,t)\varphi_0(t)\,dt$$
$$\psi_i(x, \lambda) = \varphi_i(x) - \lambda \int_a^b K(x,t)\varphi_i(t)\,dt, \qquad i = 1, \ldots, n. \tag{7}$$

According to the collocation method, we require that the residual $\varepsilon[Y_n(x)]$ be zero at the given system of *the collocation points* x_1, ..., x_n on the interval $[a, b]$, i.e., we set

$$\varepsilon[Y_n(x_j)] = 0, \qquad j = 1, \ldots, n,$$

where

$$a \leq x_1 < x_2 < \cdots < x_{n-1} < x_n \leq b.$$

It is common practice to set $x_1 = a$ and $x_n = b$.

This, together with formula (6) implies the linear algebraic system

$$\sum_{i=1}^n A_i \psi_i(x_j, \lambda) = -\psi_0(x_j, \lambda), \qquad j = 1, \ldots, n, \tag{8}$$

for the coefficients A_1, ..., A_n. If the determinant of system (8) is nonzero,

$$\det[\psi_i(x_j, \lambda)] = \begin{vmatrix} \psi_1(x_1, \lambda) & \psi_1(x_2, \lambda) & \cdots & \psi_1(x_n, \lambda) \\ \psi_2(x_1, \lambda) & \psi_2(x_2, \lambda) & \cdots & \psi_2(x_n, \lambda) \\ \vdots & \vdots & \ddots & \vdots \\ \psi_n(x_1, \lambda) & \psi_n(x_2, \lambda) & \cdots & \psi_n(x_n, \lambda) \end{vmatrix} \neq 0,$$

then system (8) uniquely determines the numbers A_1, ..., A_n, and hence makes it possible to find the approximate solution $Y_n(x)$ by formula (5).

11.15-3. The Eigenfunctions of the Equation

On equating the determinant with zero, we obtain the relation

$$\det[\psi_i(x_j, \lambda)] = 0,$$

which, in general, enables us to find approximate values $\tilde{\lambda}_k$ ($k = 1, \dots, n$) for the characteristic values of the kernel $K(x, t)$.

If we set

$$f(x) \equiv 0, \quad \varphi_0(x) \equiv 0, \quad \lambda = \tilde{\lambda}_k,$$

then, instead of system (8), we obtain the homogeneous system

$$\sum_{i=1}^{n} \tilde{A}_i^{(k)} \psi_i(x_j, \tilde{\lambda}_k) = 0, \qquad j = 1, \dots, n. \tag{9}$$

On finding nonzero solutions $\tilde{A}_i^{(k)}$ ($i = 1, \dots, n$) of system (9), we obtain approximate eigenfunctions for the kernel $K(x, t)$:

$$\tilde{Y}_n^{(k)}(x) = \sum_{i=1}^{n} \tilde{A}_i^{(k)} \varphi_i(x),$$

that correspond to its characteristic value $\lambda_k \approx \tilde{\lambda}_k$.

Example. Let us solve the equation

$$y(x) - \int_0^1 \frac{t^2 y(t)}{x^2 + t^2} \, dt = x \arctan \frac{1}{x} \tag{10}$$

by the collocation method.

We set

$$Y_2(x) = A_1 + A_2 x.$$

On substituting this expression into Eq. (10), we obtain the residual

$$\varepsilon[Y_2(x)] = -A_1 x \arctan \frac{1}{x} + A_2 \left[x - \frac{1}{2} + \frac{x^2}{2} \ln \left(1 + \frac{1}{x^2} \right) \right] - x \arctan \frac{1}{x}.$$

On choosing the collocation points $x_1 = 0$ and $x_2 = 1$ and taking into account the relations

$$\lim_{x \to 0} x \arctan \frac{1}{x} = 0, \quad \lim_{x \to 0} x^2 \ln \left(1 + \frac{1}{x^2} \right) = 0,$$

we obtain the following system for the coefficients A_1 and A_2:

$$0 \times A_1 - \tfrac{1}{2} A_2 = 0,$$
$$-\tfrac{\pi}{4} A_1 + \tfrac{1}{2}(1 + \ln 2) A_2 = \tfrac{\pi}{4}.$$

This implies $A_2 = 0$ and $A_1 = -1$. Thus,

$$Y_2(x) = -1. \tag{11}$$

We can readily verify that the approximate solution (11) thus obtained is exact.

⊙ References for Section 11.15: L. Collatz (1960), B. P. Demidovich, I. A. Maron, and E. Z. Shuvalova (1963), A. F. Verlan' and V. S. Sizikov (1986).

11.16. The Method of Least Squares

11.16-1. Description of the Method

By analogy with the collocation method, for the equation

$$\varepsilon[y(x)] \equiv y(x) - \lambda \int_a^b K(x,t)y(t)\,dt - f(x) = 0 \tag{1}$$

we set

$$Y_n(x) = \varphi_0(x) + \sum_{i=1}^n A_i\varphi_i(x), \tag{2}$$

where $\varphi_0(x)$, $\varphi_1(x)$, ..., $\varphi_n(x)$ are given functions, $A_1, \ldots, A_n$ are indeterminate coefficients, and the $\varphi_i(x)$ $(i = 1, \ldots, n)$ are linearly independent.

On substituting (2) into the left-hand side of Eq. (1), we obtain the residual

$$\varepsilon[Y_n(x)] = \psi_0(x, \lambda) + \sum_{i=1}^n A_i\psi_i(x, \lambda), \tag{3}$$

where $\psi_0(x, \lambda)$ and the $\psi_i(x, \lambda)$ $(i = 1, \ldots, n)$ are defined by formulas (7) of Subsection 11.15-2.

According to the method of least squares, the coefficients A_i $(i = 1, \ldots, n)$ can be found from the condition for the minimum of the integral

$$I = \int_a^b \{\varepsilon[Y_n(x)]\}^2\,dx = \int_a^b \left[\psi_0(x, \lambda) + \sum_{i=1}^n A_i\psi_i(x, \lambda)\right]^2 dx. \tag{4}$$

This requirement leads to the algebraic system of equations

$$\frac{\partial I}{\partial A_j} = 0, \qquad j = 1, \ldots, n, \tag{5}$$

and hence, on the basis of (4), by differentiating with respect to the parameters $A_1, \ldots, A_n$ under the integral sign, we obtain

$$\frac{1}{2}\frac{\partial I}{\partial A_j} = \int_a^b \psi_j(x, \lambda)\left[\psi_0(x, \lambda) + \sum_{i=1}^n A_i\psi_i(x, \lambda)\right] dx = 0, \qquad j = 1, \ldots, n. \tag{6}$$

Using the notation

$$c_{ij}(\lambda) = \int_a^b \psi_i(x, \lambda)\psi_j(x, \lambda)\,dx, \tag{7}$$

we can rewrite system (6) in the form of the *normal system of the method of least squares*:

$$\begin{aligned}
c_{11}(\lambda)A_1 + c_{12}(\lambda)A_2 + \cdots + c_{1n}(\lambda)A_n &= -c_{10}(\lambda), \\
c_{21}(\lambda)A_1 + c_{22}(\lambda)A_2 + \cdots + c_{2n}(\lambda)A_n &= -c_{20}(\lambda), \\
\cdots\cdots\cdots\cdots\cdots\cdots\cdots\cdots\cdots\cdots\cdots\cdots\cdots \\
c_{n1}(\lambda)A_1 + c_{n2}(\lambda)A_2 + \cdots + c_{nn}(\lambda)A_n &= -c_{n0}(\lambda).
\end{aligned} \tag{8}$$

Note that if $\varphi_0(x) \equiv 0$, then $\psi_0(x) = -f(x)$. Moreover, since $c_{ij}(\lambda) = c_{ji}(\lambda)$, the matrix of system (8) is symmetric.

11.16-2. The Construction of Eigenfunctions

The method of least squares can also be applied for the approximate construction of characteristic values and eigenfunctions of the kernel $K(x, s)$, similarly to the way in which it can be done in the collocation method. Namely, by setting $f(x) \equiv 0$ and $\varphi_0(x) \equiv 0$, which implies $\psi_0(x) \equiv 0$, we determine approximate values of the characteristic values from the algebraic equation

$$\det[c_{ij}(\lambda)] = 0. \tag{9}$$

After this, approximate eigenfunctions can be found from the homogeneous system of the form (8), where, instead of λ, the corresponding approximate value is substituted.

Example. Let us find an approximate solution of the equation

$$y(x) = x^2 + \int_{-1}^{1} \sinh(x + t)y(t)\, dt \tag{10}$$

by the method of least squares.

For the form of an approximate solution we take $Y_2(x) = x^2 + A_2 x + A_1$. This implies

$$\varphi_1(x) = 1, \quad \varphi_2(x) = x, \quad \varphi_0(x) = x^2.$$

Taking into account the relations

$$\int_{-1}^{1} \sinh(x + t)\, dt = a \sinh x, \quad \int_{-1}^{1} t \sinh(x + t)\, dt = b \sinh x, \quad \int_{-1}^{1} t^2 \sinh(x + t)\, dt = c \sinh x,$$

$$a = 2 \sinh 1 = 2.3504, \quad b = 2e^{-1} = 0.7358, \quad c = 6 \sinh 1 - 4 \cosh 1 = 0.8788,$$

on the basis of formulas (7) of Subsection 11.15-2 we have

$$\psi_1 = 1 - a \sinh x, \quad \psi_2 = x - b \cosh x, \quad \psi_0 = -c \sinh x.$$

Furthermore, we see that (to the fourth decimal place)

$$c_{11} = 2 + a^2 \left(\tfrac{1}{2} \sinh 2 - 1 \right) = 6.4935, \quad c_{22} = \tfrac{2}{3} + b^2 \left(\tfrac{1}{2} \sinh 2 + 1 \right) = 2.1896,$$

$$c_{12} = -4(ae^{-1} + b \sinh 1) = -8e^{-1} \sinh 1 = -3.4586, \quad c_{10} = ac\left(\tfrac{1}{2} \sinh 2 - 1 \right) = 1.6800, \quad c_{20} = -2ce^{-1} = -0.6466,$$

and obtain the following system for the coefficients A_1 and A_2:

$$6.4935 A_1 - 3.4586 A_2 = -1.6800,$$
$$-3.4586 A_1 + 2.1896 A_2 = 0.6466.$$

Hence, we have $A_1 = -0.5423$ and $A_2 = -0.5613$. Thus,

$$Y_2(x) = x^2 - 0.5613x - 0.5423. \tag{11}$$

Since the kernel

$$K(x, t) = \sinh(x + t) = \sinh x \cosh t + \cosh x \sinh t$$

of Eq. (10) is degenerate, we can readily obtain the exact solution

$$y(x) = x^2 + \alpha \sinh x + \beta \cosh x, \tag{12}$$

$$\alpha = \frac{6 \sinh 1 - 4 \cosh 1}{2 - \left(\tfrac{1}{2} \sinh 2 \right)^2} = -0.6821, \quad \beta = \alpha\left(\tfrac{1}{2} \sinh 2 - 1 \right) = -0.5548.$$

On comparing formulas (11) and (12) we conclude that the approximate solution $Y_2(x)$ is close to the exact solution $y(x)$ if $|x|$ is small. At the endpoints $x = \pm 1$, the discrepancy $|y(x) - Y_2(x)|$ is rather significant.

⊙ References for Section 11.16: L. V. Kantorovich and V. I. Krylov (1958), B. P. Demidovich, I. A. Maron, and E. Z. Shuvalova (1963), M. L. Krasnov, A. I. Kiselev, and G. I. Makarenko (1971).

11.17. The Bubnov–Galerkin Method

11.17-1. Description of the Method

Let

$$\varepsilon[y(x)] \equiv y(x) - \lambda \int_a^b K(x,t)y(t)\,dt - f(x) = 0. \tag{1}$$

Similarly to the above reasoning, we seek an approximate solution of Eq. (1) in the form of a finite sum

$$Y_n(x) = f(x) + \sum_{i=1}^n A_i \varphi_i(x), \qquad i = 1, \ldots, n, \tag{2}$$

where the $\varphi_i(x)$ $(i = 1, \ldots, n)$ are some given linearly independent functions (*coordinate functions*) and $A_1, \ldots, A_n$ are indeterminate coefficients. On substituting the expression (2) into the left-hand side of Eq. (1), we obtain the residual

$$\varepsilon[Y_n(x)] = \sum_{j=1}^n A_j \left[\varphi_j(x) - \lambda \int_a^b K(x,t)\varphi_j(t)\,dt \right] - \lambda \int_a^b K(x,t)f(t)\,dt. \tag{3}$$

According to the Bubnov–Galerkin method, the coefficients A_i $(i = 1, \ldots, n)$ are defined from the condition that the residual is orthogonal to all coordinate functions $\varphi_1(x), \ldots, \varphi_n(x)$. This gives the system of equations

$$\int_a^b \varepsilon[Y_n(x)]\varphi_i(x)\,dx = 0, \qquad i = 1, \ldots, n,$$

or, by virtue of (3),

$$\sum_{j=1}^n (\alpha_{ij} - \lambda \beta_{ij})A_j = \lambda \gamma_i, \qquad i = 1, \ldots, n, \tag{4}$$

where

$$\alpha_{ij} = \int_a^b \varphi_i(x)\varphi_j(x)\,dx, \quad \beta_{ij} = \int_a^b \int_a^b K(x,t)\varphi_i(x)\varphi_j(t)\,dt\,dx, \quad \gamma_i = \int_a^b \int_a^b K(x,t)\varphi_i(x)f(t)\,dt\,dx.$$

If the determinant of system (4)

$$D(\lambda) = \det[\alpha_{ij} - \lambda \beta_{ij}]$$

is nonzero, then this system uniquely determines the coefficients $A_1, \ldots, A_n$. In this case, formula (2) gives an approximate solution of the integral equation (1).

11.17-2. Characteristic Values

The equation $D(\lambda) = 0$ gives approximate characteristic values $\tilde{\lambda}_1, \ldots, \tilde{\lambda}_n$ of the integral equation. On finding nonzero solutions of the homogeneous linear system

$$\sum_{j=1}^n (\alpha_{ij} - \tilde{\lambda}_k \beta_{ij})\tilde{A}_j^{(k)} = 0, \qquad i = 1, \ldots, n,$$

we can construct approximate eigenfunctions $\tilde{Y}_n^{(k)}(x)$ corresponding to characteristic values $\tilde{\lambda}_k$:

$$\tilde{Y}_n^{(k)}(x) = \sum_{i=1}^n \tilde{A}_i^{(k)} \varphi(x).$$

It can be shown that the Bubnov–Galerkin method is equivalent to the replacement of the kernel $K(x,t)$ by some degenerate kernel $K_{(n)}(x,t)$. Therefore, for the approximate solution $Y_n(x)$ we have an error estimate similar to that presented in Subsection 11.13-2.

Example. Let us find the first two characteristic values of the integral equation

$$\varepsilon[y(x)] \equiv y(x) - \lambda \int_0^1 K(x,t)y(t)\,dt = 0,$$

where

$$K(x,t) = \begin{cases} t & \text{for } t \le x, \\ x & \text{for } t > x. \end{cases} \tag{5}$$

On the basis of (5), we have

$$\varepsilon[y(x)] = y(x) - \lambda \left\{ \int_0^x ty(t)\,dt + \int_x^1 xy(t)\,dt \right\}.$$

We set $Y_2(x) = A_1 x + A_2 x^2$. In this case

$$\varepsilon[Y_2(x)] = A_1 x + A_2 x^2 - \lambda \left[\tfrac{1}{3}A_1 x^3 + \tfrac{1}{4}A_2 x^4 + x\left(\tfrac{1}{2}A_1 + \tfrac{1}{3}A_2\right) - \left(\tfrac{1}{2}A_1 x^3 + \tfrac{1}{3}A_2 x^4\right) \right] =$$
$$= A_1 \left[\left(1 - \tfrac{1}{2}\lambda\right)x + \tfrac{1}{6}\lambda x^3 \right] + A_2 \left(-\tfrac{1}{3}\lambda x + x^2 + \tfrac{1}{12}\lambda x^4 \right).$$

On orthogonalizing the residual $\varepsilon[Y_2(x)]$, we obtain the system

$$\int_0^1 \varepsilon[Y_2(x)]x\,dx = 0,$$

$$\int_0^1 \varepsilon[Y_2(x)]x^2\,dx = 0,$$

or the following homogeneous system of two algebraic equations with two unknowns:

$$A_1(120 - 48\lambda) + A_2(90 - 35\lambda) = 0$$
$$A_1(630 - 245\lambda) + A_2(504 - 180\lambda) = 0. \tag{6}$$

On equating the determinant of system (6) with zero, we obtain the following equation for the characteristic values:

$$D(\lambda) \equiv \begin{vmatrix} 120 - 48\lambda & 90 - 35\lambda \\ 630 - 245\lambda & 504 - 180\lambda \end{vmatrix} = 0.$$

Hence,

$$\lambda^2 - 26.03\lambda + 58.15 = 0. \tag{7}$$

Equations (7) imply

$$\bar{\lambda}_1 = 2.462\ldots \quad \text{and} \quad \bar{\lambda}_2 = 23.568\ldots$$

For comparison we present the exact characteristic values:

$$\lambda_1 = \tfrac{1}{4}\pi^2 = 2.467\ldots \quad \text{and} \quad \lambda_2 = \tfrac{9}{4}\pi^2 = 22.206\ldots,$$

which can be obtained from the solution of the following boundary value problem equivalent to the original equation:

$$y''_{xx}(x) + \lambda y(x) = 0; \qquad y(0) = 0, \quad y'_x(1) = 0.$$

Thus, the error of $\bar{\lambda}_1$ is approximately equal to 0.2% and that of $\bar{\lambda}_2$, to 6%.

⊙ References for Section 11.17: L. V. Kantorovich and V. I. Krylov (1958), B. P. Demidovich, I. A. Maron, and E. Z. Shuvalova (1963), A. F. Verlan' and V. S. Sizikov (1986).

11.18. The Quadrature Method

11.18-1. The General Scheme for Fredholm Equations of the Second Kind

In the solution of an integral equation, the reduction to the solution of systems of algebraic equations obtained by replacing the integrals with finite sums is one of the most effective tools. The method of quadratures is related to the approximation methods. It is widespread in practice because it is rather universal with respect to the principle of constructing algorithms for solving both linear and nonlinear equations.

Just as in the case of Volterra equations, the method is based on a quadrature formula (see Subsection 8.7-1):

$$\int_a^b \varphi(x)\,dx = \sum_{j=1}^n A_j \varphi(x_j) + \varepsilon_n[\varphi], \tag{1}$$

where the x_j are the nodes of the quadrature formula, the A_j are given coefficients that do not depend on the function $\varphi(x)$, and $\varepsilon_n[\varphi]$ is the error of replacement of the integral by the sum (the truncation error).

If in the Fredholm integral equation of the second kind,

$$y(x) - \lambda \int_a^b K(x,t)y(t)\,dt = f(x), \qquad a \le x \le b, \tag{2}$$

we set $x = x_i$ ($i = 1, \ldots, n$), then we obtain the following relation that is the basic formula for the method under consideration:

$$y(x_i) - \lambda \int_a^b K(x_i,t)y(t)\,dt = f(x_i), \qquad i = 1, \ldots, n. \tag{3}$$

Applying the quadrature formula (1) to the integral in (3), we arrive at the following system of equations:

$$y(x_i) - \lambda \sum_{j=1}^n A_j K(x_i, x_j)y(x_j) = f(x_i) + \lambda \varepsilon_n[y]. \tag{4}$$

By neglecting the small term $\lambda \varepsilon_n[y]$ in this formula, we obtain the system of linear algebraic equations for approximate values y_i of the solution $y(x)$ at the nodes $x_1, \ldots, x_n$:

$$y_i - \lambda \sum_{j=1}^n A_j K_{ij} y_j = f_i, \qquad i = 1, \ldots, n, \tag{5}$$

where $K_{ij} = K(x_i, x_j)$, $f_i = f(x_i)$.

The solution of system (5) gives the values $y_1, \ldots, y_n$, which determine an approximate solution of the integral equation (2) on the entire interval $[a, b]$ by interpolation. Here for the approximate solution we can take the function obtained by linear interpolation, i.e., the function that coincides with y_i at the points x_i and is linear on each of the intervals $[x_i, x_{i+1}]$. Moreover, for an analytic expression of the approximate solution to the equation, a function

$$\tilde{y}(x) = f(x) + \lambda \sum_{j=1}^n A_j K(x, x_j)y_j \tag{6}$$

can be chosen, which also takes the values $y_1, \ldots, y_n$ at the points $x_1, \ldots, x_n$.

11.18-2. Construction of the Eigenfunctions

The method of quadratures can also be applied for solutions of homogeneous Fredholm equations of the second kind. In this case, system (5) becomes homogeneous ($f_i = 0$) and has a nontrivial solution only if its determinant $D(\lambda)$ is equal to zero. The algebraic equation $D(\lambda) = 0$ of degree n for λ makes it possible to find the roots $\tilde{\lambda}_1, \ldots, \tilde{\lambda}_n$, which are approximate values of n characteristic values of the equation. The substitution of each value $\tilde{\lambda}_k$ ($k = 1, \ldots, n$) into (5) for $f_i \equiv 0$ leads to the system of equations

$$y_i^{(k)} - \tilde{\lambda}_k \sum_{j=1}^n A_j K_{ij} y_j^{(k)} = 0, \qquad i = 1, \ldots, n,$$

whose nonzero solutions $y_i^{(k)}$ make it possible to obtain approximate expressions for the eigenfunctions of the integral equation:

$$\tilde{y}_k(x) = \tilde{\lambda}_k \sum_{j=1}^n A_j K(x, x_j)y_j^{(k)}.$$

If λ differs from each of the roots $\tilde{\lambda}_k$, then the nonhomogeneous system of linear algebraic equations (5) has a unique solution. In the same case, the homogeneous system of equations (5) has only the trivial solution.

11.18-3. Specific Features of the Application of Quadrature Formulas

The accuracy of the resulting solutions essentially depends on the smoothness of the kernel and the constant term. When choosing the quadrature formula, it is necessary to take into account that the more accurate an applied formula is, the more serious requirements must be imposed on the smoothness of the kernel, the solution, and the right-hand side.

If the right-hand side or the kernel have singularities, then it is reasonable to perform a preliminary transform of the original equation to obtain a more accurate approximate solution. Here the following methods can be applied.

If the right-hand side $f(x)$ has singularities and the kernel is smooth, then we can introduce the new unknown function $z(x) = y(x) - f(x)$ instead of $y(x)$, and the substitution of $z(x)$ in the original equation leads to the equation

$$z(x) - \lambda \int_a^b K(x,t)z(t)\,dt = \lambda \int_a^b K(x,t)f(t)\,dt,$$

in which the right-hand side is smoothed, and hence a solution $z(x)$ is smoother. From the function $z(x)$ thus obtained we can readily find the desired solution $y(x)$.

For the cases in which the kernel $K(x,t)$ or its derivatives with respect to t have discontinuities on the diagonal $x = t$, it is useful to rewrite the equation under consideration in the equivalent form

$$y(x)\left[1 - \lambda \int_a^b K(x,t)\,dt\right] - \lambda \int_a^b K(x,t)[y(t) - y(x)]\,dt = f(x),$$

where the integrand in the second integral has no singularities because the difference $y(t) - y(x)$ vanishes on the diagonal $x = t$, and the calculation of the integral $\int_a^b K(x,t)\,dt$ is performed without unknown functions and is possible in the explicit form.

Example. Consider the equation

$$y(x) - \frac{1}{2}\int_0^1 xty(t)\,dt = \frac{5}{6}x.$$

Let us choose the nodes $x_1 = 0$, $x_2 = \frac{1}{2}$, $x_3 = 1$ and calculate the values of the right-hand side $f(x) = \frac{5}{6}x$ and of the kernel $K(x,t) = xt$ at these nodes:

$$f(0) = 0, \quad f\left(\tfrac{1}{2}\right) = \tfrac{5}{12}, \quad f(1) = \tfrac{5}{6},$$

$$K(0,0) = 0, \quad K\left(0, \tfrac{1}{2}\right) = 0, \quad K(0,1) = 0, \quad K\left(\tfrac{1}{2},0\right) = 0, \quad K\left(\tfrac{1}{2}, \tfrac{1}{2}\right) = \tfrac{1}{4},$$

$$K\left(\tfrac{1}{2},1\right) = \tfrac{1}{2}, \quad K(1,0) = 0, \quad K\left(1, \tfrac{1}{2}\right) = \tfrac{1}{2}, \quad K(1,1) = 1.$$

On applying Simpson's rule (see Subsection 8.7-1)

$$\int_0^1 F(x)\,dx \approx \frac{1}{6}\left[F(0) + 4F\left(\tfrac{1}{2}\right) + F(1)\right]$$

to determine the approximate values y_i ($i = 1, 2, 3$) of the solution $y(x)$ at the nodes x_i we obtain the system

$$y_1 = 0,$$
$$\tfrac{11}{12}y_2 - \tfrac{1}{24}y_3 = \tfrac{5}{12},$$
$$-\tfrac{2}{12}y_2 + \tfrac{11}{12}y_3 = \tfrac{5}{6},$$

whose solution is $y_1 = 0$, $y_2 = \frac{1}{2}$, $y_3 = 1$. In accordance with the expression (6), the approximate solution can be presented in the form

$$\bar{y}(x) = \tfrac{5}{6}x + \tfrac{1}{2} \times \tfrac{1}{6}\left(0 + 4 \times \tfrac{1}{2} \times \tfrac{1}{2}x + 1 \times 1 \times x\right) = x.$$

We can readily verify that it coincides with the exact solution.

⊙ References for Section 11.18: N. S. Bakhvalov (1973), V. I. Krylov, V. V. Bobkov, and P. I. Monastyrnyi (1984), A. F. Verlan' and V. S. Sizikov (1986).

11.19. Systems of Fredholm Integral Equations of the Second Kind

11.19-1. Some Remarks

A system of Fredholm integral equations of the second kind has the form

$$y_i(x) - \lambda \sum_{j=1}^{n} \int_a^b K_{ij}(x,t)y_j(t)\,dt = f_i(x), \qquad a \le x \le b, \quad i = 1, \ldots, n. \tag{1}$$

Assume that the kernels $K_{ij}(x,t)$ are continuous or square integrable on the square $S = \{a \le x \le b, a \le t \le b\}$ and the right-hand sides $f_i(x)$ are continuous or square integrable on $[a,b]$. We also assume that the functions $y_i(x)$ to be defined are continuous or square integrable on $[a,b]$ as well. The theory developed above for Fredholm equations of the second kind can be completely extended to such systems. In particular, it can be shown that for systems (1), the successive approximations converge in mean-square to the solution of the system if λ satisfies the inequality

$$|\lambda| < \frac{1}{B_*}, \tag{2}$$

where

$$\sum_{i=1}^{n} \sum_{j=1}^{n} \int_a^b \int_a^b |K_{ij}(x,t)|^2\,dx\,dt = B_*^2 < \infty. \tag{3}$$

If the kernel $K_{ij}(x,t)$ satisfies the additional condition

$$\int_a^b K_{ij}^2(x,t)\,dt \le A_{ij}, \qquad a \le x \le b, \tag{4}$$

where A_{ij} are some constants, then the successive approximations converge absolutely and uniformly.

If all kernels $K_{ij}(x,t)$ are degenerate, then system (1) can be reduced to a linear algebraic system. It can be established that for a system of Fredholm integral equations, all Fredholm theorems are satisfied.

11.19-2. The Method of Reducing a System of Equations to a Single Equation

System (1) can be transformed into a single Fredholm integral equation of the second kind. Indeed, let us introduce the functions $Y(x)$ and $F(x)$ on $[a,\ nb - (n-1)a]$ by setting

$$Y(x) = y_i\big(x - (i-1)(b-a)\big), \quad F(x) = f_i\big(x - (i-1)(b-a)\big),$$

for

$$(i-1)b - (i-2)a \le x \le ib - (i-1)a.$$

Let us define a kernel $K(x,t)$ on the square $\{a \le x \le nb - (n-1)a,\ a \le t \le nb - (n-1)a\}$ as follows:

$$K(x,t) = K_{ij}\big(x - (i-1)(b-a),\ t - (j-1)(b-a)\big)$$

for

$$(i-1)b - (i-2)a \le x \le ib - (i-1)a, \quad (j-1)b - (j-2)a \le t \le jb - (j-1)a.$$

Now system (1) can be rewritten as the single Fredholm equation

$$Y(x) - \lambda \int_a^{nb-(n-1)a} K(x,t)Y(t)\,dt = F(x), \qquad a \le x \le nb-(n-1)a.$$

If the kernels $K_{ij}(x,t)$ are square integrable on the square $S = \{a \le x \le b,\ a \le t \le b\}$ and the right-hand sides $f_i(x)$ are square integrable on $[a, b]$, then the kernel $K(x,t)$ is square integrable on the new square

$$S_n = \{a < x < nb-(n-1)a,\ a < t < nb-(n-1)a\},$$

and the right-hand side $F(x)$ is square integrable on $[a,\ nb - (n-1)a]$.

If condition (4) is satisfied, then the kernel $K(x,t)$ satisfies the inequality

$$\int_a^b K^2(x,t)\,dt \le A_*, \qquad a < x < nb-(n-1)a,$$

where A_* is a constant.

⊙ Reference for Section 11.19: S. G. Mikhlin (1960).

11.20. Regularization Method for Equations With Infinite Limits of Integration

11.20-1. Basic Equation and Fredholm Theorems

Consider an integral equation of the second kind in the form

$$y(x) + \frac{1}{\sqrt{2\pi}} \int_0^\infty K_1(x-t)y(t)\,dt + \frac{1}{\sqrt{2\pi}} \int_{-\infty}^0 K_2(x-t)y(t)\,dt + \int_{-\infty}^\infty M(x,t)y(t)\,dt = f(x), \quad (1)$$

where $-\infty < x < \infty$. We assume that the functions $y(x)$ and $f(x)$ and the kernels $K_1(x)$ and $K_2(x)$ are such that their Fourier transforms belong to $L_2(-\infty, \infty)$ and satisfy the Hölder condition. We also assume that the Fourier transforms of the kernel $M(x,t)$ with respect to each variable belong to $L_2(-\infty, \infty)$ and satisfy the Hölder condition and, in addition,

$$\int_{-\infty}^\infty \int_{-\infty}^\infty |M(x,t)|^2\,dx\,dt < \infty.$$

It should be noted that Eq. (1) with $M(x,t) \equiv 0$ is the convolution-type integral equation with two kernels which was discussed in Subsection 11.9-2.

The transposed homogeneous equation has the form

$$\varphi(x) + \frac{1}{\sqrt{2\pi}} \int_0^\infty K_1(t-x)\varphi(t)\,dt + \frac{1}{\sqrt{2\pi}} \int_{-\infty}^0 K_2(t-x)\varphi(t)\,dt + \int_{-\infty}^\infty M(t,x)\varphi(t)\,dt = 0, \quad (2)$$

where $-\infty < x < \infty$.

Assume that the normality conditions (see Subsection 11.9-2) hold, that is,

$$1 + \mathcal{K}_1(u) \ne 0, \quad 1 + \mathcal{K}_2(u) \ne 0, \qquad -\infty < u < \infty. \tag{3}$$

THEOREM 1. *The number of linearly independent solutions of the homogeneous ($f(x) \equiv 0$) equation (1) and that of the transposed homogeneous ($g(x) \equiv 0$) equation (2) are finite.*

THEOREM 2. *For the nonhomogeneous equation (1) to be solvable, it is necessary and sufficient that*

$$\int_{-\infty}^{\infty} f(t)\varphi_k(t)\,dt = 0, \qquad k = 1,\dots,N, \tag{4}$$

where $\varphi_k(x)$ is a complete finite set of linearly independent solutions to the transposed homogeneous equation (2).

THEOREM 3. *The difference between the number of linearly independent solutions to the homogeneous equation (1) and the number of linearly independent solutions to the homogeneous transposed equation (2) is equal to the index*

$$\nu = \operatorname{Ind} \frac{1+\mathcal{K}_2(u)}{1+\mathcal{K}_1(u)} = \frac{1}{2\pi}\left[\arg\frac{1+\mathcal{K}_2(u)}{1+\mathcal{K}_1(u)}\right]_{-\infty}^{\infty}. \tag{5}$$

11.20-2. Regularizing Operators

An important method for the theoretical investigation and practical solution of the integral equations in question is a regularization of these equations, i.e., their reduction to a Fredholm equation of the second kind.

Let us denote by **K** the operator determined by the left-hand side of Eq. (1):

$$\mathbf{K}[y(x)] \equiv y(x) + \frac{1}{\sqrt{2\pi}}\int_0^{\infty} K_1(x-t)y(t)\,dt + \frac{1}{\sqrt{2\pi}}\int_{-\infty}^0 K_2(x-t)y(t)\,dt + \int_{-\infty}^{\infty} M(x,t)y(t)\,dt \tag{6}$$

and introduce the similar operator

$$\mathbf{L}[\omega(x)] \equiv \omega(x) + \frac{1}{\sqrt{2\pi}}\int_0^{\infty} L_1(x-t)\omega(t)\,dt + \frac{1}{\sqrt{2\pi}}\int_{-\infty}^0 L_2(x-t)\omega(t)\,dt + \int_{-\infty}^{\infty} Q(x,t)\omega(t)\,dt. \tag{7}$$

Let us find an operator **L** such that the product **LK** is determined by the left-hand side of a Fredholm equation of the second kind with a kernel $K(x,t)$:

$$\mathbf{LK}[y(x)] \equiv y(x) + \int_{-\infty}^{\infty} K(x,t)y(t)\,dt, \qquad \int_{-\infty}^{\infty}\int_{-\infty}^{\infty}|K(x,t)|^2\,dx\,dt < \infty. \tag{8}$$

The operator **L** is called a *left regularizer*.

For the operator **K** of the integral equation (1) to have a left regularizer **L** of the form (7), it is necessary and sufficient that the normality conditions (3) hold.

If conditions (3) are satisfied, then the left regularizer **L** has the form

$$\mathbf{L}\omega(x) \equiv \omega(x) - \frac{1}{\sqrt{2\pi}}\int_0^{\infty} R_1(x-t)\omega(t)\,dt - \frac{1}{\sqrt{2\pi}}\int_{-\infty}^0 R_2(x-t)\omega(t)\,dt + \int_{-\infty}^{\infty} Q(x,t)\omega(t)\,dt, \tag{9}$$

where the resolvents $R_1(x-t)$ and $R_2(x-t)$ of the kernels $K_1(x-t)$ and $K_2(x-t)$ are given by (see Subsection 11.8-1)

$$R_j(x) = \frac{1}{\sqrt{2\pi}}\int_{-\infty}^{\infty}\frac{\mathcal{K}_j(u)}{1+\mathcal{K}_j(u)}e^{-iux}\,du, \quad \mathcal{K}_j(u) = \frac{1}{\sqrt{2\pi}}\int_{-\infty}^{\infty} K_j(x)e^{iux}\,dx, \quad j=1,2,$$

and $Q(x,t)$ is any function such that

$$\int_{-\infty}^{\infty}\int_{-\infty}^{\infty}|Q(x,t)|^2\,dx\,dt < \infty.$$

If condition (3) is satisfied, then the operator **L** given by formula (9) is simultaneously a *right regularizer* of the operator **K**:

$$\mathbf{KL}[y(x)] \equiv y(x) + \int_{-\infty}^{\infty} K_*(x, t) y(t)\, dt, \tag{10}$$

where the function $K_*(x, t)$ satisfies the condition

$$\int_{-\infty}^{\infty} \int_{-\infty}^{\infty} |K_*(x, t)|^2\, dx\, dt < \infty. \tag{11}$$

11.20-3. The Regularization Method

Consider the equation of the form

$$\mathbf{K}[y(x)] = f(x), \qquad -\infty < x < \infty, \tag{12}$$

where the operator **K** is defined by (6).

There are several ways of regularizing this equation, i.e., of its reduction to a Fredholm equation. First, this equation can be reduced to an equation with a Cauchy kernel. On regularizing the last equation by a method presented in Section 13.4, we can achieve our aim. This approach can be applied if we can find, for given functions $K_1(x)$, $K_2(x)$, $M(x, t)$, and $f(x)$, simple expressions for their Fourier integrals. Otherwise it is natural to perform the regularization of Eq. (12) directly, without passing to the inverse transforms.

A left regularization of Eq. (12) involves the application of the regularizer **L** constructed in the previous subsection to both its sides:

$$\mathbf{LK}[y(x)] = \mathbf{L}[f(x)]. \tag{13}$$

It follows from (8) that Eq. (13) is a Fredholm equation

$$y(x) + \int_{-\infty}^{\infty} K(x, t) y(t)\, dt = \mathbf{L}[f(x)]. \tag{14}$$

Thus, Eq. (12) can be transformed by left regularization to a Fredholm equation with the same unknown function $y(x)$ and the known right-hand side $\mathbf{L}[f(x)]$. Left regularization is known to imply no loss of solutions: all solutions of the original equation (12) are solutions of the regularized equation. However, in the general case, a solution of the regularized equation need not be a solution of the original equation.

The right regularization consists in the substitution of the expression

$$y(x) = \mathbf{L}[\omega(x)] \tag{15}$$

for the desired function into Eq. (12), where $\omega(x)$ is a new unknown function. We finally arrive at the following integral equation:

$$\mathbf{KL}[\omega(x)] = f(x), \tag{16}$$

which is a Fredholm equation as well by virtue of (10):

$$\mathbf{KL}[\omega(x)] \equiv \omega(x) + \int_{-\infty}^{\infty} K_*(x, t) \omega(t)\, dt = f(x), \qquad -\infty < x < \infty. \tag{17}$$

Thus, we have passed from Eq. (12) for the unknown function $y(x)$ to a Fredholm integral equation for a new unknown function $\omega(x)$. On solving the Fredholm equation (17), we find a solution of the original equation (12) by formula (15). Right regularization can give no irrelevant solutions, but it is known that it can lead to a loss of a solution.

A solution of the problem on an equivalent regularization, for which neither the loss of solutions nor the appearance of irrelevant "solutions" occur, is of significant theoretical and practical interest.

For Eq. (12) with an arbitrary right-hand side $f(x)$ to admit an equivalent left regularization, it is necessary and sufficient that the index ν given by formula (5) be nonnegative. For an equivalently regularizing operator we can take the operator

$$\mathbf{L}^\circ[\omega(x)] \equiv \omega(x) - \frac{1}{\sqrt{2\pi}} \int_0^\infty R_1(x-t)\omega(t)\,dt - \frac{1}{\sqrt{2\pi}} \int_{-\infty}^0 R_2(x-t)\omega(t)\,dt.$$

Thus, the Fredholm equation

$$\mathbf{L}^\circ\mathbf{K}[y(x)] = \mathbf{L}^\circ[f(x)], \tag{18}$$

for the case $\nu \geq 0$, has those and only those solutions that are solutions to Eq. (12).

For the case in which the index ν is nonpositive, the operator $\mathbf{L}^\circ$ performs an equivalent right regularization of Eq. (12) for an arbitrary right-hand side $f(x)$. In other words, for $\nu \leq 0$, on finding the solution to the Fredholm equation

$$\mathbf{K}\mathbf{L}^\circ[\omega(x)] = f(x),$$

we can obtain all solutions of the original equation (12) by the formula $y(x) = \mathbf{L}^\circ[\omega(x)]$.

Another method of regularization is known, the so-called Carleman–Vekua regularization, which is based on the solution of the corresponding characteristic equation. Equation (12) can formally be rewritten as a convolution type equation with two kernels:

$$y(x) + \frac{1}{\sqrt{2\pi}} \int_0^\infty K_1(x-t)y(t)\,dt + \frac{1}{\sqrt{2\pi}} \int_{-\infty}^0 K_2(x-t)y(t)\,dt = f_1(x), \tag{19}$$

where

$$f_1(x) = f(x) - \int_{-\infty}^\infty M(x,t)y(t)\,dt.$$

Next, the function $f_1(x)$ is provisionally assumed to be known, and Eq. (19) is solved (see Subsection 11.9-2). The analysis of the resulting formula for the function $y(x)$ shows that, for $\nu = 0$, this is a Fredholm integral equation with the unknown function $y(x)$. For the case in which $\nu > 0$, the resulting equation contains ν arbitrary constants. For a negative index ν, solvability conditions must be added to the equation.

⊙ Reference for Section 11.20: F. D. Gakhov and Yu. I. Cherskii (1978).

Chapter 12

Methods for Solving Singular Integral Equations of the First Kind

12.1. Some Definitions and Remarks

12.1-1. Integral Equations of the First Kind With Cauchy Kernel

A *singular integral equation of the first kind with Cauchy kernel* has the form

$$\frac{1}{\pi i} \int_L \frac{\varphi(\tau)}{\tau - t} \, d\tau = f(t), \qquad i^2 = -1, \tag{1}$$

where L is a smooth closed or nonclosed contour in the complex plane of the variable $z = x + iy$, t and τ are the complex coordinates on L, $\varphi(t)$ is the unknown function, $\dfrac{1}{\tau - t}$ is the Cauchy kernel, and $f(t)$ is a given function, which is called the right-hand side of Eq. (1). The integral on the left-hand side only exists in the sense of the Cauchy principal value (see Subsection 12.2-5).

A singular integral equation in which L is a smooth closed contour, as well as an equation of the form

$$\frac{1}{\pi} \int_{-\infty}^{\infty} \frac{\varphi(t)}{t - x} \, dt = f(x), \quad -\infty < x < \infty, \tag{2}$$

on the real axis and an equation with Cauchy kernel

$$\frac{1}{\pi} \int_a^b \frac{\varphi(t)}{t - x} \, dt = f(x), \qquad a \le x \le b, \tag{3}$$

on a finite interval, are special cases of Eq. (1).

A general singular integral equation of the first kind with Cauchy kernel has the form

$$\frac{1}{\pi i} \int_L \frac{M(t, \tau)}{\tau - t} \varphi(\tau) \, d\tau = f(t), \tag{4}$$

where $M(t, \tau)$ is a given function. This equation can also be rewritten in a different (equivalent) form, which is given in Subsection 12.4-4.

Assume that all functions in Eqs. (1)–(4) satisfy the Hölder condition (Subsection 12.2-2) and the function $M(t, \tau)$ satisfies this condition with respect to both variables.

12.1-2. Integral Equations of the First Kind With Hilbert Kernel

The simplest *singular integral equation of the first kind with Hilbert kernel* has the form

$$\frac{1}{2\pi} \int_0^{2\pi} \cot\left(\frac{\xi - x}{2}\right) \varphi(\xi) \, d\xi = f(x), \tag{5}$$

where $\varphi(x)$ is the unknown function ($0 \le x \le 2\pi$), $\cot\left[\frac{1}{2}(\xi - x)\right]$ is the Hilbert kernel, and $f(x)$ is the given right-hand side of the equation ($0 \le x \le 2\pi$).

A general singular integral equation of the first kind with Hilbert kernel has the form

$$-\frac{1}{2\pi} \int_0^{2\pi} N(x, \xi) \cot\left(\frac{\xi - x}{2}\right) \varphi(\xi)\, d\xi = f(x), \tag{6}$$

where $N(x, \xi)$ is a given function. Equation (6) can often be rewritten in an equivalent form, which is presented in Subsection 12.4-5.

Assume that all functions in Eqs. (5) and (6) also satisfy the Hölder condition (see Subsection 12.2-2) and the function $N(x, \xi)$ satisfies this condition with respect to both variables.

If the right-hand sides of Eqs. (1)–(6) are identically zero, then the equations are said to be *homogeneous*, otherwise they are said to be *nonhomogeneous*.

⊙ References for Section 12.1: F. D. Gakhov (1977), S. G. Mikhlin and S. Prössdorf (1986), S. Prössdorf and B. Silbermann (1991), A. Dzhuraev (1992), N. I. Muskhelishvili (1992), I. K. Lifanov (1996).

12.2. The Cauchy Type Integral

12.2-1. Definition of the Cauchy Type Integral

Let L be a smooth closed contour* on the plane of a complex variable $z = x + iy$. The domain inside the contour L is called the *interior domain* and is denoted by Ω^+, and the complement of $\Omega^+ \cup L$, which contains the point at infinity, is called the *exterior domain* and is denoted by Ω^-.

If a function $f(z)$ is analytic in Ω^+ and continuous in $\Omega^+ \cup L$, then according to the familiar Cauchy formula in the theory of functions of a complex variable we have

$$\frac{1}{2\pi i} \int_L \frac{f(\tau)}{\tau - z}\, d\tau = \begin{cases} f(z) & \text{for } z \in \Omega^+, \\ 0 & \text{for } z \in \Omega^-. \end{cases} \tag{1}$$

If a function $f(z)$ is analytic in Ω^- and continuous in $\Omega^- \cup L$, then

$$\frac{1}{2\pi i} \int_L \frac{f(\tau)}{\tau - z}\, d\tau = \begin{cases} f(\infty) & \text{for } z \in \Omega^+, \\ -f(z) + f(\infty) & \text{for } z \in \Omega^-. \end{cases} \tag{2}$$

As usual, the positive direction on L is defined as the direction for which the domain Ω^+ remains to the left of the contour.

The Cauchy formula permits one to calculate the values of a function at any point of the domain provided that the values on the boundary of the domain are known, i.e., the Cauchy formula solves the boundary value problem for analytic functions. The integral on the left-hand side in (1) and (2) is called the *Cauchy integral*.

Assume that L is a smooth closed or nonclosed contour that entirely belongs to the finite part of the complex plane. Let τ be the complex coordinate on L, and let $\varphi(\tau)$ be a continuous function of a point of the contour. In this case the integral

$$\Phi(z) = \frac{1}{2\pi i} \int_L \frac{\varphi(\tau)}{\tau - z}\, d\tau, \tag{3}$$

which is constructed in the same way as the Cauchy integral, is called a *Cauchy type integral*. The function $\varphi(\tau)$ is called its *density* and $1/(\tau - z)$ its *kernel*.

---

* By a *smooth contour* we mean a simple curve (i.e., a curve without points of self-intersection) that is either closed or nonclosed, has a continuous tangent, and has no cuspidal points.

For a Cauchy type integral with continuous density $\varphi(\tau)$, the only points at which the integrand is not analytic with respect to z are the points of the integration curve L. This curve is singular for the function $\Phi(z)$.

If L is a nonclosed contour, then $\Phi(z)$ is an analytic function on the entire plane with the singularity curve L. Assume that L is a closed contour. In this case, $\Phi(z)$ splits into two independent functions: a function $\Phi^+(z)$ defined on the domain Ω^+ and a function $\Phi^-(z)$ defined on the domain Ω^-. In general, these functions are not analytic continuations of each other.

By a *piecewise analytic function* we mean an analytic function $\Phi(z)$ defined by two independent expressions $\Phi^+(z)$ and $\Phi^-(z)$ on two complementary domains Ω^+ and Ω^- of the complex plane.

We note an important property of a Cauchy type integral. The function $\Phi(z)$ expressed by a Cauchy type integral of the form (3) vanishes at infinity, i.e., $\Phi^-(\infty) = 0$. This condition is also sufficient for the representability of a piecewise analytic function by a Cauchy type integral.

12.2-2. The Hölder Condition

Let L be a smooth curve in the complex plane $z = x + iy$, and let $\varphi(t)$ be a function on this curve. We say that $\varphi(t)$ satisfies the *Hölder condition* on L if for any two points t_1, $t_2 \in L$ we have

$$|\varphi(t_2) - \varphi(t_1)| < A|t_2 - t_1|^\lambda, \tag{4}$$

where A and λ are positive constants. The number A is called the *Hölder constant* and λ is called the *Hölder exponent*. If $\lambda > 1$, then by condition (4) the derivative $\varphi'_t(t)$ vanishes everywhere, and $\varphi(t)$ must be constant. Therefore, we assume that $0 < \lambda \le 1$. For $\lambda = 1$, the Hölder condition is often called the *Lipschitz condition*. Sometimes the Hölder condition is called the *Lipschitz condition of order λ*.

If t_1 and t_2 are sufficiently close to each other and if the Hölder condition holds for some exponent λ_1, then this condition certainly holds for each exponent $\lambda < \lambda_1$. In general, the converse assertion fails. The smaller λ, the broader the class of Hölder continuous functions is. The narrowest class is that of functions satisfying the Lipschitz condition.

It follows from the last property that if functions $\varphi_1(t)$ and $\varphi_2(t)$ satisfy the Hölder condition with exponents λ_1 and λ_2, respectively, then their sum and the product, as well as their ratio provided that the denominator is nonzero, satisfy the Hölder condition with exponent $\lambda = \min(\lambda_1, \lambda_2)$.

If $\varphi(t)$ is differentiable and has a bounded derivative, then $\varphi(t)$ satisfies the Lipschitz condition. In general, the converse assertion fails.

12.2-3. The Principal Value of a Singular Integral

Consider the integral

$$\int_a^b \frac{dx}{x - c}, \qquad a < c < b.$$

Evaluating this integral as an improper integral, we obtain

$$\int_a^b \frac{dx}{x - c} = \lim_{\substack{\varepsilon_1 \to 0 \\ \varepsilon_2 \to 0}} \left(-\int_a^{c-\varepsilon_1} \frac{dx}{c - x} + \int_{c+\varepsilon_2}^b \frac{dx}{x - c} \right) = \ln \frac{b - c}{c - a} + \lim_{\substack{\varepsilon_1 \to 0 \\ \varepsilon_2 \to 0}} \ln \frac{\varepsilon_1}{\varepsilon_2}. \tag{5}$$

The limit of the last expression obviously depends on the way in which ε_1 and ε_2 tend to zero. Hence, the improper integral does not exist. This integral is called a *singular integral*. However, this integral can be assigned a meaning if we assume that there is some relationship between ε_1 and ε_2. For example, if the deleted interval is symmetric with respect to the point c, i.e.,

$$\varepsilon_1 = \varepsilon_2 = \varepsilon, \tag{6}$$

we arrive at the notion of the Cauchy principal value of a singular integral.

The *Cauchy principal value* of the singular integral

$$\int_a^b \frac{dx}{x-c}, \qquad a < c < b$$

is the number

$$\lim_{\varepsilon \to 0} \left(\int_a^{c-\varepsilon} \frac{dx}{x-c} + \int_{c+\varepsilon}^b \frac{dx}{x-c} \right).$$

With regard to formula (5), we have

$$\int_a^b \frac{dx}{x-c} = \ln \frac{b-c}{c-a}. \tag{7}$$

Consider the more general integral

$$\int_a^b \frac{\varphi(x)}{x-c} \, dx, \tag{8}$$

where $\varphi(x) \in [a, b]$ is a function satisfying the Hölder condition. Let us understand this integral in the sense of the Cauchy principal value, which we define as follows:

$$\int_a^b \frac{\varphi(x)}{x-c} \, dx = \lim_{\varepsilon \to 0} \left(\int_a^{c-\varepsilon} \frac{\varphi(x)}{x-c} \, dx + \int_{c+\varepsilon}^b \frac{\varphi(x)}{x-c} \, dx \right).$$

We have the identity

$$\int_a^b \frac{\varphi(x)}{x-c} \, dx = \int_a^b \frac{\varphi(x) - \varphi(c)}{x-c} \, dx + \varphi(c) \int_a^b \frac{dx}{x-c};$$

moreover, the first integral on the right-hand side is convergent as an improper integral, because it follows from the Hölder condition that

$$\left| \frac{\varphi(x) - \varphi(c)}{x-c} \right| < \frac{A}{|x-c|^{1-\lambda}}, \qquad 0 < \lambda \le 1,$$

and the second integral coincides with (7).

Thus, we see that the singular integral (8), where $\varphi(x)$ satisfies the Hölder condition, exists in the sense of the Cauchy principal value and is equal to

$$\int_a^b \frac{\varphi(x)}{x-c} \, dx = \int_a^b \frac{\varphi(x) - \varphi(c)}{x-c} \, dx + \varphi(c) \ln \frac{b-c}{c-a}.$$

Some authors denote singular integrals by special symbols like v.p. $\int$ (valeur principale). However, this is not necessary because, on one hand, if an integral of the form (8) exists as a proper or an improper integral, then it exists in the sense of the Cauchy principal value, and their values coincide; on the other hand, we shall always understand a singular integral in the sense of the Cauchy principal value. For this reason, we denote a singular integral by the usual integral sign.

12.2-4. Multivalued Functions

In the representation $z = \rho e^{i\theta}$ of a complex number, the modulus ρ is determined uniquely, whereas the argument θ is only defined modulo 2π. This does not make the representation of a number ambiguous, because the argument enters this representation via the function $e^{i\theta}$, which is 2π-periodic. However, if the dependence of an analytic function on the argument θ is not 2π-periodic, then this function turns out to be multivalued. Of the elementary functions, the logarithm and the power function with noninteger exponent have this property:

$$\ln(z - z_0) = \ln|z - z_0| + i\arg(z - z_0) = \ln\rho + i\Theta, \tag{9}$$

$$(z - z_0)^\gamma = \rho^\gamma e^{i\gamma\Theta} = \rho^\alpha[\cos(\beta\ln\rho) + i\sin(\beta\ln\rho)]e^{i\gamma\theta}e^{i2\pi k\gamma}, \qquad \gamma = \alpha + i\beta. \tag{10}$$

In our reasoning, the logarithm of the modulus of a complex number is always understood as a real number, according to the usual definition. The general representation of the argument Θ has the form

$$\Theta = \theta + 2\pi k,$$

where k ranges over all integers ($k = 0, \pm1, \pm2, \dots$) and θ is the argument with the least absolute value.

To any k, there corresponds a branch of the multivalued function. The logarithmic function has infinitely many branches. The same holds for the power function with an irrational or nonreal exponent. However, if the exponent is rational, $\gamma = p/q$, with $\gcd(p, q) = 1$, then the power function has q branches. The branches of the logarithm differ by a constant of the form $i2\pi m$, and the branches of a power function differ by a factor of the form $e^{i2\pi m\gamma}$ (m is an integer). Obviously, to define a multivalued function, it is necessary to indicate which branch is chosen. However, in contrast to the case of functions of a real variable, this is not sufficient for the complete definition of a multivalued function of a complex variable. For the latter functions, there are points on the plane with the following property: as the independent variable goes along a closed contour surrounding this point and returns to the initial value, the chosen branch of the function changes to some other branch. Such points are called the *branching points* of the multivalued function. For the functions (9) and (10), the branching points are z_0 and the point at infinity. If the variable is going along a contour surrounding the point z_0 counterclockwise or clockwise, then the argument Θ is changed by 2π or by -2π, respectively.

Accordingly, the logarithm is increased or decreased by $i2\pi$, and the power function is multiplied by $e^{i2\pi\gamma}$ or $e^{-i2\pi\gamma}$. Hence, the branch corresponding to the value $k = n$ passes to the neighboring branch corresponding to $k = n+1$ or $k = n-1$. As usual, the study of the point at infinity is performed by the substitution $z = 1/\zeta$ with the subsequent investigation at the point $\zeta = 0$.

We can preserve a chosen branch of a function only if we forbid going around an arbitrary branching point. To this end, we may use *cuts* joining the branching points. In the above cases of the logarithmic and the power function, we can make a cut along a curve issuing from the point z_0 and passing to infinity. A multivalued function is defined uniquely if the branch is chosen and the cut is given.

The range of Θ is determined by the position of the cut. For example, if the cut passes along the ray that forms an angle θ_0 with the real axis, then for the principal branch ($k = 0$) we have $\theta_0 \leq \Theta \leq \theta_0 + 2\pi$. In particular, for the cut that passes along the positive real axis, we have $0 \leq \Theta \leq 2\pi$; and for cut along the negative real axis, we obtain $-\pi \leq \Theta \leq \pi$. If the cut is curvilinear, then the range of the argument depends on the functions of a point. The initial value of the argument corresponds to the left edge of the cut (with respect to z_0) and the final value corresponds to the right edge. Let us denote the value of the argument on the left and on the right edge of the cut by Θ^+ and Θ^-, respectively. Then we have

$$\Theta^- - \Theta^+ = 2\pi.$$

For the chosen branch, the cut is a curve of discontinuity. On the edges of the cut we have

$$\ln(z^- - z_0) = \ln(z^+ - z_0) + i2\pi,$$
$$(z^- - z_0)^\gamma = e^{i2\pi\gamma}(z^+ - z_0)^\gamma.$$

This discontinuity property of branches of multivalued functions on the edges of a cut is widely used in the solution of boundary value problems with discontinuous boundary conditions. The logarithm is applied for the case in which a discontinuous function enters the boundary condition as a summand, and the power function corresponds to the case of a discontinuous factor in the boundary conditions.

12.2-5. The Principal Value of a Singular Curvilinear Integral

Let L be a smooth contour and let τ and t be complex coordinates of its points. Consider the singular curvilinear integral

$$\int_L \frac{\varphi(\tau)}{\tau - t}\, d\tau. \tag{11}$$

Let us take a circle of some radius ρ centered at the point t on the contour. Let t_1 and t_2 be the points of intersection of this circle with the curve. Assume that the radius is so small that the circle has no other points of intersection with L. Let l be the part of the contour L cut out by the circle. Consider the integral over the remaining arc,

$$\int_{L-l} \frac{\varphi(\tau)}{\tau - t}\, d\tau. \tag{12}$$

The limit of the integral (12) as $\rho \to 0$ is called the *principal value* of the singular integral (11). Using the representation

$$\int_L \frac{\varphi(\tau)}{\tau - t}\, d\tau = \int_L \frac{\varphi(\tau) - \varphi(t)}{\tau - t}\, d\tau + \varphi(t) \int_L \frac{d\tau}{\tau - t}$$

and the same reasoning as above, we see that the singular integral (11) exists in the sense of the Cauchy principal value for any function $\varphi(\tau)$ satisfying the Hölder condition.

At any point of smoothness, this integral can be presented in two forms:

$$\int_L \frac{\varphi(\tau)}{\tau - t}\, d\tau = \int_L \frac{\varphi(\tau) - \varphi(t)}{\tau - t}\, d\tau + \varphi(t)\left(\ln\frac{b - t}{a - t} + i\pi\right)$$
$$\int_L \frac{\varphi(\tau)}{\tau - t}\, d\tau = \int_L \frac{\varphi(\tau) - \varphi(t)}{\tau - t}\, d\tau + \varphi(t)\ln\frac{b - t}{t - a},$$

where a and b are the endpoints of L.

In particular, if the contour is closed, then by setting $a = b$ we obtain

$$\int_L \frac{\varphi(\tau)}{\tau - t}\, d\tau = \int_L \frac{\varphi(\tau) - \varphi(t)}{\tau - t}\, d\tau + i\pi\varphi(t).$$

Throughout the following, any singular integral will be understood in the sense of the Cauchy principal value.

Let L be a smooth contour (closed or nonclosed) and let $\varphi(\tau)$ be a Hölder function of a point on the contour. Then the Cauchy type integral

$$\Phi(z) = \frac{1}{2\pi i} \int_L \frac{\varphi(\tau)}{\tau - z}\, d\tau \tag{13}$$

has limit values $\Phi^+(t)$ and $\Phi^-(t)$ at any point of $t \in L$ other than the endpoints of the contour, as $z \to t$ from the left or from the right along any path; and these limit values can be expressed via the density $\varphi(t)$ of the integral and via the singular integral (13) by the *Sokhotski–Plemelj formulas*

$$\Phi^+(t) = \frac{1}{2}\varphi(t) + \frac{1}{2\pi i} \int_L \frac{\varphi(\tau)}{\tau - t}\, d\tau, \quad \Phi^-(t) = -\frac{1}{2}\varphi(t) + \frac{1}{2\pi i} \int_L \frac{\varphi(\tau)}{\tau - t}\, d\tau. \tag{14}$$

The sum and the difference of formulas (14) give the equivalent formulas

$$\Phi^+(t) - \Phi^-(t) = \varphi(t), \tag{15}$$

$$\Phi^+(t) + \Phi^-(t) = \frac{1}{\pi i} \int_L \frac{\varphi(\tau)}{\tau - t}\, d\tau, \tag{16}$$

which are often used instead of (14).

The *Sokhotski–Plemelj formulas for the real axis* have the form

$$\Phi^+(x) = \frac{1}{2}\varphi(x) + \frac{1}{2\pi i} \int_{-\infty}^{\infty} \frac{\varphi(\tau)}{\tau - x}\, d\tau, \quad \Phi^-(x) = -\frac{1}{2}\varphi(x) + \frac{1}{2\pi i} \int_{-\infty}^{\infty} \frac{\varphi(\tau)}{\tau - x}\, d\tau. \tag{17}$$

Moreover, we have

$$\Phi^+(\infty) = \tfrac{1}{2}\varphi(\infty), \quad \Phi^-(\infty) = -\tfrac{1}{2}\varphi(\infty).$$

This, together with (17), implies

$$\Phi^+(\infty) + \Phi^-(\infty) = 0, \tag{18}$$

$$\lim_{x \to \infty} \int_{-\infty}^{\infty} \frac{\varphi(\tau)}{\tau - x}\, d\tau = 0. \tag{19}$$

Any function representable by a Cauchy type integral on the real axis necessarily satisfies condition (18). This condition is also sufficient for the representability of a piecewise analytic function in the upper and the lower half-plane by an integral over the real axis.

Consider a Cauchy type integral over the real axis and assume that z is not real:

$$\Phi(z) = \frac{1}{2\pi i} \int_{-\infty}^{\infty} \frac{\varphi(x)}{x - z}\, dx, \tag{20}$$

where $\varphi(x)$ is a complex function of a real variable x satisfying the Hölder condition on the real axis.

If a function $\varphi(z)$ is analytic in the upper half-plane, is continuous in the closed upper half-plane, and satisfies the Hölder condition on the real axis, then

$$\frac{1}{2\pi i} \int_{-\infty}^{\infty} \frac{\varphi(x)}{x - z}\, dx = \begin{cases} \varphi(z) - \tfrac{1}{2}\varphi(\infty) & \text{for } \operatorname{Im} z > 0, \\ -\tfrac{1}{2}\varphi(\infty) & \text{for } \operatorname{Im} z < 0. \end{cases} \tag{21}$$

We also have the formula

$$\frac{1}{2\pi i} \int_{-\infty}^{\infty} \frac{\varphi(x) - \varphi(\infty)}{x - z}\, dx = \begin{cases} \tfrac{1}{2}\varphi(\infty) & \text{for } \operatorname{Im} z > 0, \\ -\varphi(z) + \tfrac{1}{2}\varphi(\infty) & \text{for } \operatorname{Im} z < 0 \end{cases} \tag{22}$$

provided that $\varphi(z)$ is analytic in the lower half-plane, continuous in the closed lower half-plane and satisfies the Hölder condition on the real axis.

12.2-6. The Poincaré–Bertrand Formula

Consider the following pair of iterated singular integrals:

$$N(t) = \frac{1}{\pi i} \int_L \frac{d\tau}{\tau - t} \frac{1}{\pi i} \int_L \frac{K(\tau, \tau_1)}{\tau_1 - \tau} \, d\tau_1, \tag{23}$$

$$M(t) = \frac{1}{\pi i} \int_L d\tau_1 \frac{1}{\pi i} \int_L \frac{K(\tau, \tau_1)}{(\tau - t)(\tau_1 - \tau)} \, d\tau, \tag{24}$$

where L is a smooth contour and the function $K(\tau, \tau_1)$ satisfies the Hölder condition with respect to both variables.

Both integrals make sense, and although N differs from M only by the order of integration, they are not equal, as shown by the following *Poincaré–Bertrand formula*

$$\frac{1}{\pi i} \int_L \frac{d\tau}{\tau - t} \frac{1}{\pi i} \int_L \frac{K(\tau, \tau_1)}{\tau_1 - \tau} \, d\tau_1 = K(t, t) + \frac{1}{\pi i} \int_L d\tau_1 \frac{1}{\pi i} \int_L \frac{K(\tau, \tau_1)}{(\tau - t)(\tau_1 - \tau)} \, d\tau, \tag{25}$$

which can also be rewritten in the form

$$\int_L \frac{d\tau}{\tau - t} \int_L \frac{K(\tau, \tau_1)}{\tau_1 - \tau} \, d\tau_1 = -\pi^2 K(t, t) + \int_L d\tau_1 \int_L \frac{K(\tau, \tau_1)}{(\tau - t)(\tau_1 - \tau)} \, d\tau. \tag{26}$$

Example. Let us evaluate the Cauchy type integral over the unit circle $|z| = 1$ with density $\varphi(\tau) = 2/[\tau(\tau - 2)]$, i.e.,

$$\Phi(z) = \frac{1}{2\pi i} \int_L \frac{1}{\tau - 2} \frac{d\tau}{\tau - z} - \frac{1}{2\pi i} \int_L \frac{1}{\tau} \frac{d\tau}{\tau - z}.$$

The function $1/(z - 2)$ is analytic in Ω^+, and $1/z$ is analytic in Ω^- and vanishes at infinity. By formula (1), the first integral is equal to $1/(z - 2)$ for $z \in \Omega^+$ and is zero for $z \in \Omega^-$. By formula (2), the second integral is equal to $-1/z$ for $z \in \Omega^-$ and is zero for $z \in \Omega^+$. Hence,

$$\Phi^+(z) = \frac{1}{z - 2}, \quad \Phi^-(z) = \frac{1}{z}.$$

⊙ References for Section 12.2: F. D. Gakhov (1977), S. G. Mikhlin and S. Prössdorf (1986), N. I. Muskhelishvili (1992).

12.3. The Riemann Boundary Value Problem

12.3-1. The Principle of Argument. The Generalized Liouville Theorem

THE THEOREM ON THE ANALYTIC CONTINUATION (THE PRINCIPLE OF CONTINUITY). *Assume that a domain Ω_1 borders a domain Ω_2 along a smooth curve L. Let analytic functions $f_1(z)$ and $f_2(z)$ be given in Ω_1 and Ω_2. Assume that, as the point z tends to L, both functions tend to the same continuous limit function on the curve L. Under these assumptions, the functions $f_1(z)$ and $f_2(z)$ are analytic continuations of each other.*

Assume that a function $f(z)$ is analytic in a domain Ω bounded by a contour L except for finitely many points, where it may have poles. Let us write out the power series expansion of $f(z)$ around some point z_0:

$$f(z) = c_n(z - z_0)^n + c_{n+1}(z - z_0)^{n+1} + \cdots = (z - z_0)^n f_1(z), \qquad f_1(z_0) = c_n \neq 0.$$

The number n is called the *order of the function $f(z)$ at the point z_0*. If $n > 0$, then the order of the function is the order of zero; if $n < 0$, then the order of the function is minus the order of the pole. If the order of a function at z_0 is zero, then at z_0 the function has a finite nonzero value at z_0.

When considering the point at infinity, we must replace the difference $z - z_0$ by $1/z$. If $z_0 \in L$, then we define the order of the function to be equal to $\frac{1}{2}n$.

Let N_Ω and P_Ω (N_L and P_L) be the numbers of zeros and poles on the domain (on the contour, respectively), where each zero and pole is taken according to its multiplicity. Let $[\delta]_L$ denote the increment of the variable δ when going around the contour in the positive direction. As usual, by the positive direction we mean the direction the domain under consideration remains to the left of the contour.

THE PRINCIPLE OF ARGUMENT. *Let $f(z)$ be a single-valued analytic function in a multiply connected domain Ω bounded by a smooth contour $L = L_0 + L_1 + \cdots + L_m$ except for finitely many points at which $f(z)$ may have poles, and let $f(z)$ be continuous in the closed domain $\Omega \cup L$ (except for these poles) and have at most finitely many zeros of integer order on the contour. In this case, the following formula holds:*

$$N_\Omega - P_\Omega + \frac{1}{2}(N_L - P_L) = \frac{1}{2\pi}[\arg f(z)]_L.$$

THE GENERALIZED LIOUVILLE THEOREM. *Assume that a function $f(z)$ is analytic on the entire complex plane except for points $a_0 = \infty$, a_k ($k = 1, \ldots, n$), where it has poles, and that the principal parts of the Laurent series expansions of $f(z)$ at the poles have the form*

$$Q_0(z) = c_1^0 z + c_2^0 z^2 + \cdots + c_{m_0}^0 z^{m_0} \qquad \textit{at the point } a_0,$$

$$Q_k\left(\frac{1}{z - a_k}\right) = \frac{c_1^k}{z - a_k} + \frac{c_2^k}{(z - a_k)^2} + \cdots + \frac{c_{m_k}^k}{(z - a_k)^{m_k}} \qquad \textit{at the points } a_k.$$

Then $f(z)$ is a rational function, and can be represented by the formula

$$f(z) = C + Q_0(z) + \sum_{k=1}^{n} Q_k\left(\frac{1}{z - a_k}\right),$$

where C is a constant. In particular, if the only singularity of $f(z)$ is a pole of order m at infinity, then $f(z)$ is a polynomial of degree m,

$$f(z) = c_0 + c_1 z + \cdots + c_m z^m.$$

The following notation is customary:
(a) $\overline{f(z)}$ is the function conjugate to a given function $f(z)$;
(b) $f(\bar{z})$ is the function obtained from $f(z)$ by replacing z by $\bar{z}$, i.e., y by $-y$ in $f(z)$;
(c) $\bar{f}(z)$ is the function defined by the condition $\bar{f}(z) = \overline{f(\bar{z})}$.
  If $z = x + iy$ and $f(z) = u(x, y) + iv(x, y)$, then

$$\overline{f(z)} = u(x, y) - iv(x, y), \quad f(\bar{z}) = u(x, -y) + iv(x, -y), \quad \bar{f}(z) = u(x, -y) - iv(x, -y).$$

In particular, if $f(z)$ is given by a series $f(z) = \sum_{k=0}^{n} c_k z^k$, then

$$\overline{f(z)} = \sum_{k=0}^{n} \bar{c}_k \bar{z}^k, \quad f(\bar{z}) = \sum_{k=0}^{n} c_k \bar{z}^k, \quad \bar{f}(z) = \sum_{k=0}^{n} \bar{c}_k z^k.$$

For a function represented by a Cauchy type integral

$$f(z) = \frac{1}{2\pi i} \int_L \frac{\varphi(\tau)}{\tau - z} \, d\tau,$$

we have

$$\overline{f(z)} = -\frac{1}{2\pi i} \int_L \frac{\overline{\varphi(\tau)}}{\bar{\tau} - \bar{z}} \, \overline{d\tau}, \quad f(\bar{z}) = \frac{1}{2\pi i} \int_L \frac{\varphi(\tau)}{\tau - \bar{z}} \, d\tau, \quad \bar{f}(z) = -\frac{1}{2\pi i} \int_L \frac{\overline{\varphi(\tau)}}{\bar{\tau} - z} \, \overline{d\tau}.$$

Note that if a function satisfies the condition $\bar{f}(z) = f(z)$, then it takes real values for all real values of z. The converse assertion also holds.

12.3-2. The Hermite Interpolation Polynomial

The Hermite interpolation polynomial is used for the construction of the canonical function of the nonhomogeneous Riemann problem in Subsections 10.4-7 and 12.3-9.

Let distinct points z_k ($k = 1, \ldots, m$) be given, and a number $\Delta_k^{(j)}$ ($j = 0, 1, \ldots, n_k - 1$) be assigned to each point z_k, where the n_k are given positive integers. It is required to construct a polynomial $\mathcal{U}_p(z)$ of the least possible degree such that

$$\mathcal{U}_p^{(j)}(z_k) = \Delta_k^{(j)}, \quad k = 1, \ldots, m, \quad j = 0, 1, \ldots, n_k - 1,$$

where the $\mathcal{U}_p^{(j)}(z_k)$ are the values of the jth-order derivatives of the polynomial at the points z_k. The numbers z_k are called the *interpolation nodes* and n_k the *interpolation multiplicities at the nodes* z_k.

There exists a unique polynomial with these properties. It has the form (e.g., see V. I. Smirnov and N. A. Lebedev (1964))

$$\mathcal{U}_p(z) = \sum_{k=1}^{m} \frac{\zeta(z)}{(z - z_k)^{n_k}} \sum_{r=0}^{n_k-1} A_{k,r}(z - z_k)^r, \quad p = \sum_{k=1}^{m} n_k - 1,$$

$$\zeta(z) = \prod_{k=1}^{m} (z - z_k)^{n_k}, \quad A_{k,r} = \sum_{j=0}^{r} \frac{\Delta_k^{(j)}}{j!\,(r-j)!} \left[\frac{d^{r-j}}{dz^{r-j}} \frac{(z - z_k)^{n_k}}{\zeta(z)} \right]_{z=z_k},$$

$$k = 1, \ldots, m, \quad r = 0, 1, \ldots, n_k - 1;$$

and this polynomial is unique.

The interpolation polynomial $\mathcal{U}_p(z)$ constructed for some function $f(z)$ must satisfy the following conditions at the points z_k:

$$\mathcal{U}_p^{(j)}(z_k) = \Delta_k^{(j)} = f^{(j)}(z_k), \quad k = 1, \ldots, m, \quad j = 0, 1, \ldots, n_k - 1,$$

where $f^{(j)}(z_k)$ is the value of the jth-order derivative of $f(z)$ at the point z_k.

12.3-3. Notion of the Index

Let L be a smooth closed contour, and let $D(t)$ be a continuous nowhere vanishing function on this contour.

The *index* ν of the function $D(t)$ with respect to the contour L is the increment of the argument of $D(t)$ along L (traversed in the positive direction) divided by 2π:

$$\nu = \operatorname{Ind} D(t) = \frac{1}{2\pi}[\arg D(t)]_L. \tag{1}$$

Since $\ln D(t) = \ln |D(t)| + i \arg D(t)$ and since after the traverse the function $|D(t)|$ returns to its original value, it follows that $[\ln D(t)]_L = i[\arg D(t)]_L$, and hence

$$\nu = \frac{1}{2\pi i}[\ln D(t)]_L. \tag{2}$$

The index can be expressed in the form of an integral as follows:

$$\nu = \operatorname{Ind} D(t) = \frac{1}{2\pi i} \int_L d \ln D(t) = \frac{1}{2\pi} \int_L d \arg D(t). \tag{3}$$

If the function $D(t)$ is not differentiable but has bounded variation, then the integral is regarded as the Stieltjes integral. Since $D(t)$ is continuous, the image $\check{\Gamma}$ of the closed contour L is a closed contour as well, and the increment of the argument $D(t)$ along L is a multiple of 2π. Hence, the following assertions hold.

1°. The index of a function that is continuous on a closed contour and vanishes nowhere is an integer (possibly zero).

2°. The index of the product of two functions is equal to the sum of the indexes of the factors. The index of a ratio is equal to the difference of the indexes of the numerator and the denominator.

We now assume that $D(t)$ is differentiable and is the boundary value of a function analytic in the interior or exterior of L. In this case, the number

$$\nu = \frac{1}{2\pi i} \int_L d\ln D(t) = \frac{1}{2\pi i} \int_L \frac{D'_t(t)}{D(t)} \, dt \tag{4}$$

is equal to the logarithmic residue of the function $D(t)$. The principle of argument (see Subsection 12.3-1) implies the following properties of the index:

3°. If $D(t)$ is the boundary value of a function analytic in the interior or exterior of the contour, then its index is equal to the number of zeros inside the contour or minus the number of zeros outside the contour, respectively.

4°. If a function $D(z)$ is analytic in the interior of the contour except for finitely many points at which it may have poles, then the number of zeros must be replaced by the difference of the number of zeros and the number of poles.

Here the zeros and the poles are counted according to their multiplicities. We also note that the indexes of complex conjugate functions have opposite signs.

Let

$$t = t_1(s) + it_2(s) \qquad (0 \le s \le l)$$

be the equation of the contour L. On substituting the expression of the complex coordinate t into the function $D(t)$, we obtain

$$D(t) = D\big(t_1(s) + it_2(s)\big) = \xi(s) + i\eta(s). \tag{5}$$

Let us regard ξ and η as Cartesian coordinates. Then

$$\xi = \xi(s), \quad \eta = \eta(s)$$

is a parametric equation of some curve Γ. Since the function $D(t)$ is continuous and the contour L is closed, it follows that the curve Γ is closed as well.

The number of turns of the curve Γ around the origin, i.e., the number of full rotations of the radius vector as the variable s varies from 0 to l, is obviously the index of the function $D(t)$. This number is often called the *winding number of the curve* Γ with respect to the origin.

If the curve Γ is successfully constructed, then the winding number can be observed directly. There are many examples for which the index can be found by analyzing the shape of the curve Γ. For instance, if $D(t)$ is a real or a pure imaginary function that does not vanish, then Γ is a line segment (traversed an even number of times), and the index $D(t)$ is equal to zero. If the real part $\xi(s)$ or the imaginary part $\eta(s)$ preserves its sign, then the index is obviously zero, and so on. If the function $D(t)$ can be represented as the product or the ratio of functions that are limit values of functions analytic in the interior or exterior of the contour, then the index can be calculated on the basis of properties 2°, 3° and 4°.

In the general case, the calculation of the index can be performed by formula (3). On the basis of formula (5) we substitute the expression

$$d\arg D(t) = d\arctan \frac{\eta(s)}{\xi(s)}$$

into (3) and assume that ξ and η are differentiable. Then we obtain

$$\nu = \frac{1}{2\pi} \int_\Gamma \frac{\xi \, d\eta - \eta \, d\xi}{\xi^2 + \eta^2} = \frac{1}{2\pi} \int_0^l \frac{\xi(s)\eta_s'(s) - \eta(s)\xi_s'(s)}{\xi^2(s) + \eta^2(s)} \, ds. \tag{6}$$

Example. Let us calculate the index of $D(t) = t^n$ with respect to an arbitrary contour L surrounding the origin.

First method. The function t^n is the boundary value of the function z^n, which has precisely one zero of order n inside the contour. Hence

$$\nu = \operatorname{Ind} t^n = n.$$

Second method. If the argument of t is φ, then the argument of t^n is $n\varphi$. As the point t traverses the contour L and returns to the original value, the argument φ obtains the increment 2π. Hence,

$$\operatorname{Ind} t^n = n.$$

The index can also be found numerically. Since the index is integer-valued, an approximate value whose error is less than $\frac{1}{2}$ can be rounded off to the nearest integer to obtain the exact value.

12.3-4. Statement of the Riemann Problem

Let L be a simple smooth closed contour which divides the complex plane into the interior domain Ω^+ and the exterior domain Ω^-, and let two functions of points of the contours $D(t)$ and $H(t)$ satisfying the Hölder condition (see Subsection 12.2-2) be given; moreover, suppose that $D(t)$ does not vanish.

The Riemann Problem. Find two functions (or a single piecewise analytic function), namely, a function $\Phi^+(z)$ analytic in Ω^+ and a function $\Phi^-(z)$ analytic in the domain Ω^- including $z = \infty$, so that the following linear relation is satisfied on the contour L:

$$\Phi^+(t) = D(t)\Phi^-(t) \qquad \text{(the homogeneous problem)} \tag{7}$$

or

$$\Phi^+(t) = D(t)\Phi^-(t) + H(t) \qquad \text{(the nonhomogeneous problem)}. \tag{8}$$

The function $D(t)$ is called the *coefficient* of the Riemann problem, and the function $H(t)$ is called the *right-hand side*.

We first consider a Riemann problem of special form that is called the *jump problem*. Let a function $\varphi(t)$ defined on a closed contour L satisfy the Hölder condition. The problem is to find a piecewise analytic function $\Phi(z)$ ($\Phi(z) = \Phi^+(z)$ for $z \in \Omega^+$ and $\Phi(z) = \Phi^-(z)$ for $z \in \Omega^-$) that vanishes at infinity and has a jump of magnitude $\varphi(t)$ on L, i.e., such that

$$\Phi^+(t) - \Phi^-(t) = \varphi(t).$$

It follows from the Sokhotski–Plemelj formulas (see Subsection 12.2-5) that the function

$$\Phi(z) = \frac{1}{2\pi i} \int_L \frac{\varphi(\tau)}{\tau - z} \, d\tau$$

is the unique solution to the above problem.

Thus, an arbitrary function $\varphi(t)$ given on the closed contour and satisfying the Hölder condition can be uniquely represented as the difference of functions $\Phi^+(t)$ and $\Phi^-(t)$ that are the boundary values of analytic functions $\Phi^+(z)$ and $\Phi^-(z)$ under the additional condition $\Phi^-(\infty) = 0$.

If we neglect the additional condition $\Phi^-(\infty) = 0$, then the solution will be given by the formula

$$\Phi(z) = \frac{1}{2\pi i} \int_L \frac{\varphi(\tau)}{\tau - z} \, d\tau + \text{const}. \tag{9}$$

Let us seek a particular solution of the homogeneous problem (7) in the class of functions that do not vanish on the contour. Let N_+ and N_- be the numbers of zeros of the desired functions in the domains Ω^+ and Ω^-, respectively. Taking the index of both parts of Eq. (7), on the basis of properties 2° and 3° we obtain

$$N_+ + N_- = \operatorname{Ind} D(t) = \nu. \tag{10}$$

We call the index ν of the coefficient $D(t)$ the *index of the Riemann problem.*

Let $\nu = 0$. Under this condition, $\ln D(t)$ is a single-valued function. It follows from (10) that $N_+ = N_- = 0$, i.e., the solution has no zeros on the entire plane. Therefore, the functions $\ln \Phi^{\pm}(z)$ are analytic in their domains and hence single-valued together with the boundary values $\ln \Phi^{\pm}(t)$.

Taking the logarithm of the boundary condition (7), we obtain

$$\ln \Phi^+(t) - \ln \Phi^-(t) = \ln D(t). \tag{11}$$

We can choose an arbitrary branch of $\ln D(t)$ because the final result is independent of the choice of this branch. Thus, we must find a piecewise analytic function $\ln \Phi(z)$ with a prescribed jump on L. The solution of this problem under the additional condition $\ln \Phi^-(\infty) = 0$ is given by the formula

$$\ln \Phi(z) = \frac{1}{2\pi i} \int_L \frac{\ln D(\tau)}{\tau - z} \, d\tau. \tag{12}$$

For brevity, we write

$$\frac{1}{2\pi i} \int_L \frac{\ln D(\tau)}{\tau - z} \, d\tau = G(z). \tag{13}$$

It readily follows from the Sokhotski–Plemelj formulas that the functions

$$\Phi^+(z) = e^{G^+(z)} \quad \text{and} \quad \Phi^-(z) = e^{G^-(z)} \tag{14}$$

are the solution of the boundary value problem (7) with the condition $\Phi^-(\infty) = 1$.

If we neglect the additional condition $\Phi^-(\infty) = 1$, then in formula (12) we must add an arbitrary constant, and the solution becomes

$$\Phi^+(z) = C e^{G^+(z)}, \quad \Phi^-(z) = C e^{G^-(z)}, \tag{15}$$

where C is an arbitrary constant. Since $G^-(\infty) = 0$, it follows that C is the value of $\Phi^-(z)$ at infinity.

Thus, in the case $\nu = 0$ and for arbitrary $\Phi^-(\infty) \neq 0$, the solution contains a single arbitrary constant, and hence there is a unique linearly independent solution. If $\Phi^-(\infty) = 0$, then $C = 0$, and the problem has only the trivial solution (which is identically zero), which is natural because $N_- = 0$.

This gives an important corollary. An arbitrary function $D(t) \neq 0$ on L that satisfies the Hölder condition and has zero index can be represented as the ratio of the boundary values $\Phi^+(t)$ and $\Phi^-(t)$ of functions that are analytic in Ω^+ and Ω^- and have no zeros in these domains. These functions are determined modulo an arbitrary constant factor and are given by formulas (15).

On passing to the general case, we seek a piecewise analytic function satisfying the homogeneous boundary condition (7) and having zero order on the entire plane except for the point at infinity, where the order of the function is equal to the index of the problem.

By the *canonical function* (of the homogeneous Riemann problem) $X(z)$ we mean the function satisfying the boundary condition (7) and piecewise analytic on the entire plane except for the point at infinity, where the order of this function is equal to the index of the problem.

This function can be constructed by reducing the problem to the case of zero index. Indeed, let us rewrite the boundary condition (7) in the form

$$\Phi^+(t) = t^{-\nu} D(t) t^{\nu} \Phi^-(t).$$

On representing the function $t^{-\nu}D(t)$ with zero index as the ratio of boundary values of analytic functions,

$$t^{-\nu}D(t) = \frac{e^{G^+(t)}}{e^{G^-(t)}}, \qquad G(z) = \frac{1}{2\pi i}\int_L \frac{\ln[\tau^{-\nu}D(\tau)]}{\tau - z}\, d\tau, \tag{16}$$

we obtain the following expression for the canonical function:

$$X^+(z) = e^{G^+(z)}, \qquad X^-(z) = z^{-\nu}e^{G^-(z)}. \tag{17}$$

Since $X^+(t) = D(t)X^-(t)$, it follows that the coefficient of the Riemann problem can be represented as the ratio of canonical functions:

$$D(t) = \frac{X^+(t)}{X^-(t)}. \tag{18}$$

The representation (18) is often called a *factorization*.

For $\nu \geq 0$, the canonical function, which has a zero of order ν at infinity, is a particular solution of the boundary value problem (7). For $\nu < 0$, the canonical function has a pole of order $|\nu|$ at infinity and is not a solution, but in this case it is still used as an auxiliary function in the solution of the nonhomogeneous problem.

12.3-5. The Solution of the Homogeneous Problem

Let $\nu = \operatorname{Ind} D(t)$ be an arbitrary integer. On representing $D(t)$ by formula (18), we reduce the boundary condition (7) to the form

$$\frac{\Phi^+(t)}{X^+(t)} = \frac{\Phi^-(t)}{X^-(t)}.$$

The left-hand side of the last relation contains the boundary value of a function that is analytic in Ω^+, and the right-hand side contains the boundary value of a function that has at least the order $-\nu$ at infinity. By the principle of continuity (see Subsection 12.3-1), the functions on the left-hand side and on the right-hand side are analytic continuations of each other to the entire plane possibly except for the point at infinity at which, in the case $\nu > 0$, a pole of order $\leq \nu$ can occur. Hence, for $\nu > 0$, by the generalized Liouville theorem (see Subsection 12.3-1), this single analytic function is a polynomial of degree $\leq \nu$ with arbitrary coefficients. For $\nu < 0$, it follows from the Liouville theorem that this function is constant. However, since this function must vanish at infinity, it follows that it is identically zero. Hence, for $\nu < 0$, the homogeneous problem has only the trivial solution (which is identically zero). A problem that has no nontrivial solutions is said to be *unsolvable*. Thus, for a negative index, the homogeneous problem (7) is unsolvable.

Let $\nu > 0$. Let $P_\nu(z)$ stand for a polynomial of degree ν with arbitrary coefficients. In this case, we obtain a solution in the form

$$\Phi(z) = P_\nu(z)X(z),$$

or

$$\Phi^+(z) = P_\nu(z)e^{G^+(z)}, \qquad \Phi^-(z) = z^{-\nu}P_\nu(z)e^{G^-(z)}, \tag{19}$$

where $G(z)$ is determined by formula (16).

Thus, if the index ν of the Riemann boundary value problem is nonnegative, then the homogeneous problem (7) has $\nu + 1$ linearly independent solutions

$$\Phi_k^+(z) = z^k e^{G^+(z)}, \qquad \Phi_k^-(z) = z^{k-\nu}e^{G^-(z)} \qquad (k = 0, 1, \ldots, \nu). \tag{20}$$

The general solution contains $\nu + 1$ arbitrary constants and is given by formula (19). For a negative index, problem (7) is unsolvable.

The polynomial $P_\nu(z)$ has exactly ν zeros in the complex plane. It follows from formulas (19) that the number of all zeros of a solution to the homogeneous Riemann boundary value problem is equal to the index ν. Depending on the choice of the coefficients of the polynomial, these zeros can occur in each of the domains $\Omega^\pm$ and also on the contour itself. Just as above, we denote by $N_\pm$ the number of zeros in the domains $\Omega^\pm$ and by N_0 the number of zeros on the contour L. We can see that in the general case (without the condition that there are no zeros on the contour), formula (10) becomes

$$N_+ + N_- + N_0 = \nu.$$ (21)

12.3-6. The Solution of the Nonhomogeneous Problem

On replacing the coefficient $D(t)$ in the boundary condition (8) by the ratio of the boundary values of the canonical functions by formula (18), we reduce (8) to the form

$$\frac{\Phi^+(t)}{X^+(t)} = \frac{\Phi^-(t)}{X^-(t)} + \frac{H(t)}{X^+(t)}.$$ (22)

The function $H(t)/X^+(t)$ satisfies the Hölder condition. Let us replace it by the difference of the boundary values of analytic functions (see the jump problem in Subsection 12.3-4):

$$\frac{H(t)}{X^+(t)} = \Psi^+(t) - \Psi^-(t),$$

where

$$\Psi(z) = \frac{1}{2\pi i} \int_L \frac{H(\tau)}{X^+(\tau)} \frac{d\tau}{\tau - z}.$$ (23)

Then the boundary condition (22) can be rewritten in the form

$$\frac{\Phi^+(t)}{X^+(t)} - \Psi^+(t) = \frac{\Phi^-(t)}{X^-(t)} - \Psi^-(t).$$

Note that for $\nu \geq 0$ the function $\Phi^-(z)/X^-(z)$ has a pole at infinity, and for $\nu < 0$ it has a zero of order ν.

By the same reasoning as in the solution of the homogeneous problem, we obtain the following results.

Let $\nu \geq 0$. In this case,

$$\frac{\Phi^+(t)}{X^+(t)} - \Psi^+(t) = \frac{\Phi^-(t)}{X^-(t)} - \Psi^-(t) = P_\nu(t).$$

This gives the solution

$$\Phi(z) = X(z)[\Psi(z) + P_\nu(z)],$$ (24)

where the functions $X(z)$ and $\Psi(z)$ are expressed by formulas (17) and (23) and P_ν is a polynomial of degree ν with arbitrary coefficients.

We can readily see that formula (24) gives the general solution of the nonhomogeneous problem because it contains the general solution $X(z)P_\nu(z)$ of the homogeneous problem as a summand.

Let $\nu < 0$. In this case, $\Phi^-(z)/X^-(z)$ vanishes at infinity and

$$\frac{\Phi^+(t)}{X^+(t)} - \Psi^+(t) = \frac{\Phi^-(t)}{X^-(t)} - \Psi^-(t) = 0,$$

so that

$$\Phi(z) = X(z)\Psi(z).$$ (25)

In the expression for the function $\Phi^-(z)$, the first factor has a pole of order $-\nu$ at infinity by virtue of formula (17), and the second factor is the Cauchy type integral (23) and, in general, has a first-order zero at infinity. Hence, $\Phi^-(z)$ has a pole of order $\leq -\nu - 1$ at infinity. Thus, if $\nu < -1$, then the nonhomogeneous problem is unsolvable in general. It is solvable only if the constant term satisfies some additional conditions. To find these conditions, we expand the Cauchy type integral (23) in a series in a neighborhood of the point at infinity:

$$\Psi^-(z) = \sum_{k=1}^{\infty} c_k z^{-k}, \qquad \text{where} \quad c_k = -\frac{1}{2\pi i} \int_L \frac{H(\tau)}{X^+(\tau)} \tau^{k-1}\, d\tau.$$

For $\Phi^-(z)$ to be analytic at the point at infinity, it is necessary that the first $-\nu - 1$ coefficients of the expansion of $\Psi^-(z)$ be zero. This means that for the solvability of the nonhomogeneous problem in the case of negative index ($\nu < -1$), it is necessary and sufficient that the following $-\nu - 1$ conditions hold:

$$\int_L \frac{H(\tau)}{X^+(\tau)} \tau^{k-1}\, d\tau = 0, \qquad k = 1, 2, \ldots, -\nu - 1. \tag{26}$$

Thus, in the case $\nu \geq 0$, the nonhomogeneous Riemann problem is solvable for an arbitrary right-hand side, and the general solution is given by the formula

$$\Phi(z) = \frac{X(z)}{2\pi i} \int_L \frac{H(\tau)}{X^+(\tau)} \frac{d\tau}{\tau - z} + X(z) P_\nu(z), \tag{27}$$

where the canonical function $X(z)$ is given by (17) and $P_\nu(z)$ is a polynomial of degree ν with arbitrary complex coefficients. If $\nu = -1$, then the nonhomogeneous problem is also solvable and has a unique solution.

In the case $\nu < -1$, the nonhomogeneous problem is unsolvable in general. For this problem to be solvable, it is necessary and sufficient that the right-hand side of the problem satisfy $-\nu - 1$ conditions (26). If these conditions are satisfied, then the solution of the problem is unique and is given by formula (27), where we must set $P_\nu(z) \equiv 0$.

The solution with the additional condition of vanishing at infinity has important applications. In this case, instead of a polynomial of degree ν, we must take a polynomial of degree $\nu - 1$. For the solvability of the problem in the case of negative index, it is necessary that the coefficient $c_{-\nu}$ be zero as well.

Hence, under the assumption that $\Phi^-(\infty) = 0$, the solution is given for $\nu \geq 0$ by the formula

$$\Phi(z) = X(z)[\Psi(z) + P_{\nu-1}(z)], \tag{28}$$

where, for $\nu = 0$, we must set $P_{\nu-1}(z) \equiv 0$.

If $\nu < 0$, then the solution is still given by formula (28) with $P_{\nu-1}(z) \equiv 0$ under the following $-\nu$ solvability conditions:

$$\int_L \frac{H(\tau)}{X^+(\tau)} \tau^{k-1}\, d\tau = 0, \qquad k = 1, 2, \ldots, -\nu. \tag{29}$$

In this case, the assertion on the solvability of the nonhomogeneous problem acquires a more symmetric form. For $\nu \geq 0$, the general solution of the nonhomogeneous problem linearly depends on ν arbitrary constants. For $\nu < 0$, the number of the solvability conditions is equal to $-\nu$. Note that for $\nu = 0$ the nonhomogeneous problem is unconditionally solvable, and the solution is unique.

On the basis of the above reasoning, the solution of the Riemann boundary value problem is mainly reduced to the following two operations:

1°. A representation of an arbitrary function given on the contour in the form of the difference of boundary values of analytic functions in the domains Ω^+ and Ω^- (the jump problem).

2°. A representation of a nonvanishing function in the form of the ratio of boundary values of analytic functions (factorization).

Here the second operation can be reduced to the first by taking the logarithm. Some complications related to the case of a nonzero index are due to the multivaluedness of the logarithm only. The first operation for arbitrary functions is equivalent to the calculation of a Cauchy type integral. In this connection, the solution to the problem by formulas (17) and (23)–(25) is explicitly expressed (in the closed form) via Cauchy type integrals.

12.3-7. The Riemann Problem With Rational Coefficients

Consider the Riemann boundary value problem with a contour that consists of finitely many simple curves and with coefficient $D(t)$ a rational function that has neither zeros nor poles on the contour. Note that an arbitrary continuous function (and all the functions satisfying the Hölder condition) can be approximated with arbitrary accuracy by rational functions, and the solution of problems with rational coefficients can serve as a basis for the approximate solution in the general case. Assume that the Riemann problem has the form

$$\Phi^+(t) = \frac{p(t)}{q(t)} \Phi^-(t) + H(t), \tag{30}$$

and the polynomials $p(z)$ and $q(z)$ can be factorized as follows:

$$p(z) = p_+(z)p_-(z), \quad q(z) = q_+(z)q_-(z), \tag{31}$$

where $p_+(z)$ and $q_+(z)$ are polynomials whose roots belong to Ω^+ and $p_-(z)$ and $q_-(z)$ are polynomials with roots in Ω^-. It readily follows from property 4° of the index (Subsection 12.3-3) that $\nu = m_+ - n_+$, where m_+ and n_+ are the numbers of zeros of the polynomials $p_+(z)$ and $q_+(z)$.

Since the coefficient of the problem is a function that can be analytically continued to the domain $\Omega^\pm$, it follows that in this case it is reasonable to avoid using the general formulas and obtain a solution directly by analytic continuation; here the role of the standard function of the type t^ν that is used in the reduction of the index to zero can be played by the product $\prod_{j=1}^{\nu}(t-a_j)$, where $a_1, \ldots, a_\nu$ are arbitrary points of the domain Ω^+. On representing the boundary condition in the form

$$\frac{q_-(t)}{p_-(t)} \Phi^+(t) - \frac{p_+(t)}{q_+(t)} \Phi^-(t) = \frac{q_-(t)}{p_-(t)} H(t),$$

where the canonical function is determined by the expressions

$$X^+(z) = \frac{p_-(z)}{q_-(z)}, \quad X^-(z) = \frac{q_+(z)}{p_+(z)}, \tag{32}$$

we obtain the solution by the same reasoning as in Subsection 12.3-6 in the following form:

$$\Phi^+(z) = \frac{p_-(z)}{q_-(z)}[\Psi(z) + P_{\nu-1}(z)], \quad \Phi^-(z) = \frac{q_+(z)}{p_+(z)}[\Psi(z) + P_{\nu-1}(z)], \tag{33}$$

where

$$\Psi(z) = \frac{1}{2\pi i} \int_L \frac{q_-(\tau)}{p_-(\tau)} \frac{H(\tau)\,d\tau}{\tau - z}, \quad \Phi^-(\infty) = 0.$$

If the index is negative, then we must set $P_{\nu-1}(z) \equiv 0$ and add the solvability conditions

$$\int_L \frac{q_-(\tau)}{p_-(\tau)} H(\tau)\tau^{k-1}\,d\tau = 0, \quad k = 1, 2, \ldots, -\nu, \tag{34}$$

which agree with the general formula (29), because the canonical function has the form (32).

Note that for the general case, in the practical solution of the Riemann problem, it can also be convenient to express the coefficient in the form

$$D(t) = \frac{p_+(t)p_-(t)}{q_+(t)q_-(t)} D_1(t),$$

where $D_1(t)$ is a function with zero index and the polynomials $p_\pm(t)$ and $q_\pm(t)$ are chosen for a given coefficient in a special way. For an appropriate choice of such polynomials, the solution can be obtained in the simplest possible way.

Example. Consider the Riemann problem

$$\Phi^+(t) = \frac{t}{t^2 - 1} \Phi^-(t) + \frac{t^3 - t^2 + 1}{t^3 - t}$$

under the assumption that $\Phi^-(\infty) = 0$ and L is an arbitrary smooth closed contour of one of the following forms:

1°. The interior of the contour L contains the point $z_1 = 0$ and does not contain the points $z_2 = 1$ and $z_3 = -1$.
2°. The interior of the contour L contains the points $z_1 = 0$ and $z_2 = 1$ and does not contain the point $z_3 = -1$.
3°. The interior of the contour L contains the points $z_1 = 0$, $z_2 = 1$, and $z_3 = -1$.
4°. The interior of the contour L contains the points $z_2 = 1$ and $z_3 = -1$ and does not contain the point $z_1 = 0$.

Consider cases 1°–4° in order. In the solution we apply the method of Subsection 12.3-7.

1°. We have

$$p_+(t) = t, \quad p_-(t) = 1, \quad q_+(t) = 1, \quad q_-(t) = t^2 - 1; \quad m_+ = 1, \quad n_+ = 0, \quad \nu = m_+ - n_+ = 1.$$

Let us rewrite the boundary condition in the form

$$(t^2 - 1)\Phi^+(t) - t\Phi^-(t) = \frac{1}{t}(t^3 - t^2 + 1)(t + 1).$$

Hence,

$$\Psi(z) = \frac{1}{2\pi i} \int_L \frac{q_-(\tau)}{p_-(\tau)} H(\tau) \frac{d\tau}{\tau - z} = \frac{1}{2\pi i} \int_L \frac{\tau^3 - \tau + 1}{\tau - z} d\tau + \frac{1}{2\pi i} \int_L \frac{1/\tau}{\tau - z} d\tau,$$

and the formulas for the Cauchy integral (see Subsection 12.2-1) imply

$$\Psi^+(z) = z^3 - z + 1, \quad \Psi^-(z) = -\frac{1}{z}.$$

The general solution of the problem contains a single (arbitrary) constant. By formula (33), we obtain

$$\Phi^+(z) = \frac{1}{z^2 - 1}(z^3 - z + 1 + C) = \frac{z^3 - z + 1}{z^2 - 1} + \frac{C}{z^2 - 1}, \qquad \Phi^-(z) = \frac{1}{z}\left(-\frac{1}{z} + C\right) = -\frac{1}{z^2} + \frac{C}{z},$$

where C is an arbitrary constant. On replacing C by $C - 1$ we can rewrite the solution in the form

$$\Phi^+(z) = z + \frac{C}{z^2 - 1}, \quad \Phi^-(z) = -\frac{z + 1}{z^2} + \frac{C}{z}.$$

2°. We have

$$p_+(t) = t, \quad p_-(t) = 1, \quad q_+(t) = t - 1, \quad q_-(t) = t + 1, \quad m_+ = n_+ = 1, \quad \nu = 0,$$

$$(t + 1)\Phi^+(t) - \frac{t}{t - 1}\Phi^-(t) = \frac{(t + 1)(t^3 - t^2 + 1)}{t(t - 1)},$$

$$\Psi(z) = \frac{1}{2\pi i} \int_L \frac{\tau^2 + \tau}{\tau - z} d\tau + \frac{1}{2\pi i} \int_L \frac{(\tau + 1)/[\tau(\tau - 1)]}{\tau - z} d\tau = \begin{cases} z^2 + z & \text{for } z \in \Omega^+, \\ -\dfrac{z + 1}{z(z - 1)} & \text{for } z \in \Omega^-. \end{cases}$$

The problem has the unique solution

$$\Phi^+(z) = \frac{p_-(z)}{q_-(z)} \Phi^+(z) = \frac{1}{z + 1}(z^2 + z) = z,$$

$$\Phi^-(z) = \frac{q_+(z)}{p_+(z)} \Phi^-(z) = \frac{z - 1}{z}\left(-\frac{z + 1}{z(z - 1)}\right) = -\frac{z + 1}{z^2}.$$

3°. We have

$$p_+(t) = t, \quad p_-(t) = 1, \quad q_+(t) = t^2 - 1, \quad q_-(t) = 1, \quad m_+ = 1, \quad n_+ = 2, \quad \nu = -1,$$

$$\Psi(z) = \frac{1}{2\pi i} \int_L \frac{\tau}{\tau - z} \, d\tau + \frac{1}{2\pi i} \int_L \frac{1/[\tau(\tau - 1)]}{\tau - z} \, d\tau = \begin{cases} z & \text{for } z \in \Omega^+, \\ -\dfrac{1}{z(z-1)} & \text{for } z \in \Omega^-. \end{cases}$$

The solution of the problem exists only under the solvability conditions (34) or, for the case in question, under the single condition

$$\int_L \frac{q_-(\tau)}{p_-(\tau)} H(\tau) \, d\tau = 0.$$

On calculating this integral, we obtain

$$\int_L \frac{\tau^3 - \tau^2 + 1}{\tau^2 - \tau} \, d\tau = \int_L \tau \, d\tau + \int_L \frac{d\tau}{\tau - 1} - \int_L \frac{d\tau}{\tau} = 0 + 2\pi i - 2\pi i = 0.$$

Thus, the solvability condition holds, and the unique solution of the problem is

$$\Phi^+(z) = z, \quad \Phi^-(z) = -\frac{z+1}{z^2}.$$

4°. We have

$$p_+(t) = 1, \quad p_-(t) = t, \quad q_+(t) = t^2 - 1, \quad q_-(t) = 1, \quad \nu = m_+ - n_+ = -2 < 0.$$

For the solvability of the problem, the following two conditions are necessary:

$$\int_L \frac{q_-(\tau)}{p_-(\tau)} H(\tau) \tau^{k-1} \, d\tau = 0, \qquad k = 1, 2.$$

On calculating the last integral for $k = 1$, we obtain

$$\int_L \frac{\tau^3 - \tau^2 + 1}{\tau(\tau^2 - \tau)} \, d\tau = \int_L \left(1 - \frac{1}{\tau} - \frac{1}{\tau^2} + \frac{1}{\tau - 1} \right) d\tau = 2\pi i \neq 0.$$

Thus, the solvability condition fails, and hence the problem has no solution.

Note that if we formally calculate the function $\Phi(z)$, then it has a pole at infinity, and hence cannot be a solution of the problem.

12.3-8. The Riemann Problem for a Half-Plane

Let the contour L be the real axis. Just as above, the Riemann problem is to find two bounded analytic functions $\Phi^+(z)$ and $\Phi^-(z)$ in the upper and the lower half-plane, respectively (or a single piecewise analytic function $\Phi(z)$ on the plane), whose limit values on the contour satisfy the boundary condition

$$\Phi^+(x) = D(x)\Phi^-(x) + H(x), \qquad -\infty < x < \infty. \tag{35}$$

The given functions $D(x)$ and $H(x)$ satisfy the Hölder condition both at the endpoints and in a neighborhood of the point at infinity on the contour. We also assume that $D(x) \neq 0$.

The main difference from the above case of a finite curve is that here the point at infinity and the origin belong to the contour itself, and therefore cannot be taken as exceptional points at which the canonical function can have a nonzero order. Instead of the auxiliary function t which was used in the above discussion (and has the unit index with respect to L), we use the linear-fractional function on the real axis with the same property:

$$\frac{x-i}{x+i}.$$

The argument of this function

$$\arg \frac{x-i}{x+i} = \arg \frac{(x-i)^2}{x^2 + i} = 2 \arg(x - i)$$

increases by 2π as x ranges over the real axis in the positive direction. Thus,

$$\text{Ind } \frac{x-i}{x+i} = 1.$$

If Ind $D(x) = \nu$, then the function

$$\left(\frac{x-i}{x+i}\right)^{-\nu} D(x)$$

has zero index. Its logarithm is single-valued on the real axis.

We construct the canonical function for which the point $-i$ is the exceptional point as follows:

$$X^+(z) = e^{G^+(z)}, \quad X^-(z) = \left(\frac{z-i}{z+i}\right)^{-\nu} e^{G^-(z)}, \tag{36}$$

where

$$G(z) = \frac{1}{2\pi i} \int_{-\infty}^{\infty} \ln\left[\left(\frac{\tau-i}{\tau+i}\right)^{-\nu} D(\tau)\right] \frac{d\tau}{\tau-z}.$$

Using the limit values of this function, we transform the boundary condition (35) to the form

$$\frac{\Phi^+(x)}{X^+(x)} = \frac{\Phi^-(x)}{X^-(x)} + \frac{H(x)}{X^+(x)}.$$

Next, introducing the analytic function

$$\Psi(z) = \frac{1}{2\pi i} \int_{-\infty}^{\infty} \frac{H(\tau)}{X^+(\tau)} \frac{d\tau}{\tau-z}, \tag{37}$$

we represent the boundary condition in the form

$$\frac{\Phi^+(x)}{X^+(x)} - \Psi^+(x) = \frac{\Phi^-(x)}{X^-(x)} - \Psi^-(x).$$

Note that, in contrast with the case of a finite contour, here we have $\Psi^-(\infty) \neq 0$ in general. On applying the theorem on analytic continuation and taking into account the fact that the only possible singularity of the function under consideration is a pole at the point $z = -i$ of order $\leq \nu$ (for $\nu > 0$), on the basis of the generalized Liouville theorem we obtain (see Subsection 12.3-1)

$$\frac{\Phi^+(z)}{X^+(z)} - \Psi^+(z) = \frac{\Phi^-(z)}{X^-(z)} - \Psi^-(z) = \frac{P_\nu(z)}{(z+i)^\nu}, \quad \nu \geq 0,$$

where $P_\nu(z)$ is a polynomial of degree $\leq \nu$ with arbitrary coefficients. This gives the general solution of the problem:

$$\Phi(z) = X(z)\left[\Psi(z) + \frac{P_\nu(z)}{(z+i)^\nu}\right] \qquad \text{for} \quad \nu \geq 0, \tag{38}$$

$$\Phi(z) = X(z)[\Psi(z) + C] \qquad \text{for} \quad \nu < 0, \tag{39}$$

where C is an arbitrary constant. For $\nu < 0$, the function $X(z)$ has a pole of order $-\nu$ at the point $z = -i$, and therefore for the solvability of the problem we must set $C = -\Psi^-(-i)$. For $\nu < -1$, the following conditions must additionally hold:

$$\int_{-\infty}^{\infty} \frac{H(x)}{X^+(x)} \frac{dx}{(x+i)^k} = 0, \qquad k = 2, 3, \ldots, -\nu. \tag{40}$$

Thus, we obtained results similar to those for a finite contour.

Indeed, for $\nu \geq 0$, the homogeneous and nonhomogeneous Riemann boundary value problems for the half-plane are unconditionally solvable, and their solution linearly depends on $\nu + 1$ arbitrary constants. For $\nu < 0$, the homogeneous problem is unsolvable. For $\nu < 0$, the nonhomogeneous problem is uniquely solvable; moreover, in the case $\nu = -1$ the problem is unconditionally solvable, and in the case $\nu < -1$, it is solvable under $-\nu - 1$ solvability conditions (40) only.

Let us also discuss the case of solutions vanishing at infinity (see also Subsection 10.3-4). On substituting the relation $\Phi^+(\infty) = \Phi^-(\infty) = 0$ into the boundary condition, we obtain $H(\infty) = 0$. Hence, for a Riemann problem to have a solution that vanishes at infinity, the right-hand side of the boundary condition must vanish at infinity. Assume that this condition is satisfied. To obtain a solution for the case under consideration, we must replace the expression $P_\nu(z)$ in (38) by $P_{\nu-1}(z)$ and equate the constant C in (39) with zero. Thus,

$$\Phi(z) = X(z)\left[\Psi(z) + \frac{P_{\nu-1}(z)}{(z+i)^\nu}\right]. \tag{41}$$

For $\nu \leq 0$, we must set $P_{\nu-1}(z) \equiv 0$ in this formula. We must add another condition to the solvability conditions (40), namely, $\Psi(-i) = 0$, and finally we obtain the following solvability conditions:

$$\int_{-\infty}^{\infty} \frac{H(x)}{X^+(x)} \frac{dx}{(x+i)^k} = 0, \qquad k = 1, 2, \ldots, -\nu. \tag{42}$$

Now, for $\nu > 0$ we have a solution that depends on ν arbitrary constants. For $\nu \leq 0$, a solution is unique, and for $\nu < 0$, a solution exists if and only if $-\nu$ conditions hold.

12.3-9. Exceptional Cases of the Riemann Problem

In the statement of the Riemann boundary value problem it was required that the coefficient $D(t)$ satisfies the Hölder condition (this prevents infinite values of this coefficient) and vanishes nowhere. As can be observed from the solution (the use of $\ln D(t)$), these restrictions are essential. Now we assume that $D(t)$ vanishes or tends to infinity, with an integer order, at some points of the contour. We assume that the contour L consists of a single closed curve.

Consider the homogeneous problem. We rewrite the boundary condition of the homogeneous Riemann problem in the form

$$\Phi^+(t) = \frac{\displaystyle\prod_{k=1}^{\mu}(t-\alpha_k)^{m_k}}{\displaystyle\prod_{j=1}^{\kappa}(t-\beta_j)^{p_j}} D_1(t)\Phi^-(t). \tag{43}$$

Here α_k $(k = 1, \ldots, \mu)$ and β_j $(j = 1, \ldots, \kappa)$ are some points of the contour, m_k and p_j are positive integers, and $D_1(t)$ is a function that is everywhere nonzero and satisfies the Hölder condition. The points α_k are zeros of the function $D(t)$. The points β_j will be called the *poles* of this function. The use of the term "pole" is not completely rigorous because the function $D(t)$ is not analytic. We shall use this term for brevity for a point at which a function (not analytic) tends to infinity with some integer order. We write

$$\text{Ind}\, D_1(t) = \nu, \qquad \sum_{j=1}^{\kappa} p_j = p, \qquad \sum_{k=1}^{\mu} m_k = m.$$

We seek the solution in the class of functions bounded on the contour.

Let $X(z)$ be the canonical function of the Riemann problem with coefficient $D_1(t)$. Let us substitute the expression $D_1(t) = X^+(t)/X^-(t)$ into (43) and rewrite the boundary condition in the form

$$\frac{\Phi^+(t)}{X^+(t)\prod\limits_{k=1}^{\mu}(t-\alpha_k)^{m_k}} = \frac{\Phi^-(t)}{X^-(t)\prod\limits_{j=1}^{\kappa}(t-\beta_j)^{p_j}}. \tag{44}$$

To the last relation we apply the theorem on analytic continuation and the generalized Liouville theorem (see Subsection 12.3-1). The points α_k and β_j cannot be singular points of the same analytic function because this would contradict the assumption that $\Phi^+(t)$ or $\Phi^-(t)$ be bounded. Hence, the only possible singularity is the point at infinity. The order at infinity of $X^-(z)$ is ν, and the order of $\prod\limits_{j=1}^{\kappa}(z-\beta_j)^{p_j}$ is equal to $-p$. Hence, the order at infinity of the function $\Phi^-(z)/\left[X^-(z)\prod\limits_{j=1}^{\kappa}(z-\beta_j)^{p_j}\right]$ is $-\nu + p$. For $\nu - p \geq 0$ it follows from the generalized Liouville theorem that

$$\frac{\Phi^+(z)}{X^+(z)\prod\limits_{k=1}^{\mu}(z-\alpha_k)^{m_k}} = \frac{\Phi^-(z)}{X^-(z)\prod\limits_{j=1}^{\kappa}(z-\beta_j)^{p_j}} = P_{\nu-p}(z),$$

and hence

$$\Phi^+(z) = X^+(z)\prod\limits_{k=1}^{\mu}(z-\alpha_k)^{m_k} P_{\nu-p}(z), \quad \Phi^-(z) = X^-(z)\prod\limits_{j=1}^{\kappa}(z-\beta_j)^{p_j} P_{\nu-p}(z). \tag{45}$$

If $\nu - p < 0$, then we must set $P_{\nu-p}(z) \equiv 0$, and hence the problem has no solutions.

The boundary value problem with coefficient $D_1(t)$ is called the *reduced problem*. The index ν of the reduced problem will be called the *index* of the original problem. Formulas (45) show that the degree of the occurring polynomial is less by p than the index ν of the problem.

Hence, the number of solutions of problem (43) in the class of functions bounded on the contour is independent of the number of zeros of the coefficient and is diminished by the total number of all poles. In particular, if the index is less than the total order of the poles, then the problem is unsolvable. If the problem is solvable, then its solution can be expressed by formulas (45) in which the canonical function $X(z)$ of the reduced problem can be found by formulas (16) and (17) after replacing $D(t)$ by $D_1(t)$ in these formulas. Under the additional condition $\Phi^-(\infty) = 0$, the number of solutions is diminished by one, and the degree of the polynomial in (45) must be at most $\nu - p - 1$.

Now let us extend the class of solutions by assuming that one of the desired functions $\Phi^+(z)$ and $\Phi^-(z)$ can tend to infinity with integral order at some points of the contour, and at the same time another function remains bounded at these points. We can readily see that this assumption implies no modifications at nonexceptional points. Here the boundedness of one of the functions automatically implies the boundedness of the other. This is not the case for the exceptional points. Let us rewrite the boundary condition (43) in the form

$$\frac{\prod\limits_{j=1}^{\kappa}(t-\beta_j)^{p_j}\,\Phi^+(t)}{X^+(t)} = \frac{\prod\limits_{k=1}^{\mu}(t-\alpha_k)^{m_k}\,\Phi^-(t)}{X^-(t)}. \tag{46}$$

Applying the above reasoning and taking into account the fact that the right-hand side has a pole of order $\nu + m$ at infinity, we obtain the general solution in the form

$$\Phi^+(z) = X^+(z)\prod\limits_{k=1}^{\mu}(z-\alpha_k)^{-m_k} P_{\nu+m}(z), \quad \Phi^-(z) = X^-(z)\prod\limits_{j=1}^{\kappa}(z-\beta_j)^{-p_j} P_{\nu+m}(z). \tag{47}$$

Formulas (47) show that in the class of solutions with admissible polar singularity for one of the functions, the number of solutions is greater than that in the class of functions bounded on the contour (for $\nu > 0$) by the total order of all zeros and poles of the coefficient.

We now consider the nonhomogeneous problem. Let us write out the boundary condition in the form

$$\Phi^+(t) = \frac{\displaystyle\prod_{k=1}^{\mu}(t-\alpha_k)^{m_k}}{\displaystyle\prod_{j=1}^{\kappa}(t-\beta_j)^{p_j}} D_1(t)\Phi^-(t) + H(t). \tag{48}$$

We can readily see that the boundary condition cannot be satisfied by finite functions $\Phi^+(t)$ and $\Phi^-(t)$ if we assume that $H(t)$ has poles at points that differ from β_j or if at these points, the orders of the poles of $H(t)$ exceed p_j. Hence, we assume that $H(t)$ can have poles at the points β_j only and that their orders do not exceed p_j. To perform the subsequent reasoning, we must also assume that the functions $D_1(t)$ and $\prod_{j=1}^{\kappa}(t-\beta_j)^{p_j} H(t)$ at the exceptional points are differentiable sufficiently many times.

Just as in the homogeneous problem, we replace $D_1(t)$ by the ratio of the canonical functions $X^+(t)/X^-(t)$ and rewrite the boundary condition (48) in the form

$$\prod_{j=1}^{\kappa}(t-\beta_j)^{p_j}\frac{\Phi^+(t)}{X^+(t)} = \prod_{k=1}^{\mu}(t-\alpha_k)^{m_k}\frac{\Phi^-(t)}{X^-(t)} + \prod_{j=1}^{\kappa}(t-\beta_j)^{p_j}\frac{H(t)}{X^+(t)}. \tag{49}$$

On replacing the function defined by the second summand on the right-hand side in (49) by the difference of the boundary values of analytic functions

$$\prod_{j=1}^{\kappa}(t-\beta_j)^{p_j}\frac{H(t)}{X^+(t)} = \Psi^+(t) - \Psi^-(t),$$

where

$$\Psi(z) = \frac{1}{2\pi i}\int_L \prod_{j=1}^{\kappa}(\tau-\beta_j)^{p_j}\frac{H(\tau)}{X^+(\tau)}\frac{d\tau}{\tau-z}, \tag{50}$$

we reduce the boundary condition to the form

$$\prod_{j=1}^{\kappa}(t-\beta_j)^{p_j}\frac{\Phi^+(t)}{X^+(t)} - \Psi^+(t) = \prod_{k=1}^{\mu}(t-\alpha_k)^{m_k}\frac{\Phi^-(t)}{X^-(t)} - \Psi^-(t).$$

On applying the theorem on analytic continuation and the generalized Liouville theorem (see Subsection 12.3-1), we obtain

$$\Phi^+(z) = \frac{X^+(z)}{\displaystyle\prod_{j=1}^{\kappa}(z-\beta_j)^{p_j}}[\Psi^+(z) + P_{\nu+m}(z)], \quad \Phi^-(z) = \frac{X^-(z)}{\displaystyle\prod_{k=1}^{\mu}(z-\alpha_k)^{m_k}}[\Psi^-(z) + P_{\nu+m}(z)]. \tag{51}$$

In general, the last formulas give solutions that can tend to infinity at the points α_k and β_k. For a solution to be bounded it is necessary that the function $\Psi^+(z) + P_{\nu+m}(z)$ have zeros of orders p_j at the points β_j and the function $\Psi^-(z) + P_{\nu+m}(z)$ have zeros of orders m_k at the points α_k. These requirements form $m+p$ conditions for the coefficients of the polynomial $P_{\nu+m}(z)$. If the coefficients of the polynomial $P_{\nu+m}(z)$ are chosen in accordance with the above conditions, then formulas (51) give a solution of the nonhomogeneous problem (48) in the class of bounded functions.

Consider another way of constructing a solution, which is more convenient and based on the construction of a special particular solution.

By the *canonical function* $Y(z)$ *of the nonhomogeneous problem* we mean a piecewise analytic function that satisfies the boundary condition (48), has zero order everywhere in the finite part of the domain (including the points α_k and β_j), and has the least possible order at infinity.

In the construction of the canonical function, we start from the solution given by formulas (51). Let us construct a polynomial $\mathcal{U}_n(z)$ that satisfies the following conditions:

$$\mathcal{U}_n^{(i)}(\beta_j) = \Psi^{+(i)}(\beta_j), \qquad i = 0, 1, \ldots, p_j - 1, \qquad j = 1, \ldots, \kappa,$$
$$\mathcal{U}_n^{(l)}(\alpha_k) = \Psi^{-(l)}(\alpha_k), \qquad l = 0, 1, \ldots, m_k - 1, \qquad k = 1, \ldots, \mu,$$

where $\Psi^{+(i)}(\beta_j)$ and $\Psi^{-(l)}(\alpha_k)$ are the values of the ith and the lth derivatives at the corresponding points. Thus, $\mathcal{U}_n(z)$ is the Hermite interpolation polynomial for the functions

$$\Psi(z) = \begin{cases} \Psi^+(z) & \text{at the points } \beta_j, \\ \Psi^-(z) & \text{at the points } \alpha_k \end{cases}$$

with interpolation nodes β_j and α_k of multiplicities p_j and m_k, respectively (see Subsection 12.3-2). Such a polynomial is uniquely determined, and its degree is at most $n = m + p - 1$.

The canonical function of the nonhomogeneous problem can be expressed via the interpolation polynomial as follows:

$$Y^+(z) = X^+(z) \frac{\Psi^+(z) - \mathcal{U}_n(z)}{\displaystyle\prod_{j=1}^{\kappa}(z - \beta_j)^{p_j}}, \qquad Y^-(z) = X^-(z) \frac{\Psi^-(z) - \mathcal{U}_n(z)}{\displaystyle\prod_{k=1}^{\mu}(z - \alpha_k)^{m_k}}. \tag{52}$$

To construct the general solution of the nonhomogeneous problem (48), we use the fact that this general solution is the sum of a particular solution of the nonhomogeneous problem and of the general solution of the homogeneous problem. Applying formulas (47) and (52), we obtain

$$\Phi^+(z) = Y^+(z) + X^+(z) \prod_{k=1}^{\mu}(z - \alpha_k)^{m_k} P_{\nu-p}(z),$$

$$\Phi^-(z) = Y^-(z) + X^-(z) \prod_{j=1}^{\kappa}(z - \beta_j)^{p_j} P_{\nu-p}(z). \tag{53}$$

For the case in which $\nu - p < 0$, we must set $P_{\nu-p}(z) \equiv 0$. Applying formula (52), we readily find that the order of $Y^-(z)$ at infinity is equal to $\nu - p + 1$. If $\nu < p - 1$, then $Y^-(z)$ has a pole at infinity, and the canonical function is no longer a solution of the nonhomogeneous problem.

However, on subjecting the constant term $H(t)$ to $p - \nu - 1$ conditions, we can increase the order of the functions $Y(z)$ at infinity by $p - \nu - 1$ and thus again make the canonical function $Y(z)$ a solution of the nonhomogeneous problem. Obviously, to this end it is necessary and sufficient that in the expansion of the function $\Psi(z) - \mathcal{U}_n(z)$ in a neighborhood of the point at infinity, the first $p - \nu - 1$ coefficients be zero. This gives just $p - \nu - 1$ solvability conditions of the problem for the case under consideration. Let us clarify the character of these conditions. The expansion of $\Psi(z) - \mathcal{U}_n(z)$ can be represented in the form

$$\Psi(z) - \mathcal{U}_n(z) = -a_n z^n - a_{n-1} z^{n-1} - \cdots - a_0 + a_{-1} z^{-1} + a_{-2} z^{-2} + \cdots + a_{-k} z^{-k} + \cdots,$$

where $a_0, a_1, \ldots, a_n$ are the coefficients of the polynomial $\mathcal{U}_n(z)$, and the a_{-k} are the coefficients of the expansion of the function $\Psi(z)$, which are given by the obvious formula

$$a_{-k} = -\frac{1}{2\pi i} \int_L \prod_{j=1}^{\kappa} (\tau - \beta_j)^{p_j} \frac{H(\tau)\tau^{k-1}}{X^+(\tau)} \, d\tau.$$

The solvability conditions acquire the form

$$a_n = a_{n-1} = \cdots = a_{n-p+\nu+2} = 0.$$

If a solution must satisfy the additional condition $\Phi^-(\infty) = 0$, then, for $\nu - p > 0$, in formulas (53) we must take the polynomial $P_{\nu-p-1}(z)$, and for $\nu - p < 0$, $p - \nu$ conditions must be satisfied.

12.3-10. The Riemann Problem for a Multiply Connected Domain

Let $L = L_0 + L_1 + \cdots + L_m$ be a collection of $m + 1$ disjoint contours, and let the interior of the contour L_0 contain the other contours. By Ω^+ we denote the $(m + 1)$-connected domain interior for L_0 and exterior for $L_1, \ldots, L_m$. By Ω^- we denote the complement of $\Omega^+ + L$ in the entire complex plane. To be definite, we assume that the origin lies in Ω^+. The positive direction of the contour L is that for which the domain Ω^+ remains to the left, i.e., the contour L_0 must be traversed counterclockwise and the contours $L_1, \ldots, L_m$, clockwise.

We first note that the jump problem

$$\Phi^+(t) - \Phi^-(t) = H(t)$$

is solved by the same formula

$$\Phi(z) = \frac{1}{2\pi i} \int_L \frac{H(\tau)\, d\tau}{\tau - z}$$

as in the case of a simply connected domain. This follows from the Sokhotski–Plemelj formulas, which have the same form for a multiply connected domain as for a simply connected domain.

The Riemann problem (homogeneous and nonhomogeneous) can be posed in the same way as for a simply connected domain.

We write $\nu_k = \frac{1}{2\pi}[\arg D(t)]_{L_k}$ (all contours are passed in the positive direction). By the *index of the problem* we mean the number

$$\nu = \sum_{k=0}^{m} \nu_k. \tag{54}$$

If ν_k $(k = 1, \ldots, m)$ are zero for the inner contours, then the solution of the problem has just the same form as for a simply connected domain.

To reduce the general case to the simplest one, we introduce the function

$$\prod_{k=1}^{m} (t - z_k)^{\nu_k},$$

where the z_k are some points inside the contours L_k $(k = 1, \ldots, m)$. Taking into account the fact that $[\arg(t - z_k)]_{L_j} = 0$ for $k \neq j$ and $[\arg(t - z_j)]_{L_j} = -2\pi$, we obtain

$$\frac{1}{2\pi}\left[\arg \prod_{k=1}^{m}(t - z_k)^{\nu_k}\right]_{L_j} = \frac{1}{2\pi}\left[\arg(t - z_j)^{\nu_j}\right]_{L_j} = -\nu_j, \qquad j = 1, \ldots, m.$$

Hence,

$$\left[\arg \left(D(t) \prod_{k=1}^{m} (t - z_k)^{\nu_k} \right) \right]_{L_j} = 0, \qquad j = 1, \ldots, m.$$

Let us calculate the increment of the argument of the function $D(t) \prod_{k=1}^{m} (t - z_k)^{\nu_k}$ with respect to the contour L_0:

$$\frac{1}{2\pi} \left[\arg \left(D(t) \prod_{k=1}^{m} (t - z_k)^{\nu_k} \right) \right]_{L_0} = \frac{1}{2\pi} \left[\arg D(t) \right]_{L_0} + \frac{1}{2\pi} \sum_{k=1}^{m} [\nu_k \arg(t - z_k)]_{L_0} = \nu_0 + \sum_{k=1}^{m} \nu_k = \nu.$$

Since the origin belongs to the domain Ω^+, it follows that

$$[\arg t]_{L_k} = 0, \quad k = 1, \ldots, m, \qquad [\arg t]_{L_0} = 2\pi.$$

Therefore,

$$\left[\arg \left(t^{-\nu} \prod_{k=1}^{m} (t - z_k)^{\nu_k} D(t) \right) \right]_{L} = 0. \tag{55}$$

$1°$. *The Homogeneous Problem.* Let us rewrite the boundary condition

$$\Phi^+(t) = D(t)\Phi^-(t) \tag{56}$$

in the form

$$\Phi^+(t) = \frac{t^\nu}{\displaystyle\prod_{k=1}^{m} (t - z_k)^{\nu_k}} \left(t^{-\nu} \prod_{k=1}^{m} (t - z_k)^{\nu_k} D(t) \right) \Phi^-(t). \tag{57}$$

The function $t^{-\nu} \prod_{k=1}^{m} (t - z_k)^{\nu_k} D(t)$ has zero index on each of the contours L_k $(k = 1, \ldots, m)$, and hence it can be expressed as the ratio

$$t^{-\nu} \prod_{k=1}^{m} (t - z_k)^{\nu_k} D(t) = \frac{e^{G^+(t)}}{e^{G^-(t)}}, \tag{58}$$

where

$$G(z) = \frac{1}{2\pi i} \int_L \ln \left(\tau^{-\nu} \prod_{k=1}^{m} (\tau - z_k)^{\nu_k} D(\tau) \right) \frac{d\tau}{\tau - z}. \tag{59}$$

The canonical function of the problem is given by the formulas

$$X^+(z) = \prod_{k=1}^{m} (z - z_k)^{-\nu_k} e^{G^+(z)}, \quad X^-(z) = z^{-\nu} e^{G^-(z)}. \tag{60}$$

Now the boundary condition (57) can be rewritten in the form

$$\frac{\Phi^+(t)}{X^+(t)} = \frac{\Phi^-(t)}{X^-(t)}.$$

As usual, by applying the theorem on analytic continuation and the generalized Liouville theorem (see Subsection 12.3-1), we obtain

$$\Phi^+(z) = \prod_{k=1}^{m}(z - z_k)^{-\nu_k} e^{G^+(z)} P_\nu(z), \quad \Phi^-(z) = z^{-\nu} e^{G^-(z)} P_\nu(z). \tag{61}$$

We can see that this solution differs from the above solution of the problem for a simply connected domain only in that the function $\Phi^+(z)$ has the factor $\prod_{k=1}^{m}(z - z_k)^{-\nu_k}$. Under the additional condition $\Phi^-(\infty) = 0$, in formulas (61) we must take the polynomial $P_{\nu-1}(z)$.

Applying the Sokhotski–Plemelj formulas, we obtain

$$G^\pm(t) = \pm\tfrac{1}{2}\ln[t^{-\nu}\Pi(t)D(t)] + G(t),$$

where $G(t)$ is the Cauchy principal value of the integral (59) and

$$\Pi(t) = \prod_{k=1}^{m}(t - z_k)^{\nu_k}.$$

On passing to the limit as $z \to t$ in formulas (60) we obtain

$$X^+(t) = \sqrt{\frac{D(t)}{t^\nu \Pi(t)}}\, e^{G(t)}, \quad X^-(t) = \frac{1}{\sqrt{t^\nu \Pi(t)D(t)}}\, e^{G(t)}. \tag{62}$$

The sign of the root is determined by the (arbitrary) choice of a branch of the function $\ln[t^{-\nu}\Pi(t)D(t)]$.

2°. *The Nonhomogeneous Problem.* By the same reasoning as above, we represent the boundary condition

$$\Phi^+(t) = D(t)\Phi^-(t) + H(t) \tag{63}$$

in the form

$$\frac{\Phi^+(t)}{X^+(t)} - \Psi^+(t) = \frac{\Phi^-(t)}{X^-(t)} - \Psi^-(t),$$

where $\Psi(z)$ is defined by the formula

$$\Psi(z) = \frac{1}{2\pi i}\int_L \frac{H(\tau)}{X^+(\tau)}\frac{d\tau}{\tau - z}.$$

This gives the general solution

$$\Phi(z) = X(z)[\Psi(z) + P_\nu(z)] \tag{64}$$

or

$$\Phi(z) = X(z)[\Psi(z) + P_{\nu-1}(z)], \tag{65}$$

if the solution satisfies the condition $\Phi^-(\infty) = 0$.

For $\nu < 0$, the nonhomogeneous problem is solvable if and only if the following conditions are satisfied:

$$\int_L \frac{H(t)}{X^+(t)} t^{k-1}\, dt = 0, \tag{66}$$

where k ranges from 1 to $-\nu - 1$ if we seek solutions bounded at infinity and from 1 to $-\nu$ if we assume that $\Phi^-(\infty) = 0$.

Under conditions (66), the solution can also be found from formulas (64) or (65) by setting $P_\nu \equiv 0$.

If the external contour L_0 is absent and the domain Ω^+ is the plane with holes, then the main difference from the preceding case is that here the zero index with respect to all contours L_k ($k = 1, \dots, m$) is attained by the function $\prod_{k=1}^{m}(t - z_k)^{\nu_k} D(t)$ that does not involve the factor $t^{-\nu}$.

Therefore, to obtain a solution to the problem, it suffices to repeat the above reasoning on omitting this factor.

12.3-11. The Cases of Discontinuous Coefficients and Nonclosed Contours

Assume that the functions $D(t)$ and $H(t)$ in the boundary condition of the Riemann problem (63) satisfy the Hölder condition everywhere on L except for points $t_1, \dots, t_m$ at which these functions have jumps, and assume that L is a closed curve. None of the limit values vanishes, and the boundary condition holds everywhere except for the discontinuity points at which it makes no sense.

A solution to the problem is sought in the class of functions that are integrable on the contour. Therefore, a solution is everywhere continuous, in the sense of the Hölder condition, possibly except for the points t_k. For these points, there are different possibilities.

$1°$. We can assume boundedness at all discontinuity points, and thus seek a solution that is everywhere bounded.

$2°$. We can assume that a solution is bounded at some discontinuity points and admit an integrable singularity at the other discontinuity points.

$3°$. We can admit integrable singularity at all points which are admitted by the conditions of the problem.

The first class of solutions is the narrowest, the second class is broader, and the third class is the largest. The number of solutions depends on the class in which it is sought, and it can turn out that a problem that is solvable in a broader class is unsolvable in a narrower class.

We make a few remarks on the Riemann problem for nonclosed contours. Assume that a contour L consists of a collection of m simple closed disjoint curves $L_1, \dots, L_m$ whose endpoints are a_k and b_k (the positive direction is from a_k to b_k). Assume that $D(t)$ and $H(t)$ are functions given on L and satisfy the Hölder condition, and $D(t) \neq 0$ everywhere.

It is required to find a function $\Phi(z)$ that is analytic on the entire plane except for the points of the contour L, and whose boundary values $\Phi^+(t)$ and $\Phi^-(t)$, when tending to L from the left and from the right, are integrable functions satisfying the boundary condition (63).

As can be seen from the setting, the Riemann problem for a nonclosed contour principally differs from the problem for a closed contour in that the entire plane with the cut along the curve L forms a single domain, and instead of two independent analytic functions $\Phi^+(z)$ and $\Phi^-(z)$, we must find a single analytic function $\Phi(z)$ for which the contour L is the line of jumps. The problem posed above can be reduced to that for a closed contour with discontinuous coefficients.

The details on the Riemann boundary value problem with discontinuous coefficients and non-closed contours can be found in the references cited below.

12.3-12. The Hilbert Boundary Value Problem

Let a simple smooth closed contour L and real Hölder functions $a(s)$, $b(s)$, and $c(s)$ of the arc length s on the contour be given.

By the *Hilbert boundary value problem* we mean the following problem. Find a function

$$f(z) = u(x, y) + iv(x, y)$$

that is analytic on the domain Ω^+ and continuous on the contour for which the limit values of the real and the imaginary part on the contour satisfy the linear relation

$$a(s)u(s) + b(s)v(s) = c(s). \tag{67}$$

For $c(s) \equiv 0$ we obtain the *homogeneous* problem and, for nonzero $c(s)$, a *nonhomogeneous*. The Hilbert boundary value problem can be reduced to the Riemann boundary value problem. The methods of this reduction can be found in the references cited at the end of the section.

⊙ References for Section 12.3: F. D. Gakhov (1977), N. I. Muskhelishvili (1992).

12.4. Singular Integral Equations of the First Kind

12.4-1. The Simplest Equation With Cauchy Kernel

Consider the singular integral equation of the first kind

$$\frac{1}{\pi i} \int_L \frac{\varphi(\tau)}{\tau - t} \, d\tau = f(t), \tag{1}$$

where L is a closed contour. Let us construct the solution. In this relation we replace the variable t by τ_1, multiply by $\frac{1}{\pi i} \frac{d\tau_1}{\tau_1 - t}$, integrate along the contour L, and change the order of integration according to the Poincaré–Bertrand formula (see Subsection 12.2-6). Then we obtain

$$\frac{1}{\pi i} \int_L \frac{f(\tau_1)}{\tau_1 - t} \, d\tau_1 = \varphi(t) + \frac{1}{\pi i} \int_L \varphi(\tau) \, d\tau \, \frac{1}{\pi i} \int_L \frac{d\tau_1}{(\tau_1 - t)(\tau - \tau_1)}. \tag{2}$$

Let us calculate the second integral on the right-hand side of (2):

$$\int_L \frac{d\tau_1}{(\tau_1 - t)(\tau - \tau_1)} = \frac{1}{\tau - t} \left(\int_L \frac{d\tau_1}{\tau_1 - t} - \int_L \frac{d\tau_1}{\tau_1 - \tau} \right) = \frac{1}{\tau - t} (i\pi - i\pi) = 0.$$

Thus,

$$\varphi(t) = \frac{1}{\pi i} \int_L \frac{f(\tau)}{\tau - t} \, d\tau. \tag{3}$$

The last formula gives the solution of the singular integral equation of the first kind (1) for a closed contour L.

12.4-2. An Equation With Cauchy Kernel on the Real Axis

Consider the following singular integral equation of the first kind on the real axis:

$$\frac{1}{\pi i} \int_{-\infty}^{\infty} \frac{\varphi(t)}{t - x} \, dt = f(x), \qquad -\infty < x < \infty. \tag{4}$$

Equation (4) is a special case of the characteristic integral equation on the real axis (see Subsection 13.2-4). In the class of functions vanishing at infinity, Eq. (4) has the solution

$$\varphi(x) = \frac{1}{\pi i} \int_{-\infty}^{\infty} \frac{f(t)}{t - x} \, dt, \qquad -\infty < x < \infty. \tag{5}$$

Denoting $f(x) = F(x)i^{-1}$, we rewrite Eqs. (4) and (5) in the form

$$\frac{1}{\pi} \int_{-\infty}^{\infty} \frac{\varphi(t)}{t - x} \, dt = F(x), \quad \varphi(x) = -\frac{1}{\pi} \int_{-\infty}^{\infty} \frac{F(t)}{t - x} \, dt. \qquad -\infty < x < \infty. \tag{6}$$

The two formulas (6) are called the Hilbert transform pair (see Subsection 7.6-3).

12.4-3. An Equation of the First Kind on a Finite Interval

Consider the singular integral equation of the first kind

$$\frac{1}{\pi} \int_a^b \frac{\varphi(t)}{t-x} \, dt = f(x), \qquad a \le x \le b, \tag{7}$$

on a finite interval. Its solutions can be constructed by using the theory of the Riemann boundary value problem for a nonclosed contour (see Subsection 12.3-11). Let us present the final results.

1°. A solution that is unbounded at both endpoints:

$$\varphi(x) = -\frac{1}{\pi} \frac{1}{\sqrt{(x-a)(b-x)}} \left(\int_a^b \frac{\sqrt{(t-a)(b-t)}}{t-x} f(t) \, dt + C \right), \tag{8}$$

where C is an arbitrary constant and

$$\int_a^b \varphi(t) \, dt = C. \tag{9}$$

2°. A solution bounded at the endpoint a and unbounded at the endpoint b:

$$\varphi(x) = -\frac{1}{\pi} \sqrt{\frac{x-a}{b-x}} \int_a^b \sqrt{\frac{b-t}{t-a}} \frac{f(t)}{t-x} \, dt. \tag{10}$$

3°. A solution bounded at both endpoints:

$$\varphi(x) = -\frac{1}{\pi} \sqrt{(x-a)(b-x)} \int_a^b \frac{f(t)}{\sqrt{(t-a)(b-t)}} \frac{dt}{t-x}, \tag{11}$$

under the condition that

$$\int_a^b \frac{f(t) \, dt}{\sqrt{(t-a)(b-t)}} = 0. \tag{12}$$

Solutions that have a singularity point s inside the interval $[a, b]$ can also be constructed. These solutions have the following form:

4°. A singular solution that is unbounded at both endpoints:

$$\varphi(x) = -\frac{1}{\pi} \frac{1}{\sqrt{(x-a)(b-x)}} \left(\int_a^b \frac{\sqrt{(t-a)(b-t)}}{t-x} f(t) \, dt + C_1 + \frac{C_2}{x-s} \right), \tag{13}$$

where C_1 and C_2 are arbitrary constants.

5°. A singular solution bounded at one endpoint:

$$\varphi(x) = -\frac{1}{\pi} \sqrt{(x-a)(b-x)} \left(\int_a^b \sqrt{\frac{b-t}{t-a}} \frac{f(t)}{t-x} \, dt + \frac{C}{x-s} \right), \tag{14}$$

where C is an arbitrary constants.

6°. A singular solution bounded at both endpoints:

$$\varphi(x) = -\frac{1}{\pi} \sqrt{(x-a)(b-x)} \left(\int_a^b \frac{f(t)}{\sqrt{(t-a)(b-t)}} \frac{dt}{t-x} + \frac{A}{x-s} \right), \qquad A = \int_{-1}^1 \frac{f(t) \, dt}{\sqrt{(t-a)(b-t)}}. \tag{15}$$

12.4-4. The General Equation of the First Kind With Cauchy Kernel

Consider the general equation of the first kind with Cauchy kernel

$$\frac{1}{\pi i} \int_L \frac{M(t,\tau)}{\tau - t} \varphi(\tau)\, d\tau = f(t), \tag{16}$$

where the integral is understood in the sense of the Cauchy principal value and is taken over a closed or nonclosed contour L. As usual, the functions $a(t)$, $f(t)$, and $M(t,\tau)$ on L are assumed to satisfy the Hölder condition, where the last function satisfies this condition with respect to both variables.

We perform the following manipulation with the kernel:

$$\frac{M(t,\tau)}{\tau - t} = \frac{M(t,\tau) - M(t,t)}{\tau - t} + \frac{M(t,t)}{\tau - t}$$

and write

$$M(t,t) = b(t), \qquad \frac{1}{\pi i} \frac{M(t,\tau) - M(t,t)}{\tau - t} = K(t,\tau). \tag{17}$$

We can rewrite Eq. (16) in the form

$$\frac{b(t)}{\pi i} \int_L \frac{\varphi(\tau)}{\tau - t}\, d\tau + \int_L K(t,\tau)\varphi(\tau)\, d\tau = f(t). \tag{18}$$

It follows from formulas (17) that the function $b(t)$ satisfies the Hölder condition on the entire contour L and $K(t,\tau)$ satisfies this condition everywhere except for the points with $\tau = t$ at which this function satisfies the estimate

$$|K(t,\tau)| < \frac{A}{|\tau - t|^\lambda}, \qquad 0 \le \lambda < 1.$$

The general singular integral equation of the first kind with Cauchy kernel is frequently written in the form (18).

The general singular integral equation of the first kind is a special case of the complete singular integral equation whose theory is treated in Chapter 13. In general, it cannot be solved in a closed form. However, there are some cases in which such a solution is possible.

Let the function $M(t,\tau)$ in Eq. (16), which satisfies the Hölder condition with respect to both variables on the smooth closed contour L by assumption, have an analytic continuation to the domain Ω^+ with respect to each of the variables. If $M(t,t) \equiv 1$, then the solution of Eq. (16) can be obtained by means of the Poincaré–Bertrand formula (see Subsection 12.2-6). This solution is given by the relation

$$\varphi(t) = \frac{1}{\pi i} \int_L \frac{M(t,\tau)}{\tau - t} f(\tau)\, d\tau. \tag{19}$$

Eq. (16) can be solved without the assumption that the function $M(t,\tau)$ satisfies the condition $M(t,t) \equiv 1$. Namely, assume that the function $M(t,\tau)$ has the analytic continuation to Ω^+ with respect to each of the variables and that $M(z,z) \ne 0$ for $z \in \overline{\Omega^+}$. In this case, the solution of Eq. (16) has the form

$$\varphi(t) = \frac{1}{\pi i} \frac{1}{M(t,t)} \int_L \frac{M(t,\tau)}{M(\tau,\tau)} \frac{f(\tau)}{\tau - t}\, d\tau. \tag{20}$$

In Section 12.5, a numerical method for solving a special case of the general equation of the first kind is given, which is of independent interest from the viewpoint of applications.

Remark 1. The solutions of complete singular integral equations that are constructed in Subsection 12.4-4 can also be applied for the case in which the contour L is a collection of finitely many disjoint smooth closed contours.

12.4-5. **Equations of the First Kind With Hilbert Kernel**

$1°$. Consider the simplest singular integral equation of the first kind with Hilbert kernel

$$\frac{1}{2\pi} \int_0^{2\pi} \cot\left(\frac{\xi - x}{2}\right) \varphi(\xi)\,d\xi = f(x), \qquad 0 \le x \le 2\pi, \tag{21}$$

under the additional assumption

$$\int_0^{2\pi} \varphi(x)\,dx = 0. \tag{22}$$

Equation (21) can have a solution only if a solvability condition is satisfied. This condition is obtained by integrating Eq. (21) with respect to x from zero to 2π and, with regard for the relation

$$\int_0^{2\pi} \cot\left(\frac{\xi - x}{2}\right) dx = 0,$$

becomes

$$\int_0^{2\pi} f(x)\,dx = 0. \tag{23}$$

To construct a solution of Eq. (21), we apply the solution of the simplest singular integral equation of the first kind with Cauchy kernel by assuming that the contour L is the circle of unit radius centered at the origin (see Subsection 12.4-1). We rewrite the equation with Cauchy kernel and its solution in the form

$$\frac{1}{\pi} \int_L \frac{\varphi_1(\tau)}{\tau - t}\,d\tau = f_1(t), \tag{24}$$

$$\varphi_1(t) = -\frac{1}{\pi} \int_L \frac{f_1(\tau)}{\tau - t}\,d\tau, \tag{25}$$

which is obtained by substituting the function $\varphi_1(t)$ instead of $\varphi(t)$ and the function $f_1(t)i^{-1}$ instead of $f(t)$ into the relations of 12.4-1.

We set $t = e^{ix}$ and $\tau = e^{i\xi}$ and find the relationship between the Cauchy kernel and the Hilbert kernel:

$$\frac{d\tau}{\tau - t} = \frac{1}{2} \cot\left(\frac{\xi - x}{2}\right) d\xi + \frac{i}{2}\,d\xi. \tag{26}$$

On substituting relation (26) into Eq. (24) and into the solution (25), with regard to the change of variables $\varphi(x) = \varphi_1(t)$ and $f(x) = f_1(t)$ we obtain

$$\frac{1}{2\pi} \int_0^{2\pi} \cot\left(\frac{\xi - x}{2}\right) \varphi(\xi)\,d\xi + \frac{i}{2\pi} \int_0^{2\pi} \varphi(\xi)\,d\xi = f(x), \tag{27}$$

$$\varphi(x) = -\frac{1}{2\pi} \int_0^{2\pi} \cot\left(\frac{\xi - x}{2}\right) f(\xi)\,d\xi - \frac{i}{2\pi} \int_0^{2\pi} f(\xi)\,d\xi. \tag{28}$$

Equation (21), under the additional assumption (22), coincides with Eq. (27), and hence its solution is given by the expression (28). Taking into account the solvability conditions (23), on the basis of (28) we rewrite a solution of Eq. (21) in the form

$$\varphi(x) = -\frac{1}{2\pi} \int_0^{2\pi} \cot\left(\frac{\xi - x}{2}\right) f(\xi)\,d\xi. \tag{29}$$

Formulas (21) and (29), together with conditions (22) and (23), are called the *Hilbert inversion formula*.

Remark 2. Equation (21) is a special case of the characteristic singular integral equation with Hilbert kernel (see Subsections 13.1-2 and 13.2-5).

2°. Consider the general singular integral equation of the first kind with Hilbert kernel

$$\frac{1}{2\pi} \int_0^{2\pi} N(x, \xi) \cot\left(\frac{\xi - x}{2}\right) \varphi(\xi) \, d\xi = f(x). \tag{30}$$

Let us represent its kernel in the form

$$N(x, \xi) \cot\frac{\xi - x}{2} = \left[N(x, \xi) - N(x, x)\right] \cot\frac{\xi - x}{2} + N(x, x) \cot\frac{\xi - x}{2}.$$

We introduce the notation

$$N(x, x) = -b(x), \quad \frac{1}{2\pi}\left[N(x, \xi) - N(x, x)\right] \cot\frac{\xi - x}{2} = K(x, \xi), \tag{31}$$

and rewrite Eq. (30) as follows:

$$-\frac{b(x)}{2\pi} \int_0^{2\pi} \cot\left(\frac{\xi - x}{2}\right) \varphi(\xi) \, d\xi + \int_0^{2\pi} K(x, \xi)\varphi(\xi) \, d\xi = f(x), \tag{32}$$

It follows from formulas (31) that the function $b(x)$ satisfies the Hölder condition, whereas the kernel $K(x, \xi)$ satisfies the Hölder condition everywhere except possibly for the points $x = \xi$, at which the following estimate holds:

$$|K(x, \xi)| < \frac{A}{|\xi - x|^\lambda}, \quad A = \text{const} < \infty, \quad 0 \le \lambda < 1.$$

The general singular integral equation of the first kind with Hilbert kernel is frequently written in the form (32). It is a special case of the complete singular integral equation with Hilbert kernel, which is treated in Subsections 13.1-2 and 13.4-8.

⊙ References for Section 12.4: F. D. Gakhov (1977), F. D. Gakhov and Yu. I. Cherskii (1978), S. G. Mikhlin and S. Prössdorf (1986), N. I. Muskhelishvili (1992), I. K. Lifanov (1996).

12.5. Multhopp–Kalandiya Method

Consider a general singular integral equation of the first kind with Cauchy kernel on the finite interval $[-1, 1]$ of the form

$$\frac{1}{\pi} \int_{-1}^1 \frac{\varphi(t) \, dt}{t - x} + \frac{1}{\pi} \int_{-1}^1 K(x, t)\varphi(t) \, dt = f(x). \tag{1}$$

This equation frequently occurs in applications, especially in aerodynamics and 2D elasticity.

We present here a method of approximate solution of Eq. (1) under the assumption that this equation has a solution in the classes indicated below.

12.5-1. A Solution That is Unbounded at the Endpoints of the Interval

According to the general theory of singular integral equations (e.g., see N. I. Muskhelishvili (1992)), such a solution can be represented in the form

$$\varphi(x) = \frac{\psi(x)}{\sqrt{1 - x^2}}, \tag{2}$$

where $\psi(x)$ is a bounded function on $[-1, 1]$. Let us substitute the expression (2) into Eq. (1) and introduce new variables θ and τ by the relations $x = \cos\theta$ and $t = \cos\tau$, $0 \leq \theta \leq \pi$, $0 \leq \tau \leq \pi$. In this case, Eq. (1) becomes

$$\frac{1}{\pi} \int_0^\pi \frac{\psi(\cos\tau)\, d\tau}{\cos\tau - \cos\theta} + \frac{1}{\pi} \int_0^\pi K(\cos\theta, \cos\tau)\psi(\cos\tau)\, d\tau = f(\cos x). \tag{3}$$

Let us construct the *Lagrange interpolation polynomial* for the desired function $\psi(x)$ with the *Chebyshev nodes*

$$x_m = \cos\theta_m, \quad \theta_m = \frac{2m-1}{2n}\pi, \quad m = 1, \ldots, n.$$

This polynomial is known to have the form

$$L_n(\psi; \cos\theta) = \frac{1}{n} \sum_{l=1}^n (-1)^{l+1} \psi(\cos\theta_l) \frac{\cos n\theta\, \sin\theta_l}{\cos\theta - \cos\theta_l}. \tag{4}$$

Note that for each l the fraction on the right-hand side in (4) is an even trigonometric polynomial of degree $\leq n - 1$. We define the coefficients of this polynomial by means of the known relations

$$\frac{1}{\pi} \int_0^\pi \frac{\cos n\tau\, d\tau}{\cos\tau - \cos\theta} = \frac{\sin n\theta}{\sin\theta}, \quad 0 \leq \theta \leq \pi, \quad n = 0, 1, 2, \ldots \tag{5}$$

and rewrite (4) in the form

$$L_n(\psi; \cos\theta) = \frac{2}{n} \sum_{l=1}^n \psi(\cos\theta_l) \sum_{m=0}^{n-1} \cos m\theta_l \cos m\theta - \frac{1}{n} \sum_{l=1}^n \psi(\cos\theta_l). \tag{6}$$

On the basis of the above two relations we write out the following quadrature formula for the singular integral:

$$\frac{1}{\pi} \int_{-1}^1 \frac{\varphi(t)\, dt}{t - x} = \frac{2}{n\sin\theta} \sum_{l=1}^n \psi(\cos\theta_l) \sum_{m=1}^{n-1} \cos m\theta_l \sin m\theta. \tag{7}$$

This formula is exact for the case in which $\psi(t)$ is a polynomial of order $\leq n - 1$ in t.

To the second integral on the left-hand side of Eq. (1), we apply the formula

$$\frac{1}{\pi} \int_{-1}^1 \frac{P(x)\, dx}{\sqrt{1-x^2}} = \frac{1}{n} \sum_{l=1}^n P(\cos\theta_l), \tag{8}$$

which holds for any polynomial $P(x)$ of degree $\leq 2n - 1$. In this case, by (8) we have

$$\frac{1}{\pi} \int_{-1}^1 K(x, t)\varphi(t)\, dt = \frac{1}{n} \sum_{l=1}^n K(\cos\theta, \cos\theta_l)\psi(\cos\theta_l). \tag{9}$$

On substituting relations (7) and (9) into Eq. (1), we obtain

$$\frac{2}{n\sin\theta} \sum_{l=1}^n \psi(\cos\theta_l) \sum_{m=1}^{n-1} \cos m\theta_l \sin m\theta + \frac{1}{n} \sum_{l=1}^n K(\cos\theta, \cos\theta_l)\psi(\cos\theta_l) = f(\cos\theta). \tag{10}$$

By setting $\theta = \theta_k$ $(k = 1, \ldots, n)$ and with regard to the formula

$$\sum_{m=1}^{n-1} \cos m\theta_l \sin m\theta_k = \frac{1}{2} \cot \frac{\theta_k \pm \theta_l}{2}, \tag{11}$$

where the sign "plus" is taken for the case in which $|k - l|$ is even and "minus" if $|k - l|$ odd, we obtain the following system of linear algebraic equations for the approximate values ψ_l of the desired function $\psi(x)$ at the nodes:

$$\sum_{l=1}^{n} a_{kl}\psi_l = f_k, \quad f_k = f(\cos\theta_k), \quad k = 1, \ldots, n,$$

$$a_{kl} = \frac{1}{n}\left[\frac{1}{\sin\theta_k}\cot\frac{\theta_k \pm \theta_l}{2} + K(\cos\theta_k, \cos\theta_l)\right]. \tag{12}$$

After solving the system (12), the corresponding approximate solution to Eq. (1) can be found by formulas (2) and (4).

12.5-2. A Solution Bounded at One Endpoint of the Interval

In this case we set

$$\varphi(x) = \sqrt{\frac{1-x}{1+x}}\,\zeta(x), \tag{13}$$

where $\zeta(x)$ is a bounded function on $[-1, 1]$.

We take the same interpolation nodes as in Section 12.5-1, replace $\zeta(x)$ by the polynomial

$$L_n(\zeta; \cos\theta) = \frac{1}{n}\sum_{l=1}^{n}(-1)^{l+1}\zeta(\cos\theta_l)\frac{\cos n\theta\,\sin\theta_l}{\cos\theta - \cos\theta_l}, \tag{14}$$

and substitute the result into the singular integral that enters the expression (1). Just as above, we obtain the following quadrature formula:

$$\frac{1}{\pi}\int_{-1}^{1}\frac{\varphi(t)\,dt}{t-x} = 2\frac{1-\cos\theta}{n\sin\theta}\sum_{l=1}^{n}\zeta(\cos\theta_l)\sum_{m=1}^{n-1}\cos m\theta_l\sin m\theta - \frac{1}{n}\sum_{l=1}^{n}\zeta(\cos\theta_l). \tag{15}$$

This formula is exact for the case in which $\zeta(t)$ is a polynomial of order $\leq n - 1$ in t.

The formula for the second summand on the left-hand side of the equation becomes

$$\frac{1}{\pi}\int_{-1}^{1}K(x,t)\varphi(t)\,dt = \frac{1}{n}\sum_{l=1}^{n}(1 - \cos\theta_l)K(\cos\theta, \cos\theta_l)\zeta(\cos\theta_l). \tag{16}$$

This formula is exact if the integrand is a polynomial in t of degree $\leq 2n - 2$.

On substituting relations (15) and (16) into Eq. (1) and on setting $\theta = \theta_k$ $(k = 1, \ldots, n)$, with regard to formula (11), we obtain a system of linear algebraic equations for the approximate values ζ_l of the desired function $\zeta(x)$ at the nodes:

$$\sum_{l=1}^{n} b_{kl}\zeta_l = f_k, \quad f_k = f(\cos\theta_k), \quad k = 1, \ldots, n,$$

$$b_{kl} = \frac{1}{n}\left[\tan\frac{\theta_k}{2}\cot\frac{\theta_k \pm \theta_l}{2} - 1 + 2\sin^2\frac{\theta_l}{2}K(\cos\theta_k, \cos\theta_l)\right]. \tag{17}$$

After solving the system (17), the corresponding approximate solution to Eq. (1) can be found by formulas (13) and (14).

12.5-3. Solution Bounded at Both Endpoints of the Interval

A solution of Eq. (1) that is bounded at the endpoints of the interval vanishes at the endpoints,

$$\varphi(1) = \varphi(-1) = 0. \tag{18}$$

Let us approximate the function $\varphi(x)$ by an even trigonometric polynomial of θ constructed for the interpolation nodes that are the roots of the corresponding *Chebyshev polynomial of the second kind*:

$$x_k = \cos\theta_k, \quad \theta_k = \frac{k\pi}{n+1}, \quad k = 1, \dots, n. \tag{19}$$

This polynomial has the form

$$M_n(\varphi; \cos\theta) = \frac{2}{n+1} \sum_{l=1}^{n} \varphi(\cos\theta_l) \sum_{m=1}^{n} \sin m\theta_l \sin m\theta. \tag{20}$$

We thus obtain the following quadrature formula:

$$\frac{1}{\pi} \int_{-1}^{1} \frac{\varphi(t)\,dt}{t-x} = -\frac{2}{n+1} \sum_{l=1}^{n} \varphi(\cos\theta_l) \sum_{m=1}^{n} \sin m\theta_l \cos m\theta. \tag{21}$$

This formula holds for any odd trigonometric polynomial $\varphi(x)$ of degree $\leq n$.

To the regular integral in Eq. (1) we apply the formula

$$\int_{-1}^{1} \sqrt{1-x^2}\, P(x)\,dx = \frac{\pi}{n+1} \sum_{l=1}^{n} \sin^2\theta_l\, P(\cos\theta_l), \tag{22}$$

whose accuracy coincides with that of formula (8). On the basis of (22), we have

$$\frac{1}{\pi} \int_{-1}^{1} K(x,t)\varphi(t)\,dt = \frac{1}{n+1} \sum_{l=1}^{n} \sin\theta_l\, K(\cos\theta, \cos\theta_l)\varphi(\cos\theta_l). \tag{23}$$

On substituting relations (21) and (23) into Eq. (1) and on setting $\theta = \theta_k$ ($k = 1, \dots, n$), we obtain a system of linear algebraic equations in the form

$$\sum_{l=1}^{n} c_{kl}\varphi_l = f_k, \quad k = 1, \dots, n,$$

$$c_{kl} = \frac{\sin\theta_l}{n+1}\left[\frac{2\varepsilon_{kl}}{\cos\theta_l - \cos\theta_k} + K(\cos\theta_k, \cos\theta_l)\right], \quad \varepsilon_{kl} = \begin{cases} 0 & \text{for even } |k-l|, \\ 1 & \text{for odd } |k-l|, \end{cases} \tag{24}$$

where $f_k = f(\cos\theta_k)$ and φ_l are approximate values of the unknown function $\varphi(x)$ at the nodes.

After solving system (24), the corresponding approximate solution is defined by formula (20).

When solving a singular integral equation by the Multhopp–Kalandiya method, it is important that the desired solutions have a representation

$$\varphi(x) = (1-x)^\alpha (1+x)^\beta \chi(x), \tag{25}$$

where $\alpha = \pm\frac{1}{2}$, $\beta = \pm\frac{1}{2}$, and $\chi(x)$ is a bounded function on the interval with well-defined values at the endpoints. If the representation (25) holds, then the method can be applied to the complete singular integral equation, which is treated in Chapter 13.

In the literature cited below, some other methods of numerical solution of singular integral equations are discussed as well.

⊙ References for Section 12.5: A. I. Kalandiya (1973), N. I. Muskhelishvili (1992), S. M. Belotserkovskii and I. K. Lifanov (1993), and I. K. Lifanov (1996).

Chapter 13

Methods for Solving Complete Singular Integral Equations

13.1. Some Definitions and Remarks

13.1-1. Integral Equations With Cauchy Kernel

A complete singular integral equation with Cauchy kernel has the form

$$a(t)\varphi(t) + \frac{1}{\pi i} \int_L \frac{M(t,\tau)}{\tau - t} \varphi(\tau)\, d\tau = f(t), \qquad i^2 = -1, \tag{1}$$

where the integral, which is understood in the sense of the Cauchy principal value, is taken over a closed or nonclosed contour L and t and τ are the complex coordinates of points of the contour. It is assumed that the functions $a(t)$, $f(t)$, and $M(t,\tau)$ given on L and the unknown function $\varphi(t)$ satisfy the Hölder condition (see Subsection 12.2-2), and $M(t,\tau)$ satisfies this condition with respect to both variables.

The integral in Eq. (1) can also be written in a frequently used equivalent form. To this end, we consider the following transformation of the kernel:

$$\frac{M(t,\tau)}{\tau - t} = \frac{M(t,\tau) - M(t,t)}{\tau - t} + \frac{M(t,t)}{\tau - t}, \tag{2}$$

where we set

$$M(t,t) = b(t), \qquad \frac{1}{\pi i} \frac{M(t,\tau) - M(t,t)}{\tau - t} = K(t,\tau). \tag{3}$$

In this case Eq. (1), with regard to (2) and (3), becomes

$$a(t)\varphi(t) + \frac{b(t)}{\pi i} \int_L \frac{\varphi(\tau)}{\tau - t}\, d\tau + \int_L K(t,\tau)\varphi(\tau)\, d\tau = f(t). \tag{4}$$

It follows from formulas (3) that the function $b(t)$ satisfies the Hölder condition on the entire contour L and $K(t,\tau)$ satisfies the Hölder condition everywhere except for the points $\tau = t$, at which one has the estimate

$$|K(t,\tau)| < \frac{A}{|\tau - t|^\lambda}, \qquad A = \text{const} < \infty, \qquad 0 \le \lambda < 1.$$

Naturally, Eq. (4) is also called a *complete singular integral equation with Cauchy kernel*. The functions $a(t)$ and $b(t)$ are called the *coefficients* of Eq. (4), $\dfrac{1}{\tau - t}$ is called the *Cauchy kernel*, and the known function $f(t)$ is called the *right-hand side* of the equation. The first and the second terms on the left-hand side of Eq. (4) form the *characteristic part* or the *characteristic* of the complete singular equation and the third summand is called the *regular part*, and the function $K(t, \tau)$ is called the *kernel of the regular part*. It follows from the above estimate for the kernel of the regular part that $K(t, \tau)$ is a Fredholm kernel.

For Eqs. (1) and (4) we shall use the operator notation

$$\mathbf{K}[\varphi(t)] = f(t), \tag{5}$$

where the operator $\mathbf{K}$ is called a *singular operator.*

The equation

$$\mathbf{K}^\circ[\varphi(t)] \equiv a(t)\varphi(t) + \frac{b(t)}{\pi i} \int_L \frac{\varphi(\tau)}{\tau - t}\, d\tau = f(t) \tag{6}$$

is called the *characteristic equation* corresponding to the complete equation (4), and the operator $\mathbf{K}^\circ$ is called the *characteristic operator.*

For the regular part of the equation we introduce the notation

$$\mathbf{K}_r[\varphi(t)] \equiv \int_L K(t, \tau)\varphi(\tau)\, d\tau,$$

where the operator $\mathbf{K}_r$ is called a *regular (Fredholm) operator*, and we rewrite the complete singular equation in another operator form:

$$\mathbf{K}[\varphi(t)] \equiv \mathbf{K}^\circ[\varphi(t)] + \mathbf{K}_r[\varphi(t)] = f(t), \tag{7}$$

which will be used in what follows.

The equation

$$\mathbf{K}^*[\psi(t)] \equiv a(t)\psi(t) - \frac{1}{\pi i} \int_L \frac{b(\tau)\psi(\tau)}{\tau - t}\, d\tau + \int_L K(\tau, t)\psi(\tau)\, d\tau = g(t), \tag{8}$$

obtained from Eq. (4) by transposing the variables in the kernel is said to be *transposed* to (4). The operator $\mathbf{K}^*$ is said to be *transposed* to the operator $\mathbf{K}$.

In particular, the equation

$$\mathbf{K}^{\circ*}[\psi(t)] \equiv a(t)\psi(t) - \frac{1}{\pi i} \int_L \frac{b(\tau)}{\tau - t}\psi(\tau)\, d\tau = g(t) \tag{9}$$

is the equation transposed to the characteristic equation (6). It should be noted that the operator $\mathbf{K}^{\circ*}$ transposed to the characteristic operator $\mathbf{K}^\circ$ differs from the operator $\mathbf{K}^{*\circ}$ that is characteristic for the transposed equation (9). The latter is defined by the formula

$$\mathbf{K}^{*\circ}[\psi(t)] \equiv a(t)\psi(t) - \frac{b(t)}{\pi i} \int_L \frac{\psi(\tau)}{\tau - t}\, d\tau. \tag{10}$$

Throughout the following we assume that in the general case the contour L consists of $m + 1$ closed smooth curves $L = L_0 + L_1 + \cdots + L_m$. For equations with nonclosed contours, see, for example, the books by F. D. Gakhov (1977) and N. I. Muskhelishvili (1992).

Remark 1. The above relationship between Eqs. (1) and (4) that involves the properties of these equations is violated if we modify the condition and assume that in Eq. (1) the function $M(t, \tau)$ satisfies the Hölder condition everywhere on the contour except for finitely many points at which M has jump discontinuities. In this case, the complete singular integral equation must be represented in the form (4) with separated characteristic and regular parts in some way that differs from the transformation (2) and (3) because the above transformation of Eq. (1) does not lead to the desired decomposition. For equations with discontinuous coefficients, see the cited books.

13.1-2. Integral Equations With Hilbert Kernel

A *complete singular integral equation with Hilbert kernel* has the form

$$a(x)\varphi(x) + \frac{1}{2\pi} \int_0^{2\pi} N(x,\xi) \cot \frac{\xi-x}{2} \, \varphi(\xi) \, d\xi = f(x), \tag{11}$$

where the real functions $a(x)$, $f(x)$, and $N(x,\xi)$ and the unknown function $\varphi(x)$ satisfy the Hölder condition (see Subsection 12.2-2), with the function $N(x,\xi)$ satisfying the condition with respect to both variables.

The integral equation (11) can also be written in the following equivalent form, which is frequently used. We transform the kernel as follows:

$$N(x,\xi) \cot \frac{\xi-x}{2} = \big[N(x,\xi) - N(x,x)\big] \cot \frac{\xi-x}{2} + N(x,x) \cot \frac{\xi-x}{2}, \tag{12}$$

where we write

$$N(x,x) = -b(x), \quad \frac{1}{2\pi} \big[N(x,\xi) - N(x,x)\big] \cot \frac{\xi-x}{2} = K(x,\xi). \tag{13}$$

In this case, Eq. (11) with regard to (12) and (13) becomes

$$a(x)\varphi(x) - \frac{b(x)}{2\pi} \int_0^{2\pi} \cot \frac{\xi-x}{2} \, \varphi(\xi) \, d\xi + \int_0^{2\pi} K(x,\xi)\varphi(\xi) \, d\xi = f(x). \tag{14}$$

It follows from formulas (13) that the function $b(x)$ satisfies the Hölder condition, and the kernel $K(x,\xi)$ satisfies the Hölder condition everywhere except possibly for the points $x = \xi$ at which the following estimate holds:

$$|K(x,\xi)| < \frac{A}{|\xi-x|^\lambda}, \quad A = \text{const} < \infty, \quad 0 \le \lambda < 1.$$

The equation in the form (14) is also called a complete singular integral equation with Hilbert kernel. The functions $a(x)$ and $b(x)$ are called the *coefficients* of Eq. (14), $\cot\big[\frac{1}{2}(\xi-x)\big]$ is called the *Hilbert kernel*, and the known function $f(x)$ is called the *right-hand side* of the equation. The first and second summands in Eq. (14) form the so-called *characteristic part* or the *characteristic* of the complete singular equation, and the third summand is called its *regular part*; the function $K(x,\xi)$ is called the *kernel of the regular part*.

The equation

$$a(x)\varphi(x) - \frac{b(x)}{2\pi} \int_0^{2\pi} \cot \frac{\xi-x}{2} \, \varphi(\xi) \, d\xi = f(x), \tag{15}$$

is called the *characteristic equation* corresponding to the complete equation (14).

As usual, the above and the forthcoming equations whose right-hand sides are zero everywhere on their domains are said to be *homogeneous*, and otherwise they are said to be *nonhomogeneous*.

13.1-3. Fredholm Equations of the Second Kind on a Contour

Fredholm theory and methods for solving Fredholm integral equations of the second kind presented in Chapter 11 remain valid if all functions and parameters in the equations are treated as complex ones and an interval of the real axis is replaced by a contour L. Here we present only some information and write the Fredholm integral equation of the second kind in the form that is convenient for the purposes of this chapter.

Consider the Fredholm integral equation

$$\varphi(t) + \lambda \int_L K(t, \tau)\varphi(\tau)\, d\tau = f(t), \tag{16}$$

where L is a smooth contour, t and τ are complex coordinates of its points, $\varphi(t)$ is the desired function, $f(t)$ is the right-hand side of the equation, and $K(t, \tau)$ is the kernel.

If for some λ, the homogeneous Fredholm equation has a nontrivial solution (or nontrivial solutions), then λ is called a *characteristic value*, and the nontrivial solutions themselves are called *eigenfunctions* of the kernel $K(t, \tau)$ or of Eq. (16).

The set of characteristic values of Eq. (16) is at most countable. If this set is infinite, then its only limit point is the point at infinity. To each characteristic value, there are corresponding finitely many linearly independent eigenfunctions. The set of characteristic values of an integral equation is called its *spectrum*. The spectrum of a Fredholm integral equation is a discrete set.

If λ does not coincide with any characteristic value (in this case the value λ is said to be regular), i.e., the homogeneous equation has only the trivial solution, then the nonhomogeneous equation (16) is solvable for any right-hand side $f(t)$.

The general solution is given by the formula

$$\varphi(t) = f(t) - \int_L R(t, \tau;\ \lambda) f(\tau)\, d\tau, \tag{17}$$

where the function $R(t, \tau; \lambda)$ is called the *resolvent of the equation* or the *resolvent of the kernel* $K(t, \tau)$ and can be expressed via $K(t, \tau)$.

If a value of the parameter λ is characteristic for Eq. (16), then the homogeneous integral equation

$$\varphi(t) + \lambda \int_L K(t, \tau)\varphi(\tau)\, d\tau = 0, \tag{18}$$

as well as the transposed homogeneous equation

$$\psi(t) + \lambda \int_L K(\tau, t)\varphi(\tau)\, d\tau = 0, \tag{19}$$

has nontrivial solutions, and the number of solutions of Eq. (18) is finite and is equal to the number of linearly independent solutions of Eq. (19).

The general solution of the homogeneous equation can be represented in the form

$$\varphi(t) = \sum_{k=1}^{n} C_k \varphi_k(t), \tag{20}$$

where $\varphi_1(t), \ldots, \varphi_n(t)$ is a (complete) finite set of linearly independent eigenfunctions that correspond to the characteristic value λ, and C_k are arbitrary constants.

If the homogeneous equation (18) is solvable, then the nonhomogeneous equation (16) is, in general, unsolvable. This equation is solvable if and only if the following conditions hold:

$$\int_L f(t)\psi_k(t)\, dt = 0, \tag{21}$$

where $\{\psi_k(t)\}$ $(k = 1, \ldots, n)$ is a (complete) finite set of linearly independent eigenfunctions of the transposed equation that correspond to the characteristic value λ.

If conditions (21) are satisfied, then the general solution of the nonhomogeneous equation (16) can be given by the formula (e.g., see Subsection 11.6-5)

$$\varphi(t) = f(t) - \int_L R_g(t, \tau; \lambda) f(\tau) \, d\tau + \sum_{k=1}^{n} C_k \varphi_k(t), \tag{22}$$

where $R_g(t, \tau; \lambda)$ is called the *generalized resolvent* and the sum on the right-hand side of (22) is the general solution of the corresponding homogeneous equation.

Now we consider an equation of the second kind with weak singularity on the contour:

$$\varphi(t) + \int_L \frac{M(t, \tau)}{|\tau - t|^{\alpha}} \varphi(\tau) \, d\tau = f(t), \tag{23}$$

where $M(t, \tau)$ is a continuous function and $0 < \alpha < 1$. By iterating we can reduce this equation to a Fredholm integral equation of the second kind (e.g., see Remark 1 in Section 11.3). It has all properties of a Fredholm equation.

For the above reasons, in the theory of singular integral equations it is customary to make no difference between Fredholm equations and equations with weak singularity and use for them the same notation

$$\varphi(t) + \lambda \int_L K(t, \tau) \varphi(\tau) \, d\tau = 0, \quad K(t, \tau) = \frac{M(t, \tau)}{|\tau - t|^{\alpha}}, \quad 0 \le \alpha < 1. \tag{24}$$

The integral equation (24) is called simply a *Fredholm equation*, and its kernel is called a *Fredholm kernel*.

If in Eq. (24) the known functions satisfy the Hölder condition, and $M(t, \tau)$ satisfies this condition with respect to both variables, then each bounded integrable solution of Eq. (24) also satisfies the Hölder condition.

Remark 2. By the above estimates, the kernels of the regular parts of the above singular integral equations are Fredholm kernels.

Remark 3. The complete and characteristic singular integral equations are sometimes called singular integral equations of the second kind.

⊙ References for Section 13.1: F. D. Gakhov (1977), F. G. Tricomi (1985), S. G. Mikhlin and S. Prössdorf (1986), A. Dzhuraev (1992), N. I. Muskhelishvili (1992), I. K. Lifanov (1996).

13.2. The Carleman Method for Characteristic Equations

13.2-1. A Characteristic Equation With Cauchy Kernel

Consider a characteristic equation with Cauchy kernel:

$$\mathbf{K}^{\circ}[\varphi(t)] \equiv a(t)\varphi(t) + \frac{b(t)}{\pi i} \int_L \frac{\varphi(\tau)}{\tau - t} \, d\tau = f(t), \tag{1}$$

where the contour L consists of $m + 1$ closed smooth curves $L = L_0 + L_1 + \cdots + L_m$.

Solving Eq. (1) can be reduced to solving a Riemann boundary value problem (see Subsection 12.3-10), and the solution of the equation can be presented in a closed form.

Let us introduce the piecewise analytic function given by the Cauchy integral whose density is the desired solution of the characteristic equation:

$$\Phi(z) = \frac{1}{2\pi i} \int_L \frac{\varphi(\tau)}{\tau - z} \, d\tau. \tag{2}$$

According to the Sokhotski–Plemelj formulas (see Subsection 12.2-5), we have

$$\varphi(t) = \Phi^+(t) - \Phi^-(t),$$

$$\frac{1}{\pi i} \int_L \frac{\varphi(\tau)}{\tau - z} \, d\tau = \Phi^+(t) + \Phi^-(t). \tag{3}$$

On substituting (3) into (1) and solving the resultant equation for $\Phi^+(t)$, we see that the piecewise analytic function $\Phi(z)$ must be a solution of the Riemann boundary value problem

$$\Phi^+(t) = D(t)\Phi^-(t) + H(t), \tag{4}$$

where

$$D(t) = \frac{a(t) - b(t)}{a(t) + b(t)}, \quad H(t) = \frac{f(t)}{a(t) + b(t)}. \tag{5}$$

Since the function $\Phi(z)$ is represented by a Cauchy type integral, it follows that this function must satisfy the additional condition

$$\Phi^-(\infty) = 0. \tag{6}$$

The index ν of the coefficient $D(t)$ of the Riemann problem (4) is called the *index of the integral equation* (1). On solving the boundary value problem (4), we find the solution of Eq. (1) by the first formula in (3).

Thus, the integral equation (1) is reduced to the Riemann boundary value problem (4). To establish the equivalence of the equation to the boundary value problem we note that, conversely, the function $\varphi(t)$ that is found by the above-mentioned method from the solution of the boundary value problem necessarily satisfies Eq. (1).

We first consider the following normal (nonexceptional) case in which the coefficient $D(t)$ of the Riemann problem (4) admits no zero or infinite values, which amounts to the condition

$$a(t) \pm b(t) \neq 0 \tag{7}$$

for Eq. (1). To simplify the subsequent formulas, we assume that the coefficients of Eq. (1) satisfy the condition

$$a^2(t) - b^2(t) = 1. \tag{8}$$

This can always be achieved by dividing the equation by $\sqrt{a^2(t) - b^2(t)}$.

Let us write out the solution of the Riemann boundary value problem (4) under the assumption $\nu \geq 0$ and then use the Sokhotski–Plemelj formulas to find the limit values of the corresponding functions (see Subsections 12.2-5, 13.3-6, and 12.3-10):

$$\Phi^+(t) = X^+(t)\left[\frac{1}{2}\frac{H(t)}{X^+(t)} + \Psi(t) - \frac{1}{2}P_{\nu-1}(t)\right], \qquad \Phi^-(t) = X^-(t)\left[-\frac{1}{2}\frac{H(t)}{X^+(t)} + \Psi(t) - \frac{1}{2}P_{\nu-1}(t)\right], \tag{9}$$

where

$$\Psi(t) = \frac{1}{2\pi i} \int_L \frac{H(\tau)}{X^+(\tau)} \frac{d\tau}{\tau - t}. \tag{10}$$

The arbitrary polynomial is taken in the form $-\frac{1}{2}P_{\nu-1}(t)$, which is convenient for the subsequent notation.

Hence, by formula (3) we have

$$\varphi(t) = \frac{1}{2}\left[1 + \frac{X^-(t)}{X^+(t)}\right]H(t) + X^+(t)\left[1 - \frac{X^-(t)}{X^+(t)}\right]\left[\Psi(t) - \frac{1}{2}P_{\nu-1}(t)\right].$$

Representing the coefficient of the Riemann problem in the form $D(t) = X^+(t)/X^-(t)$ and replacing the function $\Psi(t)$ by the expression on the right-hand side in (10), we obtain

$$\varphi(t) = \frac{1}{2}\left[1 + \frac{1}{D(t)}\right]H(t) + X^+(t)\left[1 - \frac{1}{D(t)}\right]\left[\frac{1}{2\pi i}\int_L \frac{H(\tau)}{X^+(\tau)}\frac{d\tau}{\tau - t} - \frac{1}{2}P_{\nu-1}(t)\right].$$

Finally, on replacing $X^+(t)$ by the expression (62) in Subsection 12.3-10 and substituting the expressions for $D(t)$ and $H(t)$ given in (5), we obtain

$$\varphi(t) = a(t)f(t) - \frac{b(t)Z(t)}{\pi i}\int_L \frac{f(\tau)}{Z(\tau)}\frac{d\tau}{\tau - t} + b(t)Z(t)P_{\nu-1}(t), \tag{11}$$

where

$$Z(t) = [a(t) + b(t)]X^+(t) = [a(t) - b(t)]X^-(t) = \frac{e^{G(t)}}{\sqrt{t^\nu\Pi(t)}},$$

$$G(t) = \frac{1}{2\pi i}\int_L \ln\left[\tau^{-\nu}\Pi(\tau)\frac{a(\tau) - b(\tau)}{a(\tau) + b(\tau)}\right]\frac{d\tau}{\tau - t}, \qquad \Pi(t) = \sum_{k=1}^m (t - z_k)^{\nu_k}, \tag{12}$$

and the coefficients $a(t)$ and $b(t)$ satisfy condition (7). Here $\Pi(t) \equiv 1$ for the case in which L is a simple contour enclosing a simply connected domain. Since the functions $a(t)$, $b(t)$, and $f(t)$ satisfy the Hölder condition, it follows from the properties of the limit values of the Cauchy type integral that the function $\varphi(t)$ also satisfies the Hölder condition.

The last term in formula (11) is the general solution of the homogeneous equation ($f(t) \equiv 0$), and the first two terms form a particular solution of the nonhomogeneous equation.

The particular solution of Eq. (1) can be represented in the form $\mathbf{R}[f(t)]$, where $\mathbf{R}$ is the operator defined by

$$\mathbf{R}[f(t)] = a(t)f(t) - \frac{b(t)Z(t)}{\pi i}\int_L \frac{f(\tau)}{Z(\tau)}\frac{d\tau}{\tau - t}.$$

In this case, the general solution of Eq. (1) becomes

$$\varphi(t) = \mathbf{R}[f(t)] + \sum_{k=1}^\nu c_k\varphi_k(t), \tag{13}$$

where $\varphi_k(t) = b(t)Z(t)t^{k-1}$ ($k = 1, 2, \ldots, \nu$) are the linearly independent eigenfunctions of the characteristic equation.

If $\nu < 0$, then the Riemann problem (4) is in general unsolvable. The solvability conditions

$$\int_L \frac{H(\tau)}{X^+(\tau)}\tau^{k-1}\,d\tau = 0, \qquad k = 1, 2, \ldots, -\nu, \tag{14}$$

for problem (4) are the solvability conditions for Eq. (1) as well.

Replacing $H(\tau)$ and $X^+(\tau)$ by their expressions from (5) and (12), we can rewrite the solvability conditions in the form

$$\int_L \frac{f(\tau)}{Z(\tau)}\tau^{k-1}\,d\tau = 0, \qquad k = 1, 2, \ldots, -\nu. \tag{15}$$

If the solvability conditions hold, then the solution of the nonhomogeneous equation (4) is given by formula (11) for $P_{\nu-1} \equiv 0$.

1.° If $\nu > 0$, then the homogeneous equation $\mathbf{K}°[\varphi(t)] = 0$ has ν linearly independent solutions

$$\varphi_k(t) = b(t)Z(t)t^{k-1}, \qquad k = 1, 2, \ldots, \nu.$$

2.° If $\nu \leq 0$, then the homogeneous equation is unsolvable (has only the trivial solution).

3.° If $\nu \geq 0$, then the nonhomogeneous equation is solvable for an arbitrary right-hand side $f(t)$, and its general solution linearly depends on ν arbitrary constants.

4.° If $\nu < 0$, then the nonhomogeneous equation is solvable if and only if its right-hand side f satisfies the $-\nu$ conditions,

$$\int_L \psi_k(t)f(t)\,dt = 0, \qquad \psi_k(t) = \frac{t^{k-1}}{Z(t)}. \tag{16}$$

The above properties of characteristic singular integral equations are essentially different from the properties of Fredholm integral equations (see Subsection 13.1-3). With Fredholm equations, if the homogeneous equation is solvable, then the nonhomogeneous equation is in general unsolvable, and conversely, if the homogeneous equation is unsolvable, then the nonhomogeneous equation is solvable. However, for a singular equation, if the homogeneous equation is solvable, then the nonhomogeneous equation is unconditionally solvable, and if the homogeneous equation is unsolvable, then the nonhomogeneous equation is in general unsolvable as well.

By analogy with the case of Fredholm equations, we introduce a parameter λ into the kernel of the characteristic equation and consider the equation

$$a(t)\varphi(t) + \frac{\lambda b(t)}{\pi i} \int_L \frac{\varphi(\tau)}{\tau - t}\,d\tau = 0.$$

As shown above, the last equation is solvable if

$$\nu = \text{Ind}\,\frac{a(t) - \lambda b(t)}{a(t) + \lambda b(t)} > 0.$$

The index of a continuous function changes by jumps and only for the values of λ such that $a(t) \mp \lambda b(t) = 0$. If in the complex plane $\lambda = \lambda_1 + i\lambda_2$ we draw the curves $\lambda = \pm a(t)/b(t)$, then these curves divide the plane into domains in each of which the index is constant. Thus, the characteristic values of the characteristic integral equation occupy entire domains, and hence the spectrum is continuous, in contrast with the spectrum of a Fredholm equation.

13.2-2. The Transposed Equation of a Characteristic Equation

The equation

$$\mathbf{K}^{\circ*}[\psi(t)] \equiv a(t)\psi(t) - \frac{1}{\pi i} \int_L \frac{b(\tau)\psi(\tau)}{\tau - t}\,d\tau = g(t), \tag{17}$$

which is transposed to the characteristic equation $\mathbf{K}^\circ[\varphi(t)] = f(t)$, is not characteristic. However, the substitution

$$b(t)\psi(t) = \omega(t) \tag{18}$$

reduces it to a characteristic equation for the function $\omega(t)$:

$$a(t)\omega(t) - \frac{b(t)}{\pi i} \int_L \frac{\omega(\tau)}{\tau - t}\,d\tau = b(t)g(t). \tag{19}$$

From the last equation we find $\omega(t)$, by the formula obtained by adding (17) to (18), and determine the desired function $\psi(t)$:

$$\psi(t) = \frac{1}{a(t) + b(t)}\left[\omega(t) + \frac{1}{\pi i} \int_L \frac{\omega(\tau)}{\tau - t}\,d\tau + g(t)\right].$$

Introducing the piecewise analytic function

$$\Phi_*(z) = \frac{1}{2\pi i} \int_L \frac{\omega(\tau)}{\tau - z} \, d\tau, \tag{20}$$

we arrive at the Riemann boundary value problem

$$\Phi_*^+(t) = \frac{a(t) + b(t)}{a(t) - b(t)} \Phi_*^-(t) + \frac{b(t)g(t)}{a(t) - b(t)}. \tag{21}$$

The coefficient of the boundary value problem (21) is the inverse of the coefficient of the Riemann problem (4) corresponding to the equation $\mathbf{K}^\circ[\varphi(t)] = f(t)$. Hence,

$$\nu^* = \operatorname{Ind} \frac{a(t) + b(t)}{a(t) - b(t)} = -\operatorname{Ind} \frac{a(t) - b(t)}{a(t) + b(t)} = -\nu. \tag{22}$$

Note that it follows from formulas (17) in Subsection 12.3-4 that the canonical function $X^*(z)$ for Eq. (21) and the canonical function $X(z)$ for (4) are reciprocal:

$$X^*(z) = \frac{1}{X(z)}.$$

By analogy with the reasoning in Subsection 13.2-1, we obtain a solution of the singular integral equation (17) for $\nu^* = -\nu \geq 0$ in the form

$$\psi(t) = a(t)g(t) + \frac{1}{\pi i Z(t)} \int_L \frac{b(\tau)Z(\tau)g(\tau)}{\tau - t} \, d\tau + \frac{1}{Z(t)} Q_{\nu^*-1}(t), \tag{23}$$

where $Z(t)$ is given by formula (12) and $Q_{\nu^*-1}(t)$ is a polynomial of degree at most $\nu^* - 1$ with arbitrary coefficients. If $\nu^* = 0$, then we must set $Q_{\nu^*-1}(t) \equiv 0$.

If $\nu^* = -\nu < 0$, then for the solvability of Eq. (17) it is necessary and sufficient that

$$\int_L b(t)Z(t)g(t)t^{k-1} \, dt = 0, \qquad k = 1, 2, \ldots, -\nu^*, \tag{24}$$

and if these conditions hold, then the solution is given by formula (23), where we must set $Q_{\nu^*-1}(t) \equiv 0$.

The results of simultaneous investigation of a characteristic equation and the transposed equation show another essential difference from the properties of Fredholm equations (see Subsection 13.1-3). Transposed homogeneous characteristic equations cannot be solvable simultaneously. Either they are both unsolvable ($\nu = 0$), or, for a nonzero index, only the equation with a positive index is solvable.

We point out that the difference between the numbers of solutions of a characteristic homogeneous equation and the transposed equation is equal to the index ν.

Assertions 1° and 2° and assertions 3° and 4° in Subsection 13-2.1 are called, respectively, the *first Fredholm theorem* and the *second Fredholm theorem* for a characteristic equation, and the relationship between the index of an equation and the number of solutions of the homogeneous equations $\mathbf{K}^\circ[\varphi(t)] = 0$ and $\mathbf{K}^{\circ*}[\psi(t)] = 0$ is called the *third Fredholm theorem*.

13.2-3. The Characteristic Equation on the Real Axis

The theory of the Cauchy type integral (see Section 12.2) shows that if the density of the Cauchy type integral taken over an infinite curve vanishes at infinity, then the properties of the integral for the cases in which the contour is finite and infinite are essentially the same. Therefore, the theory of singular integral equations on an infinite contour in the class of functions that vanish at infinity coincides with the theory of equations on a finite contour.

Just as for the case of a finite contour, the characteristic integral equation

$$a(x)\varphi(x) + \frac{b(x)}{\pi i} \int_{-\infty}^{\infty} \frac{\varphi(\tau)}{\tau - x} \, d\tau = f(x) \tag{25}$$

can be reduced by means of the Cauchy type integral

$$\Phi(z) = \frac{1}{2\pi i} \int_{-\infty}^{\infty} \frac{\varphi(\tau)}{\tau - z} \, d\tau \tag{26}$$

and the Sokhotski–Plemelj formulas (see Subsection 12.2-5), to the following Riemann boundary value problem for the real axis (see Subsection 12.3-8):

$$\Phi^{+}(x) = \frac{a(x) - b(x)}{a(x) + b(x)} \Phi^{-}(x) + \frac{f(x)}{a(x) + b(x)}, \qquad -\infty < x < \infty. \tag{27}$$

We assume that

$$a^{2}(x) - b^{2}(x) = 1, \tag{28}$$

because Eq. (25) can always be reduced to case (28) by the division by $\sqrt{a^{2}(t) - b^{2}(t)}$. Note that the index ν of the integral equation (25) is given by the formula

$$\nu = \text{Ind} \, \frac{a(x) - b(x)}{a(x) + b(x)}. \tag{29}$$

In this case for $\nu \geq 0$ we obtain

$$\varphi(x) = a(x)f(x) - \frac{b(x)Z(x)}{\pi i} \int_{-\infty}^{\infty} \frac{f(\tau)}{Z(\tau)} \frac{d\tau}{\tau - x} + b(x)Z(x) \frac{P_{\nu-1}(x)}{(x + i)^{\nu}}, \tag{30}$$

where

$$Z(x) = [a(x) + b(x)]X^{+}(x) = [a(x) - b(x)]X^{-}(x) = \left(\frac{x - i}{x + i} \right)^{-\nu/2} e^{G(x)},$$

$$G(x) = \frac{1}{2\pi i} \int_{-\infty}^{\infty} \ln \left[\left(\frac{\tau - i}{\tau + i} \right)^{-\nu} \frac{a(\tau) - b(\tau)}{a(\tau) + b(\tau)} \right] \frac{d\tau}{\tau - x}.$$

For the case in which $\nu \leq 0$ we must set $P_{\nu-1}(x) \equiv 0$. For $\nu < 0$, we must also impose the solvability conditions

$$\int_{-\infty}^{\infty} \frac{f(x)}{Z(x)} \frac{dx}{(x + i)^{k}} = 0, \qquad k = 1, 2, \ldots, -\nu. \tag{31}$$

For the solution of Eq. (25) in the class of functions bounded at infinity, see F. D. Gakhov (1977).

The analog of the characteristic equation on the real axis is the equation of the form

$$a(x)\varphi(x) + \frac{b(x)}{\pi i} \int_{-\infty}^{\infty} \frac{x - z_0}{\tau - z_0} \frac{\varphi(\tau)}{\tau - x} \, d\tau = f(x), \tag{32}$$

where z_0 is a point that does not belong to the contour. For this equation, all qualitative results obtained for the characteristic equation with finite contour are still valid together with the formulas. In particular, the following inversion formulas for the Cauchy type integral hold:

$$\psi(x) = \frac{1}{\pi i} \int_{-\infty}^{\infty} \frac{x - z_0}{\tau - z_0} \frac{\varphi(\tau)}{\tau - x} \, d\tau, \qquad \varphi(x) = \frac{1}{\pi i} \int_{-\infty}^{\infty} \frac{x - z_0}{\tau - z_0} \frac{\psi(\tau)}{\tau - x} \, d\tau. \tag{33}$$

13.2-4. The Exceptional Case of a Characteristic Equation

In the study of the characteristic equation in Subsection 13-2.1, the case in which the functions $a(t) \pm b(t)$ can vanish on the contour L was excluded. The reason was that the coefficient $D(t)$ of the Riemann problem to which the characteristic equation can be reduced has in the exceptional case zeros and poles on the contour, and hence this problem is outside the framework of the general theory. Let us perform an investigation of the above exceptional case.

We assume that the coefficients of the singular equations under consideration have properties that provide the additional differentiability requirements that were introduced in the consideration of exceptional cases of the Riemann problem (see 12.3-9).

Consider a characteristic equation with Cauchy kernel (1) under the assumption that the functions $a(t)-b(t)$ and $a(t)+b(t)$ have zeros on the contour at the points $\alpha_1, \ldots, \alpha_\mu$ and $\beta_1 \ldots, \beta_\eta$, respectively, of integral orders, and hence are representable in the form

$$a(t) - b(t) = \prod_{k=1}^{\mu} (t - \alpha_k)^{m_k} r(t), \quad a(t) + b(t) = \prod_{j=1}^{\eta} (t - \beta_j)^{p_j} s(t),$$

where $r(t)$ and $s(t)$ vanish nowhere. We assume that all points α_k and β_j are different.

Assume that the coefficients of Eq. (1) satisfy the relation

$$a^2(t) - b^2(t) = \prod_{k=1}^{\mu} (t - \alpha_k)^{m_k} \prod_{j=1}^{\eta} (t - \beta_j)^{p_j} = A_0(t). \tag{34}$$

The equation under consideration can be reduced to the above case by dividing it by $\sqrt{s(t)r(t)}$.

In the exceptional case, by analogy with the case studied in Subsection 13.2-1, Eq. (1) can be reduced to the Riemann problem

$$\Phi^+(t) = \frac{\displaystyle\prod_{k=1}^{\mu} (t - \alpha_k)^{m_k}}{\displaystyle\prod_{j=1}^{\eta} (t - \beta_j)^{p_j}} D_1(t) \Phi^-(t) + \frac{f(t)}{\displaystyle\prod_{j=1}^{\eta} (t - \beta_j)^{p_j} s(t)}, \tag{35}$$

where $D_1(t) = r(t)/s(t)$. The solution of this problem in the class of functions that satisfy the condition $\Phi(\infty) = 0$ is given by the formulas

$$\Phi^+(z) = \frac{X^+(z)}{\displaystyle\prod_{j=1}^{\eta} (z - \beta_j)^{p_j}} [\Psi^+(z) - \mathcal{U}_\rho(z) + A_0(z)P_{\nu-p-1}(z)],$$

$$\Phi^-(z) = \frac{X^-(z)}{\displaystyle\prod_{k=1}^{\mu} (z - \alpha_k)^{m_k}} [\Psi^-(z) - \mathcal{U}_\rho(z) + A_0(z)P_{\nu-p-1}(z)], \tag{36}$$

where

$$\Psi(z) = \frac{1}{2\pi i} \int_L \frac{f(\tau)}{s(\tau) X^+(\tau)} \frac{d\tau}{\tau - z}, \tag{37}$$

and $\mathcal{U}_\rho(z)$ is the Hermite interpolation polynomial (see Subsection 12.3-2) for the function $\Psi(z)$ of degree $\rho = m + p - 1$ with nodes at the points α_k and β_j, respectively, and of the multiplicities m_k and p_j, respectively, where $m = \sum m_k$ and $p = \sum p_j$.

We regard the polynomial $\mathcal{U}_\rho(z)$ as an operator that maps the right-hand side $f(t)$ of Eq. (1) to the polynomial that interpolates the Cauchy type integral (37) as above. Let us denote this operator by

$$\tfrac{1}{2}\mathbf{T}[f(t)] = \mathcal{U}_\rho(z). \tag{38}$$

Here the coefficient $\frac{1}{2}$ is taken for the convenience of the subsequent manipulations.

Furthermore, by analogy with the normal case, from (36) we can find

$$\Phi^+(t) = \frac{X^+(t)}{\displaystyle\prod_{j=1}^{\eta}(t-\beta_j)^{p_j}} \left[\frac{1}{2}\frac{f(t)}{s(t)X^+(t)} + \frac{1}{2\pi i}\int_L \frac{f(\tau)}{s(\tau)X^+(\tau)}\frac{d\tau}{\tau-t} - \frac{1}{2}\mathbf{T}[f(t)] - \frac{1}{2}A_0(t)P_{\nu-p-1}(t) \right],$$

$$\Phi^-(t) = \frac{X^-(t)}{\displaystyle\prod_{k=1}^{\mu}(t-\alpha_k)^{m_k}} \left[-\frac{1}{2}\frac{f(t)}{s(t)X^+(t)} + \frac{1}{2\pi i}\int_L \frac{f(\tau)}{s(\tau)X^+(\tau)}\frac{d\tau}{\tau-t} - \frac{1}{2}\mathbf{T}[f(t)] - \frac{1}{2}A_0(t)P_{\nu-p-1}(t) \right].$$

We introduced the coefficient $-\frac{1}{2}$ in the last summands of these formulas using the fact that the coefficients of the polynomial $P_{\nu-p-1}(t)$ are arbitrary. Hence,

$$\varphi(t) = \Phi^+(t) - \Phi^-(t) = \frac{\Delta_1(t)f(t)}{s(t)X^+(t)} + \Delta_2(t)\left[\frac{1}{\pi i}\int_L \frac{f(\tau)\,d\tau}{s(\tau)X^+(\tau)(\tau-t)} - \mathbf{T}[f(t)] - A_0(t)P_{\nu-p-1}(t) \right], \tag{39}$$

where

$$\Delta_1(t) = \frac{X^+(t)}{2\displaystyle\prod_{j=1}^{\eta}(t-\beta_j)^{p_j}} + \frac{X^-(t)}{2\displaystyle\prod_{k=1}^{\mu}(t-\alpha_k)^{m_k}}, \qquad \Delta_2(t) = \frac{X^+(t)}{2\displaystyle\prod_{j=1}^{\eta}(t-\beta_j)^{p_j}} - \frac{X^-(t)}{2\displaystyle\prod_{k=1}^{\mu}(t-\alpha_k)^{m_k}}.$$

We write

$$Z(t) = s(t)X^+(t) = r(t)X^-(t), \tag{40}$$

and, applying relation (34), represent formula (39) as follows:

$$\varphi(t) = \frac{1}{A_0(t)}\left[a(t)f(t) - \frac{b(t)Z(t)}{\pi i}\int_L \frac{f(\tau)}{Z(\tau)}\frac{d\tau}{\tau-t} + b(t)Z(t)\mathbf{T}[f(t)] \right] + b(t)Z(t)P_{\nu-p-1}(t).$$

Let us introduce the operator $\mathbf{R}_1[f(t)]$ by the formula

$$\mathbf{R}_1[f(t)] \equiv \frac{1}{A_0(t)}\left[a(t)f(t) - \frac{b(t)Z(t)}{\pi i}\int_L \frac{f(\tau)}{Z(\tau)}\frac{d\tau}{\tau-t} + b(t)Z(t)\mathbf{T}[f(t)] \right], \tag{41}$$

and finally obtain

$$\varphi(t) = \mathbf{R}_1[f(t)] + b(t)Z(t)P_{\nu-p-1}(t). \tag{42}$$

Formula (42) gives a solution of Eq. (1) for the exceptional case in which $\nu - p > 0$. This solution linearly depends on $\nu - p$ arbitrary constants. If $\nu - p < 0$, then the solution exists only under $p - \nu$ special solvability conditions imposed on $f(t)$, which follow from the solvability conditions for the Riemann problem (35) corresponding to this case.

13.2-5. The Characteristic Equation With Hilbert Kernel

Consider the characteristic equation with Hilbert kernel

$$a(x)\varphi(x) - \frac{b(x)}{2\pi} \int_0^{2\pi} \cot\frac{\xi - x}{2}\, \varphi(\xi)\, d\xi = f(x). \tag{43}$$

Just as the characteristic integral equation with Cauchy kernel is related to the Riemann boundary value problem, so the characteristic equation (43) with Hilbert kernel can be analytically reduced to a Hilbert problem in a straightforward manner. In turn, the Hilbert problem can be reduced to the Riemann problem (see Subsection 12.3-12), and hence the solution of Eq. (43) can be constructed in a closed form.

For $\nu > 0$, the homogeneous equation (43) ($f(x) \equiv 0$) has 2ν linearly independent solutions, and the nonhomogeneous problem is unconditionally solvable and linearly depends on 2ν real constants.

For $\nu < 0$, the homogeneous equation is unsolvable, and the nonhomogeneous equation is solvable only under -2ν real solvability conditions.

Taking into account the fact that any complex parameter contains two real parameters, and a complex solvability condition is equivalent to two real conditions, we see that, for $\nu \neq 0$, the qualitative results of investigating the characteristic equation with Hilbert kernel completely agree with the corresponding results for the characteristic equation with Cauchy kernel.

13.2-6. The Tricomi Equation

The singular integral Tricomi equation has the form

$$\varphi(x) - \lambda \int_0^1 \left(\frac{1}{\xi - x} - \frac{1}{x + \xi - 2x\xi} \right) \varphi(\xi)\, d\xi = f(x), \quad 0 \le x \le 1. \tag{44}$$

The kernel of this equation consists of two terms. The first term is the Cauchy kernel. The second term is continuous if at least one of the variables x and ξ varies strictly inside the interval $[0, 1]$; however, for $x = \xi = 0$ and for $x = \xi = 1$, this kernel becomes infinite and is nonintegrable in the square $\{0 \le x \le 1, 0 \le \xi \le 1\}$.

By using the function

$$\Phi(z) = \frac{1}{2\pi i} \int_0^1 \left(\frac{1}{\xi - z} - \frac{1}{z + \xi - 2z\xi} \right) \varphi(\xi)\, d\xi,$$

which is piecewise analytic in the upper and the lower half-plane, we can reduce Eq. (44) to the Riemann problem with boundary condition on the real axis. The solution of the Tricomi equation has the form

$$y(x) = \frac{1}{1 + \lambda^2 \pi^2} \left[f(x) + \int_0^1 \frac{\xi^\alpha (1-x)^\alpha}{x^\alpha (1-\xi)^\alpha} \left(\frac{1}{\xi - x} - \frac{1}{x + \xi - 2x\xi} \right) f(\xi)\, d\xi \right] + \frac{C(1-x)^\beta}{x^{1+\beta}},$$

$$\alpha = \frac{2}{\pi} \arctan(\lambda\pi) \ (-1 < \alpha < 1), \quad \tan\frac{\beta\pi}{2} = \lambda\pi \ (-2 < \beta < 0),$$

where C is an arbitrary constant.

⊙ References for Section 13.2: P. P. Zabreyko, A. I. Koshelev, et al. (1975), F. D. Gakhov (1977), F. G. Tricomi (1985), N. I. Muskhelishvili (1992).

13.3. Complete Singular Integral Equations Solvable in a Closed Form

In contrast with characteristic equations and their transposed equations, complete singular integral equations cannot be solved in the closed form in general. However, there are some cases in which complete equations can be solved in a closed form.

13.3-1. Closed-Form Solutions in the Case of Constant Coefficients

Consider the complete singular integral equation with Cauchy kernel in the form (see Subsection 13.1-1)

$$a(t)\varphi(t) + \frac{b(t)}{\pi i} \int_L \frac{\varphi(\tau)}{\tau - t} d\tau + \int_L K(t, \tau)\varphi(\tau) d\tau = f(t), \tag{1}$$

where L is an arbitrary closed contour. Let us show that Eq. (1) can be solved in a closed form if $a(t) = a$ and $b(t) = b$ are constants and $K(t, \tau)$ is an arbitrary function that has an analytic continuation to the domain Ω^+ with respect to each variable.

Under the above assumptions, Eq. (1) has the form

$$a\varphi(t) + \frac{1}{\pi i} \int_L \frac{M(t, \tau)}{\tau - t}\varphi(\tau) d\tau = f(t), \tag{2}$$

where $M(t, \tau) = b + \pi i(t - \tau)K(t, \tau)$, so that $M(t, t) = b = \text{const}$. Let $b \neq 0$. We write

$$\psi(t) = \frac{1}{b\pi i} \int_L \frac{M(t, \tau)}{\tau - t}\varphi(\tau) d\tau. \tag{3}$$

According to Subsection 12.4-4, the function $\varphi(t)$ can be expressed via $\psi(t)$ and $\psi(t)$ can be expressed via $\varphi(t)$. Then we rewrite Eq. (2) as follows:

$$a\varphi(t) + b\psi(t) = f(t). \tag{4}$$

On applying the operation (3) to this equation, we obtain

$$a\psi(t) + b\varphi(t) = w(t), \tag{5}$$

where

$$w(t) = \frac{1}{b\pi i} \int_L \frac{M(t, \tau)}{\tau - t} f(\tau) d\tau.$$

By solving system (4), (5) we find $\varphi(t)$:

$$\varphi(t) = \frac{1}{a^2 - b^2} \left[af(t) - \frac{1}{\pi i} \int_L \frac{M(t, \tau)}{\tau - t} f(\tau) d\tau \right] \tag{6}$$

under the assumption that $a \neq \pm b$.

Thus, for $a \neq \pm b$ and for a kernel $K(t, \tau)$ that can be analytically continued, Eq. (1) or (2) is solvable and has the unique solution given by formula (6).

Equation (1) was studied above for $b \neq 0$. This assumption is natural because, for $b \equiv 0$, Eq. (1) is no longer singular. However, the Fredholm equation obtained for $b = 0$, that is,

$$a\varphi(t) + \int_L K(t, \tau)\varphi(\tau) d\tau = f(t), \qquad a = \text{const}, \tag{7}$$

is solvable in a closed form for a kernel $K(t, \tau)$ that has analytic continuation.

Let a function $K(t, \tau)$ have an analytic continuation to the domain Ω^+ with respect to each of the variables and continuous for $t, \tau \in L$. In this case, the following assertions hold.

1°. The function

$$\Phi^+(t) = \int_L K(t, \tau)\varphi(\tau)\, d\tau$$

has an analytic continuation to the domain Ω^+ for any function $\varphi(t)$ satisfying the Hölder condition.

2°. If a function $\varphi^+(t)$ satisfying the Hölder condition has an analytic continuation to the domain Ω^+, then

$$\int_L K(t, \tau)\varphi^+(\tau)\, d\tau = 0. \tag{8}$$

This implies the relation

$$\int_L K(t, \tau) \int_L K(\tau, \tau_1)\varphi(\tau_1)\, d\tau_1\, d\tau = 0 \tag{9}$$

for each function $\varphi(t)$ (satisfying the Hölder condition). Therefore, it follows from (7) that

$$a \int_L K(t, \tau)\varphi(\tau)\, d\tau = \int_L K(t, \tau)f(\tau)\, d\tau,$$

and hence

$$\varphi(t) = \frac{1}{a^2}\left[af(t) - \int_L K(t, \tau)f(\tau)\, d\tau\right]. \tag{10}$$

Therefore, if a kernel $K(t, \tau)$ is analytic in the domain Ω^+ with respect to each of the variables and continuous for $t, \tau \in L$, then Eq. (7) is solvable for each right-hand side, and the solution is given by formula (10).

13.3-2. Closed-Form Solutions in the General Case

Let us pass to the general case of the solvability of Eq. (1) in a closed form under the condition that a function $K(t, \tau)[a(t) + b(t)]^{-1}$ is analytic with respect to τ and meromorphic with respect to t in the domain Ω^+.

For brevity, we write

$$\mathbf{K}_r[\varphi(t)] = \int_L K(t, \tau)\varphi(\tau)\, d\tau$$

and note that

$$\mathbf{K}_r[\varphi^+(t)] = 0 \tag{11}$$

for each function $\varphi^+(t)$ that has an analytic continuation to the domain Ω^+. By setting $\varphi(t) = \varphi^+(t) - \varphi^-(t)$ and with regard to (11), we reduce Eq. (1) to a relation similar to that of the Riemann problem:

$$\varphi^+(t) - \frac{1}{a(t) + b(t)}\mathbf{K}_r[\varphi^-(t)] = D(t)\varphi^-(t) + H(t), \tag{12}$$

where

$$D(t) = \frac{a(t) - b(t)}{a(t) + b(t)}, \quad H(t) = \frac{f(t)}{a(t) + b(t)}.$$

By assumption, we have

$$\frac{K(t, \tau)}{a(t) + b(t)} = \frac{A^+(t, \tau)}{\Pi^+(t)}, \quad \Pi^+(t) = \prod_{k=1}^{n}(t - z_k)^{m_k}, \tag{13}$$

where $z_k \in \Omega^+$ and m_k are positive integers and the function $A^+(t, \tau)$ is analytic with respect to t and with respect to τ on Ω^+.

Relation (12) becomes

$$\Pi^+(t)\varphi^+(t) + \mathbf{A}^+[\varphi^-(t)] = \Pi^+(t)[D(t)\varphi^-(t) + H(t)], \tag{14}$$

where $\mathbf{A}^+$ is the integral operator with kernel $A^+(t, \tau)$. Since the function $\mathbf{A}^+[\varphi^-(t)]$ is analytic on Ω^+, it follows that the last relation is an ordinary Riemann problem for which the functions $\Pi^+(t)\varphi^+(t) + \mathbf{A}^+[\varphi^-(t)]$ and $\varphi^-(t)$ can be defined in a closed form, and hence the same holds for $\varphi(t)$. Namely, let us rewrite the function $D(t)$ in the form $D(t) = X^+(t)/X^-(t)$, where $X^\pm(z)$ is the canonical function of the Riemann problem, and reduce relation (14) to the form in which the generalized Liouville theorem can be applied (see Subsection 12.3-1). We arrive at a polynomial of degree at most $\nu - 1 + \sum_{k=1}^{n} m_k$ with arbitrary coefficients (for the case in which $\nu + \sum_{k=1}^{n} m_k > 0$). However, the presence of the factor $\Pi^+(t)$ (on $\varphi^+(t)$), which vanishes in Ω^+ with total order of zeros $\sum_{k=1}^{n} m_k$, clearly reduces the number of arbitrary constants in the general solution.

Remark 1. Following the lines of the discussion in Subsection 13.3-2 we can treat the case in which the kernel $K(t, \tau)$ is meromorphic with respect to τ as well. In this case, Eq. (1) can be reduced to a Riemann problem of the type (12) and a linear algebraic system.

Remark 2. The solutions of a complete singular integral equation that are constructed in Section 13.3 can be applied for the case in which the contour L is a collection of finitely many disjoint smooth closed contours.

Example 1. Consider the equation

$$\lambda\varphi(t) + \frac{1}{\pi i} \int_L \frac{\cos(\tau - t)}{\tau - t} \varphi(\tau)\, d\tau = f(t), \tag{15}$$

where L is an arbitrary closed contour.

Note that the function $M(t, \tau) = \cos(\tau - t)$ has the property $M(t, t) \equiv 1$. Therefore, it remains to apply formula (6), and thus for (15) we have

$$\varphi(t) = \frac{1}{\lambda^2 - 1} \left[\lambda f(t) - \frac{1}{\pi i} \int_L \frac{\cos(\tau - t)}{\tau - t} f(\tau)\, d\tau \right], \qquad \lambda \neq \pm 1.$$

Example 2. Consider the equation

$$\lambda\varphi(t) + \frac{1}{\pi i} \int_L \frac{\sin(\tau - t)}{(\tau - t)^2} \varphi(\tau)\, d\tau = f(t), \tag{16}$$

where L is an arbitrary closed contour.

The function $M(t, \tau) = \sin(\tau - t)/(\tau - t)$ has the property $M(t, t) \equiv 1$. Therefore, applying formula (6), for (16) we obtain

$$\varphi(t) = \frac{1}{\lambda^2 - 1} \left[\lambda f(t) - \frac{1}{\pi i} \int_L \frac{\sin(\tau - t)}{(\tau - t)^2} f(\tau)\, d\tau \right], \qquad \lambda \neq \pm 1.$$

⊙ Reference for Section 13.3: F. D. Gakhov (1977).

13.4. The Regularization Method for Complete Singular Integral Equations

13.4-1. Certain Properties of Singular Operators

Let $\mathbf{K}_1$ and $\mathbf{K}_2$ be singular operators,

$$\mathbf{K}_1[\varphi(t)] \equiv a_1(t)\varphi(t) + \frac{1}{\pi i} \int_L \frac{M_1(t, \tau)}{\tau - t} \varphi(\tau)\, d\tau, \tag{1}$$

$$\mathbf{K}_2[\omega(t)] \equiv a_2(t)\omega(t) + \frac{1}{\pi i} \int_L \frac{M_2(t, \tau)}{\tau - t} \omega(\tau)\, d\tau. \tag{2}$$

The operator $\mathbf{K} = \mathbf{K}_2\mathbf{K}_1$ defined by the formula $\mathbf{K}[\varphi(t)] = \mathbf{K}_2\big[\mathbf{K}_1[\varphi(t)]\big]$ is called the *composition* or the *product* of the operators $\mathbf{K}_1$ and $\mathbf{K}_2$.

Let us form the expression for the operator $\mathbf{K}$,

$$\mathbf{K}[\varphi(t)] = \mathbf{K}_2\mathbf{K}_1[\varphi(t)] \equiv a_2(t)\left[a_1(t)\varphi(t) + \frac{1}{\pi i}\int_L \frac{M_1(t,\tau)}{\tau - t}\varphi(\tau)\,d\tau\right]$$
$$+ \frac{1}{\pi i}\int_L \frac{M_2(t,\tau)}{\tau - t}\left[a_1(\tau)\varphi(\tau) + \frac{1}{\pi i}\int_L \frac{M_1(\tau,\tau_1)}{\tau_1 - \tau}\varphi(\tau_1)\,d\tau_1\right]d\tau, \qquad (3)$$

and select its characteristic part. To this end, we perform the following manipulations:

$$\int_L \frac{M_1(t,\tau)}{\tau - t}\varphi(\tau)\,d\tau = M_1(t,t)\int_L \frac{\varphi(\tau)}{\tau - t}\,d\tau + \int_L \frac{M_1(t,\tau) - M_1(t,t)}{\tau - t}\varphi(\tau)\,d\tau,$$
$$\int_L \frac{a_1(\tau)M_2(t,\tau)}{\tau - t}\varphi(\tau)\,d\tau = a_1(t)M_2(t,t)\int_L \frac{\varphi(\tau)}{\tau - t}\,d\tau + \int_L \frac{a_1(\tau)M_2(t,\tau) - a_1(t)M_2(t,t)}{\tau - t}\varphi(\tau)\,d\tau, \qquad (4)$$
$$\int_L \frac{M_2(t,\tau)}{\tau - t}\,d\tau \int_L \frac{M_1(\tau,\tau_1)}{\tau_1 - \tau}\varphi(\tau_1)\,d\tau_1 = -\pi^2 M_2(t,t)M_1(t,t)\varphi(t) + \int_L \varphi(\tau_1)\,d\tau_1 \int_L \frac{M_2(t,\tau)M_1(\tau,\tau_1)}{(\tau_1 - \tau)(\tau - t)}\,d\tau.$$

Here we applied the Poincaré–Bertrand formula (see Subsection 12.2-6). We can see that all kernels of the integrals of the last summands on the right-hand sides in (4) are Fredholm kernels.

We write

$$M_1(t,t) = b_1(t), \qquad M_2(t,t) = b_2(t) \qquad (5)$$

and see that the characteristic operator $\mathbf{K}^\circ$ of the composition (product) $\mathbf{K}$ of two singular operators $\mathbf{K}_1$ and $\mathbf{K}_2$ can be expressed by the formula

$$\mathbf{K}^\circ[\varphi(t)] = (\mathbf{K}_2\mathbf{K}_1)^\circ[\varphi(t)] = [a_2(t)a_1(t) + b_2(t)b_1(t)]\varphi(t) + \frac{a_2(t)b_1(t) + b_2(t)a_1(t)}{\pi i}\int_L \frac{\varphi(\tau)}{\tau - t}\,d\tau. \quad (6)$$

Let us write out the operator $\mathbf{K}_1$ and $\mathbf{K}_2$ in the form (3) with explicitly expressed characteristic parts:

$$\mathbf{K}_1[\varphi(t)] \equiv a_1(t)\varphi(t) + \frac{b_1(t)}{\pi i}\int_L \frac{\varphi(\tau)}{\tau - t}\,d\tau + \int_L K_1(t,\tau)\varphi(\tau)\,d\tau, \qquad (7)$$

$$\mathbf{K}_2[\omega(t)] \equiv a_2(t)\omega(t) + \frac{b_2(t)}{\pi i}\int_L \frac{\omega(\tau)}{\tau - t}\,d\tau + \int_L K_2(t,\tau)\omega(\tau)\,d\tau. \qquad (8)$$

Thus, the coefficients $a(t)$ and $b(t)$ of the characteristic part of the product of the operators $\mathbf{K}_1$ and $\mathbf{K}_2$ can be expressed by the formulas

$$a(t) = a_2(t)a_1(t) + b_2(t)b_1(t), \qquad b(t) = a_2(t)b_1(t) + b_2(t)a_1(t). \qquad (9)$$

These formulas do not contain regular kernels k_1 and k_2 and are symmetric with respect to the indices 1 and 2. This means that the characteristic part of the product of singular operators depends neither on their regular parts nor on the order of these operators in the product.

Thus, any change of order of the factors, as well as a change of the regular parts of the factors, influences the regular part of the product of the operators only and preserves the characteristic part of the product.

Let us calculate the coefficient of the Riemann problem that corresponds to the characteristic operator $(\mathbf{K}_2\mathbf{K}_1)^\circ$:

$$D(t) = \frac{a(t) - b(t)}{a(t) + b(t)} = \frac{[a_2(t) - b_2(t)][a_1(t) - b_1(t)]}{[a_2(t) + b_2(t)][a_1(t) + b_1(t)]} = D_2(t)D_1(t), \qquad (10)$$

where we denote by

$$D_1(t) = \frac{a_1(t) - b_1(t)}{a_1(t) + b_1(t)}, \qquad D_2(t) = \frac{a_2(t) - b_2(t)}{a_2(t) + b_2(t)} \qquad (11)$$

the coefficients of the Riemann problems that correspond to the operators $\mathbf{K}_1^\circ$ and $\mathbf{K}_2^\circ$. This means that the coefficient of the Riemann problem for the operator $(\mathbf{K}_2\mathbf{K}_1)^\circ$ is equal to the product of the coefficients of the Riemann problems for the operators $\mathbf{K}_1^\circ$ and $\mathbf{K}_2^\circ$, and hence the index of the product of singular operators is equal to the sum of indices of the factors:

$$\nu = \nu_1 + \nu_2. \tag{12}$$

In its complete form, the operator $\mathbf{K}_2\mathbf{K}_1$ is defined by the expression

$$\mathbf{K}_2\mathbf{K}_1[\varphi(t)] \equiv a(t)\varphi(t) + \frac{b(t)}{\pi i} \int_L \frac{\varphi(\tau)}{\tau - t}\, d\tau + \int_L K(t,\tau)\varphi(\tau)\, d\tau,$$

where $a(t)$ and $b(t)$ are defined by formulas (9). For a regular kernel $K(t,\tau)$, on the basis of formulas (4) we can write out the explicit expression.

For a singular operator $\mathbf{K}$ and its transposed operator $\mathbf{K}^*$ (see Subsection 13.1-1), the following relations hold:

$$\int_L \psi(t)\mathbf{K}[\varphi(t)]\, dt = \int_L \varphi\mathbf{K}^*[\psi(t)]\, dt$$

for any functions $\varphi(t)$ and $\psi(t)$ that satisfy the Hölder condition, and

$$(\mathbf{K}_2\mathbf{K}_1)^* = \mathbf{K}_1^*\mathbf{K}_2^*.$$

13.4-2. The Regularizer

The regularization method is a reduction of a singular integral equation to a Fredholm equation. The reduction process itself is known as *regularization*.

If a singular operator $\mathbf{K}_2$ is such that the operator $\mathbf{K}_2\mathbf{K}_1$ is regular (Fredholm), i.e., contains no singular integral ($b(t) \equiv 0$), then $\mathbf{K}_2$ is called the *regularizing operator* with respect to the singular operator $\mathbf{K}_1$ or, briefly, a *regularizer*. Note that if $\mathbf{K}_2$ is a regularizer, then the operator $\mathbf{K}_1\mathbf{K}_2$ is regular as well.

Let us find the general form of a regularizer. By definition, the following relation must hold:

$$b(t) = a_2(t)b_1(t) + b_2(t)a_1(t) = 0, \tag{13}$$

which implies that

$$a_2(t) = g(t)a_1(t), \quad b_2(t) = -g(t)b_1(t), \tag{14}$$

where $g(t)$ is an arbitrary function that vanishes nowhere and satisfies the Hölder condition.

Hence, if $\mathbf{K}$ is a singular operator,

$$\mathbf{K}[\varphi(t)] \equiv a(t)\varphi(t) + \frac{b(t)}{\pi i} \int_L \frac{\varphi(\tau)}{\tau - t}\, d\tau + \int_L K(t,\tau)\varphi(\tau)\, d\tau, \tag{15}$$

then, in general, the regularizer $\tilde{\mathbf{K}}$ can be expressed as follows:

$$\tilde{\mathbf{K}}[\omega(t)] \equiv g(t)a(t)\omega(t) - \frac{g(t)b(t)}{\pi i} \int_L \frac{\omega(\tau)}{\tau - t}\, d\tau + \int_L \tilde{K}(t,\tau)\omega(\tau)\, d\tau, \tag{16}$$

where $\tilde{K}(t,\tau)$ is an arbitrary Fredholm kernel and $g(t)$ is an arbitrary function satisfying the Hölder condition.

Since the index of a regular operator ($b(t) \equiv 0$) is clearly equal to zero, it follows from the property of the product of operators that the index of the regularizer has the same modulus as the index of the original operator and the opposite sign. The same fact can be established directly by the form of a regularizer (16) from the formula

$$\tilde{D}(t) = \frac{\tilde{a}(t) - \tilde{b}(t)}{\tilde{a}(t) + \tilde{b}(t)} = \frac{a(t) + b(t)}{a(t) - b(t)} = \frac{1}{D(t)}.$$

Thus, for any singular operator with Cauchy kernel (15) of the normal type $(a(t) \pm b(t) \neq 0)$, there exist infinitely many regularizers (16) whose characteristic part depends on an arbitrary function $g(t)$ that contains an arbitrary regular kernel $\tilde{K}(t, \tau)$.

Since the elements $g(t)$ and $\tilde{K}(t, \tau)$ are arbitrary, we can choose them so that the regularizer will satisfy some additional conditions. For instance, we can make the coefficient of $\varphi(t)$ in the regularized equation be normalized, i.e., equal to one. To this end we must set $g(t) = [a^2(t) - b^2(t)]^{-1}$. If no conditions are imposed, then it is natural to apply the simplest regularizers. These can be obtained by setting $g(t) \equiv 1$ and $\tilde{K}(t, \tau) \equiv 0$ in formula (16), which gives the regularizer

$$\tilde{\mathbf{K}}[\omega(t)] = \mathbf{K}^{*\circ}[\omega(t)] \equiv a(t)\omega(t) - \frac{b(t)}{\pi i} \int_L \frac{\omega(\tau)}{\tau - t} \, d\tau, \tag{17}$$

or we can set $g(t) \equiv 1$ and $\tilde{K}(t, \tau) = -\frac{1}{\pi i} \frac{b(\tau) - b(t)}{\tau - t}$ and obtain

$$\tilde{\mathbf{K}}[\omega(t)] = \mathbf{K}^{\circ *}[\omega(t)] \equiv a(t)\omega(t) - \frac{1}{\pi i} \int_L \frac{b(\tau)\omega(\tau)}{\tau - t} \, d\tau. \tag{18}$$

The simplest operators $\mathbf{K}^{*\circ}$ and $\mathbf{K}^{\circ *}$ are most frequently used as regularizers.

Since the multiplication of operators is not commutative, we must distinguish two forms of regularization: left regularization, which gives the operator $\tilde{\mathbf{K}}\mathbf{K}$, and right regularization which leads to the operator $\mathbf{K}\tilde{\mathbf{K}}$. On the basis of the above remark we can claim that a right regularizer is simultaneously a left regularizer, and vice versa. Thus, the operation of regularization is commutative.

If an operator $\tilde{\mathbf{K}}$ is a regularizer for an operator $\mathbf{K}$, then, in turn, the operator $\mathbf{K}$ is a regularizer for the operator $\tilde{\mathbf{K}}$. The operators $\mathbf{K}_1\mathbf{K}_2$ and $\mathbf{K}_2\mathbf{K}_1$ can differ by a regular part only.

13.4-3. The Methods of Left and Right Regularization

Let a complete singular integral equation be given:

$$\mathbf{K}[\varphi(t)] \equiv a(t)\varphi(t) + \frac{b(t)}{\pi i} \int_L \frac{\varphi(\tau)}{\tau - t} \, d\tau + \int_L K(t, \tau)\varphi(\tau) \, d\tau = f(t). \tag{19}$$

Three methods of regularization are used. The first two methods are based on the composition of a given singular operator and its regularizer (left and right regularization). The third method differs essentially from the first two, namely, the elimination of the singular integral is performed by solving the corresponding characteristic equation.

1°. *Left regularization.* Let us take the regularizer (16):

$$\tilde{\mathbf{K}}[\omega(t)] \equiv g(t)a(t)\omega(t) - \frac{g(t)b(t)}{\pi i} \int_L \frac{\omega(\tau)}{\tau - t} \, d\tau + \int_L \tilde{K}(t, \tau)\omega(\tau) \, d\tau. \tag{20}$$

On replacing the function $\omega(t)$ in $\tilde{\mathbf{K}}[\omega(t)]$ with the expression $\mathbf{K}[\varphi(t)] - f(t)$ we arrive at the integral equation

$$\tilde{\mathbf{K}}\mathbf{K}[\varphi(t)] = \tilde{\mathbf{K}}[f(t)]. \tag{21}$$

By definition, $\tilde{\mathbf{K}}\mathbf{K}$ is a Fredholm operator, because $\tilde{\mathbf{K}}$ is a regularizer. Hence, Eq. (21) is a Fredholm equation. Thus, we have transformed the singular integral equation (19) into the Fredholm integral equation (21) for the same unknown function $\varphi(t)$.

This is the first regularization method, which is called *left regularization*.

$2°$. *Right Regularization.* On replacing in Eq. (19) the desired function by the expression (20),

$$\varphi(t) = \tilde{\mathbf{K}}[\omega(t)], \tag{22}$$

where $\omega(t)$ is a new unknown function, we arrive at the integral equation

$$\mathbf{K}\tilde{\mathbf{K}}[\omega(t)] = f(t), \tag{23}$$

which is a Fredholm equation as well. Thus, from the singular integral equation (19) for the unknown function $\varphi(t)$ we passed to the Fredholm integral equation for the new unknown function $\omega(t)$.

On solving the Fredholm equation (23), we find a solution of the original equation (19) by formula (22). The application of formula (22) requires integration only (a proper integral and a singular integral must be found).

This is the second method of the regularization, which is called *right regularization.*

13.4-4. The Problem of Equivalent Regularization

In the reduction of a singular integral equation to a regular one we perform a functional transformation over the corresponding equation. In general, this transformation can either introduce new irrelevant solutions that do not satisfy the original equation or imply a loss of some solutions. Therefore, in general, the resultant equation is not equivalent to the original equation. Consider the relationship between the solutions of these equations and find out in what cases these equations are equivalent.

$1°$. *Left Regularization.* Consider a singular equation

$$\mathbf{K}[\varphi(t)] = f(t) \tag{24}$$

and the corresponding regular equation

$$\tilde{\mathbf{K}}\mathbf{K}[\varphi(t)] = \tilde{\mathbf{K}}[f(t)]. \tag{25}$$

Let us write out Eq. (25) in the form

$$\tilde{\mathbf{K}}\big[\mathbf{K}[\varphi(t)] - f(t)\big] = 0. \tag{26}$$

Since the operator $\tilde{\mathbf{K}}$ is homogeneous, it follows that each solution of the original equation (24) (a function that vanishes the expression $\mathbf{K}[\varphi(t)] - f(t)$) satisfies Eq. (26) as well. Hence, the left regularization implies no loss of solutions. However, a solution of the regularized equation need not be a solution of the original equation.

Consider the singular integral equation corresponding to the regularizer

$$\tilde{\mathbf{K}}[\omega(t)] = 0. \tag{27}$$

Let $\omega_1(t), \ldots, \omega_p(t)$ be a complete system of its solutions, i.e., a maximal collection of linearly independent eigenfunctions of the regularizer $\tilde{\mathbf{K}}$.

We regard Eq. (26) as a singular equation of the form (27) with the unknown function $\omega(t) = \mathbf{K}[\varphi(t)] - f(t)$. We obtain

$$\mathbf{K}[\varphi(t)] - f(t) = \sum_{j=1}^{p} \alpha_j \omega_j(t), \tag{28}$$

where the α_j are some constants.

We see that the regularized equation is equivalent to Eq. (28) rather than the original equation (24).

Thus, Eq. (25) is equivalent to Eq. (28) in which α_j are arbitrary or definite constants. It may occur that Eq. (28) is solvable only under the assumption that all α_j satisfy the condition $\alpha_j = 0$. In this case, Eq. (25) is equivalent to the original equation (24), and the regularizer defines an equivalent transformation. In particular, if the regularizer has no eigenfunctions, then the right-hand side of Eq. (28) is identically zero, and it must be equivalent. This operator certainly exists for $\nu \geq 0$. For instance, we can take the regularizer $\mathbf{K}^{*\circ}$, which has no eigenfunctions for the case under consideration because the index of the regularizer $\mathbf{K}^{*\circ}$ is equal to $-\nu \leq 0$.

$2°$. *Right Regularization.* Consider Eq. (24) and the corresponding regularized equation

$$\mathbf{K}\tilde{\mathbf{K}}[\omega(t)] = f(t), \tag{29}$$

which is obtained by substitution

$$\tilde{\mathbf{K}}[\omega(t)] = \varphi(t). \tag{30}$$

If $\omega_j(t)$ is a solution of Eq. (29), then formula (30) gives the corresponding solution of the original equation

$$\varphi_j(t) = \tilde{\mathbf{K}}[\omega_j(t)].$$

Hence, the right regularization cannot lead to irrelevant solutions.

Conversely, assume that $\varphi_k(t)$ is a solution of the original equation. In this case a solution of the regularized equation (29) can be obtained as a solution of the nonhomogeneous singular equation

$$\tilde{\mathbf{K}}[\omega(t)] = \varphi_k(t);$$

however, this solution may be unsolvable. Thus, the right regularization can lead to loss of solutions. We have no loss of solutions if Eq. (30) is solvable for each right-hand side. In this case the operator $\tilde{\mathbf{K}}$ will be an equivalent right regularizer.

$3°$. *The Equivalent Regularization.* The operator $\tilde{\mathbf{K}} = \mathbf{K}^{*\circ}$ is an equivalent regularizer for any index; for $\nu \geq 0$, we must apply left regularization, while for $\nu \leq 0$ we must use right regularization.

In the latter case we obtain an equation for a new function $\omega(t)$, and if it is determined, then we can construct all solutions to the original equation in antiderivatives, and it follows from the properties of the right regularization that no irrelevant solutions can occur.

For the other methods of equivalent regularization, see the references at the end of this section.

13.4-5. Fredholm Theorems

Let a complete singular integral equation be given:

$$\mathbf{K}[\varphi(t)] = f(t). \tag{31}$$

THEOREM 1. *The number of solutions of the singular integral equation (31) is finite.*

THEOREM 2. *A necessary and sufficient solvability condition for the singular equation (31) is*

$$\int_L f(t)\psi_j(t)\,dt = 0, \qquad j = 1,\dots,m, \tag{32}$$

where $\psi_1(t), \dots, \psi_m(t)$ is a maximal finite set of linearly independent solutions of the transposed homogeneous equation $\mathbf{K}^*[\psi(t)] = 0$. (Since the functions under consideration are complex, it follows that condition (32) is not the orthogonality condition for the functions $f(t)$ and $\psi_j(t)$.)

THEOREM 3. *The difference between the number n of linearly independent solutions of the singular equation $\mathbf{K}[\varphi(t)] = 0$ and the number m of linearly independent solutions of the transposed equation $\mathbf{K}^*[\psi(t)] = 0$ depends on the characteristic part of the operator $\mathbf{K}$ only and is equal to its index, i.e.,*

$$n - m = \nu. \tag{33}$$

Corollary. The number of linearly independent solutions of characteristic equations is minimal among all singular equations with given index ν.

13.4-6. The Carleman–Vekua Approach to the Regularization

Let us transfer the regular part of a singular equation to the right-hand side and rewrite the equation as follows:

$$a(t)\varphi(t) + \frac{b(t)}{\pi i} \int_L \frac{\varphi(\tau)}{\tau - t} d\tau = f(t) - \int_L K(t, \tau)\varphi(\tau) d\tau, \tag{34}$$

or, in the operator form,

$$\mathbf{K}^\circ[\varphi(t)] = f(t) - \mathbf{K}_r[\varphi(t)]. \tag{35}$$

We regard the last equation as a characteristic one and solve it by temporarily assuming that the right-hand side is a known function. In this case (see Subsection 13.2-1)

$$\varphi(t) = \left[a(t)f(t) - \frac{b(t)Z(t)}{\pi i} \int_L \frac{f(\tau)}{Z(\tau)} \frac{d\tau}{\tau - t} + b(t)Z(t)P_{\nu-1}(t) \right]$$
$$- \left[a(t) \int_L K(t, \tau)\varphi(\tau) d\tau - \frac{b(t)Z(t)}{\pi i} \int_L \frac{d\tau_1}{Z(\tau_1)(\tau_1 - t)} \int_L K(\tau_1, \tau)\varphi(\tau) d\tau \right], \tag{36}$$

where for $\nu \le 0$ we must set $P_{\nu-1}(t) \equiv 0$. Let us reverse the order of integration in the iterated integral and rewrite the expression in the last parentheses as follows:

$$\int_L \left[a(t)K(t, \tau) - \frac{b(t)Z(t)}{\pi i} \int_L \frac{K(\tau_1, \tau)}{Z(\tau_1)(\tau_1 - t)} d\tau_1 \right] \varphi(\tau) d\tau.$$

Since $Z(t)$ satisfies the Hölder condition (and hence is bounded) and does not vanish and since $K(\tau_1, \tau)$ satisfies the estimate $|K(\tau_1, \tau)| < A|\tau_1 - \tau|^{-\lambda}$ (with $0 \le \lambda < 1$) near the point $\tau_1 = \tau$, we can see that the entire integral

$$\int_L \frac{K(\tau_1, \tau)}{Z(\tau_1)(\tau_1 - t)} d\tau_1$$

satisfies an estimate similar to that for $K(\tau_1, \tau)$. Hence, the kernel

$$N(t, \tau) = a(t)K(t, \tau) - \frac{b(t)Z(t)}{\pi i} \int_L \frac{K(\tau_1, \tau)}{Z(\tau_1)(\tau_1 - t)} d\tau_1 \tag{37}$$

is a Fredholm kernel. On transferring the terms with $\varphi(t)$ to the right-hand side, we obtain

$$\varphi(t) + \int_L N(t, \tau)\varphi(\tau) d\tau = f_1(t), \tag{38}$$

where $N(t, \tau)$ is the Fredholm kernel defined by formula (37) and $f_1(t)$ has the form

$$f_1(t) = a(t)f(t) - \frac{b(t)Z(t)}{\pi i} \int_L \frac{f(\tau)}{Z(\tau)} \frac{d\tau}{\tau - t} + b(t)Z(t)P_{\nu-1}(t). \tag{39}$$

If the index of Eq. (34) ν is negative, then the function must satisfy not only the Fredholm equation (38) but also the relations

$$\int_L \left[\int_L \frac{K(t, \tau)}{Z(t)} t^{k-1} dt \right] \varphi(\tau) d\tau = \int_L \frac{f(t)}{Z(t)} t^{k-1} dt, \qquad k = 1, 2, \ldots, -\nu. \tag{40}$$

Thus, if $\nu \ge 0$, then the solution of a complete singular integral equation (34) is reduced to the solution of the Fredholm integral equation (38). If $\nu < 0$, then Eq. (34) can be reduced to Eq. (38) (where we must set $P_{\nu-1}(t) \equiv 0$) together with conditions (40), which can be rewritten in the form

$$\int_L \rho_k(\tau)\varphi(\tau) d\tau = f_k, \qquad k = 1, 2, \ldots, -\nu,$$
$$\rho_k(\tau) = \int_L \frac{K(t, \tau)}{Z(t)} t^{k-1} dt, \qquad f_k = \int_L \frac{f(t)}{Z(t)} t^{k-1} dt, \tag{41}$$

where the $\rho_k(\tau)$ are known functions and the f_k are known constants.

Relations (41) are the solvability conditions for the regularized equation (38). However, they need not be the solvability conditions for the original Cauchy singular integral equation (34). Some of them can be the equivalence conditions for these two equations. Let us select the conditions of these two types.

Assume that among the functions $\rho_k(t)$ there are precisely h linearly independent functions. We can choose the numbering so that these are the functions $\rho_1(t), \ldots, \rho_h(t)$. In this case we have

$$\int_L \rho_k(t)\varphi(t)\,dt = f_k, \qquad k = 1, 2, \ldots, h. \tag{42}$$

Moreover, the following $\eta = |\nu| - h$ linearly independent relations must hold:

$$\alpha_{j1}\rho_1(t) + \cdots + \alpha_{j|\nu|}\rho_{|\nu|}(t) = 0, \qquad j = 1, 2, \ldots, \eta.$$

Let us multiply the relations in (40) successively by $\alpha_{j1}, \ldots, \alpha_{j|\nu|}$ and sum the products. Taking into account the last relations, we have

$$\int_L f(t)\psi_j(t)\,dt = 0, \qquad \psi_j(t) = \frac{1}{Z(t)}\sum_{k=1}^{|\nu|}\alpha_{jk}t^{k-1}; \qquad j = 1, 2, \ldots, \eta. \tag{43}$$

These relations, which do not involve the desired function $\varphi(t)$, are the necessary solvability conditions on the right-hand side $f(t)$ for the original singular equation and the regularized equation to be solvable. Relations (42) are the equivalence conditions for the original singular equation and the regularized equation. The solution of the Fredholm equation (38) satisfies the original singular equation (34) if and only if it satisfies conditions (42).

Thus, for $\nu \geq 0$, the regularized equation (38) is equivalent to the original singular equation. For $\nu < 0$, the original equation is equivalent to the regularized equation (with common solvability conditions (43)) together with conditions (42).

Remark 1. If the kernel of the regular part of a complete singular integral equation with Cauchy kernel is degenerate, then by the Carleman–Vekua regularization this equation can be reduced to the investigation of a system of linear algebraic equations (see, e.g., S. G. Mikhlin and K. L. Smolitskiy (1967)).

Remark 2. The Carleman–Vekua regularization is sometimes called the regularization by solving the characteristic equation.

13.4-7. Regularization in Exceptional Cases

Consider the complete singular equation with Cauchy kernel

$$\mathbf{K}[\varphi(t)] \equiv a(t)\varphi(t) + \frac{b(t)}{\pi i}\int_L \frac{\varphi(\tau)}{\tau - t}\,d\tau + \int_L K(t, \tau)\varphi(\tau)\,d\tau = f(t) \tag{44}$$

under the same conditions on the functions $a(t) \pm b(t)$ as above in Subsection 13.2-4.

We represent this equation in the form

$$\mathbf{K}^\circ[\varphi(t)] = f(t) - \int_L K(t, \tau)\varphi(\tau)\,d\tau,$$

and apply the Carleman–Vekua regularization. In this case by formula (42) of Subsection 13.2-4 we obtain the equation

$$\varphi(t) + \mathbf{R}_1\left[\int_L K(t, \tau)\varphi(\tau)\,d\tau\right] = \mathbf{R}_1[f(t)] + b(t)Z(t)P_{\nu-p-1}(t), \tag{45}$$

where the operator $\mathbf{R}_1$ is defined by formula (41) of Subsection 13.2-4.

In the expression for the second summand on the left-hand side in (45), the operation $\mathbf{R}_1$ with respect to the variable t commutes with the operation of integration with respect to τ. Therefore, Eq. (45) can be rewritten in the form

$$\varphi(t) + \int_L \mathbf{R}_1^t \left[K(t,\tau)\right]\varphi(\tau)\,d\tau = \mathbf{R}_1[f(t)] + b(t)Z(t)P_{\nu-p-1}(t), \tag{46}$$

where the superscript t at the symbol of the operator $\mathbf{R}_1^t$ means that the operation is performed with respect to the variable t.

Since the operator $\mathbf{R}_1$ is bounded, it follows that the resulting integral equation (46) is a Fredholm equation, and hence the regularization problem for the singular equation (44) is solved.

It follows from the general theory of the regularization that Eq. (44) is equivalent to Eq. (46) for $\nu - p \geq 0$ and to Eq. (46) and a system of functional equations for $\nu - p < 0$.

In conclusion we note that for the above cases of singular integral equations, the Fredholm theorems fail in general.

Remark 3. Exceptional cases of singular integral equations with Cauchy kernel can be reduced to equations of the normal type.

13.4-8. The Complete Equation With Hilbert Kernel

Consider the complete singular integral equation with Hilbert kernel (see Subsection 13.1-2)

$$a(x)\varphi(x) - \frac{b(x)}{2\pi} \int_0^{2\pi} \cot\left(\frac{\xi-x}{2}\right)\varphi(\xi)\,d\xi + \int_0^{2\pi} K(x,\xi)\varphi(\xi)\,d\xi = f(x). \tag{47}$$

Let us show that Eq. (47) can be reduced to a complete singular integral equation with a kernel of the Cauchy type, and in this connection, the theory of the latter equation can be directly extended to Eq. (47). Since the regular parts of these two types of equations have the same character, it follows that it suffices to apply the relationship between the Hilbert kernel and the Cauchy kernel (see Subsection 12.4-5):

$$\frac{d\tau}{\tau-t} = \frac{1}{2}\cot\left(\frac{\xi-x}{2}\right)d\xi + \frac{i}{2}\,d\xi. \tag{48}$$

Hence,

$$\frac{1}{2}\cot\left(\frac{\xi-x}{2}\right)d\xi = \frac{d\tau}{\tau-t} - \frac{1}{2}\frac{d\tau}{\tau}, \tag{49}$$

where $t = e^{ix}$ and $\tau = e^{i\xi}$ are the complex coordinates of points of the contour L, that is, the unit circle.

On replacing the Hilbert kernel in Eq. (47) with the expression (49) and on substituting $x = -i\ln t$, $\xi = -i\ln\tau$, and $d\xi = -i\tau^{-1}\,d\tau$, after obvious manipulations we reduce Eq. (47) to a complete singular integral equation with Cauchy kernel of the form

$$a_1(t)\varphi_1(t) - \frac{ib_1(t)}{\pi i}\int_L \frac{\varphi_1(\tau)}{\tau-t}\,d\tau + \int_L K_1(t,\tau)\,d\tau = f_1(t). \tag{50}$$

The coefficient of the Riemann problem corresponding to Eq. (50) is

$$D(t) = \frac{a_1(t) + ib_1(t)}{a_1(t) - ib_1(t)} = \frac{a(x) + ib(x)}{a(x) - ib(x)}, \tag{51}$$

and the index is expressed by the formula

$$\operatorname{Ind} D(t) = 2 \operatorname{Ind}[a(x) + ib(x)]. \tag{52}$$

Example. Let us perform the regularization of the following singular integral equations in different ways:

$$\mathbf{K}[\varphi(t)] \equiv (t + t^{-1})\varphi(t) + \frac{t - t^{-1}}{\pi i} \int_L \frac{\varphi(\tau)}{\tau - t} \, d\tau - \frac{1}{2\pi i} \int_L (t + t^{-1})(\tau + \tau^{-1})\varphi(\tau) \, d\tau = 2t^2, \tag{53}$$

where L is the unit circle.

The regular part of the kernel is degenerate. Therefore, in the same way as was applied in the solution of Fredholm equations with degenerate kernel (see Section 11.2), the equation can be reduced to the investigation of the characteristic equation and a linear algebraic equation, and hence it can be solved in a closed form. Thus, we need no regularization. However, the equation under consideration is useful in the illustration of general methods because all calculations can be performed to the very end.

For convenience of the subsequent discussion, we first solve this equation. We write

$$\frac{1}{2\pi i} \int_L (\tau + \tau^{-1})\varphi(\tau) \, d\tau = A, \tag{54}$$

and write out the equation in the characteristic form:

$$(t + t^{-1})\varphi(t) + \frac{t - t^{-1}}{\pi i} \int_L \frac{\varphi(\tau)}{\tau - t} \, d\tau = 2t^2 + A(t + t^{-1}).$$

For the corresponding Riemann boundary value problem

$$\Phi^+(t) = t^{-2}\Phi^-(t) + t + \tfrac{1}{2}A(1 + t^{-2}), \tag{55}$$

we have the index $\nu = -2$, and the solvability conditions (see Subsection 13.2-1) hold for $A = 0$ only. In this case, $\Phi^+(z) = z$ and $\Phi^-(z) = 0$. This gives a solution to Eq. (53) in the form $\varphi(t) = \Phi^+(t) - \Phi^-(t) = t$. On substituting the last expression into Eq. (54) we see that this relation holds for $A = 0$. Hence, the given equation is solvable and has a unique solution of the form

$$\varphi(t) = t.$$

$1°$. *Left Regularization.* Since the index of the equation $\nu = -2 < 0$ is negative, it follows that any its regularizer has eigenfunctions (at least two linearly independent), and hence the left regularization leads, in general, to an equation that is not equivalent to the original one.

We first consider the left regularization by means of the simplest regularizer $\mathbf{K}^{*\circ}$. Let us find the linearly independent eigenfunctions of the equation

$$\mathbf{K}^{*\circ}[\omega(t)] \equiv (t + t^{-1})\omega(t) - \frac{t - t^{-1}}{\pi i} \int_L \frac{\omega(\tau)}{\tau - t} \, d\tau = 0.$$

The corresponding Riemann boundary value problem

$$\Phi^+(t) = t^2 \Phi^-(t)$$

now has the index $\nu = 2$. We can find the eigenfunctions of the operator $\mathbf{K}^{*\circ}$ by the formulas of Subsection 13.2-1 and obtain

$$\omega_1(t) = 1 - t^{-2}, \quad \omega_2(t) = t - t^{-1}.$$

On the basis of the general theory (see Subsection 13.4-4), the regular equation $\mathbf{K}^{*\circ}\mathbf{K}[\varphi(t)] = \mathbf{K}^{*\circ}[f(t)]$ is equivalent to the singular equation:

$$\mathbf{K}[\varphi(t)] = f(t) + \alpha_1\omega_1(t) + \alpha_2\omega_2(t), \tag{56}$$

where α_1 and α_2 are constants that can be either arbitrary or definite. Taking into account Eq. (54), we write out Eq. (56) in the form of a characteristic equation:

$$(t + t^{-1})\varphi(t) + \frac{t - t^{-1}}{\pi i} \int_L \frac{\varphi(\tau)}{\tau - t} \, d\tau = 2t^2 + A(t + t^{-1}) + \alpha_1(1 - t^{-2}) + \alpha_2(t - t^{-1}).$$

The corresponding Riemann boundary value problem has the form

$$\Phi^+(t) = t^{-2}\Phi^-(t) + t + \tfrac{1}{2}A(1 + t^{-2}) + \tfrac{1}{2}\alpha_1(t^{-1} + t^{-3}) + \tfrac{1}{2}\alpha_2(1 - t^{-2}).$$

Its solution can be represented as follows:

$$\Phi^+(z) = z + \tfrac{1}{2}A + \tfrac{1}{2}\alpha_2, \quad \Phi^-(z) = \tfrac{1}{2}z^2[\alpha_1 z^{-3} + (\alpha_2 - A)z^{-2} - \alpha_1 z^{-1}].$$

The solvability conditions give $\alpha_1 = 0$ and $\alpha_2 = A$. In this case, the solution of Eq. (56) is defined by the formula

$$\varphi(t) = \Phi^+(t) - \Phi^-(t) = t + A.$$

On substituting the above expression for $\varphi(t)$ into Eq. (54) we obtain the identity $A = A$. Hence, the constant $\alpha_2 = A$ remains arbitrary, and the regularized equation is equivalent not to the original equation but to the equation

$$\mathbf{K}[\varphi(t)] = f(t) + \alpha_2\omega_2(t),$$

which has the solution $\varphi(t) = t + A$, where A is an arbitrary constant. The last function φ satisfies the original equation only for $A = 0$.

$2°$. *Right Regularization.* For a right regularizer we take the simplest operator $\mathbf{K}^{*\circ}$. By setting

$$\varphi(t) = \mathbf{K}^{*\circ}[\omega(t)] \equiv (t + t^{-1})\omega(t) - \frac{t - t^{-1}}{\pi i} \int_L \frac{\omega(\tau)}{\tau - t}\, d\tau, \tag{57}$$

we obtain the following Fredholm equation with respect to the function $\omega(t)$:

$$\mathbf{KK}^{*\circ}[\omega(t)] \equiv \omega(t) - \frac{1}{4\pi i}\int_L [t(\tau^2 - 1 + \tau^{-2}) + 2\tau^{-1} + t^{-1}(\tau^2 + 3 + \tau^{-2}) - 2\tau^{-2}\tau^{-1}]\omega(\tau)\, d\tau = \tfrac{1}{2}t^2. \tag{58}$$

The last equation is degenerate. On solving it we obtain

$$\omega(t) = \tfrac{1}{2}t^2 + \alpha(t - t^{-1}) + \beta(1 - t^{-2}),$$

where α and β are arbitrary constants.

Thus, the regularized equation for $\omega(t)$ has two linearly independent solutions, while the original equation (53) has a unique solution. On substituting the above expression for $\omega(t)$ into formula (57) we obtain

$$\varphi(t) = \mathbf{K}^{*\circ}\left[\tfrac{1}{2}t^2 + \alpha(t - t^{-1}) + \beta(1 - t^{-2})\right] = t,$$

where $\varphi(t)$ is the (unique) solution of the original singular equation. The result agrees with the general theory because, for a negative index, the right regularization by means of the operator $\mathbf{K}^{*\circ}$ is an equivalent regularization.

$3°$. *The Carleman–Vekua Regularization.* This method of regularization is performed by formulas (36)–(39). However, we must recall that these formulas can be applied only for an equation such that $a^2(t) - b^2(t) = 1$. Therefore, we must first divide Eq. (53) by two. In this case, we have

$$a = \tfrac{1}{2}(t + t^{-1}), \quad b = \tfrac{1}{2}(t - t^{-1}), \quad f(t) = t^2, \quad K(t, \tau) = -\frac{1}{4\pi i}(t + t^{-1})(\tau + \tau^{-1}), \quad X^+(z) = 1, \quad Z(t) = (a + b)X^+ = t,$$

$$f_1(t) = \tfrac{1}{2}(t + t^{-1})t^2 - \frac{(t - t^{-1})t}{2\pi i}\int_L \frac{\tau^2}{\tau}\frac{d\tau}{\tau - t} = t,$$

$$N(t, \tau) = -\frac{1}{2}(t + t^{-1})\frac{1}{4\pi i}(t + t^{-1})(\tau + \tau^{-1}) + \frac{(t - t^{-1})\,t\,(\tau + \tau^{-1})}{2\pi i \cdot 4\pi i}\int_L \frac{\tau_1 + \tau_1^{-1}}{\tau_1}\frac{d\tau_1}{\tau_1 - t} = -\frac{1}{2\pi i}(\tau + \tau^{-1}).$$

The regularized equation has the form

$$\varphi(t) - \frac{1}{2\pi i}\int_L (\tau + \tau^{-1})\varphi(\tau)\, d\tau = t. \tag{59}$$

To this equation we must add conditions (41) for $k = 1, 2$. This equation is degenerate, and on solving it we find the general solution $\varphi(t) = t + A$, where A is an arbitrary constant. Let us write out conditions (42) and (43). Here we have

$$\rho_k(\tau) = \int_L \frac{K(t, \tau)}{Z(t)}t^{k-1}\, dt = -\frac{\tau + \tau^{-1}}{4\pi i}\int_L (1 + t^{-2})t^{k-1}\, dt, \quad k = 1, 2,$$

$$\rho_1(\tau) = 0, \quad \rho_2(\tau) = -\tfrac{1}{2}(\tau + \tau^{-1}), \quad f_k = \int_L \frac{f(t)}{Z(t)}t^{k-1}\, dt = \int_L t^k\, dt, \quad f_1 = f_2 = 0.$$

The functions $\rho_1(t)$ and $\rho_2(t)$ are linearly dependent. The dependence $\alpha_{j1}\rho_1(t) + \cdots + \alpha_{j|\nu|}\rho_{|\nu|}(t) = 0$ (see Subsection 13.4-6) has the form

$$\alpha_1 \rho_1(t) + 0 \cdot \rho_2(t) = 0.$$

Hence, the solvability condition (43) holds identically. The equivalence condition (42)

$$\int_L \rho_2(\tau)\varphi(\tau)\, d\tau = -\tfrac{1}{2}\int_L (\tau + \tau^{-1})(\tau + A)\, d\tau = 0$$

holds for $A = 0$ only. Hence, among the solutions to the regularized equation, $\varphi(t) = t + A$, only the function $\varphi(t) = t$ satisfies the original equation.

⊙   References for Section 13.4: F. D. Gakhov (1977), S. G. Mikhlin and S. Prössdorf (1986), N. I. Muskhelishvili (1992).

Chapter 14

Methods for Solving Nonlinear Integral Equations

14.1. Some Definitions and Remarks

14.1-1. Nonlinear Volterra Integral Equations

Nonlinear Volterra integral equations can be represented in the form

$$\int_a^x K\big(x,t,y(t)\big)\,dt = F\big(x,y(x)\big), \tag{1}$$

where $K\big(x,t,y(t)\big)$ is the kernel of the integral equation and $y(x)$ is the unknown function ($a \le x \le b$). All functions in (1) are usually assumed to be continuous.

The form (1) does not cover all possible forms of nonlinear Volterra integral equations; however, it includes the types of nonlinear equations which are most frequently used and studied. A nonlinear integral equation (1) is called a *Volterra integral equation in the Urysohn form*.

In some cases, Eq. (1) can be rewritten in the form

$$\int_a^x K\big(x,t,y(t)\big)\,dt = f(x). \tag{2}$$

Equation (2) is called a *Volterra equation of the first kind in the Urysohn form*. Similarly, the equation

$$y(x) - \int_a^x K\big(x,t,y(t)\big)\,dt = f(x), \tag{3}$$

is called a *Volterra equation of the second kind in the Urysohn form*.

By the substitution $u(x) = y(x) - f(x)$, Eq. (3) can be reduced to the canonical form

$$u(x) = \int_a^x \mathcal{K}\big(x,t,u(t)\big)\,dt, \tag{4}$$

where $\mathcal{K}\big(x,t,u(t)\big)$ is the kernel* of the canonical integral equation.

The kernel $K\big(x,t,y(t)\big)$ is said to be *degenerate* if

$$K\big(x,t,y(t)\big) = \sum_{k=1}^{n} g_k(x)h_k\big(t,y(t)\big).$$

---

* There are another ways of reducing Eq. (3) to the form (4) for which the form of the function $\mathcal{K}$ may be different.

If in Eq. (1) the kernel is $K\big(x, t, y(t)\big) = Q(x, t)\Phi\big(t, y(t)\big)$, where $Q(x, t)$ and $\Phi(t, y)$ are known functions, then we obtain the Volterra integral equation in the Hammerstein form:

$$\int_a^x Q(x, t)\Phi\big(t, y(t)\big)\, dt = F\big(x, y(x)\big). \tag{5}$$

In some cases Eq. (5) can be rewritten in the form

$$\int_a^x Q(x, t)\Phi\big(t, y(t)\big)\, dt = f(x). \tag{6}$$

Equation (6) is called a *Volterra equation of the first kind in the Hammerstein form*. Similarly, an equation of the form

$$y(x) - \int_a^x Q(x, t)\Phi\big(t, y(t)\big)\, dt = f(x), \tag{7}$$

is called a *Volterra equation of the second kind in the Hammerstein form*.

It is possible to reduce Eq. (7) to the canonical form

$$u(x) = \int_a^x Q(x, t)\Phi_*\big(t, u(t)\big)\, dt, \tag{8}$$

where $u(x) = y(x) - f(x)$.

Remark 1. Since a Volterra equation in the Hammerstein form is a special case of a Volterra equation in the Urysohn form, the methods discussed below for the latter are certainly applicable to the former.

Remark 2. Some other types of nonlinear integral equations with variable limits of integration are considered in Chapter 5.

14.1-2. Nonlinear Equations With Constant Integration Limits

Nonlinear integral equations with constant integration limits can be represented in the form

$$\int_a^b K\big(x, t, y(t)\big)\, dt = F(x, y(x)), \qquad \alpha \le x \le \beta, \tag{9}$$

where $K\big(x, t, y(t)\big)$ is the kernel of the integral equation and $y(x)$ is the unknown function. Usually, all functions in (9) are assumed to be continuous and the case of $\alpha = a$ and $\beta = b$ is considered.

The form (9) does not cover all possible forms of nonlinear integral equations with constant integration limits; however, just as the form (1) for the Volterra equations, it includes the most frequently used and most studied types of these equations. A nonlinear integral equation (1) with constant limits of integration is called an *integral equation of the Urysohn type*.

If Eq. (9) can be rewritten in the form

$$\int_a^b K\big(x, t, y(t)\big)\, dt = f(x), \tag{10}$$

then (10) is called an *Urysohn equation of the first kind*. Similarly, the equation

$$y(x) - \int_a^b K\big(x, t, y(t)\big)\, dt = f(x), \tag{11}$$

is called an *Urysohn equation of the second kind*.

An Urysohn equation of the second kind can be rewritten in the canonical form

$$u(x) = \int_a^b \mathcal{K}(x, t, u(t)) \, dt. \tag{12}$$

Remark 3. Conditions for existence and uniqueness of the solution of an Urysohn equation are discussed below in Subsections 14.3-4 and 14.3-5.

If in Eq. (9) the kernel is $K(x, t, y(t)) = Q(x, t)\Phi(t, y(t))$, and $Q(x, t)$ and $\Phi(t, y)$ are given functions, then we obtain an *integral equation of the Hammerstein type*:

$$\int_a^b Q(x, t)\Phi(t, y(t)) \, dt = F(x, y(x)), \tag{13}$$

where, as usual, all functions in the equation are assumed to be continuous.

If Eq. (13) can be rewritten in the form

$$\int_a^b Q(x, t)\Phi(t, y(t)) \, dt = f(x), \tag{14}$$

then (14) is called a *Hammerstein equation of the first kind*. Similarly, an equation of the form

$$y(x) - \int_a^b Q(x, t)\Phi(t, y(t)) \, dt = f(x), \tag{15}$$

is called a *Hammerstein equation of the second kind*.

A Hammerstein equation of the second kind can be rewritten in the *canonical form*

$$u(x) = \int_a^b Q(x, t)\Phi_*(t, u(t)) \, dt. \tag{16}$$

The existence of the canonical forms (4), (8), (12) and (16) means that the distinction between the inhomogeneous and homogeneous nonlinear integral equations is unessential, unlike the case of linear equations. Another specific feature of a nonlinear equation is that it frequently has several solutions.

Remark 4. Since a Hammerstein equation is a special case of an Urysohn equation, the methods discussed below for the latter are certainly applicable to the former.

Remark 5. Some other types of nonlinear integral equations with constant limits of integration are considered in Chapter 6.

⊙ References for Section 14.1: N. S. Smirnov (1951), M. A. Krasnosel'skii (1964), M. L. Krasnov, A. I. Kiselev, and G. I. Makarenko (1971), P. P. Zabreyko, A. I. Koshelev, et al. (1975), F. G. Tricomi (1985), A. F. Verlan' and V. S. Sizikov (1986).

14.2. Nonlinear Volterra Integral Equations

14.2-1. The Method of Integral Transforms

Consider a Volterra integral equation with quadratic nonlinearity

$$\mu y(x) - \lambda \int_0^x y(x - t)y(t) \, dt = f(x). \tag{1}$$

To solve this equation, the Laplace transform can be applied, which, with regard to the convolution theorem (see Section 7.2), leads to a quadratic equation for the transform $\tilde{y}(p) = \mathfrak{L}\{y(x)\}$:

$$\mu\tilde{y}(p) - \lambda\tilde{y}^2(p) = \tilde{f}(p).$$

This implies

$$\tilde{y}(p) = \frac{\mu \pm \sqrt{\mu^2 - 4\lambda\tilde{f}(p)}}{2\lambda}. \tag{2}$$

The inverse Laplace transform $y(x) = \mathfrak{L}^{-1}\{\tilde{y}(p)\}$ (if it exists) is a solution to Eq. (1). Note that for the two different signs in formula (2), there are two corresponding solutions of the original equation.

Example 1. Consider the integral equation

$$\int_0^x y(x-t)y(t)\,dt = Ax^m, \qquad m > -1.$$

Applying the Laplace transform to the equation under consideration with regard to the relation $\mathcal{L}\{x^m\} = \Gamma(m+1)p^{-m-1}$, we obtain

$$\tilde{y}^2(p) = A\Gamma(m+1)p^{-m-1},$$

where $\Gamma(m)$ is the gamma function. On extracting the square root of both sides of the equation, we obtain

$$\tilde{y}(p) = \pm\sqrt{A\Gamma(m+1)}\, p^{-\frac{m+1}{2}}.$$

Applying the Laplace inversion formula, we obtain two solutions to the original integral equation

$$y_1(x) = -\frac{\sqrt{A\Gamma(m+1)}}{\Gamma\left(\dfrac{m+1}{2}\right)} x^{\frac{m-1}{2}}, \qquad y_2(x) = \frac{\sqrt{A\Gamma(m+1)}}{\Gamma\left(\dfrac{m+1}{2}\right)} x^{\frac{m-1}{2}}.$$

14.2-2. The Method of Differentiation for Integral Equations

Sometimes, differentiation (possibly multiple) of a nonlinear integral equation with subsequent elimination of the integral terms by means of the original equation makes it possible to reduce this equation to a nonlinear ordinary differential equation. Below we briefly list some equations of this type.

$1°$. The equation

$$y(x) + \int_a^x f\big(t, y(t)\big)\,dt = g(x) \tag{3}$$

can be reduced by differentiation to the nonlinear first-order equation

$$y_x' + f(x, y) - g_x'(x) = 0$$

with the initial condition $y(a) = g(a)$.

$2°$. The equation

$$y(x) + \int_a^x (x-t)f\big(t, y(t)\big)\,dt = g(x) \tag{4}$$

can be reduced by double differentiation (with the subsequent elimination of the integral term by using the original equation) to the nonlinear second-order equation:

$$y_{xx}'' + f(x, y) - g_{xx}''(x) = 0. \tag{5}$$

The initial conditions for the function $y = y(x)$ have the form

$$y(a) = g(a), \qquad y_x'(a) = g_x'(a). \tag{6}$$

$3°$. The equation

$$y(x) + \int_a^x e^{\lambda(x-t)} f\big(t, y(t)\big)\,dt = g(x) \tag{7}$$

can be reduced by differentiation to the nonlinear first-order equation

$$y_x' + f(x, y) - \lambda y + \lambda g(x) - g_x'(x) = 0. \tag{8}$$

The desired function $y = y(x)$ must satisfy the initial condition $y(a) = g(a)$.

4°. Equations of the form

$$y(x) + \int_a^x \cosh\left[\lambda(x-t)\right] f\left(t, y(t)\right) dt = g(x), \tag{9}$$

$$y(x) + \int_a^x \sinh\left[\lambda(x-t)\right] f\left(t, y(t)\right) dt = g(x), \tag{10}$$

$$y(x) + \int_a^x \cos\left[\lambda(x-t)\right] f\left(t, y(t)\right) dt = g(x), \tag{11}$$

$$y(x) + \int_a^x \sin\left[\lambda(x-t)\right] f\left(t, y(t)\right) dt = g(x) \tag{12}$$

can also be reduced to second-order ordinary differential equations by double differentiation. For these equations, see Section 5.8 in the first part of the book (Eqs. 22, 23, 24, and 25, respectively).

14.2-3. The Successive Approximation Method

1°. In many cases, the successive approximation method can be successfully applied to solve various types of integral equations. The principles of constructing the iteration process are the same as in the case of linear equations. For Volterra equations of the second kind in the Urysohn form

$$y(x) - \int_a^x K\left(x, t, y(t)\right) dt = f(x), \qquad a \le x \le b, \tag{13}$$

the corresponding recurrent expression has the form

$$y_{k+1}(x) = f(x) + \int_a^x K\left(x, t, y_k(t)\right) dt, \qquad k = 0, 1, 2, \ldots \tag{14}$$

It is customary to take the initial approximation either in the form $y_0(x) \equiv 0$ or in the form $y_0(x) = f(x)$.

In contrast to the case of linear equations, the successive approximation method has a smaller domain of convergence. Let us present the convergence conditions for the iteration process (14) that are simultaneously the existence conditions for a solution of Eq. (13). To be definite, we assume that $y_0(x) = f(x)$.

If for any z_1 and z_2 we have the relation

$$|K(x, t, z_1) - K(x, t, z_2)| \le \varphi(x, t)|z_1 - z_2|$$

and the relation

$$\left| \int_a^x K\left(x, t, f(t)\right) dt \right| \le \psi(x)$$

holds, where

$$\int_a^x \psi^2(t)\, dt \le N^2, \qquad \int_a^b \int_a^x \varphi^2(x, t)\, dt\, dx \le M^2,$$

for some constants N and M, then the successive approximations converge to a unique solution of Eq. (13) almost everywhere absolutely and uniformly.

Example 2. Let us apply the successive approximation method to solve the equation

$$y(x) = \int_0^x \frac{1 + y^2(t)}{1 + t^2}\, dt.$$

If $y_0(x) \equiv 0$, then

$$y_1(x) = \int_0^x \frac{dt}{1 + t^2} = \arctan x,$$

$$y_2(x) = \int_0^x \frac{1 + \arctan^2 t}{1 + t^2} \, dt = \arctan x + \tfrac{1}{3} \arctan^3 x,$$

$$y_3(x) = \int_0^x \frac{1 + \arctan t + \tfrac{1}{3} \arctan^3 t}{1 + t^2} \, dt = \arctan x + \tfrac{1}{3} \arctan^3 x + \tfrac{2}{3 \cdot 5} \arctan^5 x + \tfrac{1}{7 \cdot 9} \arctan^7 x.$$

On continuing this process, we can observe that $y_k(x) \to \tan(\arctan x) = x$ as $k \to \infty$, i.e., $y(x) = x$. The substitution of this result into the original equation shows the validity of the result.

Example 3. For the nonlinear equation

$$y(x) = \int_0^x [ty^2(t) - 1] \, dt$$

we must obtain the first three approximations. If we set $y_0(x) = 0$, then

$$y_1(x) = \int_0^x (-1) \, dt = -x,$$

$$y_2(x) = \int_0^x (t^3 - 1) \, dt = -x + \tfrac{1}{4} x^4,$$

$$y_3(x) = \int_0^x \left[t \left(\tfrac{1}{16} t^8 - \tfrac{1}{2} t^5 + t^2 \right) - 1 \right] dt = -x + \tfrac{1}{4} x^4 - \tfrac{1}{14} x^7 + \tfrac{1}{160} x^{10}.$$

$2°$. The successive approximation method can be applied to solve other forms of nonlinear equations, for instance, equations of the form

$$y(x) = F\left(x, \int_a^x K(x, t) y(t) \, dt \right)$$

solved for $y(x)$ in which the integral has x as the upper integration limit. This makes it possible to obtain a numerical solution by applying small steps with respect to x and by linearization at each step, which usually provides the uniqueness of the result of the iterations for an arbitrary initial approximation.

$3°$. The initial approximation substantially influences the number of iterations necessary to obtain the result with prescribed accuracy, and therefore when choosing this approximation, some additional arguments are usually applied. Namely, for the equation

$$Ay(x) - \int_0^x Q(x - t) \Phi\big(y(t)\big) \, dt = f(x),$$

where A is a constant, a good initial approximation $y_0(x)$ can sometimes be found from the solution of the following (in general, transcendental) equation for $\tilde{y}_0(p)$:

$$A\tilde{y}_0(p) - \tilde{Q}(p) \Phi\big(\tilde{y}_0(p)\big) = \tilde{f}(p),$$

where $\tilde{y}_0(p)$, $\tilde{Q}(p)$, and $\tilde{f}(p)$ are the transforms of the corresponding functions obtained by means of the Laplace transform. If $\tilde{y}_0(p)$ is defined, then the initial approximation can be found by applying the Laplace inversion formula: $y_0(x) = \mathcal{L}^{-1}\{\tilde{y}_0(p)\}$.

14.2-4. The Newton–Kantorovich Method

A merit of the iteration methods when applied to Volterra linear equations of the second kind is their unconditional convergence under weak restrictions on the kernel and the right-hand side. When solving nonlinear equations, the applicability domain of the method of simple iterations is smaller, and if the process is still convergent, then, in many cases, the rate of convergence can be very low. An effective method that makes it possible to overcome the indicated complications is the Newton–Kantorovich method. The main objective of this method is the solution of nonlinear integral equations of the second kind with constant limits of integration. Nevertheless, this method is useful in the solution of many problems for the Volterra equations and makes it possible to significantly increase the rate of convergence compared with the successive approximation method.

Let us apply the Newton–Kantorovich method to solve a Volterra equation of the second kind in the Urysohn form

$$y(x) = f(x) + \int_a^x K\big(x, t, y(t)\big)\, dt. \tag{15}$$

We obtain the following iteration process:

$$y_k(x) = y_{k-1}(x) + \varphi_{k-1}(x), \qquad k = 1, 2, \ldots, \tag{16}$$

$$\varphi_{k-1}(x) = \varepsilon_{k-1}(x) + \int_a^x K_y'\big(x, t, y_{k-1}(t)\big)\varphi_{k-1}(t)\, dt, \tag{17}$$

$$\varepsilon_{k-1}(x) = f(x) + \int_a^x K\big(x, t, y_{k-1}(t)\big)\, dt - y_{k-1}(x). \tag{18}$$

The algorithm is based on the solution of the linear integral equation (17) for the correction $\varphi_{k-1}(x)$ with the kernel and right-hand side that vary from step to step. This process has a high rate of convergence, but it is rather complicated because we must solve a new equation at each step of iteration. To simplify the problem, we can replace Eq. (17) by the equation

$$\varphi_{k-1}(x) = \varepsilon_{k-1}(x) + \int_a^x K_y'\big(x, t, y_0(t)\big)\varphi_{k-1}(t)\, dt \tag{19}$$

or by the equation

$$\varphi_{k-1}(x) = \varepsilon_{k-1}(x) + \int_a^x K_y'\big(x, t, y_m(t)\big)\varphi_{k-1}(t)\, dt, \tag{20}$$

whose kernels do not vary. In Eq. (20), m is fixed and satisfies the condition $m < k - 1$.

It is reasonable to apply Eq. (19) with an appropriately chosen initial approximation. Otherwise we can stop at some mth approximation and, beginning with this approximation, apply the simplified equation (20). The iteration process thus obtained is the modified Newton–Kantorovich method. In principle, it converges somewhat slower than the original process (16)–(18); however, it is not so cumbersome in the calculations.

Example 4. Let us apply the Newton–Kantorovich method to solve the equation

$$y(x) = \int_0^x [ty^2(t) - 1]\, dt.$$

The derivative of the integrand with respect to y has the form

$$K_y'\big(t, y(t)\big) = 2ty(t).$$

For the zero approximation we take $y_0(x) \equiv 0$. According to (17) and (18) we obtain $\varphi_0(x) = -x$ and $y_1(x) = -x$. Furthermore, $y_2(x) = y_1(x) + \varphi_1(x)$. By (18) we have

$$\varepsilon_1(x) = \int_0^x [t(-t)^2 - 1]\, dt + x = \tfrac{1}{4}x^4.$$

The equation for the correction has the form

$$\varphi_1(x) = -2 \int_0^x t^2 \varphi_1(t)\, dt + \frac{1}{4} x^4$$

and can be solved by any of the known methods for Volterra linear equations of the second kind. In the case under consideration, we apply the successive approximation method, which leads to the following results (the number of the step is indicated in the superscript):

$$\varphi_1^{(0)} = \frac{1}{4} x^4,$$

$$\varphi_1^{(1)} = \frac{1}{4} x^4 - 2 \int_0^x \frac{1}{4} t^6\, dt = \frac{1}{4} x^4 - \frac{1}{14} x^7,$$

$$\varphi_1^{(2)} = \frac{1}{4} x^4 - 2 \int_0^x t^2 \left(\frac{1}{4} t^4 - \frac{1}{14} t^7 \right) dt = \frac{1}{4} x^4 - \frac{1}{14} x^7 + \frac{1}{70} x^{10}.$$

We restrict ourselves to the second approximation and obtain

$$y_2(x) = -x + \frac{1}{4} x^4 - \frac{1}{14} x^7 + \frac{1}{70} x^{10}$$

and then pass to the third iteration step of the Newton–Kantorovich method:

$$y_3(x) = y_2(x) + \varphi_2(x),$$

$$\varepsilon_2(x) = \frac{1}{160} x^{10} - \frac{1}{1820} x^{13} - \frac{1}{7840} x^{16} + \frac{1}{9340} x^{19} + \frac{1}{107800} x^{22},$$

$$\varphi_2(x) = \varepsilon_2(x) + 2 \int_0^x t \left(-t + \frac{1}{4} t^4 - \frac{1}{14} t^7 + \frac{1}{70} t^{10} \right) \varphi_2(t)\, dt.$$

When solving the last equation, we restrict ourselves to the zero approximation and obtain

$$y_3(x) = -x + \frac{1}{4} x^4 - \frac{1}{14} x^7 + \frac{23}{112} x^{10} - \frac{1}{1820} x^{13} - \frac{1}{7840} x^{16} + \frac{1}{9340} x^{19} + \frac{1}{107800} x^{22}.$$

The application of the successive approximation method to the original equation leads to the same result at the fourth step.

As usual, in the numerical solution the integral is replaced by a quadrature formula. The main difficulty of the implementation of the method in this case is in evaluating the derivative of the kernel. The problem can be simplified if the kernel is given as an analytic expression that can be differentiated in the analytic form. However, if the kernel is given by a table, then the evaluation must be performed numerically.

14.2-5. The Collocation Method

When applied to the solution of a Volterra equation of the first kind in the Urysohn form

$$\int_a^x K\big(x, t, y(t)\big)\, dt = f(x), \qquad a \le x \le b, \tag{21}$$

the *collocation method* is as follows. The interval $[a, b]$ is divided into N parts on each of which the desired solution can be presented by a function of a certain form

$$\tilde{y}(x) = \Phi(x, A_1, \dots, A_m), \tag{22}$$

involving free parameters A_i, $i = 1, \dots, m$.

On the $(k+1)$st part $x_k \le x \le x_{k+1}$, where $k = 0, 1, \dots, N-1$, the solution can be written in the form

$$\int_{x_k}^x K\big(x, t, \tilde{y}(t)\big)\, dt = f(x) - \Psi_k(x), \tag{23}$$

where the integral

$$\Psi_k(x) = \int_a^{x_k} K\big(x, t, \tilde{y}(t)\big)\, dt, \tag{24}$$

can always be calculated for the approximate solution $\tilde{y}(x)$, which is known on the interval $a \le x \le x_k$ and was previously obtained for $k - 1$ parts. The initial value $y(a)$ of the desired solution can be found by an auxiliary method or is assumed to be given.

To solve Eq. (23), representation (22) is applied, and the free parameters A_i $(i = 1, \ldots, m)$ can be defined from the condition that the residuals vanish:

$$\varepsilon(A_i, x_{k,j}) = \int_{x_k}^{x_{k,j}} K\big(x_{k,j}, t, \Phi(t, A_1, \ldots, A_m)\big)\, dt - f(x_{k,j}) - \Psi_k(x_{k,j}), \tag{25}$$

where the $x_{k,j}$ $(j = 1, \ldots, m)$ are the nodes that correspond to the partition of the interval $[x_k, x_{k+1}]$ into m parts (subintervals). System (25) is a system of m equations for $A_1, \ldots, A_m$.

For convenience of the calculations, it is reasonable to present the desired solution on any part as a polynomial

$$\tilde{y}(x) = \sum_{i=1}^{m} A_i \varphi_i(x), \tag{26}$$

where the $\varphi_i(x)$ are linearly independent coordinate functions. For the functions $\varphi_i(x)$, power and trigonometric polynomials are frequently used; for instance, $\varphi_i(x) = x^{i-1}$.

In applications, the concrete form of the functions $\varphi_i(x)$ in formula (26), as well as the form of the functions Φ in (20), can sometimes be given on the basis of physical reasoning or defined by the structure of the solution of a simpler model equation.

14.2-6. The Quadrature Method

To solve a nonlinear Volterra equation, we can apply the method based on the use of quadrature formulas. The procedure of constructing the approximate system of equations is the same as in the linear case (see Subsection 9.10-1).

$1°$. We consider the nonlinear Volterra equation of the second kind in the Urysohn form

$$y(x) - \int_a^x K\big(x, t, y(t)\big)\, dt = f(x) \tag{27}$$

on an interval $a \le x \le b$. Assume that $K\big(x, t, y(t)\big)$ and $f(x)$ are continuous functions.

From Eq. (27) we find that $y(a) = f(a)$. Let us choose a constant integration step h and consider the discrete set of points $x_i = a + h(i - 1)$, where $i = 1, \ldots, n$. For $x = x_i$, Eq. (27) becomes

$$y(x_i) - \int_a^{x_i} K\big(x_i, t, y(t)\big)\, dt = f(x_i). \tag{28}$$

Applying the quadrature formula (see Subsection 8.7-1) to the integral in (28), choosing x_j $(j = 1, \ldots, i)$ to be the nodes in t, and neglecting the truncation error, we arrive at the following system of nonlinear algebraic (or transcendental) equations:

$$y_1 = f_1, \quad y_i - \sum_{j=1}^{i} A_{ij} K_{ij}(y_j) = f_i, \qquad i = 2, \ldots, n, \tag{29}$$

where the A_{ij} are the coefficients of the quadrature formula on the interval $[a, x_i]$, the y_i are the approximate values of the solution $y(x)$ at the nodes x_i, $f_i = f(x_i)$, and $K_{ij}(y_j) = K(x_i, t_j, y_j)$.

Relations (29) can be rewritten as a sequence of recurrent nonlinear equations,

$$y_1 = f_1, \quad y_i - A_{ii} K_{ii}(y_i) = f_i + \sum_{j=1}^{i-1} A_{ij} K_{ij}(y_j), \qquad i = 2, \ldots, n, \tag{30}$$

for the approximate values of the desired solution at the nodes.

$2°$. When applied to the Volterra equation of the second kind in the Hammerstein form

$$y(x) - \int_a^x Q(x,t)\Phi\big(t,y(t)\big)\,dt = f(x), \tag{31}$$

the main relations of the quadrature method have the form ($x_1 = a$)

$$y_1 = f_1, \quad y_i - \sum_{j=1}^{i} A_{ij}Q_{ij}\Phi_j(y_j) = f_i, \qquad i = 2,\ldots,n, \tag{32}$$

where $Q_{ij} = Q(x_i, t_j)$ and $\Phi_j(y_j) = \Phi(t_j, y_j)$. These relations lead to the sequence of nonlinear recurrent equations

$$y_1 = f_1, \quad y_i - A_{ii}Q_{ii}\Phi_i(y_i) = f_i + \sum_{j=1}^{i-1} A_{ij}Q_{ij}\Phi_j(y_j), \qquad i = 2,\ldots,n, \tag{33}$$

whose solutions give approximate values of the desired function.

Example 5. In the solution of the equation

$$y(x) - \int_0^x e^{-(x-t)}y^2(t)\,dt = e^{-x}, \qquad 0 \le x \le 0.1,$$

where $Q(x,t) = e^{-(x-t)}$, $\Phi\big(t,y(t)\big) = y^2(t)$, and $f(x) = e^{-x}$, the approximate expression has the form

$$y(x_i) - \int_0^{x_i} e^{-(x_i-t)}y^2(t)\,dt = e^{-x_i}.$$

On applying the trapezoidal rule to evaluate the integral (with step $h = 0.02$) and finding the solution at the nodes $x_i = 0$, 0.02, 0.04, 0.06, 0.08, 0.1, we obtain, according to (33), the following system of computational relations:

$$y_1 = f_1, \quad y_i - 0.01\,Q_{ii}y_i^2 = f_i + \sum_{j=1}^{i-1} 0.02\,Q_{ij}y_j^2, \qquad i = 2,\ldots,6.$$

Thus, to find an approximate solution, we must solve a quadratic equation for each value y_i, which makes it possible to write out the answer

$$y_i = 50 \pm 50\left[1 - 0.04\left(f_i + \sum_{j=1}^{i-1} 0.02\,Q_{ij}y_j^2\right)\right]^{1/2}, \qquad i = 2,\ldots,6.$$

⊙ References for Section 14.2: M. L. Krasnov, A. I. Kiselev, and G. I. Makarenko (1971), P. P. Zabreyko, A. I. Koshelev, et al. (1975), A. F. Verlan' and V. S. Sizikov (1986).

14.3. Equations With Constant Integration Limits

14.3-1. Nonlinear Equations With Degenerate Kernels

$1°$. Consider a Hammerstein equation of the second kind in the canonical form

$$y(x) = \int_a^b Q(x,t)\Phi\big(t,y(t)\big)\,dt, \tag{1}$$

where $Q(x,t)$ and $\Phi(t,y)$ are given functions and $y(x)$ is the unknown function.

Let the kernel $Q(x,t)$ be degenerate, i.e.,

$$Q(x,t) = \sum_{k=1}^{m} g_k(x)h_k(t). \tag{2}$$

In this case Eq. (1) becomes

$$y(x) = \sum_{k=1}^{m} g_k(x) \int_a^b h_k(t)\Phi\big(t, y(t)\big)\, dt. \tag{3}$$

We write

$$A_k = \int_a^b h_k(t)\Phi\big(t, y(t)\big)\, dt, \qquad k = 1, \ldots, m, \tag{4}$$

where the constants A_k are yet unknown. Then it follows from (3) that

$$y(x) = \sum_{k=1}^{m} A_k g_k(x). \tag{5}$$

On substituting the expression (5) for $y(x)$ into relations (4), we obtain (in the general case) m transcendental equations of the form

$$A_k = \Psi_k(A_1, \ldots, A_m), \qquad k = 1, \ldots, m, \tag{6}$$

which contain m unknown numbers $A_1, \ldots, A_m$.

For the case in which $\Phi(t, y)$ is a polynomial in y, i.e.,

$$\Phi(t, y) = p_0(t) + p_1(t)y + \cdots + p_n(t)y^n, \tag{7}$$

where $p_0(t), \ldots, p_n(t)$ are, for instance, continuous functions of t on the interval $[a, b]$, system (6) becomes a system of nonlinear algebraic equations for $A_1, \ldots, A_m$.

The number of solutions of the integral equation (3) is equal to the number of solutions of system (6). Each solution of system (6) generates a solution (5) of the integral equation.

2°. Consider the Urysohn equation of the second kind with the simplified degenerate kernel of the following form:

$$y(x) + \int_a^b \left\{ \sum_{k=1}^{n} g_k(x)f_k\big(t, y(t)\big) \right\} dt = h(x). \tag{8}$$

Its solution has the form

$$y(x) = h(x) + \sum_{k=1}^{n} \lambda_k g_k(x), \tag{9}$$

where the constants λ_k can be defined by solving the algebraic (or transcendental) system of equations

$$\lambda_m + \int_a^b f_m\left(t, h(t) + \sum_{k=1}^{n} \lambda_k g_k(t) \right) dt = 0, \qquad m = 1, \ldots, n. \tag{10}$$

To different roots of this system, there are different corresponding solutions of the nonlinear integral equation. It may happen that (real) solutions are absent.

A solution of an Urysohn equation of the second kind with degenerate kernel in the general form

$$f\big(x, y(x)\big) + \int_a^b \left\{ \sum_{k=1}^{n} g_k\big(x, y(x)\big) h_k\big(t, y(t)\big) \right\} dt = 0. \tag{11}$$

can be represented in the implicit form

$$f\big(x, y(x)\big) + \sum_{k=1}^{n} \lambda_k g_k\big(x, y(x)\big) = 0, \tag{12}$$

where the parameters λ_k are determined from the system of algebraic (or transcendental) equations:

$$\lambda_k - H_k(\vec{\lambda}) = 0, \qquad k = 1, \dots, n,$$
$$H_k(\vec{\lambda}) = \int_a^b h_k\big(t, y(t)\big)\, dt, \quad \vec{\lambda} = \{\lambda_1, \dots, \lambda_n\}. \tag{13}$$

Into system (13), we must substitute the function $y(x) = y(x, \vec{\lambda})$, which can be obtained by solving Eq. (12).

The number of solutions of the integral equation is defined by the number of solutions obtained from (12) and (13). It can occur that there is no solution.

Example 1. Let us solve the integral equation

$$y(x) = \lambda \int_0^1 xty^3(t)\, dt \tag{14}$$

with parameter λ. We write

$$A = \int_0^1 ty^3(t)\, dt. \tag{15}$$

In this case, it follows from (14) that

$$y(x) = \lambda Ax. \tag{16}$$

On substituting $y(x)$ in the form (16) into relation (15), we obtain

$$A = \int_0^1 t\lambda^3 A^3 t^3\, dt.$$

Hence,

$$A = \tfrac{1}{5}\lambda^3 A^3. \tag{17}$$

For $\lambda > 0$, Eq. (17) has three solutions:

$$A_1 = 0, \quad A_2 = \left(\frac{5}{\lambda^3}\right)^{1/2}, \quad A_3 = -\left(\frac{5}{\lambda^3}\right)^{1/2}.$$

Hence, the integral equation (14) also has three solutions for any $\lambda > 0$:

$$y_1(x) \equiv 0, \quad y_2(x) = \left(\frac{5}{\lambda^3}\right)^{1/2} x, \quad y_3(x) = -\left(\frac{5}{\lambda^3}\right)^{1/2} x.$$

For $\lambda \leq 0$, Eq. (17) has only the trivial solution $y(x) \equiv 0$.

14.3-2. The Method of Integral Transforms

$1°$. Consider the following nonlinear integral equation with quadratic nonlinearity on a semi-axis:

$$\mu y(x) - \lambda \int_0^\infty \frac{1}{t} y\left(\frac{x}{t}\right) y(t)\, dt = f(x). \tag{18}$$

To solve this equation, the Mellin transform can be applied, which, with regard to the convolution theorem (see Section 7.3), leads to a quadratic equation for the transform $\hat{y}(s) = \mathfrak{M}\{y(x)\}$:

$$\mu \hat{y}(s) - \lambda \hat{y}^2(s) = \hat{f}(s).$$

This implies

$$\hat{y}(s) = \frac{\mu \pm \sqrt{\mu^2 - 4\lambda \hat{f}(s)}}{2\lambda}. \tag{19}$$

The inverse transform $y(x) = \mathfrak{M}^{-1}\{\hat{y}(s)\}$ obtained by means of the Mellin inversion formula (if it exists) is a solution of Eq. (18). To different signs in the formula for the images (19), there are two corresponding solutions of the original equation.

2°. By applying the Mellin transform, one can solve nonlinear integral equations of the form

$$y(x) - \lambda \int_0^\infty t^\beta y(xt) y(t) \, dt = f(x). \tag{20}$$

The Mellin transform (see Table 2 in Section 7.3) reduces (20) to the following functional equation for the transform $\hat{y}(s) = \mathfrak{M}\{y(x)\}$:

$$\hat{y}(s) - \lambda \hat{y}(s) \hat{y}(1 - s + \beta) = \hat{f}(s). \tag{21}$$

On replacing s by $1 - s + \beta$ in (21), we obtain the relationship

$$\hat{y}(1 - s + \beta) - \lambda \hat{y}(s) \hat{y}(1 - s + \beta) = \hat{f}(1 - s + \beta). \tag{22}$$

On eliminating the quadratic term from (21) and (22), we obtain

$$\hat{y}(s) - \hat{f}(s) = \hat{y}(1 - s + \beta) - \hat{f}(1 - s + \beta). \tag{23}$$

We express $\hat{y}(1 - s + \beta)$ from this relation and substitute it into (21). We arrive at the quadratic equation

$$\lambda \hat{y}^2(s) - \left[1 + \hat{f}(s) - \hat{f}(1 - s + \beta)\right] \hat{y}(s) + \hat{f}(s) = 0. \tag{24}$$

On solving (24) for $\hat{y}(s)$, by means of the Mellin inversion formula we can find a solution of the original integral equation (20).

14.3-3. The Method of Differentiating for Integral Equations

1°. The equation

$$y(x) + \int_a^b |x - t| f\left(t, y(t)\right) dt = g(x). \tag{25}$$

can be reduced to a nonlinear second-order equation by double differentiation (with subsequent elimination of the integral term by using the original equation):

$$y''_{xx} + 2f(x, y) = g''_{xx}(x). \tag{26}$$

For the boundary conditions for this equation, see Section 6.8 in the first part of the book (Eq. 35).

2°. The equation

$$y(x) + \int_a^b e^{\lambda|x - t|} f\left(t, y(t)\right) dt = g(x). \tag{27}$$

can also be reduced to a nonlinear second-order equation by double differentiation (with subsequent elimination of the integral term by using the original equation):

$$y''_{xx} + 2\lambda f(x, y) - \lambda^2 y = g''_{xx}(x) - \lambda^2 g(x). \tag{28}$$

For the boundary conditions for this equation, see Section 6.8 of the first part of the book (Eq. 36).

3°. The equations

$$y(x) + \int_a^b \sinh\left(\lambda|x - t|\right) f\left(t, y(t)\right) dt = g(x), \tag{29}$$

$$y(x) + \int_a^b \sin\left(\lambda|x - t|\right) f\left(t, y(t)\right) dt = g(x), \tag{30}$$

can also be reduced to second-order ordinary differential equations by means of the differentiation. For these equations, see Section 6.8 of the first part of the book (Eqs. 37 and 38).

14.3-4. The Successive Approximation Method

Consider the nonlinear Urysohn integral equation in the canonical form:

$$y(x) = \int_a^b \mathcal{K}\big(x, t, y(t)\big)\, dt, \qquad a \le x \le b. \tag{31}$$

The iteration process for this equation is constructed by the formula

$$y_k(x) = \int_a^b \mathcal{K}\big(x, t, y_{k-1}(t)\big)\, dt, \qquad k = 1, 2, \dots. \tag{32}$$

If the function $\mathcal{K}(x, t, y)$ is jointly continuous together with the derivative $\mathcal{K}'_y(x, t, y)$ (with respect to the variables x, t, and ρ, $a \le x \le b$, $a \le t \le b$, and $|y| \le \rho$) and if

$$\int_a^b \sup_y |\mathcal{K}(x, t, y)|\, dt \le \rho, \quad \int_a^b \sup_y |\mathcal{K}'_y(x, t, y)|\, dt \le \beta < 1, \tag{33}$$

then for any continuous function $y_0(x)$ of the initial approximation from the domain $\{|y| \le \rho,\ a \le x \le b\}$, the successive approximations (32) converge to a continuous solution $y^*(x)$, which lies in the same domain and is unique in this domain. The rate of convergence is defined by the inequality

$$|y^*(x) - y_k(x)| \le \frac{\beta^k}{1 - \beta} \sup_x |y_1(x) - y_0(x)|, \qquad a \le x \le b, \tag{34}$$

which gives an *a priori* estimate for the error of the kth approximation. The *a posteriori* estimate (which is, in general, more precise) has the form

$$|y^*(x) - y_k(x)| \le \frac{\beta}{1 - \beta} \sup_x |y_k(x) - y_{k-1}(x)|, \qquad a \le x \le b. \tag{35}$$

A solution of an equation of the form (31) with an additional term $f(x)$ on the right-hand side can be constructed in a similar manner.

Example 2. Let us apply the successive approximation method to solve the equation

$$y(x) = \int_0^1 xt y^2(t)\, dt - \tfrac{5}{12}x + 1.$$

The recurrent formula has the form

$$y_k(x) = \int_0^1 xt y_{k-1}^2(t)\, dt - \tfrac{5}{12}x + 1, \qquad k = 1, 2, \dots$$

For the initial approximation we take $y_0(x) = 1$. The calculation yields

$$y_1(x) = 1 + 0.083\, x, \quad y_2(x) = 1 + 0.14\, x, \quad y_3(x) = 1 + 0.18\, x, \quad \dots$$
$$y_8(x) = 1 + 0.27\, x, \quad y_9(x) = 1 + 0.26\, x, \quad y_{10}(x) = 1 + 0.29\, x, \quad \dots$$
$$y_{16}(x) = 1 + 0.318\, x, \quad y_{17}(x) = 1 + 0.321\, x, \quad y_{18}(x) = 1 + 0.323\, x, \quad \dots$$

Thus, the approximations tend to the exact solution $y(x) = 1 + \tfrac{1}{3}x$. We see that the rate of convergence of the iteration process is fairly small.

Note that in Subsection 14.3-5, the equation in question is solved by a more efficient method.

14.3-5. The Newton–Kantorovich Method

The solution of nonlinear integral equations is a complicated problem of computational mathematics, which is related to difficulties of both a principal and computational character. In this connection, methods are developed that are especially designed for solving nonlinear equations, including the Newton–Kantorovich method, which makes it possible to provide and accelerate the convergence of iteration processes in many cases.

We consider this method in connection with the Urysohn equation in the canonical form (31). The iteration process is constructed as follows:

$$y_k(x) = y_{k-1}(x) + \varphi_{k-1}(x), \qquad k = 1, 2, \ldots, \tag{36}$$

$$\varphi_{k-1}(x) = \varepsilon_{k-1}(x) + \int_a^b \mathcal{K}_y'\big(x, t, y_{k-1}(t)\big)\varphi_{k-1}(t)\,dt, \tag{37}$$

$$\varepsilon_{k-1}(x) = \int_a^b \mathcal{K}\big(x, t, y_{k-1}(t)\big)\,dt - y_{k-1}(x). \tag{38}$$

At each step of the algorithm, a linear integral equation for the correction $\varphi_{k-1}(x)$ is solved. Under some conditions, the process (36) has high rate of convergence; however, it is rather complicated, because at each iteration we must obtain the new kernel $\mathcal{K}_y'\big(x, t, y_{k-1}(t)\big)$ for Eqs. (37).

The algorithm can be simplified by using the equation

$$\varphi_{k-1}(x) = \varepsilon_{k-1}(x) + \int_a^b \mathcal{K}_y'\big(x, t, y_0(t)\big)\varphi_{k-1}(t)\,dt \tag{39}$$

instead of (37). If the initial approximation is chosen successfully, then the difference between the integral operators in (37) and (39) is small, and the kernel in (39) remains the same in the course of the solution.

The successive approximation method that consists in the application of formulas (36), (38), and (39) is called the *modified Newton–Kantorovich method*. In principle, its rate of convergence is less than that of the original (nonmodified) method; however, this version of the method is less complicated in calculations, and therefore it is frequently preferable.

Let the function $\mathcal{K}(x, t, y)$ be jointly continuous together with the derivatives $\mathcal{K}_y'(x, t, y)$ and $\mathcal{K}_{yy}''(x, t, y)$ with respect to the variables x, t, y, where $a \le x \le b$ and $a \le t \le b$, and let the following conditions hold:

1°. For the initial approximation $y_0(x)$, the resolvent $\mathcal{R}(x, t)$ of the linear integral equation (37) with the kernel $\mathcal{K}_y'\big(x, t, y_0(t)\big)$ satisfies the condition

$$\int_a^b |\mathcal{R}(x, t)|\,dt \le A < \infty, \qquad a \le x \le b.$$

2°. The residual $\varepsilon_0(x)$ of Eq. (38) for the approximation $y_0(x)$ satisfies the inequality

$$|\varepsilon_0(x)| = \left| \int_a^b \mathcal{K}\big(x, t, y_0(t)\big)\,dt - y_0(x) \right| \le B < \infty.$$

3°. In the domain $|y(x) - y_0(x)| \le 2(1 + A)B$, the following relation holds:

$$\int_a^b \sup_y \big|\mathcal{K}_{yy}''(x, t, y)\big|\,dt \le D < \infty.$$

4°. The constants A, B, and D satisfy the condition

$$H = (1 + A)^2 BD \le \tfrac{1}{2}.$$

In this case, under assumptions $1°$–$4°$, the process (36) converges to a solution $y^*(x)$ of Eq. (31) in the domain

$$|y(x) - y_0(x)| \leq (1 - \sqrt{1 - 2H})H^{-1}(1 - A)B, \qquad a \leq x \leq b.$$

This solution is unique in the domain

$$|y(x) - y_0(x)| \leq 2(1 + A)B, \qquad a \leq x \leq b.$$

The rate of convergence is determined by the estimate

$$|y^*(x) - y_k(x)| \leq 2^{1-k}(2H)^{2^k - 1}(1 - A)B, \qquad a \leq x \leq b.$$

Thus, the above conditions establish the convergence of the algorithm and the existence, the position, and the uniqueness domain of a solution of the nonlinear equation (31). These conditions impose certain restrictions on the initial approximation $y_0(x)$ whose choice is an important independent problem that has no unified approach. As usual, the initial approximation is determined either by more detailed *a priori* analysis of the equation under consideration or by physical reasoning implied by the essence of the problem described by this equation. Under a successful choice of the initial approximation, the Newton–Kantorovich method provides a high rate of convergence of the iteration process to obtain an approximate solution with given accuracy.

Remark. Let the right-hand side of Eq. (31) contain an additional term $f(x)$. Then such an equation can be represented in the form (31), where the integrand is $K\big(x, t, y(t)\big) + (b - a)^{-1}f(x)$.

Example 3. Let us apply the Newton–Kantorovich method to solve the equation

$$y(x) = \int_0^1 xty^2(t)\, dt - \tfrac{5}{12}x + 1. \qquad (40)$$

For the initial approximation we take $y_0(x) = 1$. According to (38), we find the residual

$$\varepsilon_0(x) = \int_0^1 xty_0^2(t)\, dt - \tfrac{5}{12}x + 1 - y_0(x) = x\int_0^1 t\, dt - \tfrac{5}{12}x + 1 - 1 = \tfrac{1}{12}x.$$

The y-derivative of the kernel $\mathcal{K}(x, t, y) = xty^2(t)$, which is needed in the calculations, has the form $\mathcal{K}'_y(x, t, y) = 2xty(t)$. According to (37), we form the following equation for $\varphi_0(x)$:

$$\varphi_0(x) = \tfrac{1}{12}x + 2x\int_0^1 ty_0(t)\varphi_0(t)\, dt,$$

where the kernel turns out to be degenerate, which makes it possible to obtain the solution $\varphi_0(x) = \tfrac{1}{4}x$ directly. Now we define the first approximation to the desired function:

$$y_1(x) = y_0(x) + \varphi_0(x) = 1 + \tfrac{1}{4}x.$$

We continue the iteration process and obtain

$$\varepsilon_1(x) = \int_0^1 xt\big(1 + \tfrac{1}{4}t\big)\, dt + \big(1 - \tfrac{5}{12}x\big) - \big(1 + \tfrac{1}{4}x\big) = \tfrac{1}{64}x.$$

The equation for $\varphi_1(x)$ has the form

$$\varphi_1(x) = \tfrac{1}{64}x + 2x\int_0^1 t\big(1 + \tfrac{1}{4}t\big)\, dt + \big(1 - \tfrac{5}{12}x\big) - \big(1 + \tfrac{1}{4}x\big),$$

and the solution is $\varphi_1(x) = \tfrac{3}{40}x$. Hence, $y_2(x) = 1 + \tfrac{1}{4}x + \tfrac{3}{40}x = 1 + 0.325\,x$. The maximal difference between the exact solution $y(x) = 1 + \tfrac{1}{3}x$ and the approximate solution $y_2(x)$ is observed at $x = 1$ and is less than 0.5%.

This solution is not unique. The other solution can be obtained by taking the function $y_0(x) = 1 + 0.8\,x$ for the initial approximation. In this case we can repeat the above sequence of approximations and obtain the following results (the numerical coefficient of x is rounded):

$$y_1(x) = 1 + 0.82\,x, \quad y_2(x) = 1 + 1.13\,x, \quad y_3(x) = 1 + 0.98\,x, \quad \ldots,$$

and the subsequent approximations tend to the exact solution $y(x) = 1 + x$.

We see that the rate of convergence of the iteration process performed by the Newton–Kantorovich method is significantly higher than that performed by the method of successive approximations (see Example 2 in Subsection 14.4-4).

To estimate the rate of convergence of the performed iteration process, we can compare the above results with the realization of the modified Newton–Kantorovich method. In connection with the latter, for the above versions of the approximations we can obtain

$y_n(x) = 1 + k_n x;$

k_0	k_1	k_2	k_3	k_4	k_5	k_6	k_7	k_8	$\ldots$
0	0.25	0.69	0.60	0.51	0.44	0.38	0.36	0.345	$\ldots$

The iteration process converges to the exact solution $y(x) = 1 + \frac{1}{3}x$.

We see that the modified Newton–Kantorovich method is less efficient than the Newton–Kantorovich method, but more efficient than the method of successive approximations (see Example 2 in Subsection 14.4-4).

14.3-6. The Quadrature Method

To solve an arbitrary nonlinear equation, we can apply the method based on the application of quadrature formulas. The procedure of composing the approximating system of equations is the same as in the linear case (see Subsection 11.18-1). We consider this procedure for an example of the Urysohn equation of the second kind:

$$y(x) - \int_a^b K\big(x, t, y(t)\big)\, dt = f(x), \qquad a \le x \le b. \tag{41}$$

We set $x = x_i$ $(i = 1, \ldots, n)$. Then we obtain

$$y(x_i) - \int_a^b K\big(x_i, t, y(t)\big)\, dt = f(x_i). \qquad i = 1, \ldots, n, \tag{42}$$

On applying the quadrature formula from Subsection 11.18-1 and neglecting the approximation error, we transform relations (42) into the system of nonlinear equations

$$y_i - \sum_{j=1}^n A_j K_{ij}(y_j) = f_i, \qquad i = 1, \ldots, n, \tag{43}$$

for the approximate values y_i of the solution $y(x)$ at the nodes $x_1, \ldots, x_n$, where $f_i = f(x_i)$ and $K_{ij}(y_j) = K(x_i, t_j, y_j)$, and A_j are the coefficients of the quadrature formula.

The solution of the nonlinear system (43) gives values $y_1, \ldots, y_n$ for which by interpolation we find an approximate solution of the integral equation (41) on the entire interval $[a, b]$. For the analytic expression of an approximate solution, we can take the function

$$\tilde{y}(x) = f(x) + \sum_{j=1}^n A_j K(x, x_j, y_j). \tag{44}$$

14.3-7. The Tikhonov Regularization Method

In connection with the nonlinear Urysohn integral equation of the first kind

$$\int_a^b K\big(x, t, y(t)\big)\, dt = f(x), \qquad c \le x \le d, \tag{45}$$

where $f(x) \in L_2(c, d)$ and $y(t) \in L_2(a, b)$, the Tikhonov regularization method leads to a regularized nonlinear integral equation in the form

$$\alpha y_\alpha(x) + \int_a^b M\big(t, x, y_\alpha(t), y_\alpha(x)\big) \, dt = F\big(x, y_\alpha(x)\big), \qquad a \leq x \leq b, \tag{46}$$

$$M\big(t, x, y(t), y(x)\big) = \int_c^d K\big(s, t, y(t)\big) K_y'\big(s, x, y(x)\big) \, ds, \tag{47}$$

$$F\big(x, y(x)\big) = \int_c^d K_y'\big(t, x, y(x)\big) f(t) \, dt, \tag{48}$$

where α is a regularization parameter.

For instance, by applying the quadrature method on the basis of the trapezoidal rule, we can reduce Eq. (46) to a system of nonlinear algebraic equations. An approximate solution of (45) is constructed by the principle described above for linear equations (see Section 10.8).

⊙ References for Section 14.3: N. S. Smirnov (1951), P. P. Zabreyko, A. I. Koshelev, et al. (1975), F. G. Tricomi (1985), A. F. Verlan' and V. S. Sizikov (1986).

Supplements

Supplement 1

Elementary Functions and Their Properties

Throughout Supplement 1 it is assumed that n is a positive integer, unless otherwise specified.

1.1. Trigonometric Functions

▶ **Simplest relations**

$$\sin^2 x + \cos^2 x = 1, \quad \sin(-x) = -\sin x, \quad \cos(-x) = \cos x,$$

$$\tan x = \frac{\sin x}{\cos x}, \quad \cot x = \frac{\cos x}{\sin x}, \quad 1 + \tan^2 x = \frac{1}{\cos^2 x}, \quad 1 + \cot^2 x = \frac{1}{\sin^2 x},$$

$$\tan x \cot x = 1, \quad \tan(-x) = -\tan x, \quad \cot(-x) = -\cot x.$$

▶ **Relations between trigonometric functions of single argument**

$$\sin x = \pm\sqrt{1 - \cos^2 x} = \pm\frac{\tan x}{\sqrt{1 + \tan^2 x}} = \pm\frac{1}{\sqrt{1 + \cot^2 x}},$$

$$\cos x = \pm\sqrt{1 - \sin^2 x} = \pm\frac{1}{\sqrt{1 + \tan^2 x}} = \pm\frac{\cot x}{\sqrt{1 + \cot^2 x}},$$

$$\tan x = \pm\frac{\sin x}{\sqrt{1 - \sin^2 x}} = \pm\frac{\sqrt{1 - \cos^2 x}}{\cos x} = \frac{1}{\cot x},$$

$$\cot x = \pm\frac{\sqrt{1 - \sin^2 x}}{\sin x} = \pm\frac{\cos x}{\sqrt{1 - \cos^2 x}} = \frac{1}{\tan x}.$$

▶ **Reduction formulas**

$$\sin(x \pm n\pi) = (-1)^n \sin x, \qquad \cos(x \pm n\pi) = (-1)^n \cos x,$$

$$\sin\left(x \pm \frac{2n+1}{2}\pi\right) = \pm(-1)^n \cos x, \qquad \cos\left(x \pm \frac{2n+1}{2}\pi\right) = \mp(-1)^n \sin x,$$

$$\tan(x \pm n\pi) = \tan x, \qquad \cot(x \pm n\pi) = \cot x,$$

$$\tan\left(x \pm \frac{2n+1}{2}\pi\right) = -\cot x, \qquad \cot\left(x \pm \frac{2n+1}{2}\pi\right) = -\tan x,$$

$$\sin\left(x \pm \frac{\pi}{4}\right) = \frac{\sqrt{2}}{2}(\sin x \pm \cos x), \qquad \cos\left(x \pm \frac{\pi}{4}\right) = \frac{\sqrt{2}}{2}(\cos x \mp \sin x),$$

$$\tan\left(x \pm \frac{\pi}{4}\right) = \frac{\tan x \pm 1}{1 \mp \tan x}, \qquad \cot\left(x \pm \frac{\pi}{4}\right) = \frac{\cot x \mp 1}{1 \pm \cot x}.$$

▶ **Addition formulas**

$$\sin(x \pm y) = \sin x \cos y \pm \cos x \sin y, \quad \cos(x \pm y) = \cos x \cos y \mp \sin x \sin y,$$

$$\tan(x \pm y) = \frac{\tan x \pm \tan y}{1 \mp \tan x \tan y}, \quad \cot(x \pm y) = \frac{1 \mp \tan x \tan y}{\tan x \pm \tan y}.$$

▶ **Addition and subtraction of trigonometric functions**

$$\sin x + \sin y = 2 \sin\left(\frac{x+y}{2}\right) \cos\left(\frac{x-y}{2}\right),$$

$$\sin x - \sin y = 2 \sin\left(\frac{x-y}{2}\right) \cos\left(\frac{x+y}{2}\right),$$

$$\cos x + \cos y = 2 \cos\left(\frac{x+y}{2}\right) \cos\left(\frac{x-y}{2}\right),$$

$$\cos x - \cos y = -2 \sin\left(\frac{x+y}{2}\right) \sin\left(\frac{x-y}{2}\right),$$

$$\sin^2 x - \sin^2 y = \cos^2 y - \cos^2 x = \sin(x+y) \sin(x-y),$$

$$\sin^2 x - \cos^2 y = -\cos(x+y) \cos(x-y),$$

$$\tan x \pm \tan y = \frac{\sin(x \pm y)}{\cos x \cos y}, \quad \cot x \pm \cot y = \frac{\sin(y \pm x)}{\sin x \sin y},$$

$$a \cos x + b \sin x = r \sin(x + \varphi) = r \cos(x - \psi).$$

Here $r = \sqrt{a^2 + b^2}$, $\sin \varphi = a/r$, $\cos \varphi = b/r$, $\sin \psi = b/r$, and $\cos \psi = a/r$.

▶ **Products of trigonometric functions**

$$\sin x \sin y = \tfrac{1}{2}[\cos(x-y) - \cos(x+y)],$$

$$\cos x \cos y = \tfrac{1}{2}[\cos(x-y) + \cos(x+y)],$$

$$\sin x \cos y = \tfrac{1}{2}[\sin(x-y) + \sin(x+y)].$$

▶ **Powers of trigonometric functions**

$$\cos^2 x = \tfrac{1}{2} \cos 2x + \tfrac{1}{2}, \qquad\qquad \sin^2 x = -\tfrac{1}{2} \cos 2x + \tfrac{1}{2},$$

$$\cos^3 x = \tfrac{1}{4} \cos 3x + \tfrac{3}{4} \cos x, \qquad\qquad \sin^3 x = -\tfrac{1}{4} \sin 3x + \tfrac{3}{4} \sin x,$$

$$\cos^4 x = \tfrac{1}{8} \cos 4x + \tfrac{1}{2} \cos 2x + \tfrac{3}{8}, \qquad \sin^4 x = \tfrac{1}{8} \cos 4x - \tfrac{1}{2} \cos 2x + \tfrac{3}{8},$$

$$\cos^5 x = \tfrac{1}{16} \cos 5x + \tfrac{5}{16} \cos 3x + \tfrac{5}{8} \cos x, \quad \sin^5 x = \tfrac{1}{16} \sin 5x - \tfrac{5}{16} \sin 3x + \tfrac{5}{8} \sin x,$$

$$\cos^{2n} x = \frac{1}{2^{2n-1}} \sum_{k=0}^{n-1} C_{2n}^k \cos[2(n-k)x] + \frac{1}{2^{2n}} C_{2n}^n,$$

$$\cos^{2n+1} x = \frac{1}{2^{2n}} \sum_{k=0}^{n} C_{2n+1}^k \cos[(2n-2k+1)x],$$

$$\sin^{2n} x = \frac{1}{2^{2n-1}} \sum_{k=0}^{n-1} (-1)^{n-k} C_{2n}^k \cos[2(n-k)x] + \frac{1}{2^{2n}} C_{2n}^n,$$

$$\sin^{2n+1} x = \frac{1}{2^{2n}} \sum_{k=0}^{n} (-1)^{n-k} C_{2n+1}^k \sin[(2n-2k+1)x].$$

Here $C_m^k = \dfrac{m!}{k!\,(m-k)!}$ are binomial coefficients ($0! = 1$).

▶ **Trigonometric functions of multiple arguments**

$$\cos 2x = 2\cos^2 x - 1 = 1 - 2\sin^2 x, \qquad \sin 2x = 2\sin x \cos x,$$

$$\cos 3x = -3\cos x + 4\cos^3 x, \qquad \sin 3x = 3\sin x - 4\sin^3 x,$$

$$\cos 4x = 1 - 8\cos^2 x + 8\cos^4 x, \qquad \sin 4x = 4\cos x\,(\sin x - 2\sin^3 x),$$

$$\cos 5x = 5\cos x - 20\cos^3 x + 16\cos^5 x, \qquad \sin 5x = 5\sin x - 20\sin^3 x + 16\sin^5 x,$$

$$\cos(2nx) = 1 + \sum_{k=1}^{n}(-1)^k \frac{n^2(n^2-1)\dots[n^2-(k-1)^2]}{(2k)!}4^k \sin^{2k} x,$$

$$\cos[(2n+1)x] = \cos x \left\{ 1 + \sum_{k=1}^{n}(-1)^k \frac{[(2n+1)^2-1][(2n+1)^2-3^2]\dots[(2n+1)^2-(2k-1)^2]}{(2k)!} \sin^{2k} x \right\},$$

$$\sin(2nx) = 2n\cos x \left[\sin x + \sum_{k=1}^{n}(-4)^k \frac{(n^2-1)(n^2-2^2)\dots(n^2-k^2)}{(2k-1)!} \sin^{2k-1} x \right],$$

$$\sin[(2n+1)x] = (2n+1)\left\{ \sin x + \sum_{k=1}^{n}(-1)^k \frac{[(2n+1)^2-1][(2n+1)^2-3^2]\dots[(2n+1)^2-(2k-1)^2]}{(2k+1)!} \sin^{2k+1} x \right\},$$

$$\tan 2x = \frac{2\tan x}{1-\tan^2 x}, \qquad \tan 3x = \frac{3\tan x - \tan^3 x}{1-3\tan^2 x}, \qquad \tan 4x = \frac{4\tan x - 4\tan^3 x}{1-6\tan^2 x + \tan^4 x}.$$

▶ **Trigonometric functions of half argument**

$$\sin^2 \frac{x}{2} = \frac{1-\cos x}{2}, \qquad \cos^2 \frac{x}{2} = \frac{1+\cos x}{2},$$

$$\tan \frac{x}{2} = \frac{\sin x}{1+\cos x} = \frac{1-\cos x}{\sin x}, \qquad \cot \frac{x}{2} = \frac{\sin x}{1-\cos x} = \frac{1+\cos x}{\sin x},$$

$$\sin x = \frac{2\tan \frac{x}{2}}{1+\tan^2 \frac{x}{2}}, \qquad \cos x = \frac{1-\tan^2 \frac{x}{2}}{1+\tan^2 \frac{x}{2}}, \qquad \tan x = \frac{2\tan \frac{x}{2}}{1-\tan^2 \frac{x}{2}}.$$

▶ **Euler and de Moivre formulas. Relationship with hyperbolic functions**

$$e^{y+ix} = e^y(\cos x + i\sin x), \qquad (\cos x + i\sin x)^n = \cos(nx) + i\sin(nx), \qquad i^2 = -1,$$

$$\sin(ix) = i\sinh x, \qquad \cos(ix) = \cosh x, \qquad \tan(ix) = i\tanh x, \qquad \cot(ix) = -i\coth x.$$

▶ **Differentiation formulas**

$$\frac{d\sin x}{dx} = \cos x, \qquad \frac{d\cos x}{dx} = -\sin x, \qquad \frac{d\tan x}{dx} = \frac{1}{\cos^2 x}, \qquad \frac{d\cot x}{dx} = -\frac{1}{\sin^2 x}.$$

▶ **Expansion into power series**

$$\cos x = 1 - \frac{x^2}{2!} + \frac{x^4}{4!} - \frac{x^6}{6!} + \cdots \qquad (|x| < \infty),$$

$$\sin x = x - \frac{x^3}{3!} + \frac{x^5}{5!} - \frac{x^7}{7!} + \cdots \qquad (|x| < \infty),$$

$$\tan x = x + \frac{x^3}{3} + \frac{2x^5}{15} + \frac{17x^7}{315} + \cdots \qquad (|x| < \pi/2),$$

$$\cot x = \frac{1}{x} - \frac{x}{3} - \frac{x^3}{45} - \frac{2x^5}{945} - \cdots \qquad (|x| < \pi).$$

1.2. Hyperbolic Functions

▶ **Definitions**

$$\sinh x = \frac{e^x - e^{-x}}{2}, \quad \cosh x = \frac{e^x + e^{-x}}{2}, \quad \tanh x = \frac{e^x - e^{-x}}{e^x + e^{-x}}, \quad \coth x = \frac{e^x + e^{-x}}{e^x - e^{-x}}.$$

▶ **Simplest relations**

$$\cosh^2 x - \sinh^2 x = 1, \qquad \tanh x \coth x = 1,$$
$$\sinh(-x) = -\sinh x, \qquad \cosh(-x) = \cosh x,$$
$$\tanh x = \frac{\sinh x}{\cosh x}, \qquad \coth x = \frac{\cosh x}{\sinh x},$$
$$\tanh(-x) = -\tanh x, \qquad \coth(-x) = -\coth x,$$
$$1 - \tanh^2 x = \frac{1}{\cosh^2 x}, \qquad \coth^2 x - 1 = \frac{1}{\sinh^2 x}.$$

▶ **Relations between hyperbolic functions of single argument** ($x \geq 0$)

$$\sinh x = \sqrt{\cosh^2 x - 1} = \frac{\tanh x}{\sqrt{1 - \tanh^2 x}} = \frac{1}{\sqrt{\coth^2 x - 1}},$$
$$\cosh x = \sqrt{\sinh^2 x + 1} = \frac{1}{\sqrt{1 - \tanh^2 x}} = \frac{\coth x}{\sqrt{\coth^2 x - 1}},$$
$$\tanh x = \frac{\sinh x}{\sqrt{\sinh^2 x + 1}} = \frac{\sqrt{\cosh^2 x - 1}}{\cosh x} = \frac{1}{\coth x},$$
$$\coth x = \frac{\sqrt{\sinh^2 x + 1}}{\sinh x} = \frac{\cosh x}{\sqrt{\cosh^2 x - 1}} = \frac{1}{\tanh x}.$$

▶ **Addition formulas**

$$\sinh(x \pm y) = \sinh x \cosh y \pm \sinh y \cosh x, \qquad \cosh(x \pm y) = \cosh x \cosh y \pm \sinh x \sinh y,$$
$$\tanh(x \pm y) = \frac{\tanh x \pm \tanh y}{1 \pm \tanh x \tanh y}, \qquad \coth(x \pm y) = \frac{\coth x \coth y \pm 1}{\coth y \pm \coth x}.$$

▶ **Addition and subtraction of hyperbolic functions**

$$\sinh x \pm \sinh y = 2 \sinh\left(\frac{x \pm y}{2}\right) \cosh\left(\frac{x \mp y}{2}\right),$$
$$\cosh x + \cosh y = 2 \cosh\left(\frac{x + y}{2}\right) \cosh\left(\frac{x - y}{2}\right),$$
$$\cosh x - \cosh y = 2 \sinh\left(\frac{x + y}{2}\right) \sinh\left(\frac{x - y}{2}\right),$$
$$\sinh^2 x - \sinh^2 y = \cosh^2 x - \cosh^2 y = \sinh(x + y) \sinh(x - y),$$
$$\sinh^2 x + \cosh^2 y = \cosh(x + y) \cosh(x - y),$$
$$\tanh x \pm \tanh y = \frac{\sinh(x \pm y)}{\cosh x \cosh y}, \qquad \coth x \pm \coth y = \pm\frac{\sinh(x \pm y)}{\sinh x \sinh y}.$$

▶ **Products of hyperbolic functions**

$$\sinh x \sinh y = \tfrac{1}{2}[\cosh(x + y) - \cosh(x - y)],$$
$$\cosh x \cosh y = \tfrac{1}{2}[\cosh(x + y) + \cosh(x - y)],$$
$$\sinh x \cosh y = \tfrac{1}{2}[\sinh(x + y) + \sinh(x - y)].$$

▶ **Powers of hyperbolic functions**

$\cosh^2 x = \frac{1}{2}\cosh 2x + \frac{1}{2}$,

$\sinh^2 x = \frac{1}{2}\cosh 2x - \frac{1}{2}$,

$\cosh^3 x = \frac{1}{4}\cosh 3x + \frac{3}{4}\cosh x$,

$\sinh^3 x = \frac{1}{4}\sinh 3x - \frac{3}{4}\sinh x$,

$\cosh^4 x = \frac{1}{8}\cosh 4x + \frac{1}{2}\cosh 2x + \frac{3}{8}$,

$\sinh^4 x = \frac{1}{8}\cosh 4x - \frac{1}{2}\cosh 2x + \frac{3}{8}$,

$\cosh^5 x = \frac{1}{16}\cosh 5x + \frac{5}{16}\cosh 3x + \frac{5}{8}\cosh x$,

$\sinh^5 x = \frac{1}{16}\sinh 5x - \frac{5}{16}\sinh 3x + \frac{5}{8}\sinh x$,

$$\cosh^{2n} x = \frac{1}{2^{2n-1}}\sum_{k=0}^{n-1} C_{2n}^k \cosh[2(n-k)x] + \frac{1}{2^{2n}}C_{2n}^n,$$

$$\cosh^{2n+1} x = \frac{1}{2^{2n}}\sum_{k=0}^{n} C_{2n+1}^k \cosh[(2n-2k+1)x],$$

$$\sinh^{2n} x = \frac{1}{2^{2n-1}}\sum_{k=0}^{n-1}(-1)^k C_{2n}^k \cosh[2(n-k)x] + \frac{(-1)^n}{2^{2n}}C_{2n}^n,$$

$$\sinh^{2n+1} x = \frac{1}{2^{2n}}\sum_{k=0}^{n}(-1)^k C_{2n+1}^k \sinh[(2n-2k+1)x].$$

Here C_m^k are binomial coefficients.

▶ **Hyperbolic functions of multiple arguments**

$\cosh 2x = 2\cosh^2 x - 1$,

$\sinh 2x = 2\sinh x \cosh x$,

$\cosh 3x = -3\cosh x + 4\cosh^3 x$,

$\sinh 3x = 3\sinh x + 4\sinh^3 x$,

$\cosh 4x = 1 - 8\cosh^2 x + 8\cosh^4 x$,

$\sinh 4x = 4\cosh x(\sinh x + 2\sinh^3 x)$,

$\cosh 5x = 5\cosh x - 20\cosh^3 x + 16\cosh^5 x$,

$\sinh 5x = 5\sinh x + 20\sinh^3 x + 16\sinh^5 x$.

$$\cosh(nx) = 2^{n-1}\cosh^n x + \frac{n}{2}\sum_{k=0}^{[n/2]}\frac{(-1)^{k+1}}{k+1}C_{n-k-2}^{k-2}2^{n-2k-2}(\cosh x)^{n-2k-2},$$

$$\sinh(nx) = \sinh x\sum_{k=0}^{[(n-1)/2]} 2^{n-k-1}C_{n-k-1}^k(\cosh x)^{n-2k-1}.$$

Here C_m^k are binomial coefficients and $[A]$ stands for the integer part of a number A.

▶ **Relationship with trigonometric functions**

$\sinh(ix) = i\sin x$, $\quad \cosh(ix) = \cos x$, $\quad \tanh(ix) = i\tan x$, $\quad \coth(ix) = -i\cot x$, $\qquad i^2 = -1$.

▶ **Differentiation formulas**

$$\frac{d\sinh x}{dx} = \cosh x, \qquad \frac{d\cosh x}{dx} = \sinh x, \qquad \frac{d\tanh x}{dx} = \frac{1}{\cosh^2 x}, \qquad \frac{d\coth x}{dx} = -\frac{1}{\sinh^2 x}.$$

▶ **Expansion into power series**

$$\cosh x = 1 + \frac{x^2}{2!} + \frac{x^4}{4!} + \frac{x^6}{6!} + \cdots \qquad (|x| < \infty),$$

$$\sinh x = x + \frac{x^3}{3!} + \frac{x^5}{5!} + \frac{x^7}{7!} + \cdots \qquad (|x| < \infty),$$

$$\tanh x = x - \frac{x^3}{3} + \frac{2x^5}{15} - \frac{17x^7}{315} + \cdots \qquad (|x| < \pi/2),$$

$$\coth x = \frac{1}{x} + \frac{x}{3} - \frac{x^3}{45} + \frac{2x^5}{945} - \cdots \qquad (|x| < \pi).$$

1.3. Inverse Trigonometric Functions

▶ **Definitions and some properties**

$$\sin(\arcsin x) = x, \qquad \cos(\arccos x) = x,$$
$$\tan(\arctan x) = x, \qquad \cot(\text{arccot}\, x) = x.$$

Principal values of inverse trigonometric functions are defined by the inequalities

$$-\tfrac{\pi}{2} \le \arcsin x \le \tfrac{\pi}{2}, \quad 0 \le \arccos x \le \pi \quad (-1 \le x \le 1),$$
$$-\tfrac{\pi}{2} < \arctan x < \tfrac{\pi}{2}, \quad 0 < \text{arccot}\, x < \pi \quad (-\infty < x < \infty).$$

▶ **Simplest formulas**

$$\arcsin(-x) = -\arcsin x, \qquad \arccos(-x) = \pi - \arccos x,$$
$$\arctan(-x) = -\arctan x, \qquad \text{arccot}(-x) = \pi - \text{arccot}\, x,$$

$$\arcsin(\sin x) = \begin{cases} x - 2n\pi & \text{if } 2n\pi - \tfrac{\pi}{2} \le x \le 2n\pi + \tfrac{\pi}{2}, \\ -x + 2(n+1)\pi & \text{if } (2n+1)\pi - \tfrac{\pi}{2} \le x \le 2(n+1)\pi + \tfrac{\pi}{2}, \end{cases}$$

$$\arccos(\cos x) = \begin{cases} x - 2n\pi & \text{if } 2n\pi \le x \le (2n+1)\pi, \\ -x + 2(n+1)\pi & \text{if } (2n+1)\pi \le x \le 2(n+1)\pi, \end{cases}$$

$$\arctan(\tan x) = x - n\pi \quad \text{if } n\pi - \tfrac{\pi}{2} < x < n\pi + \tfrac{\pi}{2},$$
$$\text{arccot}(\cot x) = x - n\pi \quad \text{if } n\pi < x < (n+1)\pi.$$

▶ **Relations between inverse trigonometric functions**

$$\arcsin x + \arccos x = \tfrac{\pi}{2}, \qquad \arctan x + \text{arccot}\, x = \tfrac{\pi}{2};$$

$$\arcsin x = \begin{cases} \arccos \sqrt{1-x^2} & \text{if } 0 \le x \le 1, \\ -\arccos \sqrt{1-x^2} & \text{if } -1 \le x \le 0, \\ \arctan \dfrac{x}{\sqrt{1-x^2}} & \text{if } -1 < x < 1, \\ \text{arccot} \dfrac{\sqrt{1-x^2}}{x} - \pi & \text{if } -1 \le x < 0; \end{cases} \qquad \arccos x = \begin{cases} \arcsin \sqrt{1-x^2} & \text{if } 0 \le x \le 1, \\ \pi - \arcsin \sqrt{1-x^2} & \text{if } -1 \le x \le 0, \\ \arctan \dfrac{\sqrt{1-x^2}}{x} & \text{if } 0 < x \le 1, \\ \text{arccot} \dfrac{x}{\sqrt{1-x^2}} & \text{if } -1 < x < 1; \end{cases}$$

$$\arctan x = \begin{cases} \arcsin \dfrac{x}{\sqrt{1+x^2}} & \text{for any } x, \\ \arccos \dfrac{1}{\sqrt{1+x^2}} & \text{if } x \ge 0, \\ -\arccos \dfrac{1}{\sqrt{1+x^2}} & \text{if } x \le 0, \\ \text{arccot} \dfrac{1}{x} & \text{if } x > 0; \end{cases} \qquad \text{arccot}\, x = \begin{cases} \arcsin \dfrac{1}{\sqrt{1+x^2}} & \text{if } x > 0, \\ \pi - \arcsin \dfrac{1}{\sqrt{1+x^2}} & \text{if } x < 0, \\ \arctan \dfrac{1}{x} & \text{if } x > 0, \\ \pi + \arctan \dfrac{1}{x} & \text{if } x < 0. \end{cases}$$

▶ **Addition and subtraction of inverse trigonometric functions**

$$\arcsin x + \arcsin y = \arcsin\left(x\sqrt{1-y^2} + y\sqrt{1-x^2}\right) \qquad \text{for } x^2 + y^2 \le 1,$$
$$\arccos x \pm \arccos y = \pm \arccos\left[xy \mp \sqrt{(1-x^2)(1-y^2)}\right] \quad \text{for } x \pm y \ge 0,$$
$$\arctan x + \arctan y = \arctan \frac{x+y}{1-xy} \qquad \text{for } xy < 1,$$
$$\arctan x - \arctan y = \arctan \frac{x-y}{1+xy} \qquad \text{for } xy > -1.$$

▶ **Differentiation formulas**

$$\frac{d}{dx}\arcsin x = \frac{1}{\sqrt{1-x^2}}, \quad \frac{d}{dx}\arccos x = -\frac{1}{\sqrt{1-x^2}}, \quad \frac{d}{dx}\arctan x = \frac{1}{1+x^2}, \quad \frac{d}{dx}\operatorname{arccot} x = -\frac{1}{1+x^2}.$$

▶ **Expansion into power series**

$$\arcsin x = x + \frac{1}{2}\frac{x^3}{3} + \frac{1\cdot 3}{2\cdot 4}\frac{x^5}{5} + \frac{1\cdot 3\cdot 5}{2\cdot 4\cdot 6}\frac{x^7}{7} + \cdots \quad (|x| < 1),$$

$$\arctan x = x - \frac{x^3}{3} + \frac{x^5}{5} - \frac{x^7}{7} + \cdots \quad (|x| \le 1),$$

$$\arctan x = \frac{\pi}{2} - \frac{1}{x} + \frac{1}{3x^3} - \frac{1}{5x^5} + \cdots \quad (|x| > 1).$$

The expansions for $\arccos x$ and $\operatorname{arccot} x$ can be obtained with the aid of the formulas $\arccos x = \frac{\pi}{2} - \arcsin x$ and $\operatorname{arccot} x = \frac{\pi}{2} - \arctan x$.

1.4. Inverse Hyperbolic Functions

▶ **Relationship with logarithmic function**

$$\operatorname{Arsinh} x = \ln\left(x + \sqrt{x^2 + 1}\right), \qquad \operatorname{Artanh} x = \frac{1}{2}\ln\frac{1+x}{1-x},$$

$$\operatorname{Arcosh} x = \ln\left(x + \sqrt{x^2 - 1}\right), \qquad \operatorname{Arcoth} x = \frac{1}{2}\ln\frac{1+x}{x-1};$$

$$\operatorname{Arsinh}(-x) = -\operatorname{Arsinh} x, \qquad \operatorname{Artanh}(-x) = -\operatorname{Artanh} x,$$

$$\operatorname{Arcosh}(-x) = \operatorname{Arcosh} x, \qquad \operatorname{Arcoth}(-x) = -\operatorname{Arcoth} x.$$

▶ **Relations between inverse hyperbolic functions**

$$\operatorname{Arsinh} x = \operatorname{Arcosh}\sqrt{x^2 + 1} = \operatorname{Artanh}\frac{x}{\sqrt{x^2 + 1}},$$

$$\operatorname{Arcosh} x = \operatorname{Arsinh}\sqrt{x^2 - 1} = \operatorname{Artanh}\frac{\sqrt{x^2 - 1}}{x},$$

$$\operatorname{Artanh} x = \operatorname{Arsinh}\frac{x}{\sqrt{1 - x^2}} = \operatorname{Arcosh}\frac{1}{\sqrt{1 - x^2}} = \operatorname{Arcoth}\frac{1}{x}.$$

▶ **Addition and subtraction of inverse hyperbolic functions**

$$\operatorname{Arsinh} x \pm \operatorname{Arsinh} y = \operatorname{Arsinh}\left(x\sqrt{1 + y^2} \pm y\sqrt{1 + x^2}\right),$$

$$\operatorname{Arcosh} x \pm \operatorname{Arcosh} y = \operatorname{Arcosh}\left[xy \pm \sqrt{(x^2 - 1)(y^2 - 1)}\right],$$

$$\operatorname{Arsinh} x \pm \operatorname{Arcosh} y = \operatorname{Arsinh}\left[xy \pm \sqrt{(x^2 + 1)(y^2 - 1)}\right],$$

$$\operatorname{Artanh} x \pm \operatorname{Artanh} y = \operatorname{Artanh}\frac{x \pm y}{1 \pm xy}, \quad \operatorname{Artanh} x \pm \operatorname{Arcoth} y = \operatorname{Artanh}\frac{xy \pm 1}{y \pm x}.$$

▶ **Differentiation formulas**

$$\frac{d}{dx}\operatorname{Arsinh} x = \frac{1}{\sqrt{x^2 + 1}}, \qquad \frac{d}{dx}\operatorname{Arcosh} x = \frac{1}{\sqrt{x^2 - 1}},$$

$$\frac{d}{dx}\operatorname{Artanh} x = \frac{1}{1 - x^2} \quad (x^2 < 1), \qquad \frac{d}{dx}\operatorname{Arcoth} x = \frac{1}{1 - x^2} \quad (x^2 > 1).$$

▶ **Expansion into power series**

$$\text{Arsinh}\, x = x - \frac{1}{2}\frac{x^3}{3} + \frac{1 \cdot 3}{2 \cdot 4}\frac{x^5}{5} - \frac{1 \cdot 3 \cdot 5}{2 \cdot 4 \cdot 6}\frac{x^7}{7} + \cdots \quad (|x| < 1),$$

$$\text{Artanh}\, x = x + \frac{x^3}{3} + \frac{x^5}{5} + \frac{x^7}{7} + \cdots \qquad\qquad (|x| < 1).$$

⊙ References for Supplement 1: H. B. Dwight (1961), M. Arbramowitz and I. A. Stegun (1964), G. A. Korn and T. M. Korn (1968), W. H. Beyer (1991).

Supplement 2

Tables of Indefinite Integrals

2.1. Integrals Containing Rational Functions

▶ **Integrals containing $a + bx$.** *

1. $\displaystyle \int \frac{dx}{a + bx} = \frac{1}{b} \ln|a + bx|.$

2. $\displaystyle \int (a + bx)^n dx = \frac{(a + bx)^{n+1}}{b(n + 1)}, \quad n \neq -1.$

3. $\displaystyle \int \frac{x\, dx}{a + bx} = \frac{1}{b^2} \left(a + bx - a \ln|a + bx| \right).$

4. $\displaystyle \int \frac{x^2\, dx}{a + bx} = \frac{1}{b^3} \left[\frac{1}{2}(a + bx)^2 - 2a(a + bx) + a^2 \ln|a + bx| \right].$

5. $\displaystyle \int \frac{dx}{x(a + bx)} = -\frac{1}{a} \ln \left| \frac{a + bx}{x} \right|.$

6. $\displaystyle \int \frac{dx}{x^2(a + bx)} = -\frac{1}{ax} + \frac{b}{a^2} \ln \left| \frac{a + bx}{x} \right|.$

7. $\displaystyle \int \frac{x\, dx}{(a + bx)^2} = \frac{1}{b^2} \left(\ln|a + bx| + \frac{a}{a + bx} \right).$

8. $\displaystyle \int \frac{x^2\, dx}{(a + bx)^2} = \frac{1}{b^3} \left(a + bx - 2a \ln|a + bx| - \frac{a^2}{a + bx} \right).$

9. $\displaystyle \int \frac{dx}{x(a + bx)^2} = \frac{1}{a(a + bx)} - \frac{1}{a^2} \ln \left| \frac{a + bx}{x} \right|.$

10. $\displaystyle \int \frac{x\, dx}{(a + bx)^3} = \frac{1}{b^2} \left[-\frac{1}{a + bx} + \frac{a}{2(a + bx)^2} \right].$

▶ **Integrals containing $a + x$ and $b + x$.**

11. $\displaystyle \int \frac{a + x}{b + x}\, dx = x + (a - b) \ln|b + x|.$

12. $\displaystyle \int \frac{dx}{(a + x)(b + x)} = \frac{1}{a - b} \ln \left| \frac{b + x}{a + x} \right|, \quad a \neq b.$ For $a = b$, see integral 2 with $n = -2$.

13. $\displaystyle \int \frac{x\, dx}{(a + x)(b + x)} = \frac{1}{a - b} \left(a \ln|a + x| - b \ln|b + x| \right).$

14. $\displaystyle \int \frac{dx}{(a + x)(b + x)^2} = \frac{1}{(b - a)(b + x)} + \frac{1}{(a - b)^2} \ln \left| \frac{a + x}{b + x} \right|.$

---

* Throughout this section, the integration constant C is omitted for brevity.

15. $\displaystyle\int \frac{x\,dx}{(a+x)(b+x)^2} = \frac{b}{(a-b)(b+x)} - \frac{a}{(a-b)^2}\ln\left|\frac{a+x}{b+x}\right|.$

16. $\displaystyle\int \frac{x^2\,dx}{(a+x)(b+x)^2} = \frac{b^2}{(b-a)(b+x)} + \frac{a^2}{(a-b)^2}\ln|a+x| + \frac{b^2-2ab}{(b-a)^2}\ln|b+x|.$

17. $\displaystyle\int \frac{dx}{(a+x)^2(b+x)^2} = -\frac{1}{(a-b)^2}\left(\frac{1}{a+x} + \frac{1}{b+x}\right) + \frac{2}{(a-b)^3}\ln\left|\frac{a+x}{b+x}\right|.$

18. $\displaystyle\int \frac{x\,dx}{(a+x)^2(b+x)^2} = \frac{1}{(a-b)^2}\left(\frac{a}{a+x} + \frac{b}{b+x}\right) + \frac{a+b}{(a-b)^3}\ln\left|\frac{a+x}{b+x}\right|.$

19. $\displaystyle\int \frac{x^2\,dx}{(a+x)^2(b+x)^2} = -\frac{1}{(a-b)^2}\left(\frac{a^2}{a+x} + \frac{b^2}{b+x}\right) + \frac{2ab}{(a-b)^3}\ln\left|\frac{a+x}{b+x}\right|.$

▶ **Integrals containing $a^2 + x^2$.**

20. $\displaystyle\int \frac{dx}{a^2+x^2} = \frac{1}{a}\arctan\frac{x}{a}.$

21. $\displaystyle\int \frac{dx}{(a^2+x^2)^2} = \frac{x}{2a^2(a^2+x^2)} + \frac{1}{2a^3}\arctan\frac{x}{a}.$

22. $\displaystyle\int \frac{dx}{(a^2+x^2)^3} = \frac{x}{4a^2(a^2+x^2)^2} + \frac{3x}{8a^4(a^2+x^2)} + \frac{3}{8a^5}\arctan\frac{x}{a}.$

23. $\displaystyle\int \frac{dx}{(a^2+x^2)^{n+1}} = \frac{x}{2na^2(a^2+x^2)^n} + \frac{2n-1}{2na^2}\int \frac{dx}{(a^2+x^2)^n}; \quad n = 1, 2, \ldots$

24. $\displaystyle\int \frac{x\,dx}{a^2+x^2} = \frac{1}{2}\ln(a^2+x^2).$

25. $\displaystyle\int \frac{x\,dx}{(a^2+x^2)^2} = -\frac{1}{2(a^2+x^2)}.$

26. $\displaystyle\int \frac{x\,dx}{(a^2+x^2)^3} = -\frac{1}{4(a^2+x^2)^2}.$

27. $\displaystyle\int \frac{x\,dx}{(a^2+x^2)^{n+1}} = -\frac{1}{2n(a^2+x^2)^n}; \quad n = 1, 2, \ldots$

28. $\displaystyle\int \frac{x^2\,dx}{a^2+x^2} = x - a\arctan\frac{x}{a}.$

29. $\displaystyle\int \frac{x^2\,dx}{(a^2+x^2)^2} = -\frac{x}{2(a^2+x^2)} + \frac{1}{2a}\arctan\frac{x}{a}.$

30. $\displaystyle\int \frac{x^2\,dx}{(a^2+x^2)^3} = -\frac{x}{4(a^2+x^2)^2} + \frac{x}{8a^2(a^2+x^2)} + \frac{1}{8a^3}\arctan\frac{x}{a}.$

31. $\displaystyle\int \frac{x^2\,dx}{(a^2+x^2)^{n+1}} = -\frac{x}{2n(a^2+x^2)^n} + \frac{1}{2n}\int \frac{dx}{(a^2+x^2)^n}; \quad n = 1, 2, \ldots$

32. $\displaystyle\int \frac{x^3\,dx}{a^2+x^2} = \frac{x^2}{2} - \frac{a^2}{2}\ln(a^2+x^2).$

33. $\displaystyle\int \frac{x^3\,dx}{(a^2+x^2)^2} = \frac{a^2}{2(a^2+x^2)} + \frac{1}{2}\ln(a^2+x^2).$

34. $\displaystyle\int \frac{x^3\,dx}{(a^2+x^2)^{n+1}} = -\frac{1}{2(n-1)(a^2+x^2)^{n-1}} + \frac{a^2}{2n(a^2+x^2)^n}; \quad n = 2, 3, \ldots$

35. $\displaystyle\int \frac{dx}{x(a^2+x^2)} = \frac{1}{2a^2}\ln\frac{x^2}{a^2+x^2}.$

36. $\displaystyle\int \frac{dx}{x(a^2+x^2)^2} = \frac{1}{2a^2(a^2+x^2)} + \frac{1}{2a^4}\ln\frac{x^2}{a^2+x^2}.$

37. $\displaystyle\int \frac{dx}{x(a^2+x^2)^3} = \frac{1}{4a^2(a^2+x^2)^2} + \frac{1}{2a^4(a^2+x^2)} + \frac{1}{2a^6}\ln\frac{x^2}{a^2+x^2}$.

38. $\displaystyle\int \frac{dx}{x^2(a^2+x^2)} = -\frac{1}{a^2x} - \frac{1}{a^3}\arctan\frac{x}{a}$.

39. $\displaystyle\int \frac{dx}{x^2(a^2+x^2)^2} = -\frac{1}{a^4x} - \frac{x}{2a^4(a^2+x^2)} - \frac{3}{2a^5}\arctan\frac{x}{a}$.

40. $\displaystyle\int \frac{dx}{x^3(a^2+x^2)^2} = -\frac{1}{2a^4x^2} - \frac{1}{2a^4(a^2+x^2)} - \frac{1}{a^6}\ln\frac{x^2}{a^2+x^2}$.

41. $\displaystyle\int \frac{dx}{x^2(a^2+x^2)^3} = -\frac{1}{a^6x} - \frac{x}{4a^4(a^2+x^2)^2} - \frac{7x}{8a^6(a^2+x^2)} - \frac{15}{8a^7}\arctan\frac{x}{a}$.

42. $\displaystyle\int \frac{dx}{x^3(a^2+x^2)^3} = -\frac{1}{2a^6x^2} - \frac{1}{a^6(a^2+x^2)} - \frac{1}{4a^4(a^2+x^2)^2} - \frac{3}{2a^8}\ln\frac{x^2}{a^2+x^2}$.

▶ **Integrals containing $a^2 - x^2$.**

43. $\displaystyle\int \frac{dx}{a^2-x^2} = \frac{1}{2a}\ln\left|\frac{a+x}{a-x}\right|$.

44. $\displaystyle\int \frac{dx}{(a^2-x^2)^2} = \frac{x}{2a^2(a^2-x^2)} + \frac{1}{4a^3}\ln\left|\frac{a+x}{a-x}\right|$.

45. $\displaystyle\int \frac{dx}{(a^2-x^2)^3} = \frac{x}{4a^2(a^2-x^2)^2} + \frac{3x}{8a^4(a^2-x^2)} + \frac{3}{16a^5}\ln\left|\frac{a+x}{a-x}\right|$.

46. $\displaystyle\int \frac{dx}{(a^2-x^2)^{n+1}} = \frac{x}{2na^2(a^2-x^2)^n} + \frac{2n-1}{2na^2}\int \frac{dx}{(a^2-x^2)^n}; \quad n = 1, 2, \ldots$

47. $\displaystyle\int \frac{x\,dx}{a^2-x^2} = -\frac{1}{2}\ln|a^2-x^2|$.

48. $\displaystyle\int \frac{x\,dx}{(a^2-x^2)^2} = \frac{1}{2(a^2-x^2)}$.

49. $\displaystyle\int \frac{x\,dx}{(a^2-x^2)^3} = \frac{1}{4(a^2-x^2)^2}$.

50. $\displaystyle\int \frac{x\,dx}{(a^2-x^2)^{n+1}} = \frac{1}{2n(a^2-x^2)^n}; \quad n = 1, 2, \ldots$

51. $\displaystyle\int \frac{x^2\,dx}{a^2-x^2} = -x + \frac{a}{2}\ln\left|\frac{a+x}{a-x}\right|$.

52. $\displaystyle\int \frac{x^2\,dx}{(a^2-x^2)^2} = \frac{x}{2(a^2-x^2)} - \frac{1}{4a}\ln\left|\frac{a+x}{a-x}\right|$.

53. $\displaystyle\int \frac{x^2\,dx}{(a^2-x^2)^3} = \frac{x}{4(a^2-x^2)^2} - \frac{x}{8a^2(a^2-x^2)} - \frac{1}{16a^3}\ln\left|\frac{a+x}{a-x}\right|$.

54. $\displaystyle\int \frac{x^2\,dx}{(a^2-x^2)^{n+1}} = \frac{x}{2n(a^2-x^2)^n} - \frac{1}{2n}\int \frac{dx}{(a^2-x^2)^n}; \quad n = 1, 2, \ldots$

55. $\displaystyle\int \frac{x^3\,dx}{a^2-x^2} = -\frac{x^2}{2} - \frac{a^2}{2}\ln|a^2-x^2|$.

56. $\displaystyle\int \frac{x^3\,dx}{(a^2-x^2)^2} = \frac{a^2}{2(a^2-x^2)} + \frac{1}{2}\ln|a^2-x^2|$.

57. $\displaystyle\int \frac{x^3\,dx}{(a^2-x^2)^{n+1}} = -\frac{1}{2(n-1)(a^2-x^2)^{n-1}} + \frac{a^2}{2n(a^2-x^2)^n}; \quad n = 2, 3, \ldots$

58. $\displaystyle\int \frac{dx}{x(a^2-x^2)} = \frac{1}{2a^2}\ln\left|\frac{x^2}{a^2-x^2}\right|$.

59. $\displaystyle\int \frac{dx}{x(a^2 - x^2)^2} = \frac{1}{2a^2(a^2 - x^2)} + \frac{1}{2a^4} \ln\left|\frac{x^2}{a^2 - x^2}\right|.$

60. $\displaystyle\int \frac{dx}{x(a^2 - x^2)^3} = \frac{1}{4a^2(a^2 - x^2)^2} + \frac{1}{2a^4(a^2 - x^2)} + \frac{1}{2a^6} \ln\left|\frac{x^2}{a^2 - x^2}\right|.$

▶ **Integrals containing $a^3 + x^3$.**

61. $\displaystyle\int \frac{dx}{a^3 + x^3} = \frac{1}{6a^2} \ln \frac{(a + x)^2}{a^2 - ax + x^2} + \frac{1}{a^2\sqrt{3}} \arctan \frac{2x - a}{a\sqrt{3}}.$

62. $\displaystyle\int \frac{dx}{(a^3 + x^3)^2} = \frac{x}{3a^3(a^3 + x^3)} + \frac{2}{3a^3} \int \frac{dx}{a^3 + x^3}.$

63. $\displaystyle\int \frac{x\,dx}{a^3 + x^3} = \frac{1}{6a} \ln \frac{a^2 - ax + x^2}{(a + x)^2} + \frac{1}{a\sqrt{3}} \arctan \frac{2x - a}{a\sqrt{3}}.$

64. $\displaystyle\int \frac{x\,dx}{(a^3 + x^3)^2} = \frac{x^2}{3a^3(a^3 + x^3)} + \frac{1}{3a^3} \int \frac{x\,dx}{a^3 + x^3}.$

65. $\displaystyle\int \frac{x^2\,dx}{a^3 + x^3} = \frac{1}{3} \ln|a^3 + x^3|.$

66. $\displaystyle\int \frac{dx}{x(a^3 + x^3)} = \frac{1}{3a^3} \ln\left|\frac{x^3}{a^3 + x^3}\right|.$

67. $\displaystyle\int \frac{dx}{x(a^3 + x^3)^2} = \frac{1}{3a^3(a^3 + x^3)} + \frac{1}{3a^6} \ln\left|\frac{x^3}{a^3 + x^3}\right|.$

68. $\displaystyle\int \frac{dx}{x^2(a^3 + x^3)} = -\frac{1}{a^3 x} - \frac{1}{a^3} \int \frac{x\,dx}{a^3 + x^3}.$

69. $\displaystyle\int \frac{dx}{x^2(a^3 + x^3)^2} = -\frac{1}{a^6 x} - \frac{x^2}{3a^6(a^3 + x^3)} - \frac{4}{3a^6} \int \frac{x\,dx}{a^3 + x^3}.$

▶ **Integrals containing $a^3 - x^3$.**

70. $\displaystyle\int \frac{dx}{a^3 - x^3} = \frac{1}{6a^2} \ln \frac{a^2 + ax + x^2}{(a - x)^2} + \frac{1}{a^2\sqrt{3}} \arctan \frac{2x + a}{a\sqrt{3}}.$

71. $\displaystyle\int \frac{dx}{(a^3 - x^3)^2} = \frac{x}{3a^3(a^3 - x^3)} + \frac{2}{3a^3} \int \frac{dx}{a^3 - x^3}.$

72. $\displaystyle\int \frac{x\,dx}{a^3 - x^3} = \frac{1}{6a} \ln \frac{a^2 + ax + x^2}{(a - x)^2} - \frac{1}{a\sqrt{3}} \arctan \frac{2x + a}{a\sqrt{3}}.$

73. $\displaystyle\int \frac{x\,dx}{(a^3 - x^3)^2} = \frac{x^2}{3a^3(a^3 - x^3)} + \frac{1}{3a^3} \int \frac{x\,dx}{a^3 - x^3}.$

74. $\displaystyle\int \frac{x^2\,dx}{a^3 - x^3} = -\frac{1}{3} \ln|a^3 - x^3|.$

75. $\displaystyle\int \frac{dx}{x(a^3 - x^3)} = \frac{1}{3a^3} \ln\left|\frac{x^3}{a^3 - x^3}\right|.$

76. $\displaystyle\int \frac{dx}{x(a^3 - x^3)^2} = \frac{1}{3a^3(a^3 - x^3)} + \frac{1}{3a^6} \ln\left|\frac{x^3}{a^3 - x^3}\right|.$

77. $\displaystyle\int \frac{dx}{x^2(a^3 - x^3)} = -\frac{1}{a^3 x} + \frac{1}{a^3} \int \frac{x\,dx}{a^3 - x^3}.$

78. $\displaystyle\int \frac{dx}{x^2(a^3 - x^3)^2} = -\frac{1}{a^6 x} - \frac{x^2}{3a^6(a^3 - x^3)} + \frac{4}{3a^6} \int \frac{x\,dx}{a^3 - x^3}.$

▶ **Integrals containing $a^4 \pm x^4$.**

79. $\displaystyle \int \frac{dx}{a^4 + x^4} = \frac{1}{4a^3\sqrt{2}} \ln \frac{a^2 + ax\sqrt{2} + x^2}{a^2 - ax\sqrt{2} + x^2} + \frac{1}{2a^3\sqrt{2}} \arctan \frac{ax\sqrt{2}}{a^2 - x^2}.$

80. $\displaystyle \int \frac{x\, dx}{a^4 + x^4} = \frac{1}{2a^2} \arctan \frac{x^2}{a^2}.$

81. $\displaystyle \int \frac{x^2\, dx}{a^4 + x^4} = -\frac{1}{4a\sqrt{2}} \ln \frac{a^2 + ax\sqrt{2} + x^2}{a^2 - ax\sqrt{2} + x^2} + \frac{1}{2a\sqrt{2}} \arctan \frac{ax\sqrt{2}}{a^2 - x^2}.$

82. $\displaystyle \int \frac{dx}{a^4 - x^4} = \frac{1}{4a^3} \ln \left| \frac{a + x}{a - x} \right| + \frac{1}{2a^3} \arctan \frac{x}{a}.$

83. $\displaystyle \int \frac{x\, dx}{a^4 - x^4} = \frac{1}{4a^2} \ln \left| \frac{a^2 + x^2}{a^2 - x^2} \right|.$

84. $\displaystyle \int \frac{x^2\, dx}{a^4 - x^4} = \frac{1}{4a} \ln \left| \frac{a + x}{a - x} \right| - \frac{1}{2a} \arctan \frac{x}{a}.$

85. $\displaystyle \int \frac{dx}{x(a + bx^m)} = \frac{1}{am} \ln \left| \frac{x^m}{a + bx^m} \right|.$

2.2. Integrals Containing Irrational Functions

▶ **Integrals containing $x^{1/2}$.**

1. $\displaystyle \int \frac{x^{1/2}\, dx}{a^2 + b^2 x} = \frac{2}{b^2} x^{1/2} - \frac{2a}{b^3} \arctan \frac{bx^{1/2}}{a}.$

2. $\displaystyle \int \frac{x^{3/2}\, dx}{a^2 + b^2 x} = \frac{2x^{3/2}}{3b^2} - \frac{2a^2 x^{1/2}}{b^4} + \frac{2a^3}{b^5} \arctan \frac{bx^{1/2}}{a}.$

3. $\displaystyle \int \frac{x^{1/2}\, dx}{(a^2 + b^2 x)^2} = -\frac{x^{1/2}}{b^2(a^2 + b^2 x)} + \frac{1}{ab^3} \arctan \frac{bx^{1/2}}{a}.$

4. $\displaystyle \int \frac{x^{3/2}\, dx}{(a^2 + b^2 x)^2} = \frac{2x^{3/2}}{b^2(a^2 + b^2 x)} + \frac{3a^2 x^{1/2}}{b^4(a^2 + b^2 x)} - \frac{3a}{b^5} \arctan \frac{bx^{1/2}}{a}.$

5. $\displaystyle \int \frac{dx}{(a^2 + b^2 x)x^{1/2}} = \frac{2}{ab} \arctan \frac{bx^{1/2}}{a}.$

6. $\displaystyle \int \frac{dx}{(a^2 + b^2 x)x^{3/2}} = -\frac{2}{a^2 x^{1/2}} - \frac{2b}{a^3} \arctan \frac{bx^{1/2}}{a}.$

7. $\displaystyle \int \frac{dx}{(a^2 + b^2 x)^2 x^{1/2}} = \frac{x^{1/2}}{a^2(a^2 + b^2 x)} + \frac{1}{a^3 b} \arctan \frac{bx^{1/2}}{a}.$

8. $\displaystyle \int \frac{x^{1/2}\, dx}{a^2 - b^2 x} = -\frac{2}{b^2} x^{1/2} + \frac{2a}{b^3} \ln \left| \frac{a + bx^{1/2}}{a - bx^{1/2}} \right|.$

9. $\displaystyle \int \frac{x^{3/2}\, dx}{a^2 - b^2 x} = -\frac{2x^{3/2}}{3b^2} - \frac{2a^2 x^{1/2}}{b^4} + \frac{a^3}{b^5} \ln \left| \frac{a + bx^{1/2}}{a - bx^{1/2}} \right|.$

10. $\displaystyle \int \frac{x^{1/2}\, dx}{(a^2 - b^2 x)^2} = \frac{x^{1/2}}{b^2(a^2 - b^2 x)} - \frac{1}{2ab^3} \ln \left| \frac{a + bx^{1/2}}{a - bx^{1/2}} \right|.$

11. $\displaystyle \int \frac{x^{3/2}\, dx}{(a^2 - b^2 x)^2} = \frac{3a^2 x^{1/2} - 2b^2 x^{3/2}}{b^4(a^2 - b^2 x)} - \frac{3a}{2b^5} \ln \left| \frac{a + bx^{1/2}}{a - bx^{1/2}} \right|.$

12. $\displaystyle \int \frac{dx}{(a^2 - b^2 x)x^{1/2}} = \frac{1}{ab} \ln \left| \frac{a + bx^{1/2}}{a - bx^{1/2}} \right|.$

13. $\displaystyle\int \frac{dx}{(a^2 - b^2x)x^{3/2}} = -\frac{2}{a^2x^{1/2}} + \frac{b}{a^3}\ln\left|\frac{a + bx^{1/2}}{a - bx^{1/2}}\right|.$

14. $\displaystyle\int \frac{dx}{(a^2 - b^2x)^2x^{1/2}} = \frac{x^{1/2}}{a^2(a^2 - b^2x)} + \frac{1}{2a^3b}\ln\left|\frac{a + bx^{1/2}}{a - bx^{1/2}}\right|.$

▶ **Integrals containing $(a + bx)^{p/2}$.**

15. $\displaystyle\int (a + bx)^{p/2}\, dx = \frac{2}{b(p + 2)}(a + bx)^{(p+2)/2}.$

16. $\displaystyle\int x(a + bx)^{p/2}\, dx = \frac{2}{b^2}\left[\frac{(a + bx)^{(p+4)/2}}{p + 4} - \frac{a(a + bx)^{(p+2)/2}}{p + 2}\right].$

17. $\displaystyle\int x^2(a + bx)^{p/2}\, dx = \frac{2}{b^3}\left[\frac{(a + bx)^{(p+6)/2}}{p + 6} - \frac{2a(a + bx)^{(p+4)/2}}{p + 4} + \frac{a^2(a + bx)^{(p+2)/2}}{p + 2}\right].$

▶ **Integrals containing $(x^2 + a^2)^{1/2}$.**

18. $\displaystyle\int (x^2 + a^2)^{1/2}\, dx = \frac{1}{2}x(a^2 + x^2)^{1/2} + \frac{a^2}{2}\ln\left[x + (x^2 + a^2)^{1/2}\right].$

19. $\displaystyle\int x(x^2 + a^2)^{1/2}\, dx = \frac{1}{3}(a^2 + x^2)^{3/2}.$

20. $\displaystyle\int (x^2 + a^2)^{3/2}\, dx = \frac{1}{4}x(a^2 + x^2)^{3/2} + \frac{3}{8}a^2x(a^2 + x^2)^{1/2} + \frac{3}{8}a^4\ln\left|x + (x^2 + a^2)^{1/2}\right|.$

21. $\displaystyle\int \frac{1}{x}(x^2 + a^2)^{1/2}\, dx = (a^2 + x^2)^{1/2} - a\ln\left|\frac{a + (x^2 + a^2)^{1/2}}{x}\right|.$

22. $\displaystyle\int \frac{dx}{\sqrt{x^2 + a^2}} = \ln\left[x + (x^2 + a^2)^{1/2}\right].$

23. $\displaystyle\int \frac{x\, dx}{\sqrt{x^2 + a^2}} = (x^2 + a^2)^{1/2}.$

24. $\displaystyle\int (x^2 + a^2)^{-3/2}\, dx = a^{-2}x(x^2 + a^2)^{-1/2}.$

▶ **Integrals containing $(x^2 - a^2)^{1/2}$.**

25. $\displaystyle\int (x^2 - a^2)^{1/2}\, dx = \frac{1}{2}x(x^2 - a^2)^{1/2} - \frac{a^2}{2}\ln\left|x + (x^2 - a^2)^{1/2}\right|.$

26. $\displaystyle\int x(x^2 - a^2)^{1/2}\, dx = \frac{1}{3}(x^2 - a^2)^{3/2}.$

27. $\displaystyle\int (x^2 - a^2)^{3/2}\, dx = \frac{1}{4}x(x^2 - a^2)^{3/2} - \frac{3}{8}a^2x(x^2 - a^2)^{1/2} + \frac{3}{8}a^4\ln\left|x + (x^2 - a^2)^{1/2}\right|.$

28. $\displaystyle\int \frac{1}{x}(x^2 - a^2)^{1/2}\, dx = (x^2 - a^2)^{1/2} - a\arccos\left|\frac{a}{x}\right|.$

29. $\displaystyle\int \frac{dx}{\sqrt{x^2 - a^2}} = \ln\left|x + (x^2 - a^2)^{1/2}\right|.$

30. $\displaystyle\int \frac{x\, dx}{\sqrt{x^2 - a^2}} = (x^2 - a^2)^{1/2}.$

31. $\displaystyle\int (x^2 - a^2)^{-3/2}\, dx = -a^{-2}x(x^2 - a^2)^{-1/2}.$

▶ **Integrals containing $(a^2 - x^2)^{1/2}$.**

32. $\displaystyle \int (a^2 - x^2)^{1/2}\, dx = \frac{1}{2}x(a^2 - x^2)^{1/2} + \frac{a^2}{2}\arcsin\frac{x}{a}.$

33. $\displaystyle \int x(a^2 - x^2)^{1/2}\, dx = -\frac{1}{3}(a^2 - x^2)^{3/2}.$

34. $\displaystyle \int (a^2 - x^2)^{3/2}\, dx = \frac{1}{4}x(a^2 - x^2)^{3/2} + \frac{3}{8}a^2 x(a^2 - x^2)^{1/2} + \frac{3}{8}a^4 \arcsin\frac{x}{a}.$

35. $\displaystyle \int \frac{1}{x}(a^2 - x^2)^{1/2}\, dx = (a^2 - x^2)^{1/2} - a\ln\left|\frac{a + (a^2 - x^2)^{1/2}}{x}\right|.$

36. $\displaystyle \int \frac{dx}{\sqrt{a^2 - x^2}} = \arcsin\frac{x}{a}.$

37. $\displaystyle \int \frac{x\, dx}{\sqrt{a^2 - x^2}} = -(a^2 - x^2)^{1/2}.$

38. $\displaystyle \int (a^2 - x^2)^{-3/2}\, dx = a^{-2}x(a^2 - x^2)^{-1/2}.$

▶ **Reduction formulas.** The parameters a, b, p, m, and n below can assume arbitrary values, except for those at which denominators vanish in successive applications of a formula. Notation: $w = ax^n + b$.

39. $\displaystyle \int x^m(ax^n + b)^p\, dx = \frac{1}{m + np + 1}\left(x^{m+1}w^p + npb\int x^m w^{p-1}\, dx\right).$

40. $\displaystyle \int x^m(ax^n + b)^p\, dx = \frac{1}{bn(p+1)}\left[-x^{m+1}w^{p+1} + (m + n + np + 1)\int x^m w^{p+1}\, dx\right].$

41. $\displaystyle \int x^m(ax^n + b)^p\, dx = \frac{1}{b(m+1)}\left[x^{m+1}w^{p+1} - a(m + n + np + 1)\int x^{m+n}w^p\, dx\right].$

42. $\displaystyle \int x^m(ax^n + b)^p\, dx = \frac{1}{a(m + np + 1)}\left[x^{m-n+1}w^{p+1} - b(m - n + 1)\int x^{m-n}w^p\, dx\right].$

2.3. Integrals Containing Exponential Functions

1. $\displaystyle \int e^{ax}\, dx = \frac{1}{a}e^{ax}.$

2. $\displaystyle \int a^x\, dx = \frac{a^x}{\ln a}.$

3. $\displaystyle \int xe^{ax}\, dx = e^{ax}\left(\frac{x}{a} - \frac{1}{a^2}\right).$

4. $\displaystyle \int x^2 e^{ax}\, dx = e^{ax}\left(\frac{x^2}{a} - \frac{2x}{a^2} + \frac{2}{a^3}\right).$

5. $\displaystyle \int x^n e^{ax}\, dx = e^{ax}\left[\frac{1}{a}x^n - \frac{n}{a^2}x^{n-1} + \frac{n(n-1)}{a^3}x^{n-2} - \cdots + (-1)^{n-1}\frac{n!}{a^n}x + (-1)^n\frac{n!}{a^{n+1}}\right], \quad n = 1, 2, \ldots$

6. $\displaystyle \int P_n(x)e^{ax}\, dx = e^{ax}\sum_{k=0}^{n}\frac{(-1)^k}{a^{k+1}}\frac{d^k}{dx^k}P_n(x),$ where $P_n(x)$ is an arbitrary nth-degree polynomial.

7. $\displaystyle \int \frac{dx}{a + be^{px}} = \frac{x}{a} - \frac{1}{ap}\ln|a + be^{px}|.$

8. $\displaystyle\int \frac{dx}{ae^{px} + be^{-px}} = \begin{cases} \dfrac{1}{p\sqrt{ab}} \arctan\left(e^{px}\sqrt{\dfrac{a}{b}}\right) & \text{if } ab > 0, \\[3mm] \dfrac{1}{2p\sqrt{-ab}} \ln\left(\dfrac{b + e^{px}\sqrt{-ab}}{b - e^{px}\sqrt{-ab}}\right) & \text{if } ab < 0. \end{cases}$

9. $\displaystyle\int \frac{dx}{\sqrt{a + be^{px}}} = \begin{cases} \dfrac{1}{p\sqrt{a}} \ln \dfrac{\sqrt{a + be^{px}} - \sqrt{a}}{\sqrt{a + be^{px}} + \sqrt{a}} & \text{if } a > 0, \\[3mm] \dfrac{2}{p\sqrt{-a}} \arctan \dfrac{\sqrt{a + be^{px}}}{\sqrt{-a}} & \text{if } a < 0. \end{cases}$

2.4. Integrals Containing Hyperbolic Functions

▶ **Integrals containing cosh x.**

1. $\displaystyle\int \cosh(a + bx)\, dx = \frac{1}{b} \sinh(a + bx).$

2. $\displaystyle\int x \cosh x\, dx = x \sinh x - \cosh x.$

3. $\displaystyle\int x^2 \cosh x\, dx = (x^2 + 2) \sinh x - 2x \cosh x.$

4. $\displaystyle\int x^{2n} \cosh x\, dx = (2n)! \sum_{k=1}^{n} \left[\frac{x^{2k}}{(2k)!} \sinh x - \frac{x^{2k-1}}{(2k-1)!} \cosh x \right].$

5. $\displaystyle\int x^{2n+1} \cosh x\, dx = (2n+1)! \sum_{k=0}^{n} \left[\frac{x^{2k+1}}{(2k+1)!} \sinh x - \frac{x^{2k}}{(2k)!} \cosh x \right].$

6. $\displaystyle\int x^p \cosh x\, dx = x^p \sinh x - px^{p-1} \cosh x + p(p-1) \int x^{p-2} \cosh x\, dx.$

7. $\displaystyle\int \cosh^2 x\, dx = \frac{1}{2}x + \frac{1}{4} \sinh 2x.$

8. $\displaystyle\int \cosh^3 x\, dx = \sinh x + \frac{1}{3} \sinh^3 x.$

9. $\displaystyle\int \cosh^{2n} x\, dx = C_{2n}^n \frac{x}{2^{2n}} + \frac{1}{2^{2n-1}} \sum_{k=0}^{n-1} C_{2n}^k \frac{\sinh[2(n-k)x]}{2(n-k)}, \quad n = 1, 2, \ldots$

10. $\displaystyle\int \cosh^{2n+1} x\, dx = \frac{1}{2^{2n}} \sum_{k=0}^{n} C_{2n+1}^k \frac{\sinh[(2n - 2k + 1)x]}{2n - 2k + 1} = \sum_{k=0}^{n} C_n^k \frac{\sinh^{2k+1} x}{2k + 1}, \quad n = 1, 2, \ldots$

11. $\displaystyle\int \cosh^p x\, dx = \frac{1}{p} \sinh x \cosh^{p-1} x + \frac{p-1}{p} \int \cosh^{p-2} x\, dx.$

12. $\displaystyle\int \cosh ax \cosh bx\, dx = \frac{1}{a^2 - b^2} \left[a \cosh bx \sinh ax - b \cosh ax \sinh bx \right].$

13. $\displaystyle\int \frac{dx}{\cosh ax} = \frac{2}{a} \arctan\left(e^{ax}\right).$

14. $\displaystyle\int \frac{dx}{\cosh^{2n} x} = \frac{\sinh x}{2n - 1} \left[\frac{1}{\cosh^{2n-1} x} + \sum_{k=1}^{n-1} \frac{2^k (n-1)(n-2) \ldots (n-k)}{(2n-3)(2n-5) \ldots (2n - 2k - 1)} \frac{1}{\cosh^{2n-2k-1} x} \right],$
$n = 1, 2, \ldots$

15. $\displaystyle\int \frac{dx}{\cosh^{2n+1} x} = \frac{\sinh x}{2n} \left[\frac{1}{\cosh^{2n} x} + \sum_{k=1}^{n-1} \frac{(2n-1)(2n-3) \ldots (2n - 2k + 1)}{2^k (n-1)(n-2) \ldots (n-k)} \frac{1}{\cosh^{2n-2k} x} \right]$
$+ \dfrac{(2n-1)!!}{(2n)!!} \arctan \sinh x, \quad n = 1, 2, \ldots$

16. $\displaystyle\int \frac{dx}{a + b\cosh x} = \begin{cases} -\dfrac{\operatorname{sign} x}{\sqrt{b^2 - a^2}} \arcsin \dfrac{b + a\cosh x}{a + b\cosh x} & \text{if } a^2 < b^2, \\[4mm] \dfrac{1}{\sqrt{a^2 - b^2}} \ln \dfrac{a + b + \sqrt{a^2 - b^2}\,\tanh(x/2)}{a + b - \sqrt{a^2 - b^2}\,\tanh(x/2)} & \text{if } a^2 > b^2. \end{cases}$

▶ **Integrals containing sinh x.**

17. $\displaystyle\int \sinh(a + bx)\,dx = \frac{1}{b}\cosh(a + bx).$

18. $\displaystyle\int x\sinh x\,dx = x\cosh x - \sinh x.$

19. $\displaystyle\int x^2 \sinh x\,dx = (x^2 + 2)\cosh x - 2x\sinh x.$

20. $\displaystyle\int x^{2n}\sinh x\,dx = (2n)!\left[\sum_{k=0}^{n}\frac{x^{2k}}{(2k)!}\cosh x - \sum_{k=1}^{n}\frac{x^{2k-1}}{(2k-1)!}\sinh x\right].$

21. $\displaystyle\int x^{2n+1}\sinh x\,dx = (2n+1)!\sum_{k=0}^{n}\left[\frac{x^{2k+1}}{(2k+1)!}\cosh x - \frac{x^{2k}}{(2k)!}\sinh x\right].$

22. $\displaystyle\int x^p \sinh x\,dx = x^p\cosh x - px^{p-1}\sinh x + p(p-1)\int x^{p-2}\sinh x\,dx.$

23. $\displaystyle\int \sinh^2 x\,dx = -\frac{1}{2}x + \frac{1}{4}\sinh 2x.$

24. $\displaystyle\int \sinh^3 x\,dx = -\cosh x + \frac{1}{3}\cosh^3 x.$

25. $\displaystyle\int \sinh^{2n} x\,dx = (-1)^n C_{2n}^n \frac{x}{2^{2n}} + \frac{1}{2^{2n-1}}\sum_{k=0}^{n-1}(-1)^k C_{2n}^k \frac{\sinh[2(n-k)x]}{2(n-k)}, \quad n = 1, 2, \ldots$

26. $\displaystyle\int \sinh^{2n+1} x\,dx = \frac{1}{2^{2n}}\sum_{k=0}^{n}(-1)^k C_{2n+1}^k \frac{\cosh[(2n-2k+1)x]}{2n-2k+1} = \sum_{k=0}^{n}(-1)^{n+k}C_n^k \frac{\cosh^{2k+1}x}{2k+1},$
$n = 1, 2, \ldots$

27. $\displaystyle\int \sinh^p x\,dx = \frac{1}{p}\sinh^{p-1}x\cosh x - \frac{p-1}{p}\int \sinh^{p-2}x\,dx.$

28. $\displaystyle\int \sinh ax\sinh bx\,dx = \frac{1}{a^2 - b^2}\left[a\cosh ax\sinh bx - b\cosh bx\sinh ax\right].$

29. $\displaystyle\int \frac{dx}{\sinh ax} = \frac{1}{a}\ln\left|\tanh\frac{ax}{2}\right|.$

30. $\displaystyle\int \frac{dx}{\sinh^{2n}x} = \frac{\cosh x}{2n-1}\left[-\frac{1}{\sinh^{2n-1}x}\right.$
$\left.+\sum_{k=1}^{n-1}(-1)^{k-1}\frac{2^k(n-1)(n-2)\ldots(n-k)}{(2n-3)(2n-5)\ldots(2n-2k-1)}\frac{1}{\sinh^{2n-2k-1}x}\right], \quad n = 1, 2, \ldots$

31. $\displaystyle\int \frac{dx}{\sinh^{2n+1}x} = \frac{\cosh x}{2n}\left[\frac{1}{\sinh^{2n}x} + \sum_{k=1}^{n-1}(-1)^{k-1}\frac{(2n-1)(2n-3)\ldots(2n-2k+1)}{2^k(n-1)(n-2)\ldots(n-k)}\frac{1}{\sinh^{2n-2k}x}\right]$
$+(-1)^n\frac{(2n-1)!!}{(2n)!!}\ln\tanh\frac{x}{2}, \quad n = 1, 2, \ldots$

32. $\displaystyle\int \frac{dx}{a + b\sinh x} = \frac{1}{\sqrt{a^2 + b^2}}\ln\frac{a\tanh(x/2) - b + \sqrt{a^2 + b^2}}{a\tanh(x/2) - b - \sqrt{a^2 + b^2}}.$

33. $\displaystyle\int \frac{Ax + B\sinh x}{a + b\sinh x}\,dx = \frac{B}{b}x + \frac{Ab - Ba}{b\sqrt{a^2 + b^2}}\ln\frac{a\tanh(x/2) - b + \sqrt{a^2 + b^2}}{a\tanh(x/2) - b - \sqrt{a^2 + b^2}}.$

▶ **Integrals containing tanh x or coth x.**

34. $\displaystyle\int \tanh x \, dx = \ln \cosh x.$

35. $\displaystyle\int \tanh^2 x \, dx = x - \tanh x.$

36. $\displaystyle\int \tanh^3 x \, dx = -\tfrac{1}{2} \tanh^2 x + \ln \cosh x.$

37. $\displaystyle\int \tanh^{2n} x \, dx = x - \sum_{k=1}^{n} \frac{\tanh^{2n-2k+1} x}{2n - 2k + 1}, \qquad n = 1, 2, \ldots$

38. $\displaystyle\int \tanh^{2n+1} x \, dx = \ln \cosh x - \sum_{k=1}^{n} \frac{(-1)^k C_n^k}{2k \cosh^{2k} x} = \ln \cosh x - \sum_{k=1}^{n} \frac{\tanh^{2n-2k+2} x}{2n - 2k + 2}, \qquad n = 1, 2, \ldots$

39. $\displaystyle\int \tanh^p x \, dx = -\frac{1}{p-1} \tanh^{p-1} x + \int \tanh^{p-2} x \, dx.$

40. $\displaystyle\int \coth x \, dx = \ln |\sinh x|.$

41. $\displaystyle\int \coth^2 x \, dx = x - \coth x.$

42. $\displaystyle\int \coth^3 x \, dx = -\tfrac{1}{2} \coth^2 x + \ln |\sinh x|.$

43. $\displaystyle\int \coth^{2n} x \, dx = x - \sum_{k=1}^{n} \frac{\coth^{2n-2k+1} x}{2n - 2k + 1}, \qquad n = 1, 2, \ldots$

44. $\displaystyle\int \coth^{2n+1} x \, dx = \ln |\sinh x| - \sum_{k=1}^{n} \frac{C_n^k}{2k \sinh^{2k} x} = \ln |\sinh x| - \sum_{k=1}^{n} \frac{\coth^{2n-2k+2} x}{2n - 2k + 2}, \qquad n = 1, 2, \ldots$

45. $\displaystyle\int \coth^p x \, dx = -\frac{1}{p-1} \coth^{p-1} x + \int \coth^{p-2} x \, dx.$

2.5. Integrals Containing Logarithmic Functions

1. $\displaystyle\int \ln ax \, dx = x \ln ax - x.$

2. $\displaystyle\int x \ln x \, dx = \tfrac{1}{2} x^2 \ln x - \tfrac{1}{4} x^2.$

3. $\displaystyle\int x^p \ln ax \, dx = \begin{cases} \dfrac{1}{p+1} x^{p+1} \ln ax - \dfrac{1}{(p+1)^2} x^{p+1} & \text{if } p \neq -1, \\ \tfrac{1}{2} \ln^2 ax & \text{if } p = -1. \end{cases}$

4. $\displaystyle\int (\ln x)^2 \, dx = x(\ln x)^2 - 2x \ln x + 2x.$

5. $\displaystyle\int x(\ln x)^2 \, dx = \tfrac{1}{2} x^2 (\ln x)^2 - \tfrac{1}{2} x^2 \ln x + \tfrac{1}{4} x^2.$

6. $\displaystyle\int x^p (\ln x)^2 \, dx = \begin{cases} \dfrac{x^{p+1}}{p+1} (\ln x)^2 - \dfrac{2x^{p+1}}{(p+1)^2} \ln x + \dfrac{2x^{p+1}}{(p+1)^3} & \text{if } p \neq -1, \\ \tfrac{1}{3} \ln^3 x & \text{if } p = -1. \end{cases}$

7. $\displaystyle\int (\ln x)^n \, dx = \frac{x}{n+1} \sum_{k=0}^{n} (-1)^k (n+1)n \ldots (n-k+1)(\ln x)^{n-k}, \qquad n = 1, 2, \ldots$

8. $\int (\ln x)^q \, dx = x(\ln x)^q - q \int (\ln x)^{q-1} \, dx, \ q \neq -1.$

9. $\int x^n (\ln x)^m \, dx = \dfrac{x^{n+1}}{m+1} \sum\limits_{k=0}^{m} \dfrac{(-1)^k}{(n+1)^{k+1}} (m+1)m \ldots (m-k+1)(\ln x)^{m-k}, \ n, m = 1, 2, \ldots$

10. $\int x^p (\ln x)^q \, dx = \dfrac{1}{p+1} x^{p+1} (\ln x)^q - \dfrac{q}{p+1} \int x^p (\ln x)^{q-1} \, dx, \quad p, \, q \neq -1.$

11. $\int \ln(a + bx) \, dx = \dfrac{1}{b}(ax + b)\ln(ax + b) - x.$

12. $\int x \ln(a + bx) \, dx = \dfrac{1}{2}\left(x^2 - \dfrac{a^2}{b^2}\right)\ln(a + bx) - \dfrac{1}{2}\left(\dfrac{x^2}{2} - \dfrac{a}{b}x\right).$

13. $\int x^2 \ln(a + bx) \, dx = \dfrac{1}{3}\left(x^3 - \dfrac{a^3}{b^3}\right)\ln(a + bx) - \dfrac{1}{3}\left(\dfrac{x^3}{3} - \dfrac{ax^2}{2b} + \dfrac{a^2 x}{b^2}\right).$

14. $\int \dfrac{\ln x \, dx}{(a + bx)^2} = -\dfrac{\ln x}{b(a + bx)} + \dfrac{1}{ab} \ln \dfrac{x}{a + bx}.$

15. $\int \dfrac{\ln x \, dx}{(a + bx)^3} = -\dfrac{\ln x}{2b(a + bx)^2} + \dfrac{1}{2ab(a + bx)} + \dfrac{1}{2a^2 b} \ln \dfrac{x}{a + bx}.$

16. $\int \dfrac{\ln x \, dx}{\sqrt{a + bx}} = \begin{cases} \dfrac{2}{b}\left[(\ln x - 2)\sqrt{a + bx} + \sqrt{a}\, \ln \dfrac{\sqrt{a + bx} + \sqrt{a}}{\sqrt{a + bx} - \sqrt{a}}\right] & \text{if } a > 0, \\[4mm] \dfrac{2}{b}\left[(\ln x - 2)\sqrt{a + bx} + 2\sqrt{-a}\, \arctan \dfrac{\sqrt{a + bx}}{\sqrt{-a}}\right] & \text{if } a < 0. \end{cases}$

17. $\int \ln(x^2 + a^2) \, dx = x \ln(x^2 + a^2) - 2x + 2a \arctan(x/a).$

18. $\int x \ln(x^2 + a^2) \, dx = \tfrac{1}{2}\left[(x^2 + a^2)\ln(x^2 + a^2) - x^2\right].$

19. $\int x^2 \ln(x^2 + a^2) \, dx = \tfrac{1}{3}\left[x^3 \ln(x^2 + a^2) - \tfrac{2}{3}x^3 + 2a^2 x - 2a^3 \arctan(x/a)\right].$

2.6. Integrals Containing Trigonometric Functions

▶ **Integrals containing $\cos x$.** Notation: $n = 1, 2, \ldots$

1. $\int \cos(a + bx) \, dx = \dfrac{1}{b} \sin(a + bx).$

2. $\int x \cos x \, dx = \cos x + x \sin x.$

3. $\int x^2 \cos x \, dx = 2x \cos x + (x^2 - 2) \sin x.$

4. $\int x^{2n} \cos x \, dx = (2n)! \left[\sum\limits_{k=0}^{n}(-1)^k \dfrac{x^{2n-2k}}{(2n-2k)!} \sin x + \sum\limits_{k=0}^{n-1}(-1)^k \dfrac{x^{2n-2k-1}}{(2n-2k-1)!} \cos x\right].$

5. $\int x^{2n+1} \cos x \, dx = (2n+1)! \sum\limits_{k=0}^{n}\left[(-1)^k \dfrac{x^{2n-2k+1}}{(2n-2k+1)!} \sin x + \dfrac{x^{2n-2k}}{(2n-2k)!} \cos x\right].$

6. $\int x^p \cos x \, dx = x^p \sin x + px^{p-1} \cos x - p(p-1) \int x^{p-2} \cos x \, dx.$

7. $\int \cos^2 x \, dx = \tfrac{1}{2}x + \tfrac{1}{4} \sin 2x.$

8. $\displaystyle\int \cos^3 x \, dx = \sin x - \frac{1}{3}\sin^3 x.$

9. $\displaystyle\int \cos^{2n} x \, dx = \frac{1}{2^{2n}} C_{2n}^n x + \frac{1}{2^{2n-1}} \sum_{k=0}^{n-1} C_{2n}^k \frac{\sin[(2n-2k)x]}{2n-2k}.$

10. $\displaystyle\int \cos^{2n+1} x \, dx = \frac{1}{2^{2n}} \sum_{k=0}^{n} C_{2n+1}^k \frac{\sin[(2n-2k+1)x]}{2n-2k+1}.$

11. $\displaystyle\int \frac{dx}{\cos x} = \ln\left|\tan\left(\frac{x}{2}+\frac{\pi}{4}\right)\right|.$

12. $\displaystyle\int \frac{dx}{\cos^2 x} = \tan x.$

13. $\displaystyle\int \frac{dx}{\cos^3 x} = \frac{\sin x}{2\cos^2 x} + \frac{1}{2}\ln\left|\tan\left(\frac{x}{2}+\frac{\pi}{4}\right)\right|.$

14. $\displaystyle\int \frac{dx}{\cos^n x} = \frac{\sin x}{(n-1)\cos^{n-1} x} + \frac{n-2}{n-1}\int \frac{dx}{\cos^{n-2} x}, \quad n > 1.$

15. $\displaystyle\int \frac{x\,dx}{\cos^{2n} x} = \sum_{k=0}^{n-1} \frac{(2n-2)(2n-4)\ldots(2n-2k+2)}{(2n-1)(2n-3)\ldots(2n-2k+3)} \frac{(2n-2k)x\sin x - \cos x}{(2n-2k+1)(2n-2k)\cos^{2n-2k+1} x}$

$\displaystyle\qquad\qquad + \frac{2^{n-1}(n-1)!}{(2n-1)!!}\left(x\tan x + \ln|\cos x|\right).$

16. $\displaystyle\int \cos ax \cos bx \, dx = \frac{\sin\big[(b-a)x\big]}{2(b-a)} + \frac{\sin\big[(b+a)x\big]}{2(b+a)}, \quad a \neq \pm b.$

17. $\displaystyle\int \frac{dx}{a+b\cos x} = \begin{cases} \dfrac{2}{\sqrt{a^2-b^2}}\arctan\dfrac{(a-b)\tan(x/2)}{\sqrt{a^2-b^2}} & \text{if } a^2 > b^2, \\[4mm] \dfrac{1}{\sqrt{b^2-a^2}}\ln\left|\dfrac{\sqrt{b^2-a^2}+(b-a)\tan(x/2)}{\sqrt{b^2-a^2}-(b-a)\tan(x/2)}\right| & \text{if } b^2 > a^2. \end{cases}$

18. $\displaystyle\int \frac{dx}{(a+b\cos x)^2} = \frac{b\sin x}{(b^2-a^2)(a+b\cos x)} - \frac{a}{b^2-a^2}\int \frac{dx}{a+b\cos x}.$

19. $\displaystyle\int \frac{dx}{a^2+b^2\cos^2 x} = \frac{1}{a\sqrt{a^2+b^2}}\arctan\frac{a\tan x}{\sqrt{a^2+b^2}}.$

20. $\displaystyle\int \frac{dx}{a^2-b^2\cos^2 x} = \begin{cases} \dfrac{1}{a\sqrt{a^2-b^2}}\arctan\dfrac{a\tan x}{\sqrt{a^2-b^2}} & \text{if } a^2 > b^2, \\[4mm] \dfrac{1}{2a\sqrt{b^2-a^2}}\ln\left|\dfrac{\sqrt{b^2-a^2}-a\tan x}{\sqrt{b^2-a^2}+a\tan x}\right| & \text{if } b^2 > a^2. \end{cases}$

21. $\displaystyle\int e^{ax}\cos bx \, dx = e^{ax}\left[\frac{b}{a^2+b^2}\sin bx + \frac{a}{a^2+b^2}\cos bx\right].$

22. $\displaystyle\int e^{ax}\cos^2 x \, dx = \frac{e^{ax}}{a^2+4}\left(a\cos^2 x + 2\sin x\cos x + \frac{2}{a}\right).$

23. $\displaystyle\int e^{ax}\cos^n x \, dx = \frac{e^{ax}\cos^{n-1} x}{a^2+n^2}(a\cos x + n\sin x) + \frac{n(n-1)}{a^2+n^2}\int e^{ax}\cos^{n-2} x \, dx.$

▶ **Integrals containing $\sin x$.** Notation: $n = 1, 2, \ldots$

24. $\displaystyle\int \sin(a+bx) \, dx = -\frac{1}{b}\cos(a+bx).$

25. $\displaystyle\int x\sin x \, dx = \sin x - x\cos x.$

26. $\displaystyle\int x^2 \sin x \, dx = 2x \sin x - (x^2 - 2) \cos x.$

27. $\displaystyle\int x^3 \sin x \, dx = (3x^2 - 6) \sin x - (x^3 - 6x) \cos x.$

28. $\displaystyle\int x^{2n} \sin x \, dx = (2n)! \left[\sum_{k=0}^{n} (-1)^{k+1} \frac{x^{2n-2k}}{(2n-2k)!} \cos x + \sum_{k=0}^{n-1} (-1)^k \frac{x^{2n-2k-1}}{(2n-2k-1)!} \sin x \right].$

29. $\displaystyle\int x^{2n+1} \sin x \, dx = (2n+1)! \sum_{k=0}^{n} \left[(-1)^{k+1} \frac{x^{2n-2k+1}}{(2n-2k+1)!} \cos x + (-1)^k \frac{x^{2n-2k}}{(2n-2k)!} \sin x \right].$

30. $\displaystyle\int x^p \sin x \, dx = -x^p \cos x + px^{p-1} \sin x - p(p-1) \int x^{p-2} \sin x \, dx.$

31. $\displaystyle\int \sin^2 x \, dx = \tfrac{1}{2}x - \tfrac{1}{4} \sin 2x.$

32. $\displaystyle\int x \sin^2 x \, dx = \tfrac{1}{4}x^2 - \tfrac{1}{4}x \sin 2x - \tfrac{1}{8} \cos 2x.$

33. $\displaystyle\int \sin^3 x \, dx = -\cos x + \tfrac{1}{3} \cos^3 x.$

34. $\displaystyle\int \sin^{2n} x \, dx = \frac{1}{2^{2n}} C_{2n}^n x + \frac{(-1)^n}{2^{2n-1}} \sum_{k=0}^{n-1} (-1)^k C_{2n}^k \frac{\sin[(2n-2k)x]}{2n-2k},$

    where $C_m^k = \dfrac{m!}{k!\,(m-k)!}$ are binomial coefficients $(0! = 1)$.

35. $\displaystyle\int \sin^{2n+1} x \, dx = \frac{1}{2^{2n}} \sum_{k=0}^{n} (-1)^{n+k+1} C_{2n+1}^k \frac{\cos[(2n-2k+1)x]}{2n-2k+1}.$

36. $\displaystyle\int \frac{dx}{\sin x} = \ln \left| \tan \frac{x}{2} \right|.$

37. $\displaystyle\int \frac{dx}{\sin^2 x} = -\cot x.$

38. $\displaystyle\int \frac{dx}{\sin^3 x} = -\frac{\cos x}{2 \sin^2 x} + \frac{1}{2} \ln \left| \tan \frac{x}{2} \right|.$

39. $\displaystyle\int \frac{dx}{\sin^n x} = -\frac{\cos x}{(n-1) \sin^{n-1} x} + \frac{n-2}{n-1} \int \frac{dx}{\sin^{n-2} x}, \quad n > 1.$

40. $\displaystyle\int \frac{x \, dx}{\sin^{2n} x} = -\sum_{k=0}^{n-1} \frac{(2n-2)(2n-4)\ldots(2n-2k+2)}{(2n-1)(2n-3)\ldots(2n-2k+3)} \frac{\sin x + (2n-2k)x \cos x}{(2n-2k+1)(2n-2k) \sin^{2n-2k+1} x}$

    $\displaystyle + \frac{2^{n-1}(n-1)!}{(2n-1)!!} \left(\ln |\sin x| - x \cot x \right).$

41. $\displaystyle\int \sin ax \, \sin bx \, dx = \frac{\sin[(b-a)x]}{2(b-a)} - \frac{\sin[(b+a)x]}{2(b+a)}, \quad a \neq \pm b.$

42. $\displaystyle\int \frac{dx}{a + b \sin x} = \begin{cases} \dfrac{2}{\sqrt{a^2 - b^2}} \arctan \dfrac{b + a \tan x/2}{\sqrt{a^2 - b^2}} & \text{if } a^2 > b^2, \\[3mm] \dfrac{1}{\sqrt{b^2 - a^2}} \ln \left| \dfrac{b - \sqrt{b^2 - a^2} + a \tan x/2}{b + \sqrt{b^2 - a^2} + a \tan x/2} \right| & \text{if } b^2 > a^2. \end{cases}$

43. $\displaystyle\int \frac{dx}{(a + b \sin x)^2} = \frac{b \cos x}{(a^2 - b^2)(a + b \sin x)} + \frac{a}{a^2 - b^2} \int \frac{dx}{a + b \sin x}.$

44. $\displaystyle\int \frac{dx}{a^2 + b^2 \sin^2 x} = \frac{1}{a\sqrt{a^2 + b^2}} \arctan \frac{\sqrt{a^2 + b^2} \, \tan x}{a}.$

45. $\displaystyle\int \frac{dx}{a^2 - b^2 \sin^2 x} = \begin{cases} \dfrac{1}{a\sqrt{a^2-b^2}} \arctan \dfrac{\sqrt{a^2-b^2}\,\tan x}{a} & \text{if } a^2 > b^2, \\[3mm] \dfrac{1}{2a\sqrt{b^2-a^2}} \ln\left|\dfrac{\sqrt{b^2-a^2}\,\tan x + a}{\sqrt{b^2-a^2}\,\tan x - a}\right| & \text{if } b^2 > a^2. \end{cases}$

46. $\displaystyle\int \frac{\sin x\, dx}{\sqrt{1+k^2\sin^2 x}} = -\frac{1}{k}\arcsin\frac{k\cos x}{\sqrt{1+k^2}}.$

47. $\displaystyle\int \frac{\sin x\, dx}{\sqrt{1-k^2\sin^2 x}} = -\frac{1}{k}\ln\left|k\cos x + \sqrt{1-k^2\sin^2 x}\right|.$

48. $\displaystyle\int \sin x\sqrt{1+k^2\sin^2 x}\, dx = -\frac{\cos x}{2}\sqrt{1+k^2\sin^2 x} - \frac{1+k^2}{2k}\arcsin\frac{k\cos x}{\sqrt{1+k^2}}.$

49. $\displaystyle\int \sin x\sqrt{1-k^2\sin^2 x}\, dx = -\frac{\cos x}{2}\sqrt{1-k^2\sin^2 x} - \frac{1-k^2}{2k}\ln\left|k\cos x + \sqrt{1-k^2\sin^2 x}\right|.$

50. $\displaystyle\int e^{ax}\sin bx\, dx = e^{ax}\left[\frac{a}{a^2+b^2}\sin bx - \frac{b}{a^2+b^2}\cos bx\right].$

51. $\displaystyle\int e^{ax}\sin^2 x\, dx = \frac{e^{ax}}{a^2+4}\left(a\sin^2 x - 2\sin x\cos x + \frac{2}{a}\right).$

52. $\displaystyle\int e^{ax}\sin^n x\, dx = \frac{e^{ax}\sin^{n-1} x}{a^2+n^2}(a\sin x - n\cos x) + \frac{n(n-1)}{a^2+n^2}\int e^{ax}\sin^{n-2} x\, dx.$

▶ **Integrals containing** $\sin x$ **and** $\cos x$.

53. $\displaystyle\int \sin ax\cos bx\, dx = -\frac{\cos[(a+b)x]}{2(a+b)} - \frac{\cos[(a-b)x]}{2(a-b)}, \quad a \neq \pm b.$

54. $\displaystyle\int \frac{dx}{b^2\cos^2 ax + c^2\sin^2 ax} = \frac{1}{abc}\arctan\left(\frac{c}{b}\tan ax\right).$

55. $\displaystyle\int \frac{dx}{b^2\cos^2 ax - c^2\sin^2 ax} = \frac{1}{2abc}\ln\left|\frac{c\tan ax + b}{c\tan ax - b}\right|.$

56. $\displaystyle\int \frac{dx}{\cos^{2n} x\,\sin^{2m} x} = \sum_{k=0}^{n+m-1} C_{n+m-1}^k \frac{\tan^{2k-2m+1} x}{2k-2m+1}, \quad n, m = 1, 2, \ldots$

57. $\displaystyle\int \frac{dx}{\cos^{2n+1} x\,\sin^{2m+1} x} = C_{n+m}^m \ln|\tan x| + \sum_{k=0}^{n+m} C_{n+m}^k \frac{\tan^{2k-2m} x}{2k-2m}, \quad n, m = 1, 2, \ldots$

▶ **Reduction formulas.** The parameters p and q below can assume any values, except for those at which the denominators on the right-hand side vanish.

58. $\displaystyle\int \sin^p x\cos^q x\, dx = -\frac{\sin^{p-1} x\cos^{q+1} x}{p+q} + \frac{p-1}{p+q}\int \sin^{p-2} x\cos^q x\, dx.$

59. $\displaystyle\int \sin^p x\cos^q x\, dx = \frac{\sin^{p+1} x\cos^{q-1} x}{p+q} + \frac{q-1}{p+q}\int \sin^p x\cos^{q-2} x\, dx.$

60. $\displaystyle\int \sin^p x\cos^q x\, dx = \frac{\sin^{p-1} x\cos^{q-1} x}{p+q}\left(\sin^2 x - \frac{q-1}{p+q-2}\right)$
$\displaystyle\qquad + \frac{(p-1)(q-1)}{(p+q)(p+q-2)}\int \sin^{p-2} x\cos^{q-2} x\, dx.$

61. $\displaystyle\int \sin^p x\cos^q x\, dx = \frac{\sin^{p+1} x\cos^{q+1} x}{p+1} + \frac{p+q+2}{p+1}\int \sin^{p+2} x\cos^q x\, dx.$

62. $\displaystyle\int \sin^p x\cos^q x\, dx = -\frac{\sin^{p+1} x\cos^{q+1} x}{q+1} + \frac{p+q+2}{q+1}\int \sin^p x\cos^{q+2} x\, dx.$

63. $\displaystyle\int \sin^p x \cos^q x \, dx = -\frac{\sin^{p-1} x \cos^{q+1} x}{q+1} + \frac{p-1}{q+1} \int \sin^{p-2} x \cos^{q+2} x \, dx.$

64. $\displaystyle\int \sin^p x \cos^q x \, dx = \frac{\sin^{p+1} x \cos^{q-1} x}{p+1} + \frac{q-1}{p+1} \int \sin^{p+2} x \cos^{q-2} x \, dx.$

▶ **Integrals containing $\tan x$ and $\cot x$.**

65. $\displaystyle\int \tan x \, dx = -\ln|\cos x|.$

66. $\displaystyle\int \tan^2 x \, dx = \tan x - x.$

67. $\displaystyle\int \tan^3 x \, dx = \frac{1}{2} \tan^2 x + \ln|\cos x|.$

68. $\displaystyle\int \tan^{2n} x \, dx = (-1)^n x - \sum_{k=1}^{n} \frac{(-1)^k (\tan x)^{2n-2k+1}}{2n-2k+1}, \quad n = 1, 2, \ldots$

69. $\displaystyle\int \tan^{2n+1} x \, dx = (-1)^{n+1} \ln|\cos x| - \sum_{k=1}^{n} \frac{(-1)^k (\tan x)^{2n-2k+2}}{2n-2k+2}, \quad n = 1, 2, \ldots$

70. $\displaystyle\int \frac{dx}{a+b\tan x} = \frac{1}{a^2+b^2} \left(ax + b \ln|a \cos x + b \sin x| \right).$

71. $\displaystyle\int \frac{\tan x \, dx}{\sqrt{a+b\tan^2 x}} = \frac{1}{\sqrt{b-a}} \arccos\left(\sqrt{1-\frac{a}{b}} \cos x \right), \quad b > a, \ b > 0.$

72. $\displaystyle\int \cot x \, dx = \ln|\sin x|.$

73. $\displaystyle\int \cot^2 x \, dx = -\cot x - x.$

74. $\displaystyle\int \cot^3 x \, dx = -\frac{1}{2} \cot^2 x - \ln|\sin x|.$

75. $\displaystyle\int \cot^{2n} x \, dx = (-1)^n x + \sum_{k=1}^{n} \frac{(-1)^k (\cot x)^{2n-2k+1}}{2n-2k+1}, \quad n = 1, 2, \ldots$

76. $\displaystyle\int \cot^{2n+1} x \, dx = (-1)^n \ln|\sin x| + \sum_{k=1}^{n} \frac{(-1)^k (\cot x)^{2n-2k+2}}{2n-2k+2}, \quad n = 1, 2, \ldots$

77. $\displaystyle\int \frac{dx}{a+b\cot x} = \frac{1}{a^2+b^2} \left(ax - b \ln|a \sin x + b \cos x| \right).$

2.7. Integrals Containing Inverse Trigonometric Functions

1. $\displaystyle\int \arcsin \frac{x}{a} \, dx = x \arcsin \frac{x}{a} + \sqrt{a^2 - x^2}.$

2. $\displaystyle\int \left(\arcsin \frac{x}{a} \right)^2 dx = x \left(\arcsin \frac{x}{a} \right)^2 - 2x + 2\sqrt{a^2 - x^2} \, \arcsin \frac{x}{a}.$

3. $\displaystyle\int x \arcsin \frac{x}{a} \, dx = \frac{1}{4}(2x^2 - a^2) \arcsin \frac{x}{a} + \frac{x}{4} \sqrt{a^2 - x^2}.$

4. $\displaystyle\int x^2 \arcsin \frac{x}{a} \, dx = \frac{x^3}{3} \arcsin \frac{x}{a} + \frac{1}{9}(x^2 + 2a^2)\sqrt{a^2 - x^2}.$

5. $\displaystyle\int \arccos \frac{x}{a}\, dx = x \arccos \frac{x}{a} - \sqrt{a^2 - x^2}.$

6. $\displaystyle\int \left(\arccos \frac{x}{a}\right)^2 dx = x \left(\arccos \frac{x}{a}\right)^2 - 2x - 2\sqrt{a^2 - x^2}\, \arccos \frac{x}{a}.$

7. $\displaystyle\int x \arccos \frac{x}{a}\, dx = \frac{1}{4}(2x^2 - a^2) \arccos \frac{x}{a} - \frac{x}{4}\sqrt{a^2 - x^2}.$

8. $\displaystyle\int x^2 \arccos \frac{x}{a}\, dx = \frac{x^3}{3} \arccos \frac{x}{a} - \frac{1}{9}(x^2 + 2a^2)\sqrt{a^2 - x^2}.$

9. $\displaystyle\int \arctan \frac{x}{a}\, dx = x \arctan \frac{x}{a} - \frac{a}{2}\ln(a^2 + x^2).$

10. $\displaystyle\int x \arctan \frac{x}{a}\, dx = \frac{1}{2}(x^2 + a^2) \arctan \frac{x}{a} - \frac{ax}{2}.$

11. $\displaystyle\int x^2 \arctan \frac{x}{a}\, dx = \frac{x^3}{3} \arctan \frac{x}{a} - \frac{ax^2}{6} + \frac{a^3}{6}\ln(a^2 + x^2).$

12. $\displaystyle\int \operatorname{arccot} \frac{x}{a}\, dx = x \operatorname{arccot} \frac{x}{a} + \frac{a}{2}\ln(a^2 + x^2).$

13. $\displaystyle\int x \operatorname{arccot} \frac{x}{a}\, dx = \frac{1}{2}(x^2 + a^2) \operatorname{arccot} \frac{x}{a} + \frac{ax}{2}.$

14. $\displaystyle\int x^2 \operatorname{arccot} \frac{x}{a}\, dx = \frac{x^3}{3} \operatorname{arccot} \frac{x}{a} + \frac{ax^2}{6} - \frac{a^3}{6}\ln(a^2 + x^2).$

⊙ References for Supplement 2: H. B. Dwight (1961), I. S. Gradshteyn and I. M. Ryzhik (1980), A. P. Prudnikov, Yu. A. Brychkov, and O. I. Marichev (1986, 1988).

Supplement 3

Tables of Definite Integrals

Throughout Supplement 3 it is assumed that n is a positive integer, unless otherwise specified.

3.1. Integrals Containing Power-Law Functions

1. $\displaystyle\int_0^\infty \frac{dx}{ax^2 + b} = \frac{\pi}{2\sqrt{ab}}$.

2. $\displaystyle\int_0^\infty \frac{dx}{x^4 + 1} = \frac{\pi\sqrt{2}}{4}$.

3. $\displaystyle\int_0^1 \frac{x^n\, dx}{x + 1} = (-1)^n \left[\ln 2 + \sum_{k=1}^n \frac{(-1)^k}{k} \right]$.

4. $\displaystyle\int_0^\infty \frac{x^{a-1}\, dx}{x + 1} = \frac{\pi}{\sin(\pi a)}, \quad 0 < a < 1$.

5. $\displaystyle\int_0^\infty \frac{x^{\lambda-1}\, dx}{(1 + ax)^2} = \frac{\pi(1 - \lambda)}{a^\lambda \sin(\pi\lambda)}, \quad 0 < \lambda < 2$.

6. $\displaystyle\int_0^1 \frac{dx}{x^2 + 2x \cos\beta + 1} = \frac{\beta}{2 \sin\beta}$.

7. $\displaystyle\int_0^1 \frac{\left(x^a + x^{-a}\right) dx}{x^2 + 2x \cos\beta + 1} = \frac{\pi \sin(a\beta)}{\sin(\pi a) \sin\beta}, \quad |a| < 1,\ \beta \neq (2n + 1)\pi$.

8. $\displaystyle\int_0^\infty \frac{x^{\lambda-1}\, dx}{(x + a)(x + b)} = \frac{\pi(a^{\lambda-1} - b^{\lambda-1})}{(b - a) \sin(\pi\lambda)}, \quad 0 < \lambda < 2$.

9. $\displaystyle\int_0^\infty \frac{x^{\lambda-1}(x + c)\, dx}{(x + a)(x + b)} = \frac{\pi}{\sin(\pi\lambda)} \left(\frac{a - c}{a - b} a^{\lambda-1} + \frac{b - c}{b - a} b^{\lambda-1} \right), \quad 0 < \lambda < 1$.

10. $\displaystyle\int_0^\infty \frac{x^\lambda\, dx}{(x + 1)^3} = \frac{\pi\lambda(1 - \lambda)}{2 \sin(\pi\lambda)}, \quad -1 < \lambda < 2$.

11. $\displaystyle\int_0^\infty \frac{x^{\lambda-1}\, dx}{(x^2 + a^2)(x^2 + b^2)} = \frac{\pi\left(b^{\lambda-2} - a^{\lambda-2}\right)}{2\left(a^2 - b^2\right) \sin(\pi\lambda/2)}, \quad 0 < \lambda < 4$.

12. $\displaystyle\int_0^1 x^a(1 - x)^{1-a}\, dx = \frac{\pi a(1 - a)}{2 \sin(\pi a)}, \quad -1 < a < 1$.

13. $\displaystyle\int_0^1 \frac{dx}{x^a(1 - x)^{1-a}} = \frac{\pi}{\sin(\pi a)}, \quad 0 < a < 1$.

14. $\displaystyle\int_0^1 \frac{x^a\,dx}{(1-x)^a} = \frac{\pi a}{\sin(\pi a)}, \quad -1 < a < 1.$

15. $\displaystyle\int_0^1 x^{p-1}(1-x)^{q-1}\,dx \equiv B(p,q) = \frac{\Gamma(p)\Gamma(q)}{\Gamma(p+q)}, \quad p,q > 0.$

16. $\displaystyle\int_0^1 x^{p-1}(1-x^q)^{-p/q}\,dx = \frac{\pi}{q\sin(\pi p/q)}, \quad q > p > 0.$

17. $\displaystyle\int_0^1 x^{p+q-1}(1-x^q)^{-p/q}\,dx = \frac{\pi p}{q^2\sin(\pi p/q)}, \quad q > p.$

18. $\displaystyle\int_0^1 x^{q/p-1}(1-x^q)^{-1/p}\,dx = \frac{\pi}{q\sin(\pi/p)}, \quad p > 1,\ q > 0.$

19. $\displaystyle\int_0^1 \frac{x^{p-1}-x^{-p}}{1-x}\,dx = \pi\cot(\pi p), \quad |p| < 1.$

20. $\displaystyle\int_0^1 \frac{x^{p-1}-x^{-p}}{1+x}\,dx = \frac{\pi}{\sin(\pi p)}, \quad |p| < 1.$

21. $\displaystyle\int_0^1 \frac{x^p-x^{-p}}{x-1}\,dx = \frac{1}{p} - \pi\cot(\pi p), \quad |p| < 1.$

22. $\displaystyle\int_0^1 \frac{x^p-x^{-p}}{1+x}\,dx = \frac{1}{p} - \frac{\pi}{\sin(\pi p)}, \quad |p| < 1.$

23. $\displaystyle\int_0^1 \frac{x^{1+p}-x^{1-p}}{1-x^2}\,dx = \frac{\pi}{2}\cot\left(\frac{\pi p}{2}\right) - \frac{1}{p}, \quad |p| < 1.$

24. $\displaystyle\int_0^1 \frac{x^{1+p}-x^{1-p}}{1+x^2}\,dx = \frac{1}{p} - \frac{\pi}{2\sin(\pi p/2)}, \quad |p| < 1.$

25. $\displaystyle\int_0^\infty \frac{x^{p-1}-x^{q-1}}{1-x}\,dx = \pi[\cot(\pi p) - \cot(\pi q)], \quad p,q > 0.$

26. $\displaystyle\int_0^1 \frac{dx}{\sqrt{(1+a^2x)(1-x)}} = \frac{2}{a}\arctan a.$

27. $\displaystyle\int_0^1 \frac{dx}{\sqrt{(1-a^2x)(1-x)}} = \frac{1}{a}\ln\frac{1+a}{1-a}.$

28. $\displaystyle\int_{-1}^1 \frac{dx}{(a-x)\sqrt{1-x^2}} = \frac{\pi}{\sqrt{a^2-1}}, \quad 1 < a.$

29. $\displaystyle\int_0^1 \frac{x^n\,dx}{\sqrt{1-x}} = \frac{2\,(2n)!!}{(2n+1)!!}, \quad n = 1, 2, \ldots$

30. $\displaystyle\int_0^1 \frac{x^{n-1/2}\,dx}{\sqrt{1-x}} = \frac{\pi\,(2n-1)!!}{(2n)!!}, \quad n = 1, 2, \ldots$

31. $\displaystyle\int_0^1 \frac{x^{2n}\,dx}{\sqrt{1-x^2}} = \frac{\pi}{2}\,\frac{1\cdot 3\ldots(2n-1)}{2\cdot 4\ldots(2n)}, \quad n = 1, 2, \ldots$

32. $\displaystyle\int_0^1 \frac{x^{2n+1}\,dx}{\sqrt{1-x^2}} = \frac{2\cdot 4\ldots(2n)}{1\cdot 3\ldots(2n+1)}, \quad n = 1, 2, \ldots$

33. $\displaystyle\int_0^\infty \frac{x^{\lambda-1}\,dx}{(1+ax)^{n+1}} = (-1)^n\,\frac{\pi C_{\lambda-1}^n}{a^\lambda\sin(\pi\lambda)}, \quad 0 < \lambda < n+1.$

34. $\displaystyle\int_0^\infty \frac{x^m\,dx}{(a+bx)^{n+1/2}} = 2^{m+1}m!\,\frac{(2n-2m-3)!!}{(2n-1)!!}\,\frac{a^{m-n+1/2}}{b^{m+1}}, \quad a,b>0,$
$n,m=1,2,\ldots, \quad m < b - \frac{1}{2}.$

35. $\displaystyle\int_0^\infty \frac{dx}{(x^2+a^2)^n} = \frac{\pi}{2}\frac{(2n-3)!!}{(2n-2)!!}\frac{1}{a^{2n-1}}, \quad n=1,2,\ldots$

36. $\displaystyle\int_0^\infty \frac{(x+1)^{\lambda-1}}{(x+a)^{\lambda+1}}\,dx = \frac{1-a^{-\lambda}}{\lambda(a-1)}, \quad a>0.$

37. $\displaystyle\int_0^1 \frac{x^{\lambda-1}\,dx}{(1+ax)(1-x)^\lambda} = \frac{\pi}{(1+a)^\lambda\sin(\pi\lambda)}, \quad 0<\lambda<1,\, a>-1.$

38. $\displaystyle\int_0^1 \frac{x^{\lambda-1/2}\,dx}{(1+ax)^\lambda(1-x)^\lambda} = 2\pi^{-1/2}\Gamma\!\left(\lambda+\tfrac{1}{2}\right)\Gamma\!\left(1-\lambda\right)\cos^{2\lambda}k\,\frac{\sin[(2\lambda-1)k]}{(2\lambda-1)\sin k}, \quad k=\arctan\sqrt{a};$
$-\frac{1}{2}<\lambda<1,\, a>0.$

39. $\displaystyle\int_0^\infty \frac{x^{a-1}\,dx}{x^b+1} = \frac{\pi}{b\sin(\pi a/b)}, \quad 0<a\le b.$

40. $\displaystyle\int_0^\infty \frac{x^{a-1}\,dx}{(x^b+1)^2} = \frac{\pi(a-b)}{b^2\sin[\pi(a-b)/b]}, \quad a<2b.$

41. $\displaystyle\int_0^\infty \frac{x^{\lambda-1/2}\,dx}{(x+a)^\lambda(x+b)^\lambda} = \sqrt{\pi}\left(\sqrt{a}+\sqrt{b}\right)^{1-2\lambda}\frac{\Gamma(\lambda-1/2)}{\Gamma(\lambda)}, \quad \lambda>0.$

42. $\displaystyle\int_0^\infty \frac{1-x^a}{1-x^b}\,x^{c-1}\,dx = \frac{\pi\sin A}{b\sin C\sin(A+C)}, \quad A=\frac{\pi a}{b}, \quad C=\frac{\pi c}{b}; \quad a+c<b, \quad c>0.$

43. $\displaystyle\int_0^\infty \frac{x^{a-1}\,dx}{(1+x^2)^{1-b}} = \tfrac{1}{2}B\!\left(\tfrac{1}{2}a, 1-b-\tfrac{1}{2}a\right), \quad \tfrac{1}{2}a+b<1, \quad a>0.$

44. $\displaystyle\int_0^\infty \frac{x^{2m}\,dx}{(ax^2+b)^n} = \frac{\pi(2m-1)!!\,(2n-2m-3)!!}{2\,(2n-2)!!\,a^m b^{n-m-1}\sqrt{ab}}, \quad a,b>0, \quad n>m+1.$

45. $\displaystyle\int_0^\infty \frac{x^{2m+1}\,dx}{(ax^2+b)^n} = \frac{m!\,(n-m-2)!}{2(n-1)!a^{m+1}b^{n-m-1}}, \quad ab>0, \quad n>m+1\ge 1.$

46. $\displaystyle\int_0^\infty \frac{x^{\mu-1}\,dx}{(1+ax^p)^\nu} = \frac{1}{pa^{\mu/p}}B\!\left(\frac{\mu}{p}, \nu-\frac{\mu}{p}\right), \quad p>0, \quad 0<\mu<p\nu.$

47. $\displaystyle\int_0^\infty \left(\sqrt{x^2+a^2}-x\right)^n dx = \frac{na^{n+1}}{n^2-1}, \quad n=2,3,\ldots$

48. $\displaystyle\int_0^\infty \frac{dx}{\left(x+\sqrt{x^2+a^2}\,\right)^n} = \frac{n}{a^{n-1}(n^2-1)}, \quad n=2,3,\ldots$

49. $\displaystyle\int_0^\infty x^m\left(\sqrt{x^2+a^2}-x\right)^n dx = \frac{n\cdot m!\,a^{n+m+1}}{(n-m-1)(n-m+1)\ldots(n+m+1)},$
$n,m=1,2,\ldots, \quad 0\le m\le n-2$

50. $\displaystyle\int_0^\infty \frac{x^m\,dx}{\left(x+\sqrt{x^2+a^2}\,\right)^n} = \frac{n\cdot m!}{(n-m-1)(n-m+1)\ldots(n+m+1)a^{n-m-1}}, \quad n=2,3,\ldots$

3.2. Integrals Containing Exponential Functions

1. $\displaystyle\int_0^\infty e^{-ax}\,dx = \frac{1}{a}, \quad a>0.$

2. $\displaystyle\int_0^1 x^n e^{-ax}\,dx = \frac{n!}{a^{n+1}} - e^{-a}\sum_{k=0}^n \frac{n!}{k!}\frac{1}{a^{n-k+1}}, \quad a > 0, \ n = 1, 2, \ldots$

3. $\displaystyle\int_0^\infty x^n e^{-ax}\,dx = \frac{n!}{a^{n+1}}, \quad a > 0, \ n = 1, 2, \ldots$

4. $\displaystyle\int_0^\infty \frac{e^{-ax}}{\sqrt{x}}\,dx = \sqrt{\frac{\pi}{a}}, \quad a > 0.$

5. $\displaystyle\int_0^\infty x^{\nu-1} e^{-\mu x}\,dx = \frac{\Gamma(\nu)}{\mu^\nu}, \quad \mu, \nu > 0.$

6. $\displaystyle\int_0^\infty \frac{dx}{1+e^{ax}} = \frac{\ln 2}{a}.$

7. $\displaystyle\int_0^\infty \frac{x^{2n-1}\,dx}{e^{px}-1} = (-1)^{n-1}\left(\frac{2\pi}{p}\right)^{2n}\frac{B_{2n}}{4n}, \quad n = 1, 2, \ldots \quad (B_m \text{ are the Bernoulli numbers}).$

8. $\displaystyle\int_0^\infty \frac{x^{2n-1}\,dx}{e^{px}+1} = (1-2^{1-2n})\left(\frac{2\pi}{p}\right)^{2n}\frac{|B_{2n}|}{4n}, \quad n = 1, 2, \ldots$

9. $\displaystyle\int_{-\infty}^\infty \frac{e^{-px}\,dx}{1+e^{-qx}} = \frac{\pi}{q\sin(\pi p/q)}, \quad q > p > 0 \text{ or } 0 > p > q.$

10. $\displaystyle\int_0^\infty \frac{e^{ax}+e^{-ax}}{e^{bx}+e^{-bx}}\,dx = \frac{\pi}{2b\cos\left(\dfrac{\pi a}{2b}\right)}, \quad b > a.$

11. $\displaystyle\int_0^\infty \frac{e^{-px}-e^{-qx}}{1-e^{-(p+q)x}}\,dx = \frac{\pi}{p+q}\cot\frac{\pi p}{p+q}, \quad p, q > 0.$

12. $\displaystyle\int_0^\infty \left(1-e^{-\beta x}\right)^\nu e^{-\mu x}\,dx = \frac{1}{\beta}B\left(\frac{\mu}{\beta}, \nu+1\right).$

13. $\displaystyle\int_0^\infty \exp\left(-ax^2\right)dx = \frac{1}{2}\sqrt{\frac{\pi}{a}}, \quad a > 0.$

14. $\displaystyle\int_0^\infty x^{2n+1}\exp\left(-ax^2\right)dx = \frac{n!}{2a^{n+1}}, \quad a > 0, \ n = 1, 2, \ldots$

15. $\displaystyle\int_0^\infty x^{2n}\exp\left(-ax^2\right)dx = \frac{1\cdot 3\ldots(2n-1)\sqrt{\pi}}{2^{n+1}a^{n+1/2}}, \quad a > 0, \ n = 1, 2, \ldots$

16. $\displaystyle\int_{-\infty}^\infty \exp\left(-a^2x^2 \pm bx\right)dx = \frac{\sqrt{\pi}}{|a|}\exp\left(\frac{b^2}{4a^2}\right).$

17. $\displaystyle\int_0^\infty \exp\left(-ax^2 - \frac{b}{x^2}\right)dx = \frac{1}{2}\sqrt{\frac{\pi}{a}}\exp(-2\sqrt{ab}), \quad a, b > 0.$

18. $\displaystyle\int_0^\infty \exp\left(-x^a\right)dx = \frac{1}{a}\Gamma\left(\frac{1}{a}\right), \quad a > 0.$

3.3. Integrals Containing Hyperbolic Functions

1. $\displaystyle\int_0^\infty \frac{dx}{\cosh ax} = \frac{\pi}{2|a|}.$

2. $\displaystyle\int_0^\infty \frac{dx}{a+b\cosh x} = \begin{cases} \dfrac{2}{\sqrt{b^2-a^2}}\arctan\dfrac{\sqrt{b^2-a^2}}{a+b} & \text{if } |b| > |a|, \\[2ex] \dfrac{1}{\sqrt{a^2-b^2}}\ln\dfrac{a+b+\sqrt{a^2-b^2}}{a+b-\sqrt{a^2+b^2}} & \text{if } |b| < |a|. \end{cases}$

3. $\displaystyle\int_0^\infty \frac{x^{2n}\,dx}{\cosh ax} = \left(\frac{\pi}{2a}\right)^{2n+1}|E_{2n}|, \quad a > 0.$

4. $\displaystyle\int_0^\infty \frac{x^{2n}}{\cosh^2 ax}\,dx = \frac{\pi^{2n}(2^{2n}-2)}{a(2a)^{2n}}|B_{2n}|, \quad a > 0.$

5. $\displaystyle\int_0^\infty \frac{\cosh ax}{\cosh bx}\,dx = \frac{\pi}{2b\cos\left(\dfrac{\pi a}{2b}\right)}, \quad b > |a|.$

6. $\displaystyle\int_0^\infty x^{2n}\frac{\cosh ax}{\cosh bx}\,dx = \frac{\pi}{2b}\frac{d^{2n}}{da^{2n}}\frac{1}{\cos\left(\frac{1}{2}\pi a/b\right)}, \quad b > |a|, \quad n = 1, 2, \ldots$

7. $\displaystyle\int_0^\infty \frac{\cosh ax \cosh bx}{\cosh(cx)}\,dx = \frac{\pi}{c}\frac{\cos\left(\dfrac{\pi a}{2c}\right)\cos\left(\dfrac{\pi b}{2c}\right)}{\cos\left(\dfrac{\pi a}{c}\right)+\cos\left(\dfrac{\pi b}{c}\right)}, \quad c > |a| + |b|.$

8. $\displaystyle\int_0^\infty \frac{x\,dx}{\sinh ax} = \frac{\pi^2}{2a^2}, \quad a > 0.$

9. $\displaystyle\int_0^\infty \frac{dx}{a + b\sinh x} = \frac{1}{\sqrt{a^2+b^2}}\ln\frac{a+b+\sqrt{a^2+b^2}}{a+b-\sqrt{a^2+b^2}}, \quad ab \neq 0.$

10. $\displaystyle\int_0^\infty \frac{\sinh ax}{\sinh bx}\,dx = \frac{\pi}{2b}\tan\left(\frac{\pi a}{2b}\right), \quad b > |a|.$

11. $\displaystyle\int_0^\infty x^{2n}\frac{\sinh ax}{\sinh bx}\,dx = \frac{\pi}{2b}\frac{d^{2n}}{dx^{2n}}\tan\left(\frac{\pi a}{2b}\right), \quad b > |a|, \quad n = 1, 2, \ldots$

12. $\displaystyle\int_0^\infty \frac{x^{2n}}{\sinh^2 ax}\,dx = \frac{\pi^{2n}}{a^{2n+1}}|B_{2n}|, \quad a > 0.$

3.4. Integrals Containing Logarithmic Functions

1. $\displaystyle\int_0^1 x^{a-1}\ln^n x\,dx = (-1)^n n!\,a^{-n-1}, \quad a > 0, \quad n = 1, 2, \ldots$

2. $\displaystyle\int_0^1 \frac{\ln x}{x+1}\,dx = -\frac{\pi^2}{12}.$

3. $\displaystyle\int_0^1 \frac{x^n \ln x}{x+1}\,dx = (-1)^{n+1}\left[\frac{\pi^2}{12} + \sum_{k=1}^n \frac{(-1)^k}{k^2}\right], \quad n = 1, 2, \ldots$

4. $\displaystyle\int_0^1 \frac{x^{\mu-1}\ln x}{x+a}\,dx = \frac{\pi a^{\mu-1}}{\sin(\pi\mu)}\left[\ln a - \pi\cot(\pi\mu)\right], \quad 0 < \mu < 1.$

5. $\displaystyle\int_0^1 |\ln x|^\mu\,dx = \Gamma(\mu+1), \quad \mu > -1.$

6. $\displaystyle\int_0^\infty x^{\mu-1}\ln(1+ax)\,dx = \frac{\pi}{\mu a^\mu \sin(\pi\mu)}, \quad -1 < \mu < 0.$

7. $\displaystyle\int_0^1 x^{2n-1}\ln(1+x)\,dx = \frac{1}{2n}\sum_{k=1}^{2n}\frac{(-1)^{k-1}}{k}, \quad n = 1, 2, \ldots$

8. $\displaystyle\int_0^1 x^{2n}\ln(1+x)\,dx = \frac{1}{2n+1}\left[\ln 4 + \sum_{k=1}^{2n+1}\frac{(-1)^k}{k}\right], \quad n = 0, 1, \ldots$

9. $\displaystyle\int_0^1 x^{n-1/2} \ln(1+x)\,dx = \frac{2\ln 2}{2n+1} + \frac{4(-1)^n}{2n+1}\left[\pi - \sum_{k=0}^n \frac{(-1)^k}{2k+1}\right], \quad n = 1, 2, \ldots$

10. $\displaystyle\int_0^\infty \ln\frac{a^2+x^2}{b^2+x^2}\,dx = \pi(a-b), \quad a, b > 0.$

11. $\displaystyle\int_0^\infty \frac{x^{p-1}\ln x}{1+x^q}\,dx = -\frac{\pi^2 \cos(\pi p/q)}{q^2 \sin^2(\pi p/q)}, \quad 0 < p < q.$

12. $\displaystyle\int_0^\infty e^{-\mu x} \ln x\,dx = -\frac{1}{\mu}(C + \ln\mu), \quad \mu > 0, \quad C = 0.5772\ldots$

3.5. Integrals Containing Trigonometric Functions

1. $\displaystyle\int_0^{\pi/2} \cos^{2n} x\,dx = \frac{\pi}{2}\frac{1\cdot 3\ldots(2n-1)}{2\cdot 4\ldots(2n)}, \quad n = 1, 2, \ldots$

2. $\displaystyle\int_0^{\pi/2} \cos^{2n+1} x\,dx = \frac{2\cdot 4\ldots(2n)}{1\cdot 3\ldots(2n+1)}, \quad n = 1, 2, \ldots$

3. $\displaystyle\int_0^{\pi/2} x \cos^n x\,dx = -\sum_{k=0}^{m-1} \frac{(n-2k+1)(n-2k+3)\ldots(n-1)}{(n-2k)(n-2k+2)\ldots n}\frac{1}{n-2k}$

$+ \begin{cases} \dfrac{\pi}{2}\dfrac{(2m-2)!!}{(2m-1)!!} & \text{if } n = 2m-1, \\[2mm] \dfrac{\pi^2}{8}\cdot\dfrac{(2m-1)!!}{(2m)!!} & \text{if } n = 2m, \end{cases} \quad m = 1, 2, \ldots$

4. $\displaystyle\int_0^\pi \frac{dx}{(a+b\cos x)^{n+1}} = \frac{\pi}{2^n(a+b)^n\sqrt{a^2-b^2}}\sum_{k=0}^n \frac{(2n-2k-1)!!\,(2k-1)!!}{(n-k)!\,k!}\left(\frac{a+b}{a-b}\right)^k, \quad a > |b|.$

5. $\displaystyle\int_0^\infty \frac{\cos ax}{\sqrt{x}}\,dx = \sqrt{\frac{\pi}{2a}}, \quad a > 0.$

6. $\displaystyle\int_0^\infty \frac{\cos ax - \cos bx}{x}\,dx = \ln\left|\frac{b}{a}\right|, \quad ab \neq 0.$

7. $\displaystyle\int_0^\infty \frac{\cos ax - \cos bx}{x^2}\,dx = \tfrac{1}{2}\pi(b-a), \quad a, b \geq 0.$

8. $\displaystyle\int_0^\infty x^{\mu-1}\cos ax\,dx = a^{-\mu}\Gamma(\mu)\cos\left(\tfrac{1}{2}\pi\mu\right), \quad a > 0, \quad 0 < \mu < 1.$

9. $\displaystyle\int_0^\infty \frac{\cos ax}{b^2+x^2}\,dx = \frac{\pi}{2b}e^{-ab}, \quad a, b > 0.$

10. $\displaystyle\int_0^\infty \frac{\cos ax}{b^4+x^4}\,dx = \frac{\pi\sqrt{2}}{4b^3}\exp\left(-\frac{ab}{\sqrt{2}}\right)\left[\cos\left(\frac{ab}{\sqrt{2}}\right) + \sin\left(\frac{ab}{\sqrt{2}}\right)\right], \quad a, b > 0.$

11. $\displaystyle\int_0^\infty \frac{\cos ax}{(b^2+x^2)^2}\,dx = \frac{\pi}{4b^3}(1+ab)e^{-ab}, \quad a, b > 0.$

12. $\displaystyle\int_0^\infty \frac{\cos ax\,dx}{(b^2+x^2)(c^2+x^2)} = \frac{\pi\left(be^{-ac} - ce^{-ab}\right)}{2bc\left(b^2-c^2\right)}, \quad a, b, c > 0.$

13. $\displaystyle\int_0^\infty \cos\left(ax^2\right)dx = \frac{1}{2}\sqrt{\frac{\pi}{2a}}, \quad a > 0.$

14. $\displaystyle\int_0^\infty \cos(ax^p)\,dx = \frac{\Gamma(1/p)}{pa^{1/p}}\cos\frac{\pi}{2p}, \quad a > 0, \quad p > 1.$

15. $\displaystyle\int_0^{\pi/2} \sin^{2n} x\,dx = \frac{\pi}{2}\,\frac{1\cdot 3\ldots(2n-1)}{2\cdot 4\ldots(2n)}, \quad n = 1, 2, \ldots$

16. $\displaystyle\int_0^{\pi/2} \sin^{2n+1} x\,dx = \frac{2\cdot 4\ldots(2n)}{1\cdot 3\ldots(2n+1)}, \quad n = 1, 2, \ldots$

17. $\displaystyle\int_0^\infty \frac{\sin ax}{x}\,dx = \frac{\pi}{2}\,\mathrm{sign}\,a.$

18. $\displaystyle\int_0^\infty \frac{\sin^2 ax}{x^2}\,dx = \frac{\pi}{2}|a|.$

19. $\displaystyle\int_0^\infty \frac{\sin ax}{\sqrt{x}}\,dx = \sqrt{\frac{\pi}{2a}}, \quad a > 0.$

20. $\displaystyle\int_0^\pi x\sin^\mu x\,dx = \frac{\pi^2}{2^{\mu+1}}\,\frac{\Gamma(\mu+1)}{\left[\Gamma\left(\mu+\frac{1}{2}\right)\right]^2}, \quad \mu > -1.$

21. $\displaystyle\int_0^\infty x^{\mu-1}\sin ax\,dx = a^{-\mu}\Gamma(\mu)\sin\left(\tfrac{1}{2}\pi\mu\right), \quad a > 0, \quad 0 < \mu < 1.$

22. $\displaystyle\int_0^{\pi/2} \frac{\sin x\,dx}{\sqrt{1-k^2\sin^2 x}} = \frac{1}{2k}\ln\frac{1+k}{1-k}.$

23. $\displaystyle\int_0^\infty \sin(ax^2)\,dx = \frac{1}{2}\sqrt{\frac{\pi}{2a}}, \quad a > 0.$

24. $\displaystyle\int_0^\infty \sin(ax^p)\,dx = \frac{\Gamma(1/p)}{pa^{1/p}}\sin\frac{\pi}{2p}, \quad a > 0, \quad p > 1.$

25. $\displaystyle\int_0^{\pi/2} \sin^{2n+1} x\cos^{2m+1} x\,dx = \frac{n!\,m!}{2(n+m+1)!}, \quad n, m = 1, 2, \ldots$

26. $\displaystyle\int_0^{\pi/2} \sin^{p-1} x\cos^{q-1} x\,dx = \tfrac{1}{2}B\left(\tfrac{1}{2}p, \tfrac{1}{2}q\right).$

27. $\displaystyle\int_0^{2\pi} (a\sin x + b\cos x)^{2n}\,dx = 2\pi\frac{(2n-1)!!}{(2n)!!}\left(a^2+b^2\right)^n, \quad n = 1, 2, \ldots$

28. $\displaystyle\int_0^\infty \frac{\sin x\cos ax}{x}\,dx = \begin{cases} \dfrac{\pi}{2} & \text{if } |a| < 1, \\ \dfrac{\pi}{4} & \text{if } |a| = 1, \\ 0 & \text{if } 1 < |a|. \end{cases}$

29. $\displaystyle\int_0^\pi \frac{\sin x\,dx}{\sqrt{a^2+1-2a\cos x}} = \begin{cases} 2 & \text{if } 0 \le a \le 1, \\ 2/a & \text{if } 1 < a. \end{cases}$

30. $\displaystyle\int_0^\infty \frac{\tan ax}{x}\,dx = \frac{\pi}{2}\,\mathrm{sign}\,a.$

31. $\displaystyle\int_0^{\pi/2} (\tan x)^{\pm\lambda}\,dx = \frac{\pi}{2\cos\left(\tfrac{1}{2}\pi\lambda\right)}, \quad |\lambda| < 1.$

32. $\displaystyle\int_0^\infty e^{-ax}\sin bx\,dx = \frac{b}{a^2+b^2}, \quad a > 0.$

33. $\displaystyle\int_0^\infty e^{-ax}\cos bx\,dx = \frac{a}{a^2+b^2}, \quad a > 0.$

34. $\displaystyle\int_0^\infty \exp(-ax^2)\cos bx\,dx = \frac{1}{2}\sqrt{\frac{\pi}{a}}\,\exp\!\left(-\frac{b^2}{4a}\right).$

35. $\displaystyle\int_0^\infty \cos(ax^2)\cos bx\,dx = \sqrt{\frac{\pi}{8a}}\left[\cos\!\left(\frac{b^2}{4a}\right) + \sin\!\left(\frac{b^2}{4a}\right)\right], \quad a, b > 0.$

36. $\displaystyle\int_0^\infty (\cos ax + \sin ax)\cos(b^2 x^2)\,dx = \frac{1}{b}\sqrt{\frac{\pi}{8}}\,\exp\!\left(-\frac{a^2}{2b}\right), \quad a, b > 0.$

37. $\displaystyle\int_0^\infty \left[\cos ax + \sin ax\right]\sin(b^2 x^2)\,dx = \frac{1}{b}\sqrt{\frac{\pi}{8}}\,\exp\!\left(-\frac{a^2}{2b}\right), \quad a, b > 0.$

⊙ References for Supplement 3: H. B. Dwight (1961), I. S. Gradshteyn and I. M. Ryzhik (1980), A. P. Prudnikov, Yu. A. Brychkov, and O. I. Marichev (1986, 1988).

Supplement 4

Tables of Laplace Transforms

4.1. General Formulas

No	Original function, $f(x)$	Laplace transform, $\tilde{f}(p) = \int_0^\infty e^{-px} f(x)\,dx$
1	$af_1(x) + bf_2(x)$	$a\tilde{f}_1(p) + b\tilde{f}_2(p)$
2	$f(x/a), \quad a > 0$	$a\tilde{f}(ap)$
3	$\begin{cases} 0 & \text{if } 0 < x < a, \\ f(x-a) & \text{if } a < x, \end{cases}$	$e^{-ap}\tilde{f}(p)$
4	$x^n f(x); \quad n = 1, 2, \ldots$	$(-1)^n \dfrac{d^n}{dp^n} \tilde{f}(p)$
5	$\dfrac{1}{x} f(x)$	$\displaystyle\int_p^\infty \tilde{f}(q)\,dq$
6	$e^{ax} f(x)$	$\tilde{f}(p-a)$
7	$\sinh(ax) f(x)$	$\frac{1}{2}\left[\tilde{f}(p-a) - \tilde{f}(p+a)\right]$
8	$\cosh(ax) f(x)$	$\frac{1}{2}\left[\tilde{f}(p-a) + \tilde{f}(p+a)\right]$
9	$\sin(\omega x) f(x)$	$-\frac{i}{2}\left[\tilde{f}(p-i\omega) - \tilde{f}(p+i\omega)\right], \quad i^2 = -1$
10	$\cos(\omega x) f(x)$	$\frac{1}{2}\left[\tilde{f}(p-i\omega) + \tilde{f}(p+i\omega)\right], \quad i^2 = -1$
11	$f(x^2)$	$\dfrac{1}{\sqrt{\pi}} \displaystyle\int_0^\infty \exp\left(-\dfrac{p^2}{4t^2}\right) \tilde{f}(t^2)\,dt$
12	$x^{a-1} f\left(\dfrac{1}{x}\right), \quad a > -1$	$\displaystyle\int_0^\infty (t/p)^{a/2} J_a\left(2\sqrt{pt}\right) \tilde{f}(t)\,dt$
13	$f(a \sinh x), \quad a > 0$	$\displaystyle\int_0^\infty J_p(at) \tilde{f}(t)\,dt$
14	$f(x+a) = f(x) \ \text{(periodic function)}$	$\dfrac{1}{1-e^{ap}} \displaystyle\int_0^a f(x) e^{-px}\,dx$
15	$f(x+a) = -f(x)$ (antiperiodic function)	$\dfrac{1}{1+e^{-ap}} \displaystyle\int_0^a f(x) e^{-px}\,dx$
16	$f_x'(x)$	$p\tilde{f}(p) - f(+0)$
17	$f_x^{(n)}(x)$	$p^n \tilde{f}(p) - \displaystyle\sum_{k=1}^n p^{n-k} f_x^{(k-1)}(+0)$

No	Original function, $f(x)$	Laplace transform, $\tilde{f}(p) = \int_0^\infty e^{-px} f(x)\, dx$
18	$x^m f_x^{(n)}(x), \quad m \geq n$	$\left(-\dfrac{d}{dp}\right)^m \left[p^n \tilde{f}(p)\right]$
19	$\dfrac{d^n}{dx^n}\left[x^m f(x)\right], \quad m \geq n$	$(-1)^m p^n \dfrac{d^m}{dp^m} \tilde{f}(p)$
20	$\displaystyle\int_0^x f(t)\, dt$	$\dfrac{\tilde{f}(p)}{p}$
21	$\displaystyle\int_0^x (x-t) f(t)\, dt$	$\dfrac{1}{p^2}\tilde{f}(p)$
22	$\displaystyle\int_0^x (x-t)^\nu f(t)\, dt, \qquad \nu > -1$	$\Gamma(\nu+1) p^{-\nu-1} \tilde{f}(p)$
23	$\displaystyle\int_0^x e^{-a(x-t)} f(t)\, dt$	$\dfrac{1}{p+a}\tilde{f}(p)$
24	$\displaystyle\int_0^x \sinh\left[a(x-t)\right] f(t)\, dt$	$\dfrac{a\tilde{f}(p)}{p^2 - a^2}$
25	$\displaystyle\int_0^x \sin\left[a(x-t)\right] f(t)\, dt$	$\dfrac{a\tilde{f}(p)}{p^2 + a^2}$
26	$\displaystyle\int_0^x f_1(t) f_2(x-t)\, dt$	$\tilde{f}_1(p)\tilde{f}_2(p)$
27	$\displaystyle\int_0^x \dfrac{1}{t} f(t)\, dt$	$\dfrac{1}{p}\displaystyle\int_p^\infty \tilde{f}(q)\, dq$
28	$\displaystyle\int_x^\infty \dfrac{1}{t} f(t)\, dt$	$\dfrac{1}{p}\displaystyle\int_0^p \tilde{f}(q)\, dq$
29	$\displaystyle\int_0^\infty \dfrac{1}{\sqrt{t}} \sin\left(2\sqrt{xt}\,\right) f(t)\, dt$	$\dfrac{\sqrt{\pi}}{p\sqrt{p}} \tilde{f}\!\left(\dfrac{1}{p}\right)$
30	$\dfrac{1}{\sqrt{x}}\displaystyle\int_0^\infty \cos\left(2\sqrt{xt}\,\right) f(t)\, dt$	$\dfrac{\sqrt{\pi}}{\sqrt{p}} \tilde{f}\!\left(\dfrac{1}{p}\right)$
31	$\displaystyle\int_0^\infty \dfrac{1}{\sqrt{\pi x}} \exp\!\left(-\dfrac{t^2}{4x}\right) f(t)\, dt$	$\dfrac{1}{\sqrt{p}} \tilde{f}\!\left(\sqrt{p}\,\right)$
32	$\displaystyle\int_0^\infty \dfrac{t}{2\sqrt{\pi x^3}} \exp\!\left(-\dfrac{t^2}{4x}\right) f(t)\, dt$	$\tilde{f}\!\left(\sqrt{p}\,\right)$
33	$f(x) - a\displaystyle\int_0^x f\left(\sqrt{x^2 - t^2}\,\right) J_1(at)\, dt$	$\tilde{f}\!\left(\sqrt{p^2 + a^2}\,\right)$
34	$f(x) + a\displaystyle\int_0^x f\left(\sqrt{x^2 - t^2}\,\right) I_1(at)\, dt$	$\tilde{f}\!\left(\sqrt{p^2 - a^2}\,\right)$

4.2. Expressions With Power-Law Functions

No	Original function, $f(x)$	Laplace transform, $\tilde{f}(p) = \int_0^\infty e^{-px} f(x)\,dx$
1	1	$\dfrac{1}{p}$
2	$\begin{cases} 0 & \text{if } 0 < x < a, \\ 1 & \text{if } a < x < b, \\ 0 & \text{if } b < x. \end{cases}$	$\dfrac{1}{p}\left(e^{-ap} - e^{-bp}\right)$
3	x	$\dfrac{1}{p^2}$
4	$\dfrac{1}{x+a}$	$-e^{ap}\,\mathrm{Ei}(-ap)$
5	$x^n, \qquad n = 1, 2, \ldots$	$\dfrac{n!}{p^{n+1}}$
6	$x^{n-1/2}, \qquad n = 1, 2, \ldots$	$\dfrac{1 \cdot 3 \ldots (2n-1)\sqrt{\pi}}{2^n p^{n+1/2}}$
7	$\dfrac{1}{\sqrt{x+a}}$	$\sqrt{\dfrac{\pi}{p}}\,e^{ap}\,\mathrm{erfc}\left(\sqrt{ap}\right)$
8	$\dfrac{\sqrt{x}}{x+a}$	$\sqrt{\dfrac{\pi}{p}} - \pi\sqrt{a}\,e^{ap}\,\mathrm{erfc}\left(\sqrt{ap}\right)$
9	$(x+a)^{-3/2}$	$2a^{-1/2} - 2(\pi p)^{1/2}e^{ap}\,\mathrm{erfc}\left(\sqrt{ap}\right)$
10	$x^{1/2}(x+a)^{-1}$	$(\pi/p)^{1/2} - \pi a^{1/2}e^{ap}\,\mathrm{erfc}\left(\sqrt{ap}\right)$
11	$x^{-1/2}(x+a)^{-1}$	$\pi a^{-1/2}e^{ap}\,\mathrm{erfc}\left(\sqrt{ap}\right)$
12	$x^\nu, \qquad \nu > -1$	$\Gamma(\nu+1)p^{-\nu-1}$
13	$(x+a)^\nu, \qquad \nu > -1$	$p^{-\nu-1}e^{-ap}\Gamma(\nu+1, ap)$
14	$x^\nu(x+a)^{-1}, \qquad \nu > -1$	$ke^{ap}\Gamma(-\nu, ap), \qquad k = a^\nu\Gamma(\nu+1)$
15	$(x^2 + 2ax)^{-1/2}(x+a)$	$ae^{ap}K_1(ap)$

4.3. Expressions With Exponential Functions

No	Original function, $f(x)$	Laplace transform, $\tilde{f}(p) = \int_0^\infty e^{-px} f(x)\,dx$
1	e^{-ax}	$(p+a)^{-1}$
2	xe^{-ax}	$(p+a)^{-2}$
3	$x^{\nu-1}e^{-ax}, \qquad \nu > 0$	$\Gamma(\nu)(p+a)^{-\nu}$
4	$\dfrac{1}{x}\left(e^{-ax} - e^{-bx}\right)$	$\ln(p+b) - \ln(p+a)$

No	Original function, $f(x)$	Laplace transform, $\tilde{f}(p) = \int_0^\infty e^{-px} f(x)\, dx$
5	$\dfrac{1}{x^2}\left(1 - e^{-ax}\right)^2$	$(p+2a)\ln(p+2a) + p\ln p - 2(p+a)\ln(p+a)$
6	$\exp(-ax^2),\qquad a > 0$	$(\pi b)^{1/2}\exp(bp^2)\operatorname{erfc}(p\sqrt{b}),\qquad a = \dfrac{1}{4b}$
7	$x\exp(-ax^2)$	$2b - 2\pi^{1/2}b^{3/2}p\operatorname{erfc}(p\sqrt{b}),\qquad a = \dfrac{1}{4b}$
8	$\exp(-a/x),\qquad a \geq 0$	$2\sqrt{a/p}\,K_1\!\left(2\sqrt{ap}\right)$
9	$\sqrt{x}\exp(-a/x),\qquad a \geq 0$	$\tfrac{1}{2}\sqrt{\pi/p^3}\left(1 + 2\sqrt{ap}\right)\exp\!\left(-2\sqrt{ap}\right)$
10	$\dfrac{1}{\sqrt{x}}\exp(-a/x),\qquad a \geq 0$	$\sqrt{\pi/p}\exp\!\left(-2\sqrt{ap}\right)$
11	$\dfrac{1}{x\sqrt{x}}\exp(-a/x),\qquad a > 0$	$\sqrt{\pi/a}\exp\!\left(-2\sqrt{ap}\right)$
12	$x^{\nu-1}\exp(-a/x),\qquad a > 0$	$2(a/p)^{\nu/2}K_\nu\!\left(2\sqrt{ap}\right)$
13	$\exp\!\left(-2\sqrt{ax}\right)$	$p^{-1} - (\pi a)^{1/2}p^{-3/2}e^{a/p}\operatorname{erfc}\!\left(\sqrt{a/p}\right)$
14	$\dfrac{1}{\sqrt{x}}\exp\!\left(-2\sqrt{ax}\right)$	$(\pi/p)^{1/2}e^{a/p}\operatorname{erfc}\!\left(\sqrt{a/p}\right)$

4.4. Expressions With Hyperbolic Functions

No	Original function, $f(x)$	Laplace transform, $\tilde{f}(p) = \int_0^\infty e^{-px} f(x)\, dx$
1	$\sinh(ax)$	$\dfrac{a}{p^2 - a^2}$
2	$\sinh^2(ax)$	$\dfrac{2a^2}{p^3 - 4a^2 p}$
3	$\dfrac{1}{x}\sinh(ax)$	$\dfrac{1}{2}\ln\dfrac{p+a}{p-a}$
4	$x^{\nu-1}\sinh(ax),\qquad \nu > -1$	$\tfrac{1}{2}\Gamma(\nu)\left[(p-a)^{-\nu} - (p+a)^{-\nu}\right]$
5	$\sinh\!\left(2\sqrt{ax}\right)$	$\dfrac{\sqrt{\pi a}}{p\sqrt{p}}e^{a/p}$
6	$\sqrt{x}\sinh\!\left(2\sqrt{ax}\right)$	$\pi^{1/2}p^{-5/2}\left(\tfrac{1}{2}p + a\right)e^{a/p}\operatorname{erf}\!\left(\sqrt{a/p}\right) - a^{1/2}p^{-2}$
7	$\dfrac{1}{\sqrt{x}}\sinh\!\left(2\sqrt{ax}\right)$	$\pi^{1/2}p^{-1/2}e^{a/p}\operatorname{erf}\!\left(\sqrt{a/p}\right)$
8	$\dfrac{1}{\sqrt{x}}\sinh^2\!\left(\sqrt{ax}\right)$	$\tfrac{1}{2}\pi^{1/2}p^{-1/2}\left(e^{a/p} - 1\right)$
9	$\cosh(ax)$	$\dfrac{p}{p^2 - a^2}$

No	Original function, $f(x)$	Laplace transform, $\tilde{f}(p) = \int_0^\infty e^{-px} f(x)\, dx$
10	$\cosh^2(ax)$	$\dfrac{p^2 - 2a^2}{p^3 - 4a^2 p}$
11	$x^{\nu-1} \cosh(ax), \qquad \nu > 0$	$\frac{1}{2}\Gamma(\nu)\big[(p-a)^{-\nu} + (p+a)^{-\nu}\big]$
12	$\cosh\big(2\sqrt{ax}\,\big)$	$\dfrac{1}{p} + \dfrac{\sqrt{\pi a}}{p\sqrt{p}} e^{a/p} \operatorname{erf}\big(\sqrt{a/p}\,\big)$
13	$\sqrt{x} \cosh\big(2\sqrt{ax}\,\big)$	$\pi^{1/2} p^{-5/2}\big(\frac{1}{2}p + a\big) e^{a/p}$
14	$\dfrac{1}{\sqrt{x}} \cosh\big(2\sqrt{ax}\,\big)$	$\pi^{1/2} p^{-1/2} e^{a/p}$
15	$\dfrac{1}{\sqrt{x}} \cosh^2\big(\sqrt{ax}\,\big)$	$\frac{1}{2}\pi^{1/2} p^{-1/2}\big(e^{a/p} + 1\big)$

4.5. Expressions With Logarithmic Functions

No	Original function, $f(x)$	Laplace transform, $\tilde{f}(p) = \int_0^\infty e^{-px} f(x)\, dx$
1	$\ln x$	$-\dfrac{1}{p}(\ln p + \mathcal{C})$, $\mathcal{C} = 0.5772\ldots$ is the Euler constant
2	$\ln(1 + ax)$	$-\dfrac{1}{p} e^{p/a} \operatorname{Ei}(-p/a)$
3	$\ln(x + a)$	$\dfrac{1}{p}\big[\ln a - e^{ap} \operatorname{Ei}(-ap)\big]$
4	$x^n \ln x, \qquad n = 1, 2, \ldots$	$\dfrac{n!}{p^{n+1}}\big(1 + \frac{1}{2} + \frac{1}{3} + \cdots + \frac{1}{n} - \ln p - \mathcal{C}\big)$, $\mathcal{C} = 0.5772\ldots$ is the Euler constant
5	$\dfrac{1}{\sqrt{x}} \ln x$	$-\sqrt{\pi/p}\,\big[\ln(4p) + \mathcal{C}\big]$
6	$x^{n-1/2} \ln x, \qquad n = 1, 2, \ldots$	$\dfrac{k_n}{p^{n+1/2}}\big[2 + \frac{2}{3} + \frac{2}{5} + \cdots + \frac{2}{2n-1} - \ln(4p) - \mathcal{C}\big]$, $k_n = 1 \cdot 3 \cdot 5 \ldots (2n-1)\dfrac{\sqrt{\pi}}{2^n}, \quad \mathcal{C} = 0.5772\ldots$
7	$x^{\nu-1} \ln x, \quad \nu > 0$	$\Gamma(\nu) p^{-\nu}\big[\psi(\nu) - \ln p\big], \quad \psi(\nu)$ is the logarithmic derivative of the gamma function
8	$(\ln x)^2$	$\dfrac{1}{p}\big[(\ln x + \mathcal{C})^2 + \frac{1}{6}\pi^2\big], \quad \mathcal{C} = 0.5772\ldots$
9	$e^{-ax} \ln x$	$-\dfrac{\ln(p+a) + \mathcal{C}}{p + a}, \qquad \mathcal{C} = 0.5772\ldots$

4.6. Expressions With Trigonometric Functions

No	Original function, $f(x)$	Laplace transform, $\tilde{f}(p) = \int_0^\infty e^{-px} f(x)\, dx$
1	$\sin(ax)$	$\dfrac{a}{p^2 + a^2}$
2	$|\sin(ax)|, \qquad a > 0$	$\dfrac{a}{p^2 + a^2} \coth\left(\dfrac{\pi p}{2a}\right)$
3	$\sin^{2n}(ax), \qquad n = 1, 2, \ldots$	$\dfrac{a^{2n}(2n)!}{p\left[p^2 + (2a)^2\right]\left[p^2 + (4a)^2\right] \ldots \left[p^2 + (2na)^2\right]}$
4	$\sin^{2n+1}(ax), \qquad n = 1, 2, \ldots$	$\dfrac{a^{2n+1}(2n+1)!}{\left[p^2 + a^2\right]\left[p^2 + 3^2 a^2\right] \ldots \left[p^2 + (2n+1)^2 a^2\right]}$
5	$x^n \sin(ax), \qquad n = 1, 2, \ldots$	$\dfrac{n!\, p^{n+1}}{\left(p^2 + a^2\right)^{n+1}} \displaystyle\sum_{0 \le 2k \le n} (-1)^k C_{n+1}^{2k+1} \left(\dfrac{a}{p}\right)^{2k+1}$
6	$\dfrac{1}{x} \sin(ax)$	$\arctan\left(\dfrac{a}{p}\right)$
7	$\dfrac{1}{x} \sin^2(ax)$	$\tfrac{1}{4} \ln\left(1 + 4a^2 p^{-2}\right)$
8	$\dfrac{1}{x^2} \sin^2(ax)$	$a \arctan(2a/p) - \tfrac{1}{4} p \ln\left(1 + 4a^2 p^{-2}\right)$
9	$\sin\left(2\sqrt{ax}\right)$	$\dfrac{\sqrt{\pi a}}{p\sqrt{p}} e^{-a/p}$
10	$\dfrac{1}{x} \sin\left(2\sqrt{ax}\right)$	$\pi \operatorname{erf}\left(\sqrt{a/p}\right)$
11	$\cos(ax)$	$\dfrac{p}{p^2 + a^2}$
12	$\cos^2(ax)$	$\dfrac{p^2 + 2a^2}{p\left(p^2 + 4a^2\right)}$
13	$x^n \cos(ax), \qquad n = 1, 2, \ldots$	$\dfrac{n!\, p^{n+1}}{\left(p^2 + a^2\right)^{n+1}} \displaystyle\sum_{0 \le 2k \le n+1} (-1)^k C_{n+1}^{2k} \left(\dfrac{a}{p}\right)^{2k}$
14	$\dfrac{1}{x}\left[1 - \cos(ax)\right]$	$\tfrac{1}{2} \ln\left(1 + a^2 p^{-2}\right)$
15	$\dfrac{1}{x}\left[\cos(ax) - \cos(bx)\right]$	$\dfrac{1}{2} \ln \dfrac{p^2 + b^2}{p^2 + a^2}$
16	$\sqrt{x} \cos\left(2\sqrt{ax}\right)$	$\tfrac{1}{2}\pi^{1/2} p^{-5/2}(p - 2a) e^{-a/p}$
17	$\dfrac{1}{\sqrt{x}} \cos\left(2\sqrt{ax}\right)$	$\sqrt{\pi/p}\, e^{-a/p}$
18	$\sin(ax) \sin(bx)$	$\dfrac{2abp}{\left[p^2 + (a+b)^2\right]\left[p^2 + (a-b)^2\right]}$
19	$\cos(ax) \sin(bx)$	$\dfrac{b\left(p^2 - a^2 + b^2\right)}{\left[p^2 + (a+b)^2\right]\left[p^2 + (a-b)^2\right]}$

No	Original function, $f(x)$	Laplace transform, $\tilde{f}(p) = \int_0^\infty e^{-px} f(x)\,dx$
20	$\cos(ax)\cos(bx)$	$\dfrac{p\left(p^2 + a^2 + b^2\right)}{\left[p^2 + (a+b)^2\right]\left[p^2 + (a-b)^2\right]}$
21	$\dfrac{ax\cos(ax) - \sin(ax)}{x^2}$	$p\arctan\dfrac{a}{x} - a$
22	$e^{bx}\sin(ax)$	$\dfrac{a}{(p-b)^2 + a^2}$
23	$e^{bx}\cos(ax)$	$\dfrac{p-b}{(p-b)^2 + a^2}$
24	$\sin(ax)\sinh(ax)$	$\dfrac{2a^2 p}{p^4 + 4a^4}$
25	$\sin(ax)\cosh(ax)$	$\dfrac{a\left(p^2 + 2a^2\right)}{p^4 + 4a^4}$
26	$\cos(ax)\sinh(ax)$	$\dfrac{a\left(p^2 - 2a^2\right)}{p^4 + 4a^4}$
27	$\cos(ax)\cosh(ax)$	$\dfrac{p^3}{p^4 + 4a^4}$

4.7. Expressions With Special Functions

No	Original function, $f(x)$	Laplace transform, $\tilde{f}(p) = \int_0^\infty e^{-px} f(x)\,dx$
1	$\operatorname{erf}(ax)$	$\dfrac{1}{p}\exp\left(b^2 p^2\right)\operatorname{erfc}(bp), \qquad b = \dfrac{1}{2a}$
2	$\operatorname{erf}\left(\sqrt{ax}\right)$	$\dfrac{\sqrt{a}}{p\sqrt{p+a}}$
3	$e^{ax}\operatorname{erf}\left(\sqrt{ax}\right)$	$\dfrac{\sqrt{a}}{\sqrt{p}\,(p-a)}$
4	$\operatorname{erf}\left(\frac{1}{2}\sqrt{a/x}\right)$	$\dfrac{1}{p}\left[1 - \exp\left(-\sqrt{ap}\right)\right]$
5	$\operatorname{erfc}\left(\sqrt{ax}\right)$	$\dfrac{\sqrt{p+a} - \sqrt{a}}{p\sqrt{p+a}}$
6	$e^{ax}\operatorname{erfc}\left(\sqrt{ax}\right)$	$\dfrac{1}{p + \sqrt{ap}}$
7	$\operatorname{erfc}\left(\frac{1}{2}\sqrt{a/x}\right)$	$\dfrac{1}{p}\exp\left(-\sqrt{ap}\right)$
8	$\operatorname{Ci}(x)$	$\dfrac{1}{2p}\ln(p^2 + 1)$

No	Original function, $f(x)$	Laplace transform, $\tilde{f}(p) = \int_0^\infty e^{-px} f(x)\, dx$
9	$\mathrm{Si}(x)$	$\dfrac{1}{p}\operatorname{arccot} p$
10	$\mathrm{Ei}(-x)$	$-\dfrac{1}{p}\ln(p+1)$
11	$J_0(ax)$	$\dfrac{1}{\sqrt{p^2+a^2}}$
12	$J_\nu(ax), \qquad \nu > -1$	$\dfrac{a^\nu}{\sqrt{p^2+a^2}\,\left(p+\sqrt{p^2+a^2}\,\right)^\nu}$
13	$x^n J_n(ax), \qquad n = 1, 2, \dots$	$1\cdot 3\cdot 5\dots(2n-1)a^n\left(p^2+a^2\right)^{-n-1/2}$
14	$x^\nu J_\nu(ax), \qquad \nu > -\tfrac{1}{2}$	$2^\nu \pi^{-1/2}\Gamma\!\left(\nu+\tfrac{1}{2}\right)a^\nu\left(p^2+a^2\right)^{-\nu-1/2}$
15	$x^{\nu+1} J_\nu(ax), \qquad \nu > -1$	$2^{\nu+1}\pi^{-1/2}\Gamma\!\left(\nu+\tfrac{3}{2}\right)a^\nu p\left(p^2+a^2\right)^{-\nu-3/2}$
16	$J_0\!\left(2\sqrt{ax}\,\right)$	$\dfrac{1}{p}e^{-a/p}$
17	$\sqrt{x}\,J_1\!\left(2\sqrt{ax}\,\right)$	$\dfrac{\sqrt{a}}{p^2}e^{-a/p}$
18	$x^{\nu/2} J_\nu\!\left(2\sqrt{ax}\,\right), \qquad \nu > -1$	$a^{\nu/2}p^{-\nu-1}e^{-a/p}$
19	$I_0(ax)$	$\dfrac{1}{\sqrt{p^2-a^2}}$
20	$I_\nu(ax), \qquad \nu > -1$	$\dfrac{a^\nu}{\sqrt{p^2-a^2}\,\left(p+\sqrt{p^2-a^2}\,\right)^\nu}$
21	$x^\nu I_\nu(ax), \qquad \nu > -\tfrac{1}{2}$	$2^\nu \pi^{-1/2}\Gamma\!\left(\nu+\tfrac{1}{2}\right)a^\nu\left(p^2-a^2\right)^{-\nu-1/2}$
22	$x^{\nu+1} I_\nu(ax), \qquad \nu > -1$	$2^{\nu+1}\pi^{-1/2}\Gamma\!\left(\nu+\tfrac{3}{2}\right)a^\nu p\left(p^2-a^2\right)^{-\nu-3/2}$
23	$I_0\!\left(2\sqrt{ax}\,\right)$	$\dfrac{1}{p}e^{a/p}$
24	$\dfrac{1}{\sqrt{x}}I_1\!\left(2\sqrt{ax}\,\right)$	$\dfrac{1}{\sqrt{a}}\left(e^{a/p}-1\right)$
25	$x^{\nu/2} I_\nu\!\left(2\sqrt{ax}\,\right), \qquad \nu > -1$	$a^{\nu/2}p^{-\nu-1}e^{a/p}$
26	$Y_0(ax)$	$-\dfrac{2}{\pi}\dfrac{\operatorname{Arsinh}(p/a)}{\sqrt{p^2+a^2}}$
27	$K_0(ax)$	$\dfrac{\ln\!\left(p+\sqrt{p^2-a^2}\,\right)-\ln a}{\sqrt{p^2-a^2}}$

⊙ References for Supplement 4: G. Doetsch (1950, 1956, 1958), H. Bateman and A. Erdélyi (1954), V. A. Ditkin and A. P. Prudnikov (1965).

Supplement 5

Tables of Inverse Laplace Transforms

5.1. General Formulas

No	Laplace transform, $\tilde{f}(p)$	Inverse transform, $f(x) = \dfrac{1}{2\pi i} \displaystyle\int_{c-i\infty}^{c+i\infty} e^{px} \tilde{f}(p)\, dp$
1	$\tilde{f}(p+a)$	$e^{-ax} f(x)$
2	$\tilde{f}(ap), \quad a > 0$	$\dfrac{1}{a} f\left(\dfrac{x}{a}\right)$
3	$\tilde{f}(ap+b), \quad a > 0$	$\dfrac{1}{a} \exp\left(-\dfrac{b}{a}x\right) f\left(\dfrac{x}{a}\right)$
4	$\tilde{f}(p-a) + \tilde{f}(p+a)$	$2f(x)\cosh(ax)$
5	$\tilde{f}(p-a) - \tilde{f}(p+a)$	$2f(x)\sinh(ax)$
6	$e^{-ap} \tilde{f}(p), \quad a \geq 0$	$\begin{cases} 0 & \text{if } 0 \leq x < a, \\ f(x-a) & \text{if } a < x. \end{cases}$
7	$p\tilde{f}(p)$	$\dfrac{df(x)}{dx}, \quad \text{if } f(+0) = 0$
8	$\dfrac{1}{p}\tilde{f}(p)$	$\displaystyle\int_0^x f(t)\, dt$
9	$\dfrac{1}{p+a}\tilde{f}(p)$	$e^{-ax} \displaystyle\int_0^x e^{at} f(t)\, dt$
10	$\dfrac{1}{p^2}\tilde{f}(p)$	$\displaystyle\int_0^x (x-t)f(t)\, dt$
11	$\dfrac{\tilde{f}(p)}{p(p+a)}$	$\dfrac{1}{a}\displaystyle\int_0^x \left[1 - e^{a(x-t)}\right] f(t)\, dt$
12	$\dfrac{\tilde{f}(p)}{(p+a)^2}$	$\displaystyle\int_0^x (x-t)e^{-a(x-t)} f(t)\, dt$
13	$\dfrac{\tilde{f}(p)}{(p+a)(p+b)}$	$\dfrac{1}{b-a}\displaystyle\int_0^x \left[e^{-a(x-t)} - e^{-b(x-t)}\right] f(t)\, dt$
14	$\dfrac{\tilde{f}(p)}{(p+a)^2 + b^2}$	$\dfrac{1}{b}\displaystyle\int_0^x e^{-a(x-t)} \sin\left[b(x-t)\right] f(t)\, dt$
15	$\dfrac{1}{p^n}\tilde{f}(p), \quad n = 1, 2, \ldots$	$\dfrac{1}{(n-1)!}\displaystyle\int_0^x (x-t)^{n-1} f(t)\, dt$
16	$\tilde{f}_1(p)\tilde{f}_2(p)$	$\displaystyle\int_0^x f_1(t)f_2(x-t)\, dt$

No	Laplace transform, $\tilde{f}(p)$	Inverse transform, $f(x) = \dfrac{1}{2\pi i}\displaystyle\int_{c-i\infty}^{c+i\infty} e^{px}\,\tilde{f}(p)\,dp$
17	$\dfrac{1}{\sqrt{p}}\tilde{f}\!\left(\dfrac{1}{p}\right)$	$\displaystyle\int_0^\infty \dfrac{\cos\left(2\sqrt{xt}\right)}{\sqrt{\pi x}}\,f(t)\,dt$
18	$\dfrac{1}{p\sqrt{p}}\tilde{f}\!\left(\dfrac{1}{p}\right)$	$\displaystyle\int_0^\infty \dfrac{\sin\left(2\sqrt{xt}\right)}{\sqrt{\pi t}}\,f(t)\,dt$
19	$\dfrac{1}{p^{2\nu+1}}\tilde{f}\!\left(\dfrac{1}{p}\right)$	$\displaystyle\int_0^\infty (x/t)^\nu\, J_{2\nu}\left(2\sqrt{xt}\right) f(t)\,dt$
20	$\dfrac{1}{p}\tilde{f}\!\left(\dfrac{1}{p}\right)$	$\displaystyle\int_0^\infty J_0\left(2\sqrt{xt}\right) f(t)\,dt$
21	$\dfrac{1}{p}\tilde{f}\!\left(p+\dfrac{1}{p}\right)$	$\displaystyle\int_0^x J_0\left(2\sqrt{xt-t^2}\right) f(t)\,dt$
22	$\dfrac{1}{p^{2\nu+1}}\tilde{f}\!\left(p+\dfrac{a}{p}\right)$	$\displaystyle\int_0^x \left(\dfrac{x-t}{at}\right)^\nu J_{2\nu}\left(2\sqrt{axt-at^2}\right) f(t)\,dt$
23	$\tilde{f}\left(\sqrt{p}\right)$	$\displaystyle\int_0^\infty \dfrac{t}{2\sqrt{\pi x^3}}\exp\!\left(-\dfrac{t^2}{4x}\right) f(t)\,dt$
24	$\dfrac{1}{\sqrt{p}}\tilde{f}\left(\sqrt{p}\right)$	$\dfrac{1}{\sqrt{\pi x}}\displaystyle\int_0^\infty \exp\!\left(-\dfrac{t^2}{4x}\right) f(t)\,dt$
25	$\tilde{f}\left(p+\sqrt{p}\right)$	$\dfrac{1}{2\sqrt{\pi}}\displaystyle\int_0^x \dfrac{t}{(x-t)^{3/2}}\exp\!\left[-\dfrac{t^2}{4(x-t)}\right] f(t)\,dt$
26	$\tilde{f}\left(\sqrt{p^2+a^2}\right)$	$f(x) - a\displaystyle\int_0^x f\left(\sqrt{x^2-t^2}\right) J_1(at)\,dt$
27	$\tilde{f}\left(\sqrt{p^2-a^2}\right)$	$f(x) + a\displaystyle\int_0^x f\left(\sqrt{x^2-t^2}\right) I_1(at)\,dt$
28	$\dfrac{\tilde{f}\left(\sqrt{p^2+a^2}\right)}{\sqrt{p^2+a^2}}$	$\displaystyle\int_0^x J_0\left(a\sqrt{x^2-t^2}\right) f(t)\,dt$
29	$\dfrac{\tilde{f}\left(\sqrt{p^2-a^2}\right)}{\sqrt{p^2-a^2}}$	$\displaystyle\int_0^x I_0\left(a\sqrt{x^2-t^2}\right) f(t)\,dt$
30	$\tilde{f}\left(\sqrt{(p+a)^2-b^2}\right)$	$e^{-ax} f(x) + b e^{-ax}\displaystyle\int_0^x f\left(\sqrt{x^2-t^2}\right) I_1(bt)\,dt$
31	$\tilde{f}(\ln p)$	$\displaystyle\int_0^\infty \dfrac{x^{t-1}}{\Gamma(t)} f(t)\,dt$
32	$\dfrac{1}{p}\tilde{f}(\ln p)$	$\displaystyle\int_0^\infty \dfrac{x^t}{\Gamma(t+1)} f(t)\,dt$
33	$\tilde{f}(p-ia) + \tilde{f}(p+ia),\ \ i^2=-1$	$2 f(x)\cos(ax)$
34	$i\left[\tilde{f}(p-ia) - \tilde{f}(p+ia)\right],\ \ i^2=-1$	$2 f(x)\sin(ax)$
35	$\dfrac{d\tilde{f}(p)}{dp}$	$-x f(x)$
36	$\dfrac{d^n \tilde{f}(p)}{dp^n}$	$(-x)^n f(x)$

No	Laplace transform, $\tilde{f}(p)$	Inverse transform, $f(x) = \dfrac{1}{2\pi i} \displaystyle\int_{c-i\infty}^{c+i\infty} e^{px}\tilde{f}(p)\,dp$
37	$p^n \dfrac{d^m \tilde{f}(p)}{dp^m}, \quad m \geq n$	$(-1)^m \dfrac{d^n}{dx^n}\left[x^m f(x)\right]$
38	$\displaystyle\int_p^\infty \tilde{f}(q)\,dq$	$\dfrac{1}{x} f(x)$
39	$\dfrac{1}{p}\displaystyle\int_0^p \tilde{f}(q)\,dq$	$\displaystyle\int_x^\infty \dfrac{f(t)}{t}\,dt$
40	$\dfrac{1}{p}\displaystyle\int_p^\infty \tilde{f}(q)\,dq$	$\displaystyle\int_0^x \dfrac{f(t)}{t}\,dt$

5.2. Expressions With Rational Functions

No	Laplace transform, $\tilde{f}(p)$	Inverse transform, $f(x) = \dfrac{1}{2\pi i} \displaystyle\int_{c-i\infty}^{c+i\infty} e^{px}\tilde{f}(p)\,dp$
1	$\dfrac{1}{p}$	1
2	$\dfrac{1}{p+a}$	e^{-ax}
3	$\dfrac{1}{p^2}$	x
4	$\dfrac{1}{p(p+a)}$	$\dfrac{1}{a}\left(1 - e^{-ax}\right)$
5	$\dfrac{1}{(p+a)^2}$	xe^{-ax}
6	$\dfrac{p}{(p+a)^2}$	$(1 - ax)e^{-ax}$
7	$\dfrac{1}{p^2 - a^2}$	$\dfrac{1}{a}\sinh(ax)$
8	$\dfrac{p}{p^2 - a^2}$	$\cosh(ax)$
9	$\dfrac{1}{(p+a)(p+b)}$	$\dfrac{1}{a-b}\left(e^{-bx} - e^{-ax}\right)$
10	$\dfrac{p}{(p+a)(p+b)}$	$\dfrac{1}{a-b}\left(ae^{-ax} - be^{-bx}\right)$
11	$\dfrac{1}{p^2 + a^2}$	$\dfrac{1}{a}\sin(ax)$
12	$\dfrac{p}{p^2 + a^2}$	$\cos(ax)$
13	$\dfrac{1}{(p+b)^2 + a^2}$	$\dfrac{1}{a}e^{-bx}\sin(ax)$
14	$\dfrac{p}{(p+b)^2 + a^2}$	$e^{-bx}\left[\cos(ax) - \dfrac{b}{a}\sin(ax)\right]$

No	Laplace transform, $\tilde{f}(p)$	Inverse transform, $f(x) = \dfrac{1}{2\pi i}\displaystyle\int_{c-i\infty}^{c+i\infty} e^{px}\tilde{f}(p)\,dp$
15	$\dfrac{1}{p^3}$	$\frac{1}{2}x^2$
16	$\dfrac{1}{p^2(p+a)}$	$\dfrac{1}{a^2}\left(e^{-ax}+ax-1\right)$
17	$\dfrac{1}{p(p+a)(p+b)}$	$\dfrac{1}{ab(a-b)}\left(a-b+be^{-ax}-ae^{-bx}\right)$
18	$\dfrac{1}{p(p+a)^2}$	$\dfrac{1}{a^2}\left(1-e^{-ax}-axe^{-ax}\right)$
19	$\dfrac{1}{(p+a)(p+b)(p+c)}$	$\dfrac{(c-b)e^{-ax}+(a-c)e^{-bx}+(b-a)e^{-cx}}{(a-b)(b-c)(c-a)}$
20	$\dfrac{p}{(p+a)(p+b)(p+c)}$	$\dfrac{a(b-c)e^{-ax}+b(c-a)e^{-bx}+c(a-b)e^{-cx}}{(a-b)(b-c)(c-a)}$
21	$\dfrac{p^2}{(p+a)(p+b)(p+c)}$	$\dfrac{a^2(c-b)e^{-ax}+b^2(a-c)e^{-bx}+c^2(b-a)e^{-cx}}{(a-b)(b-c)(c-a)}$
22	$\dfrac{1}{(p+a)(p+b)^2}$	$\dfrac{1}{(a-b)^2}\left[e^{-ax}-e^{-bx}+(a-b)xe^{-bx}\right]$
23	$\dfrac{p}{(p+a)(p+b)^2}$	$\dfrac{1}{(a-b)^2}\left\{-ae^{-ax}+\left[a+b(b-a)x\right]e^{-bx}\right\}$
24	$\dfrac{p^2}{(p+a)(p+b)^2}$	$\dfrac{1}{(a-b)^2}\left[a^2e^{-ax}+b(b-2a-b^2x+abx)e^{-bx}\right]$
25	$\dfrac{1}{(p+a)^3}$	$\frac{1}{2}x^2e^{-ax}$
26	$\dfrac{p}{(p+a)^3}$	$x\left(1-\frac{1}{2}ax\right)e^{-ax}$
27	$\dfrac{p^2}{(p+a)^3}$	$\left(1-2ax+\frac{1}{2}a^2x^2\right)e^{-ax}$
28	$\dfrac{1}{p(p^2+a^2)}$	$\dfrac{1}{a^2}\left[1-\cos(ax)\right]$
29	$\dfrac{1}{p\left[(p+b)^2+a^2\right]}$	$\dfrac{1}{a^2+b^2}\left\{1-e^{-bx}\left[\cos(ax)+\dfrac{b}{a}\sin(ax)\right]\right\}$
30	$\dfrac{1}{(p+a)(p^2+b^2)}$	$\dfrac{1}{a^2+b^2}\left[e^{-ax}+\dfrac{a}{b}\sin(bx)-\cos(bx)\right]$
31	$\dfrac{p}{(p+a)(p^2+b^2)}$	$\dfrac{1}{a^2+b^2}\left[-ae^{-ax}+a\cos(bx)+b\sin(bx)\right]$
32	$\dfrac{p^2}{(p+a)(p^2+b^2)}$	$\dfrac{1}{a^2+b^2}\left[a^2e^{-ax}-ab\sin(bx)+b^2\cos(bx)\right]$
33	$\dfrac{1}{p^3+a^3}$	$\dfrac{1}{3a^2}e^{-ax}-\dfrac{1}{3a^2}e^{ax/2}\left[\cos(kx)-\sqrt{3}\sin(kx)\right],$ $k=\frac{1}{2}a\sqrt{3}$
34	$\dfrac{p}{p^3+a^3}$	$-\dfrac{1}{3a}e^{-ax}+\dfrac{1}{3a}e^{ax/2}\left[\cos(kx)+\sqrt{3}\sin(kx)\right],$ $k=\frac{1}{2}a\sqrt{3}$

No	Laplace transform, $\tilde{f}(p)$	Inverse transform, $f(x) = \dfrac{1}{2\pi i}\displaystyle\int_{c-i\infty}^{c+i\infty} e^{px}\,\tilde{f}(p)\,dp$
35	$\dfrac{p^2}{p^3 + a^3}$	$\frac{1}{3}e^{-ax} + \frac{2}{3}e^{ax/2}\cos(kx), \quad k = \frac{1}{2}a\sqrt{3}$
36	$\dfrac{1}{(p+a)\left[(p+b)^2 + c^2\right]}$	$\dfrac{e^{-ax} - e^{-bx}\cos(cx) + ke^{-bx}\sin(cx)}{(a-b)^2 + c^2}, \quad k = \dfrac{a-b}{c}$
37	$\dfrac{p}{(p+a)\left[(p+b)^2 + c^2\right]}$	$\dfrac{-ae^{-ax} + ae^{-bx}\cos(cx) + ke^{-bx}\sin(cx)}{(a-b)^2 + c^2},$ $k = \dfrac{b^2 + c^2 - ab}{c}$
38	$\dfrac{p^2}{(p+a)\left[(p+b)^2 + c^2\right]}$	$\dfrac{a^2 e^{-ax} + (b^2 + c^2 - 2ab)e^{-bx}\cos(cx) + ke^{-bx}\sin(cx)}{(a-b)^2 + c^2},$ $k = -ac - bc + \dfrac{ab^2 - b^3}{c}$
39	$\dfrac{1}{p^4}$	$\frac{1}{6}x^3$
40	$\dfrac{1}{p^3(p+a)}$	$\dfrac{1}{a^3} - \dfrac{1}{a^2}x + \dfrac{1}{2a}x^2 - \dfrac{1}{a^3}e^{-ax}$
41	$\dfrac{1}{p^2(p+a)^2}$	$\dfrac{1}{a^2}x\left(1 + e^{-ax}\right) + \dfrac{2}{a^3}\left(e^{-ax} - 1\right)$
42	$\dfrac{1}{p^2(p+a)(p+b)}$	$-\dfrac{a+b}{a^2b^2} + \dfrac{1}{ab}x + \dfrac{1}{a^2(b-a)}e^{-ax} + \dfrac{1}{b^2(a-b)}e^{-bx}$
43	$\dfrac{1}{(p+a)^2(p+b)^2}$	$\dfrac{1}{(a-b)^2}\left[e^{-ax}\left(x + \dfrac{2}{a-b}\right) + e^{-bx}\left(x - \dfrac{2}{a-b}\right)\right]$
44	$\dfrac{1}{(p+a)^4}$	$\frac{1}{6}x^3 e^{-ax}$
45	$\dfrac{p}{(p+a)^4}$	$\frac{1}{2}x^2 e^{-ax} - \frac{1}{6}ax^3 e^{-ax}$
46	$\dfrac{1}{p^2(p^2 + a^2)}$	$\dfrac{1}{a^3}\left[ax - \sin(ax)\right]$
47	$\dfrac{1}{p^4 - a^4}$	$\dfrac{1}{2a^3}\left[\sinh(ax) - \sin(ax)\right]$
48	$\dfrac{p}{p^4 - a^4}$	$\dfrac{1}{2a^2}\left[\cosh(ax) - \cos(ax)\right]$
49	$\dfrac{p^2}{p^4 - a^4}$	$\dfrac{1}{2a}\left[\sinh(ax) + \sin(ax)\right]$
50	$\dfrac{p^3}{p^4 - a^4}$	$\dfrac{1}{2}\left[\cosh(ax) + \cos(ax)\right]$
51	$\dfrac{1}{p^4 + a^4}$	$\dfrac{1}{a^3\sqrt{2}}\left(\cosh\xi\sin\xi - \sinh\xi\cos\xi\right), \quad \xi = \dfrac{ax}{\sqrt{2}}$
52	$\dfrac{p}{p^4 + a^4}$	$\dfrac{1}{a^2}\sin\left(\dfrac{ax}{\sqrt{2}}\right)\sinh\left(\dfrac{ax}{\sqrt{2}}\right)$
53	$\dfrac{p^2}{p^4 + a^4}$	$\dfrac{1}{a\sqrt{2}}\left(\cos\xi\sinh\xi + \sin\xi\cosh\xi\right), \quad \xi = \dfrac{ax}{\sqrt{2}}$

No	Laplace transform, $\tilde{f}(p)$	Inverse transform, $f(x) = \dfrac{1}{2\pi i} \displaystyle\int_{c-i\infty}^{c+i\infty} e^{px}\tilde{f}(p)\,dp$
54	$\dfrac{1}{(p^2+a^2)^2}$	$\dfrac{1}{2a^3}\big[\sin(ax)-ax\cos(ax)\big]$
55	$\dfrac{p}{(p^2+a^2)^2}$	$\dfrac{1}{2a}x\sin(ax)$
56	$\dfrac{p^2}{(p^2+a^2)^2}$	$\dfrac{1}{2a}\big[\sin(ax)+ax\cos(ax)\big]$
57	$\dfrac{p^3}{(p^2+a^2)^2}$	$\cos(ax)-\tfrac{1}{2}ax\sin(ax)$
58	$\dfrac{1}{\big[(p+b)^2+a^2\big]^2}$	$\dfrac{1}{2a^3}e^{-bx}\big[\sin(ax)-ax\cos(ax)\big]$
59	$\dfrac{1}{(p^2-a^2)(p^2-b^2)}$	$\dfrac{1}{a^2-b^2}\Big[\dfrac{1}{a}\sinh(ax)-\dfrac{1}{b}\sinh(bx)\Big]$
60	$\dfrac{p}{(p^2-a^2)(p^2-b^2)}$	$\dfrac{\cosh(ax)-\cosh(bx)}{a^2-b^2}$
61	$\dfrac{p^2}{(p^2-a^2)(p^2-b^2)}$	$\dfrac{a\sinh(ax)-b\sinh(bx)}{a^2-b^2}$
62	$\dfrac{p^3}{(p^2-a^2)(p^2-b^2)}$	$\dfrac{a^2\cosh(ax)-b^2\cosh(bx)}{a^2-b^2}$
63	$\dfrac{1}{(p^2+a^2)(p^2+b^2)}$	$\dfrac{1}{b^2-a^2}\Big[\dfrac{1}{a}\sin(ax)-\dfrac{1}{b}\sin(bx)\Big]$
64	$\dfrac{p}{(p^2+a^2)(p^2+b^2)}$	$\dfrac{\cos(ax)-\cos(bx)}{b^2-a^2}$
65	$\dfrac{p^2}{(p^2+a^2)(p^2+b^2)}$	$\dfrac{-a\sin(ax)+b\sin(bx)}{b^2-a^2}$
66	$\dfrac{p^3}{(p^2+a^2)(p^2+b^2)}$	$\dfrac{-a^2\cos(ax)+b^2\cos(bx)}{b^2-a^2}$
67	$\dfrac{1}{p^n},\quad n=1,2,\dots$	$\dfrac{1}{(n-1)!}x^{n-1}$
68	$\dfrac{1}{(p+a)^n},\quad n=1,2,\dots$	$\dfrac{1}{(n-1)!}x^{n-1}e^{-ax}$
69	$\dfrac{1}{p(p+a)^n},\quad n=1,2,\dots$	$a^{-n}\big[1-e^{-ax}e_n(ax)\big],\quad e_n(z)=1+\dfrac{z}{1!}+\cdots+\dfrac{z^n}{n!}$
70	$\dfrac{1}{p^{2n}+a^{2n}},\quad n=1,2,\dots$	$-\dfrac{1}{na^{2n}}\displaystyle\sum_{k=1}^{n}\exp(a_k x)\big[a_k\cos(b_k x)-b_k\sin(b_k x)\big],$ $a_k=a\cos\varphi_k,\ b_k=a\sin\varphi_k,\ \varphi_k=\dfrac{\pi(2k-1)}{2n}$
71	$\dfrac{1}{p^{2n}-a^{2n}},\quad n=1,2,\dots$	$\dfrac{1}{na^{2n-1}}\sinh(ax)+\dfrac{1}{na^{2n}}\displaystyle\sum_{k=2}^{n}\exp(a_k x)$ $\times\big[a_k\cos(b_k x)-b_k\sin(b_k x)\big],$ $a_k=a\cos\varphi_k,\ b_k=a\sin\varphi_k,\ \varphi_k=\dfrac{\pi(k-1)}{n}$

No	Laplace transform, $\tilde{f}(p)$	Inverse transform, $f(x) = \dfrac{1}{2\pi i}\displaystyle\int_{c-i\infty}^{c+i\infty} e^{px}\tilde{f}(p)\,dp$
6	$\dfrac{1}{p\sqrt{p}}$	$2\sqrt{\dfrac{x}{\pi}}$
7	$\dfrac{1}{(p+a)\sqrt{p+b}}$	$(b-a)^{-1/2}e^{-ax}\,\mathrm{erf}\!\left[(b-a)^{1/2}x^{1/2}\right]$
8	$\dfrac{1}{\sqrt{p}\,(p-a)}$	$\dfrac{1}{\sqrt{a}}e^{ax}\,\mathrm{erf}\!\left(\sqrt{ax}\right)$
9	$\dfrac{1}{p^{3/2}(p-a)}$	$a^{-3/2}e^{ax}\,\mathrm{erf}\!\left(\sqrt{ax}\right)-2a^{-1}\pi^{-1/2}x^{1/2}$
10	$\dfrac{1}{\sqrt{p}+a}$	$\pi^{-1/2}x^{-1/2}-ae^{a^2x}\,\mathrm{erfc}\!\left(a\sqrt{x}\right)$
11	$\dfrac{a}{p(\sqrt{p}+a)}$	$1-e^{a^2x}\,\mathrm{erfc}\!\left(a\sqrt{x}\right)$
12	$\dfrac{1}{p+a\sqrt{p}}$	$e^{a^2x}\,\mathrm{erfc}\!\left(a\sqrt{x}\right)$
13	$\dfrac{1}{\left(\sqrt{p}+\sqrt{a}\right)^2}$	$1-\dfrac{2}{\sqrt{\pi}}(ax)^{1/2}+(1-2ax)e^{ax}\left[\mathrm{erf}\!\left(\sqrt{ax}\right)-1\right]$
14	$\dfrac{1}{p\left(\sqrt{p}+\sqrt{a}\right)^2}$	$\dfrac{1}{a}+\left(2x-\dfrac{1}{a}\right)e^{ax}\,\mathrm{erfc}\!\left(\sqrt{ax}\right)-\dfrac{2}{\sqrt{\pi a}}\sqrt{x}$
15	$\dfrac{1}{\sqrt{p}\left(\sqrt{p}+a\right)^2}$	$2\pi^{-1/2}x^{1/2}-2axe^{a^2x}\,\mathrm{erfc}\!\left(a\sqrt{x}\right)$
16	$\dfrac{1}{\left(\sqrt{p}+a\right)^3}$	$\dfrac{2}{\sqrt{\pi}}(a^2x+1)\sqrt{x}-ax(2a^2x+3)e^{a^2x}\,\mathrm{erfc}\!\left(a\sqrt{x}\right)$
17	$p^{-n-1/2},\quad n=1,2,\ldots$	$\dfrac{2^n}{1\cdot3\ldots(2n-1)\sqrt{\pi}}\,x^{n-1/2}$
18	$(p+a)^{-n-1/2}$	$\dfrac{2^n}{1\cdot3\ldots(2n-1)\sqrt{\pi}}\,x^{n-1/2}e^{-ax}$
19	$\dfrac{1}{\sqrt{p^2+a^2}}$	$J_0(ax)$
20	$\dfrac{1}{\sqrt{p^2-a^2}}$	$I_0(ax)$
21	$\dfrac{1}{\sqrt{p^2+ap+b}}$	$\exp\!\left(-\tfrac{1}{2}ax\right)J_0\!\left[(b-\tfrac{1}{4}a^2)^{1/2}x\right]$
22	$\left(\sqrt{p^2+a^2}-p\right)^{1/2}$	$\dfrac{1}{\sqrt{2\pi x^3}}\sin(ax)$
23	$\dfrac{1}{\sqrt{p^2+a^2}}\left(\sqrt{p^2+a^2}+p\right)^{1/2}$	$\dfrac{\sqrt{2}}{\sqrt{\pi x}}\cos(ax)$
24	$\dfrac{1}{\sqrt{p^2-a^2}}\left(\sqrt{p^2-a^2}+p\right)^{1/2}$	$\dfrac{\sqrt{2}}{\sqrt{\pi x}}\cosh(ax)$

No	Laplace transform, $\tilde{f}(p)$	Inverse transform, $f(x) = \dfrac{1}{2\pi i}\displaystyle\int_{c-i\infty}^{c+i\infty} e^{px}\,\tilde{f}(p)\,dp$
72	$\dfrac{1}{p^{2n+1} + a^{2n+1}},\quad n = 0,\,1,\,\ldots$	$\dfrac{e^{-ax}}{(2n+1)a^{2n}} - \dfrac{2}{(2n+1)a^{2n+1}}\displaystyle\sum_{k=1}^{n}\exp(a_k x)$ $\times\big[a_k\cos(b_k x) - b_k\sin(b_k x)\big],$ $a_k = a\cos\varphi_k,\ b_k = a\sin\varphi_k,\ \varphi_k = \dfrac{\pi(2k-1)}{2n+1}$
73	$\dfrac{1}{p^{2n+1} - a^{2n+1}},\quad n = 0,\,1,\,\ldots$	$\dfrac{e^{ax}}{(2n+1)a^{2n}} + \dfrac{2}{(2n+1)a^{2n+1}}\displaystyle\sum_{k=1}^{n}\exp(a_k x)$ $\times\big[a_k\cos(b_k x) - b_k\sin(b_k x)\big],$ $a_k = a\cos\varphi_k,\ b_k = a\sin\varphi_k,\ \varphi_k = \dfrac{2\pi k}{2n+1}$
74	$\dfrac{Q(p)}{P(p)},$ $P(p) = (p - a_1)\ldots(p - a_n);$ $Q(p)$ is a polynomial of degree $\leq n - 1;\ a_i \neq a_j$ if $i \neq j$	$\displaystyle\sum_{k=1}^{n}\dfrac{Q(a_k)}{P'(a_k)}\exp(a_k x),$ (the prime stand for the differentiation)
75	$\dfrac{Q(p)}{P(p)},$ $P(p) = (p - a_1)^{m_1}\ldots(p - a_n)^{m_n};$ $Q(p)$ is a polynomial of degree $< m_1 + m_2 + \cdots + m_n - 1;$ $a_i \neq a_j$ if $i \neq j$	$\displaystyle\sum_{k=1}^{n}\sum_{l=1}^{m_k}\dfrac{\Phi_{kl}(a_k)}{(m_k - l)!\,(l-1)!}x^{m_k-l}\exp(a_k x),$ $\Phi_{kl}(p) = \dfrac{d^{l-1}}{dp^{l-1}}\left[\dfrac{Q(p)}{P_k(p)}\right],\quad P_k(p) = \dfrac{P(p)}{(p - a_k)^m}$
76	$\dfrac{Q(p) + pR(p)}{P(p)},$ $P(p) = (p^2 + a_1^2)\ldots(p^2 + a_n^2);$ $Q(p)$ and $R(p)$ are polynomials of degree $\leq 2n - 2;\ a_l \neq a_j,\ l \neq j$	$\displaystyle\sum_{k=1}^{n}\dfrac{Q(ia_k)\sin(a_k x) + a_k R(ia_k)\cos(a_k x)}{a_k P_k(ia_k)},$ $P_m(p) = \dfrac{P(p)}{p^2 + a_m^2},\quad i^2 = -1$

5.3. Expressions With Square Roots

No	Laplace transform, $\tilde{f}(p)$	Inverse transform, $f(x) = \dfrac{1}{2\pi i}\displaystyle\int_{c-i\infty}^{c+i\infty} e^{px}\,\tilde{f}(p)\,dp$
1	$\dfrac{1}{\sqrt{p}}$	$\dfrac{1}{\sqrt{\pi x}}$
2	$\sqrt{p-a} - \sqrt{p-b}$	$\dfrac{e^{bx} - e^{ax}}{2\sqrt{\pi x^3}}$
3	$\dfrac{1}{\sqrt{p+a}}$	$\dfrac{1}{\sqrt{\pi x}}e^{-ax}$
4	$\sqrt{\dfrac{p+a}{p}} - 1$	$\tfrac{1}{2}ae^{-ax/2}\big[I_1\big(\tfrac{1}{2}ax\big) + I_0\big(\tfrac{1}{2}ax\big)\big]$
5	$\dfrac{\sqrt{p+a}}{p+b}$	$\dfrac{e^{-ax}}{\sqrt{\pi x}} + (a-b)^{1/2}e^{-bx}\,\mathrm{erf}\big[(a-b)^{1/2}x^{1/2}\big]$

No	Laplace transform, $\tilde{f}(p)$	Inverse transform, $f(x) = \dfrac{1}{2\pi i} \displaystyle\int_{c-i\infty}^{c+i\infty} e^{px}\,\tilde{f}(p)\,dp$
25	$\left(\sqrt{p^2+a^2}+p\right)^{-n}$	$na^{-n}x^{-1}J_n(ax)$
26	$\left(\sqrt{p^2-a^2}+p\right)^{-n}$	$na^{-n}x^{-1}I_n(ax)$
27	$\left(p^2+a^2\right)^{-n-1/2}$	$\dfrac{(x/a)^n J_n(ax)}{1\cdot 3\cdot 5\ldots(2n-1)}$
28	$\left(p^2-a^2\right)^{-n-1/2}$	$\dfrac{(x/a)^n I_n(ax)}{1\cdot 3\cdot 5\ldots(2n-1)}$

5.4. Expressions With Arbitrary Powers

No	Laplace transform, $\tilde{f}(p)$	Inverse transform, $f(x) = \dfrac{1}{2\pi i} \displaystyle\int_{c-i\infty}^{c+i\infty} e^{px}\,\tilde{f}(p)\,dp$
1	$(p+a)^{-\nu},\ \ \nu>0$	$\dfrac{1}{\Gamma(\nu)}x^{\nu-1}e^{-ax}$
2	$\left[(p+a)^{1/2}+(p+b)^{1/2}\right]^{-2\nu},\ \ \nu>0$	$\dfrac{\nu}{(a-b)^\nu}x^{-1}\exp\left[-\tfrac{1}{2}(a+b)x\right]I_\nu\left[\tfrac{1}{2}(a-b)x\right]$
3	$\left[(p+a)(p+b)\right]^{-\nu},\ \ \nu>0$	$\dfrac{\sqrt{\pi}}{\Gamma(\nu)}\left(\dfrac{x}{a-b}\right)^{\nu-1/2}\exp\left(-\dfrac{a+b}{2}x\right)I_{\nu-1/2}\left(\dfrac{a-b}{2}x\right)$
4	$\left(p^2+a^2\right)^{-\nu-1/2},\ \ \nu>-\tfrac{1}{2}$	$\dfrac{\sqrt{\pi}}{(2a)^\nu\Gamma\left(\nu+\tfrac{1}{2}\right)}x^\nu J_\nu(ax)$
5	$\left(p^2-a^2\right)^{-\nu-1/2},\ \ \nu>-\tfrac{1}{2}$	$\dfrac{\sqrt{\pi}}{(2a)^\nu\Gamma\left(\nu+\tfrac{1}{2}\right)}x^\nu I_\nu(ax)$
6	$p\left(p^2+a^2\right)^{-\nu-1/2},\ \ \nu>0$	$\dfrac{a\sqrt{\pi}}{(2a)^\nu\Gamma\left(\nu+\tfrac{1}{2}\right)}x^\nu J_{\nu-1}(ax)$
7	$p\left(p^2-a^2\right)^{-\nu-1/2},\ \ \nu>0$	$\dfrac{a\sqrt{\pi}}{(2a)^\nu\Gamma\left(\nu+\tfrac{1}{2}\right)}x^\nu I_{\nu-1}(ax)$
8	$\left[(p^2+a^2)^{1/2}+p\right]^{-\nu}=$ $a^{-2\nu}\left[(p^2+a^2)^{1/2}-p\right]^\nu,\ \ \nu>0$	$\nu a^{-\nu}x^{-1}J_\nu(ax)$
9	$\left[(p^2-a^2)^{1/2}+p\right]^{-\nu}=$ $a^{-2\nu}\left[p-(p^2-a^2)^{1/2}\right]^\nu,\ \ \nu>0$	$\nu a^{-\nu}x^{-1}I_\nu(ax)$
10	$p\left[(p^2+a^2)^{1/2}+p\right]^{-\nu},\ \ \nu>1$	$\nu a^{1-\nu}x^{-1}J_{\nu-1}(ax)-\nu(\nu+1)a^{-\nu}x^{-2}J_\nu(ax)$
11	$p\left[(p^2-a^2)^{1/2}+p\right]^{-\nu},\ \ \nu>1$	$\nu a^{1-\nu}x^{-1}I_{\nu-1}(ax)-\nu(\nu+1)a^{-\nu}x^{-2}I_\nu(ax)$
12	$\dfrac{\left(\sqrt{p^2+a^2}+p\right)^{-\nu}}{\sqrt{p^2+a^2}},\ \ \nu>-1$	$a^{-\nu}J_\nu(ax)$
13	$\dfrac{\left(\sqrt{p^2-a^2}+p\right)^{-\nu}}{\sqrt{p^2-a^2}},\ \ \nu>-1$	$a^{-\nu}I_\nu(ax)$

5.5. Expressions With Exponential Functions

No	Laplace transform, $\tilde{f}(p)$	Inverse transform, $f(x) = \dfrac{1}{2\pi i} \displaystyle\int_{c-i\infty}^{c+i\infty} e^{px} \tilde{f}(p)\, dp$
1	$p^{-1}e^{-ap}, \quad a > 0$	$\begin{cases} 0 & \text{if } 0 < x < a, \\ 1 & \text{if } a < x. \end{cases}$
2	$p^{-1}\left(1 - e^{-ap}\right), \quad a > 0$	$\begin{cases} 1 & \text{if } 0 < x < a, \\ 0 & \text{if } a < x. \end{cases}$
3	$p^{-1}\left(e^{-ap} - e^{-bp}\right), \quad 0 \le a < b$	$\begin{cases} 0 & \text{if } 0 < x < a, \\ 1 & \text{if } a < x < b, \\ 0 & \text{if } b < x. \end{cases}$
4	$p^{-2}\left(e^{-ap} - e^{-bp}\right), \quad 0 \le a < b$	$\begin{cases} 0 & \text{if } 0 < x < a, \\ x - a & \text{if } a < x < b, \\ b - a & \text{if } b < x. \end{cases}$
5	$(p + b)^{-1}e^{-ap}, \quad a > 0$	$\begin{cases} 0 & \text{if } 0 < x < a, \\ e^{-b(x-a)} & \text{if } a < x. \end{cases}$
6	$p^{-\nu}e^{-ap}, \quad \nu > 0$	$\begin{cases} 0 & \text{if } 0 < x < a, \\ \dfrac{(x-a)^{\nu-1}}{\Gamma(\nu)} & \text{if } a < x. \end{cases}$
7	$p^{-1}\left(e^{ap} - 1\right)^{-1}, \quad a > 0$	$f(x) = n \ \text{ if }\ na < x < (n+1)a; \quad n = 0, 1, 2, \ldots$
8	$e^{a/p} - 1$	$\sqrt{\dfrac{a}{x}}\, I_1\left(2\sqrt{ax}\right)$
9	$p^{-1/2}e^{a/p}$	$\dfrac{1}{\sqrt{\pi x}} \cosh\left(2\sqrt{ax}\right)$
10	$p^{-3/2}e^{a/p}$	$\dfrac{1}{\sqrt{\pi a}} \sinh\left(2\sqrt{ax}\right)$
11	$p^{-5/2}e^{a/p}$	$\sqrt{\dfrac{x}{\pi a}} \cosh\left(2\sqrt{ax}\right) - \dfrac{1}{2\sqrt{\pi a^3}} \sinh\left(2\sqrt{ax}\right)$
12	$p^{-\nu-1}e^{a/p}, \quad \nu > -1$	$(x/a)^{\nu/2} I_\nu\left(2\sqrt{ax}\right)$
13	$1 - e^{-a/p}$	$\sqrt{\dfrac{a}{x}}\, J_1\left(2\sqrt{ax}\right)$
14	$p^{-1/2}e^{-a/p}$	$\dfrac{1}{\sqrt{\pi x}} \cos\left(2\sqrt{ax}\right)$
15	$p^{-3/2}e^{-a/p}$	$\dfrac{1}{\sqrt{\pi a}} \sin\left(2\sqrt{ax}\right)$
16	$p^{-5/2}e^{-a/p}$	$\dfrac{1}{2\sqrt{\pi a^3}} \sin\left(2\sqrt{ax}\right) - \sqrt{\dfrac{x}{\pi a}} \cos\left(2\sqrt{ax}\right)$
17	$p^{-\nu-1}e^{-a/p}, \quad \nu > -1$	$(x/a)^{\nu/2} J_\nu\left(2\sqrt{ax}\right)$
18	$\exp\left(-\sqrt{ap}\right), \quad a > 0$	$\dfrac{\sqrt{a}}{2\sqrt{\pi}}\, x^{-3/2} \exp\left(-\dfrac{a}{4x}\right)$
19	$p\exp\left(-\sqrt{ap}\right), \quad a > 0$	$\dfrac{\sqrt{a}}{8\sqrt{\pi}}\, (a - 6x)x^{-7/2} \exp\left(-\dfrac{a}{4x}\right)$
20	$\dfrac{1}{p}\exp\left(-\sqrt{ap}\right), \quad a \ge 0$	$\mathrm{erfc}\left(\dfrac{\sqrt{a}}{2\sqrt{x}}\right)$

No	Laplace transform, $\tilde{f}(p)$	Inverse transform, $f(x) = \dfrac{1}{2\pi i} \displaystyle\int_{c-i\infty}^{c+i\infty} e^{px}\,\tilde{f}(p)\,dp$
21	$\sqrt{p}\exp(-\sqrt{ap}\,), \quad a > 0$	$\dfrac{1}{4\sqrt{\pi}}\,(a - 2x)x^{-5/2}\exp\!\left(-\dfrac{a}{4x}\right)$
22	$\dfrac{1}{\sqrt{p}}\exp(-\sqrt{ap}\,), \quad a \geq 0$	$\dfrac{1}{\sqrt{\pi x}}\exp\!\left(-\dfrac{a}{4x}\right)$
23	$\dfrac{1}{p\sqrt{p}}\exp(-\sqrt{ap}\,), \quad a \geq 0$	$\dfrac{2\sqrt{x}}{\sqrt{\pi}}\exp\!\left(-\dfrac{a}{4x}\right) - \sqrt{a}\,\mathrm{erfc}\!\left(\dfrac{\sqrt{a}}{2\sqrt{x}}\right)$
24	$\dfrac{\exp\!\left(-k\sqrt{p^2 + a^2}\,\right)}{\sqrt{p^2 + a^2}}, \quad k > 0$	$\begin{cases} 0 & \text{if } 0 < x < k, \\ J_0\!\left(a\sqrt{x^2 - k^2}\,\right) & \text{if } k < x. \end{cases}$
25	$\dfrac{\exp\!\left(-k\sqrt{p^2 - a^2}\,\right)}{\sqrt{p^2 - a^2}}, \quad k > 0$	$\begin{cases} 0 & \text{if } 0 < x < k, \\ I_0\!\left(a\sqrt{x^2 - k^2}\,\right) & \text{if } k < x. \end{cases}$

5.6. Expressions With Hyperbolic Functions

No	Laplace transform, $\tilde{f}(p)$	Inverse transform, $f(x) = \dfrac{1}{2\pi i} \displaystyle\int_{c-i\infty}^{c+i\infty} e^{px}\,\tilde{f}(p)\,dp$
1	$\dfrac{1}{p\sinh(ap)}, \quad a > 0$	$f(x) = 2n$ if $a(2n-1) < x < a(2n+1)$; $n = 0, 1, 2, \dots \;(x > 0)$
2	$\dfrac{1}{p^2\sinh(ap)}, \quad a > 0$	$f(x) = 2n(x - an)$ if $a(2n-1) < x < a(2n+1)$; $n = 0, 1, 2, \dots \;(x > 0)$
3	$\dfrac{\sinh(a/p)}{\sqrt{p}}$	$\dfrac{1}{2\sqrt{\pi x}}\left[\cosh\!\left(2\sqrt{ax}\,\right) - \cos\!\left(2\sqrt{ax}\,\right)\right]$
4	$\dfrac{\sinh(a/p)}{p\sqrt{p}}$	$\dfrac{1}{2\sqrt{\pi a}}\left[\sinh\!\left(2\sqrt{ax}\,\right) - \sin\!\left(2\sqrt{ax}\,\right)\right]$
5	$p^{-\nu-1}\sinh(a/p), \quad \nu > -2$	$\tfrac{1}{2}(x/a)^{\nu/2}\left[I_\nu\!\left(2\sqrt{ax}\,\right) - J_\nu\!\left(2\sqrt{ax}\,\right)\right]$
6	$\dfrac{1}{p\cosh(ap)}, \quad a > 0$	$f(x) = \begin{cases} 0 & \text{if } a(4n-1) < x < a(4n+1), \\ 2 & \text{if } a(4n+1) < x < a(4n+3), \end{cases}$ $n = 0, 1, 2, \dots \;(x > 0)$
7	$\dfrac{1}{p^2\cosh(ap)}, \quad a > 0$	$x - (-1)^n(x - 2an)$ if $2n - 1 < x/a < 2n + 1$; $n = 0, 1, 2, \dots \;(x > 0)$
8	$\dfrac{\cosh(a/p)}{\sqrt{p}}$	$\dfrac{1}{2\sqrt{\pi x}}\left[\cosh\!\left(2\sqrt{ax}\,\right) + \cos\!\left(2\sqrt{ax}\,\right)\right]$
9	$\dfrac{\cosh(a/p)}{p\sqrt{p}}$	$\dfrac{1}{2\sqrt{\pi a}}\left[\sinh\!\left(2\sqrt{ax}\,\right) + \sin\!\left(2\sqrt{ax}\,\right)\right]$
10	$p^{-\nu-1}\cosh(a/p), \quad \nu > -1$	$\tfrac{1}{2}(x/a)^{\nu/2}\left[I_\nu\!\left(2\sqrt{ax}\,\right) + J_\nu\!\left(2\sqrt{ax}\,\right)\right]$
11	$\dfrac{1}{p}\tanh(ap), \quad a > 0$	$f(x) = (-1)^{n-1}$ if $2a(n-1) < x < 2an$; $n = 1, 2, \dots$

No	Laplace transform, $\tilde{f}(p)$	Inverse transform, $f(x) = \dfrac{1}{2\pi i}\displaystyle\int_{c-i\infty}^{c+i\infty} e^{px}\tilde{f}(p)\,dp$
12	$\dfrac{1}{p}\coth(ap), \quad a > 0$	$f(x) = (2n-1)$ if $2a(n-1) < x < 2an;$ $n = 1, 2, \ldots$
13	$\operatorname{Arcoth}(p/a)$	$\dfrac{1}{x}\sinh(ax)$

5.7. Expressions With Logarithmic Functions

No	Laplace transform, $\tilde{f}(p)$	Inverse transform, $f(x) = \dfrac{1}{2\pi i}\displaystyle\int_{c-i\infty}^{c+i\infty} e^{px}\tilde{f}(p)\,dp$
1	$\dfrac{1}{p}\ln p$	$-\ln x - \mathcal{C},$ $\mathcal{C} = 0.5772\ldots$ is the Euler constant
2	$p^{-n-1}\ln p$	$\left(1 + \frac{1}{2} + \frac{1}{3} + \cdots + \frac{1}{n} - \ln x - \mathcal{C}\right)\dfrac{x^n}{n!},$ $\mathcal{C} = 0.5772\ldots$ is the Euler constant
3	$p^{-n-1/2}\ln p$	$k_n\left[2 + \frac{2}{3} + \frac{2}{5} + \cdots + \frac{2}{2n-1} - \ln(4x) - \mathcal{C}\right]x^{n-1/2},$ $k_n = \dfrac{2^n}{1\cdot 3\cdot 5\ldots(2n-1)\sqrt{\pi}}, \quad \mathcal{C} = 0.5772\ldots$
4	$p^{-\nu}\ln p, \quad \nu > 0$	$\dfrac{1}{\Gamma(\nu)}x^{\nu-1}\big[\psi(\nu) - \ln x\big], \quad \psi(\nu)$ is the logarithmic derivative of the gamma function
5	$\dfrac{1}{p}(\ln p)^2$	$(\ln x + \mathcal{C})^2 - \frac{1}{6}\pi^2, \quad \mathcal{C} = 0.5772\ldots$
6	$\dfrac{1}{p^2}(\ln p)^2$	$x\big[(\ln x + \mathcal{C} - 1)^2 + 1 - \frac{1}{6}\pi^2\big]$
7	$\dfrac{\ln(p+b)}{p+a}$	$e^{-ax}\big\{\ln(b-a) - \operatorname{Ei}\big[(a-b)x\big]\big\}$
8	$\dfrac{\ln p}{p^2 + a^2}$	$\dfrac{1}{a}\cos(ax)\operatorname{Si}(ax) + \dfrac{1}{a}\sin(ax)\big[\ln a - \operatorname{Ci}(ax)\big]$
9	$\dfrac{p\ln p}{p^2 + a^2}$	$\cos(ax)\big[\ln a - \operatorname{Ci}(ax)\big] - \sin(ax)\operatorname{Si}(ax)$
10	$\ln\dfrac{p+b}{p+a}$	$\dfrac{1}{x}\left(e^{-ax} - e^{-bx}\right)$
11	$\ln\dfrac{p^2 + b^2}{p^2 + a^2}$	$\dfrac{2}{x}\big[\cos(ax) - \cos(bx)\big]$
12	$p\ln\dfrac{p^2 + b^2}{p^2 + a^2}$	$\dfrac{2}{x}\big[\cos(bx) + bx\sin(bx) - \cos(ax) - ax\sin(ax)\big]$
13	$\ln\dfrac{(p+a)^2 + k^2}{(p+b)^2 + k^2}$	$\dfrac{2}{x}\cos(kx)(e^{-bx} - e^{-ax})$
14	$p\ln\left(\dfrac{1}{p}\sqrt{p^2 + a^2}\right)$	$\dfrac{1}{x^2}\big[\cos(ax) - 1\big] + \dfrac{a}{x}\sin(ax)$
15	$p\ln\left(\dfrac{1}{p}\sqrt{p^2 - a^2}\right)$	$\dfrac{1}{x^2}\big[\cosh(ax) - 1\big] - \dfrac{a}{x}\sinh(ax)$

5.8. Expressions With Trigonometric Functions

No	Laplace transform, $\tilde{f}(p)$	Inverse transform, $f(x) = \dfrac{1}{2\pi i}\displaystyle\int_{c-i\infty}^{c+i\infty} e^{px}\,\tilde{f}(p)\,dp$
1	$\dfrac{\sin(a/p)}{\sqrt{p}}$	$\dfrac{1}{\sqrt{\pi x}}\sinh\!\left(\sqrt{2ax}\,\right)\sin\!\left(\sqrt{2ax}\,\right)$
2	$\dfrac{\sin(a/p)}{p\sqrt{p}}$	$\dfrac{1}{\sqrt{\pi a}}\cosh\!\left(\sqrt{2ax}\,\right)\sin\!\left(\sqrt{2ax}\,\right)$
3	$\dfrac{\cos(a/p)}{\sqrt{p}}$	$\dfrac{1}{\sqrt{\pi x}}\cosh\!\left(\sqrt{2ax}\,\right)\cos\!\left(\sqrt{2ax}\,\right)$
4	$\dfrac{\cos(a/p)}{p\sqrt{p}}$	$\dfrac{1}{\sqrt{\pi a}}\sinh\!\left(\sqrt{2ax}\,\right)\cos\!\left(\sqrt{2ax}\,\right)$
5	$\dfrac{1}{\sqrt{p}}\exp\!\left(-\sqrt{ap}\,\right)\sin\!\left(\sqrt{ap}\,\right)$	$\dfrac{1}{\sqrt{\pi x}}\sin\!\left(\dfrac{a}{2x}\right)$
6	$\dfrac{1}{\sqrt{p}}\exp\!\left(-\sqrt{ap}\,\right)\cos\!\left(\sqrt{ap}\,\right)$	$\dfrac{1}{\sqrt{\pi x}}\cos\!\left(\dfrac{a}{2x}\right)$
7	$\arctan\dfrac{a}{p}$	$\dfrac{1}{x}\sin(ax)$
8	$\dfrac{1}{p}\arctan\dfrac{a}{p}$	$\mathrm{Si}(ax)$
9	$p\arctan\dfrac{a}{p}-a$	$\dfrac{1}{x^2}\big[ax\cos(ax)-\sin(ax)\big]$
10	$\arctan\dfrac{2ap}{p^2+b^2}$	$\dfrac{2}{x}\sin(ax)\cos\!\left(x\sqrt{a^2+b^2}\,\right)$

5.9. Expressions With Special Functions

No	Laplace transform, $\tilde{f}(p)$	Inverse transform, $f(x) = \dfrac{1}{2\pi i}\displaystyle\int_{c-i\infty}^{c+i\infty} e^{px}\,\tilde{f}(p)\,dp$
1	$\exp\!\left(ap^2\right)\mathrm{erfc}\!\left(p\sqrt{a}\,\right)$	$\dfrac{1}{\sqrt{\pi a}}\exp\!\left(-\dfrac{x^2}{4a}\right)$
2	$\dfrac{1}{p}\exp\!\left(ap^2\right)\mathrm{erfc}\!\left(p\sqrt{a}\,\right)$	$\mathrm{erf}\!\left(\dfrac{x}{2\sqrt{a}}\right)$
3	$\mathrm{erfc}\!\left(\sqrt{ap}\,\right),\quad a>0$	$\begin{cases} 0 & \text{if } 0<x<a, \\[2mm] \dfrac{\sqrt{a}}{\pi x\sqrt{x-a}} & \text{if } a<x. \end{cases}$
4	$e^{ap}\,\mathrm{erfc}\!\left(\sqrt{ap}\,\right)$	$\dfrac{\sqrt{a}}{\pi\sqrt{x}\,(x+a)}$
5	$\dfrac{1}{\sqrt{p}}e^{ap}\,\mathrm{erfc}\!\left(\sqrt{ap}\,\right)$	$\dfrac{1}{\sqrt{\pi(x+a)}}$
6	$\mathrm{erf}\!\left(\sqrt{a/p}\,\right)$	$\dfrac{1}{\pi x}\sin\!\left(2\sqrt{ax}\,\right)$

No	Laplace transform, $\tilde{f}(p)$	Inverse transform, $f(x) = \dfrac{1}{2\pi i}\displaystyle\int_{c-i\infty}^{c+i\infty} e^{px}\tilde{f}(p)\,dp$
7	$\dfrac{1}{\sqrt{p}}\exp(a/p)\operatorname{erf}\!\left(\sqrt{a/p}\right)$	$\dfrac{1}{\sqrt{\pi x}}\sinh\!\left(2\sqrt{ax}\right)$
8	$\dfrac{1}{\sqrt{p}}\exp(a/p)\operatorname{erfc}\!\left(\sqrt{a/p}\right)$	$\dfrac{1}{\sqrt{\pi x}}\exp\!\left(-2\sqrt{ax}\right)$
9	$p^{-a}\gamma(a,bp),\quad a,b>0$	$\begin{cases} x^{a-1} & \text{if } 0<x<b, \\ 0 & \text{if } b<x. \end{cases}$
10	$\gamma(a,b/p),\quad a>0$	$b^{a/2}x^{a/2-1}J_a\!\left(2\sqrt{bx}\right)$
11	$a^{-p}\gamma(p,a)$	$\exp\!\left(-ae^{-x}\right)$
12	$K_0(ap),\quad a>0$	$\begin{cases} 0 & \text{if } 0<x<a, \\ (x^2-a^2)^{-1/2} & \text{if } a<x. \end{cases}$
13	$K_\nu(ap),\quad a>0$	$\begin{cases} 0 & \text{if } 0<x<a, \\ \dfrac{\cosh\!\left[\nu\operatorname{Arcosh}(x/a)\right]}{\sqrt{x^2-a^2}} & \text{if } a<x. \end{cases}$
14	$K_0\!\left(a\sqrt{p}\right)$	$\dfrac{1}{2x}\exp\!\left(-\dfrac{a^2}{4x}\right)$
15	$\dfrac{1}{\sqrt{p}}K_1\!\left(a\sqrt{p}\right)$	$\dfrac{1}{a}\exp\!\left(-\dfrac{a^2}{4x}\right)$

⊙ References for Supplement 5: G. Doetsch (1950, 1956, 1958), H. Bateman and A. Erdélyi (1954), I. I. Hirschman and D. V. Widder (1955), V. A. Ditkin and A. P. Prudnikov (1965).

Supplement 6

Tables of Fourier Cosine Transforms

6.1. General Formulas

No	Original function, $f(x)$	Cosine transform, $\check{f}_c(u) = \int_0^\infty f(x)\cos(ux)\,dx$
1	$af_1(x) + bf_2(x)$	$a\check{f}_{1c}(u) + b\check{f}_{2c}(u)$
2	$f(ax), \quad a > 0$	$\dfrac{1}{a}\check{f}_c\left(\dfrac{u}{a}\right)$
3	$x^{2n}f(x), \quad n = 1, 2, \ldots$	$(-1)^n \dfrac{d^{2n}}{du^{2n}}\check{f}_c(u)$
4	$x^{2n+1}f(ax), \quad n = 0, 1, \ldots$	$(-1)^n \dfrac{d^{2n+1}}{du^{2n+1}}\check{f}_s(u), \quad \check{f}_s(u) = \int_0^\infty f(x)\sin(xu)\,dx$
5	$f(ax)\cos(bx), \quad a, b > 0$	$\dfrac{1}{2a}\left[\check{f}_c\left(\dfrac{u+b}{a}\right) + \check{f}_c\left(\dfrac{u-b}{a}\right)\right]$

6.2. Expressions With Power-Law Functions

No	Original function, $f(x)$	Cosine transform, $\check{f}_c(u) = \int_0^\infty f(x)\cos(ux)\,dx$
1	$\begin{cases} 1 & \text{if } 0 < x < a, \\ 0 & \text{if } a < x \end{cases}$	$\dfrac{1}{u}\sin(au)$
2	$\begin{cases} x & \text{if } 0 < x < 1, \\ 2-x & \text{if } 1 < x < 2, \\ 0 & \text{if } 2 < x \end{cases}$	$\dfrac{4}{u^2}\cos u \sin^2\dfrac{u}{2}$
3	$\dfrac{1}{a+x}, \quad a > 0$	$-\sin(au)\,\text{si}(au) - \cos(au)\,\text{Ci}(au)$
4	$\dfrac{1}{a^2+x^2}, \quad a > 0$	$\dfrac{\pi}{2a}e^{-au}$ (the integral is understood in the sense of Cauchy principal value)
5	$\dfrac{1}{a^2-x^2}, \quad a > 0$	$\dfrac{\pi\sin(au)}{2u}$
6	$\dfrac{a}{a^2+(b+x)^2} + \dfrac{a}{a^2+(b-x)^2}$	$\pi e^{-au}\cos(bu)$
7	$\dfrac{b+x}{a^2+(b+x)^2} + \dfrac{b-x}{a^2+(b-x)^2}$	$\pi e^{-au}\sin(bu)$
8	$\dfrac{1}{a^4+x^4}, \quad a > 0$	$\tfrac{1}{2}\pi a^{-3}\exp\left(-\dfrac{au}{\sqrt{2}}\right)\sin\left(\dfrac{\pi}{4} + \dfrac{au}{\sqrt{2}}\right)$

No	Original function, $f(x)$	Cosine transform, $\check{f}_c(u) = \int_0^\infty f(x)\cos(ux)\,dx$
9	$\dfrac{1}{(a^2+x^2)(b^2+x^2)}, \quad a,b>0$	$\dfrac{\pi}{2}\dfrac{ae^{-bu}-be^{-au}}{ab(a^2-b^2)}$
10	$\dfrac{x^{2m}}{(x^2+a)^{n+1}},$ $n,m=1,2,\ldots;\ \ n+1>m\geq 0$	$(-1)^{n+m}\dfrac{\pi}{2n!}\dfrac{\partial^n}{\partial a^n}\left(a^{1/\sqrt{m}}e^{-u\sqrt{a}}\right)$
11	$\dfrac{1}{\sqrt{x}}$	$\sqrt{\dfrac{\pi}{2u}}$
12	$\begin{cases}\dfrac{1}{\sqrt{x}} & \text{if } 0<x<a,\\ 0 & \text{if } a<x\end{cases}$	$2\sqrt{\dfrac{\pi}{2u}}\,C(au),\ \ C(u)$ is the Fresnel integral
13	$\begin{cases}0 & \text{if } 0<x<a,\\ \dfrac{1}{\sqrt{x}} & \text{if } a<x\end{cases}$	$\sqrt{\dfrac{\pi}{2u}}\left[1-2C(au)\right],\ \ C(u)$ is the Fresnel integral
14	$\begin{cases}0 & \text{if } 0<x<a,\\ \dfrac{1}{\sqrt{x-a}} & \text{if } a<x\end{cases}$	$\sqrt{\dfrac{\pi}{2u}}\left[\cos(au)-\sin(au)\right]$
15	$\dfrac{1}{\sqrt{a^2+x^2}}$	$K_0(au)$
16	$\begin{cases}\dfrac{1}{\sqrt{a^2-x^2}} & \text{if } 0<x<a,\\ 0 & \text{if } a<x\end{cases}$	$\dfrac{\pi}{2}\,J_0(au)$
17	$x^{-\nu},\quad 0<\nu<1$	$\sin\left(\tfrac{1}{2}\pi\nu\right)\Gamma(1-\nu)u^{\nu-1}$

6.3. Expressions With Exponential Functions

No	Original function, $f(x)$	Cosine transform, $\check{f}_c(u) = \int_0^\infty f(x)\cos(ux)\,dx$
1	e^{-ax}	$\dfrac{a}{a^2+u^2}$
2	$\dfrac{1}{x}\left(e^{-ax}-e^{-bx}\right)$	$\dfrac{1}{2}\ln\dfrac{b^2+u^2}{a^2+u^2}$
3	$\sqrt{x}\,e^{-ax}$	$\tfrac{1}{2}\sqrt{\pi}\,(a^2+u^2)^{-3/4}\cos\left(\tfrac{3}{2}\arctan\dfrac{u}{a}\right)$
4	$\dfrac{1}{\sqrt{x}}e^{-ax}$	$\sqrt{\dfrac{\pi}{2}}\left[\dfrac{a+(a^2+u^2)^{1/2}}{a^2+u^2}\right]^{1/2}$
5	$x^n e^{-ax},\quad n=1,2,\ldots$	$\dfrac{a^{n+1}n!}{(a^2+u^2)^{n+1}}\sum_{0\leq 2k\leq n+1}(-1)^k C_{n+1}^{2k}\left(\dfrac{u}{a}\right)^{2k}$
6	$x^{n-1/2}e^{-ax},\quad n=1,2,\ldots$	$k_n u\dfrac{\partial^n}{\partial a^n}\dfrac{1}{r\sqrt{r-a}},$ where $\ r=\sqrt{a^2+u^2},\ k_n=(-1)^n\sqrt{\pi/2}$
7	$x^{\nu-1}e^{-ax}$	$\Gamma(\nu)(a^2+u^2)^{-\nu/2}\cos\left(\nu\arctan\dfrac{u}{a}\right)$

No	Original function, $f(x)$	Cosine transform, $\breve{f}_c(u) = \int_0^\infty f(x)\cos(ux)\,dx$
8	$\dfrac{x}{e^{ax}-1}$	$\dfrac{1}{2u^2} - \dfrac{\pi^2}{2a^2\sinh^2(\pi a^{-1}u)}$
9	$\dfrac{1}{x}\left(\dfrac{1}{2} - \dfrac{1}{x} + \dfrac{1}{e^x - 1}\right)$	$-\dfrac{1}{2}\ln\left(1 - e^{-2\pi u}\right)$
10	$\exp\left(-ax^2\right)$	$\dfrac{1}{2}\sqrt{\dfrac{\pi}{a}}\exp\left(-\dfrac{u^2}{4a}\right)$
11	$\dfrac{1}{\sqrt{x}}\exp\left(-\dfrac{a}{x}\right)$	$\sqrt{\dfrac{\pi}{2u}}\,e^{-\sqrt{2au}}\left[\cos\left(\sqrt{2au}\right) - \sin\left(\sqrt{2au}\right)\right]$
12	$\dfrac{1}{x\sqrt{x}}\exp\left(-\dfrac{a}{x}\right)$	$\sqrt{\dfrac{\pi}{a}}\,e^{-\sqrt{2au}}\cos\left(\sqrt{2au}\right)$

6.4. Expressions With Hyperbolic Functions

No	Original function, $f(x)$	Cosine transform, $\breve{f}_c(u) = \int_0^\infty f(x)\cos(ux)\,dx$
1	$\dfrac{1}{\cosh(ax)}, \quad a > 0$	$\dfrac{\pi}{2a\cosh\left(\frac{1}{2}\pi a^{-1}u\right)}$
2	$\dfrac{1}{\cosh^2(ax)}, \quad a > 0$	$\dfrac{\pi u}{2a^2\sinh\left(\frac{1}{2}\pi a^{-1}u\right)}$
3	$\dfrac{\cosh(ax)}{\cosh(bx)}, \quad |a| < b$	$\dfrac{\pi}{b}\left[\dfrac{\cos\left(\frac{1}{2}\pi ab^{-1}\right)\cosh\left(\frac{1}{2}\pi b^{-1}u\right)}{\cos\left(\pi ab^{-1}\right) + \cosh\left(\pi b^{-1}u\right)}\right]$
4	$\dfrac{1}{\cosh(ax) + \cos b}$	$\dfrac{\pi\sinh\left(a^{-1}bu\right)}{a\sin b\,\sinh\left(\pi a^{-1}u\right)}$
5	$\exp\left(-ax^2\right)\cosh(bx), \quad a > 0$	$\dfrac{1}{2}\sqrt{\dfrac{\pi}{a}}\exp\left(\dfrac{b^2 - u^2}{4a}\right)\cos\left(\dfrac{abu}{2}\right)$
6	$\dfrac{x}{\sinh(ax)}$	$\dfrac{\pi^2}{4a^2\cosh^2\left(\frac{1}{2}\pi a^{-1}u\right)}$
7	$\dfrac{\sinh(ax)}{\sinh(bx)}, \quad |a| < b$	$\dfrac{\pi}{2b}\dfrac{\sin\left(\pi ab^{-1}\right)}{\cos\left(\pi ab^{-1}\right) + \cosh\left(\pi b^{-1}u\right)}$
8	$\dfrac{1}{x}\tanh(ax), \quad a > 0$	$\ln\left[\coth\left(\frac{1}{4}\pi a^{-1}u\right)\right]$

6.5. Expressions With Logarithmic Functions

No	Original function, $f(x)$	Cosine transform, $\breve{f}_c(u) = \int_0^\infty f(x)\cos(ux)\,dx$
1	$\begin{cases} \ln x & \text{if } 0 < x < 1, \\ 0 & \text{if } 1 < x \end{cases}$	$-\dfrac{1}{u}\operatorname{Si}(u)$

No	Original function, $f(x)$	Cosine transform, $\breve{f}_c(u) = \int_0^\infty f(x)\cos(ux)\,dx$
2	$\dfrac{\ln x}{\sqrt{x}}$	$-\sqrt{\dfrac{\pi}{2u}}\left[\ln(4u) + C + \dfrac{\pi}{2}\right],$ $C = 0.5772\ldots$ is the Euler constant
3	$x^{\nu-1}\ln x, \quad 0 < \nu < 1$	$\Gamma(\nu)\cos\left(\dfrac{\pi\nu}{2}\right)u^{-\nu}\left[\psi(\nu) - \dfrac{\pi}{2}\tan\left(\dfrac{\pi\nu}{2}\right) - \ln u\right]$
4	$\ln\left\vert\dfrac{a+x}{a-x}\right\vert, \quad a > 0$	$\dfrac{2}{u}\left[\cos(au)\,\mathrm{Si}(au) - \sin(au)\,\mathrm{Ci}(au)\right]$
5	$\ln\left(1 + a^2/x^2\right), \quad a > 0$	$\dfrac{\pi}{u}\left(1 - e^{-au}\right)$
6	$\ln\dfrac{a^2+x^2}{b^2+x^2}, \quad a,b > 0$	$\dfrac{\pi}{u}\left(e^{-bu} - e^{-au}\right)$
7	$e^{-ax}\ln x, \quad a > 0$	$-\dfrac{aC + \frac{1}{2}a\ln(u^2+a^2) + u\arctan(u/a)}{u^2+a^2}$
8	$\ln\left(1 + e^{-ax}\right), \quad a > 0$	$\dfrac{a}{2u^2} - \dfrac{\pi}{2u\sinh\left(\pi a^{-1}u\right)}$
9	$\ln\left(1 - e^{-ax}\right), \quad a > 0$	$\dfrac{a}{2u^2} - \dfrac{\pi}{2u}\coth\left(\pi a^{-1}u\right)$

6.6. Expressions With Trigonometric Functions

No	Original function, $f(x)$	Cosine transform, $\breve{f}_c(u) = \int_0^\infty f(x)\cos(ux)\,dx$
1	$\dfrac{\sin(ax)}{x}, \quad a > 0$	$\begin{cases} \frac{1}{2}\pi & \text{if } u < a, \\ \frac{1}{4}\pi & \text{if } u = a, \\ 0 & \text{if } u > a \end{cases}$
2	$x^{\nu-1}\sin(ax), \quad a > 0,\ \vert\nu\vert < 1$	$\pi\,\dfrac{(u+a)^{-\nu} - \vert u+a\vert^{-\nu}\,\mathrm{sign}(u-a)}{4\Gamma(1-\nu)\cos\left(\frac{1}{2}\pi\nu\right)}$
3	$\dfrac{x\sin(ax)}{x^2+b^2}, \quad a,b > 0$	$\begin{cases} \frac{1}{2}\pi e^{-ab}\cosh(bu) & \text{if } u < a, \\ -\frac{1}{2}\pi e^{-bu}\sinh(ab) & \text{if } u > a \end{cases}$
4	$\dfrac{\sin(ax)}{x(x^2+b^2)}, \quad a,b > 0$	$\begin{cases} \frac{1}{2}\pi b^{-2}\left[1 - e^{-ab}\cosh(bu)\right] & \text{if } u < a, \\ \frac{1}{2}\pi b^{-2}e^{-bu}\sinh(ab) & \text{if } u > a \end{cases}$
5	$e^{-bx}\sin(ax), \quad a,b > 0$	$\dfrac{1}{2}\left[\dfrac{a+u}{(a+u)^2+b^2} + \dfrac{a-u}{(a-u)^2+b^2}\right]$
6	$\dfrac{1}{x}\sin^2(ax), \quad a > 0$	$\dfrac{1}{4}\ln\left\vert 1 - 4\dfrac{a^2}{u^2}\right\vert$
7	$\dfrac{1}{x^2}\sin^2(ax), \quad a > 0$	$\begin{cases} \frac{1}{4}\pi(2a - u) & \text{if } u < 2a, \\ 0 & \text{if } u > 2a \end{cases}$
8	$\dfrac{1}{x}\sin\left(\dfrac{a}{x}\right), \quad a > 0$	$\dfrac{\pi}{2}J_0\left(2\sqrt{au}\right)$

No	Original function, $f(x)$	Cosine transform, $\check{f}_c(u) = \int_0^\infty f(x)\cos(ux)\,dx$
9	$\dfrac{1}{\sqrt{x}}\sin(a\sqrt{x})\sin(b\sqrt{x})$, $a,b>0$	$\sqrt{\dfrac{\pi}{u}}\,\sin\left(\dfrac{ab}{2u}\right)\sin\left(\dfrac{a^2+b^2}{4u}-\dfrac{\pi}{4}\right)$
10	$\sin(ax^2)$, $a>0$	$\sqrt{\dfrac{\pi}{8a}}\left[\cos\left(\dfrac{u^2}{4a}\right)-\sin\left(\dfrac{u^2}{4a}\right)\right]$
11	$\exp(-ax^2)\sin(bx^2)$, $a>0$	$\dfrac{\sqrt{\pi}}{(A^2+B^2)^{1/4}}\exp\left(-\dfrac{Au^2}{A^2+B^2}\right)\sin\left(\varphi-\dfrac{Bu^2}{A^2+B^2}\right)$, $A=4a$, $B=4b$, $\varphi=\tfrac{1}{2}\arctan(b/a)$
12	$\dfrac{1-\cos(ax)}{x}$, $a>0$	$\dfrac{1}{2}\ln\left\|1-\dfrac{a^2}{u^2}\right\|$
13	$\dfrac{1-\cos(ax)}{x^2}$, $a>0$	$\begin{cases}\tfrac{1}{2}\pi(a-u) & \text{if } u<a,\\ 0 & \text{if } u>a\end{cases}$
14	$x^{\nu-1}\cos(ax)$, $a>0$, $0<\nu<1$	$\tfrac{1}{2}\Gamma(\nu)\cos\left(\tfrac{1}{2}\pi\nu\right)\left[\|u-a\|^{-\nu}+(u+a)^{-\nu}\right]$
15	$\dfrac{\cos(ax)}{x^2+b^2}$, $a,b>0$	$\begin{cases}\tfrac{1}{2}\pi b^{-1}e^{-ab}\cosh(bu) & \text{if } u<a,\\ \tfrac{1}{2}\pi b^{-1}e^{-bu}\cosh(ab) & \text{if } u>a\end{cases}$
16	$e^{-bx}\cos(ax)$, $a,b>0$	$\dfrac{b}{2}\left[\dfrac{1}{(a+u)^2+b^2}+\dfrac{1}{(a-u)^2+b^2}\right]$
17	$\dfrac{1}{\sqrt{x}}\cos(a\sqrt{x})$	$\sqrt{\dfrac{\pi}{u}}\,\sin\left(\dfrac{a^2}{4u}+\dfrac{\pi}{4}\right)$
18	$\dfrac{1}{\sqrt{x}}\cos(a\sqrt{x})\cos(b\sqrt{x})$	$\sqrt{\dfrac{\pi}{u}}\,\cos\left(\dfrac{ab}{2u}\right)\sin\left(\dfrac{a^2+b^2}{4u}+\dfrac{\pi}{4}\right)$
19	$\exp(-bx^2)\cos(ax)$, $b>0$	$\dfrac{1}{2}\sqrt{\dfrac{\pi}{b}}\exp\left(-\dfrac{a^2+u^2}{4b}\right)\cosh\left(\dfrac{au}{2b}\right)$
20	$\cos(ax^2)$, $a>0$	$\sqrt{\dfrac{\pi}{8a}}\left[\cos\left(\tfrac{1}{4}a^{-1}u^2\right)+\sin\left(\tfrac{1}{4}a^{-1}u^2\right)\right]$
21	$\exp(-ax^2)\cos(bx^2)$, $a>0$	$\dfrac{\sqrt{\pi}}{(A^2+B^2)^{1/4}}\exp\left(-\dfrac{Au^2}{A^2+B^2}\right)\cos\left(\varphi-\dfrac{Bu^2}{A^2+B^2}\right)$, $A=4a$, $B=4b$, $\varphi=\tfrac{1}{2}\arctan(b/a)$

6.7. Expressions With Special Functions

No	Original function, $f(x)$	Cosine transform, $\check{f}_c(u) = \int_0^\infty f(x)\cos(ux)\,dx$
1	$\mathrm{Ei}(-ax)$	$-\dfrac{1}{u}\arctan\left(\dfrac{u}{a}\right)$
2	$\mathrm{Ci}(ax)$	$\begin{cases}0 & \text{if } 0<u<a,\\ -\dfrac{\pi}{2u} & \text{if } a<u\end{cases}$
3	$\mathrm{si}(ax)$	$-\dfrac{1}{2u}\ln\left\|\dfrac{u+a}{u-a}\right\|$, $u\neq a$

No	Original function, $f(x)$	Cosine transform, $\check{f}_c(u) = \int_0^\infty f(x)\cos(ux)\,dx$
4	$J_0(ax), \quad a > 0$	$\begin{cases} \dfrac{1}{\sqrt{a^2 - u^2}} & \text{if } 0 < u < a, \\ 0 & \text{if } a < u \end{cases}$
5	$J_\nu(ax), \quad a > 0,\ \nu > -1$	$\begin{cases} \dfrac{\cos\left[\nu\arcsin(u/a)\right]}{\sqrt{a^2 - u^2}} & \text{if } 0 < u < a, \\ -\dfrac{a^\nu \sin(\pi\nu/2)}{\xi(u + \xi)^\nu} & \text{if } a < u, \end{cases}$ where $\xi = \sqrt{u^2 - a^2}$
6	$\dfrac{1}{x}J_\nu(ax), \quad a > 0,\ \nu > 0$	$\begin{cases} \nu^{-1}\cos\left[\nu\arcsin(u/a)\right] & \text{if } 0 < u < a, \\ \dfrac{a^\nu \cos(\pi\nu/2)}{\nu\left(u + \sqrt{u^2 - a^2}\right)^\nu} & \text{if } a < u \end{cases}$
7	$x^{-\nu}J_\nu(ax), \quad a > 0,\ \nu > -\frac{1}{2}$	$\begin{cases} \dfrac{\sqrt{\pi}\left(a^2 - u^2\right)^{\nu-1/2}}{(2a)^\nu \Gamma\left(\nu + \frac{1}{2}\right)} & \text{if } 0 < u < a, \\ 0 & \text{if } a < u \end{cases}$
8	$x^{\nu+1}J_\nu(ax),$ $a > 0,\ -1 < \nu < -\frac{1}{2}$	$\begin{cases} 0 & \text{if } 0 < u < a, \\ \dfrac{2^{\nu+1}\sqrt{\pi}\,a^\nu u}{\Gamma\left(-\nu - \frac{1}{2}\right)(u^2 - a^2)^{\nu+3/2}} & \text{if } a < u \end{cases}$
9	$J_0\left(a\sqrt{x}\right), \quad a > 0$	$\dfrac{1}{u}\sin\left(\dfrac{a^2}{4u}\right)$
10	$\dfrac{1}{\sqrt{x}}J_1\left(a\sqrt{x}\right), \quad a > 0$	$\dfrac{4}{a}\sin^2\left(\dfrac{a^2}{8u}\right)$
11	$x^{\nu/2}J_\nu\left(a\sqrt{x}\right),\ a > 0,\ -1 < \nu < \frac{1}{2}$	$\left(\dfrac{a}{2}\right)^\nu u^{-\nu-1}\sin\left(\dfrac{a^2}{4u} - \dfrac{\pi\nu}{2}\right)$
12	$J_0\left(a\sqrt{x^2 + b^2}\right)$	$\begin{cases} \dfrac{\cos\left(b\sqrt{a^2 - u^2}\right)}{\sqrt{a^2 - u^2}} & \text{if } 0 < u < a, \\ 0 & \text{if } a < u \end{cases}$
13	$Y_0(ax), \quad a > 0$	$\begin{cases} 0 & \text{if } 0 < u < a, \\ -\dfrac{1}{\sqrt{u^2 - a^2}} & \text{if } a < u \end{cases}$
14	$x^\nu Y_\nu(ax), \quad a > 0,\ |\nu| < \frac{1}{2}$	$\begin{cases} 0 & \text{if } 0 < u < a, \\ -\dfrac{(2a)^\nu\sqrt{\pi}}{\Gamma\left(\frac{1}{2} - \nu\right)(u^2 - a^2)^{\nu+1/2}} & \text{if } a < u \end{cases}$
15	$K_0\left(a\sqrt{x^2 + b^2}\right), \quad a, b > 0$	$\dfrac{\pi}{2\sqrt{u^2 + a^2}}\exp\left(-b\sqrt{u^2 + a^2}\right)$

⊙ References for Supplement 6: G. Doetsch (1950, 1956, 1958), H. Bateman and A. Erdélyi (1954), V. A. Ditkin and A. P. Prudnikov (1965).

Supplement 7

Tables of Fourier Sine Transforms

7.1. General Formulas

No	Original function, $f(x)$	Sine transform, $\check{f}_s(u) = \int_0^\infty f(x)\sin(ux)\,dx$
1	$af_1(x) + bf_2(x)$	$a\check{f}_{1s}(u) + b\check{f}_{2s}(u)$
2	$f(ax), \quad a > 0$	$\dfrac{1}{a}\check{f}_s\left(\dfrac{u}{a}\right)$
3	$x^{2n}f(x), \quad n = 1, 2, \ldots$	$(-1)^n\dfrac{d^{2n}}{du^{2n}}\check{f}_s(u)$
4	$x^{2n+1}f(ax), \quad n = 0, 1, \ldots$	$(-1)^{n+1}\dfrac{d^{2n+1}}{du^{2n+1}}\check{f}_c(u), \quad \check{f}_c(u) = \int_0^\infty f(x)\cos(xu)\,dx$
5	$f(ax)\cos(bx), \quad a, b > 0$	$\dfrac{1}{2a}\left[\check{f}_s\left(\dfrac{u+b}{a}\right) + F_s\left(\dfrac{u-b}{a}\right)\right]$

7.2. Expressions With Power-Law Functions

No	Original function, $f(x)$	Sine transform, $\check{f}_s(u) = \int_0^\infty f(x)\sin(ux)\,dx$
1	$\begin{cases} 1 & \text{if } 0 < x < a, \\ 0 & \text{if } a < x \end{cases}$	$\dfrac{1}{u}[1 - \cos(au)]$
2	$\begin{cases} x & \text{if } 0 < x < 1, \\ 2-x & \text{if } 1 < x < 2, \\ 0 & \text{if } 2 < x \end{cases}$	$\dfrac{4}{u^2}\sin u \sin^2\dfrac{u}{2}$
3	$\dfrac{1}{x}$	$\dfrac{\pi}{2}$
4	$\dfrac{1}{a+x}, \quad a > 0$	$\sin(au)\operatorname{Ci}(au) - \cos(au)\operatorname{si}(au)$
5	$\dfrac{x}{a^2 + x^2}, \quad a > 0$	$\dfrac{\pi}{2}e^{-au}$
6	$\dfrac{1}{x(a^2 + x^2)}, \quad a > 0$	$\dfrac{\pi}{2a^2}\left(1 - e^{-au}\right)$
7	$\dfrac{a}{a^2 + (x-b)^2} - \dfrac{a}{a^2 + (x+b)^2}$	$\pi e^{-au}\sin(bu)$
8	$\dfrac{x+b}{a^2 + (x+b)^2} - \dfrac{x-b}{a^2 + (x-b)^2}$	$\pi e^{-au}\cos(bu)$

No	Original function, $f(x)$	Sine transform, $\check{f}_s(u) = \int_0^\infty f(x)\sin(ux)\,dx$
9	$\dfrac{x}{(x^2+a^2)^n}$, $\quad a>0$, $n=1,2,\ldots$	$\dfrac{\pi u e^{-au}}{2^{2n-2}(n-1)!\,a^{2n-3}}\displaystyle\sum_{k=0}^{n-2}\dfrac{(2n-k-4)!}{k!\,(n-k-2)!}(2au)^k$
10	$\dfrac{x^{2m+1}}{(x^2+a)^{n+1}}$, $\;n,m=0,1,\ldots;\;0\le m\le n$	$(-1)^{n+m}\dfrac{\pi}{2n!}\dfrac{\partial^n}{\partial a^n}\left(a^m e^{-u\sqrt{a}}\right)$
11	$\dfrac{1}{\sqrt{x}}$	$\sqrt{\dfrac{\pi}{2u}}$
12	$\dfrac{1}{x\sqrt{x}}$	$\sqrt{2\pi u}$
13	$x(a^2+x^2)^{-3/2}$	$uK_0(au)$
14	$\dfrac{\left(\sqrt{a^2+x^2}-a\right)^{1/2}}{\sqrt{a^2+x^2}}$	$\sqrt{\dfrac{\pi}{2u}}\,e^{-au}$
15	$x^{-\nu}$, $\quad 0<\nu<2$	$\cos\left(\tfrac{1}{2}\pi\nu\right)\Gamma(1-\nu)u^{\nu-1}$

7.3. Expressions With Exponential Functions

No	Original function, $f(x)$	Sine transform, $\check{f}_s(u) = \int_0^\infty f(x)\sin(ux)\,dx$
1	e^{-ax}, $\quad a>0$	$\dfrac{u}{a^2+u^2}$
2	$x^n e^{-ax}$, $\quad a>0$, $n=1,2,\ldots$	$n!\left(\dfrac{a}{a^2+u^2}\right)^{n+1}\displaystyle\sum_{k=0}^{[n/2]}(-1)^k C_{n+1}^{2k+1}\left(\dfrac{u}{a}\right)^{2k+1}$
3	$\dfrac{1}{x}e^{-ax}$, $\quad a>0$	$\arctan\dfrac{u}{a}$
4	$\sqrt{x}\,e^{-ax}$, $\quad a>0$	$\dfrac{\sqrt{\pi}}{2}(a^2+u^2)^{-3/4}\sin\left(\dfrac{3}{2}\arctan\dfrac{u}{a}\right)$
5	$\dfrac{1}{\sqrt{x}}e^{-ax}$, $\quad a>0$	$\sqrt{\dfrac{\pi}{2}}\dfrac{(\sqrt{a^2+u^2}-a)^{1/2}}{\sqrt{a^2+u^2}}$
6	$\dfrac{1}{x\sqrt{x}}e^{-ax}$, $\quad a>0$	$\sqrt{2\pi}\left(\sqrt{a^2+u^2}-a\right)^{1/2}$
7	$x^{n-1/2}e^{-ax}$, $\quad a>0$, $n=1,2,\ldots$	$(-1)^n\sqrt{\dfrac{\pi}{2}}\dfrac{\partial^n}{\partial a^n}\left[\dfrac{(\sqrt{a^2+u^2}-a)^{1/2}}{\sqrt{a^2+u^2}}\right]$
8	$x^{\nu-1}e^{-ax}$, $\quad a>0$, $\nu>-1$	$\Gamma(\nu)(a^2+u^2)^{-\nu/2}\sin\left(\nu\arctan\dfrac{u}{a}\right)$
9	$x^{-2}\left(e^{-ax}-e^{-bx}\right)$, $\quad a,b>0$	$\dfrac{u}{2}\ln\left(\dfrac{u^2+b^2}{u^2+a^2}\right)+b\arctan\left(\dfrac{u}{b}\right)-a\arctan\left(\dfrac{u}{a}\right)$

No	Original function, $f(x)$	Sine transform, $\check{f}_s(u) = \int_0^\infty f(x) \sin(ux)\, dx$
10	$\dfrac{1}{e^{ax}+1}, \quad a > 0$	$\dfrac{1}{2u} - \dfrac{\pi}{2a \sinh(\pi u/a)}$
11	$\dfrac{1}{e^{ax}-1}, \quad a > 0$	$\dfrac{\pi}{2a} \coth\left(\dfrac{\pi u}{a}\right) - \dfrac{1}{2u}$
12	$\dfrac{e^{x/2}}{e^x - 1}$	$-\tfrac{1}{2}\tanh(\pi u)$
13	$x\exp\left(-ax^2\right)$	$\dfrac{\sqrt{\pi}}{4a^{3/2}}\, u \exp\left(-\dfrac{u^2}{4a}\right)$
14	$\dfrac{1}{x}\exp\left(-ax^2\right)$	$\dfrac{\pi}{2}\operatorname{erf}\left(\dfrac{u}{2\sqrt{a}}\right)$
15	$\dfrac{1}{\sqrt{x}}\exp\left(-\dfrac{a}{x}\right)$	$\sqrt{\dfrac{\pi}{2u}}\, e^{-\sqrt{2au}}\left[\cos\left(\sqrt{2au}\right) + \sin\left(\sqrt{2au}\right)\right]$
16	$\dfrac{1}{x\sqrt{x}}\exp\left(-\dfrac{a}{x}\right)$	$\sqrt{\dfrac{\pi}{a}}\, e^{-\sqrt{2au}}\sin\left(\sqrt{2au}\right)$

7.4. Expressions With Hyperbolic Functions

No	Original function, $f(x)$	Sine transform, $\check{f}_s(u) = \int_0^\infty f(x) \sin(ux)\, dx$
1	$\dfrac{1}{\sinh(ax)}, \quad a > 0$	$\dfrac{\pi}{2a}\tanh\left(\tfrac{1}{2}\pi a^{-1}u\right)$
2	$\dfrac{x}{\sinh(ax)}, \quad a > 0$	$\dfrac{\pi^2 \sinh\left(\tfrac{1}{2}\pi a^{-1}u\right)}{4a^2 \cosh^2\left(\tfrac{1}{2}\pi a^{-1}u\right)}$
3	$\dfrac{1}{x}e^{-bx}\sinh(ax), \quad b > |a|$	$\tfrac{1}{2}\arctan\left(\dfrac{2au}{u^2 + b^2 - a^2}\right)$
4	$\dfrac{1}{x\cosh(ax)}, \quad a > 0$	$\arctan\left[\sinh\left(\tfrac{1}{2}\pi a^{-1}u\right)\right]$
5	$1 - \tanh\left(\tfrac{1}{2}ax\right), \quad a > 0$	$\dfrac{1}{u} - \dfrac{\pi}{a \sinh\left(\pi a^{-1}u\right)}$
6	$\coth\left(\tfrac{1}{2}ax\right) - 1, \quad a > 0$	$\dfrac{\pi}{a}\coth\left(\pi a^{-1}u\right) - \dfrac{1}{u}$
7	$\dfrac{\cosh(ax)}{\sinh(bx)}, \quad |a| < b$	$\dfrac{\pi}{2b}\dfrac{\sinh\left(\pi b^{-1}u\right)}{\cos\left(\pi ab^{-1}\right) + \cosh\left(\pi b^{-1}u\right)}$
8	$\dfrac{\sinh(ax)}{\cosh(bx)}, \quad |a| < b$	$\dfrac{\pi}{b}\dfrac{\sin\left(\tfrac{1}{2}\pi ab^{-1}\right)\sinh\left(\tfrac{1}{2}\pi b^{-1}u\right)}{\cos\left(\pi ab^{-1}\right) + \cosh\left(\pi b^{-1}u\right)}$

7.5. Expressions With Logarithmic Functions

No	Original function, $f(x)$	Sine transform, $\check{f}_s(u) = \int_0^\infty f(x)\sin(ux)\,dx$
1	$\begin{cases} \ln x & \text{if } 0 < x < 1, \\ 0 & \text{if } 1 < x \end{cases}$	$\dfrac{1}{u}\left[\mathrm{Ci}(u) - \ln u - \mathcal{C}\right],$ $\mathcal{C} = 0.5772\ldots$ is the Euler constant
2	$\dfrac{\ln x}{x}$	$-\tfrac{1}{2}\pi(\ln u + \mathcal{C})$
3	$\dfrac{\ln x}{\sqrt{x}}$	$-\sqrt{\dfrac{\pi}{2u}}\left[\ln(4u) + \mathcal{C} - \dfrac{\pi}{2}\right]$
4	$x^{\nu-1}\ln x, \quad \|\nu\| < 1$	$\dfrac{\pi u^{-\nu}\left[\psi(\nu) + \frac{\pi}{2}\cot\left(\frac{\pi\nu}{2}\right) - \ln u\right]}{2\Gamma(1-\nu)\cos\left(\frac{\pi\nu}{2}\right)}$
5	$\ln\left\|\dfrac{a+x}{a-x}\right\|, \quad a > 0$	$\dfrac{\pi}{u}\sin(au)$
6	$\ln\dfrac{(x+b)^2 + a^2}{(x-b)^2 + a^2}, \quad a, b > 0$	$\dfrac{2\pi}{u}e^{-au}\sin(bu)$
7	$e^{-ax}\ln x, \quad a > 0$	$\dfrac{a\arctan(u/a) - \frac{1}{2}u\ln(u^2+a^2) - e^{\mathcal{C}}u}{u^2 + a^2}$
8	$\dfrac{1}{x}\ln\left(1 + a^2 x^2\right), \quad a > 0$	$-\pi\,\mathrm{Ei}\left(-\dfrac{u}{a}\right)$

7.6. Expressions With Trigonometric Functions

No	Original function, $f(x)$	Sine transform, $\check{f}_s(u) = \int_0^\infty f(x)\sin(ux)\,dx$
1	$\dfrac{\sin(ax)}{x}, \quad a > 0$	$\dfrac{1}{2}\ln\left\|\dfrac{u+a}{u-a}\right\|$
2	$\dfrac{\sin(ax)}{x^2}, \quad a > 0$	$\begin{cases} \frac{1}{2}\pi u & \text{if } 0 < u < a, \\ \frac{1}{2}\pi a & \text{if } u > a \end{cases}$
3	$x^{\nu-1}\sin(ax), \quad a > 0, \; -2 < \nu < 1$	$\pi\,\dfrac{\|u-a\|^{-\nu} - \|u+a\|^{-\nu}}{4\Gamma(1-\nu)\sin\left(\frac{1}{2}\pi\nu\right)}, \quad \nu \neq 0$
4	$\dfrac{\sin(ax)}{x^2 + b^2}, \quad a, b > 0$	$\begin{cases} \frac{1}{2}\pi b^{-1}e^{-ab}\sinh(bu) & \text{if } 0 < u < a, \\ \frac{1}{2}\pi b^{-1}e^{-bu}\sinh(ab) & \text{if } u > a \end{cases}$
5	$\dfrac{\sin(\pi x)}{1 - x^2}$	$\begin{cases} \sin u & \text{if } 0 < u < \pi, \\ 0 & \text{if } u > \pi \end{cases}$
6	$e^{-ax}\sin(bx), \quad a > 0$	$\dfrac{a}{2}\left[\dfrac{1}{a^2 + (b-u)^2} - \dfrac{1}{a^2 + (b+u)^2}\right]$
7	$x^{-1}e^{-ax}\sin(bx), \quad a > 0$	$\dfrac{1}{4}\ln\dfrac{(u+b)^2 + a^2}{(u-b)^2 + a^2}$
8	$\dfrac{1}{x}\sin^2(ax), \quad a > 0$	$\begin{cases} \frac{1}{4}\pi & \text{if } 0 < u < 2a, \\ \frac{1}{8}\pi & \text{if } u = 2a, \\ 0 & \text{if } u > 2a \end{cases}$

No	Original function, $f(x)$	Sine transform, $\check{f}_s(u) = \int_0^\infty f(x) \sin(ux)\, dx$
9	$\dfrac{1}{x^2} \sin^2(ax), \quad a > 0$	$\frac{1}{4}(u + 2a) \ln|u + 2a| + \frac{1}{4}(u - 2a) \ln|u - 2a|$ $-\frac{1}{2} u \ln u$
10	$\exp(-ax^2) \sin(bx), \quad a > 0$	$\dfrac{1}{2} \sqrt{\dfrac{\pi}{a}} \exp\left(-\dfrac{u^2 + b^2}{4a}\right) \sinh\left(\dfrac{bu}{2a}\right)$
11	$\dfrac{1}{x} \sin(ax) \sin(bx), \quad a \geq b > 0$	$\begin{cases} 0 & \text{if } 0 < u < a - b, \\ \frac{\pi}{4} & \text{if } a - b < u < a + b, \\ 0 & \text{if } a + b < u \end{cases}$
12	$\sin\left(\dfrac{a}{x}\right), \quad a > 0$	$\dfrac{\pi \sqrt{a}}{2\sqrt{u}} J_1\left(2\sqrt{au}\,\right)$
13	$\dfrac{1}{\sqrt{x}} \sin\left(\dfrac{a}{x}\right), \quad a > 0$	$\sqrt{\dfrac{\pi}{8u}} \left[\sin\left(2\sqrt{au}\,\right) - \cos\left(2\sqrt{au}\,\right) + \exp\left(-2\sqrt{au}\,\right)\right]$
14	$\exp(-a\sqrt{x}\,) \sin\left(a\sqrt{x}\,\right), \quad a > 0$	$a\sqrt{\dfrac{\pi}{8}}\, u^{-3/2} \exp\left(-\dfrac{a^2}{2u}\right)$
15	$\dfrac{\cos(ax)}{x}, \quad a > 0$	$\begin{cases} 0 & \text{if } 0 < u < a, \\ \frac{1}{4}\pi & \text{if } u = a, \\ \frac{1}{2}\pi & \text{if } a < u \end{cases}$
16	$x^{\nu - 1} \cos(ax), \quad a > 0, \; |\nu| < 1$	$\dfrac{\pi(u + a)^{-\nu} - \text{sign}(u - a)|u - a|^{-\nu}}{4\Gamma(1 - \nu) \cos\left(\frac{1}{2}\pi\nu\right)}$
17	$\dfrac{x \cos(ax)}{x^2 + b^2}, \quad a, b > 0$	$\begin{cases} -\frac{1}{2}\pi e^{-ab} \sinh(bu) & \text{if } u < a, \\ \frac{1}{2}\pi e^{-bu} \cosh(ab) & \text{if } u > a \end{cases}$
18	$\dfrac{1 - \cos(ax)}{x^2}, \quad a > 0$	$\dfrac{u}{2} \ln\left|\dfrac{u^2 - a^2}{u^2}\right| + \dfrac{a}{2} \ln\left|\dfrac{u + a}{u - a}\right|$
19	$\dfrac{1}{\sqrt{x}} \cos\left(a\sqrt{x}\,\right)$	$\sqrt{\dfrac{\pi}{u}} \cos\left(\dfrac{a^2}{4u} + \dfrac{\pi}{4}\right)$
20	$\dfrac{1}{\sqrt{x}} \cos\left(a\sqrt{x}\,\right) \cos\left(b\sqrt{x}\,\right), \; a, b > 0$	$\sqrt{\dfrac{\pi}{u}} \cos\left(\dfrac{ab}{2u}\right) \cos\left(\dfrac{a^2 + b^2}{4u} + \dfrac{\pi}{4}\right)$

7.7. Expressions With Special Functions

No	Original function, $f(x)$	Sine transform, $\check{f}_s(u) = \int_0^\infty f(x) \sin(ux)\, dx$
1	$\text{erfc}(ax), \quad a > 0$	$\dfrac{1}{u}\left[1 - \exp\left(-\dfrac{u^2}{4a^2}\right)\right]$
2	$\text{ci}(ax), \quad a > 0$	$-\dfrac{1}{2u} \ln\left|1 - \dfrac{u^2}{a^2}\right|$
3	$\text{si}(ax), \quad a > 0$	$\begin{cases} 0 & \text{if } 0 < u < a, \\ -\frac{1}{2}\pi u^{-1} & \text{if } a < u \end{cases}$

No	Original function, $f(x)$	Sine transform, $\breve{f}_s(u) = \int_0^\infty f(x)\sin(ux)\,dx$
4	$J_0(ax), \quad a > 0$	$\begin{cases} 0 & \text{if } 0 < u < a, \\ \dfrac{1}{\sqrt{u^2 - a^2}} & \text{if } a < u \end{cases}$
5	$J_\nu(ax), \quad a > 0, \ \nu > -2$	$\begin{cases} \dfrac{\sin[\nu \arcsin(u/a)]}{\sqrt{a^2 - u^2}} & \text{if } 0 < u < a, \\ \dfrac{a^\nu \cos(\pi\nu/2)}{\xi(u + \xi)^\nu} & \text{if } a < u, \end{cases}$ where $\ \xi = \sqrt{u^2 - a^2}$
6	$\dfrac{1}{x} J_0(ax), \quad a > 0, \ \nu > 0$	$\begin{cases} \arcsin(u/a) & \text{if } 0 < u < a, \\ \pi/2 & \text{if } a < u \end{cases}$
7	$\dfrac{1}{x} J_\nu(ax), \quad a > 0, \ \nu > -1$	$\begin{cases} \nu^{-1} \sin[\nu \arcsin(u/a)] & \text{if } 0 < u < a, \\ \dfrac{a^\nu \sin(\pi\nu/2)}{\nu(u + \sqrt{u^2 - a^2})^\nu} & \text{if } a < u \end{cases}$
8	$x^\nu J_\nu(ax), \quad a > 0, \ -1 < \nu < \tfrac{1}{2}$	$\begin{cases} 0 & \text{if } 0 < u < a, \\ \dfrac{\sqrt{\pi}(2a)^\nu}{\Gamma(\tfrac{1}{2} - \nu)(u^2 - a^2)^{\nu + 1/2}} & \text{if } a < u \end{cases}$
9	$x^{-1} e^{-ax} J_0(bx), \quad a > 0$	$\arcsin\left(\dfrac{2u}{\sqrt{(u+b)^2 + a^2} + \sqrt{(u-b)^2 + a^2}}\right)$
10	$\dfrac{J_0(ax)}{x^2 + b^2}, \quad a, b > 0$	$\begin{cases} b^{-1} \sinh(bu) K_0(ab) & \text{if } 0 < u < a, \\ 0 & \text{if } a < u \end{cases}$
11	$\dfrac{x J_0(ax)}{x^2 + b^2}, \quad a, b > 0$	$\begin{cases} 0 & \text{if } 0 < u < a, \\ \tfrac{1}{2}\pi e^{-bu} I_0(ab) & \text{if } a < u \end{cases}$
12	$\dfrac{\sqrt{x} J_{2n+1/2}(ax)}{x^2 + b^2}, \\ a, b > 0, \quad n = 0, 1, 2, \ldots$	$\begin{cases} (-1)^n \sinh(bu) K_{2n+1/2}(ab) & \text{if } 0 < u < a, \\ 0 & \text{if } a < u \end{cases}$
13	$\dfrac{x^\nu J_\nu(ax)}{x^2 + b^2}, \\ a, b > 0, \quad -1 < \nu < \tfrac{5}{2}$	$\begin{cases} b^{\nu-1} \sinh(bu) K_\nu(ab) & \text{if } 0 < u < a, \\ 0 & \text{if } a < u \end{cases}$
14	$\dfrac{x^{1-\nu} J_\nu(ax)}{x^2 + b^2}, \\ a, b > 0, \quad \nu > -\tfrac{3}{2}$	$\begin{cases} 0 & \text{if } 0 < u < a, \\ \tfrac{1}{2}\pi b^{-\nu} e^{-bu} I_\nu(ab) & \text{if } a < u \end{cases}$
15	$J_0(a\sqrt{x}), \quad a > 0$	$\dfrac{1}{u} \cos\left(\dfrac{a^2}{4u}\right)$
16	$\dfrac{1}{\sqrt{x}} J_1(a\sqrt{x}), \quad a > 0$	$\dfrac{2}{a} \sin\left(\dfrac{a^2}{4u}\right)$
17	$x^{\nu/2} J_\nu(a\sqrt{x}), \\ a > 0, \quad -2 < \nu < \tfrac{1}{2}$	$\dfrac{a^\nu}{2^\nu u^{\nu+1}} \cos\left(\dfrac{a^2}{4u} - \dfrac{\pi\nu}{2}\right)$

No	Original function, $f(x)$	Sine transform, $\check{f}_s(u) = \int_0^\infty f(x)\sin(ux)\,dx$
18	$Y_0(ax), \quad a > 0$	$\begin{cases} \dfrac{2\arcsin(u/a)}{\pi\sqrt{a^2-u^2}} & \text{if } 0 < u < a, \\[2mm] \dfrac{2\left[\ln\left(u-\sqrt{u^2-a^2}\right)-\ln a\right]}{\pi\sqrt{u^2-a^2}} & \text{if } a < u \end{cases}$
19	$Y_1(ax), \quad a > 0$	$\begin{cases} 0 & \text{if } 0 < u < a, \\[2mm] -\dfrac{u}{a\sqrt{u^2-a^2}} & \text{if } a < u \end{cases}$
20	$K_0(ax), \quad a > 0$	$\dfrac{\ln\left(u+\sqrt{u^2+a^2}\right)-\ln a}{\sqrt{u^2+a^2}}$
21	$xK_0(ax), \quad a > 0$	$\dfrac{\pi u}{2(u^2+a^2)^{3/2}}$
22	$x^{\nu+1}K_\nu(ax), \quad a > 0,\ \nu > -\tfrac{3}{2}$	$\sqrt{\pi}\,(2a)^\nu\Gamma\left(\nu+\tfrac{3}{2}\right)u(u^2+a^2)^{-\nu-3/2}$

⊙ References for Supplement 7: G. Doetsch (1950, 1956, 1958), H. Bateman and A. Erdélyi (1954), I. I. Hirschman and D. V. Widder (1955), V. A. Ditkin and A. P. Prudnikov (1965).

Supplement 8

Tables of Mellin Transforms

8.1. General Formulas

No	Original function, $f(x)$	Mellin transform, $\hat{f}(s) = \int_0^\infty f(x)x^{s-1}\,dx$
1	$af_1(x) + bf_2(x)$	$a\hat{f}_1(s) + b\hat{f}_2(s)$
2	$f(ax),\ a > 0$	$a^{-s}\hat{f}(s)$
3	$x^a f(x)$	$\hat{f}(s+a)$
4	$f(1/x)$	$\hat{f}(-s)$
5	$f(x^\beta),\ \beta > 0$	$\dfrac{1}{\beta}\hat{f}\!\left(\dfrac{s}{\beta}\right)$
6	$f(x^{-\beta}),\ \beta > 0$	$\dfrac{1}{\beta}\hat{f}\!\left(-\dfrac{s}{\beta}\right)$
7	$x^\lambda f(ax^\beta),\ a,\beta > 0$	$\dfrac{1}{\beta}a^{-\frac{s+\lambda}{\beta}}\hat{f}\!\left(\dfrac{s+\lambda}{\beta}\right)$
8	$x^\lambda f(ax^{-\beta}),\ a,\beta > 0$	$\dfrac{1}{\beta}a^{\frac{s+\lambda}{\beta}}\hat{f}\!\left(-\dfrac{s+\lambda}{\beta}\right)$
9	$f'_x(x)$	$-(s-1)\hat{f}(s-1)$
10	$xf'_x(x)$	$-s\hat{f}(s)$
11	$f_x^{(n)}(x)$	$(-1)^n\dfrac{\Gamma(s)}{\Gamma(s-n)}\hat{f}(s-n)$
12	$\left(x\dfrac{d}{dx}\right)^n f(x)$	$(-1)^n s^n \hat{f}(s)$
13	$\left(\dfrac{d}{dx}x\right)^n f(x)$	$(-1)^n (s-1)^n \hat{f}(s)$
14	$x^\alpha \displaystyle\int_0^\infty t^\beta f_1(xt)f_2(t)\,dt$	$\hat{f}_1(s+\alpha)\hat{f}_2(1-s-\alpha+\beta)$
15	$x^\alpha \displaystyle\int_0^\infty t^\beta f_1\!\left(\dfrac{x}{t}\right)f_2(t)\,dt$	$\hat{f}_1(s+\alpha)\hat{f}_2(s+\alpha+\beta+1)$

8.2. Expressions With Power-Law Functions

No	Original function, $f(x)$	Mellin transform, $\hat{f}(s) = \int_0^\infty f(x)x^{s-1}\,dx$
1	$\begin{cases} x & \text{if } 0 < x < 1, \\ 2-x & \text{if } 1 < x < 2, \\ 0 & \text{if } 2 < x \end{cases}$	$\begin{cases} \dfrac{2(2^s - 1)}{s(s+1)} & \text{if } s \neq 0, \\ 2\ln 2 & \text{if } s = 0, \end{cases}$ $\operatorname{Re} s > -1$
2	$\dfrac{1}{x+a}, \quad a > 0$	$\dfrac{\pi a^{s-1}}{\sin(\pi s)}, \quad 0 < \operatorname{Re} s < 1$
3	$\dfrac{1}{(x+a)(x+b)}, \quad a, b > 0$	$\dfrac{\pi\left(a^{s-1} - b^{s-1}\right)}{(b-a)\sin(\pi s)}, \quad 0 < \operatorname{Re} s < 2$
4	$\dfrac{x+a}{(x+b)(x+c)}, \quad b, c > 0$	$\dfrac{\pi}{\sin(\pi s)}\left[\left(\dfrac{b-a}{b-c}\right)b^{s-1} + \left(\dfrac{c-a}{c-b}\right)c^{s-1}\right],$ $0 < \operatorname{Re} s < 1$
5	$\dfrac{1}{x^2 + a^2}, \quad a > 0$	$\dfrac{\pi a^{s-2}}{2\sin\left(\frac{1}{2}\pi s\right)}, \quad 0 < \operatorname{Re} s < 2$
6	$\dfrac{1}{x^2 + 2ax\cos\beta + a^2}, \quad a > 0, \ |\beta| < \pi$	$-\dfrac{\pi a^{s-2}\sin\left[\beta(s-1)\right]}{\sin\beta\,\sin(\pi s)}, \quad 0 < \operatorname{Re} s < 2$
7	$\dfrac{1}{(x^2 + a^2)(x^2 + b^2)}, \quad a, b > 0$	$\dfrac{\pi\left(a^{s-2} - b^{s-2}\right)}{2(b^2 - a^2)\sin\left(\frac{1}{2}\pi s\right)}, \quad 0 < \operatorname{Re} s < 4$
8	$\dfrac{1}{(1+ax)^{n+1}}, \quad a > 0, \ n = 1, 2, \ldots$	$\dfrac{(-1)^n \pi}{a^s \sin(\pi s)}C_{s-1}^n, \quad 0 < \operatorname{Re} s < n+1$
9	$\dfrac{1}{x^n + a^n}, \quad a > 0, \ n = 1, 2, \ldots$	$\dfrac{\pi a^{s-n}}{n\sin(\pi s/n)}, \quad 0 < \operatorname{Re} s < n$
10	$\dfrac{1-x}{1-x^n}, \quad n = 2, 3, \ldots$	$\dfrac{\pi\sin(\pi/n)}{n\sin(\pi s/n)\sin\left[\pi(s+1)/n\right]}, \quad 0 < \operatorname{Re} s < n-1$
11	$\begin{cases} x^\nu & \text{if } 0 < x < 1, \\ 0 & \text{if } 1 < x \end{cases}$	$\dfrac{1}{s+\nu}, \quad \operatorname{Re} s > -\nu$
12	$\dfrac{1-x^\nu}{1-x^{n\nu}}, \quad n = 2, 3, \ldots$	$\dfrac{\pi\sin(\pi/n)}{n\nu\sin\left(\frac{\pi s}{n\nu}\right)\sin\left[\frac{\pi(s+\nu)}{n\nu}\right]}, \quad 0 < \operatorname{Re} s < (n-1)\nu$

8.3. Expressions With Exponential Functions

No	Original function, $f(x)$	Mellin transform, $\hat{f}(s) = \int_0^\infty f(x)x^{s-1}\,dx$
1	$e^{-ax}, \quad a > 0$	$a^{-s}\Gamma(s), \quad \operatorname{Re} s > 0$
2	$\begin{cases} e^{-bx} & \text{if } 0 < x < a, \\ 0 & \text{if } a < x, \end{cases} \quad b > 0$	$b^{-s}\gamma(s, ab), \quad \operatorname{Re} s > 0$
3	$\begin{cases} 0 & \text{if } 0 < x < a, \\ e^{-bx} & \text{if } a < x, \end{cases} \quad b > 0$	$b^{-s}\Gamma(s, ab)$
4	$\dfrac{e^{-ax}}{x+b}, \quad a, b > 0$	$e^{ab}b^{s-1}\Gamma(s)\Gamma(1-s, ab), \quad \operatorname{Re} s > 0$
5	$\exp\left(-ax^\beta\right), \quad a, \beta > 0$	$\beta^{-1}a^{-s/\beta}\Gamma(s/\beta), \quad \operatorname{Re} s > 0$

No	Original function, $f(x)$	Mellin transform, $\hat{f}(s) = \int_0^\infty f(x)x^{s-1}\,dx$
6	$\exp\left(-ax^{-\beta}\right), \quad a, \beta > 0$	$\beta^{-1}a^{s/\beta}\Gamma(-s/\beta), \quad \operatorname{Re} s < 0$
7	$1 - \exp\left(-ax^{\beta}\right), \quad a, \beta > 0$	$-\beta^{-1}a^{-s/\beta}\Gamma(s/\beta), \quad -\beta < \operatorname{Re} s < 0$
8	$1 - \exp\left(-ax^{-\beta}\right), \quad a, \beta > 0$	$-\beta^{-1}a^{s/\beta}\Gamma(-s/\beta), \quad 0 < \operatorname{Re} s < \beta$

8.4. Expressions With Logarithmic Functions

No	Original function, $f(x)$	Mellin transform, $\hat{f}(s) = \int_0^\infty f(x)x^{s-1}\,dx$
1	$\begin{cases} \ln x & \text{if } 0 < x < a, \\ 0 & \text{if } a < x \end{cases}$	$\dfrac{s \ln a - 1}{s^2 a^s}, \quad \operatorname{Re} s > 0$
2	$\ln(1 + ax), \quad a > 0$	$\dfrac{\pi}{sa^s \sin(\pi s)}, \quad -1 < \operatorname{Re} s < 0$
3	$\ln|1 - x|$	$\dfrac{\pi}{s}\cot(\pi s), \quad -1 < \operatorname{Re} s < 0$
4	$\dfrac{\ln x}{x + a}, \quad a > 0$	$\dfrac{\pi a^{s-1}\left[\ln a - \pi \cot(\pi s)\right]}{\sin(\pi s)}, \quad 0 < \operatorname{Re} s < 1$
5	$\dfrac{\ln x}{(x + a)(x + b)}, \quad a, b > 0$	$\dfrac{\pi\left[a^{s-1}\ln a - b^{s-1}\ln b - \pi \cot(\pi s)(a^{s-1} - b^{s-1})\right]}{(b - a)\sin(\pi s)},$ $0 < \operatorname{Re} s < 1$
6	$\begin{cases} x^\nu \ln x & \text{if } 0 < x < 1, \\ 0 & \text{if } 1 < x \end{cases}$	$-\dfrac{1}{(s + \nu)^2}, \quad \operatorname{Re} s > -\nu$
7	$\dfrac{\ln^2 x}{x + 1}$	$\dfrac{\pi^3\left[2 - \sin^2(\pi s)\right]}{\sin^3(\pi s)}, \quad 0 < \operatorname{Re} s < 1$
8	$\begin{cases} \ln^{\nu-1} x & \text{if } 0 < x < 1, \\ 0 & \text{if } 1 < x \end{cases}$	$\Gamma(\nu)(-s)^{-\nu}, \quad \operatorname{Re} s < 0, \ \nu > 0$
9	$\ln\left(x^2 + 2x \cos \beta + 1\right), \quad |\beta| < \pi$	$\dfrac{2\pi \cos(\beta s)}{s \sin(\pi s)}, \quad -1 < \operatorname{Re} s < 0$
10	$\ln\left|\dfrac{1 + x}{1 - x}\right|$	$\dfrac{\pi}{s}\tan\left(\tfrac{1}{2}\pi s\right), \quad -1 < \operatorname{Re} s < 1$
11	$e^{-x}\ln^n x, \quad n = 1, 2, \ldots$	$\dfrac{d^n}{ds^n}\Gamma(s), \quad \operatorname{Re} s > 0$

8.5. Expressions With Trigonometric Functions

No	Original function, $f(x)$	Mellin transform, $\hat{f}(s) = \int_0^\infty f(x)x^{s-1}\,dx$
1	$\sin(ax), \quad a > 0$	$a^{-s}\Gamma(s)\sin\left(\tfrac{1}{2}\pi s\right), \quad -1 < \operatorname{Re} s < 1$
2	$\sin^2(ax), \quad a > 0$	$-2^{-s-1}a^{-s}\Gamma(s)\cos\left(\tfrac{1}{2}\pi s\right), \quad -2 < \operatorname{Re} s < 0$
3	$\sin(ax)\sin(bx), \quad a, b > 0, \ a \neq b$	$\tfrac{1}{2}\Gamma(s)\cos\left(\tfrac{1}{2}\pi s\right)\left[|b - a|^{-s} - (b + a)^{-s}\right],$ $-2 < \operatorname{Re} s < 1$

No	Original function, $f(x)$	Mellin transform, $\hat{f}(s) = \int_0^\infty f(x)x^{s-1}\,dx$
4	$\cos(ax), \quad a > 0$	$a^{-s}\Gamma(s)\cos\left(\tfrac{1}{2}\pi s\right), \quad 0 < \operatorname{Re} s < 1$
5	$\sin(ax)\cos(bx), \quad a, b > 0$	$\dfrac{\Gamma(s)}{2}\sin\left(\dfrac{\pi s}{2}\right)\left[(a+b)^{-s} + \lvert a-b\rvert^{-s}\operatorname{sign}(a-b)\right],$ $-1 < \operatorname{Re} s < 1$
6	$e^{-ax}\sin(bx), \quad a > 0$	$\dfrac{\Gamma(s)\sin\left[s\arctan(b/a)\right]}{(a^2+b^2)^{s/2}}, \quad -1 < \operatorname{Re} s$
7	$e^{-ax}\cos(bx), \quad a > 0$	$\dfrac{\Gamma(s)\cos\left[s\arctan(b/a)\right]}{(a^2+b^2)^{s/2}}, \quad 0 < \operatorname{Re} s$
8	$\begin{cases}\sin(a\ln x) & \text{if } 0 < x < 1, \\ 0 & \text{if } 1 < x\end{cases}$	$-\dfrac{a}{s^2+a^2}, \quad \operatorname{Re} s > 0$
9	$\begin{cases}\cos(a\ln x) & \text{if } 0 < x < 1, \\ 0 & \text{if } 1 < x\end{cases}$	$\dfrac{s}{s^2+a^2}, \quad \operatorname{Re} s > 0$
10	$\arctan x$	$-\dfrac{\pi}{2s\cos\left(\tfrac{1}{2}\pi s\right)}, \quad -1 < \operatorname{Re} s < 0$
11	$\operatorname{arccot} x$	$\dfrac{\pi}{2s\cos\left(\tfrac{1}{2}\pi s\right)}, \quad 0 < \operatorname{Re} s < 1$

8.6. Expressions With Special Functions

No	Original function, $f(x)$	Mellin transform, $\hat{f}(s) = \int_0^\infty f(x)x^{s-1}\,dx$
1	$\operatorname{erfc} x$	$\dfrac{\Gamma\left(\tfrac{1}{2}s + \tfrac{1}{2}\right)}{\sqrt{\pi}\,s}, \quad \operatorname{Re} s > 0$
2	$\operatorname{Ei}(-x)$	$-s^{-1}\Gamma(s), \quad \operatorname{Re} s > 0$
3	$\operatorname{Si}(x)$	$-s^{-1}\sin\left(\tfrac{1}{2}\pi s\right)\Gamma(s), \quad -1 < \operatorname{Re} s < 0$
4	$\operatorname{si}(x)$	$-4s^{-1}\sin\left(\tfrac{1}{2}\pi s\right)\Gamma(s), \quad -1 < \operatorname{Re} s < 0$
5	$\operatorname{Ci}(x)$	$-s^{-1}\cos\left(\tfrac{1}{2}\pi s\right)\Gamma(s), \quad 0 < \operatorname{Re} s < 1$
6	$J_\nu(ax), \quad a > 0$	$\dfrac{2^{s-1}\Gamma\left(\tfrac{1}{2}\nu + \tfrac{1}{2}s\right)}{a^s\Gamma\left(\tfrac{1}{2}\nu - \tfrac{1}{2}s + 1\right)}, \quad -\nu < \operatorname{Re} s < \tfrac{3}{2}$
7	$Y_\nu(ax), \quad a > 0$	$-\dfrac{2^{s-1}}{\pi a^s}\Gamma\left(\dfrac{s}{2} + \dfrac{\nu}{2}\right)\Gamma\left(\dfrac{s}{2} - \dfrac{\nu}{2}\right)\cos\left[\dfrac{\pi(s-\nu)}{2}\right],$ $\lvert\nu\rvert < \operatorname{Re} s < \tfrac{3}{2}$
8	$e^{-ax}I_\nu(ax), \quad a > 0$	$\dfrac{\Gamma(1/2 - s)\Gamma(s + \nu)}{\sqrt{\pi}\,(2a)^s\Gamma(1 + \nu - s)}, \quad -\nu < \operatorname{Re} s < \tfrac{1}{2}$
9	$K_\nu(ax), \quad a > 0$	$\dfrac{2^{s-2}}{a^s}\Gamma\left(\dfrac{s}{2} + \dfrac{\nu}{2}\right)\Gamma\left(\dfrac{s}{2} - \dfrac{\nu}{2}\right), \quad \lvert\nu\rvert < \operatorname{Re} s$
10	$e^{-ax}K_\nu(ax), \quad a > 0$	$\dfrac{\sqrt{\pi}\,\Gamma(s - \nu)\Gamma(s + \nu)}{(2a)^s\Gamma(s + 1/2)}, \quad \lvert\nu\rvert < \operatorname{Re} s$

⊙ References for Supplement 8: H. Bateman and A. Erdélyi (1954), V. A. Ditkin and A. P. Prudnikov (1965).

Supplement 9

Tables of Inverse Mellin Transforms

See Section 8.1 of Supplement 8 for general formulas.

9.1. Expressions With Power-Law Functions

No	Direct transform, $\hat{f}(s)$	Inverse transform, $f(x) = \dfrac{1}{2\pi i} \displaystyle\int_{\sigma-i\infty}^{\sigma+i\infty} \hat{f}(s)x^{-s}\,ds$
1	$\dfrac{1}{s}$, $\quad \mathrm{Re}\,s > 0$	$\begin{cases} 1 & \text{if } 0 < x < 1, \\ 0 & \text{if } 1 < x \end{cases}$
2	$\dfrac{1}{s}$, $\quad \mathrm{Re}\,s < 0$	$\begin{cases} 0 & \text{if } 0 < x < 1, \\ -1 & \text{if } 1 < x \end{cases}$
3	$\dfrac{1}{s+a}$, $\quad \mathrm{Re}\,s > -a$	$\begin{cases} x^a & \text{if } 0 < x < 1, \\ 0 & \text{if } 1 < x \end{cases}$
4	$\dfrac{1}{s+a}$, $\quad \mathrm{Re}\,s < -a$	$\begin{cases} 0 & \text{if } 0 < x < 1, \\ -x^a & \text{if } 1 < x \end{cases}$
5	$\dfrac{1}{(s+a)^2}$, $\quad \mathrm{Re}\,s > -a$	$\begin{cases} -x^a \ln x & \text{if } 0 < x < 1, \\ 0 & \text{if } 1 < x \end{cases}$
6	$\dfrac{1}{(s+a)^2}$, $\quad \mathrm{Re}\,s < -a$	$\begin{cases} 0 & \text{if } 0 < x < 1, \\ x^a \ln x & \text{if } 1 < x \end{cases}$
7	$\dfrac{1}{(s+a)(s+b)}$, $\quad \mathrm{Re}\,s > -a, -b$	$\begin{cases} \dfrac{x^a - x^b}{b - a} & \text{if } 0 < x < 1, \\ 0 & \text{if } 1 < x \end{cases}$
8	$\dfrac{1}{(s+a)(s+b)}$, $\quad -a < \mathrm{Re}\,s < -b$	$\begin{cases} \dfrac{x^a}{b-a} & \text{if } 0 < x < 1, \\ \dfrac{x^b}{b-a} & \text{if } 1 < x \end{cases}$
9	$\dfrac{1}{(s+a)(s+b)}$, $\quad \mathrm{Re}\,s < -a, -b$	$\begin{cases} 0 & \text{if } 0 < x < 1, \\ \dfrac{x^b - x^a}{b - a} & \text{if } 1 < x \end{cases}$
10	$\dfrac{1}{(s+a)^2 + b^2}$, $\quad \mathrm{Re}\,s > -a$	$\begin{cases} \dfrac{1}{b} x^a \sin\left(b \ln \dfrac{1}{x}\right) & \text{if } 0 < x < 1, \\ 0 & \text{if } 1 < x \end{cases}$
11	$\dfrac{s+a}{(s+a)^2 + b^2}$, $\quad \mathrm{Re}\,s > -a$	$\begin{cases} x^a \cos(b \ln x) & \text{if } 0 < x < 1, \\ 0 & \text{if } 1 < x \end{cases}$

No	Direct transform, $\hat{f}(s)$	Inverse transform, $f(x) = \dfrac{1}{2\pi i}\displaystyle\int_{\sigma-i\infty}^{\sigma+i\infty}\hat{f}(s)x^{-s}\,ds$
12	$\sqrt{s^2-a^2}-s,\quad \mathrm{Re}\,s>\lvert a\rvert$	$\begin{cases}-\dfrac{a}{\ln x}I_1(-a\ln x) & \text{if } 0<x<1,\\[2mm] 0 & \text{if } 1<x\end{cases}$
13	$\sqrt{\dfrac{s+a}{s-a}}-1,\quad \mathrm{Re}\,s>\lvert a\rvert$	$\begin{cases}aI_0(-a\ln x)+aI_1(-a\ln x) & \text{if } 0<x<1,\\[2mm] 0 & \text{if } 1<x\end{cases}$
14	$(s+a)^{-\nu},\quad \mathrm{Re}\,s>-a,\ \nu>0$	$\begin{cases}\dfrac{1}{\Gamma(\nu)}x^a(-\ln x)^{\nu-1} & \text{if } 0<x<1,\\[2mm] 0 & \text{if } 1<x\end{cases}$
15	$s^{-1}(s+a)^{-\nu},$ $\mathrm{Re}\,s>0,\ \mathrm{Re}\,s>-a,\ \nu>0$	$\begin{cases}a^{-\nu}\left[\Gamma(\nu)\right]^{-1}\gamma(\nu,-a\ln x) & \text{if } 0<x<1,\\[2mm] 0 & \text{if } 1<x\end{cases}$
16	$s^{-1}(s+a)^{-\nu},$ $-a<\mathrm{Re}\,s<0,\ \nu>0$	$\begin{cases}-a^{-\nu}\left[\Gamma(\nu)\right]^{-1}\Gamma(\nu,-a\ln x) & \text{if } 0<x<1,\\[2mm] -a^{-\nu} & \text{if } 1<x\end{cases}$
17	$(s^2-a^2)^{-\nu},\quad \mathrm{Re}\,s>\lvert a\rvert,\ \nu>0$	$\begin{cases}\dfrac{\sqrt{\pi}\,(-\ln x)^{\nu-1/2}I_{\nu-1/2}(-a\ln x)}{\Gamma(\nu)(2a)^{\nu-1/2}} & \text{if } 0<x<1,\\[3mm] 0 & \text{if } 1<x\end{cases}$
18	$(a^2-s^2)^{-\nu},\quad \mathrm{Re}\,s<\lvert a\rvert,\ \nu>0$	$\begin{cases}\dfrac{(-\ln x)^{\nu-1/2}K_{\nu-1/2}(-a\ln x)}{\sqrt{\pi}\,\Gamma(\nu)(2a)^{\nu-1/2}} & \text{if } 0<x<1,\\[3mm] \dfrac{(\ln x)^{\nu-1/2}K_{\nu-1/2}(a\ln x)}{\sqrt{\pi}\,\Gamma(\nu)(2a)^{\nu-1/2}} & \text{if } 1<x\end{cases}$

9.2. Expressions With Exponential and Logarithmic Functions

No	Direct transform, $\hat{f}(s)$	Inverse transform, $f(x) = \dfrac{1}{2\pi i}\displaystyle\int_{\sigma-i\infty}^{\sigma+i\infty}\hat{f}(s)x^{-s}\,ds$
1	$\exp(as^2),\quad a>0$	$\dfrac{1}{2\sqrt{\pi a}}\exp\left(-\dfrac{\ln^2 x}{4a}\right)$
2	$s^{-\nu}e^{-a/s},\quad \mathrm{Re}\,s>0;\ a,\nu>0$	$\begin{cases}\left\lvert\dfrac{a}{\ln x}\right\rvert^{\frac{1-\nu}{2}}J_{\nu-1}\left(2\sqrt{a\lvert\ln x\rvert}\right) & \text{if } 0<x<1,\\[2mm] 0 & \text{if } 1<x\end{cases}$
3	$\exp\left(-\sqrt{as}\right),\quad \mathrm{Re}\,s>0,\ a>0$	$\begin{cases}\dfrac{(a/\pi)^{1/2}}{2\lvert\ln x\rvert^{3/2}}\exp\left(-\dfrac{a}{4\lvert\ln x\rvert}\right) & \text{if } 0<x<1,\\[2mm] 0 & \text{if } 1<x\end{cases}$
4	$\dfrac{1}{s}\exp\left(-a\sqrt{s}\right),\quad \mathrm{Re}\,s>0$	$\begin{cases}\mathrm{erfc}\left(\dfrac{a}{2\sqrt{\lvert\ln x\rvert}}\right) & \text{if } 0<x<1,\\[2mm] 0 & \text{if } 1<x\end{cases}$
5	$\dfrac{1}{s}\left[\exp\left(-a\sqrt{s}\right)-1\right],\quad \mathrm{Re}\,s>0$	$\begin{cases}-\mathrm{erf}\left(\dfrac{a}{2\sqrt{\lvert\ln x\rvert}}\right) & \text{if } 0<x<1,\\[2mm] 0 & \text{if } 1<x\end{cases}$

No	Direct transform, $\hat{f}(s)$	Inverse transform, $f(x) = \dfrac{1}{2\pi i}\displaystyle\int_{\sigma-i\infty}^{\sigma+i\infty}\hat{f}(s)x^{-s}\,ds$
6	$\sqrt{s}\exp(-\sqrt{as})$, $\quad \operatorname{Re}s>0$	$\begin{cases}\dfrac{a-2|\ln x|}{4\sqrt{\pi}|\ln x|^5}\exp\!\left(-\dfrac{a}{4|\ln x|}\right) & \text{if } 0<x<1,\\[2mm] 0 & \text{if } 1<x\end{cases}$
7	$\dfrac{1}{\sqrt{s}}\exp(-\sqrt{as})$, $\quad \operatorname{Re}s>0$	$\begin{cases}\dfrac{1}{\sqrt{\pi}|\ln x|}\exp\!\left(-\dfrac{a}{4|\ln x|}\right) & \text{if } 0<x<1,\\[2mm] 0 & \text{if } 1<x\end{cases}$
8	$\ln\dfrac{s+a}{s+b}$, $\quad \operatorname{Re}s>-a,-b$	$\begin{cases}\dfrac{x^a-x^b}{\ln x} & \text{if } 0<x<1,\\[2mm] 0 & \text{if } 1<x\end{cases}$
9	$s^{-\nu}\ln s$, $\quad \operatorname{Re}s>0,\ \nu>0$	$\begin{cases}|\ln x|^{\nu-1}\dfrac{\psi(\nu)-\ln|\ln x|}{\Gamma(\nu)} & \text{if } 0<x<1,\\[2mm] 0 & \text{if } 1<x\end{cases}$

9.3. Expressions With Trigonometric Functions

No	Direct transform, $\hat{f}(s)$	Inverse transform, $f(x) = \dfrac{1}{2\pi i}\displaystyle\int_{\sigma-i\infty}^{\sigma+i\infty}\hat{f}(s)x^{-s}\,ds$
1	$\dfrac{\pi}{\sin(\pi s)}$, $\quad 0<\operatorname{Re}s<1$	$\dfrac{1}{x+1}$
2	$\dfrac{\pi}{\sin(\pi s)}$, $\quad -n<\operatorname{Re}s<1-n,$ $n=\ldots,-1,0,1,2,\ldots$	$(-1)^n\dfrac{x^n}{x+1}$
3	$\dfrac{\pi^2}{\sin^2(\pi s)}$, $\quad 0<\operatorname{Re}s<1$	$\dfrac{\ln x}{x-1}$
4	$\dfrac{\pi^2}{\sin^2(\pi s)}$, $\quad n<\operatorname{Re}s<n+1,$ $n=\ldots,-1,0,1,2,\ldots$	$\dfrac{\ln x}{x^n(x-1)}$
5	$\dfrac{2\pi^3}{\sin^3(\pi s)}$, $\quad 0<\operatorname{Re}s<1$	$\dfrac{\pi^2+\ln^2 x}{x+1}$
6	$\dfrac{2\pi^3}{\sin^3(\pi s)}$, $\quad n<\operatorname{Re}s<n+1,$ $n=\ldots,-1,0,1,2,\ldots$	$\dfrac{\pi^2+\ln^2 x}{(-x)^n(x+1)}$
7	$\sin(s^2/a)$, $\quad a>0$	$\dfrac{1}{2}\sqrt{\dfrac{a}{\pi}}\,\sin\!\left(\tfrac{1}{4}a|\ln x|^2-\tfrac{1}{4}\pi\right)$
8	$\dfrac{\pi}{\cos(\pi s)}$, $\quad -\tfrac{1}{2}<\operatorname{Re}s<\tfrac{1}{2}$	$\dfrac{\sqrt{x}}{x+1}$
9	$\dfrac{\pi}{\cos(\pi s)}$, $\quad n-\tfrac{1}{2}<\operatorname{Re}s<n+\tfrac{1}{2}$ $n=\ldots,-1,0,1,2,\ldots$	$(-1)^n\dfrac{x^{1/2-n}}{x+1}$
10	$\dfrac{\cos(\beta s)}{s\cos(\pi s)}$, $\quad -1<\operatorname{Re}s<0,\ |\beta|<\pi$	$\dfrac{1}{2\pi}\ln(x^2+2x\cos\beta+1)$

No	Direct transform, $\hat{f}(s)$	Inverse transform, $f(x) = \dfrac{1}{2\pi i}\displaystyle\int_{\sigma-i\infty}^{\sigma+i\infty}\hat{f}(s)x^{-s}\,ds$
11	$\cos(s^2/a),\quad a>0$	$\dfrac{1}{2}\sqrt{\dfrac{a}{\pi}}\,\cos\!\left(\tfrac{1}{4}a|\ln x|^2-\tfrac{1}{4}\pi\right)$
12	$\arctan\!\left(\dfrac{a}{s+b}\right),\quad \operatorname{Re}s>-b$	$\begin{cases}\dfrac{x^b}{|\ln x|}\sin(a|\ln x|) & \text{if } 0<x<1,\\ 0 & \text{if } 1<x\end{cases}$

9.4. Expressions With Special Functions

No	Direct transform, $\hat{f}(s)$	Inverse transform, $f(x) = \dfrac{1}{2\pi i}\displaystyle\int_{\sigma-i\infty}^{\sigma+i\infty}\hat{f}(s)x^{-s}\,ds$
1	$\Gamma(s),\quad \operatorname{Re}s>0$	e^{-x}
2	$\Gamma(s),\quad -1<\operatorname{Re}s<0$	$e^{-x}-1$
3	$\sin\!\left(\tfrac{1}{2}\pi s\right)\Gamma(s),\quad -1<\operatorname{Re}s<1$	$\sin x$
4	$\sin(as)\Gamma(s),$ $\operatorname{Re}s>-1,\ \|a\|<\dfrac{\pi}{2}$	$\exp(-x\cos a)\sin(x\sin a)$
5	$\cos\!\left(\tfrac{1}{2}\pi s\right)\Gamma(s),\quad 0<\operatorname{Re}s<1$	$\cos x$
6	$\cos\!\left(\tfrac{1}{2}\pi s\right)\Gamma(s),\quad -2<\operatorname{Re}s<0$	$-2\sin^2(x/2)$
7	$\cos(as)\Gamma(s),\quad \operatorname{Re}s>0,\ \|a\|<\dfrac{\pi}{2}$	$\exp(-x\cos a)\cos(x\sin a)$
8	$\dfrac{\Gamma(s)}{\cos(\pi s)},\quad 0<\operatorname{Re}s<\dfrac{1}{2}$	$e^x\operatorname{erfc}\!\left(\sqrt{x}\right)$
9	$\Gamma(a+s)\Gamma(b-s),$ $-a<\operatorname{Re}s<b,\ a+b>0$	$\Gamma(a+b)x^a(x+1)^{-a-b}$
10	$\Gamma(a+s)\Gamma(b+s),$ $\operatorname{Re}s>-a,-b$	$2x^{(a+b)/2}K_{a-b}\!\left(2\sqrt{x}\right)$
11	$\dfrac{\Gamma(s)}{\Gamma(s+\nu)},\quad \operatorname{Re}s>0,\ \nu>0$	$\begin{cases}\dfrac{(1-x)^{\nu-1}}{\Gamma(\nu)} & \text{if } 0<x<1,\\ 0 & \text{if } 1<x\end{cases}$
12	$\dfrac{\Gamma(1-\nu-s)}{\Gamma(1-s)},$ $\operatorname{Re}s<1-\nu,\ \nu>0$	$\begin{cases}0 & \text{if } 0<x<1,\\ \dfrac{(x-1)^{\nu-1}}{\Gamma(\nu)} & \text{if } 1<x\end{cases}$
13	$\dfrac{\Gamma(s)}{\Gamma(\nu-s+1)},$ $0<\operatorname{Re}s<\dfrac{\nu}{2}+\dfrac{3}{4}$	$x^{-\nu/2}J_\nu\!\left(2\sqrt{x}\right)$
14	$\dfrac{\Gamma(s+\nu)\Gamma(s-\nu)}{\Gamma(s+1/2)},\quad \operatorname{Re}s>\|\nu\|$	$\pi^{-1/2}e^{-x/2}K_\nu(x/2)$

No	Direct transform, $\hat{f}(s)$	Inverse transform, $f(x) = \dfrac{1}{2\pi i}\displaystyle\int_{\sigma-i\infty}^{\sigma+i\infty} \hat{f}(s)x^{-s}\,ds$
15	$\dfrac{\Gamma(s+\nu)\Gamma(1/2-s)}{\Gamma(1+\nu-s)}$, $-\nu < \operatorname{Re}s < \tfrac{1}{2}$	$\pi^{1/2}e^{-x/2}I_\nu(x/2)$
16	$\psi(s+a)-\psi(s+b)$, $\operatorname{Re}s > -a,-b$	$\begin{cases} \dfrac{x^b - x^a}{1-x} & \text{if } 0 < x < 1, \\ 0 & \text{if } 1 < x \end{cases}$
17	$\Gamma(s)\psi(s)$, $\operatorname{Re}s > 0$	$e^{-x}\ln x$
18	$\Gamma(s,a)$, $a > 0$	$\begin{cases} 0 & \text{if } 0 < x < a, \\ e^{-x} & \text{if } a < x \end{cases}$
19	$\Gamma(s)\Gamma(1-s,a)$, $\operatorname{Re}s > 0,\ a > 0$	$(x+1)^{-1}e^{-a(x+1)}$
20	$\gamma(s,a)$, $\operatorname{Re}s > 0,\ a > 0$	$\begin{cases} e^{-x} & \text{if } 0 < x < a, \\ 0 & \text{if } a < x \end{cases}$
21	$J_0\!\left(a\sqrt{b^2-s^2}\right)$, $a > 0$	$\begin{cases} 0 & \text{if } 0 < x < e^{-a}, \\ \dfrac{\cos\!\left(b\sqrt{a^2-\ln^2 x}\right)}{\pi\sqrt{a^2-\ln^2 x}} & \text{if } e^{-a} < x < e^{a}, \\ 0 & \text{if } e^{a} < x \end{cases}$
22	$s^{-1}I_0(s)$, $\operatorname{Re}s > 0$	$\begin{cases} 1 & \text{if } 0 < x < e^{-1}, \\ \pi^{-1}\arccos(\ln x) & \text{if } e^{-1} < x < e, \\ 0 & \text{if } e < x \end{cases}$
23	$I_\nu(s)$, $\operatorname{Re}s > 0$	$\begin{cases} -\dfrac{2^\nu \sin(\pi\nu)}{\pi F(x)\sqrt{\ln^2 x - 1}} & \text{if } 0 < x < e^{-1}, \\ \dfrac{\cos\!\left[\nu\arccos(\ln x)\right]}{\pi\sqrt{1-\ln^2 x}} & \text{if } e^{-1} < x < e, \\ 0 & \text{if } e < x, \end{cases}$ $F(x) = \left(\sqrt{-1-\ln x} + \sqrt{1-\ln x}\,\right)^{2\nu}$
24	$s^{-1}I_\nu(s)$, $\operatorname{Re}s > 0$	$\begin{cases} \dfrac{2^\nu \sin(\pi\nu)}{\pi\nu F(x)} & \text{if } 0 < x < e^{-1}, \\ \dfrac{\sin\!\left[\nu\arccos(\ln x)\right]}{\pi\nu} & \text{if } e^{-1} < x < e, \\ 0 & \text{if } e < x, \end{cases}$ $F(x) = \left(\sqrt{-1-\ln x} + \sqrt{1-\ln x}\,\right)^{2\nu}$
25	$s^{-\nu}I_\nu(s)$, $\operatorname{Re}s > -\tfrac{1}{2}$	$\begin{cases} 0 & \text{if } 0 < x < e^{-1}, \\ \dfrac{(1-\ln^2 x)^{\nu-1/2}}{\sqrt{\pi}\,2^\nu\Gamma(\nu+1/2)} & \text{if } e^{-1} < x < e, \\ 0 & \text{if } e < x \end{cases}$
26	$s^{-1}K_0(s)$, $\operatorname{Re}s > 0$	$\begin{cases} \operatorname{Arcosh}(-\ln x) & \text{if } 0 < x < e^{-1}, \\ 0 & \text{if } e^{-1} < x \end{cases}$
27	$s^{-1}K_1(s)$, $\operatorname{Re}s > 0$	$\begin{cases} \sqrt{\ln^2 x - 1} & \text{if } 0 < x < e^{-1}, \\ 0 & \text{if } e^{-1} < x \end{cases}$

No	Direct transform, $\hat{f}(s)$	Inverse transform, $f(x) = \dfrac{1}{2\pi i} \displaystyle\int_{\sigma-i\infty}^{\sigma+i\infty} \hat{f}(s) x^{-s}\, ds$
28	$K_\nu(s), \quad \mathrm{Re}\, s > 0$	$\begin{cases} \dfrac{\cosh\left[\nu \operatorname{Arcosh}(-\ln x)\right]}{\sqrt{\ln^2 x - 1}} & \text{if } 0 < x < e^{-1}, \\ 0 & \text{if } e^{-1} < x \end{cases}$
29	$s^{-1} K_\nu(s), \quad \mathrm{Re}\, s > 0$	$\begin{cases} \dfrac{1}{\nu} \sinh\left[\nu \operatorname{Arcosh}(-\ln x)\right] & \text{if } 0 < x < e^{-1}, \\ 0 & \text{if } e^{-1} < x \end{cases}$
30	$s^{-\nu} K_\nu(s), \quad \mathrm{Re}\, s > 0, \ \nu > -\tfrac{1}{2}$	$\begin{cases} \dfrac{\sqrt{\pi}\,(\ln^2 x - 1)^{\nu-1/2}}{2^\nu \Gamma(\nu + 1/2)} & \text{if } 0 < x < e^{-1}, \\ 0 & \text{if } e^{-1} < x \end{cases}$

⊙ References for Supplement 9: H. Bateman and A. Erdélyi (1954), V. A. Ditkin and A. P. Prudnikov (1965).

Supplement 10

Special Functions and Their Properties

Throughout Supplement 10 it is assumed that n is a positive integer, unless otherwise specified.

10.1. Some Symbols and Coefficients

▶ **Factorial**

$$0! = 1! = 1, \quad n! = 1 \cdot 2 \cdot 3 \dots (n-1)n, \quad n = 2, 3, \dots,$$

$$(2n)!! = 2 \cdot 4 \cdot 6 \dots (2n-2)(2n) = 2^n n!,$$

$$(2n+1)!! = 1 \cdot 3 \cdot 5 \dots (2n-1)(2n+1) = \frac{2^{n+1}}{\sqrt{\pi}} \Gamma\left(n + \frac{3}{2}\right),$$

$$n!! = \begin{cases} (2k)!! & \text{if } n = 2k, \\ (2k+1)!! & \text{if } n = 2k+1, \end{cases} \quad 0!! = 1.$$

▶ **Binomial coefficients**

$$C_n^k = \frac{n!}{k!(n-k)!}, \qquad \text{where} \quad k = 1, \dots, n,$$

$$C_a^k = (-1)^k \frac{(-a)_k}{k!} = \frac{a(a-1) \dots (a-k+1)}{k!}, \qquad \text{where} \quad k = 1, 2, \dots$$

General case:

$$C_a^b = \frac{\Gamma(a+1)}{\Gamma(b+1)\Gamma(a-b+1)}, \qquad \text{where} \quad \Gamma(x) \text{ is the gamma function.}$$

Properties:

$$C_a^0 = 1, \quad C_n^k = 0 \quad \text{for } k = -1, -2, \dots \text{ or } k > n,$$

$$C_a^{b+1} = \frac{a}{b+1} C_{a-1}^b = \frac{a-b}{b+1} C_a^b, \quad C_a^b + C_a^{b+1} = C_{a+1}^{b+1},$$

$$C_{-1/2}^n = \frac{(-1)^n}{2^{2n}} C_{2n}^n = (-1)^n \frac{(2n-1)!!}{(2n)!!},$$

$$C_{1/2}^n = \frac{(-1)^{n-1}}{n2^{2n-1}} C_{2n-2}^{n-1} = \frac{(-1)^{n-1}}{n} \frac{(2n-3)!!}{(2n-2)!!},$$

$$C_{n+1/2}^{2n+1} = (-1)^n 2^{-4n-1} C_{2n}^n, \quad C_{2n+1/2}^n = 2^{-2n} C_{4n+1}^{2n},$$

$$C_n^{1/2} = \frac{2^{2n+1}}{\pi C_{2n}^n}, \quad C_n^{n/2} = \frac{2^{2n}}{\pi} C_n^{(n-1)/2}.$$

▶ **Pochhammer symbol** $(k = 1, 2, \dots)$

$$(a)_n = a(a + 1) \dots (a + n - 1) = \frac{\Gamma(a + n)}{\Gamma(a)} = (-1)^n \frac{\Gamma(1 - a)}{\Gamma(1 - a - n)},$$

$$(a)_0 = 1, \quad (a)_{n+k} = (a)_n (a + n)_k, \quad (n)_k = \frac{(n + k - 1)!}{(n - 1)!},$$

$$(a)_{-n} = \frac{\Gamma(a - n)}{\Gamma(a)} = \frac{(-1)^n}{(1 - a)_n}, \quad \text{where } a \neq 1, \dots, n;$$

$$(1)_n = n!, \quad (1/2)_n = 2^{-2n} \frac{(2n)!}{n!}, \quad (3/2)_n = 2^{-2n} \frac{(2n + 1)!}{n!},$$

$$(a + mk)_{nk} = \frac{(a)_{mk+nk}}{(a)_{mk}}, \quad (a + n)_n = \frac{(a)_{2n}}{(a)_n}, \quad (a + n)_k = \frac{(a)_k (a + k)_n}{(a)_n}.$$

▶ **Bernoulli numbers, B_n**
Definition:

$$\frac{x}{e^x - 1} = \sum_{n=0}^{\infty} B_n \frac{x^n}{n!}.$$

The numbers:

$$B_0 = 1, \quad B_1 = -\frac{1}{2}, \quad B_2 = \frac{1}{6}, \quad B_4 = -\frac{1}{30}, \quad B_6 = \frac{1}{42}, \quad B_8 = -\frac{1}{30}, \quad B_{10} = \frac{5}{66}, \quad \dots,$$
$$B_{2m+1} = 0 \quad \text{for} \quad m = 1, 2, \dots$$

10.2. Error Functions and Integral Exponent

▶ **Error function and complementary error function (probability integrals)**
Definitions:

$$\text{erf } x = \frac{2}{\sqrt{\pi}} \int_0^x \exp(-t^2)\, dt, \qquad \text{erfc } x = 1 - \text{erf } x = \frac{2}{\sqrt{\pi}} \int_x^{\infty} \exp(-t^2)\, dt.$$

Expansion of erf x into series in powers of x as $x \to 0$:

$$\text{erf } x = \frac{2}{\sqrt{\pi}} \sum_{k=0}^{\infty} (-1)^k \frac{x^{2k+1}}{(k)!(2k + 1)} = \frac{2}{\sqrt{\pi}} \exp(-x^2) \sum_{k=0}^{\infty} \frac{2^k x^{2k+1}}{2k + 1)!!}.$$

Asymptotic expansion of erfc x as $x \to \infty$:

$$\text{erfc } x = \frac{1}{\sqrt{\pi}} \exp(-x^2) \left[\sum_{m=0}^{M-1} (-1)^m \frac{\left(\frac{1}{2}\right)_m}{x^{2m+1}} + O\left(|x|^{-2M-1}\right) \right], \qquad M = 1, 2, \dots$$

▶ **Integral exponent**
Definition:

$$\text{Ei}(x) = \int_{-\infty}^{x} \frac{e^t}{t}\, dt \qquad\qquad \text{for} \quad x < 0,$$

$$\text{Ei}(x) = \lim_{\varepsilon \to +0} \left(\int_{-\infty}^{-\varepsilon} \frac{e^t}{t}\, dt + \int_{\varepsilon}^{x} \frac{e^t}{t}\, dt \right) \quad \text{for} \quad x > 0.$$

Other integral representations:

$$\text{Ei}(-x) = -e^{-x} \int_0^\infty \frac{x \sin t + t \cos t}{x^2 + t^2}\, dt \quad \text{for} \quad x > 0,$$

$$\text{Ei}(-x) = e^{-x} \int_0^\infty \frac{x \sin t - t \cos t}{x^2 + t^2}\, dt \quad \text{for} \quad x < 0,$$

$$\text{Ei}(-x) = -x \int_1^\infty e^{-xt} \ln t\, dt \quad \text{for} \quad x > 0.$$

Expansion into series in powers of x as $x \to 0$:

$$\text{Ei}(x) = \begin{cases} \mathcal{C} + \ln(-x) + \displaystyle\sum_{k=1}^\infty \frac{x^k}{k \cdot k!} & \text{if } x < 0, \\[2mm] \mathcal{C} + \ln x + \displaystyle\sum_{k=1}^\infty \frac{x^k}{k \cdot k!} & \text{if } x > 0, \end{cases}$$

where $\mathcal{C} = 0.5572\ldots$ is the Euler constant.

Asymptotic expansion as $x \to \infty$:

$$\text{Ei}(-x) = e^{-x} \sum_{k=1}^n (-1)^k \frac{(k-1)!}{x^k} + R_n, \qquad R_n < \frac{n!}{x^n}.$$

▶ **Integral logarithm**

Definition:

$$\text{li}(x) = \begin{cases} \displaystyle\int_0^x \frac{dt}{\ln t} = \text{Ei}(\ln x) & \text{if } 0 < x < 1, \\[2mm] \displaystyle\lim_{\varepsilon \to +0}\left(\int_0^{1-\varepsilon} \frac{dt}{\ln t} + \int_{1+\varepsilon}^x \frac{dt}{\ln t} \right) & \text{if } x > 1. \end{cases}$$

For small x,

$$\text{li}(x) \approx \frac{x}{\ln(1/x)}.$$

Asymptotic expansion as $x \to 1$:

$$\text{li}(x) = \mathcal{C} + \ln|\ln x| + \sum_{k=1}^\infty \frac{\ln^k x}{k \cdot k!}.$$

10.3. Integral Sine and Integral Cosine. Fresnel Integrals

▶ **Integral sine**

Definition:

$$\text{Si}(x) = \int_0^x \frac{\sin t}{t}\, dt, \qquad \text{si}(x) = -\int_x^\infty \frac{\sin t}{t}\, dt = \text{Si}(x) - \frac{\pi}{2}.$$

Specific values:

$$\text{Si}(0) = 0, \quad \text{Si}(\infty) = \frac{\pi}{2}, \quad \text{si}(\infty) = 0.$$

Properties:

$$\text{Si}(-x) = -\text{Si}(x), \quad \text{si}(x) + \text{si}(-x) = -\pi, \quad \lim_{x \to -\infty} \text{si}(x) = -\pi.$$

Expansion into series in powers of x as $x \to 0$:

$$\mathrm{Si}(x) = \sum_{k=1}^{\infty} \frac{(-1)^{k+1} x^{2k-1}}{(2k-1)(2k-1)!}.$$

Asymptotic expansion as $x \to \infty$:

$$\mathrm{si}(x) = -\cos x \left[\sum_{m=0}^{M-1} \frac{(-1)^m (2m)!}{x^{2m+1}} + O\left(|x|^{-2M-1}\right) \right] + \sin x \left[\sum_{m=1}^{N-1} \frac{(-1)^m (2m-1)!}{x^{2m}} + O\left(|x|^{-2N}\right) \right],$$

where $M, N = 1, 2, \ldots$

▶ **Integral cosine**

Definition:

$$\mathrm{Ci}(x) = -\int_x^{\infty} \frac{\cos t}{t} \, dt = \mathcal{C} + \ln x + \int_0^x \frac{\cos t - 1}{t} \, dt, \qquad \mathcal{C} = 0.5572\ldots$$

Expansion into series in powers of x as $x \to 0$:

$$\mathrm{Ci}(x) = \mathcal{C} + \ln x + \sum_{k=1}^{\infty} \frac{(-1)^k x^{2k}}{2k \, (2k)!}.$$

Asymptotic expansion as $x \to \infty$:

$$\mathrm{Ci}(x) = \cos x \left[\sum_{m=1}^{M-1} \frac{(-1)^m (2m-1)!}{x^{2m}} + O\left(|x|^{-2M}\right) \right] + \sin x \left[\sum_{m=0}^{N-1} \frac{(-1)^m (2m)!}{x^{2m+1}} + O\left(|x|^{-2N-1}\right) \right],$$

where $M, N = 1, 2, \ldots$

▶ **Fresnel integrals**

Definitions:

$$S(x) = \frac{1}{\sqrt{2\pi}} \int_0^x \frac{\sin t}{\sqrt{t}} \, dt = \sqrt{\frac{2}{\pi}} \int_0^{\sqrt{x}} \sin t^2 \, dt,$$

$$C(x) = \frac{1}{\sqrt{2\pi}} \int_0^x \frac{\cos t}{\sqrt{t}} \, dt = \sqrt{\frac{2}{\pi}} \int_0^{\sqrt{x}} \cos t^2 \, dt.$$

Expansion into series in powers of x as $x \to 0$:

$$S(x) = \sqrt{\frac{2}{\pi}} \, x \sum_{k=0}^{\infty} \frac{(-1)^k x^{2k+1}}{(4k+3)(2k+1)!},$$

$$C(x) = \sqrt{\frac{2}{\pi}} \, x \sum_{k=0}^{\infty} \frac{(-1)^k x^{2k}}{(4k+1)(2k)!}.$$

Asymptotic expansion as $x \to \infty$:

$$S(x) = \frac{1}{2} - \frac{\cos x}{\sqrt{2\pi x}} P(x) - \frac{\sin x}{\sqrt{2\pi x}} Q(x),$$

$$C(x) = \frac{1}{2} + \frac{\sin x}{\sqrt{2\pi x}} P(x) - \frac{\cos x}{\sqrt{2\pi x}} Q(x),$$

$$P(x) = 1 - \frac{1 \cdot 3}{(2x)^2} + \frac{1 \cdot 3 \cdot 5 \cdot 7}{(2x)^4} - \cdots, \qquad Q(x) = \frac{1}{2x} - \frac{1 \cdot 3 \cdot 5}{(2x)^3} + \cdots.$$

10.4. Gamma Function. Beta Function

▶ **Definition. Integral representations**

The gamma function, $\Gamma(z)$, is an analytic function of the complex argument z everywhere, except for the points $z = 0, -1, -2, \ldots$

For Re $z > 0$,

$$\Gamma(z) = \int_0^\infty t^{z-1} e^{-t}\, dt.$$

For $-(n + 1) < \text{Re}\, z < -n$, where $n = 0, 1, 2, \ldots$,

$$\Gamma(z) = \int_0^\infty \left[e^{-t} - \sum_{m=0}^n \frac{(-1)^m}{m!} \right] t^{z-1}\, dt.$$

▶ **Euler formula**

$$\Gamma(z) = \lim_{n\to\infty} \frac{n!\, n^z}{z(z+1)\ldots(z+n)} \qquad (z \neq 0, -1, -2, \ldots).$$

▶ **Simplest properties**

$$\Gamma(z + 1) = z\Gamma(z), \quad \Gamma(n + 1) = n!, \quad \Gamma(1) = \Gamma(2) = 1.$$

▶ **Symmetry formulas**

$$\Gamma(z)\Gamma(-z) = -\frac{\pi}{z \sin(\pi z)}, \quad \Gamma(z)\Gamma(1 - z) = \frac{\pi}{\sin(\pi z)},$$

$$\Gamma\left(\frac{1}{2} + z\right)\Gamma\left(\frac{1}{2} - z\right) = \frac{\pi}{\cos(\pi z)}.$$

▶ **Multiple argument formulas**

$$\Gamma(2z) = \frac{2^{2z-1}}{\sqrt{\pi}}\Gamma(z)\Gamma\left(z + \frac{1}{2}\right),$$

$$\Gamma(3z) = \frac{3^{3z-1/2}}{2\pi}\Gamma(z)\Gamma\left(z + \frac{1}{3}\right)\Gamma\left(z + \frac{2}{3}\right),$$

$$\Gamma(nz) = (2\pi)^{(1-n)/2} n^{nz-1/2} \prod_{k=0}^{n-1} \Gamma\left(z + \frac{k}{n}\right).$$

▶ **Fractional values of the argument**

$$\Gamma\left(\frac{1}{2}\right) = \sqrt{\pi}, \qquad \Gamma\left(n + \frac{1}{2}\right) = \frac{\sqrt{\pi}}{2^n}(2n - 1)!!,$$

$$\Gamma\left(-\frac{1}{2}\right) = -2\sqrt{\pi}, \quad \Gamma\left(\frac{1}{2} - n\right) = (-1)^n \frac{2^n \sqrt{\pi}}{(2n - 1)!!}.$$

▶ **Asymptotic expansion (Stirling formula)**

$$\Gamma(z) = \sqrt{2\pi}\, e^{-z} z^{z-1/2}\left[1 + \tfrac{1}{12} z^{-1} + \tfrac{1}{288} z^{-2} + O(z^{-3})\right] \qquad (|\arg|z < \pi).$$

▶ **Logarithmic derivative of the gamma function**

Definition:

$$\psi(z) = \frac{\ln \Gamma(z)}{dz} = \frac{\Gamma'_z(z)}{\Gamma(z)}.$$

Functional relations:

$$\psi(z) - \psi(1 + z) = -\frac{1}{z},$$

$$\psi(z) - \psi(1 - z) = -\pi \cot(\pi z),$$

$$\psi(z) - \psi(-z) = -\pi \cot(\pi z) - \frac{1}{z},$$

$$\psi\left(\tfrac{1}{2} + z\right) - \psi\left(\tfrac{1}{2} - z\right) = \pi \tan(\pi z),$$

$$\psi(mz) = \ln m + \frac{1}{m} \sum_{k=0}^{m-1} \psi\left(z + \frac{k}{m}\right).$$

Integral representations (Re $z > 0$):

$$\psi(z) = \int_0^\infty \left[e^{-t} - (1 + t)^{-z}\right] t^{-1} \, dt,$$

$$\psi(z) = \ln z + \int_0^\infty \left[t^{-1} - (1 - e^{-t})^{-1}\right] e^{-tz} \, dt,$$

$$\psi(z) = -\mathcal{C} + \int_0^1 \frac{1 - t^{z-1}}{1 - t} \, dt,$$

where $\mathcal{C} = -\psi(1) = 0.5572\ldots$ is the Euler constant.

Values for integer argument:

$$\psi(1) = -\mathcal{C}, \qquad \psi(n) = -\mathcal{C} + \sum_{k=1}^{n-1} k^{-1} \quad (n = 2, 3, \ldots)$$

▶ **Beta function**

Definition:

$$B(x, y) = \int_0^1 t^{x-1}(1 - t)^{y-1} \, dt,$$

where Re $x > 0$ and Re $y > 0$.

Relationship with the gamma function:

$$B(x, y) = \frac{\Gamma(x)\Gamma(y)}{\Gamma(x + y)}.$$

10.5. Incomplete Gamma Function

▶ **Definitions. Integral representations**

$$\gamma(\alpha, x) = \int_0^x e^{-t} t^{\alpha-1} \, dt, \qquad \text{Re } \alpha > 0,$$

$$\Gamma(\alpha, x) = \int_x^\infty e^{-t} t^{\alpha-1} \, dt = \Gamma(\alpha) - \gamma(\alpha, x).$$

▶ **Recurrent formulas**

$$\gamma(\alpha + 1, x) = \alpha\gamma(\alpha, x) - x^{\alpha}e^{-x},$$

$$\Gamma(\alpha + 1, x) = \alpha\Gamma(\alpha, x) + x^{\alpha}e^{-x}.$$

▶ **Asymptotic expansions as $x \to 0$:**

$$\gamma(\alpha, x) = \sum_{n=0}^{\infty} \frac{(-1)^n x^{\alpha+n}}{n!\,(\alpha + n)},$$

$$\Gamma(\alpha, x) = \Gamma(\alpha) - \sum_{n=0}^{\infty} \frac{(-1)^n x^{\alpha+n}}{n!\,(\alpha + n)}.$$

▶ **Asymptotic expansions as $x \to \infty$:**

$$\gamma(\alpha, x) = \Gamma(\alpha) - x^{\alpha-1}e^{-x}\left[\sum_{m=0}^{M-1} \frac{(1-\alpha)_m}{(-x)^m} + O\left(|x|^{-M}\right)\right],$$

$$\Gamma(\alpha, x) = x^{\alpha-1}e^{-x}\left[\sum_{m=0}^{M-1} \frac{(1-\alpha)_m}{(-x)^m} + O\left(|x|^{-M}\right)\right] \quad \left(-\tfrac{3}{2}\pi < \arg x < \tfrac{3}{2}\right).$$

▶ **Integral functions related to the gamma function:**

$$\operatorname{erf} x = \frac{1}{\sqrt{\pi}}\gamma\left(\frac{1}{2}, x^2\right), \quad \operatorname{erfc} x = \frac{1}{\sqrt{\pi}}\Gamma\left(\frac{1}{2}, x^2\right), \quad \operatorname{Ei}(-x) = -\Gamma(0, x).$$

▶ **Incomplete beta function:**

$$B_x(p, q) = \int_0^1 t^{p-1}(1-t)^{q-1}\,dt,$$

where $\operatorname{Re} x > 0$ and $\operatorname{Re} y > 0$.

10.6. Bessel Functions

▶ **Definition and basic formulas**

The Bessel function of the first kind, $J_\nu(x)$, and the Bessel function of the second kind, $Y_\nu(x)$ (also called the Neumann function), are solutions of the Bessel equation

$$x^2 y''_{xx} + x y'_x + (x^2 - \nu^2)y = 0$$

and are defined by the formulas

$$J_\nu(x) = \sum_{k=0}^{\infty} \frac{(-1)^k (x/2)^{\nu+2k}}{k!\,\Gamma(\nu + k + 1)}, \quad Y_\nu(x) = \frac{J_\nu(x)\cos\pi\nu - J_{-\nu}(x)}{\sin\pi\nu}. \tag{1}$$

The formula for $Y_\nu(x)$ is valid for $\nu \neq 0, \pm 1, \pm 2, \ldots$ (the cases $\nu \neq 0, \pm 1, \pm 2, \ldots$ are discussed in what follows).

The general solution of the Bessel equation has the form $Z_\nu(x) = C_1 J_\nu(x) + C_2 Y_\nu(x)$ and is called the cylinder function.

The Bessel functions possess the properties

$$2\nu Z_\nu(x) = x[Z_{\nu-1}(x) + Z_{\nu+1}(x)],$$

$$\frac{d}{dx} Z_\nu(x) = \frac{1}{2}[Z_{\nu-1}(x) - Z_{\nu+1}(x)] = \pm\left[\frac{\nu}{x} Z_\nu(x) - Z_{\nu\pm1}(x)\right],$$

$$\frac{d}{dx}[x^\nu Z_\nu(x)] = x^\nu Z_{\nu-1}(x), \qquad \frac{d}{dx}[x^{-\nu} Z_\nu(x)] = -x^{-\nu} Z_{\nu+1}(x),$$

$$\left(\frac{1}{x}\frac{d}{dx}\right)^n [x^\nu J_\nu(x)] = x^{\nu-n} J_{\nu-n}(x), \qquad \left(\frac{1}{x}\frac{d}{dx}\right)^n [x^{-\nu} J_\nu(x)] = (-1)^n x^{-\nu-n} J_{\nu+n}(x),$$

$$J_{-n}(x) = (-1)^n J_n(x), \quad Y_{-n}(x) = (-1)^n Y_n(x), \qquad n = 0, 1, 2, \ldots$$

▶ **The Bessel functions for $\nu = \pm n \pm \frac{1}{2}$; $n = 0, 1, \ldots$**

$$J_{1/2}(x) = \sqrt{\frac{2}{\pi x}} \sin x, \qquad\qquad J_{-1/2}(x) = \sqrt{\frac{2}{\pi x}} \cos x,$$

$$J_{3/2}(x) = \sqrt{\frac{2}{\pi x}}\left(\frac{1}{x} \sin x - \cos x\right), \qquad J_{-3/2}(x) = \sqrt{\frac{2}{\pi x}}\left(-\frac{1}{x} \cos x - \sin x\right),$$

$$J_{n+1/2}(x) = \sqrt{\frac{2}{\pi x}}\left[\sin\left(x - \frac{n\pi}{2}\right) \sum_{k=0}^{[n/2]} \frac{(-1)^k (n+2k)!}{(2k)!\,(n-2k)!\,(2x)^{2k}}\right.$$

$$\left. + \cos\left(x - \frac{n\pi}{2}\right) \sum_{k=0}^{[(n-1)/2]} \frac{(-1)^k (n+2k+1)!}{(2k+1)!\,(n-2k-1)!\,(2x)^{2k+1}}\right],$$

$$J_{-n-1/2}(x) = \sqrt{\frac{2}{\pi x}}\left[\cos\left(x + \frac{n\pi}{2}\right) \sum_{k=0}^{[n/2]} \frac{(-1)^k (n+2k)!}{(2k)!\,(n-2k)!\,(2x)^{2k}}\right.$$

$$\left. - \sin\left(x + \frac{n\pi}{2}\right) \sum_{k=0}^{[(n-1)/2]} \frac{(-1)^k (n+2k+1)!}{(2k+1)!\,(n-2k-1)!\,(2x)^{2k+1}}\right],$$

$$Y_{1/2}(x) = -\sqrt{\frac{2}{\pi x}} \cos x, \qquad\qquad Y_{-1/2}(x) = \sqrt{\frac{2}{\pi x}} \sin x,$$

$$Y_{n+1/2}(x) = (-1)^{n+1} J_{-n-1/2}(x), \qquad Y_{-n-1/2}(x) = (-1)^n J_{n+1/2}(x).$$

▶ **The Bessel functions for $\nu = \pm n$; $n = 0, 1, 2, \ldots$**
Let $\nu = n$ be an arbitrary integer. The relations

$$J_{-n}(x) = (-1)^n J_n(x), \quad Y_{-n}(x) = (-1)^n Y_n(x)$$

are valid. The function $J_n(x)$ is given by the first formula in (1) with $\nu = n$, and $Y_n(x)$ can be obtained from the second formula in (1) by proceeding to the limit $\nu \to n$. For nonnegative n, $Y_n(x)$ can be represented in the form

$$Y_n(x) = \frac{2}{\pi} J_n(x) \ln\frac{x}{2} - \frac{1}{\pi} \sum_{k=0}^{n-1} \frac{(n-k-1)!}{k!}\left(\frac{2}{x}\right)^{n-2k} - \frac{1}{\pi} \sum_{k=0}^{\infty} (-1)^k \left(\frac{x}{2}\right)^{n+2k} \frac{\psi(k+1) + \psi(n+k+1)}{k!\,(n+k)!},$$

where $\psi(1) = -\mathcal{C}$, $\psi(n) = -\mathcal{C} + \sum_{k=1}^{n-1} k^{-1}$, $\mathcal{C} = 0.5572\ldots$ is the Euler constant, $\psi(x) = [\ln \Gamma(x)]'_x$ is the logarithmic derivative of the gamma function.

▶ **Wronskians and similar formulas**

$$W(J_\nu, J_{-\nu}) = -\frac{2}{\pi x}\sin(\pi\nu), \qquad W(J_\nu, Y_\nu) = \frac{2}{\pi x},$$

$$J_\nu(x)J_{-\nu+1}(x) + J_{-\nu}(x)J_{\nu-1}(x) = \frac{2\sin(\pi\nu)}{\pi x}, \qquad J_\nu(x)Y_{\nu+1}(x) - J_{\nu+1}(x)Y_\nu(x) = -\frac{2}{\pi x}.$$

Here the notation $W(f, g) = fg'_x - f'_x g$ is used.

▶ **Integral representations**
The functions J_ν and Y_ν can be represented in the form of definite integrals (for $x > 0$):

$$\pi J_\nu(x) = \int_0^\pi \cos(x\sin\theta - \nu\theta)\,d\theta - \sin\pi\nu \int_0^\infty \exp(-x\sinh t - \nu t)\,dt,$$

$$\pi Y_\nu(x) = \int_0^\pi \sin(x\sin\theta - \nu\theta)\,d\theta - \int_0^\infty (e^{\nu t} + e^{-\nu t}\cos\pi\nu)e^{-x\sinh t}\,dt.$$

For $|\nu| < \frac{1}{2}$, $x > 0$,

$$J_\nu(x) = \frac{2^{1+\nu}x^{-\nu}}{\pi^{1/2}\Gamma(\frac{1}{2} - \nu)} \int_1^\infty \frac{\sin(xt)\,dt}{(t^2 - 1)^{\nu+1/2}},$$

$$Y_\nu(x) = -\frac{2^{1+\nu}x^{-\nu}}{\pi^{1/2}\Gamma(\frac{1}{2} - \nu)} \int_1^\infty \frac{\cos(xt)\,dt}{(t^2 - 1)^{\nu+1/2}}.$$

For $\nu > -\frac{1}{2}$,

$$J_\nu(x) = \frac{2(x/2)^\nu}{\pi^{1/2}\Gamma(\frac{1}{2} + \nu)} \int_0^{\pi/2} \cos(x\cos t)\sin^{2\nu} t\,dt \quad \text{(Poisson's formula)}.$$

For $\nu = 0$, $x > 0$,

$$J_0(x) = \frac{2}{\pi}\int_0^\infty \sin(x\cosh t)\,dt, \qquad Y_0(x) = -\frac{2}{\pi}\int_0^\infty \cos(x\cosh t)\,dt.$$

For integer $\nu = n = 0, 1, 2, \ldots$,

$$J_n(x) = \frac{1}{\pi}\int_0^\pi \cos(nt - x\sin t)\,dt \quad \text{(Bessel's formula)},$$

$$J_{2n}(x) = \frac{2}{\pi}\int_0^{\pi/2} \cos(x\sin t)\cos(2nt)\,dt,$$

$$J_{2n+1}(x) = \frac{2}{\pi}\int_0^{\pi/2} \sin(x\sin t)\sin[(2n + 1)t]\,dt.$$

▶ **Integrals with Bessel functions**

$$\int_0^x x^\lambda J_\nu(x)\,dx = \frac{x^{\lambda+\nu+1}}{2^\nu(\lambda + \nu + 1)\Gamma(\nu + 1)} F\left(\frac{\lambda + \nu + 1}{2}, \frac{\lambda + \nu + 3}{2}, \nu+1; -\frac{x^2}{4}\right), \qquad \mathrm{Re}(\lambda+\nu) > -1,$$

where $F(a, b, c; x)$ is the hypergeometric series (see Section 10.9 of this supplement),

$$\int_0^x x^\lambda Y_\nu(x)\,dx = -\frac{\cos(\nu\pi)\Gamma(-\nu)}{2^\nu\pi(\lambda + \nu + 1)} x^{\lambda+\nu+1} F\left(\frac{\lambda + \nu + 1}{2}, \nu + 1, \frac{\lambda + \nu + 3}{2}, -\frac{x^2}{4}\right)$$

$$- \frac{2^\nu\Gamma(\nu)}{\lambda - \nu + 1} x^{\lambda-\nu+1} F\left(\frac{\lambda - \nu + 1}{2}, 1 - \nu, \frac{\lambda - \nu + 3}{2}, -\frac{x^2}{4}\right), \qquad \mathrm{Re}\,\lambda > |\mathrm{Re}\,\nu| - 1.$$

▶ **Asymptotic expansions as $|x| \to \infty$**

$$J_\nu(x) = \sqrt{\frac{2}{\pi x}} \left\{ \cos\left(\frac{4x - 2\nu\pi - \pi}{4}\right) \left[\sum_{m=0}^{M-1} (-1)^m (\nu, 2m)(2x)^{-2m} + O(|x|^{-2M})\right] \right.$$

$$\left. - \sin\left(\frac{4x - 2\nu\pi - \pi}{4}\right) \left[\sum_{m=0}^{M-1} (-1)^m (\nu, 2m+1)(2x)^{-2m-1} + O(|x|^{-2M-1})\right] \right\},$$

$$Y_\nu(x) = \sqrt{\frac{2}{\pi x}} \left\{ \sin\left(\frac{4x - 2\nu\pi - \pi}{4}\right) \left[\sum_{m=0}^{M-1} (-1)^m (\nu, 2m)(2x)^{-2m} + O(|x|^{-2M})\right] \right.$$

$$\left. + \cos\left(\frac{4x - 2\nu\pi - \pi}{4}\right) \left[\sum_{m=0}^{M-1} (-1)^m (\nu, 2m+1)(2x)^{-2m-1} + O(|x|^{-2M-1})\right] \right\},$$

where $(\nu, m) = \dfrac{1}{2^{2m}m!}(4\nu^2 - 1)(4\nu^2 - 3^2)\dots[4\nu^2 - (2m-1)^2] = \dfrac{\Gamma(\frac{1}{2} + \nu + m)}{m!\,\Gamma(\frac{1}{2} + \nu - m)}$.

For nonnegative integer n and large x,

$$\sqrt{\pi x}\, J_{2n}(x) = (-1)^n(\cos x + \sin x) + O(x^{-2}),$$
$$\sqrt{\pi x}\, J_{2n+1}(x) = (-1)^{n+1}(\cos x - \sin x) + O(x^{-2}).$$

▶ **Asymptotic for large ν ($\nu \to \infty$).**

$$J_\nu(x) \to \frac{1}{\sqrt{2\pi\nu}}\left(\frac{ex}{2\nu}\right)^\nu, \quad Y_\nu(x) \to -\sqrt{\frac{2}{\pi\nu}}\left(\frac{ex}{2\nu}\right)^{-\nu},$$

where x is fixed,

$$J_\nu(\nu) \to \frac{2^{1/3}}{3^{2/3}\Gamma(2/3)}\frac{1}{\nu^{1/3}}, \quad Y_\nu(\nu) \to -\frac{2^{1/3}}{3^{1/6}\Gamma(2/3)}\frac{1}{\nu^{1/3}}.$$

▶ **Zeros of Bessel functions**

Each of the functions $J_\nu(x)$ and $Y_\nu(x)$ has infinitely many real zeros (for real ν). All zeros are simple, possibly except for the point $x = 0$.

The zeros γ_m of $J_0(x)$, i.e., the roots of the equation $J_0(\gamma_m) = 0$, are approximately given by

$$\gamma_m = 2.4 + 3.13\,(m - 1) \qquad (m = 1, 2, \dots),$$

with maximum error 0.2%.

▶ **Hankel functions (Bessel functions of the third kind)**

$$H_\nu^{(1)}(z) = J_\nu(z) + iY_\nu(z), \quad H_\nu^{(2)}(z) = J_\nu(z) - iY_\nu(z), \qquad i^2 = -1.$$

10.7. Modified Bessel Functions

▶ **Definitions. Basic formulas**

The modified Bessel functions of the first kind, $I_\nu(x)$, and the second kind, $K_\nu(x)$ (also called the Macdonald function), of order ν are solutions of the modified Bessel equation

$$x^2 y''_{xx} + xy'_x - (x^2 + \nu^2)y = 0$$

and are defined by the formulas

$$I_\nu(x) = \sum_{k=0}^\infty \frac{(x/2)^{2k+\nu}}{k!\,\Gamma(\nu+k+1)}, \qquad K_\nu(x) = \frac{\pi}{2} \frac{I_{-\nu} - I_\nu}{\sin \pi\nu},$$

(see below for $K_\nu(x)$ with $\nu = 0, 1, 2, \ldots$).

The modified Bessel functions possess the properties

$$K_{-\nu}(x) = K_\nu(x); \qquad I_{-n}(x) = (-1)^n I_n(x), \quad n = 0, 1, 2, \ldots$$

$$2\nu I_\nu(x) = x[I_{\nu-1}(x) - I_{\nu+1}(x)], \qquad 2\nu K_\nu(x) = -x[K_{\nu-1}(x) - K_{\nu+1}(x)],$$

$$\frac{d}{dx} I_\nu(x) = \frac{1}{2}[I_{\nu-1}(x) + I_{\nu+1}(x)], \qquad \frac{d}{dx} K_\nu(x) = -\frac{1}{2}[K_{\nu-1}(x) + K_{\nu+1}(x)].$$

▶ **Modified Bessel functions for $\nu = \pm n \pm \frac{1}{2}$, where $n = 0, 1, 2, \ldots$**

$$I_{1/2}(x) = \sqrt{\frac{2}{\pi x}} \sinh x, \qquad I_{-1/2}(x) = \sqrt{\frac{2}{\pi x}} \cosh x,$$

$$I_{3/2}(x) = \sqrt{\frac{2}{\pi x}} \left(-\frac{1}{x} \sinh x + \cosh x \right), \qquad I_{-3/2}(x) = \sqrt{\frac{2}{\pi x}} \left(-\frac{1}{x} \cosh x + \sinh x \right),$$

$$I_{n+1/2}(x) = \frac{1}{\sqrt{2\pi x}} \left[e^x \sum_{k=0}^n \frac{(-1)^k (n+k)!}{k!\,(n-k)!\,(2x)^k} - (-1)^n e^{-x} \sum_{k=0}^n \frac{(n+k)!}{k!\,(n-k)!\,(2x)^k} \right],$$

$$I_{-n-1/2}(x) = \frac{1}{\sqrt{2\pi x}} \left[e^x \sum_{k=0}^n \frac{(-1)^k (n+k)!}{k!\,(n-k)!\,(2x)^k} + (-1)^n e^{-x} \sum_{k=0}^n \frac{(n+k)!}{k!\,(n-k)!\,(2x)^k} \right],$$

$$K_{\pm 1/2}(x) = \sqrt{\frac{\pi}{2x}} e^{-x}, \qquad K_{\pm 3/2}(x) = \sqrt{\frac{\pi}{2x}} \left(1 + \frac{1}{x} \right) e^{-x},$$

$$K_{n+1/2}(x) = K_{-n-1/2}(x) = \sqrt{\frac{\pi}{2x}} e^{-x} \sum_{k=0}^n \frac{(n+k)!}{k!\,(n-k)!\,(2x)^k}.$$

▶ **Modified Bessel functions $\nu = n$, where $n = 0, 1, 2, \ldots$**

If $\nu = n$ is a nonnegative integer, then

$$K_n(x) = (-1)^{n+1} I_n(x) \ln \frac{x}{2} + \frac{1}{2} \sum_{m=0}^{n-1} (-1)^m \left(\frac{x}{2} \right)^{2m-n} \frac{(n-m-1)!}{m!}$$

$$+ \frac{1}{2} (-1)^n \sum_{m=0}^\infty \left(\frac{x}{2} \right)^{n+2m} \frac{\psi(n+m+1) + \psi(m+1)}{m!\,(n+m)!}; \quad n = 0, 1, 2, \ldots,$$

where $\psi(z)$ is the logarithmic derivative of the gamma function; for $n = 0$, the first sum is dropped.

▶ **Wronskians and similar formulas.**

$$W(I_\nu, I_{-\nu}) = -\frac{2}{\pi x} \sin(\pi\nu), \quad W(I_\nu, K_\nu) = -\frac{1}{x},$$

$$I_\nu(x) I_{-\nu+1}(x) - I_{-\nu}(x) I_{\nu-1}(x) = -\frac{2\sin(\pi\nu)}{\pi x}, \quad I_\nu(x) K_{\nu+1}(x) + I_{\nu+1}(x) K_\nu(x) = \frac{1}{x},$$

where $W(f, g) = f g'_x - f'_x g$.

▶ **Integral representations.**

The functions $I_\nu(x)$ and $K_\nu(x)$ can be represented in terms of definite integrals:

$$I_\nu(x) = \frac{x^\nu}{\pi^{1/2} 2^\nu \Gamma(\nu + \frac{1}{2})} \int_{-1}^{1} \exp(-xt)(1 - t^2)^{\nu-1/2}\, dt \qquad (x > 0,\ \nu > -\tfrac{1}{2}),$$

$$K_\nu(x) = \int_{0}^{\infty} \exp(-x \cosh t) \cosh(\nu t)\, dt \qquad (x > 0),$$

$$K_\nu(x) = \frac{1}{\cos(\frac{1}{2}\pi\nu)} \int_{0}^{\infty} \cos(x \sinh t) \cosh(\nu t)\, dt \qquad (x > 0,\ -1 < \nu < 1),$$

$$K_\nu(x) = \frac{1}{\sin(\frac{1}{2}\pi\nu)} \int_{0}^{\infty} \sin(x \sinh t) \sinh(\nu t)\, dt \qquad (x > 0,\ -1 < \nu < 1).$$

For integer $\nu = n$,

$$I_n(x) = \frac{1}{\pi} \int_{0}^{\pi} \exp(x \cos t) \cos(nt)\, dt \qquad (n = 0, 1, 2, \dots),$$

$$K_0(x) = \int_{0}^{\infty} \cos(x \sinh t)\, dt = \int_{0}^{\infty} \frac{\cos(xt)}{\sqrt{t^2 + 1}}\, dt \qquad (x > 0).$$

▶ **Integrals with modified Bessel functions**

$$\int_{0}^{x} x^\lambda I_\nu(x)\, dx = \frac{x^{\lambda+\nu+1}}{2^\nu (\lambda + \nu + 1)\Gamma(\nu + 1)} F\left(\frac{\lambda + \nu + 1}{2}, \frac{\lambda + \nu + 3}{2}, \nu + 1; \frac{x^2}{4}\right), \qquad \mathrm{Re}(\lambda + \nu) > -1,$$

where $F(a, b, c; x)$ is the hypergeometric series (see Section 10.9 of this supplement),

$$\int_{0}^{x} x^\lambda K_\nu(x)\, dx = \frac{2^{\nu-1}\Gamma(\nu)}{\lambda - \nu + 1} x^{\lambda-\nu+1} F\left(\frac{\lambda - \nu + 1}{2}, 1 - \nu, \frac{\lambda - \nu + 3}{2}, \frac{x^2}{4}\right)$$
$$+ \frac{2^{-\nu-1}\Gamma(-\nu)}{\lambda + \nu + 1} x^{\lambda+\nu+1} F\left(\frac{\lambda + \nu + 1}{2}, 1 + \nu, \frac{\lambda + \nu + 3}{2}, \frac{x^2}{4}\right), \qquad \mathrm{Re}\,\lambda > |\mathrm{Re}\,\nu| - 1.$$

▶ **Asymptotic expansions as $x \to \infty$**

$$I_\nu(x) = \frac{e^x}{\sqrt{2\pi x}}\left\{1 + \sum_{m=1}^{M}(-1)^m \frac{(4\nu^2 - 1)(4\nu^2 - 3^2)\dots[4\nu^2 - (2m-1)^2]}{m!\,(8x)^m}\right\},$$

$$K_\nu(x) = \sqrt{\frac{\pi}{2x}}\, e^{-x}\left\{1 + \sum_{m=1}^{M} \frac{(4\nu^2 - 1)(4\nu^2 - 3^2)\dots[4\nu^2 - (2m-1)^2]}{m!\,(8x)^m}\right\}.$$

The terms of the order of $O(x^{-M-1})$ are omitted in the braces.

10.8. Degenerate Hypergeometric Functions

▶ **Definitions. Basic Formulas**

The degenerate hypergeometric functions $\Phi(a, b; x)$ and $\Psi(a, b; x)$ are solutions of the degenerate hypergeometric equation

$$xy''_{xx} + (b - x)y'_x - ay = 0.$$

TABLE S1
Special cases of the Kummer function $\Phi(a, b; z)$

a	b	z	Φ	Conventional notation
a	a	x	e^x	
1	2	$2x$	$\dfrac{1}{x}e^x \sinh x$	
a	$a+1$	$-x$	$ax^{-a}\gamma(a, x)$	Incomplete gamma function $\gamma(a, x) = \displaystyle\int_0^x e^{-t}t^{a-1}\, dt$
$\dfrac{1}{2}$	$\dfrac{3}{2}$	$-x^2$	$\dfrac{\sqrt{\pi}}{2}\,\mathrm{erf}\,x$	Error function $\mathrm{erf}\,x = \dfrac{2}{\sqrt{\pi}}\displaystyle\int_0^x \exp(-t^2)\, dt$
$-n$	$\dfrac{1}{2}$	$\dfrac{x^2}{2}$	$\dfrac{n!}{(2n)!}\left(-\dfrac{1}{2}\right)^{-n}H_{2n}(x)$	Hermite polynomials $H_n = (-1)^n e^{x^2}\dfrac{d^n}{dx^n}\left(e^{-x^2}\right),$ $n = 0, 1, 2, \ldots$
$-n$	$\dfrac{3}{2}$	$\dfrac{x^2}{2}$	$\dfrac{n!}{(2n+1)!}\left(-\dfrac{1}{2}\right)^{-n}H_{2n+1}(x)$	
$-n$	b	x	$\dfrac{n!}{(b)_n}L_n^{(b-1)}(x)$	Laguerre polynomials $L_n^{(\alpha)}(x) = \dfrac{e^x x^{-\alpha}}{n!}\dfrac{d^n}{dx^n}\left(e^{-x}x^{n+\alpha}\right),$ $\alpha = b-1,$ $(b)_n = b(b+1)\ldots(b+n-1)$
$\nu+\dfrac{1}{2}$	$2\nu+1$	$2x$	$\Gamma(1+\nu)e^x\left(\dfrac{x}{2}\right)^{-\nu}I_\nu(x)$	Modified Bessel functions $I_\nu(x)$
$n+1$	$2n+2$	$2x$	$\Gamma\left(n+\dfrac{3}{2}\right)e^x\left(\dfrac{x}{2}\right)^{-n-\frac{1}{2}}I_{n+\frac{1}{2}}(x)$	

In the case $b \neq 0, -1, -2, -3, \ldots$, the function $\Phi(a, b; x)$ can be represented as Kummer's series:

$$\Phi(a, b; x) = 1 + \sum_{k=1}^{\infty} \frac{(a)_k}{(b)_k}\frac{x^k}{k!},$$

where $(a)_k = a(a+1)\ldots(a+k-1)$, $(a)_0 = 1$.

Table S1 presents some special cases when Φ can be expressed in terms of simpler functions. The function $\Psi(a, b; x)$ is defined as follows:

$$\Psi(a, b; x) = \frac{\Gamma(1-b)}{\Gamma(a-b+1)}\Phi(a, b; x) + \frac{\Gamma(b-1)}{\Gamma(a)}x^{1-b}\Phi(a-b+1, 2-b; x).$$

▶ **Some transformations and linear relations**
Kummer transformation:

$$\Phi(a, b; x) = e^x\Phi(b-a, b; -x), \qquad \Psi(a, b; x) = x^{1-b}\Psi(1+a-b, 2-b; x).$$

Linear relations for Φ:

$$(b-a)\Phi(a-1,b;x) + (2a-b+x)\Phi(a,b;x) - a\Phi(a+1,b;x) = 0,$$
$$b(b-1)\Phi(a,b-1;x) - b(b-1+x)\Phi(a,b;x) + (b-a)x\Phi(a,b+1;x) = 0,$$
$$(a-b+1)\Phi(a,b;x) - a\Phi(a+1,b;x) + (b-1)\Phi(a,b-1;x) = 0,$$
$$b\Phi(a,b;x) - b\Phi(a-1,b;x) - x\Phi(a,b+1;x) = 0,$$
$$b(a+x)\Phi(a,b;x) - (b-a)x\Phi(a,b+1;x) - ab\Phi(a+1,b;x) = 0,$$
$$(a-1+x)\Phi(a,b;x) + (b-a)\Phi(a-1,b;x) - (b-1)\Phi(a,b-1;x) = 0.$$

Linear relations for Ψ:

$$\Psi(a-1,b;x) - (2a-b+x)\Psi(a,b;x) + a(a-b+1)\Psi(a+1,b;x) = 0,$$
$$(b-a-1)\Psi(a,b-1;x) - (b-1+x)\Psi(a,b;x) + x\Psi(a,b+1;x) = 0,$$
$$\Psi(a,b;x) - a\Psi(a+1,b;x) - \Psi(a,b-1;x) = 0,$$
$$(b-a)\Psi(a,b;x) - x\Psi(a,b+1;x) + \Psi(a-1,b;x) = 0,$$
$$(a+x)\Psi(a,b;x) + a(b-a-1)\Psi(a+1,b;x) - x\Psi(a,b+1;x) = 0,$$
$$(a-1+x)\Psi(a,b;x) - \Psi(a-1,b;x) + (a-c+1)\Psi(a,b-1;x) = 0.$$

▶ **Differentiation formulas and Wronskian**

Differentiation formulas:

$$\frac{d}{dx}\Phi(a,b;x) = \frac{a}{b}\Phi(a+1,b+1;x), \qquad \frac{d^n}{dx^n}\Phi(a,b;x) = \frac{(a)_n}{(b)_n}\Phi(a+n,b+n;x),$$

$$\frac{d}{dx}\Psi(a,b;x) = -a\Psi(a+1,b+1;x), \qquad \frac{d^n}{dx^n}\Psi(a,b;x) = (-1)^n(a)_n\Psi(a+n,b+n;x).$$

Wronskian:

$$W(\Phi,\Psi) = \Phi\Psi'_x - \Phi'_x\Psi = -\frac{\Gamma(b)}{\Gamma(a)}x^{-b}e^x.$$

▶ **Degenerate hypergeometric functions for $n = 0, 1, \ldots$**

$$\Psi(a,n+1;x) = \frac{(-1)^{n-1}}{n!\,\Gamma(a-n)}\left\{ \Phi(a,n+1;x)\ln x \right.$$

$$\left. + \sum_{r=0}^{\infty}\frac{(a)_r}{(n+1)_r}\left[\psi(a+r) - \psi(1+r) - \psi(1+n+r)\right]\frac{x^r}{r!} \right\} + \frac{(n-1)!}{\Gamma(a)}\sum_{r=0}^{n-1}\frac{(a-n)_r}{(1-n)_r}\frac{x^{r-n}}{r!},$$

where $n = 0, 1, 2, \ldots$ (the last sum is dropped for $n = 0$), $\psi(z) = [\ln\Gamma(z)]'_z$ is the logarithmic derivative of the gamma function,

$$\psi(1) = -\mathcal{C}, \quad \psi(n) = -\mathcal{C} + \sum_{k=1}^{n-1}k^{-1},$$

where $\mathcal{C} = 0.5572\ldots$ is the Euler constant.

If $b < 0$, then the formula

$$\Psi(a,b;x) = x^{1-b}\Psi(a-b+1, 2-b; x)$$

is valid for any x.

For $b \neq 0, -1, -2, -3, \ldots$, the general solution of the degenerate hypergeometric equation can be represented in the form

$$y = C_1\Phi(a,b;x) + C_2\Psi(a,b;x),$$

and for $b = 0, -1, -2, -3, \ldots$, in the form

$$y = x^{1-b}\left[C_1\Phi(a-b+1, 2-b; x) + C_2\Psi(a-b+1, 2-b; x)\right].$$

▶ **Integral representations**

$$\Phi(a, b; x) = \frac{\Gamma(b)}{\Gamma(a)\,\Gamma(b-a)} \int_0^1 e^{xt} t^{a-1} (1-t)^{b-a-1}\, dt \qquad (\text{for } b > a > 0),$$

$$\Psi(a, b; x) = \frac{1}{\Gamma(a)} \int_0^\infty e^{-xt} t^{a-1} (1+t)^{b-a-1}\, dt \qquad (\text{for } a > 0,\ x > 0),$$

where $\Gamma(a)$ is the gamma function.

▶ **Integrals with degenerate hypergeometric functions**

$$\int \Phi(a, b; x)\, dx = \frac{b-1}{a-1} \Psi(a-1, b-1; x) + C,$$

$$\int \Psi(a, b; x)\, dx = \frac{1}{1-a} \Psi(a-1, b-1; x) + C,$$

$$\int x^n \Phi(a, b; x)\, dx = n! \sum_{k=1}^{n+1} \frac{(-1)^{k+1}(1-b)_k x^{n-k+1}}{(1-a)_k (n-k+1)!} \Phi(a-k, b-k; x) + C,$$

$$\int x^n \Psi(a, b; x)\, dx = n! \sum_{k=1}^{n+1} \frac{(-1)^{k+1} x^{n-k+1}}{(1-a)_k (n-k+1)!} \Psi(a-k, b-k; x) + C.$$

▶ **Asymptotic expansion as $|x| \to \infty$.**

$$\Phi(a, b; x) = \frac{\Gamma(b)}{\Gamma(a)} e^x x^{a-b} \left[\sum_{n=0}^N \frac{(b-a)_n(1-a)_n}{n!} x^{-n} + \varepsilon \right], \quad x > 0,$$

$$\Phi(a, b; x) = \frac{\Gamma(b)}{\Gamma(b-a)} (-x)^{-a} \left[\sum_{n=0}^N \frac{(a)_n(a-b+1)_n}{n!} (-x)^{-n} + \varepsilon \right], \quad x < 0,$$

$$\Psi(a, b; x) = x^{-a} \left[\sum_{n=0}^N (-1)^n \frac{(a)_n(a-b+1)_n}{n!} x^{-n} + \varepsilon \right], \quad -\infty < x < \infty,$$

where $\varepsilon = O(x^{-N-1})$.

10.9. Hypergeometric Functions

▶ **Definition**

The hypergeometric functions $F(\alpha, \beta, \gamma; x)$ is a solution the Gaussian hypergeometric equation

$$x(x-1)y''_{xx} + [(\alpha + \beta + 1)x - \gamma]y'_x + \alpha\beta y = 0.$$

For $\gamma \neq 0, -1, -2, -3, \ldots$, the function $F(\alpha, \beta, \gamma; x)$ can be expressed in terms of the hypergeometric series:

$$F(\alpha, \beta, \gamma; x) = 1 + \sum_{k=1}^\infty \frac{(\alpha)_k(\beta)_k}{(\gamma)_k} \frac{x^k}{k!}, \qquad (\alpha)_k = \alpha(\alpha+1)\ldots(\alpha+k-1),$$

which certainly converges for $|x| < 1$.

Table S2 shows some special cases when F can be expressed in term of elementary functions.

TABLE S2
Some special cases when the hypergeometric function $F(\alpha, \beta, \gamma; z)$
can be expressed in terms of elementary functions

α	β	γ	z	F
$-n$	β	γ	x	$\displaystyle\sum_{k=0}^{n} \frac{(-n)_k (\beta)_k}{(\gamma)_k} \frac{x^k}{k!},$ where $n = 1, 2, \ldots$
$-n$	β	$-n-m$	x	$\displaystyle\sum_{k=0}^{n} \frac{(-n)_k (\beta)_k}{(-n-m)_k} \frac{x^k}{k!},$ where $n = 1, 2, \ldots$
α	β	β	x	$(1-x)^{-\alpha}$
α	$\alpha + \frac{1}{2}$	$\frac{1}{2}$	x^2	$\frac{1}{2}\left[(1+x)^{-2\alpha} + (1-x)^{-2\alpha}\right]$
α	$\alpha + \frac{1}{2}$	$\frac{3}{2}$	x^2	$\dfrac{(1+x)^{1-2\alpha} - (1-x)^{1-2\alpha}}{2x(1-2\alpha)}$
α	$-\alpha$	$\frac{1}{2}$	$-x^2$	$\frac{1}{2}\left[\left(\sqrt{1+x^2}+x\right)^{2\alpha} + \left(\sqrt{1+x^2}-x\right)^{2\alpha}\right]$
α	$1-\alpha$	$\frac{1}{2}$	$-x^2$	$\dfrac{\left(\sqrt{1+x^2}+x\right)^{2\alpha-1} + \left(\sqrt{1+x^2}-x\right)^{2\alpha-1}}{2\sqrt{1+x^2}}$
α	$\alpha - \frac{1}{2}$	$2\alpha - 1$	x	$2^{2\alpha-2}\left(1 + \sqrt{1-x}\right)^{2-2\alpha}$
α	$1-\alpha$	$\frac{3}{2}$	$\sin^2 x$	$\dfrac{\sin[(2\alpha-1)x]}{(\alpha-1)\sin(2x)}$
α	$2-\alpha$	$\frac{3}{2}$	$\sin^2 x$	$\dfrac{\sin[(2\alpha-2)x]}{(\alpha-1)\sin(2x)}$
α	$1-\alpha$	$\frac{1}{2}$	$\sin^2 x$	$\dfrac{\cos[(2\alpha-1)x]}{\cos x}$
α	$\alpha + 1$	$\frac{1}{2}\alpha$	x	$(1+x)(1-x)^{-\alpha-1}$
α	$\alpha + \frac{1}{2}$	$2\alpha + 1$	x	$\left(\dfrac{1+\sqrt{1-x}}{2}\right)^{-2\alpha}$
α	$\alpha + \frac{1}{2}$	2α	x	$\dfrac{1}{\sqrt{1-x}}\left(\dfrac{1+\sqrt{1-x}}{2}\right)^{1-2\alpha}$
$\frac{1}{2}$	$\frac{1}{2}$	$\frac{3}{2}$	x^2	$\dfrac{1}{x}\arcsin x$
$\frac{1}{2}$	1	$\frac{3}{2}$	$-x^2$	$\dfrac{1}{x}\arctan x$
1	1	2	$-x$	$\dfrac{1}{x}\ln(x+1)$
$\frac{1}{2}$	1	$\frac{3}{2}$	x^2	$\dfrac{1}{2x}\ln\dfrac{1+x}{1-x}$
$n+1$	$n+m+1$	$n+m+l+2$	x	$\dfrac{(-1)^m (n+m+l+1)!}{n!\, l!\, (n+m)!\, (m+l)!} \dfrac{d^{n+m}}{dx^{n+m}}\left\{(1-x)^{m+l}\dfrac{d^l F}{dx^l}\right\},$ $\quad F = -\dfrac{\ln(1-x)}{x}, \quad n, m, l = 0, 1, 2, \ldots$

▶ **Basic properties**

The function F possesses the following properties:

$$F(\alpha, \beta, \gamma; x) = F(\beta, \alpha, \gamma; x),$$

$$F(\alpha, \beta, \gamma; x) = (1 - x)^{\gamma-\alpha-\beta} F(\gamma - \alpha, \ \gamma - \beta, \ \gamma; \ x),$$

$$F(\alpha, \beta, \gamma; x) = (1 - x)^{-\alpha} F\left(\alpha, \ \gamma - \beta, \ \gamma; \ \frac{x}{x-1}\right),$$

$$\frac{d^n}{dx^n} F(\alpha, \beta, \gamma; x) = \frac{(\alpha)_n (\beta)_n}{(\gamma)_n} F(\alpha + n, \ \beta + n, \ \gamma + n; \ x).$$

If γ is not an integer, then the general solution of the hypergeometric equation can be written in the form

$$y = C_1 F(\alpha, \beta, \gamma; x) + C_2 x^{1-\gamma} F(\alpha - \gamma + 1, \ \beta - \gamma + 1, \ 2 - \gamma; \ x).$$

▶ **Integral representations**

For $\gamma > \beta > 0$, the hypergeometric function can be expressed in terms of a definite integral:

$$F(\alpha, \beta, \gamma; x) = \frac{\Gamma(\gamma)}{\Gamma(\beta)\,\Gamma(\gamma-\beta)} \int_0^1 t^{\beta-1}(1-t)^{\gamma-\beta-1}(1-tx)^{-\alpha}\, dt,$$

where $\Gamma(\beta)$ is the gamma function.

See M. Abramowitz and I. Stegun (1979) and H. Bateman and A. Erdélyi (1973, Vol. 1) for more detailed information about hypergeometric functions.

10.10. Legendre Functions

▶ **Definitions. Basic formulas**

The associated Legendre functions $P_\nu^\mu(z)$ and $Q_\nu^\mu(z)$ of the first and the second kind are linearly independent solutions of the Legendre equation:

$$(1 - z^2)y_{zz}'' - 2zy_z' + [\nu(\nu + 1) - \mu^2(1 - z^2)^{-1}]y = 0,$$

where the parameters ν and μ and the variable z can assume arbitrary real or complex values.

For $|1 - z| < 2$, the formulas

$$P_\nu^\mu(z) = \frac{1}{\Gamma(1-\mu)} \left(\frac{z+1}{z-1}\right)^{\mu/2} F\left(-\nu, \ 1 + \nu, \ 1 - \mu, \ \frac{1-z}{2}\right),$$

$$Q_\nu^\mu(z) = A\left(\frac{z-1}{z+1}\right)^{\frac{\mu}{2}} F\left(-\nu, \ 1 + \nu, \ 1 + \mu, \ \frac{1-z}{2}\right) + B\left(\frac{z+1}{z-1}\right)^{\frac{\mu}{2}} F\left(-\nu, \ 1 + \nu, \ 1 - \mu, \ \frac{1-z}{2}\right),$$

$$A = e^{i\mu\pi} \frac{\Gamma(-\mu)\,\Gamma(1+\nu+\mu)}{2\,\Gamma(1+\nu-\mu)}, \quad B = e^{i\mu\pi} \frac{\Gamma(\mu)}{2}, \quad i^2 = -1,$$

are valid, where $F(a, b, c; z)$ is the hypergeometric series (see (see Section 10.9 of this supplement).

For $|z| > 1$,

$$P_\nu^\mu(z) = \frac{2^{-\nu-1}\Gamma(-\frac{1}{2}-\nu)}{\sqrt{\pi}\,\Gamma(-\nu-\mu)} z^{-\nu+\mu-1}(z^2-1)^{-\mu/2} F\left(\frac{1+\nu-\mu}{2}, \ \frac{2+\nu-\mu}{2}, \ \frac{2\nu+3}{2}, \ \frac{1}{z^2}\right)$$

$$+ \frac{2^\nu \Gamma(\frac{1}{2}+\nu)}{\sqrt{\pi}\,\Gamma(1+\nu-\mu)} z^{\nu+\mu}(z^2-1)^{-\mu/2} F\left(-\frac{\nu+\mu}{2}, \ \frac{1-\nu-\mu}{2}, \ \frac{1-2\nu}{2}, \ \frac{1}{z^2}\right),$$

$$Q_\nu^\mu(z) = e^{i\pi\mu} \frac{\sqrt{\pi}\,\Gamma(\nu+\mu+1)}{2^{\nu+1}\Gamma(\nu+\frac{3}{2})} z^{-\nu-\mu-1}(z^2-1)^{\mu/2} F\left(\frac{2+\nu+\mu}{2}, \ \frac{1+\nu+\mu}{2}, \ \frac{2\nu+3}{2}, \ \frac{1}{z^2}\right).$$

The functions $P_\nu(z) \equiv P_\nu^0(z)$ and $Q_\nu(z) \equiv Q_\nu^0(z)$ are called the *Legendre functions*.

The modified associated Legendre functions, on the cut $z = x$, $-1 < x < 1$, of the real axis are defined by the formulas

$$P_\nu^\mu(x) = \tfrac{1}{2} \left[e^{\frac{1}{2} i \mu \pi} P_\nu^\mu(x + i0) + e^{-\frac{1}{2} i \mu \pi} P_\nu^\mu(x - i0) \right],$$

$$Q_\nu^\mu(x) = \tfrac{1}{2} e^{-i\mu\pi} \left[e^{-\frac{1}{2} i \mu \pi} Q_\nu^\mu(x + i0) + e^{\frac{1}{2} i \mu \pi} Q_\nu^\mu(x - i0) \right].$$

▶ **Trigonometric expansions**

For $-1 < x < 1$, the modified associated Legendre functions can be represented in the form trigonometric series:

$$P_\nu^\mu(\cos\theta) = \frac{2^{\mu+1}}{\sqrt{\pi}} \frac{\Gamma(\nu+\mu+1)}{\Gamma(\nu+\frac{3}{2})} (\sin\theta)^\mu \sum_{k=0}^\infty \frac{(\frac{1}{2}+\mu)_k (1+\nu+\mu)_k}{k!\,(\nu+\frac{3}{2})_k} \sin[(2k+\nu+\mu+1)\theta],$$

$$Q_\nu^\mu(\cos\theta) = \sqrt{\pi}\, 2^\mu \frac{\Gamma(\nu+\mu+1)}{\Gamma(\nu+\frac{3}{2})} (\sin\theta)^\mu \sum_{k=0}^\infty \frac{(\frac{1}{2}+\mu)_k (1+\nu+\mu)_k}{k!\,(\nu+\frac{3}{2})_k} \cos[(2k+\nu+\mu+1)\theta],$$

where $0 < \theta < \pi$.

▶ **Some relations**

$$P_\nu^\mu(z) = P_{-\nu-1}^\mu(z), \quad P_\nu^n(z) = \frac{\Gamma(\nu+n+1)}{\Gamma(\nu-n+1)} P_\nu^{-n}(z), \quad n = 0, 1, 2, \ldots$$

$$Q_\nu^\mu(z) = \frac{\pi}{2 \sin(\mu\pi)} e^{i\pi\mu} \left[P_\nu^\mu(z) - \frac{\Gamma(1+\nu+\mu)}{\Gamma(1+\nu-\mu)} P_\nu^{-\mu}(z) \right].$$

For $0 < x < 1$,

$$P_\nu^\mu(-x) = P_\nu^\mu(x) \cos[\pi(\nu+\mu)] - 2\pi^{-1} Q_\nu^\mu(x) \sin[\pi(\nu+\mu)],$$

$$Q_\nu^\mu(-x) = -Q_\nu^\mu(x) \cos[\pi(\nu+\mu)] - \tfrac{1}{2}\pi P_\nu^\mu(x) \sin[\pi(\nu+\mu)].$$

For $-1 < x < 1$,

$$P_{\nu+1}^\mu(x) = \frac{2\nu+1}{\nu-\mu+1} x P_\nu^\mu(x) - \frac{\nu+\mu}{\nu-\mu+1} P_{\nu-1}^\mu(x).$$

Wronskians:

$$W(P_\nu, Q_\nu) = \frac{1}{1-x^2}, \quad W(P_\nu^\mu, Q_\nu^\mu) = \frac{k}{1-x^2}, \quad k = 2^{2\mu} \frac{\Gamma\left(\frac{\nu+\mu+1}{2}\right) \Gamma\left(\frac{\nu+\mu+2}{2}\right)}{\Gamma\left(\frac{\nu-\mu+1}{2}\right) \Gamma\left(\frac{\nu-\mu+2}{2}\right)}.$$

For $n = 0, 1, 2, \ldots$,

$$P_\nu^n(x) = (-1)^n (1-x^2)^{n/2} \frac{d^n}{dx^n} P_\nu(x), \quad Q_\nu^n(x) = (-1)^n (1-x^2)^{n/2} \frac{d^n}{dx^n} Q_\nu(x).$$

▶ **Legendre polynomials**

The Legendre polynomials $P_n(x)$ and the Legendre functions $Q_n(x)$ are defined by the formulas

$$P_n(x) = \frac{1}{n!\,2^n} \frac{d^n}{dx^n} (x^2 - 1)^n, \quad Q_n(x) = \frac{1}{2} P_n(x) \ln \frac{1+x}{1-x} - \sum_{m=1}^n \frac{1}{m} P_{m-1}(x) P_{n-m}(x).$$

The polynomials $P_n = P_n(x)$ can be calculated recursively using the relations

$$P_0(x) = 1, \quad P_1(x) = x, \quad P_2(x) = \frac{1}{2}(3x^2 - 1), \quad \ldots, \quad P_{n+1}(x) = \frac{2n+1}{n+1}xP_n(x) - \frac{n}{n+1}P_{n-1}(x).$$

The first three functions $Q_n = Q_n(x)$ have the form

$$Q_0(x) = \frac{1}{2}\ln\frac{1+x}{1-x}, \quad Q_1(x) = \frac{x}{2}\ln\frac{1+x}{1-x} - 1, \quad Q_2(x) = \frac{3x^2-1}{4}\ln\frac{1+x}{1-x} - \frac{3}{2}x.$$

The polynomials $P_n(x)$ have the implicit representation

$$P_n(x) = 2^{-n}\sum_{m=0}^{[n/2]}(-1)^m C_n^m C_{2n-2m}^n x^{n-2m},$$

where $[A]$ is the integer part of a number A.

All zeros of $P_n(x)$ are real and lie on the interval $-1 < x < +1$; the functions $P_n(x)$ form an orthogonal system on the interval $-1 \leq x \leq +1$, with

$$\int_{-1}^{+1} P_n(x)P_m(x)\,dx = \begin{cases} 0 & \text{if } n \neq m, \\ \dfrac{2}{2n+1} & \text{if } n = m. \end{cases}$$

The generating function is

$$\frac{1}{\sqrt{1 - 2sx + s^2}} = \sum_{n=0}^{\infty} P_n(x)s^n \qquad (|s| < 1).$$

▶ **Integral representations**
For $n = 0, 1, 2, \ldots,$

$$P_\nu^n(z) = \frac{\Gamma(\nu+n+1)}{\pi\Gamma(\nu+1)}\int_0^\pi \left(z + \cos t\sqrt{z^2-1}\right)^\nu \cos(nt)\,dt, \quad \text{Re } z > 0,$$

$$Q_\nu^n(z) = (-1)^n\frac{\Gamma(\nu+n+1)}{2^{\nu+1}\Gamma(\nu+1)}(z^2-1)^{-n/2}\int_0^\pi (z+\cos t)^{n-\nu-1}(\sin t)^{2\nu+1}\,dt, \quad \text{Re }\nu > -1,$$

Note that $z \neq x$, $-1 < x < 1$, in the latter formula.

10.11. Orthogonal Polynomials

All zeros of each of the orthogonal polynomials $\mathcal{P}_n(x)$ considered in this section are real and simple. The zeros of the polynomials $\mathcal{P}_n(x)$ and $\mathcal{P}_{n+1}(x)$ are alternating.

▶ **Legendre polynomials**
The Legendre polynomials $P_n = P_n(x)$ satisfy the equation

$$(1 - x^2)y_{xx}'' - 2xy_x' + n(n+1)y = 0.$$

They are outlined in Section 10.10 of this supplement.

▶ **Laguerre polynomials**

The Laguerre polynomials $L_n = L_n(x)$ satisfy the equation

$$xy''_{xx} + (1 - x)y'_x + ny = 0$$

and are defined by the formulas

$$L_n(x) = e^x \frac{d^n}{dx^n}\left(x^n e^{-x}\right)(-1)^n\left[x^n - n^2 x^{n-1} + \frac{n^2(n-1)^2}{2!}x^{n-2} + \cdots\right].$$

The first four polynomials have the form

$$L_0 = 1, \quad L_1 = -x + 1, \quad L_2 = x^2 - 4x + 2, \quad L_3 = -x^3 + 9x^2 - 18x + 6.$$

To calculate L_n for $n \geq 2$, one can use the recurrent formulas

$$L_{n+1}(x) = (2n + 1 - x)L_n(x) - n^2 L_{n-1}(x).$$

The functions $L_n(x)$ form an orthogonal system on the interval $0 < x < \infty$, with

$$\int_0^\infty e^{-x} L_n(x)L_m(x)\,dx = \begin{cases} 0 & \text{if } n \neq m, \\ (n!)^2 & \text{if } n = m. \end{cases}$$

The associated Laguerre polynomials of degree $n - k$ and order k are given by

$$L_n^k(x) = \frac{d^k}{dx^k}L_n(x).$$

These satisfy the differential equation

$$xy''_{xx} + (k + 1 - x)y'_x + (n - k)y = 0,$$

where $n = 1, 2, \ldots$ and $k = 0, 1, 2, \ldots$

The generating function is

$$\frac{1}{1-s}\exp\left(-\frac{sx}{1-s}\right) = \sum_{n=0}^\infty L_n(x)\frac{s^n}{n!}.$$

▶ **Chebyshev polynomials**

The Chebyshev polynomials $T_n = T_n(x)$ satisfy the equation

$$(1 - x^2)y''_{xx} - xy'_x + n^2 y = 0 \tag{1}$$

and are defined by the formulas

$$T_n(x) = \cos(n \arccos x) = \frac{(-2)^n n!}{(2n)!}\sqrt{1-x^2}\,\frac{d^n}{dx^n}\left[(1-x^2)^{n-\frac{1}{2}}\right]$$

$$= \frac{n}{2}\sum_{m=0}^{[n/2]}(-1)^m \frac{(n-m-1)!}{m!\,(n-2m)!}(2x)^{n-2m} \quad (n = 0, 1, 2, \ldots),$$

where $[A]$ stands for the integer part of a number A.

The first four polynomials are
$$T_0 = 1, \quad T_1 = x, \quad T_2 = 2x^2 - 1, \quad T_3 = 4x^3 - 3x.$$
The recurrent formulas:
$$T_{n+1}(x) = 2xT_n(x) - T_{n-1}(x), \qquad n \geq 2.$$
The functions $T_n(x)$ form an orthogonal system on the interval $-1 < x < +1$, with
$$\int_{-1}^{+1} \frac{T_n(x)T_m(x)}{\sqrt{1-x^2}}\, dx = \begin{cases} 0 & \text{if } n \neq m, \\ \frac{1}{2}\pi & \text{if } n = m \neq 0, \\ \pi & \text{if } n = m = 0. \end{cases}$$
The Chebyshev functions of the second kind,
$$U_0(x) = \arcsin x,$$
$$U_n(x) = \sin(n \arcsin x) = \frac{\sqrt{1-x^2}}{n}\frac{dT_n(x)}{dx} \qquad (n = 1, 2, \dots),$$
just as the Chebyshev polynomials, also satisfy the differential equation (1).

The generating function is
$$\frac{1-sx}{1-2sx+s^2} = \sum_{n=0}^{\infty} T_n(x)s^n \qquad (|s| < 1).$$

▶ **Hermite polynomial**
The Hermite polynomial $H_n = H_n(x)$ satisfies the equation
$$y''_{xx} - 2xy'_x + 2ny = 0$$
and is defined by the formulas
$$H_n(x) = (-1)^n \exp(x^2)\frac{d^n}{dx^n}\exp(-x^2).$$
The first four polynomials are
$$H_0 = 1, \quad H_1 = x, \quad H_2 = 4x^2 - 2, \quad H_3 = 8x^3 - 12x.$$
The recurrent formulas:
$$H_{n+1}(x) = 2xH_n(x) - 2nH_{n-1}(x), \qquad n \geq 2.$$
The functions $H_n(x)$ form an orthogonal system on the interval $-\infty < x < \infty$, with
$$\int_{-\infty}^{\infty} \exp(-x^2)H_n(x)H_m(x)\, dx = \begin{cases} 0 & \text{if } n \neq m, \\ \sqrt{\pi}\, 2^n n! & \text{if } n = m. \end{cases}$$
The Hermite functions $\psi_n(x)$ are introduced by the formula $\psi_n(x) = \exp(-\frac{1}{2}x^2)H_n(x)$, where $n = 0, 1, 2, \dots$

The generating function:
$$\exp(-s^2 + 2sx) = \sum_{n=0}^{\infty} H_n(x)\frac{s^n}{n!}.$$

▶ **Jacobi polynomials**
The Jacobi polynomials $P_n^{\alpha,\beta} = P_n^{\alpha,\beta}(x)$ satisfy the equation
$$(1-x^2)y''_{xx} + [\beta - \alpha - (\alpha+\beta+2)x]y'_x + n(n+\alpha+\beta+1)y = 0$$
and are defined by the formulas
$$P_n^{\alpha,\beta} = \frac{(-1)^n}{2^n n!}(1-x)^{-\alpha}(1+x)^{-\beta}\frac{d^n}{dx^n}\left[(1-x)^{\alpha+n}(1+x)^{\beta+n}\right] = 2^{-n}\sum_{m=0}^{n} C_{n+\alpha}^m C_{n+\beta}^{n-m}(x-1)^{n-m}(x+1)^m,$$
where C_b^a are binomial coefficients.

⊙ References for Supplement 10: H. Bateman and A. Erdélyi (1953, 1955), M. Abramowitz and I. A. Stegun (1964).

References

Abramowitz, M. and Stegun, I. A. (Editors), *Handbook of Mathematical Functions With Formulas, Graphs and Mathematical Tables,* National Bureau of Standards, Washington, 1964.

Aleksandrov, V. M., *Asymptotic methods in the mechanics of continuous media: problems with mixed boundary conditions,* Appl. Math. and Mech. (PMM), Vol. 57, No. 2, pp. 321–327, 1993.

Aleksandrov, V. M. and Kovalenko, E. V., *Problems With Mixed Boundary Conditions in Continuum Mechanics* [in Russian], Nauka, Moscow, 1986.

Arutyunyan, N. Kh., *A plane contact problem of creep theory,* Appl. Math. and Mech. (PMM), Vol. 23, No. 5, pp. 901–924, 1959.

Arutyunyan, N. Kh., *Some Problems in the Theory of Creep,* Pergamon Press, Oxford, 1966.

Babenko, Yu. I., *Heat and Mass Transfer: A Method for Computing Heat and Diffusion Flows* [in Russian], Khimiya, Moscow, 1986.

Bakhvalov, N. S., *Numerical Methods* [in Russian], Nauka, Moscow, 1973.

Bateman, H., *On the numerical solution of linear integral equations,* Proc. Roy. Soc. (A), Vol. 100, No. 705, pp. 441–449, 1922.

Bateman, H. and Erdélyi, A., *Higher Transcendental Functions. Vol. 1,* McGraw-Hill Book Co., New York, 1953.

Bateman, H. and Erdélyi, A., *Higher Transcendental Functions. Vol. 2,* McGraw-Hill Book Co., New York, 1953.

Bateman, H. and Erdélyi, A., *Higher Transcendental Functions. Vol. 3,* McGraw-Hill Book Co., New York, 1955.

Bateman, H. and Erdélyi, A., *Tables of Integral Transforms. Vol. 1,* McGraw-Hill Book Co., New York, 1954.

Bateman, H. and Erdélyi, A., *Tables of Integral Transforms. Vol. 2,* McGraw-Hill Book Co., New York, 1954.

Bellman, R. and Cooke, K. L., *Differential–Difference Equations,* Academic Press, New York, 1963.

Belotserkovskii, S. M. and Lifanov I. K., *Method of Discrete Vortices,* CRC Press, Boca Raton–New York, 1993.

Beyer, W. H., *CRC Standard Mathematical Tables and Formulae,* CRC Press, Boca Raton, 1991.

Brakhage, H., Nickel, K., and Rieder, P., *Auflösung der Abelschen Integralgleichung 2. Art,* ZAMP, Vol. 16, Fasc. 2, S. 295–298, 1965.

Brychkov, Yu. A. and Prudnikov, A. P., *Integral Transforms of Generalized Functions,* Gordon & Breach Sci. Publ., New York, 1989.

Butkovskii, A. G., *Characteristics of Systems With Distributed Parameters* [in Russian], Nauka, Moscow, 1979.

Cochran, J. A., *The Analysis of Linear Integral Equations,* McGraw-Hill Book Co., New York, 1972.

Collatz, L., *The Numerical Treatment of Differential Equations,* Springer-Verlag, Berlin, 1960.

Corduneanu, C., *Integral Equations and Stability of Feedback Systems,* Academic Press, New York, 1973.

Corduneanu, C., *Integral Equations and Applications*, Cambridge Univ. Press, Cambridge–New York, 1991.

Courant, R. and Hilbert, D., *Methods of Mathematical Physics. Vol. 1.*, Interscience Publ., New York, 1953.

Davenport, W. B. and Root, W. L., *An Introduction to the Theory of Random Signals and Noise*, McGraw-Hill Book Co., New York, 1958.

Davis, B., *Integral Transforms and Their Applications*, Springer-Verlag, New York, 1978.

Delves, L. M. and Mohamed J. L., *Computational Methods for Integral Equations*, Cambridge Univ. Press, Cambridge–New York, 1985.

Demidovich, B. P., Maron, I. A., and Shuvalova E. Z., *Numerical Methods. Approximation of Functions and Differential and Integral Equations* [in Russian], Fizmatgiz, Moscow, 1963.

Ditkin, V. A. and Prudnikov, A. P., *Integral Transforms and Operational Calculus*, Pergamon Press, New York, 1965.

Doetsch, G., *Handbuch der Laplace-Transformation. Theorie der Laplace-Transformation*, Birkhäuser Verlag, Basel–Stuttgart, 1950.

Doetsch, G., *Handbuch der Laplace-Transformation. Anwendungen der Laplace-Transformation*, Birkhäuser Verlag, Basel–Stuttgart, 1956.

Doetsch, G., *Einführung in Theorie und Anwendung der Laplace-Transformation*, Birkhäuser Verlag, Basel–Stuttgart, 1958.

Dwight, H. B., *Tables of Integrals and Other Mathematical Data*, Macmillan, New York, 1961.

Dzhuraev, A., *Methods of Singular Integral Equations*, J. Wiley, New York, 1992.

Fock, V. A., *Some integral equations of mathematical physics*, Doklady AN SSSR, Vol. 26, No. 4–5, pp. 147–151, 1942.

Gakhov, F. D., *Boundary Value Problems* [in Russian], Nauka, Moscow, 1977.

Gakhov, F. D. and Cherskii, Yu. I., *Equations of Convolution Type* [in Russian], Nauka, Moscow, 1978.

Gohberg, I. C. and Krein, M. G., *The Theory of Volterra Operators in a Hilbert Space and Its Applications* [in Russian], Nauka, Moscow, 1967.

Golberg, A. (Editor), *Numerical Solution of Integral Equations*, Plenum Press, New York, 1990.

Gorenflo, R. and Vessella, S., *Abel Integral Equations: Analysis and Applications*, Springer-Verlag, Berlin–New York, 1991.

Goursat, E., *Cours d'Analyse Mathématique, III*, 3me éd., Gauthier–Villars, Paris, 1923.

Gradshteyn, I. S. and Ryzhik, I. M., *Tables of Integrals, Series, and Products*, Academic Press, New York, 1980.

Gripenberg, G., Londen, S.-O., and Staffans, O., *Volterra Integral and Functional Equations*, Cambridge Univ. Press, Cambridge–New York, 1990.

Hackbusch, W., *Integral Equations: Theory and Numerical Treatment*, Birkhäuser Verlag, Boston, 1995.

Hirschman, I. I. and Widder, D. V., *The Convolution Transform*, Princeton Univ. Press, Princeton, New Jersey, 1955.

Jerry, A. J., *Introduction to Integral Equations With Applications*, Marcel Dekker, New York–Basel, 1985.

Kalandiya, A. I., *Mathematical Methods of Two-Dimensional Elasticity*, Mir Publ., Moscow, 1973.

Kamke, E., *Differentialgleichungen: Lösungsmethoden und Lösungen, Bd. 1*, B. G. Teubner, Leipzig, 1977.

Kantorovich, L. V. and Akilov, G. P., *Functional Analysis in Normed Spaces*, Macmillan, New York, 1964.

Kantorovich, L. V. and Krylov, V. I., *Approximate Methods of Higher Analysis*, Interscience Publ., New York, 1958.

Kanwal, R. P., *Linear Integral Equations*, Birkhäuser, Boston, 1997.

Kolmogorov, A. N. and Fomin, S. V., *Introductory Real Analysis*, Prentice-Hall, Englewood Cliffs, 1970.

Kondo, J., *Integral Equations*, Clarendon Press, Oxford, 1991.

Korn, G. A. and Korn, T. M., *Mathematical Handbook for Scientists and Engineers*, McGraw-Hill Book Co., New York, 1968.

Krasnosel'skii, M. A., *Topological Methods in the Theory of Nonlinear Integral Equations*, Macmillan, New York, 1964.

Krasnov, M. L., Kiselev, A. I., and Makarenko, G. I., *Problems and Exercises in Integral Equations*, Mir Publ., Moscow, 1971.

Krein, M. G., *Integral equations on a half-line with kernels depending upon the difference of the arguments* [in Russian], Uspekhi Mat. Nauk, Vol. 13, No. 5 (83), pp. 3–120, 1958.

Krylov, V. I., Bobkov, V. V., and Monastyrnyi, P. I., *Introduction to the Theory of Numerical Methods. Integral Equations, Ill-Posed Problems, and Improvement of Convergence* [in Russian], Nauka i Tekhnika, Minsk, 1984.

Lavrentiev, M. M., *Some Improperly Posed Problems of Mathematical Physics*, Springer-Verlag, New York, 1967.

Lavrentiev, M. M., Romanov, V. G., and Shishatskii, S. P., *Ill-Posed Problems of Mathematical Physics and Analysis* [in Russian], Nauka, Moscow, 1980.

Lifanov, I. K., *Singular Integral Equations and Discrete Vortices*, VSP, Amsterdam, 1996.

Lovitt, W. V., *Linear Integral Equations*, Dover Publ., New York, 1950.

Mikhailov, L. G., *Integral Equations With Homogeneous Kernel of Degree* -1 [in Russian], Donish, Dushanbe, 1966.

Mikhlin, S. G., *Linear Integral Equations*, Hindustan Publ. Corp., Delhi, 1960.

Mikhlin, S. G. and Prössdorf, S., *Singular Integral Operators*, Springer-Verlag, Berlin–New York, 1986.

Mikhlin, S. G. and Smolitskiy K. L., *Approximate Methods for Solution of Differential and Integral Equations*, American Elsevier Publ. Co., New York, 1967.

Miles, J. W., *Integral Transforms in Applied Mathematics*, Cambridge Univ. Press, Cambridge, 1971.

Müntz, H. M., *Integral Equations* [in Russian], GTTI, Leningrad, 1934.

Murphy, G. M., *Ordinary Differential Equations and Their Solutions*, D. Van Nostrand, New York, 1960.

Muskhelishvili N. I., *Singular Integral Equations: Boundary Problems of Function Theory and Their Applications to Mathematical Physics*, Dover Publ., New York, 1992.

Naylor, D., *On an integral transform*, Int. J. Math. & Math. Sci., Vol. 9, No. 2, pp. 283–292, 1986.

Nikol'skii S. M., *Quadrature Formulas* [in Russian], Nauka, Moscow, 1979.

Noble, B., *Methods Based on Wiener–Hopf Technique for the Solution of Partial Differential Equations*, Pergamon Press, London, 1958.

Oldham, K. B. and Spanier, J., *The Fractional Calculus*, Academic Press, London, 1974.

Petrovskii, I. G., *Lectures on the Theory of Integral Equations*, Graylock Press, Rochester, 1957.

Pipkin, A. C., *A Course on Integral Equations*, Springer-Verlag, New York, 1991.

Polyanin, A. D. and Manzhirov, A. V., *Method of model solutions in the theory of linear integral equations* [in Russian], Doklady AN, Vol. 354, No. 1, pp. 30–34, 1997.

Polyanin, A. D. and Manzhirov, A. V., *Handbuch der Integralgleichungen*, Spectrum Akad. Verlag, Heidelberg–Berlin, 1998.

Polyanin, A. D. and Zaitsev, V. F., *Handbook of Exact Solutions for Ordinary Differential Equations*, CRC Press, Boca Raton–New York, 1995.

Polyanin, A. D. and Zaitsev, V. F., *Handbuch der linearen Differentialgleichungen*, Spectrum Akad. Verlag, Heidelberg–Berlin, 1996.

Porter, D. and Stirling, D. S. G., *Integral Equations: A Practical Treatment, from Spectral Theory to Applications*, Cambridge Univ. Press, Cambridge–New York, 1990.

Privalov, I. I., *Integral Equations* [in Russian], ONTI, Moscow–Leningrad, 1935.

Prössdorf, S. and Silbermann, B., *Numerical Analysis for Integral and Related Operator Equations*, Birkhäuser Verlag, Basel–Boston, 1991.

Prudnikov, A. P., Brychkov, Yu. A., and Marichev, O. I., *Integrals and Series, Vol. 1, Elementary Functions*, Gordon & Breach Sci. Publ., New York, 1986.

Prudnikov, A. P., Brychkov, Yu. A., and Marichev, O. I., *Integrals and Series, Vol. 2, Special Functions*, Gordon & Breach Sci. Publ., New York, 1986.

Prudnikov, A. P., Brychkov, Yu. A., and Marichev, O. I., *Integrals and Series, Vol. 3, More Special Functions*, Gordon & Breach Sci. Publ., New York, 1988.

Riesz, F. and Sz.-Nagy, B., *Functional Analysis*, Ungar, New York, 1955.

Sakhnovich, L. A., *Integral Equations With Difference Kernels on Finite Intervals*, Birkhäuser Verlag, Basel–Boston, 1996.

Samko, S. G., Kilbas, A. A., and Marichev, O. I. (Editors), *Fractional Integrals and Derivatives. Theory and Applications*, Gordon & Breach Sci. Publ., New York, 1993.

Smirnov, N. S., *Introduction to the Theory of Nonlinear Integral Equations* [in Russian], Gostekhizdat, Moscow, 1951.

Smirnov, V. I., *A Course in Higher Mathematics. Vol. 4. Part 1* [in Russian], Nauka, Moscow, 1974.

Smirnov, V. I. and Lebedev, N. A., *Functions of a Complex Variable: Constructive Theory* [in Russian], Nauka, Moscow–Leningrad, 1964.

Sneddon, I., *Fourier Transforms*, McGraw-Hill Book Co., New York–Toronto–London, 1951.

Sveshnikov, A. G. and Tikhonov, A. N., *Theory of Functions of a Complex Variable* [in Russian], Nauka, Moscow, 1970.

Tikhonov, A. N. and Arsenin, V. Ya., *Methods for the Solution of Ill-Posed Problems* [in Russian], Nauka, Moscow, 1979.

Titchmarsh, E. C., *Introduction to the Theory of Fourier Integrals*, Clarendon Press, Oxford, 1948.

Tricomi, F. G., *Integral Equations*, Dover Publ., New York, 1985.

Tslaf, L. Ya., *Variational Calculus and Integral Equations* [in Russian], Nauka, Moscow, 1970.

Uflyand, Ya. S., *Dual Integral Equations Method in Problems of Mathematical Physics* [in Russian], Nauka, Leningrad, 1977.

Verlan', A. F. and Sizikov V. S., *Integral Equations: Methods, Algorithms, and Programs* [in Russian], Naukova Dumka, Kiev, 1986.

Vinogradov, I. M. (Editor), *Encyclopedia of Mathematics. Vol. 2* [in Russian], Sovetskaya Entsiklopediya, Moscow, 1979.

Volterra, V., *Theory of Functionals and of Integral and Integro-Differential Equations*, Dover Publ., New York, 1959.

Vorovich, I. I., Aleksandrov, V. M., and Babeshko, V. A., *Nonclassical Mixed Problems in the Theory of Elasticity* [in Russian], Nauka, Moscow, 1974.

Whittaker, E. T. and Watson, G. N., *A Course of Modern Analysis*, Cambridge Univ. Press, Cambridge, 1958.

Wiarda, G., *Integralgleichungen unter besonderer Berücksichtigung der Anwendungen*, Verlag und Druck von B. G. Teubner, Leipzig und Berlin, 1930.

Zabreyko, P. P., Koshelev, A. I., et al., *Integral Equations: A Reference Text*, Noordhoff Int. Publ., Leyden, 1975.

Zaitsev, V. F. and Polyanin, A. D., *Discrete-Group Methods for Integrating Equations of Nonlinear Mechanics*, CRC Press–Begel House, Boca Raton, 1994.

Zwillinger, D., *Handbook of Differential Equations*, Academic Press, San Diego, 1989.

Index